INTRODUCTION TO OPTIMUM DESIGN

THIRD EDITION

JASBIR S. ARORA
The University of Iowa
College of Engineering
Iowa City, Iowa

AMSTERDAM • BOSTON • HEIDELBERG • LONDON
NEW YORK • OXFORD • PARIS • SAN DIEGO
SAN FRANCISCO • SINGAPORE • SYDNEY • TOKYO

Academic Press is an imprint of Elsevier

Academic Press is an imprint of Elsevier
225 Wyman Street, Waltham, MA 02451, USA
The Boulevard, Langford Lane, Kidlington, Oxford, OX5 1GB, UK

Notices

Knowledge and best practice in this field are constantly changing. As new research and experience broaden our understanding, changes in research methods, professional practices, or medical treatment may become necessary.

Practitioners and researchers must always rely on their own experience and knowledge in evaluating and using any information, methods, compounds, or experiments described herein. In using such information or methods they should be mindful of their own safety and the safety of others, including parties for whom they have a professional responsibility.

To the fullest extent of the law, neither the Publisher nor the authors, contributors, or editors, assume any liability for any injury and/or damage to persons or property as a matter of products liability, negligence or otherwise, or from any use or operation of any methods, products, instructions, or ideas contained in the material herein.

MATLAB® is a trademark of TheMathWorks, Inc., and is used with permission. TheMathWorks does not warrant the accuracy of the text or exercises in this book. This book's use or discussion of the MATLAB® software or related products does not constitute endorsement or sponsorship by TheMathWorks of a particular pedagogical approach or particular use of the MATLAB® software.

MATLAB® and Handle Graphics® are registered trademarks of TheMathWorks, Inc.

Library of Congress Cataloging-in-Publication Data
Arora, Jasbir S.
   Introduction to optimum design / Jasbir Arora. — 3rd ed.
     p. cm.
   Includes bibliographical references and index.
   ISBN 978-0-12-381375-6 (hardback)
1. Engineering design—Mathematical models. I. Title.
   TA174.A76 2011
   620'.0042015118—dc23   2011026976

British Library Cataloguing-in-Publication Data
A catalogue record for this book is available from the British Library

For information on all Academic Press publications
visit our Web site at *www.elsevierdirect.com*

Printed in the United States

12 13 14 15   10 9 8 7 6 5 4 3

To
Ruhee
Rita
and
in memory of my parents
Balwant Kaur
Wazir Singh

Contents

Preface to Third Edition   xiii
Acknowledgments   xv
Key Symbols and Abbreviations   xvi

I
THE BASIC CONCEPTS

1 Introduction to Design Optimization   1

1.1 The Design Process   2
1.2 Engineering Design versus Engineering
    Analysis   4
1.3 Conventional versus Optimum Design
    Process   4
1.4 Optimum Design versus Optimal Control   6
1.5 Basic Terminology and Notation   6
    1.5.1 Points and Sets   6
    1.5.2 Notation for Constraints   8
    1.5.3 Superscripts/Subscripts and Summation
          Notation   9
    1.5.4 Norm/Length of a Vector   10
    1.5.5 Functions   11
    1.5.6 Derivatives of Functions   12
    1.5.7 U.S.–British versus SI Units   13

2 Optimum Design Problem
    Formulation   17

2.1 The Problem Formulation Process   18
    2.1.1 Step 1: Project/Problem Description   18
    2.1.2 Step 2: Data and Information
          Collection   19
    2.1.3 Step 3: Definition of Design Variables   20
    2.1.4 Step 4: Optimization Criterion   21
    2.1.5 Step 5: Formulation of Constraints   22
2.2 Design of a Can   25
2.3 Insulated Spherical Tank Design   26

2.4 Sawmill Operation   28
2.5 Design of a Two-Bar Bracket   30
2.6 Design of a Cabinet   37
    2.6.1 Formulation 1 for Cabinet Design   37
    2.6.2 Formulation 2 for Cabinet Design   38
    2.6.3 Formulation 3 for Cabinet Design   39
2.7 Minimum-Weight Tubular Column Design   40
    2.7.1 Formulation 1 for Column Design   41
    2.7.2 Formulation 2 for Column Design   41
2.8 Minimum-Cost Cylindrical Tank Design   42
2.9 Design of Coil Springs   43
2.10 Minimum-Weight Design of a Symmetric
     Three-Bar Truss   46
2.11 A General Mathematical Model for Optimum
     Design   50
    2.11.1 Standard Design Optimization
           Model   50
    2.11.2 Maximization Problem Treatment   51
    2.11.3 Treatment of "Greater Than Type"
           Constraints   51
    2.11.4 Application to Different Engineering
           Fields   52
    2.11.5 Important Observations about the
           Standard Model   52
    2.11.6 Feasible Set   53
    2.11.7 Active/Inactive/Violated
           Constraints   53
    2.11.8 Discrete and Integer Design
           Variables   54
    2.11.9 Types of Optimization Problems   55
Exercises for Chapter 2   56

3 Graphical Optimization and Basic
    Concepts   65

3.1 Graphical Solution Process   65
    3.1.1 Profit Maximization Problem   65
    3.1.2 Step-by-Step Graphical Solution
          Procedure   67

3.2 Use of *Mathematica* for Graphical
Optimization   71
  3.2.1 Plotting Functions   72
  3.2.2 Identification and Shading of Infeasible
    Region for an Inequality   73
  3.2.3 Identification of Feasible Region   73
  3.2.4 Plotting of Objective Function
    Contours   74
  3.2.5 Identification of Optimum Solution   74
3.3 Use of MATLAB for Graphical Optimization   75
  3.3.1 Plotting of Function Contours   75
  3.3.2 Editing of Graph   77
3.4 Design Problem with Multiple Solutions   77
3.5 Problem with Unbounded Solution   79
3.6 Infeasible Problem   79
3.7 Graphical Solution for the Minimum-Weight
  Tubular Column   80
3.8 Graphical Solution for a Beam Design
  Problem   82
Exercises for Chapter 3   83

4 Optimum Design Concepts: Optimality
  Conditions   95

4.1 Definitions of Global and Local Minima   96
  4.1.1 Minimum   97
  4.1.2 Existence of a Minimum   102
4.2 Review of Some Basic Calculus Concepts   103
  4.2.1 Gradient Vector: Partial Derivatives
    of a Function   103
  4.2.2 Hessian Matrix: Second-Order Partial
    Derivatives   105
  4.2.3 Taylor's Expansion   106
  4.2.4 Quadratic Forms and Definite
    Matrices   109
4.3 Concept of Necessary and Sufficient
  Conditions   115
4.4 Optimality Conditions: Unconstrained
  Problem   116
  4.4.1 Concepts Related to Optimality
    Conditions   116
  4.4.2 Optimality Conditions for Functions
    of a Single Variable   117
  4.4.3 Optimality Conditions for Functions
    of Several Variables   122
4.5 Necessary Conditions: Equality-Constrained
  Problem   130
  4.5.1 Lagrange Multipliers   131

  4.5.2 Lagrange Multiplier Theorem   135
4.6 Necessary Conditions for a General Constrained
  Problem   137
  4.6.1 The Role of Inequalities   137
  4.6.2 Karush-Kuhn-Tucker Necessary
    Conditions   139
  4.6.3 Summary of the KKT Solution
    Approach   152
4.7 Postoptimality Analysis: The Physical Meaning
  of Lagrange Multipliers   153
  4.7.1 Effect of Changing Constraint
    Limits   153
  4.7.2 Effect of Cost Function Scaling on the
    Lagrange Multipliers   156
  4.7.3 Effect of Scaling a Constraint on Its
    Lagrange Multiplier   158
  4.7.4 Generalization of Constraint Variation
    Sensitivity Result   159
4.8 Global Optimality   159
  4.8.1 Convex Sets   160
  4.8.2 Convex Functions   162
  4.8.3 Convex Programming Problem   164
  4.8.4 Transformation of a Constraint   168
  4.8.5 Sufficient Conditions for Convex
    Programming Problems   169
4.9 Engineering Design Examples   171
  4.9.1 Design of a Wall Bracket   171
  4.9.2 Design of a Rectangular
    Beam   174
Exercises for Chapter 4   178

5 More on Optimum Design Concepts:
  Optimality Conditions   189

5.1 Alternate Form of KKT Necessary
  Conditions   189
5.2 Irregular Points   192
5.3 Second-Order Conditions for Constrained
  Optimization   194
5.4 Second-Order Conditions for Rectangular
  Beam Design Problem   199
5.5 Duality in Nonlinear Programming   201
  5.5.1 Local Duality: Equality Constraints
    Case   201
  5.5.2 Local Duality: The Inequality Constraints
    Case   206
Exercises for Chapter 5   208

II

NUMERICAL METHODS FOR CONTINUOUS VARIABLE OPTIMIZATION

6 Optimum Design with Excel Solver 213

6.1 Introduction to Numerical Methods for Optimum Design 213
   6.1.1 Classification of Search Methods 214
   6.1.2 What to Do If the Solution Process Fails 215
   6.1.3 Simple Scaling of Variables 217
6.2 Excel Solver: An Introduction 218
   6.2.1 Excel Solver 218
   6.2.2 Roots of a Nonlinear Equation 219
   6.2.3 Roots of a Set of Nonlinear Equations 222
6.3 Excel Solver for Unconstrained Optimization Problems 224
6.4 Excel Solver for Linear Programming Problems 225
6.5 Excel Solver for Nonlinear Programming: Optimum Design of Springs 227
6.6 Optimum Design of Plate Girders Using Excel Solver 231
6.7 Optimum Design of Tension Members 238
6.8 Optimum Design of Compression Members 243
   6.8.1 Formulation of the Problem 243
   6.8.2 Formulation of the Problem for Inelastic Buckling 247
   6.8.3 Formulation of the Problem for Elastic Buckling 249
6.9 Optimum Design of Members for Flexure 250
6.10 Optimum Design of Telecommunication Poles 263
Exercises for Chapter 6 271

7 Optimum Design with MATLAB 275

7.1 Introduction to the Optimization Toolbox 275
   7.1.1 Variables and Expressions 275
   7.1.2 Scalar, Array, and Matrix Operations 276
   7.1.3 Optimization Toolbox 276

7.2 Unconstrained Optimum Design Problems 278
7.3 Constrained Optimum Design Problems 281
7.4 Optimum Design Examples with MATLAB 284
   7.4.1 Location of Maximum Shear Stress for Two Spherical Bodies in Contact 284
   7.4.2 Column Design for Minimum Mass 286
   7.4.3 Flywheel Design for Minimum Mass 290
Exercises for Chapter 7 294

8 Linear Programming Methods for Optimum Design 299

8.1 Linear Functions 300
8.2 Definition of a Standard Linear Programming Problem 300
   8.2.1 Standard LP Definition 300
   8.2.2 Transcription to Standard LP 302
8.3 Basic Concepts Related to Linear Programming Problems 305
   8.3.1 Basic Concepts 305
   8.3.2 LP Terminology 310
   8.3.3 Optimum Solution to LP Problems 313
8.4 Calculation of Basic Solutions 314
   8.4.1 The Tableau 314
   8.4.2 The Pivot Step 316
   8.4.3 Basic Solutions to $\mathbf{Ax} = \mathbf{b}$ 317
8.5 The Simplex Method 321
   8.5.1 The Simplex 321
   8.5.2 Basic Steps in the Simplex Method 321
   8.5.3 Basic Theorems of Linear Programming 326
8.6 The Two-Phase Simplex Method—Artificial Variables 334
   8.6.1 Artificial Variables 334
   8.6.2 Artificial Cost Function 336
   8.6.3 Definition of the Phase I Problem 336
   8.6.4 Phase I Algorithm 337
   8.6.5 Phase II Algorithm 339
   8.6.6 Degenerate Basic Feasible Solution 345
8.7 Postoptimality Analysis 348
   8.7.1 Changes in Constraint Limits 348
   8.7.2 Ranging Right-Side Parameters 354
   8.7.3 Ranging Cost Coefficients 359
   8.7.4 Changes in the Coefficient Matrix 361
Exercises for Chapter 8 363

9 More on Linear Programming Methods
   for Optimum Design   377

9.1 Derivation of the Simplex Method   377
     9.1.1 General Solution to $\mathbf{Ax} = \mathbf{b}$ 377
     9.1.2 Selection of a Nonbasic Variable that
            Should Become Basic   379
     9.1.3 Selection of a Basic Variable that Should
            Become Nonbasic   381
     9.1.4 Artificial Cost Function   382
     9.1.5 The Pivot Step   384
     9.1.6 Simplex Algorithm   384
9.2 An Alternate Simplex Method   385
9.3 Duality in Linear Programming   387
     9.3.1 Standard Primal LP Problem   387
     9.3.2 Dual LP Problem   388
     9.3.3 Treatment of Equality Constraints   389
     9.3.4 Alternate Treatment of Equality
            Constraints   391
     9.3.5 Determination of the Primal Solution
            from the Dual Solution   392
     9.3.6 Use of the Dual Tableau to Recover
            the Primal Solution   395
     9.3.7 Dual Variables as Lagrange
            Multipliers   398
9.4 KKT Conditions for the LP Problem   400
     9.4.1 KKT Optimality Conditions   400
     9.4.2 Solution to the KKT Conditions   400
9.5 Quadratic Programming Problems   402
     9.5.1 Definition of a QP Problem   402
     9.5.2 KKT Necessary Conditions for the QP
            Problem   403
     9.5.3 Transformation of KKT Conditions   404
     9.5.4 The Simplex Method for Solving QP
            Problem   405
Exercises for Chapter 9   409

10 Numerical Methods for Unconstrained
    Optimum Design   411

10.1 Gradient-Based and Direct Search
      Methods   411
10.2 General Concepts: Gradient-Based
      Methods   412
      10.2.1 General Concepts   413
      10.2.2 A General Iterative Algorithm   413
10.3 Descent Direction and Convergence of
      Algorithms   415
      10.3.1 Descent Direction and Descent
              Step   415

10.3.2 Convergence of Algorithms   417
10.3.3 Rate of Convergence   417
10.4 Step Size Determination: Basic Ideas   418
      10.4.1 Definition of the Step Size
              Determination Subproblem   418
      10.4.2 Analytical Method to Compute Step
              Size   419
10.5 Numerical Methods to Compute Step Size   421
      10.5.1 General Concepts   421
      10.5.2 Equal-Interval Search   423
      10.5.3 Alternate Equal-Interval Search   425
      10.5.4 Golden Section Search   425
10.6 Search Direction Determination: The
      Steepest-Descent Method   431
10.7 Search Direction Determination: The
      Conjugate Gradient Method   434
10.8 Other Conjugate Gradient Methods   437
Exercises for Chapter 10   438

11 More on Numerical Methods for
    Unconstrained Optimum Design   443

11.1 More on Step Size Determination   444
      11.1.1 Polynomial Interpolation   444
      11.1.2 Inexact Line Search: Armijo's
              Rule   448
      11.1.3 Inexact Line Search: Wolfe
              Conditions   449
      11.1.4 Inexact Line Search: Goldstein Test   450
11.2 More on the Steepest-Descent Method   451
      11.2.1 Properties of the Gradient Vector   451
      11.2.2 Orthogonality of Steepest-Descent
              Directions   454
11.3 Scaling of Design Variables   456
11.4 Search Direction Determination: Newton's
      Method   459
      11.4.1 Classical Newton's Method   460
      11.4.2 Modified Newton's Method   461
      11.4.3 Marquardt Modification   465
11.5 Search Direction Determination: Quasi-Newton
      Methods   466
      11.5.1 Inverse Hessian Updating: The DFP
              Method   467
      11.5.2 Direct Hessian Updating: The BFGS
              Method   470
11.6 Engineering Applications of Unconstrained
      Methods   472
      11.6.1 Data Interpolation   472
      11.6.2 Minimization of Total Potential
              Energy   473

11.6.3 Solutions of Nonlinear Equations   475
11.7 Solutions to Constrained Problems Using
     Unconstrained Optimization Methods   477
     11.7.1 Sequential Unconstrained Minimization
            Techniques   478
     11.7.2 Augmented Lagrangian (Multiplier)
            Methods   479
11.8 Rate of Convergence of Algorithms   481
     11.8.1 Definitions   481
     11.8.2 Steepest-Descent Method   482
     11.8.3 Newton's Method   483
     11.8.4 Conjugate Gradient Method   484
     11.8.5 Quasi-Newton Methods   484
11.9 Direct Search Methods   485
     11.9.1 Univariate Search   485
     11.9.2 Hooke-Jeeves Method   486
Exercises for Chapter 11   487

12 Numerical Methods for Constrained
   Optimum Design   491

12.1 Basic Concepts Related to Numerical
     Methods   492
     12.1.1 Basic Concepts Related to Algorithms
            for Constrained Problems   492
     12.1.2 Constraint Status at a Design
            Point   495
     12.1.3 Constraint Normalization   496
     12.1.4 The Descent Function   498
     12.1.5 Convergence of an Algorithm   498
12.2 Linearization of the Constrained Problem   499
12.3 The Sequential Linear Programming
     Algorithm   506
     12.3.1 Move Limits in SLP   506
     12.3.2 An SLP Algorithm   508
     12.3.3 The SLP Algorithm: Some
            Observations   512
12.4 Sequential Quadratic Programming   513
12.5 Search Direction Calculation: The QP
     Subproblem   514
     12.5.1 Definition of the QP Subproblem   514
     12.5.2 Solving of the QP Subproblem   518
12.6 The Step Size Calculation Subproblem   520
     12.6.1 The Descent Function   520
     12.6.2 Step Size Calculation: Line
            Search   522
12.7 The Constrained Steepest-Descent
     Method   525
     12.7.1 The CSD Algorithm   526

12.7.2 The CSD Algorithm: Some
       Observations   527
Exercises for Chapter 12   527

13 More on Numerical Methods
   for Constrained Optimum
   Design   533

13.1 Potential Constraint Strategy   534
13.2 Inexact Step Size Calculation   537
     13.2.1 Basic Concept   537
     13.2.2 Descent Condition   538
     13.2.3 CSD Algorithm with Inexact Step
            Size   542
13.3 Bound-Constrained Optimization   549
     13.3.1 Optimality Conditions   549
     13.3.2 Projection Methods   550
     13.3.3 Step Size Calculation   552
13.4 Sequential Quadratic Programming: SQP
     Methods   553
     13.4.1 Derivation of the Quadratic
            Programming Subproblem   554
     13.4.2 Quasi-Newton Hessian
            Approximation   557
     13.4.3 SQP Algorithm   558
     13.4.4 Observations on SQP Methods   561
     13.4.5 Descent Functions   563
13.5 Other Numerical Optimization Methods   564
     13.5.1 Method of Feasible Directions   564
     13.5.2 Gradient Projection Method   566
     13.5.3 Generalized Reduced Gradient
            Method   567
13.6 Solution to the Quadratic Programming
     Subproblem   569
     13.6.1 Solving the KKT Necessary
            Conditions   570
     13.6.2 Direct Solution to the QP
            Subproblem   571
Exercises for Chapter 13   572

14 Practical Applications
   of Optimization   575

14.1 Formulation of Practical Design Optimization
     Problems   576
     14.1.1 General Guidelines   576
     14.1.2 Example of a Practical Design
            Optimization Problem   577

14.2 Gradient Evaluation of Implicit Functions   582
14.3 Issues in Practical Design Optimization   587
 14.3.1 Selection of an Algorithm   587
 14.3.2 Attributes of a Good Optimization
  Algorithm   588
14.4 Use of General-Purpose Software   589
 14.4.1 Software Selection   589
 14.4.2 Integration of an Application into
  General-Purpose Software   589
14.5 Optimum Design of Two-Member Frame with
 Out-of-Plane Loads   590
14.6 Optimum Design of a Three-Bar Structure for
 Multiple Performance Requirements   592
 14.6.1 Symmetric Three-Bar Structure   592
 14.6.2 Asymmetric Three-Bar Structure   594
 14.6.3 Comparison of Solutions   598
14.7 Optimal Control of Systems by Nonlinear
 Programming   598
 14.7.1 A Prototype Optimal Control
  Problem   598
 14.7.2 Minimization of Error in State
  Variable   602
 14.7.3 Minimum Control Effort Problem   608
 14.7.4 Minimum Time Control Problem   609
 14.7.5 Comparison of Three Formulations
  for the Optimal Control of System
  Motion   611
14.8 Alternative Formulations for Structural
 Optimization Problems   612
14.9 Alternative Formulations for Time-Dependent
 Problems   613
Exercises for Chapter 14   615

III

ADVANCED AND MODERN TOPICS ON OPTIMUM DESIGN

15 Discrete Variable Optimum Design Concepts and Methods   619

15.1 Basic Concepts and Definitions   620
 15.1.1 Definition of Mixed Variable Optimum
  Design Problem: MV-OPT   620
 15.1.2 Classification of Mixed Variable
  Optimum Design Problems   621

15.1.3 Overview of Solution Concepts   622
15.2 Branch-and-Bound Methods   623
 15.2.1 Basic BBM   623
 15.2.2 BBM with Local Minimization   625
 15.2.3 BBM for General MV-OPT   627
15.3 Integer Programming   628
15.4 Sequential Linearization Methods   629
15.5 Simulated Annealing   630
15.6 Dynamic Rounding-Off Method   632
15.7 Neighborhood Search Method   633
15.8 Methods for Linked Discrete Variables   633
15.9 Selection of a Method   635
15.10 Adaptive Numerical Method for Discrete
 Variable Optimization   636
 15.10.1 Continuous Variable
  Optimization   636
 15.10.2 Discrete Variable Optimization   637
Exercises for Chapter 15   639

16 Genetic Algorithms for Optimum Design   643

16.1 Basic Concepts and Definitions   644
16.2 Fundamentals of Genetic Algorithms   646
16.3 Genetic Algorithm for Sequencing-Type
 Problems   651
16.4 Applications   653
Exercises for Chapter 16   653

17 Multi-objective Optimum Design Concepts and Methods   657

17.1 Problem Definition   658
17.2 Terminology and Basic Concepts   660
 17.2.1 Criterion Space and Design Space   660
 17.2.2 Solution Concepts   662
 17.2.3 Preferences and Utility Functions   665
 17.2.4 Vector Methods and Scalarization
  Methods   666
 17.2.5 Generation of Pareto Optimal Set   666
 17.2.6 Normalization of Objective
  Functions   667
 17.2.7 Optimization Engine   667
17.3 Multi-objective Genetic Algorithms   667
17.4 Weighted Sum Method   671
17.5 Weighted Min-Max Method   672
17.6 Weighted Global Criterion Method   673
17.7 Lexicographic Method   674

17.8 Bounded Objective Function Method 675
17.9 Goal Programming 676
17.10 Selection of Methods 677
Exercises for Chapter 17 678

18 Global Optimization Concepts and Methods 681

18.1 Basic Concepts of Solution Methods 682
    18.1.1 Basic Solution Concepts 682
    18.1.2 Overview of Methods 683
18.2 Overview of Deterministic Methods 684
    18.2.1 Covering Methods 684
    18.2.2 Zooming Method 685
    18.2.3 Methods of Generalized Descent 686
    18.2.4 Tunneling Method 688
18.3 Overview of Stochastic Methods 689
    18.3.1 Pure Random Search Method 690
    18.3.2 Multistart Method 691
    18.3.3 Clustering Methods 691
    18.3.4 Controlled Random Search: Nelder-Mead Method 694
    18.3.5 Acceptance-Rejection Methods 697
    18.3.6 Stochastic Integration 698
18.4 Two Local-Global Stochastic Methods 699
    18.4.1 Conceptual Local-Global Algorithm 699
    18.4.2 Domain Elimination Method 700
    18.4.3 Stochastic Zooming Method 702
    18.4.4 Operations Analysis of Methods 702
18.5 Numerical Performance of Methods 705
    18.5.1 Summary of Features of Methods 705
    18.5.2 Performance of Some Methods with Unconstrained Problems 706
    18.5.3 Performance of Stochastic Zooming and Domain Elimination Methods 707
    18.5.4 Global Optimization of Structural Design Problems 708
Exercises for Chapter 18 710

19 Nature-Inspired Search Methods 713

19.1 Differential Evolution Algorithm 714
    19.1.1 Generation of an Initial Population 715
    19.1.2 Generation of a Donor Design 716

19.1.3 Crossover Operation to Generate the Trial Design 716
    19.1.4 Acceptance/Rejection of the Trial Design 717
    19.1.5 DE Algorithm 717
19.2 Ant Colony Optimization 718
    19.2.1 Ant Behavior 718
    19.2.2 ACO Algorithm for the Traveling Salesman Problem 721
    19.2.3 ACO Algorithm for Design Optimization 724
19.3 Particle Swarm Optimization 727
    19.3.1 Swarm Behavior and Terminology 727
    19.3.2 Particle Swarm Optimization Algorithm 728
Exercises for Chapter 19 729

20 Additional Topics on Optimum Design 731

20.1 Meta-Models for Design Optimization 731
    20.1.1 Meta-Model 731
    20.1.2 Response Surface Method 733
    20.1.3 Normalization of Variables 737
20.2 Design of Experiments for Response Surface Generation 741
20.3 Discrete Design with Orthogonal Arrays 749
20.4 Robust Design Approach 754
    20.4.1 Robust Optimization 754
    20.4.2 The Taguchi Method 761
20.5 Reliability-Based Design Optimization—Design under Uncertainty 767
    20.5.1 Review of Background Material for RBDO 768
    20.5.2 Calculation of the Reliability Index 774
    20.5.3 Formulation of Reliability-Based Design Optimization 784

Appendix A: Vector and Matrix Algebra 785

A.1 Definition of Matrices 785
A.2 Types of Matrices and Their Operations 787
    A.2.1 Null Matrix 787
    A.2.2 Vector 787
    A.2.3 Addition of Matrices 787

A.2.4 Multiplication of Matrices 788

A.2.5 Transpose of a Matrix 790

A.2.6 Elementary Row—Column Operations 790

A.2.7 Equivalence of Matrices 790

A.2.8 Scalar Product—Dot Product of Vectors 790

A.2.9 Square Matrices 791

A.2.10 Partitioning of Matrices 791

A.3 Solving n Linear Equations in n Unknowns 792

A.3.1 Linear Systems 792

A.3.2 Determinants 793

A.3.3 Gaussian Elimination Procedure 796

A.3.4 Inverse of a Matrix: Gauss-Jordan Elimination 800

A.4 Solution to m Linear Equations in n Unknowns 803

A.4.1 Rank of a Matrix 803

A.4.2 General Solution of $m \times n$ Linear Equations 804

A.5 Concepts Related to a Set of Vectors 810

A.5.1 Linear Independence of a Set of Vectors 810

A.5.2 Vector Spaces 814

A.6 Eigenvalues and Eigenvectors 816

A.7 Norm and Condition Number of a Matrix 818

A.7.1 Norm of Vectors and Matrices 818

A.7.2 Condition Number of a Matrix 819

Exercises for Appendix A 819

Appendix B: Sample Computer Programs 823

B.1 Equal Interval Search 823

B.2 Golden Section Search 826

B.3 Steepest-Descent Method 829

B.4 Modified Newton's Method 829

Bibliography 841

Answers to Selected Exercises 851

Index 861

Preface to Third Edition

The philosophy of this third edition of *Introduction to Optimum Design* is to provide readers with an organized approach to engineering design optimization that is both rigorous and simple, that illustrates basic concepts and procedures with simple examples, and that demonstrates the applicability of these concepts and procedures to engineering design problems. The key step in the optimum design process is the formulation of a design problem as an optimization problem, which is emphasized and illustrated with examples. In addition, insights into, and interpretations of, optimality conditions are discussed and illustrated.

Two main objectives were set for the third edition: (1) to enhance the presentation of the book's content and (2) to include advanced topics so that the book will be suitable for higher-level courses on design optimization. The first objective is achieved by making the material more concise, organizing it with more second-, third-, and fourth-level headings, and using illustrations in example problems that have more details. The second objective is achieved by including several new topics suitable for both alternate basic and advanced courses.

New topics include duality in nonlinear programming, optimality conditions for the Simplex method, the rate of convergence of iterative algorithms, solution methods for quadratic programming problems, direct search methods, nature-inspired search methods, response surface methods, design of experiments, robust design optimization, and reliability-based design optimization.

This edition can be broadly divided into three parts. Part I, Chapters 1 through 5, presents the basic concepts related to optimum design and optimality conditions. Part II, Chapters 6 through 14, treats numerical methods for continuous variable optimization problems and their applications. Finally, Part III, Chapters 15 through 20, offers advanced and modern topics on optimum design, including methods that do not require derivatives of the problem functions.

Introduction to Optimum Design, Third Edition, can be used to construct several types of courses depending on the instructor's preference and learning objectives for students. Three course types are suggested, although several variations are possible.

Undergraduate/First-Year Graduate Course

Topics for an undergraduate and/or first-year graduate course include

- Formulation of optimization problems (Chapters 1 and 2)
- Optimization concepts using the graphical method (Chapter 3)
- Optimality conditions for unconstrained and constrained problems (Chapter 4)
- Use of Excel and MATLAB® illustrating optimum design of practical problems (Chapters 6 and 7)
- Linear programming (Chapter 8)
- Numerical methods for unconstrained and constrained problems (Chapters 10 and 12)

The use of Excel and MATLAB is to be introduced mid-semester so that students have a chance to formulate and solve more challenging project-type problems by semester's end. Note that advanced project-type exercises and sections with advanced material are marked with an asterisk (*) next to section headings, which means that they may be omitted for this course.

First Graduate-Level Course

Topics for a first graduate-level course include

- Theory and numerical methods for unconstrained optimization (Chapters 1 through 4 and 10 and 11)
- Theory and numerical methods for constrained optimization (Chapters 4, 5, 12, and 13)
- Linear and quadratic programming (Chapters 8 and 9)

The pace of material coverage should be faster for this course type. Students can code some of the algorithms into computer programs and solve practical problems.

Second Graduate-Level Course

This course presents advanced topics on optimum design:

- Duality theory in nonlinear programming, rate of convergence of iterative algorithms, derivation of numerical methods, and direct search methods (Chapters 1 through 14)
- Methods for discrete variable problems (Chapter 15)
- Nature-inspired search methods (Chapters 16 and 19)
- Multi-objective optimization (Chapter 17)
- Global optimization (Chapter 18)
- Response surface methods, robust design, and reliability-based design optimization (Chapter 20)

During this course, students write computer programs to implement some of the numerical methods and to solve practical problems.

Acknowledgments

I would like to give special thanks to my colleague, Professor Karim Abdel-Malek, Director of the Center for Computer-Aided Design at The University of Iowa, for his enthuastic support for this project and for getting me involved with the very exciting research taking place in the area of digital human modeling under the Virtual Soldier Research Program.

I would also like to acknowledge the contributions of the following colleagues: Professor Tae Hee Lee provided me with a first draft of the material for Chapter 7; Dr. Tim Marler provided me with a first draft of the material for Chapter 17; Professor G. J. Park provided me with a first draft of the material for Chapter 20; and Drs. Marcelo A. da Silva and Qian Wang provided me with a first draft of some of the material for Chapter 6. Their contributions were invaluable in the polishing of these chapters. In addition, Dr. Tim Marler, Dr. Yujiang Xiang, Dr. Rajan Bhatt, Dr. Hyun Joon Chung, and John Nicholson provided me with valuable input for improving the presentation of material in some chapters. I would also like to acknowledge the help of Jun Choi, Hyun-Jung Kwon, and John Nicholson with parts of the book's solutions manual.

I am grateful to numerous colleagues and friends around the globe for their fruitful associations with me and for discussions on the subject of optimum design. I appreciate my colleagues at The University of Iowa who used the previous editions of the book to teach an undergraduate course on optimum design: Professors Karim Abdel-Malek, Asghar Bhatti, Kyung Choi, Vijay Goel, Ray Han, Harry Kane, George Lance, and Emad Tanbour. Their input and suggestions greatly helped me improve the presentation of material in the first 12 chapters of this edition. I would also like to acknowledge all of my former graduate students whose thesis work on various topics of optimization contributed to the broadening of my horizons on the subject.

I would like to thank Bob Canfield, Hamid Torab, Jingang Yi, and others for reviewing various parts the third edition. Their suggestions helped me greatly in its fine-tuning. I would also like to thank Steve Merken and Marilyn Rash at Elsevier for their superb handling of the manuscript and production of the book. I also thank Melanie Laverman for help with the editing of some of the book's chapters.

I am grateful to the Department of Civil and Environmental Engineering, Center for Computer-Aided Design, College of Engineering, and The University of Iowa for providing me with time, resources, and support for this very satisfying endeavor. Finally, I would like to thank my family and friends for their love and support.

Key Symbols and Abbreviations

$(\mathbf{a} \cdot \mathbf{b})$	Dot product of vectors $\mathbf{a}$ and $\mathbf{b}$; $\mathbf{a}^T\mathbf{b}$	ACO	Ant colony optimization
$\mathbf{c}(\mathbf{x})$	Gradient of cost function, $\nabla f(\mathbf{x})$	BBM	Branch-and-bound method
$f(\mathbf{x})$	Cost function to be minimized	CDF	Cumulative distribution function
$g_j(\mathbf{x})$	jth inequality constraint	CSD	Constrained steepest descent
$h_i(\mathbf{x})$	ith equality constraint	DE	Differential evolution; Domain elimination
m	Number of inequality constraints	GA	Genetic algorithm
n	Number of design variables	ILP	Integer linear programming
p	Number of equality constraints	KKT	Karush-Kuhn-Tucker
$\mathbf{x}$	Design variable vector of dimension n	LP	Linear programming
x_i	ith component of design variable vector $\mathbf{x}$	MV-OPT	Mixed variable optimization problem
$\mathbf{x}^{(k)}$	kth design variable vector	NLP	Nonlinear programming

Note: A superscript (i) indicates optimum value for a variable, (ii) indicates advanced material section, and (iii) indicates a project-type exercise.

PSO	Particle swarm optimization
QP	Quadratic programming
RBDO	Reliability-based design optimization
SA	Simulated annealing
SLP	Sequential linear programming
SQP	Sequential quadratic programming
TS	Traveling salesman (salesperson)

1

Introduction to Design Optimization

Upon completion of this chapter, you will be able to

- Describe the overall process of designing systems
- Distinguish between engineering design and engineering analysis activities
- Distinguish between the conventional design process and the optimum design process
- Distinguish between optimum design and optimal control problems
- Understand the notations used for operations with vectors, matrices, and functions and their derivatives

Engineering consists of a number of well-established activities, including analysis, design, fabrication, sales, research, and development of systems. The subject of this text—the design of systems—is a major field in the engineering profession. The process of designing and fabricating systems has been developed over centuries. The existence of many complex systems, such as buildings, bridges, highways, automobiles, airplanes, space vehicles, and others, is an excellent testimonial to its long history. However, the evolution of such systems has been slow and the entire process is both time-consuming and costly, requiring substantial human and material resources. Therefore, the procedure has been to design, fabricate, and use a system regardless of whether it is the *best one*. Improved systems have been designed only after a substantial investment has been recovered.

The preceding discussion indicates that several systems can usually accomplish the same task, and that some systems are better than others. For example, the purpose of a bridge is to provide continuity in traffic from one side of the river to the other side. Several types of bridges can serve this purpose. However, to analyze and design all possibilities can be time-consuming and costly. Usually one type is selected based on some preliminary analyses and is designed in detail.

The design of a system can be *formulated as problems of optimization* in which a performance measure is optimized while all other requirements are satisfied. Many numerical methods of optimization have been developed and used to design better systems. This text

1

describes the basic concepts of optimization and numerical methods for the design of engineering systems. Design process, rather than optimization theory, is emphasized. Various theorems are stated as results without rigorous proofs; however, their implications from an engineering point of view are discussed.

Any problem in which certain parameters need to be determined to satisfy constraints can be formulated as one optimization problem. Once this has been done, the concepts and methods described in this text can be used to solve it. For this reason, the optimization techniques are quite general, having a wide range of applicability in diverse fields. It is impossible to discuss every application of optimization concepts and techniques in this introductory text. However, using simple applications, we discuss concepts, fundamental principles, and basic techniques that are used in numerous applications. The student should understand them without becoming bogged down with the notation, terminology, and details of the particular area of application.

1.1 THE DESIGN PROCESS

How Do I Begin to Design a System?

The design of many engineering systems can be a complex process. Assumptions must be made to develop realistic models that can be subjected to mathematical analysis by the available methods, and the models must be verified by experiments. Many possibilities and factors must be considered during problem formulation. *Economic considerations* play an important role in designing cost-effective systems. To complete the design of an engineering system, designers from different fields of engineering usually must cooperate. For example, the design of a high-rise building involves designers from architectural, structural, mechanical, electrical, and environmental engineering as well as construction management experts. Design of a passenger car requires cooperation among structural, mechanical, automotive, electrical, chemical, hydraulics design, and human factors engineers. Thus, in an *interdisciplinary environment* considerable interaction is needed among various design teams to complete the project. For most applications the entire design project must be broken down into several subproblems, which are then treated somewhat independently. Each of the subproblems can be posed as a problem of optimum design.

The design of a system begins with the analysis of various options. Subsystems and their components are identified, designed, and tested. This process results in a set of drawings, calculations, and reports by which the system can be fabricated. We use a systems engineering model to describe the *design process*. Although a complete discussion of this subject is beyond the scope of this text, some basic concepts are discussed using a simple block diagram.

Design is an *iterative process. Iterative* implies analyzing several *trial designs* one after another until an acceptable design is obtained. It is important to understand the concept of trial design. In the design process, the designer estimates a trial design of the system based on experience, intuition, or some simple mathematical analyses. The trial design is then analyzed to determine if it is acceptable. If it is, the design process is terminated. In the optimization process, the trial design is analyzed to determine if it is the best. Depending

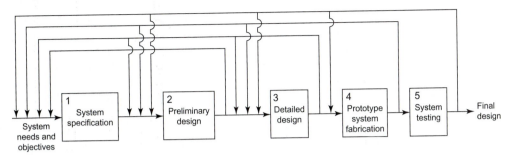

FIGURE 1.1   System evolution model.

on the specifications, "best" can have different connotations for different systems. In general, it implies that a system is cost-effective, efficient, reliable, and durable. The basic concepts are described in this text to aid the engineer in designing systems at the minimum cost and in the shortest amount of time.

The design process should be well organized. To discuss it, we consider a *system evolution model*, shown in Figure 1.1, where the process begins with the identification of a need that may be conceived by engineers or non-engineers. The five steps of the model in the figure are described in the following paragraphs.

The *first step* in the evolutionary process is to precisely define the specifications for the system. Considerable interaction between the engineer and the sponsor of the project is usually necessary to quantify the *system specifications*.

The *second step* in the process is to develop a *preliminary design* of the system. Various system concepts are studied. Since this must be done in a relatively short time, *simplified models* are used at this stage. Various subsystems are identified and their preliminary designs estimated. Decisions made at this stage generally influence the system's final appearance and performance. At the end of the preliminary design phase, a few promising concepts that need further analysis are identified.

The *third step* in the process is a *detailed design* for all subsystems using the iterative process described earlier. To evaluate various possibilities, this must be done for all previously identified promising concepts. The design parameters for the subsystems must be identified. The system performance requirements must be identified and satisfied. The subsystems must be designed to maximize system worth or to minimize a measure of the cost. Systematic optimization methods described in this text aid the designer in accelerating the detailed design process. At the end of the process, a description of the system is available in the form of reports and drawings.

The *fourth and fifth steps* shown in Figure 1.1 may or may not be necessary for all systems. They involve fabrication of a prototype system and testing, and are necessary when the system must be mass-produced or when human lives are involved. These steps may appear to be the final ones in the design process, but they are not because the system may not perform according to specifications during the testing phase. Therefore, the specifications may have to be modified or other concepts may have to be studied. In fact, this re-examination may be necessary at any point during the design process. It is for this reason that *feedback loops* are placed at every stage of the system evolution process, as shown in

Figure 1.1. The iterative process must be continued until the best system evolves. Depending on the complexity of the system, the process may take a few days or several months.

The model described in Figure 1.1 is a simplified block diagram for system evolution. In actual practice, each block may have to be broken down into several sub-blocks to carry out the studies properly and arrive at rational decisions. *The important point is that optimization concepts and methods are helpful at every stage of the process.* Such methods, along with the appropriate software, can be useful in studying various design possibilities rapidly. Therefore, in this text we discuss optimization methods and their use in the design process.

1.2 ENGINEERING DESIGN VERSUS ENGINEERING ANALYSIS

Can I Design without Analysis?
No, You Must Analyze!

It is important to recognize the differences between *engineering analysis* and *design activities*. The analysis problem is concerned with determining the behavior of an existing system or a trial system being designed for a given task. Determination of the behavior of the system implies calculation of its response to specified inputs. For this reason, the sizes of various parts and their configurations are given for the analysis problem; that is, the design of the system is known. On the other hand, the design process calculates the sizes and shapes of various parts of the system to meet performance requirements. The design of a system is an iterative process; we estimate a design and analyze it to see if it performs according to given specifications. If it does, we have an *acceptable (feasible) design*, although we may still want to change it to improve its performance. If the trial design does not work, we need to change it to come up with an acceptable system. In both cases, we must be able to *analyze designs* to make further decisions. Thus, analysis capability must be available in the design process.

This book is intended for use in all branches of engineering. It is assumed throughout that students understand the analysis methods covered in undergraduate engineering statics and physics courses. However, *we will not let the lack of analysis capability hinder understanding of the systematic process of optimum design.* Equations for analysis of the system are given wherever feasible.

1.3 CONVENTIONAL VERSUS OPTIMUM DESIGN PROCESS

Why Do I Want to Optimize?
Because You Want to Beat the Competition and Improve Your Bottom Line!

It is a challenge for engineers to design efficient and cost-effective systems without compromising their integrity. Figure 1.2(a) presents a self-explanatory flowchart for a conventional design method; Figure 1.2(b) presents a similar flowchart for the optimum design method. It is important to note that both methods are iterative, as indicated by a loop between blocks 6 and 3. Both methods have some blocks that require similar

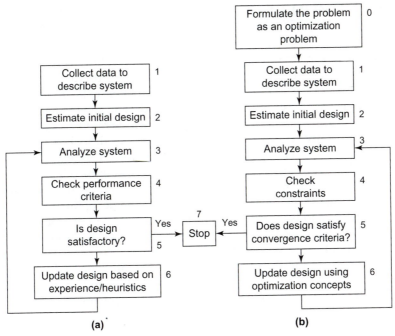

FIGURE 1.2 Comparison of (a) conventional design method and (b) optimum design method.

calculations and others that require different calculations. The key features of the two processes are these:

1. The optimum design method has block 0, where the problem is formulated as one of optimization (discussed in more detail in Chapter 2). An objective function is defined that measures the merits of different designs.
2. Both methods require data to describe the system in block 1.
3. Both methods require an initial design estimate in block 2.
4. Both methods require analysis of the system in block 3.
5. In block 4, the conventional design method checks to ensure that the performance criteria are met, whereas the optimum design method checks for satisfaction of all of the constraints for the problem formulated in block 0.
6. In block 5, stopping criteria for the two methods are checked, and the iteration is stopped if the specified stopping criteria are met.
7. In block 6, the conventional design method updates the design based on the designer's experience and intuition and other information gathered from one or more trial designs; the optimum design method uses optimization concepts and procedures to update the current design.

The foregoing distinction between the two design approaches indicates that the conventional design process is less formal. An objective function that measures a design's merit is not identified. Trend information is usually not calculated; nor is it used in block 6 to make design decisions for system improvement. In contrast, the optimization process is more formal, using trend information to make design changes.

1.4 OPTIMUM DESIGN VERSUS OPTIMAL CONTROL

What Is Optimal Control?

Optimum design and optimal control of systems are separate activities. There are numerous applications in which methods of optimum design are useful in designing systems. There are many other applications where optimal control concepts are needed. In addition, there are some applications in which both optimum design and optimal control concepts must be used. Sample applications of both techniques include *robotics* and *aerospace structures*. In this text, optimal control problems and methods are not described in detail. However, the fundamental differences between the two activities are briefly explained in the sequel. It turns out that optimal control problems can be transformed into optimum design problems and treated by the methods described in this text. Thus, methods of optimum design are very powerful and should be clearly understood. A simple optimal control problem is described in Chapter 14 and is solved by the methods of optimum design.

The optimal control problem consists of finding feedback controllers for a system to produce the desired output. The system has active elements that sense output fluctuations. System controls are automatically adjusted to correct the situation and optimize a measure of performance. Thus, control problems are usually dynamic in nature. In optimum design, on the other hand, we design the system and its elements to optimize an objective function. The system then remains fixed for its entire life.

As an example, consider the cruise control mechanism in passenger cars. The idea behind this feedback system is to control fuel injection to maintain a constant speed. Thus, the system's output (i.e., the vehicle's cruising speed) is known. The job of the control mechanism is to sense fluctuations in speed depending on road conditions and to adjust fuel injection accordingly.

1.5 BASIC TERMINOLOGY AND NOTATION

Which Notation Do I Need to Know?

To understand and to be comfortable with the methods of optimum design, the student must be familiar with linear algebra (vector and matrix operations) and basic calculus. Operations of *linear algebra* are described in Appendix A. Students who are not comfortable with this material need to review it thoroughly. Calculus of functions of single and multiple variables must also be understood. Calculus concepts are reviewed wherever they are needed. In this section, the *standard terminology* and *notations* used throughout the text are defined. It is important to understand and to memorize these notations and operations.

1.5.1 Points and Sets

Because realistic systems generally involve several variables, it is necessary to define and use some convenient and compact notations. *Set* and *vector notations* serve this purpose quite well.

Vectors and Points

A point is an ordered list of numbers. Thus, (x_1, x_2) is a point consisting of two numbers whereas $(x_1, x_2, \ldots, x_n)$ is a point consisting of n numbers. Such a point is often called an *n-tuple*. The n components $x_1, x_2, \ldots, x_n$ are collected into a column vector as

$$\mathbf{x} = \begin{bmatrix} x_1 \\ x_2 \\ \vdots \\ x_n \end{bmatrix} = [x_1 \ x_2 \ \cdots \ x_n]^T \tag{1.1}$$

where the superscript T denotes the *transpose* of a vector or a matrix. This is called an *n-vector*. Each number x_i is called a component of the (point) vector. Thus, x_1 is the first component, x_2 is the second, and so on.

We also use the following notation to represent a point or a vector in the n-dimensional space:

$$\mathbf{x} = (x_1, x_2, \ldots, x_n) \tag{1.2}$$

In 3-dimensional space, the vector $\mathbf{x} = [x_1 \ x_2 \ x_3]^T$ represents a point P, as shown in Figure 1.3. Similarly, when there are n components in a vector, as in Eqs. (1.1) and (1.2), $\mathbf{x}$ is interpreted as a point in the n-dimensional space, denoted as R^n. The space R^n is simply the collection of all n-dimensional vectors (points) of real numbers. For example, the real line is R^1, the plane is R^2, and so on.

The terms *vector* and *point* are used interchangeably, and lowercase letters in roman boldface are used to denote them. Uppercase letters in roman boldface represent matrices.

Sets

Often we deal with *sets* of points satisfying certain conditions. For example, we may consider a set S of all points having three components, with the last having a fixed value of 3, which is written as

$$S = \big\{\mathbf{x} = (x_1, x_2, x_3) \,|\, x_3 = 3\big\} \tag{1.3}$$

FIGURE 1.3 Vector representation of a point P that is in 3-dimensional space.

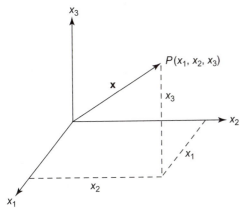

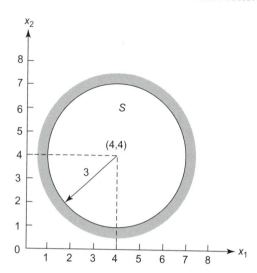

FIGURE 1.4 Image of a geometrical representation for the set $S = \{x \mid (x_1 - 4)^2 + (x_2 - 4)^2 \leq 9\}$.

Information about the set is contained in braces ({}). Equation (1.3) reads as "S equals the set of all points (x_1, x_2, x_3) with $x_3 = 3$." The vertical bar divides information about the set S into two parts: To the left of the bar is the dimension of points in the set; to the right are the properties that distinguish those points from others not in the set (for example, properties a point must possess to be in the set S).

Members of a set are sometimes called *elements*. If a point x is an element of the set S, then we write $x \in S$. The expression $x \in S$ is read as "x is an element of (belongs to) S." Conversely, the expression "$y \notin S$" is read as "y is not an element of (does not belong to) S."

If all the elements of a set S are also elements of another set T, then S is said to be a *subset* of T. Symbolically, we write $S \subset T$, which is read as "S is a subset of T" or "S is contained in T." Alternatively, we say "T is a superset of S," which is written as $T \supset S$.

As an example of a set S, consider a domain of the $x_1 - x_2$ plane enclosed by a circle of radius 3 with the center at the point (4, 4), as shown in Figure 1.4. Mathematically, all points within and on the circle can be expressed as

$$S = \{x \in R^2 \mid (x_1 - 4)^2 + (x_2 - 4)^2 \leq 9\} \tag{1.4}$$

Thus, the center of the circle (4, 4) is in the set S because it satisfies the inequality in Eq. (1.4). We write this as $(4, 4) \in S$. The origin of coordinates (0, 0) does not belong to the set because it does not satisfy the inequality in Eq. (1.4). We write this as $(0, 0) \notin S$. It can be verified that the following points belong to the set: (3, 3), (2, 2), (3, 2), (6, 6). In fact, set S has an infinite number of points. Many other points are not in the set. It can be verified that the following points are not in the set: (1, 1), (8, 8), and (−1, 2).

1.5.2 Notation for Constraints

Constraints arise naturally in optimum design problems. For example, the material of the system must not fail, the demand must be met, resources must not be exceeded, and so on. We shall discuss the constraints in more detail in Chapter 2. Here we discuss the terminology and notations for the constraints.

We encountered a constraint in Figure 1.4 that shows a set S of points within and on the circle of radius 3. The set S is defined by the following constraint:

$$(x_1 - 4)^2 + (x_2 - 4)^2 \leq 9 \tag{1.5}$$

A constraint of this form is a "less than or equal to type" *constraint* and is abbreviated as "$\leq$ type." Similarly, there are *greater than or equal to type constraints*, abbreviated as "$\geq$ type." Both are called *inequality constraints*.

1.5.3 Superscripts/Subscripts and Summation Notation

Later we will discuss a set of vectors, components of vectors, and multiplication of matrices and vectors. To write such quantities in a convenient form, consistent and compact notations must be used. We define these notations here. *Superscripts are used to represent different vectors and matrices.* For example, $\mathbf{x}^{(i)}$ represents the ith vector of a set and $\mathbf{A}^{(k)}$ represents the kth matrix. *Subscripts are used to represent components of vectors and matrices.* For example, x_j is the jth component of $\mathbf{x}$ and a_{ij} is the $i-j$th element of matrix $\mathbf{A}$. Double subscripts are used to denote elements of a matrix.

To indicate the *range of a subscript or superscript* we use the notation

$$x_i; \quad i = 1 \text{ to } n \tag{1.6}$$

This represents the numbers $x_1, x_2, \ldots, x_n$. Note that "$i = 1$ to n" represents the range for the index i and is read, "i goes from 1 to n." Similarly, a set of k vectors, each having n components, is represented by the *superscript* notation as

$$\mathbf{x}^{(j)}; \quad j = 1 \text{ to } k \tag{1.7}$$

This represents the k vectors $\mathbf{x}^{(1)}, \mathbf{x}^{(2)}, \ldots, \mathbf{x}^{(k)}$. It is important to note that subscript i in Eq. (1.6) and superscript j in Eq. (1.7) are *free indices*; that is, they can be replaced by any other variable. For example, Eq. (1.6) can also be written as x_j, $j = 1$ to n and Eq. (1.7) can be written as $\mathbf{x}^{(i)}$, $i = 1$ to k. Note that the superscript j in Eq. (1.7) does not represent the power of $\mathbf{x}$. It is an index that represents the jth vector of a set of vectors.

We also use the *summation notation* quite frequently. For example,

$$c = x_1 y_1 + x_2 y_2 + \ldots + x_n y_n \tag{1.8}$$

is written as

$$c = \sum_{i=1}^{n} x_i y_i \tag{1.9}$$

Also, multiplication of an n-dimensional vector $\mathbf{x}$ by an $m \times n$ matrix $\mathbf{A}$ to obtain an m-dimensional vector $\mathbf{y}$ is written as

$$\mathbf{y} = \mathbf{A}\mathbf{x} \tag{1.10}$$

Or, in summation notation, the ith component of $\mathbf{y}$ is

$$y_i = \sum_{j=1}^{n} a_{ij}x_j = a_{i1}x_1 + a_{i2}x_2 + \ldots + a_{in}x_n; \quad i = 1 \text{ to } m \tag{1.11}$$

There is another way of writing the matrix multiplication of Eq. (1.10). Let m-dimensional vectors $\mathbf{a}^{(i)}$; $i = 1$ to n represent columns of the matrix $\mathbf{A}$. Then $\mathbf{y} = \mathbf{A}\mathbf{x}$ is also given as

$$\mathbf{y} = \sum_{j=1}^{n} \mathbf{a}^{(j)}x_j = \mathbf{a}^{(1)}x_1 + \mathbf{a}^{(2)}x_2 + \ldots + \mathbf{a}^{(n)}x_n \tag{1.12}$$

The sum on the right side of Eq. (1.12) is said to be a *linear combination* of columns of matrix $\mathbf{A}$ with x_j, $j = 1$ to n as its multipliers. Or $\mathbf{y}$ is given as a linear combination of columns of $\mathbf{A}$ (refer to Appendix A for further discussion of the linear combination of vectors).

Occasionally, we must use the double summation notation. For example, assuming $m = n$ and substituting y_i from Eq. (1.11) into Eq. (1.9), we obtain the double sum as

$$c = \sum_{i=1}^{n} x_i \left(\sum_{j=1}^{n} a_{ij}x_j \right) = \sum_{i=1}^{n} \sum_{j=1}^{n} a_{ij}x_i x_j \tag{1.13}$$

Note that the indices i and j in Eq. (1.13) can be interchanged. This is possible because c is a *scalar quantity*, so its value is not affected by whether we sum first on i or on j. Equation (1.13) can also be written in the matrix form, as we will see later.

1.5.4 Norm/Length of a Vector

If we let $\mathbf{x}$ and $\mathbf{y}$ be two n-dimensional vectors, then their *dot product* is defined as

$$(\mathbf{x} \cdot \mathbf{y}) = \mathbf{x}^T\mathbf{y} = \sum_{i=1}^{n} x_i y_i \tag{1.14}$$

Thus, the dot product is a sum of the product of corresponding elements of the vectors $\mathbf{x}$ and $\mathbf{y}$. Two vectors are said to be *orthogonal* (*normal*) if their dot product is 0; that is, $\mathbf{x}$ and $\mathbf{y}$ are orthogonal if $\mathbf{x} \cdot \mathbf{y} = 0$. If the vectors are not orthogonal, the angle between them can be calculated from the definition of the dot product:

$$\mathbf{x} \cdot \mathbf{y} = \|\mathbf{x}\| \, \|\mathbf{y}\| \, \cos\theta \tag{1.15}$$

where θ is the angle between vectors $\mathbf{x}$ and $\mathbf{y}$ and $||\mathbf{x}||$ represents the *length of vector* $\mathbf{x}$. This is also called the *norm of the vector*. The length of vector $\mathbf{x}$ is defined as the square root of the sum of squares of the components:

$$\|\mathbf{x}\| = \sqrt{\sum_{i=1}^{n} x_i^2} = \sqrt{\mathbf{x} \cdot \mathbf{x}} \tag{1.16}$$

The double sum of Eq. (1.13) can be written in the matrix form as follows:

$$c = \sum_{i=1}^{n} \sum_{j=1}^{n} a_{ij} x_i x_j = \sum_{i=1}^{n} x_i \left(\sum_{j=1}^{n} a_{ij} x_j \right) = \mathbf{x}^T \mathbf{A} \mathbf{x} \tag{1.17}$$

Since $\mathbf{Ax}$ represents a vector, the triple product of Eq. (1.17) is also written as a dot product:

$$c = \mathbf{x}^T \mathbf{A} \mathbf{x} = (\mathbf{x} \cdot \mathbf{A} \mathbf{x}) \tag{1.18}$$

1.5.5 Functions

Just as a function of a single variable is represented as $f(x)$, a function of n independent variables $x_1, x_2, \ldots, x_n$ is written as

$$f(\mathbf{x}) = f(x_1, x_2, \ldots, x_n) \tag{1.19}$$

We deal with many functions of vector variables. To distinguish between functions, subscripts are used. Thus, the ith function is written as

$$g_i(\mathbf{x}) = g_i(x_1, x_2, \ldots, x_n) \tag{1.20}$$

If there are m functions $g_i(\mathbf{x})$, $i = 1$ to m, these are represented in the vector form

$$\mathbf{g}(\mathbf{x}) = \begin{bmatrix} g_1(\mathbf{x}) \\ g_2(\mathbf{x}) \\ \vdots \\ g_m(\mathbf{x}) \end{bmatrix} = [g_1(\mathbf{x}) \ g_2(\mathbf{x}) \ \cdots \ g_m(\mathbf{x})]^T \tag{1.21}$$

Throughout the text it is *assumed* that all functions are *continuous* and at least *twice continuously differentiable*. A function $f(\mathbf{x})$ of n variables is called *continuous* at a point $\mathbf{x}^*$ if, for any $\varepsilon > 0$, there is a $\delta > 0$ such that

$$|f(\mathbf{x}) - f(\mathbf{x}^*)| < \varepsilon \tag{1.22}$$

whenever $||\mathbf{x} - \mathbf{x}^*|| < \delta$. Thus, for all points $\mathbf{x}$ in a small neighborhood of point $\mathbf{x}^*$, a change in the function value from $\mathbf{x}^*$ to $\mathbf{x}$ is small when the function is continuous. A continuous function need not be differentiable. *Twice-continuous differentiability* of a function implies not only that it is differentiable two times, but also that its second derivative is continuous.

Figures 1.5(a) and 1.5(b) show continuous and discontinuous functions. The function in Figure 1.5(a) is differentiable everywhere, whereas the function in Figure 1.5(b) is not differentiable at points x_1, x_2, and x_3. Figure 1.5(c) is an example in which f is not a function because it has infinite values at x_1. Figure 1.5(d) is an example of a discontinuous function. As examples, functions $f(x) = x^3$ and $f(x) = \sin x$ are continuous everywhere and are also continuously differentiable. However, function $f(x) = |x|$ is continuous everywhere but not differentiable at $x = 0$.

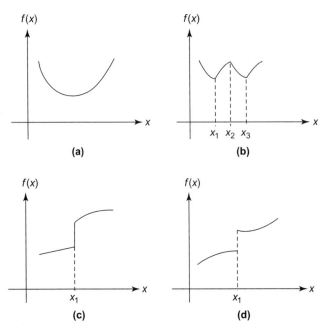

FIGURE 1.5 Continuous and discontinuous functions: (a) and (b) continuous functions; (c) not a function; (d) discontinuous function.

1.5.6 Derivatives of Functions

Often in this text we must calculate derivatives of functions of several variables. Here we introduce some of the basic notations used to represent the partial derivatives of functions of several variables.

First Partial Derivatives

For a function $f(\mathbf{x})$ of n variables, the first partial derivatives are written as

$$\frac{\partial f(\mathbf{x})}{\partial x_i}; \quad i = 1 \text{ to } n \tag{1.23}$$

The n partial derivatives in Eq. (1.23) are usually arranged in a column vector known as the *gradient* of the function $f(\mathbf{x})$. The gradient is written as $\partial f / \partial \mathbf{x}$ or $\nabla f(\mathbf{x})$. Therefore,

$$\nabla f(\mathbf{x}) = \frac{\partial f(\mathbf{x})}{\partial \mathbf{x}} = \begin{bmatrix} \dfrac{\partial f(\mathbf{x})}{\partial x_1} \\[2mm] \dfrac{\partial f(\mathbf{x})}{\partial x_2} \\[2mm] \vdots \\[2mm] \dfrac{\partial f(\mathbf{x})}{\partial x_n} \end{bmatrix} \tag{1.24}$$

Note that each component of the gradient in Eq. (1.23) or (1.24) is a function of vector $\mathbf{x}$.

Second Partial Derivatives

Each component of the gradient vector in Eq. (1.24) can be differentiated again with respect to a variable to obtain the second partial derivatives for the function $f(\mathbf{x})$:

$$\frac{\partial^2 f(\mathbf{x})}{\partial x_i \partial x_j}; \quad i, j = 1 \text{ to } n \tag{1.25}$$

We see that there are n^2 partial derivatives in Eq. (1.25). These can be arranged in a matrix known as the *Hessian matrix*, written as $\mathbf{H}(\mathbf{x})$, or simply the matrix of second partial derivatives of $f(\mathbf{x})$, written as $\nabla^2 f(\mathbf{x})$:

$$\mathbf{H}(\mathbf{x}) = \nabla^2 f(\mathbf{x}) = \left[\frac{\partial^2 f(\mathbf{x})}{\partial x_i \partial x_j} \right]_{n \times n} \tag{1.26}$$

Note that if $f(\mathbf{x})$ is continuously differentiable two times, then Hessian matrix $\mathbf{H}(\mathbf{x})$ in Eq. (1.26) is *symmetric*.

Partial Derivatives of Vector Functions

On several occasions we must differentiate a vector function of n variables, such as the vector $\mathbf{g}(\mathbf{x})$ in Eq. (1.21), with respect to the n variables in vector $\mathbf{x}$. Differentiation of each component of the vector $\mathbf{g}(\mathbf{x})$ results in a gradient vector, such as $\nabla g_i(\mathbf{x})$. Each of these gradients is an n-dimensional vector. They can be arranged as columns of a matrix of dimension $m \times n$, referred to as the gradient matrix of $\mathbf{g}(\mathbf{x})$. This is written as

$$\nabla \mathbf{g}(\mathbf{x}) = \frac{\partial \mathbf{g}(\mathbf{x})}{\partial \mathbf{x}} = [\nabla g_1(\mathbf{x}) \ \nabla g_2(\mathbf{x}) \ \dots \ \nabla g_m(\mathbf{x})]_{n \times m} \tag{1.27}$$

This gradient matrix is usually written as matrix $\mathbf{A}$:

$$\mathbf{A} = [a_{ij}]_{n \times m}; \quad a_{ij} = \frac{\partial g_j}{\partial x_i}; \quad i = 1 \text{ to } n; \quad j = 1 \text{ to } m \tag{1.28}$$

1.5.7 U.S.−British versus SI Units

The formulation of the design problem and the methods of optimization do not depend on the units of measure used. Thus, it does not matter which units are used in defining the problem. However, the final form of some of the analytical expressions for the problem does depend on the units used. In the text, we use both U.S.−British and SI units in examples and exercises. Readers unfamiliar with either system should not feel at a disadvantage when reading and understanding the material since it is simple to switch from one system to the other. To facilitate the conversion from U.S.−British to SI units or vice versa, Table 1.1 gives conversion factors for the most commonly used quantities. For a complete list of conversion factors, consult the IEEE ASTM (1997) publication.

TABLE 1.1　Conversion factors for U.S.–British and SI units

To convert from U.S.–British	To SI units	Multiply by
Acceleration		
foot/second^2 (ft/s^2)	meter/second^2 (m/s^2)	0.3048*
inch/second^2 (in/s^2)	meter/second^2 (m/s^2)	0.0254*
Area		
foot^2 (ft^2)	meter^2 (m^2)	0.09290304*
inch^2 (in^2)	meter^2 (m^2)	6.4516E−04*
Bending Moment or Torque		
pound force inch (lbf · in)	Newton meter (N · m)	0.1129848
pound force foot (lbf · ft)	Newton meter (N · m)	1.355818
Density		
pound mass/inch^3 (lbm/in^3)	kilogram/meter^3 (kg/m^3)	27,679.90
pound mass/foot^3 (lbm/ft^3)	kilogram/meter^3 (kg/m^3)	16.01846
Energy or Work		
British thermal unit (BTU)	Joule (J)	1055.056
foot-pound force (ft · lbf)	Joule (J)	1.355818
kilowatt-hour (KWh)	Joule (J)	3,600,000*
Force		
kip (1000 lbf)	Newton (N)	4448.222
pound force (lbf)	Newton (N)	4.448222
Length		
foot (ft)	meter (m)	0.3048*
inch (in)	meter (m)	0.0254*
mile (mi), U.S. statute	meter (m)	1609.347
mile (mi), International, nautical	meter (m)	1852*
Mass		
pound mass (lbm)	kilogram (kg)	0.4535924
slug (lbf · s^{2}ft)	kilogram (kg)	14.5939
ton (short, 2000 lbm)	kilogram (kg)	907.1847
ton (long, 2240 lbm)	kilogram (kg)	1016.047
tonne (t, metric ton)	kilogram (kg)	1000*

TABLE 1.1   (*Continued*)

To convert from U.S.–British	To SI units	Multiply by
Power		
foot-pound/minute (ft · lbf/min)	Watt (W)	0.02259697
horsepower (550 ft · lbf/s)	Watt (W)	745.6999
Pressure or Stress		
atmosphere (std) (14.7 lbf/in^2)	Newton/meter^2 (N/m^2 or Pa)	101,325*
one bar (b)	Newton/meter^2 (N/m^2 or Pa)	100,000*
pound/foot^2 (lbf/ft^2)	Newton/meter^2 (N/m^2 or Pa)	47.88026
pound/inch^2 (lbf/in^2 or psi)	Newton/meter^2 (N/m^2 or Pa)	6894.757
Velocity		
foot/minute (ft/min)	meter/second (m/s)	0.00508*
foot/second (ft/s)	meter/second (m/s)	0.3048*
knot (nautical mi/h), international	meter/second (m/s)	0.5144444
mile/hour (mi/h), international	meter/second (m/s)	0.44704*
mile/hour (mi/h), international	kilometer/hour (km/h)	1.609344*
mile/second (mi/s), international	kilometer/second (km/s)	1.609344*
Volume		
foot^3 (ft^3)	meter^3 (m^3)	0.02831685
inch^3 (in^3)	meter^3 (m^3)	1.638706E−05
gallon (Canadian liquid)	meter^3 (m^3)	0.004546090
gallon (U.K. liquid)	meter^3 (m^3)	0.004546092
gallon (U.S. dry)	meter^3 (m^3)	0.004404884
gallon (U.S. liquid)	meter^3 (m^3)	0.003785412
one liter (L)	meter^3 (m^3)	0.001*
ounce (U.K. fluid)	meter^3 (m^3)	2.841307E−05
ounce (U.S. fluid)	meter^3 (m^3)	2.957353E−05
pint (U.S. dry)	meter^3 (m^3)	5.506105E−04
pint (U.S. liquid)	meter^3 (m^3)	4.731765E−04
quart (U.S. dry)	meter^3 (m^3)	0.001101221
quart (U.S. liquid)	meter^3 (m^3)	9.463529E−04

* Exact conversion factor.

2

Optimum Design Problem Formulation

Upon completion of this chapter, you will be able to

- Translate a descriptive statement of the design problem into a mathematical statement for optimization
- Identify and define the problem's design variables
- Identify and define an optimization criterion for the problem
- Identify and define the design problem's constraints
- Transcribe the problem formulation into a standard model for design optimization

It is generally accepted that the *proper definition and formulation of a problem* take roughly 50 percent of the total effort needed to solve it. Therefore, it is critical to follow well-defined procedures for formulating design optimization problems. In this chapter, we describe the process of transforming the design of a selected system and/or subsystem into an optimum design problem.

Several simple and moderately complex applications are discussed in this chapter to illustrate the problem formulation process. More advanced applications are discussed in Chapters 6 and 7 and 14 through 19.

The *importance of properly formulating* a design optimization problem must be stressed because the optimum solution will be only as good as the formulation. For example, if we forget to include a critical constraint in the formulation, the optimum solution will most likely violate it. Also, if we have too many constraints, or if they are inconsistent, there may be no solution. However, once the problem is properly formulated, good software is usually available to deal with it. For most design optimization problems, we will use the following *five-step* formulation procedure:

Step 1: Project/problem description
Step 2: Data and information collection
Step 3: Definition of design variables
Step 4: Optimization criterion
Step 5: Formulation of constraints

2.1 THE PROBLEM FORMULATION PROCESS

The formulation of an optimum design problem involves translating a descriptive statement of it into a well-defined mathematical statement. We will describe the tasks to be performed in each of the foregoing five steps to develop a mathematical formulation for the design optimization problem. These steps are illustrated with some examples in this section and in later sections.

At this stage, it is also important to understand the solution process for optimization of a design problem. As illustrated earlier in Figure 1.2(b), optimization methods are iterative where the solution process is started by selecting a *trial design* or a *set of trial designs*. The trial designs are analyzed and evaluated, and a new trial design is generated. This iterative process is continued until an optimum solution is reached.

2.1.1 Step 1: Project/Problem Description

Are the Project Goals Clear?

The formulation process begins by developing a descriptive statement for the project/problem, usually by the project's owner/sponsor. The statement describes the overall *objectives* of the project and the *requirements* to be met. This is also called the *statement of work*.

EXAMPLE 2.1 DESIGN OF A CANTILEVER BEAM—PROBLEM DESCRIPTION

Cantilever beams are used in many practical applications in civil, mechanical, and aerospace engineering. To illustrate the step of problem description, we consider the design of a hollow square-cross-section *cantilever beam* to support a load of 20 kN at its end. The beam, made of steel, is 2 m long, as shown in Figure 2.1. The failure conditions for the beam are as follows: (1) the material should not fail under the action of the load, and (2) the deflection of the free end should be no more than 1 cm. The width-to-thickness ratio for the beam should be no more than 8. A minimum-mass beam is desired. The width and thickness of the beam must be within the following limits:

$$60 \le width \le 300 \text{ mm} \tag{a}$$

$$10 \le thickness \le 40 \text{ mm} \tag{b}$$

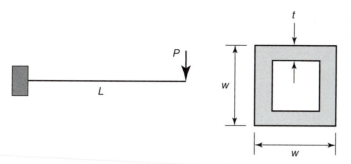

FIGURE 2.1 Cantilever beam of a hollow square cross-section.

2.1.2 Step 2: Data and Information Collection

Is All the Information Available to Solve the Problem?

To develop a mathematical formulation for the problem, we need to gather information on material properties, performance requirements, resource limits, cost of raw materials, and so forth. In addition, most problems require the capability to *analyze trial designs*. Therefore, *analysis procedures* and *analysis tools* must be identified at this stage. For example, the finite-element method is commonly used for analysis of structures, so the software tool available for such an analysis needs to be identified. In many cases, the project statement is vague, and assumptions about modeling of the problem need to be made in order to formulate and solve it.

EXAMPLE 2.2 DATA AND INFORMATION COLLECTION FOR A CANTILEVER BEAM

The information needed for the *cantilever beam design problem* of Example 2.1 includes expressions for bending and shear stresses, and the expression for the deflection of the free end. The notation and data for this purpose are defined in the table that follows.

Useful expressions for the beam are

$$A = w^2 - (w - 2t)^2 = 4t(w - t), \text{ mm}^2 \tag{c}$$

$$I = \frac{8}{3}wt^3 + \frac{2}{3}w^3t - 2w^2t^2 - \frac{4}{3}t^4, \text{ mm}^4 \tag{d}$$

$$Q = \frac{3}{4}w^2t - \frac{3}{2}wt^2 + t^3, \text{ mm}^3 \tag{e}$$

$$M = PL, \text{ N} \cdot \text{mm} \tag{f}$$

$$V = P, \text{ N} \tag{g}$$

$$\sigma = \frac{Mw}{2I}, \text{ N} \cdot \text{mm}^{-2} \tag{h}$$

$$\tau = \frac{VQ}{2It}, \ \mathrm{N \cdot mm^{-2}} \qquad\qquad (i)$$

$$q = \frac{PL^3}{3EI}, \ \mathrm{mm} \qquad\qquad (j)$$

Notation	Data
A	cross-sectional area, $\mathrm{mm^2}$
E	modulus of elasticity, $21 \times 10^4 \ \mathrm{N \cdot mm^{-2}}$
G	shear modulus, $8 \times 10^4 \ \mathrm{N \cdot mm^{-2}}$
I	moment of inertia, $\mathrm{mm^4}$
L	length of the member, 2000 mm
M	bending moment, $\mathrm{N \cdot mm}$
P	load at the free end, 20,000 N
Q	moment about the neutral axis of the area above the neutral axis, $\mathrm{mm^3}$
q	vertical deflection of the free end, mm
q_a	allowable vertical deflection of the free end, 10 mm
V	shear force, N
w	width (depth) of the section, mm
t	wall thickness, mm
σ	bending stress, $\mathrm{N \cdot mm^{-2}}$
σ_a	allowable bending stress, $165 \ \mathrm{N \cdot mm^{-2}}$
τ	shear stress, $\mathrm{N \cdot mm^{-2}}$
τ_a	allowable shear stress, $90 \ \mathrm{N \cdot mm^{-2}}$

2.1.3 Step 3: Definition of Design Variables

What Are These Variables?
How Do I Identify Them?

The next step in the formulation process is to identify a set of variables that describe the system, called the *design variables*. In general, these are referred to as optimization variables and are regarded as *free* because we should be able to assign any value to them. Different values for the variables produce different designs. The design variables should be independent of each other as far as possible. If they are dependent, their values cannot be specified independently because there are constraints between them. The number of independent design variables gives the *design degrees of freedom* for the problem.

For some problems, different sets of variables can be identified to describe the same system. Problem formulation will depend on the selected set. We will present some examples later in this chapter to elaborate on this point.

Once the design variables are given numerical values, we have a *design of the system*. Whether this design *satisfies all requirements* is another question. We will introduce a number of concepts to investigate such questions in later chapters.

If proper design variables are not selected for a problem, the formulation will be either incorrect or not possible. At the initial stage of problem formulation, all options for specification of design variables should be investigated. Sometimes it may be desirable to designate more design variables than apparent design degrees of freedom. This gives added flexibility to problem formulation. Later, it will be possible to assign a fixed numerical value to any variable and thus eliminate it from the formulation.

At times it is difficult to clearly identify a problem's design variables. In such a case, a complete list of all variables may be prepared. Then, by considering each variable individually, we can decide whether or not to treat it as an optimization variable. If it is a valid design variable, the designer should be able to specify a numerical value for it to select a trial design.

We will use the term "design variables" to indicate all optimization variables for the optimization problem and will represent them in the vector **x**. To summarize, the following considerations should be given in identifying design variables for a problem:

- Design variables should be independent of each other as far as possible. If they are not, there must be some equality constraints between them (explained later).
- A minimum number of design variables required to properly formulate a design optimization problem must exist.
- As many independent parameters as possible should be designated as design variables at the problem formulation phase. Later on, some of the variables can be assigned fixed values.
- A numerical value should be given to each identified design variable to determine if a trial design of the system is specified.

EXAMPLE 2.3 DESIGN VARIABLES FOR A CANTILEVER BEAM

Only dimensions of the cross-section are identified as design variables for the *cantilever beam design problem* of Example 2.1; all other parameters are specified:

w = width (depth) of the section, mm
t = wall thickness, mm

2.1.4 Step 4: Optimization Criterion

How Do I Know that My Design Is the Best?

There can be many feasible designs for a system, and some are better than others. The question is how we compare designs and designate one as better than another. For this, we must have a criterion that associates a number with each design. Thus, the merit of a given design is specified. The criterion must be a scalar function whose numerical value can be obtained once a design is specified; that is, it must be a *function of the design variable vector* **x**. Such a criterion is usually called an *objective function* for the optimum design problem, and it needs to be *maximized* or *minimized* depending on problem requirements. A criterion that is to be minimized is usually called a *cost function* in engineering literature, which is the term used throughout this text. It is emphasized that a *valid objective function must be influenced directly or indirectly by the variables of the design problem*; otherwise, it is not a meaningful objective function.

The selection of a proper objective function is an important decision in the design process. Some objective functions are cost (to be minimized), profit (to be maximized), weight (to be minimized), energy expenditure (to be minimized), and, for example, ride quality of a vehicle (to be maximized). In many situations an obvious objective function can be identified. For example, we always want to minimize the cost of manufacturing goods or maximize return on investment. In some situations, two or more objective functions may be identified. For example, we may want to minimize the weight of a structure and at the same time minimize the deflection or stress at a certain point. These are called *multiobjective design optimization problems* and are discussed in a later chapter.

For some design problems, it is not obvious what the objective function should be or how it should relate to the design variables. Some insight and experience may be needed to identify a proper objective function. For example, consider the optimization of a passenger car. What are the design variables? What is the objective function, and what is its functional form in terms of the design variables? Although this is a very practical problem, it is quite complex. Usually, such problems are divided into several smaller subproblems and each one is formulated as an optimum design problem. For example, design of a passenger car can be divided into a number of optimization subproblems involving the trunk lid, doors, side panels, roof, seats, suspension system, transmission system, chassis, hood, power plant, bumpers, and so on. Each subproblem is now manageable and can be formulated as an optimum design problem.

EXAMPLE 2.4 OPTIMIZATION CRITERION FOR A CANTILEVER BEAM

For the *design problem in Example 2.1*, the objective is to design a minimum-mass cantilever beam. Since the mass is proportional to the cross-sectional area of the beam, the objective function for the problem is taken as the cross-sectional area:

$$f(w, t) = A = 4t(w - t), \text{ mm}^2 \tag{k}$$

2.1.5 Step 5: Formulation of Constraints

What Restrictions Do I Have on My Design?

All restrictions placed on the design are collectively called *constraints*. The final step in the formulation process is to identify all constraints and develop expressions for them. Most realistic systems must be designed and fabricated with the given *resources* and must meet *performance requirements*. For example, structural members should not fail under normal operating loads. The vibration frequencies of a structure must be different from the operating frequency of the machine it supports; otherwise, resonance can occur and cause catastrophic failure. Members must fit into the available space.

These constraints, as well as others, must depend on the design variables, since only then do their values change with different trial designs; that is, a meaningful constraint must be a function of at least one design variable. Several concepts and terms related to constraints are explained next.

Linear and Nonlinear Constraints

Many constraint functions have only first-order terms in design variables. These are called *linear constraints*. *Linear programming problems* have only linear constraints and objective functions. More general problems have nonlinear cost and/or constraint functions. These are called *nonlinear programming problems*. Methods to treat both linear and nonlinear constraints and objective functions are presented in this text.

Feasible Design

The design of a system is a set of numerical values assigned to the design variables (i.e., a particular design variable vector x). Even if this design is absurd (e.g., negative radius) or inadequate in terms of its function, it can still be called a design. Clearly, some designs are useful and others are not. A design meeting all requirements is called a *feasible design* (*acceptable* or *workable*). An *infeasible design* (*unacceptable*) does not meet one or more of the requirements.

Equality and Inequality Constraints

Design problems may have equality as well as inequality constraints. The problem description should be studied carefully to determine which requirements need to be formulated as equalities and which ones as inequalities. For example, a machine component may be required to move precisely by Δ to perform the desired operation, so we must treat this as an equality constraint. A feasible design must satisfy precisely all equality constraints. Also, most design problems have inequality constraints, sometimes called *unilateral* or *one-sided constraints*. Note that the *feasible region* with respect to an inequality constraint is much larger than that with respect to the same constraint expressed as equality.

To illustrate the difference between equality and inequality constraints, we consider a constraint written in both equality and inequality forms. Figure 2.2(a) shows the equality constraint $x_1 = x_2$. Feasible designs with respect to the constraint must lie on the straight line A−B. However, if the constraint is written as an inequality $x_1 \le x_2$, the feasible region is much larger, as shown in Figure 2.2(b). Any point on the line A−B or above it gives a feasible design.

Implicit Constraints

Some constraints are quite simple, such as the smallest and largest allowable values for the design variables, whereas more complex ones may be indirectly influenced by the design variables. For example, deflection at a point in a large structure depends on its design. However, it is impossible to express deflection as an explicit function of the design variables except for very simple structures. These are called *implicit constraints*. When there are implicit functions in the problem formulation, it is not possible to formulate the problem functions explicitly in terms of design variables alone. Instead, we must use some intermediate variables in the problem formulation. We will discuss formulations having implicit functions in Chapter 14.

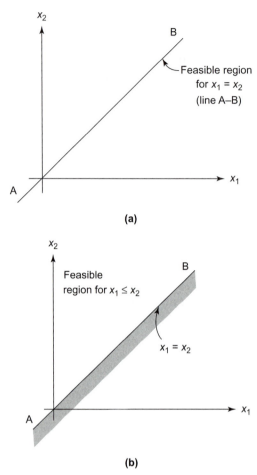

FIGURE 2.2   Shown here is the distinction between equality and inequality constraints: (a) Feasible region for constraint $x_1 = x_2$ (line A−B); (b) feasible region for constraint $x_1 \le x_2$ (line A−B and the region above it).

EXAMPLE 2.5 CONSTRAINTS FOR A CANTILEVER BEAM

Using various expressions given in Eqs. (c) through (j), we formulate the constraints for the *cantilever beam design problem* from Example 2.1 as follows:

Bending stress constraint: $\sigma \le \sigma_a$

$$\frac{PLw}{2I} - \sigma_a \le 0 \tag{1}$$

Shear stress constraint: $\tau \le \tau_a$

$$\frac{PQ}{2It} - \tau_a \le 0 \tag{m}$$

Deflection constraint: $q \le q_a$

$$\frac{PL^3}{3EI} - q_a \le 0 \tag{n}$$

Width−thickness restriction: $\frac{w}{t} \le 8$

$$w - 8t \le 0 \tag{o}$$

Dimension restrictions

$$60 - w \le 0, \text{ mm}; \quad w - 300 \le 0, \text{ mm} \tag{p}$$

$$3 - t \le 0, \text{ mm}; \quad t - 15 \le 0, \text{ mm} \tag{q}$$

Thus the optimization problem is to find w and t to minimize the cost function of Eq. (k) subject to the eight inequality constraints of Eqs. (l) through (q). Note that the constraints of Eqs. (l) through (n) are nonlinear functions and others are linear functions of the design variables. There are eight inequality constraints and no equality constraints for this problem. Substituting various expressions, Eqs. (l) through (n) can be expressed explicitly in terms of the design variables, if desired.

2.2 DESIGN OF A CAN

STEP 1: PROJECT/PROBLEM DESCRIPTION The purpose of this project is to design a can, shown in Figure 2.3, to hold *at least* 400 ml of liquid (1 ml = 1 cm³), as well as to meet other design requirements. The cans will be produced in the billions, so it is desirable to minimize their manufacturing costs. Since cost can be directly related to the surface area of the sheet metal used, it is reasonable to minimize the amount of sheet metal required. Fabrication, handling, aesthetics, and shipping considerations impose the following restrictions on the size of the can: The diameter should be no more than 8 cm and no less than 3.5 cm, whereas the height should be no more than 18 cm and no less than 8 cm.

STEP 2: DATA AND INFORMATION COLLECTION Data for the problem are given in the project statement.

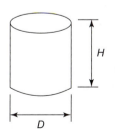

FIGURE 2.3 Can.

STEP 3: DEFINITION OF DESIGN VARIABLES   The two design variables are defined as

D = diameter of the can, cm
H = height of the can, cm

STEP 4: OPTIMIZATION CRITERION   The design objective is to minimize the total surface area S of the sheet metal for the three parts of the cylindrical can: the surface area of the cylinder (circumference × height) and the surface area of the two ends. Therefore, the optimization criterion, or *cost function* (the total area of sheet metal), is given as

$$S = \pi DH + 2\left(\frac{\pi}{4}D^2\right), \ \ \text{cm}^2 \tag{a}$$

STEP 5: FORMULATION OF CONSTRAINTS   The first constraint is that the can must hold *at least* 400 cm^3 of fluid, which is written as

$$\frac{\pi}{4}D^2H \geq 400, \ \ \text{cm}^3 \tag{b}$$

If it had been stated that "the can must hold 400 ml of fluid," then the preceding volume constraint would be an equality. The other constraints on the size of the can are

$$3.5 \leq D \leq 8, \ \ \text{cm}$$
$$8 \leq H \leq 18, \ \ \text{cm} \tag{c}$$

The explicit constraints on design variables in Eqs. (c) have many different names in the literature, such as *side constraints, technological constraints, simple bounds, sizing constraints,* and *upper and lower limits on the design variables.* Note that for the present problem there are really four constraints in Eqs. (c). Thus, the problem has two design variables and a total of five inequality constraints.

Note also that the cost function and the first constraint are nonlinear in variables; the remaining constraints are linear.

2.3 INSULATED SPHERICAL TANK DESIGN

STEP 1: PROJECT/PROBLEM DESCRIPTION   The goal of this project is to choose an insulation thickness t to minimize the life-cycle cooling cost for a spherical tank. The cooling costs include installing and running the refrigeration equipment, and installing the insulation. Assume a 10-year life, a 10 percent annual interest rate, and no salvage value. The tank has already been designed having r (m) as its radius.

STEP 2: DATA AND INFORMATION COLLECTION   To formulate this design optimization problem, we need some data and expressions. To calculate the volume of the insulation material, we require the surface area of the spherical tank, which is given as

$$A = 4\pi r^2, \ \ \text{m}^2 \tag{a}$$

To calculate the capacity of the refrigeration equipment and the cost of its operation, we need to calculate the annual heat gain G, which is given as

$$G = \frac{(365)(24)(\Delta T)A}{c_1 t}, \quad \text{Watt-hours} \tag{b}$$

where ΔT is the average difference between the internal and external temperatures in Kelvin, c_1 is the thermal resistivity per unit thickness in Kelvin-meter per Watt, and t is the insulation thickness in meters. ΔT can be estimated from the historical data for temperatures in the region in which the tank is to be used. Let c_2 = the insulation cost per cubic meter ($\$/m^3$), c_3 = the cost of the refrigeration equipment per Watt-hour of capacity ($\$/Wh$), and c_4 = the annual cost of running the refrigeration equipment per Watt-hour ($\$/Wh$).

STEP 3: DEFINITION OF DESIGN VARIABLES　　Only one design variable is identified for this problem:

t = insulation thickness, m

STEP 4: OPTIMIZATION CRITERION　　The goal is to minimize the life-cycle cooling cost of refrigeration for the spherical tank over 10 years. The life-cycle cost has three components: insulation, refrigeration equipment, and operations for 10 years. Once the annual operations cost has been converted to the present cost, the total cost is given as

$$Cost = c_2 At + c_3 G + c_4 G \tag{c}$$

where *uspwf* $(0.1, 10) = 6.14457$ is the uniform series present worth factor, calculated using the equation

$$uspwf(i,n) = \frac{1}{i}\left[1 - (1-i)^{-n}\right] \tag{d}$$

where i is the rate of return per dollar per period and n is the number of periods. Note that to calculate the volume of the insulation as At, it is assumed that the insulation thickness is much smaller than the radius of the spherical tank; that is, $t \ll r$.

STEP 5: FORMULATION OF CONSTRAINTS　　Although no constraints are indicated in the problem statement, it is important to require that the insulation thickness be non-negative (i.e., $t \geq 0$). Although this may appear obvious, it is important to include the constraint explicitly in the mathematical formulation of the problem. Without its explicit inclusion, the mathematics of optimization may assign negative values to thickness, which is, of course, meaningless. Note also that in reality t cannot be zero because it appears in the denominator of the expression for G. Therefore, the constraint should really be expressed as $t > 0$. However, *strict inequalities* cannot be treated mathematically or numerically in the solution process because they give an open feasible set. We must allow the possibility of satisfying inequalities as equalities; that is, we must allow the possibility that $t = 0$ in the solution process. Therefore, a more realistic constraint is $t \geq t_{min}$, where t_{min} is the smallest insulation thickness available on the market.

EXAMPLE 2.6 FORMULATION OF THE SPHERICAL TANK PROBLEM WITH INTERMEDIATE VARIABLES

A summary of the problem formulation for the design optimization of insulation for a spherical tank with intermediate variables is as follows:

Specified data: r, ΔT, c_1, c_2, c_3, c_4, t_{min}
Design variable: t, m
Intermediate variables:

$$A = 4\pi r^2$$

$$G = \frac{(365)(24)(\Delta T)A}{c_1 t} \tag{e}$$

Cost function: Minimize the life-cycle cooling cost of refrigeration of the spherical tank,

$$Cost = c_2 At + c_3 G + 6.14457 c_4 G \tag{f}$$

Constraint:

$$t \geq t_{min} \tag{g}$$

Note that A and G may also be treated as design variables in this formulation. In such a case, A must be assigned a fixed numerical value since r has already been determined, and the expression for G must be treated as an equality constraint.

EXAMPLE 2.7 FORMULATION OF THE SPHERICAL TANK PROBLEM WITH THE DESIGN VARIABLE ONLY

Following is a summary of the problem formulation for the design optimization of insulation for a spherical tank in terms of the design variable only:

Specified data: r, ΔT, c_1, c_2, c_3, c_4, t_{min}
Design variable: t, m
Cost function: Minimize the life-cycle cooling cost of refrigeration of the spherical tank,

$$Cost = at + \frac{b}{t}, \quad a = 4c_2 \pi r^2,$$

$$b = \frac{(c_3 + 6.14457 c_4)}{c_1}(365)(24)(\Delta T)(4\pi r^2) \tag{h}$$

Constraint:

$$t \geq t_{min} \tag{i}$$

2.4 SAWMILL OPERATION

STEP 1: PROJECT/PROBLEM DESCRIPTION A company owns two sawmills and two forests. Table 2.1 shows the capacity of each of the mills (logs/day) and the distances

TABLE 2.1   Data for sawmills

Mill	Distance from Mill 1	Distance from Mill 2	Mill capacity per day
A	24.0 km	20.5 km	240 logs
B	17.2 km	18.0 km	300 logs

between the forests and the mills (km). Each forest can yield up to 200 logs/day for the duration of the project, and the cost to transport the logs is estimated at $10/km/log. At least 300 logs are needed daily. The goal is to minimize the total daily cost of transporting the logs.

STEP 2: DATA AND INFORMATION COLLECTION   Data are given in Table 2.1 and in the problem statement.

STEP 3: DEFINITION OF DESIGN VARIABLES   The design problem is to determine how many logs to ship from Forest i to Mill j, as shown in Figure 2.4. Therefore, the design variables are identified and defined as follows:

x_1 = number of logs shipped from Forest 1 to Mill A
x_2 = number of logs shipped from Forest 2 to Mill A
x_3 = number of logs shipped from Forest 1 to Mill B
x_4 = number of logs shipped from Forest 2 to Mill B

Note that if we assign numerical values to these variables, an operational plan for the project is specified and the cost of daily log transportation can be calculated. The selected design may or may not satisfy all constraints.

STEP 4: OPTIMIZATION CRITERION   The design objective is to minimize the daily cost of transporting the logs to the mills. The cost of transportation, which depends on the distance between the forests and the mills, is

$$Cost = 24(10)x_1 + 20.5(10)x_2 + 17.2(10)x_3 + 18(10)x_4$$
$$= 240.0x_1 + 205.0x_2 + 172.0x_3 + 180.0x_4$$

(a)

FIGURE 2.4   Sawmill operation.

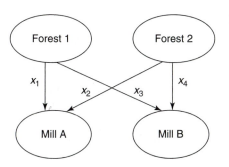

STEP 5: FORMULATION OF CONSTRAINTS The constraints for the problem are based on mill capacity and forest yield:

$$x_1 + x_2 \leq 240 \quad \text{(Mill A capacity)}$$
$$x_3 + x_4 \leq 300 \quad \text{(Mill B capacity)}$$
$$x_1 + x_3 \leq 200 \quad \text{(Forest 1 yield)} \tag{b}$$
$$x_2 + x_4 \leq 200 \quad \text{(Forest 2 yield)}$$

The constraint on the number of logs needed for each day is expressed as

$$x_1 + x_2 + x_3 + x_4 \geq 300 \quad \text{(demand for logs)} \tag{c}$$

For a realistic problem formulation, all design variables must be non-negative; that is,

$$x_i \geq 0; \quad i = 1 \text{ to } 4 \tag{d}$$

The problem has four design variables, five inequality constraints, and four non-negativity constraints on the design variables. Note that all problem functions are linear in design variables, so this is a *linear programming problem*. Note also that for a meaningful solution, all design variables must have *integer* values. Such problems are called *integer programming problems* and require special solution methods. Some such methods are discussed in Chapter 15.

It is also noted that the problem of sawmill operation falls into a class known as *transportation problems*. For such problems, we would like to ship items from several distribution centers to several retail stores to meet their demand at a minimum cost of transportation.

2.5 DESIGN OF A TWO-BAR BRACKET

STEP 1: PROJECT/PROBLEM DESCRIPTION The objective of this project is to design a two-bar bracket (shown in Figure 2.5) to support a load W without structural failure. The load is applied at an angle θ, which is between 0 and $90°$, h is the height, and s is the bracket's base width. The bracket will be produced in large quantities. It has also been determined that its total cost (material, fabrication, maintenance, and so on) is directly related to the size of the two bars. Thus, the design objective is to minimize the total mass of the bracket while satisfying performance, fabrication, and space limitations.

STEP 2: DATA AND INFORMATION COLLECTION First, the load W and its angle of application θ need to be specified. Since the bracket may be used in several applications, it may not be possible to specify just one angle for W. It is possible to formulate the design optimization problem such that a range is specified for angle θ (i.e., load W may be applied at any angle within that specified range). In this case, the formulation will be slightly more complex because performance requirements will need to be satisfied for each angle of application. In the present formulation, it is assumed that angle θ is specified.

Second, the material to be used for the bars must be specified because the material properties are needed to formulate the optimization criterion and performance

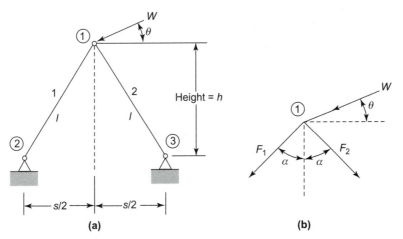

FIGURE 2.5   Two-bar bracket: (a) structure and (b) free-body diagram for node 1.

requirements. Whether the two bars are to be fabricated using the same material also needs to be determined. In the present formulation, it is assumed that they are, although it may be prudent to assume otherwise for some advanced applications. In addition, we need to determine the fabrication and space limitations for the bracket (e.g., on the size of the bars, height, and base width).

In formulating the design problem, we also need to define *structural performance* more precisely. Forces F_1 and F_2 carried by bars 1 and 2, respectively, can be used to define failure conditions for the bars. To compute these forces, we use the principle of *static equilibrium*. Using the *free-body diagram* for node 1 (shown in Figure 2.5(b)), equilibrium of forces in the horizontal and vertical directions gives

$$-F_1\sin\alpha + F_2\sin\alpha = W\ \cos\theta$$
$$-F_1\cos\alpha - F_2\cos\alpha = W\ \sin\theta$$

(a)

From the geometry of Figure 2.5, $\sin\alpha = 0.5\ s/l$ and $\cos\alpha = h/l$, where l is the length of members given as $l = \sqrt{h^2 + (0.5s)^2}$. Note that F_1 and F_2 are shown as tensile forces in the free-body diagram. The solution to Eqs. (a) will determine the magnitude and direction of the forces. In addition, the *tensile force will be taken as positive*. Thus, the bar will be in compression if the force carried by it has negative value. By solving the two equations simultaneously for the unknowns F_1 and F_2, we obtain

$$F_1 = -0.5Wl\left[\frac{\sin\theta}{h} + \frac{2\ \cos\theta}{s}\right]$$
$$F_2 = -0.5Wl\left[\frac{\sin\theta}{h} - \frac{2\ \cos\theta}{s}\right]$$

(b)

To avoid bar failure due to overstressing, we need to calculate bar stress. If we know the force carried by a bar, then the stress σ can be calculated as the force divided by the bar's cross-sectional area (stress = force/area). The SI unit for stress is Newton/meter^2 (N/m^2), also called Pascal (Pa), whereas the U.S.–British unit is pound/in^2 (written as psi). The expression for the cross-sectional area depends on the cross-sectional shape used for the bars and selected design variables. Therefore, a structural shape for the bars and associated design variables must be selected. This is illustrated later in the formulation process.

In addition to analysis equations, we need to define the properties of the selected material. Several formulations for optimum design of the bracket are possible depending on the application's requirements. To illustrate, a material with known properties is assumed for the bracket. However, the structure can be optimized using other materials along with their associated fabrication costs. Solutions can then be compared to select the best possible one for the structure.

For the selected material, let ρ be the mass density and $\sigma_a > 0$ be the allowable design stress. As a performance requirement, it is assumed that if the stress exceeds this allowable value, the bar is considered to have failed. The *allowable stress* is defined as the material failure stress (a property of the material) divided by a factor of safety greater than one. In addition, it is assumed that the allowable stress is calculated in such a way that the buckling failure of a bar in compression is avoided.

STEP 3: DEFINITION OF DESIGN VARIABLES Several sets of design variables may be identified for the two-bar structure. The height h and span s can be treated as design variables in the initial formulation. Later, they may be assigned numerical values, if desired, to eliminate them from the formulation. Other design variables will depend on the cross-sectional shape of bars 1 and 2. Several cross-sectional shapes are possible, as shown in Figure 2.6, where design variables for each shape are also identified.

Note that for many cross-sectional shapes, different design variables can be selected. For example, in the case of the circular tube in Figure 2.6(a), the outer diameter d_o and the ratio between the inner and outer diameters $r = d_i/d_o$ may be selected as the design variables. Or d_o and d_i may be selected. However, it is not desirable to designate d_o, d_i, and r as the design variables because they are not independent of each other. If they are selected, then a relationship between them must be specified as an equality constraint. Similar remarks can be made for the design variables associated with other cross-sections, also shown in Figure 2.6.

As an example of problem formulation, consider the design of a bracket with hollow circular tubes, as shown in Figure 2.6(a). The inner and outer diameters d_i and d_o and wall thickness t may be identified as the design variables, although they are not all independent of each other. For example, we cannot specify $d_i = 10$, $d_o = 12$, and $t = 2$ because it violates the physical condition $t = 0.5(d_o - d_i)$. Therefore, if we formulate the problem with d_i, d_o, and t as design variables, we must also impose the constraint $t = 0.5(d_o - d_i)$. To illustrate a formulation of the problem, let the design variables be defined as

x_1 = height h of the bracket
x_2 = span s of the bracket
x_3 = outer diameter of bar 1
x_4 = inner diameter of bar 1
x_5 = outer diameter of bar 2
x_6 = inner diameter of bar 2

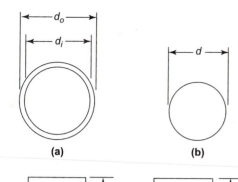

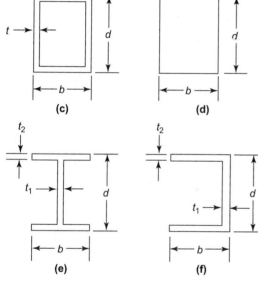

FIGURE 2.6 Bar cross-sectional shapes: (a) circular tube; (b) solid circular; (c) rectangular tube; (d) solid rectangular; (e) I-section; (f) channel section.

In terms of these variables, the cross-sectional areas A_1 and A_2 of bars 1 and 2 are given as

$$A_1 = \frac{\pi}{4}(x_3^2 - x_4^2); \quad A_2 = \frac{\pi}{4}(x_5^2 - x_6^2) \qquad (c)$$

Once the problem is formulated in terms of the six selected design variables, it is always possible to modify it to meet more specialized needs. For example, the height x_1 may be assigned a fixed numerical value, thus eliminating it from the problem formulation. In addition, complete symmetry of the structure may be required to make its fabrication easier; that is, it may be necessary for the two bars to have the same cross-section, size, and material. In such a case, we set $x_3 = x_5$ and $x_4 = x_6$ in all expressions of the problem formulation. Such modifications are left as exercises.

STEP 4: OPTIMIZATION CRITERION The structure's mass is identified as the objective function in the problem statement. Since it is to be minimized, it is called the *cost function* for the problem. An expression for the mass is determined by the cross-sectional shape of the bars and associated design variables. For the hollow circular tubes and selected design variables, the total mass of the structure is calculated as (density × material volume):

$$Mass = \rho[l(A_1 + A_2)] = \left[\rho\sqrt{x_1^2 + (0.5x_2)^2}\right]\frac{\pi}{4}(x_3^2 - x_4^2 + x_5^2 - x_6^2) \qquad \text{(d)}$$

Note that if the outer diameter and the ratio between the inner and outer diameters are selected as design variables, the form of the mass function changes. Thus, the *final form* depends on the design variables selected for the problem.

STEP 5: FORMULATION OF CONSTRAINTS   It is important to include all constraints in the problem formulation because the final solution depends on them. For the two-bar structure, the constraints are on the stress in the bars and on the design variables themselves. These constraints will be formulated for hollow circular tubes using the previously defined design variables. They can be similarly formulated for other sets of design variables and cross-sectional shapes.

To avoid overstressing a bar, the calculated stress σ (tensile or compressive) must not exceed the material allowable stress $\sigma_a > 0$. The stresses σ_1 and σ_2 in the two bars are calculated as force/area:

$$\sigma_1 = \frac{F_1}{A_1} \quad \text{(stress in bar 1)}$$
$$\sigma_2 = \frac{F_2}{A_2} \quad \text{(stress in bar 2)} \qquad \text{(e)}$$

Note that to treat positive and negative stresses (tension and compression), we must use the absolute value of the calculated stress in writing the constraints (e.g., $|\sigma| \leq \sigma_a$). The absolute-value constraints can be treated by different approaches in optimization methods. Here we split each absolute-value constraint into two constraints. For example, the stress constraint for bar 1 is written as the following two constraints:

$$\sigma_1 \leq \sigma_a \quad \text{(tensile stress in bar 1)}$$
$$-\sigma_1 \leq \sigma_a \quad \text{(compressive stress in bar 1)} \qquad \text{(f)}$$

With this approach, the second constraint is satisfied automatically if bar 1 is in tension, and the first constraint is automatically satisfied if bar 1 is in compression. Similarly, the stress constraint for bar 2 is written as

$$\sigma_2 \leq \sigma_a \quad \text{(tensile stress in bar 2)}$$
$$-\sigma_2 \leq \sigma_a \quad \text{(compressive stress in bar 2)} \qquad \text{(g)}$$

Finally, to impose fabrication and space limitations, constraints on the design variables are imposed as

$$x_{iL} \leq x_i \leq x_{iU}; \quad i = 1 \text{ to } 6 \qquad \text{(h)}$$

where x_{iL} and x_{iU} are the minimum and maximum allowed values for the ith design variable. Their numerical values must be specified before the problem can be solved.

Note that the expression for bar stress changes if different design variables are chosen for circular tubes, or if a different cross-sectional shape is chosen for the bars. For example, inner and outer radii, mean radius and wall thickness, or outside diameter and the ratio of inside to outside diameter as design variables will all produce different expressions for the cross-sectional areas and stresses. *These results show that the choice of design variables greatly influences the problem formulation.*

Note also that we had to first *analyze* the structure (calculate its response to given inputs) to write the constraints properly. It was only after we had calculated the forces in the bars that we were able to write the constraints. This is an important step in any engineering design problem formulation: *We must be able to analyze the system before we can formulate the design optimization problem.*

In the following examples, we summarize two formulations of the problem. The first uses several intermediate variables, which is useful when the problem is transcribed into a computer program. Because this formulation involves simpler expressions of various quantities, it is easier to write and debug a computer program. In the second formulation, all intermediate variables are eliminated to obtain the formulation exclusively in terms of design variables. This formulation has slightly more complex expressions. It is important to note that the second formulation may not be possible for all applications because some problem functions may only be implicit functions of the design variables. One such formulation is presented in Chapter 14.

EXAMPLE 2.8 FORMULATION OF THE TWO-BAR BRACKET PROBLEM WITH INTERMEDIATE VARIABLES

A summary of the problem formulation for optimum design of the two-bar bracket using intermediate variables is as follows:

Specified data: W, θ, $\sigma_a > 0$, x_{iL} and x_{iU}, $i = 1$ to 6
Design variables: x_1, x_2, x_3, x_4, x_5, x_6
Intermediate variables:

   Bar cross-sectional areas: $A_1 = \dfrac{\pi}{4}(x_3^2 - x_4^2);\quad A_2 = \dfrac{\pi}{4}(x_5^2 - x_6^2)$ (a)

   Length of bars:

$$l = \sqrt{x_1^2 + (0.5x_2)^2}$$ (b)

   Forces in bars:

$$F_1 = -0.5Wl\left[\frac{\sin\theta}{x_1} + \frac{2\cos\theta}{x_2}\right]$$

     (c)

$$F_2 = -0.5Wl\left[\frac{\sin\theta}{x_1} - \frac{2\cos\theta}{x_2}\right]$$

   Bar stresses:

$$\sigma_1 = \frac{F_1}{A_1};\quad \sigma_2 = \frac{F_2}{A_2}$$ (d)

Cost function: Minimize the total mass of the bars,

$$\text{Mass} = \rho l(A_1 + A_2)$$ (e)

Constraints:

Bar stress:

$$-\sigma_1 \le \sigma_a; \quad \sigma_1 \le \sigma_a; \quad -\sigma_2 \le \sigma_a; \quad \sigma_2 \le \sigma_a \tag{f}$$

Design variable limits:

$$x_{iL} \le x_i \le x_{iU}; \quad i = 1 \text{ to } 6 \tag{g}$$

Note that the intermediate variables, such as A_1, A_2, F_1, F_2, σ_1, and σ_2, may also be treated as optimization variables. However, in that case, we have six equality constraints between the variables, in addition to the other constraints.

EXAMPLE 2.9 FORMULATION OF THE TWO-BAR BRACKET WITH DESIGN VARIABLES ONLY

A summary of the problem formulation for optimum design of the two-bar bracket in terms of design variables only is obtained by eliminating the intermediate variables from all the expressions as follows:

Specified data: W, θ, $\sigma_a > 0$, x_{iL} and x_{iU}, $i = 1$ to 6
Design variables: x_1, x_2, x_3, x_4, x_5, x_6
Cost function: Minimize total mass of the bars,

$$Mass = \frac{\pi\rho}{4}\sqrt{x_1^2 + (0.5x_2)^2}\,(x_3^2 - x_4^2 + x_5^2 - x_6^2) \tag{a}$$

Constraints:

Bar stress:

$$\frac{2W\sqrt{x_1^2 + (0.5x_2)^2}}{\pi(x_3^2 - x_4^2)}\left[\frac{\sin\theta}{x_1} + \frac{2\cos\theta}{x_2}\right] \le \sigma_a \tag{b}$$

$$\frac{-2W\sqrt{x_1^2 + (0.5x_2)^2}}{\pi(x_3^2 - x_4^2)}\left[\frac{\sin\theta}{x_1} + \frac{2\cos\theta}{x_2}\right] \le \sigma_a \tag{c}$$

$$\frac{2W\sqrt{x_1^2 + (0.5x_2)^2}}{\pi(x_5^2 - x_6^2)}\left[\frac{\sin\theta}{x_1} - \frac{2\cos\theta}{x_2}\right] \le \sigma_a \tag{d}$$

$$\frac{-2W\sqrt{x_1^2 + (0.5x_2)^2}}{\pi(x_5^2 - x_6^2)}\left[\frac{\sin\theta}{x_1} - \frac{2\cos\theta}{x_2}\right] \le \sigma_a \tag{e}$$

Design variable limits:

$$x_{iL} \le x_i \le x_{iU}; \quad i = 1 \text{ to } 6 \tag{f}$$

2.6 DESIGN OF A CABINET

STEP 1: PROJECT/PROBLEM DESCRIPTION A cabinet is assembled from components C_1, C_2, and C_3. Each cabinet requires 8 C_1, 5 C_2, and 15 C_3 components. The assembly of C_1 requires either 5 bolts or 5 rivets, whereas C_2 requires 6 bolts or 6 rivets, and C_3 requires 3 bolts or 3 rivets. The cost of installing a bolt, including the cost of the bolt itself, is $0.70 for C_1, $1.00 for C_2, and $0.60 for C_3. Similarly, riveting costs are $0.60 for C_1, $0.80 for C_2, and $1.00 for C_3. Bolting and riveting capacities per day are 6000 and 8000, respectively. To minimize the cost for the 100 cabinets that must be assembled each day, we wish to determine the number of components to be bolted and riveted (after Siddall, 1972).

STEP 2: DATA AND INFORMATION COLLECTION All data for the problem are given in the project statement. This problem can be formulated in several different ways depending on the assumptions made and the definition of the design variables. Three formulations are presented, and for each one, the design variables are identified and expressions for the cost and constraint functions are derived; that is, steps 3 through 5 are presented.

2.6.1 Formulation 1 for Cabinet Design

STEP 3: DEFINITION OF DESIGN VARIABLES In the first formulation, the following design variables are identified for 100 cabinets:

x_1 = number of C_1 to be bolted for 100 cabinets
x_2 = number of C_1 to be riveted for 100 cabinets
x_3 = number of C_2 to be bolted for 100 cabinets
x_4 = number of C_2 to be riveted for 100 cabinets
x_5 = number of C_3 to be bolted for 100 cabinets
x_6 = number of C_3 to be riveted for 100 cabinets

STEP 4: OPTIMIZATION CRITERION The design objective is to minimize the total cost of cabinet fabrication, which is obtained from the specified costs for bolting and riveting each component:

$$\text{Cost} = 0.70(5)x_1 + 0.60(5)x_2 + 1.00(6)x_3 + 0.80(6)x_4 + 0.60(3)x_5 + 1.00(3)x_6$$
$$= 3.5x_1 + 3.0x_2 + 6.0x_3 + 4.8x_4 + 1.8x_5 + 3.0x_6 \tag{a}$$

STEP 5: FORMULATION OF CONSTRAINTS The constraints for the problem consist of riveting and bolting capacities and the number of cabinets fabricated each day. Since 100 cabinets must be fabricated, the required numbers of C_1, C_2, and C_3 are given in the following constraints:

$$x_1 + x_2 = 8 \times 100 \quad \text{(number of } C_1 \text{ needed)}$$
$$x_3 + x_4 = 5 \times 100 \quad \text{(number of } C_2 \text{ needed)} \tag{b}$$
$$x_5 + x_6 = 15 \times 100 \quad \text{(number of } C_3 \text{ needed)}$$

Bolting and riveting capacities must not be exceeded. Thus,

$$5x_1 + 6x_3 + 3x_5 \leq 6000 \quad \text{(bolting capacity)}$$
$$5x_2 + 6x_4 + 3x_6 \leq 8000 \quad \text{(riveting capacity)}$$

(c)

Finally, all design variables must be non-negative to find a meaningful solution:

$$x_i \geq 0; \quad i = 1 \text{ to } 6$$

(d)

2.6.2 Formulation 2 for Cabinet Design

STEP 3: DEFINITION OF DESIGN VARIABLES   If we relax the constraint that each component must be bolted or riveted, then the following design variables can be defined:

x_1 = total number of bolts required for all C_1
x_2 = total number of bolts required for all C_2
x_3 = total number of bolts required for all C_3
x_4 = total number of rivets required for all C_1
x_5 = total number of rivets required for all C_2
x_6 = total number of rivets required for all C_3

STEP 4: OPTIMIZATION CRITERION   The objective is still to minimize the total cost of fabricating 100 cabinets, given as

$$Cost = 0.70x_1 + 1.00x_2 + 0.60x_3 + 0.60x_4 + 0.80x_5 + 1.00x_6$$

(e)

STEP 5: FORMULATION OF CONSTRAINTS   Since 100 cabinets must be built every day, it will be necessary to have 800 C_1, 500 C_2, and 1500 C_3 components. The total number of bolts and rivets needed for all C_1, C_2, and C_3 components is indicated by the following equality constraints:

$$x_1 + x_4 = 5 \times 800 \quad \text{(bolts and rivets needed for } C_1)$$
$$x_2 + x_5 = 6 \times 500 \quad \text{(bolts and rivets needed for } C_2)$$
$$x_3 + x_6 = 3 \times 1500 \quad \text{(bolts and rivets needed for } C_3)$$

(f)

Constraints on capacity for bolting and riveting are

$$x_1 + x_2 + x_3 \leq 6000 \quad \text{(bolting capacity)}$$
$$x_4 + x_5 + x_6 \leq 8000 \quad \text{(riveting capacity)}$$

(g)

Finally, all design variables must be non-negative:

$$x_i \geq 0; \quad i = 1 \text{ to } 6$$

(h)

Thus, this formulation also has six design variables, three equality constraints, and two inequality constraints. After an optimum solution has been obtained, we can decide how many components to bolt and how many to rivet.

2.6.3 Formulation 3 for Cabinet Design

STEP 3: DEFINITION OF DESIGN VARIABLES Another formulation of the problem is possible if we require that all cabinets be identical. The following design variables can be identified:

x_1 = number of C_1 to be bolted on one cabinet
x_2 = number of C_1 to be riveted on one cabinet
x_3 = number of C_2 to be bolted on one cabinet
x_4 = number of C_2 to be riveted on one cabinet
x_5 = number of C_3 to be bolted on one cabinet
x_6 = number of C_3 to be riveted on one cabinet

STEP 4: OPTIMIZATION CRITERION With these design variables, the cost of fabricating 100 cabinets each day is given as

$$Cost = 100[0.70(5)x_1 + 0.60(5)x_2 + 1.00(6)x_3 + 0.80(6)x_4 + 0.60(3)x_5 + 1.00(3)x_6]$$
$$= 350x_1 + 300x_2 + 600x_3 + 480x_4 + 180x_5 + 300x_6 \tag{i}$$

STEP 5: FORMULATION OF CONSTRAINTS Since each cabinet needs 8 C_1, 5 C_2, and 15 C_3 components, the following equality constraints can be identified:

$$x_1 + x_2 = 8 \quad \text{(number of } C_1 \text{ needed)}$$
$$x_3 + x_4 = 5 \quad \text{(number of } C_2 \text{ needed)} \tag{j}$$
$$x_5 + x_6 = 15 \quad \text{(number of } C_3 \text{ needed)}$$

Constraints on the capacity to rivet and bolt are expressed as the following inequalities:

$$(5x_1 + 6x_3 + 3x_5)100 \leq 6000 \quad \text{(bolting capacity)}$$
$$(5x_2 + 6x_4 + 3x_6)100 \leq 8000 \quad \text{(riveting capacity)} \tag{k}$$

Finally, all design variables must be non-negative:

$$x_i \geq 0; \quad i = 1 \text{ to } 6 \tag{l}$$

The following points are noted for the three formulations:

1. Because cost and constraint functions are *linear* in all three formulations, they are linear programming problems. It is conceivable that each formulation will yield a different optimum solution. After solving the problems, the designer can select the best strategy for fabricating cabinets.
2. All formulations have *three equality constraints*, each involving two design variables. Using these constraints, we can eliminate three variables from the problem and thus reduce its dimension. This is desirable from a computational standpoint because the number of variables and constraints is reduced. However, because the elimination of variables is not possible for many complex problems, we must develop and use methods to treat both equality and inequality constraints.

3. For a meaningful solution for these formulations, all design variables must have integer values. These are called *integer programming problems*. Some numerical methods to treat this class of problem are discussed in Chapter 15.

2.7 MINIMUM-WEIGHT TUBULAR COLUMN DESIGN

STEP 1: PROJECT/PROBLEM DESCRIPTION   Straight columns are used as structural elements in civil, mechanical, aerospace, agricultural, and automotive structures. Many such applications can be observed in daily life—for example, a street light pole, a traffic light post, a flag pole, a water tower support, a highway sign post, a power transmission pole. It is important to optimize the design of a straight column since it may be mass-produced. The objective of this project is to design a minimum-mass *tubular* column of length l supporting a load P without buckling or overstressing. The column is fixed at the base and free at the top, as shown in Figure 2.7. This type of structure is called a cantilever column.

STEP 2: DATA AND INFORMATION COLLECTION   The *buckling load* (also called the *critical load*) for a cantilever column is given as

$$P_{cr} = \frac{\pi^2 EI}{4l^2} \tag{a}$$

The buckling load formula for a column with other support conditions is different from this formula (Crandall, Dahl, and Lardner, 1999). Here, I is the moment of inertia for the cross-section of the column and E is the material property, called the modulus of elasticity (Young's modulus). Note that the buckling load depends on the design of the column (i.e., the moment of inertia I). It imposes a limit on the applied load; that is, the column fails if the applied load exceeds the buckling load. The material stress σ for the column is defined as P/A, where A is the cross-sectional area of the column. The material allowable stress under the axial load is σ_a, and the material mass density is ρ (mass per unit volume).

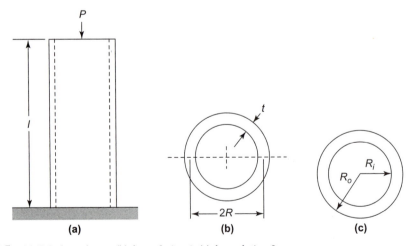

FIGURE 2.7   (a) Tubular column; (b) formulation 1; (c) formulation 2.

A cross-section of the tubular column is shown in Figure 2.7. Many formulations for the design problem are possible depending on how the design variables are defined. Two such formulations are described here.

2.7.1 Formulation 1 for Column Design

STEP 3: DEFINITION OF DESIGN VARIABLES   For the first formulation, the following design variables are defined:

R = mean radius of the column
t = wall thickness

Assuming that the column wall is thin ($R \gg t$), the material cross-sectional area and moment of inertia are

$$A = 2\pi Rt; \quad I = \pi R^3 t \tag{b}$$

STEP 4: OPTIMIZATION CRITERION   The total mass of the column to be minimized is given as

$$Mass = \rho(lA) = 2\rho l\pi Rt \tag{c}$$

STEP 5: FORMULATION OF CONSTRAINTS   The first constraint is that the stress (P/A) should not exceed the material allowable stress σ_a to avoid material failure. This is expressed as the inequality $\sigma \leq \sigma_a$. Replacing σ with P/A and then substituting for A, we obtain

$$\frac{P}{2\pi Rt} \leq \sigma_a \tag{d}$$

The column should not buckle under the applied load P, which implies that the applied load should not exceed the buckling load (i.e., $P \leq P_{cr}$). Using the given expression for the buckling load in Eq. (a) and substituting for I, we obtain

$$P \leq \frac{\pi^3 ER^3 t}{4l^2} \tag{e}$$

Finally, the design variables R and t must be within the specified minimum and maximum values:

$$R_{min} \leq R \leq R_{max}; \quad t_{min} \leq t \leq t_{max} \tag{f}$$

2.7.2 Formulation 2 for Column Design

STEP 3: DEFINITION OF DESIGN VARIABLES   Another formulation of the design problem is possible if the following design variables are defined:

R_o = outer radius of the column
R_i = inner radius of the column

In terms of these design variables, the cross-sectional area A and the moment of inertia I are

$$A = \pi(R_o^2 - R_i^2); \quad I = \frac{\pi}{4}(R_o^4 - R_i^4).$$ (g)

STEP 4: OPTIMIZATION CRITERION   Minimize the total mass of the column:

$$Mass = \rho(lA) = \pi\rho l(R_o^2 - R_i^2)$$ (h)

STEP 5: FORMULATION OF THE CONSTRAINTS   The material crushing constraint is $(P/A \le \sigma_a)$:

$$\frac{P}{\pi(R_o^2 - R_i^2)} \le \sigma_a$$ (i)

Using the foregoing expression for I, the buckling load constraint is $(P \le P_{cr})$:

$$P \le \frac{\pi^3 E}{16l^3}(R_o^4 - R_i^4)$$ (j)

Finally, the design variables R_o and R_i must be within specified limits:

$$R_{o\ min} \le R_o \le R_{o\ max}; \quad R_{i\ min} \le R_i \le R_{o\ max}$$ (k)

When this problem is solved using a numerical method, a constraint $R_o > R_i$ must also be imposed. Otherwise, some methods may take the design to the point where $R_o < R_i$. This situation is not physically possible and must be explicitly excluded to numerically solve the design problem.

In addition to the foregoing constraints, local buckling of the column wall needs to be considered for both formulations. Local buckling can occur if the wall thickness becomes too small. This can be avoided if the ratio of mean radius to wall thickness is required to be smaller than a limiting value, that is,

$$\frac{(R_o + R_i)}{2(R_o - R_i)} \le k \quad or \quad \frac{R}{t} \le k$$ (l)

where k is a specified value that depends on Young's modulus and the yield stress of the material. For steel with $E = 29{,}000$ ksi and a yield stress of 50 ksi, k is given as 32 (AISC, 2005).

2.8 MINIMUM-COST CYLINDRICAL TANK DESIGN

STEP 1: PROJECT/PROBLEM DESCRIPTION   Design a minimum-cost cylindrical tank closed at both ends to contain a fixed volume of fluid V. The cost is found to depend directly on the area of sheet metal used.

STEP 2: DATA AND INFORMATION COLLECTION   Let c be the dollar cost per unit area of the sheet metal. Other data are given in the project statement.

STEP 3: DEFINITION OF DESIGN VARIABLES   The design variables for the problem are identified as

R = radius of the tank
H = height of the tank

STEP 4: OPTIMIZATION CRITERION   The cost function for the problem is the dollar cost of the sheet metal for the tank. Total surface area of the sheet metal consisting of the end plates and cylinder is given as

$$A = 2\pi R^2 + 2\pi RH \tag{a}$$

Therefore, the cost function for the problem is given as

$$f = c(2\pi R^2 + 2\pi RH) \tag{b}$$

STEP 5: FORMULATION OF CONSTRAINTS   The volume of the tank ($\pi R^2 H$) is required to be V. Therefore,

$$\pi R^2 H = V \tag{c}$$

Also, both of the design variables R and H must be within some minimum and maximum values:

$$R_{min} \le R \le R_{max}; \quad H_{min} \le H \le H_{max} \tag{d}$$

This problem is quite similar to the can problem discussed in Section 2.2. The only difference is in the volume constraint. There the constraint is an inequality and here it is an equality.

2.9   DESIGN OF COIL SPRINGS

STEP 1: PROJECT/PROBLEM DESCRIPTION   Coil springs are used in numerous practical applications. Detailed methods for analyzing and designing such mechanical components have been developed over the years (e.g., Spotts, 1953; Wahl, 1963; Shigley, Mischke, and Budynas, 2004; Haug and Arora, 1979). The purpose of this project is to design a minimum-mass spring (shown in Figure 2.8) to carry a given axial load (called a tension-compression spring) without material failure and while satisfying two performance requirements: The spring must deflect by at least Δ (in) and the frequency of surge waves must not be less than ω_0 (Hertz, Hz).

STEP 2: DATA AND INFORMATION COLLECTION   To formulate the problem of designing a coil spring, see the notation and data defined in Table 2.2.

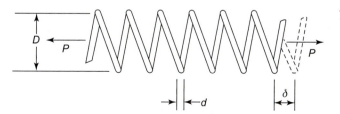

FIGURE 2.8   Coil spring.

TABLE 2.2   Information to design a coil spring

Notation	Data
Deflection along the axis of spring	δ, in
Mean coil diameter	D, in
Wire diameter	d, in
Number of active coils	N
Gravitational constant	$g = 386$ in/s^2
Frequency of surge waves	ω, Hz
Weight density of spring material	$\gamma = 0.285$ lb/in^3
Shear modulus	$G = (1.15 \times 10^7)$ lb/in^2
Mass density of material ($\rho = \gamma/g$)	$\rho = (7.38342 \times 10^{-4})$ lb-s^2/in^4
Allowable shear stress	$\tau_a = 80{,}000$ lb/in^2
Number of inactive coils	$Q = 2$
Applied load	$P = 10$ lb
Minimum spring deflection	$\Delta = 0.5$ in
Lower limit on surge wave frequency	$\omega_0 = 100$ Hz
Limit on outer diameter of coil	$D_o = 1.5$ in

The wire twists when the spring is subjected to a tensile or a compressive load. Therefore, shear stress needs to be calculated so that a constraint on it can be included in the formulation. In addition, surge wave frequency needs to be calculated. These and other design equations for the spring are given as

Load deflection equation:
$$P = K\delta \qquad\qquad\qquad (a)$$

Spring constant:
$$K = \frac{d^4 G}{8D^3 N} \qquad\qquad\qquad (b)$$

Shear stress:
$$\tau = \frac{8kPD}{\pi d^3} \qquad\qquad\qquad (c)$$

Wahl stress concentration factor:
$$k = \frac{(4D - d)}{4(D - d)} + \frac{0.615d}{D} \qquad\qquad\qquad (d)$$

Frequency of surge waves:
$$\omega = \frac{d}{2\pi ND^2} \sqrt{\frac{G}{2\rho}} \qquad\qquad\qquad (e)$$

The expression for the Wahl stress concentration factor k in Eq. (d) has been determined experimentally to account for unusually high stresses at certain points on the spring. These analysis equations are used to define the constraints.

STEP 3: DEFINITION OF DESIGN VARIABLES   The three design variables for the problem are defined as

d = wire diameter, in
D = mean coil diameter, in
N = number of active coils, integer

STEP 4: OPTIMIZATION CRITERION   The problem is to *minimize the mass* of the spring, given as volume × mass density:

$$Mass = \left(\frac{\pi}{4}d^2\right)[(N + Q)\pi D]\rho = \frac{1}{4}(N + Q)\pi^2 Dd^2 \rho \tag{f}$$

STEP 5: FORMULATION OF CONSTRAINTS
Deflection constraint. It is often a requirement that *deflection* under a load P be at least Δ. Therefore, the constraint is that the calculated deflection δ must be greater than or equal to Δ. Such a constraint is common to spring design. The function of the spring in many applications is to provide a modest restoring force as parts undergo large displacement in carrying out kinematic functions. Mathematically, this performance requirement ($\delta \geq \Delta$) is stated in an inequality form, using Eq. (a), as

$$\frac{P}{K} \geq \Delta \tag{g}$$

Shear stress constraint. To prevent material overstressing, *shear stress* in the wire must be no greater than τ_a, which is expressed in mathematical form as

$$\tau \leq \tau_a \tag{h}$$

Constraint on the frequency of surge waves. We also wish to avoid resonance in dynamic applications by making the *frequency of surge waves* (along the spring) as great as possible. For the present problem, we require the frequency of surge waves for the spring to be at least ω_0 (Hz). The constraint is expressed in mathematical form as

$$\omega \geq \omega_0 \tag{i}$$

Diameter constraint. The *outer diameter* of the spring should not be greater than D_o, so

$$D + d \leq D_0 \tag{j}$$

Explicit bounds on design variables. To avoid fabrication and other practical difficulties, we put *minimum and maximum size limits* on the wire diameter, coil diameter, and number of turns:

$$d_{min} \leq d \leq d_{max}$$
$$D_{min} \leq D \leq D_{max} \tag{k}$$
$$N_{min} \leq N \leq N_{max}$$

Thus, the purpose of the minimum-mass spring design problem is to select the design variables d, D, and N to minimize the mass of Eq. (f), while satisfying the ten inequality constraints of Eqs. (g) through (k). If the intermediate variables are eliminated, the problem formulation can be summarized in terms of the design variables only.

EXAMPLE 2.10 FORMULATION OF THE SPRING DESIGN PROBLEM WITH DESIGN VARIABLES ONLY

A summary of the problem formulation for the optimum design of coil springs is as follows:

Specified data: Q, P, ρ, γ, τ_a, G, Δ, ω_0, D_0, d_{min}, d_{max}, D_{min}, D_{max}, N_{min}, N_{max}
Design variables: d, D, N
Cost function: Minimize the mass of the spring given in Eq. (f).
Constraints:

Deflection limit:
$$\frac{8PD^3N}{d^4G} \geq \Delta \tag{l}$$

Shear stress:
$$\frac{8PD}{\pi d^3}\left[\frac{(4D-d)}{4(D-d)} + \frac{0.615d}{D}\right] \leq \tau_a \tag{m}$$

Frequency of surge waves:
$$\frac{d}{2\pi ND^2}\sqrt{\frac{G}{2\rho}} \geq \omega_0 \tag{n}$$

Diameter constraint: Given in Eq. (j).
Design variable bounds: Given in Eqs. (k).

2.10 MINIMUM-WEIGHT DESIGN OF A SYMMETRIC THREE-BAR TRUSS

STEP 1: PROJECT/PROBLEM DESCRIPTION As an example of a slightly more complex design problem, consider the three-bar structure shown in Figure 2.9 (Schmit, 1960; Haug and Arora, 1979). This is a statically indeterminate structure for which the member forces cannot be calculated solely from equilibrium equations. The structure is to be designed for minimum volume (or, equivalently, minimum mass) to support a force P. It must satisfy various performance and technological constraints, such as member crushing, member buckling, failure by excessive deflection of node 4, and failure by resonance when the natural frequency of the structure is below a given threshold.

STEP 2: DATA AND INFORMATION COLLECTION Needed to solve the problem are geometry data, properties of the material used, and loading data. In addition, since the structure is statically indeterminate, the static equilibrium equations alone are not enough to analyze it. We need to use advanced analysis procedures to obtain expressions for member forces,

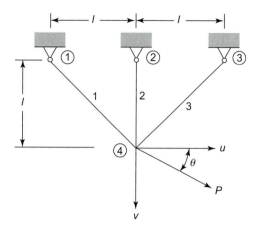

FIGURE 2.9   Three-bar truss.

nodal displacements, and the natural frequency to formulate constraints for the problem. Here we will give such expressions.

Since the structure must be symmetric, members 1 and 3 will have the same cross-sectional area, say A_1. Let A_2 be the cross-sectional area of member 2. Using analysis procedures for statically indeterminate structures, horizontal and vertical displacements u and v of node 4 are calculated as

$$u = \frac{\sqrt{2}lP_u}{A_1 E}; \quad v = \frac{\sqrt{2}lP_v}{(A_1 + \sqrt{2}A_2)E} \tag{a}$$

where E is the modulus of elasticity for the material, P_u and P_v are the horizontal and vertical components of the applied load P given as $P_u = P\cos\theta$ and $P_v = P\sin\theta$, and l is the height of the truss as shown in Figure 2.9. Using the displacements, forces carried by the members of the truss can be calculated. Then the stresses σ_1, σ_2, and σ_3 in members 1, 2, and 3 under the applied load P can be computed from member forces as (stress = force/area; $\sigma_i = F_i/A_i$):

$$\sigma_1 = \frac{1}{\sqrt{2}}\left[\frac{P_u}{A_1} + \frac{P_v}{(A_1 + \sqrt{2}A_2)}\right] \tag{b}$$

$$\sigma_2 = \frac{\sqrt{2}P_v}{(A_1 + \sqrt{2}A_2)} \tag{c}$$

$$\sigma_3 = \frac{1}{\sqrt{2}}\left[-\frac{P_u}{A_1} + \frac{P_v}{(A_1 + \sqrt{2}A_2)}\right] \tag{d}$$

Note that the member forces, and hence stresses, are dependent on the design of the structure, that is, the member areas.

Many structures support moving machinery and other dynamic loads. These structures vibrate with a certain frequency known as *natural frequency*. This is an intrinsic dynamic property of a structural system. There can be several modes of vibration, each having its own frequency. *Resonance* causes catastrophic failure of the structure, which occurs when any one of its vibration frequencies coincides with the frequency of the operating machinery it supports.

Therefore, it is reasonable to demand that no structural frequency be close to the frequency of the operating machinery. The mode of vibration corresponding to the lowest natural frequency is important because that mode is excited first. It is important to make the lowest (fundamental) natural frequency of the structure as high as possible to avoid any possibility of resonance. This also makes the structure stiffer. Frequencies of a structure are obtained by solving an eigenvalue problem involving the structure's stiffness and mass properties. The lowest eigenvalue ζ related to the lowest natural frequency of the symmetric three-bar truss is computed using a consistent-mass model:

$$\zeta = \frac{3EA_1}{\rho l^2 \left(4A_1 + \sqrt{2}A_2\right)} \tag{e}$$

where ρ is the material mass per unit volume (mass density). This completes the analysis of the structure.

STEP 3: DEFINITION OF DESIGN VARIABLES   The following design variables are defined for the symmetric structure:

A_1 = cross-sectional area of material for members 1 and 3
A_2 = cross-sectional area of material for member 2

Other design variables for the problem are possible depending on the cross-sectional shape of members, as shown in Figure 2.6.

STEP 4: OPTIMIZATION CRITERION   The relative merit of any design for the problem is measured in its material weight. Therefore, the total weight of the structure serves as a cost function (weight of member = cross-sectional area × length × weight density):

$$Volume = l\gamma\left(2\sqrt{2}A_1 + A_2\right) \tag{f}$$

where γ is the weight density.

STEP 5: FORMULATION OF CONSTRAINTS   The structure is designed for use in two applications. In each application, it supports different loads. These are called loading conditions for the structure. In the present application, a symmetric structure is obtained if the following two loading conditions are considered. The first load is applied at an angle θ and the second one, of same magnitude, at an angle $(\pi - \theta)$, where the angle θ ($0° \leq \theta \leq 90°$) is shown earlier in Figure 2.9. If we let member 1 be the same as member 3, then the second loading condition can be ignored. Therefore, we consider only one load applied at an angle θ ($0° \leq \theta \leq 90°$).

Note from Eqs. (b) and (c) that the stresses σ_1 and σ_2 are always positive (tensile). If $\sigma_a > 0$ is an allowable stress for the material, then the *stress constraints* for members 1 and 2 are

$$\sigma_1 \leq \sigma_a; \quad \sigma_2 \leq \sigma_a \tag{g}$$

However, from Eq. (c), stress in member 3 can be positive (tensile) or negative (compressive) depending on the load angle. Therefore, both possibilities need to be considered in formulating the stress constraint for member 3. One way to formulate such a constraint was explained in Section 2.5. Another way is as follows:

$$\text{IF } (\sigma_3 < 0) \text{ THEN } -\sigma_3 \le \sigma_a \text{ ELSE } \sigma_3 \le \sigma_a \tag{h}$$

Since the sign of the stress does not change with design, if the member is in compression, it remains in compression throughout the optimization process. Therefore, the constraint function remains continuous and differentiable.

A similar procedure can be used for stresses in bars 1 and 2 if the stresses can reverse their sign (e.g., when the load direction is reversed). Horizontal and vertical deflections of node 4 must be within the specified limits Δ_u and Δ_v, respectively. Using Eq. (a), the *deflection constraints* are

$$u \le \Delta_u; \quad v \le \Delta_v \tag{i}$$

As discussed previously, the *fundamental natural frequency* of the structure should be higher than a specified frequency ω_0 (Hz). This constraint can be written in terms of the lowest eigenvalue for the structure. The eigenvalue corresponding to a frequency of ω_0 (Hz) is given as $(2\pi\omega_0)^2$. The lowest eigenvalue ζ for the structure given in Eq. (e) should be higher than $(2\pi\omega_0)^2$, that is,

$$\zeta \ge (2\pi\omega_0)^2 \tag{j}$$

To impose *buckling constraints* for members under compression, an expression for the moment of inertia of the cross-section is needed. This expression cannot be obtained because the cross-sectional shape and dimensions are not specified. However, the moment of inertia I can be related to the cross-sectional area of the members as $I = \beta A^2$, where A is the cross-sectional area and β is a nondimensional constant. This relation follows if the shape of the cross-section is fixed and all of its dimensions are varied in the same proportion.

The axial force for the ith member is given as $F_i = A_i\sigma_i$, where $i = 1, 2, 3$ with tensile force taken as positive. Members of the truss are considered columns with pin ends. Therefore, the buckling load for the ith member is given as $\pi^2 EI/l_i^2$, where l_i is the length of the ith member (Crandall, Dahl, and Lardner, 1999). Buckling constraints are expressed as $-F_i \le \pi^2 EI/l_i^2$, where $i = 1, 2, 3$. The negative sign for F_i is used to make the left side of the constraints positive when the member is in compression. Also, there is no need to impose buckling constraints for members in tension. With the foregoing formulation, the buckling constraint for tensile members is automatically satisfied. Substituting various quantities, member buckling constraints are

$$-\sigma_1 \le \frac{\pi^2 E\beta A_1}{2l^2} \le \sigma_a; \quad -\sigma_2 \le \frac{\pi^2 E\beta A_2}{l^2} \le \sigma_a; \quad -\sigma_3 \le \frac{\pi^2 E\beta A_1}{2l^2} \le \sigma_a \tag{k}$$

Note that the buckling load has been divided by the member area to obtain the buckling stress in Eqs. (k). The buckling stress is required not to exceed the material allowable stress σ_a. It is additionally noted that with the foregoing formulation, the load P in Figure 2.9 can be applied in the positive or negative direction. When the load is applied in the opposite direction, the member forces are also reversed. The foregoing formulation for the buckling constraints can treat both positive and negative load in the solution process.

Finally, A_1 and A_2 must both be non-negative, that is, $A_1, A_2 \geq 0$. Most practical design problems require each member to have a certain minimum area, A_{min}. The minimum area constraints can be written as

$$A_1, \quad A_2 \geq A_{min} \tag{1}$$

The optimum design problem, then, is to find cross-sectional areas $A_1, A_2 \geq A_{min}$ to minimize the volume of Eq. (f) subject to the constraints of Eqs. (g) through (l). This small-scale problem has 11 inequality constraints and 2 design variables.

2.11 A GENERAL MATHEMATICAL MODEL FOR OPTIMUM DESIGN

To describe optimization concepts and methods, we need a general mathematical statement for the optimum design problem. Such a mathematical model is defined as the minimization of a cost function while satisfying all equality and inequality constraints. The inequality constraints in the model are always transformed as "$\leq$ types." This will be called the *standard design optimization model* that is treated throughout this text. It will be shown that all design problems can easily be transcribed into the standard form.

2.11.1 Standard Design Optimization Model

In previous sections, several design problems were formulated. All problems have an optimization criterion that can be used to compare various designs and to determine an optimum or the best one. Most design problems must also satisfy certain constraints. Some design problems have only inequality constraints, others have only equality constraints, and some have both. We can define a general mathematical model for optimum design to encompass all of the possibilities. A standard form of the model is first stated, and then transformation of various problems into the standard form is explained.

Standard Design Optimization Model

Find an n-vector $\mathbf{x} = (x_1, x_2, \ldots, x_n)$ of design variables to

Minimize a cost function:

$$f(\mathbf{x}) = f(x_1, x_2, \ldots, x_n) \tag{2.1}$$

subject to the p equality constraints:

$$h_j(\mathbf{x}) = h_j(x_1, x_2, \ldots, x_n) = 0; \quad j = 1 \text{ to } p \tag{2.2}$$

and the m inequality constraints:

$$g_i(\mathbf{x}) = g_i(x_1, x_2, \ldots, x_n) \leq 0; \quad i = 1 \text{ to } m \tag{2.3}$$

Note that the simple bounds on design variables, such as $x_i \geq 0$, or $x_{iL} \leq x_i \leq x_{iU}$, where x_{iL} and x_{iU} are the smallest and largest allowed values for x_i, are assumed to be included in the inequalities of Eq. (2.3).

In numerical methods, these constraints are treated explicitly to take advantage of their simple form to achieve efficiency. However, in discussing the basic optimization concepts, we assume that the inequalities in Eq. (2.3) include these constraints as well.

2.11.2 Maximization Problem Treatment

The general design model treats only minimization problems. This is no restriction, as maximization of a function $F(\mathbf{x})$ is the same as minimization of a transformed function $f(\mathbf{x}) = -F(\mathbf{x})$. To see this graphically, consider a plot of the function of one variable $F(x)$, shown in Figure 2.10(a). The function $F(x)$ takes its maximum value at the point x^*. Next consider a graph of the function $f(x) = -F(x)$, shown in Figure 2.10(b). It is clear that $f(x)$ is a reflection of $F(x)$ about the x-axis. It is also clear from the graph that $f(x)$ takes on a minimum value at the same point x^* where the maximum of $F(x)$ occurs. Therefore, minimization of $f(x)$ is equivalent to maximization of $F(x)$.

2.11.3 Treatment of "Greater Than Type" Constraints

The standard design optimization model treats only "$\le$ type" inequality constraints. Many design problems may also have "$\ge$ type" inequalities. Such constraints can be

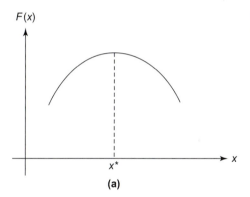

$F(x)$

x^*

(a)

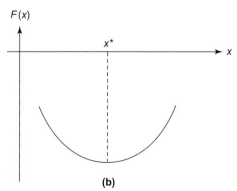

$F(x)$

x^*

(b)

FIGURE 2.10 Point maximizing $F(x)$ equals point minimizing $-F(x)$: (a) plot of $F(x)$; (b) plot of $f(x) = -F(x)$.

converted to the standard form without much difficulty. The " $\geq$ type" constraint $G_j(x) \geq 0$ is equivalent to the " $\leq$ type" inequality $g_j(x) = -G_j(x) \leq 0$. Therefore, we can multiply any " $\geq$ type" constraint by -1 to convert it to a " $\leq$ type."

2.11.4 Application to Different Engineering Fields

Design optimization problems from different fields of engineering can be transcribed into the standard model. It must be realized that *the overall process of designing different engineering systems is the same.* Analytical and numerical methods for analyzing systems can differ. Formulation of the design problem can contain terminology that is specific to the particular domain of application. For example, in the fields of structural, mechanical, and aerospace engineering, we are concerned with the integrity of the structure and its components. The performance requirements involve constraints on member stresses, strains, deflections at key points, frequencies of vibration, buckling failure, and so on. These concepts are specific to each field, and designers working in the particular field understand their meaning and the constraints.

Other fields of engineering also have their own terminology to describe design optimization problems. However, once the problems from different fields have been transcribed into mathematical statements using a standard notation, they have the same mathematical form. They are contained in the standard design optimization model defined in Eqs. (2.1) through (2.3). For example, all of the problems formulated earlier in this chapter can be transformed into the form of Eqs. (2.1) through (2.3). Therefore, the optimization concepts and methods described in the text are quite general and can be used to solve problems from diverse fields. *The methods can be developed without reference to any design application.* This is a key point and must be kept in mind while studying the optimization concepts and methods.

2.11.5 Important Observations about the Standard Model

Several points must be clearly understood about the standard model:

1. *Dependence of functions on design variables:* First of all, the functions $f(x)$, $h_j(x)$, and $g_i(x)$ must *depend*, explicitly or implicitly, on some of the *design variables.* Only then are they valid for the design problem. Functions that do not depend on any variable have no relation to the problem and can be safely ignored.
2. *Number of equality constraints:* The number of *independent equality constraints* must be less than, or at the most equal to, the number of design variables (i.e., $p \leq n$). When $p > n$, we have an *overdetermined system* of equations. In that case, either some *equality constraints* are *redundant* (linearly dependent on other constraints) or they are *inconsistent*. In the former case, redundant constraints can be deleted and, if $p < n$, the optimum solution for the problem is possible. In the latter case, no solution for the design problem is possible and the problem formulation needs to be closely reexamined. When $p = n$, no optimization of the system is necessary because the roots of the equality constraints are the only candidate points for optimum design.
3. *Number of inequality constraints:* While there is a restriction on the number of independent equality constraints, *there is no restriction on the number of inequality*

constraints. However, the total number of active constraints (satisfied at equality) must, at the optimum, be less than or at the most equal to the number of design variables.

4. *Unconstrained problems:* Some design problems may not have any constraints. These are called *unconstrained*; those with constraints are called *constrained*.

5. *Linear programming problems:* If all of the functions $f(x)$, $h_j(x)$, and $g_i(x)$ are linear in design variables x, then the problem is called a *linear programming problem*. If any of these functions is nonlinear, the problem is called a *nonlinear programming problem*.

6. *Scaling of problem functions:* It is important to note that if the *cost function is scaled* by multiplying it with a positive constant, the optimum design does not change. However, the optimum cost function value does change. Also, any constant can be added to the cost function without affecting the optimum design. Similarly, the inequality *constraints can be scaled* by any positive constant and the equalities by any constant. This will not affect the feasible region and hence the optimum solution. All the foregoing transformations, however, affect the values of the *Lagrange multipliers* (defined in Chapter 4). Also, performance of the numerical algorithms for a solution to the optimization problem may be affected by these transformations.

2.11.6 Feasible Set

The term *feasible set* will be used throughout the text. *A feasible set for the design problem is a collection of all feasible designs*. The terms *constraint set* and *feasible design space* are also used to represent the feasible set of designs. The letter S is used to represent the feasible set. Mathematically, the set S is a collection of design points satisfying all constraints:

$$S = \{x \mid h_j(x) = 0, \ j = 1 \text{ to } p; \ g_i(x) \leq 0, \ i = 1 \text{ to } m\} \tag{2.4}$$

The *set of feasible designs* is sometimes referred to as the *feasible region*, especially for optimization problems with two design variables. It is important to note that the *feasible region usually shrinks when more constraints are added to the design model and expands when some constraints are deleted*. When the feasible region shrinks, the number of possible designs that can optimize the cost function is reduced; that is, there are fewer feasible designs. In this event, the minimum value of the cost function is likely to increase. The effect is completely the opposite when some constraints are dropped. This observation is significant for practical design problems and should be clearly understood.

2.11.7 Active/Inactive/Violated Constraints

We will quite frequently refer to a constraint as *active, tight, inactive,* or *violated*. We define these terms precisely. An inequality constraint $g_j(x) \leq 0$ is said to be *active* at a design point x^* if it is satisfied at equality (i.e., $g_j(x^*) = 0$). This is also called a *tight* or *binding* constraint. For a feasible design, an inequality constraint may or may not be active. However, all equality constraints are active for all feasible designs.

An inequality constraint $g_j(x) \leq 0$ is said to be *inactive* at a design point x^* if it is strictly satisfied (i.e., $g_j(x^*) < 0$). It is said to be *violated* at a design point x^* if its value is positive (i.e., $g_j(x^*) > 0$). An *equality constraint* $h_i(x) = 0$ is violated at a design point x^* if $h_i(x^*)$ is not identically zero. Note that by these definitions, an equality constraint is either active or violated at a given design point.

2.11.8 Discrete and Integer Design Variables

So far, we have assumed in the standard model that variables x_i can have any numerical value within the feasible region. Many times, however, some variables are required to have discrete or integer values. Such variables appear quite often in engineering design problems. We encountered problems in Sections 2.4, 2.6, and 2.9 that have integer design variables. Before describing how to treat them, let us define what we mean by discrete and integer variables.

A design variable is called *discrete* if its value must be selected from a given finite set of values. For example, a plate thickness must be one that is available commercially: 1/8, 1/4, 3/8, 1/2, 5/8, 3/4, 1 in, and so on. Similarly, structural members must be selected from a catalog to reduce fabrication cost. Such variables must be treated as discrete in the standard formulation.

An *integer variable*, as the name implies, must have an integer value, for example, the number of logs to be shipped, the number of bolts used, the number of coils in a spring, the number of items to be shipped, and so on. Problems with such variables are called *discrete* and *integer programming problems*. Depending on the type of problem functions, the problems can be classified into five different categories. These classifications and the methods to solve them are discussed in Chapter 15.

In some sense, discrete and integer variables impose additional constraints on the design problem. Therefore, as noted before, the optimum value of the cost function is likely to increase with these variables compared with the same problem that is solved with continuous variables. If we treat all design variables as continuous, the minimum value of the cost function represents a lower bound on the true minimum value when discrete or integer variables are used. This gives some idea of the "best" optimum solution if all design variables are continuous. The optimum cost function value is likely to increase when discrete values are assigned to variables. Thus, the first suggested procedure is to solve the problem assuming continuous design variables if possible. Then the nearest discrete/integer values are assigned to the variables and the design is checked for feasibility. With a few trials, the best feasible design close to the continuous optimum can be obtained.

As a second approach for solving such problems, an *adaptive numerical optimization procedure* may be used. An optimum solution with continuous variables is first obtained if possible. Then only the variables that are close to their discrete or integer value are assigned that value. They are held fixed and the problem is optimized again. The procedure is continued until all variables have been assigned discrete or integer values. A few further trials may be carried out to improve the optimum cost function value. This procedure has been demonstrated by Arora and Tseng (1988).

The foregoing procedures require additional computational effort and do not guarantee a true minimum solution. However, they are quite straightforward and do not require any additional methods or software for solution of discrete/integer variable problems.

2.11.9 Types of Optimization Problems

The standard design optimization model can represent many different problem types. We saw that it can be used to represent unconstrained, constrained, linear programming, and nonlinear programming optimization problems. It is important to understand other optimization problems that are encountered in practical applications. Many times these problems can be transformed into the standard model and solved by the optimization methods presented and discussed in this text. Here we present an overview of the types of optimization problems.

Continuous/Discrete-Variable Optimization Problems

When the design variables can have any numerical value within their allowable range, the problem is called a *continuous-variable* optimization problem. When the problem has only discrete/integer variables, it is called a *discrete/integer-variable* optimization problem. When the problem has both continuous and discrete variables, it is called a mixed-variable optimization problem. Numerical methods for these types of problems have been developed, as discussed in later chapters.

Smooth/Nonsmooth Optimization Problems

When its functions are continuous and differentiable, the problem is referred to as smooth (*differentiable*). There are numerous practical optimization problems in which the functions can be formulated as continuous and differentiable. There are also many practical applications where the problem functions are not differentiable or even discontinuous. Such problems are called nonsmooth (*nondifferentiable*).

Numerical methods to solve these two classes of problems can be different. Theory and numerical methods for smooth problems are well developed. Therefore, it is most desirable to formulate the problem with continuous and differentiable functions as far as possible. Sometimes, a problem with discontinuous or nondifferentiable functions can be transformed into one that has continuous and differentiable functions so that optimization methods for smooth problems can be used. Such applications are discussed in Chapter 6.

Network Optimization Problems

A network or a graph consists of points and lines connecting pairs of points. Network models are used to represent many practical problems and processes from different branches of engineering, computer science, operations research, transportation, telecommunication, decision support, manufacturing, airline scheduling, and many other disciplines. Depending on the application type, network optimization problems have been classified as transportation problems, assignment problems, shortest-path problems, maximum-flow problems, minimum-cost-flow problems, and critical path problems.

To understand the concept of network problems, let us describe the transportation problem in more detail. Transportation models play an important role in logistics and supply chain management for reducing cost and improving service. Therefore the goal is to find the most effective way to transport goods. A shipper having m warehouses with supply s_i of goods at the ith warehouse must ship goods to n geographically dispersed retail centers, each with a customer demand d_j that must be met. The objective is to determine the minimum cost distribution system, given that the unit cost of transportation between the ith warehouse and the jth retail center is c_{ij}.

This problem can be formulated as one of linear programming. Since such network optimization problems are encountered in diverse fields, special methods have been developed to solve them more efficiently and perhaps in real time. Many textbooks are available on this subject. We do not address these problems in any detail, although some of the methods presented in Chapters 15 through 20 can be used to solve them.

Dynamic-Response Optimization Problems

Many practical systems are subjected to transient dynamic inputs. In such cases, some of the problem constraints are time-dependent. Each of these constraints must be imposed for the entire time interval of interest. Therefore each represents an infinite set of constraints because the constraint must be imposed at each time point in the given interval. The usual approach to treating such a constraint is to impose it at a finite number of time points in the given interval. This way the problem is transformed into the standard form and treated with the methods presented in this textbook.

Design Variables as Functions

In some applications, the design variables are not parameters but functions of one, two, or even three variables. Such design variables arise in optimal control problems where the input needs to be determined over the desired range of time to control the behavior of the system. The usual treatment of design functions is to parameterize them. In other words, each function is represented in terms of some known functions, called the *basis functions*, and the parameters multiplying them. The parameters are then treated as design variables. In this way the problem is transformed into the standard form and the methods presented in this textbook can be used to solve it.

EXERCISES FOR CHAPTER 2

2.1 A 100×100-m lot is available to construct a multistory office building. At least $20{,}000\,\text{m}^2$ of total floor space is needed. According to a zoning ordinance, the maximum height of the building can be only 21 m, and the parking area outside the building must be at least 25 percent of the total floor area. It has been decided to fix the height of each story at 3.5 m. The cost of the building in millions of dollars is estimated at $0.6h + 0.001A$, where A is the cross-sectional area of the building per floor and h is the height of the building. Formulate the minimum-cost design problem.

2.2 A refinery has two crude oils:
   1. Crude A costs \$120/barrel (bbl) and 20,000 bbl are available.
   2. Crude B costs \$150/bbl and 30,000 bbl are available.
The company manufactures gasoline and lube oil from its crudes. Yield and sale price per barrel and markets are shown in Table E2.2. How much crude oil should the company use to maximize its profit? Formulate the optimum design problem.

TABLE E2.2   Data for refinery operations

	Yield/bbl		Sale price	Market
Product	Crude A	Crude B	per bbl ($)	(bbl)
Gasoline	0.6	0.8	200	20,000
Lube oil	0.4	0.2	450	10,000

2.3 Design a beer mug, shown in Figure E2.3, to hold as much beer as possible. The height and radius of the mug should be no more than 20 cm. The mug must be at least 5 cm in radius. The surface area of the sides must be no greater than 900 cm² (ignore the bottom area of the mug and mug handle). Formulate the optimum design problem.

2.4 A company is redesigning its parallel-flow heat exchanger of length l to increase its heat transfer. An end view of the unit is shown in Figure E2.4. There are certain limitations on the design problem. The smallest available conducting tube has a radius of 0.5 cm, and all tubes must be of the same size. Further, the total cross-sectional area of all of the tubes cannot exceed 2000 cm² to ensure adequate space inside the outer shell. Formulate the problem to determine the number of tubes and the radius of each one to maximize the surface area of the tubes in the exchanger.

2.5 Proposals for a parking ramp have been defeated, so we plan to build a parking lot in the downtown urban renewal section. The cost of land is $200W + 100D$, where W is the width along the street and D is the depth of the lot in meters. The available width along the street is 100 m, while the maximum depth available is 200 m. We want the size of the lot to be at least 10,000 m². To avoid unsightliness, the city requires that the longer dimension of any

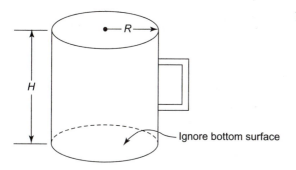

FIGURE E2.3   Beer mug.

Ignore bottom surface

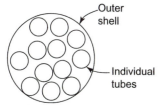

FIGURE E2.4   Cross-section of a heat exchanger.

Outer shell

Individual tubes

I. THE BASIC CONCEPTS

lot be no more than twice the shorter dimension. Formulate the minimum-cost design problem.

2.6 A manufacturer sells products A and B. Profit from A is $10/kg and is $8/kg from B. Available raw materials for the products are 100 kg of C and 80 kg of D. To produce 1 kg of A, we need 0.4 kg of C and 0.6 kg of D. To produce 1 kg of B, we need 0.5 kg of C and 0.5 kg of D. The markets for the products are 70 kg for A and 110 kg for B. How much of A and B should be produced to maximize profit? Formulate the design optimization problem.

2.7 Design a diet of bread and milk to get at least 5 units of vitamin A and 4 units of vitamin B daily. The amount of vitamins A and B in 1 kg of each food and the cost per kilogram of the food are given in Table E2.7. Formulate the design optimization problem so that we get at least the basic requirements of vitamins at the minimum cost.

TABLE E2.7 Data for the diet problem

Vitamin	Bread	Milk
A	1	2
B	3	2
Cost/kg, $	2	1

2.8 Enterprising chemical engineering students have set up a still in a bathtub. They can produce 225 bottles of pure alcohol each week. They bottle two products from alcohol: (1) wine, at 20 proof, and (2) whiskey, at 80 proof. Recall that pure alcohol is 200 proof. They have an unlimited supply of water, but can only obtain 800 empty bottles per week because of stiff competition. The weekly supply of sugar is enough for either 600 bottles of wine or 1200 bottles of whiskey. They make a $1.00 profit on each bottle of wine and a $2.00 profit on each bottle of whiskey. They can sell whatever they produce. How many bottles of wine and whiskey should they produce each week to maximize profit? Formulate the design optimization problem (created by D. Levy).

2.9 Design a can closed at one end using the smallest area of sheet metal for a specified interior volume of 600 cm^3. The can is a right-circular cylinder with interior height h and radius r. The ratio of height to diameter must not be less than 1.0 nor greater than 1.5. The height cannot be more than 20 cm. Formulate the design optimization problem.

2.10 Design a shipping container closed at both ends with dimensions $b \times b \times h$ to minimize the ratio: (round-trip cost of shipping container only)/(one-way cost of shipping contents only). Use the data in the following table. Formulate the design optimization problem.

Mass of container/surface area	80 kg/m^2
Maximum b	10 m
Maximum h	18 m
One-way shipping cost, full or empty	$18/kg gross mass
Mass of contents	150 kg/m^3

2.11 Certain mining operations require an open-top rectangular container to transport materials. The data for the problem are as follows:

Construction costs:
- Sides: $50/m^2
- Ends: $60/m^2
- Bottom: $90/m^2
Minimum volume needed: 150 m^3

Formulate the problem of determining the container dimensions at a minimum cost.

2.12 Design a circular tank closed at both ends to have a volume of 250 m^3. The fabrication cost is proportional to the surface area of the sheet metal and is $400/m^2. The tank is to be housed in a shed with a sloping roof. Therefore, height H of the tank is limited by the relation $H \leq (10 - D/2)$, where D is the tank's diameter. Formulate the minimum-cost design problem.

2.13 Design the steel framework shown in Figure E2.13 at a minimum cost. The cost of a horizontal member in one direction is $20 w and in the other direction it is $30 d. The cost of a vertical column is $50 h. The frame must enclose a total volume of at least 600 m^3. Formulate the design optimization problem.

2.14 Two electric generators are interconnected to provide total power to meet the load. Each generator's cost is a function of the power output, as shown in Figure E2.14. All costs and power are expressed on a per-unit basis. The total power needed is at least 60 units. Formulate a minimum-cost design problem to determine the power outputs P_1 and P_2.

2.15 Transportation problem. A company has m manufacturing facilities. The facility at the ith location has capacity to produce b_i units of an item. The product should be shipped to n distribution centers. The distribution center at the jth location requires at least a_j units of the item to satisfy demand. The cost of shipping an item from the ith plant to the jth distribution center is c_{ij}. Formulate a minimum-cost transportation system to meet each of the distribution center's demands without exceeding the capacity of any manufacturing facility.

2.16 Design of a two-bar truss. Design a symmetric two-bar truss (both members have the same cross-section), as shown in Figure E2.16, to support a load W. The truss consists of two steel tubes pinned together at one end and supported on the ground at the other. The span of

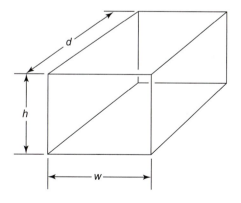

FIGURE E2.13   Steel frame.

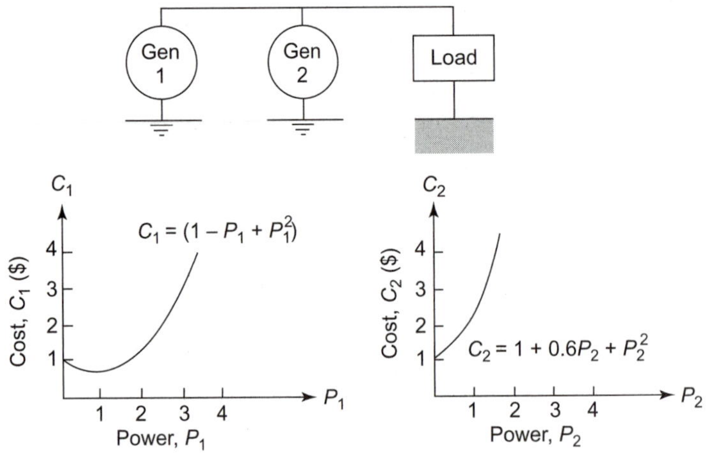

FIGURE E2.14 Graphic of a power generator.

$$C_1 = (1 - P_1 + P_1^2)$$

$$C_2 = 1 + 0.6P_2 + P_2^2$$

the truss is fixed at s. Formulate the minimum-mass truss design problem using height and cross-sectional dimensions as design variables. The design should satisfy the following constraints:

1. Because of space limitations, the height of the truss must not exceed b_1 and must not be less than b_2.
2. The ratio of mean diameter to thickness of the tube must not exceed b_3.
3. The compressive stress in the tubes must not exceed the allowable stress σ_a for steel.
4. The height, diameter, and thickness must be chosen to safeguard against member buckling.

Use the following data: $W = 10$ kN; span $s = 2$ m; $b_1 = 5$ m; $b_2 = 2$ m; $b_3 = 90$; allowable stress $\sigma_a = 250$ MPa; modulus of elasticity $E = 210$ GPa; mass density $\rho = 7850$ kg/m^3; factor of safety against buckling $FS = 2$; $0.1 \leq D \leq 2$ (m); and $0.01 \leq t \leq 0.1$ (m).

2.17 A beam of rectangular cross-section (Figure E2.17) is subjected to a maximum bending moment of M and a maximum shear of V. The allowable bending and shearing stresses are σ_a and τ_a, respectively. The bending stress in the beam is calculated as

$$\sigma = \frac{6M}{bd^2}$$

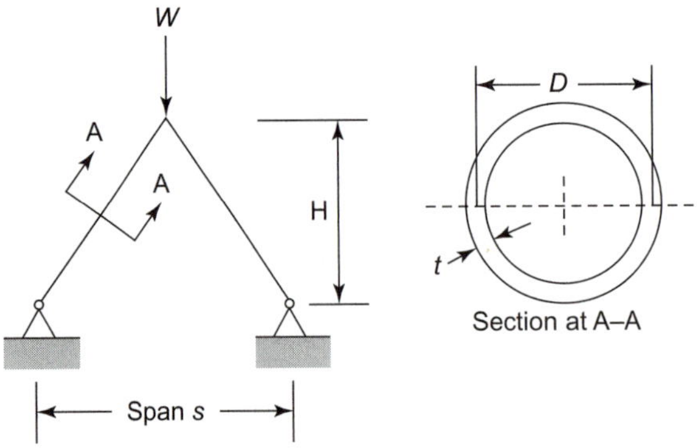

FIGURE E2.16 Two-bar structure.

Section at A–A

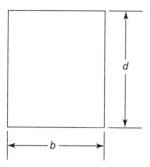

FIGURE E2.17   Cross-section of a rectangular beam.

and the average shear stress in the beam is calculated as

$$\tau = \frac{3V}{2bd}$$

where d is the depth and b is the width of the beam. It is also desirable to have the depth of the beam not exceed twice its width. Formulate the design problem for minimum cross-sectional area using this data: $M = 140$ kN $\cdot$ m, $V = 24$ kN, $\sigma_a = 165$ MPa, $\tau_a = 50$ MPa.

2.18 A vegetable oil processor wishes to determine how much shortening, salad oil, and margarine to produce to optimize the use its current oil stock supply. At the present time, he has 250,000 kg of soybean oil, 110,000 kg of cottonseed oil, and 2000 kg of milk-base substances. The milk-base substances are required only in the production of margarine. There are certain processing losses associated with each product: 10 percent for shortening, 5 percent for salad oil, and no loss for margarine. The producer's back orders require him to produce at least 100,000 kg of shortening, 50,000 kg of salad oil, and 10,000 kg of margarine. In addition, sales forecasts indicate a strong demand for all products in the near future. The profit per kilogram and the base stock required per kilogram of each product are given in Table E2.18. Formulate the problem to maximize profit over the next production scheduling period. (created by J. Liittschwager)

TABLE E2.18   Data for the vegetable oil processing problem

Product	Profit per kg	Parts per kg of base stock requirements		
		Soybean	Cottonseed	Milk base
Shortening	1.00	2	1	0
Salad oil	0.80	0	1	0
Margarine	0.50	3	1	1

Section 2.11: A General Mathematical Model for Optimum Design

2.19 *Answer True or False:*

1. Design of a system implies specification of the design variable values.
2. All design problems have only linear inequality constraints.
3. All design variables should be independent of each other as far as possible.

4. If there is an equality constraint in the design problem, the optimum solution must satisfy it.
5. Each optimization problem must have certain parameters called the design variables.
6. A feasible design may violate equality constraints.
7. A feasible design may violate " ≥ type" constraints.
8. A "≤ type" constraint expressed in the standard form is active at a design point if it has zero value there.
9. The constraint set for a design problem consists of all feasible points.
10. The number of independent equality constraints can be larger than the number of design variables for the problem.
11. The number of "≤ type" constraints must be less than the number of design variables for a valid problem formulation.
12. The feasible region for an equality constraint is a subset of that for the same constraint expressed as an inequality.
13. Maximization of $f(x)$ is equivalent to minimization of $1/f(x)$.
14. A lower minimum value for the cost function is obtained if more constraints are added to the problem formulation.
15. Let f_n be the minimum value for the cost function with n design variables for a problem. If the number of design variables for the same problem is increased to, say, $m = 2n$, then $f_m > f_n$, where f_m is the minimum value for the cost function with m design variables.

*2.20 A trucking company wants to purchase several new trucks. It has $2 million to spend. The investment should yield a maximum of trucking capacity for each day in tonnes × kilometers. Data for the three available truck models are given in Table E2.20: truck load capacity, average speed, crew required per shift, hours of operation for three shifts, and cost of each truck. There are some limitations on the operations that need to be considered. The labor market is such that the company can hire at most 150 truck drivers. Garage and maintenance facilities can handle at the most 25 trucks. How many trucks of each type should the company purchase? Formulate the design optimization problem.

TABLE E2.20   Data for available trucks

Truck model	Truck load capacity (tonnes)	Average truck speed (km/h)	Crew required per shift	No. of hours of operations per day (3 shifts)	Cost of each truck ($)
A	10	55	1	18	40,000
B	20	50	2	18	60,000
C	18	50	2	21	70,000

*2.21 A large steel corporation has two iron ore reduction plants. Each plant processes iron ore into two different ingot stocks, which are shipped to any of three fabricating plants where they are made into either of two finished products. In total, there are two reduction plants, two ingot stocks, three fabricating plants, and two finished products. For the upcoming season, the company wants to minimize total tonnage of iron ore processed in its reduction plants, subject to production and demand constraints. Formulate the design optimization problem and transcribe it into the standard model.

Nomenclature

$a(r, s)$ = tonnage yield of ingot stock s from 1 ton of iron ore processed at reduction plant r
$b(s, f, p)$ = total yield from 1 ton of ingot stock s shipped to fabricating plant f and
        manufactured into product p
$c(r)$ = ore-processing capacity in tonnage at reduction plant r
$k(f)$ = capacity of fabricating plant f in tonnage for all stocks
$D(p)$ = tonnage demand requirement for product p

Production and Demand Constraints

1. The total tonnage of iron ore processed by both reduction plants must equal the total tonnage processed into ingot stocks for shipment to the fabricating plants.
2. The total tonnage of iron ore processed by each reduction plant cannot exceed its capacity.
3. The total tonnage of ingot stock manufactured into products at each fabricating plant must equal the tonnage of ingot stock shipped to it by the reduction plants.
4. The total tonnage of ingot stock manufactured into products at each fabricating plant cannot exceed the plant's available capacity.
5. The total tonnage of each product must equal its demand.

Constants for the Problem

$a(1,1) = 0.39$	$c(1) = 1,200,000$	$k(1) = 190,000$	$D(1) = 330,000$
$a(1,2) = 0.46$	$c(2) = 1,000,0\,00$	$k(2) = 240,000$	$D(2) = 125,000$
$a(2,1) = 0.44$		$k(3) = 290,000$	
$a(2,2) = 0.48$			

$b(1,1,1) = 0.79$	$b(1,1,2) = 0.84$
$b(2,1,1) = 0.68$	$b(2,1,2) = 0.81$
$b(1,2,1) = 0.73$	$b(1,2,2) = 0.85$
$b(2,2,1) = 0.67$	$b(2,2,2) = 0.77$
$b(1,3,1) = 0.74$	$b(1,3,2) = 0.72$
$b(2,3,1) = 0.62$	$b(2,3,2) = 0.78$

2.22 Optimization of a water canal. Design a water canal having a cross-sectional area of 150 m². The lowest construction costs occur when the volume of the excavated material equals the amount of material required for the dykes, as shown in Figure E2.22. Formulate the problem to minimize the dug-out material A_1. Transcribe the problem into the standard design optimization model.

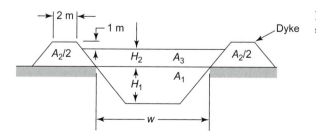

FIGURE E2.22 Graphic of a cross-section of a canal. *(Created by V. K. Goel.)*

Beam

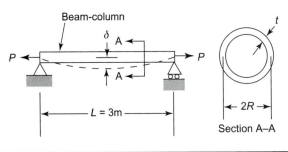

FIGURE E2.23 Cantilever beam.

Section A–A

2.23 A cantilever beam is subjected to the point load P (kN), as shown in Figure E2.23. The maximum bending moment in the beam is PL (kN · m) and the maximum shear is P (kN). Formulate the minimum-mass design problem using a hollow circular cross-section. The material should not fail under bending or shear stress. The maximum bending stress is calculated as

$$\sigma = \frac{PL}{I}R_o$$

where I = moment of inertia of the cross-section. The maximum shearing stress is calculated as

$$\tau = \frac{P}{3I}(R_o^2 + R_oR_i + R_i^2)$$

Transcribe the problem into the standard design optimization model (also use $R_o \leq$ 40.0 cm, $R_i \leq$ 40.0 cm). Use this data: $P = 14$ kN; $L = 10$ m; mass density $\rho = 7850$ kg/m³; allowable bending stress $\sigma_b = 165$ MPa; allowable shear stress $\tau_a = 50$ MPa.

2.24 Design a hollow circular beam-column, shown in Figure E2.24, for two conditions: When $P = 50$ (kN), the axial stress σ must not exceed an allowable value σ_a, and when $P = 0$, deflection δ due to self-weight should satisfy $\delta \leq 0.001L$. The limits for dimensions are $t = 0.10$ to 1.0 cm, $R = 2.0$ to 20.0 cm, and $R/t \leq 20$ (AISC, 2005). Formulate the minimum-weight design problem and transcribe it into the standard form. Use the following data: $\delta = 5wL^4/384EI$; w = self-weight force/length (N/m); $\sigma_a = 250$ MPa; modulus of elasticity $E = 210$ GPa; mass density $\rho = 7800$ kg/m³; $\sigma = P/A$; gravitational constant $g = 9.80$ m/s²; moment of inertia $I = \pi R^3t$ (m⁴).

Beam-column

δ

P ⟶ ⟵ P

$L = 3$m

t

FIGURE E2.24 Graphic of a hollow circular beam-column.

← 2R →

Section A–A

Graphical Optimization and Basic Concepts

- Graphically solve any optimization problem having two design variables
- Plot constraints and identify their feasible/infeasible side
- Identify the feasible region (feasible set) for a problem
- Plot objective function contours through the feasible region
- Graphically locate the optimum solution for a problem and identify active/inactive constraints
- Identify problems that may have multiple, unbounded, or infeasible solutions
- Explain basic concepts and terms associated with optimum design

Optimization problems having only two design variables can be solved by observing how they are graphically represented. All constraint functions are plotted, and a set of feasible designs (the feasible set) for the problem is identified. Objective function contours are then drawn, and the optimum design is determined by visual inspection. In this chapter, we illustrate the graphical solution process and introduce several concepts related to optimum design problems. In Section 3.1, a design optimization problem is formulated and used to describe the solution process. Several more example problems are solved in later sections to illustrate concepts and the procedure.

3.1 GRAPHICAL SOLUTION PROCESS

3.1.1 Profit Maximization Problem

STEP 1: PROJECT/PROBLEM DESCRIPTION A company manufactures two machines, A and B. Using available resources, either 28 A or 14 B can be manufactured daily. The sales department can sell up to 14 A machines or 24 B machines. The shipping facility can handle no more than 16 machines per day. The company makes a profit of $400 on each A machine and $600 on each B machine. How many A and B machines should the company manufacture every day to maximize its profit?

STEP 2: DATA AND INFORMATION COLLECTION   Data and information are defined in the project statement.

STEP 3: DEFINITION OF DESIGN VARIABLES   The following two design variables are identified in the problem statement:

x_1 = number of A machines manufactured each day
x_2 = number of B machines manufactured each day

STEP 4: OPTIMIZATION CRITERION   The objective is to maximize daily profit, which can be expressed in terms of design variables as

$$P = 400x_1 + 600x_2 \tag{a}$$

STEP 5: FORMULATION OF CONSTRAINTS   Design constraints are placed on manufacturing capacity, on sales personnel, and on the shipping and handling facility. The constraint on the shipping and handling facility is quite straightforward:

$$x_1 + x_2 \leq 16 \text{ (shipping and handling constraint)} \tag{b}$$

Constraints on manufacturing and sales facilities are a bit tricky. First, consider the manufacturing limitation. It is assumed that if the company is manufacturing x_1 A machines per day, then the remaining resources and equipment can be proportionately used to manufacture x_2 B machines, and vice versa. Therefore, noting that $x_1/28$ is the fraction of resources used to produce A and $x_2/14$ is the fraction used to produce B, the constraint is expressed as

$$\frac{x_1}{28} + \frac{x_2}{14} \leq 1 \text{ (manufacturing constraint)} \tag{c}$$

Similarly, the constraint on sales department resources is given as

$$\frac{x_1}{14} + \frac{x_2}{24} \leq 1 \text{ (limitation on sales department)} \tag{d}$$

Finally, the design variables must be non-negative as

$$x_1, \ x_2 \geq 0 \tag{e}$$

Note that for this problem, the formulation remains valid even when a design variable has zero value. The problem has two design variables and five inequality constraints. All functions of the problem are linear in variables x_1 and x_2. Therefore, it is a *linear programming problem*. Note also that for a meaningful solution, both design variables must have integer values at the optimum point.

3.1.2 Step-by-Step Graphical Solution Procedure

STEP 1: COORDINATE SYSTEM SET-UP   The first step in the solution process is to set up an origin for the *x-y* coordinate system and scales along the *x-* and *y*-axes. By looking at the constraint functions, a coordinate system for the profit maximization problem can be set up using a range of 0 to 25 along both the *x* and *y* axes. In some cases, the scale may need to be adjusted after the problem has been graphed because the original scale may provide too small or too large a graph for the problem.

STEP 2: INEQUALITY CONSTRAINT BOUNDARY PLOT   To illustrate the graphing of a constraint, let us consider the inequality $x_1 + x_2 \leq 16$ given in Eq. (b). To represent the constraint graphically, we first need to plot the constraint boundary; that is, the points that satisfy the constraint as an equality $x_1 + x_2 = 16$. This is a linear function of the variables x_1 and x_2. To plot such a function, we need two points that satisfy the equation $x_1 + x_2 = 16$. Let these points be calculated as (16,0) and (0,16). Locating them on the graph and joining them by a straight line produces the line F–J, as shown in Figure 3.1. Line F–J then represents the boundary of the feasible region for the inequality constraint $x_1 + x_2 \leq 16$. Points on one side of this line violate the constraint, while those on the other side satisfy it.

STEP 3: IDENTIFICATION OF THE FEASIBLE REGION FOR AN INEQUALITY   The next task is to determine which side of constraint boundary F–J is feasible for the constraint $x_1 + x_2 \leq 16$. To accomplish this, we select a point on either side of F–J and evaluate the constraint function there. For example, at point (0,0), the left side of the constraint has a value of 0. Because the value is less than 16, the constraint is satisfied and the region

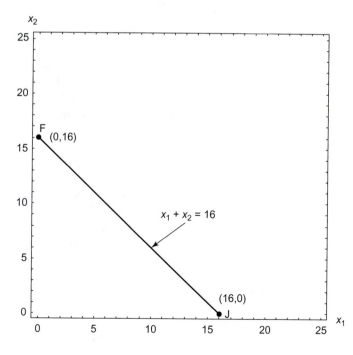

FIGURE 3.1   Constraint boundary for the inequality $x_1 + x_2 \leq 16$ in the profit maximization problem.

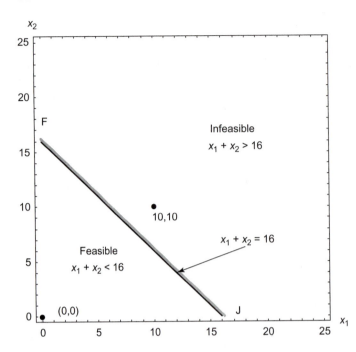

FIGURE 3.2   Feasible/infeasible side for the inequality $x_1 + x_2 \leq 16$ in the profit maximization problem.

below F–J is feasible. We can test the constraint at another point on the opposite side of F–J, say at point (10,10). At this point the constraint is violated because the left side of the constraint function is 20, which is larger than 16. Therefore, the region above F–J is infeasible with respect to the constraint, as shown in Figure 3.2. The infeasible region is "shaded-out," *a convention that is used throughout this text*.

Note that if this were an *equality* constraint $x_1 + x_2 = 16$, the feasible region for it would only be the points on line F–J. Although there are infinite points on F–J, the feasible region for the equality constraint is much smaller than that for the same constraint written as an inequality. This shows the importance of properly formulating all the constraints of the problem.

STEP 4: IDENTIFICATION OF THE FEASIBLE REGION   By following the procedure that is described in step 3, all inequalities are plotted on the graph and the feasible side of each one is identified (if equality constraints were present, they would also be plotted at this stage). Note that the constraints $x_1, x_2 \geq 0$ restrict the feasible region to the first quadrant of the coordinate system. The intersection of feasible regions for all constraints provides the feasible region for the profit maximization problem, indicated as ABCDE in Figure 3.3. Any point in this region or on its boundary provides a feasible solution to the problem.

STEP 5: PLOTTING OF OBJECTIVE FUNCTION CONTOURS   The next task is to plot the objective function on the graph and locate its optimum points. For the present problem, the objective is to maximize the profit $P = 400x_1 + 600x_2$, which involves three variables: P, x_1, and x_2. The function needs to be represented on the graph so that the value of P can be

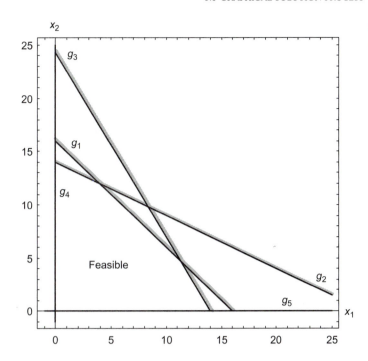

FIGURE 3.3 Feasible region for the profit maximization problem.

compared for different feasible designs to locate the best design. However, because there are infinite feasible points, it is not possible to evaluate the objective function at every point. One way of overcoming this impasse is to plot the contours of the objective function.

A *contour* is a curve on the graph that connects all points having the same objective function value. A collection of points on a contour is also called the *level set*. If the objective function is to be minimized, the contours are also called *isocost curves*. To plot a contour through the feasible region, we need to assign it a value. To obtain this value, consider a point in the feasible region and evaluate the profit function there. For example, at point $(6,4)$, P is $P = 6 \times 400 + 4 \times 600 = 4800$. To plot the $P = 4800$ contour, we plot the function $400x_1 + 600x_2 = 4800$. This contour is a straight line, as shown in Figure 3.4.

STEP 6: IDENTIFICATION OF THE OPTIMUM SOLUTION   To locate an optimum point for the objective function, we need at least two contours that pass through the feasible region. We can then observe trends for the values of the objective function at different feasible points to locate the best solution point. Contours for $P = 2400$, 4800, and 7200 are plotted in Figure 3.5. We now observe the following trend: As the contours move up toward point D, feasible designs can be found with larger values for P. It is clear from observation that point D has the largest value for P in the feasible region. We now simply read the coordinates of point D $(4, 12)$ to obtain the optimum design, having a maximum value for the profit function as $P = 8800$.

Thus, the best strategy for the company is to manufacture 4 A and 12 B machines to maximize its daily profit. The inequality constraints in Eqs. (b) and (c) are *active* at the optimum; that is, they are satisfied at equality. These represent limitations on shipping and handling

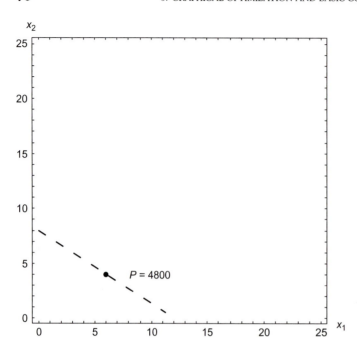

FIGURE 3.4 Plot of $P = 4800$ objective function contour for the profit maximization problem.

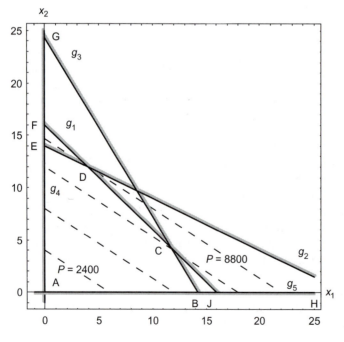

FIGURE 3.5 Graphical solution to the profit maximization problem: optimum point D = (4, 12); maximum profit, $P = 8800$.

facilities, and on manufacturing. The company can think about relaxing these constraints to improve its profit. All other inequalities are strictly satisfied and therefore *inactive*.

Note that in this example the design variables must have integer values. Fortunately, the optimum solution has integer values for the variables. If this had not been the case, we would have used the procedure suggested in Section 2.11.8 or in Chapter 15 to solve this problem. Note also that for this example all functions are linear in design variables. Therefore, all curves in Figures 3.1 through 3.5 are *straight lines*. In general, the functions of a design problem may not be linear, in which case curves must be plotted to identify the feasible region, and contours or *isocost curves* must be drawn to identify the optimum design. To *plot a nonlinear function*, a table of numerical values for x_1 and x_2 must be generated. These points must be then plotted on a graph and connected by a smooth curve.

3.2 USE OF MATHEMATICA FOR GRAPHICAL OPTIMIZATION

It turns out that good programs, such as Mathematica and MATLAB®, are available to implement the step-by-step procedure of the previous section and obtain a graphical solution for the problem on the computer screen. Mathematica is an interactive software package with many capabilities; however, we will explain its use to solve a two-variable optimization problem by plotting all functions on the computer screen. Although other commands for plotting functions are available, the most convenient for working with inequality constraints and objective function contours is the *ContourPlot* command. As with most Mathematica commands, this one is followed by what we call subcommands as "arguments" that define the nature of the plot. All Mathematica commands are case-sensitive, so it is important to pay attention to which letters are capitalized.

Mathematica input is organized into what is called a *notebook*. A notebook is divided into *cells*, with each cell containing input that can be executed independently. To explain the graphical optimization capability of Mathematica, we will again use the profit maximization problem. (*Note that the commands used here may change in future releases of the program.*) We start by entering in the notebook the problem functions as follows (the first two commands are for initialization of the program):

```
<<Graphics`Arrow`
Clear[x1,x2];

P=400*x1+600*x2;
g1=x1+x2-16;          (*shipping and handling constraint*)
g2=x1/28+x2/14-1;     (*manufacturing constraint*)
g3=x1/14+x2/24-1;     (*limitation on sales department*)
g4=-x1;               (*non-negativity*)
g5=-x2;               (*non-negativity*)
```

This input illustrates some basic features concerning Mathematica format. Note that the ENTER key acts simply as a carriage return, taking the blinking cursor to the next line. Pressing SHIFT and ENTER actually inputs the typed information into Mathematica. When no immediate output from Mathematica is desired, the input line must end with a semicolon (;). If the semicolon is omitted, Mathematica will simplify the input and display it on the screen or execute an arithmetic expression and display the result. Comments are bracketed as (*Comment*). Note also that all constraints are assumed to be in the standard "≤" form. This helps in identifying the infeasible region for constraints on the screen using the *ContourPlot* command.

3.2.1 Plotting Functions

The Mathematica command used to plot the contour of a function, say g1=0, is entered as follows:

```
Plotg1=ContourPlot[g1,{x1,0,25},{x2,0,25}, ContourShading→False, Contours→{0},
ContourStyle→{{Thickness[.01]}}, Axes→True, AxesLabel→{"x1","x2"},
PlotLabel→"Profit Maximization Problem", Epilog→{Disk[{0,16},{.4,.4}],
Text["(0,16)",{2,16}], Disk[{16,0},{.4,.4}], Text["(16,0)",{17,1.5}],
Text["F",{0,17}], Text["J",{17,0}], Text["x1+x2=16",{13,9}], Arrow[{13,8.3},{10,6}]},
DefaultFont→{"Times",12}, ImageSize→72.5];
```

Plotg1 is simply an arbitrary name referring to the data points for the function g1 determined by the *ContourPlot* command; it is used in future commands to refer to this particular plot. This *ContourPlot* command plots a contour defined by the equation g1=0 as shown earlier in Figure 3.1. Arguments of the *ContourPlot* command containing various subcommands are explained as follows (note that the arguments are separated by commas and are enclosed in square brackets ([ ]):

g1: function to be plotted.

{x1,0,25}, {x2,0,25}: ranges for the variables x1 and x2; 0 to 25.

ContourShading→False: indicates that shading will not be used to plot contours, whereas *ContourShading*→True would indicate that shading will be used. Note that most subcommands are followed by an arrow (→) or (->) and a set of parameters enclosed in braces ({}).

Contours→{0}: contour values for g1; one contour is requested having 0 value.

ContourStyle→{{Thickness[.01]}}: defines characteristics of the contour such as thickness and color. Here, the thickness of the contour is specified as ".01". It is given as a fraction of the total width of the graph and needs to be determined by trial and error.

Axes→True: indicates whether axes should be drawn at the origin; in the present case, where the origin (0,0) is located at the bottom left corner of the graph, the *Axes* subcommand is irrelevant except that it allows for the use of the *AxesLabel* command.

AxesLabel→{"x1","x2"}: allows one to indicate labels for each axis.

PlotLabel→"Profit Maximization Problem": places a label at the top of the graph.

Epilog→{...}: allows insertion of additional graphics primitives and text in the figure on the screen figure on the screen; Disk [{0,16}, {.4,.4}] allows insertion of a dot at the location (0,16) of radius .4 in both directions; Text ["(0,16)", (2,16)] allows "(0,16)" to be placed at the location (2,16).

ImageSize → *72 5*: indicates that the width of the plot should be 5 inches; the size of the plot can be adjusted by selecting the image and dragging one of the black square control points; the images in Mathematica can be copied and pasted to a word processor file.

DefaultFont → {"Times",12}: specifies the preferred font and size for the text.

3.2.2 Identification and Shading of Infeasible Region for an Inequality

Figure 3.2 is created using a slightly modified *ContourPlot* command used earlier for Figure 3.1:

```
Plotg1=ContourPlot[g1,{x1,0,25},{x2,0,25}, ContourShading→False, Contours→{0,.65},
ContourStyle→{{Thickness[.01]}, {GrayLevel[.8],Thickness[.025]}}, Axes→True,
AxesLabel→{"x1","x2"}, PlotLabel→"Profit Maximization Problem",
Epilog→{Disk[{10,10},{.4,.4}], Text["(10,10)",{11,9}], Disk[{0,0},{.4,.4}],
Text["(0,0)",{2,.5}], Text["x1+x2=16",{18,7}], Arrow[{18,6.3},{12,4}],
Text["Infeasible",{17,17}], Text["x1+x2>16",{17,15.5}], Text["Feasible",{5,6}],
Text["x1+x2<16",{5,4.5}]}, DefaultFont→{"Times",12}, ImageSize→72.5];
```

Here, two contour lines are specified, the second one having a small positive value. This is indicated by the command: *Contours* → {0,.65}. The constraint boundary is represented by the contour g1=0. The contour g1=0.65 will pass through the infeasible region, where the positive number 0.65 is determined by trial and error.

To shade the infeasible region, the characteristics of the contour are changed. Each set of brackets {} with the *ContourStyle* subcommand corresponds to a specific contour. In this case, {Thickness[.01]} provides characteristics for the first contour g1=0, and {GrayLevel[.8],Thickness[0.025]} provides characteristics for the second contour g1=0.65. *GrayLevel* specifies a color for the contour line. A gray level of 0 yields a black line, whereas a gray level of 1 yields a white line. Thus, this *ContourPlot* command essentially draws one thin, black line and one thick, gray line. This way the infeasible side of an inequality is shaded out.

3.2.3 Identification of Feasible Region

By using the foregoing procedure, all constraint functions for the problem are plotted and their feasible sides are identified. The plot functions for the five constraints g1 through g5 are named Plotg1, Plotg2, Plotg3, Plotg4, and Plotg5. All of these functions are quite similar to the one that was created using the *ContourPlot* command explained earlier. As an example, the Plotg4 function is given as

```
Plotg4=ContourPlot[g4,{x1,−1,25},{x2,−1,25}, ContourShading→False, Contours→{0,.35},
ContourStyle→{{Thickness[.01]}, {GrayLevel[.8],Thickness[.02]}},
DisplayFunction→Identity];
```

The *DisplayFunction* → *Identity* subcommand is added to the *ContourPlot* command to suppress display of output from each Plotgi function; without that, Mathematica executes each `Plotgi` function and displays the results. Next, with the following *Show* command, the five plots are combined to display the complete feasible set in Figure 3.3:

```
Show[{Plotg1,Plotg2,Plotg3,Plotg4,Plotg5}, Axes→True,AxesLabel→{"x1","x2"},
PlotLabel→"Profit Maximization Problem", DefaultFont→{"Times",12}, Epilog→
{Text["g1",{2.5,16.2}], Text["g2",{24,4}], Text["g3",{2,24}], Text["g5",{21,1}],
Text["g4",{1,10}], Text["Feasible",{5,6}]}, DefaultFont→{"Times",12},
ImageSize→72.5,DisplayFunction→ $DisplayFunction];
```

The *Text* subcommands are included to add text to the graph at various locations. The *DisplayFunction→$DisplayFunction* subcommand is added to display the final graph; without that it is not displayed.

3.2.4 Plotting of Objective Function Contours

The next task is to plot the objective function contours and locate its optimum point. The objective function contours of values 2400, 4800, 7200, and 8800, shown in Figure 3.4, are drawn by using the *ContourPlot* command as follows:

```
PlotP=ContourPlot[P,{x1,0,25},{x2,0,25}, ContourShading→False, Contours→{4800},
ContourStyle→{{Dashing[{.03,.04}], Thickness[.007]}}, Axes→True,
AxesLabel→{"x1","x2"}, PlotLabel→"Profit Maximization Problem",
DefaultFont→{"Times",12}, Epilog→{Disk[{6,4},{.4,.4}], Text["P= 4800",{9.75,4}]},
ImageSize→72.5];
```

The *ContourStyle* subcommand provides four sets of characteristics, one for each contour. *Dashing*[{a,b}] yields a dashed line with "a" as the length of each dash and "b" as the space between dashes. These parameters represent a fraction of the total width of the graph.

3.2.5 Identification of Optimum Solution

The *Show* command used to plot the feasible region for the problem in Figure 3.3 can be extended to plot the profit function contours as well. Figure 3.5 contains the graphical representation of the problem, obtained using the following *Show* command:

```
Show[{Plotg1,Plotg2,Plotg3,Plotg4,Plotg5, PlotP}, Axes→True, AxesLabel→{"x1","x2"},
PlotLabel→"Profit Maximization Problem", DefaultFont→{"Times",12},
Epilog→{Text["g1",{2.5,16.2}], Text["g2",{24,4}], Text["g3",{3,23}], Text["g5",{23,1}],
Text["g4",{1,10}], Text["P= 2400",{3.5,2}], Text["P= 8800",{17,3.5}],Text["G",{1,24.5}],
Text["C",{10.5,4}], Text["D",{3.5,11}], Text["A",{1,1}], Text["B",{14,−1}],
Text["J",{16,−1}], Text["H",{25,−1}], Text["E",{−1,14}], Text["F",{−1,16}]},
DefaultFont→{"Times",12}, ImageSize→72.5, DisplayFunction→ $DisplayFunction];
```

Additional *Text* subcommands have been added to label different objective function contours and different points. The final graph is used to obtain the graphical solution. The *Disk* subcommand can be added to the *Epilog* command to put a dot at the optimum point.

3.3 USE OF MATLAB FOR GRAPHICAL OPTIMIZATION

MATLAB has many capabilities for solving engineering problems. For example, it can plot problem functions and graphically solve a two-variable optimization problem. In this section, we explain how to use the program for this purpose; other uses of the program for solving optimization problems are explained in Chapter 7.

There are two modes of input with MATLAB. We can enter commands interactively, one at a time, with results displayed immediately after each one. Alternatively, we can create an input file, called an *m-file* that is executed in batch mode. The m-file can be created using the text editor in MATLAB. To access this editor, select "File," "New," and "m-file." When saved, this file will have the suffix ".m" (dot m). To submit or run the file, after starting MATLAB, we simply type the name of the file we wish to run in the command window, without the suffix (the *current directory* in the MATLAB program must be one where the file is located). In this section, we will solve the profit maximization problem of the previous section using MATLAB7.6. It is important to note that with future releases, the commands we will discuss may change.

3.3.1 Plotting of Function Contours

For plotting all of the constraints with MATLAB and identifying the feasible region, it is assumed that all inequality constraints are written in the standard "$\leq$" form. The M-file for the profit maximization problem with explanatory comments is displayed in Table 3.1. Note that the file comments are preceded by the percent sign, %. The comments are ignored during MATLAB execution. For contour plots, the first command in the input file is entered as follows:

```
[x1,x2]=meshgrid(-1.0:0.5:25.0, -1.0:0.5:25.0);
```

This command creates a grid or array of points where all functions to be plotted are evaluated. The command indicates that x1 and x2 will start at -1.0 and increase in increments of 0.5 up to 25.0. These variables now represent two-dimensional arrays and require special attention in operations using them. "*" (star) and "/" (slash) indicate scalar multiplication and division, respectively, whereas ".*" (dot star) and "./" (dot slash) indicate element-by-element multiplication and division. The ".^" (dot hat) is used to apply an exponent to each element of a vector or a matrix. The semicolon ";" after a command prevents MATLAB from displaying the numerical results immediately (i.e., all of the values for x1 and x2).

This use of a semicolon is a convention in MATLAB for most commands. Note that matrix division and multiplication capabilities are not used in the present example, as the variables in the problem functions are only multiplied or divided by a scalar rather than another variable. If, for instance, a term such as x_1x_2 is present, then the element-by-element operation x1.*x2 is necessary. The "contour" command is used for plotting all problem functions on the screen.

The procedure for identifying the infeasible side of an inequality is to plot two contours for the inequality: one of value 0 and the other of a small positive value. The second

TABLE 3.1   MATLAB file for the profit maximization problem

---

m-file with explanatory comments

```
%Create a grid from −1 to 25 with an increment of 0.5 for the variables x1 and x2
[x1,x2]=meshgrid(−1:0.5:25.0,−1:0.5:25.0);
%Enter functions for the profit maximization problem
f=400*x1+600*x2;
g1=x1+x2−16;
g2=x1/28+x2/14−1;
g3=x1/14+x2/24−1;
g4=−x1;
g5=−x2;
%Initialization statements; these need not end with a semicolon
    cla reset
    axis auto    %Minimum and maximum values for axes are determined automatically
                 %Limits for x- and y-axes may also be specified with the command
                 %axis ([xmin xmax ymin ymax])
    xlabel('x1'),ylabel('x2') %Specifies labels for x- and y-axes
    title ('Profit Maximization Problem') %Displays a title for the problem
    hold on    %retains the current plot and axes properties for all subsequent plots
%Use the "contour" command to plot constraint and cost functions
    cv1=[0 .5];    %Specifies two contour values, 0 and .5
    const1=contour(x1,x2,g1,cv1,'k');    %Plots two specified contours of g1; k=black
                                    color
    clabel(const1)    %Automatically puts the contour value on the graph
    text(1,16,'g1')    %Writes g1 at the location (1, 16)
    cv2=[0 .03];
    const2=contour(x1,x2,g2,cv2,'k');
    clabel(const2)
    text(23,3,'g2')
    const3=contour(x1,x2,g3,cv2,'k');
    clabel(const3)
    text(1,23,'g3')
    cv3=[0 .5];
    const4=contour(x1,x2,g4,cv3,'k');
    clabel(const4)
    text(.25,20,'g4')
    const5=contour(x1,x2,g5,cv3,'k');
    clabel(const5)
    text(19,.5,'g5')
    text(1.5,7,'Feasible Region')
    fv=[2400, 4800, 7200, 8800];    %Defines 4 contours for the profit function
    fs=contour(x1,x2,f,fv,'k−');    %'k−' specifies black dashed lines for profit
                                    function contours
    clabel(fs)
    hold off    %Indicates end of this plotting sequence
                %Subsequent plots will appear in separate windows
```

---

contour will pass through the problem's infeasible region. The thickness of the infeasible contour is changed to indicate the infeasible side of the inequality using the graph-editing capability, which is explained in the following subsection.

In this way all constraint functions are plotted and the problem's feasible region is identified. By observing the trend of the objective function contours, the optimum point for the problem is identified.

3.3.2 Editing of Graph

Once the graph has been created using the commands just described, we can edit it before printing it or copying it to a text editor. In particular, we may need to modify the appearance of the constraints' infeasible contours and edit any text. To do this, first select "Current Object Properties..." under the "Edit" tab on the graph window. Then double-click on any item in the graph to edit its properties. For instance, we can increase the thickness of the infeasible contours to shade out the infeasible region. In addition, text may be added, deleted, or moved as desired. Note that if MATLAB is re-run, any changes made directly to the graph are lost. For this reason, it is a good idea to save the graph as a ".fig" file, which may be recalled with MATLAB.

Another way to shade out the infeasible region is to plot several closely spaced contours in it using the following commands:

```
cv1=[0:0.01:0.5];    %[Starting contour: Increment: Final contour]
const1=contour(x1,x2,g1,cv1,'g');    % g = green color
```

There are two ways to transfer the graph to a text document. First, select "Copy Figure" under the "Edit" tab so that the figure can be pasted as a bitmap into a document. Alternatively, select "Export..." under the "File" tab. The figure is exported as the specified file type and can be inserted into another document through the "Insert" command. The final MATLAB graph for the profit maximization problem is shown in Figure 3.6.

3.4 DESIGN PROBLEM WITH MULTIPLE SOLUTIONS

A situation can arise in which a constraint is parallel to the cost function. If the constraint is active at the optimum, there are multiple solutions to the problem. To illustrate this situation, consider the following design problem:

Minimize
$$f(\mathbf{x}) = -x_1 - 0.5x_2 \tag{a}$$

subject to
$$2x_1 + 3x_2 \le 12, \quad 2x_1 + x_2 \le 8, \quad -x_1 \le 0, \quad -x_2 \le 0 \tag{b}$$

In this problem, the second constraint is parallel to the cost function. Therefore, there is a possibility of *multiple optimum designs*. Figure 3.7 provides a graphical solution to the problem. It is seen that any point on the line B—C gives an optimum design, giving the problem infinite optimum solutions.

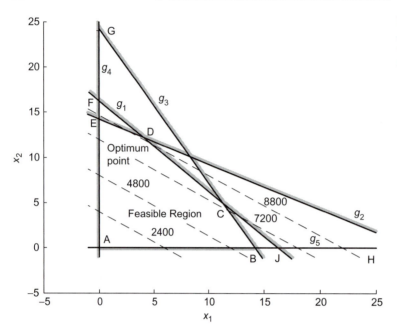

FIGURE 3.6 This shows a graphical representation of the profit maximization problem with MATLAB.

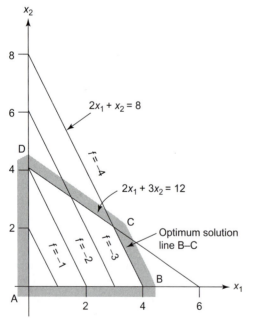

FIGURE 3.7 Example problem with multiple solutions.

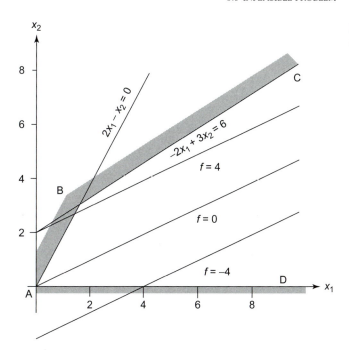

FIGURE 3.8 Example problem with an unbounded solution.

3.5 PROBLEM WITH UNBOUNDED SOLUTIONS

Some design problems may not have a bounded solution. This situation can arise if we forget a constraint or incorrectly formulate the problem. To illustrate such a situation, consider the following design problem:

Minimize
$$f(\mathbf{x}) = -x_1 + 2x_2 \qquad (c)$$

subject to
$$-2x_1 + x_2 \le 0, \quad -2x_1 + 3x_2 \le 6, \quad -x_1 \le 0, \quad -x_2 \le 0 \qquad (d)$$

The feasible set for the problem is shown in Figure 3.8 with several cost function contours. It is seen that the feasible set is unbounded. Therefore, there is no finite optimum solution, and we must re-examine the way the problem was formulated to correct the situation. Figure 3.8 shows that the problem is underconstrained.

3.6 INFEASIBLE PROBLEM

If we are not careful in formulating it, a design problem may not have a solution, which happens when there are conflicting requirements or inconsistent constraint equations. There may also be no solution when we put *too many constraints* on the system; that is, the

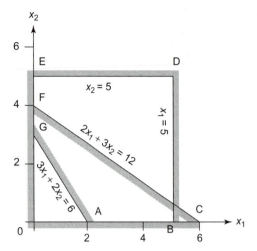

FIGURE 3.9 Infeasible design optimization problem.

constraints are so restrictive that no feasible solution is possible. These are called *infeasible problems.* To illustrate them, consider the following:

Minimize

$$f(\mathbf{x}) = x_1 + 2x_2 \qquad (e)$$

subject to

$$3x_1 + 2x_2 \le 6, \quad 2x_1 + 3x_2 \ge 12, \quad x_1, \, x_2 \le 5, \; x_1, \, x_2 \ge 0 \qquad (f)$$

Constraints for the problem are plotted in Figure 3.9 and their infeasible side is shaded-out. It is evident that there is no region within the design space that satisfies all constraints; that is, there is no feasible region for the problem. Thus, the problem is infeasible. Basically, the first two constraints impose conflicting requirements. The first requires the feasible design to be below the line A–G, whereas the second requires it to be above the line C–F. Since the two lines do not intersect in the first quadrant, the problem has no feasible region.

3.7 GRAPHICAL SOLUTION FOR THE MINIMUM-WEIGHT TUBULAR COLUMN

The design problem formulated in Section 2.7 will now be solved by the graphical method using the following data: $P = 10$ MN, $E = 207$ GPa, $\rho = 7833$ kg/m^3, $l = 5.0$ m, and $\sigma_a = 248$ MPa. Using these data, formulation 1 for the problem is defined as "Find mean radius R (m) and thickness t (m) to minimize the mass function":

$$f(R,t) = 2\rho l \pi R t = 2(7833)(5)\pi R t = 2.4608 \times 10^5 R t, \text{ kg} \qquad (a)$$

subject to the four inequality constraints

$$g_1(R,t) = \frac{P}{2\pi Rt} - \sigma_a = \frac{10 \times 10^6}{2\pi Rt} - 248 \times 10^6 \leq 0 \quad \text{(stress constraint)} \tag{b}$$

$$g_2(R,t) = P - \frac{\pi^3 ER^3 t}{4l^2} = 10 \times 10^6 - \frac{\pi^3(207 \times 10^9)R^3 t}{4(5)(5)} \leq 0 \quad \text{(buckling load constraint)} \tag{c}$$

$$g_3(R,t) = -R \leq 0 \tag{d}$$

$$g_4(R,t) = -t \leq 0 \tag{e}$$

Note that the explicit bound constraints discussed in Section 2.7 are simply replaced by the non-negativity constraints g_3 and g_4. The constraints for the problem are plotted in Figure 3.10, and the feasible region is indicated. Cost function contours for $f = 1000$ kg, 1500 kg, and 1579 kg are also shown. In this example the cost function contours run parallel to the stress constraint g_1. Since g_1 is active at the optimum, the problem has infinite optimum designs, that is, the entire curve A–B in Figure 3.10. We can read the coordinates of any point on the curve A–B as an optimum solution. In particular, point A, where constraints g_1 and g_2 intersect, is also an optimum point where $R^* = 0.1575$ m and $t^* = 0.0405$ m.

The superscript "*" on a variable indicates its optimum value, a notation that will be used throughout this text.

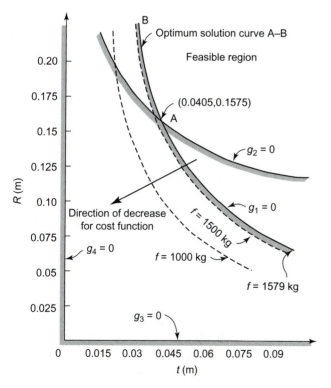

FIGURE 3.10 A graphical solution to the problem of designing a minimum-weight tubular column.

3.8 GRAPHICAL SOLUTION FOR A BEAM DESIGN PROBLEM

STEP 1: PROJECT/PROBLEM DESCRIPTION   A beam of rectangular cross-section is subjected to a bending moment M (N · m) and a maximum shear force V (N). The bending stress in the beam is calculated as $\sigma = 6M/bd^2$ (Pa), and average shear stress is calculated as $\tau = 3V/2bd$ (Pa), where b is the width and d is the depth of the beam. The allowable stresses in bending and shear are 10 MPa and 2 MPa, respectively. It is also desirable that the depth of the beam not exceed twice its width and that the cross-sectional area of the beam be minimized. In this section, we formulate and solve the problem using the graphical method.

STEP 2: DATA AND INFORMATION COLLECTION   Let bending moment $M = 40$ kN · m and the shear force $V = 150$ kN. All other data and necessary equations are given in the project statement. We shall formulate the problem using a consistent set of units, N and mm.

STEP 3: DEFINITION OF DESIGN VARIABLES   The two design variables are

d = depth of beam, mm
b = width of beam, mm

STEP 4: OPTIMIZATION CRITERION   The cost function for the problem is the cross-sectional area, which is expressed as

$$f(b,d) = bd \tag{a}$$

STEP 5: FORMULATION OF CONSTRAINTS   Constraints for the problem consist of bending stress, shear stress, and depth-to-width ratio. Bending and shear stresses are calculated as

$$\sigma = \frac{6M}{bd^2} = \frac{6(40)(1000)(1000)}{bd^2}, \ \ \text{N/mm}^2 \tag{b}$$

$$\tau = \frac{3V}{2bd} = \frac{3(150)(1000)}{2bd}, \ \ \text{N/mm}^2 \tag{c}$$

Allowable bending stress σ_a and allowable shear stress τ_a are given as

$$\sigma_a = 10 \ \text{MPa} = 10 \times 10^6 \ \text{N/m}^2 = 10 \ \text{N/mm}^2 \tag{d}$$

$$\tau_a = 2 \ \text{MPa} = 2 \times 10^6 \ \text{N/m}^2 = 2 \ \text{N/mm}^2 \tag{e}$$

Using Eqs. (b) through (e), we obtain the bending and shear stress constraints as

$$g_1 = \frac{6(40)(1000)(1000)}{bd^2} - 10 \le 0 \ \ \text{(bending stress)} \tag{f}$$

$$g_2 = \frac{3(150)(1000)}{2bd} - 2 \le 0 \ \ \text{(shear stress)} \tag{g}$$

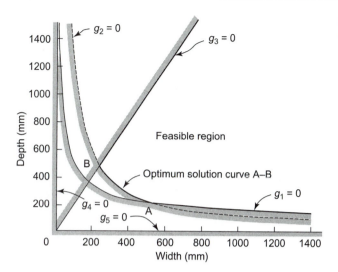

FIGURE 3.11 Graphical solution to the minimum-area beam design problem.

The constraint that requires that the depth be no more than twice the width can be expressed as

$$g_3 = d - 2b \leq 0 \tag{h}$$

Finally, both design variables should be non-negative:

$$g_4 = -b \leq 0; \quad g_5 = -d \leq 0 \tag{i}$$

In reality, b and d cannot both have zero value, so we should use some minimum value as a lower bound on them (i.e., $b \geq b_{min}$ and $d \geq d_{min}$)

Graphical Solution

Using MATLAB, the constraints for the problem are plotted in Figure 3.11, and the feasible region is identified. Note that the cost function is parallel to the constraint g_2 (both functions have the same form: $bd = $ constant). Therefore, any point along the curve A–B represents an optimum solution, so there are infinite optimum designs. This is a desirable situation since a wide choice of optimum solutions is available to meet a designer's needs.

The optimum cross-sectional area is 112,500 mm². Point B corresponds to an optimum design of $b = 237$ mm and $d = 474$ mm. Point A corresponds to $b = 527.3$ mm and $d = 213.3$ mm. These points represent the two extreme optimum solutions; all other solutions lie between these two points on the curve A–B.

EXERCISES FOR CHAPTER 3

Solve the following problems using the graphical method.

3.1 Minimize $f(x_1, x_2) = (x_1 - 3)^2 + (x_2 - 3)^2$
  subject to $x_1 + x_2 \leq 4$
  $x_1, x_2 \geq 0$

3.2 Maximize $F(x_1, x_2) = x_1 + 2x_2$
    subject to $2x_1 + x_2 \leq 4$
                 $x_1, x_2 \geq 0$

3.3 Minimize $f(x_1, x_2) = x_1 + 3x_2$
    subject to $x_1 + 4x_2 \geq 48$
              $5x_1 + x_2 \geq 50$
                $x_1, x_2 \geq 0$

3.4 Maximize $F(x_1, x_2) = x_1 + x_2 + 2x_3$
    subject to $1 \leq x_1 \leq 4$
              $3x_2 - 2x_3 = 6$
             $-1 \leq x_3 \leq 2$
             $x_2 \geq 0$

3.5 Maximize $F(x_1, x_2) = 4x_1x_2$
    subject to $x_1 + x_2 \leq 20$
              $x_2 - x_1 \leq 10$
             $x_1, x_2 \geq 0$

3.6 Minimize $f(x_1, x_2) = 5x_1 + 10x_2$
    subject to $10x_1 + 5x_2 \leq 50$
             $5x_1 - 5x_2 \geq -20$
             $x_1, x_2 \geq 0$

3.7 Minimize $f(x_1, x_2) = 3x_1 + x_2$
    subject to $2x_1 + 4x_2 \leq 21$
             $5x_1 + 3x_2 \leq 18$
             $x_1, x_2 \geq 0$

3.8 Minimize $f(x_1, x_2) = x_1^2 - 2x_2^2 - 4x_1$
    subject to $x_1 + x_2 \leq 6$
              $x_2 \leq 3$
             $x_1, x_2 \geq 0$

3.9 Minimize $f(x_1, x_2) = x_1x_2$
    subject to $x_1 + x_2^2 \leq 0$
            $x_1^2 + x_2^2 \leq 9$

3.10 Minimize $f(x_1, x_2) = 3x_1 + 6x_2$
    subject to $-3x_1 + 3x_2 \leq 2$
             $4x_1 + 2x_2 \leq 4$
            $-x_1 + 3x_2 \geq 1$

Develop an appropriate graphical representation for the following problems and determine the minimum and the maximum points for the objective function.

3.11 $f(x, y) = 2x^2 + y^2 - 2xy - 3x - 2y$
    subject to $y - x \leq 0$
          $x^2 + y^2 - 1 = 0$

3.12 $f(x, y) = 4x^2 + 3y^2 - 5xy - 8x$
    subject to $x + y = 4$

3.13 $f(x, y) = 9x^2 + 13y^2 + 18xy - 4$
subject to $x^2 + y^2 + 2x = 16$

3.14 $f(x, y) = 2x + 3y - x^3 - 2y^2$
subject to $x + 3y \leq 6$
$$5x + 2y \leq 10$$
$$x, y \geq 0$$

3.15 $f(r, t) = (r - 8)^2 + (t - 8)^2$
subject to $12 \geq r + t$
$$t \leq 5$$
$$r, t \geq 0$$

3.16 $f(x_1, x_2) = x_1^3 - 16x_1 + 2x_2 - 3x_2^2$
subject to $x_1 + x_2 \leq 3$

3.17 $f(x, y) = 9x^2 + 13y^2 + 18xy - 4$
subject to $x^2 + y^2 + 2x \geq 16$

3.18 $f(r, t) = (r - 4)^2 + (t - 4)^2$
subject to $10 - r - t \geq 0$
$$5 \geq r$$
$$r, t \geq 0$$

3.19 $f(x, y) = -x + 2y$
subject to $-x^2 + 6x + 3y \leq 27$
$$18x - y^2 \geq 180$$
$$x, y \geq 0$$

3.20 $f(x_1, x_2) = (x_1 - 4)^2 + (x_2 - 2)^2$
subject to $10 \geq x_1 + 2x_2$
$$0 \leq x_1 \leq 3$$
$$x_2 \geq 0$$

3.21 Solve the rectangular beam problem of Exercise 2.17 graphically for the following data: $M = 80$ kN $\cdot$ m, $V = 150$ kN, $\sigma_a = 8$ MPa, and $\tau_a = 3$ MPa.

3.22 Solve the cantilever beam problem of Exercise 2.23 graphically for the following data: $P = 10$ kN; $l = 5.0$ m; modulus of elasticity, $E = 210$ Gpa; allowable bending stress, $\sigma_a = 250$ MPa; allowable shear stress, $\tau_a = 90$ MPa; mass density, $\rho = 7850$ kg/m^3; $R_o \leq 20.0$ cm; $R_i \leq 20.0$ cm.

3.23 For the minimum-mass tubular column design problem formulated in Section 2.7, consider the following data: $P = 50$ kN; $l = 5.0$ m; modulus of elasticity, $E = 210$ Gpa; allowable stress, $\sigma_a = 250$ MPa; mass density $\rho = 7850$ kg/m^3.

Treating mean radius R and wall thickness t as design variables, solve the design problem graphically, imposing an additional constraint $R/t \leq 50$. This constraint is needed to avoid local crippling of the column. Also impose the member size constraints as

$$0.01 \leq R \leq 1.0 \text{ m}; \quad 5 \leq t \leq 200 \text{ mm}$$

3.24 For Exercise 3.23, treat outer radius R_o and inner radius R_i as design variables, and solve the design problem graphically. Impose the same constraints as in Exercise 3.23.

3.25 Formulate the minimum-mass column design problem of Section 2.7 using a hollow square cross-section with outside dimension w and thickness t as design variables. Solve the problem graphically using the constraints and the data given in Exercise 3.23.

3.26 Consider the symmetric (members are identical) case of the two-bar truss problem discussed in Section 2.5 with the following data: $W = 10$ kN; $\theta = 30°$; height $h = 1.0$ m; span $s = 1.5$ m; allowable stress, $\sigma_a = 250$ MPa; modulus of elasticity, $E = 210$ GPa.

Formulate the minimum-mass design problem with constraints on member stresses and bounds on design variables. Solve the problem graphically using circular tubes as members.

3.27 Formulate and solve the problem of Exercise 2.1 graphically.

3.28 In the design of the closed-end, thin-walled cylindrical pressure vessel shown in Figure E3.28, the design objective is to select the mean radius R and wall thickness t to minimize the total mass. The vessel should contain at least 25.0 m³ of gas at an internal pressure of 3.5 MPa. It is required that the circumferential stress in the pressure vessel not exceed 210 MPa and the circumferential strain not exceed (1.0E−03). The circumferential stress and strain are calculated from the equations

$$\sigma_c = \frac{PR}{t}, \quad \varepsilon_c = \frac{PR(2 - \nu)}{2Et}$$

where ρ = mass density (7850 kg/m³), σ_c = circumferential stress (Pa), ε_c = circumferential strain, P = internal pressure (Pa), E = Young's modulus (210 GPa), and ν = Poisson's ratio (0.3). Formulate the optimum design problem, and solve it graphically.

3.29 Consider the symmetric three-bar truss design problem formulated in Section 2.10. Formulate and solve the problem graphically for the following data: $l = 1.0$ m; $P = 100$ kN; $\theta = 30°$; mass density, $\rho = 2800$ kg/m³; modulus of elasticity, $E = 70$ GPa; allowable stress, $\sigma_a = 140$ MPa; $\Delta_u = 0.5$ cm; $\Delta_v = 0.5$ cm; $\omega_o = 50$ Hz; $\beta = 1.0$; $A_1, A_2 \geq 2$ cm².

3.30 Consider the cabinet design problem in Section 2.6. Use the equality constraints to eliminate three design variables from the problem. Restate the problem in terms of the remaining three variables, transcribing it into the standard form.

3.31 Graphically solve the insulated spherical tank design problem formulated in Section 2.3 for the following data: $r = 3.0$ m, $c_1 = \$10,000$, $c_2 = \$1000$, $c_3 = \$1$, $c_4 = \$0.1$, $\Delta T = 5$.

3.32 Solve the cylindrical tank design problem given in Section 2.8 graphically for the following data: $c = \$1500/m^2$, $V = 3000$ m³.

3.33 Consider the minimum-mass tubular column problem formulated in Section 2.7. Find the optimum solution for it using the graphical method for the data: load, $P = 100$ kN; length, $l = 5.0$ m; Young's modulus, $E = 210$ GPa; allowable stress, $\sigma_a = 250$ MPa; mass density, $\rho = 7850$ kg/m³; $R \leq 0.4$ m; $t \leq 0.1$ m; $R, t \geq 0$.

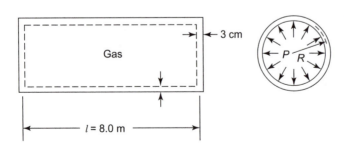

FIGURE E3.28 Graphic of a cylindrical pressure vessel.

*3.34 Design a hollow torsion rod, shown in Figure E3.34, to satisfy the following requirements (created by J. M. Trummel):

1. The calculated shear stress τ shall not exceed the allowable shear stress τ_a under the normal operating torque T_o (N · m).

2. The calculated angle of twist, θ, shall not exceed the allowable twist, θ_a (radians).

3. The member shall not buckle under a short duration torque of T_{max} (N · m).

Requirements for the rod and material properties are given in Tables E3.34 (select a material for one rod). Use the following design variables: x_1 = outside diameter of the rod; x_2 = ratio of inside/outside diameter, d_i/d_o.

Using graphical optimization, determine the inside and outside diameters for a minimum-mass rod to meet the preceding design requirements. Compare the hollow rod

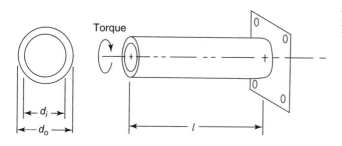

FIGURE E3.34 Graphic of a hollow torsion rod.

TABLE E3.34(a) Rod requirements

Torsion rod no.	Length l (m)	Normal torque T_o (kN · m)	Maximum T_{max} (kN · m)	Allowable twist θ_a (degrees)
1	0.50	10.0	20.0	2
2	0.75	15.0	25.0	2
3	1.00	20.0	30.0	2

TABLE E3.34(b) Materials and properties for the torsion rod

Material	Density, ρ (kg/m^3)	Allowable shear stress, τ_a (MPa)	Elastic modulus, E (GPa)	Shear modulus, G (GPa)	Poisson ratio (ν)
1. 4140 alloy steel	7850	275	210	80	0.30
2. Aluminum alloy 24 ST4	2750	165	75	28	0.32
3. Magnesium alloy A261	1800	90	45	16	0.35
4. Berylium	1850	110	300	147	0.02
5. Titanium	4500	165	110	42	0.30

with an equivalent solid rod ($d_i/d_o = 0$). Use a consistent set of units (e.g., Newtons and millimeters) and let the minimum and maximum values for design variables be given as

$$0.02 \le d_o \le 0.5 \,\text{m}, \quad 0.60 \le \frac{d_i}{d_o} \le 0.999$$

Useful expressions

Mass $M = \dfrac{\pi}{4}\rho l(d_o^2 - d_i^2), \text{ kg}$

Calculated shear stress $\tau = \dfrac{c}{J}T_o, \text{ Pa}$

Calculated angle of twist $\theta = \dfrac{l}{GJ}T_o, \text{ radians}$

Critical buckling torque $T_{cr} = \dfrac{\pi d_o^3 E}{12\sqrt{2}(1 - \nu^2)^{0.75}}\left(1 - \dfrac{d_i}{d_o}\right)^{2.5}, \text{ N} \cdot \text{m}$

Notation

M mass (kg)
d_o outside diameter (m)
d_i inside diameter (m)
ρ mass density of material (kg/m^3)
l length (m)
T_o normal operating torque (N · m)
c distance from rod axis to extreme fiber (m)
J polar moment of inertia (m^4)
θ angle of twist (radians)
G modulus of rigidity (Pa)
T_{cr} critical buckling torque (N · m)
E modulus of elasticity (Pa)
ν Poisson's ratio

---

*3.35 Formulate and solve Exercise 3.34 using the outside diameter d_o and the inside diameter d_i as design variables.

*3.36 Formulate and solve Exercise 3.34 using the mean radius R and wall thickness t as design variables. Let the bounds on design variables be given as $5 \le R \le 20$ cm and $0.2 \le t \le 4$ cm.

3.37 Formulate the problem in Exercise 2.3 and solve it using the graphical method.

3.38 Formulate the problem in Exercise 2.4 and solve it using the graphical method.

3.39 Solve Exercise 3.23 for a column pinned at both ends. The buckling load for such a column is given as $\pi^2 EI/l^2$. Use the graphical method.

3.40 Solve Exercise 3.23 for a column fixed at both ends. The buckling load for such a column is given as $4\pi^2 EI/l^2$. Use the graphical method.

3.41 Solve Exercise 3.23 for a column fixed at one end and pinned at the other. The buckling load for such a column is given as $2\pi^2 EI/l^2$. Use the graphical method.

3.42 Solve Exercise 3.24 for a column pinned at both ends. The buckling load for such a column is given as $\pi^2 EI/l^2$. Use the graphical method.

3.43 Solve Exercise 3.24 for a column fixed at both ends. The buckling load for such a column is given as $4\pi^2 EI/l^2$. Use the graphical method.

3.44 Solve Exercise 3.24 for a column fixed at one end and pinned at the other. The buckling load for such a column is given as $2\pi^2 EI/l^2$. Use the graphical method.

3.45 Solve the can design problem formulated in Section 2.2 using the graphical method.

3.46 Consider the two-bar truss shown in Figure 2.5. Using the given data, design a minimum-mass structure where $W = 100$ kN; $\theta = 30°$; $h = 1$ m; $s = 1.5$ m; modulus of elasticity $E = 210$ GPa; allowable stress $\sigma_a = 250$ MPa; mass density $\rho = 7850$ kg/m^3. Use Newtons and millimeters as units. The members should not fail in stress and their buckling should be avoided. Deflection at the top in either direction should not be more than 5 cm.

Use cross-sectional areas A_1 and A_2 of the two members as design variables and let the moment of inertia of the members be given as $I = A^2$. Areas must also satisfy the constraint $1 \leq A_i \leq 50$ cm^2.

3.47 For Exercise 3.46, use hollow circular tubes as members with mean radius R and wall thickness t as design variables. Make sure that $R/t \leq 50$. Design the structure so that member 1 is symmetric with member 2. The radius and thickness must also satisfy the constraints $2 \leq t \leq 40$ mm and $2 \leq R \leq 40$ cm.

3.48 Design a symmetric structure defined in Exercise 3.46, treating cross-sectional area A and height h as design variables. The design variables must also satisfy the constraints $1 \leq A \leq 50$ cm^2 and $0.5 \leq h \leq 3$ m.

3.49 Design a symmetric structure defined in Exercise 3.46, treating cross-sectional area A and span s as design variables. The design variables must also satisfy the constraints $1 \leq A \leq 50$ cm^2 and $0.5 \leq s \leq 4$ m.

3.50 Design a minimum-mass symmetric three-bar truss (the area of member 1 and that of member 3 are the same) to support a load P, as was shown in Figure 2.9. The following notation may be used: $P_u = P \cos \theta$, $P_v = P \sin \theta$, $A_1 = $ cross-sectional area of members 1 and 3, $A_2 = $ cross-sectional area of member 2.

The members must not fail under the stress, and the deflection at node 4 must not exceed 2 cm in either direction. Use Newtons and millimeters as units. The data is given as $P = 50$ kN; $\theta = 30°$; mass density, $\rho = 7850$ kg/m^3; $l = 1$ m; modulus of elasticity, $E = 210$ GPa; allowable stress, $\sigma_a = 150$ MPa. The design variables must also satisfy the constraints $50 \leq A_i \leq 5000$ mm^2.

***3.51** **Design of a water tower support column.** As an employee of ABC Consulting Engineers, you have been asked to design a cantilever cylindrical support column of minimum mass for a new water tank. The tank itself has already been designed in the teardrop shape, shown in Figure E3.51. The height of the base of the tank (H), the diameter of the tank (D), and the wind pressure on the tank (w) are given as $H = 30$ m, $D = 10$ m, and $w = 700$ N/m^2. Formulate the design optimization problem and then solve it graphically (created by G. Baenziger).

In addition to designing for combined axial and bending stresses and buckling, several limitations have been placed on the design. The support column must have an inside diameter of at least 0.70 m (d_i) to allow for piping and ladder access to the interior

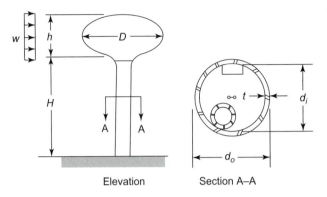

FIGURE E3.51 Graphic of a water tower support column.

Elevation   Section A–A

of the tank. To prevent local buckling of the column walls, the diameter/thickness ratio (d_o/t) cannot be greater than 92. The large mass of water and steel makes deflections critical, as they add to the bending moment. The deflection effects, as well as an assumed construction eccentricity (e) of 10 cm, must be accounted for in the design process. Deflection at the center of gravity (C.G.) of the tank should not be greater than Δ.

Limits on the inner radius and wall thickness are $0.35 \le R \le 2.0$ m and $1.0 \le t \le 20$ cm.

Pertinent constants and formulas

Height of water tank	$h = 10$ m
Allowable deflection	$\Delta = 20$ cm
Unit weight of water	$\gamma_w = 10$ kN/m^3
Unit weight of steel	$\gamma_s = 80$ kN/m^3
Modulus of elasticity	$E = 210$ GPa
Moment of inertia of the column	$I = \frac{\pi}{64}\left[d_o^4 - (d_o - 2t)^4\right]$
Cross-sectional area of column material	$A = \pi t(d_o - t)$
Allowable bending stress	$\sigma_b = 165$ MPa
Allowable axial stress	$\sigma_a = \dfrac{12\pi^2 E}{92(H/r)^2}$ (calculated using the critical buckling load with a factor of safety of 23/12)
Radius of gyration	$r = \sqrt{I/A}$
Average thickness of tank wall	$t_t = 1.5$ cm
Volume of tank	$V = 1.2\pi D^2 h$
Surface area of tank	$A_s = 1.25\pi D^2$
Projected area of tank, for wind loading	$A_p = \dfrac{2Dh}{3}$
Load on the column due to weight of water and steel tank	$P = V\gamma_w + A_s t_t \gamma_s$
Lateral load at the tank C.G. due to wind pressure	$W = wA_p$

Deflection at C.G. of tank	$\delta = \delta_1 + \delta_2$, where
	$\delta_1 = \dfrac{WH^2}{12EI}(4H + 3h)$
	$\delta_2 = \dfrac{H}{2EI}(0.5Wh + Pe)(H + h)$
Moment at base	$M = W(H + 0.5h) + (\delta + e)P$
Bending stress	$f_b = \frac{M}{2I}d_o$
Axial stress	$f_a(= P/A) = \dfrac{V\gamma_w + A_s\gamma_s t_t}{\pi t(d_o - t)}$
Combined stress constraint	$\dfrac{f_a}{\sigma_a} + \dfrac{f_b}{\sigma_b} \leq 1$
Gravitational acceleration	$g = 9.81 \text{ m/s}^2$

*3.52 Design of a flag pole. Your consulting firm has been asked to design a minimum-mass flag pole of height H. The pole will be made of uniform hollow circular tubing with d_o and d_i as outer and inner diameters, respectively. The pole must not fail under the action of high winds.

For design purposes, the pole will be treated as a cantilever that is subjected to a uniform lateral wind load of w (kN/m). In addition to the uniform load, the wind induces a concentrated load of P (kN) at the top of the pole, as shown in Figure E3.52. The flag pole must not fail in bending or shear. The deflection at the top should not exceed 10 cm. The ratio of mean diameter to thickness must not exceed 60. The pertinent data are given in the table that follows. Assume any other data if needed. The minimum and maximum values of design variables are $5 \leq d_o \leq 50$ cm and $4 \leq d_i \leq 45$ cm.

Formulate the design problem and solve it using the graphical optimization technique.

Pertinent constants and equations

Cross-sectional area	$A = \dfrac{\pi}{4}(d_o^2 - d_i^2)$
Moment of inertia	$I = \dfrac{\pi}{64}(d_o^4 - d_i^4)$
Modulus of elasticity	$E = 210$ GPa
Allowable bending stress	$\sigma_b = 165$ MPa
Allowable shear stress	$\tau_s = 50$ MPa
Mass density of pole material	$\rho = 7800 \text{ kg/m}^3$
Wind load	$w = 2.0$ kN/m
Height of flag pole	$H = 10$ m
Concentrated load at top	$P = 4.0$ kN
Moment at base	$M = (PH + 0.5wH^2)$, kN $\cdot$ m
Bending stress	$\sigma = \dfrac{M}{2I}d_o$, kPa
Shear at base	$S = (P + wH)$, kN
Shear stress	$\tau = \dfrac{S}{12I}(d_o^2 + d_od_i + d_i^2)$, kPa
Deflection at top	$\delta = \dfrac{PH^3}{3EI} + \dfrac{wH^4}{8EI}$
Minimum and maximum thickness	0.5 and 2 cm

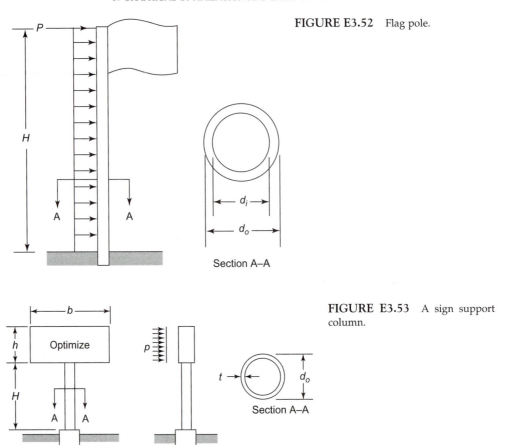

FIGURE E3.52   Flag pole.

Section A–A

FIGURE E3.53   A sign support column.

*3.53 **Design of a sign support column.** A company's design department has been asked to design a support column of minimum weight for the sign shown in Figure E3.53. The height to the bottom of the sign H, the width b, and the wind pressure p on the sign are as follows: $H = 20$ m, $b = 8$ m, $p = 800$ N/m^2.

   The sign itself weighs 2.5 kN/m^2(w). The column must be safe with respect to combined axial and bending stresses. The allowable axial stress includes a factor of safety with respect to buckling. To prevent local buckling of the plate, the diameter/thickness ratio d_o/t must not exceed 92. Note that the bending stress in the column will increase as a result of the deflection of the sign under the wind load. The maximum deflection at the sign's center of gravity should not exceed 0.1 m. The minimum and maximum values of design variables are $25 \leq d_o \leq 150$ cm and $0.5 \leq t \leq 10$ cm (created by H. Kane).

Pertinent constants and equations

Height of sign	$h = 4.0$ m
Cross-sectional area	$A = \dfrac{\pi}{4}\left[d_o^2 - (d_o - 2t)^2\right]$
Moment of inertia	$I = \dfrac{\pi}{64}(d_o^4 - (d_o - 2t)^4)$
Radius of gyration	$r = \sqrt{I/A}$
Young's modulus (aluminum alloy)	$E = 75$ GPa
Unit weight of aluminum	$\gamma = 27$ kN/m^3
Allowable bending stress	$\sigma_b = 140$ MPa
Allowable axial stress	$\sigma_a = \dfrac{12\pi^2 E}{92(H/r)^2}$
Wind force	$F = pbh$
Weight of sign	$W = wbh$
Deflection at center of gravity of sign	$\delta = \dfrac{F}{EI}\left(\dfrac{H^3}{3} + \dfrac{H^2 h}{2} + \dfrac{Hh^2}{4}\right)$
Bending stress in column	$f_b = \dfrac{M}{2I}d_o$
Axial stress	$f_a = \dfrac{W}{A}$
Moment at base	$M = F\left(H + \dfrac{h}{2}\right) + W\delta$
Combined stress requirement	$\dfrac{f_a}{\sigma_a} + \dfrac{f_b}{\sigma_b} \le 1$

*3.54 **Design of a tripod.** Design a minimum mass tripod of height H to support a vertical load $W = 60$ kN. The tripod base is an equilateral triangle with sides $B = 1200$ mm. The struts have a solid circular cross-section of diameter D (Figure E3.54).

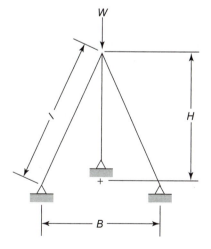

FIGURE E3.54   Tripod.

The axial stress in the struts must not exceed the allowable stress in compression, and the axial load in the strut P must not exceed the critical buckling load P_{cr} divided by a safety factor FS $= 2$. Use consistent units of Newtons and centimeters. The minimum and maximum values for the design variables are $0.5 \leq H \leq 5$ m and $0.5 \leq D \leq 50$ cm. Material properties and other relationships are given next:

Material	aluminum alloy 2014-T6
Allowable compressive stress	$\sigma_a = 150$ MPa
Young's modulus	$E = 75$ GPa
Mass density	$\rho = 2800$ kg/m^3
Strut length	$l = \left(H^2 + \frac{1}{3}B^2 \right)^{0.5}$
Critical buckling load	$P_{cr} = \frac{\pi^2 EI}{l^2}$
Moment of inertia	$I = \frac{\pi}{64}D^4$
Strut load	$P = \frac{Wl}{3H}$

4

Optimum Design Concepts
Optimality Conditions

Upon completion of this chapter, you will be able to

- Define local and global minima (maxima) for unconstrained and constrained optimization problems

- Write optimality conditions for unconstrained problems

- Write optimality conditions for constrained problems

- Check optimality of a given point for unconstrained and constrained problems

- Solve first-order optimality conditions for candidate minimum points

- Check convexity of a function and the design optimization problem

- Use Lagrange multipliers to study changes to the optimum value of the cost function due to variations in a constraint

---

In this chapter, we discuss *basic ideas, concepts and theories used for design optimization* (the minimization problem). Theorems on the subject are stated without proofs. Their implications and use in the optimization process are discussed. Useful insights regarding the optimality conditions are presented and illustrated. It is assumed that the variables are continuous and that all problem functions are continuous and at least twice continuously differentiable. Methods for discrete variable problems that may or may not need derivatives of the problem functions are presented in later chapters.

The student is reminded to review the basic terminology and notation explained in Section 1.5 as they are used throughout the present chapter and the remaining text.

Figure 4.1 shows a broad classification of the optimization approaches for continuous variable constrained and unconstrained optimization problems. The following two philosophically different viewpoints are shown: optimality criteria (or indirect) methods and search (or direct) methods.

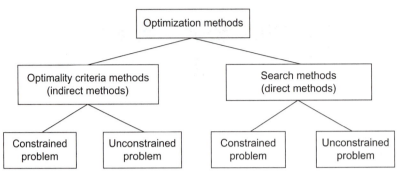

FIGURE 4.1   Classification of optimization methods.

Optimality Criteria Methods—Optimality criteria are the conditions a function must satisfy at its minimum point. Optimization methods seeking solutions (perhaps using numerical methods) to the optimality conditions are often called optimality criteria or indirect methods. In this chapter and the next, we describe methods based on this approach.

Search Methods—Search (direct) methods are based on a different philosophy. Here we start with an estimate of the optimum design for the problem. Usually the starting design will not satisfy the optimality criteria; therefore, it is improved iteratively until they are satisfied. Thus, in the direct approach we search the design space for optimum points. Methods based on this approach are described in later chapters.

A thorough knowledge of *optimality conditions* is important for an understanding of the performance of various *numerical (search) methods* discussed later in the text. This chapter and the next one focus on optimality conditions and the solution methods based on them. Simple examples are used to explain the underlying concepts and ideas. The examples will also show practical limitations of the methods.

The *search methods* are presented in Chapters 8 through 13 and refer to the results discussed in this chapter. Therefore, the *material in the present chapter should be understood thoroughly*. We will first discuss the concept of the local optimum of a function and the conditions that characterize it. The problem of global optimality of a function will be discussed later in this chapter.

4.1   DEFINITIONS OF GLOBAL AND LOCAL MINIMA

Optimality conditions for a minimum point of the function are discussed in later sections. In this section, concepts of *local* and *global minima* are defined and illustrated using the *standard mathematical model for design optimization,* defined in Chapter 2. The design optimization problem is always converted to minimization of a cost function subject to equality and inequality constraints. The problem is restated as follows:

Find design variable vector x to minimize a *cost function* f(x) subject to the *equality constraints* $h_j(x) = 0$, $j = 1$ to p and the *inequality constraints* $g_i(x) \leq 0$, $i = 1$ to m.

4.1.1 Minimum

In Section 2.11, we defined the *feasible set* S (also called the *constraint set, feasible region*, or *feasible design space*) for a design problem as a collection of feasible designs:

$$S = \{x | h_j(x) = 0, \ j = 1 \text{ to } p; \quad g_i(x) \le 0; \quad i = 1 \text{ to } m\} \tag{4.1}$$

Since there are no constraints in unconstrained problems, the entire design space is feasible for them. The *optimization problem* is to find a point in the feasible design space that gives a minimum value to the cost function. Methods to locate optimum designs are discussed throughout the text. We must first carefully define what is meant by an optimum. In the following discussion, x* is used to designate a particular point of the feasible set.

Global (Absolute) Minimum

A function $f(x)$ of n variables has a global (absolute) minimum at x* if the value of the function at x* is less than or equal to the value of the function at any other point x in the feasible set S. That is,

$$f(x^*) \le f(x) \tag{4.2}$$

for all x in the feasible set S. If strict inequality holds for all x other than x* in Eq. (4.2), then x* is called a *strong (strict) global minimum*; otherwise, it is called a *weak global minimum*.

Local (Relative) Minimum

A function $f(x)$ of n variables has a local (relative) minimum at x* if Inequality (4.2) holds for all x in a small *neighborhood* N (vicinity) of x* in the feasible set S. If strict inequality holds, then x* is called a *strong (strict) local minimum*; otherwise, it is called a *weak local minimum*.

Neighborhood N of point x* is defined as the set of points

$$N = \{x | x \in S \text{ with } ||x - x^*|| < \delta\} \tag{4.3}$$

for some small $\delta > 0$. Geometrically, it is a small feasible region around point x*.

Note that a function $f(x)$ can have *strict global minimum* at only one point. It may, however, have a global minimum at several points if it has the same value at each of those points. Similarly, a function $f(x)$ can have a *strict local minimum* at only one point in the neighborhood N (vicinity) of x*. It may, however, have a local minimum at several points in N if the function value is the same at each of those points.

Note also that *global* and *local maxima* are defined in a similar manner by simply reversing the inequality in Eq. (4.2). We also note here that these definitions do not provide a method for locating minimum points. Based on them, however, we can develop analyses and computational procedures to locate them. Also, we can use these definitions to check the optimality of points in the graphical solution process presented in Chapter 3.

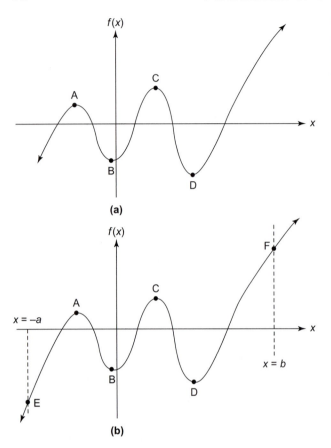

FIGURE 4.2 Representation of optimum points. (a) The unbounded domain and function (no global optimum). (b) The bounded domain and function (global minimum and maximum exist).

To understand the *graphical significance* of global and local minima, consider graphs of a function $f(x)$ of one variable, as shown in Figure 4.2. In Part (a) of the figure, where x is between $-\infty$ and ∞ ($-\infty \leq x \leq \infty$), points B and D are local minima since the function has its smallest value in their neighborhood. Similarly, both A and C are points of local maxima for the function. There is, however, no global minimum or maximum for the function since the domain and the function $f(x)$ are unbounded; that is, x and $f(x)$ are allowed to have any value between $-\infty$ and ∞. If we restrict x to lie between $-a$ and b, as in Part (b) of Figure 4.2, then point E gives the global minimum and F the global maximum for the function. Both of these points have active constraints, while points A, B, C, and D are unconstrained.

- A global minimum point is the one where there are no other feasible points with better cost function values.
- A local minimum point is the one where there are no other feasible points "in the vicinity" with better cost function values.

We will further illustrate these concepts for constrained problems with Examples 4.1 through 4.3.

EXAMPLE 4.1 USE OF THE DEFINITION OF MINIMUM POINT: UNCONSTRAINED MINIMUM FOR A CONSTRAINED PROBLEM

An optimum design problem is formulated and transcribed into the standard form in terms of the variables x and y:

Minimize

$$f(x, y) = (x - 4)^2 + (y - 6)^2 \tag{a}$$

subject to

$$g_1 = x + y - 12 \leq 0 \tag{b}$$

$$g_2 = x - 8 \leq 0 \tag{c}$$

$$g_3 = -x \leq 0 \ (x \geq 0) \tag{d}$$

$$g_4 = -y \leq 0 \ (y \geq 0) \tag{e}$$

Find the local and global minima for the function $f(x,y)$ using the graphical method.

Solution

Using the procedure for graphical optimization described in Chapter 3, the constraints for the problem are plotted and the feasible set S is identified as ABCD in Figure 4.3. The contours of the cost function $f(x,y)$, which is an equation of a circle with center at (4, 6), are also shown.

Unconstrained points To locate the minimum points, we use the definition of *local minimum* and check the inequality $f(x^*,y^*) \leq f(x,y)$ at a candidate feasible point (x^*,y^*) in its small feasible neighborhood. Note that the cost function always has a non-negative value at any point with the smallest value as zero at the center of the circle. Since the center of the circle at E(4, 6) is feasible, it is a *local minimum point* with a cost function value of zero.

Constrained points We check the local minimum condition at some other points as follows:

Point A(0, 0): $f(0, 0) = 52$ is not a local minimum point because the inequality $f(0,0) \leq f(x,y)$ is violated for any small feasible move away from point A; that is, the cost function reduces as we move away from the point A in the feasible region.

Point F(4, 0): $f(4, 0) = 36$ is also not a local minimum point since there are feasible moves from the point for which the cost function can be reduced.

It can be checked that points B, C, D, and G are also not local minimum points. In fact, there is no other local minimum point. Thus, point E is a local, as well as a *global*, minimum point for the function. It is important to note that at the minimum point no constraints are active; that is, *constraints play no role in determining the minimum points for this problem*. However, this is not always true, as we will see in Example 4.2.

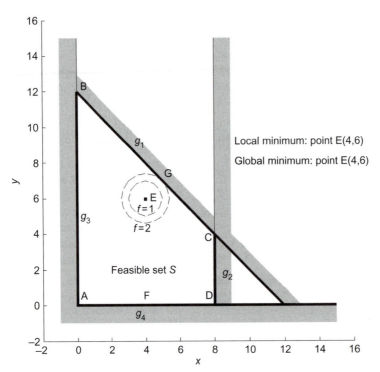

FIGURE 4.3 Representation of unconstrained minimum for Example 4.1.

EXAMPLE 4.2 USE OF THE DEFINITION OF MINIMUM POINT: CONSTRAINED MINIMUM

Solve the optimum design problem formulated in terms of variables x and y as

Minimize

$$f(x, y) = (x - 10)^2 + (y - 8)^2 \qquad \text{(a)}$$

subject to the same constraints as in Example 4.1.

Solution

The feasible region for the problem is the same as for Example 4.1, as ABCD in Figure 4.4. The cost function is an equation of a circle with the center at point E(10,8), which is an infeasible point. Two cost contours are shown in the figure. The problem now is to find a point in the feasible region that is closest to point E, that is, with the smallest value for the cost function. It is seen that point G, with coordinates (7, 5) and $f = 18$, has the smallest distance from point E. At this point, the constraint g_1 is active. Thus, *for the present objective function, the constraints play a prominent role in determining the minimum point for the problem.*

Use of the definition of minimum Use of the definition of a local minimum point also indicates that point G is indeed a local minimum for the function since any *feasible* move from G results in an increase in the cost function. The use of the definition also indicates that there is no other local minimum point. Thus, point G is a global minimum point as well.

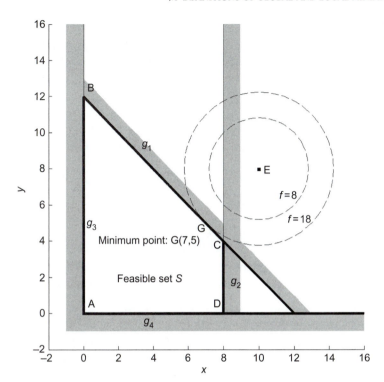

FIGURE 4.4 Representation of constrained minimum for Example 4.2.

EXAMPLE 4.3 USE OF THE DEFINITION OF MAXIMUM POINT

Solve the optimum design problem formulated in terms of variables x and y as

Maximize

$$f(x, y) = (x - 4)^2 + (y - 6)^2 \qquad \text{(a)}$$

subject to the same constraints as in Example 4.1.

Solution

The feasible region for the problem is ABCD, as shown in Figure 4.3. The objective function is an equation of a circle with center at point E(4, 6). Two objective function contours are shown in the figure. It is seen that point D(8, 0) is a local maximum point because any feasible move away from the point results in reduction of the objective function. Point C(8, 4) is not a local maximum point since a feasible move along the line CD results in an increase in the objective function, thus violating the definition of a local max-point [$f(x^*,y^*) \geq f(x,y)$]. It can be verified that points A and B are also local maximum points that and point G is not. Thus, this problem has the following three local maximum points:

Point A(0, 0): f(0, 0) = 52
Point B(0, 12): f(0, 12) = 52
Point D(8, 0): f(8, 0) = 52

It is seen that the objective function has the same value at all three points. Therefore, all of the points are global maximum points. There is no strict global minimum point. This example shows that *an objective function can have several global optimum points in the feasible region.*

4.1.2 Existence of a Minimum

In general, we do not know before attempting to solve a problem if a minimum even exists. In certain cases we can *ensure existence of a minimum*, even though we may not know how to find it. The Weierstrass theorem guarantees this when certain conditions are satisfied.

THEOREM 4.1

Weierstrass Theorem—Existence of a Global Minimum If $f(\mathbf{x})$ is continuous on a non-empty feasible set S that is closed and bounded, then $f(\mathbf{x})$ has a global minimum in S.

To use the theorem, we must understand the meaning of a *closed and bounded set*. A set S is *closed* if it includes all of its boundary points and every sequence of points has a subsequence that converges to a point in the set. This implies that there cannot be strictly "< type" inequality constraints in the formulation of the problem.

A set is bounded if for any point, $\mathbf{x} \in S$, $\mathbf{x}^T\mathbf{x} < c$, where c is a finite number. Since the domain of the function in Figure 4.2(a) is not closed, and the function is also unbounded, a global minimum or maximum for the function is not ensured. Actually, there is no global minimum or maximum for the function. However, in Figure 4.2(b), since the feasible set is closed and bounded with $-a \leq x \leq b$ and the function is continuous, it has global minimum as well as maximum points.

It is important to note that in general it is difficult to check the boundedness condition $\mathbf{x}^T\mathbf{x} < c$ since this condition must be checked for the infinite points in S. The foregoing examples are simple, where a graphical representation of the problem is available and it is easy to check the conditions. Nevertheless, it is important to keep the theorem in mind while using a numerical method to solve an optimization problem. If the numerical process is not converging to a solution, then perhaps some conditions of this theorem are not met and the problem formulation needs to be re-examined carefully. Example 4.4 further illustrates the use of the Weierstrass theorem.

EXAMPLE 4.4 EXISTENCE OF A GLOBAL MINIMUM USING THE WEIERSTRASS THEOREM

Consider a function $f(x) = -1/x$ defined on the set $S = \{x \mid 0 < x \leq 1\}$. Check the existence of a global minimum for the function.

Solution

The feasible set S is not closed since it does not include the boundary point $x = 0$. The conditions of the Weierstrass theorem are not satisfied, although f is continuous on S. The existence of a global minimum is not guaranteed, and indeed there is no point x^* satisfying $f(x^*) \le f(x)$ for all $x \in S$. If we define $S = \{x \mid 0 \le x \le 1\}$, then the feasible set is closed and bounded. However, f is not defined at $x = 0$ (hence not continuous), so the conditions of the theorem are still not satisfied and there is no guarantee of a global minimum for f in the set S.

Note that when conditions of the Weierstrass theorem are satisfied, the existence of a global optimum is guaranteed. It is important, however, to realize that when they are not satisfied, a global solution may still exist; that is, it is not an "if-and-only-if" theorem. The theorem does not rule out the possibility of a global minimum . The difference is that we cannot guarantee its existence. For example, consider the problem of minimizing $f(x) = x^2$ subject to the constraints $-1 < x < 1$. Since the feasible set is not closed, conditions of the Weierstrass theorem are not met; therefore, it cannot be used. However, the function has a global minimum at the point $x = 0$.

Note also that the theorem does not provide a method for finding a global minimum point even if its conditions are satisfied; it is only an existence theorem.

4.2 REVIEW OF SOME BASIC CALCULUS CONCEPTS

Optimality conditions for a minimum point are discussed in later sections. Since most optimization problems involve functions of several variables, these conditions use ideas from vector calculus. Therefore, in this section, *we review basic concepts from calculus using the vector and matrix notations*. Basic material related to vector and matrix algebra (linear algebra) is described in Appendix A. It is important to be comfortable with these materials in order to understand the optimality conditions.

The *partial differentiation notation* for functions of several variables is introduced. The *gradient vector* for a function of several variables requiring first partial derivatives of the function is defined. The *Hessian matrix* for the function requiring second partial derivatives of the function is then defined. *Taylor's expansions* for functions of single and multiple variables are discussed. The concept of *quadratic forms* is needed to discuss sufficiency conditions for optimality. Therefore, notation and analyses related to quadratic forms are described.

The topics from this review material may be covered all at once or reviewed on an "as needed" basis at an appropriate time during coverage of various topics in this chapter.

4.2.1 Gradient Vector: Partial Derivatives of a Function

Consider a function $f(\mathbf{x})$ of n variables $x_1, x_2, \ldots, x_n$. The partial derivative of the function with respect to x_1 at a given point $\mathbf{x}^*$ is defined as $\partial f(\mathbf{x}^*)/\partial x_1$, with respect to x_2 as

$\partial f(\mathbf{x}^*)/\partial x_2$, and so on. Let c_i *represent the partial derivative* of $f(\mathbf{x})$ with respect to x_i at the point $\mathbf{x}^*$. Then, using the index notation of Section 1.5, we can represent all partial derivatives of $f(\mathbf{x})$ as follows:

$$c_i = \frac{\partial f(\mathbf{x}^*)}{\partial x_i}; \quad i = 1 \text{ to } n \tag{4.4}$$

For convenience and compactness of notation, we arrange the partial derivatives $\partial f(\mathbf{x}^*)/\partial x_1$, $\partial f(\mathbf{x}^*)/\partial x_2$, ..., $\partial f(\mathbf{x}^*)/\partial x_n$ into a column vector called the *gradient vector* and represent it by any of the following symbols: $\mathbf{c}$, ∇f, $\partial f/\partial \mathbf{x}$, or grad f, as

$$\mathbf{c} = \nabla f(\mathbf{x}^*) = \begin{bmatrix} \dfrac{\partial f(\mathbf{x}^*)}{\partial x_1} \\[6pt] \dfrac{\partial f(\mathbf{x}^*)}{\partial x_2} \\[6pt] \vdots \\[6pt] \dfrac{\partial f(\mathbf{x}^*)}{\partial x_n} \end{bmatrix} = \begin{bmatrix} \dfrac{\partial f(\mathbf{x}^*)}{\partial x_1} & \dfrac{\partial f(\mathbf{x}^*)}{\partial x_2} & \cdots & \dfrac{\partial f(\mathbf{x}^*)}{\partial x_n} \end{bmatrix}^T \tag{4.5}$$

where superscript T denotes transpose of a vector or a matrix. Note that all partial derivatives are calculated at the given point $\mathbf{x}^*$. That is, each component of the gradient vector is a function in itself which must be evaluated at the given point $\mathbf{x}^*$.

Geometrically, the gradient vector is normal to the tangent plane at the point $\mathbf{x}^*$, as shown in Figure 4.5 for a function of three variables. Also, it points in the direction of maximum increase in the function. These properties are quite important, and will be proved and discussed in Chapter 11. They will be used in developing optimality conditions and numerical methods for optimum design. In Example 4.5 the gradient vector for a function is calculated.

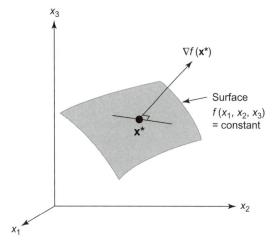

FIGURE 4.5 Gradient vector for $f(x_1, x_2, x_3)$ at the point $\mathbf{x}^*$.

EXAMPLE 4.5 CALCULATION OF A GRADIENT VECTOR

Calculate the gradient vector for the function $f(\mathbf{x}) = (x_1 - 1)^2 + (x_2 - 1)^2$ at the point $\mathbf{x}^* = (1.8, 1.6)$.

Solution

The given function is the equation for a circle with its center at the point $(1, 1)$. Since $f(1.8, 1.6) = (1.8 - 1)^2 + (1.6 - 1)^2 = 1$, the point $(1.8, 1.6)$ lies on a circle of radius 1, shown as point A in Figure 4.6. The partial derivatives for the function at the point $(1.8, 1.6)$ are calculated as

$$\frac{\partial f}{\partial x_1}(1.8, 1.6) = 2(x_1 - 1) = 2(1.8 - 1) = 1.6 \tag{a}$$

$$\frac{\partial f}{\partial x_2}(1.8, 1.6) = 2(x_2 - 1) = 2(1.6 - 1) = 1.2 \tag{b}$$

Thus, the gradient vector for $f(\mathbf{x})$ at point $(1.8, 1.6)$ is given as $\mathbf{c} = (1.6, 1.2)$. This is shown in Figure 4.6. It is seen that vector $\mathbf{c}$ is normal to the circle at point $(1.8, 1.6)$. This is consistent with the observation that the gradient is normal to the surface.

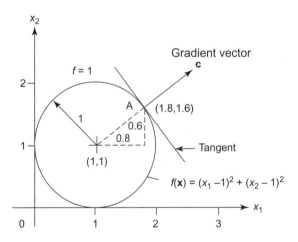

FIGURE 4.6 Gradient vector (that is not to scale) for the function $f(\mathbf{x})$ of Example 4.5 at the point $(1.8, 1.6)$.

4.2.2 Hessian Matrix: Second-Order Partial Derivatives

Differentiating the gradient vector once again, we obtain a matrix of second partial derivatives for the function $f(\mathbf{x})$ called the *Hessian matrix* or, simply, the Hessian. That is, differentiating each component of the gradient vector given in Eq. (4.5) with respect to x_1, $x_2, \ldots, x_n$, we obtain

$$\frac{\partial^2 f}{\partial \mathbf{x} \partial \mathbf{x}} = \begin{bmatrix} \dfrac{\partial^2 f}{\partial x_1^2} & \dfrac{\partial^2 f}{\partial x_1 \partial x_2} & \cdots & \dfrac{\partial^2 f}{\partial x_1 \partial x_n} \\[2mm] \dfrac{\partial^2 f}{\partial x_2 \partial x_1} & \dfrac{\partial^2 f}{\partial x_2^2} & \cdots & \dfrac{\partial^2 f}{\partial x_2 \partial x_n} \\[2mm] \vdots & \vdots & & \vdots \\[2mm] \dfrac{\partial^2 f}{\partial x_n \partial x_1} & \dfrac{\partial^2 f}{\partial x_n \partial x_2} & \cdots & \dfrac{\partial^2 f}{\partial x_n^2} \end{bmatrix} \tag{4.6}$$

where all derivatives are calculated at the given point $\mathbf{x}^*$. The Hessian is an $n \times n$ matrix, also denoted as $\mathbf{H}$ or $\nabla^2 f$. *It is important to note that each element of the Hessian is a function in itself that is evaluated at the given point* $\mathbf{x}^*$. Also, since $f(\mathbf{x})$ is assumed to be twice continuously differentiable, the cross partial derivatives are equal; that is,

$$\frac{\partial^2 f}{\partial x_i \partial x_j} = \frac{\partial^2 f}{\partial x_j \partial x_i}; \quad i = 1 \text{ to } n, \; j = 1 \text{ to } n \tag{4.7}$$

Therefore, *the Hessian is always a symmetric matrix.* It plays a prominent role in the sufficiency conditions for optimality as discussed later in this chapter. It will be written as

$$\mathbf{H} = \left[\frac{\partial^2 f}{\partial x_j \partial x_i} \right]; \quad i = 1 \text{ to } n, \; j = 1 \text{ to } n \tag{4.8}$$

The gradient and Hessian of a function are calculated in Example 4.6.

EXAMPLE 4.6 EVALUATION OF THE GRADIENT AND HESSIAN OF A FUNCTION

For the following function, calculate the gradient vector and the Hessian matrix at the point $(1, 2)$:

$$f(\mathbf{x}) = x_1^3 + x_2^3 + 2x_1^2 + 3x_2^2 - x_1 x_2 + 2x_1 + 4x_2 \tag{a}$$

Solution

The first partial derivatives of the function are given as

$$\frac{\partial f}{\partial x_1} = 3x_1^2 + 4x_1 - x_2 + 2; \quad \frac{\partial f}{\partial x_2} = 3x_2^2 + 6x_2 - x_1 + 4 \tag{b}$$

Substituting the point $x_1 = 1$, $x_2 = 2$, the gradient vector is given as $\mathbf{c} = (7, 27)$.

The second partial derivatives of the function are calculated as

$$\frac{\partial^2 f}{\partial x_1^2} = 6x_1 + 4; \quad \frac{\partial^2 f}{\partial x_1 \partial x_2} = -1; \quad \frac{\partial^2 f}{\partial x_2 \partial x_1} = -1; \quad \frac{\partial^2 f}{\partial x_2^2} = 6x_2 + 6. \tag{c}$$

Therefore, the Hessian matrix is given as

$$\mathbf{H}(\mathbf{x}) = \begin{bmatrix} 6x_1 + 4 & -1 \\ -1 & 6x_2 + 6 \end{bmatrix} \tag{d}$$

The Hessian matrix at the point $(1, 2)$ is given as

$$\mathbf{H}(1, 2) = \begin{bmatrix} 10 & -1 \\ -1 & 18 \end{bmatrix} \tag{e}$$

4.2.3 Taylor's Expansion

The idea of Taylor's expansion is fundamental to the development of optimum design concepts and numerical methods, so it is explained here. A function can be approximated by polynomials in a neighborhood of any point in terms of its value and derivatives using Taylor's expansion. Consider first a function $f(x)$ of one variable. Taylor's expansion for $f(x)$ about the point x^* is

$$f(x) = f(x^*) + \frac{df(x^*)}{dx}(x - x^*) + \frac{1}{2}\frac{d^2f(x^*)}{dx^2}(x - x^*)^2 + R \tag{4.9}$$

where R is the *remainder* term that is smaller in magnitude than the previous terms if x is sufficiently close to x^*. If we let $x - x^* = d$ (a small change in the point x^*), Taylor's expansion of Eq. (4.9) becomes a quadratic polynomial in d:

$$f(x^* + d) = f(x^*) + \frac{df(x^*)}{dx}d + \frac{1}{2}\frac{d^2f(x^*)}{dx^2}d^2 + R \tag{4.10}$$

For a function of two variables $f(x_1, x_2)$, Taylor's expansion at the point (x_1^*, x_2^*) is

$$f(x_1, x_2) = f(x_1^*, x_2^*) + \frac{\partial f}{\partial x_1}d_1 + \frac{\partial f}{\partial x_2}d_2 + \frac{1}{2}\left[\frac{\partial^2 f}{\partial x_1^2}d_1^2 + 2\frac{\partial^2 f}{\partial x_1 \partial x_2}d_1 d_2 + \frac{\partial^2 f}{\partial x_2^2}d_2^2\right] \tag{4.11}$$

where $d_1 = x_1 - x_1^*, d_2 = x_2 - x_2^*$, and all partial derivatives are calculated at the given point (x_1^*, x_2^*). For notational compactness, the arguments of these partial derivatives are omitted in Eq. (4.11) and in all subsequent discussions. Taylor's expansion in Eq. (4.11) can be written using the *summation notation* defined in Section 1.5 as

$$f(x_1, x_2) = f(x_1^*, x_2^*) + \sum_{i=1}^{2}\frac{\partial f}{\partial x_i}d_i + \frac{1}{2}\sum_{i=1}^{2}\sum_{j=1}^{2}\frac{\partial^2 f}{\partial x_i \partial x_j}d_i d_j \tag{4.12}$$

It is seen that by expanding the summations in Eq. (4.12), Eq. (4.11) is obtained. Recognizing the quantities $\partial f/\partial x_i$ as components of the gradient of the function given in Eq. (4.5) and $\partial^2 f/\partial x_i \partial x_j$ as the Hessian of Eq. (4.8) evaluated at the given point $\mathbf{x}^*$, Taylor's expansion can also be written in *matrix notation* as

$$f(\mathbf{x}^* + \mathbf{d}) = f(\mathbf{x}^*) + \nabla f^T \mathbf{d} + \frac{1}{2}\mathbf{d}^T \mathbf{H} \mathbf{d} + R \tag{4.13}$$

where $\mathbf{x} = (x_1, x_2)$, $\mathbf{x}^* = (x_1^*, x_2^*)$, $\mathbf{x} - \mathbf{x}^* = \mathbf{d}$, and $\mathbf{H}$ is the 2×2 Hessian matrix. Note that with matrix notation, Taylor's expansion in Eq. (4.13) can be generalized to functions of n variables. In that case, $\mathbf{x}$, $\mathbf{x}^*$, and ∇f are n-dimensional vectors and $\mathbf{H}$ is the $n \times n$ Hessian matrix.

Often a change in the function is desired when $\mathbf{x}^*$ moves to a neighboring point $\mathbf{x}$. Defining the change as $\Delta f = f(\mathbf{x}) - f(\mathbf{x}^*)$, Eq. (4.13) gives

$$\Delta f = \nabla f^T \mathbf{d} + \frac{1}{2}\mathbf{d}^T \mathbf{H} \mathbf{d} + R \tag{4.14}$$

A *first-order change* in $f(\mathbf{x})$ at $\mathbf{x}^*$ (denoted as δf) is obtained by retaining only the first term in Eq. (4.14):

$$\delta f = \nabla f^T \delta \mathbf{x} = \nabla f \cdot \delta \mathbf{x} \tag{4.15}$$

where δx is a small change in x^* ($\delta x = x - x^*$). Note that the first-order change in the function given in Eq. (4.15) is simply a dot product of the vectors ∇f and δx. A first-order change is an acceptable approximation for change in the original function when x is near x^*.

In Examples 4.7 through 4.9, we now consider some functions and approximate them at the given point x^* using Taylor's expansion. The remainder R will be dropped while using Eq. (4.13).

EXAMPLE 4.7 TAYLOR'S EXPANSION OF A FUNCTION OF ONE VARIABLE

Approximate $f(x) = \cos x$ around the point $x^* = 0$.

Solution

Derivatives of the function $f(x)$ are given as

$$\frac{df}{dx} = -\sin x, \quad \frac{d^2f}{dx^2} = -\cos x \tag{a}$$

Therefore, using Eq. (4.9), the second-order Taylor's expansion for $\cos x$ at the point $x^* = 0$ is given as

$$\cos x \approx \cos 0 - \sin 0(x - 0) + \frac{1}{2}(-\cos 0)(x - 0)^2 = 1 - \frac{1}{2}x^2 \tag{b}$$

EXAMPLE 4.8 TAYLOR'S EXPANSION OF A FUNCTION OF TWO VARIABLES

Obtain a second-order Taylor's expansion for the function $f(x) = 3x_1^3 x_2$ at the point $x^* = (1, 1)$.

Solution

The gradient and Hessian of the function $f(x)$ at the point $x^* = (1,1)$ using Eqs. (4.5) and (4.8) are

$$\nabla f(x) = \begin{bmatrix} \dfrac{\partial f}{\partial x_1} \\[6pt] \dfrac{\partial f}{\partial x_2} \end{bmatrix} = \begin{bmatrix} 9x_1^2 x_2 \\ 3x_1^3 \end{bmatrix} = \begin{bmatrix} 9 \\ 3 \end{bmatrix}; \quad H = \begin{bmatrix} 18x_1 x_2 & 9x_1^2 \\ 9x_1^2 & 0 \end{bmatrix} = \begin{bmatrix} 18 & 9 \\ 9 & 0 \end{bmatrix} \tag{a}$$

Substituting these in the matrix form of Taylor's expression given in Eq. (4.12), and using $d = x - x^*$, we obtain an approximation $\bar{f}(x)$ for $f(x)$ as

$$\bar{f}(x) = 3 + \begin{bmatrix} 9 \\ 3 \end{bmatrix}^T \begin{bmatrix} (x_1 - 1) \\ (x_2 - 1) \end{bmatrix} + \frac{1}{2}\begin{bmatrix} (x_1 - 1) \\ (x_2 - 1) \end{bmatrix}^T \begin{bmatrix} 18 & 9 \\ 9 & 0 \end{bmatrix} \begin{bmatrix} (x_1 - 1) \\ (x_2 - 1) \end{bmatrix} \tag{b}$$

where $f(x^*) = 3$ has been used. Simplifying the expression by expanding vector and matrix products, we obtain Taylor's expansion for $f(x)$ about the point $(1, 1)$ as

$$\bar{f}(x) = 9x_1^2 + 9x_1 x_2 - 18x_1 - 6x_2 + 9 \tag{c}$$

This expression is a second-order approximation of the function $3x_1{}^3x_2$ about the point $\mathbf{x}^* = (1, 1)$. That is, in a small neighborhood of $\mathbf{x}^*$, the expression will give almost the same value as the original function $f(\mathbf{x})$. To see how accurately $\bar{f}(\mathbf{x})$ approximates $f(\mathbf{x})$, we evaluate these functions for a 30 percent change in the given point $(1, 1)$, that is, at the point $(1.3, 1.3)$ as $\bar{f}(\mathbf{x}) = 8.2200$ and $f(\mathbf{x}) = 8.5683$. Therefore, the approximate function underestimates the original function by only 4 percent. This is quite a reasonable approximation for many practical applications.

EXAMPLE 4.9 A LINEAR TAYLOR'S EXPANSION OF A FUNCTION

Obtain a linear Taylor's expansion for the function

$$f(\mathbf{x}) = x_1^2 + x_2^2 - 4x_1 - 2x_2 + 4 \tag{a}$$

at the point $\mathbf{x}^* = (1, 2)$. Compare the approximate function with the original function in a neighborhood of the point $(1, 2)$.

Solution

The gradient of the function at the point $(1, 2)$ is given as

$$\nabla f(\mathbf{x}) = \begin{bmatrix} \dfrac{\partial f}{\partial x_1} \\[2mm] \dfrac{\partial f}{\partial x_2} \end{bmatrix} = \begin{bmatrix} (2x_1 - 4) \\ (2x_2 - 2) \end{bmatrix} = \begin{bmatrix} -2 \\ 2 \end{bmatrix} \tag{b}$$

Since $f(1, 2) = 1$, Eq. (4.13) gives a linear Taylor's approximation for $f(\mathbf{x})$ as

$$\bar{f}(\mathbf{x}) = 1 + [-2\ 2]\begin{bmatrix} (x_1 - 1) \\ (x_2 - 2) \end{bmatrix} = -2x_1 + 2x_2 - 1 \tag{c}$$

To see how accurately $\bar{f}(\mathbf{x})$ approximates the original $f(\mathbf{x})$ in the neighborhood of $(1, 2)$, we calculate the functions at the point $(1.1, 2.2)$, a 10 percent change in the point as $\bar{f}(\mathbf{x}) = 1.20$ and $f(\mathbf{x}) = 1.25$. We see that the approximate function underestimates the real function by 4 percent. An error of this magnitude is quite acceptable in many applications. Note, however, that the errors will be different for different functions and can be larger for highly nonlinear functions.

4.2.4 Quadratic Forms and Definite Matrices

Quadratic Form

The quadratic form is a special nonlinear function having only second-order terms (either the square of a variable or the product of two variables). For example, the function

$$F(\mathbf{x}) = x_1^2 + 2x_2^2 + 3x_3^2 + 2x_1x_2 + 2x_2x_3 + 2x_3x_1 \tag{4.16}$$

Quadratic forms play a prominent role in optimization theory and methods. Therefore, in this subsection, we discuss some results related to them. Generalizing the quadratic form of three variables in Eq. (4.16) to n variables and writing it in the double summation notation (refer to Section 1.5 for the summation notation), we obtain

$$F(\mathbf{x}) = \sum_{i=1}^{n}\sum_{j=1}^{n} p_{ij}x_ix_j \tag{4.17}$$

where p_{ij} are constants related to the coefficients of various terms in Eq. (4.16). It is observed that the quadratic form of Eq. (4.17) matches the second-order term of Taylor's expansion in Eq. (4.12) for n variables, except for the factor of $1/2$.

The Matrix of the Quadratic Form

The quadratic form can be written in the matrix notation. Let $\mathbf{P} = [p_{ij}]$ be an $n \times n$ matrix and $\mathbf{x} = (x_1, x_2, \ldots, x_n)$ be an n-dimensional vector. Then the quadratic form of Eq. (4.17) is given as

$$F(\mathbf{x}) = \mathbf{x}^T\mathbf{P}\mathbf{x} \tag{4.18}$$

$\mathbf{P}$ is called the *matrix of the quadratic form $F(\mathbf{x})$*. Elements of $\mathbf{P}$ are obtained from the coefficients of the terms in the function $F(\mathbf{x})$.

There are many matrices associated with the given quadratic form; in fact, there are infinite such matrices. All of the matrices are asymmetric except one. The symmetric matrix $\mathbf{A}$ associated with the quadratic form can be obtained from any asymmetric matrix $\mathbf{P}$ as

$$\mathbf{A} = \frac{1}{2}(\mathbf{P} + \mathbf{P}^T) \text{ or } a_{ij} = \frac{1}{2}(p_{ij} + p_{ji}), \ i,j = 1 \text{ to } n \tag{4.19}$$

Using this definition, the matrix $\mathbf{P}$ can be replaced with the symmetric matrix $\mathbf{A}$ and the quadratic form of Eq. (4.18) becomes

$$F(\mathbf{x}) = \mathbf{x}^T\mathbf{A}\mathbf{x} \tag{4.20}$$

The value or expression of the quadratic form does not change with $\mathbf{P}$ replaced by $\mathbf{A}$. The symmetric matrix $\mathbf{A}$ is useful in determining the nature of the quadratic form, which will be discussed later in this section. Example 4.10 illustrates identification of matrices associated with a quadratic form.

There are many matrices associated with a quadratic form. However, there is only one symmetric matrix associated with it.

EXAMPLE 4.10 A MATRIX OF THE QUADRATIC FORM

Identify a matrix associated with the quadratic form:

$$F(x_1, x_2, x_3) = 2x_1^2 + 2x_1x_2 + 4x_1x_3 - 6x_2^2 - 4x_2x_3 + 5x_3^2 \tag{a}$$

Solution

Writing F in the matrix form ($F(\mathbf{x}) = \mathbf{x}^T\mathbf{P}\mathbf{x}$), we obtain

$$F(\mathbf{x}) = [x_1 \quad x_2 \quad x_3] \begin{bmatrix} 2 & 2 & 4 \\ 0 & -6 & -4 \\ 0 & 0 & 5 \end{bmatrix} \begin{bmatrix} x_1 \\ x_2 \\ x_3 \end{bmatrix} \tag{b}$$

The matrix $\mathbf{P}$ of the quadratic form can be easily identified by comparing the above expression with Eq. (4.18). The ith diagonal element p_{ii} is the coefficient of x_i^2. Therefore, $p_{11} = 2$, the coefficient of x_1^2; $p_{22} = -6$, the coefficient of x_2^2; and $p_{33} = 5$, the coefficient of x_3^2. The coefficient of $x_i x_j$ can be divided in any proportion between the elements p_{ij} and p_{ji} of the matrix $\mathbf{P}$ as long as the sum $p_{ij} + p_{ji}$ is equal to the coefficient of $x_i x_j$. In the above matrix, $p_{12} = 2$ and $p_{21} = 0$, giving $p_{12} + p_{21} = 2$, which is the coefficient of $x_1 x_2$. Similarly, we can calculate the elements p_{13}, p_{31}, p_{23}, and p_{32}.

Since the coefficient of $x_i x_j$ can be divided between p_{ij} and p_{ji} in any proportion, there are many matrices associated with a quadratic form. For example, the following matrices are also associated with the same quadratic form:

$$\mathbf{P} = \begin{bmatrix} 2 & 0.5 & 1 \\ 1.5 & -6 & -6 \\ 3 & 2 & 5 \end{bmatrix}; \quad \mathbf{P} = \begin{bmatrix} 2 & 4 & 5 \\ -2 & -6 & 4 \\ -1 & -8 & 5 \end{bmatrix} \tag{c}$$

Dividing the coefficients equally between the off-diagonal terms, we obtain the symmetric matrix associated with the quadratic form in Eq. (a) as

$$\mathbf{A} = \begin{bmatrix} 2 & 1 & 2 \\ 1 & -6 & -2 \\ 2 & -2 & 5 \end{bmatrix} \tag{d}$$

The diagonal elements of the symmetric matrix $\mathbf{A}$ are obtained from the coefficient of x_i^2 as before. The off-diagonal elements are obtained by dividing the coefficient of the term $x_i x_j$ equally between a_{ij} and a_{ji}. Any of the matrices in Eqs. (b) through (d) give a matrix associated with the quadratic form.

Form of a Matrix

Quadratic form $F(\mathbf{x}) = \mathbf{x}^T\mathbf{A}\mathbf{x}$ may be either positive, negative, or zero for any $\mathbf{x}$. The following are the possible forms for the function $F(\mathbf{x})$ and the associated symmetric matrix $\mathbf{A}$:

1. *Positive Definite.* $F(\mathbf{x}) > 0$ for all $\mathbf{x} \neq \mathbf{0}$. The matrix A is called positive definite.
2. *Positive Semidefinite.* $F(\mathbf{x}) > 0$ for all $\mathbf{x} \neq \mathbf{0}$. The matrix A is called positive semidefinite.
3. *Negative Definite.* $F(\mathbf{x}) > 0$ for all $\mathbf{x} \neq \mathbf{0}$. The matrix $\mathbf{A}$ is called negative definite.
4. *Negative Semidefinite.* $F(\mathbf{x}) > 0$ for all $\mathbf{x} \neq \mathbf{0}$. The matrix $\mathbf{A}$ is called negative semidefinite.
5. *Indefinite.* The quadratic form is called indefinite if it is positive for some values of $\mathbf{x}$ and negative for some others. In that case, the matrix $\mathbf{A}$ is also called indefinite.

EXAMPLE 4.11 DETERMINATION OF THE FORM OF A MATRIX

Determine the form of the following matrices:

$$(i)\ \mathbf{A} = \begin{bmatrix} 2 & 0 & 0 \\ 0 & 4 & 0 \\ 0 & 0 & 3 \end{bmatrix} \qquad (ii)\ \mathbf{A} = \begin{bmatrix} -1 & 1 & 0 \\ 1 & -1 & 0 \\ 0 & 0 & -1 \end{bmatrix} \tag{a}$$

Solution

The quadratic form associated with the matrix (i) is always positive because

$$\mathbf{x}^T \mathbf{A} \mathbf{x} = (2x_1^2 + 4x_2^2 + 3x_3^2) > 0 \tag{b}$$

unless $x_1 = x_2 = x_3 = 0$ ($\mathbf{x} = 0$). Thus, the matrix is positive definite.

The quadratic form associated with the matrix (ii) is negative semidefinite, since

$$\mathbf{x}^T \mathbf{A} \mathbf{x} = (-x_1^2 - x_2^2 + 2x_1 x_2 - x_3^2) = \{-x_3^2 - (x_1 - x_2)^2\} \le 0 \tag{c}$$

for all $\mathbf{x}$, and $\mathbf{x}^T \mathbf{A} \mathbf{x} = 0$ when $x_3 = 0$, and $x_1 = x_2$ (e.g., $\mathbf{x} = (1, 1, 0)$). The quadratic form is not negative definite but is negative semidefinite since it can have a zero value for nonzero $\mathbf{x}$. Therefore, the matrix associated with it is also negative semidefinite.

We will now discuss *methods for checking* positive definiteness or semidefiniteness (form) of a quadratic form or a matrix. Since this involves calculation of eigenvalues or principal minors of a matrix, Sections A.3 and A.6 in Appendix A should be reviewed at this point.

THEOREM 4.2

Eigenvalue Check for the Form of a Matrix Let λ_i, $i = 1$ to n be the eigenvalues of a symmetric $n \times n$ matrix $\mathbf{A}$ associated with the quadratic form $F(\mathbf{x}) = \mathbf{x}^T \mathbf{A} \mathbf{x}$ (since $\mathbf{A}$ is symmetric, all eigenvalues are real). The following results can be stated regarding the quadratic form $F(\mathbf{x})$ or the matrix $\mathbf{A}$:

1. $F(\mathbf{x})$ is *positive definite if and only if* all eigenvalues of $\mathbf{A}$ are strictly positive; i.e., $\lambda_i > 0$, $i = 1$ to n.
2. $F(\mathbf{x})$ is *positive semidefinite if and only if* all eigenvalues of $\mathbf{A}$ are non-negative; i.e., $\lambda_i \ge 0$, $i = 1$ to n (note that at least one

eigenvalue must be zero for it to be called positive semidefinite).
3. $F(\mathbf{x})$ is *negative definite if and only if* all eigenvalues of $\mathbf{A}$ are strictly negative; i.e., $\lambda_i < 0$, $i = 1$ to n.
4. $F(\mathbf{x})$ is *negative semidefinite if and only if* all eigenvalues of $\mathbf{A}$ are nonpositive; i.e., $\lambda_i \le 0$, $i = 1$ to n (note that at least one eigenvalue must be zero for it to be called negative semidefinite).
5. $F(\mathbf{x})$ is *indefinite* if some $\lambda_i < 0$ and some other $\lambda_j > 0$.

Another way of checking the form of a matrix is provided in Theorem 4.3.

THEOREM 4.3

Check for the Form of a Matrix Using Principal Minors Let M_k be the kth leading principal minor of the $n \times n$ symmetric matrix $\mathbf{A}$ defined as the determinant of a $k \times k$ submatrix obtained by deleting the last $(n - k)$ rows and columns of $\mathbf{A}$ (Section A.3). Assume that *no two consecutive principal minors are zero.* Then

1. $\mathbf{A}$ is positive definite if and only if all $M_k > 0$, $k = 1$ to n.

2. $\mathbf{A}$ is positive semidefinite if and only if $M_k > 0$, $k = 1$ to r, where $r < n$ is the rank of $\mathbf{A}$ (refer to Section A.4 for a definition of the rank of a matrix).

3. $\mathbf{A}$ is negative definite if and only if $M_k < 0$ for k odd and $M_k > 0$ for k even, $k = 1$ to n.

4. $\mathbf{A}$ is negative semidefinite if and only if $M_k < 0$ for k odd and $M_k > 0$ for k even, $k = 1$ to $r < n$.

5. $\mathbf{A}$ is indefinite if it does not satisfy any of the preceding criteria.

This theorem is applicable only if the assumption of no two consecutive principal minors being zero is satisfied. When there are consecutive zero principal minors, we may resort to the eigenvalue check of Theorem 4.2. Note also that a *positive definite matrix cannot have negative or zero diagonal elements.* The form of a matrix is determined in Example 4.12.

The theory of quadratic forms is used in the second-order conditions for a local optimum point in Section 4.4. Also, it is used to determine the convexity of functions of the optimization problem. Convex functions play a role in determining the global optimum point in Section 4.8.

EXAMPLE 4.12 DETERMINATION OF THE FORM OF A MATRIX

Determine the form of the matrices given in Example 4.11.

Solution

For a given matrix $\mathbf{A}$, the eigenvalue problem is defined as $\mathbf{A}\mathbf{x} = \lambda\mathbf{x}$, where λ is an eigenvalue and $\mathbf{x}$ is the corresponding eigenvector (refer to Section A.6 for more details). To determine the eigenvalues, we set the so-called characteristic determinant to zero $|(\mathbf{A} - \lambda\mathbf{I})| = 0$. Since the matrix *(i)* is diagonal, its eigenvalues are the diagonal elements (i.e., $\lambda_1 = 2$, $\lambda_2 = 3$, and $\lambda_3 = 4$). Since all eigenvalues are strictly positive, the matrix is positive definite. The principal minor check of Theorem 4.3 also gives the same conclusion.

For the matrix *(ii)*, the characteristic determinant of the eigenvalue problem is

$$\begin{vmatrix} -1-\lambda & 1 & 0 \\ 1 & -1-\lambda & 0 \\ 0 & 0 & -1-\lambda \end{vmatrix} = 0 \tag{a}$$

Expanding the determinant by the third row, we obtain

$$(-1-\lambda)[(-1-\lambda)^2 - 1] = 0 \tag{b}$$

Therefore, the three roots give the eigenvalues as $\lambda_1 = -2$, $\lambda_2 = -1$, and $\lambda_3 = 0$. Since all eigenvalues are nonpositive, the matrix is negative semidefinite.

To use Theorem 4.3, we calculate the three leading principal minors as

$$M_1 = -1, \quad M_2 = \begin{vmatrix} -1 & 1 \\ 1 & -1 \end{vmatrix} = 0, \quad M_3 = \begin{vmatrix} -1 & 1 & 0 \\ 1 & -1 & 0 \\ 0 & 0 & -1 \end{vmatrix} = 0 \qquad \text{(c)}$$

Since there are two consecutive zero leading principal minors, we cannot use Theorem 4.3.

Differentiation of a Quadratic Form

Often we want to find the gradient and Hessian matrix for the quadratic form. We consider the quadratic form of Eq. (4.17) with the coefficients p_{ij} replaced by their symmetric counterparts a_{ij}. To calculate the derivatives of $F(\mathbf{x})$, we first expand the summations, differentiate the expression with respect to x_i, and then write it back in the summation or matrix notation:

$$\frac{\partial F(\mathbf{x})}{\partial x_i} = 2 \sum_{j=1}^{n} a_{ij} x_j; \quad \text{or} \quad \nabla F(\mathbf{x}) = 2\mathbf{A}\mathbf{x} \qquad (4.21)$$

Differentiating Eq. (4.21) once again with respect to x_i we get

$$\frac{\partial^2 F(\mathbf{x})}{\partial x_j \partial x_i} = 2a_{ij}; \quad \text{or} \quad \mathbf{H} = 2\mathbf{A} \qquad (4.22)$$

Example 4.13 shows the calculations for the gradient and the Hessian of the quadratic form.

EXAMPLE 4.13 CALCULATIONS FOR THE GRADIENT AND HESSIAN OF THE QUADRATIC FORM

Calculate the gradient and Hessian of the following quadratic form:

$$F(\mathbf{x}) = 2x_1^2 + 2x_1 x_2 + 4x_1 x_3 - 6x_2^2 - 4x_2 x_3 + 5x_3^2 \qquad \text{(a)}$$

Solution

Differentiating $F(\mathbf{x})$ with respect to x_1, x_2, and x_3, we get gradient components as

$$\frac{\partial F}{\partial x_1} = (4x_1 + 2x_2 + 4x_3); \quad \frac{\partial F}{\partial x_2} = (2x_1 - 12x_2 - 4x_3); \quad \frac{\partial F}{\partial x_3} = (4x_1 - 4x_2 + 10x_3) \qquad \text{(b)}$$

Differentiating the gradient components once again, we get the Hessian components as

$$\frac{\partial^2 F}{\partial x_1^2} = 4, \quad \frac{\partial^2 F}{\partial x_1 \partial x_2} = 2, \quad \frac{\partial^2 F}{\partial x_1 \partial x_3} = 4$$

$$\frac{\partial^2 F}{\partial x_2 \partial x_1} = 2, \quad \frac{\partial^2 F}{\partial x_2^2} = -12, \quad \frac{\partial^2 F}{\partial x_2 \partial x_3} = -4 \qquad \text{(c)}$$

$$\frac{\partial^2 F}{\partial x_3 \partial x_1} = 4, \quad \frac{\partial^2 F}{\partial x_3 \partial x_2} = -4, \quad \frac{\partial^2 F}{\partial x_3^2} = 10$$

Writing the given quadratic form in a matrix form, we identify matrix $\mathbf{A}$ as

$$\mathbf{A} = \begin{bmatrix} 2 & 1 & 2 \\ 1 & -6 & -2 \\ 2 & -2 & 5 \end{bmatrix}$$

Comparing elements of the matrix $\mathbf{A}$ with second partial derivatives of F, we observe that the Hessian $\mathbf{H} = 2\mathbf{A}$. Using Eq. (4.21), the gradient of the quadratic form is also given as

$$\nabla F(\mathbf{x}) = 2 \begin{bmatrix} 2 & 1 & 2 \\ 1 & -6 & -2 \\ 2 & -2 & 5 \end{bmatrix} \begin{bmatrix} x_1 \\ x_2 \\ x_3 \end{bmatrix} = \begin{bmatrix} (4x_1 + 2x_2 + 4x_3) \\ (2x_1 - 12x_2 - 4x_3) \\ (4x_1 - 4x_2 + 10x_3) \end{bmatrix} \tag{e}$$

4.3 CONCEPT OF NECESSARY AND SUFFICIENT CONDITIONS

In the remainder of this chapter, we will describe necessary and sufficient conditions for optimality of unconstrained and constrained optimization problems. It is important to understand the meaning of the terms *necessary* and *sufficient*. These terms have general meaning in mathematical analyses. However we will discuss them for the optimization problem only.

Necessary Conditions

The optimality conditions are derived by assuming that we are at an optimum point, and then studying the behavior of the functions and their derivatives at that point. *The conditions that must be satisfied at the optimum point are called* necessary. *Stated differently, if a point does not satisfy the necessary conditions, it cannot be optimum.* Note, however, that the satisfaction of necessary conditions does not guarantee optimality of the point; that is, there can be nonoptimum points that satisfy the same conditions. This indicates that the number of points satisfying necessary conditions can be more than the number of optima. Points satisfying the necessary conditions are called *candidate optimum points*. We must, therefore, perform further tests to distinguish between optimum and nonoptimum points, both satisfying the necessary conditions.

Sufficient Condition

If a candidate optimum point satisfies the sufficient condition, then it is indeed an optimum point. If the sufficient condition is not satisfied, however, or cannot be used, we may not be able to conclude that the candidate design is not optimum. Our conclusion will depend on the assumptions and restrictions used in deriving the sufficient condition. Further analyses of the problem or other higher-order conditions are needed to make a definite statement about optimality of the candidate point.

1. Optimum points must satisfy the necessary conditions. Points that do not satisfy them cannot be optimum.
2. A point satisfying the necessary conditions need not be optimum; that is, nonoptimum points may also satisfy the necessary conditions.
3. A candidate point satisfying a sufficient condition is indeed optimum.
4. If the sufficiency condition cannot be used or it is not satisfied, we may not be able to draw any conclusions about the optimality of the candidate point.

4.4 OPTIMALITY CONDITIONS: UNCONSTRAINED PROBLEM

We are now ready to discuss the theory and concepts of optimum design. In this section, we will discuss necessary and sufficient conditions for unconstrained optimization problems defined as "Minimize $f(\mathbf{x})$ without any constraints on $\mathbf{x}$." Such problems arise infrequently in practical engineering applications. However, we consider them here because optimality conditions for constrained problems are a logical extension of these conditions. In addition, one numerical strategy for solving a constrained problem is to convert it into a sequence of unconstrained problems. Thus, it is important to completely understand unconstrained optimization concepts.

The optimality conditions for unconstrained or constrained problems can be used in two ways:
  1. **They can be used to check whether a given point is a local optimum for the problem.**
  2. **They can be solved for local optimum points.**

We discuss only the *local optimality conditions* for unconstrained problems. Global optimality is covered in Section 4.8. First the necessary and then the sufficient conditions are discussed. As noted earlier, the necessary conditions must be satisfied at the minimum point; otherwise, it cannot be a minimum point. These conditions, however, may also be satisfied by points that are not minima. A point satisfying the necessary conditions is simply a candidate local minimum. The sufficient conditions distinguish minimum points from non-minimum points. We elaborate on these concepts with examples.

4.4.1 Concepts Related to Optimality Conditions

The basic concept for obtaining local optimality conditions is to assume that we are at a minimum point $\mathbf{x}^*$ and then examine its neighborhood to study properties of the function and its derivatives. Basically, we use the definition of a local minimum, given in Inequality (4.2), to derive the optimality conditions. Since we examine only a small neighborhood, the conditions we obtain are called *local optimality conditions*.

Let $\mathbf{x}^*$ be a *local minimum* point for $f(\mathbf{x})$. To investigate its neighborhood, let $\mathbf{x}$ be any point near $\mathbf{x}^*$. Define increments $\mathbf{d}$ and Δf in $\mathbf{x}^*$ and $f(\mathbf{x}^*)$ as $\mathbf{d} = \mathbf{x} - \mathbf{x}^*$ and $\Delta f = f(\mathbf{x}) - f(\mathbf{x}^*)$.

Since $f(x)$ has a local minimum at x^*, it will not reduce any further if we move a small distance away from x^*. Therefore, a change in the function for any move in a small neighborhood of x^* must be non-negative; that is, the function value must either remain constant or increase. This condition, also obtained directly from the definition of local minimum given in Eq. (4.2), can be expressed as the following inequality:

$$\Delta f = f(\mathbf{x}) - f(\mathbf{x}^*) \geq 0 \tag{4.23}$$

for all small changes $\mathbf{d}$. The inequality in Eq. (4.23) can be used to derive necessary and sufficient conditions for a local minimum point. Since $\mathbf{d}$ is small, we can approximate Δf by Taylor's expansion at $\mathbf{x}^*$ and derive optimality conditions using it.

4.4.2 Optimality Conditions for Functions of a Single Variable

First-Order Necessary Condition

Let us first consider a *function of only one variable*. Taylor's expansion of $f(x)$ at the point x^* gives

$$f(x) = f(x^*) + f'(x^*)d + \frac{1}{2}f''(x^*)d^2 + R \tag{4.24}$$

where R is the remainder containing higher-order terms in d and "primes" indicate the order of the derivatives. From this equation, the change in the function at x^* (i.e., $\Delta f = f(x) - f(x^*)$) is given as

$$\Delta f(x) = f'(x^*)d + \frac{1}{2}f''(x^*)d^2 + R \tag{4.25}$$

Inequality (4.23) shows that the expression for Δf must be non-negative (≥ 0) because x^* is a local minimum for $f(x)$. Since d is small, the first-order term $f'(x^*)d$ dominates other terms, and therefore Δf can be approximated as $\Delta f = f'(x^*)d$. Note that Δf in this equation can be positive or negative depending on the sign of the term $f'(x^*)d$. Since d is arbitrary (a small increment in x^*), it may be positive or negative. Therefore, if $f'(x^*) \neq 0$, the term $f'(x^*)d$ (and hence Δf) can be negative.

To see this more clearly, let the term be positive for some increment d_1 that satisfies Inequality (4.23)(i.e., $\Delta f = f'(x^*)d_1 > 0$). Since the increment d is arbitrary, it is reversible, so $d_2 = -d_1$ is another possible increment. For d_2, Δf becomes negative, which violates Inequality (4.23). Thus, the quantity $f'(x^*)d$ can have a negative value regardless of the sign of $f'(x^*)$, unless it is zero. The only way it can be non-negative for all d in a neighborhood of x^* is when

$$f'(x^*) = 0 \tag{4.26}$$

STATIONARY POINTS

Equation (4.26) is a *first-order necessary condition* for the local minimum of $f(x)$ at x^*. It is called "first-order" because it only involves the first derivative of the function. Note that the preceding arguments can be used to show that the condition of Eq. (4.26) is also necessary for local maximum points. Therefore, since the points satisfying Eq. (4.26) can be local minima or maxima, or neither minimum nor maximum (*inflection points*), they are called *stationary points*.

Sufficient Condition

Now we need a *sufficient condition* to determine which of the stationary points are actually minimum for the function. Since stationary points satisfy the necessary condition $f'(x^*) = 0$, the change in function Δf of Eq. (4.24) becomes

$$\Delta f(x) = \frac{1}{2} f''(x^*)d^2 + R \qquad (4.27)$$

Since the second-order term dominates all other higher-order terms, we need to focus on it. Note that the term can be positive for all $d \neq 0$ if

$$f''(x^*) > 0 \qquad (4.28)$$

Stationary points satisfying Inequality (4.28) must be at least local minima because they satisfy Inequality (4.23) ($\Delta f > 0$). That is, the function has positive curvature at the minimum points. Inequality (4.28) is then the *sufficient condition* for x^* to be a local minimum. Thus, if we have a point x^* satisfying both conditions in Eqs. (4.26) and (4.28), then any small move away from it will either increase the function value or keep it unchanged. This indicates that $f(x^*)$ has the smallest value in a small neighborhood (local minimum) of point x^*. Note that the foregoing conditions can be stated in terms of the *curvature of the function* since the second derivative is the curvature.

Second-Order Necessary Condition

If $f''(x^*) = 0$, we cannot conclude that x^* is not a minimum point. Note, however, from Eqs. (4.23) and (4.25), that $f(x^*)$ cannot be a minimum unless

$$f''(x^*) \geq 0 \qquad (4.29)$$

That is, if f'' evaluated at the candidate point x^* is less than zero, then x^* is not a local minimum point. Inequality (4.29) is known as a *second-order necessary condition*, so any point violating it (e.g., $f''(x^*) < 0$) cannot be a local minimum (actually, it is a local maximum point for the function).

It is important to note that if the sufficiency condition in Eq. (4.28) is satisfied, then the second-order necessary condition in Eq. (4.29) is satisfied automatically.

If $f''(x^*) = 0$, we need to evaluate higher-order derivatives to determine if the point is a local minimum (see Examples 4.14 through 4.18). By the arguments used to derive Eq. (4.26), $f'''(x^*)$ must be zero for the stationary point (necessary condition) and $f^{IV}(x^*) > 0$ for x^* must be a local minimum.

In general, the lowest nonzero derivative must be even-ordered for stationary points (necessary conditions), and it must be positive for local minimum points (sufficiency conditions). All odd-ordered derivatives lower than the nonzero even-ordered derivative must be zero as the necessary condition.

1. **The necessary conditions must be satisfied at the minimum point; otherwise, it cannot be a minimum.**
2. **The necessary conditions may also be satisfied by points that are not minima. A point satisfying the necessary conditions is simply a candidate local minimum.**
3. **If the sufficient condition is satisfied at a candidate point, then it is indeed a minimum point.**

EXAMPLE 4.14 DETERMINATION OF LOCAL MINIMUM POINTS USING NECESSARY CONDITIONS

Find the local minima for the function

$$f(x) = \sin x \tag{a}$$

Solution

Differentiating the function twice,

$$f' = \cos x; \quad f'' = -\sin x; \tag{b}$$

Stationary points are obtained as roots of $f''(x) = 0$ ($\cos x = 0$). These are

$$x = \pm \pi/2, \ \pm 3\pi/2, \ \pm 5\pi/2, \ \pm 7\pi/2, \ \dots \tag{c}$$

Local minima are identified as

$$x^* = 3\pi/2, 7\pi/2, \dots; \quad -\pi/2, -5\pi/2, \dots \tag{d}$$

since these points satisfy the sufficiency condition of Eq. (4.28) ($f'' = -\sin x > 0$ at these points). The value of $\sin x$ at the points x^* is -1. This is true from the graph of the function $\sin x$. There are infinite minimum points, and they are all actually global minima.

The points $\pi/2, 5\pi/2, \dots$, and $-3\pi/2, -7\pi/2, \dots$ are global maximum points where $\sin x$ has a value of 1. At these points, $f'(x) = 0$ and $f''(x) < 0$.

EXAMPLE 4.15 DETERMINATION OF LOCAL MINIMUM POINTS USING NECESSARY CONDITIONS

Find the local minima for the function

$$f(x) = x^2 - 4x + 4 \tag{a}$$

Solution

Figure 4.7 shows a graph for the function $f(x) = x^2 - 4x + 4$. It can be seen that the function always has a positive value except at $x = 2$, where it is zero. Therefore, this is a local as well as a global minimum point for the function. Let us see how this point will be determined using the necessary and sufficient conditions.

Differentiating the function twice,

$$f' = 2x - 4; \quad f'' = 2 \tag{b}$$

The necessary condition $f' = 0$ implies that $x^* = 2$ is a stationary point. Since $f'' > 0$ at $x^* = 2$ (actually for all x), the sufficiency condition of Eq. (4.28) is satisfied. Therefore $x^* = 2$ is a local minimum for $f(x)$. The minimum value of f is 0 at $x^* = 2$.

Note that at $x^* = 2$, the second-order necessary condition for a *local maximum* $f'' \leq 0$ is violated since $f''(2) = 2 > 0$. Therefore the point $x^* = 2$ cannot be a local maximum point. In fact the graph of the function shows that there is no local or global maximum point for the function.

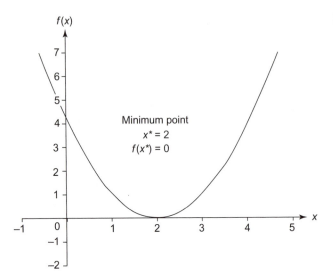

FIGURE 4.7   Representation of $f(x)$ $= x^2 - 4x + 4$ of Example 4.15.

EXAMPLE 4.16 DETERMINATION OF LOCAL MINIMUM POINTS USING NECESSARY CONDITIONS

Find the local minima for the function

$$f(x) = x^3 - x^2 - 4x + 4 \tag{a}$$

Solution

Figure 4.8 shows the graph of the function. It can be seen that point A is a local minimum point and point B is a local maximum point. We will use the necessary and sufficient conditions to prove that this is indeed true. Differentiating the function, we obtain

$$f' = 3x^2 - 2x - 4; \quad f'' = 6x - 2 \tag{b}$$

For this example there are two points satisfying the necessary condition of Eq. (4.26), that is, stationary points. These are obtained as roots of the equation $f'(x) = 0$ in Eqs. (b):

$$x_1^* = \frac{1}{6}(2 + 7.211) = 1.535 \,(\text{Point A}) \tag{c}$$

$$x_2^* = \frac{1}{6}(2 - 7.211) = -0.8685 \,(\text{Point B}) \tag{d}$$

Evaluating f'' at these points,

$$f''(1.535) = 7.211 > 0 \tag{e}$$

$$f''(-0.8685) = -7.211 < 0 \tag{f}$$

We see that only x_1^* satisfies the sufficiency condition ($f'' > 0$) of Eq. (4.28). Therefore, it is a local minimum point. From the graph in Figure 4.8, we can see that the local minimum $f(x_1^*)$ is

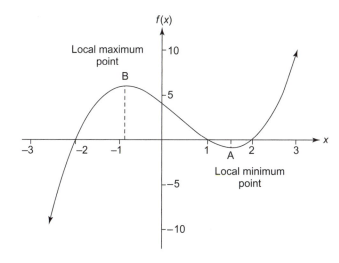

FIGURE 4.8 Representation of $f(x)$ $= x^3 - x^2 - 4x + 4$ of Example 4.16.

not the global minimum. *A global minimum for f(x) does not exist since the domain as well as the function is not bounded* (Theorem 4.1). The value of the function at the local minimum is obtained as -0.88 by substituting $x_1^* = 1.535$ in $f(x)$.

Note that $x_2^* = -0.8685$ is a local maximum point since $f''(x_2^*) < 0$. The value of the function at the maximum point is 6.065. There is no global maximum point for the function.

Checking Second-Order Necessary Conditions. Note that the second-order necessary condition for a local minimum ($f''(x^*) \geq 0$) is violated at $x_2^* = -0.8685$. Therefore, this stationary point cannot be a local minimum point. Similarly, the stationary point $x_1^* = 1.535$ cannot be a local maximum point.

Checking the Optimality of a Given Point. As noted earlier, the optimality conditions can also be used to check the optimality of a given point. To illustrate this, let us check optimality of the point $x = 1$. At this point, $f' = 3(1)^2 - 2(1) - 4 = -3 \neq 0$. Therefore $x = 1$ is not a stationary point and thus cannot be a local minimum or maximum for the function.

EXAMPLE 4.17 DETERMINATION OF LOCAL MINIMUM POINTS USING NECESSARY CONDITIONS

Find the minimal for the function

$$f(x) = x^4 \tag{a}$$

Solution

Differentiating the function twice,

$$f' = 4x^3; \quad f'' = 12x^2 \tag{b}$$

The necessary condition gives $x^* = 0$ as a stationary point. Since $f''(x^*) = 0$, we cannot conclude from the sufficiency condition of Eq. (4.28) that x^* is a minimum point. However, the second-order necessary condition of Eq. (4.29) is satisfied, so we cannot rule out the possibility of x^* being a minimum point. In fact, a graph of $f(x)$ versus x will show that x^* is indeed the global minimum point. $f''' = 24x$, which is zero at $x^* = 0$. $f^{IV}(x^*) = 24$, which is strictly greater than zero. Therefore, the fourth-order sufficiency condition is satisfied, and $x^* = 0$ is indeed a local minimum point. It is actually a global minimum point with $f(0) = 0$.

EXAMPLE 4.18 MINIMUM-COST SPHERICAL TANK DESIGN USING NECESSARY CONDITION

The result of a problem formulation in Section 2.3 is a cost function that represents the life-time cooling-related cost of an insulated spherical tank as

$$f(x) = ax + b/x, \quad a, b > 0 \tag{a}$$

where x is the thickness of insulation, and a and b are positive constants.

Solution

To minimize f, we solve the equation (necessary condition)

$$f' = a - b/x^2 = 0 \tag{b}$$

The solution is $x^* = \sqrt{b/a}$. Note that the root $x^* = -\sqrt{b/a}$ is rejected for this problem because thickness x of the insulation cannot be negative. If negative values for x are allowed, then $x^* = -\sqrt{b/a}$ satisfies the sufficiency condition for a local maximum for $f(x)$ since $f''(x^*) < 0$.

To check if the stationary point $x^* = \sqrt{b/a}$ is a local minimum, evaluate

$$f''(x^*) = 2b/x^{*3} \tag{c}$$

Since b and x^* are positive, $f''(x^*)$ is positive, and x^* is a local minimum point. The value of the function at x^* is $2\sqrt{b/a}$. Note that since the function cannot have a negative value because of the physics of the problem, x^* represents a global minimum for the problem as well.

4.4.3 Optimality Conditions for Functions of Several Variables

For the general case of a function of several variables $f(\mathbf{x})$ where $\mathbf{x}$ is an n-vector, we can repeat the derivation of the *necessary and sufficient* conditions using the multidimensional form of Taylor's expansion:

$$f(\mathbf{x}) = f(\mathbf{x}^*) + \nabla f(\mathbf{x}^*)^T \mathbf{d} + \frac{1}{2}\mathbf{d}^T \mathbf{H}(\mathbf{x}^*)\mathbf{d} + R \tag{4.30}$$

Alternatively, a change in the function $\Delta f = f(\mathbf{x}) - f(\mathbf{x}^*)$ is given as

$$\Delta f = \nabla f(\mathbf{x}^*)^T \mathbf{d} + \frac{1}{2}\mathbf{d}^T \mathbf{H}(\mathbf{x}^*)\mathbf{d} + R \tag{4.31}$$

If we assume a local minimum at $\mathbf{x}^*$ then Δf must be non-negative due to the definition of a local minimum given in Inequality (4.2), that is, $\Delta f \geq 0$. Concentrating only on the first-order term in Eq. (4.31), we observe (as before) that Δf can be non-negative for all possible $\mathbf{d}$ unless

$$\nabla f(\mathbf{x}^*) = \mathbf{0} \tag{4.32}$$

In other words, the gradient of the function at $\mathbf{x}^*$ must be zero. In the component form, this necessary condition becomes

$$\frac{\partial f(\mathbf{x}^*)}{\partial x_i} = 0; \quad i = 1 \text{ to } n \tag{4.33}$$

Points satisfying Eq. (4.33) are called *stationary points*.

Considering the second term in Eq. (4.31) evaluated at a stationary point, the positivity of Δf is assured if

$$\mathbf{d}^T \mathbf{H}(\mathbf{x}^*)\mathbf{d} > 0 \tag{4.34}$$

for all $\mathbf{d} \neq \mathbf{0}$. This is true if the Hessian $\mathbf{H}(\mathbf{x}^*)$ is a positive definite matrix (see Section 4.2), which is then the sufficient condition for a local minimum of $f(\mathbf{x})$ at $\mathbf{x}^*$. Conditions (4.33) and (4.34) are the multidimensional equivalent of Conditions (4.26) and (4.28), respectively. We summarize the development of this section in Theorem 4.4.

THEOREM 4.4

Necessary and Sufficient Conditions for Local Minimum

Necessary condition. If $f(\mathbf{x})$ has a local minimum at $\mathbf{x}^*$ then

$$\frac{\partial f(\mathbf{x}^*)}{\partial x_i} = 0; \quad i = 1 \text{ to } n \tag{a}$$

Second-order necessary condition. If $f(\mathbf{x})$ has a local minimum at $\mathbf{x}^*$, then the Hessian matrix of Eq. (4.8)

$$\mathbf{H}(\mathbf{x}^*) = \left[\frac{\partial^2 f}{\partial x_i \partial x_j} \right]_{(n \times n)} \tag{b}$$

is positive semidefinite or positive definite at the point $\mathbf{x}^*$.

Second-order sufficiency condition. If the matrix $\mathbf{H}(\mathbf{x}^*)$ is positive definite at the stationary point $\mathbf{x}^*$, then $\mathbf{x}^*$ is a local minimum point for the function $f(\mathbf{x})$.

Note that if $\mathbf{H}(\mathbf{x}^*)$ at the stationary point $\mathbf{x}^*$ is indefinite, then $\mathbf{x}^*$ is neither a local minimum nor a local maximum point because the second-order necessary condition is violated for both cases. Such stationary points are called *inflection points*. Also if $\mathbf{H}(\mathbf{x}^*)$ is at least positive semidefinite, then $\mathbf{x}^*$ cannot be a local maximum since it violates the second-order necessary condition for a local maximum of $f(\mathbf{x})$. In other words, a point cannot be a local minimum and a local maximum simultaneously. The optimality conditions for a function of a single variable and a function of several variables are summarized in Table 4.1.

In addition, note that these conditions involve derivatives of $f(\mathbf{x})$ and not the value of the function. If we *add a constant* to $f(\mathbf{x})$, the solution $\mathbf{x}^*$ of the minimization problem remains unchanged. Similarly, if we multiply $f(\mathbf{x})$ by any positive constant, the minimum point $\mathbf{x}^*$ is unchanged but the value $f(\mathbf{x}^*)$ is altered.

TABLE 4.1   Optimality conditions for unconstrained problems

Function of one variable minimize $f(x)$	Function of several variable minimize $f(x)$
First-order necessary condition: $f' = 0$. Any point satisfying this condition is called a stationary point; it can be a local maximum, local minimum, or neither of the two (inflection point)	*First-order necessary condition*: $\nabla f = \mathbf{0}$. Any point satisfying this condition is called a stationary point; it can be a local minimum, local maximum, or neither of the two (inflection point)
Second-order necessary condition for a local minimum: $f'' \geq 0$	*Second-order necessary condition* for a local minimum: **H** must be at least positive semidefinite
Second-order necessary condition for a local maximum: $f'' \leq 0$	*Second-order necessary condition* for a local maximum: **H** must be at least negative semidefinite
Second-order sufficient condition for a local minimum: $f'' > 0$	*Second-order sufficient condition* for a local minimum: **H** must be positive definite
Second-order sufficient condition for a local maximum: $f'' < 0$	*Second-order sufficient condition* for a local maximum: **H** must be negative definite
Higher-order necessary conditions for a local minimum or local maximum: Calculate a higher-ordered derivative that is not 0; all odd-ordered derivatives below this one must be 0	
Higher-order sufficient condition for a local minimum Highest nonzero derivative must be even-ordered and positive	

In a graph of $f(x)$ versus **x**, adding a constant to $f(x)$ changes the origin of the coordinate system but leaves the shape of the surface unchanged. Similarly, if we multiply $f(x)$ by any positive constant, the minimum point $\mathbf{x}^*$ is unchanged but the value $f(\mathbf{x}^*)$ is altered. In a graph of $f(\mathbf{x})$ versus **x** this is equivalent to a uniform change of the scale of the graph along the $f(\mathbf{x})$-axis, which again leaves the shape of the surface unaltered. Multiplying $f(\mathbf{x})$ by a negative constant changes the minimum at $\mathbf{x}^*$ to a maximum. We may use this property to convert maximization problems to minimization problems by multiplying $f(\mathbf{x})$ by -1, as explained earlier in Section 2.11. The effect of scaling and adding a constant to a function is shown in Example 4.19. In Examples 4.20 and 4.23, the local minima for a function are found using optimality conditions, while in Examples 4.21 and 4.22, use of the necessary conditions is explored.

EXAMPLE 4.19 EFFECTS OF SCALING OR ADDING A CONSTANT TO THE COST FUNCTION

Discuss the effect of the preceding variations for the function $f(x) = x^2 - 2x + 2$.

Solution

Consider the graphs in Figure 4.9. Part (a) represents the function $f(x) = x^2 - 2x + 2$, which has a minimum at $x^* = 1$. Parts (b), (c), and (d) show the effect of adding a constant to the function $[f(x) + 1]$, multiplying $f(x)$ by a positive number $[2f(x)]$, and multiplying it by a negative number $[-f(x)]$. In all cases, the stationary point remains unchanged.

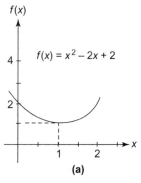

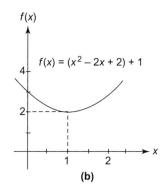

FIGURE 4.9 Graphs for Example 4.19. Effects of scaling or of adding a constant to a function. (a) A graph of $f(x) = x^2 - 2x + 2$. (b) The effect of addition of a constant to $f(x)$. (c) The effect of multiplying $f(x)$ by a positive constant. (d) Effect of multiplying $f(x)$ by -1.

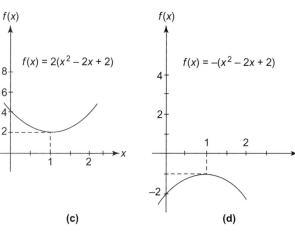

EXAMPLE 4.20 LOCAL MINIMA FOR A FUNCTION OF TWO VARIABLES USING OPTIMALITY CONDITIONS

Find local minimum points for the function

$$f(\mathbf{x}) = x_1^2 + 2x_1x_2 + 2x_2^2 - 2x_1 + x_2 + 8 \tag{a}$$

Solution

The necessary conditions for the problem give

$$\frac{\partial f}{\partial \mathbf{x}} = \begin{bmatrix} (2x_1 + 2x_2 - 2) \\ (2x_1 + 4x_2 + 1) \end{bmatrix} = \begin{bmatrix} 0 \\ 0 \end{bmatrix} \tag{b}$$

These equations are linear in variables x_1 and x_2. Solving the equations simultaneously, we get the stationary point as $\mathbf{x}^* = (2.5, -1.5)$. To check if the stationary point is a local minimum, we evaluate $\mathbf{H}$ at $\mathbf{x}^*$:

$$\mathbf{H}(2.5, -1.5) = \begin{bmatrix} \dfrac{\partial^2 f}{\partial x_1^2} & \dfrac{\partial^2 f}{\partial x_1 \partial x_2} \\[2mm] \dfrac{\partial^2 f}{\partial x_2 \partial x_1} & \dfrac{\partial^2 f}{\partial x_2^2} \end{bmatrix} = \begin{bmatrix} 2 & 2 \\ 2 & 4 \end{bmatrix} \tag{c}$$

By either of the tests of Theorems 4.2 and 4.3 or ($M_1 = 2 > 0$, $M_2 = 4 > 0$) or ($\lambda_1 = 5.236 > 0$, $\lambda_2 = 0.764 > 0$), $\mathbf{H}$ is positive definite at the stationary point $\mathbf{x}^*$. Thus, it is a local minimum with $f(\mathbf{x}^*) = 4.75$. Figure 4.10 shows a few isocost curves for the function of this problem. It is seen that the point $(2.5, -1.5)$ is the minimum for the function.

Checking the optimality of a given point As noted earlier, the optimality conditions can also be used to check the optimality of a given point. To illustrate this, let us check the optimality of the point $(1, 2)$. At this point, the gradient vector is calculated as $(4, 11)$, which is not zero. Therefore, the first-order necessary condition for a local minimum or a local maximum is violated and the point is not a stationary point.

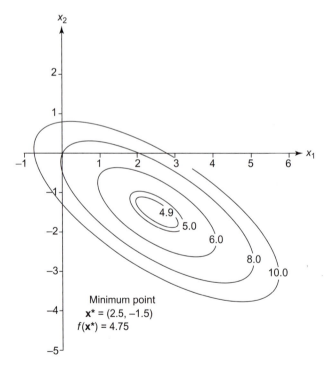

FIGURE 4.10   Isocost curves for the function of Example 4.20.

EXAMPLE 4.21 CYLINDRICAL TANK DESIGN USING NECESSARY CONDITIONS

In Section 2.8, a minimum-cost cylindrical storage tank problem was formulated. The tank is closed at both ends and is required to have volume V. The radius R and height H are selected as design variables. It is desired to design the tank having minimum surface area.

Solution

For the problem, we may simplify the cost function as

$$\bar{f} = R^2 + RH \tag{a}$$

The volume constraint is an equality,

$$h = \pi R^2 H - V = 0 \tag{b}$$

This constraint cannot be satisfied if either R or H is zero. We may then neglect the non-negativity constraints on R and H if we agree to choose only the positive value for them. We may further use the equality constraint (b) to eliminate H from the cost function:

$$H = \frac{V}{\pi R^2} \tag{c}$$

Therefore, the cost function of Eq. (a) becomes

$$\bar{\bar{f}} = R^2 + \frac{V}{\pi R} \tag{d}$$

This is an unconstrained problem in terms of R for which the necessary condition gives

$$\frac{d\bar{\bar{f}}}{dR} = 2R - \frac{V}{\pi R^2} = 0 \tag{e}$$

The solution to the necessary condition gives

$$R^* = \left(\frac{V}{2\pi}\right)^{1/3} \tag{f}$$

Using Eq. (c), we obtain

$$H^* = \left(\frac{4V}{\pi}\right)^{1/3} \tag{g}$$

Using Eq. (e), the second derivative of $\bar{\bar{f}}$ with respect to R at the stationary point is

$$\frac{d^2\bar{\bar{f}}}{dR^2} = \frac{2V}{\pi R^3} + 2 = 6 \tag{h}$$

Since the second derivative is positive for all positive R, the solution in Eqs. (f) and (g) is a local minimum point. Using Eq. (a) or (d), the cost function at the optimum is given as

$$\bar{f}(R^*, H^*) = 3\left(\frac{V}{2\pi}\right)^{2/3} \tag{i}$$

EXAMPLE 4.22 NUMERICAL SOLUTION TO THE NECESSARY CONDITIONS

Find stationary points for the following function and check the sufficiency conditions for them:

$$f(x) = \frac{1}{3}x^2 + \cos x \tag{a}$$

Solution

The function is plotted in Figure 4.11. It is seen that there are three stationary points: $x = 0$ (point A), x between 1 and 2 (point C), and x between -1 and -2 (point B). The point $x = 0$ is a local maximum for the function, and the other two are local minima.

The necessary condition is

$$f'(x) = \frac{2}{3}x - \sin x = 0 \tag{b}$$

It is seen that $x = 0$ satisfies Eq. (b), so it is a stationary point. We must find other roots of Eq. (b). Finding an analytical solution for the equation is difficult, so we must use numerical methods.

We can either plot $f'(x)$ versus x and locate the point where $f'(x) = 0$, or use a numerical method for solving nonlinear equations, such as the *Newton-Raphson method*. By either of the two methods, we find that $x^* = 1.496$ and $x^* = -1.496$ satisfy $f'(x) = 0$ in Eq. (b). Therefore, these are additional stationary points.

To determine whether they are local minimum, maximum, or inflection points, we must determine f'' at the stationary points and use the sufficient conditions of Theorem 4.4. Since $f'' = 2/3 - \cos x$, we have

1. $x^* = 0$; $f'' = -1/3 < 0$, so this is a local maximum with $f(0) = 1$.
2. $x^* = 1.496$; $f'' = 0.592 > 0$, so this is a local minimum with $f(1.496) = 0.821$.
3. $x^* = -1.496$; $f'' = 0.592 > 0$, so this is a local minimum with $f(-1.496) = 0.821$.

These results agree with the graphical solutions observed in Figure 4.11.

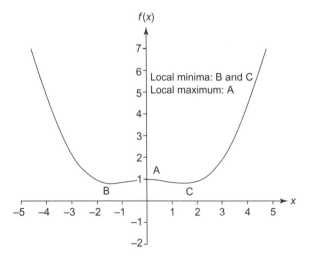

FIGURE 4.11 Graph of $f(x) = \frac{1}{3}x^2 + \cos x$ of Example 4.22.

Local minima: B and C
Local maximum: A

Global optima Note that $x^* = 1.496$ and $x^* = -1.496$ are actually global minimum points for the function, although the function is unbounded and the feasible set is not closed. Therefore, although the conditions of the Weierstrass Theorem 4.1 are not met, the function has global minimum points. This shows that Theorem 4.1 is not an "if-and-only-if" theorem. Note also that there is no global maximum point for the function since the function is unbounded and x is allowed to have any value.

EXAMPLE 4.23 LOCAL MINIMA FOR A FUNCTION OF TWO VARIABLES USING OPTIMALITY CONDITIONS

Find a local minimum point for the function

$$f(\mathbf{x}) = x_1 + \frac{(4 \times 10^6)}{x_1 x_2} + 250x_2 \tag{a}$$

Solution

The necessary conditions for optimality are

$$\frac{\partial f}{\partial x_1} = 0; \quad 1 - \frac{(4 \times 10^6)}{x_1^2 x_2} = 0 \tag{b}$$

$$\frac{\partial f}{\partial x_2} = 0; \quad 250 - \frac{(4 \times 10^6)}{x_1 x_2^2} = 0 \tag{c}$$

Equations (b) and (c) give

$$x_1^2 x_2 - (4 \times 10^6) = 0; \quad 250x_1 x_2^2 - (4 \times 10^6) = 0 \tag{d}$$

These equations give

$$x_1^2 x_2 = 250x_1 x_2^2, \quad \text{or} \quad x_1 x_2(x_1 - 250x_2) = 0 \tag{e}$$

Since neither x_1 nor x_2 can be zero (the function has singularity at $x_1 = 0$ or $x_2 = 0$), the preceding equation gives $x_1 = 250x_2$. Substituting this into Eq. (c), we obtain $x_2 = 4$. Therefore, $x_1^* = 1000$, and $x_2^* = 4$ is a stationary point for the function $f(\mathbf{x})$. Using Eqs. (b) and (c), the Hessian matrix for $f(\mathbf{x})$ at the point $\mathbf{x}^*$ is given as

$$\mathbf{H} = \frac{(4 \times 10^6)}{x_1^2 x_2^2} \begin{bmatrix} \dfrac{2x_2}{x_1} & 1 \\ 1 & \dfrac{2x_1}{x_2} \end{bmatrix}; \quad \mathbf{H}(1000, 4) = \frac{(4 \times 10^6)}{(4000)^2} \begin{bmatrix} 0.008 & 1 \\ 1 & 500 \end{bmatrix} \tag{f}$$

Eigenvalues of the Hessian are: $\lambda_1 = 0.0015$ and $\lambda_2 = 125$. Since both eigenvalues are positive: the Hessian of $f(\mathbf{x})$ at the point $\mathbf{x}^*$ is positive definite. Therefore: $\mathbf{x}^* = (1000, 4)$ satisfies the sufficiency condition for a local minimum point with $f(\mathbf{x}^*) = 3000$. Figure 4.12 shows some isocost curves for the function of this problem. It is seen that $x_1 = 1000$ and $x_2 = 4$ is the minimum point. (Note that the horizontal and vertical scales are quite different in Figure 4.12; this is done to obtain reasonable isocost curves.)

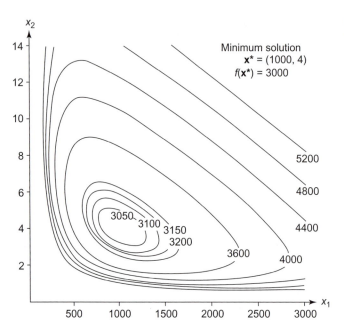

FIGURE 4.12　Isocost curves for the function of Example 4.23 (the horizontal and the vertical scales are different).

4.5 NECESSARY CONDITIONS: EQUALITY-CONSTRAINED PROBLEM

We saw in Chapter 2 that most design problems include *constraints on variables and on performance* of a system. Therefore, constraints must be included in discussing the optimality conditions. The standard design optimization model introduced in Section 2.11 needs to be considered. As a reminder, this model is restated in Table 4.2.

We begin the discussion of optimality conditions for the constrained problem by including only the equality constraints in the formulation in this section; that is, inequalities in Eq. (4.37) are ignored. The reason is that the nature of equality constraints is quite different from that of inequality constraints. Equality constraints are always active for any feasible design, whereas an inequality constraint may not be active at a feasible point. This changes the nature of the necessary conditions for the problem when inequalities are included, as we will see in Section 4.6.

The necessary conditions for an equality-constrained problem are discussed and illustrated with examples. These conditions are contained in the *Lagrange Multiplier Theorem*

TABLE 4.2　General design optimization model

Design variable vector	$\mathbf{x} = (x_1, x_2, \ldots, x_n)$	
Cost function	$f(\mathbf{x}) = f(x_1, x_2, \ldots, x_n)$	(4.35)
Equality constraints	$h_i(\mathbf{x}) = 0; \quad i = 1 \text{ top}$	(4.36)
Inequality constraints	$g_i(\mathbf{x}) \leq 0; \quad i = 1 \text{ to } m$	(4.37)

generally discussed in textbooks on calculus. The necessary conditions for the general constrained optimization problem are obtained as an extension of the Lagrange Multiplier Theorem in the next section.

4.5.1 Lagrange Multipliers

It turns out that a scalar multiplier is associated with each constraint, called the *Lagrange multiplier*. These multipliers play a prominent role in optimization theory as well as in numerical methods. Their values depend on the form of the cost and constraint functions. If these functions change, the values of the Lagrange multipliers also change. We will discuss this aspect later in Section 4.7.

Here we will introduce the idea of Lagrange multipliers by considering a simple example problem. The example outlines the development of the Lagrange Multiplier theorem. Before presenting the example problem, however, we consider an important concept of a *regular point* of the feasible set.

REGULAR POINT   Consider the constrained optimization problem of minimizing $f(\mathbf{x})$ subject to the constraints $h_i(\mathbf{x}) = 0$, $i = 1$ to p. A point $\mathbf{x}^*$ satisfying the constraints $\mathbf{h}(\mathbf{x}^*) = \mathbf{0}$ is said to be a *regular point* of the feasible set if $f(\mathbf{x}^*)$ is differentiable and gradient vectors of all constraints at the point $\mathbf{x}^*$ are linearly independent. *Linear independence* means that no two gradients are parallel to each other, and no gradient can be expressed as a linear combination of the others (refer to Appendix A for more discussion on the linear independence of a set of vectors). When inequality constraints are also included in the problem definition, then for a point to be regular, gradients of all of the active constraints must also be linearly independent.

EXAMPLE 4.24 LAGRANGE MULTIPLIERS AND THEIR GEOMETRICAL MEANING

Minimize

$$f(x_1, x_2) = (x_1 - 1.5)^2 + (x_2 - 1.5)^2 \tag{a}$$

subject to an equality constraint:

$$h(x_1, x_2) = x_1 + x_2 - 2 = 0 \tag{b}$$

Solution

The problem has two variables and can be solved easily by the graphical procedure. Figure 4.13 is a graphical representation of it. The straight line A−B represents the problem's equality constraint and its feasible region. Therefore, the optimum solution must lie on the line A−B. The cost function is an equation of a circle with its center at point (1.5, 1.5). Also shown are the isocost curves, having values of 0.5 and 0.75. It is seen that point C, having coordinates (1, 1), gives the optimum solution for the problem. The cost function contour of value 0.5 just touches the line A−B, so this is the minimum value for the cost function.

Lagrange multipliers Now let us see what mathematical conditions are satisfied at the minimum point C. Let the optimum point be represented as (x_1^*, x_2^*). To derive the conditions and to

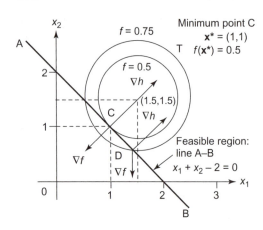

FIGURE 4.13   Graphical solution for Example 4.24. Geometrical interpretation of necessary conditions (the vectors are not to scale).

introduce the Lagrange multiplier, we first assume that the equality constraint can be used to solve for one variable in terms of the other (at least symbolically); that is, assume that we can write

$$x_2 = \phi(x_1) \tag{c}$$

where ϕ is an appropriate function of x_1. In many problems, it may not be possible to explicitly write the function $\phi(x_1)$, but for derivation purposes, we assume its existence. It will be seen later that the explicit form of this function is not needed. For the present example, $\phi(x_1)$ from Eq. (b) is given as

$$x_2 = \phi(x_1) = -x_1 + 2 \tag{d}$$

Substituting Eq. (c) into Eq. (a), we eliminate x_2 from the cost function and obtain the unconstrained minimization problem in terms of x_1 only:

Minimize

$$f(x_1, \phi(x_1)) \tag{e}$$

For the present example, substituting Eq. (d) into Eq. (a), we eliminate x_2 and obtain the minimization problem in terms of x_1 alone:

$$f(x_1) = (x_1 - 1.5)^2 + (-x_1 + 2 - 1.5)^2 \tag{f}$$

The necessary condition $df/dx_1 = 0$ gives $x_1^* = 1$. Then Eq. (d) gives $x_2^* = 1$, and the cost function at the point (1, 1) is 0.5. It can be checked that the sufficiency condition $d^2f/dx_1^2 > 0$ is also satisfied, and so the point is indeed a local minimum, as seen in Figure 4.13.

If we assume that the explicit form of the function $\phi(x_1)$ cannot be obtained (which is generally the case), then some alternate procedure must be developed to obtain the optimum solution. We will derive such a procedure and see that the Lagrange multiplier for the constraint is introduced naturally in the process. Using the chain rule of differentiation, we write the necessary condition $df/dx_1 = 0$ for the problem defined in Eq. (e) as

$$\frac{df(x_1, x_2)}{dx_1} = \frac{\partial f(x_1, x_2)}{\partial x_1} + \frac{\partial f(x_1, x_2)}{\partial x_2}\frac{dx_2}{dx_1} = 0 \tag{g}$$

Substituting Eq. (c), Eq. (g) can be written at the optimum point (x_1^*, x_2^*) as

$$\frac{\partial f(x_1^*, x_2^*)}{\partial x_1} + \frac{\partial f(x_1^*, x_2^*)}{\partial x_2} \frac{d\phi}{dx_1} = 0 \tag{h}$$

Since ϕ is not known, we need to eliminate $d\phi/dx_1$ from Eq. (h). To accomplish this, we differentiate the constraint equation $h(x_1, x_2) = 0$ at the point (x_1^*, x_2^*) as

$$\frac{dh(x_1^*, x_2^*)}{dx_1} = \frac{\partial h(x_1^*, x_2^*)}{\partial x_1} + \frac{\partial h(x_1^*, x_2^*)}{\partial x_2} \frac{d\phi}{dx_1} = 0 \tag{i}$$

Or, solving for $d\phi/dx_1$, we obtain (assuming $\partial h/\partial x_2 \neq 0$)

$$\frac{d\phi}{dx_1} = -\frac{\partial h(x_1^*, x_2^*)/\partial x_1}{\partial h(x_1^*, x_2^*)/\partial x_2} \tag{j}$$

Now, substituting for $d\phi/dx_1$ from Eq. (j) into Eq. (h), we obtain

$$\frac{\partial f(x_1^*, x_2^*)}{\partial x_1} - \frac{\partial f(x_1^*, x_2^*)}{\partial x_2} \left(\frac{\partial h(x_1^*, x_2^*)/\partial x_1}{\partial h(x_1^*, x_2^*/\partial x_2)} \right) = 0 \tag{k}$$

If we define a quantity v as

$$v = -\frac{\partial f(x_1^*, x_2^*)/\partial x_2}{\partial h(x_1^*, x_2^*)/\partial x_2} \tag{l}$$

and substitute it into Eq. (k), we obtain

$$\frac{\partial f(x_1^*, x_2^*)}{\partial x_1} + v \frac{\partial h(x_1^*, x_2^*)}{\partial x_1} = 0 \tag{m}$$

Also, rearranging Eq. (l), which defines v, we obtain

$$\frac{\partial f(x_1^*, x_2^*)}{\partial x_2} + v \frac{\partial h(x_1^*, x_2^*)}{\partial x_2} = 0 \tag{n}$$

Equations (m) and (n) along with the equality constraint $h(x_1, x_2) = 0$ are the *necessary conditions of optimality for the problem. Any point that violates these conditions cannot be a minimum point for the problem.* The scalar quantity v defined in Eq. (l) is called the *Lagrange multiplier*. If the minimum point is known, Eq. (l) can be used to calculate its value. For the present example, $\partial f(1,1)/\partial x_2 = -1$ and $\partial h(1,1)/\partial x_2 = 1$; therefore, Eq. (l) gives $v^* = 1$ as the Lagrange multiplier at the optimum point.

Recall that the necessary conditions can be used to solve for the candidate minimum points; that is, Eqs. (m), (n), and $h(x_1, x_2) = 0$ can be used to solve for x_1, x_2, and v. For the present example, these equations give

$$2(x_1 - 1.5) + v = 0; \quad 2(x_2 - 1.5) + v = 0; \quad x_1 + x_2 - 2 = 0 \tag{o}$$

The solution to these equations is indeed, $x_1^* = 1$, $x_2^* = 1$, and $v^* = 1$.

Geometrical meaning of the Lagrange multipliers It is customary to use what is known as the *Lagrange function* in writing the necessary conditions. The Lagrange function is denoted as L and defined using cost and constraint functions as

$$L(x_1, x_2, v) = f(x_1, x_2) + vh(x_1, x_2) \tag{p}$$

It is seen that the necessary conditions of Eqs. (m) and (n) are given in terms of L as

$$\frac{\partial L(x_1^*, x_2^*)}{\partial x_1} = 0, \quad \frac{\partial L(x_1^*, x_2^*)}{\partial x_2} = 0 \tag{q}$$

Or, in the vector notation, we see that the gradient of L is zero at the candidate minimum point, that is, $\nabla L(x_1^*, x_2^*) = \mathbf{0}$. Writing this condition, using Eq. (m), or writing Eqs. (m) and (n) in the vector form, we obtain

$$\nabla f(\mathbf{x}^*) + v \nabla h(\mathbf{x}^*) = \mathbf{0} \tag{r}$$

where gradients of the cost and constraint functions are given as

$$\nabla f(\mathbf{x}^*) = \begin{bmatrix} \dfrac{\partial f(x_1^*, x_2^*)}{\partial x_1} \\ \dfrac{\partial f(x_1^*, x_2^*)}{\partial x_2} \end{bmatrix}, \quad \nabla h = \begin{bmatrix} \dfrac{\partial h(x_1^*, x_2^*)}{\partial x_1} \\ \dfrac{\partial h(x_1^*, x_2^*)}{\partial x_2} \end{bmatrix} \tag{s}$$

Equation (r) can be rearranged as

$$\nabla f(\mathbf{x}^*) = -v \nabla h(\mathbf{x}^*) \tag{t}$$

The preceding equation brings out the geometrical meaning of the necessary conditions. It shows that at the candidate minimum point for the present example: gradients of the cost and constraint functions are along the same line and proportional to each other, and the Lagrange multiplier v is the proportionality constant.

For the present example, the gradients of cost and constraint functions at the candidate optimum point are given as

$$\nabla f(1, 1) = \begin{bmatrix} -1 \\ -1 \end{bmatrix}, \quad \nabla h(1, 1) = \begin{bmatrix} 1 \\ 1 \end{bmatrix} \tag{u}$$

These vectors are shown at point C in Figure 4.13. Note that they are along the same line. For any other feasible point on line A–B, say (0.4,1.6), the gradients of cost and constraint functions will not be along the same line, as seen in the following:

$$\nabla f(0.4, 1.6) = \begin{bmatrix} -2.2 \\ 0.2 \end{bmatrix}, \quad \nabla h(0.4, 1.6) = \begin{bmatrix} 1 \\ 1 \end{bmatrix} \tag{v}$$

As another example, point D in Figure 4.13 is not a candidate minimum since gradients of cost and constraint functions are not along the same line. Also, the cost function has a higher value at these points compared to the one at the minimum point; that is, we can move away from point D toward point C and reduce the cost function.

It is interesting to note that the equality constraint can be multiplied by -1 without affecting the minimum point; that is, the constraint can be written as $-x_1 - x_2 + 2 = 0$. The minimum solution is still the same: $x_1^* = 1$, $x_2^* = 1$, and $f(\mathbf{x}^*) = 0.5$; however, the sign of the Lagrange multiplier is reversed (i.e., $v^* = -1$). *This shows that the Lagrange multiplier for the equality constraint is free in sign*; that is, the sign is determined by the form of the constraint function.

It is also interesting to note that any small move from point C in the feasible region (i.e., along the line A–B) increases the cost function value, and *any further reduction in the cost function is accompanied by violation of the constraint*. Thus, point C satisfies the sufficiency condition for a local minimum point because it has the smallest value in a neighborhood of point C (note that we have used the definition of a local minimum given in Eq. (4.2)). Thus, it is indeed a local minimum point.

4.5.2 Lagrange Multiplier Theorem

The concept of Lagrange multipliers is quite general. It is encountered in many engineering applications other than optimum design. *The Lagrange multiplier for a constraint can be interpreted as the force required to impose the constraint.* We will discuss a physical meaning of Lagrange multipliers in Section 4.7. The idea of a Lagrange multiplier for an equality constraint, introduced in Example 4.24, can be generalized to many equality constraints. It can also be extended to inequality constraints.

We will first discuss the necessary conditions with multiple equality constraints in Theorem 4.5 and then describe in the next section their extensions to include the inequality constraints. Just as for the unconstrained case, solutions to the necessary conditions give candidate minimum points. The sufficient conditions—discussed in Chapter 5—can be used to determine if a candidate point is indeed a local minimum.

THEOREM 4.5

Lagrange Multiplier Theorem Consider the optimization problem defined in Eqs. (4.35) and (4.36):

Minimize $f(\mathbf{x})$

subject to equality constraints

$$h_i(\mathbf{x}) = 0, \quad i = 1 \text{ to } p$$

Let $\mathbf{x}^*$ be a regular point that is a local minimum for the problem. Then there exist unique Lagrange multipliers v_j^*, $j = 1$ to p such that

$$\frac{\partial f(\mathbf{x}^*)}{\partial x_i} + \sum_{j=1}^{p} v_j^* \frac{\partial h_j(\mathbf{x}^*)}{\partial x_i} = 0; \quad i = 1 \text{ to } n \quad (4.38)$$

$$h_j(\mathbf{x}^*) = 0; \quad j = 1 \text{ to } p \quad (4.39)$$

It is convenient to write these conditions in terms of a *Lagrange function*, defined as

$$L(\mathbf{x}, \mathbf{v}) = f(\mathbf{x}) + \sum_{j=1}^{p} v_j h_j(\mathbf{x})$$
$$= f(\mathbf{x}) + \mathbf{v}^T \mathbf{h}(\mathbf{x}) \quad (4.40)$$

Then Eq. (4.38) becomes

$$\nabla L(\mathbf{x}^*, \mathbf{v}^*) = \mathbf{0}, \quad \text{or} \quad \frac{\partial L(\mathbf{x}^*, \mathbf{v}^*)}{\partial x_i} = 0;$$
$$i = 1 \text{ to } n \quad (4.41)$$

Differentiating $L(\mathbf{x}, \mathbf{v})$ with respect to v_j, we recover the equality constraints as

$$\frac{\partial L(\mathbf{x}^*, \mathbf{v}^*)}{\partial v_j} = 0 \Rightarrow h_j(\mathbf{x}^*) = 0; \quad j = 1 \text{ to } p \quad (4.42)$$

The gradient conditions of Eqs. (4.41) and (4.42) show that the Lagrange function is stationary with respect to both $\mathbf{x}$ and $\mathbf{v}$. Therefore, it may be treated as an unconstrained function in the variables $\mathbf{x}$ and $\mathbf{v}$ to determine the stationary points. Note that any point that does not satisfy the conditions of the theorem cannot be a local minimum point. However, a point satisfying the conditions need not be a minimum point either. It is simply a candidate minimum point, which can actually be an inflection or maximum point. The second-order necessary and sufficient conditions given in Chapter 5 can distinguish between the minimum, maximum, and inflection points.

The n variables x and the p multipliers v are the unknowns, and the necessary conditions of Eqs. (4.41) and (4.42) provide enough equations to solve for them. Note also that the Lagrange multipliers v_i are free in sign; that is, they can be positive, negative, or zero. This is in contrast to the Lagrange multipliers for the inequality constraints, which are required to be non-negative, as discussed in the next section.

The gradient condition of Eq. (4.38) can be rearranged as

$$\frac{\partial f(x^*)}{\partial x_i} = -\sum_{j=1}^{p} v_j^* \frac{\partial h_j(x^*)}{\partial x_i}; \quad i = 1 \text{ to } n \tag{4.43}$$

This form shows that the *gradient of the cost function is a linear combination of the gradients of the constraints at the candidate minimum point*. The Lagrange multipliers v_j^* act as the scalars of the linear combination. This linear combination interpretation of the necessary conditions is a generalization of the concept discussed in Example 4.24 for one constraint: "... at the candidate minimum point gradients of the cost and constraint functions are along the same line." Example 4.28 illustrates the necessary conditions for an equality-constrained problem.

EXAMPLE 4.25 CYLINDRICAL TANK DESIGN—USE OF LAGRANGE MULTIPLIERS

We will re-solve the cylindrical storage tank problem (Example 4.21) using the Lagrange multiplier approach. The problem is to find radius R and length H of the cylinder to

Minimize
$$\bar{f} = R^2 + RH \tag{a}$$

subject to
$$h = \pi R^2 H - V = 0 \tag{b}$$

Solution
The Lagrange function L of Eq. (4.40) for the problem is given as

$$L = R^2 + RH + v(\pi R^2 H - V) \tag{c}$$

The necessary conditions of the Lagrange Multiplier Theorem 4.5 give

$$\frac{\partial L}{\partial R} = 2R + H + 2\pi v RH = 0 \tag{d}$$

$$\frac{\partial L}{\partial H} = R + \pi v R^2 = 0 \tag{e}$$

$$\frac{\partial L}{\partial v} = \pi R^2 H - V = 0 \tag{f}$$

These are three equations in three unknowns v, R, and H. Note that they are nonlinear. However, they can be easily solved by the elimination process, giving the solution to the necessary conditions as

$$R^* = \left(\frac{V}{2\pi}\right)^{1/3}; \quad H^* = \left(\frac{4V}{\pi}\right)^{1/3}; \quad v^* = -\frac{1}{\pi R} = -\left(\frac{2}{\pi^2 V}\right)^{1/3}; \quad f^* = 3\left(\frac{V}{2\pi}\right)^{2/3} \tag{g}$$

This is the same solution as obtained for Example 4.21, treating it as an unconstrained problem. It can be also verified that the gradients of the cost and constraint functions are along the same line at the optimum point.

Note that this problem has only one equality constraint. Therefore, the question of linear dependence of the gradients of active constraints does not arise; that is, the regularity condition for the solution point is satisfied.

Often, the necessary conditions of the Lagrange Multiplier Theorem lead to a nonlinear set of equations that cannot be solved analytically. In such cases, we must use a numerical algorithm, such as the Newton-Raphson method, to solve for their roots and the candidate minimum points. Several commercial software packages, such as Excel, MATLAB, and Mathematica, are available to find roots of nonlinear equations. We will describe the use of Excel in Chapter 6 and MATLAB in Chapter 7 for this purpose.

4.6 NECESSARY CONDITIONS FOR A GENERAL CONSTRAINED PROBLEM

4.6.1 The Role of Inequalities

In this section we will extend the Lagrange Multiplier Theorem to include inequality constraints. However, it is important to understand the role of inequalities in the necessary conditions and the solution process. As noted earlier, the inequality constraint may be active or inactive at the minimum point. However, the minimum points are not known a priori; we are trying to find them. Therefore, it is not known a priori if an inequality is active or inactive at the solution point. The question is, How do we determine the status of an inequality at the minimum point? The answer is that the determination of the status of an inequality is a part of the necessary conditions for the problem.

Examples 4.26 through 4.28 illustrate the role of inequalities in determining the minimum points.

EXAMPLE 4.26 ACTIVE INEQUALITY

Minimize
$$f(x) = (x_1 - 1.5)^2 + (x_2 - 1.5)^2 \tag{a}$$

subject to
$$g_1(\mathbf{x}) = x_1 + x_2 - 2 \le 0 \tag{b}$$

$$g_2(\mathbf{x}) = -x_1 \le 0; \quad g_3(\mathbf{x}) = -x_2 \le 0 \tag{c}$$

Solution

The feasible set S for the problem is a triangular region, shown in Figure 4.14. If constraints are ignored, $f(x)$ has a minimum at the point (1.5, 1.5) with the cost function as $f^* = 0$, which violates the constraint g_1 and therefore is an infeasible point for the problem. Note that contours of $f(x)$ are circles. They increase in diameter as $f(x)$ increases. It is clear that the minimum value for $f(x)$ corresponds to a circle with the smallest radius intersecting the feasible set. This is the point (1, 1) at which $f(x) = 0.5$. The point is on the boundary of the feasible region where the inequality constraint $g_1(x)$ is active (i.e., $g_1(x) = 0$). Thus, the location of the optimum point is governed by the constraint for this problem as well as the cost function contours.

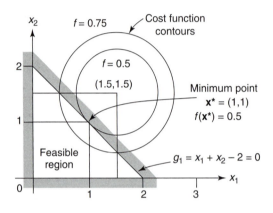

FIGURE 4.14   Example 4.26 graphical representation. Constrained optimum point.

EXAMPLE 4.27 INACTIVE INEQUALITY

Minimize

$$f(x) = (x_1 - 0.5)^2 + (x_2 - 0.5)^2 \tag{a}$$

subject to the same constraints as in Example 4.26.

Solution

The feasible set S is the same as in Example 4.26. The cost function, however, has been modified. If constraints are ignored, $f(x)$ has a minimum at (0.5, 0.5). Since the point also satisfies all of the constraints, it is the optimum solution. The solution to this problem therefore occurs in the interior of the feasible region and the constraints play no role in its location; all inequalities are inactive.

Note that a *solution to a constrained optimization problem may not exist.* **This can happen if we over-constrain the system. The requirements can be conflicting such that it is impossible to build a system to satisfy them. In such a case we must re-examine the problem formulation and relax constraints. Example 4.28 illustrates the situation.**

EXAMPLE 4.28 INFEASIBLE PROBLEM

Minimize

$$f(\mathbf{x}) = (x_1 - 2)^2 + (x_2 - 2)^2 \tag{a}$$

subject to the constraints:

$$g_1(\mathbf{x}) = x_1 + x_2 - 2 \le 0 \tag{b}$$

$$g_2(\mathbf{x}) = -x_1 + x_2 + 3 \le 0 \tag{c}$$

$$g_3(\mathbf{x}) = -x_1 \le 0; \quad g_4(\mathbf{x}) = -x_2 \le 0 \tag{d}$$

Solution

Figure 4.15 shows a plot of the constraints for the problem. It is seen that there is no design satisfying all of the constraints. The feasible set S for the problem is empty and there is no solution (i.e., no feasible design). Basically, the constraints $g_1(\mathbf{x})$ and $g_2(\mathbf{x})$ conflict with each other and need to be modified to obtain feasible solutions to the problem.

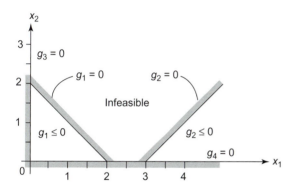

FIGURE 4.15 Plot of constraints for Example 4.28 infeasible problem.

4.6.2 Karush-Kuhn-Tucker Necessary Conditions

We now include inequality constraints $g_i(\mathbf{x}) \le 0$ and consider the general design optimization model defined in Eqs. (4.35) through (4.37). We can transform an inequality constraint into an equality constraint by adding a new variable to it, called the *slack variable*. Since the constraint is of the form "$\le$", its value is either negative or zero at a feasible point. Thus, the slack variable must always be non-negative (i.e., positive or zero) to make the inequality an equality.

An inequality constraint $g_i(\mathbf{x}) \le 0$ is equivalent to the equality constraint $g_i(\mathbf{x}) + s_i = 0$, where $s_i \ge 0$ is a slack variable. The variables s_i are treated as unknowns of the design problem along with the original variables. Their values are determined as a part of the solution. When the variable s_i has zero value, the corresponding inequality constraint is satisfied at equality. Such inequality is called an *active (tight) constraint*; that is, there is no "slack" in the constraint. For any $s_i > 0$, the corresponding constraint is a strict inequality. It is called an *inactive constraint*, and has slack given by s_i. Thus, the status of an inequality constraint is determined as a part of the solution to the problem.

Note that with the preceding procedure, we must introduce one additional variable s_i and an additional constraint $s_i \geq 0$ to treat each inequality constraint. This increases the dimension of the design problem. The constraint $s_i \geq 0$ can be avoided if we use s_i^2 as the slack variable instead of just s_i. Therefore, the inequality $g_i \leq 0$ is converted to an equality as

$$g_i + s_i^2 = 0 \tag{4.44}$$

where s_i can have any real value. This form can be used in the Lagrange Multiplier Theorem to treat inequality constraints and to derive the corresponding necessary conditions. The m new equations needed for determining the slack variables are obtained by requiring the Lagrangian L to be stationary with respect to the slack variables as well $\left(\frac{\partial L}{\partial \mathbf{s}} = \mathbf{0}\right)$.

Note that once a design point is specified, Eq. (4.44) can be used to calculate the slack variable s_i^2. If the constraint is satisfied at the point (i.e., $g_i \leq 0$), then $s_i^2 \geq 0$. If it is violated, then s_i^2 is negative, which is not acceptable; that is, the point is not a candidate minimum point.

There is an *additional necessary condition* for the Lagrange multipliers of "$\leq$ type" constraints given as

$$u_j^* \geq 0; \quad j = 1 \text{ to } m \tag{4.45}$$

where u_j^* is the Lagrange multiplier for the jth inequality constraint. Thus, *the Lagrange multiplier for each "$\leq$" inequality constraint must be non-negative.* If the constraint is inactive at the optimum, its associated Lagrange multiplier is zero. If it is active ($g_i = 0$), then the associated multiplier must be non-negative. We will explain the condition of Eq. (4.45) from a physical point of view in Section 4.7. Example 4.29 illustrates the use of necessary conditions in an inequality-constrained problem.

EXAMPLE 4.29 INEQUALITY-CONSTRAINED PROBLEM— USE OF NECESSARY CONDITIONS

We will re-solve Example 4.24 by treating the constraint as an inequality. The problem is to

Minimize

$$f(x_1, x_2) = (x_1 - 1.5)^2 + (x_2 - 1.5)^2 \tag{a}$$

subject to

$$g(\mathbf{x}) = x_1 + x_2 - 2 \leq 0 \tag{b}$$

Solution

The graphical representation of the problem remains the same as earlier in Figure 4.13 for Example 4.24, except that the feasible region is enlarged; it is line A−B and the region below it. The minimum point for the problem is same as before: $x_1^* = 1$, $x_2^* = 1$, $f(\mathbf{x}^*) = 0.5$.

Introducing a slack variable s^2 for the inequality, the Lagrange function of Eq. (4.40) for the problem is defined as

$$L = (x_1 - 1.5)^2 + (x_2 - 1.5)^2 + u(x_1 + x_2 - 2 + s^2) \tag{c}$$

where u is the Lagrange multiplier for the inequality constraint. The necessary conditions of the Lagrange Theorem give (treating x_1, x_2, u and s as unknowns):

$$\frac{\partial L}{\partial x_1} = 2(x_1 - 1.5) + u = 0 \qquad \text{(d)}$$

$$\frac{\partial L}{\partial x_2} = 2(x_2 - 1.5) + u = 0 \qquad \text{(e)}$$

$$\frac{\partial L}{\partial u} = x_1 + x_2 - 2 + s^2 = 0 \qquad \text{(f)}$$

$$\frac{\partial L}{\partial s} = 2us = 0 \qquad \text{(g)}$$

These are four equations for four unknowns x_1, x_2, u, and s. The equations must be solved simultaneously for all of the unknowns. Note that the equations are nonlinear. Therefore, they can have multiple roots.

One solution can be obtained by setting s *to zero* to satisfy the condition $2us = 0$ in Eq. (g). Equations (d) through (f) are solved to obtain $x_1^* = x_2^* = 1$, $u^* = 1$, $s = 0$. When $s = 0$, the inequality constraint is active. x_1, x_2, and u are solved from the remaining three equations, (d) through (f), which are linear in the variables. This is a stationary point of L, so it is a candidate minimum point. Note from Figure 4.13 that it is actually a minimum point, since any move away from $\mathbf{x}^*$ either violates the constraint or increases the cost function.

The second stationary point is obtained by setting $u = 0$ to satisfy the condition of Eq. (g) and solving the remaining equations for x_1, x_2, and s. This gives $x_1^* = x_2^* = 1.5$, $u^* = 0$, $s^2 = -1$. This is not a valid solution, as the constraint is violated at the point $\mathbf{x}^*$ because $g = -s^2 = 1 > 0$.

It is interesting to observe the geometrical representation of the necessary conditions for inequality-constrained problems. The gradients of the cost and constraint functions at the candidate point (1, 1) are calculated as

$$\nabla f = \begin{bmatrix} 2(x_1 - 1.5) \\ 2(x_2 - 1.5) \end{bmatrix} = \begin{bmatrix} -1 \\ -1 \end{bmatrix}; \quad \nabla g = \begin{bmatrix} 1 \\ 1 \end{bmatrix} \qquad \text{(h)}$$

These gradients are along the same line but in opposite directions, as shown in Figure 4.13. Observe also that any small move from point C either increases the cost function or takes the design into the infeasible region to reduce the cost function any further (i.e., the condition for a local minimum given in Eq. (4.2) is violated). Thus, point (1, 1) is indeed a local minimum point. This geometrical condition is called the *sufficient condition* for a local minimum point.

It turns out that the necessary condition $u \geq 0$ *ensures that the gradients of the cost and the constraint functions point in opposite directions.* This way f cannot be reduced any further by stepping in the negative gradient direction for the cost function without violating the constraint. That is, any further reduction in the cost function leads to leaving the earlier feasible region at the candidate minimum point. This can be observed in Figure 4.13.

The necessary conditions for the equality- and inequality-constrained problem written in the standard form, as in Eqs. (4.35) through (4.37), can be summed up in what are commonly known as the *Karush-Kuhn-Tucker (KKT) first-order necessary conditions*, displayed in Theorem 4.6.

THEOREM 4.6

Karush-Kuhn-Tucker Optimality Conditions
Let $\mathbf{x}^*$ be a regular point of the feasible set that is a local minimum for $f(\mathbf{x})$, subject to $h_i(\mathbf{x}) = 0$; $i = 1$ to p; $g_j(\mathbf{x}) \le 0$; $j = 1$ to m. Then there exist Lagrange multipliers $\mathbf{v}^*$ (a p-vector) and $\mathbf{u}^*$ (an m-vector) such that the Lagrangian function is stationary with respect to x_j, v_i, u_j, and s_j at the point $\mathbf{x}^*$.

1. *Lagrangian Function for the Problem Written in the Standard Form:*

$$L(\mathbf{x}, \mathbf{v}, \mathbf{u}, \mathbf{s}) = f(\mathbf{x}) + \sum_{i=1}^{p} v_i h_i(\mathbf{x})$$

$$+ \sum_{j=1}^{m} u_j (g_j(\mathbf{x}) + s_j^2)$$

$$= f(\mathbf{x}) + \mathbf{v}^T \mathbf{h}(\mathbf{x}) + \mathbf{u}^T (\mathbf{g}(\mathbf{x}) + \mathbf{s}^2)$$

$$(4.46)$$

2. *Gradient Conditions:*

$$\frac{\partial L}{\partial x_k} = \frac{\partial f}{\partial x_k} + \sum_{i=1}^{p} v_i^* \frac{\partial h_i}{\partial x_k} \qquad (4.47)$$

$$+ \sum_{j=1}^{m} u_j^* \frac{\partial g_j}{\partial x_k} = 0; \quad k = 1 \text{ to } n$$

$$\frac{\partial L}{\partial v_i} = 0 \Rightarrow h_i(\mathbf{x}^*) = 0; \quad i = 1 \text{ to } p \qquad (4.48)$$

$$\frac{\partial L}{\partial u_j} = 0 \Rightarrow (g_j(\mathbf{x}^*) + s_j^2) = 0; \quad j = 1 \text{ to } m \quad (4.49)$$

3. *Feasibility Check for Inequalities:*

$$s_j^2 \ge 0; \text{ or equivalently } g_j \le 0;$$

$$j = 1 \text{ to } m \qquad (4.50)$$

4. *Switching Conditions:*

$$\frac{\partial L}{\partial s_j} = 0 \Rightarrow 2u_j^* s_j = 0; \quad j = 1 \text{ to } m \qquad (4.51)$$

5. *Non-negativity of Lagrange Multipliers for Inequalities:*

$$u_j^* \ge 0; \quad j = 1 \text{ to } m \qquad (4.52)$$

6. *Regularity Check:* Gradients of the active constraints must be linearly independent. In such a case the Lagrange multipliers for the constraints are unique.

Geometrical Meaning of the Gradient Condition

It is important to understand the use of KKT conditions to (i) check the possible optimality of a given point, and (ii) determine the candidate local minimum points. Note first from Eqs. (4.48) through (4.50) that *the candidate minimum point must be feasible*, so we must check all of the constraints to ensure their satisfaction. The gradient conditions of Eq. (4.47) must also be satisfied simultaneously. These conditions have a *geometrical meaning*. To see this, rewrite Eq. (4.47) as

$$-\frac{\partial f}{\partial x_j} = \sum_{i=1}^{p} v_i^* \frac{\partial h_i}{\partial x_j} + \sum_{i=1}^{m} u_i^* \frac{\partial g_i}{\partial x_j}; \quad j = 1 \text{ to } n \qquad (4.53)$$

which shows that at the stationary point, the negative gradient direction on the left side (*steepest-descent direction*) for the cost function is a linear combination of the gradients of the constraints, with Lagrange multipliers that are the scalar parameters of the linear combination.

Switching Conditions

The m conditions in Eq. (4.51) *are known as the switching conditions or the complementary slackness conditions.* They can be satisfied by setting either $s_i = 0$ (zero slack implies active inequality, $g_i = 0$) or $u_i = 0$ (in this case g_i must be ≤ 0 to satisfy feasibility). These conditions determine several solution cases, and their use must be clearly understood. As discussed in Example 4.29, there was only one switching condition, which gave two possible cases: Case 1, where the slack variable was zero, and Case 2, where the Lagrange multiplier u for the inequality constraint was zero. Each of the two cases was solved for the unknowns.

For general problems, there is more than one switching condition in Eq. (4.51); the number of switching conditions is equal to the number of inequality constraints for the problem. Various combinations of these conditions can give many solution cases. In general, with m inequality constraints, the switching conditions lead to 2^m distinct *normal* solution cases (the abnormal case is the one where both $u_i = 0$ and $s_i = 0$). For each case, we need to solve the remaining necessary conditions for candidate local minimum points. Also each case may give several candidate minimum points. Depending on the functions of the problem, it may or may not be possible to solve analytically the necessary conditions of each case. If the functions are nonlinear, we must use numerical methods to find their roots. This will be discussed in Chapters 6 and 7.

KKT Conditions

To write the KKT conditions, the optimum design problem must be written in the standard form as displayed in Eqs. (4.35) through (4.37). *We will illustrate the use of the KKT conditions in several example problems.* In Example 4.29, there were only two variables, one Lagrange multiplier and one slack variable. For general problems, the unknowns are $\mathbf{x}$, $\mathbf{u}$, $\mathbf{s}$, and $\mathbf{v}$. These are n-, m-, m-, and p-dimensional vectors. Thus, there are $(n + 2m + p)$ unknown variables and we need $(n + 2m + p)$ equations to determine them. The equations needed for their solution are available in the KKT necessary conditions. The number of equations is $(n + 2m + p)$ in Eqs. (4.47) through (4.51). These equations must be solved simultaneously for the candidate minimum points. The remaining necessary conditions of Eqs. (4.50) and (4.52) must then be checked for the candidate minimum points. The conditions of Eq. (4.50) ensure feasibility of the candidate local minimum points with respect to the inequality constraints $g_i(\mathbf{x}) \leq 0$; $i = 1$ to m. And the conditions of Eq. (4.52) say that the Lagrange multipliers of the "$\leq$ type" inequality constraints must be non-negative.

Note that evaluation of s_i^2 essentially implies evaluation of the constraint function $g_i(\mathbf{x})$, since $s_i^2 = -g_i(\mathbf{x})$. This allows us to check the feasibility of the candidate points with respect to the constraint $g_i(\mathbf{x}) \leq 0$. It is also important to note the following conditions:

- If the inequality constraint $g_i(\mathbf{x}) \leq 0$ is inactive at the candidate minimum point $\mathbf{x}^*$ (i.e., $g_i(\mathbf{x}^*) < 0$, or $s_i^2 > 0$), then the corresponding *Lagrange multiplier* $u_i^* = 0$ to satisfy the switching condition of Eq. (4.51).
- If the inequality constraint $g_i(\mathbf{x}) \leq 0$ is active at the candidate minimum point $\mathbf{x}^*$ (i.e., $g_i(\mathbf{x}^*) = 0$), then the Lagrange multiplier must be non-negative, $u_i^* \geq 0$.

These conditions ensure that there are no feasible directions with respect to the ith constraint $g_i(\mathbf{x}^*) \le 0$ at the candidate point $\mathbf{x}^*$ along which the cost function can reduce any further. Stated differently, the condition ensures that any reduction in the cost function at $\mathbf{x}^*$ can occur only by stepping into the infeasible region for the constraint $g_i(\mathbf{x}^*) \le 0$.

Note further that the necessary conditions of Eqs. (4.47) through (4.52) are generally a nonlinear system of equations in the variables $\mathbf{x}, \mathbf{u}, \mathbf{s},$ *and* $\mathbf{v}$. It may not be easy to solve this system analytically. Therefore, we may have to use numerical methods such as the Newton-Raphson method to find the roots of the system. Fortunately, programs such as Excel, MATLAB, Mathematica, and others are available to solve a nonlinear set of equations. Use of Excel is discussed in Chapters 6 and use of MATLAB is discussed later in this chapter and in Chapter 7.

Important Observations about KKT Conditions

The following important points should be noted relative to the KKT first-order necessary conditions for the problem written in the standard form as is displayed in Eqs. (4.35) through (4.37):

1. The KKT conditions are *not applicable* at the points that are not *regular*. In those cases their use may yield candidate minimum points; however, the Lagrange multipliers may not be unique. This is illustrated with an example in Chapter 5.
2. Any point that *does not satisfy the KKT conditions cannot be a local minimum point* unless it is an irregular point (in that case the KKT conditions are not applicable). Points satisfying the conditions are called *KKT points*.
3. *The points satisfying the KKT conditions can be constrained or unconstrained.* They are unconstrained when there are no equalities and all inequalities are inactive. If the candidate point is unconstrained, it can be a local minimum, maximum, or inflection point depending on the form of the Hessian matrix of the cost function (refer to Section 4.4 for the necessary and sufficient conditions for unconstrained problems).
4. If there are equality constraints and no inequalities are active (i.e., $\mathbf{u} = \mathbf{0}$), then the points satisfying the KKT conditions are only stationary. They can be minimum, maximum, or inflection points.
5. If some inequality constraints are active and their multipliers are positive, then the points satisfying the KKT conditions cannot be local maxima for the cost function (they may be local maximum points if active inequalities have zero multipliers). They may not be local minima either; this will depend on the second-order necessary and sufficient conditions discussed in Chapter 5.
6. It is important to note that the value of the *Lagrange multiplier* for each constraint depends on the functional form for the constraint. For example, the Lagrange multiplier for the constraint $x/y - 10 \le 0$ ($y > 0$) is different for the same constraint expressed as $x - 10y \le 0$ or $0.1x/y - 1 \le 0$. The optimum solution for the problem does not change by changing the form of the constraint, but its Lagrange multiplier is changed. This is further explained in Section 4.7.

Examples 4.30 through 4.32 illustrate various solutions of the KKT necessary conditions for candidate local minimum points.

EXAMPLE 4.30 SOLUTION TO KKT NECESSARY CONDITIONS

Write the KKT necessary conditions and solve them for the problem

Minimize

$$f(\mathbf{x}) = \frac{1}{3}x^3 - \frac{1}{2}(b+c)x^2 + bcx + f_0 \tag{a}$$

subject to

$$a \le x \le d \tag{b}$$

where $0 < a < b < c < d$ and f_0 are specified constants (created by Y. S. Ryu).

Solution

A graph for the cost function and constraints is shown in Figure 4.16. It is seen that point A is a constrained minimum: point B is an unconstrained maximum, point C is an unconstrained minimum, and point D is a constrained maximum. We will show how the KKT conditions distinguish between these points. Note that since only one constraint can be active at the candidate minimum point (x cannot be at points A and D simultaneously), all of the feasible points are regular.

There are two inequality constraints in Eq. (b) that are written in the standard form as

$$g_1 = a - x \le 0; \quad g_2 = x - d \le 0 \tag{c}$$

The Lagrangian function of Eq. (4.46) for the problem is given as

$$L = \frac{1}{3}x^3 - \frac{1}{2}(b+c)x^2 + bcx + f_0 + u_1(a - x + s_1^2) + u_2(x - d + s_2^2) \tag{d}$$

where u_1 and u_2 are the Lagrange multipliers and s_1 and s_2 are the slack variables for the two inequalities in Eq. (c). The KKT conditions give

$$\frac{\partial L}{\partial x} = x^2 - (b+c)x + bc - u_1 + u_2 = 0 \tag{e}$$

$$(a - x) + s_1^2 = 0, \quad s_1^2 \ge 0; \quad (x - d) + s_2^2 = 0, s_2^2 \ge 0 \tag{f}$$

$$u_1 s_1 = 0; \quad u_2 s_2 = 0 \tag{g}$$

$$u_1 \ge 0; \quad u_2 \ge 0 \tag{h}$$

The switching conditions in Eq. (g) give four cases for the solution to the KKT conditions. Each case will be considered separately and solved.

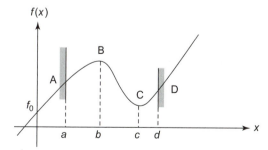

FIGURE 4.16   Example 4.30 graphical representation. Point A, constrained local minimum; B, unconstrained local maximum; C, unconstrained local minimum; D, constrained local maximum.

Case 1: $u_1 = 0$, $u_2 = 0$. For this case, Eq. (e) gives two solutions as $x = b$ and $x = c$. For these points both the inequalities are strictly satisfied because slack variables calculated from Eq. (f) are

$$\text{for } x = b: \quad s_1^2 = b - a > 0; \quad s_2^2 = d - b > 0 \tag{i}$$

$$\text{for } x = c: \quad s_1^2 = c - a > 0; \quad s_2^2 = d - c > 0 \tag{j}$$

Thus, all of the KKT conditions are satisfied, and these are the candidate minimum points. Since the points are unconstrained, they are actually stationary points. We can check the sufficient conditions by calculating the *curvature* of the cost function at the two candidate points:

$$x = b; \quad \frac{d^2 f}{dx^2} = 2x - (b + c) = b - c < 0 \tag{k}$$

Since $b < c$, d^2f/dx^2 is negative. Therefore, the sufficient condition for a local minimum is not met. Actually, the second-order necessary condition of Eq. (4.29) is violated, so the point cannot be a local minimum for the function. It is actually a local maximum point because it satisfies the sufficient condition for that, as also seen in Figure 4.16.

$$x = c; \quad \frac{d^2 f}{dx^2} = c - b > 0 \tag{l}$$

Since $b < c$, d^2f/dx^2 is positive. Therefore, the second-order sufficient condition of Eq. (4.28) is satisfied, and this is a local minimum point, as also seen in Figure 4.16. It cannot be a local maximum point since second order necessary condition for local maximum is violated.

Case 2: $u_1 = 0$, $s_2 = 0$. g_2 is active for this case, and because $s_2 = 0$, , $x = d$. Equation (e) gives

$$u_2 = -[d^2 - (b + c)d + bc] = -(d - c)(d - b) \tag{m}$$

Since $d > c > b$, u_2 is < 0. Actually the term within the square brackets is also the slope of the function at $x = d$, which is positive, so $u_2 < 0$ in Eq. (m). The KKT necessary condition is violated, so there is no solution for this case; that is, $x = d$ is not a candidate minimum point. This is true, as can be observed for point D in Figure 4.16. Actually, it can be checked that the point satisfies the KKT necessary conditions for the local maximum point.

Case 3: $s_1 = 0$, $u_2 = 0$. $s_1 = 0$ implies that g_1 is active and therefore $x = a$. Equation (c) gives

$$u_1 = a^2 - (b + c)a + bc = (a - b)(a - c) > 0 \tag{n}$$

Also, since $u_1 = $ the slope of the function at $x = a$, it is positive and all of the KKT conditions are satisfied. Thus, $x = a$ is a candidate minimum point. Actually $x = a$ is a local minimum point because a feasible move from the point increases the cost function. This is a sufficient condition, which we will discuss in Chapter 5.

Case 4: $s_1 = 0$, $s_2 = 0$. This case, for which both constraints are active, does not give any valid solution since x cannot be simultaneously equal to a and d.

EXAMPLE 4.31 SOLUTION TO THE KKT NECESSARY CONDITIONS

Solve the KKT condition for the problem:

Minimize

$$f(\mathbf{x}) = x_1^2 + x_2^2 - 3x_1 x_2 \tag{a}$$

subject to

$$g = x_1^2 + x_2^2 - 6 \leq 0 \tag{b}$$

Solution

The feasible region for the problem is a circle with its center at $(0,0)$ and its radius as $\sqrt{6}$. This is plotted in Figure 4.17. Several cost function contours are shown there. It can be seen that points A and B give minimum value for the cost function. The gradients of cost and constraint functions at these points are along the same line but in opposite directions, so the KKT necessary conditions are satisfied. We will verify this by writing these conditions and solving them for the minimum points.

The Lagrange function of Eq. (4.46) for the problem, which is already written in the standard form of Eqs. (4.35) and (4.37), is

$$L = x_1^2 + x_2^2 - 3x_1x_2 + u(x_1^2 + x_2^2 - 6 + s^2) \tag{c}$$

Since there is only one constraint for the problem, all points of the feasible region are *regular*, so the KKT necessary conditions are applicable. They are given as

$$\frac{\partial L}{\partial x_1} = 2x_1 - 3x_2 + 2ux_1 = 0 \tag{d}$$

$$\frac{\partial L}{\partial x_2} = 2x_2 - 3x_1 + 2ux_2 = 0 \tag{e}$$

$$x_1^2 + x_2^2 - 6 + s^2 = 0, \; s^2 \geq 0, \; u \geq 0 \tag{f}$$

$$us = 0 \tag{g}$$

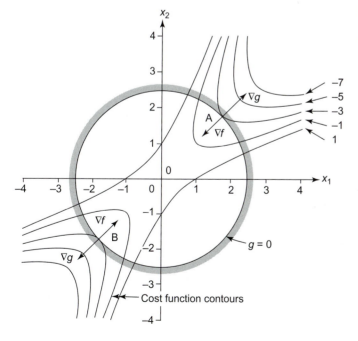

FIGURE 4.17 Graphical solution for Example 4.31. Local minimum points, A and B (the vectors are not to scale).

Equations (d) through (g) are the four equations for four unknowns, x_1, x_2, s, and u. Thus, in principle, we have enough equations to solve for all of the unknowns. The system of equations is nonlinear; however, it is possible to analytically solve for all of the roots.

There are three possible ways of satisfying the switching condition of Eq. (g): (i) $u = 0$, (ii) $s = 0$ (implying that g is active), or (iii) $u = 0$ and $s = 0$. We will consider each case separately and solve for the roots of the necessary conditions.

Case 1: u = 0. In this case, the inequality constraint is considered as inactive at the solution point. We will solve for x_1 and x_2 and then check the constraint. Equations (d) and (e) reduce to

$$2x_1 - 3x_2 = 0; \quad -3x_1 + 2x_2 = 0 \tag{h}$$

This is a 2×2 homogeneous system of linear equations (the right side is zero). Such a system has a nontrivial solution only if the determinant of the coefficient matrix is zero. However, since the determinant of the matrix is -5, the system has only a trivial solution, $x_1 = x_2 = 0$. This solution gives $s^2 = 6$ from Eq. (f), so the inequality is not active. Thus, the candidate minimum point for this case is

$$x_1^* = 0, \quad x_2^* = 0, \quad u^* = 0, \quad f(0,0) = 0 \tag{i}$$

Case 2: s = 0. In this case, $s = 0$ implies that the inequality is active. We must solve Eqs. (d) through (f) simultaneously for x_1, x_2, and u. Note that this is a nonlinear set of equations, so there can be multiple roots. Equation (d) gives $u = -1 + 3x_2/2x_1$. Substituting for u in Eq. (e), we obtain $x_1^2 = x_2^2$. Using this in Eq. (f), solving for x_1 and x_2 and then solving for u, we obtain the four roots of Eqs. (d), (e), and (f) as

$$x_1 = x_2 = \sqrt{3}, \quad u = \frac{1}{2}; \quad x_1 = x_2 = -\sqrt{3}, \quad u = \frac{1}{2} \tag{j}$$

$$x_1 = -x_2 = \sqrt{3}, \quad u = -\frac{5}{2}; \quad x_1 = -x_2 = -\sqrt{3}, \quad u = -\frac{5}{2} \tag{k}$$

The last two roots violate the KKT necessary condition, $u \geq 0$. Therefore, there are two candidate minimum points for this case. The first corresponds to point A and the second to point B in Figure 4.17.

Case 3: u = 0, s = 0. With these conditions, Eqs. (d) and (e) give $x_1 = 0$, $x_2 = 0$. Substituting these into Eq. (f), we obtain $s^2 = 6 \neq 0$. Therefore, since $s \neq 0$, this case does not give any solution for the KKT conditions.

The case where both u and s are zero usually does not give any additional KKT points and may be ignored. Finally, the points satisfying the KKT necessary conditions for the problem are summarized:

1. $x_1^* = 0$, $x_2^* = 0$, $u^* = 0$, $f^*(0, 0) = 0$, point O in Figure 4.17
2. $x_1^* = x_2^* = \sqrt{3}$, $u^* = \frac{1}{2}$, $f^*(\sqrt{3}, \sqrt{3}) = -3$, point A in Figure 4.17
3. $x_1^* = x_2^* = -\sqrt{3}$, $u^* = \frac{1}{2}$, $f^*(-\sqrt{3}, -\sqrt{3}) = -3$, point B in Figure 4.17

It is interesting to note that points A and B satisfy the sufficient condition for local minima. As can be observed from Figure 4.17, any feasible move from the points results in an increase in the cost, and any further reduction in the cost results in violation of the constraint. It can also be observed that point O does not satisfy the sufficient condition because there are feasible directions that result in a decrease in the cost function. So point O is only a stationary point. We will check the sufficient conditions for this problem in Chapter 5.

Solution to optimality conditions using MATLAB MATLAB, introduced in Chapter 3 for graphical optimization, has many capabilities for engineering calculations and analyses. It can also be used to solve a set of nonlinear equations as well. The primary command used for this purpose is fsolve. This command is part of the MATLAB Optimization Toolbox (Chapter 7), which must be installed in the computer. We will discuss use of this capability by solving the KKT conditions for the problem in Example 4.31. Use of the Optimization Toolbox to solve design optimization problems is explained in Chapter 7.

When using MATLAB, it is necessary to first create a separate m-file containing the equations in the form $\mathbf{F}(\mathbf{x}) = \mathbf{0}$. For the present example, components of the vector $\mathbf{x}$ are defined as $x(1) = x_1$, $x(2) = x_2$, $x(3) = u$, and $x(4) - s$. In terms of these variables, the KKT conditions of Eqs. (d) through (g) are given as

$$2^*x(1) - 3^*x(2) + 2^*x(3)^*x(1) = 0 \tag{l}$$

$$2^*x(2) - 3^*x(1) + 2^*x(3)^*x(2) = 0 \tag{m}$$

$$x(1)^2 + x(2)^2 - 6 + x(4)^2 = 0 \tag{n}$$

$$x(3)^*x(4) = 0 \tag{o}$$

The file defining the equations is prepared as follows:

```
Function F=kktsystem(x)
F=[2*x(1) - 3*x(2)+2*x(3)*x(1);
2*x(2) - 3*x(1)+2*x(3)*x(2);
x(1)^2+x(2)^2 - 6+x(4)^2;
x(3)*x(4)];
```

The first line defines a function, kktsystem, that accepts a vector of variable x and returns a vector of function values F. This file should be named kktsystem (the same name as the function itself), and as with other MATLAB files, it should be saved with a suffix of ".m". Next the main commands are entered interactively or in a separate file as follows:

```
x0=[1;1;1;1];
options=optimset('Display','iter')
x=fsolve(@kktsystem,x0,options)
```

x0 is the starting point or the initial guess for the root of the nonlinear equations. The options command displays output for each iteration. If the command options=optimset ('Display','off'") is used, then only the final solution is provided. The command fsolve finds a root of the system of equations provided in the function kktsystem. Although there may be many potential solutions, the solution closest to the initial guess is obtained and provided. Consequently, different starting points must be used to find different points that satisfy the KKT conditions. Starting with the given point, the solution is obtained as (1.732, 1.732, 0.5, 0).

The foregoing two examples illustrate the procedure of solving the KKT necessary conditions for candidate local minimum points. It is important to understand the procedure clearly. Example 4.31 had only one inequality constraint. The switching condition of Eq. (g) gave only two normal cases—either $u = 0$ or $s = 0$ (the abnormal case, where $u = 0$ and $s = 0$, rarely gives additional candidate points, so it can be ignored).

I. THE BASIC CONCEPTS

Each of the cases gave the candidate minimum point x^*. For Case 1 ($u = 0$), there was only one point x^* satisfying Eqs. (d), (e), and (f). However, for Case 2 ($s = 0$), there were four roots for Eqs. (d), (e), and (f). Two of the four roots did not satisfy the non-negativity conditions on the Lagrange multipliers. Therefore, the corresponding two roots were not candidate local minimum points.

The preceding procedure is valid for more general nonlinear optimization problems.

In Example 4.32, we illustrate the procedure for a problem with two design variables and two inequality constraints.

EXAMPLE 4.32 SOLUTION TO THE KKT NECESSARY CONDITIONS

Maximize

$$F(x_1, x_2) = 2x_1 + x_2 - x_1^2 - x_2^2 - 2 \tag{a}$$

subject to

$$2x_1 + x_2 \geq 4, \; x_1 + 2x_2 \geq 4 \tag{b}$$

Solution

First we write the problem in the standard form of Eqs. (4.35) and (4.37) as

Minimize

$$f(x_1, x_2) = x_1^2 + x_2^2 - 2x_1 - 2x_2 + 2 \tag{c}$$

subject to

$$g_1 = -2x_1 - x_2 + 4 \leq 0, \; g_2 = -x_1 - 2x_2 + 4 \leq 0 \tag{d}$$

Figure 4.18 is a graphical representation of the problem. The two constraint functions are plotted and the feasible region is identified. It is seen that point A(4/3, 4/3), where both of the inequality constraints are active, is the optimum solution to the problem. Since it is a two-variable problem, only two vectors can be linearly independent. It can be seen in Figure 4.18 that the constraint gradients ∇g_1 and ∇g_2 are linearly independent (hence, the *optimum point is regular*), so any other vector can be expressed as a linear combination of them.

In particular, $-\nabla f$ (the negative gradient of the cost function) can be expressed as linear combination of ∇g_1 and ∇g_2, with positive scalars as the multipliers of the linear combination, which is precisely the KKT necessary condition of Eq. (4.47). In the following, we will write these conditions and solve them to verify the graphical solution.

The Lagrange function of Eq. (4.46) for the problem defined in Eqs. (a) and (b) is given as

$$L = x_1^2 + x_2^2 - 2x_1 - 2x_2 + 2 + u_1(-2x_1 - x_2 + 4 + s_1^2) + u_2(-x_1 - 2x_2 + 4 + s_2^2) \tag{e}$$

The KKT necessary conditions are

$$\frac{\partial L}{\partial x_1} = 2x_1 - 2 - 2u_1 - u_2 = 0 \tag{f}$$

$$\frac{\partial L}{\partial x_2} = 2x_2 - 2 - u_1 - 2u_2 = 0 \tag{g}$$

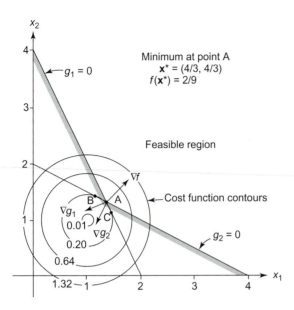

$$g_1 = -2x_1 - x_2 + 4 + s_1^2 = 0; \quad s_1^2 \geq 0, \ u_1 \geq 0 \tag{h}$$

$$g_2 = -x_1 - 2x_2 + 4 + s_2^2 = 0; \quad s_2^2 \geq 0, \ u_2 \geq 0 \tag{i}$$

$$u_i s_i = 0; \quad i = 1, 2 \tag{j}$$

Equations (f) through (j) are the six equations for six unknowns: x_1, x_2, s_1, s_2, u_1, and u_2. We must solve them simultaneously for candidate local minimum points. One way to satisfy the switching conditions of Eq. (j) is to identify various cases and then solve them for the roots. There are four cases, and we consider each separately and solve for all of the unknowns:

1. $u_1 = 0, u_2 = 0$
2. $u_1 = 0, s_2 = 0 \ (g_2 = 0)$
3. $s_1 = 0 \ (g_1 = 0), u_2 = 0$
4. $s_1 = 0 \ (g_1 = 0), s_2 = 0 \ (g_2 = 0)$

Case 1: $u_1 = 0, u_2 = 0$. Equations (f) and (g) give $x_1 = x_2 = 1$. This is not a valid solution as it gives $s_1^2 = -1 \ (g_1 = 1)$, $s_2^2 = -1 \ (g_2 = 1)$ from Eqs. (h) and (i), which implies that both inequalities are violated. Thus, the design $x_1 = 1$ and $x_2 = 1$ is not feasible.

Case 2: $u_1 = 0, s_2 = 0$. With these conditions, Eqs. (f), (g), and (i) become

$$2x_1 - 2 - u_2 = 0, \ 2x_2 - 2 - 2u_2 = 0, \ -x_1 - 2x_2 + 4 = 0 \tag{k}$$

These are three linear equations for the three unknowns x_1, x_2, and u_2. Any method of solving a linear system of equations, such as Gaussian elimination or method of determinants (Cramer's rule), can be used to find roots. Using the elimination procedure, we obtain $x_1 = 1.2$, $x_2 = 1.4$, and $u_2 = 0.4$. Therefore, the solution for this case is

$$x_1 = 1.2, \ x_2 = 1.4; \ u_1 = 0, \ u_2 = 0.4; \ f = 0.2 \tag{l}$$

We need to check for the feasibility of this design point with respect to constraint g_1 before it can be claimed as a candidate local minimum point. Substituting $x_1 = 1.2$ and $x_2 = 1.4$ into Eq. (h), we find that $s_1^2 = -0.2 < 0$ ($g_1 = 0.2$), which is a violation of constraint g_1. Therefore, Case 2 also does not give any candidate local minimum point. It can be seen in Figure 4.18 that point (1.2,1.4) corresponds to point B, which is not in the feasible set.

Case 3: $s_1 = 0$, $u_2 = 0$. With these conditions, Eqs. (f), (g), and (h) give

$$2x_1 - 2 - 2u_1 = 0; \quad 2x_2 - 2 - u_1 = 0; \quad -2x_1 - x_2 + 4 = 0 \tag{m}$$

This is again a linear system of equations for the variables x_1, x_2, and u_1. Solving the system, we obtain the solution as

$$x_1 = 1.4, \quad x_2 = 1.2; \quad u_1 = 0.4, \quad u_2 = 0; \quad f = 0.2 \tag{n}$$

Checking this design for feasibility with respect to constraint g_2, we find from Eq. (i) that $s_2^2 = -0.2 < 0$ ($g_2 = 0.2$). This is not a feasible design. Therefore, Case 3 also does not give any candidate local minimum point. It can be observed in Figure 4.18 that point (1.4, 1.2) corresponds to point C, which is not in the feasible set.

Case 4: $s_1 = 0$, $s_2 = 0$. For this case, Eqs. (f) through (i) must be solved for the four unknowns x_1, x_2, u_1, and u_2. This system of equations is again linear and can be solved easily. Using the elimination procedure as before, we obtain x_1 and x_2 from Eqs. (h) and (i), and u_1 and u_2 from Eqs. (f) and (g):

$$x_1 = 4/3, \quad x_2 = 4/3, \quad u_1 = 2/9 > 0, \quad u_2 = 2/9 > 0 \tag{o}$$

To check regularity condition for the point, we evaluate the gradients of the active constraints and define the constraint gradient matrix $\mathbf{A}$ as

$$\nabla g_1 = \begin{bmatrix} -2 \\ -1 \end{bmatrix}, \quad \nabla g_2 = \begin{bmatrix} -1 \\ -2 \end{bmatrix}, \quad \mathbf{A} = \begin{bmatrix} -2 & -1 \\ -1 & -2 \end{bmatrix} \tag{p}$$

Since rank ($\mathbf{A}$) = the number of active constraints, the gradients ∇g_1 and ∇g_2 are linearly independent. Thus, all of the KKT conditions are satisfied and the preceding solution is a candidate local minimum point. The solution corresponds to point A in Figure 4.18. The cost function at the point has a value of $2/9$.

It can be observed from Figure 4.18 that the vector $-\nabla f$ can be expressed as a linear combination of the vectors ∇g_1 and ∇g_2 at point A. This satisfies the necessary condition of Eq. (4.53). It is also seen from the figure that point A is indeed a local minimum because any further reduction in the cost function is possible only if we go into the infeasible region. Any feasible move from point A results in an increase in the cost function.

4.6.3 Summary of the KKT Solution Approach

Note the following points regarding the KKT first-order necessary conditions:

1. *The conditions can be used to check* whether a given point is a candidate minimum; the point must be feasible; the gradient of the Lagrangian with respect to the design variables must be zero; and the Lagrange multipliers for the inequality constraints must be non-negative.

2. For a given problem, *the conditions can be used to find* candidate minimum points. Several cases defined by the switching conditions must be considered and solved. Each case can provide multiple solutions.

3. For each solution case, *remember* to

    (i) Check all inequality constraints for feasibility (e.g., $g_i \leq 0$ or $s_i^2 \geq 0$).

    (ii) Calculate all of the Lagrange multipliers.

    (iii) Ensure that the Lagrange multipliers for all of the inequality constraints are non-negative.

Limitation of the KKT Solution Approach

Note that addition of an inequality to the problem formulation doubles the number of KKT solution cases. With 2 inequalities, we had 4 KKT cases; with 3 inequalities, we have 8 cases; and with 4 inequalities, we have 16 cases. Therefore, the number of cases quickly becomes out of hand, and thus this solution procedure cannot be used to solve most practical problems. Based on these conditions, however, numerical methods have been developed that can handle any number of equality and inequality constraints. In Section 4.9, we will solve two problems having 16 and 32 cases, respectively.

4.7 POSTOPTIMALITY ANALYSIS: THE PHYSICAL MEANING OF LAGRANGE MULTIPLIERS

The study of variations in the optimum solution when some of the original problem parameters are changed is known as *postoptimality* or *sensitivity* analysis. This is an important topic for optimum design of engineering systems, as we can gain some insights regarding the optimum solution. Variations in the optimum cost function and design variables due to changes in certain problem parameters can be studied. Since sensitivity of the cost function to the variations in the constraint limit values can be studied without any further calculations, we will focus on this aspect of sensitivity analysis only.

4.7.1 Effect of Changing Constraint Limits

We will assume that the minimization problem has been solved with $h_i(\mathbf{x}) = 0$ and $g_j(\mathbf{x}) \leq 0$, that is, with the current limit values for the constraints as zero. We want to know what happens to the optimum cost function value if more resources become available (a constraint is relaxed) or if resources are reduced (a constraint needs to be tightened).

It turns out that the Lagrange multipliers $(\mathbf{v}^*, \mathbf{u}^*)$ at the optimum design provide information to answer the foregoing sensitivity question. The investigation of this question also leads to a physical interpretation of the Lagrange multipliers that can be very useful in practical applications. The interpretation will also show why the Lagrange multipliers for the "$\leq$ type" constraints have to be non-negative.

To discuss changes in the cost function due to changes in the constraint limits, we consider the modified problem:

Minimize

$$f(\mathbf{x}) \tag{4.54}$$

subject to

$$h_i(\mathbf{x}) = b_i; \quad i = 1 \text{ to } p \tag{4.55}$$

$$g_j(\mathbf{x}) \leq e_j; \quad j = 1 \text{ to } m \tag{4.56}$$

where b_i and e_j are small variations in the neighborhood of zero. It is clear that the optimum point for the perturbed problem depends on vectors $\mathbf{b}$ and $\mathbf{e}$; that is, it is a function of $\mathbf{b}$ and $\mathbf{e}$ that can be written as $\mathbf{x}^* = \mathbf{x}^*(\mathbf{b},\mathbf{e})$. Also, optimum cost function value depends on $\mathbf{b}$ and $\mathbf{e}$: $f^* = f^*(\mathbf{b},\mathbf{e})$. However, explicit dependence of the cost function on $\mathbf{b}$ and $\mathbf{e}$ is not known; an expression for f^* in terms of b_i and e_j cannot be obtained. The following theorem gives a way of obtaining the partial derivatives $\partial f^*/\partial b_i$ and $\partial f^*/\partial e_j$.

THEOREM 4.7

Constraint Variation Sensitivity Theorem Let $f(\mathbf{x})$, $h_i(\mathbf{x})$, $i = 1$ to p, and $g_j(\mathbf{x})$, $j = 1$ to m have two continuous derivatives. Let $\mathbf{x}^*$ be a regular point that, together with the multipliers v_i^* and u_j^*, satisfies both the KKT necessary conditions and the sufficient conditions (presented in the next chapter) for an isolated local minimum point for the problem defined in Eqs. (4.35) through (4.37).

If, for each $g_j(\mathbf{x}^*)$, it is true that $u_j^* > 0$, then the solution $\mathbf{x}^*(\mathbf{b},\mathbf{e})$ of the modified

optimization problem defined in Eqs. (4.54) through (4.56) is a continuously differentiable function of $\mathbf{b}$ and $\mathbf{e}$ in some neighborhood of $\mathbf{b} = \mathbf{0}$, $\mathbf{e} = \mathbf{0}$. Furthermore,

$$\frac{\partial f^*}{\partial b_i} = \frac{\partial f(\mathbf{x}^*(0,0))}{\partial b_i} = -v_i^*; \quad i = 1 \text{ to } p \tag{4.57}$$

$$\frac{\partial f^*}{\partial e_j} = \frac{\partial f(\mathbf{x}^*(0,0))}{\partial e_j} = -u_j^*; \quad j = 1 \text{ to } m \tag{4.58}$$

First-Order Changes in the Cost Function

The theorem gives values for implicit first-order derivatives of the cost function f^* with respect to the right side parameters of the constraints b_i and e_j. The derivatives can be used to calculate changes in the cost function as b_i and e_j are changed. Note that the theorem is applicable only when the inequality constraints are written in the "$\leq$" form. Using the theorem we can estimate changes in the cost function if we decide to adjust the right side of the constraints in the neighborhood of zero. For this purpose, Taylor's expansion for the cost function in terms of b_i and e_j can be used. Let us assume that we want to vary the right sides, b_i and e_j, of the ith equality and jth inequality constraints. First-order Taylor's expansion for the cost function $f(b_i, e_j)$ about the point $b_i = 0$ and $e_j = 0$ is given as

$$f(b_i, e_j) = f^*(0, 0) + \frac{\partial f^*(0, 0)}{\partial b_i} b_i + \frac{\partial f^*(0, 0)}{\partial e_j} e_j \tag{4.59}$$

Or, substituting from Eqs. (4.57) and (4.58), we obtain

$$f(b_i, e_j) = f^*(0,0) - v_i^* b_i - u_j^* e_j \tag{4.60}$$

where $f(0, 0)$ is the optimum cost function value that is obtained with $b_i = 0$ and $e_j = 0$. Using Eq. (4.60), a first-order change in the cost function δf due to small changes in b_i and e_j is given as

$$\delta f^* = f(b_i, e_j) - f^*(0, 0) = -v_i^* b_i - u_j^* e_j \tag{4.61}$$

For given values of b_i and e_j we can estimate the new value of the cost function from Eq. (4.60). Also, Eqs. (4.60) and (4.61) can be used for physical interpretation of the Lagrange multipliers. It is seen that the multipliers give the *benefit of relaxing a constraint* or *the penalty associated with tightening it;* relaxation enlarges the feasible set, while tightening contracts it.

If we want to change the right side of more constraints, we simply include them in Eq. (4.61) and obtain the change in cost function as

$$\delta f^* = -\sum v_i^* b_i - \sum u_j^* e_j \tag{4.62}$$

It is useful to note that if the conditions of Theorem 4.7 are not satisfied, the existence of implicit derivatives of Eqs. (4.57) and (4.58) is not ruled out by the theorem. That is, the derivatives may still exist but their existence cannot be guaranteed by Theorem 4.7.

This observation will be verified in an example problem in Section 4.9.2.

Non-negativity of Lagrange Multipliers

Equation (4.61) can also be used to show that the *Lagrange multiplier corresponding to a "≤ type" constraint must be non-negative.* To see this, let us assume that we want to relax an inequality constraint $g_j \leq 0$ that is active ($g_j = 0$) at the optimum point; that is, we select $e_j > 0$ in Eq. (4.56). When a constraint is relaxed, the feasible set for the design problem expands. We allow more feasible designs to be candidate minimum points. Therefore, with the expanded feasible set we expect the optimum cost function to reduce further or at the most remain unchanged (Example 4.33). We observe from Eq. (4.61) that if $u_j^* < 0$, then relaxation of the constraint ($e_j > 0$) results in an increase in cost (that is, $\delta f^* = -u_j^* e_j > 0$). This is a contradiction, as it implies that there is a penalty for relaxing the constraint. Therefore, the *Lagrange multiplier for a "≤ type" constraint must be non-negative.*

EXAMPLE 4.33 EFFECT OF VARIATIONS OF CONSTRAINT LIMITS ON THE OPTIMUM COST FUNCTION

To illustrate the use of constraint variation sensitivity theorem, we consider the following problem solved as Example 4.31 and discuss the effect of changing the limit for the constraint:

Minimize

$$f(x_1, x_2) = x_1^2 + x_2^2 - 3x_1 x_2 \tag{a}$$

subject to

$$g(x_1, x_2) = x_1^2 + x_2^2 - 6 \leq 0. \tag{b}$$

Solution

The graphical solution to the problem is given in Figure 4.17. A point satisfying both necessary and sufficient conditions is

$$x_1^* = x_2^* = \sqrt{3}, \ u^* = \frac{1}{2}, \ f(\mathbf{x}^*) = -3 \tag{c}$$

We want to see what happens if we change the right side of the constraint equation to a value "e" from zero. Note that the constraint $g(x_1, x_2) \leq 0$ gives a circular feasible region with its center at (0,0) and its radius as $\sqrt{6}$, as shown earlier in Figure 4.17. Therefore, changing the right side of the constraint changes the radius of the circle.

From Theorem 4.7, we have

$$\frac{\partial f(\mathbf{x}^*)}{\partial e} = -u^* = -\frac{1}{2} \tag{d}$$

If we set $e = 1$, the new value of the cost function will be approximately $-3 + (-1/2)(1) = -3.5$ using Eq. (4.60). This is consistent with the new feasible set because with $e = 1$, the radius of the circle becomes $\sqrt{7}$ and the feasible region is expanded (as can be seen in Figure 4.17). We should expect some reduction in the cost function.

If we set $e = -1$, then the effect is the opposite. The feasible set becomes smaller and the cost function increases to -2.5 using Eq. (4.60).

Practical Use of Lagrange Multipliers

From the foregoing discussion and example, we see that *optimum Lagrange multipliers give very useful information about the problem.* The designer can compare the magnitude of the multipliers for the active constraints. The multipliers with relatively larger values will have a significant effect on the optimum cost if the corresponding constraints are changed. *The larger the value of the Lagrange multiplier, the larger the dividend to relax the constraint, or the larger the penalty to tighten the constraint.* Knowing this, the designer can select a few critical constraints having the greatest influence on the cost function, and then analyze to see if they can be relaxed to further reduce the optimum cost function value.

4.7.2 Effect of Cost Function Scaling on Lagrange Multipliers

On many occasions, a cost function for the problem is multiplied by a positive constant. As noted in Section 4.3, any scaling of the cost function does not alter the optimum point. It does, however, change the optimum value for the cost function. The scaling should also influence the implicit derivatives of Eqs. (4.57) and (4.58) for the cost function with respect to the right side parameters of the constraints. *We observe from these equations that all the Lagrange multipliers are also multiplied by the same constant.*

Let u_j^* and v_i^* be the Lagrange multipliers for the inequality and equality constraints, respectively, and $f(\mathbf{x}^*)$ be the optimum value of the cost function at the solution point $\mathbf{x}^*$. Let the cost function be scaled as $\bar{f}(\mathbf{x}) = Kf(\mathbf{x})$, where $K > 0$ is a given constant, and $\bar{u}_j^*$ and $\bar{v}_i^*$ are the optimum Lagrange multipliers for the inequality and equality constraints, respectively, for the changed problem. Then the optimum design variable vector for the perturbed problem is $\mathbf{x}^*$ and the relationship between optimum Lagrange multipliers is derived using the KKT necessary conditions for the original and the changed problems as follows:

$$\bar{u}_j^* = Ku_j^* \quad \text{and} \quad \bar{v}_i^* = Kv_i^* \tag{4.63}$$

Thus all the Lagrange multipliers are scaled by the factor K. Example 4.34 shows the effect of scaling the cost function on the Lagrange multipliers.

EXAMPLE 4.34 EFFECT OF SCALING THE COST FUNCTION ON THE LAGRANGE MULTIPLIERS

Consider Example 4.31 written in the standard form:

Minimize

$$f(\mathbf{x}) = x_1^2 + x_2^2 - 3x_1x_2 \tag{a}$$

subject to

$$g(\mathbf{x}) = x_1^2 + x_2^2 - 6 \leq 0 \tag{b}$$

Study the effect on the optimum solution of scaling the cost function by a constant $K > 0$.

Solution

A graphical solution to the problem is given in Figure 4.17. A point satisfying both the necessary and sufficient conditions is

$$x_1^* = x_2^* = \sqrt{3}, \ u^* = \frac{1}{2}, \ f(\mathbf{x}^*) = -3 \tag{c}$$

$$x_1^* = x_2^* = -\sqrt{3}, \ u^* = \frac{1}{2}, \ f(\mathbf{x}^*) = -3 \tag{d}$$

Let us solve the scaled problem by writing KKT conditions. The Lagrangian for the problem is given as (quantities with an over bar are for the perturbed problem):

$$L = K(x_1^2 + x_2^2 - 3x_1x_2) + \bar{u}(x_1^2 + x_2^2 - 6 + \bar{s}^2) \tag{e}$$

The necessary conditions give

$$\frac{\partial L}{\partial x_1} = 2Kx_1 - 3Kx_2 + 2\bar{u}x_1 = 0 \tag{f}$$

$$\frac{\partial L}{\partial x_2} = 2Kx_2 - 3Kx_1 + 2\bar{u}x_2 = 0 \tag{g}$$

$$x_1^2 + x_2^2 - 6 + \bar{s}^2 = 0; \ \bar{s}^2 \geq 0 \tag{h}$$

$$\bar{u}\bar{s} = 0, \quad \bar{u} \geq 0 \tag{i}$$

As in Example 4.31, the case where $\bar{s} = 0$ gives candidate minimum points. Solving Eqs. (f) through (h), we get the two KKT points as

$$x_1^* = x_2^* = \sqrt{3}, \ \bar{u}^* = K/2, \ \bar{f}(\mathbf{x}^*) = -3K \tag{j}$$

$$x_1^* = x_2^* = -\sqrt{3}, \ \bar{u}^* = K/2, \ \bar{f}(\mathbf{x}^*) = -3K \tag{k}$$

Therefore, comparing the solutions with those obtained in Example 4.31, we observe that $\bar{u}^* = Ku^*$.

4.7.3 Effect of Scaling a Constraint on Its Lagrange Multiplier

Many times, a constraint is scaled by a positive constant. We want to know the effect of this scaling on the Lagrange multiplier for the constraint.

It should be noted that the scaling of a constraint does not change the constraint boundary, so it has no effect on the optimum solution. Only the Lagrange multiplier for the scaled constraint is affected. Looking at the implicit derivatives of the cost function with respect to the constraint's right-side parameters, we observe that the Lagrange multiplier for the scaled constraint is divided by the scaling parameter.

Let $M_j > 0$ and P_i be the scale parameters for the jth inequality and ith equality constraints $(\bar{g}_j = M_j g_j;\ \bar{h}_i = P_i h_i)$, with u_j^* and and v_i^* and $\bar{u}_j^*$ and $\bar{v}_j^*$ as the corresponding Lagrange multipliers for the original and the scaled constraints, respectively. Then the following relations hold for the Lagrange multipliers:

$$\bar{u}_j^* = u_j^*/M_j \quad \text{and} \quad \bar{v}_i^* = v_i^*/P_i \tag{4.64}$$

Example 4.35 illustrates the effect of scaling a constraint on its Lagrange multiplier.

EXAMPLE 4.35 EFFECT OF SCALING A CONSTRAINT ON ITS LAGRANGE MULTIPLIER

Consider Example 4.31 and study the effect of multiplying the inequality by a constant $M > 0$.

Solution

The Lagrange function for the problem with a scaled constraint is given as

$$L = x_1^2 + x_2^2 - 3x_1x_2 + \bar{u}[M(x_1^2 + x_2^2 - 6) + \bar{s}^2] \tag{a}$$

The KKT conditions give

$$\frac{\partial L}{\partial x_1} = 2x_1 - 3x_2 + 2\bar{u}Mx_1 = 0 \tag{b}$$

$$\frac{\partial L}{\partial x_2} = 2x_2 - 3x_1 + 2\bar{u}Mx_2 = 0 \tag{c}$$

$$M(x_1^2 + x_2^2 - 6) + \bar{s}^2 = 0; \quad \bar{s}^2 \ge 0 \tag{d}$$

$$\bar{u}\bar{s} = 0, \quad \bar{u} \ge 0 \tag{e}$$

As in Example 4.31, only the case with $\bar{s} = 0$ gives candidate optimum points. Solving this case, we get the two KKT points:

$$x_1^* = x_2^* = \sqrt{3}, \ \bar{u}^* = \frac{1}{2M}, \ f(\mathbf{x}^*) = -3 \tag{f}$$

$$x_1^* = x_2^* = -\sqrt{3}, \ \bar{u}^* = \frac{1}{2M}, \ f(\mathbf{x}^*) = -3 \tag{g}$$

Therefore, comparing these solutions with the ones for Example 4.31, we observe that $\bar{u}^* = \bar{u}^*/M$.

4.7.4 Generalization of the Constraint Variation Sensitivity Result

Many times variations are desired with respect to parameters that are embedded in the constraint expression in a complex way. Therefore, the sensitivity expressions given in Eqs. (4.57) and (4.58) cannot be used directly and need to be generalized. We will pursue these generalizations for the inequality constraints only in the following paragraphs; equality constraints can be treated in similar ways. It turns out that the sensitivity of the optimum cost function with respect to an inequality constraint can be written as

$$\frac{\partial f(\mathbf{x}^*)}{\partial g_j} = u_j^*, \quad j = 1 \text{ to } m \tag{4.65}$$

If the constraint function depends on a parameter s as $g_j(s)$, then variations with respect to the parameter s can be written using the chain rule of differentiation as

$$\frac{df(\mathbf{x}^*)}{ds} = \frac{\partial f(\mathbf{x}^*)}{\partial g_j}\frac{dg_j}{ds} = u_j^*\frac{dg_j}{ds} \tag{4.66}$$

Therefore, a change in the cost function due to a small change δs in the parameter s is given as

$$\delta f^* = \frac{df}{ds}\delta s = u_j^*\frac{dg_j}{ds}\delta s \tag{4.67}$$

Another way of writing this small change to the cost function is to express it in terms of changes to the constraint function itself, using Eq. (4.65) as

$$\delta f^* = \frac{\partial f}{\partial g_j}\delta g_j = u_j^*\delta g_j \tag{4.68}$$

Sometimes the right side e_j is dependent on a parameter s. In that case the sensitivity of the cost function f with respect to s (the derivative of f with respect to s) can be obtained directly from Eq. (4.57) using the chain rule of differentiation as

$$\frac{df(\mathbf{x}^*)}{ds} = \frac{\partial f(\mathbf{x}^*)}{\partial e_j}\frac{de_j}{ds} = -u_j^*\frac{de_j}{ds} \tag{4.69}$$

4.8 GLOBAL OPTIMALITY

In the optimum design of systems, the question of the global optimality of a solution always arises. In general, it is difficult to answer the question satisfactorily. However, an answer can be attempted in the following two ways:

1. If the cost function $f(\mathbf{x})$ is continuous on a closed and bounded feasible set, then the Weierstrauss Theorem 4.1 guarantees the existence of a global minimum. Therefore, if we calculate all the local minimum points, the point that gives the least value to the cost function can be selected as a global minimum for the function. This is called *exhaustive search*.
2. If the optimization problem can be shown to be convex, then any local minimum is also a global minimum. Also the KKT necessary conditions are sufficient for the minimum point.

Both of these procedures can involve substantial computations. Methods based on the first procedure are described in Chapter 18. In this section we pursue the second approach and discuss topics of *convexity* and *convex programming problems*. Such problems are defined in terms of *convex sets* and *convex functions*, specifically, convexity of the feasible set and the cost function. Therefore, we introduce these concepts and discuss results regarding global optimum solutions.

4.8.1 Convex Sets

A convex set S is a collection of points (vectors $\mathbf{x}$) having the following property: If P_1 and P_2 are any points in S, then the entire line segment P_1-P_2 is also in S. This is a necessary and sufficient condition for convexity of the set S. Figure 4.19 shows some examples of convex and nonconvex sets.

To explain convex sets further, let us consider points on a real line along the *x*-axis (Figure 4.20). Points in any interval on the line represent a convex set. Consider an interval between points a and b as shown in Figure 4.20. To show that it is a convex set, let x_1 and x_2 be two points in the interval.

The line segment between the points can be written as

$$x = \alpha x_2 + (1 - \alpha)x_1; \quad 0 \le \alpha \le 1 \tag{4.70}$$

FIGURE 4.19   (a) Convex sets. (b) Nonconvex sets.

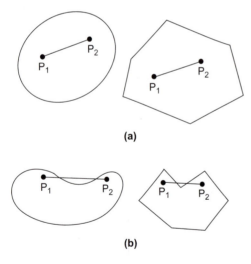

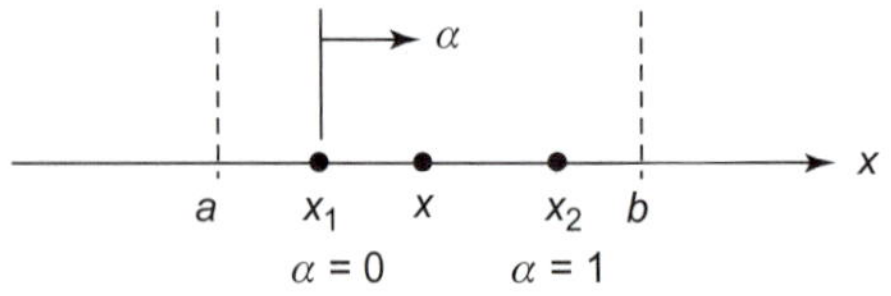

(a)

(b)

FIGURE 4.20   Convex interval between a and b on a real line.

In this equation, if $\alpha = 0$, $x = x_1$; if $\alpha = 1$, $x = x_2$. It is clear that the line defined in Eq. (4.70) is in the interval $[a, b]$. The entire line segment is on the line between a and b. Therefore, the set of points between a and b is a convex set.

In general, for the n-dimensional space, the *line segment* between any two points $\mathbf{x}^{(1)}$ and $\mathbf{x}^{(2)}$ is written as

$$\mathbf{x} = \alpha \mathbf{x}^{(2)} + (1 - \alpha)\mathbf{x}^{(1)}; \quad 0 \le \alpha \le 1 \tag{4.71}$$

Equation (4.71) is a generalization of Eq. (4.70) and is called the *parametric representation of a line segment* between the points $\mathbf{x}^{(1)}$ and $\mathbf{x}^{(2)}$. If the entire line segment of Eq. (4.71) is in the set S, then it is a convex set. A check of the convexity of a set is demonstrated in Example 4.36.

EXAMPLE 4.36 CHECK FOR CONVEXITY OF A SET

Show the convexity of the set

$$S = \{\mathbf{x} \mid x_1^2 + x_2^2 - 1.0 \le 0\} \tag{a}$$

Solution

To show the set S graphically, we first plot the constraint as an equality that represents a circle of radius 1 centered at $(0, 0)$, as shown in Figure 4.21. Points inside or on the circle are in S. Geometrically we see that for any two points inside the circle, the line segment between them is also inside the circle.

Therefore, S is a convex set. We can also use Eq. (4.71) to show convexity of S. To do this, take any two points $\mathbf{x}^{(1)}$ and $\mathbf{x}^{(2)}$ in the set S. Use of Eq. (4.71) to calculate $\mathbf{x}$ and the condition that the distance between $\mathbf{x}^{(1)}$ and $\mathbf{x}^{(2)}$ is non-negative (that is, $||\mathbf{x}^{(1)} - \mathbf{x}^{(2)}|| \ge 0$) will show $\mathbf{x} \in S$. This will prove the convexity of S and is left as an exercise.

Note that if the foregoing set S is defined by reversing the inequality as $x_1^2 + x_2^2 - 1.0 \ge 0$, then the feasible set S will consist of points outside the circle. Such a set is clearly nonconvex because it violates the condition that the line segment of Eq. (4.71) defined by any two points in the set is not entirely in the set.

FIGURE 4.21 Convex set S for Example 4.36.

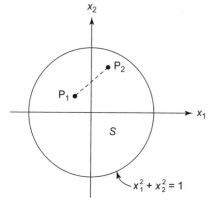

4.8.2 Convex Functions

Consider a function of a single variable $f(x) = x^2$. A graph of the function is shown in Figure 4.22. If a straight line is constructed between any two points $(x_1, f(x_1))$ and $(x_2, f(x_2))$ on the curve, the line lies above the graph of $f(x)$ at all points between x_1 and x_2. This property characterizes convex functions.

The convex function of a single variable $f(x)$ is defined on a convex set; that is, the independent variable x must lie in a convex set. A function $f(x)$ is called convex on the convex set S if the graph of the function lies below the line joining any two points on the curve $f(x)$. Figure 4.23 is a geometrical representation of a convex function. Using the geometry, the foregoing definition of a convex function can be expressed by the inequality

$$f(x) \le \alpha f(x_2) + (1 - \alpha) f(x_1) \tag{4.72}$$

Since $x = \alpha x_2 + (1 - \alpha)x_1$, the above inequality becomes

$$f(\alpha x_2 + (1 - \alpha)x_1) \le \alpha f(x_2) + (1 - \alpha)f(x_1) \quad \text{for} \quad 0 \le \alpha \le 1 \tag{4.73}$$

The foregoing definition of a convex function of one variable can be generalized to functions of n variables. A function $f(\mathbf{x})$ defined on a convex set S is convex if it satisfies the inequality

$$f(\alpha \mathbf{x}^{(2)} + (1 - \alpha)\mathbf{x}^{(1)}) \le \alpha f(\mathbf{x}^{(2)}) + (1 - \alpha)f(\mathbf{x}^{(1)}) \quad \text{for} \quad 0 \le \alpha \le 1 \tag{4.74}$$

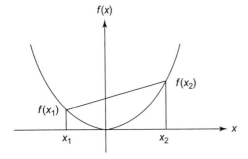

FIGURE 4.22   Convex function $f(x) = x^2$.

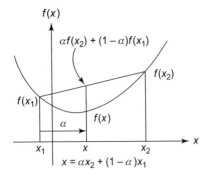

FIGURE 4.23   Characterization of a convex function.

for any two points $\mathbf{x}^{(1)}$ and $\mathbf{x}^{(2)}$ in S. Note that convex set S is a region in the n-dimensional space satisfying the convexity condition. Equations (4.73) and (4.74) give *necessary and sufficient conditions for the convexity of a function*. However, they are difficult to use in practice because we have to check an infinite number of pairs of points. Fortunately, the following theorem gives an easier way of checking the convexity of a function.

THEOREM 4.8

Check for the Convexity of a Function A function of n variables $f(x_1, x_2, \ldots, x_n)$ defined on a *convex set* S is convex *if and only if* the Hessian matrix of the function is *positive semidefinite* or positive definite at all points in the set S. If the Hessian matrix is positive definite for all points in the feasible set, then f is called a *strictly convex function*. (*Note:* The converse of this is not true: A strictly convex function may have only a positive semidefinite Hessian at some points; for example, $f(x) = x^4$ is a strictly convex function but its second derivative is zero at $x = 0$.)

Note that the Hessian condition of Theorem 4.8 is both necessary and sufficient; that is, the function is not convex if the Hessian is not at least positive semidefinite for all points in the set S. Therefore, if it can be shown that the Hessian is not positive definite or positive semidefinite at some points in the set S, then the function is not convex because the condition of Theorem 4.8 is violated.

In one dimension, the convexity check of the theorem reduces to the condition that the second derivative (curvature) of the function be non-negative. The graph of such a function has non-negative curvature, as for the functions in Figures 4.22 and 4.23. The theorem can be proved by writing a Taylor's expansion for the function $f(\mathbf{x})$ and then using the definitions of Eqs. (4.73) and (4.74). Examples 4.37 and 4.38 illustrate the check for convexity of functions.

EXAMPLE 4.37 CHECK FOR CONVEXITY OF A FUNCTION

$$f(\mathbf{x}) = x_1^2 + x_2^2 - 1 \tag{a}$$

Solution

The domain for the function (which is all values of x_1 and x_2) is convex. The gradient and Hessian of the function are given as

$$\nabla f = \begin{bmatrix} 2x_1 \\ 2x_2 \end{bmatrix}, \quad \mathbf{H} = \begin{bmatrix} 2 & 0 \\ 0 & 2 \end{bmatrix} \tag{b}$$

By either of the tests given in Theorems 4.2 and 4.3 ($M_1 = 2$, $M_2 = 4$, $\lambda_1 = 2$, $\lambda_2 = 2$), we see that $\mathbf{H}$ is positive definite everywhere. Therefore, f is a strictly convex function.

EXAMPLE 4.38 CHECK FOR THE CONVEXITY OF A FUNCTION

$$f(x) = 10 - 4x + 2x^2 - x^3 \tag{a}$$

Solution

The second derivative of the function is $d^2f/dx^2 = 4 - 6x$. For the function to be convex, $d^2f/dx^2 \geq 0$. Thus, the function is convex only if $4 - 6x \geq 0$ or $x \leq 2/3$. *The convexity check actually defines a domain for the function over which it is convex.* The function $f(x)$ is plotted in Figure 4.24. It can be seen that the function is convex for $x \leq 2/3$ and concave for $x \geq 2/3$ (a function $f(x)$ is called *concave* if $-f(x)$ is convex).

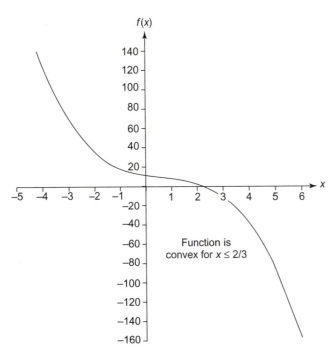

FIGURE 4.24   Example 4.38 graph for the function $f(x) = 10 - 4x + 2x2 - x3$.

4.8.3 Convex Programming Problem

If a function $g_i(x)$ is convex, then the set $g_i(x) \leq e_i$ is convex, where e_i is any constant. If functions $g_i(x)$ for $i = 1$ to m are convex, then the set defined by $g_i(x) \leq e_i$ for $i = 1$ to m is also convex. The set $g_i(x) \leq e_i$ for $i = 1$ to m is the intersection of sets defined by the individual constraints $g_i(x) \leq e_i$. Therefore, the intersection of convex sets is itself a convex set. We can relate convexity of functions and sets by using Theorem 4.9.

THEOREM 4.9

Convex Functions and Convex Sets Let the feasible set S be defined, with the constraints of the general optimization problem defined in the standard form in Eqs. (4.35) through (4.37), as

$$S = \{\mathbf{x} \mid h_i(\mathbf{x}) = 0, i = 1 \text{ to } p;$$
$$g_j(\mathbf{x}) \leq 0, \quad j = 1 \text{ to } m\} \quad (4.75)$$

Then S is a convex set if functions g_j are convex and functions h_i are linear.

The set S of Example 4.36 is convex because it is defined by a convex function. It is important to realize that if we have a nonlinear equality constraint $h_i(\mathbf{x}) = 0$, then the feasible set S is always nonconvex. This can be seen from the definition of a convex set. For an equality constraint, the set S is a collection of points lying on the surface $h_i(\mathbf{x}) = 0$. If we take any two points on the surface, the straight line joining them cannot be on the surface, unless it is a plane (linear equality). Therefore, a feasible set defined by any nonlinear equality constraint is always nonconvex. On the contrary, a feasible set defined by a linear equality or inequality is always convex.

If all inequality constraint functions for an optimum design problem are convex, and all equality constraint are linear, then the feasible set S is convex by Theorem 4.9. If the cost function is also convex over the set S, then we have what is known as a *convex programming problem*. Such problems have a very useful property, which is that KKT necessary conditions are also sufficient and any local minimum is also a global minimum.

It is important to note that Theorem 4.9 does not say that the feasible set S cannot be convex if a constraint function fails the convexity check (i.e., it is not an "if and only if" theorem). There are some problems where the constraint functions fail the convexity check, but the feasible set is still convex. Thus, *the conditions of the theorem are only sufficient but not necessary for the convexity of the problem.*

THEOREM 4.10

Global Minimum If $f(\mathbf{x}^*)$ is a local minimum for a convex function $f(\mathbf{x})$ that is defined on a convex feasible set S, then it is also a global minimum.

It is important to note that the theorem does not say that $\mathbf{x}^*$ cannot be a global minimum point if functions of the problem fail the convexity test.

The point may indeed be a global minimum; however, we cannot claim global optimality using Theorem 4.10. We will have to use some other procedure, such as exhaustive search. Note also that the theorem does not say that the global minimum is unique; that is, there can be multiple minimum points in the feasible set, all having the same cost function value. The convexity of several problems is checked in Examples 4.39 to 4.41.

EXAMPLE 4.39 CHECK FOR THE CONVEXITY OF A PROBLEM

Minimize

$$f(x_1, x_2) = x_1^3 - x_2^3 \tag{a}$$

subject to

$$x_1 \geq 0, \quad x_2 \leq 0 \tag{b}$$

Solution

The constraints actually define the domain for the function $f(\mathbf{x})$, which is the fourth quadrant of a plane (shown in Figure 4.25). This domain is convex. The Hessian of f is given as

$$\mathbf{H} = \begin{bmatrix} 6x_1 & 0 \\ 0 & -6x_2 \end{bmatrix} \tag{c}$$

The Hessian is positive semidefinite or positive definite over the domain defined by the constraints $(x_1 \geq 0, x_2 \leq 0)$. Therefore, the cost function is convex and the problem is convex. Note that if constraints $x_1 \geq 0$ and $x_2 \leq 0$ are not imposed, then the cost function will not be convex for all feasible $\mathbf{x}$. This can be observed in Figure 4.25, where several cost function contours are shown. Thus, the condition of positive semidefiniteness of the Hessian ($6x_1 \geq 0$, $-6x_2 \geq 0$) can define the domain for the function over which it is convex.

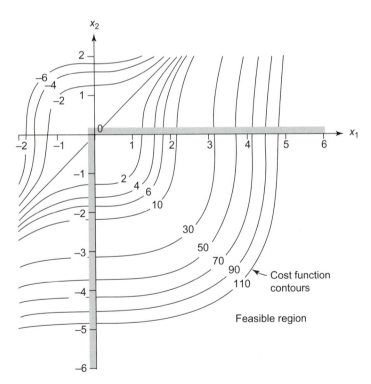

FIGURE 4.25 Example 4.39 graphical representation.

EXAMPLE 4.40 CHECK FOR THE CONVEXITY OF A PROBLEM

Minimize

$$f(x_1, x_2) = 2x_1 + 3x_2 - x_1^3 - 2x_2^2 \tag{a}$$

subject to

$$x_1 + 3x_2 \leq 6, \quad 5x_1 + 2x_2 \leq 10, \quad x_1, x_2 \geq 0 \tag{b}$$

Solution

Since all of the constraint functions are linear in the variables x_1 and x_2, the feasible set for the problem is convex. If the cost function f is also convex, then the problem is convex. The Hessian of the cost function is

$$\mathbf{H} = \begin{bmatrix} -6x_1 & 0 \\ 0 & -4 \end{bmatrix} \tag{c}$$

The eigenvalues of $\mathbf{H}$ are $-6x_1$ and -4. Since the first eigenvalue is nonpositive for $x_1 \geq 0$, and the second eigenvalue is negative, the function is not convex (Theorem 4.8), so the problem cannot be classified as a convex programming problem. Global optimality of a local minimum is not guaranteed.

Figure 4.26 shows the feasible set for the problem along with several isocost curves. It is seen that the feasible set is convex but the cost function is not. Thus the problem can have multiple local minima having different values for the cost function.

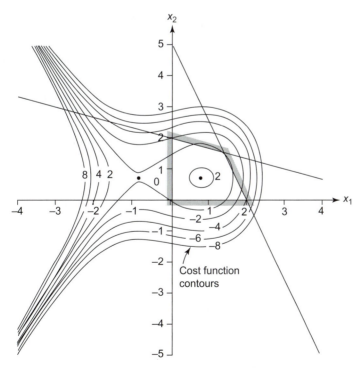

FIGURE 4.26 Example 4.40 graphical representation.

EXAMPLE 4.41 CHECK FOR THE CONVEXITY OF A PROBLEM

Minimize

$$f(x_1, x_2) = 9x_1^2 - 18x_1x_2 + 13x_2^2 - 4 \tag{a}$$

subject to

$$x_1^2 + x_2^2 + 2x_1 \geq 16 \tag{b}$$

Solution

To check for the convexity of the problem, we need to write the constraint in the standard form as

$$g(\mathbf{x}) = -x_1^2 - x_2^2 - 2x_1 + 16 \leq 0 \tag{c}$$

The Hessian of $g(\mathbf{x})$ is

$$\mathbf{H} = \begin{bmatrix} -2 & 0 \\ 0 & -2 \end{bmatrix} \tag{d}$$

Eigenvalues of the Hessian are -2 and -2. Since, the Hessian is neither positive definite nor positive semidefinite, $g(\mathbf{x})$ is not convex (in fact, the Hessian is negative definite, so $g(\mathbf{x})$ is concave). Therefore, the problem cannot be classified as a convex programming problem, and global optimality for the solution cannot be guaranteed by Theorem 4.10.

4.8.4 Transformation of a Constraint

A constraint function can be transformed into a different form that is equivalent to the original function; that is, the constraint boundary and the feasible set for the problem do not change but the form of the function changes. Transformation of a constraint function, however, may affect its convexity check: *A transformed constraint function may fail the convexity check*. The convexity of the feasible set is, however, not affected by the transformation.

In order to illustrate the effect of transformations, let us consider the following inequality constraint:

$$g_1 = \frac{a}{x_1x_2} - b \leq 0 \tag{a}$$

with $x_1 > 0$, $x_2 > 0$, and a and b as the given positive constants. To check the convexity of the constraint, we calculate the Hessian matrix as

$$\nabla^2 g_1 = \frac{2a}{x_1^2x_2^2} \begin{bmatrix} \dfrac{x_2}{x_1} & 0.5 \\ 0.5 & \dfrac{x_1}{x_2} \end{bmatrix} \tag{b}$$

Both eigenvalues, as well as the two leading principal minors of the preceding matrix, are strictly positive, so the matrix is positive definite and the constraint function g_1 is convex. The feasible set for g_1 is convex.

Now let us transform the constraint by multiplying throughout by $x_1 x_2$ (since $x_1 > 0$, $x_2 > 0$, the sense of the inequality is not changed) to obtain

$$g_2 = a - bx_1 x_2 \leq 0 \tag{c}$$

The constraints g_1 and g_2 are equivalent and will give the same optimum solution for the problem. To check convexity of the constraint function, we calculate the Hessian matrix as

$$\nabla^2 g_2 = \begin{bmatrix} 0 & -b \\ -b & 0 \end{bmatrix} \tag{d}$$

The eigenvalues of the preceding matrix are: $\lambda_1 = -b$ and $\lambda_2 = b$. Therefore, the matrix is indefinite by Theorem 4.2 and, by Theorem 4.8, the constraint function g_2 is not convex. Thus, we lose the convexity of the constraint function and we cannot claim convexity of the feasible set by Theorem 4.9. Since the problem cannot be shown to be convex, we cannot use results related to convex programming problems.

4.8.5 Sufficient Conditions for Convex Programming Problems

If we can show convexity of a problem, any solution to the necessary conditions will automatically satisfy the sufficient conditions (see Example 4.42). In addition, the solution will be a global minimum. Following the procedure of Section 4.4, we consider various cases defined by the switching conditions of Eq. (4.51) until a solution is found. We can stop there, as the solution is a *global optimum design*.

THEOREM 4.11

Sufficient Conditions for Convex Programming Problems If $f(x)$ is a convex cost function defined on a convex feasible set, then the first-order KKT conditions are necessary as well as sufficient for a global minimum.

EXAMPLE 4.42 CHECK FOR THE CONVEXITY OF A PROBLEM

Let us consider Example 4.29 again and check for its convexity:

Minimize

$$f(x) = (x_1 - 1.5)^2 + (x_2 - 1.5)^2 \tag{a}$$

subject to

$$g(x) = x_1 + x_2 - 2 \leq 0 \tag{b}$$

Solution

The KKT necessary conditions give the candidate local minimum as $x_1^* = 1$, $x_2^* = 1$, and $u^* = 1$. The constraint function $g(x)$ is linear, so it is convex. Since the inequality constraint function is convex and there is no equality constraint, the feasible set S is convex. The Hessian matrix for the cost function is

$$H = \begin{bmatrix} 2 & 0 \\ 0 & 2 \end{bmatrix} \qquad (c)$$

Since H is positive definite everywhere by Theorem 4.2 or Theorem 4.3, the cost function $f(x)$ is strictly convex by Theorem 4.8. Therefore, the problem is convex and the solution $x_1^* = 1, x_2^* = 1$ satisfies the sufficiency condition of Theorem 4.11. It is a strict global minimum point for the problem.

The convexity results are summarized in Table 4.3.

TABLE 4.3   Convex programming problem—summary of results

Problem must be written in standard form: Minimize $f(x)$, subject to $h_i(x) = 0$, $g_j(x) \leq 0$

1. Convex set	The geometrical condition, that a line joining two points in the set is to be in the set, is an *"if-and-only-if"* condition for the convexity of the set.
2. Convexity of feasible set S	All of the constraint functions should be convex. *This condition is sufficient but not necessary*; that is, functions failing the convexity check may still define convex sets. — Nonlinear equality constraints always give nonconvex sets. — Linear equalities or inequalities always give convex sets.
3. Convex functions	A function is convex *if and only if* its Hessian is at least *positive semidefinite* everywhere. A function is *strictly convex* if its Hessian is *positive definite* everywhere; however, the *converse is not true*: A strictly convex function may not have a positive definite Hessian everywhere. Thus this condition is *sufficient* but not necessary.
4. Form of constraint function	Changing the form of a constraint function can result in failure of the convexity check for the new constraint or vice versa.
5. Convex programming problem	$f(x)$ is convex over the convex feasible set S. — KKT first-order conditions are necessary as well as sufficient for global minimums — Any local minimum point is also a global minimum point.
6. Nonconvex programming problem	If a problem fails a convexity check, it does not imply that there is no global minimum for the problem. It could have only one local minimum in the feasible set S, which would then be a global minimum as well.

4.9 ENGINEERING DESIGN EXAMPLES

The procedures described in the previous sections are used to solve two engineering design examples. The problems are formulated, convexity is verified, KKT necessary conditions are written and solved, and the Constraint Variation Sensitivity Theorem is illustrated and discussed.

4.9.1 Design of a Wall Bracket

The wall bracket that is shown in Figure 4.27 is to be designed to support a load of $W = 1.2$ MN. The material for the bracket should not fail under the action of forces in the bars. These are expressed as the following stress constraints:

$$\text{Bar 1: } \sigma_1 \leq \sigma_a \tag{a}$$

$$\text{Bar 2: } \sigma_2 \leq \sigma_a \tag{b}$$

where

$\sigma_a = $ allowable stress for the material ($16{,}000$ N/cm^2)
$\sigma_1 = $ stress in Bar 1 which is given as F_1/A_1, N/cm^2
$\sigma_2 = $ stress in Bar 2 which is given as F_2/A_2, N/cm^2
$A_1 = $ cross-sectional area of Bar 1, cm^2
$A_2 = $ cross-sectional area of Bar 2, cm^2
$F_1 = $ force due to load W in Bar 1, N
$F_2 = $ force due to load W in Bar 2, N

Total volume of the bracket is to be minimized.

Problem Formulation

The cross-sectional areas A_1 and A_2 are the two design variables, and the cost function for the problem is the volume, which is given as

$$f(A_1, A_2) = l_1 A_1 + l_2 A_2, \text{ cm}^3 \tag{c}$$

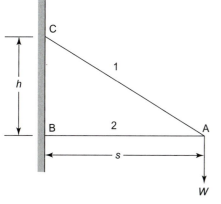

FIGURE 4.27   Graphic of a wall bracket. $h = 30$ cm, $s = 40$ cm, and $W = 1.2$ MN.

where $l_1 = \sqrt{30^2 + 40^2} = 50$ cm is the length of member 1 and $l_2 = 40$ cm is the length of member 2. To write the stress constraints, we need forces in the members, which are obtained using static equilibrium of node A as follows: $F_1 = 2.0 \times 10^6$ N, $F_2 = 1.6 \times 10^6$ N. Therefore, the stress constraints of Eqs. (a) and (b) are given as

$$g_1 = \frac{(2.0 \times 10^6)}{A_1} - 16,000 \leq 0 \tag{d}$$

$$g_2 = \frac{(1.6 \times 10^6)}{A_2} - 16,000 \leq 0 \tag{e}$$

The cross-sectional areas must both be non-negative:

$$g_3 = -A_1 \leq 0, \quad g_4 = -A_2 \leq 0 \tag{f}$$

Constraints for the problem are plotted in Figure 4.28, and the feasible region is identified. A few cost function contours are also shown. It is seen that the optimum solution is at the point A with $A_1^* = 125$ cm^2, $A_2^* = 100$ cm^2, and $f = 10{,}250$ cm^3.

Convexity

Since the cost function of Eq. (c) is linear in terms of design variables, it is convex. The Hessian matrix for the constraint g_1 is

$$\nabla^2 g_1 = \begin{bmatrix} \dfrac{(4.0 \times 10^6)}{A_1^3} & 0 \\ 0 & 0 \end{bmatrix} \tag{g}$$

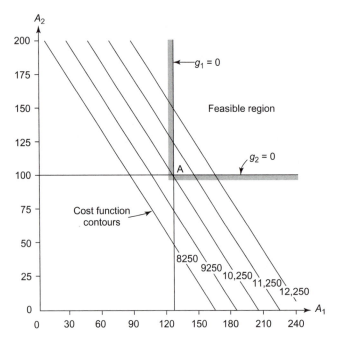

FIGURE 4.28   Graphical solution for the wall bracket problem.

which is a positive semidefinite matrix for $A_1 > 0$, so g_1 is convex. Similarly, g_2 is convex, and since g_3 and g_4 are linear, they are convex. Thus the problem is *convex*, the KKT necessary conditions are *sufficient* and any design satisfying the KKT conditions is a global minimum.

KKT Necessary Conditions

To use the KKT conditions, we introduce slack variables into the constraints and define the Lagrange function of Eq. (4.46) for the problem as

$$L = (l_1 A_1 + l_2 A_2) + u_1 \left[\frac{(2.0 \times 10^6)}{A_1} - 16{,}000 + s_1^2 \right] + u_2 \left[\frac{(1.6 \times 10^6)}{A_2} - 16{,}000 + s_2^2 \right] \qquad \text{(h)}$$
$$+ u_3(-A_1 + s_3^2) + u_4(-A_2 + s_4^2)$$

The necessary conditions become

$$\frac{\partial L}{\partial A_1} = l_1 - u_1 \frac{(2.0 \times 10^6)}{A_1^2} - u_3 = 0 \qquad \text{(i)}$$

$$\frac{\partial L}{\partial A_2} = l_2 - u_2 \frac{(1.6 \times 10^6)}{A_2^2} - u_4 = 0 \qquad \text{(j)}$$

$$u_i s_i = 0, \quad u_i \geq 0, \quad g_i + s_i^2 = 0, \quad s_i^2 \geq 0; \quad i = 1 \text{ to } 4 \qquad \text{(k)}$$

The switching conditions in Eqs. (k) give 16 solution cases. These cases can be identified using a systematic procedure, as shown in Table 4.4. Note that any case that requires $s_3 = 0$ (that is, $g_3 = 0$) makes the area $A_1 = 0$. For such a case the constraint g_1 of Eq. (d) is violated, so it does not give a candidate solution. Similarly, $s_4 = 0$ makes $A_2 = 0$, which violates the constraint of Eq. (e). In addition, A_1 and A_2 cannot be negative because the corresponding solution has no physical meaning. Therefore, all of the cases requiring $s_3 = 0$ and/or $s_4 = 0$ do not give any candidate solution.

These cases need not be considered any further. This leaves only Cases 1 through 3 and Case 6 for further consideration, and we solve them as follows. (Any of the cases giving $A_1 < 0$ or $A_2 < 0$ will also be discarded.)

> *Case 1:* $u_1 = 0$, $u_2 = 0$, $u_3 = 0$, $u_4 = 0$. This gives $l_1 = 0$ and $l_2 = 0$ in Eqs. (i) and (j) which is *not acceptable*.
> *Case 2:* $s_1 = 0$, $u_2 = 0$, $u_3 = 0$, $u_4 = 0$. This gives $l_1 = 0$ in Eq. (i), which is *not acceptable*.
> *Case 3:* $u_1 = 0$, $s_2 = 0$, $u_3 = 0$, $u_4 = 0$. This gives $l_2 = 0$ in Eq. (j), which is *not acceptable*.

In Case 6, $s_1 = 0$, $s_2 = 0$, $u_3 = 0$, $u_4 = 0$, and Equations (d) and (e) give $A_1^* = 125 \text{ cm}^2$, $A_2^* = 100 \text{ cm}^2$. Equations (i) and (j) give the Lagrange multipliers as $u_1 = 0.391$ and $u_2 = 0.25$, and since both are non-negative, all of the KKT conditions are satisfied. The cost function at optimum is obtained as $f^* = 50(125) + 40(100)$, or $f^* = 10{,}250 \text{ cm}^3$. The gradients of the active constraints are $[(-2.0 \times 10^6)/A_1^2, 0)$, $(0, (-1.0 \times 10^6)/A_2^2]$. These vectors are linearly independent, and so the minimum point is a regular point of the feasible set. By convexity, this point is a global minimum point for f.

TABLE 4.4   Definition of KKT cases with four inequalities

No.	Case	Active constraints
1	$u_1 = 0$, $u_2 = 0$, $u_3 = 0$, $u_4 = 0$	No inequality active
2	$s_1 = 0$, $u_2 = 0$, $u_3 = 0$, $u_4 = 0$	One inequality active at a time
3	$u_1 = 0$, $s_2 = 0$, $u_3 = 0$, $u_4 = 0$	
4	$u_1 = 0$, $u_2 = 0$, $s_3 = 0$, $u_4 = 0$	
5	$u_1 = 0$, $u_2 = 0$, $u_3 = 0$, $s_4 = 0$	
6	$s_1 = 0$, $s_2 = 0$, $u_3 = 0$, $u_4 = 0$	Two inequalities active at a time
7	$u_1 = 0$, $s_2 = 0$, $s_3 = 0$, $u_4 = 0$	
8	$u_1 = 0$, $u_2 = 0$, $s_3 = 0$, $s_4 = 0$	
9	$s_1 = 0$, $u_2 = 0$, $u_3 = 0$, $s_4 = 0$	
10	$s_1 = 0$, $u_2 = 0$, $s_3 = 0$, $u_4 = 0$	
11	$u_1 = 0$, $s_2 = 0$, $u_3 = 0$, $s_4 = 0$	
12	$s_1 = 0$, $s_2 = 0$, $s_3 = 0$, $u_4 = 0$	Three inequalities active at a time
13	$u_1 = 0$, $s_2 = 0$, $s_3 = 0$, $s_4 = 0$	
14	$s_1 = 0$, $u_2 = 0$, $s_3 = 0$, $s_4 = 0$	
15	$s_1 = 0$, $s_2 = 0$, $u_3 = 0$, $s_4 = 0$	
16	$s_1 = 0$, $s_2 = 0$, $s_3 = 0$, $s_4 = 0$	Four inequalities active at a time

Sensitivity Analysis

If the allowable stress changes to 16,500 N/cm^2 from 16,000 N/cm^2, we need to know how the cost function will change. Using Eq. (4.61) we get the change in the cost function as $\delta f^* = -u_1 e_1 - u_2 e_2$, where $e_1 = e_2 = 16,500 - 16,000 = 500$ N/cm^2. Therefore, the change in the cost function is $\delta f^* = -0.391(500) - 0.25(500) = -320.5$ cm^3. The volume of the bracket thus reduces by 320.5 cm^3.

4.9.2 Design of a Rectangular Beam

In Section 3.8, a rectangular beam design problem was formulated and solved graphically. We will solve the same problem using the KKT necessary conditions. The problem is formulated as follows. Find b and d to minimize

$$f(b,d) = bd \tag{a}$$

subject to the inequality constraints

$$g_1 = \frac{(2.40 \times 10^8)}{bd^2} - 10 \le 0 \tag{b}$$

$$g_2 = \frac{(2.25 \times 10^5)}{bd} - 2 \leq 0 \tag{c}$$

$$g_3 = -2b + d \leq 0 \tag{d}$$

$$g_4 = -b \leq 0, \quad g_5 = -d \leq 0 \tag{e}$$

Convexity

Constraints g_3, g_4, and g_5 are linear in terms of b and d, and are therefore convex. The Hessian for the constraint g_1 is given as

$$\nabla^2 g_1 = \frac{(4.80 \times 10^8)}{b^3 d^4} \begin{bmatrix} d^2 & bd \\ bd & 3b^2 \end{bmatrix} \tag{f}$$

Since this matrix is positive definite for $b > 0$ and $d > 0$, g_1 is a strictly convex function. The Hessian for the constraint g_2 is given as

$$\nabla^2 g_2 = \frac{(2.25 \times 10^5)}{b^3 d^3} \begin{bmatrix} 2d^2 & bd \\ bd & 2b^2 \end{bmatrix} \tag{g}$$

This matrix is positive definite, so the constraint g_2 is also strictly convex. Because all of the constraints of the problem are convex, the feasible convex.

It is interesting to note that constraints g_1 and g_2 can be *transformed* as (since $b > 0$ and $d > 0$, the sense of inequality is not changed)

$$\bar{g}_1 = (2.40 \times 10^8) - 10bd^2 \leq 0 \tag{h}$$

$$\bar{g}_2 = (2.25 \times 10^5) - 2bd \leq 0 \tag{i}$$

Hessians of the functions $\bar{g}_1$ and $\bar{g}_2$ are given as

$$\nabla^2 \bar{g}_1 = \begin{bmatrix} 0 & -20d \\ -20d & -20b \end{bmatrix}; \quad \nabla^2 \bar{g}_2 = \begin{bmatrix} 0 & -2 \\ -2 & 0 \end{bmatrix} \tag{j}$$

Both of the preceding matrices are not positive semidefinite. Therefore, the constraint functions $\bar{g}_1$ and $\bar{g}_2$ given in Eqs. (f) and (g) are not convex. *This goes to show that convexity of a function can be lost if it is changed into another form.* This is an important observation, and it shows that we should be careful in transformation of constraint functions. Note, however, that the transformation of constraints does not change the optimum solution. It does change the values of the Lagrange multipliers for the constraints, however, as discussed in Section 4.7.

To check convexity of the cost function, we write its Hessian as

$$\nabla^2 f = \begin{bmatrix} 0 & 1 \\ 1 & 0 \end{bmatrix} \tag{k}$$

This matrix is indefinite, so the cost function is nonconvex. The problem fails the convexity check of Theorem 4.9, and we cannot guarantee global optimality of the solution by

Theorem 4.10. *Note that this does not say that a local minimum cannot be a global minimum. It may still be a global minimum, but cannot be guaranteed by Theorem 4.10.*

KKT Necessary Conditions

To use the KKT conditions, we introduce slack variables into the constraints and define the Lagrange function for the problem as

$$L = bd + u_1\left(\frac{(2.40 \times 10^8)}{bd^2} - 10 + s_1^2\right) + u_2\left(\frac{(2.25 \times 10^5)}{bd} - 2 + s_2^2\right) + u_3(d - 2b + s_3^2) \tag{1}$$
$$+ u_4(-b + s_4^2) + u_5(-d + s_5^2)$$

The necessary conditions give

$$\frac{\partial L}{\partial b} = d + u_1\frac{(-2.40 \times 10^8)}{b^2 d^2} + u_2\frac{(-2.25 \times 10^5)}{b^2 d} - 2u_3 - u_4 = 0 \tag{m}$$

$$\frac{\partial L}{\partial d} = b + u_1\frac{(-4.80 \times 10^8)}{bd^3} + u_2\frac{(-2.25 \times 10^5)}{bd^2} + u_3 - u_5 = 0 \tag{n}$$

$$u_i s_i = 0, \quad u_i \geq 0, \quad g_i + s_i^2 = 0; \quad s_i^2 \geq 0; \quad i = 1 \text{ to } 5 \tag{o}$$

The switching conditions in Eq. (o) give 32 cases for the necessary conditions. However, note that the cases requiring either $s_4 = 0$ or $s_5 = 0$, or both as zero, do not give any candidate optimum points because they violate the constraint of either Eqs. (b) and (c) or of Eq. (d). Therefore, these cases will not be considered, which can be done by setting $u_4 = 0$ and $u_5 = 0$ in the remaining cases. This leaves the following eight cases for further consideration:

1. $u_1 = 0$, $u_2 = 0$, $u_3 = 0$, $u_4 = 0$, $u_5 = 0$
2. $u_1 = 0$, $u_2 = 0$, $s_3 = 0$, $u_4 = 0$, $u_5 = 0$
3. $u_1 = 0$, $s_2 = 0$, $u_3 = 0$, $u_4 = 0$, $u_5 = 0$
4. $s_1 = 0$, $u_2 = 0$, $u_3 = 0$, $u_4 = 0$, $u_5 = 0$
5. $u_1 = 0$, $s_2 = 0$, $s_3 = 0$, $u_4 = 0$, $u_5 = 0$
6. $s_1 = 0$, $s_2 = 0$, $u_3 = 0$, $u_4 = 0$, $u_5 = 0$
7. $s_1 = 0$, $u_2 = 0$, $s_3 = 0$, $u_4 = 0$, $u_5 = 0$
8. $s_1 = 0$, $s_2 = 0$, $s_3 = 0$, $u_4 = 0$, $u_5 = 0$

We consider each case one at a time and solve for the candidate optimum points. Note that any solution having $b < 0$ or $d < 0$ violates constraint g_4 or g_5 and will be discarded.

Case 1: $u_1 = 0$, $u_2 = 0$, $u_3 = 0$, $u_4 = 0$, $u_5 = 0$. This case gives $d = 0$, $b = 0$ in Eqs. (k) and (l). Therefore, this case does not give a solution.

Case 2: $u_1 = 0$, $u_2 = 0$, $s_3 = 0$, $u_4 = 0$, $u_5 = 0$. Equation (d) gives $d = 2b$. Equations (k) and (l) give $d - 2u_3 = 0$ and $d + u_3 = 0$. These three equations give $b = 0$ and $d = 0$, which is not feasible.

Case 3: $u_1 = 0$, $s_2 = 0$, $u_3 = 0$, $u_4 = 0$, $u_5 = 0$. Equations (k), (l), and (c) give

$$d - u_2\frac{(2.25 \times 10^5)}{b^2 d} = 0 \tag{p}$$

$$b - u_2 \frac{(2.25 \times 10^5)}{bd^2} = 0 \tag{q}$$

$$\frac{(2.25 \times 10^5)}{bd} - 2 = 0 \tag{r}$$

These equations give a solution as $u_2 = (5.625 \times 10^4)$ and $bd = (1.125 \times 10^5)$. Since $u_2 > 0$, this is a valid solution. Actually, there is a family of solutions given by $bd = (1.125 \times 10^5)$; for any $d > 0$, b can be found from this equation. However, there must be some limits on the values of b and d for which this family of solutions is valid. These ranges are provided by requiring $s_1^2 \geq 0$ and $s_3^2 \geq 0$, or $g_1 \leq 0$ and $g_3 \leq 0$.

Substituting $b = (1.125 \times 10^5)/d$ from Eq. (p) into g_1 (Eq. b),

$$\frac{(2.40 \times 10^8)}{(1.125 \times 10^5)\, d} - 10 \leq 0; \quad \text{or} \quad d \geq 213.33 \text{ mm} \tag{s}$$

Substituting $b = (1.125 \times 10^5)/d$ from Eq. (p) into g_3 (Eq. d),

$$d - \frac{(2.25 \times 10^5)}{bd} \leq 0; \quad \text{or} \quad d \leq 474.34 \text{ mm} \tag{t}$$

This gives limits on the depth d. We can find limits on the width b by substituting Eqs. (s) and (t) into $bd = (1.125 \times 10^5)$ from Eq. (r):

$$d \geq 213.33, \quad b \leq 527.34 \tag{u}$$

$$d \leq 474.33, \quad b \geq 237.17 \tag{v}$$

Therefore, there are infinite solutions for this case as

$$237.17 \leq b \leq 527.34 \text{ mm}; \quad 213.33 \leq d \leq 474.33 \text{ mm} \tag{w}$$

$$bd = (1.125 \times 10^5) \text{ mm}^2 \tag{x}$$

Case 4: $s_1 = 0$, $u_2 = 0$, $u_3 = 0$, $u_4 = 0$, $u_5 = 0$. Equations (m) and (n) reduce to

$$d - \frac{(2.40 \times 10^8)}{b^2 d^2} = 0; \quad \text{or} \quad b^2 d^3 = (2.40 \times 10^8) \tag{y}$$

$$b - \frac{(4.80 \times 10^8)}{bd^3} = 0; \quad \text{or} \quad b^2 d^3 = \left(4.80 \times 10^8\right) \tag{z}$$

Since these two equations are inconsistent, there is no solution for this case.

Case 5: $u_1 = 0$, $s_2 = 0$, $s_3 = 0$, $u_4 = 0$, $u_5 = 0$. Equations (c) and (d) can be solved for b and d—for example, by substituting $b = 2d$ from Eq. (d) into Eq. (c), we get $b = 237.17$ mm. Therefore, $d = 2(237.17) = 474.34$ mm. We can calculate u_2 and u_3 from Eqs. (m) and (n) as $u_2 = (5.625 \times 10^4)$, $u_3 = 0$. Substituting values of b and d into Eq. (b), we get $g_1 = -5.5 < 0$, so the constraint is satisfied (i.e., $s_1^2 > 0$). It can be verified that the gradients of g_2 and g_3 at the candidate point are linearly independent, and so the regularity condition is satisfied. Since all of the necessary conditions are satisfied, this is a valid solution. The Constraint Sensitivity Theorem 4.7 and Eq. (4.57) tell us that, because $u_3 = 0$, we can move away from that constraint toward the feasible region without affecting the optimum cost function value. This can also be observed from Figure 3.11 where the graphical solution for the problem was given. In the figure, point B represents the solution for this case. We can leave point B toward point A and remain on the constraint $g_2 = 0$ for optimum designs.

Case 6: $s_1 = 0$, $s_2 = 0$, $u_3 = 0$, $u_4 = 0$, $u_5 = 0$. Equations (b) and (c) can be solved for the band d as $b = 527.34$ mm and $d = 213.33$ mm. We can solve for u_1 and u_2 from Eqs. (m) and (n) as $u_1 = 0$ and $u_2 = (5.625 \times 10^4)$. Substituting values of b and d into Eq. (d), we get $g_3 = -841.35 < 0$, so the constraint is satisfied (i.e., $s_3^2 \geq 0$). It can also be verified that the point also satisfies the regularity condition. Since all of the KKT conditions are satisfied, this is a valid solution. It is quite similar to the one for Case 5. The solution corresponds to point A in Figure 3.11. If we leave constraint $g_1 = 0$ (point A) and remain on the curve A − B, we obtain other optimum designs near point A.

Case 7: $s_1 = 0$, $u_2 = 0$, $s_3 = 0$, $u_4 = 0$, $u_5 = 0$. Equations (b) and (d) can be solved as $b = 181.71$ mm and $d = 363.42$ mm. Equations (m) and (n) give the Lagrange multipliers $u_1 = 4402.35$ and $u_3 = -60.57$. Since $u_3 < 0$, this case does not give a valid solution.

Case 8: $s_1 = 0$, $s_2 = 0$, $s_3 = 0$, $u_4 = 0$, $u_5 = 0$. This case gives three equations in two unknowns (an over-determined system), which has no solution.

Sensitivity Analysis

It should be observed that none of the candidate minimum points (A and B and curve A−B in Figure 3.11) satisfies the sufficiency conditions presented in the next chapter. Therefore, *the existence of partial derivatives of the cost function with respect to the right-side parameters in Eq. (4.58) is not guaranteed by Theorem 4.7.* However, since we have a graphical solution to the problem in Figure 3.11, we can check what happens if we do use the sensitivity theorem.

For point A in Figure 3.11 (Case 6), constraints g_1 and g_2 are active, $b = 527.34$ mm, $d = 213.33$ mm, $u_1 = 0$, and $u_2 = (5.625 \times 10^4)$. Since $u_1 = 0$, Eq. (4.58) gives $\partial f / \partial e_1 = 0$. This means that any small change in the constraint limit does not change the optimum cost function value. That this is true can be observed from Figure 3.11. The optimum point A is changed but constraint g_2 remains active; that is, $bd = (1.125 \times 10^5)$ must be satisfied since point A stays on the g_2 boundary. Any change in g_2 moves the constraint parallel to itself, changing the optimum solution (the design variables and the cost function). Because $u_2 = (5.625 \times 10^4)$, Eq. (4.58) gives $\partial f / \partial e_2 = (-5.625 \times 10^4)$. It can be verified that the sensitivity coefficient predicts correct changes in the cost function.

It can also be verified that the other two solution cases (Cases 3 and 5) also give correct values for the sensitivity coefficients.

EXERCISES FOR CHAPTER 4

Section 4.2 Review of Some Basic Calculus Concepts

4.1 *Answer True or False.*

   1. A function can have several local minimum points in a small neighborhood of x^*.

   2. A function cannot have more than one global minimum point.

   3. The value of the function having a global minimum at several points must be the same.

4. A function defined on an open set cannot have a global minimum.
5. The gradient of a function $f(\mathbf{x})$ at a point is normal to the surface defined by the level surface $f(\mathbf{x}) = $ constant.
6. The gradient of a function at a point gives a local direction of maximum decrease in the function.
7. The Hessian matrix of a continuously differentiable function can be asymmetric.
8. The Hessian matrix for a function is calculated using only the first derivatives of the function.
9. Taylor series expansion for a function at a point uses the function value and its derivatives.
10. Taylor series expansion can be written at a point where the function is discontinuous.
11. Taylor series expansion of a complicated function replaces it with a polynomial function at the point.
12. Linear Taylor series expansion of a complicated function at a point is only a good local approximation for the function.
13. A quadratic form can have first-order terms in the variables.
14. For a given $\mathbf{x}$, the quadratic form defines a vector.
15. Every quadratic form has a symmetric matrix associated with it.
16. A symmetric matrix is positive definite if its eigenvalues are non-negative.
17. A matrix is positive semidefinite if some of its eigenvalues are negative and others are non-negative.
18. All eigenvalues of a negative definite matrix are strictly negative.
19. The quadratic form appears as one of the terms in Taylor's expansion of a function.
20. A positive definite quadratic form must have positive value for any $\mathbf{x} \neq \mathbf{0}$.

Write the Taylor's expansion for the following functions up to quadratic terms.

4.2 $\cos x$ about the point $x^* = \pi/4$
4.3 $\cos x$ about the point $x^* = \pi/3$
4.4 $\sin x$ about the point $x^* = \pi/6$
4.5 $\sin x$ about the point $x^* = \pi/4$
4.6 e^x about the point $x^* = 0$
4.7 e^x about the point $x^* = 2$
4.8 $f(x_1, x_2) = 10x_1^4 - 20x_1^2 x_2 + 10x_2^2 + x_1^2 - 2x_1 + 5$ about the point (1, 1). Compare approximate and exact values of the function at the point (1.2, 0.8).

Determine the nature of the following quadratic forms.

4.9 $F(\mathbf{x}) = x_1^2 + 4x_1 x_2 + 2x_1 x_3 - 7x_2^2 - 6x_2 x_3 + 5x_3^2$
4.10 $F(\mathbf{x}) = 2x_1^2 + 2x_2^2 - 5x_1 x_2$
4.11 $F(\mathbf{x}) = x_1^2 + x_2^2 + 3x_1 x_2$
4.12 $F(\mathbf{x}) = 3x_1^2 + x_2^2 - x_1 x_2$
4.13 $F(\mathbf{x}) = x_1^2 - x_2^2 + 4x_1 x_2$
4.14 $F(\mathbf{x}) = x_1^2 - x_2^2 + x_3^2 - 2x_2 x_3$
4.15 $F(\mathbf{x}) = x_1^2 - 2x_1 x_2 + 2x_2^2$
4.16 $F(\mathbf{x}) = x_1^2 - x_1 x_2 - x_2^2$

4.17 $F(\mathbf{x}) = x_1^2 + 2x_1x_3 - 2x_2^2 + 4x_3^2 - 2x_2x_3$

4.18 $F(\mathbf{x}) = 2x_1^2 + x_1x_2 + 2x_2^2 + 3x_3^2 - 2x_1x_3$

4.19 $F(\mathbf{x}) = x_1^2 + 2x_2x_3 + x_2^2 + 4x_3^2$

4.20 $F(\mathbf{x}) = 4x_1^2 + 2x_1x_3 - x_2^2 + 4x_3^2$

Section 4.4 Optimality Conditions: Unconstrained Problems

4.21 *Answer True or False.*

1. If the first-order necessary condition at a point is satisfied for an unconstrained problem, it can be a local maximum point for the function.
2. A point satisfying first-order necessary conditions for an unconstrained function may not be a local minimum point.
3. A function can have a negative value at its maximum point.
4. If a constant is added to a function, the location of its minimum point is changed.
5. If a function is multiplied by a positive constant, the location of the function's minimum point is unchanged.
6. If curvature of an unconstrained function of a single variable at the point x* is zero, then it is a local maximum point for the function.
7. The curvature of an unconstrained function of a single variable at its local minimum point is negative.
8. The Hessian of an unconstrained function at its local minimum point must be positive semidefinite.
9. The Hessian of an unconstrained function at its minimum point is negative definite.
10. If the Hessian of an unconstrained function is indefinite at a candidate point, the point may be a local maximum or minimum.

Find stationary points for the following functions (use a numerical method or a software package like Excel, MATLAB, and Mathematica, if needed). Also determine the local minimum, local maximum, and inflection points for the functions (inflection points are those stationary points that are neither minimum nor maximum).

4.22 $f(x_1, x_2) = 3x_1^2 + 2x_1x_2 + 2x_2^2 + 7$

4.23 $f(x_1, x_2) = x_1^2 + 4x_1x_2 + x_2^2 + 3$

4.24 $f(x_1, x_2) = x_1^3 + 12x_1x_2^2 + 2x_2^2 + 5x_1^2 + 3x_2$

4.25 $f(x_2, x_2) = 5x_1 - \frac{1}{16}x_1^2x_2 + \frac{1}{4x_1}x_2^2$

4.26 $f(x) = \cos x$

4.27 $f(x_1, x_2) = x_1^2 + x_1x_2 + x_2^2$

4.28 $f(x) = x^2 e^{-x}$

4.29 $f(x_1, x_2) = x_1 - \frac{10}{x_1x_2} + 5x_2$

4.30 $f(x_1, x_2) = x_1^2 - 2x_1 + 4x_2^2 - 8x_2 + 6$

4.31 $f(x_1, x_2) = 3x_1^2 - 2x_1x_2 + 5x_2^2 + 8x_2$

4.32 The annual operating cost U for an electrical line system is given by the following expression

$$U = \frac{(21.9 \times 10^7)}{V^2C} + (3.9 \times 10^6)C + 1000 \, V$$

where $V =$ line voltage in kilovolts and $C =$ line conductance in mhos. Find stationary points for the function, and determine V and C to minimize the operating cost.

4.33 $f(x_1, x_2) = x_1^2 + 2x_2^2 - 4x_1 - 2x_1x_2$

4.34 $f(x_1, x_2) = 12x_1^2 + 22x_2^2 - 1.5x_1 - x_2$

4.35 $f(x_1, x_2) = 7x_1^2 + 12x_2^2 - x_1$

4.36 $f(x_1, x_2) = 12x_1^2 + 21x_2^2 - x_2$

4.37 $f(x_1, x_2) = 25x_1^2 + 20x_2^2 - 2x_1 - x_2$

4.38 $f(x_1, x_2, x_3) = x_1^2 + 2x_2^2 + 2x_3^2 + 2x_1x_2 + 2x_2x_3$

4.39 $f(x_1, x_2) = 8x_1^2 + 8x_2^2 - 80\sqrt{x_1^2 + x_2^2 - 20x_2 + 100} - 80\sqrt{x_1^2 + x_2^2 + 20x_2 + 100} - 5x_1 - 5x_2$

4.40 $f(x_1, x_2) = 9x_1^2 + 9x_2^2 - 100\sqrt{x_1^2 + x_2^2 - 20x_2 + 100} - 64\sqrt{x_1^2 + x_2^2 + 16x_2 + 64} - 5x_1 - 41x_2$

4.41 $f(x_1, x_2) = 100(x_2 - x_1^2)^2 + (1 - x_1)^2$

4.42 $f(x_1, x_2, x_3, x_4) = (x_1 - 10x_2)^2 + 5(x_3 - x_4)^2 + (x_2 - 2x_3)^4 + 10(x_1 - x_4)^4$

Section 4.5 Necessary Conditions: Equality Constrained Problem

Find points satisfying the necessary conditions for the following problems; check if they are optimum points using the graphical method (if possible).

4.43 Minimize $f(x_1, x_2) = 4x_1^2 + 3x_2^2 - 5x_1x_2 - 8x_1$

subject to $x_1 + x_2 = 4$

4.44 Maximizte $f(x_1, x_2) = 4x_1^2 + 3x_2^2 - 5x_1x_2 - 8x_1$

subject to $x_1 + x_2 = 4$

4.45 Minimize $f(x_1, x_2) = (x_1 - 2)^2 + (x_2 + 1)^2$

subject to $2x_1 + 3x_2 - 4 = 0$

4.46 Minimize $f(x_1, x_2) = 4x_1^2 + 9x_2^2 + 6x_2 - 4x_1 + 13$

subject to $x_1 - 3x_2 + 3 = 0$

4.47 Minimize $f(\mathbf{x}) = (x_1 - 1)^2 + (x_2 + 2)^2 + (x_3 - 2)^2$

subject to $2x_1 + 3x_2 - 1 = 0$

$x_1 + x_2 + 2x_3 - 4 = 0$

4.48 Minimize $f(x_1, x_2) = 9x_1^2 + 18x_1x_2 + 13x_2^2 - 4$

subject to $x_1^2 + x_2^2 + 2x_1 = 16$

4.49 Minimize $f(x_1, x_2) = (x_1 - 1)^2 + (x_2 - 1)^2$

subject to $x_1 + x_2 - 4 = 0$

4.50 Consider the following problem with equality constraints:

Minimize $(x_1 - 1)^2 + (x_2 - 1)^2$

subject to $x_1 + x_2 - 4 = 0$

$x_1 - x_2 - 2 = 0$

1. Is it a valid optimization problem? Explain.

2. Explain how you would solve the problem? Are necessary conditions needed to find the optimum solution?

4.51 Minimize $f(x_1, x_2) = 4x_1^2 + 3x_2^2 - 5x_1x_2 - 8$

subject to $x_1 + x_2 = 4$

4.52 Maximize $F(x_1, x_2) = 4x_1^2 + 3x_2^2 - 5x_1x_2 - 8$

subject to $x_1 + x_2 = 4$

Section 4.6 Necessary Conditions for General Constrained Problem

4.53 *Answer True or False.*

1. A regular point of the feasible region is defined as a point where the cost function gradient is independent of the gradients of active constraints.
2. A point satisfying KKT conditions for a general optimum design problem can be a local max-point for the cost function.
3. At the optimum point, the number of active independent constraints is always more than the number of design variables.
4. In the general optimum design problem formulation, the number of independent equality constraints must be "$\leq$" to the number of design variables.
5. In the general optimum design problem formulation, the number of inequality constraints cannot exceed the number of design variables.
6. At the optimum point, Lagrange multipliers for the "$\leq$ type" inequality constraints must be non-negative.
7. At the optimum point, the Lagrange multiplier for a "$\leq$ type" constraint can be zero.
8. While solving an optimum design problem by KKT conditions, each case defined by the switching conditions can have multiple solutions.
9. In optimum design problem formulation, "$\geq$ type" constraints cannot be treated.
10. Optimum design points for constrained optimization problems give stationary value to the Lagrange function with respect to design variables.
11. Optimum design points having at least one active constraint give stationary value to the cost function.
12. At a constrained optimum design point that is regular, the cost function gradient is linearly dependent on the gradients of the active constraint functions.
13. If a slack variable has zero value at the optimum, the inequality constraint is inactive.
14. Gradients of inequality constraints that are active at the optimum point must be zero.
15. Design problems with equality constraints have the gradient of the cost function as zero at the optimum point.

Find points satisfying KKT necessary conditions for the following problems; check if they are optimum points using the graphical method (if possible).

4.54 Maximize $F(x_1, x_2) = 4x_1^2 + 3x_2^2 - 5x_1x_2 - 8$
subject to $x_1 + x_2 \leq 4$

4.55 Minimize $f(x_1, x_2) = 4x_1^2 + 3x_2^2 - 5x_1x_2 - 8$
subject to $x_1 + x_2 \leq 4$

4.56 Maximize $F(x_1, x_2) = 4x_1^2 + 3x_2^2 - 5x_1x_2 - 8x_1$
subject to $x_1 + x_2 \leq 4$

4.57 Minimize $f(x_1, x_2) = (x_1 - 1)^2 + (x_2 - 1)^2$
subject to $x_1 + x_2 \geq 4$
$$x_1 - x_2 - 2 = 0$$

4.58 Minimize $f(x_1, x_2) = (x_1 - 1)^2 + (x_2 - 1)^2$
subject to $x_1 + x_2 = 4$
$$x_1 - x_2 - 2 \geq 0$$

4.59 Minimize $f(x_1, x_2) = (x_1 - 1)^2 + (x_2 - 1)^2$

subject to $x_1 + x_2 \geq 4$

$x_1 - x_2 \geq 2$

4.60 Minimize $f(x, y) = (x - 4)^2 + (y - 6)^2$

subject to $12 \geq x + y$

$x \geq 6, y \geq 0$

4.61 Minimize $f(x_1, x_2) = 2x_1 + 3x_2 - x_1^3 - 2x_2^2$

subject to $x_1 + 3x_2 \leq 6$

$5x_1 + 2x_2 \leq 10$

$x_1, x_2 \geq 0$

4.62 Minimize $f(x_1, x_2) = 4x_1^2 + 3x_2^2 - 5x_1x_2 - 8x_1$

subject to $x_1 + x_2 \leq 4$

4.63 Minimize $f(x_1, x_2) = x_1^2 + x_2^2 - 4x_1 - 2x_2 + 6$

subject to $x_1 + x_2 \geq 4$

4.64 Minimize $f(x_1, x_2) = 2x_1^2 - 6x_1x_2 + 9x_2^2 - 18x_1 + 9x_2$

subject to $x_1 + 2x_2 \leq 10$

$4x_1 - 3x_2 \leq 20; x_i \geq 0; i = 1, 2$

4.65 Minimize $f(x_1, x_2) = (x_1 - 1)^2 + (x_2 - 1)^2$

subject to $x_1 + x_2 - 4 \leq 0$

4.66 Minimize $f(x_1, x_2) = (x_1 - 1)^2 + (x_2 - 1)^2$

subject to $x_1 + x_2 - 4 \leq 0$

$x_1 - x_2 - 2 \leq 0$

4.67 Minimize $f(x_1, x_2) = (x_1 - 1)^2 + (x_2 - 1)^2$

subject to $x_1 + x_2 - 4 \leq 0$

$2 - x_1 \leq 0$

4.68 Minimize $f(x_1, x_2) = 9x_1^2 - 18x_1x_2 + 13x_2^2 - 4$

subject to $x_1^2 + x_2^2 + 2x_1 \geq 16$

4.69 Minimize $f(x_1, x_2) = (x_1 - 3)^2 + (x_2 - 3)^2$

subject to $x_1 + x_2 \leq 4$

$x_1 - 3x_2 = 1$

4.70 Minimize $f(x_1, x_2) = x_1^3 - 16x_1 + 2x_2 - 3x_2^2$

subject to $x_1 + x_2 \leq 3$

4.71 Minimize $f(x_1, x_2) = 3x_1^2 - 2x_1x_2 + 5x_2^2 + 8x_2$

subject to $x_1^2 - x_2^2 + 8x_2 \leq 16$

4.72 Minimize $f(x, y) = (x - 4)^2 + (y - 6)^2$

subject to $x + y \leq 12$

$x \leq 6$

$x, y \geq 0$

4.73 Minimize $f(x, y) = (x - 8)^2 + (y - 8)^2$

subject to $x + y \leq 12$

$x \leq 6$

$x, y \geq 0$

I. THE BASIC CONCEPTS

4.74 Maximize $F(x, y) = (x - 4)^2 + (y - 6)^2$
  subject to $x + y \leq 12$
  $\qquad 6 \geq x$
  $\qquad x, y \geq 0$

4.75 Maximize $F(r, t) = (r - 8)^2 + (t - 8)^2$
  subject to $10 \geq r + t$
  $\qquad t \leq 5$
  $\qquad r, t \geq 0$

4.76 Maximize $F(r, t) = (r - 3)^2 + (t - 2)^2$
  subject to $10 \geq r + t$
  $\qquad t \leq 5$
  $\qquad r, t \geq 0$

4.77 Maximize $F(r, t) = (r - 8)^2 + (t - 8)^2$
  subject to $r + t \leq 10$
  $\qquad t \geq 0$
  $\qquad r \leq 0$

4.78 Maximize $F(r, t) = (r - 3)^2 + (t - 2)^2$
  subject to $10 \geq r + t$
  $\qquad t \geq 5$
  $\qquad r, t \geq 0$

4.79 Consider the problem of designing the "can" formulated in Section 2.2. Write KKT conditions and solve them. Interpret the necessary conditions at the solution point graphically.

4.80 A minimum weight tubular column design problem is formulated in Section 2.7 using mean radius R and thickness t as design variables. Solve the KKT conditions for the problem imposing an additional constraint $R/t \leq 50$ for this data: $P = 50$ kN, $l = 5.0$ m, $E = 210$ GPa, $\sigma_a = 250$ MPa and $\rho = 7850$ kg/m^3. Interpret the necessary conditions at the solution point graphically.

4.81 A minimum weight tubular column design problem is formulated in Section 2.7 using outer radius R_o and inner radius R_i as design variables. Solve the KKT conditions for the problem imposing an additional constraint $0.5(R_o + R_i)/(R_o - R_i) \leq 50$ Use the same data as in Exercise 4.80. Interpret the necessary conditions at the solution point graphically.

4.82 An engineering design problem is formulated as
  Minimize $f(x_1, x_2) = x_1^2 + 320x_1x_2$
  subject to $\frac{1}{100}(x_1 - 60x_2) \leq 0$

$$1 - \frac{1}{3600}x_1(x_1 - x_2) \leq 0$$

  $\qquad x_1, x_2 \geq 0$

Write KKT necessary conditions and solve for the candidate minimum designs. Verify the solutions graphically. Interpret the KKT conditions on the graph for the problem.

Formulate and solve the following problems graphically. Verify the KKT conditions at the solution point and show gradients of the cost function and active constraints on the graph.

4.83 Exercise 2.1	**4.84** Exercise 2.2	**4.85** Exercise 2.3
4.86 Exercise 2.4	**4.87** Exercise 2.5	**4.88** Exercise 2.6
4.89 Exercise 2.7	**4.90** Exercise 2.8	**4.91** Exercise 2.9
4.92 Exercise 2.10	**4.93** Exercise 2.11	**4.94** Exercise 2.12
4.95 Exercise 2.13	**4.96** Exercise 2.14	

Section 4.7 Physical Meaning of Lagrange Multipliers

Solve the following problems graphically, verify the KKT necessary conditions for the solution points and study the effect on the cost function of changing the boundary of the active constraint(s) by one unit.

4.97 Exercise 4.43	**4.98** Exercise 4.44	**4.99** Exercise 4.45
4.100 Exercise 4.46	**4.101** Exercise 4.47	**4.102** Exercise 4.48
4.103 Exercise 4.49	**4.104** Exercise 4.50	**4.105** Exercise 4.51
4.106 Exercise 4.52	**4.107** Exercise 4.54	**4.108** Exercise 4.55
4.109 Exercise 4.56	**4.110** Exercise 4.57	**4.111** Exercise 4.58
4.112 Exercise 4.59	**4.113** Exercise 4.60	**4.114** Exercise 4.61
4.115 Exercise 4.62	**4.116** Exercise 4.63	**4.117** Exercise 4.64
4.118 Exercise 4.65	**4.119** Exercise 4.66	**4.120** Exercise 4.67
4.121 Exercise 4.68	**4.122** Exercise 4.69	**4.123** Exercise 4.70
4.124 Exercise 4.71	**4.125** Exercise 4.72	**4.126** Exercise 4.73
4.127 Exercise 4.74	**4.128** Exercise 4.75	**4.129** Exercise 4.76
4.130 Exercise 4.77	**4.131** Exercise 4.78	

Section 4.8 Global Optimality

4.132 *Answer True or False.*

1. A linear inequality constraint always defines a convex feasible region.
2. A linear equality constraint always defines a convex feasible region.
3. A nonlinear equality constraint cannot give a convex feasible region.
4. A function is convex if and only if its Hessian is positive definite everywhere.
5. An optimum design problem is convex if all constraints are linear and the cost function is convex.
6. A convex programming problem always has an optimum solution.
7. An optimum solution for a convex programming problem is always unique.
8. A nonconvex programming problem cannot have global optimum solution.
9. For a convex design problem, the Hessian of the cost function must be positive semidefinite everywhere.
10. Checking for the convexity of a function can actually identify a domain over which the function may be convex.

4.133 Using the definition of a line segment given in Eq. (4.71), show that the following set is convex
$$S = \{\mathbf{x} \mid x_1^2 + x_2^2 - 1.0 \le 0\}$$

4.134 Find the domain for which the following functions are convex: (i) $\sin x$, (ii) $\cos x$.

Check for convexity of the following functions. If the function is not convex everywhere, then determine the domain (feasible set S) over which the function is convex.

4.135 $f(x_1, x_2) = 3x_1^2 + 2x_1x_2 + 2x_2^2 + 7$

4.136 $f(x_1, x_2) = x_1^2 + 4x_1x_2 + x_2^2 + 3$

4.137 $f(x_1, x_2) = x_1^3 + 12x_1x_2^2 + 2x_2^2 + 5x_1^2 + 3x_2$

4.138 $f(x_1, x_2) = 5x_1 - \frac{1}{16}x_1^2x_2^2 + \frac{1}{4x_1}x_2^2$

4.139 $f(x_1, x_2) = x_1^2 + x_1x_2 + x_2^2$

4.140 $U = \frac{(21.9 \times 10^7)}{V^2C} + (3.9 \times 10^6)C + 1000 V$

4.141 Consider the problem of designing the "can" formulated in Section 2.2. Check convexity of the problem. Solve the problem graphically and check the KKT conditions at the solution point.

Formulate and check convexity of the following problems; solve the problems graphically and verify the KKT conditions at the solution point.

4.142 Exercise 2.1	**4.143** Exercise 2.3	**4.144** Exercise 2.4
4.145 Exercise 2.5	**4.146** Exercise 2.9	**4.147** Exercise 2.10
4.148 Exercise 2.12	**4.149** Exercise 2.14	

Section 4.9 Engineering Design Examples

4.150 The problem of minimum weight design of the symmetric three-bar truss of Figure 2.6 is formulated as follows:

Minimize $f(x_1, x_2) = 2x_1 + x_2$

subject to the constraints

$$g_1 = \frac{1}{\sqrt{2}}\left[\frac{P_u}{x_1} + \frac{P_v}{(x_1 + \sqrt{2}x_2)}\right] - 20{,}000 \le 0$$

$$g_2 = \frac{\sqrt{2}P_v}{(x_1 + \sqrt{2}x_2)} - 20{,}000 \le 0$$

$$g_3 = -x_1 \le 0$$

$$g_4 = -x_2 \le 0$$

where x_1 is the cross-sectional area of members 1 and 3 (symmetric structure) and x_2 is the cross-sectional area of member 2, $P_u = P\cos\theta$, $P_v = P\sin\theta$, with $P > 0$ and $0 \le \theta \le 90$. Check for convexity of the problem for $\theta = 60°$.

4.151 For the three-bar truss problem of Exercise 4.150, consider the case of KKT conditions with g_1 as the only active constraint. Solve the conditions for optimum solution and determine the range for the load angle θ for which the solution is valid.

4.152 For the three-bar truss problem of Exercise 4.150, consider the case of KKT conditions with only g_1 and g_2 as active constraints. Solve the conditions for optimum solution and determine the range for the load angle θ for which the solution is valid.

4.153 For the three-bar truss problem of Exercise 4.150, consider the case of KKT conditions with g_2 as the only active constraint. Solve the conditions for optimum solution and determine the range for the load angle θ for which the solution is valid.

4.154 For the three-bar truss problem of Exercise 4.150, consider the case of KKT conditions with g_1 and g_4 as active constraints. Solve the conditions for optimum solution and determine the range for the load angle θ for which the solution is valid.

More on Optimum Design Concepts
Optimality Conditions

- Write and use an alternate form of optimality conditions for constrained problems

- Determine if the candidate minimum points are irregular

- Check the second-order optimality conditions at the candidate minimum points for general constrained problems

- Describe duality theory in nonlinear programming

In this chapter, we discuss some additional topics related to the optimality condition for constrained problems. Implications of the regularity requirement in the Karush-Kuhn-Tucker (KKT) necessary conditions are discussed. Second-order optimality conditions for the problem are presented and discussed. These topics are usually not covered in a first course on optimization. Also, they may be omitted in a first reading of this book. They are more suitable for a second course or a graduate level course on the subject.

5.1 ALTERNATE FORM OF KKT NECESSARY CONDITIONS

There is an alternate but entirely equivalent form for the KKT necessary conditions. In this form, the slack variables are not added to the inequality constraints and the conditions of Eqs. (4.46) through (4.52) are written without them. It can be seen that in the necessary conditions of Eqs. (4.46) through (4.52), the slack variable s_i appears in only two equations: Eq. (4.49) as $g_i(\mathbf{x}^*) + s_i^2 = 0$, and Eq. (4.51) as $u_i^* s_i = 0$. We will show that both the equations can be written in an equivalent form without the slack variable s_i.

Consider first Eq. (4.49): $g_i(\mathbf{x}^*) + s_i^2 = 0$ for $i = 1$ to m. The purpose of this equation is to ensure that all the inequalities remain satisfied at the candidate minimum point. The equation can be written as $s_i^2 = -g_i(\mathbf{x}^*)$ and, since $s_i^2 \geq 0$ ensures satisfaction of the constraint,

TABLE 5.1 Alternate form of KKT necessary conditions

Problem: Minimize $f(x)$ subject to $h_i(x) = 0$, $i = 1$ to p; $g_j(x) \leq 0$, $j = 1$ to m

1. Lagrangian function definition	$L = f + \sum_{i=1}^{p} v_i h_i + \sum_{j=1}^{m} u_j g_j$	(5.1)
2. Gradient conditions	$\dfrac{\partial L}{\partial x_k} = 0; \quad \dfrac{\partial f}{\partial x_k} + \sum_{i=1}^{p} v_i^* \dfrac{\partial h_i}{\partial x_k} + \sum_{j=1}^{m} u_j^* \dfrac{\partial g_j}{\partial x_k} = 0; \quad k = 1$ to n	(5.2)
3. Feasibility check	$h_i(\mathbf{x}^*) = 0; \quad i = 1$ to $p; \quad g_j(\mathbf{x}^*) \leq 0; \quad j = 1$ to m	(5.3)
4. Switching conditions	$u_j^* g_j(\mathbf{x}^*) = 0; \quad j = 1$ to m	(5.4)
5. Non-negativity of Lagrange multipliers for inequalities	$u_j^* \geq 0; \quad j = 1$ to m	(5.5)
6. Regularity check: Gradients of active constraints must be linearly independent. In such a case, the Lagrange multipliers for the constraints are unique.		

we get $-g_i(\mathbf{x}^*) \geq 0$, or $g_i(\mathbf{x}^*) \leq 0$ for $i = 1$ to m. Thus, Eq. (4.49), $g_i(\mathbf{x}^*) + s_i^2 = 0$, can be simply replaced by $g_i(\mathbf{x}^*) \leq 0$.

The second equation involving the slack variable is Eq. (4.51), $u_i^* s_i = 0$, $i = 1$ to m. Multiplying the equation by s_i, we get $u_i^* s_i^2 = 0$. Now substituting $s_i^2 = -g_i(\mathbf{x}^*)$, we get $u_i^* g_i(\mathbf{x}^*) = 0$, $i = 1$ to m. This way the slack variable is eliminated from the equation and the switching condition of Eq. (4.51) can be written as $u_i^* g_i(\mathbf{x}^*) = 0$, $i = 1$ to m. These conditions can be used to define various cases as $u_i^* = 0$ or $g_i = 0$ (instead of $s_i = 0$). Table 5.1 gives the KKT conditions of Theorem 4.6 in the alternate form without the slack variables, and Examples 5.1 and 5.2 provide an illustration of its use.

EXAMPLE 5.1 USE OF THE ALTERNATE FORM OF THE KKT CONDITIONS

Minimize

$$f(x, y) = (x - 10)^2 + (y - 8)^2 \tag{a}$$

subject to

$$g_1 = x + y - 12 \leq 0 \tag{b}$$

$$g_2 = x - 8 \leq 0 \tag{c}$$

Solution

The problem is already expressed in the standard form. The KKT conditions are

1. Lagrangian function definition of Eq. (5.1):

$$L = (x - 10)^2 + (y - 8)^2 + u_1(x + y - 12) + u_2(x - 8) \tag{d}$$

2. Gradient condition of Eq. (5.2):

$$\frac{\partial L}{\partial x} = 2(x - 10) + u_1 + u_2 = 0$$

$$\frac{\partial L}{\partial y} = 2(y - 8) + u_1 = 0$$

(e)

3. Feasibility check of Eq. (5.3):

$$g_1 \leq 0, \quad g_2 \leq 0$$

(f)

4. Switching conditions of Eq. (5.4):

$$u_1 g_1 = 0, \quad u_2 g_2 = 0$$

(g)

5. Non-negativity of Lagrange multipliers of Eq. (5.5):

$$u_1, u_2 \geq 0$$

(h)

6. Regularity check.

The switching conditions of Eq. (g) give the following four cases:

1. $u_1 = 0$, $u_2 = 0$ (both g_1 and g_2 inactive)
2. $u_1 = 0$, $g_2 = 0$ (g_1 inactive, g_2 active)
3. $g_1 = 0$, $u_2 = 0$ (g_1 active, g_2 inactive)
4. $g_1 = 0$, $g_2 = 0$ (both g_1 and g_2 active)

Case 1: $u_1 = 0$, $u_2 = 0$ (both g_1 and g_2 inactive). Equations (a) give the solution as, $x = 10$, $y = 8$. Checking feasibility of this point gives $g_1 = 6 > 0$, $g_2 = 2 > 0$; thus both constraints are violated and so this case does not give any feasible candidate minimum point.

Case 2: $u_1 = 0$, $g_2 = 0$ (g_1 inactive, g_2 active). $g_2 = 0$ gives $x = 8$. Equations (a) give $y = 8$ and $u_2 = 4$. At the point (8, 8), $g_1 = 4 > 0$, which is a violation. Thus the point (8, 8) is infeasible and this case also does not give any feasible candidate minimum points.

Case 3: $g_1 = 0$, $u_2 = 0$ (g_1 active, g_2 inactive). Equations (a) and $g_1 = 0$ give $x = 7$, $y = 5$, $u_1 = 6 > 0$. Checking feasibility, $g_2 = -1 < 0$, which is satisfied. Since there is only one active constraint, the question of linear dependence of gradients of active constraints does not arise; therefore, regularity condition is satisfied. Thus point (7, 5) satisfies all the KKT necessary conditions.

Case 4: $g_1 = 0$, $g_2 = 0$ (both g_1 and g_2 active). The case $g_1 = 0$, $g_2 = 0$ gives $x = 8$, $y = 4$. Equations (a) give $u_1 = 8$, $u_2 = -4 < 0$, which is a violation of the necessary conditions. Therefore, this case also does not give any candidate minimum points.

It may be checked that this is a convex programming problem since constraints are linear and the cost function is convex. Therefore, the point obtained in Case 3 is indeed a global minimum point according to the convexity results of Section 4.8.

EXAMPLE 5.2 CHECK FOR KKT NECESSARY CONDITIONS

An optimization problem has one equality constraint h and one inequality constraint g. Check the KKT necessary conditions at what is believed to be the minimum point using the following information:

$$h = 0, \quad g = 0, \quad \nabla f = (2, 3, 2), \quad \nabla h = (1, -1, 1), \quad \nabla g = (-1, -2, -1)$$

(a)

Solution

At the candidate minimum point, the gradients of h and g are linearly independent, so the given point is regular. The KKT conditions are

$$\nabla L = \nabla f + v\nabla h + u\nabla g = 0$$
$$h = 0, \quad g \leq 0, \quad ug = 0, \quad u \geq 0 \tag{b}$$

Substituting for ∇f, ∇h and ∇g, we get the following three equations:

$$2 + v - u = 0, \quad 3 - v - 2u = 0, \quad 2 + v - u = 0 \tag{c}$$

These are three equations in two unknowns; however, only two of them are linearly independent. Solving for u and v, we get $u = 5/3 \geq 0$ and $v = -1/3$. Thus, all of the KKT necessary conditions are satisfied.

5.2 IRREGULAR POINTS

In all of the examples that have been considered thus far it is implicitly assumed that conditions of the KKT Theorem 4.6 or the Lagrange Theorem 4.5 are satisfied. In particular; we have assumed that $\mathbf{x}^*$ is a *regular point* of the feasible design space. That is, gradients of all the active constraints at $\mathbf{x}^*$ are linearly independent (i.e., they are not parallel to each other, nor can any gradient be expressed as a linear combination of others). It must be realized that necessary conditions are *applicable only if the assumption of the regularity of* $\mathbf{x}^*$ is satisfied. To show that the necessary conditions are not applicable if $\mathbf{x}^*$ is not a regular point, we consider Example 5.3.

EXAMPLE 5.3   CHECK FOR KKT CONDITIONS AT IRREGULAR POINTS

Minimize

$$f(x_1, x_2) = x_1^2 + x_2^2 - 4x_1 + 4 \tag{a}$$

subject to

$$g_1 = -x_1 \leq 0 \tag{b}$$

$$g_2 = -x_2 \leq 0 \tag{c}$$

$$g_3 = x_2 - (1 - x_1)^3 \leq 0 \tag{d}$$

Check if the minimum point (1, 0) satisfies the KKT necessary conditions (McCormick, 1967).

Solution

The graphical solution, shown in Figure 5.1, gives the global minimum for the problem at $\mathbf{x}^* = (1, 0)$. Let us see if the solution satisfies the KKT necessary conditions:

1. Lagrangian function definition of Eq. (5.1):

$$L = x_1^2 + x_2^2 - 4x_1 + 4 + u_1(-x_1) + u_2(-x_2) + u_3(x_2 - [1 - x_1]^3) \tag{e}$$

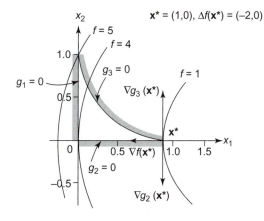

$\mathbf{x}^* = (1,0),\ \Delta f(\mathbf{x}^*) = (-2,0)$

FIGURE 5.1   Graphic solution for Example 5.3: irregular optimum point.

2. Gradient condition of Eq. (5.2):

$$\frac{\partial L}{\partial x_1} = 2x_1 - 4 - u_1 + u_3(3)(1 - x_1)^2 = 0$$

$$\frac{\partial L}{\partial x_2} = 2x_2 - u_2 + u_3 = 0 \tag{f}$$

3. Feasibility check of Eq. (5.3):

$$g_i \leq 0, \quad i = 1, 2, 3 \tag{g}$$

4. Switching conditions of Eq. (5.4):

$$u_i g_i = 0, \quad i = 1, 2, 3 \tag{h}$$

5. Non-negativity of Lagrange multipliers of Eq. (5.5):

$$u_i \geq 0, \quad i = 1, 2, 3 \tag{i}$$

6. Regularity check. At $\mathbf{x}^* = (1, 0)$ the first constraint (g_1) is inactive and the second and third constraints are active. The switching conditions in Eq. (h) identify the case as $u_1 = 0$, $g_2 = 0$, $g_3 = 0$. Substituting the solution into Eqs. (f), we find that the first equation gives $-2 = 0$ and therefore, it is not satisfied. Thus, the KKT necessary conditions are not satisfied at the minimum point.

This apparent *contradiction* can be resolved by checking the regularity condition at the minimum point $\mathbf{x}^* = (1, 0)$. The gradients of the active constraints g_2 and g_3 are given as

$$\nabla g_2 = \begin{bmatrix} 0 \\ -1 \end{bmatrix}; \quad \nabla g_3 = \begin{bmatrix} 0 \\ 1 \end{bmatrix} \tag{j}$$

These vectors are not linearly independent. They are along the same line but in opposite directions, as shown in Figure 5.1. Thus $\mathbf{x}^*$ is not a regular point of the feasible set. Since this is assumed in the KKT conditions, their use is invalid here. Note also that the geometrical interpretation of the KKT conditions of Eq. (4.53) is violated; that is, for the present example, ∇f at (1, 0) cannot be written as a linear combination of the gradients of the active constraints g_2 and g_3. Actually, ∇f is normal to both ∇g_2 and ∇g_3, as shown in the figure.

Note that for some problems, irregular points can be obtained as a solution to the KKT conditions; however, in such cases, the Lagrange multipliers of the active constraints cannot be guaranteed to be unique. Also, the constraint variation sensitivity result of Section 4.7 may or may not be applicable to some values of the Lagrange multipliers.

5.3 SECOND-ORDER CONDITIONS FOR CONSTRAINED OPTIMIZATION

Solutions to the first-order necessary conditions are candidate local minimum designs. In this section, we will discuss second-order necessary and sufficiency conditions for constrained optimization problems. As in the unconstrained case, *second-order information* about the functions at the candidate point x^* will be used to determine if the point is indeed a local minimum. Recall for the unconstrained problem that the local sufficiency of Theorem 4.4 requires the quadratic part of Taylor's expansion for the function at x^* to be positive for all nonzero design changes d. *In the constrained case, we must also consider active constraints at x^* to determine feasible changes d.* We will consider only the points $x = x^* + d$ in the neighborhood of x^* that satisfy the active constraint equations.

Any $d \neq 0$ satisfying active constraints to the first order must be in the constraint tangent hyperplane (Figure 5.2). Such d's are then orthogonal to the gradients of the active constraints since constraint gradients are normal to the constraint tangent hyperplane. Therefore, the dot product of d with each of the constraint gradients ∇h_i and ∇g_i must be zero; that is, $\nabla h_i^T d = 0$ and $\nabla g_i^T d = 0$. These equations are used to determine directions d that define a feasible region around the point x^*. Note that only active inequality constraints ($g_i = 0$) are used in determining d. The situation is depicted in Figure 5.2 for one inequality constraint.

To derive the second-order conditions, we write Taylor's expansion of the Lagrange function and consider only those d that satisfy the preceding conditions. x^* is then a *local minimum point if the second-order term of Taylor's expansion is positive for all d in the constraint tangent hyperplane.* This is then the sufficient condition for an isolated local minimum point. As a necessary condition the second-order term must be nonnegative. We summarize these results in Theorems 5.1 and 5.2.

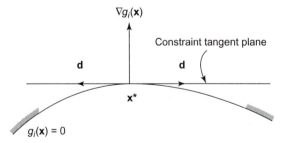

FIGURE 5.2   Directions d used in second-order conditions.

THEOREM 5.1

Second-Order Necessary Conditions for General Constrained Problems Let x^* satisfy the first-order KKT necessary conditions for the general optimum design problem. Define the Hessian of the Lagrange function L at x^* as

$$\nabla^2 L = \nabla^2 f + \sum_{i=1}^{p} v_i^* \nabla^2 h_i + \sum_{i=1}^{m} u_i^* \nabla^2 g_i \quad (5.6)$$

Let there be nonzero feasible directions, $\mathbf{d} \neq \mathbf{0}$, satisfying the following linear systems at the point x^*:

$$\nabla h_i^T \mathbf{d} = 0; \quad i = 1 \text{ to } p \quad (5.7)$$

$$\nabla g_i^T \mathbf{d} = 0;$$

for all active inequalities (i.e., for those i with $g_i(x^*) = 0$)   (5.8)

Then, if x^* is a local minimum point for the optimum design problem, it must be true that

$$Q \geq 0 \quad \text{where} \quad Q = \mathbf{d}^T \nabla^2 L(x^*) \mathbf{d} \quad (5.9)$$

Note that any point that does not satisfy the second-order necessary conditions cannot be a local minimum point.

THEOREM 5.2

Sufficient Conditions for General Constrained Problems Let x^* satisfy the first-order KKT necessary conditions for the general optimum design problem. Define the Hessian of the Lagrange function L at x^* as shown in Eq. (5.6). Define nonzero feasible directions, $\mathbf{d} \neq \mathbf{0}$, as solutions to the linear systems:

$$\nabla h_i^T \mathbf{d} = 0; \quad i = 1 \text{ to } p \quad (5.10)$$

$$\nabla g_i^T \mathbf{d} = 0 \text{ for all active}$$

inequalities with $u_i^* > 0$ (5.11)

Also let $\nabla g_i^T \mathbf{d} \leq 0$ for those active inequalities with $u_i^* = 0$. If

$$Q > 0, \quad \text{where} \quad Q = \mathbf{d}^T \nabla^2 L(x^*) \mathbf{d} \quad (5.12)$$

then x^* is an *isolated local minimum* point (*isolated* means that there are no other local minimum points in the neighborhood of x^*).

Insights for Second-Order Conditions

Note first the difference in the conditions for the directions $\mathbf{d}$ in Eq. (5.8) for the necessary condition and Eq. (5.11) for the sufficient condition. In Eq. (5.8) all active inequalities with non-negative multipliers are included, whereas in Eq. (5.11) only those active inequalities with a positive multiplier are included. Equations (5.10) and (5.11) simply say that the dot product of vectors ∇h_i and $\mathbf{d}$ and ∇g_i (having $u_i^* > 0$) and $\mathbf{d}$ should be zero. Thus, only the $\mathbf{d}$ orthogonal to the gradients of equality and active inequality constraints with $u_i^* > 0$ are considered. Stated differently, only $\mathbf{d}$ in the tangent hyperplane to the active constraints at the candidate minimum point are considered. Equation (5.12) says that the Hessian of the Lagrangian must be positive definite for all $\mathbf{d}$ lying in the constraint tangent hyperplane. Note that ∇h_i, ∇g_i, and $\nabla^2 L$ are calculated at the candidate local minimum points $\mathbf{x}^*$ satisfying the KKT necessary conditions.

It should also be emphasized that if the inequality in Eq. (5.12) is not satisfied (i.e., $Q \not> 0$), we cannot conclude that $\mathbf{x}^*$ is not a local minimum. It may still be a local minimum but not an isolated one. Note also that the theorem cannot be used for any $\mathbf{x}^*$ if its assumptions are not satisfied. In that case, we cannot draw any conclusions for the point $\mathbf{x}^*$.

It is important to note that if matrix $\nabla^2 L(\mathbf{x}^*)$ is negative definite or negative semidefinite then the second-order necessary condition in Eq. (5.9) for a local minimum is violated and $\mathbf{x}^*$ cannot be a local minimum point. Also if $\nabla^2 L(\mathbf{x}^*)$ is positive definite (i.e., Q in Eq. (5.12) is positive for any $\mathbf{d} \neq 0$) then $\mathbf{x}^*$ satisfies the sufficiency condition for an isolated local minimum and no further checks are needed. The reason is that if $\nabla^2 L(\mathbf{x}^*)$ is positive definite, then it is positive definite for those $\mathbf{d}$ that satisfy Eqs. (5.10) and (5.11). However, *if $\nabla^2 L(\mathbf{x}^*)$ is not positive definite (i.e., it is positive semidefinite or indefinite), then we cannot conclude that $\mathbf{x}^*$ is not an isolated local minimum.* We must calculate $\mathbf{d}$ to satisfy Eqs. (5.10) and (5.11) and carry out the sufficiency test given in Theorem 5.2. This result is summarized in Theorem 5.3.

THEOREM 5.3

Strong Sufficient Condition   Let $\mathbf{x}^*$ satisfy the first-order Karush-Kuhn-Tucker necessary conditions for the general optimum design problem. Define Hessian $\nabla^2 L$ ($\mathbf{x}^*$) for the Lagrange function at $\mathbf{x}^*$ as shown in Eq. (5.6). Then, if $\nabla^2 L(\mathbf{x}^*)$ is positive definite, $\mathbf{x}^*$ is an isolated minimum point.

One case arising in some applications needs special mention. This occurs when the total number of active constraints (with at least one inequality) at the candidate minimum point $\mathbf{x}^*$ is equal to the number of independent design variables; that is, there are no design degrees of freedom at the candidate minimum point. Since $\mathbf{x}^*$ satisfies the KKT necessary conditions, the gradients of all the active constraints are linearly independent. Thus, the only solution for the system of Eqs. (5.10) and (5.11) is $\mathbf{d} = 0$ and Theorem 5.2 cannot be used. However, since $\mathbf{d} = 0$ is the only solution, there are no feasible directions in the neighborhood that can reduce the cost function any further. Thus, the point $\mathbf{x}^*$ is indeed a local minimum for the cost function (see also the definition of a local minimum in Section 4.1.1). We consider Examples 5.4 through 5.6 to illustrate the use of second-order conditions of optimality.

EXAMPLE 5.4   CHECK FOR SECOND-ORDER CONDITIONS

Check the second-order condition for Example 4.30:

Minimize

$$f(x) = \frac{1}{3}x^3 - \frac{1}{2}(b+c)x^2 + bcx + f_0 \tag{a}$$

subject to

$$a \leq x \leq d \tag{b}$$

where $0 < a < b < c < d$ and f_0 are specified constants.

Solution

There is only one constrained candidate local minimum point, $x = a$. Since there is only one design variable and one active constraint, the condition $\nabla g_1 \bar{d} = 0$ of Eq. (5.11) gives $\bar{d} = 0$ as the only solution (note that $\bar{d}$ is used as a direction for sufficiency check since d is used as a constant in the example). Therefore, Theorem 5.2 cannot be used for a sufficiency check. Also note that at $x = a$, $d^2L/dx^2 = 2a - b - c$, which can be positive, negative, or zero depending on the values of a, b, and c, so we cannot use curvature of the function to check the sufficiency condition (Strong Sufficient Theorem 5.3). However, from Figure 4.16 we observe that $x = a$ is indeed an isolated local minimum point. From this example we can conclude that if the number of active inequality constraints is equal to the number of independent design variables and all other KKT conditions are satisfied, then the candidate point is indeed a local minimum point.

EXAMPLE 5.5   CHECK FOR SECOND-ORDER CONDITIONS

Consider the optimization problem of Example 4.31:

Minimize

$$f(\mathbf{x}) = x_1{}^2 + x_2{}^2 - 3x_1 x_2 \tag{a}$$

subject to

$$g(\mathbf{x}) = x_1{}^2 + x_2{}^2 - 6 \leq 0 \tag{b}$$

Check for sufficient conditions for the candidate minimum points.

Solution

The points satisfying KKT necessary conditions are

$$\text{(i) } \mathbf{x}^* = (0,0), \ u^* = 0; \quad \text{(ii) } \mathbf{x}^* = (\sqrt{3}, \sqrt{3}), \ u^* = \frac{1}{2}; \quad \text{(iii) } \mathbf{x}^* = (-\sqrt{3}, -\sqrt{3}), \ u^* = \frac{1}{2} \tag{c}$$

It was observed in Example 4.31 and Figure 4.17 that the point $(0, 0)$ did not satisfy the sufficiency condition and that the other two points did satisfy it. Those geometrical observations will be mathematically verified using the second-order optimality conditions.

The Hessian matrices for the cost and constraint functions are

$$\nabla^2 f = \begin{bmatrix} 2 & -3 \\ -3 & 2 \end{bmatrix}, \quad \nabla^2 g = \begin{bmatrix} 2 & 0 \\ 0 & 2 \end{bmatrix} \tag{d}$$

By the method of Appendix A, eigenvalues of $\nabla^2 g$ are $\lambda_1 = 2$ and $\lambda_2 = 2$. Since both eigenvalues are positive, the function g is convex, and so the feasible set defined by $g(\mathbf{x}) \leq 0$ is convex by Theorem 4.9. However, since eigenvalues of $\nabla^2 f$ are -1 and 5, f is not convex. Therefore, it cannot be classified as a convex programming problem and sufficiency cannot be shown by Theorem 4.11. We must resort to the General Sufficiency Theorem 5.2.

The Hessian of the Lagrangian function is given as

$$\nabla^2 L = \nabla^2 f + u \nabla^2 g = \begin{bmatrix} 2+2u & -3 \\ -3 & 2+2u \end{bmatrix} \tag{e}$$

(i) For the first point $\mathbf{x}^* = (0, 0)$, $u^* = 0$, $\nabla^2 L$ becomes $\nabla^2 f$ (the constraint $g(\mathbf{x}) \leq 0$ is inactive). In this case the problem is unconstrained and the local sufficiency requires $\mathbf{d}^T \nabla^2 f(\mathbf{x}^*)\mathbf{d} > 0$ for all $\mathbf{d}$. Or $\nabla^2 f$ should be positive definite at $\mathbf{x}^*$. Since both eigenvalues of $\nabla^2 f$ are not positive, we conclude that the above condition is not satisfied. Therefore, $\mathbf{x}^* = (0, 0)$ does not satisfy the second-order sufficiency condition.

Note that since $\lambda_1 = -1$ and $\lambda_2 = 5$, the matrix $\nabla^2 f$ is indefinite at $\mathbf{x}^*$. The point $\mathbf{x}^* = (0, 0)$, then, *violates the second-order necessary condition* of Theorem 4.4 requiring $\nabla^2 f$ to be at least positive semidefinite at the candidate local minimum point. Thus, $\mathbf{x}^* = (0, 0)$ cannot be a local minimum point. This agrees with graphical observation made in Example 4.31.

(ii) At points $\mathbf{x}^* = (\sqrt{3}, \sqrt{3})$, $u^* = \dfrac{1}{2}$ and $\mathbf{x}^* = (-\sqrt{3}, -\sqrt{3})$, $u^* = \dfrac{1}{2}$,

$$\nabla^2 L = \nabla^2 f + u \nabla^2 g = \begin{bmatrix} 2+2u & -3 \\ -3 & 2+2u \end{bmatrix} = \begin{bmatrix} 3 & -3 \\ -3 & 3 \end{bmatrix} \tag{f}$$

$$\nabla g = \pm (2\sqrt{3}, 2\sqrt{3}) = \pm 2\sqrt{3}\,(1,\ 1) \tag{g}$$

It may be checked that $\nabla^2 L$ is not positive definite at either of the two points. Therefore, we cannot use Theorem 5.3 to conclude that $\mathbf{x}^*$ is a minimum point. We must find $\mathbf{d}$ satisfying Eqs. (5.10) and (5.11). If we let $\mathbf{d} = (d_1, d_2)$, then $\nabla g^T \mathbf{d} = 0$ gives

$$\pm 2\sqrt{3}\begin{bmatrix} 1 & 1 \end{bmatrix}\begin{bmatrix} d_1 \\ d_2 \end{bmatrix} = 0; \quad \text{or } d_1 + d_2 = 0 \tag{h}$$

Thus, $d_1 = -d_2 = c$, where $c \neq 0$ is an arbitrary constant, and a $\mathbf{d} \neq \mathbf{0}$ satisfying $\nabla g^T \mathbf{d} = 0$ is given as $\mathbf{d} = c(1, -1)$. The sufficiency condition of Eq. (5.12) gives

$$Q = \mathbf{d}^T (\nabla^2 L)\mathbf{d} = c[1\ -1]\begin{bmatrix} 3 & -3 \\ -3 & 3 \end{bmatrix} c\begin{bmatrix} 1 \\ -1 \end{bmatrix} = 12c^2 > 0 \text{ for } c \neq 0 \tag{i}$$

The points $\mathbf{x}^* = (\sqrt{3}, \sqrt{3})$ and $\mathbf{x}^* = (-\sqrt{3}, -\sqrt{3})$ satisfy the sufficiency condition of Eq. (5.12). They are therefore, isolated local minimum points, as was observed graphically in Example 4.31 and Figure 4.17. We see for this example that $\nabla^2 L$ is not positive definite at $\mathbf{x}^*$, but $\mathbf{x}^*$ is still an isolated minimum point.

Note that since f is continuous and the feasible set is closed and bounded, we are guaranteed the existence of a global minimum by the Weierstrass Theorem 4.1. Also we have examined every possible point satisfying necessary conditions. Therefore, we must conclude by elimination that $\mathbf{x}^* = (\sqrt{3}, \sqrt{3})$ and $\mathbf{x}^* = (-\sqrt{3}, -\sqrt{3})$ are global minimum points. The value of the cost function for both points is $f(\mathbf{x}^*) = -3$.

EXAMPLE 5.6 CHECK FOR SECOND-ORDER CONDITIONS

Consider Example 4.32:

Minimize

$$f(x_1, x_2) = x_1{}^2 + x^2{}_2 - 2x_1 - 2x_2 + 2 \tag{a}$$

subject to

$$g_1 = -2x_1 - x_2 + 4 \le 0 \tag{b}$$

$$g_2 = -x_1 - 2x_2 + 4 \le 0 \tag{c}$$

Check the second-order conditions for the candidate minimum point.

Solution

The KKT necessary conditions are satisfied for the point

$$x_1^* = \frac{4}{3}, \quad x_2^* = \frac{4}{3}, \quad u_1^* = \frac{2}{9}, \quad u_2^* = \frac{2}{9} \tag{d}$$

Since all the constraint functions are linear, the feasible set S is convex. The Hessian of the cost function is positive definite. Therefore, it is also convex and the problem is convex. By Theorem 4.11,

$$x_1^* = \frac{4}{3}, \; x_2^* = \frac{4}{3}$$

satisfies sufficiency conditions for a global minimum with the cost function as $f(x^*) = \frac{2}{9}$.

Note that local sufficiency cannot be shown by the method of Theorem 5.2. The reason is that the conditions of Eq. (5.11) give two equations in two unknowns:

$$-2d_1 - d_2 = 0, \; -d_1 - 2d_2 = 0 \tag{e}$$

This is a homogeneous system of equations with a nonsingular coefficient matrix. Therefore, its only solution is $d_1 = d_2 = 0$. Thus, we cannot find a $\mathbf{d} \ne \mathbf{0}$ for use in the condition of Eq. (5.12), and Theorem 5.2 cannot be used. However, we have seen in the foregoing and in Figure 4.18 that the point is actually an isolated global minimum point. Since it is a two-variable problem and two inequality constraints are active at the KKT point, the condition for a local minimum is satisfied.

5.4 SECOND-ORDER CONDITIONS FOR THE RECTANGULAR BEAM DESIGN PROBLEM

The rectangular beam problem was formulated and graphically solved in Section 3.8. The KKT necessary conditions were written and solved in Section 4.9.2. Several points that satisfy the KKT conditions are obtained. It is seen from the graphical representation of the problem that all of these points are global minima for the problem; however, none of the points is an isolated minimum. Let us show that the sufficiency condition will not be satisfied for any of these points.

Cases 3, 5, and 6 in Section 4.9.2 gave solutions that satisfy the KKT conditions. Cases 5 and 6 had two active constraints; however, only the constraint with positive multiplier needs to be considered in Eq. (5.11). The sufficiency theorem requires only constraints with $u_i > 0$ to be considered in calculating the feasible directions for use in Eq. (5.12). Therefore, only the g_2 constraint needs to be included in the check for sufficiency conditions. Thus, *all three cases have the same sufficiency check.*

We need to calculate Hessians of the cost function and the second constraint:

$$\nabla^2 f = \begin{bmatrix} 0 & 1 \\ 1 & 0 \end{bmatrix}, \quad \nabla^2 g_2 = \frac{(2.25 \times 10^5)}{b^3 d^3} \begin{bmatrix} 2d^2 & bd \\ bd & 2b^2 \end{bmatrix} \tag{a}$$

Since $bd = (1.125 \times 10^5)$, $\nabla^2 g_2$ becomes

$$\nabla^2 g_2 = 2 \begin{bmatrix} \dfrac{2}{b^2} & (1.125 \times 10^5)^{-1} \\ (1.125 \times 10^5)^{-1} & \dfrac{2}{d^2} \end{bmatrix} \tag{b}$$

The Hessian of the Lagrangian is given as

$$\nabla^2 L = \nabla^2 f + u_2 \nabla^2 g_2 = \begin{bmatrix} 0 & 1 \\ 1 & 0 \end{bmatrix} + 2(56,250) \begin{bmatrix} \dfrac{2}{b^2} & (1.125 \times 10^5)^{-1} \\ (1.125 \times 10^5)^{-1} & \dfrac{2}{d^2} \end{bmatrix} \tag{c}$$

$$\nabla^2 L = \begin{bmatrix} \dfrac{(2.25 \times 10^5)}{b^2} & 2 \\ 2 & \dfrac{(2.25 \times 10^5)}{d^2} \end{bmatrix} \tag{d}$$

The determinant of $\nabla^2 L$ is 0 for $bd = (1.125 \times 10^5)$; the matrix is only positive semidefinite. Therefore, the Strong Sufficiency Theorem 5.3 cannot be used to show the sufficiency of x^*. We must check the sufficiency condition of Eq. (5.12). In order to do that, we must find directions y satisfying Eq. (5.11). The gradient of g_2 is given as

$$\nabla g_2 = \left[\frac{-(2.25 \times 10^5)}{b^2 d}, \quad \frac{-(2.25 \times 10^5)}{bd^2} \right] \tag{e}$$

The feasible directions y are given by $\nabla g_2^T y = 0$, as

$$\frac{1}{b} y_1 + \frac{1}{d} y_2 = 0, \quad \text{or } y_2 = -\frac{d}{b} y_1 \tag{f}$$

Therefore, vector y is given as $y = (1, -d/b)c$, where $c = y_1$ is any constant. Using $\nabla^2 L$ and y, Q of Eq. (5.12) is given as

$$Q = y^T \nabla^2 L y = 0 \tag{g}$$

Thus, the sufficiency condition of Theorem 5.2 is not satisfied. The points satisfying $bd = (1.125 \times 10^5)$ are not isolated minimum points. This is, of course, true from Figure 3.11. Note, however, that since $Q = 0$, the second-order necessary condition of Theorem 5.1 is satisfied for Case 3. Theorem 5.2 cannot be used for solutions to Cases 5 and 6 since there are two active constraints for this two-variable problem; therefore, there are no nonzero **d** vectors.

It is important to note that this problem does not satisfy the condition for a convex programming problem and all of the points satisfying KKT conditions do not satisfy the sufficiency condition for an isolated local minimum. Yet all of the points are actually global minimum points. Two conclusions can be drawn from this example:

1. *Global minimum points* can be obtained for problems that cannot be classified as convex programming problems. We cannot show global optimality unless we find all of the local minimum points in the closed and bounded set (the Weierstrass Theorem 4.1).
2. *If sufficiency conditions are not satisfied*, the only conclusion we can draw is that the candidate point need not be an isolated minimum. It may have many local optima in the neighborhood, and they may all be actually global minimum points.

5.5 DUALITY IN NONLINEAR PROGRAMMING

Given a nonlinear programming problem, there is another nonlinear programming problem closely associated with it. The former is called the *primal problem*, and the latter is called the *dual problem*. Under certain convexity assumptions, the primal and dual problems have the same optimum objective function values and therefore, it is possible to solve the primal problem indirectly by solving the dual problem. As a by-product of one of the duality theorems, we obtain the *saddle point necessary conditions*.

Duality has played an important role in the development of optimization theory and numerical methods. Development of the duality theory requires *assumptions* about the convexity of the problem. However to be broadly applicable, the theory should require a minimum of convexity assumptions. This leads to the concept of local convexity and to the *local duality theory*

In this section, we will present only the local duality. The theory can be used to develop computational methods for solving optimization problems. We will see in Chapter 11 that it can be used to develop the so-called *augmented Lagrangian methods*.

5.5.1 Local Duality: Equality Constraints Case

For sake of developing the *local duality theory*, we consider the equality-constrained problem first.

Problem E

Find an *n*-vector **x** to

Minimize

$$f(\mathbf{x}) \tag{5.13}$$

subject to

$$h_i(\mathbf{x}) = 0; \quad i = 1 \text{ to } p \tag{5.14}$$

Later on we will extend the theory to both equality and inequality constrained problems. The theory we are going to present is sometimes called the *strong duality* or *Lagrangian duality*. We assume that functions f and h_i are twice continuously differentiable. We will first define a dual function associated with Problem E and study its properties. Then we will define the dual problem.

To present the duality results for Problem E the following notation is used.

The Lagrangian function:

$$L(\mathbf{x}, \mathbf{v}) = f(\mathbf{x}) + \sum_{i=1}^{p} v_i h_i = f(\mathbf{x}) + (\mathbf{v} \cdot \mathbf{h}) \tag{5.15}$$

The *Hessian of the Lagrangian function* with respect to $\mathbf{x}$:

$$\mathbf{H}_x(\mathbf{x}, \mathbf{v}) = \frac{\partial^2 L}{\partial \mathbf{x}^2} = \frac{\partial^2 f(\mathbf{x})}{\partial \mathbf{x}^2} + \sum_{i=1}^{p} v_i \frac{\partial^2 h_i}{\partial \mathbf{x}^2} \tag{5.16}$$

The *gradient matrix* of equality constraints:

$$\mathbf{N} = \left[\frac{\partial h_j}{\partial x_i} \right]_{n \times p} \tag{5.17}$$

In these equations, $\mathbf{v}$ is the p-dimensional Lagrange multiplier vector for the equality constraints.

Let $\mathbf{x}^*$ be a local minimum of Problem E that is also a regular point of the feasible set. Then there exists a unique Lagrange multiplier v_i^* for each constraint such that the first-order necessary condition is met:

$$\frac{\partial L(\mathbf{x}^*, \mathbf{v}^*)}{\partial \mathbf{x}} = \mathbf{0}, \ or \ \frac{\partial f(\mathbf{x}^*)}{\partial \mathbf{x}} + \sum_{i=1}^{p} v_i^* \frac{\partial h_i(\mathbf{x}^*)}{\partial \mathbf{x}} = \mathbf{0} \tag{5.18}$$

For development of the local duality theory, we make the assumption that the Hessian of the Lagrangian function $\mathbf{H}_x(\mathbf{x}^*, \mathbf{v}^*)$ at the minimum point $\mathbf{x}^*$ is positive definite. This assumption guarantees that the *Lagrangian* of Eq. (5.15) is *locally convex* at $\mathbf{x}^*$. This also satisfies the sufficiency condition for $\mathbf{x}^*$ to be an isolated local minimum of Problem E. With this assumption, the point $\mathbf{x}^*$ is not only a local minimum of Problem E, it is also a local minimum for the unconstrained problem:

$$\underset{\mathbf{x}}{\text{minimize}} \ L(\mathbf{x}, \mathbf{v}^*) \ or \ \underset{\mathbf{x}}{\text{minimize}} \left(f(\mathbf{x}) + \sum_{i=1}^{p} v_i^* h_i \right) \tag{5.19}$$

where $\mathbf{v}^*$ is a vector of Lagrange multipliers at $\mathbf{x}^*$. The necessary and sufficient conditions for the above unconstrained problem are the same as for the constrained Problem E (with $\mathbf{H}_x(\mathbf{x}^*, \mathbf{v}^*)$ being positive definite). In addition for any $\mathbf{v}$ *sufficiently close* to $\mathbf{v}^*$, the Lagrange function $L(\mathbf{x}, \mathbf{v})$ will have a local minimum at a point $\mathbf{x}$ near $\mathbf{x}^*$. Now we will establish the condition that $\mathbf{x}(\mathbf{v})$ exists and is a differentiable function of $\mathbf{v}$.

The necessary condition at the point $(\mathbf{x},\mathbf{v})$ in the vicinity of $(\mathbf{x}^*,\mathbf{v}^*)$ is given as

$$\frac{\partial L(\mathbf{x},\mathbf{v})}{\partial \mathbf{x}} = \frac{\partial f(\mathbf{x})}{\partial \mathbf{x}} + \sum_{i=1}^{p} v_i \frac{\partial h_i}{\partial \mathbf{x}} = 0, \; or \; \frac{\partial f(\mathbf{x})}{\partial \mathbf{x}} + \mathbf{N}\mathbf{v} = 0 \tag{5.20}$$

Since $\mathbf{H}_x(\mathbf{x}^*,\mathbf{v}^*)$ is positive definite, it is nonsingular. Also because of this positive definiteness, $\mathbf{H}_x(\mathbf{x},\mathbf{v})$ is positive definite in the vicinity of $(\mathbf{x}^*,\mathbf{v}^*)$ and thus nonsingular. This is a generalization of a theorem from calculus: *If a function is positive at a point, it is positive in a neighborhood of that point.* Note that $\mathbf{H}_x(\mathbf{x},\mathbf{v})$ is also the Jacobian of the necessary conditions of Eq. (5.20) with respect to $\mathbf{x}$. Therefore, Eq. (5.20) has a solution $\mathbf{x}$ near $\mathbf{x}^*$ when $\mathbf{v}$ is near $\mathbf{v}^*$. Thus, locally there is a unique correspondence between $\mathbf{v}$ and $\mathbf{x}$ through a solution to the unconstrained problem:

$$\underset{\mathbf{x}}{\text{minimize}} \; L(\mathbf{x},\mathbf{v}) \; or \; \underset{\mathbf{x}}{\text{minimize}} \left[f(\mathbf{x}) + \sum_{i=1}^{p} v_i h_i \right] \tag{5.21}$$

Furthermore, for a given $\mathbf{v}$, $\mathbf{x}(\mathbf{v})$ is a differentiable function of $\mathbf{v}$ (by the Implicit Functions Theorem of calculus).

Dual Function

Near $\mathbf{v}^*$, we define the dual function $\phi(\mathbf{v})$ by the equation

$$\phi(\mathbf{v}) = \underset{\mathbf{x}}{\text{minimize}} \; L(\mathbf{x}, \mathbf{v}) = \underset{\mathbf{x}}{\text{minimize}} \left[f(\mathbf{x}) + \sum_{i=1}^{p} v_i h_i \right] \tag{5.22}$$

where the minimum is taken locally with respect to $\mathbf{x}$ near $\mathbf{x}^*$.

Dual Problem

$$\underset{\mathbf{v}}{\text{maximize}} \; \phi(\mathbf{v}) \tag{5.23}$$

With this definition of the dual function we can show that locally the original constrained Problem E is equivalent to unconstrained *local maximization of the dual function* $\phi(\mathbf{v})$ with respect to $\mathbf{v}$. Thus, we can establish equivalence between a constrained problem in $\mathbf{x}$ and an unconstrained problem in $\mathbf{v}$. To establish the duality relation, we must prove two lemmas.

LEMMA 5.1

The gradient of the dual function $\phi(\mathbf{v})$ is given as

$$\frac{\partial \phi(\mathbf{v})}{\partial \mathbf{v}} = \mathbf{h}(\mathbf{x}(\mathbf{v})) \tag{5.24}$$

Proof Let $\mathbf{x}(\mathbf{v})$ represent a local minimum for the Lagrange function

$$L(\mathbf{x}, \mathbf{v}) = f(\mathbf{x}) + (\mathbf{v} \cdot \mathbf{h}) \tag{5.25}$$

Therefore, the dual function can be explicitly written from Eq. (5.22) as

$$\phi(\mathbf{x}(\mathbf{v})) = \left[f(\mathbf{x}(\mathbf{v})) + (\mathbf{v} \cdot \mathbf{h}(\mathbf{x}(\mathbf{v}))) \right] \tag{5.26}$$

where $\mathbf{x}(\mathbf{v})$ is a solution of the necessary condition in Eq. (5.20).

Now, differentiating $\phi(\mathbf{v})$ in Eq. (5.26) with respect to $\mathbf{v}$, and using the fact that $\mathbf{x}(\mathbf{v})$ is a differentiable function of $\mathbf{v}$, we get

$$\frac{\partial \phi(\mathbf{x}(\mathbf{v}))}{\partial \mathbf{v}} = \frac{\partial \phi(\mathbf{v})}{\partial \mathbf{v}} + \frac{\partial \mathbf{x}(\mathbf{v})}{\partial \mathbf{v}} \frac{\partial \phi}{\partial \mathbf{x}} = \mathbf{h}(\mathbf{x}(\mathbf{v})) + \frac{\partial \mathbf{x}(\mathbf{v})}{\partial \mathbf{v}} \frac{\partial L}{\partial \mathbf{x}} \tag{5.27}$$

where $\dfrac{\partial \mathbf{x}(\mathbf{v})}{\partial \mathbf{v}}$ is a $p \times n$ matrix. But $\partial L/\partial \mathbf{x}$ in Eq. (5.27) is zero because $\mathbf{x}(\mathbf{v})$ minimizes the Lagrange function of Eq. (5.25). This proves the result of Eq. (5.24).

Lemma 5.1 is of practical importance because it shows that the gradient of the dual function is quite simple to calculate. Once the dual function is evaluated by minimization with respect to $\mathbf{x}$, the corresponding $\mathbf{h}(\mathbf{x})$, which is the gradient of $\phi(\mathbf{v})$, can be evaluated without any further calculation.

LEMMA 5.2

The Hessian of the dual function is given as

$$\mathbf{H}_v = \frac{\partial^2 \phi(\mathbf{v})}{\partial \mathbf{v}^2} = -\mathbf{N}^T [\mathbf{H}_x(\mathbf{x})]^{-1} \mathbf{N} \tag{5.28}$$

Proof Using Lemma 5.1 in Eq. (5.24),

$$\mathbf{H}_v = \frac{\partial}{\partial \mathbf{v}} \left(\frac{\partial \phi(\mathbf{x}(\mathbf{v}))}{\partial \mathbf{v}} \right) = \frac{\partial \mathbf{h}(\mathbf{x}(\mathbf{v}))}{\partial \mathbf{v}} + \frac{\partial \mathbf{x}(\mathbf{v})}{\partial \mathbf{v}} \mathbf{N} \tag{5.29}$$

To calculate $\dfrac{\partial \mathbf{x}(\mathbf{v})}{\partial \mathbf{v}}$, we differentiate the necessary condition of Eq. (5.20) with respect to $\mathbf{v}$ to obtain

$$\mathbf{N}^T + \frac{\partial \mathbf{x}(\mathbf{v})}{\partial \mathbf{v}} \mathbf{H}_x(\mathbf{x}) = 0 \tag{5.30}$$

Solving for $\dfrac{\partial \mathbf{x}(\mathbf{v})}{\partial \mathbf{v}}$ from Eq. (5.30), we get

$$\frac{\partial \mathbf{x}(\mathbf{v})}{\partial \mathbf{v}} = -\mathbf{N}^T [\mathbf{H}_x(\mathbf{x})]^{-1} \tag{5.31}$$

Substituting Eq. (5.31) into Eq. (5.29), we obtain the result of Eq. (5.28), which was to be proved.

Since $[\mathbf{H}_x(\mathbf{x})]^{-1}$ is positive definite, and since $\mathbf{N}$ is of full column rank near $\mathbf{x}$, we have $\mathbf{H}_v(\mathbf{v})$, a $p \times p$ matrix (Hessian of $\phi(\mathbf{v})$), to be *negative definite*. This observation and the Hessian of $\phi(\mathbf{v})$ play a role in the analysis of dual methods.

THEOREM 5.4

Local Duality Theorem   For Problem E, let

(i) $\mathbf{x}^*$ be a local minimum.
(ii) $\mathbf{x}^*$ be a regular point.
(iii) $\mathbf{v}^*$ be the Lagrange multipliers at $\mathbf{x}^*$.
(iv) $\mathbf{H}_x(\mathbf{x}^*, \mathbf{v}^*)$ be positive definite.

Then for the dual problem

Maximize

$$\phi(\mathbf{v}) \tag{5.32}$$

has a local solution at $\mathbf{v}^*$ with $\mathbf{x}^* = \mathbf{x}(\mathbf{v}^*)$. The maximum value of the dual function is equal to the minimum value of $f(\mathbf{x})$; that is,

$$\phi(\mathbf{v}^*) = f(x^*) \tag{5.33}$$

Proof   It is clear that $\mathbf{x}^* = \mathbf{x}(\mathbf{v}^*)$ by definition of $\phi(\mathbf{v})$. Now, at $\mathbf{v}^*$, we have Lemma 5.1:

$$\frac{\partial \phi(\mathbf{v}^*)}{\partial \mathbf{v}} = \mathbf{h}(\mathbf{x}) = 0 \tag{a}$$

Also, by Lemma 5.2, the Hessian of $\phi(\mathbf{v})$ is negative definite. Thus, $\mathbf{v}^*$ satisfies the first-order

necessary and second-order sufficiency conditions for an unconstrained maximum point of $\phi(\mathbf{v})$.

Substituting $\mathbf{v}^*$ in the definition of $\phi(\mathbf{v})$ in Eq. (5.26), we get

$$\phi(\mathbf{v}^*) = [f(\mathbf{x}(\mathbf{v}^*)) + (\mathbf{v}^* \cdot \mathbf{h}(\mathbf{x}(\mathbf{v}^*)))]$$
$$= [f(\mathbf{x}^*) + (\mathbf{v}^* \cdot \mathbf{h}(\mathbf{x}^*))] \tag{b}$$
$$= f(\mathbf{x}^*)$$

which was to be proved.

EXAMPLE 5.7   SOLUTION TO THE DUAL PROBLEM

Consider the following problem in two variables:

Minimize

$$f = -x_1 x_2 \tag{a}$$

subject to

$$(x_1 - 3)^2 + x_2^2 = 5 \tag{b}$$

Solution

Let us first solve the primal problem using the optimality conditions. The Lagrangian for the problem is given as

$$L = -x_1 x_2 + v[(x_1 - 3)^2 + x_2^2 - 5] \tag{c}$$

The first-order necessary conditions are

$$-x_2 + (2x_1 - 6)v = 0 \tag{d}$$
$$-x_1 + 2x_2 v = 0 \tag{e}$$

Together with the equality constraint in Eq. (b), these equations have a solution:

$$x_1^* = 4, \ x_2^* = 2, \ v^* = 1, \ f^* = -8 \tag{f}$$

The Hessian of the Lagrangian function is given as

$$H_x(\mathbf{x}^*, \mathbf{v}^*) = \begin{bmatrix} 2 & -1 \\ -1 & 2 \end{bmatrix} \tag{g}$$

Since this is a positive definite matrix, we conclude that the solution obtained is an isolated local minimum.

Since $H_x(\mathbf{x}^*, \mathbf{v}^*)$ is positive definite, we can apply the local duality theory near the solution point. Define a dual function as

$$\phi(\mathbf{v}) = \underset{\mathbf{x}}{\text{minimize}}\, L(\mathbf{x}, \mathbf{v}) \tag{h}$$

Solving Eqs. (d) and (e), we get x_1 and x_2 in terms of v, provided that

$$4v^2 - 1 \neq 0 \tag{i}$$

as

$$x_1 = \frac{12v^2}{4v^2 - 1}, \quad x_2 = \frac{6v}{4v^2 - 1} \tag{j}$$

Substituting Eqs. (j) into Eq. (c), the dual function of Eq. (h) is given as

$$\phi(v) = \frac{4v + 4v^3 - 80v^5}{(4v^2 - 1)^2} \tag{k}$$

which is valid for $v \neq \pm\frac{1}{2}$. This $\phi(\mathbf{v})$ has a local maximum at $v^* = 1$. Substituting $v = 1$ in Eqs. (j), we get the same solution as in Eqs. (f). Note that $\phi(v^*) = -8$, which is the same as f^* in Eq. (f).

5.5.2 Local Duality: The Inequality Constraints Case

Consider the equality/inequality-constrained problem.

Problem P

In addition to the equality constraints in Problem E, we impose inequality constraints:

$$g_i(\mathbf{x}) \leq 0; \quad i = 1 \text{ to } m \tag{5.34}$$

The feasible set S is defined as

$$S = \{\mathbf{x} | h_i(\mathbf{x}) = 0, \; i = 1 \text{ to } p; \; g_j(\mathbf{x}) \leq 0, \; j = 1 \text{ to } m\} \tag{5.35}$$

The Lagrangian function is defined as

$$L(\mathbf{x}, \mathbf{v}, \mathbf{u}) = f(\mathbf{x}) + \sum_{i=1}^{p} v_i h_i + \sum_{j=1}^{m} u_j g_j \tag{5.36}$$

$$= f(\mathbf{x}) + (\mathbf{v} \cdot \mathbf{h}) + (\mathbf{u} \cdot \mathbf{g}); u_j \geq 0, \; j = 1 \text{ to } m$$

The dual function for Problem P is defined as

$$\phi(\mathbf{v}, \mathbf{u}) = \underset{\mathbf{x}}{\text{minimize}}\, L(\mathbf{x}, \mathbf{v}, \mathbf{u}); \quad u_j \geq 0, \; j = 1 \text{ to } m \tag{5.37}$$

The dual problem is defined as

$$\underset{\mathbf{v}, \mathbf{u}}{\text{maximize}}\, \phi(\mathbf{v}, \mathbf{u}); \quad u_j \geq 0, \; j = 1 \text{ to } m \tag{5.38}$$

THEOREM 5.5

Strong Duality Theorem Let the following apply:

(i) $\mathbf{x}^*$ be a local minimum of Problem P.

(ii) $\mathbf{x}^*$ be a regular point.

(iii) $H_x(\mathbf{x}^*, \mathbf{v}^*, \mathbf{u}^*)$ be positive definite.

(iv) $\mathbf{v}^*$, $\mathbf{u}^*$ be the Lagrange multipliers at the optimum point $\mathbf{x}^*$.

Then $\mathbf{v}^*$, $\mathbf{u}^*$ solves the dual problem that is defined in Eq. (5.38) with $f(\mathbf{x}^*) = \phi(\mathbf{v}^*, \mathbf{u}^*)$ and $\mathbf{x}^* = \mathbf{x}(\mathbf{v}^*, \mathbf{u}^*)$.

If the assumption of the positive definiteness of $H_x(\mathbf{x}^*, \mathbf{v}^*)$ is not made, we get the weak duality theorem.

THEOREM 5.6

Weak Duality Theorem Let $\mathbf{x}$ be a feasible solution for Problem P and let $\mathbf{v}$ and $\mathbf{u}$ be the feasible solution for the dual problem that is defined in Eq. (5.38); thus, $h_i(\mathbf{x}) = 0$, $i = 1$ to p, and $g_j(\mathbf{x}) \leq 0$ and $u_j \geq 0$, $j = 1$ to m. Then

$$\phi(\mathbf{v}, \mathbf{u}) \leq f(\mathbf{x}) \qquad (5.39)$$

Proof By definition

$$\phi(\mathbf{v}, \mathbf{u}) = \underset{\mathbf{x}}{\text{minimize}} \, L(\mathbf{x}, \mathbf{v}, \mathbf{u})$$

$$= \underset{\mathbf{x}}{\text{minimize}} \, (f(\mathbf{x}) + (\mathbf{v} \cdot \mathbf{h}) + (\mathbf{u} \cdot \mathbf{g}))$$

$$\leq (f(\mathbf{x}) + (\mathbf{v} \cdot \mathbf{h}) + (\mathbf{u} \cdot \mathbf{g})) \leq f(\mathbf{x})$$

since $u_i \geq 0$, $g_i(\mathbf{x}) \leq 0$, and $u_i g_i = 0$ for $i = 1$ to m; and $h_i(\mathbf{x}) = 0$, $i = 1$ to p.

From Theorem 5.5, we obtain the following results:

1. Minimum $[f(\mathbf{x})$ with $\mathbf{x} \in S] \geq$ maximum $[\phi(\mathbf{v}, \mathbf{u})$ with $u_i \geq 0, i = 1$ to $m]$.
2. If $f(\mathbf{x}^*) = \phi(\mathbf{v}^*, \mathbf{u}^*)$ with $u_i \geq 0$, $i = 1$ to m and $\mathbf{x}^* \in S$, then $\mathbf{x}^*$ and $(\mathbf{v}^*, \mathbf{u}^*)$ solve the primal and dual problems, respectively.
3. If Minimum $[f(\mathbf{x})$ with $\mathbf{x} \in S] = -\infty$, then the dual is *infeasible*, and vice versa (i.e., if dual is *infeasible*, the primal is *unbounded*).
4. If Maximum $[\phi(\mathbf{v}, \mathbf{u})$ with $u_i \geq 0$, $i = 1$ to $m] = \infty$, then the primal problem has *no feasible solution*, and vice versa (i.e., if primal is *infeasible*, the dual is *unbounded*).

LEMMA 5.3 LOWER BOUND FOR PRIMAL COST FUNCTION

For any $\mathbf{v}$ and $\mathbf{u}$ with $u_i \geq 0$, $i = 1$ to m

$$\phi(\mathbf{v}, \mathbf{u}) \leq f(\mathbf{x}^*) \qquad (5.40)$$

Proof $\phi(\mathbf{v}, \mathbf{u}) \leq$ maximum $\phi(\mathbf{v}, \mathbf{u})$; $u_i \geq 0$, $i = 1$ to m

$$= \underset{\mathbf{v}, \mathbf{u}}{\text{maximize}} \left\{ \underset{\mathbf{x}}{\text{minimize}} \, (f(\mathbf{x}) + (\mathbf{v} \cdot \mathbf{h}) + (\mathbf{u} \cdot \mathbf{g})); \quad u_i \geq 0, \quad i = 1 \text{ to } m \right\}$$

$$= \begin{array}{c} \text{maximize} \\ \mathbf{v}, \mathbf{u} \end{array} \{f(\mathbf{x}(\mathbf{v}, \mathbf{u})) + (\mathbf{v} \cdot \mathbf{h}) + (\mathbf{u} \cdot \mathbf{g})\}; \quad u_i \geq 0, \quad i = 1 \text{ to } m$$

$$= f(\mathbf{x}(\mathbf{v}^*, \mathbf{u}^*)) + (\mathbf{v}^* \cdot \mathbf{h}) + (\mathbf{u}^* \cdot \mathbf{g}) = f(\mathbf{x}^*)$$

Lemma 5.3 is quite useful for practical applications. It tells us how to find a lower bound on the optimum primal cost function. The dual cost function for arbitrary v_i, $i = 1$ to p and $u_i \geq 0$, $i = 1$ to m provides a *lower bound* for the primal cost function. For any $\mathbf{x} \in S$, $f(\mathbf{x})$ provides an *upper bound* for the optimum cost function.

SADDLE POINTS Let $L(\mathbf{x},\mathbf{v},\mathbf{u})$ be the Lagrange function. L has a saddle point at $\mathbf{x}^*,\mathbf{v}^*$, $\mathbf{u}^*$ subject to $u_i \geq 0$, $i = 1$ to m if

$$L(\mathbf{x}^*, \mathbf{v}, \mathbf{u}) \leq L(\mathbf{x}^*, \mathbf{v}^*, \mathbf{u}^*) \leq L(\mathbf{x}, \mathbf{v}^*, \mathbf{u}^*) \tag{5.41}$$

holds for all $\mathbf{x}$ near $\mathbf{x}^*$ and $(\mathbf{v}, \mathbf{u})$ near $(\mathbf{v}^*, \mathbf{u}^*)$ with $u_i \geq 0$ for $i = 1$ to m.

THEOREM 5.7

Saddle Point Theorem For Problem P let all functions be twice continuously differentiable and let $L(\mathbf{x}, \mathbf{v}, \mathbf{u})$ be defined as

$$L(\mathbf{x}, \mathbf{v}, \mathbf{u}) = f(\mathbf{x}) + (\mathbf{v} \cdot \mathbf{h}) + (\mathbf{u} \cdot \mathbf{g}); \tag{5.42}$$
$$u_j \geq 0, \quad j = 1 \text{ to } m$$

Let $L(\mathbf{x}^*, \mathbf{v}^*, \mathbf{u}^*)$ exist with $u_i^* \geq 0, i = 1$ to m. Also let $\mathbf{H}_x(\mathbf{x}^*, \mathbf{v}^*, \mathbf{u}^*)$ be *positive definite*.

Then $\mathbf{x}^*$ satisfying a suitable constraint qualification is a local minimum of Problem P if and only if $(\mathbf{x}^*, \mathbf{v}^*, \mathbf{u}^*)$ is a saddle point of the Lagrangian; that is,

$$L(\mathbf{x}^*, \mathbf{v}, \mathbf{u}) \leq L(\mathbf{x}^*, \mathbf{v}^*, \mathbf{u}^*) \leq L(\mathbf{x}, \mathbf{v}^*, \mathbf{u}^*) \tag{5.43}$$

for all $\mathbf{x}$ near $\mathbf{x}^*$ and all $(\mathbf{v}, \mathbf{u})$ near $(\mathbf{v}^*, \mathbf{u}^*)$, with $u_i \geq 0$ for $i = 1$ to m.

See Bazaraa et al. (2006) for proof of Theorem 5.7.

EXERCISES FOR CHAPTER 5

5.1 *Answer True or False.*

1. A convex programming problem always has a unique global minimum point.
2. For a convex programming problem, KKT necessary conditions are also sufficient.
3. The Hessian of the Lagrange function must be positive definite at constrained minimum points.
4. For a constrained problem, if the sufficiency condition of Theorem 5.2 is violated, the candidate point $\mathbf{x}^*$ may still be a minimum point.
5. If the Hessian of the Lagrange function at $\mathbf{x}^*$, $\nabla^2 L(\mathbf{x}^*)$, is positive definite, the optimum design problem is convex.
6. For a constrained problem, the sufficient condition at $\mathbf{x}^*$ is satisfied if there are no feasible directions in a neighborhood of $\mathbf{x}^*$ along which the cost function reduces.

5.2 Formulate the problem of Exercise 4.84. Show that the solution point for the problem is not a regular point. Write KKT conditions for the problem, and study the implication of the irregularity of the solution point.

5.3 Solve the following problem using the graphical method:

Minimize $f(x_1, x_2) = (x_1 - 10)^2 + (x_2 - 5)^2$
subject to $x_1 + x_2 \leq 12$, $x_1 \leq 8$, $x_1 - x_2 \leq 4$

Show that the minimum point does not satisfy the regularity condition. Study the implications of this situation.

Solve the following problems graphically. Check necessary and sufficient conditions for candidate local minimum points and verify them on the graph for the problem.

5.4 Minimize $f(x_1, x_2) = 4x_1^2 + 3x_2^2 - 5x_1x_2 - 8x_1$
subject to $x_1 + x_2 = 4$

5.5 Maximize $F(x_1, x_2) = 4x_1^2 + 3x_2^2 - 5x_1x_2 - 8x_1$
subject to $x_1 + x_2 = 4$

5.6 Minimize $f(x_1, x_2) = (x_1 - 2)^2 + (x_2 + 1)^2$
subject to $2x_1 + 3x_2 - 4 = 0$

5.7 Minimize $f(x_1, x_2) = 4x_1^2 + 9x_2^2 + 6x_2 - 4x_1 + 13$
subject to $x_1 - 3x_2 + 3 = 0$

5.8 Minimize $f(\mathbf{x}) = (x_1 - 1)^2 + (x_2 + 2)^2 + (x_3 - 2)^2$
subject to $2x_1 + 3x_2 - 1 = 0$
$x_1 + x_2 + 2x_3 - 4 = 0$

5.9 Minimize $f(x_1, x_2) = 9x_1^2 + 18x_1x_2 + 13x_2^2 - 4$
subject to $x_1^2 + x_2^2 + 2x_1 = 16$

5.10 Minimize $f(x_1, x_2) = (x_1 - 1)^2 + (x_2 - 1)^2$
subject to $x_1 + x_2 - 4 = 0$

5.11 Minimize $f(x_1, x_2) = 4x_1^2 + 3x_2^2 - 5x_1x_2 - 8$
subject to $x_1 + x_2 = 4$

5.12 Maximize $F(x_1, x_2) = 4x_1^2 + 3x_2^2 - 5x_1x_2 - 8$
subject to $x_1 + x_2 = 4$

5.13 Maximize $F(x_1, x_2) = 4x_1^2 + 3x_2^2 - 5x_1x_2 - 8$
subject to $x_1 + x_2 \leq 4$

5.14 Minimize $f(x_1, x_2) = 4x_1^2 + 3x_2^2 - 5x_1x_2 - 8$
subject to $x_1 + x_2 \leq 4$

5.15 Maximize $F(x_1, x_2) = 4x_1^2 + 3x_2^2 - 5x_1x_2 - 8x_1$
subject to $x_1 + x_2 \leq 4$

5.16 Minimize $f(x_1, x_2) = (x_1 - 1)^2 + (x_2 - 1)^2$
subject to $x_1 + x_2 \geq 4$
$x_1 - x_2 - 2 = 0$

5.17 Minimize $f(x_1, x_2) = (x_1 - 1)^2 + (x_2 - 1)^2$
subject to $x_1 + x_2 = 4$
$x_1 - x_2 - 2 \geq 0$

5.18 Minimize $f(x_1, x_2) = (x_1 - 1)^2 + (x_2 - 1)^2$
subject to $x_1 + x_2 \geq 4$
$x_1 - x_2 \geq 2$

5.19 Minimize $f(x, y) = (x - 4)^2 + (y - 6)^2$
subject to $12 \geq x + y$
$x \geq 6, y \geq 0$

5.20 Minimize $f(x_1, x_2) = 2x_1 + 3x_2 - x_1^3 - 2x_2^2$

subject to $x_1 + 3x_2 \leq 6$

$$5x_1 + 2x_2 \leq 10$$

$$x_1, x_2 \geq 0$$

5.21 Minimize $f(x_1, x_2) = 4x_1^2 + 3x_2^2 - 5x_1x_2 - 8x_1$

subject to $x_1 + x_2 \leq 4$

5.22 Minimize $f(x_1, x_2) = x_1^2 + x_2^2 - 4x_1 - 2x_2 + 6$

subject to $x_1 + x_2 \geq 4$

5.23 Minimize $f(x_1, x_2) = 2x_1^2 - 6x_1x_2 + 9x_2^2 - 18x_1 + 9x_2$

subject to $x_1 + 2x_2 \leq 10$

$$4x_1 - 3x_2 \leq 20; \ x_i \geq 0; \ i = 1, 2$$

5.24 Minimize $f(x_1, x_2) = (x_1 - 1)^2 + (x_2 - 1)^2$

subject to $x_1 + x_2 - 4 \leq 0$

5.25 Minimize $f(x_1, x_2) = (x_1 - 1)^2 + (x_2 - 1)^2$

subject to $x_1 + x_2 - 4 \leq 0$

$$x_1 - x_2 - 2 \leq 0$$

5.26 Minimize $f(x_1, x_2) = (x_1 - 1)^2 + (x_2 - 1)^2$

subject to $x_1 + x_2 - 4 \leq 0$

$$2 - x_1 \leq 0$$

5.27 Minimize $f(x_1, x_2) = 9x_1^2 - 18x_1x_2 + 13x_2^2 - 4$

subject to $x_1^2 + x_2^2 + 2x_1 \geq 16$

5.28 Minimize $f(x_1, x_2) = (x_1 - 3)^2 + (x_2 - 3)^2$

subject to $x_1 + x_2 \leq 4$

$$x_1 - 3x_2 = 1$$

5.29 Minimize $f(x_1, x_2) = x_1^3 - 16x_1 + 2x_2 - 3x_2^2$

subject to $x_1 + x_2 \leq 3$

5.30 Minimize $f(x_1, x_2) = 3x_1^2 - 2x_1x_2 + 5x_2^2 + 8x_2$

subject to $x_1^2 - x_2^2 + 8x_2 \leq 16$

5.31 Minimize $f(x, y) = (x - 4)^2 + (y - 6)^2$

subject to $x + y \leq 12$

$$x \leq 6$$

$$x, y \geq 0$$

5.32 Minimize $f(x, y) = (x - 8)^2 + (y - 8)^2$

subject to $x + y \leq 12$

$$x \leq 6$$

$$x, y \geq 0$$

5.33 Maximize $F(x, y) = (x - 4)^2 + (y - 6)^2$

subject to $x + y \leq 12$

$$6 \geq x$$

$$x, y \geq 0$$

5.34 Maximize $F(r, t) = (r - 8)^2 + (t - 8)^2$

subject to $10 \geq r + t$

$$t \leq 5$$

$$r, t \geq 0$$

5.35 Maximize $F(r, t) = (r - 3)^2 + (t - 2)^2$
  subject to $10 \geq r + t$
  $\qquad t \leq 5$
  $\qquad r, t \geq 0$

5.36 Maximize $F(r, t) = (r - 8)^2 + (t - 8)^2$
  subject to $r + t \leq 10$
  $\qquad t \geq 0$
  $\qquad r \geq 0$

5.37 Maximize $F(r, t) = (r - 3)^2 + (t - 2)^2$
  subject to $10 \geq r + t$
  $\qquad t \geq 5$
  $\qquad r, t \geq 0$

5.38 Formulate and graphically solve Exercise 2.23 of the design of a cantilever beam using hollow circular cross section. Check the necessary and sufficient conditions at the optimum point. The data for the problem are $P = 10$ kN; $l = 5$ m; modulus of elasticity, $E = 210$ GPa; allowable bending stress, $\sigma_a = 250$ MPa; allowable shear stress, $\tau_a = 90$ MPa; and mass density, $\rho = 7850$ kg/m^3; $0 \leq R_o \leq 20$ cm, and $0 \leq R_i \leq 20$ cm.

5.39 Formulate and graphically solve Exercise 2.24. Check the necessary and sufficient conditions for the solution points and verify them on the graph.

5.40 Formulate and graphically solve Exercise 3.28. Check the necessary and sufficient conditions for the solution points and verify them on the graph.
  Find optimum solutions for the following problems graphically. Check necessary and sufficient conditions for the solution points and verify them on the graph for the problem.

5.41 A minimum weight tubular column design problem is formulated in Section 2.7 using mean radius R and thickness t as design variables. Solve the problem by imposing an additional constraint $R/t \leq 50$ for the following data: $P = 50$ kN, $l = 5.0$ m, $E = 210$ GPa, $\sigma_a = 250$ MPa, and $\rho = 7850$ kg/m^3.

5.42 A minimum weight tubular column design problem is formulated in Section 2.7 using outer radius R_o and inner radius R_i as design variables. Solve the problem by imposing an additional constraint $0.5(R_o + R_i)/(R_o - R_i) \leq 50$. Use the same data as in Exercise 5.41.

5.43 Solve the problem of designing a "can" formulated in Section 2.2.

5.44 Exercise 2.1

*5.45 Exercise 3.34

*5.46 Exercise 3.35

*5.47 Exercise 3.36

*5.48 Exercise 3.54

5.49 *Answer True or False.*

  1. Candidate minimum points for a constrained problem that do not satisfy second-order sufficiency conditions can be global minimum designs.

  2. Lagrange multipliers may be used to calculate the sensitivity coefficient for the cost function with respect to the right side parameters even if Theorem 4.7 cannot be used.

  3. Relative magnitudes of the Lagrange multipliers provide useful information for practical design problems.

5.50 A circular tank that is closed at both ends is to be fabricated to have a volume of 250π m^3. The fabrication cost is found to be proportional to the surface area of the sheet metal needed for fabrication of the tank and is \$400/m^2. The tank is to be housed in a shed with a sloping roof which limits the height of the tank by the relation $H \le 8D$, where H is the height and D is the diameter of the tank. The problem is formulated as minimize $f(D,H) = 400$ $(0.5\pi D^2 + \pi DH)$ subject to the constraints $\frac{\pi}{4}D^2H = 250\pi$, and $H \le 8D$. Ignore any other constraints.
1. Check for convexity of the problem.
2. Write KKT necessary conditions.
3. Solve KKT necessary conditions for local minimum points. Check sufficient conditions and verify the conditions graphically.
4. What will be the change in cost if the volume requirement is changed to 255π m^3 in place of 250π m^3?

5.51 A symmetric (area of member 1 is the same as area of member 3) three-bar truss problem is described in Section 2.10.
1. Formulate the minimum mass design problem treating A_1 and A_2 as design variables.
2. Check for convexity of the problem.
3. Write KKT necessary conditions for the problem.
4. Solve the optimum design problem using the data: $P = 50$ kN, $\theta = 30°$, $\rho = 7800$ kg/m^3, $\sigma_a = 150$ MPa. Verify the solution graphically and interpret the necessary conditions on the graph for the problem.
5. What will be the effect on the cost function if σ_a is increased to 152 MPa?

Formulate and solve the following problems graphically; check necessary and sufficient conditions at the solution points, verify the conditions on the graph for the problem and study the effect of variations in constraint limits on the cost function.

5.52 Exercise 2.1	**5.53** Exercise 2.3	**5.54** Exercise 2.4
5.55 Exercise 2.5	**5.56** Exercise 2.9	**5.57** Exercise 4.92
5.58 Exercise 2.12	**5.59** Exercise 2.14	**5.60** Exercise 2.23
5.61 Exercise 2.24	**5.62** Exercise 5.41	**5.63** Exercise 5.42
5.64 Exercise 5.43	**5.65** Exercise 3.28	***5.66** Exercise 3.34
***5.67** Exercise 3.35	***5.68** Exercise 3.36	***5.69** Exercise 3.39
***5.70** Exercise 3.40	***5.71** Exercise 3.41	***5.72** Exercise 3.46
***5.73** Exercise 3.47	***5.74** Exercise 3.48	***5.75** Exercise 3.49
***5.76** Exercise 3.50	***5.77** Exercise 3.51	***5.78** Exercise 3.52
***5.79** Exercise 3.53	***5.80** Exercise 3.54	

6

Optimum Design with Excel Solver

Upon completion of this chapter, you will be able to

- Prepare an Excel worksheet for use with Excel Solver

- Solve for the roots of nonlinear equations using Solver

- Formulate practical design problems as optimization problems

- Solve unconstrained optimization problems using Solver

- Solve linear optimum design problems using Solver

- Solve nonlinear constrained optimum design problems using Solver

- Interpret the Solver solution

It turns out that several commercial computer programs, such as Excel Solver, Mathematica Optimization Tool Box, MATLAB Optimization Tool Box, and others, are available to solve an optimization problem once it has been properly formulated. In this chapter we describe how to solve optimization problems using Excel Solver. Solutions using the Solver involve (1) preparation of an Excel worksheet; (2) initialization of the Solver; and (3) interpretation of the optimum solution. The basic concepts and methods implemented in the optimization software are presented and discussed in later chapters.

The purpose of introducing a numerical optimization program early on is to allow students to start work on class projects that involve real-world optimization problems, while the instructor is covering basic concepts and numerical methods in class. In addition to starting class projects early on, students can use these programs to verify their homework solutions. This approach is believed to be conducive to learning optimization concepts and computational algorithms.

6.1 INTRODUCTION TO NUMERICAL METHODS FOR OPTIMUM DESIGN

So far we have discussed problem formulation, graphical optimization, and optimality conditions. The graphical method is applicable to two-variable problems only. The approach to solving optimality conditions described in Chapters 4 and 5 becomes difficult to use when the number of variables or the number of constraints is greater than three. Also, this approach leads to a set of nonlinear equations that needs to be solved using a numerical method anyway. Therefore, numerical methods, which can handle many variables and constraints, as well as directly search for optimum points, have been developed. In this section, we describe some basic concepts of numerical optimization methods and present an overview of different classes of search methods for nonlinear problems. Also "what to do if the solution process fails" and si mple scaling of design variables are discussed.

6.1.1 Classification of Search Methods

Derivative-Based Methods

Search methods for smooth problems are based on the assumption that all functions of the problem are continuous and at least twice continuously differentiable. Also, at least the first-order derivatives of the functions can be calculated accurately. In addition, design variables are assumed to be continuous within their allowable range. Methods in this class are also known as *gradient-based search methods*.

Most methods in this class are based on the following iterative equation:

$$x_i^{(k+1)} = x_i^{(k)} + \Delta x_i^{(k)}; \quad i = 1 \text{ to } n; \quad \text{and} \quad k = 0, 1, 2, \ldots \tag{6.1}$$

In this equation, the subscript i refers to the design variable number, and the superscript k refers to the iteration number. The iterative search process starts with an initial estimate of the design variables, $x_i^{(0)}$. By selecting different initial estimates, or starting points, different optimum solutions can be obtained. Using values of the functions, gradients of the functions, and in some cases Hessians of the functions, a change in the design, $\Delta x_i^{(k)}$, is calculated. The design is then updated using Eq. (6.1) and the process is repeated until some stopping criterion is satisfied.

It is important to note that since the methods based on Eq. (6.1) use only local information about the problem functions, the methods always converge to a local minimum point only. However, strategies based on these methods can be developed to find global solutions to smooth problems. Such strategies are presented in Chapter 18.

Based on the philosophy of Eq. (6.1), numerous methods have been developed and evaluated during the last several decades. From those works, several good algorithms have emerged that are implemented in commercial programs such as Excel, MATLAB, Mathematica, and others. We will present and discuss, in Chapters 9 through 13, some of the methods for unconstrained and constrained nonlinear optimization problems.

Direct Search Methods

Direct search methods do not use, calculate, or approximate derivatives of functions in their search strategy. The functions are assumed to be continuous and differentiable; however, their derivatives are either unavailable or not trustworthy. Only the functions' values are calculated and used in the search. Even if the numerical values for the functions are not available, the methods can still be used as long as it can be determined which point leads to a better value for the function compared to other points. Several methods and their variants are available in the literature (Lewis, et al, 2000; Kolda, et al., 2003). We will describe two prominent methods in this class: the Hooke-Jeeves pattern search method (Chapter 11) and the Nelder-Mead simplex method (Chapter 18).

Derivative-Free Methods

The term *derivative-free* refers to a class of methods that do not require explicit calculation of analytical derivatives of the functions. However, an approximation of derivatives is used to construct a local model. The functions are assumed to be continuous and differentiable. Approximation of the derivatives is generated using the function values only, such as in the finite difference approach. This class also includes the response surface methods that generate approximation for complex functions using only the function values and regression analysis (Box and Wilson, 1951). A method based on this approach is described in Chapter 20.

Nature-Inspired Search Methods

During the past several years, many methods have been proposed and evaluated that are based on observation of some natural phenomenon. These methods also use only the values of the problem functions in their solution procedure. Most of these methods use statistical concepts and random numbers to advance the search toward a solution point. One class of methods starts with an initial design and updates it to a new design using various statistical calculations (simulated annealing, described in Chapter 15). In another class, we start with an initial set of designs and generate a new set using the initial set, random numbers, and statistical concepts (genetic algorithms, described in Chapter 16). Some other methods in this class are presented in Chapter 19.

Methods in this class are quite general because they can be used to solve all kinds of optimization problems, smooth problems, nonsmooth problems, discrete/integer programming problems, network optimization problems, and problems with noisy functions for which derivatives are difficult or expensive to calculate. The methods use only the function values that may be evaluated by any means. Since no trend information about the problem functions is used, the methods typically require thousands of function evaluations. This can be quite time-consuming. However, modern computers and parallel computations can be used to reduce the wall-clock time for solving complex problems.

6.1.2 What to Do If the Solution Process Fails

Optimization methods are iterative, and each iteration can require a large number of calculations depending on the application. For smooth problems, these iterations are based on the design update procedure given in Eq. (6.1). The basic steps to implement an iteration of an optimization algorithm for smooth problems are as follows:

1. Calculation of cost and constraint functions and their gradients at the current point.
2. Definition of a subproblem, using the function values and their gradients to determine the search direction for design update, followed by solution to the subproblem for the search direction.
3. Step size determination in the search direction to update the current design point.

Each of the foregoing steps can require substantial computations depending on the complexity of the application. The round-off and truncation errors can accumulate and can cause the iterative process to fail without producing the final solution. The solution process can also fail if the formulation of the design problem is inaccurate or inconsistent. In this subsection, we discuss some steps that can be used to analyze the situation and fix the cause of failure of the optimization process.

How to Find Feasible Points

First, it is useful to determine if the problem has feasible points. There are several methods to find feasible points for a problem. However, a simple procedure that has worked well is to use the original formulation of the problem with the cost function replaced by a constant. That is, the optimization problem is redefined as

$$\text{Minimize } f = \text{constant}$$
$$\text{subject to all constraints}$$

(6.2)

In this way, the cost function is essentially ignored and the optimization iterations are performed to correct the constraint violations to determine a feasible point. The same algorithm that is to be used for the original problem can be used to solve the problem with a constant cost function. An advantage to this approach is that the original formulation for the constraints, as well as the solution algorithm, can be used to check the feasibility of the problem. Note that this process gives a single feasible point. When the process is repeated with multiple initial points, a set of feasible points can be obtained.

If a feasible point cannot be found starting from an initial point, several other starting points should be tried before declaring that the problem has no feasible solution.

A Feasible Point Cannot Be Obtained

There can be several reasons for this condition. The following steps should be followed to analyze the situation and fix the problem:

1. Check the formulation to ensure that the constraints are formulated properly and that there are no inconsistencies in them. Ensure that all of the functions are continuous and differentiable for a smooth optimization problem.
2. The problem defined in Eq. (6.2) can be used to check the feasibility of individual constraints or a subset of constraints while ignoring the remaining ones. In this way, constraints causing infeasibility can be identified.
3. Ensure that the formulation and data are properly transferred to the optimization software.
4. The constraint limits may be too severe. Relax the constraint limits, if possible, and see if that alleviates the difficulty.
5. Check the constraint feasibility tolerance; relax it to see if that resolves the difficulty.
6. Normalize the constraints if they have different orders of magnitude.
7. Check derivation and implementation of the gradients of the constraint functions. If the gradients are evaluated using finite differences, then their accuracy needs to be verified.
8. Increase precision of all calculations, if possible.

Algorithm Does Not Converge

There can be several reasons for this condition. The following steps can help in analyzing and fixing the problem:

1. Check the formulation to ensure that the constraints and the cost function are formulated properly. Ensure that all of the functions are continuous and differentiable for a smooth optimization problem.
2. Normalize the constraints and the cost function if they have different orders of magnitude.
3. Check the implementation of the cost function and the constraint functions evaluations.
4. Check the derivation and implementation of the gradients of all of the functions. If the gradients are evaluated using finite differences, then their accuracy needs to be verified.
5. Examine the final point reported by the program; it may actually be a solution point. The termination/stopping criterion may be too stringent for the algorithm to meet; try relaxing it.
6. If an overflow of calculations is reported, the problem may be unbounded. Additional constraints may be needed to obtain a bounded solution point.
7. Try different starting points.
8. Ignore some of the constraints and solve the resulting problem. If the algorithm converges, add some of the ignored constraints and solve the problem again. Continue this process until the algorithm does not converge. In this way the problematic constraint can be identified.
9. Use a smaller limit on the number of iterations and restart the algorithm with the final point of the previous run of the program as the starting point.
10. If the design variables are of differing orders of magnitude, try scaling them so that the scaled variables have the same order of magnitude. For example, if x_1 is of the order 10^5 and x_2 is of the order 10^{-5}, then define the scaled variables y_1 and y_2 using the transformations:

$$x_1 = 10^5 y_1; \quad x_2 = 10^{-5} y_2 \tag{6.3}$$

This way the scaled variables y_1 and y_2 are of the order 1. Note that this transformation of variables changes the gradients of all functions as well.
11. Ensure that the optimization algorithm has been proven to converge to a local minimum point starting from any initial guess, and that it is implemented in a robust way. This may be done by solving a few difficult problems with known solutions.
12. Increase the precision of the calculations, if possible.

6.1.3 Simple Scaling of Variables

There are several ways to define scaling of variables; Eq. (6.3) shows one such way. Some advanced procedures require knowledge of the Hessian matrix, or its approximation at the solution point for the cost function, or the Lagrangian function. Such procedures can be quite useful in accelerating the rate of convergence of the iterative procedure. However, in this subsection, we discuss some other simple scaling procedures.

Let a variable x have lower and upper bounds as a and b with $b > a$:

$$a \le x \le b \tag{6.4}$$

In the following procedures, it is assumed that a and b are realistic practical bounds; otherwise, the scaling procedure may actually be detrimental for calculations.

1. If it is desired that the scaled variable y vary between -1 and 1, then the transformation from x to y is defined as

$$y = \frac{2x}{(b-a)} - \frac{(b+a)}{(b-a)}; \quad -1 \le y \le 1 \tag{6.5}$$

2. If the largest value of the scaled variable y is desired to be 1, then the transformation from x to y is defined as

$$y = \frac{x}{b}; \quad \frac{a}{b} \le y \le 1 \tag{6.6}$$

3. If the scaled variable y is desired to be 1 in the middle of the range for x, then the transformation from x to y is defined as

$$y = \frac{2x}{(b+a)}; \quad \frac{2a}{(b+a)} \le y \le \frac{2b}{(b+a)} \tag{6.7}$$

6.2 EXCEL SOLVER: AN INTRODUCTION

In this section, we describe Excel Solver and demonstrate it for solution of nonlinear equations. Preparation of the Excel worksheet for the Solver is explained, as is the Solver Dialog Box, which sets up the solution conditions and various parameters. It turns out that the Solver can also be used for solution of linear and nonlinear programming problems. Therefore, the material in this section should be thoroughly understood because it will be used in all subsequent sections.

6.2.1 Excel Solver

Excel is a spreadsheet program that has many useful capabilities for engineering calculations. In particular, it can be used to solve a system of simultaneous equations and optimization problems. Online help is available to assist with using the program. When the program is invoked, it opens what is called a "workbook." A workbook contains several pages called "worksheets." These worksheets are used to store related data and information. Each worksheet is divided into cells that are referenced by their location (i.e., their column/row numbers). All of the information and the data and its manipulation must be organized in terms of cells. Cells contain raw data, formulas, and references to other cells.

"Solver" is the tool available in Excel to solve a nonlinear equation, a system of linear/nonlinear equations, and optimization problems. Use of this tool to solve linear and non-linear optimization problems is demonstrated later in this chapter. In this section, we introduce the use of this tool to solve for roots of nonlinear equations. Solver is invoked through the `Data` tab. If it is not visible under the `Data` tab, then it is not yet installed on your computer. To install it, use the `Add-in` capability under the `File—Excel Options` menu. Note that the location of Solver and other commands, and the procedure to install Solver, may change with later versions of Excel.

To use the Excel Solver, two major steps need to be followed:

1. An Excel worksheet for the problem needs to be prepared, identifying the cells that are the variables for the problem. Also, all equations for the problem need to be entered into different cells.
2. The Solver is then invoked, which results in the display of the `Solver Parameters` dialog box. In this box, the actual problem to be solved is defined. The cells that contain the variables for the problem are identified. The cell containing the objective function for the problem is identified. The cells defining various constraints for the problem are identified. Various `Options` for the Solver are also invoked.

In the following sections, we elaborate on these two steps and show their use to solve several different types of problems.

6.2.2 Roots of a Nonlinear Equation

Here we will use Solver to find the roots of the nonlinear equation

$$\frac{2x}{3} - \sin x = 0 \tag{6.8}$$

This equation was obtained as a necessary condition for the minimization problem of Example 4.22. We need to prepare a worksheet that defines the problem. The worksheet can be prepared in many different ways; Figure 6.1 shows one.

We rename cell C3 as x, which is the solution variable, and show a value of 2 for it. To name a cell, use the "`Define Name`" command under the "`Formulas`" tab. Defining

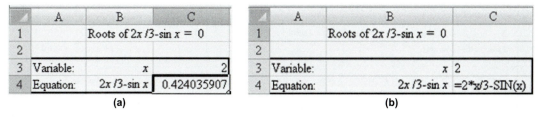

FIGURE 6.1 Excel worksheet for finding roots of $\frac{2x}{3} - \sin x = 0$: (a) worksheet; (b) worksheet with formulas showing.

meaningful names for cells allows them to be referenced by their names rather than their cell numbers. The cells in Figure 6.1 contain the following information:

> **Cell A3**: indicates that row 3 is associated with the variable x (inactive cell)
> **Cell A4**: indicates that row 4 is associated with the equation (inactive cell)
> **Cell B3**: the variable name that will appear later in the Answer Report (inactive cell)
> **Cell C3**: the starting value for the variable named x (the current value is shown as 2); will be updated later by Solver to a final value (active cell)
> **Cell B4**: the equation whose roots need to be found; it will appear later in the Answer Report (inactive cell)
> **Cell C4**: contains the expression for the equation: "= 2*x/3 − sin(x)"; currently it displays the value of the expression for $x = 2$ in cell C3; to see the expression in cell C4, use the Show Formulas command under the Formulas tab (active cell).

Solver Parameters Dialog Box

Whenever the value in cell C3 is changed, cell C4 is updated automatically. Now Solver is invoked under the Data tab to define the target cell and the cell to be varied. Figure 6.2 shows the Solver Parameters dialog box. We set the target cell as C4 because it contains the equation whose roots need to be determined. We set its value to zero in the dialog box because when the root is found its value should be zero. (Solver uses the $ symbol to identify cells; for example, $C$4 refers to cell C4. Use of $ in a cell reference is convenient when the "copy formula" capability is used. With the "copy formula" command, a reference to a cell that uses a $ symbol is not changed.)

Next, we define the cell whose value needs to be treated as the variable by inserting x in the By Changing Cells box. Then, by clicking the Options button, we can reset some of

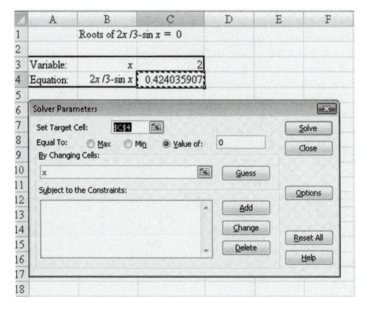

FIGURE 6.2 A Solver Parameters dialog box to define the problem.

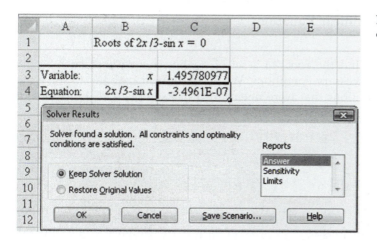

FIGURE 6.3 A Solver Results dialog box and the final worksheet.

FIGURE 6.4 A Solver Answer Report for roots of $\frac{2x}{3} - \sin x = 0$.

	A	B	C	D	E
1	Microsoft Excel 12.0 Answer Report				
2	Worksheet: [nonlin.xls]Figure 4.13				
3	Report Created: 3/8/2010 4:35:09 PM				
4					
5	Target Cell (Value Of)				
6		Cell	Name	Original Value	Final Value
7		$C$4	2x/3-sin x	0.424035907	-3.49607E-07
8					
9	Adjustable Cells				
10		Cell	Name	Original Value	Final Value
11		$C$3	x	2	1.495780977

the parameters related to the solution procedure, if desired. Otherwise, we click the Solve button to find a root starting from $x = 2$.

SOLVER OUTPUT

When Solver is finished, it produces three reports: Answer, Sensitivity, and Limits. (See the Solver Results dialog box in Figure 6.3.) We select only the Answer Report because it contains the relevant information for the roots of the nonlinear equation. When OK is clicked with the Answer option highlighted under Reports, Solver produces another worksheet that contains the final results, as shown in Figure 6.4.

Other roots for the equation can be found by changing the starting value of x in cell C3. If one starting value does not work, another value should be tried. Using this procedure, three roots of the equation are found as 0, 1.496, and -1.496, as given in Example 4.22.

6.2.3 Roots of a Set of Nonlinear Equations

Excel Worksheet

Solver can also be used to find the roots of a set of nonlinear equations, such as the ones obtained from the KKT necessary conditions. We will use this capability to solve the KKT first-order necessary conditions that were determined in Example 4.31. Equations of the necessary conditions are

$$2x_1 - 3x_2 + 2ux_1 = 0 \tag{6.9a}$$

$$2x_2 - 3x_1 + 2ux_2 = 0 \tag{6.9b}$$

$$x_1^2 + x_2^2 - 6 + s^2 = 0; \quad s^2 \geq 0; \quad u \geq 0 \tag{6.9c}$$

$$us = 0 \tag{6.9d}$$

The first step in the solution process is to prepare an Excel worksheet to describe the problem functions. Then Solver is invoked under the Data tab to define equations and constraints. As noted earlier, an Excel worksheet can be prepared in a number of different ways. One way is shown in Figure 6.5, which shows not only the completed worksheet

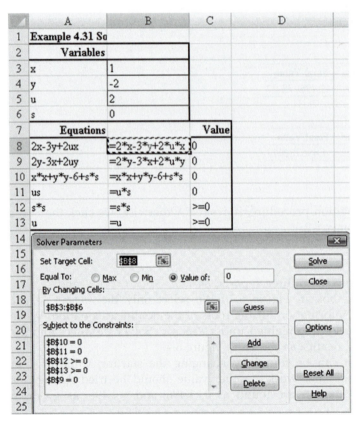

FIGURE 6.5 Worksheet and Solver Parameters dialog box for KKT conditions for Example 4.31.

but also the `Solver Parameters` dialog box for the problem. Various cells are defined as follows:

Cells A3 to A6: variable names that will appear later in the `Answer Report` worksheet (inactive cells)

Cells A8 to A13: expressions for the KKT conditions given in Eqs. (6.9a) through (6.9d); these expressions will appear later in the `Answer Report` (inactive cells)

Cells B3 to B6: renamed x, y, u, and s, respectively, containing the starting values for the four variables; note that the variables x_1 and x_2 have been changed to x and y because x_1 and x_2 are not valid names in Excel; current starting values are shown as 1, -2, 2, 0 (active cells)

Cell B8: $= 2{*}x - 3{*}y + 2{*}u{*}x$ (expression for $\partial L/\partial x$) (active cell)

Cell B9: $= 2{*}y - 3{*}x + 2{*}u{*}y$ (expression for $\partial L/\partial y$) (active cell)

Cell B10: $= x{*}x + y{*}y - 6 + s{*}s$ (constraint, $g + s^2$) (active cell)

Cell B11: $= u{*}s$ (switching condition) (active cell)

Cell B12: $= s{*}s$ (active cell)

Cell B13: $= u$ (active cell)

Cells C8 to C13: right side of expressions in Cells B8 to B13 (inactive cells)

The current values for the expressions in cells B8 through B13 at the starting values of the variables in cells B3 through B6 are given as 12, -15, -1, 0, 0, and 2. Expressions coded in cells B8 through B13 are seen by using the `Show Formulas` command under the `Data` tab.

Solver Parameters Dialog Box

Now the Solver is invoked under the `Data` tab and the root-finding problem is defined in the `Solver Parameters` dialog box. The target cell is set to B8, whose value is set to zero at the solution point, as shown in Figure 6.5. The variable cells are identified as B3 through B6. The rest of the equations are entered as constraints by clicking the `Add` button. Note that in order to solve a set of nonlinear equations, one of them is identified as the target equation, Eq. (6.9a) in the present case, and the rest are identified as equality constraints. Once the problem has been defined, the `Solve` button is clicked to solve it. Solver solves the problem and reports the final results by updating the original worksheet and opening the `Solver Results` dialog box, as shown in Figure 6.6. The final `Answer` worksheet can be generated if desired. The current starting point of (1, -2, 2, 0) gave the KKT point as (-1.732, -1.732, 0.5, 0).

Solution to KKT Cases with Solver

It is important to note that with the worksheet shown in Figure 6.5, the two KKT cases can be solved separately. These cases can be generated using starting values for the slack variable and the Lagrange multiplier in cells B5 and B6. For example, selecting $u = 0$ and $s > 0$ generates the case where the inequality constraint is inactive. This gives the solution: $x = 0$ and $y = 0$. Selecting $u > 0$ and $s = 0$ gives the case where the inequality constraint is active. Selecting different starting values for x and y gives two other points as solutions to the necessary conditions. When there are two or more inequality constraints, various KKT cases can be generated in a similar way.

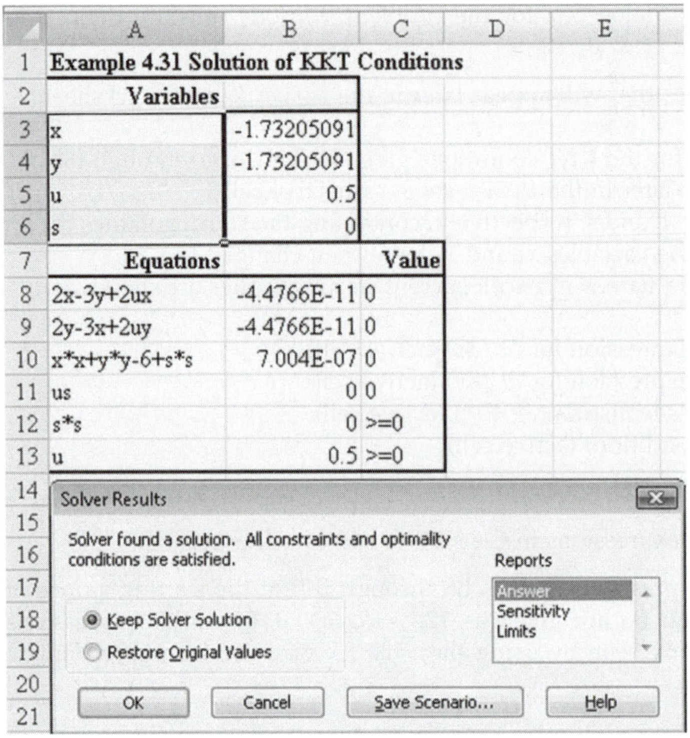

◢	A	B	C	D	E
1	**Example 4.31 Solution of KKT Conditions**				
2	**Variables**				
3	x	-1.73205091			
4	y	-1.73205091			
5	u	0.5			
6	s	0			
7	**Equations**		**Value**		
8	2x-3y+2ux	-4.4766E-11	0		
9	2y-3x+2uy	-4.4766E-11	0		
10	x*x+y*y-6+s*s	7.004E-07	0		
11	us	0	0		
12	s*s	0	>=0		
13	u	0.5	>=0		
14					

FIGURE 6.6 Solver Results for KKT conditions for Example 4.31.

Solver Results

Solver found a solution. All constraints and optimality conditions are satisfied.

Reports

Answer
Sensitivity
Limits

◉ Keep Solver Solution

○ Restore Original Values

[OK]   [Cancel]   [Save Scenario...]   [Help]

6.3 EXCEL SOLVER FOR UNCONSTRAINED OPTIMIZATION PROBLEMS

Excel Solver can be used to solve any unconstrained optimization problem. To show this, let us consider the unconstrained optimization problem:

Minimize

$$f(x, y, z) = x^2 + 2y^2 + 2z^2 + 2xy + 2yz \tag{6.10}$$

Figure 6.7 shows the worksheet and the Solver Parameters dialog box for the problem. The worksheet can be prepared in several different ways, as explained in Section 6.2. For the present example, cells B4 through B6 define the design variables for the problem. These are renamed as x, y, and z and show the starting values for the design variables as 2, 4, and 10. Cell D9 defines the final expression for the cost function. Once the worksheet has been prepared, Solver is invoked under the Data tab, and the Options button is used to invoke the conjugate gradient method. The Newton method can also be used. The forward finite difference option is selected for calculation of the gradient of the cost function. The central difference approach can also be used. The algorithm converges to the solution $(0,0,0)$ with $f^* = 0$ after five iterations.

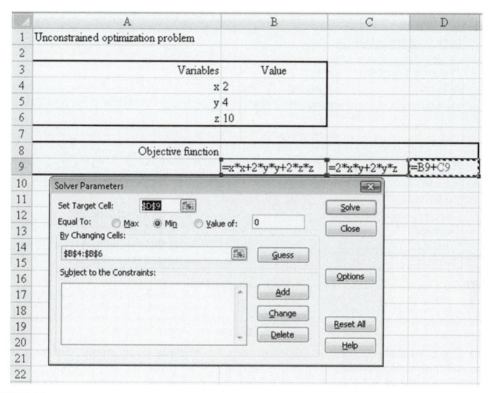

FIGURE 6.7   Excel worksheet and Solver Parameters dialog box for unconstrained problem.

6.4 EXCEL SOLVER FOR LINEAR PROGRAMMING PROBLEMS

Excel Solver can be used to solve linear programming problems as well. The procedure for solving this type of problem is basically the same as the procedure explained for solving nonlinear equations or unconstrained optimization problems in the previous two sections. To start, an Excel worksheet needs to be prepared that contains all of the data and equations for the problem. Next, the Solver dialog box is activated under the Data tab. Last, the objective function, the design variables, and the constraints are defined, and the problem is solved. We will demonstrate this process by solving the problem given as

Maximize

$$z = x_1 + 4x_2 \tag{6.11a}$$

subject to

$$x_1 + 2x_2 \le 5 \tag{6.11b}$$

$$2x_1 + x_2 = 4 \tag{6.11c}$$

$$x_1 - x_2 \ge 1 \tag{6.11d}$$

	A	B	C	D	E	F
1	Linear programming problem					
2						
3	*Problem* is to maximize:	x1+4x2				
4	subject to	x1+2x2<=5				
5		2x1+x2=4				
6		x1-x2>=1				
7		x1, x2>=0				
8						
9	*Problem set up for Solver*					
10	Variables		x1	x2	Sum of LHS	RHS Limit
11	Variable value		0	0		
12	Objective function: max		1	4	=C12*$C$11+D12*$D$11	
13	Constraint 1		1	2	=C13*$C$11+D13*$D$11	5
14	Constraint 2		2	1	=C14*$C$11+D14*$D$11	4
15	Constraint 3		1	-1	=C15*$C$11+D15*$D$11	1

FIGURE 6.8   Excel worksheet for the linear programming problem.

$$x_1, x_2 \geq 0 \tag{6.11e}$$

The Excel worksheet for the problem can be organized in many different ways. Figure 6.8 shows one possible format for setting up the problem. The original problem is entered at the top of the sheet just as a reminder. Other cells containing data about the problem are explained as follows:

A10 to A15: row designations (inactive cell)
C11, D11: starting values for the variables x_1 and x_2, respectively; currently set to 0 (active cell)
C12, D12: objective function coefficients for x_1 and x_2 (inactive cell)
C13 to D15: coefficients of constraints in Eqs. (6.11b) through (6.11d). (inactive cell)
E12: formula to calculate the objective function value using the design variable values in the cells C11 and D11 (active cell)
E13: formula to calculate left side of the "$\leq$ type" constraint in Eq. (6.11b) (active cell)
E14: formula to calculate the left side of the equality constraint in Eq. (6.11c) (active cell)
E15: formula to calculate the left side of the "$\geq$ type" constraint in Eq. (6.11d) (active cell)
F13 to F15: right-side limits for the constraints. (inactive cell)

Note that for the present example, the design variable cells C11 and D11 are not renamed. The `Show Formulas` command under the `Formulas` tab is used to display the formulas in cells E12 through E15; without that command the cells display the current evaluation of the formulas. The formula in cell E12 is entered, and the rest of the formulas are generated using the `Copy Cell` command. It is important to note the $ signs used in referencing some of the cells in the formulas entered in cell E12, as `=C12*$C$11+D12*$D$11`. The cells required to remain fixed in the formula while copying need to have a $ prefix. For example, cells C11 and D11 have design variable values that are needed with each formula; therefore these cells are entered as $C$11 and $D$11. References to these cells do not change in the formulas in cells E13 through E15. Alternatively, equations can be entered manually in each cell.

The next step is to identify the objective function, variables, and constraints for Solver by invoking Solver under the Data tab. This is shown in Figure 6.9, where cell E12 is identified as the objective function in the Target Cell. The Max button is selected to indicate that the objective is to be maximized. Next, the design variables are entered as cells C11 and D11 in the By Changing Cells text box. Excel will change the values in these cells as it determines the optimum solution. The constraints are entered by clicking the Add button; a dialog box appears in which the cells for the left and right sides of a constraint are entered. The final set-up for the present problem in the Solver Parameters dialog box is shown in Figure 6.9. Now we click the Options button and identify the problem as a Linear Model and click the Solve button to obtain the Solver results.

Figure 6.10 shows the Solver Results dialog box and the updated worksheet. Since the Keep Solver Solution option is chosen, the Solver updates the values of the cells C11, D11, and E12 through E15. Three reports are produced in separate worksheets: Answers, Sensitivity, and Limits. Any of these can be highlighted before clicking OK. Figure 6.11 shows the Answer Report. Figure 6.12 shows the Sensitivity Report;. it gives ranges for the right-side parameters and the objective function coefficients. The Limits Report (not shown) gives the lower and upper limits for each variable and the corresponding value of the objective function. Solver determines these limits by rerunning the optimizer with all variables fixed to their optimal values except one, which is optimized.

6.5 EXCEL SOLVER FOR NONLINEAR PROGRAMMING: OPTIMUM DESIGN OF SPRINGS

Now we consider a nonlinear programming problem and solve it using Excel Solver. The spring design problem was described and formulated in Section 2.9. An Excel worksheet for the problem is prepared as shown in Figure 6.13. After normalizing the constraints, the final formulation for the problem is given as follows: find d, D, and N to

Minimize

$$f = (N + 2)Dd^2 \tag{6.12a}$$

subject to the constraints

$$g_1 = 1 - \frac{P}{K\Delta} \le 0 \text{ (Deflection)} \tag{6.12b}$$

$$g_2 = \frac{\tau}{\tau_a} - 1 \le 0 \text{ (Shear Stress)} \tag{6.12c}$$

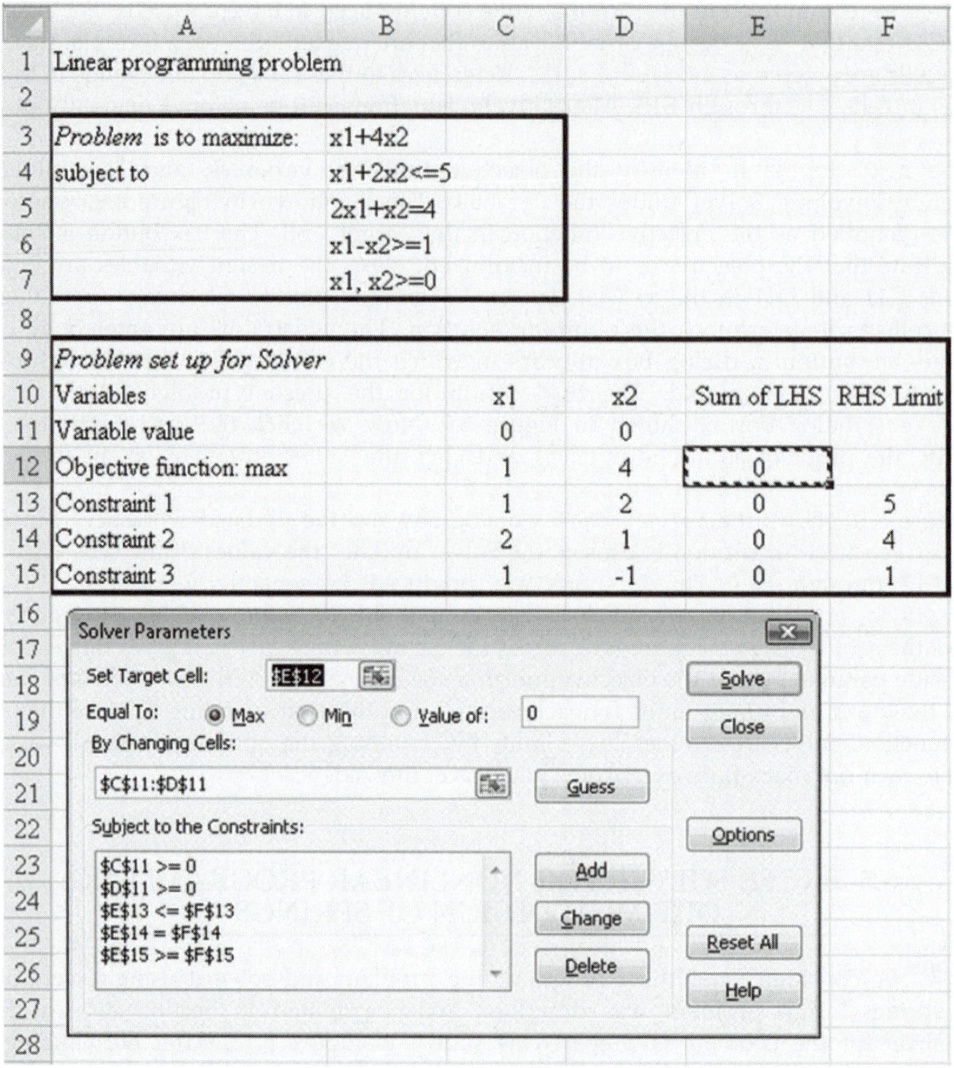

	A	B	C	D	E	F
1	Linear programming problem					
2						
3	*Problem* is to maximize:	x1+4x2				
4	subject to	x1+2x2<=5				
5		2x1+x2=4				
6		x1-x2>=1				
7		x1, x2>=0				
8						
9	*Problem set up for Solver*					
10	Variables		x1	x2	Sum of LHS	RHS Limit
11	Variable value		0	0		
12	Objective function: max		1	4	0	
13	Constraint 1		1	2	0	5
14	Constraint 2		2	1	0	4
15	Constraint 3		1	-1	0	1

Solver Parameters ☒

Set Target Cell: $E$12 🔢

Equal To: ⦿ Max   ◯ Min   ◯ Value of: 0

By Changing Cells:

$C$11:$D$11   🔢   Guess

Subject to the Constraints:

$C$11 >= 0
$D$11 >= 0
$E$13 <= $F$13
$E$14 = $F$14
$E$15 >= $F$15

[Add] [Change] [Delete]

[Solve] [Close] [Options] [Reset All] [Help]

FIGURE 6.9   Solver Parameters dialog box for the linear programming problem.

$$g_3 = 1 - \frac{\omega}{\omega_0} \leq 0 \text{ (Frequency)} \qquad (6.12d)$$

$$g_4 = \frac{D+d}{1.5} - 1 \leq 0 \text{ (Outer Diameter)} \qquad (6.12e)$$

The lower and upper bounds for the design variables are selected as shown in Figure 6.13. Note that the constant $\pi^2 \rho/4$ is ignored in the cost function of Eq. (6.12a); the mass, however, is calculated in cell C24.

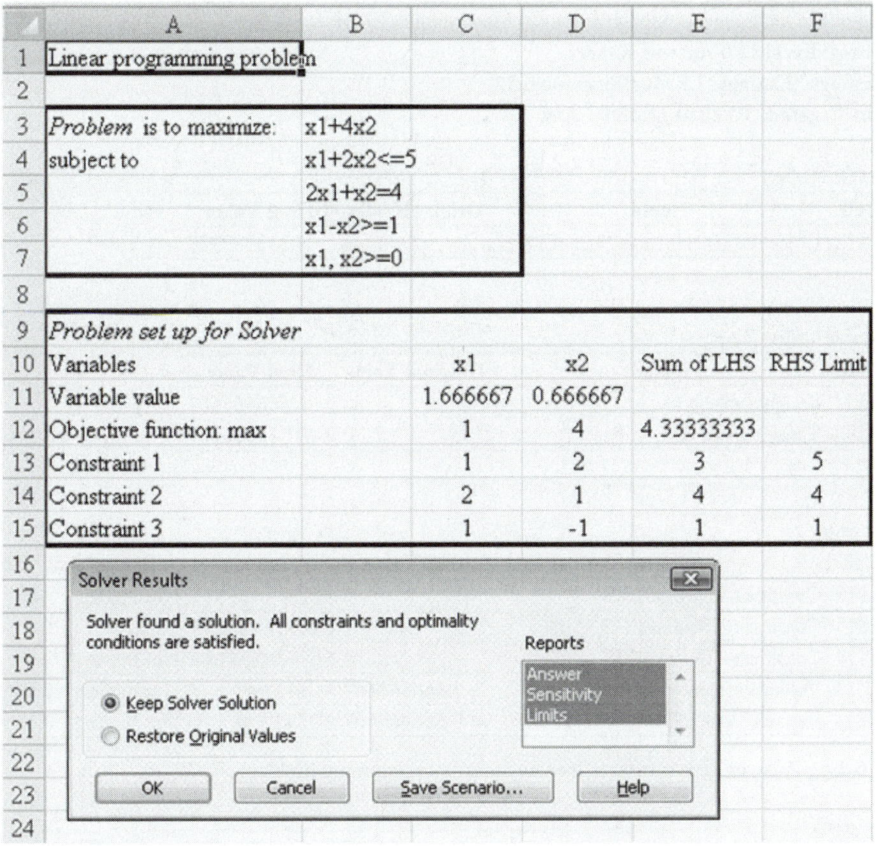

	A	B	C	D	E	F
1	Linear programming problem					
2						
3	Problem is to maximize:	x1+4x2				
4	subject to	x1+2x2<=5				
5		2x1+x2=4				
6		x1-x2>=1				
7		x1, x2>=0				
8						
9	Problem set up for Solver					
10	Variables		x1	x2	Sum of LHS	RHS Limit
11	Variable value		1.666667	0.666667		
12	Objective function: max		1	4	4.33333333	
13	Constraint 1		1	2	3	5
14	Constraint 2		2	1	4	4
15	Constraint 3		1	-1	1	1
16						
17	Solver Results					
18	Solver found a solution. All constraints and optimality					
19	conditions are satisfied.		Reports			
20	⦿ Keep Solver Solution		Answer / Sensitivity / Limits			
21	○ Restore Original Values					
22						
23	OK Cancel Save Scenario... Help					
24						

FIGURE 6.10   Solver Results dialog box for the linear programming problem.

In the worksheet of Figure 6.13, the following notation is used for some of the variables (others are the same as in the formulation):

D_coil: mean coil diameter D, in
Def: deflection δ, in
Def_min: minimum required deflection Δ, 0.5 in
omega: frequency of surge waves ω, Hz
omega_0: lower limit on surge wave frequency ω_0, 100 Hz
rho: mass density of the material ρ, lb-s^2/in
tau: shear stress τ, psi
tau_a: allowable shear stress τ_a, 80,000 psi

In Figure 6.13, rows 4 through 6 contain the design variable data, rows 8 through 15 contain the data for the problem, rows 17 through 21 contain calculations for the analysis variables, row 23 defines the objective function, and rows 26 through 29 define the constraints of Eqs. (6.12b) through (6.12e).

	A	B	C	D	E	F	G
1	Microsoft Excel 12.0 Answer Report						
2	Worksheet: [Example LP.xlsx]Example 6.12						
3	Report Created: 3/5/2010 11:43:01 AM						
4							
5	Target Cell (Max)						
6		Cell	Name	Original Value	Final Value		
7		$E$12	Objective function: max Sum of LHS	0	4.333333333		
8							
9							
10	Adjustable Cells						
11		Cell	Name	Original Value	Final Value		
12		$C$11	Variable value x1	0	1.666666667		
13		$D$11	Variable value x2	0	0.666666667		
14							
15							
16	Constraints						
17		Cell	Name	Cell Value	Formula	Status	Slack
18		$E$13	Constraint 1 Sum of LHS	3	$E$13<=$F$13	Not Binding	2
19		$E$15	Constraint 3 Sum of LHS	1	$E$15>=$F$15	Binding	0
20		$E$14	Constraint 2 Sum of LHS	4	$E$14=$F$14	Not Binding	0
21		$C$11	Variable value x1	1.666666667	$C$11>=0	Not Binding	1.666666667
22		$D$11	Variable value x2	0.666666667	$D$11>=0	Not Binding	0.666666667

FIGURE 6.11   Answer Report from Solver for linear programming problem.

	A	B	C	D	E	F	G	H
1	Microsoft Excel 12.0 Sensitivity Report							
2	Worksheet: [Example LP.xlsx]Example 6.12							
3	Report Created: 3/5/2010 11:43:02 AM							
4								
5	Adjustable Cells							
6				Final	Reduced	Objective	Allowable	Allowable
7		Cell	Name	Value	Cost	Coefficient	Increase	Decrease
8		$C$11	Variable value x1	1.666666667	0	1	7	1E+30
9		$D$11	Variable value x2	0.666666667	0	4	1E+30	3.5
10								
11	Constraints							
12				Final	Shadow	Constraint	Allowable	Allowable
13		Cell	Name	Value	Price	R.H. Side	Increase	Decrease
14		$E$13	Constraint 1 Sum of LHS	3	0	5	1E+30	2
15		$E$15	Constraint 3 Sum of LHS	1	-2.333333333	1	1	2
16		$E$14	Constraint 2 Sum of LHS	4	1.666666667	4	2	2

FIGURE 6.12   Sensitivity Report from Solver for the linear programming problem.

	A	B	C	D	E
1	**Design of Coil Springs**				
2					
3	**1. Design Variables**	Lower limit	Symbol	Value	Upper limit
4	Wire diameter	=0.05	d, in	0.2	=0.2
5	Mean coil diameter	=0.25	D, in	1.3	=1.3
6	Number of active coils	=2	N	2	
7	**2. Parameters**	Symbol	Value	Units	
8	Shear modulus	G	=11500000	lb/in^2	
9	Mass density	ρ, ro	=7.38342*10^-4	$lb\text{-}s^2/in^4$	
10	Allowable shear stress	τ_a, tau_a	=80000	lb/in^2	
11	Number of inactive coils	Q	=2		
12	Applied load	P	=10	lb	
13	Minimum spring deflection	Δ Def_min	=0.5	in	
14	Lower limit on surge wave frequency	ω_o, omega_0	=100	Hz	
15	Limit on outer diameter of the coil	D_o	=1.5	in	
16	**3. Analysis Variables**	Symbol	Equation	Units	
17	Load deflection equation	δ, Def	=P/K	in	
18	Spring Constant	K	=(d^4*G)/(8*D_coil^3*N)	lb/in	
19	Shear Stress	τ, tau	=(8*k_CF*P*D_coil)/(PI()*d^3)	lb/in^2	
20	Wahl stress concentration factor	k, k_CF	=(4*D_coil-d)/(4*(D_coil-d))+(0.615*d)/(D_coil)		
21	Frequency of surge waves	ω, omega	=((d)/(2*PI()*N*D_coil^2))*SQRT(G/(2*ro))	Hz	
22	**4. Objective Function**	Symbol	Equation	Units	
23	Minimize f	f	=(N+2)*D_coil*d^2		
24	Mass	M	=f*ro*PI()^2/4	$lb\text{-}s^2/in$	
25	**5. Constraints**	Value	$</>/=$	RS	
26	Deflection constraint	=1-P/(K*Def_min)	<	0	
27	Shear stress constraint	=tau/tau_a - 1	<	0	
28	Frequency constraint	=1-omega/omega_0	<	0	
29	Outer diameter constraint	=(D_coil+d)/(Do)-1	<	0	

FIGURE 6.13 Excel worksheet for the spring design problem.

The starting point for the design variables is selected as (0.2, 1.3, 2), where the cost function value is 0.208. The solver found an optimum solution as (0.0517, 0.3565, 11.31), with the cost function value as 0.01268, as seen in the `Answer Report` in Figure 6.14. The deflection and the shear stress constraints are active at the optimum solution with their Lagrange multipliers as (0.0108, 0.0244). The Lagrange multipliers are given in the `Sensitivity Report` (not shown).

It is observed that the number of coils is not an integer at the optimum solution. If desired, we can now fix the number of coils to 11 or 12 and reoptimize the problem. When this is done, the following solution is obtained: (0.0512, 0.3454, 12), with $f^* = 0.01268$.

6.6 OPTIMUM DESIGN OF PLATE GIRDERS USING EXCEL SOLVER

STEP 1: PROJECT/PROBLEM DESCRIPTION Welded *plate girders* are used in many practical applications, such as overhead cranes and highway and railway bridges. As an

	A	B	C	D	E	F	G
1		**Microsoft Excel 12.0 Answer Report**					
2		**Worksheet: [Spring_section2.9.xlsx]Formulation**					
3		**Report Created: 3/5/2010 2:31:41 PM**					
4							
5		Target Cell (Min)					
6		**Cell**	**Name**	**Original Value**	**Final Value**		
7		$C$23 f		0.208	0.012677954		
8							
9		Adjustable Cells					
10		**Cell**	**Name**	**Original Value**	**Final Value**		
11		$D$4 d		0.2	0.051680775		
12		$D$5 D_coil		1.3	0.356532293		
13		$D$6 N		2	11.313501		
14							
15		Constraints					
16		**Cell**	**Name**	**Cell Value**	**Formula**	**Status**	**Slack**
17		$B$4	Wire diameter Lower limit	0.05	$B$4<=$D$4	Not Binding	0.001680775
18		$B$5	Mean coil diameter Lower limit	0.25	$B$5<=$D$5	Not Binding	0.106532293
19		$B$6	Number of active coils Lower limit	2	$B$6<=$D$6	Not Binding	9.313501004
20		$B$26	Deflection constraint Value	6.45516E-07	$B$26<=$D$26	Binding	0
21		$B$27	Shear stress constraint Value	-6.36133E-08	$B$27<=$D$27	Binding	0
22		$B$28	Frequency constraint Value	-4.047304635	$B$28<=$D$28	Not Binding	4.047304635
23		$B$29	Outer diameter constraint Value	-0.727857955	$B$29<=$D$29	Not Binding	0.727857955
24		$D$4	d	0.051680775	$D$4<=$E$4	Not Binding	0.148319225
25		$D$5	D_coil	0.356532293	$D$5<=$E$5	Not Binding	0.943467707
26		$D$6	N	11.313501	$D$6<=$E$6	Not Binding	3.686498996

FIGURE 6.14   Solver "Answer Report" for the spring design problem.

example of the formulation of a practical design problem and the optimization solution process, we will present the design of a welded plate girder for a highway bridge to minimize its cost.

Other applications of plate girders can be formulated and solved in a similar way. It has been determined that the life-cycle cost of the girder is related to its total mass. Since mass is proportional to material volume, the objective of this project is to design a minimum-volume girder and at the same time satisfy the requirements of the American Association of State Highway and Transportation Officials (AASHTO) *Specifications* (Arora et al., 1997). The dead load for the girder consists of the weight of the pavement and the girder's self-weight. The live load consists of the equivalent uniform load and concentrated loads based on the HS-20(MS18) truck-loading condition. The cross-section of the girder is shown in Figure 6.15.

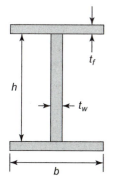

FIGURE 6.15   Cross-section of plate girder.

In this section, we present a formulation for the problem using the procedure described in Chapter 2. Preparation of the Excel worksheet to solve the problem is explained and the problem is solved using Excel Solver.

STEP 2: DATA AND INFORMATION COLLECTION   Material and loading data and other parameters for the plate girder are specified as follows:

L	=	span, 25 m
E	=	modulus of elasticity, 210 GPa
σ_y	=	yield stress, 262 MPa
σ_a	=	allowable bending stress, $0.55\sigma_y = 144.1$ MPa
τ_a	=	allowable shear stress, $0.33\sigma_y = 86.46$ MPa
σ_t	=	allowable fatigue stress, 255 MPa
D_a	=	allowable deflection, $L/800$, m
P_m	=	concentrated load for moment, 104 kN
P_s	=	concentrated load for shear, 155 kN
LLIF	=	live load impact factor, $1 + \frac{50}{(L + 125)}$

Note that the live load impact factor depends on the span length L. For $L = 25$ m, this factor is calculated as 1.33, and it is assumed that the loads P_m and P_s already incorporate this factor. The dependent variables for the problem that can be evaluated using the cross-sectional dimensions and other data are defined as

Cross-sectional area:

$$A = (ht_w + 2bt_f), \ \text{m}^2 \tag{6.13a}$$

Moment of inertia:

$$I = \frac{1}{12}t_w h^3 + \frac{2}{3}bt_f^3 + \frac{1}{2}bt_f h(h + 2t_f), \ \text{m}^4 \tag{6.13b}$$

Uniform load for the girder:

$$w = (19 + 77A), \ \text{kN/m} \tag{6.13c}$$

Bending moment:

$$M = \frac{L}{8}(2P_m + wL), \text{ kN-m} \qquad (6.13d)$$

Bending stress:

$$\sigma = \frac{M}{1000I}(0.5h + t_f), \text{ MPa} \qquad (6.13e)$$

Flange buckling stress limit:

$$\sigma_f = 72,845\left(\frac{t_f}{b}\right)^2, \text{ MPa} \qquad (6.13f)$$

Web crippling stress limit:

$$\sigma_w = 3,648,276\left(\frac{t_w}{h}\right)^2, \text{ MPa} \qquad (6.13g)$$

Shear force:

$$S = 0.5(P_s + wL), \text{ kN} \qquad (6.13h)$$

Deflection:

$$D = \frac{L^3}{384 \times 10^6 \ EI}(8P_m + 5wL), \text{ m} \qquad (6.13i)$$

Average shear stress:

$$\tau = \frac{S}{1000ht_w}, \text{ MPa} \qquad (6.13j)$$

STEP 3: DEFINITION OF DESIGN VARIABLES The cross-sectional dimensions of the plate girder are treated as four design variables for the problem:

h = web height, m
b = flange width, m
t_f = flange thickness, m
t_w = web thickness, m

STEP 4: OPTIMIZATION CRITERION The objective is to minimize the material volume of the girder:

$$\text{Vol} = AL = (ht_w + 2bt_f)L, \text{ m}^3 \qquad (6.14)$$

STEP 5: FORMULATION OF CONSTRAINTS The following constraints for the plate girder are defined:

Bending stress:

$$\sigma \leq \sigma_a \qquad (6.15)$$

Flange buckling:

$$\sigma \le \sigma_f \tag{6.16}$$

Web crippling:

$$\sigma \le \sigma_w \tag{6.17}$$

Shear stress:

$$\tau \le \tau_a \tag{6.18}$$

Deflection:

$$D \le D_a \tag{6.19}$$

Fatigue stress:

$$\sigma \le \frac{1}{2}\sigma_t \tag{6.20}$$

Size constraints:

$$0.30 \le h \le 2.5; \quad 0.30 \le b \le 2.5$$
$$0.01 \le t_f \le 0.10; \quad 0.01 \le t_w \le 0.10 \tag{6.21}$$

Note that the lower and upper limits on the design variables have been specified arbitrarily in the present example. In practice, appropriate values for the given design problem will have to be specified based on the available plate sizes. It is also important to note that the constraints of Eqs. (6.15) through (6.20) can be written explicitly in terms of the design variables h, b, t_f, and t_w by substituting into them expressions for all of the dependent variables. However, there are many applications where it is not possible or convenient to eliminate the dependent variables to obtain explicit expressions for all functions of the optimization problem in terms of the design variables alone. In such cases, the dependent variables must be kept in the problem formulation and treated in the solution process. In addition, use of dependent variables makes it easier to read and debug the program that contains the problem formulation.

SPREADSHEET LAYOUT

The layout of the spreadsheet for solving the KKT optimality conditions, linear programming problems, and unconstrained problems was explained earlier in this chapter. As previously noted, Solver is an "Add-in" to Microsoft Excel. If it does not appear under the Data tab, then it can be easily installed by following the steps outlined in Section 6.2. Figure 6.16 shows the layout of the spreadsheet showing formulas for the plate girder design problem in various cells. The spreadsheet can be organized in any convenient way. The main requirement is that the cells containing objective and constraint functions and the design variables be clearly identified. For the present problem, the spreadsheet is organized into five distinct blocks.

Block 1 contains information about the design variables. Symbols for the variables and their upper and lower limits are defined. The cells containing the starting values for the

	A	B	C	D	E	F
1	Plate Girder Design					
2	**1. Design variable name**	**Lower limit**	**Symbol**	**Value**	**Upper limit**	**Units**
3	web height	0.3	h	1	2.5	m
4	flange width	0.3	b	1	2.5	m
5	flange thickness	0.01	tf	0.1	0.1	m
6	web thickness	0.01	tw	0.1	0.1	m
7						
8	**2. Parameter name**	**Symbol**	**Value**	**Units**		
9	Span length	L	25	m		
10	Mudulus of elasticity	E	210	GPa		
11	Yield stress	sigma_y	262	MPa		
12	Allowable fatigue stress	sigma_t	255	MPa		
13	Concentrated load for moment	Pm	104	kN		
14	Concentrated load for shear	Ps	155	kN		
15	Live load impact factor	LLIF	=1+50/(L+125)	none		
16						
17	**3. Dependent Variable name**	**Symbol**	**Equation**	**Units**		
18	Cross sectional area	A	=h*tw+2*b*tf	m^2		
19	Moment of inertia	I	=(1/12)*tw*h^3+(2/3)*b*tf^3+(1/2)*b*tf*h*(h+2*tf)	m^4		
20	Uniform load	w	=19+77*A	kN/m		
21	Bending moment	M	=L*(2*Pm+w*L)/8	kN-m		
22	Bending stress	sigma	=M*(h/2+tf)/(1000*I)	MPa		
23	Shear force	S	=(Ps+w*L)/2	kN		
24	Deflection	D	=L^3*(8*Pm+5*w*L)/(384*E*I*1000000)	m		
25	Average shear stress	tau	=S/(1000*h*tw)	MPa		
26						
27	**4. Objective Function name**	**Symbol**	**Equation**	**Units**		
28	Volume of material	Vol	=A*L	m^3		
29						
30	**5. Constraints**	**Value/Eq.**	**</>/=**	**Value/Eq.**	**Name**	
31	Bending stress	=sigma	<	=0.55*sigma_y	Allowable bending stress	
32	Bending stress	=sigma	<	=72845*(tf/b)^2	Flange buckling limit	
33	Bending stress	=sigma	<	=3648276*(tw/h)^2	Web cripping limit	
34	Shear stress	=tau	<	=0.33*sigma_y	Allowable shear stress	
35	Deflection	=D	<	=L/800	Allowable deflection	
36	Bending stress	=sigma	<	=sigma_t/2	Allowable fatigue stress	

FIGURE 6.16   Layout of the spreadsheet for plate girder design problem.

variables are identified as D3 through D6. These are updated during the solution process. Also, since these cells are used in all expressions, they are given real names, such as h, b, t_f, and t_w. This is done with the Define Name command under the Formulas tab.

Block 2 defines various data and parameters for the problem. Material properties, loading data, and span length are defined.

Block 3 contains equations for the dependent variables in cells C18 through C25. Although it is not necessary to include them, it can be very useful because they can be incorporated explicitly into the constraint and objective function formulas. First, they simplify the formulation of the constraint expressions, reducing algebraic manipulation errors. Second, they provide a check of these intermediate quantities for debugging purposes and for information feedback.

Block 4 identifies the cell that contains the objective function, cell C28.

Block 5 contains information about the constraints. Cells B31 through B36 contain the left sides, and cells D31 through D36 contain the right sides of the constraints.

Constraints are implemented in Excel by relating two cells through an inequality ($\leq$ or $\geq$) or an equality (=) relationship. This is defined in the `Solver Parameters` dialog box, which is described next. Although many of the quantities appearing in the constraint section also appear elsewhere in the spreadsheet, they are simply references to other cells in the spreadsheet's variables and parameters sections (see the formulas in Figure 6.16). Thus, the only cells that need to be modified during a "what-if" analysis are those in the independent variable or parameters section. The constraints are automatically updated to reflect any changes.

SOLVER PARAMETERS DIALOG BOX

Once the spreadsheet has been created, the next step is to define the optimization problem for the Solver. Figure 6.17 shows a screen shot of the `Solver Parameters` dialog box. The objective function cell is entered as the `Target Cell`, which is to be minimized. The independent design variables are identified next under the `By Changing Cells:` heading. A range of cells has been entered here, but individual cells, separated by commas, could be entered instead. Finally, the constraints are entered under the `Subject to the Constraints`

	A	B	C	D	E	F
1	Plate Girder Design					
2	**1. Design variable name**	**Lower limit**	**Symbol**	**Value**	**Upper limit**	**Units**
3	web height	0.3	h	1	2.5	m
4	flange width	0.3	b	1	2.5	m
5	flange thickness	0.01	tf	0.1	0.1	m
6	web thickness	0.01	tw	0.1	0.1	m
7						
8	**2. Parameter name**	**Symbol**	**Value**	**Units**		
9	Span length	L	25	m		
10	Modulus of elasticity	E	210	GPa		
11	Yield stress	sigma_y	262	MPa		
12	Allowable fatigue stress	sigma_t	255	MPa		
13	Concentrated load for moment	Pm	104	kN		
14	Concentrated load for shear	Ps	155	kN		
15	Live load impact factor	LLIF	1.333333333	none		
16						
17	**3. Dependent Variable name**	**Symbol**	**Equation**	**Units**		
18	Cross sectional area	A	0.3	m²		
19	Moment of inertia	I	0.069	m⁴		
20	Uniform load	w	42.1	kN/m		
21	Bending moment	M	3939.0625	kN-m		
22	Bending stress	sigma	34.25271739	MPa		
23	Shear force	S	603.75	kN		
24	Deflection	D	0.017114275	m		
25	Average shear stress	tau	6.0375	MPa		
26						
27	**4. Objective Function name**	**Symbol**	**Equation**	**Units**		
28	Volume of material	Vol	7.5	m²		
29						
30	**5. Constraints**	**Value/Eq.**	**</>/=**	**Value/Eq.**	**Name**	
31	Bending stress	34.25271739	<	144.1	Allowable bending stress	
32	Bending stress	34.25271739	<	728.45	Flange buckling limit	
33	Bending stress	34.25271739	<	36482.76	Web crippling limit	
34	Shear stress	6.0375	<	86.46	Allowable shear stress	
35	Deflection	0.017114275	<	0.03125	Allowable deflection	
36	Bending stress	34.25271739	<	127.5	Allowable fatigue stress	

Solver Parameters

Set Target Cell: Vol

Equal To: ○ Max ● Min ○ Value of: 0

By Changing Cells:

$D$3:$D$6

Subject to the Constraints:

$B$31:$B$36 <= $D$31:$D$36
$B$3:$B$6 <= $D$3:$D$6
$D$3:$D$6 <= $E$3:$E$6

Solve · Close · Guess · Options · Add · Change · Delete · Reset All · Help

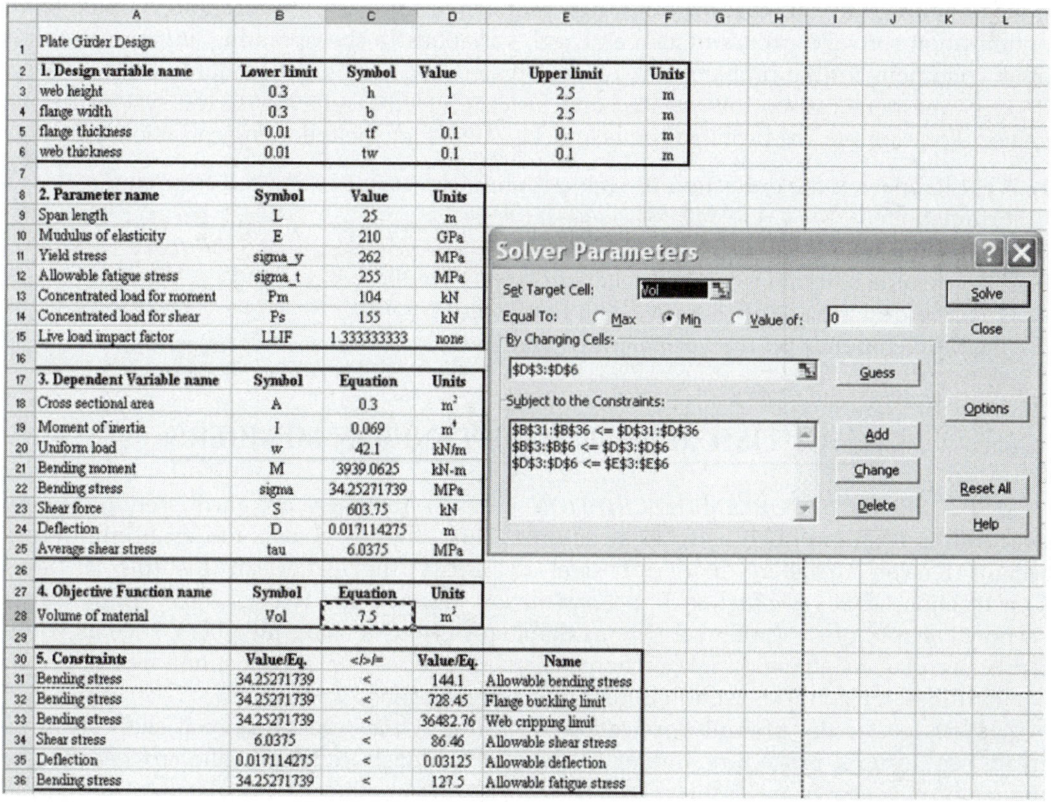

FIGURE 6.17 Solver parameters dialog box and spreadsheet for plate girder design problem.

heading. The constraints are defined by clicking the Add button and then providing the appropriate information. The constraints include not only those identified in the constraints section of the spreadsheet but also the bounds on the design variables.

SOLUTION

Once the problem has been defined in the Solver Parameters dialog box, clicking the Solve button initiates the optimization process. Once Solver has found the solution, the design variable cells (D3–D6), the dependent variable cells (C18–C25), and the constraint function cells (B31–B36 and D31–D36) are updated using the optimum values of the design variables. Solver also generates three reports in separate worksheets, Answer, Sensitivity, and Limits (as explained earlier in this chapter). The Lagrange multipliers and constraint activity can be recovered from these reports.

The optimum solution for the plate girder is obtained as follows:

$$h = 2.0753 \text{ m}, \ b = 0.3960 \text{ m}, \ t_f = 0.0156 \text{ m}, \ t_w = 0.0115 \text{ m}, \ \text{Vol} = 0.90563 \text{ m}^3$$

The flange buckling, web crippling, and deflection constraints are active at the optimum point.

It is important to note that once a design problem has been formulated and coupled to an optimization software program such as Excel, variations in the operating environment and other conditions for the problem can be investigated in a very short amount of time. "What if" type questions can be investigated and *insights* into the behavior of the system can be gained. For example, the problem can be quickly solved for the following conditions:

1. What happens if the deflection or web crippling constraint is omitted from the formulation?
2. What if the span length is changed?
3. What if some material properties change?
4. What if a design variable is assigned a fixed value?
5. What if the bounds on the variables are changed?

6.7 OPTIMUM DESIGN OF TENSION MEMBERS

STEP 1: PROJECT/PROBLEM DESCRIPTION   Tension members are encountered in many practical applications such as truss structures. This project's goal is to design minimum-weight tension members made of steel to satisfy the American Institute of Steel Construction (AISC) *Manual of Steel Construction Requirements* (AISC, 2005). Many cross-sectional shapes of steel members are available for use as tension members, such as wide-flange sections (W-shape), angle sections, channel sections, tee sections, hollow circular or square tubes, tension rods (solid circular or square sections), and cables.

A W-shape for the present application is desired. Other cross-sectional shapes can be treated by using a procedure similar to the one demonstrated here. The cross-section of the member is shown in Figure 6.18. The load for the member is calculated based on the application and loads requirement of the region where the structure is located (ASCE, 2005). The specified material is ASTM A992, grade 50 steel.

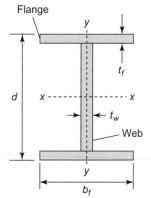

Flange

y

t_f

d x ----------x

t_w

Web

y

b_f

FIGURE 6.18   W-shape for a member.

x-x is the major (strong) axis
y-y is the minor (weak) axis

STEP 2: DATA AND INFORMATION COLLECTION   To formulate the problem of optimizing the design of tension members, the information that needs to be collected includes AISC requirements, load, and material properties. To achieve these objectives, see the notation and data defined in Table 6.1.

Some useful expressions for problem formulation are:

$$A_g = 2b_f t_f + (d - 2t_f)t_w, \text{ in}^2 \tag{6.22a}$$

$$I_y = 2\left(\frac{1}{12}t_f b_f^3\right) + \frac{1}{12}(d - 2t_f)t_w^3, \text{ in}^4 \tag{6.22b}$$

$$r_y = \sqrt{I_y/A_g}, \text{ in} \tag{6.22c}$$

STEP 3: DEFINITION OF DESIGN VARIABLES   The design variables for the member are its cross-sectional dimensions, shown in Figure 6.18. Thus, the design variable vector is $\mathbf{x} = (d, b_f, t_f, t_w)$.

STEP 4: OPTIMIZATION CRITERION   The goal is to select the lightest possible W-shape for the member. Thus we minimize the weight of the member per unit length, given as cross-sectional area × density:

$$f = 12\gamma A_g, \text{ lbs/ft} \tag{6.23}$$

STEP 5: FORMULATION OF CONSTRAINTS   AISC (2005) requires three limit states to be satisfied by the member:

1. Yielding of the gross section
2. Rupture of the net section
3. Slenderness limit state

TABLE 6.1   Data for optimizing the design of tension members

Notation	Data
A_g	Gross area of the section, in^2
A_n	Net area (gross area minus cross-sectional areas due to bolt holes), in^2
A_e	Effective net area (reduction of net area to account for stress concentrations at holes and shear lag effect where not all cross-sectional elements are connected to transmit load), $A_e = UA_n$, in^2
b_f	Width of flange, in
d	Depth of section, in
F_y	Specified minimum yield stress, 50 ksi for A992 steel, ksi
F_u	Specified minimum ultimate stress, 65 ksi for A992 steel, ksi
L	Laterally unsupported length of member, in
P_n	Nominal axial strength, kips
P_a	Required strength, kips
r_y	Least radius of gyration, in
t_f	Thickness of flange, in
t_w	Thickness of web, in
U	*Shear lag coefficient: reduction coefficient for net area* whenever tension is transmitted through some but not all cross-sectional elements of member (such as angle section, where only one leg of angle is connected to a gusset plate), $U = 1 - \bar{x}/l$. Also, connection design should be such that $U \geq 0.6$. Table D3.1 of AISC *Manual* (AISC, 2005) can be used to evaluate U for different conditions.
$\bar{x}$	Distance from plane of shear transfer (plane of connection) to centroid of tension member cross-section, in
Ω_t	Factor of safety for tension; 1.67 for yielding of gross cross-section and 2.00 for rupture (fracture) of net cross-section
γ	Density of steel, 0.283 lb/in^3

The yield limit says that the required strength should be less than or equal to the allowable yield strength (capacity of the member in tension). The allowable yield strength is obtained by dividing the nominal yield strength P_{ny} by a factor of safety. Thus the yield limit state is written as

$$P_a \leq \frac{P_{ny}}{\Omega_t}; \quad P_{ny} = F_y A_g \tag{6.24}$$

Since $\Omega_t = 5/3$ the yield limit state constraint becomes

$$P_a \leq 0.6 F_y A_g \tag{6.25}$$

The rupture limit state imposes the constraint that the required strength be no more than the allowable rupture strength of the net section. The allowable rupture strength is obtained by dividing the nominal rupture strength by a factor of safety. Thus the rupture limit state constraint is written as

$$P_a \leq \frac{P_{nr}}{\Omega_t}; \quad P_{nr} = F_u A_e \tag{6.26}$$

Since $\Omega_t = 2$ for the rupture limit state, this constraint becomes

$$P_a \leq 0.5 F_u A_e \tag{6.27}$$

Calculation of the net effective area A_e depends on the details of the connection (e.g., length of connection, number and arrangement of bolts, and distance of the shear plane from the centroid of the section). Since these details are not known at the initial design stage, assumptions are made about the effective net area A_e and are then verified once the design of the member is known. For the present application, $A_e = 0.75 A_g$ is used.

The slenderness ratio constraint is written as

$$\frac{L}{r_y} \leq 300 \tag{6.28}$$

Although this constraint is not required for tension members, AISC (2005) suggests imposing it to avoid having members that are too slender. This may also avoid buckling of the members if reversal of loads occurs.

EXAMPLE 6.1 SELECTION OF A W10 SHAPE

To solve the optimization problem formulated in the foregoing section for a member of length 25 ft, we use Excel Solver. Since the W10 shape is to be selected, the following lower and upper limits for the design variables are specified based on the data in the dimensions table in the AISC *Manual* (2005):

$$9.73 \leq d \leq 11.4 \text{ in} \tag{a}$$

$$3.96 \leq b_f \leq 10.4 \text{ in} \tag{b}$$

$$0.21 \leq t_f \leq 1.25 \text{ in} \tag{c}$$

$$0.19 \leq t_w \leq 0.755 \text{ in} \tag{d}$$

Based on an analysis of loads for the structure, it is determined that the service dead load is 50 kips and the live load is 100 kips for the tension member. Therefore, the required strength P_a for the member is 150 kips.

The Excel worksheet for the problem is prepared, and the Solver is invoked. In the implementation, the constraints are normalized with respect to their limit values; for example, the yielding limit state constraint of Eq. (6.25) is normalized as $1 \leq 0.6F_y A_g/P_a$. The initial design is selected as a W10 × 15 section, for which the constraints of Eqs. (6.25), (6.27), and (6.28) are violated; the section is therefore not adequate. The initial values of the design variables are set as (10.00, 4.00, 0.27, 0.23) for the W10 × 15 section. The initial cost function value is 14.72 lbs/ft. The Solver gives the following optimum design:

$$x^* = (d = 11.138, \quad b_f = 4.945, \quad t_f = 0.321, \quad t_w = 0.2838) \text{ in;}$$

$$\text{Weight} = 20.90 \text{ lbs/ft}; \quad A_g = 6.154 \text{ in.}^2, \quad r_y = 1.027 \text{ in;}$$

(e)

Several different starting points give basically the same optimum weight but slightly different design variable values. Analyzing the optimum design, we observe that the rupture limit state constraint is active at the optimum point. When using the optimum design, it is suggested that we use a W10 × 22 section, which has allowable strengths of 194 kips and 158 kips in the yielding and rupture limit states, respectively.

EXAMPLE 6.2   SELECTION OF A W8 SHAPE

Now we want to investigate the use of a W8 section for the member. For the W8 shape, the following lower and upper limits for the design variables are specified based on the data in the dimensions table in the AISC *Manual* (2005):

$$7.93 \leq d \leq 9.00 \text{ in} \tag{a}$$

$$3.94 \leq b_f \leq 8.28 \text{ in} \tag{b}$$

$$0.205 \leq t_f \leq 0.935 \text{ in} \tag{c}$$

$$0.17 \leq t_w \leq 0.507 \text{ in} \tag{d}$$

The initial design is selected as a W8 × 15 section. For this section, the constraints of Eqs. (6.25), (6.27), and (6.28) are violated, so the section is not adequate. Using the W8 × 15 section, the initial design variable values are set as (8.11, 4.01, 0.315, 0.245), where the cost function has a value of 14.80 lbs/ft. The Solver gives the following optimum design:

$$x^* = (d = 8.764, \quad b_f = 5.049, \quad t_f = 0.385, \quad t_w = 0.284) \text{ in}$$

$$\text{Weight} = 20.90 \text{ lbs/ft}; \quad A_g = 6.154 \text{ in.}^2, \quad r_y = 1.159 \text{ in}$$

(e)

Several different starting points give the same optimum weight but different design variable values. Analyzing the optimum design, we observe that only the rupture limit state constraint is active at the optimum point. When using the optimum design, it is suggested that we use a W8 × 24 section, which has allowable strengths of 208 kips and 169 kips in yielding and rupture limit states, respectively. A W8 × 21 section did not meet the rupture limit state constraint. Also, it appears that a W10 × 22 section is lighter in this application. If there is no particular restriction on the depth of the section, a W12 or even a W14 section could be tried for an even lighter section.

DISCUSSION

It is important to note that while designing the connection for the member, the assumption for the net effective area $A_e = 0.75A_g$ must be verified. AISC (2005) also requires checking the section for block shear failure of the member. Block shear is a failure phenomenon at the bolted connections for the member. The member or the gusset plate can shear off and/or rupture in tension. There can be several modes of block shear failure, depending on the details of the connection. All of the modes need to be guarded against shear failure by either yielding or rupture. Sections J4.2 and J4.3 of the AISC Specification (2005) should be consulted for more details.

In the present application, we have treated members made of steel. Members made of other materials, such as aluminum and wood, can be optimized in a manner similar to the one described here.

6.8 OPTIMUM DESIGN OF COMPRESSION MEMBERS

6.8.1 Formulation of the Problem

STEP 1: PROJECT/PROBLEM DESCRIPTION   Compression members are encountered in many practical applications, such as pole structures, columns in building frames and members of truss structures. In this project, the goal is to design minimum weight compression members made of steel to satisfy the AISC Manual requirements (AISC, 2005). Many cross-sectional shapes of steel members are available for use as compression members, such as wide-flange sections (W-shape), angle sections, channel sections, tee sections, hollow circular or square tubes, tension rods (solid circular or square sections) and cables. A W-shape is desired for the present application. The cross-section of the member is shown in Figure 6.18. The load for the member is calculated based on the application and load requirements of the region where the structure is located (ASCE, 2005). The specified material is ASTM A992 Grade 50 steel.

STEP 2: DATA/INFORMATION COLLECTION   To formulate the problem of optimizing the design of compression members, the information that needs to be collected includes AISC requirements, load, and material properties. To achieve these goals, see the notation and data defined in Table 6.2.

Some useful expressions for the problem formulation are:

$$A_g = 2b_f t_f + (d - 2t_f)t_w, \text{ in}^2 \tag{6.29a}$$

$$I_x = 2\left(\frac{1}{12}b_f t_f^3\right) + \frac{1}{12}t_w(d - 2t_f)^3 + 2b_f t_f \left(\frac{d}{2} - \frac{t_f}{2}\right)^2, \text{ in}^4 \tag{6.29b}$$

$$I_y = 2\left(\frac{1}{12}t_f b_f^3\right) + \frac{1}{12}(d - 2t_f)t_w^3, \text{ in}^4 \tag{6.29c}$$

$$r_x = \sqrt{I_x/A_g}, \text{ in} \tag{6.29d}$$

TABLE 6.2 Data for optimizing the design of compression members

Notation	Data
A_g	Gross area of the section, in^2
b_f	Width of the flange, in
d	Depth of the section, in
E	Modulus of elasticity; 29,000 ksi
F_e	Euler stress, ksi
F_{ex}	Euler stress for buckling with respect to the strong (x) axis, ksi
F_{ey}	Euler stress for buckling with respect to the weak (y) axis, ksi
F_y	Specified minimum yield stress; 50 ksi for A992 steel
F_{cr}	Critical stress for the member, ksi
F_{crx}	Critical stress for the member for buckling with respect to the strong (x) axis, ksi
F_{cry}	Critical stress for the member for buckling with respect to the weak (y) axis, ksi
I_x	Moment of inertia about the strong (x) axis, in^4
I_y	Moment of inertia about the weak (y) axis, in^4
K	Dimensionless coefficient called the *effective length factor*; its value depends on the end conditions for the member
K_x	Effective length factor for buckling with respect to the strong (x) axis; 1.0
K_y	Effective length factor for buckling with respect to the weak (y) axis; 1.0
L_x	Laterally unsupported length of the member for buckling with respect to strong (x) axis, 420 in
L_y	Laterally unsupported length of the member for buckling with respect to weak (y) axis, 180 in
P_n	Nominal axial compressive strength, kips
P_a	Required compressive strength; 1500 kips
r_x	Radius of gyration about the strong (x) axis, in
r_y	Radius of gyration about the weak (y) axis, in
t_f	Thickness of the flange, in
t_w	Thickness of the web, in
Ω_c	Factor of safety for compression; 5/3 for the yielding of the gross cross-section
λ	Slenderness ratio
λ_e	Limiting value of slenderness ratio for elastic/inelastic buckling
λ_x	Slenderness ratio for buckling with respect to strong (x) axis
λ_y	Slenderness ratio for buckling with respect to weak (y) axis
γ	Density of steel; 0.283 lb/in^3

$$r_y = \sqrt{I_y/A_g}, \text{ in} \qquad (6.29e)$$

$$\lambda = \frac{KL}{r} \qquad (6.29f)$$

$$\lambda_e = 4.71\sqrt{\frac{E}{F_y}} \qquad (6.29g)$$

$$\lambda_x = \frac{K_xL_x}{r_x} \qquad (6.29h)$$

$$\lambda_y = \frac{K_yL_y}{r_y} \qquad (6.29i)$$

The American Institute of Steel Construction requirements for design of compression members using the allowable strength design (ASD) approach are

(i) the required strength $\leq$ the available (allowable) strength of the section
(ii) $KL/r \leq 200$

The strength constraint is written as:

$$P_a \leq \frac{P_n}{\Omega_c} = 0.6P_n \qquad (6.30)$$

The nominal strength P_n is given as:

$$P_n = F_{cr}A_g \qquad (6.31)$$

To calculate the nominal strength P_n, the critical stress F_{cr} needs to be calculated. The critical stress for the member depends on several factors, such as buckling about the strong (x) axis or the weak (y) axis, unbraced length of the member, end conditions for the member, slenderness ratio, material properties, and elastic or inelastic buckling. When the following condition is met, buckling is *inelastic* and the critical stress is given by Eq. (6.33):

$$\lambda \leq \lambda_e \text{ (or, } F_e \geq 0.44F_y) \qquad (6.32)$$

$$F_{cr} = (0.658^{F_y/F_e})F_y \qquad (6.33)$$

When the following condition is met, buckling is *elastic* and the critical stress is given by Eq. (6.35):

$$\lambda > \lambda_e \text{ (or, } F_e < 0.44F_y) \qquad (6.34)$$

$$F_{cr} = 0.877F_e \qquad (6.35)$$

In the foregoing expressions, the Euler stress F_e is given as

$$F_e = \frac{\pi^2E}{\lambda^2} \qquad (6.36)$$

In calculating the Euler stress, it needs to be determined whether the buckling is about the strong (x) axis or about the weak (y) axis. This means that the slenderness ratios λ about the x and y axes need to be calculated. The larger of these two values determines the buckling axis.

STEP 3: DEFINITION OF DESIGN VARIABLES The design variables for the member are the cross-sectional dimensions of the member, shown in Figure 6.18. Thus the design variable vector is $\mathbf{x} = (d, b_f, t_f, t_w)$.

STEP 4: OPTIMIZATION CRITERION The goal is to select the lightest possible W-shape for the member. Thus we minimize the weight of the member per unit length given as the cross-sectional area × density:

$$f = 12\gamma A_g, \ \text{lbs/ft} \tag{6.37}$$

STEP 5: FORMULATION OF CONSTRAINTS It is challenging to formulate the constraint of Eq. (6.30) for a derivative-based optimization algorithm. The reason is that the constraint depends on two "IF THEN ELSE" conditions, which could make the constraint function discontinuous or at least non-differentiable at some points in the feasible set for the problem. The first "IF THEN ELSE" condition is for calculation of the Euler stress F_e in Eq. (6.36):

$$\text{IF } \lambda_x \leq \lambda_y, \ \text{THEN } F_e = \frac{\pi^2 E}{(\lambda_y)^2} \tag{6.38}$$

$$\text{ELSE, } F_e = \frac{\pi^2 E}{(\lambda_x)^2} \tag{6.39}$$

This condition basically says that a smaller value of F_e from Eqs. (6.38) and (6.39) should be used in calculating the critical stress in Eqs. (6.33) and (6.35). To overcome this uncertainty of buckling about the x or y axis, we use both the expressions for F_e in Eqs. (6.38) and (6.39) to calculate the critical stress and impose the strength requirement constraint for both the cases. Therefore we define:

$$F_{ey} = \frac{\pi^2 E}{(\lambda_y)^2} \tag{6.40}$$

$$F_{ex} = \frac{\pi^2 E}{(\lambda_x)^2} \tag{6.41}$$

where F_{ey} and F_{ex} are the Euler stresses with respect to the buckling about weak (y) and strong (x) axes, respectively. These expressions are used to calculate critical stresses, F_{crx} and $F_{cry,}$ from Eqs. (6.33) and (6.35).

The second major difficulty with the formulation is with the "IF THEN ELSE" condition in Eq. (6.32), which determines whether the buckling is going to be elastic or inelastic. This condition determines which of the two expressions, given in Eqs. (6.33) and (6.35), governs the critical stress value for the section. To overcome this difficulty, we design the member to remain in the inelastic buckling mode or remain in the elastic buckling mode. In other words, we impose a constraint requiring the member to remain in the inelastic buckling mode and optimize the member. Then we re-optimize the problem by imposing the condition for the member to remain in the elastic buckling mode. The better of the two solutions is then used as the final design.

6.8.2 Formulation of the Problem for Inelastic Buckling

We formulate the problem where the inelastic buckling constraint is imposed. To optimize the problem, we first assume buckling of the member about the weak (y) axis and impose the following constraints:

$$\lambda_y \leq \lambda_e \tag{6.42}$$

$$P_a \leq 0.6 F_{cry} A_g \tag{6.43}$$

where the critical stress F_{cry} is given from Eq. (6.33) as

$$F_{cry} = (0.658^{F_y/F_{ey}}) F_y \tag{6.44}$$

We monitor the slenderness ratio λ_x about the strong (x) axis. If $\lambda_x \geq \lambda_y$ at the optimum solution, then the buckling would be about the strong (x) axis. We re-optimize the problem by imposing the following constraints:

$$\lambda_x \leq \lambda_e \tag{6.45}$$

$$P_a \leq 0.6 F_{crx} A_g \tag{6.46}$$

where the critical stress F_{crx} is given from Eq. (6.33) as

$$F_{crx} = (0.658^{F_y/F_{ex}}) F_y \tag{6.47}$$

To avoid local buckling of the flange and the web, the following constraints are imposed (AISC, 2005):

$$\frac{(d - 2t_f)}{t_w} \leq 0.56 \sqrt{\frac{E}{F_y}} \tag{6.48}$$

$$\frac{b_f}{2t_f} \leq 1.49 \sqrt{\frac{E}{F_y}} \tag{6.49}$$

EXAMPLE 6.3   INELASTIC BUCKLING SOLUTION

As indicated before, we assume that the member is going to buckle about the weak (y) and optimize the problem. We also assume that a W18 shape is desired for the member and impose the following lower and upper limits for the design variables based on the data that is in the dimensions table in the AISC Manual (2005) (if other W-shapes for the member are desired, appropriate upper and lower limits for the design variables can be imposed):

$$17.7 \leq d \leq 21.1 \text{ in} \tag{a}$$

$$6.0 \leq b_f \leq 11.7 \text{ in} \tag{b}$$

$$0.425 \leq t_f \leq 2.11 \text{ in} \tag{c}$$

$$0.30 \leq t_w \leq 1.16 \text{ in} \tag{d}$$

The optimization problem becomes: find design variables d, b_f, t_f, t_w to minimize the cost function of Eq. (6.37) subject to the constraints of Eqs. (6.42), (6.43), (6.48), (6.49) and (a) through (d). The Excel sheet for the problem is prepared, and the Solver is invoked. In the implementation, the constraints are normalized with respect to their limit values, e.g., the yielding limit state constraint of Eq. (6.43) is normalized as $1 \leq 0.6 F_{cry} A_a / P_a$. The initial values of the design variables are set to their lower limits as (17.70, 6.00, 0.425, 0.30). The initial cost function value is 34.5 lbs/ft. The Solver gives the following optimum design:

$$(d = 18.48, \quad b_f = 11.70, \quad t_f = 2.11, \quad t_w = 1.16) \text{ in}$$

$$\text{Weight} = 224 \text{ lbs/ft}; \quad A_g = 65.91 \text{ in}^2 \tag{e}$$

$$\lambda_x = 56.8, \quad \lambda_y = 61.5, \quad \lambda_e = 113.4, \quad F_{crx} = 39.5 \text{ ksi}, \quad F_{cry} = 37.9 \text{ ksi}$$

Several different starting points give the same optimum solution.

Analyzing the optimum design, we observe that the strength constraint for buckling about the y axis is active along with upper bounds for flange width, flange thickness, and web thickness at the optimum point. Since $\lambda_y > \lambda_x$, the assumption of buckling about the weak axis is correct. Based on the optimum design, it is suggested to use the W18 × 234 section that has allowable strength of 1550 kips.

To gain some insights into the behavior of the column and the solution process, we assume buckling of the member to take place about the strong (x) axis and then re-optimize the column. The optimization problem becomes: find design variables d, b_f, t_f, t_w to minimize the cost function of Eq. (6.37) subject to the constraints of Eqs. (6.45), (6.46), (6.48), (6.49) and (a) through (d). The initial values of the design variables are set as (17.70, 6.00, 0.425, 0.30). The initial cost function value is 34.5 lbs/ft. The Solver gives the following optimum design:

$$(d = 21.1, \quad b_f = 11.70, \quad t_f = 2.11, \quad t_w = 0.556) \text{ in}$$

$$\text{Weight} = 199.6 \text{ lbs/ft}; \quad A_g = 58.77 \text{ in}^2 \tag{f}$$

$$\lambda_x = 47.0, \quad \lambda_y = 58.1, \quad \lambda_e = 113.4, \quad F_{crx} = 42.54 \text{ ksi}, \quad F_{cry} = 39.1 \text{ ksi}$$

Several different starting points give the same optimum solution. Analyzing the optimum design, we observe that the strength constraint for buckling about the x axis is active along with upper bounds for section depth, flange width, and flange thickness at the optimum point. However, at the optimum point, $\lambda_y > \lambda_x$, indicating buckling about the weak (y) axis. Therefore this optimum design is not acceptable.

6.8.3 Formulation of the Problem for Elastic Buckling

Now we formulate the problem for the case where buckling is assumed to be elastic. Just as for the inelastic buckling formulation, we first assume buckling of the member about the weak (y) axis and optimize the column by imposing the following constraints:

$$\lambda_e \le \lambda_y \le 200 \tag{6.50}$$

$$P_a \le 0.6 F_{cry} A_g \tag{6.51}$$

where the critical stress F_{cry} is given from Eq. (6.35) as

$$F_{cry} = 0.877 F_{ey} \tag{6.52}$$

We monitor the slenderness ratio λ_x about the strong (x) axis. If $\lambda_x \ge \lambda_y$ at the optimum point, then the buckling would be about the strong (x) axis. We then re-optimize the problem by imposing the following constraints:

$$\lambda_e \le \lambda_x \le 200 \tag{6.53}$$

$$P_a \le 0.6 F_{crx} A_g \tag{6.54}$$

where the critical stress F_{crx} is given from Eq. (6.33) as

$$F_{crx} = 0.877 F_{ex} \tag{6.55}$$

EXAMPLE 6.4 ELASTIC BUCKLING SOLUTION

As indicated before, we assume that the member is going to buckle about the weak (y) axis and optimize the problem. We also assume that a W18 shape is desired for the member. The optimization problem becomes: find design variables d, b_f, t_f, t_w to minimize the cost function of Eq. (6.37) subject to the constraints of Eqs. (6.50), (6.51), (6.48), (6.49), and (a) through (d). The Excel worksheet for the problem is prepared, and the Solver is invoked. In the implementation, the constraints are normalized with respect to their limit values, as for the previous example. The initial values of the design variables are set as (17.70, 6.00, 0.425, 0.30). The initial cost function value is 34.5 lbs/ft. The Solver could not find a feasible solution since the strength constraint of Eq. (6.51) could not be satisfied. The depth of the section reached its lower limit, and the thicknesses of the flange and the web reached their upper limits. Thus elastic buckling of a W18 shape about its weaker axis is not possible for this problem.

To investigate the possibility of buckling of the member about the strong axis, we re-optimize the column. The optimization problem becomes: find design variables d, b_f, t_f, t_w to minimize the cost function of Eq. (6.37) subject to the constraints of Eqs. (6.53), (6.54), (6.48), (6.49), and (a) through (d). The initial values of the design variables are set as (17.70, 6.00, 0.425, 0.30). The initial cost function value is 34.5 lbs/ft. Again the Solver could not find a feasible solution for the problem. The constraint on the lower limit of the slenderness ratio in Eq. (6.53) could not be satisfied. The depth of the section and its flange width reached their lower limits, the web thickness reached its upper limit, and the flange thickness almost reached its upper limit.

The foregoing two solutions indicate that for this problem, elastic buckling of the W18 shape is not possible for the required strength. Therefore, the final solution for the problem is a W18 × 234 shape for the column where the column buckles inelastically about the weak (y) axis.

DISCUSSION

In this section, we have formulated the problem of optimizing compression members to comply with the AISC Manual requirements (AISC, 2005). It turns out that the Manual requirements cannot be formulated as continuous and differentiable functions. Therefore a derivative-based optimization algorithm may not be appropriate for the problem. An approach is presented here to handle these requirements in a derivative-based optimization method. Basically the problem is solved four times where the formulation is continuous and differentiable so that a derivative-based optimization method can be used. This approach works quite well and yields an optimum design for the problem.

6.9 OPTIMUM DESIGN OF MEMBERS FOR FLEXURE

STEP 1: PROJECT/PROBLEM DESCRIPTION Beams are encountered in many practical applications, such as building frames, bridges, and other structures. In this project, the goal is to design minimum-weight steel beams to satisfy the AISC *Manual* requirements (AISC, 2005). Many cross-sectional shapes of steel members are available for use as flexural members, such as wide-flange sections (W-shape), angle sections, channel sections, tee sections, and hollow circular or square tubes.

A W-shape should be selected for the present application. The cross-section of the member is shown in Figure 6.19. The load is calculated based on the application and load requirement of the region where the structure is located (ASCE, 2005). The specified material is ASTM A992, Grade 50 steel.

We will first present a general formulation of this problem and then solve several examples.

STEP 2: DATA AND INFORMATION COLLECTION To formulate the problem of optimizing the design of flexural members, the information that needs to be collected includes AISC requirements, load, and material properties. To achieve these goals, see the notation and data defined in Table 6.3.

Some useful expressions for W-shapes for formulation of the problem of optimum design of members for flexure are as follows:

$$h = d - 2t_f, \text{ in.; } h_0 = d - t_f, \text{ in} \tag{6.55a}$$

$$A_g = 2b_f t_f + (d - 2t_f)t_w, \text{ in}^2 \tag{6.55b}$$

$$\bar{y} = \frac{1}{A_g}\left[b_f t_f^2 + h t_w(0.25h + t_f)\right], \text{ in} \tag{6.55c}$$

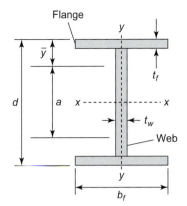
Flange

$\bar{y}$

t_f

d a

t_w

Web

b_f

y

FIGURE 6.19 W-shape for the flexural member.

x-x is the major (strong) axis
y-y is the minor (weak) axis

$$Z_x = 0.5aA_g, \text{ in}^3; \quad a = d - 2\bar{y}, \text{ in} \tag{6.55d}$$

$$M_p = \frac{1}{12}F_yZ_x, \text{ kip-ft}; \quad M_y = \frac{1}{12}F_yS_x, \text{ kip-ft} \tag{6.55e}$$

$$C_b = \frac{12.5M_{max}}{2.5M_{max} + 3M_A + 4M_B + 3M_C}R_m \le 3.0 \tag{6.55f}$$

$$C_w = \frac{1}{4}I_yh_0^2, \text{ in}^6 \text{ (for doubly symmetric W-shape)} \tag{6.55g}$$

$$F_{cr} = \frac{C_b\pi^2E}{\left(\frac{L_b}{r_{ts}}\right)^2}\sqrt{1 + 0.078\frac{Jc}{S_xh_o}\left(\frac{L_b}{r_{ts}}\right)^2}, \text{ ksi} \tag{6.55h}$$

$$I_x = 2\left(\frac{1}{12}b_ft_f^3\right) + \frac{1}{12}t_w(d - 2t_f)^3 + 2b_ft_f\left(\frac{d}{2} - \frac{t_f}{2}\right)^2, \text{ in}^4 \tag{6.55i}$$

$$I_y = 2\left(\frac{1}{12}t_fb_f^3\right) + \frac{1}{12}(d - 2t_f)t_w^3, \text{ in}^4 \tag{6.55j}$$

$$S_x = \frac{I_x}{0.5d}, \text{ in}^3 \tag{6.55k}$$

$$J = \frac{1}{3}(2b_ft_f^3 + h_0t_w^3), \text{ in}^4 \tag{6.55l}$$

$$L_p = 1.76r_y\sqrt{\frac{E}{F_y}}, \text{ in}; \quad r_y = \sqrt{I_y/A_g}, \text{ in} \tag{6.55m}$$

$$L_r = 1.95r_{ts}\frac{E}{0.7F_y}\sqrt{\frac{Jc}{S_xh_o}}\sqrt{1 + \sqrt{1 + 6.76\left(\frac{0.7F_y}{E}\frac{S_xh_o}{Jc}\right)^2}}, \text{ in} \tag{6.55n}$$

TABLE 6.3   Data for optimizing the design of flexural members

Notation	Data
A_g	Gross area of section, in^2
A_w	Area of web $\approx d t_w$, in^2
a	Distance between centroids of two half-areas of cross-section, in
b_f	Width of flange, in
C	Parameter (1 for doubly symmetric shapes)
C_b	Factor (*beam bending coefficient*) that takes into account nonuniform bending moment distribution over unbraced length L_b
C_v	Ratio of critical web stress to shear yield stress
C_w	Warping constant, in^6
d	Depth of section, in
E	Modulus of elasticity; 29,000 ksi
F_y	Specified minimum yield stress; 50 ksi for A992 grade 50 steel
F_{cr}	Critical stress for member, ksi
FLB	Flange Local Buckling
G	Shear modulus of steel; 11,200 ksi
h	Height of web, in
h_0	Distance between flange centroids, in
I_x	Moment of inertia with respect to strong (x) axis, in^4
I_y	Moment of inertia with respect to weak (y) axis, in^4
J	Torsional constant, in^4
L_b	Unbraced length; distance between points braced against lateral displacement of compression flange, in
L_p	Limiting laterally unbraced length for full plastic bending capacity (property of section), in
L_r	Limiting laterally unbraced length for inelastic lateral-torsional buckling (property of section), in
LTB	Lateral-Torsional Buckling
M_a	Required moment strength (i.e., maximum moment corresponding to controlling load combination (ASCE, 2005), kip-ft
M_A	Absolute value of moment at quarter point of unbraced segment, kip-ft
M_B	Absolute value of moment at mid-point of unbraced segment, kip-ft
M_C	Absolute value of moment at three-quarter point of unbraced segment, kip-ft

TABLE 6.3 (*Continued*)

Notation	Data
M_{max}	Absolute value of maximum moment in unbraced segment, kip-ft
M_p	Plastic moment, kip-ft
M_y	Moment that brings beam to point of yielding, kip-ft
M_n	Nominal moment strength, kip-ft
R_m	Cross-section monosymmetry parameter (1 for doubly symmetric sections)
r_x	Radius of gyration with respect to strong (x) axis, in
r_y	Radius of gyration with respect to weak (y) axis, in
r_{ts}	Property of cross-section, in
S_x	Section modulus, in^3
t_f	Thickness of flange, in
t_w	Thickness of web, in
V_a	Maximum shear based on controlling combination of loads, kips
V_n	Nominal shear strength of section, kips
WLB	Web Local Buckling
$\bar{y}$	Distance of centroid of half-area of cross-section from extreme fiber, in
Z_x	Plastic section modulus, in^3
Ω_b	Safety factor for bending; 5/3
Ω_v	Safety factor for shear
λ	Width-thickness ratio
λ_p	Upper limit for λ for compactness
λ_r	Upper limit for λ for noncompactness
λ_f	Width-thickness ratio for flange
λ_{pf}	Upper limit for λ_f for compactness of flange
λ_{rf}	Upper limit for λ_f for noncompactness of flange
λ_w	Width-thickness ratio for web
λ_{pw}	Upper limit for λ_w for compactness of web
λ_{rw}	Upper limit for λ_w for noncompactness of web
γ	Density of steel; 0.283 lb/in^3
Δ	Maximum deflection due to live loads, in

II. NUMERICAL METHODS FOR CONTINUOUS VARIABLE OPTIMIZATION

$$r_{ts}^2 = \frac{\sqrt{I_y C_w}}{S_x}, \ \text{in}^2 \tag{6.55o}$$

$$\lambda_f = \frac{b_f}{2t_f}; \quad \lambda_w = \frac{h}{t_w} \tag{6.55p}$$

$$\lambda_{pf} = 0.38\sqrt{\frac{E}{F_y}}; \quad \lambda_{rf} = \sqrt{\frac{E}{F_y}} \tag{6.55q}$$

$$\lambda_{pw} = 3.76\sqrt{\frac{E}{F_y}}; \quad \lambda_{rw} = 5.70\sqrt{\frac{E}{F_y}} \tag{6.55r}$$

The American Institute of Steel Construction requirements for design of flexural members using the allowable strength design (ASD) approach are

(i) The required moment strength for the member will not exceed the available (allowable) moment strength of the member.
(ii) The required shear strength for the member will not exceed the available (allowable) shear strength of the member.
(iii) The deflection of the beam will not exceed a specified limit

MOMENT STRENGTH REQUIREMENT

The required moment strength must not exceed the available (allowable) moment strength of the member:

$$M_a \leq \frac{M_n}{\Omega_b} = 0.6M_n \tag{6.56}$$

To determine the nominal strength M_n of the beam, several failure modes for the beam need to be considered: elastic or inelastic lateral-torsional buckling (global buckling of the beam), elastic or inelastic flange local buckling, and elastic or inelastic web local buckling. For each of these failure modes, the nominal strength of the section is calculated. Then the smallest of these values is taken as the nominal strength M_n for the beam.

For local buckling considerations, AISC (2005) classifies cross-sectional shapes as *compact*, *noncompact*, and *slender*, depending on the width-thickness ratios.

1. *Compact.* The section is classified as compact if the width-thickness ratio for the flange and the web satisfies the following condition (for compact shapes, the local buckling of web and flange is not a failure mode):

$$\lambda \leq \lambda_p \tag{6.57}$$

2. *Noncompact.* The section is classified as noncompact if the width-thickness ratio for the flange or the web satisfies the following condition (local buckling of web or flange or of both needs to be considered for a noncompact shape):

$$\lambda_p < \lambda \le \lambda_r \qquad (6.58)$$

3. *Slender.* The section is classified as slender if the width-thickness ratio for the flange or the web satisfies the following condition (the tension flange yielding state needs to be considered in addition to the global and local buckling and yielding limit states):

$$\lambda > \lambda_r \qquad (6.59)$$

The foregoing conditions are based on the worst width-thickness ratio of the elements of the cross-section. The width-thickness ratio limits for flexure of rolled I-shapes are given in Eqs. (6.55q) and (6.55r). When the shape is *compact* ($\lambda \le \lambda_p$ for both flange and web), there is no need to check flange local buckling or web local buckling. Most of the standard rolled W-shapes are compact for $F_y \le 65\,ksi$; noncompact shapes are identified with a footnote f in the AISC *Manual* (AISC, 2005). Moreover, all of the standard rolled shapes satisfy the width-thickness ratio for the web, so only the flanges of the standard rolled W-shapes identified with the footnote f are noncompact. Built-up welded I-shapes such as plate girders can have noncompact or slender flanges and/or web.

NOMINAL BENDING STRENGTH OF COMPACT SHAPES

The nominal flexural strength M_n for compact shapes will be the lower value obtained from two failure modes:

1. Limit states of yielding (plastic moment)
2. Lateral-Torsional Buckling (LTB)

For the limit state of yielding, the nominal moment is given as

$$M_n = M_p = F_y Z_x \qquad (6.60)$$

To calculate the nominal moment strength M_n for the LTB limit state, we need to first determine if the buckling is elastic or inelastic. To do that, we calculate the lengths L_p and L_r for I-shaped members using Eqs. (6.55m) and (6.55n), respectively. The nominal strength is calculated as follows:

1. The LTB is not a failure mode if

$$L_b \le L_p \qquad (6.61)$$

The limit state of yielding gives the nominal moment strength as the plastic moment in Eq. (6.60).
2. Inelastic LTB occurs if

$$L_p < L_b \le L_r \qquad (6.62)$$

In this case, the nominal moment strength is given as

$$M_n = C_b \left[M_p - (M_p - 0.7F_yS_x) \left(\frac{L_b - L_p}{L_r - L_p} \right) \right] \leq M_p \tag{6.63}$$

3. Elastic LTB occurs when (the member is classified as slender)

$$L_b > L_r \tag{6.64}$$

In this case, the nominal moment strength is given as

$$M_n = F_{cr}S_x \leq M_p \tag{6.65}$$

where the critical stress F_{cr} is calculated using Eq. (6.55h). If the bending moment is uniform, all moment values are the same in Eq. (6.55f), giving $C_b = 1$. This is also true for a conservative design.

NOMINAL BENDING STRENGTH OF NONCOMPACT SHAPES

Most standard rolled W, M, S and C shapes are compact for $F_y \leq 65$ ksi. A few are noncompact because of the flange width–thickness ratio, but none are slender. In general, a noncompact beam may fail by LTB, FLB, or web local buckling (WLB). Any of these failures can be elastic or inelastic. All of them need to be investigated to calculate the nominal moment strength of the member.

The webs of all hot-rolled shapes in the AISC *Manual* (AISC, 2005) are compact, so the noncompact shapes are subject only to the limit states of LTB and FLB. If the shape is *noncompact* ($\lambda_p < \lambda \leq \lambda_r$) because of the flange, then the nominal moment strength is the smallest of the following:

1. For LTB, the nominal moment is calculated using Eqs. (6.60), (6.63), or (6.65).
2. For FLB, the M_n is calculated as follows:
   (i) There is no FLB if

$$\lambda_f \leq \lambda_{pf} \tag{6.66}$$

   (ii) The flange is noncompact and FLB is inelastic if

$$\lambda_{pf} < \lambda_f \leq \lambda_{rf} \tag{6.67}$$

   The nominal moment strength is given as

$$M_n = \left[M_p - (M_p - 0.7F_yS_x) \left(\frac{\lambda_f - \lambda_{pf}}{\lambda_{rf} - \lambda_{pf}} \right) \right] \leq M_p \tag{6.68}$$

   (iii) The flange is slender and FLB is elastic if

$$\lambda_f > \lambda_{rf} \tag{6.69}$$

   The nominal moment strength is given as

$$M_n = \frac{0.9Ek_cS_x}{\lambda_f^2} \leq M_p \tag{6.70}$$

$$k_c = \frac{4}{\sqrt{h/t_w}} \tag{6.71}$$

The value of k_c will not be taken as less than 0.35 or greater than 0.76.

SHEAR STRENGTH REQUIREMENT

The required shear strength (applied shear) must not exceed the available (allowable) shear strength (AISC, 2005):

$$V_a \leq \frac{V_n}{\Omega_v} \tag{6.72}$$

AISC (2005) Specification covers both beams with stiffened webs and beams with unstiffened webs. The basic nominal shear strength equation is

$$V_n = (0.6F_y)A_wC_v \tag{6.73}$$

where $0.6F_y$ = shear yield stress (60% of the tensile yield stress). The value of C_v depends on whether the limit state is web shear yielding, web shear inelastic buckling, or web shear elastic buckling. For the special case of hot-rolled I-shapes with

$$\frac{h}{t_w} \leq 2.24\sqrt{\frac{E}{F_y}} \tag{6.74}$$

the limit state is shear yielding, and

$$C_v = 1.0; \quad \Omega_v = 1.50 \tag{6.75}$$

Most W-shapes with $F_y \leq 65$ ksi fall into this category (AISC, 2005). For all other doubly and singly symmetric shapes, except for round hollow-structural-sections (HSS),

$$\Omega_v = 1.67 \tag{6.76}$$

and C_v is determined as follows:

1. There is no web shear instability if

$$\frac{h}{t_w} \leq 1.10\sqrt{\frac{k_vE}{F_y}} \tag{6.77}$$

$$C_v = 1.0 \tag{6.78}$$

2. Inelastic web shear buckling occurs if

$$1.10\sqrt{\frac{k_vE}{F_y}} < \frac{h}{t_w} \leq 1.37\sqrt{\frac{k_vE}{F_y}} \tag{6.79}$$

$$C_v = \frac{1.10\sqrt{\frac{k_v E}{F_y}}}{h/t_w}$$

(6.80)

3. The limit state is elastic web shear buckling if

$$\frac{h}{t_w} > 1.37\sqrt{\frac{k_v E}{F_y}}$$

(6.81)

$$C_v = \frac{1.51 E k_v}{(h/t_w)^2 F_y}$$

(6.82)

where $k_v = 5$. This value of k_v is for unstiffened webs with $\frac{h}{t_w} \le 260$.

Equation (6.82) is based on elastic stability theory, and Eq. (6.80) is an empirical equation for the inelastic region, providing a transition between the limit states of web shear yielding and web shear elastic buckling.

DEFLECTION REQUIREMENT

Deflection of the beam should not be excessive under service loads. This is the *serviceability requirement* for a structure. It turns out that the live load deflection is more important to control because the dead load deflection is usually controlled by providing a camber to the beam. The live load deflection requirement for the beam is taken as

$$\Delta = \frac{L}{240}$$

(6.83)

where L is the length of the beam and Δ is the deflection due to live loads.

STEP 3: DEFINITION OF DESIGN VARIABLES The design variables for the member are the cross-sectional dimensions of the member, shown in Figure 6.19. Thus the design variable vector is $\mathbf{x} = (d, b_f, t_f, t_w)$

STEP 4: OPTIMIZATION CRITERION The goal is to select the lightest possible W-shape for the member. Thus we minimize the weight of the member per unit length given as cross-sectional area × density:

$$f = 12\gamma A_g, \ \text{lbs/ft}$$

(6.84)

STEP 5: FORMULATION OF CONSTRAINTS It is challenging to formulate the strength constraints for the beam design problem for a derivative-based optimization algorithm. The reason is that the constraint depends on several "IF THEN ELSE" conditions that could make the constraint function discontinuous or at least nondifferentiable at some points in the feasible set for the problem. The first "IF THEN ELSE" condition is for classifying shapes as compact, noncompact, or slender. If the shape is compact, then local buckling of flanges or the web is not a failure mode. Next, it needs to be decided for compact shapes if there is LTB. If there is, then it further needs to be decided whether the buckling is elastic or inelastic. Thus several conditions need to be checked before the nominal strength of even the compact shapes can be decided. The nominal strength for the section is given as the smallest value from Eqs. (6.60) through (6.65).

The second "IF THEN ELSE" condition is related to noncompact sections where the FLB failure mode must also be considered in addition to LTB, as discussed above. In addition, FLB can be inelastic or elastic, and this needs to be determined. Therefore, the nominal moment strength is given by the smaller value from Eqs. (6.60) through (6.70).

In design of beams, the usual procedure is to size the member for the moment strength and then check it for shear strength and the deflection requirement. We will follow this procedure. Also, we first design the member as a compact section and then as a noncompact section and compare the two designs. More detailed formulations of the problem for these two cases are presented in the following examples.

EXAMPLE 6.5 DESIGN OF A COMPACT SHAPE FOR INELASTIC LTB

We wish to design a simply supported beam of span 30 ft that is braced only at the ends. The beam is subjected to a uniform dead load of 2 kip/ft and a concentrated live load of 15 kips at the mid-span, as shown in Figure 6.20. The material for the beam is A992 grade 50 steel.

Following the procedure outlined above, we optimize the member for its flexural strength and then check its adequacy for shear and deflection requirements. Analysis of the beam gives the required moment and shear strengths as

$$M_a = 337.5 \text{ kip-ft}, \quad V_a = 37.5 \text{ kips} \tag{a}$$

Since we require the section to be compact, we impose the following constraints on the width-thickness ratios of the flange and the web:

$$\lambda_f \leq \lambda_{pf} \text{ and } \lambda_w \leq \lambda_{pw} \tag{b}$$

Since the unbraced length $L_b = 30$ ft is fairly large, the LTB failure mode must be considered. First, we assume the LTB to be inelastic (later we will consider it to be elastic) implying that the following constraints for the unbraced length need to be imposed:

$$L_p < L_b \leq L_r \tag{c}$$

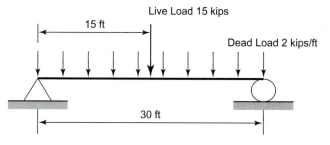

Live Load 15 kips

15 ft

Dead Load 2 kips/ft

30 ft

FIGURE 6.20   A simply supported beam subjected to dead and live loads.

Since the beam is subjected to a nonuniform bending moment, the factor C_b is calculated using Eq. (6.55f) as 1.19. The nominal strength M_n for the section is given by Eq. (6.63). This value must not exceed the plastic moment for the section, thus giving the constraint

$$M_n \leq M_p \tag{d}$$

Now the strength constraint of Eq. (6.56) is written as

$$M_a \leq 0.6 M_n \tag{e}$$

We also assume that a W14-shape is desired for the member and impose the following lower and upper limits for the design variables (if other W-shapes for the member are desired, appropriate upper and lower limits on the design variables can be imposed):

$$(13.7 \leq d \leq 16.4, \quad 5.0 \leq b_f \leq 16.0, \quad 0.335 \leq t_f \leq 1.89, \quad 0.23 \leq t_w \leq 1.18) \text{ in} \tag{f}$$

The optimization problem becomes one of finding the design variables d, b_f, t_f, t_w to minimize the cost function of Eq. (6.84) subject to the constraint in Eqs. (b) through (f). The Excel worksheet for the problem is prepared and the Solver is invoked. In the implementation, the constraints are normalized with respect to their limit values; for example, the moment strength constraint of Eq. (e) is normalized as $1 \leq 0.6 M_n / M_a$. The initial values of the design variables are set as (16.40, 16.00, 1.89, 1.18). The initial cost function value is 256 lbs/ft. The Solver gives the following optimum design:

$$(d = 16.4, \quad b_f = 12.96, \quad t_f = 0.71, \quad t_w = 0.23) \text{ in}$$

$$\text{Weight} = 74.1 \text{ lbs/ft}; \quad A_g = 21.81 \text{ in}^2, \quad Z_x = 157.0 \text{ in}^3, \quad S_x = 145.8 \text{ in}^3 \tag{g}$$

$$L_p = 12.1 \text{ ft}, \quad L_r = 34.7 \text{ ft}, \quad V_n = 113.2 \text{ kips}, \quad \Delta = 0.42 \text{ in}$$

Analyzing the optimum design, we observe the following: The moment strength constraint, the flange compactness constraint, the upper limit on depth, and the lower limit on web thickness are active at the optimum point. We also observe that, since the allowable shear strength of $\frac{V_n}{\Omega_v} = \frac{113.2}{1.5} = 75.5$ kips exceeds the required shear strength of 37.5 kips, the optimum design satisfies the shear strength constraint. Also the live load deflection constraint is satisfied since the allowable deflection of $L/240 = 1.5$ in exceeds the actual deflection of 0.42 in.

Based on the optimum weight of 74.1 lbs for the section, we select a W14 $\times$ 82 shape. However, this section has an allowable moment strength of only 270.0 kip-ft, which violates the required moment strength constraint of Eq. (e). The next two heavier sections are W14 $\times$ 90 and W14 $\times$ 99; however, both are noncompact. Therefore, we select the W14 $\times$ 109 shape, which has an allowable bending strength of 464.4 kip-ft satisfying the required moment strength constraint. All other constraints are also satisfied.

If we relax the constraints on design variables, we can select a W18-shape with the following limits on the design variables:

$$(17.7 \leq d \leq 20.7, \quad 6.0 \leq b_f \leq 11.6, \quad 0.425 \leq t_f \leq 1.91, \quad 0.30 \leq t_w \leq 1.06) \text{ in} \tag{h}$$

The initial values of the design variables are set as (16.40, 12.96, 0.71, 0.23). The initial cost function value is 74.2 lbs/ft. The Solver gives the following optimum design:

$$(d = 18.38, \quad b_f = 11.6, \quad t_f = 0.76, \quad t_w = 0.3) \text{ in}$$

$$\text{Weight} = 77.0 \text{ lbs/ft}; \quad A_g = 22.7 \text{ in}^2, \quad Z_x = 176.6 \text{ in}^3, \quad S_x = 162 \text{ in}^3 \tag{i}$$

$$L_p = 10.43 \text{ ft}, \quad L_r = 30 \text{ ft}, \quad V_n = 165.4 \text{ kips}, \quad \Delta = 0.34 \text{ in}$$

Analyzing the optimum design, we observe that the moment strength constraint, the upper limit on the unbraced length, the upper limit on the flange width, and the lower limit on web thickness are active at the optimum point. Based on the optimum design, we choose a W18 × 97 section that has an available (allowable) bending strength of 390.9 kip-ft and meets all other constraints. Other lighter sections did not meet all of the constraints.

If we further relax the upper bound constraints on design variables, we can select a W21 or a W24 shape:

$$(13.7 \le d \le 25.5, \quad 5.0 \le b_f \le 13.0, \quad 0.335 \le t_f \le 2.11, \quad 0.23 \le t_w \le 1.16) \text{ in} \tag{j}$$

The initial values of the design variables are set as (16.40, 16.0, 1.89, 1.18). The initial cost function value is 256.0 lbs/ft. The Solver gives the following optimum design:

$$(d = 19.47, \quad b_f = 12.31, \quad t_f = 0.67, \quad t_w = 0.23) \text{ in}$$

$$\text{Weight} = 70.4 \text{ lbs/ft}; \quad A_g = 20.73 \text{ in}^2, \quad Z_x = 174.55 \text{ in}^3, \quad S_x = 162.1 \text{ in}^3 \tag{k}$$

$$L_p = 11.2 \text{ ft}, \quad L_r = 30 \text{ ft}, \quad V_n = 134.38 \text{ kips}, \quad \Delta = 0.32 \text{ in}$$

Analyzing the optimum design, we observe that the moment strength constraint, the upper limit on the unbraced length, the upper limit on the flange the compactness, and the lower limit on the web thickness are active at the optimum point. Based on the optimum design, it is suggested that we use a W21 × 101 or a W24 × 117 section in order to meet all of the constraints.

EXAMPLE 6.6 DESIGN OF A COMPACT SHAPE WITH ELASTIC LTB

Next we require the LTB to be elastic. Therefore, we impose the constraint in Eq. (6.64) and calculate the nominal strength using Eq. (6.65). Lower and upper limits on the design variables are set as in Eq. (j). The initial values of the design variables are set as (24.1, 12.8, 0.75, 0.5). The initial cost function value is 103.6 lbs/ft. The Solver gives the following optimum design:

$$(d = 22.15, \quad b_f = 12.06, \quad t_f = 0.0659, \quad t_w = 0.23) \text{ in}$$

$$\text{Weight} = 70.2 \text{ lbs/ft}; \quad A_g = 20.68 \text{ in}^2, \quad Z_x = 195.61 \text{ in}^3, \quad S_x = 181.28 \text{ in}^3 \tag{l}$$

$$L_p = 10.78 \text{ ft}, \quad L_r = 28.07 \text{ ft}, \quad V_n = 152.80 \text{ kips}, \quad \Delta = 0.25 \text{ in}$$

Analyzing the optimum design, we observe that the moment strength constraint, the upper limit on web compactness, the upper limit on flange compactness, and the lower limit on web thickness are active at the optimum point. Based on the optimum design, it is suggested that we use a W18 × 97 section in order to meet all of the constraints.

Based on the solutions in Examples 6.5 and 6.6, it appears that W18 × 97 is the lightest compact shape for the problem.

EXAMPLE 6.7   DESIGN OF A NONCOMPACT SHAPE

Now we redesign the problem of Example 6.5, assuming the shape to be noncompact. The web of the section is still required to be compact; therefore, the constraint of Eq. (6.57) is imposed for the web width-thickness ratio. The flange is required to be noncompact; therefore, the constraints of Eq. (6.58) are imposed for the flange width-thickness ratio. The nominal strength for the section assuming inelastic FLB is calculated using Eq. (6.68). This value is used in the bending strength constraint of Eq. (e).

We also assume that the LTB for the member is inelastic; therefore, the constraints for the unbraced length given in Eq. (c) are imposed. The nominal strength for this failure mode is calculated using Eq. (6.63). This is used in another constraint on the required bending strength given in Eq. (e).

The Excel Solver sheet for Example 6.5 is modified for the noncompact shape. A W18-shape is desired, so the constraints bound by the design variables, Eq. (h), are imposed. The Solver gives the following optimum design:

$$(d = 18.24, \quad b_f = 11.7, \quad t_f = 0.63, \quad t_w = 0.81) \text{ in}$$
$$\text{Weight} = 96.9 \text{ lbs/ft}; \quad A_g = 28.5 \text{ in}^2, \quad Z_x = 188.6 \text{ in}^3, \quad S_x = 162.1 \text{ in}^3 \tag{m}$$
$$L_p = 8.61 \text{ ft}, \quad L_r = 30.00 \text{ ft}, \quad V_n = 442.3 \text{ kips}, \quad \Delta = 0.34 \text{ in}$$

The starting point is selected as (15.79, 11.7, 0.50, 0.41), which has a weight of 60.4 lbs/ft. Analyzing the optimum design, we observe the active constraints to be as follows: The lower limit on the flange width-thickness ratio, the inelastic LTB moment strength constraint, the upper bound on the unbraced length, and the upper limit on the flange width. Note that the optimum weight obtained for the noncompact shape is much greater than that obtained for the compact shape in Eqs. (i). Based on the solution, a W18 × 97 shape is suggested for the member. Although this is a compact shape, it is selected because no noncompact shape near the optimum weight is available in the AISC *Manual*.

DISCUSSION

In this section, we formulate the problem of optimizing beams to comply with the AISC *Manual* requirements (AISC, 2005). It turns out that these requirements cannot be formulated as continuous and differentiable functions. Therefore, a derivative-based optimization algorithm may not be appropriate for the problem; direct search methods are more appropriate.

An approach is presented here to handle these requirements with a derivative-based optimization method. Basically the problem is formulated as a compact or a noncompact shape, and either an inelastic or an elastic LTB condition is imposed. In this way, all of the problem functions are continuous and differentiable so that a derivative-based optimization method can be used. This approach works quite well and yields an optimum design for each of the formulations.

6.10 OPTIMUM DESIGN OF TELECOMMUNICATION POLES

STEP 1: PROJECT/PROBLEM DESCRIPTION Steel poles are encountered in many practical applications, such as telecommunication poles, electrical transmission line poles, and others. The goal of this project is to design a minimum-weight telecommunication steel pole based on Allowable Stress Design (ASD). The pole is tapered and has a hollow circular cross-section. It is subjected to horizontal wind loads F_h and p_h and gravity dead loads F_v and p_v, as shown in Figure 6.21. These are the main loads that act on the structure due to wind and the dead load of the platforms, antennas, cables, ladder, and the steel pole itself. The pole should perform its function safely; that is, the material should not fail when the loads F_h, p_h, F_v, and p_v are applied. Also, it should not deflect too much at the top so that it can perform its function correctly.

The height of the pole is H, the tip diameter is d_t, the taper is τ, and the wall thickness is t, as shown in Figure 6.21. Figure 6.22 shows a real structure installed in the field (created by Marcelo A. da Silva).

STEP 2: DATA AND INFORMATION COLLECTION To formulate the problem of optimizing the design of a telecommunication pole, the information that needs to be collected includes procedure for analysis of the structure, expressions for stresses, and cross-sectional properties. In addition, various data for the problem, such as material properties, wind loads, dead loads, and constraint limits, are needed. To achieve these goals, see the notation and data defined in Table 6.4.

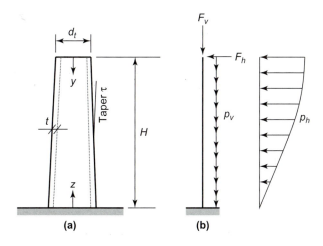

FIGURE 6.21   Pole structure and loads.

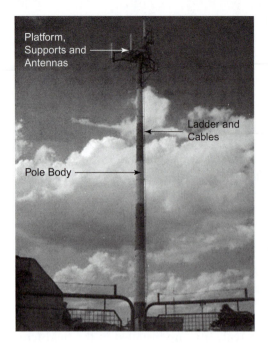

Platform,
Supports and
Antennas

Ladder and
Cables

Pole Body

FIGURE 6.22　Real telecommunication steel pole.

Some useful expressions for the structural analysis are

$$q(z) = 463.1\,z^{0.25} \tag{6.85a}$$

$$F_h = C_t A_t q(H) = C_t A_t (463.1\,H^{0.25}) \tag{6.85b}$$

$$d_e(z) = d_t + 2\tau(H - z) \tag{6.85c}$$

$$d_i(z) = d_e(z) - 2t \tag{6.85d}$$

$$A(z) = \frac{\pi}{4}\left[d_e(z)^2 - d_i(z)^2\right] \tag{6.85e}$$

$$I(z) = \frac{\pi}{64}\left[d_e(z)^4 - d_i(z)^4\right] \tag{6.85f}$$

$$S(z) = \frac{2I(z)}{d_e(z)} \tag{6.85g}$$

$$p_h(y) = [A_{lc}C_{lc} + d_e(z)C_p]q(z); \quad z = H - y \tag{6.85h}$$

$$p_v(y) = p_{lc} + A(z)\gamma; \quad z = H - y \tag{6.85i}$$

$$N(z) = F_v + \int_0^{H-z} p_v(y)dy \tag{6.85j}$$

$$M(z) = F_h(H - z) + \int_0^{H-z} p_h(y)(H - z - y)dy \tag{6.85k}$$

TABLE 6.4 Data for optimizing the design of a telecommunication pole

Notation	Data
A_{lc}	Distributed area of ladder and cables, $0.3 \, \text{m}^2/\text{m}$
A_t	Projected area on vertical plane of platforms, supports, and antennas, $10 \, \text{m}^2$
$A(z)$	Cross-sectional area of pole section at height z, m^2
C_{lc}	Drag coefficient of wind load acting on ladder and cables, taken as 1
C_p	Drag coefficient of wind load acting on pole body, taken as 0.75
C_t	Drag coefficient of wind load acting on area A_t, taken as 1
d_e	External diameter of given pole section, m
d_i	Internal diameter of given pole section, m
d_t	External diameter of pole tip, m
E	Modulus of elasticity, $210 \times 10^9 \, \text{Pa}$
F_h	Wind load at top due to platform, supports, and antennas, N
F_v	Dead load at top due to weight of platform, supports, and antennas, taken as 10,400 N
H	Height of pole, 30 m
$I(z)$	Moment of inertia of section at height z, m^4
l_{max}	Maximum allowable value for ratio d_e/t
$M(z)$	Bending moment at a height z, N/m
$N(z)$	Axial load at height z, N
p_{lc}	Self-weight of ladder and cables, taken as 400 N/m
$p_h(y)$	Horizontal distributed load at distance y from top, N/m
$p_v(y)$	Vertical distributed load at distance y from top, N/m
V_0	Basic wind velocity, also called maximum wind velocity, determined as 40 m/s
V_{oper}	Operational wind velocity, considered 55% V_0, 22 m/s
q	Effective wind velocity pressure, N/m^2
$S(z)$	Section modulus of pole at height z, m^3
t	Thickness of pole wall, m
v_a	Allowable tip deflection, taken as $2\%H = 0.60 \, \text{m}$
v'_a	Allowable rotation under operational wind loading, taken as $0° \, 30' = \frac{30}{60} \times \frac{\pi}{180} = 0.00873$ radians
$v(z)$	Horizontal displacement at height z, m
$v'(z)$	Rotation in vertical plane of pole section at height z, radians
$v'_{oper}(z)$	Rotation in vertical plane of section at height z due to operational wind loading, radians

(Continued)

II. NUMERICAL METHODS FOR CONTINUOUS VARIABLE OPTIMIZATION

TABLE 6.4 Data for optimizing the design of a telecommunication pole (*Continued*)

Notation	Data
$v''(z)$	Curvature in vertical plane of pole section at height z, m^{-1}
y	Distance from pole tip to given section, m
z	Height above ground of given pole section, m
γ	Specific weight of steel, 78,500 N/m^3
σ_a	Steel allowable stress, 150×10^6 Pa (obtained by dividing yield stress by factor of safety larger than 1)
$\sigma(z)$	Axial stress in pole section at height z, Pa
τ	Taper of pole, m/m

$$\sigma(z) = \frac{N(z)}{A(z)} + \frac{M(z)}{S(z)} \qquad (6.85l)$$

The expression for the effective wind velocity pressure $q(z)$ in Eq. (6.85a) has been obtained using several factors such as the wind velocity, which depends on the region where the pole is to be installed, a topographic factor that depends on the surrounding terrain, and an importance factor for the structure, as given by the NBR-6123 code (ABNT, 1988). The effective wind velocity (V_0) is also called the maximum wind velocity, distinguished from V_{oper} (operational wind velocity).

Therefore, $q(z)$ in Eq. (6.85a) is also called the maximum wind pressure. The internal loads, stresses, displacements, rotations, and curvatures derived from $q(z)$ are based on the maximum wind velocity. The expressions for the loads and moment in Eqs. (6.85h) through (6.85l) are derived by considering a section at height z, as shown in Figure 6.23. Note that the vertical load expression in Eq. (6.85i) contains the self-weight of the pole. This implies that the total axial load and the moment at a point are dependent on the pole's design.

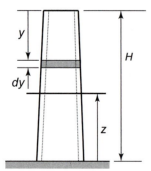

FIGURE 6.23   Pole structure — section at height z.

To develop expressions for the cost and constraint functions, considerable background information is needed. First, we need to calculate stresses in the pole section at height z, which requires analysis of the structure. The members are subjected to both bending moment and axial load, and the stresses of these internal loads are combined in Eq. (6.85l) to obtain the effective axial stress. For the cantilever pole, the maximum stress occurs at ground level, where the structure is clamped. To be safe, the effective stress constraint is imposed at section $z = 0$. In the present analysis, the effect of the shear load is neglected to calculate stress and tip deflection. However, this must be taken into account in the design of the foundation, which is not considered in the present project.

Assuming linearly elastic behavior of the structure, the displacements can be computed by integration of the differential elastic line equation (Hibbeler, 2007) given as

$$EI(z)v''(z) = -M(z) \tag{6.86}$$

with boundary conditions as $v(0) = 0$ and $v'(0) = 0$. This differential equation must be integrated twice to calculate the displacements. It is difficult to integrate this equation analytically since the moment $M(z)$ and the moment of inertia $I(z)$ vary along z. A practical procedure to accomplish this integration is the trapezoidal rule presented below. For this purpose the pole is divided into n segments by introducing $(n + 1)$ node points along the pole axis as $\{z_0, z_1, \ldots z_i, \ldots z_n\}$. At each point z_i, $i > 0$, we compute $v_i = v(z_i)$, $v'_i = v'(z_i)$ and $v''_i = v''(z_i)$ using the following integration scheme:

$$v''_i = -\frac{M(z_i)}{EI(z_i)}$$

$$v'_i = v'_{i-1} + \frac{v''_i + v''_{i-1}}{2} h \tag{6.87}$$

$$v_i = v_{i-1} + \frac{v'_i + v'_{i-1}}{2} h$$

where $h = z_i - z_{i-1}$. Once the structural dimensions, material properties, loading, and internal loads are known, the displacements are calculated from Eqs. (6.87).

One important subject that is not treated in the current formulation of the problem is the dynamic analysis of this kind of structure. When the pole's first natural frequency of vibration is smaller than 1 Hz, the dynamic effects of the wind loads cannot be neglected, and dynamic analysis of the structure must be performed. In addition, geometric nonlinear effects should be included for more accurate analysis of the pole. Verification of these effects is left as a topic of further study.

STEP 3: DEFINITION OF DESIGN VARIABLES   Since the height, loading, and material for the structure are specified, only the cross-sectional dimensions need to be determined to complete its design. Therefore, the three design variables for the pole are identified as

d_t = external diameter of tip section, m
t = thickness of section wall, m
τ = taper of pole, m/m

Thus the design variable vector is $x = (d_t, t, \tau)$.

STEP 4: OPTIMIZATION CRITERION   The objective is to minimize the weight of the pole structure, which is calculated using the following integral:

$$f = \int_0^H A(y)\gamma dy \tag{6.88}$$

Carrying out the integration, the weight function is given as

$$f = \frac{1}{24\tau}\pi\gamma\left[((d_t + 2\tau H)^3 - (d_t)^3) - ((d_t + 2\tau H - 2t)^3 - (d_t - 2t)^3)\right], \ \ N \tag{6.89}$$

An approximation of the weight function is given as

$$f = \frac{1}{3}\gamma H\left[A(H) + A(0) + \sqrt{A(H)A(0)}\right], \ \ N \tag{6.90}$$

STEP 5: FORMULATION OF CONSTRAINTS   The first constraint for the problem is on the material failure, which is written as $\sigma(0) \leq \sigma_a$. Substituting $z = 0$ in Eq. (6.85l), we obtain the stress constraint for the ground-level section (the most stressed) as follows:

$$\frac{N(0)}{A(0)} + \frac{M(0)}{S(0)} \leq \sigma_a \tag{6.91}$$

The second constraint is related to the tip deflection, which must be within its allowable value:

$$v(H) \leq v_a. \tag{6.92}$$

Note that to obtain v, it is necessary to twice integrate the elastic line equation, Eq. (6.86), as explained earlier, and $v_a = 0.60$ m.

Simple bounds on the design variables are imposed as

$$0.30 \leq d_t \leq 1.0, \ \ m \tag{6.93a}$$

$$0.0032 \leq t \leq 0.0254, \ \ m \tag{6.93b}$$

$$0 \leq \tau \leq 0.05, \ \ m/m \tag{6.93c}$$

Thus the formulation of the problem is stated as follows: Find the design variables d_t, t, and τ to minimize the cost function of Eq. (6.90), subject to the stress constraint of Eq. (6.91), the deflection constraint of Eq. (6.92), and the explicit design variable−bound constraints of Eqs. (6.93a) through (6.93c).

EXAMPLE 6.8   OPTIMUM DESIGN OF POLE

To solve this optimization problem, we use Excel Solver. The Excel worksheet is prepared to implement all equations of the formulation, and the Solver is invoked. The stress and the displacement constraints are normalized as $\frac{\sigma}{\sigma_a} \leq 1$, and $v/v_a \leq 1$ in implementation. The initial values of the design variables are set as (0.400, 0.005, 0.020), where the cost function has a value of 36,804 N, the maximum stress is 72×10^6 Pa, and the tip deflection is 0.24 m. Note that the starting design is feasible and can be improved. The Solver gives the following optimum design:

$\mathbf{x}^* = (d_t = 0.30$ m, $t = 0.0032$ m, $\tau = 0.018$ m/m)
Cost function = 19,730 N
Stress at the base, $\sigma = 141 \times 10^6$ Pa
Tip deflection, $v(30) = 0.60$ m

Analyzing the optimum design, we observe that the minimum tip diameter, the minimum wall thickness, and the tip deflection constraints are active at the optimum point. The Lagrange multiplier for the tip deflection constraint is 673.

EXAMPLE 6.9 OPTIMUM DESIGN WITH THE TIP ROTATION CONSTRAINT

In practice, the pole's antennas must not lose the link with the receiver under operational wind conditions due to wind velocity, V_{oper}. In this case, the rotation of the antennas must be smaller than a given limit, called the maximum rotation allowable for the antennas (v_a'). Figure 6.24 shows this rotational limit constraint. Antenna A is installed on pole A and has a link with antenna B on pole B. Because the antennas are fixed on the pole, if the poles rotate more than the allowable value for them, they will lose their link and the system will go off the air.

For that reason, we impose a new constraint that is related to tip rotation, which must be within its allowable value:

$$v_{oper}'(H) \leq v_a' \tag{6.94}$$

Note that to obtain v_{oper}', it is necessary to integrate the elastic line equation, Eq. (6.86), for operational wind loading, as explained earlier. It is noted that for linear analysis we have

$$v_{oper}'(H) = v'(H) \times 0.55^2 \tag{6.95}$$

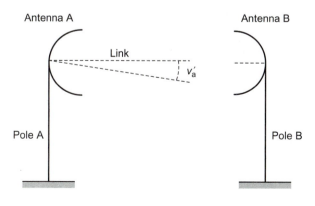

FIGURE 6.24   Geometrical view of the allowable rotation for the antenna.

where $v'(H)$ is computed using the expressions in Eqs. (6.87). With this, the constraint of Eq. (6.94) is rewritten as

$$v'(H) \times 0.55^2 \le v'_a \tag{6.96}$$

The optimization problem is now defined by Eqs. (6.90) through (6.93c) and Eq. (6.96). Starting with the same initial design as for Example 6.8, where the rotation at the top is 0.00507 rad, the Solver gives the following optimum design:

$\mathbf{x}^* = (d_t = 0.5016$ m, $t = 0.0032$ m, $\tau = 0.015$ m/m$)$
Cost function = 22,447 N
Stress at the base, $\sigma = 146 \times 10^6$ Pa
Tip deflection, $v(30) = 0.47$ m
Tip rotation, $v'_{oper}(30) = 0.00873$ rad

Analyzing the optimum design, we observe that the minimum wall thickness and the tip rotation constraints are active at the optimum point. The Lagrange multiplier for the tip rotation constraint is 806.

EXAMPLE 6.10 OPTIMUM DESIGN WITH THE LOCAL BUCKLING CONSTRAINT

If the wall of the pole is too thin, local buckling can happen, causing catastrophic failure of the pole. One way to avoid local buckling is to constrain the ratio of diameter to wall thickness in the design process. Thus the following additional constraint is imposed in the formulation:

$$d_e(0)/t \le l_{max} \tag{6.97}$$

where l_{max} is an upper limit on the diameter over the thickness ratio. For demonstration purposes, it is taken as 200; however, the value depends on the applicable design code. The optimization problem is now defined by Eqs. (6.90) through (6.93c) and Eq. (6.96) and (6.97). With the same initial design as for Examples 6.8 and 6.9, the $d_e(0)/t$ ratio is 320, which violates the local buckling constraint. The Solver gives the following optimum design:

$\mathbf{x}^* = (d_t = 0.5565$ m, $t = 0.0053$ m, $\tau = 0.0084$ m/m$)$
Cost function = 31,504 N
Stress at the base, $\sigma = 150 \times 10^6$ Pa
Tip deflection, $v(30) = 0.52$ m
Tip rotation, $v'_{oper}(30) = 0.00873$ rad
Ratio, $d_e(0)/t = 200$

Analyzing the optimum design, we observe that the constraints related to the stress at the base, the tip rotation, and the local buckling are active at the optimum point. The Lagrange multipliers for the constraints are 180, 1546, and 1438.

EXERCISES FOR CHAPTER 6

Section 6.3 Excel Solver for Unconstrained Optimization Problems

Solve the following problems using the Excel Solver (choose any reasonable starting point):

6.1 Exercise 4.32   **6.2** Exercise 4.39   **6.3** Exercise 4.40

6.4 Exercise 4.41   **6.5** Exercise 4.42

Section 6.4 Excel Solver for Linear Programming Problems

Solve the following LP problems using the Excel Solver:

6.6 Maximize $z = x_1 + 2x_2$

subject to $-x_1 + 3x_2 \leq 10$

$$x_1 + x_2 \leq 6$$
$$x_1 - x_2 \leq 2$$
$$x_1 + 3x_2 \geq 6$$
$$x_1, x_2 \geq 0$$

6.7 Maximize $z = x_1 + 4x_2$

subject to $x_1 + 2x_2 \leq 5$

$$x_1 + x_2 = 4$$
$$x_1 - x_2 \geq 3$$
$$x_1, x_2 \geq 0$$

6.8 Minimize $f = 5x_1 + 4x_2 - x_3$

subject to $x_1 + 2x_2 - x_3 \geq 1$

$$2x_1 + x_2 + x_3 \geq 4$$
$$x_1, x_2 \geq 0; x_3 \text{ is unrestricted in sign}$$

6.9 Maximize $z = 2x_1 + 5x_2 - 4.5x_3 + 1.5x_4$

subject to $5x_1 + 3x_2 + 1.5x_3 \leq 8$

$$1.8x_1 - 6x_2 + 4x_3 + x_4 \geq 3$$
$$-3.6x_1 + 8.2x_2 + 7.5x_3 + 5x_4 = 15$$
$$x_i \geq 0; i = 1 \text{ to } 4$$

6.10 Minimize $f = 8x - 3x_2 + 15x_3$

subject to $5x_1 - 1.8x_2 - 3.6x_3 \geq 2$

$$3x_1 + 6x_2 + 8.2x_3 \geq 5$$
$$1.5x_1 - 4x_2 + 7.5x_3 \geq -4.5$$
$$-x_2 + 5x_3 \geq 1.5$$
$$x_1, x_2 \geq 0; x_3 \text{ is unrestricted in sign}$$

6.11 Maximize $z = 10x_1 + 6x_2$

subject to $2x_1 + 3x_2 \leq 90$

$$4x_1 + 2x_2 \leq 80$$
$$x_2 \geq 15$$
$$5x_1 + x_2 = 25$$
$$x_1, x_2 \geq 0$$

Section 6.5 Excel Solver for Nonlinear Programming

6.12 Exercise 3.35   **6.13** Exercise 3.50

6.14 Exercise 3.51   **6.15** Exercise 3.54

6.16 Solve the spring design problem for the following data: Applied load $(P) = 20$ lb.

6.17 Solve the spring design problem for the following data: Number of active coils $(N) = 20$, limit on outer diameter of the coil $(D_0) = 1$ in, number of inactive coils $(Q) = 4$.

6.18 Solve the spring design problem for the following data: Aluminum coil with shear modulus $(G) = 4,000,000$ psi, mass density $(\rho) = 2.58920 \times 10^{-4}$ lb-s^2/in^4, and allowable shear stress $(\tau_a) = 50,000$ lb/in^2.

Section 6.6 Optimum Design of Plate Girders Using Excel Solver

6.19 Solve the plate girder design problem for the following data: Span length $(L) = 35$ ft.

6.20 Solve the plate girder design problem for the following data: A36 steel with modulus of elasticity $(E) = 200$ GPa, yield stress (sigma_y) $= 250$ MPa, allowable fatigue stress (sigma_t) $= 243$ MPa.

6.21 Solve the plate girder design problem for the following data: Web height $(h) = 1.5$ m, flange thickness $(t_f) = 0.015$ m.

Section 6.7 Optimum Design of Tension Members

6.22 Solve the problem of this section where a W14 shape is desired.

6.23 Solve the problem of this section where a W12 shape is desired.

6.24 Solve the problem of this section where a W8 shape is desired, the required strength P_a for the member is 200 kips, the length of the member is 13 ft, and the material is A992 Grade 50 steel.

6.25 Same as 6.24; select a W10 shape.

Section 6.8 Optimum Design of Compression Members

6.26 Solve the problem of this section where W14 shape is desired.

6.27 Solve the problem of this section where W12 shape is desired and required strength P_a is 1000 kips.

6.28 Design a compression member to carry a load of 400 kips. The length of the member is 26 ft, and the material is A572 Grade 50 steel. The member is not braced. Select a W18 shape.

6.29 Same as 6.24; select a W14 shape. The member is not braced.

6.30 Same as 6.24; select a W12 shape. The member is not braced.

Section 6.9 Optimum Design of Members for Flexure

6.31 Solve the problem of Example 6.5 for a beam of span 40 ft. Assume compact shape and inelastic LTB.

6.32 Solve the problem of Example 6.5 for a beam of span 40 ft. Assume compact shape and elastic LTB.

6.33 Solve the problem of Example 6.5 for a beam of span 10 ft. Assume noncompact shape and inelastic LTB.

6.34 Solve the problem of Example 6.5 for a beam of span 40 ft. Assume noncompact shape and elastic LTB.

6.35 Design a cantilever beam of span 15 ft subjected to a dead load of 3 kips/ft and a point live load of 20 kips at the end. The material of the beam is A572 Grade 50 steel. Assume compact shape and inelastic LTB.

6.36 Design a cantilever beam of span 15 ft subjected to a dead load of 3 kips/ft and a point live load of 20 kips at the end. The material of the beam is A572 Grade 50 steel. Assume compact shape and elastic LTB.

6.37 Design a cantilever beam of span 15 ft subjected to a dead load of 3 kips/ft and a point live load of 20 kips at the end. The material of the beam is A572 Grade 50 steel. Assume noncompact shape and inelastic LTB.

6.38 Design a cantilever beam of span 15 ft subjected to a dead load of 1 kips/ft and a point live load of 10 kips at the end. The material of the beam is A572 Grade 50 steel. Assume noncompact shape and elastic LTB.

Section 6.10 Optimum Design of Telecommunication Poles

6.39 Solve the problem of Example 6.8 for a pole of height 40 m.

6.40 Solve the problem of Example 6.9 for a pole of height 40 m.

6.41 Solve the problem of Example 6.10 for a pole of height 40 m.

Optimum Design with MATLAB®

Upon completion of this chapter you will be able to

- Use the capabilities of the Optimization Toolbox in MATLAB to solve both
  unconstrained and constrained optimization problems

MATLAB was used in Chapter 3 to graphically solve two variable optimization problems. In Chapter 4 it was used to solve a set of nonlinear equations obtained as Karush-Kuhn-Tucker (KKT) optimality conditions for constrained optimization problems. In this chapter, we describe the capabilities of the Optimization Toolbox in MATLAB to solve linear, quadratic, and nonlinear programming problems. We start by describing the basic capabilities of this toolbox. Some operators and syntax used to enter expressions and data are described. In subsequent sections, we illustrate the use of the program for unconstrained and constrained optimization problems. Then some engineering design optimization problems are solved using the program (created by Tae Hee Lee).

7.1 INTRODUCTION TO THE OPTIMIZATION TOOLBOX

7.1.1 Variables and Expressions

MATLAB can be considered a high-level programming language for numerical computation, data analysis, and graphics for applications in many fields. It interprets and evaluates expressions entered at the keyboard. The statements are usually in the form *variable=expression*. The variables can be scalars, arrays, or matrices. Arrays may store

Note: The original draft of this chapter was provided by Tae Hee Lee. The contribution to this book is very much appreciated.

many variables at a time. A simple way to define a scalar, array, or matrix is to use assignment statements as follows:

$$a = 1; \quad b = [1, 1]; \quad c = [1, 0, 0; \ 1, 1, 0; \ 1, -2, 1] \tag{7.1}$$

Note that several assignment statements can be entered in one row as in Eq. (7.1). A semicolon (;) at the end of a statement prevents the program from displaying the results. The variable a denotes a scalar that is assigned a value of 1; the variable b denotes a 1×2 row vector, and the variable c denotes a 3×3 matrix assigned as follows:

$$\mathbf{b} = \begin{bmatrix} 1 & 1 \end{bmatrix}; \quad \mathbf{c} = \begin{bmatrix} 1 & 0 & 0 \\ 1 & 1 & 0 \\ 1 & -2 & 1 \end{bmatrix} \tag{7.2}$$

The semicolons inside the brackets of the expression for c separate the rows, and the values in the rows can be separated by commas or blanks. MATLAB has a rule that the *variable name* must be a single word without spaces, and it must start with a letter followed by any number of letters, digits, or underscores. It is important to note that variable names are case sensitive. In addition, there are several built-in variables, for example, pi for the ratio of the circumference of a circle to its diameter; eps for the smallest number in the computer; inf for infinity, and so on.

7.1.2 Scalar, Array, and Matrix Operations

The arithmetic operators for scalars in MATALB are: addition (+), subtraction (−), multiplication (*), division (/), and exponentiation (^). Vector and matrix calculations can also be organized in a simple way using these operators. For example, multiplication of two matrices **A** and **B** is expressed as **A** * **B**. Slight modification of the standard operators with a "dot" prefix is used for element-by-element operations between vectors and matrices: (.*) for multiplication, (./) for division, and (.^) for exponentiation.

For example, element-by-element multiplication of vectors of the same dimension is accomplished using the operator .*:

$$c = a .* b = \begin{bmatrix} a_1 b_1 \\ a_2 b_2 \\ a_3 b_3 \end{bmatrix} \tag{7.3}$$

Here a, b, and c are column vectors with three elements. For addition and subtraction, element-by-element and usual matrix operations are the same. Other useful matrix operators are: $\mathbf{A}^2 = \mathbf{A}.* \mathbf{A}$, $\mathbf{A}^{-1} = \text{inv}(\mathbf{A})$, determinant as $\det(\mathbf{A})$, and transpose as $\mathbf{A}'$.

7.1.3 Optimization Toolbox

The Optimization Toolbox for MATLAB can solve unconstrained and constrained optimization problems. In addition, it has an algorithm to solve nonsmooth optimization problems. Some of the optimization algorithms implemented in the Optimization Toolbox are presented in later chapters. These algorithms are based on the basic concepts of algorithms for smooth and nonsmooth problems that are presented in Section 6.1. That material should be reviewed at this point.

TABLE 7.1 Optimization Toolbox functions

Problem Type	Formulation	Function
One-variable minimization in fixed intervals	Find $x \in [x_L \ x_U]$ to minimize $f(x)$	`fminbnd`
Unconstrained minimization	Find $\mathbf{x}$ to minimize $f(\mathbf{x})$	`fminunc, fminsearch`
Constrained minimization	Find $\mathbf{x}$ to minimize $f(\mathbf{x})$ subject to $\mathbf{Ax} \leq \mathbf{b}, \mathbf{Nx} = \mathbf{e}$ $g_i(\mathbf{x}) \leq 0, i = 1, \ldots, m$ $h_j = 0, j = 1, \ldots, p$ $x_{iL} \leq x_i \leq x_{iU}$	`fmincon`
Linear programming	Find $\mathbf{x}$ to minimize $f(\mathbf{x}) = \mathbf{c}^T\mathbf{x}$ subject to $\mathbf{Ax} \leq \mathbf{b}, \mathbf{Nx} = \mathbf{e}$	`linprog`
Quadratic programming	Find $\mathbf{x}$ to minimize $f(\mathbf{x}) = \mathbf{c}^T\mathbf{x} + \frac{1}{2}\mathbf{x}^T\mathbf{Hx}$ subject to $\mathbf{Ax} \leq \mathbf{b}, \mathbf{Nx} = \mathbf{e}$	`quadprog`

TABLE 7.2 Explanation of output from optimization function

Argument	Description
`x`	The solution vector or matrix found by the optimization function. If `ExitFlag` > 0, then `x` is a solution; otherwise, `x` is the latest value from the optimization routine.
`FunValue`	The value of the objective function, `ObjFun`, at the solution `x`.
`ExitFlag`	The exit condition for the optimization function. If `ExitFlag` is positive then the optimization routine converged to a solution `x`. If `ExitFlag` is zero, then the maximum number of function evaluations was reached. If `ExitFlag` is negative then the optimization routine did not converge to a solution.
`Output`	The `Output` structure contains several pieces of information about the optimization process. It provides the number of function evaluations (`Output.iterations`), the name of the algorithm used to solve the problem (`Output.algorithm`), and Lagrange multipliers for constraints, etc.

The Optimization Toolbox must be installed in the computer in addition to the basic MATLAB program before it can be used. Table 7.1 shows some of the functions available in the toolbox. Most of these optimization routines require m-files (stored in the current directory) containing a definition of the problem to be solved; several such files are presented and discussed later. Default optimization parameters are used extensively; however, they can be modified through an `options` command available in the program.

The syntax of invoking an optimization function is generally of the form

$$[\text{x, FunValue, ExitFlag, Output}] = \text{fminX(`ObjFun'}, \ldots, \text{options})\tag{7.4}$$

The left side of the statement represents the quantities returned by the function. These output arguments are described in Table 7.2. On the right side, `fminX` represents one of the functions given in Table 7.1. There can be several arguments for the function `fminX`—for example, starting values for the variables; upper and lower bounds for the variables; m-file names containing problem functions and their gradients; optimization algorithm—related data; and so on. Use of this function is demonstrated in subsequent sections for various types of problems and conditions. For further explanation of various functions and commands, extensive online help is available in MATLAB.

7.2 UNCONSTRAINED OPTIMUM DESIGN PROBLEMS

In this section, we first illustrate the use of the `fminbnd` function for minimization of a function of single variable $f(x)$ with bounds on x as $x_L \leq x \leq x_U$. Then the use of the function `fminunc` is illustrated for minimization of a function $f(\mathbf{x})$ of several variables. The m-files for the problems, containing extensive comments, are included to explain the use of these functions. Example 7.1 demonstrates use of the function `fminbnd` for functions of single variable, and Example 7.2 demonstrates the use of functions `fminsearch` and `fminunc` for multivariable unconstrained optimization problems.

EXAMPLE 7.1 SINGLE-VARIABLE UNCONSTRAINED MINIMIZATION

Find x to

Minimize

$$f(x) = 2 - 4x + e^x, \quad -10 \leq x \leq 10 \tag{a}$$

Solution

To solve this problem, we write an m-file that returns the objective function value. Then we invoke `fminbnd`, the single-variable minimization function, in fixed intervals through another m-file that is shown in Table 7.3. The file that evaluates the function, shown in Table 7.4, is called through `fminbnd`.

The output from the function is

x = 1.3863, FunVal = 0.4548, ExitFlag = 1 > 0 (i.e., minimum was found),
output = (iterations: 14, funcCount: 14, algorithm: golden section search, parabolic interpolation).

TABLE 7.3　m-file for calling `fminbnd` to find single variable minimizer in fixed interval for Example 7.1

% All comments start with %

% File name: Example7_1.m

% Problem: minimize f(x) = 2 − 4x + exp(x)

```
clear all
```

% Set lower and upper bound for the design variable

```
Lb = -10; Ub = 10;
```

% Invoke single variable unconstrained optimizer fminbnd;

% The argument ObjFunction7_1 refers to the m-file that

% contains expression for the objective function

```
[x,FunVal,ExitFlag,Output] = fminbnd('ObjFunction7_1',Lb,Ub)
```

TABLE 7.4 m-file for objective function for Example 7.1

% File name: ObjFunction7_1.m

% Example 7.1 Single variable unconstrained minimization

```
function f = ObjFunction7_1(x)
f = 2 - 4*x + exp(x);
```

EXAMPLE 7.2 MULTIVARIABLE UNCONSTRAINED MINIMIZATION

Consider a two-variable problem:

Minimize

$$f(\mathbf{x}) = 100(x_2 - x_1^2)^2 + (1 - x_1)^2 \text{ starting from } \mathbf{x}^{(0)} = (-1.2, \ 1.0) \tag{a}$$

Solve the problem using different algorithms available in the Optimization Toolbox.

Solution

The optimum solution for the problem is known as $\mathbf{x}^* = (1.0, 1.0)$ with $f(\mathbf{x}^*) = 0$ (Schittkowski, 1987). The syntax for the functions `fminsearch` and `fminunc` used to solve a multivariable unconstrained optimization problem is given as follows:

$$[\text{x, FunValue, ExitFlag, Output}] = \text{fminsearch (`ObjFun´, x0, options)} \tag{b}$$

$$[\text{x, FunValue, ExitFlag, Output}] = \text{fminunc (`ObjFun´, x0, options)} \tag{c}$$

where

ObjFun = the name of the m-file that returns the function value and its gradient if
         programmed
    x0 = the starting values of the design variables
options = a data structure of parameters that can be used to invoke various conditions for the
         optimization process

`fminsearch` uses the Simplex search method of Nelder-Mead, which does not require numerical or analytical gradients of the objective function. Thus it is a *nongradient-based method (direct search method)* that can be used for problems where the cost function is not differentiable.

Since `fminunc` does require the gradient value, with the option `LargeScale` set to `off`, it uses the BFGS quasi-Newton method (refer to Chapter 11 for details) with a mixed quadratic and cubic line search procedure. The DFP formula (refer to Chapter 11 for details), which approximates the inverse Hessian matrix, can be selected by setting the option `HessUpdate` to `dfp`. The steepest-descent method can be selected by setting option `HessUpdate` to `steepdesc`. `fminsearch` is generally less efficient than `fminunc`. However, it can be effective for problems for which the gradient evaluation is expensive or not possible.

To solve this problem, we write an m-file that returns the objective function value. Then, the unconstrained minimization function fminsearch or fminunc is invoked through execution of another m-file, shown in Table 7.5. The m-file for function and gradient evaluations is shown in Table 7.6.

The gradient evaluation option can be omitted if automatic evaluation of gradients by the finite difference method is desired. Three solution methods are used, as shown in Table 7.5. All methods converge to the known solution.

TABLE 7.5   m-file for unconstrained optimization routines for Example 7.2

% File name: Example7_2

% Rosenbruck valley function with analytical gradient of

% the objective function

```
clear all
x0 = [-1.2 1.0]'; % Set starting values
% Invoke unconstrained optimization routines
```

% 1. Nelder-Mead simplex method, *fminsearch*

% Set options: medium scale problem, maximum number of function evaluations

% Note that "..." indicates that the text is continued on the next line

```
options = optimset('LargeScale', 'off', 'MaxFunEvals', 300);
[x1, FunValue1, ExitFlag1, Output1] = ...
    fminsearch ('ObjAndGrad7_2', x0, options)
```

% 2. BFGS method, *fminunc*, default option

% Set options: medium scale problem, maximum number of function evaluations,

% gradient of objective function

```
options = optimset('LargeScale', 'off', 'MaxFunEvals', 300, ...
    'GradObj', 'on');
[x2, FunValue2, ExitFlag2, Output2] = ...
    fminunc ('ObjAndGrad7_2', x0, options)
```

% 3. DFP method, *fminunc*, HessUpdate = dfp

% Set options: medium scale optimization, maximum number of function evaluation,

% gradient of objective function, DFP method

```
options = optimset('LargeScale', 'off', 'MaxFunEvals', 300, ...
    'GradObj', 'on', 'HessUpdate', 'dfp');
[x3, FunValue3, ExitFlag3, Output3] = ...
    fminunc ('ObjAndGrad7_2', x0, options)
```

TABLE 7.6 m-file for objective function and gradient evaluations for Example 7.2

% File name: ObjAndGrad7_2.m

% Rosenbrock valley function

```
function [f, df]= ObjAndGrad7_2(x)
```

% Re-name design variable x

```
x1 = x(1); x2 = x(2); %
```

% Evaluate objective function

```
f = 100(x2 − x1^2)^2 + (1 − x1)^2;
```

% Evaluate gradient of the objective function

```
df(1) = −400*(x2−x1^2)*x1 − 2*(1−x1);
df(2) = 200*(x2−x1^2);
```

7.3 CONSTRAINED OPTIMUM DESIGN PROBLEMS

The general constrained optimization problem treated by the function `fmincon` is defined in Table 7.1. The procedure for invoking this function is the same as for unconstrained problems except that an m-file containing the constraint functions must also be provided. If analytical gradient expressions are programmed in the objective function and constraint functions m-files, they are declared through the `options` command. Otherwise, `fmincon` uses numerical gradient calculations based on the finite difference method. Example 7.3 shows the use of this function for an inequality constrained problem. Equalities, if present, can be included similarly.

EXAMPLE 7.3 CONSTRAINED MINIMIZATION PROBLEM USING FMINCON IN OPTIMIZATION TOOLBOX

Solve the problem:

Minimize

$$f(\mathbf{x}) = (x_1 - 10)^3 + (x_2 - 20)^3 \tag{a}$$

subject to

$$g_1(\mathbf{x}) = 100 - (x_1 - 5)^2 - (x_2 - 5)^2 \le 0 \tag{b}$$

$$g_2(\mathbf{x}) = -82.81 - (x_1 - 6)^2 - (x_2 - 5)^2 \le 0 \tag{c}$$

$$13 \le x_1 \le 100, \quad 0 \le x_2 \le 100 \tag{d}$$

Solution

The optimum solution for the problem is known as $\mathbf{x} = (14.095, 0.84296)$ and $f(\mathbf{x}^*) = -6961.8$ (Schittkowski, 1981). Three m-files for the problem are given in Tables 7.7 through 7.9. The script m-file in Table 7.7 invokes the function `fmincon` with appropriate arguments and options. The function m-file in Table 7.8 contains the cost function and its gradient expressions, and the function m-file in Table 7.9 contains the constraint functions and their gradients.

The problem is solved successfully, and the output from the function is given as

`Active constraints:` 5, 6 (i.e., g(1) and g(2))

$\mathbf{x} = (14.095, 0.843)$, `FunVal` $= -6.9618e + 003$, `ExitFlag` $= 1 > 0$ (i.e., minimum was found)

`output` = (iterations: 6, funcCount: 13, stepsize: 1, algorithm: medium scale: SQP, quasi-Newton, line-search).

Note that the `Active constraints` listed at the optimum solution are identified with their index counted as Lb, Ub, inequality constraints, and equality constraints. If `Display off` is included in the `options` command, the set of active constraints is not printed.

TABLE 7.7   m-file for constrained minimizer `fmincon` for Example 7.3

% File name: Example7_3

% Constrained minimization with gradient expressions available

% Calls ObjAndGrad7_3 and ConstAndGrad7_3

```
clear all
```

% Set options; medium scale, maximum number of function evaluation,

% gradient of objective function, gradient of constraints, tolerances

% Note that three periods "..." indicate continuation on next line

```
options = optimset ('LargeScale', 'off', 'GradObj', 'on', ...
    'GradConstr', 'on', 'TolCon', 1e-8, 'TolX', 1e-8);
```

% Set bounds for variables

```
Lb = [13; 0]; Ub = [100; 100];
```

% Set initial design

```
x0 = [20.1; 5.84];
```

% Invoke fmincon; four [ ] indicate no linear constraints in the problem

```
[x, FunVal, ExitFlag, Output] = ...
    fmincon('ObjAndGrad7_3', x0, [ ], [ ], [ ], [ ], Lb, ...
    Ub, 'ConstAndGrad7_3', options)
```

TABLE 7.8   m-file for objective function and gradient evaluations for Example 7.3

% File name: ObjAndGrad7_3.m

```
function [f, gf] = ObjAndGrad7_3(x)
```

% f returns value of objective function; gf returns objective function gradient

% Re-name design variables x

```
x1 = x(1); x2 = x(2);
```

% Evaluate objective function

```
f = (x1-10)^3 + (x2-20)^3;
```

% Compute gradient of objective function

```
if nargout > 1
    gf(1,1) = 3*(x1-10)^2;
    gf(2,1) = 3*(x2-20)^2;
end
```

TABLE 7.9   Constraint functions and their gradients evaluation m-file for Example 7.3

% File name: ConstAndGrad7_3.m

```
function [g, h, gg, gh] = ConstAndGrad7_3(x)
```

% g returns inequality constraints; h returns equality constraints

% gg returns gradients of inequalities; each column contains a gradient

% gh returns gradients of equalities; each column contains a gradient

% Re-name design variables

```
x1 = x(1); x2 = x(2);
```

% Inequality constraints

```
g(1) = 100-(x1-5)^2-(x2-5)^2;
g(2) = -82.81+(x1-6)^2 + (x2-5)^2;
```

% Equality constraints (none)

```
h = [ ];
```

% Gradients of constraints

```
if nargout > 2
    gg(1,1) = -2*(x1-5);
    gg(2,1) = -2*(x2-5);
    gg(1,2) = 2*(x1-6);
    gg(2,2) = 2*(x2-5);
    gh = [];
end
```

7.4 OPTIMUM DESIGN EXAMPLES WITH MATLAB

7.4.1 Location of Maximum Shear Stress for Two Spherical Bodies in Contact

PROJECT/PROBLEM STATEMENT   There are many practical applications where two spherical bodies come into contact with each other, as shown in Figure 7.1. We want to determine the maximum shear stress and its location along the z-axis for a given value of the Poisson's ratio of the material, $\nu = 0.3$.

DATA AND INFORMATION COLLECTION   The shear stress along the z-axis is calculated using the principal stresses as (Norton, 2000)

$$\sigma_{xz} = \frac{p_{max}}{2}\left(\frac{1-2v}{2} + \frac{(1+v)\alpha}{\sqrt{1+\alpha^2}} - \frac{3}{2}\frac{\alpha^3}{\left(\sqrt{1+\alpha^2}\right)^3}\right) \tag{a}$$

where $\alpha = z/a$ and a represent the contact-patch radius, as shown in Figure 7.1. The maximum pressure occurs at the center of the contact patch and is given as

$$p_{max} = \frac{3P}{2\pi a^2} \tag{b}$$

It is well known that the peak shear stress does not occur at the contact surface but rather at a small distance below it. The subsurface location of the maximum shear stress is believed to be a significant factor in the surface fatigue failure called *pitting*.

DEFINITION OF DESIGN VARIABLES   The only design variable for the problem is α.

OPTIMIZATION CRITERION   The objective is to locate a point along the z-axis where the shear stress is maximum. Transforming to the standard minimization form and normalizing with respect to p_{max}, the problem becomes finding α to minimize

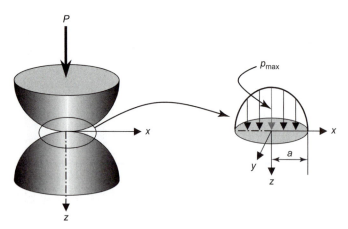

FIGURE 7.1 Graphic of spherical bodies in contact and pressure distribution on contact patch.

$$f(\alpha) = -\frac{\sigma_{xz}}{p_{max}} \qquad \text{(c)}$$

FORMULATION OF CONSTRAINTS There are no constraints for the problem except bounds on the variable α taken as $0 \le \alpha \le 5$.

SOLUTION

The exact solution for the problem is given as

$$\left.\frac{\sigma_{xz}}{p_{max}}\right|_{max} = \frac{1}{2}\left(\frac{1-2v}{2} + \frac{2}{9}(1+v)\sqrt{2(1+v)}\right) \quad \text{at} \quad \alpha = \sqrt{\frac{2+2v}{7-2v}} \qquad \text{(d)}$$

This is a single-variable optimization problem with only lower and upper bounds on the variable. Therefore, the function fminbnd in the Optimization Toolbox can be used to solve the problem. Table 7.10 shows the script m-file that invokes the function fminbnd, and Table 7.11 shows the function m-file that evaluates the function to be minimized. The script m-file also contains commands to plot the shear stress as a function of z, which is shown in Figure 7.2. The optimum solution matches the exact solution for $v = 0.3$ and is given as

$$\texttt{alpha} = 0.6374, \quad \texttt{FunVal} = -0.3329 \quad [\alpha^* = 0.6374, \ f(\alpha^*) = -0.3329] \qquad \text{(e)}$$

TABLE 7.10 m-file to invoke function fminbnd for spherical contact problem

```
% File name: sphcont_opt.m

% Design variable: ratio of the max shear stress location to

% size of the contact patch

% Find location of the maximum shear stress along the z-axis

  clear all
% Set lower and upper bound for the design variable

  Lb = 0; Ub = 5;
% Plot normalized shear stress distribution along the z-axis in spherical contact

  z = [Lb: 0.1: Ub]';
  n = size (z);
  for i = 1: n
  outz(i) = -sphcont_objf(z(i));
  end
  plot(z, outz); grid
  xlabel ('normalized depth z/a');
  ylabel ('normalized shear stress');
% Invoke the single-variable unconstrained optimizer

  [alpha, FunVal, ExitFlag, Output] = fminbnd ('sphcont_objf', Lb, Ub)
```

TABLE 7.11 m-file for evaluation of objective function for the spherical contact problem

% File name = sphcont_objf.m

% Location of max shear stress along z-axis for spherical contact problem

```
function f = sphcont_objf(alpha)
```

% f = −shear stress/max pressure

```
nu = 0.3; % Poisson's ratio
f = -0.5*((1-2*nu)/2 + (1+nu)*alpha/sqrt(1+alpha^2)-...
    1.5*(alpha/sqrt(1+alpha^2))^3);
```

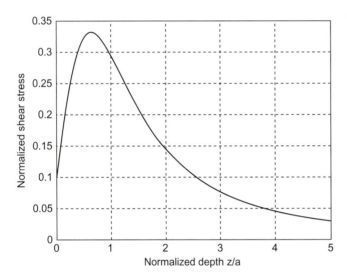

FIGURE 7.2 Normalized shear stress along the z-axis.

7.4.2 Column Design for Minimum Mass

PROJECT/PROBLEM STATEMENT As noted in Section 2.7, columns are used as structural members in many practical applications. Many times such members are subjected to eccentric loads, such as those applied by a jib crane. The problem is to design a minimum-mass tubular column that is subjected to an eccentric load, as shown in Figure 7.3. The cross-section of the column is a hollow circular tube with R and t as the mean radius and wall thickness, respectively.

DATA AND INFORMATION COLLECTION The notation and the data for the problem are given as follows:

P Load, 100 kN
L Length, 5 m
R Mean radius, m
E Young's modulus, 210 GPa
σ_a Allowable stress, 250 MPa

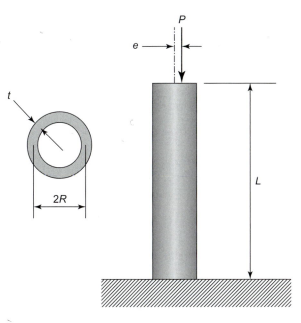

FIGURE 7.3 Configuration of vertical column with an eccentric load.

e Eccentricity (2% of radius), $0.02R$, m
Δ Allowable lateral deflection, 0.25 m
ρ Mass density, 7850 kg/m^3
A Cross-sectional area, $2\pi Rt$, m^2
I Moment of inertia, $\pi R^3 t$, m^4
C Distance to the extreme fiber, $R + \frac{1}{2}t$, m

An analysis of the structure yields the following design equations:

Normal stress:

$$\sigma = \frac{P}{A}\left[1 + \frac{ec}{k^2}\sec\left(\frac{L}{k}\sqrt{\frac{P}{EA}}\right)\right], \quad k^2 = \frac{I}{A} \tag{a}$$

Buckling load:

$$P_{cr} = \frac{\pi^2 EI}{4L^2} \tag{b}$$

Deflection:

$$\delta = e\left[\sec\left(L\sqrt{\frac{P}{EI}}\right) - 1\right] \tag{c}$$

DEFINITION OF DESIGN VARIABLES   Two design variables for the problem are defined as

R = mean radius of the tube, m
t = wall thickness, m

OPTIMIZATION CRITERION   The objective is to minimize the mass of the column, which is given as

$$f(\mathbf{x}) = \rho L A = (7850)(5)(2\pi Rt), \ \text{kg} \tag{d}$$

FORMULATION OF CONSTRAINTS   Constraints for the problem are on performance of the structure, the maximum radius-to-thickness ratio, and the bounds on the radius and thickness:

Stress constraint:

$$\sigma \le \sigma_a \tag{e}$$

Buckling load constraint:

$$P \le P_{cr} \tag{f}$$

Deflection constraint:

$$\delta \le \Delta \tag{g}$$

Radius/thickness constraint:

$$\frac{R}{t} \le 50 \tag{h}$$

Bounds on variables:

$$0.01 \le R \le 1, \ \ 0.005 \le t \le 0.2 \tag{i}$$

SOLUTION

Let us redefine the design variables and other parameters for MATLAB as

$$x_1 = R, \ \ x_2 = t \tag{j}$$

$$c = x_1 + \frac{1}{2}x_2, \ \ e = 0.02x_1 \tag{k}$$

$$A = 2\pi x_1 x_2, \ \ I = \pi x_1^3 x_2, \ \ k^2 = \frac{I}{A} = \frac{x_1^2}{2} \tag{l}$$

All constraints are normalized and rewritten using these redefined design variables. Therefore, the optimization problem is stated in the standard form as follows:

Minimize

$$f(\mathbf{x}) = 2\pi(5)(7850)x_1 x_2 \tag{m}$$

subject to

$$g_1(\mathbf{x}) = \frac{P}{2\pi x_1 x_2 \sigma_a}\left[1 + \frac{2 \times 0.02(x_1 + 0.5x_2)}{x_1}\sec\left(\frac{\sqrt{2}L}{x_1}\sqrt{\frac{P}{E(2\pi x_1 x_2)}}\right)\right] - 1 \le 0 \tag{n}$$

$$g_2(\mathbf{x}) = 1 - \frac{\pi^2 E(\pi x_1^3 x_2)}{4L^2 P} \leq 0 \tag{o}$$

$$g_3(\mathbf{x}) = \frac{0.02x_1}{\Delta} \left[\sec\left(L\sqrt{\frac{P}{E(\pi x_1^3 x_2)}} \right) - 1 \right] - 1 \leq 0 \tag{p}$$

$$g_4(\mathbf{x}) = \frac{x_1}{50x_2} - 1 \leq 0 \tag{q}$$

$$0.01 \leq x_1 \leq 1, \quad 0.005 \leq x_2 \leq 0.2 \tag{r}$$

The problem is solved using the `fmincon` function in the `Optimization Toolbox`. Table 7.12 shows the script m-file for invoking this function and setting various options for the optimization process. Tables 7.13 and 7.14 show the function m-files for the objec-

TABLE 7.12   m-file for invoking minimization function for column design problem

% File name = column_opt.m

```
clear all
```
% Set options
```
options = optimset ('LargeScale', 'off', 'TolCon', 1e-8, 'TolX', 1e – 8);
```
% Set the lower and upper bounds for design variables
```
Lb = [0.01 0.005]; Ub = [1 0.2];
```
% Set initial design
```
x0 = [1 0.2];
```
% Invoke the constrained optimization routine, fmincon
```
[x, FunVal, ExitFlag, Output] = ...
    fmincon('column_objf', x0, [], [], [], [], Lb, Ub, 'column_conf', options)
```

TABLE 7.13   m-file for objective function for the minimum-mass column design problem

% File name = column_objf.m

% Column design
```
function f = column_objf (x)
```
% Rename design variables
```
x1 = x(1); x2 = x(2);
```
% Set input parameters
```
L = 5.0; % length of column (m)
rho = 7850; % density (kg/m^3)
    f = 2*pi*L*rho*x1*x2; % mass of the column
```

TABLE 7.14   m-file for constraint functions for the column design problem

---

% File name = column_conf.m

% Column design

```
function [g, h] = column_conf (x)

x1 = x(1); x2 = x(2);
```

% Set input parameters

```
P = 50000; % loading (N)
E = 210e9; % Young's modulus (Pa)
L = 5.0; % length of the column (m)
Sy = 250e6; % allowable stress (Pa)
Delta = 0.25; % allowable deflection (m)
```

% Inequality constraints

```
g(1) = P/(2*pi*x1*x2)*(1 + ...
   2*0.02*(x1+x2/2)/x1*sec(5*sqrt(2)/x1*sqrt(P/E/(2*pi*x1*x2))))/Sy - 1;
g(2) = 1 - pi^3*E*x1^3*x2/4/L^2/P;
g(3) = 0.02*x1*(sec(L*sqrt(P/(pi*E*x1^3*x2))) - 1)/Delta - 1;
g(4) = x1/x2/50 - 1;
```

% Equality constraint (none)

```
h = [];
```

---

tive and constraint functions, respectively. Note that analytical gradients are not provided for the problem functions.

The output from the function is given as

Active Constraints: 2, 5, that is, the lower limit for thickness and g(1).

$x = (0.0537, 0.0050)$, FunVal = 66.1922, ExitFlag = 1, Output = (*iterations:* 31, funcCount: 149, stepsize: 1, algorithm: medium-scale: SQP, Quasi-Newton, line-search).

7.4.3 Flywheel Design for Minimum Mass

PROJECT/PROBLEM STATEMENT   Shafts are used in practical applications to transfer torque from a source point to another point. However, the torque to be transferred can fluctuate, causing variations in the angular speed of the shaft, which is not desirable. Flywheels are used on the shaft to smooth out these speed variations (Norton, 2000; Shigley and Mischke, 2001). The purpose of this project is to design a flywheel to smooth out variations in the speed of a solid shaft of radius r_i. The flywheel-shaft system is shown in Figure 7.4. The input torque function, which varies during a cycle, is shown in Figure 7.5. The torque variation about its average value is shown there as a function of the shaft angle from 0 to 360 degrees. The kinetic energy due to this variation is obtained by integrating the torque pulse above and below its average value (shaded area) during the cycle and is given as $E_k = 26,105$ in·lb. The shaft is rotating at a nominal angular speed of $\omega = 800$ rad/s.

DATA AND INFORMATION COLLECTION   One cycle of torque variation shown in Figure 7.5 is assumed to be repetitive and, thus, representative of the steady-state

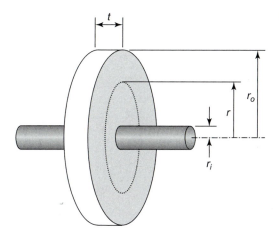

FIGURE 7.4 Flywheel-shaft system.

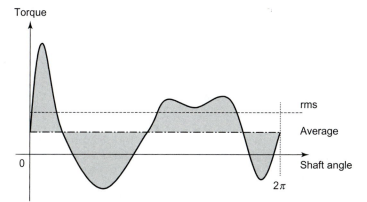

FIGURE 7.5 Fluctuation about the average value of input torque for one cycle.

condition. The desired coefficient of fluctuation is assumed to be 0.05 (C_f). The coefficient of fluctuation represents the ratio of variation of angular velocity to the nominal angular velocity: $C_f = (\omega_{max} - \omega_{min})/\omega$. The system is assumed to be in continuous operation with minimal start-stop cycles. The minimum-mass moment of inertia for the flywheel is determined using the required change in kinetic energy, E_k, specified earlier, as

$$I_s = \frac{E_k}{C_f \omega^2} = \frac{26,105}{0.05(800)^2} = 0.816 \ \text{lb} \cdot \text{in} \cdot \text{s}^2 \tag{a}$$

The design data and equations needed to formulate the minimum-mass flywheel problem are given as

γ Specific weight, 0.28 lb/in^3
G Gravitational constant, 386 in/s^2
S_y Yield stress, 62,000 psi
ω Nominal angular velocity, 800 rad/s

ν Poisson's ratio, 0.28
r_i Inner radius of flywheel, 1.0 in
r_o Outer radius of flywheel, in
t Thickness of flywheel, in

Some useful expressions for the flywheel are

Mass moment of inertia of flywheel:

$$I_m = \frac{\pi}{2}\frac{\gamma}{g}(r_o^4 - r_i^4)t, \ \ \text{lb} \cdot \text{in} \cdot \text{s}^2 \tag{b}$$

Tangential stress in flywheel at radius r:

$$\sigma_t = \frac{\gamma}{g}\omega^2\frac{3+v}{8}\left(r_i^2 + r_o^2 + \frac{r_i^2 r_o^2}{r^2} - \frac{1+3v}{3+v}r^2\right), \ \ \text{psi} \tag{c}$$

Radial stress in flywheel at radius r:

$$\sigma_r = \frac{\gamma}{g}\omega^2\frac{3+v}{8}\left(r_i^2 + r_o^2 - \frac{r_i^2 r_o^2}{r^2} - r^2\right), \ \ \text{psi} \tag{d}$$

von Mises stress:

$$\sigma' = \sqrt{\sigma_r^2 - \sigma_r\sigma_t + \sigma_t^2}, \ \ \text{psi} \tag{e}$$

DEFINITION OF DESIGN VARIABLES The two design variables for the problem are defined as

r_o = outer radius of the flywheel, in
t = thickness of the flywheel, in

OPTIMIZATION CRITERION The objective of the project is to design a flywheel of minimum mass. Since mass is proportional to material volume, we want to minimize the volume of the flywheel, which is given as

$$f = \pi(r_o^2 - r_i^2)t, \ \ \text{in}^3 \tag{f}$$

FORMULATION OF CONSTRAINTS Performance and other constraints are expressed as

Mass moment of inertia requirement:

$$I_m \geq I_s \tag{g}$$

von Mises stress constraint:

$$\sigma' \leq \frac{1}{2}S_y \tag{h}$$

Limits on design variables:

$$4.5 \leq r_o \leq 9.0, \ \ 0.25 \leq t \leq 1.25 \tag{i}$$

TABLE 7.15 m-file to invoke constrained minimization routine for flywheel design problem

% File name = flywheel_opt.m

% Flywheel design

% Design variables: outside radius (ro), and thickness (t)

```
clear all
```

% Set options

```
options = optimset ('LargeScale', 'off');
```

% Set limits for design variables

```
Lb = [4.5, 0.25]; % lower limit
Ub = [9, 1.25]; % upper limit
```

% Set initial design

```
x0 = [6, 1.0];
```

% Set radius of shaft

```
ri = 1.0;
[x, FunVal, ExitFlag, Output] = ...
    fmincon('flywheel_objf', x0, [], [], [], [], Lb, Ub, 'flywheel_conf', options, ri)
```

TABLE 7.16 m-file for objective function for flywheel design problem

% File name = flywheel_objf.m

% Objective function for flywheel design problem

```
function f = flywheel_objf(x, ri)
```

% Rename the design variables **x**

```
ro = x(1);
t = x(2);
f = pi*(ro^2 − ri^2)*t; % volume of flywheel
```

SOLUTION

The problem is solved using the `fmincon` function in the Optimization Toolbox. Table 7.15 shows the script m-file that invokes the `fmincon` function for the flywheel problem. Table 7.16 shows the function m-file that calculates the problem's objective function. Table 7.17 shows the function m-file for calculation of the constraints. Note that the von Mises stress constraint is imposed at the point of maximum stress. Therefore, this maximum is calculated using the `fminbnd` function. Table 7.18 shows the function m-file that calculates the von Mises stress. This file is called by the constraint evaluation function. Also note that all constraints are entered in the normalized " $\leq$ " form. The solution and other output from the function are given as follows:

Active constraints are 2 and 5 (i.e., lower bound on thickness and g(1))
$r_o^* = 7.3165$ in, $t^* = 0.25$ in, $f^* = 13.1328$, Output = [iterations: 8, funcCount: 37]

TABLE 7.17   m-file for constraint functions for flywheel design problem

% Constraint functions for flywheel design problem

```
function [g, h] = flywheel_conf(x, ri)
```

%Rename design variables **x**

```
ro = x(1);
t = x(2);
```

% Constraint limits

```
Is = 0.816;
Sy = 62000; % yield strength
```

% Normalized inequality constraints

```
g(1) = 1 - pi/2*(0.28/386)*(ro^4 - ri^4)*t/Is;
```

% Evaluate maximum von Mises stress

```
options = [];
[alpha, vonMS] = fminbnd('flywheel_vonMs', ri, ro, options, ri, ro);
g(2) = - vonMS/(0.5*Sy) - 1;
```

% Equality constraint (none)

```
h = [];
```

TABLE 7.18   m-file for computation of maximum von Mises stress for the current design

% File name = flywheel_vonMS.m

% von Mises stress

```
function vonMS = flywheel_vonMS (x, ri, ro)
temp = (0.28/386)*(800)^2*(3+0.28)/8;
```

% Tangential stress

```
st = temp*(ri^2 + ro^2 + ri^2*ro^2/x^2 - (1+3*0.28)/(3+0.28)*x^2 );
```

% Radial stress

```
sr = temp*(ri^2 + ro^2 - ri^2*ro^2/x^2 - x^2); % radial stress
vonMS = -sqrt(st^2 - st*sr + sr^2); % von Mises stress
```

EXERCISES FOR CHAPTER 7*

Formulate and solve the following problems.

7.1 Exercise 3.34   **7.2** Exercise 3.35   **7.3** Exercise 3.36

7.4 Exercise 3.50   **7.5** Exercise 3.51   **7.6** Exercise 3.52

7.7 Exercise 3.53   **7.8** Exercise 3.54

7.9 Consider the cantilever beam-mass system shown in Figure E7.9. Formulate and solve the minimum weight design problem for the rectangular cross section so that the fundamental

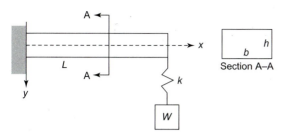

FIGURE E7.9 Cantilever beam with spring-mass at the free end.

Section A–A

vibration frequency is larger than 8rad/s and the cross-sectional dimensions satisfy the limitations

$$0.5 \le b \le 1.0, \ \ \text{in}$$
$$0.2 \le h \le 2.0, \ \ \text{in}$$

(a)

Use a nonlinear programming algorithm to solve the problem. Verify the solution graphically and trace the history of the iterative process on the graph of the problem. Let the starting point be (0.5, 0.2). The data and various equations for the problem are as shown in the following.

Fundamental vibration frequency	$\omega = \sqrt{k_e/m} \ \ \text{rad/s}$	(b)
Equivalent spring constant k_e	$\dfrac{1}{k_e} = \dfrac{1}{k} + \dfrac{L^3}{3EI}$	(c)
Mass attached to the spring	$m = W/g$	(d)
Weight attached to the spring	$W = 50 \ \text{lb}$	(e)
Length of the beam	$L = 12 \ \text{in}$	(f)
Modulus of elasticity	$E = (3 \times 10^7) \ \text{psi}$	(g)
Spring constant	$k = 10 \ \text{lb/in}$	(h)
Moment of inertia	$I, \ \text{in}^4$	(i)
Gravitational constant	$g, \ \text{in/s}^2$	(j)

7.10 A prismatic steel beam with symmetric I cross section is shown in Figure E7.10. Formulate and solve the minimum weight design problem subject to the following constraints:

1. The maximum axial stress due to combined bending and axial load effects should not exceed 100 MPa.
2. The maximum shear stress should not exceed 60 MPa.
3. The maximum deflection should not exceed 15 mm.
4. The beam should be guarded against lateral buckling.
5. Design variables should satisfy the limitations $b \ge 100 \ \text{mm}$, $t_1 \le 10 \ \text{mm}$, $t_2 \le 15 \ \text{mm}$, $h \le 150 \ \text{mm}$.

Solve the problem using a numerical optimization method, and verify the solution using KKT necessary conditions for the data shown after the figures on the next page.

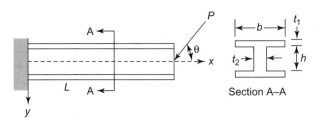

FIGURE E7.10 Graphic of a cantilever I beam. Design variables b, t_1, t_2, and h.

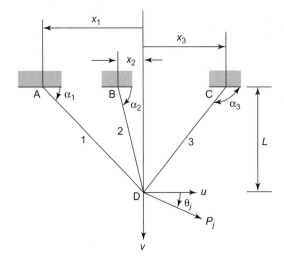

FIGURE E7.11 A three-bar structure–shape optimization graphic.

Modulus of elasticity, $E = 200$ GPa
Shear modulus, $G = 70$ GPa
Load, $P = 70$ kN
Load angle, $\theta = 45°$
Beam length, $L = 1.5$ m

7.11 Shape optimization of a structure. The design objective is to determine the shape of the three-bar structure shown in Figure E7.11 to minimize its weight (Corcoran, 1970). The design variables for the problem are the member cross-sectional areas A_1, A_2, and A_3 and the coordinates of nodes A, B, and C (note that x_1, x_2, and x_3 have positive values in the figure; the final values can be positive or negative), so that the truss is as light as possible while satisfying the stress constraints due to the following three loading conditions:

Condition no. j	Load P_j (lb)	Angle θ_j (degrees)
1	40,000	45
2	30,000	90
3	20,000	135

The stress constraints are written as

$$-5000 \quad \le \sigma_{1j} \le 5000, \text{ psi}$$
$$-20{,}000 \le \sigma_{2j} \le 20{,}000, \text{ psi}$$
$$-5000 \quad \le \sigma_{3j} \le 5000, \text{ psi}$$

where $j = 1, 2, 3$ represents the index for the three loading conditions and the stresses are calculated from the following expressions:

$$\sigma_{1j} = \frac{E}{L_1}\left[u_j \cos\alpha_1 + v_j \sin\alpha_1\right] = \frac{E}{L_1^2}(u_j x_1 + v_j L)$$

$$\sigma_{2j} = \frac{E}{L_2}\left[u_j \cos\alpha_2 + v_j \sin\alpha_2\right] = \frac{E}{L_2^2}(u_j x_2 + v_j L) \tag{a}$$

$$\sigma_{3j} = \frac{E}{L_3}\left[u_j \cos\alpha_3 + v_j \sin\alpha_3\right] = \frac{E}{L_3^2}(-u_j x_3 + v_j L)$$

where $L = 10$ in and

$$L_1 = \text{length of member } 1 = \sqrt{L^2 + x_1^2}$$

$$L_2 = \text{length of member } 2 = \sqrt{L^2 + x_2^2} \tag{b}$$

$$L_3 = \text{length of member } 3 = \sqrt{L^2 + x_3^2}$$

and u_j and v_j are the horizontal and vertical displacements for the jth loading condition determined from the following linear equations:

$$\begin{bmatrix} k_{11} & k_{12} \\ k_{21} & k_{22} \end{bmatrix}\begin{bmatrix} u_j \\ v_j \end{bmatrix} = \begin{bmatrix} P_j \cos\theta_j \\ P_j \sin\theta_j \end{bmatrix}, \quad j = 1, 2, 3 \tag{c}$$

where the stiffness coefficients are given as ($E = 3.0E + 07\text{psi}$)

$$k_{11} = E\left(\frac{A_1 x_1^2}{L_1^3} + \frac{A_2 x_2^2}{L_2^3} + \frac{A_3 x_3^2}{L_3^3}\right)$$

$$k_{12} = E\left(\frac{A_1 L x_1}{L_1^3} + \frac{A_2 L x_2}{L_2^3} - \frac{A_3 L x_3}{L_3^3}\right) = k_{21} \tag{d}$$

$$k_{22} = E\left(\frac{A_1 L^2}{L_1^3} + \frac{A_2 L^2}{L_2^3} + \frac{A_3 L^3}{L_3^3}\right)$$

Formulate the design problem and find the optimum solution starting from the point

$$A_1 = 6.0, \quad A_2 = 6.0, \quad A_3 = 6.0$$
$$x_1 = 5.0, \quad x_2 = 0.0, \quad x_3 = 5.0 \tag{e}$$

Compare the solution with that given later in Table 14.7.

II. NUMERICAL METHODS FOR CONTINUOUS VARIABLE OPTIMIZATION

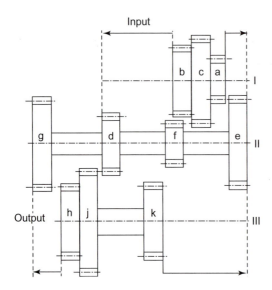

Input

FIGURE E7.12 Schematic arrangement of a nine-speed gear train.

7.12 Design synthesis of a nine-speed gear drive. The arrangement of a nine-speed gear train is shown in Figure E7.12. The objective of the synthesis is to find the size of all gears from the mesh and speed ratio equations such that the size of the largest gears are kept to a minimum (Osman et al., 1978). Because of the mesh and speed ratio equations, it is found that only the following three independent parameters need to be selected:

$$x_1 = \text{gear ratio, } d/a$$
$$x_2 = \text{gear ratio, } e/a$$
$$x_3 = \text{gear ratio, } j/a$$

Because of practical considerations, it is found that the minimization of $|x_2 - x_3|$ results in the reduction of the cost of manufacturing the gear drive.
The gear sizes must satisfy the following mesh equations:

$$\phi^2 x_1(x_1 + x_3 - x_2) - x_2 x_3 = 0$$
$$\phi^3 x_1 - x_2(1 + x_2 - x_1) = 0 \tag{a}$$

where ϕ is the step ratio in speed. Find the optimum solution for the problem for two different values of ϕ as $\sqrt{2}$ and $(2)^{1/3}$.

8

Linear Programming Methods
for Optimum Design

Upon completion of this chapter, you will be able to

- Transform a linear programming problem into the standard form

- Explain terminology and concepts related to linear programming problems

- Use the two-phase Simplex method to solve linear programming problems

- Perform postoptimality analysis for linear programming problems

An optimum design problem having linear cost and constraint functions in the design variables is called a *linear programming (LP) problem*. Linear programming problems arise in some fields of engineering such as water resources, systems engineering, traffic flow control, resources management, and transportation engineering. In the areas of aerospace, automotive, structural, or mechanical system design, most problems are not linear. However, one way of solving nonlinear programming (NLP) problems is to transform them into a sequence of linear programs (Chapter 12).

In addition, some NLP methods solve an LP problem during their iterative solution processes. Thus, linear programming methods are useful in many applications and must be clearly understood. This chapter describes the basic theory and the concepts for solving such problems.

In Section 2.11, a general mathematical model for optimum design of nonlinear problems was defined to minimize a cost function subject to equality and " ≤ " inequality constraints. In Chapter 4, a general theory of optimum design for treating the model was described. That theory can also be used to solve LP problems. However, more efficient and elegant numerical methods are available to solve the LP problem directly. Since there are numerous LP problems in the real world, it is worthwhile to discuss these methods in detail.

In this chapter, we will first define a standard LP problem that is different from the standard NLP problem defined in Chapter 2. Then details of the Simplex method are described to show the numerical steps needed to solve LP problems. Before attempting to implement the method in a computer program, however, standard packages for solving

LP problems must be investigated. Many programs are available to treat such problems—for example, Excel, MATLAB, LINDO (Schrage, 1991).

It is noted here that the subject of linear programming is well developed and several excellent full-length textbooks are available on the subject. These books may be consulted for a more in-depth treatment of the subject.

8.1 LINEAR FUNCTIONS

Cost Function

Any linear function $f(\mathbf{x})$ of k variables $\mathbf{x}$, such as the cost function, has only first-degree terms and is written in the expanded, summation or matrix form as:

$$f(\mathbf{x}) = c_1 x_1 + c_2 x_2 + \cdots + c_k x_k = \sum_{i=1}^{k} c_i x_i = \mathbf{c}^T \mathbf{x} \tag{8.1}$$

where c_i, $i = 1$ to k are constants.

Constraints

All functions of an LP problem can be represented in the form of Eq. (8.1). However, when there are multiple linear functions, the constants c_i must be represented by double subscripts rather than just one subscript. We will use the symbols a_{ij} to represent constants in the constraint expressions. The ith linear constraint involving k design variables, x_j, $j = 1$ to k has one of the following three possible forms, "$\leq$," "$=$," or "$\geq$" (written in expanded or summation notation):

$$a_{i1} x_1 + \cdots + a_{ik} x_k \leq b_i \quad \text{or} \quad \sum_{j=1}^{k} a_{ij} x_j \leq b_i \tag{8.2}$$

$$a_{i1} x_1 + \cdots + a_{ik} x_k = b_i \quad \text{or} \quad \sum_{j=1}^{k} a_{ij} x_j = b_i \tag{8.3}$$

$$a_{i1} x_1 + \cdots + a_{ik} x_k \geq b_i \quad \text{or} \quad \sum_{j=1}^{k} a_{ij} x_j \geq b_i \tag{8.4}$$

where a_{ij} and b_i are known constants. The right sides b_i of the constraints are sometimes called the *resource limits*.

8.2 DEFINITION OF A STANDARD LINEAR PROGRAMMING PROBLEM

8.2.1 Standard LP Definition

Linear programming problems may have equality as well as inequality constraints. Also, many problems require maximization of a function, whereas others require minimization. Although the *standard LP problem* can be defined and treated in several different ways, here we define it as *minimization of a cost function with equality constraints and non-negativity of design variables*. This definition will be used to describe the method (the Simplex method) to solve LP problems. The form is not as restrictive as it may appear since all other LP problems can be transcribed into it. We will explain the process of transcribing a given LP problem into the standard form.

Expanded Form of the Standard LP Problem

For notational clarity, let $\mathbf{x}$ represent an n-vector consisting of the original design variables and any additional variables used to transcribe the problem into the standard form. The standard LP problem is defined as finding the variables x_i, $i = 1$ to n to

Minimize

$$f = c_1 x_1 + c_2 x_2 + \cdots + c_n x_n \tag{8.5}$$

subject to the *m independent* equality constraints

$$a_{11}x_1 + a_{12}x_2 + \cdots + a_{1n}x_n = b_1$$
$$a_{21}x_1 + a_{22}x_2 + \cdots + a_{2n}x_n = b_2$$
$$\vdots \qquad \vdots \qquad \cdots \qquad \vdots \qquad \vdots \tag{8.6}$$
$$a_{m1}x_1 + a_{m2}x_2 + \cdots + a_{mn}x_n = b_m$$

with $b_i \geq 0$, $i = 1$ *to m*, and non-negativity constraints on the variables

$$x_j \geq 0; \quad j = 1 \text{ to } n \tag{8.7}$$

The quantities $b_i \geq 0$, c_j, and a_{ij} ($i = 1$ to m and $j = 1$ to n) are known constants, and m and n are positive integers. Note that all b_i are required to be positive or, at the most, zero. This is required while developing the algorithm to solve the LP problem, as we will see later.

Summation Form of the Standard LP Problem

The standard LP problem can also be written in the summation notation as finding the variables x_i, $i = 1$ to n to

Minimize

$$f = \sum_{i=1}^{n} c_i x_i \tag{8.8}$$

subject to m *independent* equality constraints

$$\sum_{j=1}^{n} a_{ij}x_j = b_i; \quad b_i \geq 0, \quad i = 1 \text{ to } m \tag{8.9}$$

and the non-negativity constraints of Eq. (8.7).

Matrix Form of the Standard LP Problem

Matrix notation may also be used to define the standard LP problem as finding the n-vector $\mathbf{x}$ to

Minimize

$$f = \mathbf{c}^T \mathbf{x} \tag{8.10}$$

subject to the constraints

$$\mathbf{Ax} = \mathbf{b}; \quad \mathbf{b} \geq \mathbf{0} \tag{8.11}$$
$$\mathbf{x} \geq \mathbf{0} \tag{8.12}$$

where $\mathbf{A} = [a_{ij}]$ is an $m \times n$ matrix, $\mathbf{c}$ and $\mathbf{x}$ are n-vectors, and $\mathbf{b}$ is an m-vector. Note that the vector inequalities, such as $\mathbf{b} \geq \mathbf{0}$ in Eq. (8.11), are assumed to be applied to each component of the vector throughout the text. Also, it is assumed that the matrix $\mathbf{A}$ has full row rank; that is, all constraints are linearly independent.

8.2.2 Transcription to Standard LP

The formulations given in Eqs. (8.5) through (8.12) are more general than what may appear at first sight because all LP problems can be transcribed into them. "$\leq$ type" and "$\geq$ type" inequalities can be converted to equalities using slack and surplus variables. Unrestricted variables can be decomposed into the difference of two non-negative variables. Maximization of functions can also be routinely treated. These transformations are explained in the following paragraphs.

Non-Negative Constraint Limits

The resource limits (right side of constraints) in the standard LP are assumed to be always non-negative (i.e., $b_i \geq 0$). If any b_i is negative, it can be made non-negative by multiplying both sides of the constraint by -1. Note, however, that multiplication by -1 changes the sense of the original inequality: "$\leq$ type" becomes "$\geq$ type" and vice versa. For example, a constraint $x_1 + 2x_2 \leq -2$ must be transformed as $-x_1 - 2x_2 \geq 2$ to have a non-negative right side.

Treatment of Inequalities

Since only equality constraints are treated in standard linear programming, the inequalities in Eqs. (8.2) and (8.4) must be converted to equalities. This is no real restriction since any inequality can be converted to equality by introducing a non-negative *slack* or *surplus variable*, as explained in the following paragraphs. Note also that since all b_i are required to be non-negative in Eq. (8.4), it is not always possible to convert "$\geq$" inequalities to the "$\leq$ form" and keep $b_i \geq 0$. In Chapters 2 through 5, this was done, where a standard optimization problem was defined with only "$\leq$ type" constraints. However, in this chapter, we will have to explicitly treat "$\geq$ type" linear inequalities with non-negative right sides. It will be seen later that "$\geq$ type" constraints do require a special treatment in the LP method.

Treatment of "≤ Type" Constraints

For the ith "≤ type" constraint in Eq. (8.2) with a non-negative right side, we introduce a non-negative *slack variable* $s_i \geq 0$ and convert it to an equality as

$$a_{i1}x_1 + \cdots + a_{ik}x_k + s_i = b_i; \quad b_i \geq 0; \quad s_i \geq 0 \tag{8.13}$$

We introduced the idea of slack variables in Chapter 4. There, s_i^2 was used as a slack variable instead of s_i. That was done to avoid the additional constraint $s_i \geq 0$. However, in LP problems we cannot use s_i^2 as a slack variable because it makes the problem nonlinear. Therefore, we will use s_i as a slack variable along with the additional constraint $s_i \geq 0$. For example, a constraint $2x_1 - x_2 \leq 4$ will be transformed into $2x_1 - x_2 + s_1 = 4$ with $s_1 \geq 0$ as its slack variable.

Treatment of "≥ Type" Constraints

Similarly, the ith "≥ type" constraint in Eq. (8.4) with a non-negative right side is converted to equality by subtracting a non-negative *surplus variable* $s_i \geq 0$, as

$$a_{i1}x_1 + \cdots + a_{ik}x_k - s_i = b_i; \quad b_i \geq 0; \quad s_i \geq 0 \tag{8.14}$$

The idea of a surplus variable is very similar to that of a slack variable. For the "≥ type" constraint, the left side always has to be greater than or equal to the right side, so we must subtract a non-negative variable to transform it into an equality. For example, a constraint $-x_1 + 2x_2 \geq 2$ will be transformed as $-x_1 + 2x_2 - s_1 = 2$ with $s_1 \geq 0$ as its surplus variable.

> **The slack and surplus variables are *additional unknowns* that must be determined as a part of the solution for the LP problem. At the optimum point, if the slack or surplus variable s_i is positive, the corresponding constraint is inactive; if s_i is zero, it is active.**

Unrestricted Variables

In addition to the equality constraints, we require that all design variables to be non-negative in the standard LP problem (i.e., $x_i \geq 0$, $i = 1$ to k). *If a design variable x_j is unrestricted in sign, it can always be written as the difference of two non-negative variables:*

$$x_j = x_j^+ - x_j^-; \quad x_j^+ \geq 0, \ x_j^- \geq 0 \tag{8.15}$$

This decomposition is substituted into all equations, and x_j^+ and x_j^- are treated as unknowns in the problem. At the optimum, if $x_j^+ \geq x_j^-$ then x_j is non-negative, and if $x_j^+ \leq x_j^-$, then x_j is nonpositive, as seen in Eq. (8.15).

> **Splitting each free variable into its positive and negative parts increases the dimension of the design variable vector by one.**

Maximization of a Function

Maximization of functions can be treated routinely. For example, if the objective is to maximize a function, we simply minimize its negative. That is

Maximize

$$z = (d_1 x_1 + d_2 x_2 + \ldots + d_n x_n) \Leftrightarrow \text{minimize } f = -(d_1 x_1 + d_2 x_2 + \ldots + d_n x_n) \tag{8.16}$$

Note that a function that is to be maximized is denoted as z in this chapter.

It is henceforth assumed that the LP problem has been converted into the standard form defined in Eqs. (8.5) through (8.12). Example 8.1 shows conversion to the standard LP form.

EXAMPLE 8.1 CONVERSION TO STANDARD LP FORM

Convert the following problem into the standard LP form:

Maximize

$$z = 2y_1 + 5y_2 \tag{a}$$

subject to

$$3y_1 + 2y_2 \leq 12 \tag{b}$$

$$-2y_1 - 3y_2 \leq -6 \tag{c}$$

$$y_1 \geq 0, \quad \text{and} \quad y_2 \text{ is unrestricted in sign.} \tag{d}$$

Solution

To transform the problem into the standard LP form, we take the following steps:

1. Since y_2 is *unrestricted* in sign, we split it into its positive and negative parts as

$$y_2 = y_2^+ - y_2^- \quad \text{with} \quad y_2^+ \geq 0, \; y_2^- \geq 0 \tag{e}$$

2. Substituting this definition of y_2 into the problem functions, we get:

Maximize

$$z = 2y_1 + 5(y_2^+ - y_2^-) \tag{f}$$

subject to

$$3y_1 + 2(y_2^+ - y_2^-) \leq 12 \tag{g}$$

$$-2y_1 - 3(y_2^+ - y_2^-) \leq -6 \tag{h}$$

$$y_1 \geq 0, \; y_2^+ \geq 0, \; y_2^- \geq 0 \tag{i}$$

3. The right side of the constraint in Eq. (g) is non-negative, so it conforms to the standard form. However, the right side of Eq. (h) is negative, so multiplying it by −1, we transform it into the standard form as

$$2y_1 + 3(y_2^+ - y_2^-) \geq 6 \tag{j}$$

4. Converting to a minimization problem subject to equality constraints after introduction of slack and surplus variables, we get the problem in the standard form:

Minimize

$$f = -2y_1 - 5(y_2^+ - y_2^-) \tag{k}$$

subject to

$$3y_1 + 2(y_2^+ - y_2^-) + s_1 = 12 \tag{l}$$

$$2y_1 + 3(y_2^+ - y_2^-) - s_2 = 6 \qquad \text{(m)}$$

$$y_1 \geq 0, \; y_2^+ \geq 0, \; y_2^- \geq 0, \; s_1 \geq 0, \; s_2 \geq 0 \qquad \text{(n)}$$

where s_1 = slack variable for the constraint in Eq. (g) and s_2 = surplus variable for the constraint in Eq. (j).

5. We can redefine the solution variables as

$$x_1 = y_1, \; x_2 = y_2^+, \; x_3 = y_2^-, \; x_4 = s_1, \; x_5 = s_2 \qquad \text{(o)}$$

and rewrite the problem in the standard form as:

Minimize

$$f = -2x_1 - 5x_2 + 5x_3 \qquad \text{(p)}$$

subject to

$$3x_1 + 2x_2 - 2x_3 + x_4 = 12 \qquad \text{(q)}$$

$$2x_1 + 3x_2 - 3x_3 - x_5 = 6 \qquad \text{(r)}$$

$$x_i \geq 0, \; i = 1 \text{ to } 5 \qquad \text{(s)}$$

Comparing the preceding equations with Eqs. (8.10) through (8.12), we can define the following quantities:

$m = 2$ (the number of equations); $n = 5$ (the number of variables)

$$\mathbf{x} = [x_1 \; x_2 \; x_3 \; x_4 \; x_5]^T \quad \mathbf{c} = \begin{bmatrix} -2 & -5 & 5 & 0 & 0 \end{bmatrix}^T \qquad \text{(t)}$$

$$\mathbf{b} = [12 \; 6]^T \quad \mathbf{A} = [a_{ij}]_{2 \times 5} = \begin{bmatrix} 3 & 2 & -2 & 1 & 0 \\ 2 & 3 & -3 & 0 & -1 \end{bmatrix} \qquad \text{(u)}$$

8.3 BASIC CONCEPTS RELATED TO LINEAR PROGRAMMING PROBLEMS

Several terms related to LP problems are defined and explained. Some fundamental properties of LP problems are discussed. It is shown that the optimum solution for the LP problem always lies on the boundary of the feasible set. In addition, the solution is at least at one of the vertices of the convex feasible set (called the *convex polyhedral set*). Some LP theorems are stated and their significance is discussed. The geometrical meaning of the optimum solution is explained.

8.3.1 Basic Concepts

Convexity of LP

Since all functions are linear in an LP problem, the feasible set defined by linear equalities or inequalities is *convex* (Section 4.8). Also, the cost function is linear, so it is convex.

Therefore, the LP problem is convex, and if an *optimum solution* exists, it is a *global optimum*, as stated in Theorem 4.10.

LP Solution on the Boundary of the Feasible Set

It is important to note that even when there are inequality constraints in an LP design problem, the *optimum solution, if it exists, always lies on the boundary of the feasible set*; that is, some constraints are always active at the optimum. This can be seen by writing the necessary conditions of Theorem 4.4 for an unconstrained optimum. These conditions, $\partial f / \partial x_i = 0$, when used for the cost function of Eq. (8.8), give $c_i = 0$ for $i = 1$ to n. This is not possible, as all c_i are not zero. If all c_i were zero, there would be no cost function. Therefore, by contradiction, the *optimum solution for any LP problem must lie on the boundary of the feasible set*. This is in contrast to general nonlinear problems, where the optimum point can be inside or on the boundary of the feasible set.

Infinite Roots of $\mathbf{Ax} = \mathbf{b}$

An optimum solution to the LP problem must also satisfy the equality constraints in Eq. (8.6). Only then can the solution be feasible for the problem. Therefore, to have a meaningful optimum design problem, Eq. (8.6) must have more than one solution. Only then is there a choice of feasible solutions that can have minimum cost. To have many solutions, the number of linearly independent equations in Eq. (8.6) must be less than n, the number of variables in the LP problem (refer to Section A.4 for further discussion of a general solution of m equations in n unknowns).

It is assumed in the following discussion that all of the m rows of the matrix $\mathbf{A}$ in Eq. (8.11) are linearly independent and that $m < n$. This means that there are no redundant equations. Therefore, Eq. (8.6) or (8.11) have *infinite solutions, and we seek a feasible solution that also minimizes the cost function*. A method for solving simultaneous equations (8.6) based on *Gaussian elimination* is described in Appendix A. The Simplex method for LP problems described later in the chapter uses the steps of the Gaussian elimination procedure. Therefore, that procedure must be reviewed thoroughly before studying the Simplex method.

We will use Example 8.2 to illustrate the preceding ideas and to introduce LP terminology and the basic concepts related to the Simplex method.

EXAMPLE 8.2 PROFIT MAXIMIZATION PROBLEM—INTRODUCTION TO LP TERMINOLOGY AND CONCEPTS

As an example of solving constraint equations, we consider the profit maximization problem formulated and solved graphically in Chapter 3. The problem is to find x_1 and x_2 to

Minimize

$$f = -400x_1 - 600x_2 \tag{a}$$

subject to

$$x_1 + x_2 \leq 16 \tag{b}$$

$$\frac{1}{28}x_1 + \frac{1}{14}x_2 \leq 1 \tag{c}$$

$$\frac{1}{14}x_1 + \frac{1}{24}x_2 \leq 1 \tag{d}$$

$$x_1 \geq 0, \ x_2 \geq 0 \tag{e}$$

Solution

The graphical solution to the problem is given in Figure 8.1. All constraints of Eqs. (b) through (e) are plotted and some isocost lines are shown. Each point of the region bounded by the polygon ABCDE satisfies all the constraints of Eqs. (b) through (d) and the non-negativity conditions of Eq. (e). It is seen from Figure 8.1 that the vertex D gives the optimum solution.

Conversion to standard LP form In Eq. (e), both of the variables are already required to be non-negative. The cost function in Eq. (a) is already in the standard minimization form. The right sides of constraints in Eqs. (b) through (d) are also in the standard form: $b_i \geq 0$. Therefore, only the constraints need to be converted to the equality form to transform the problem into standard LP form.

Introducing slack variables for constraints of Eqs. (b) through (d) and writing the problem in the standard LP form, we have

Minimize

$$f = -400x_1 - 600x_2 \tag{f}$$

subject to

$$x_1 + x_2 + x_3 = 16 \tag{g}$$

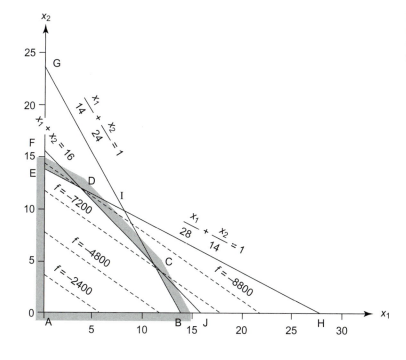

FIGURE 8.1 Solution to the profit maximization problem. Optimum point = (4, 12). Optimum cost = −8800.

$$\frac{1}{28}x_1 + \frac{1}{14}x_2 + x_4 = 1 \tag{h}$$

$$\frac{1}{14}x_1 + \frac{1}{24}x_2 + x_5 = 1 \tag{i}$$

$$x_i \geq 0, \; i = 1 \text{ to } 5 \tag{j}$$

where x_3, x_4, and x_5 are slack variables for the first, second, and third constraints, respectively.

Canonical form of $\mathbf{Ax} = \mathbf{b}$: *general solution* Note that the three equations in Eqs. (g) through (i) are linearly independent. These equations give the linear system of equations $\mathbf{Ax} = \mathbf{b}$ defined in Eq. (8.11). Since the number of variables ($n = 5$) exceeds the number of constraint equations ($m = 3$), a unique solution cannot exist for Eqs. (g) through (i) (see Appendix A). Actually there are infinite solutions. To see this, we write a *general solution* for the equations by transferring the terms associated with the variables x_1 and x_2 to the right side of Eqs. (g) through (i) as

$$x_3 = 16 - x_1 - x_2 \tag{k}$$

$$x_4 = 1 - \frac{1}{28}x_1 - \frac{1}{14}x_2 \tag{l}$$

$$x_5 = 1 - \frac{1}{14}x_1 - \frac{1}{24}x_2 \tag{m}$$

Note that the variables x_3, x_4, and x_5 appear in one and only one equation in the system $\mathbf{Ax} = \mathbf{b}$ in Eqs. (g) through (i), x_3 only in Eq. (g), x_4 only in Eq. (h), and x_5 only in Eq. (i). In addition, the coefficient of each of these variables is $+1$. Such a linear system is called a *canonical representation* of the equations $\mathbf{Ax} = \mathbf{b}$. Note that when all of the constraints in the LP problem are "$\leq$ type" with non-negative right sides, such as Eqs. (b) through (d), a general solution for the system $\mathbf{Ax} = \mathbf{b}$ is obtained without any calculations.

In Eqs. (k) through (m), x_1 and x_2 act as *independent variables* that can be assigned any value, and x_3, x_4, and x_5 are *dependent* on them. Different values for x_1 and x_2 generate different values for x_3, x_4, and x_5. Thus there are *infinite solutions* for $\mathbf{Ax} = \mathbf{b}$.

It is important to also note that the form of the general solution given in Eqs. (k) through (m) will change if a different set of variables is selected as independent variables. For example, if x_1 and x_3 are selected as independent variables, then the remaining variables x_2, x_4, and x_5 can appear in one and only one equation in Eqs. (g) through (i). This can be achieved by using the Gauss-Jordan elimination procedure described in Sections A.3 and A.4.

To determine another general solution, let us select x_2 to become a dependent variable in Eqs. (g) through (i) instead of an independent variable. This means that x_2 must appear in one and only one equation. The next question is, in which equation, Eq. (g), (h), or (i)? Each of these selections will give a different form for the general solution and hence different independent and dependent variables. For example, if x_2 is eliminated from Eqs. (g) and (i) using Eq. (h), the following canonical form of the equations is obtained:

$$\frac{1}{2}x_1 + x_3 - 14x_4 = 2 \tag{n}$$

$$\frac{1}{2}x_1 + x_2 + 14x_4 = 14 \tag{o}$$

$$\frac{17}{336}x_1 - \frac{7}{12}x_4 + x_5 = \frac{5}{12} \tag{p}$$

Thus x_3 appears only in Eq. (n), x_2 in Eq. (o), and x_5 in Eq. (p). In the canonical form of Eqs. (n) through (p), x_1 and x_4 are independent variables and x_2, x_3 and x_5 are dependent variables.

If Eq. (g) is used to eliminate x_2 from Eqs. (h) and (i), then the independent variables are x_1 and x_3 instead of x_1 and x_4. And, if Eq. (i) is used to eliminate x_2 from Eqs. (g) and (h), then the independent variables are x_1 and x_5 instead of x_1 and x_4.

Basic solution A solution of particular interest in LP problems is obtained by setting p of the variables to zero and solving for the rest, where p is the difference between the number of variables (n) and the number of independent constraint equations (m) in $\mathbf{Ax} = \mathbf{b}$ (i.e., $p = n - m$—e.g., $p = 2$ in the case of Eqs. (g) through (i)). With two variables set to zero, a unique solution to Eqs. (g) through (i) exists for the remaining three variables since there are now three independent equations in three unknowns. A solution obtained by setting p independent variables to zero is called the *basic solution*. For example, a basic solution is obtained from Eqs. (g) through (i) or (k) through (m) by setting $x_1 = 0$ and $x_2 = 0$, and solving for the remaining variables as $x_3 = 16$, $x_4 = 1$, $x_5 = 1$. The independent variables that are set to zero are called *nonbasic variables* in LP terminology (e.g., x_1 and x_2). The variables that are solved from the independent equations are called the *basic variables* (e.g., x_3, x_4, and x_5 in Eqs (k) through (m)).

Another basic solution is obtained by setting $x_1 = 0$, $x_3 = 0$, and solving the three equations in Eqs. (g) through (i) for the remaining three unknowns as $x_2 = 16$, $x_4 = -1/7$, $x_5 = 1/3$. Still another basic solution is obtained from Eqs. (n) through (p) by setting $x_1 = 0$ and $x_4 = 0$, and solving for the remaining variables as $x_3 = 2$, $x_2 = 14$, and $x_5 = 5/12$. Since there are 10 different ways in which combinations of 2 variables (out of 5) can be set to zero, there are 10 basic solutions (later a formula is given for the number of basic solutions).

Table 8.1 shows all 10 basic solutions for the present example obtained using the procedure just described. A systematic way to generate these solutions is to use the Gauss-Jordan elimination method given in Appendix A, which is demonstrated later as a way to obtain basic solutions. For each of the basic solutions given in Table 8.1, the nonbasic variables have zero value and the basic variables have nonzero values.

Basic feasible solution Note that of the 10 basic solutions in Table 8.1, exactly 5 (solutions 1, 2, 6, 8, and 9) correspond to the vertices of the feasible polygon ABCDE in Figure 8.1. The remaining 5 basic solutions violate the non-negativity condition on the variables and correspond to the infeasible vertices, F, G, H, I, and J. Therefore, 5 of the 10 basic solutions are feasible. These are called the *basic feasible solutions*.

Optimum solution By moving the isocost line parallel to itself, it is seen that the optimum solution is at point D. Note that the optimum point is at one of the vertices of the *feasible polygon*. This will be observed later as a general property of any LP problem. That is, *if an LP has a solution, it is at least at one of the vertices of the feasible set.*

The terms *canonical form* and *general solution* in reference to $\mathbf{Ax} = \mathbf{b}$ are synonymous and will be used interchangeably. They both give a basic solution to $\mathbf{Ax} = \mathbf{b}$.

TABLE 8.1 Ten basic solutions for the profit maximization problem

Solution no.	x_1	x_2	x_3	x_4	x_5	f	Vertex in Figure 8.1
1	0	0	16	1	1	0	A
2	0	14	2	0	$\frac{5}{12}$	−8400	E
3	0	16	0	$-\frac{1}{7}$	$\frac{1}{3}$	—	F (infeasible)
4	0	24	−8	$-\frac{5}{7}$	0	—	G (infeasible)
5	16	0	0	$\frac{3}{7}$	$-\frac{1}{7}$	—	J (infeasible)
6	14	0	2	$\frac{1}{2}$	0	−5600	B
7	28	0	−12	0	−1	—	H (infeasible)
8	4	12	0	0	$\frac{3}{14}$	−8800	D
9	11.2	4.8	0	$\frac{1}{5}$	0	−7360	C
10	$\frac{140}{17}$	$\frac{168}{17}$	$-\frac{36}{17}$	0	0	—	I (infeasible)

8.3.2 LP Terminology

We will now summarize various definitions and terms that are related to the LP problem. Example 8.2 and Figure 8.1 are used to illustrate the meaning of these terms. Also, the definitions of *convex set*, *convex function*, and *line segment* introduced in Section 4.8 will be used here.

- *Vertex (Extreme) Point.* This is a point of the feasible set that does not lie on a line segment joining two other points of the set. For example, every point on the circumference of a circle and each vertex of the polygon satisfy the requirements for an extreme point.
- *Feasible Solution.* Any solution of the constraint Eq. (8.6) satisfying the non-negativity conditions of Eq. (8.7) is a feasible solution. In the profit maximization example of Figure 8.1, every point bounded by the polygon ABCDE (convex set) is a feasible solution, including the boundary points.
- *Basic Solution.* A basic solution is a solution of the constraint Eq. (8.6) obtained by setting the "redundant number" $(n - m)$ of the variables to zero and solving the equations simultaneously for the remaining variables (e.g., the 10 solutions given in Table 8.1).
- *Nonbasic Variables.* The variables set to zero in the basic solution are called *nonbasic*. For example, x_1 and x_5 are set to zero in solution 4 of Table 8.1. Therefore, they are nonbasic variables.
- *Basic Variables.* The variables that are not set to zero in the basic solution are called *basic*.
- *Basic Feasible Solution.* A basic solution satisfying the non-negativity conditions on the variables in Eq. (8.7) is called a basic feasible solution. In the profit maximization example, each of the 10 solutions in Table 8.1 is basic, but only A, B, C, D, and E are basic and feasible (solutions 1, 2, 6, 8, and 9).

- *Degenerate Basic Solution.* If a basic variable has a zero value in a basic solution, the solution is a degenerate basic solution.
- *Degenerate Basic Feasible Solution.* If a basic variable has a zero value in a basic feasible solution, the solution is a degenerate basic feasible solution.
- *Optimum Solution.* A feasible solution minimizing the cost function is an optimum solution. The point D in Figure 8.1 corresponds to the optimum solution.
- *Optimum Basic Solution.* The optimum basic solution is a basic feasible solution that has an optimum cost function value. From Table 8.1 and Figure 8.1 it is clear that only solution number 8 is the optimum basic solution.
- *Convex Polyhedron.* If the feasible set for an LP problem is bounded, it is a convex polyhedron. As an example, the polygon ABCDE in Figure 8.1 represents a convex polyhedron for the problem in Example 8.2.
- *Basis.* Columns of the coefficient matrix $\mathbf{A}$ in Eq. (8.11) of the constraint equations corresponding to the basic variables are said to form a basis for the m-dimensional vector space. Any other m-dimensional vector can be expressed as a linear combination of the basis vectors.

EXAMPLE 8.3 BASIC SOLUTIONS TO AN LP PROBLEM

Find all basic solutions to the following problem and identify basic feasible solutions in a figure of the feasible set:

Maximize

$$z = 4x_1 + 5x_2 \tag{a}$$

subject to

$$-x_1 + x_2 \le 4 \tag{b}$$

$$x_1 + x_2 \le 6 \tag{c}$$

$$x_1, x_2 \ge 0 \tag{d}$$

Solution

The feasible region for the problem is shown in Figure 8.2. For conversion of the problem into the standard LP form, we observe that the right sides of the constraints are already non-negative (≥ 0) and both of the variables are required to be non-negative. Therefore, no transformations are needed for these two conditions.

Since both of the constraints are "$\le$ type", introducing the slack variables x_3 and x_4 into the constraint equations and converting maximization of z to minimization, the problem is written in the standard LP form as

Minimize

$$f = -4x_1 - 5x_2 \tag{e}$$

subject to

$$-x_1 + x_2 + x_3 = 4 \tag{f}$$

$$x_1 + x_2 + x_4 = 6 \tag{g}$$

$$x_i \geq 0; \quad i = 1 \text{ to } 4 \tag{h}$$

Since there are four variables and two constraints in Eqs. (f) and (g) ($n = 4$, $m = 2$), the problem has six basic solutions; that is, there are six different ways in which two of the four variables can be chosen as nonbasic (independent variables).

These solutions are obtained from Eqs. (f) and (g) by choosing two variables as nonbasic and the remaining two as basic. For example, x_1 and x_2 may be chosen as nonbasic (i.e., $x_1 = 0$, $x_2 = 0$). Then Eqs. (f) and (g) give $x_3 = 4$, $x_4 = 6$. Also, with $x_1 = 0$ and $x_3 = 0$, Eqs. (f) and (g) give $x_2 = 2$ and $x_4 = 2$ as another basic solution. Similarly, the remaining basic solutions are obtained by selecting two variables as nonbasic (zero) and solving for the other two from Eqs. (f) and (g).

The six basic solutions for the problem are summarized in Table 8.2 along with the corresponding cost function values. The basic feasible solutions are 1, 2, 5, and 6. These correspond to

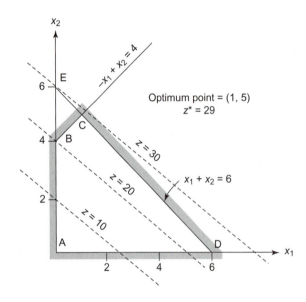

FIGURE 8.2 Graphical solution to the Example 8.3 LP problem. Optimum point = (1, 5), $z^* = 29$.

TABLE 8.2 Basic solutions for Example 8.3

Solution no.	x_1	x_2	x_3	x_4	f	Vertex in Figure 8.2
1	0	0	4	6	0	A
2	0	4	0	2	−20	B
3	0	6	−2	0	—	E (infeasible)
4	−4	0	0	10	—	Not shown (infeasible)
5	6	0	10	0	−24	D
6	1	5	0	0	−29	C

points A(0,0), B(0,4), D(6,0), and C(1,5), respectively, in Figure 8.2. The minimum value of the cost function is obtained at the point C(1,5) as $f = -29$ (maximum value of $z = 29$).

In Section 8.4, we will introduce a systematic tabular procedure based on the Gaussian-Jordan elimination method of Sections A.3 and A.4 in Appendix A to determine basic solutions of Eqs. (f) and (g).

8.3.3 Optimum Solution to LP Problems

Now two important theorems that define the optimum solution for LP problems are stated and explained.

THEOREM 8.1

Extreme Points and the Basic Feasible Solutions The collection of feasible solutions for an LP problem constitutes a convex set whose extreme points correspond to basic feasible solutions. This theorem relates *extreme points of the convex polyhedron to the basic feasible solutions.*

This is an important result, giving geometric meaning to the basic feasible solutions; they are the vertices of the polyhedron representing the feasible set for an LP problem. As an example, basic feasible solutions in Table 8.1 correspond to vertices of the feasible set in Figure 8.1.

THEOREM 8.2

The Basic Theorem of Linear Programming *This theorem establishes the importance of the basic feasible solutions.*

Let the $m \times n$ coefficient matrix **A** of the constraint equations have full row rank (i.e., rank (**A**) $= m$). Then

1. If there is a feasible solution, there is a basic feasible solution.
2. If there is an optimum feasible solution, there is an optimum basic feasible solution.

Part 1 of the theorem says that if there is any feasible solution to the LP problem, then there must be at least one extreme point or vertex of the *convex feasible set*. Part 2 of the theorem says that if the LP problem has an optimum solution, then it is at least at one of the vertices of the *convex polyhedron* representing all of the feasible solutions; that is, it is at least one of the basic feasible solutions. There can be multiple optimum solutions if the cost function is parallel to one of the active constraints, as we saw in Chapter 3.

Number of Basic Solutions

As noted earlier, the LP problem has infinite feasible designs. We seek a feasible design that minimizes the cost function. Theorem 8.2 says that such a solution must be one of the basic feasible solutions, that is, at one of the extreme points of the convex feasible set. Thus, our task of solving an LP problem is reduced to the search for an optimum only among the basic feasible solutions. For a problem having n variables and m constraints, the maximum number of basic solutions is obtained by counting the total number of combinations where m variables are nonzero out of a total of n variables. This number is given by the formula:

$$\text{\# of basic solutions} = \binom{n}{m} = \frac{n!}{m!(n - m)!} \tag{8.17}$$

This formula gives a *finite number of basic solutions*. Thus, according to Theorem 8.2, the optimum solution is at one of these basic solutions that is also feasible. We need to search this solution set systematically for the optimum.

The Simplex method of Section 8.5 is based on searching among the basic feasible solutions to reduce the cost function continuously until an optimum point is reached.

8.4 CALCULATION OF BASIC SOLUTIONS

In the last section, we observed the importance of basic solutions to the linear equations $\mathbf{Ax} = \mathbf{b}$; the optimum solution for the LP problem is at least one of the basic feasible solutions. Therefore, it is important to generate basic solutions for the problem in a systematic way. In this section we describe such a method, called the Gauss-Jordan elimination method, presented in Appendix A. The Simplex method, described in the next section, uses this procedure to search for the optimum solution among the basic feasible solutions to $\mathbf{Ax} = \mathbf{b}$.

8.4.1 The Tableau

It is customary to represent the linear system $\mathbf{Ax} = \mathbf{b}$ in a tableau. A *tableau* is defined as the representation of a scene or a picture. It is a convenient way of representing all necessary information related to an LP problem. In the Simplex method, the tableau consists of coefficients of the variables in the cost and constraint functions. In this section we will focus only on the linear system $\mathbf{Ax} = \mathbf{b}$ and its representation in a tableau; the cost function will be added in the tableau in the next section.

EXAMPLE 8.4 STRUCTURE OF THE TABLEAU

Constraints for the profit maximization problem of Example 8.2 are given in the standard form $\mathbf{Ax} = \mathbf{b}$ as follows:

$$x_1 + x_2 + x_3 = 16 \tag{a}$$

$$\frac{1}{28} x_1 + \frac{1}{14} x_2 + x_4 = 1 \tag{b}$$

$$\frac{1}{14}x_1 + \frac{1}{24}x_2 + x_5 = 1 \tag{c}$$

Write the linear system in a tableau and explain its notation.

Solution

The linear system $\mathbf{Ax} = \mathbf{b}$ is written in the tableau form in Table 8.3. It is important to understand the structure and notation of the tableau as explained in the following because the tableau is used later to develop the Simplex method.

1. *Rows of the tableau:* Each row of the tableau contains coefficients of an equation in Eqs. (a) through (c); that is, the first, second, and third rows contain coefficients of Eqs. (a) through (c), respectively.
2. *Columns of the tableau:* Columns of the tableau are associated with variables (e.g., x_1 column, x_2 column, and so on). This is because these columns contain the coefficients of each variable in $\mathbf{Ax} = \mathbf{b}$; for example, column x_1 contains coefficients of the variable x_1 in each of the equations in Eqs. (a) through (c) and column RS contains the coefficient on the right side of each equation.
3. *Identity submatrix in the tableau:* The system $\mathbf{Ax} = \mathbf{b}$ is in the canonical form; that is, there are variables that appear in one and only one equation. For example, x_3 appears in only the first equation, x_4 in the second, and x_5 in the third. Therefore, columns x_3, x_4, and x_5 have unit elements in one location and zeroes everywhere else. These columns form an *identity submatrix* in the matrix $\mathbf{A}$.
4. *Identity columns:* Columns of the identity submatrix in the tableau are called identity columns. Note that these columns can appear anywhere in the tableau. They need not be in any sequence either.
5. *Basic variable for each row:* Each row of the tableau is also associated with a variable, as indicated in the "Basic" column on the left side in Table 8.3. These variables correspond to the columns of the identity submatrix in the tableau. The row location of the unit element in an identity column determines the association of the basic variable with a row. Thus, x_3 is associated with the first row, x_4 with the second row, and x_5 with the third row. These are called the basic variables. This will become clearer later when we solve example problems.
6. *Basic solution associated with the tableau:* The tableau identifies basic and nonbasic variables and gives their values; that is, it gives a basic solution. For the tableau in Table 8.3, the basic solution is given as
7. *Nonbasic variables:* $x_1 = 0$, $x_2 = 0$
8. *Basic variables:* $x_3 = 16$, $x_4 = 1$, $x_5 = 1$
   We will see later that the tableau can be augmented with the cost function expression, and in that case, it will also immediately give the value of the cost function associated with the basic solution.
9. *Basic and nonbasic columns:* Columns associated with basic variables are called *basic columns*, and others are called *nonbasic columns*; for example, x_1 and x_2 are nonbasic columns and in Table 8.3 the others are basic. Basic columns also correspond to the identity submatrix's columns.

TABLE 8.3   Tableau form of $\mathbf{Ax} = \mathbf{b}$ for the profit maximization problem

Basic ↓	x_1	x_2	x_3	x_4	x_5	RS: b
1. x_3	1	1	1	0	0	16
2. x_4	$\frac{1}{28}$	$\frac{1}{14}$	0	1	0	1
3. x_5	$\frac{1}{14}$	$\frac{1}{24}$	0	0	1	1

8.4.2 The Pivot Step

In the Simplex method, we want to systematically search among the basic feasible solutions for the optimum design. *We must have a basic feasible solution to initiate the Simplex method.* Starting from the basic feasible solution, we want to find another one that decreases the cost function. This can be done by interchanging a current basic variable with a nonbasic variable. That is, a current basic variable is made nonbasic (i.e., reduced to 0 from a positive value), and a current nonbasic variable is made basic (i.e., increased from 0 to a positive value). The *pivot step* of the Gauss-Jordan elimination method accomplishes this task and results in a new canonical form (general solution), as explained in Example 8.5. The definitions of *pivot column*, *pivot row*, and *pivot element* are also given.

EXAMPLE 8.5 PIVOT STEP—INTERCHANGE OF BASIC AND NONBASIC VARIABLES

The problem in Example 8.3 is written as follows in the standard form with the linear system $\mathbf{Ax} = \mathbf{b}$ in the canonical form:

Minimize
$$f = -4x_1 - 5x_2 \tag{a}$$

subject to
$$-x_2 + x_2 + x_3 = 4 \tag{b}$$

$$x_1 + x_2 + x_4 = 6 \tag{c}$$

$$x_i \geq 0; \quad i - 1 \text{ to } 4 \tag{d}$$

Obtain a new canonical form by interchanging the roles of the variables x_1 and x_4 (i.e., make x_1 a basic variable and x_4 a nonbasic variable).

Solution

The given canonical form can be written as the initial tableau, as shown in Table 8.4; x_1 and x_2 are nonbasic, and x_3 and x_4 are basic ($x_1 = x_2 = 0$, $x_3 = 4$, $x_4 = 6$). This corresponds to point A in Figure 8.2. In the tableau, the basic variables are identified in the leftmost column, and their values are given in the rightmost column. Also, the basic variables can be identified by examining the columns of the tableau. The variables associated with the columns of the identity submatrix are basic (e.g., variables x_3 and x_4 in Table 8.4). The location of the positive unit element in a basic column identifies the row whose right side parameter b_i is the current value of the basic

TABLE 8.4 Pivot step to interchange basic variable x_4 with nonbasic variable x_1 for Example 8.5

Initial tableau

Basic↓	x_1	x_2	x_3	x_4	b
1. x_3	-1	1	1	0	4
2. x_4	1	1	0	1	6

Basic solution: Nonbasic variables: $x_1 = 0$, $x_2 = 0$; basic variables: $x_3 = 4$, $x_4 = 6$

To interchange x_1 with x_4, choose row 2 as the pivot row and column 1 as the pivot column. Perform elimination using a_{21} as the pivot element.

Second tableau: Result of the pivot operation

Basic↓	x_1	x_2	x_3	x_4	b
1. x_3	0	2	1	1	10
2. x_1	1	1	0	1	6

Basic solution: Nonbasic variables: $x_2 = 0$, $x_4 = 0$; basic variables: $x_1 = 6$, $x_3 = 10$

variable associated with that row. For example, the basic column x_3 has a unit element in the first row, so x_3 is the basic variable associated with the first row, having a value of 4. Similarly, x_4 is the basic variable associated with row 2 having a value of 6.

Pivot column: Since x_1 is to become a basic variable, it should become an identity column; that is, x_1 should be eliminated from all rows in the x_1 column except one. Since x_4 is to become a nonbasic variable and is associated with row 2, the unit element in the x_1 column should appear in row 2. This is achieved by eliminating x_1 for row 1 (i.e., Eq. (b)). The column in which eliminations are performed is called the pivot column.

Pivot row: The row that is used to perform elimination of a variable from various equations is called the pivot row (e.g., row 2 in the initial tableau in Table 8.4).

Pivot element: The intersection of the pivot column and the pivot row determines the pivot element (e.g., $a_{21} = 1$ for the initial tableau in Table 8.4; the pivot element is boxed).

Pivot operation: To make x_1 a basic variable and x_4 a nonbasic variable, we need to make $a_{21} = 1$ and $a_{11} = 0$. This will replace x_1 with x_4 as the basic variable and a new canonical form will be obtained. Performing Gauss-Jordan elimination in the first column with $a_{21} = 1$ as the pivot element, we obtain the second canonical form as shown in Table 8.4 (add row 2 to row 1). For this canonical form, $x_2 = x_4 = 0$ are the nonbasic variables and $x_1 = 6$ and $x_3 = 10$ are the basic variables. Thus, referring to Figure 8.2, this pivot step results in a move from the extreme point A(0, 0) to an adjacent extreme point D(6, 0).

8.4.3 Basic Solutions to Ax = b

Using the Gauss-Jordan elimination method, we can systematically generate all of the basic solutions for an LP problem. Then, evaluating the cost function for the basic feasible solutions, we can determine the optimum solution for the problem. The Simplex method described in the next section uses this approach with one exception: It searches through only the basic feasible solutions and stops once an optimum solution is obtained. Example 8.6 illustrates the process of generating basic solutions.

EXAMPLE 8.6 CALCULATION OF BASIC SOLUTIONS

For the profit maximization problem of Example 8.2, determine three basic solutions using the Gauss-Jordan elimination method.

Solution

The problem has been transcribed into the standard form in Eqs. (f) through (j) in Example 8.2 as

Minimize

$$f = -400x_1 - 600x_2 \tag{a}$$

subject to

$$x_1 + x_2 + x_3 = 16 \tag{b}$$

$$\frac{1}{28}x_1 + \frac{1}{14}x_2 + x_4 = 1 \tag{c}$$

$$\frac{1}{14}x_1 + \frac{1}{24}x_2 + x_5 = 1 \tag{d}$$

$$x_i \geq 0, \quad i = 1 \text{ to } 5 \tag{e}$$

First tableau The constraint Eqs. (b) through (d) are written in the tableau form in Table 8.5 identified as the first tableau. Columns x_3, x_4, and x_5 are the identity columns; therefore, x_3, x_4, and x_5 are the basic variables. Since the unit element in the x_3 column appears in the first row, x_3 is the basic variable associated with the first row in the tableau. Similarly, x_4 and x_5 are the basic variables associated with the second and third rows. Thus the basic solution from the first tableau is given as (corresponds to solution 1 in Table 8.1):

Basic variables: $x_3 = 16$, $x_4 = 1$, $x_5 = 1$
Nonbasic variables: $x_1 = 0$, $x_2 = 0$

Second tableau   To obtain a new basic solution, we need to obtain a new canonical form (a new general solution). This can be done if we choose different nonbasic variables. Let us select x_4 (a current basic variable) to become a nonbasic variable and x_1 (a current nonbasic variable) to become a basic variable. That is, we interchange the roles of variables x_1 and x_4; x_1 will become a basic column, and x_4 will become a nonbasic column. Since x_1 is to become a basic variable, it should appear in one and only one equation. This will be equation #2 since we have selected x_4 to become a nonbasic variable that is associated with the second row in the first tableau. In terms of the tableau terminology, we have selected the x_1 column to be the pivot column and the second row to be the pivot row. This identifies the pivot element to be $a_{21} = \frac{1}{28}$. The pivot element is boxed and the pivot column and pivot row are highlighted in the first tableau of Table 8.5.

Now we complete the pivot operation to obtain the second tableau as follows (details of the calculations for each of the following steps are displayed in Table 8.6).

TABLE 8.5 Basic solutions for the profit maximization problem

First tableau: Basic variable x_4 is selected to be replaced with the nonbasic variable x_1 to obtain a new basic solution

Basic↓	x_1	x_2	x_3	x_4	x_5	b	Remarks
1. x_3	1	1	1	0	0	16	Column x_1 is the pivot column and row 2 is the pivot row
2. x_4	$\frac{1}{28}$	$\frac{1}{14}$	0	1	0	1	
3. x_5	$\frac{1}{14}$	$\frac{1}{24}$	0	0	1	1	

Second tableau: x_3 is selected to be replaced with x_2 in the basic set

Basic↓	x_1	x_2	x_3	x_4	x_5	b	Remarks
4. x_3	0	−1	1	−28	0	−12	x_2 is the pivot column and row 4 is the pivot row
5. x_1	1	2	0	28	0	28	
6. x_5	0	$-\frac{17}{168}$	0	−2	1	−1	

Third tableau

Basic↓	x_1	x_2	x_3	x_4	x_5	b
7. x_2	0	1	−1	28	0	12
8. x_1	1	0	2	−28	0	4
9. x_5	0	0	$-\frac{17}{168}$	$\frac{5}{6}$	1	$\frac{3}{14}$

TABLE 8.6 Detailed calculation for second tableau of profit maximization problem of Example 8.6

Basic↓	x_1	x_2	x_3	x_4	x_5	b
4. x_3	$1-1=0$	$1-2=-1$	$1-0=1$	$0-28=-28$	0	$16-28=-12$
5. x_1	$\frac{1/28}{1/28}=1$	$\frac{1/14}{1/28}=2$	0	$\frac{1}{1/28}=28$	0	$\frac{1}{1/28}=28$
6. x_5	$\frac{1}{14}-1\times\frac{1}{14}=0$	$\frac{1}{24}-2\times\frac{1}{14}=-\frac{17}{168}$	0	$0-28\times\frac{1}{14}=-2$	1	$1=28\times\frac{1}{14}=-1$

Note: *Eliminations are performed in x_1 column using row 5.*

1. We divide the pivot row by the pivot element $a_{21} = \frac{1}{1/28}$; this is shown as row 5 in the second tableau of Table 8.5.
2. To eliminate x_1 from the first equation, we subtract row 5 from row 1. This is shown as row 4 in the second tableau of Table 8.5.
3. To eliminate x_1 from the third equation, we multiply row 5 by 1/14 and subtract it from row 3. This is shown as row 6 in the second tableau of Table 8.5.

Thus equations of this canonical form (general solution) are given as

$$-x_2 + x_3 - 28x_4 = -12 \tag{g}$$

$$x_1 + 2x_2 + 28x_4 = 28 \tag{h}$$

$$-\frac{17}{168}x_2 - 2x_4 + x_5 = -1 \tag{i}$$

The basic solution from the second tableau is given as

Basic variables: $\quad x_3 = -12, \ x_1 = 28, \ x_5 = -1$
Nonbasic variables: $x_2 = 0, \ x_4 = 0$

This basic solution is infeasible and corresponds to the infeasible vertex H in Figure 8.1 (solution 7 in Table 8.1).

Third tableau Next we want to obtain another basic solution by exchanging x_3 (a current basic variable) with x_2 (a current nonbasic variable). This will make x_2 an identity column and x_3 a nonbasic column. Thus x_2 is identified as the pivot column and the first row of the second tableau (row 4) is identified as the pivot row with $a_{12} = -1$ as the pivot element. (It is noted that in the Simplex method presented in the next section, a negative number is never selected as the pivot element, although here we allow that to obtain another basic solution.) The pivot operation will eliminate x_2 from all of the equations except the first one. Thus x_2 will become the identity column, similar to the x_3 column in the second tableau.

Details of the pivot operation are presented in Table 8.7, and the final results are summarized in Table 8.5. The pivot operation is carried out using the following steps:

1. We divide the pivot row by the pivot element $a_{12} = -1$; this is shown as the seventh row in the third tableau of Table 8.7.
2. To eliminate x_2 from the second equation, we multiply row 7 by 2 and subtract it from row 5. This is shown as row 8 in the third tableau of Table 8.7, and the results are summarized in Table 8.5.
3. To eliminate x_2 from the third equation, we multiply row 7 by 17/168 and add it to row 6. This is shown as row 9 in the third tableau of Table 8.7, and the results are summarized in Table 8.5.

The basic solution from the third tableau in Table 8.5 is given as

Basic variables: $\quad x_2 = 12, \ x_1 = 4, \ x_5 = \frac{3}{14}$
Nonbasic variables: $x_3 = 0, \ x_4 = 0$

This basic feasible solution corresponds to the optimum vertex D in Figure 8.1 (solution 8 in Table 8.1).

TABLE 8.7　Detailed calculation for third tableau of profit maximization problem of Example 8.6

Basic↓	x_1	x_2	x_3	x_4	x_5	b
7. x_2	0	$-1/-1 = 1$	$1/-1 = -1$	$-28/-1 = 28$	0	$-12/-1 = 12$
8. x_1	1	$2 - 1 \times 2 = 0$	$0 - (-1) \times 2 = 2$	$28 - 2 \times 28 = -28$	0	$28 - 12 \times 2 = 4$
9. x_5	0	$-\frac{17}{168} + 1 \times \frac{17}{168} = 0$	$0 - 1 \times \frac{17}{168} = -\frac{17}{168}$	$-2 + 28 \times \frac{17}{168} = \frac{5}{6}$	1	$-1 + \frac{17}{168} \times 12 = \frac{3}{14}$

Note: *Eliminations are performed in the x_2 column using row 7.*

8.5 THE SIMPLEX METHOD

8.5.1 The Simplex

A *Simplex* in two-dimensional space is formed by any three points that do not lie on a straight line. In three-dimensional space, it is formed by four points that do not lie in the same plane. Three points can lie in a plane, but the fourth has to lie outside. In general, a Simplex in the n-dimensional space is a convex hull of any $(n + 1)$ points that do not lie on one hyperplane. A *convex hull* of $(n + 1)$ points is the smallest convex set containing all of the points. Thus, the *Simplex represents a convex set*.

8.5.2 Basic Steps in the Simplex Method

The basics of the Simplex method for solving LP problems are described in this section. The ideas of canonical form, pivot row, pivot column, pivot element, and pivot step, which were introduced in the previous section, are used. The method is described as an extension of the standard Gauss-Jordan elimination procedure for solving a system of linear equations $Ax = b$, where A is an $m \times n$ $(m < n)$ matrix, x is an n-vector, and b is an m-vector. In this section, the Simplex method is developed and illustrated for "$\leq$ type" constraints, since with such constraints, the method can be developed in a straightforward manner. In the next section, "$\geq$ type" and equality constraints that require special treatment in the Simplex method are discussed. A detailed derivation of the Simplex method is presented in Chapter 9.

Basic idea of the Simplex method

Theorem 8.2 guarantees that one of the basic feasible solutions is an optimum solution for the LP problem. The basic idea of the Simplex method is to proceed from one basic feasible solution to another in a way that continually decreases the cost function until the minimum is reached. The method never calculates *basic infeasible solutions*. The *Gauss-Jordan elimination procedure*, described in the previous section, is used to systematically find basic feasible solutions of the linear system of equations $Ax = b$ until the optimum is reached.

In this subsection, we will illustrate the basic steps of the Simplex method with an example problem. The method starts with a basic feasible solution (i.e., at a vertex of the convex feasible set). A move is then made to an adjacent vertex while maintaining the feasibility of the new solution (i.e., all $x_i \geq 0$) as well as reducing the cost function. This is accomplished by replacing a basic variable with a nonbasic variable in the current basic feasible solution. Two basic questions now arise:

1. How do we choose a current nonbasic variable that should become basic?
2. Which variable from the current basic set should become nonbasic?

The Simplex method answers these questions based on theoretical considerations, which are discussed in Chapter 9. Here, we consider an example to illustrate the basic steps of the Simplex method that answer the foregoing two questions.

Reduced Cost Coefficients

Let the cost function be given only in terms of the nonbasic variables. The cost coefficients of the nonbasic variables are called the reduced cost coefficients, written as c_j'.

Cost Function in Terms of Nonbasic Variables

Before presentation of the example problem, an important requirement of the Simplex method is discussed. At the start of the Simplex iteration, the *cost function must be in terms of the nonbasic variables only.* This is readily available, as we will see in the example problems that follow. Also, at the end of each Simplex iteration, the cost function must again be available in terms of the nonbasic variables only. This can be achieved if we write the cost function $\mathbf{c}^T\mathbf{x} = f$ as the last row of the tableau and perform elimination of the basic variable in that row as well. The coefficients in the nonbasic columns of the last row are thus the *reduced cost coefficients* c_j'.

The cost function must be in terms of the nonbasic variables only in order to check the following two conditions:

Optimality of the current basic feasible solution: If all the reduced cost coefficients c_j' are non-negative (≥ 0), then the current basic feasible solution is optimum.
Determination of a nonbasic variable to become basic: If the current point is not optimum (i.e., all the reduced cost coefficients c_j' are not non-negative), then a negative coefficient c_j' determines which nonbasic variable needs to become basic to further reduce the cost function.

Example 8.7 describes and illustrates the steps of the Simplex method in a systematic way.

EXAMPLE 8.7 STEPS OF THE SIMPLEX METHOD

Solve the following LP problem using the Simplex method:

Maximize

$$z = 2x_1 + x_2 \tag{a}$$

subject to

$$-4x_1 - 3x_2 \geq -12 \tag{b}$$

$$4 \geq 2x_1 + x_2 \tag{c}$$

$$x_1 + 2x_2 \leq 4 \tag{d}$$

$$x_1, x_2 \geq 0 \tag{e}$$

Solution

The graphical solution to the problem is given in Figure 8.3. It can be seen that the problem has an infinite number of solutions along the line C–D ($z^* = 4$) because the objective function is parallel to the second constraint. The Simplex method is illustrated in the following steps.

1. *Convert the problem to the standard form.* To convert the problem to the standard LP form, we first rewrite the constraints so that their right sides are non-negative. To accomplish this, the first constraint is written as $4x_1 + 3x_2 \leq 12$ and the second constraint is written as

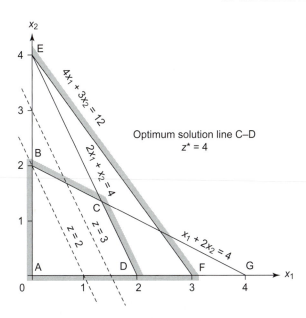

FIGURE 8.3 Graphical solution to the LP problem of Example 8.7. Optimum solution along line C–D. $z^* = 4$.

$2x_1 + x_2 \leq 4$. Maximization of z is converted to minimization of $f = -2x_1 - x_2$. Since all constraints are now in the "$\leq$ form", we add the slack variables x_3, x_4, and x_5 to the constraints and write the problem in the standard form as

Minimize

$$f = -2x_1 - x_2 \tag{f}$$

subject to

$$4x_1 + 3x_2 + x_3 = 12 \tag{g}$$

$$2x_1 + x_2 + x_4 = 4 \tag{h}$$

$$x_1 + 2x_2 + x_5 = 4 \tag{i}$$

$$x_i \geq 0; \quad i = 1 \text{ to } 5 \tag{j}$$

We use the tableau and notation of Table 8.3, augmented with the cost function expression as the last row. The initial tableau for the problem is shown in Table 8.8, where the cost function expression $-2x_1 - x_2 = f$ is written as the last row. Note also that the cost function is in terms of only the nonbasic variables x_1 and x_2. *This is one of the basic requirements of the Simplex method—that the cost function always be in terms of the nonbasic variables.* When the cost function is only in terms of the nonbasic variables, then the cost coefficients in the last row are the *reduced cost coefficients*, written as c'_j. Note that the left-most column in Table 8.8 contains the basic variables associated with each constraint.

2. *Initial basic feasible solution.* To initiate the Simplex method, a basic feasible solution is needed. This is already available in Table 8.8 and is given as:

    Basic variables: $x_3 = 12$, $x_4 = 4$, $x_5 = 4$

TABLE 8.8   Initial tableau for the LP problem of Example 8.7

Basic↓	x_1	x_2	x_3	x_4	x_5	b
1. x_3	4	3	1	0	0	12
2. x_4	2	1	0	1	0	4
3. x_5	1	2	0	0	1	4
4. Cost function	$\boxed{-2}$	**−1**	0	0	0	f

Nonbasic variables: $x_1 = 0$, $x_2 = 0$
Cost function: $f = 0$

Note that the cost row gives $0 = f$ after substituting for x_1 and x_2. This solution represents point A in Figure 8.3, where none of the constraints is active except the non-negativity constraints on the variables.

3. *Optimality check.* We scan the cost row in Table 8.8, which should have nonzero entries only in the nonbasic columns (the x_1 and x_2 columns). Therefore, $c_1' = -2$, $c_2' = -1$. If all of the nonzero entries are non-negative, then we have an optimum solution because the cost function cannot be reduced any further and the Simplex method is terminated. Since there are negative entries in the nonbasic columns of the cost row, the current basic feasible solution is not optimum.

4. *Selection of a nonbasic variable to become basic.* We select a nonbasic column having a negative cost coefficient; that is, $c_1' = -2$ in the x_1 column. This identifies the nonbasic variable (x_1) associated with the selected column that should become basic. Thus, eliminations will be performed in the x_1 column; that is, it is the pivot column. This answers question 1 posed earlier: "How does one choose a current nonbasic variable that should become basic?" Note also that when there is more than one negative entry in the cost row, the variable tapped to become basic is arbitrary among the indicated possibilities. The usual convention is to select a variable associated with the smallest value in the cost row (or the negative element with the largest absolute value).

Notation: **The reduced cost coefficients in the nonbasic columns are set in bold. The boxed negative reduced cost coefficient in Table 8.8 indicates that the nonbasic variable associated with that column is selected to become basic variable.**

5. *Selection of a basic variable to become nonbasic.* To identify which current basic variable should become nonbasic (i.e., to select the pivot row), we take ratios of the right-side parameters with the positive elements in the selected pivot column (i.e., the x_1 column), as shown in Table 8.9. We identify the row having the smallest positive ratio (the second row). This will make the current basic variable x_4 a nonbasic variable. The pivot element is $a_{21} = 2$ (the intersection of the pivot row and the pivot column). This answers question 2 posed earlier: "Which variable from the current basic set should become nonbasic?" Selection of the row with the smallest ratio as the pivot row maintains the feasibility of the new basic solution (all $x_i \geq 0$). This is justified in Chapter 9.

Notation: **The selected pivot element is boxed, and the pivot column and row are shaded.**

6. *Pivot step.* We perform eliminations in column x_1 using row 2 as the pivot row to eliminate x_1 from rows 1, 3, and the cost row. Here, we use the steps illustrated in Example 8.6 as follows:

- Divide row 2 by 2, the pivot element.
- Multiply the new row 2 by 4 and subtract it from row 1 to eliminate x_1 from row 1.
- Subtract the new row 2 from row 3 to eliminate x_1 from row 3.
- Multiply the new row 2 by 2 and add it to the cost row to eliminate x_1.

As a result of this elimination step, a new tableau is obtained, as shown in Table 8.10. The new basic feasible solution is given as

> Basic variables: $x_3 = 4$, $x_1 = 2$, $x_5 = 2$
> Nonbasic variables: $x_2 = 0$, $x_4 = 0$
> Cost function: $0 = f + 4$, $f = -4$ ($z = 4$)

7. *Optimum solution.* This solution is identified as point D in Figure 8.3. We see that the cost function has been reduced from 0 to -4. The coefficients in the nonbasic columns of the last row are non-negative $c_2' = 0$, $c_4' = 1$, so no further reduction of the cost function is possible. Thus, the foregoing solution is the optimum point. Note that for this example, only one iteration of the Simplex method gave the optimum solution. In general, more iterations are needed until all coefficients in the cost row become non-negative.

TABLE 8.9   Selection of pivot column and pivot row for Example 8.7

Basic↓	x_1	x_2	x_3	x_4	x_5	b	Ratio: b_i/a_{i1}; $a_{i1} > 0$
1. x_3	4	3	1	0	0	12	$12/4 = 3$
2. x_4	2	1	0	1	0	4	$4/2 = 2 \leftarrow$ smallest
3. x_5	1	2	0	0	1	4	$4/1 = 4$
4. Cost function	−2	−1	0	0	0	f	

The selected pivot element is boxed. Selected pivot row and column are highlighted. x_1 should become basic (pivot column). x_4 row has the smallest ratio, and so x_4 should become nonbasic (pivot row).

TABLE 8.10   Second tableau for Example 8.7 making x_1 a basic variable

Basic↓	x_1	x_2	x_3	x_4	x_5	b
1. x_3	0	1	1	−2	0	4
2. x_1	1	0.5	0	0.5	0	2
3. x_5	0	1.5	0	−0.5	1	2
4. Cost function	0	0	0	1	0	f + 4

The cost coefficients in nonbasic columns are non-negative; the tableau gives the optimum solution.

TABLE 8.11   Result of improper pivoting in Simplex method for LP problem

Basic↓	x_1	x_2	x_3	x_4	x_5	b
1. x_3	0	−5	1	0	−4	−4
2. x_4	0	−3	0	1	−2	−4
3. x_1	1	2	0	0	1	4
4. Cost function	0	3	0	0	4	$f+8$

The pivot step making x_1 basic and x_5 nonbasic in Table 8.8 gives a basic solution that is not feasible.

Consequence of zero reduced cost coefficient

The cost coefficient corresponding to the nonbasic variable x_2 in the last row is zero in the final tableau. This is an indication of multiple solutions to the problem. In general, when the reduced cost coefficient in the last row corresponding to a nonbasic variable is zero, the problem may have multiple solutions. We will discuss this point later in more detail.

Consequence of incorrect pivot selection

Let us see what happens if we do not select a row with the smallest ratio as the pivot row. Let $a_{31} = 1$ in the third row be the pivot element in Table 8.8. This will interchange the nonbasic variable x_1 with the basic variable x_5. Performing the elimination steps in the first column as explained earlier, we obtain the new tableau given in Table 8.11. From the tableau, we have

Basic variables: $x_3 = -4$, $x_4 = -4$, $x_1 = 4$
Nonbasic variables: $x_2 = 0$, $x_5 = 0$
Cost function: $0 = f + 8$, $f = -8$

The foregoing solution corresponds to point G in Figure 8.3. We see that this basic solution is not feasible because x_3 and x_4 have negative values. Thus, *we conclude that if a row with the smallest ratio (of right sides with positive elements in the pivot column) is not selected in Step 5, the new basic solution is not feasible.*

Note that a spreadsheet program, such as Excel, can be used to carry out the pivot step. Such a program can facilitate learning of the Simplex method without getting bogged down with the manual elimination process.

8.5.3 Basic Theorems of Linear Programming

In the previous subsection, the basic steps of the Simplex method were explained and illustrated with an example problem. In this subsection, the underlying principles for these steps are summarized in two theorems, the basic theorems of linear programming. We have seen that, in general, the reduced cost coefficients c_j' of the nonbasic variables may be positive, negative, or zero.

Let one c_j' be negative; then if a positive value is assigned to the associated nonbasic variable (i.e., it is made basic), the value of f will decrease. If more than one negative c_j' is present, a widely used rule of thumb is to choose the nonbasic variable associated with the smallest c_j' (i.e., the most negative) to become basic.

Thus, if any c_j' for nonbasic variables is negative, then it is possible to find a new basic feasible solution (if one exists) that will further reduce the cost function. If a c_j' is zero, then the associated nonbasic variable can be made basic without affecting the cost function value (a multiple-solution case, such as in Example 8.7). If all c_j' are non-negative, then it is not possible to reduce the cost function any further, and the current basic feasible solution is optimum. These ideas are summarized in the following Theorems 8.3 and 8.4.

THEOREM 8.3

Improvement of the Basic Feasible Solution Given a nondegenerate basic feasible solution with the corresponding cost function f_0, suppose that $c_j' < 0$ for some j.

1. *Improved basic feasible solution.* There is a feasible solution with $f < f_0$. If the jth nonbasic column associated with c_j' can be substituted for some column in the original basis, the new basic feasible solution will have $f < f_0$.

2. *Unbounded cost function.* If the jth column cannot be substituted to yield a basic feasible solution (i.e., there is no positive element in the jth column), then the feasible set is unbounded and the cost function can be made arbitrarily small (toward negative infinity).

THEOREM 8.4

Optimum Solutions to Linear Programming Problems If a basic feasible solution has reduced cost coefficients $c_j' \geq 0$ for all j, then it is optimum.

According to Theorem 8.3, the basic procedure of the Simplex method is to start with an initial basic feasible solution, that is, at the vertex of the convex polyhedron. If this solution is not optimum according to Theorem 8.4, then a move is made to an adjacent vertex to reduce the cost function. The procedure is continued until the optimum is reached.

Multiple Solutions

Note that when all reduced cost coefficients c_j' in the nonbasic columns are strictly positive, the optimum solution is unique. If at least one c_j' is zero in a nonbasic column, there is the possibility of an alternate optimum solution. If the nonbasic variable associated with a zero reduced cost coefficient can be made basic by using the foregoing pivot step procedure, the extreme point (vertex) corresponding to an alternate optimum is obtained. Since the reduced cost coefficient is zero, the optimum cost function value does not change. Any point on the line segment joining the optimum extreme points also corresponds to an optimum point. Note that all of these optima are global as opposed to local, although there is no distinct global optimum. Geometrically, multiple optima for an LP problem imply that the cost function hyperplane is parallel to one of the constraint hyperplanes.

Examples 8.8 and 8.9 show how to obtain a solution for an LP problem using the Simplex method. Example 8.10 shows how to recognize multiple solutions for the problem, and Example 8.11 shows how to recognize an unbounded cost function.

EXAMPLE 8.8 SOLUTION BY THE SIMPLEX METHOD

Using the Simplex method, in order to find the optimum (if one exists) for the LP problem of Example 8.3,

Minimize

$$f = -4x_1 - 5x_2 \tag{a}$$

subject to

$$-x_1 + x_2 + x_3 = 4 \tag{b}$$

$$x_1 + x_2 + x_4 = 6 \tag{c}$$

$$x_i \geq 0; \quad i = 1 \text{ to } 4 \tag{d}$$

Solution

Writing the problem in the Simplex tableau, we obtain the initial tableau shown in Table 8.12. From the initial tableau, the basic feasible solution is

Basic variables: $x_3 = 4$, $x_4 = 6$
Nonbasic variables: $x_1 = x_2 = 0$
Cost function: $f = 0$ from the last row of the tableau

Note that the cost function in the last row of the initial tableau is in terms of only the nonbasic variables x_1 and x_2. Thus, the coefficients in the x_1 and x_2 columns and the last row are the reduced cost coefficients c'_j. Scanning the last row, we observe that there are negative coefficients. Therefore, the current basic feasible solution is not optimum. In the last row, the most negative coefficient of -5 corresponds to the second column. Therefore, we select x_2 to become a basic variable; that is, elimination should be performed in the x_2 column. Now, taking the ratios of the right-side parameters with positive coefficients in the second column b_i/a_{i2}, we obtain a minimum ratio for the first row as 4. This identifies the first row as the pivot row according to Step 3 of the Simplex method. Therefore, the current basic variable associated with the first row, x_3, should become nonbasic.

Performing the pivot step on column 2 with a_{12} as the pivot element, we obtain the second tableau, as shown in Table 8.12. For this canonical form, the basic feasible solution is

Basic variables: $x_2 = 4$, $x_4 = 2$
Nonbasic variables: $x_1 = x_3 = 0$

The cost function is $f = -20$ $(0 = f + 20)$, which is an improvement over $f = 0$. Thus, this pivot step results in a move from (0, 0) through (0, 4) on the convex polyhedron of Figure 8.2.

The reduced cost coefficient corresponding to the nonbasic column x_1 is still negative in the second tableau. Therefore, the cost function can be reduced further. Repeating the above-mentioned process for the second tableau, we obtain $a_{21} = 2$ as the pivot element, implying that x_1 should become basic

TABLE 8.12 Solution of Example 8.8 by the Simplex method

Initial tableau: x_3 is identified to be replaced with x_2 in the basic set

Basic↓	x_1	x_2	x_3	x_4	b	Ratio: b_i/a_{i2}
1. x_3	−1	☐1	1	0	4	$\frac{4}{1}=4\leftarrow$ smallest
2. x_4	1	1	0	1	6	$\frac{6}{1}=6$
3. Cost	−4	☐−5	0	0	f	

Second tableau: x_4 is identified to be replaced with x_1 in the basic set

Basic↓	x_1	x_2	x_3	x_4	b	Ratio: b_i/a_{i1}
4. x_2	−1	1	1	0	4	Negative
5. x_4	☐2	0	−1	1	2	$\frac{2}{2}=1$
6. Cost	☐−9	0	5	0	$f+20$	

Third tableau: Reduced cost coefficients in nonbasic columns are nonnegative; the tableau gives optimum point

Basic↓	x_1	x_2	x_3	x_4	b	Ratio: b_i/a_{iq}
7. x_2	0	1	$\frac{1}{2}$	$\frac{1}{2}$	5	Not needed
8. x_1	1	0	$-\frac{1}{2}$	$\frac{1}{2}$	1	Not needed
9. Cost	0	0	$\frac{1}{2}$	$\frac{9}{2}$	$f+29$	

and x_4 should become nonbasic. After the pivot operation, the third tableau is obtained, as shown in Table 8.12. For this tableau, all the reduced cost coefficients c_j' (corresponding to the nonbasic variables) in the last row are ≥ 0. Therefore, the tableau yields the optimum solution:

Basic variables: $x_1 = 1$, $x_2 = 5$
Nonbasic variables: $x_3 = 0$, $x_4 = 0$
Cost function: $f = -29$ $(f + 29 = 0)$

This solution corresponds to the point C (1, 5) in Figure 8.2.

EXAMPLE 8.9 SOLUTION TO PROFIT MAXIMIZATION PROBLEM BY THE SIMPLEX METHOD

Use the Simplex method to find the optimum solution for the profit maximization problem in Example 8.2.

Solution

Introducing slack variables in the constraints of Eqs. (c) through (e) in Example 8.2, we get the LP problem in the standard form:

Minimize

$$f = -400x_1 - 600x_2 \tag{a}$$

subject to

$$x_1 + x_2 + x_3 = 16 \tag{b}$$

$$\frac{1}{28}x_1 + \frac{1}{14}x_2 + x_4 = 1 \tag{c}$$

$$\frac{1}{14}x_1 + \frac{1}{24}x_2 + x_5 = 1 \tag{d}$$

$$x_i \geq 0; \quad i = 1 \text{ to } 5 \tag{e}$$

Now, writing the problem in the Simplex tableau, we obtain the initial tableau as shown in Table 8.13. Thus the initial basic feasible solution is $x_1 = 0$, $x_2 = 0$, $x_3 = 16$, $x_4 = x_5 = 1$, $f = 0$, which corresponds to point A in Figure 8.1. The initial cost function is zero, and x_3, x_4, and x_5 are the basic variables.

Using the Simplex procedure, we note that $a_{22} = 1/14$ is the pivot element in the initial tableau. This implies that x_4 should be replaced by x_2 in the basic set. Carrying out the pivot operation using the second row as the pivot row, we obtain the second tableau (canonical form)

TABLE 8.13   Solution of Example 8.9 by the Simplex method

Initial tableau: x_4 is identified to be replaced with x_2 in the basic set

Basic↓	x_1	x_2	x_3	x_4	x_5	b	Ratio: b_i/a_{i2}
1. x_3	1	1	1	0	0	16	$\frac{16}{1} = 16$
2. x_4	$\frac{1}{28}$	$\frac{1}{14}$	0	1	0	1	$\frac{1}{1/14} = 14 \leftarrow$ smallest
3. x_5	$\frac{1}{14}$	$\frac{1}{24}$	0	0	1	1	$\frac{1}{1/24} = 24$
4. Cost	−400	−600	0	0	0	$f - 0$	

Second tableau: x_3 is identified to be replaced with x_1 in the basic set

Basic↓	x_1	x_2	x_3	x_4	x_5	b	Ratio: b_i/a_{i1}
5. x_3	$\frac{1}{2}$	0	1	−14	0	2	$\frac{2}{1/2} = 4 \leftarrow$ smallest
6. x_2	$\frac{1}{2}$	1	0	14	0	14	$\frac{14}{1/2} = 28$
7. x_5	$\frac{17}{336}$	0	0	$-\frac{7}{12}$	1	$\frac{5}{12}$	$\frac{5/12}{17/336} = \frac{140}{17}$
8. Cost	−100	0	0	8400	0	$f + 8400$	

Third tableau: Reduced cost coefficients in the nonbasic columns are non-negative; the tableau gives the optimum solution

Basic↓	x_1	x_2	x_3	x_4	x_5	b	Ratio
9. x_1	1	0	2	−28	0	4	Not needed
10. x_2	0	1	−1	28	0	12	Not needed
11. x_5	0	0	$-\frac{17}{168}$	$\frac{5}{6}$	1	$\frac{3}{14}$	Not needed
12. Cost	0	0	200	5600	0	$f + 8800$	

shown in Table 8.13. At this point the basic feasible solution is $x_1 = 0$, $x_2 = 14$, $x_3 = 2$, $x_4 = 0$, $x_5 = 5/12$, which corresponds to point E in Figure 8.1. The cost function is reduced to -8400.

The pivot element for the next iteration is $a_{11} = 1/2$ in the second tableau, implying that x_3 should be replaced by x_1 in the basic set. Carrying out the pivot operation, we obtain the third tableau shown in Table 8.13. At this point, all reduced cost coefficients (corresponding to nonbasic variables) are non-negative ($c_3' = 200$, $c_4' = 5600$), so according to Theorem 8.4 we have the optimum solution:

Basic variables: $x_1 = 4$, $x_2 = 12$, $x_5 = 3/14$
Nonbasic variables: $x_3 = 0$, $x_4 = 0$
Cost function: $f = -8800$

This corresponds to point D in Figure 8.1. Note that c_j', corresponding to the nonbasic variables x_3 and x_4, are positive. Therefore, the global optimum solution is unique, as may be observed in Figure 8.1 as well.

The problem in Example 8.10 has multiple solutions. The example illustrates how to recognize such solutions with the Simplex method.

EXAMPLE 8.10 LP PROBLEM WITH MULTIPLE SOLUTIONS

Solve the following problem with the Simplex method:

Maximize
$$z = x_1 + 0.5x_2 \tag{a}$$

subject to
$$2x_1 + 3x_2 \leq 12 \tag{b}$$
$$2x_1 + x_2 \leq 8 \tag{c}$$
$$x_1, x_2 \geq 0 \tag{d}$$

Solution

The problem was solved graphically in Section 3.4 3. It has *multiple solutions*, as could be seen in Figure 3.7. We will solve the problem using the Simplex method and discuss how multiple solutions can be recognized for general LP problems. The problem is converted to standard LP form:

Minimize
$$f = -x_1 - 0.5x_2 \tag{e}$$

subject to
$$2x_1 + 3x_2 + x_3 = 12 \tag{f}$$
$$2x_1 + x_2 + x_4 = 8 \tag{g}$$
$$x_i \geq 0; \quad i = 1 \text{ to } 4 \tag{h}$$

Here x_3 is the slack variable for the first constraint and x_4 is the slack variable for the second constraint. Table 8.14 contains iterations of the Simplex method. The optimum point is reached in just one iteration because all the reduced cost coefficients are non-negative in the second tableau. The optimum solution is given as

Basic variables: $x_1 = 4$, $x_3 = 4$
Nonbasic variables: $x_2 = x_4 = 0$
Optimum cost function: $f = -4$ $(z = 4)$

This solution corresponds to point B in Figure 3.7.

In the second tableau, the *reduced cost coefficient for the nonbasic variable x_2 is zero*. This means that it is possible to make x_2 basic without any change in the optimum cost function value. This suggests the existence of *multiple optimum solutions*.

Performing the pivot operation in the x_2 column, we find another solution, given in the third tableau of Table 8.14:

Basic variables: $x_1 = 3$, $x_2 = 2$
Nonbasic variables: $x_3 = x_4 = 0$
Optimum cost function: $f = -4$ $(z = 4)$

TABLE 8.14   Solution by the Simplex method for Example 8.10

Initial tableau: x_4 is identified to be replaced with x_1 in the basic set

Basic↓	x_1	x_2	x_3	x_4	b	Ratio: b_i/a_{i1}
1. x_3	2	3	1	0	12	$\frac{12}{2} = 6$
2. x_4	2	1	0	1	8	$\frac{8}{2} = 4 \leftarrow$ smallest
3. Cost	−1	−0.5	0	0	$f - 0$	

Second tableau: First optimum point; reduced cost coefficients in nonbasic columns are non-negative; the tableau gives optimum solution; $c_2' = 0$ indicates possibility of multiple solutions; x_3 is identified to be replaced with x_2 in the basic set to obtain another optimum point

Basic↓	x_1	x_2	x_3	x_4	b	Ratio: b_i/a_{i2}
4. x_3	0	2	1	−1	4	$\frac{4}{2} = 2 \leftarrow$ smallest
5. x_1	1	$\frac{1}{2}$	0	$\frac{1}{2}$	4	$\frac{4}{1/2} = 8$
6. Cost	0	0	0	$\frac{1}{2}$	$f + 4$	

Third tableau: Second optimum point

Basic↓	x_1	x_2	x_3	x_4	b	Ratio
7. x_2	0	1	$\frac{1}{2}$	$-\frac{1}{2}$	2	Not needed
8. x_1	1	0	$-\frac{1}{4}$	$\frac{3}{4}$	3	Not needed
9. Cost	0	0	0	$\frac{1}{2}$	$f + 4$	

This solution corresponds to point C on Figure 3.7. Note that any point on the line B–C also gives an optimum solution. Multiple solutions can occur when the cost function is parallel to one of the constraints. For the present example, the cost function is parallel to the second constraint, which is active at the optimum.

In general, if a reduced cost coefficient corresponding to a nonbasic variable is zero in the final tableau, multiple optimum solutions are possible. From a practical standpoint, this is not a bad situation. Actually, it may be desirable because it gives the designer options; any suitable point on the straight line joining the two optimum designs can be selected to better suit his or her needs. Note that all optimum design points are global solutions as opposed to local solutions.

Example 8.11 demonstrates how to recognize an unbounded solution for a problem.

EXAMPLE 8.11 IDENTIFICATION OF AN UNBOUNDED SOLUTION WITH THE SIMPLEX METHOD

Solve the LP problem:

Maximize

$$z = x_1 - 2x_2 \tag{a}$$

subject to

$$2x_1 - x_2 \geq 0 \tag{b}$$

$$-2x_1 + 3x_2 \leq 6 \tag{c}$$

$$x_1, x_2 \geq 0 \tag{d}$$

Solution

The problem was solved graphically in Section 3.5. It can be seen from Figure 3.8 that the solution is unbounded; the feasible set is not closed. We will solve the problem using the Simplex method and see how we can recognize unbounded problems.

Writing the problem in the standard Simplex form, we obtain the initial tableau shown in Table 8.15, where x_3 and x_4 are the slack variables (note that the first constraint has been transformed as $-2x_1 + x_2 \leq 0$). The basic feasible solution is

Basic variables: $x_3 = 0$, $x_4 = 6$

TABLE 8.15 Initial canonical form for Example 8.11 (unbounded solution)

Basic↓	x_1	x_2	x_3	x_4	b
1. x_3	−2	1	1	0	0
2. x_4	−2	3	0	1	6
3. Cost	−1	2	0	0	$f - 0$

Nonbasic variables: $x_1 = x_2 = 0$
Cost function: $f = 0$

Scanning the last row, we find that the reduced cost coefficient for the nonbasic variable x_1 is negative. Therefore, x_1 should become a basic variable. However, a pivot element cannot be selected in the x_1 column because there is no positive element. There is no other possibility of selecting another nonbasic variable to become basic; the reduced cost coefficient for x_2 (the other nonbasic variable) is positive. Therefore, no pivot steps can be performed, and yet we are not at the optimum point. Thus, the feasible set for the problem is unbounded. The foregoing observation will be true in general. For *unbounded solutions*, there will be negative reduced cost coefficients for nonbasic variables but no possibility of pivot steps.

8.6 THE TWO-PHASE SIMPLEX METHOD—ARTIFICIAL VARIABLES

The basic Simplex method of Section 8.5 is extended to handle "≥ type" and equality constraints. A basic feasible solution is needed to initiate the Simplex solution process. Such a solution is immediately available if only "≤ type" constraints are present. However, for the "≥ type" and/or equality constraints, an initial basic feasible solution is not available. To obtain such a solution, we introduce artificial variables for the "≥ type" and equality constraints, define an auxiliary minimization LP problem, and solve it. The standard Simplex method can still be used to solve the auxiliary problem. This is called Phase I of the Simplex procedure. At the end of Phase I, a basic feasible solution for the original problem becomes available. Phase II then continues to find a solution to the original LP problem. We will illustrate the method with examples.

8.6.1 Artificial Variables

When there are "≥ type" constraints in the linear programming problem, surplus variables are subtracted from them to transform the problem into the standard form. The equality constraints, if present, are not changed because they are already in the standard form. For such problems, an initial basic feasible solution cannot be obtained by selecting the original design variables as nonbasic (setting them to zero), as is the case when there are only "≤ type" constraints (e.g., for all of the examples in Section 8.5). To obtain an initial basic feasible solution, the Gauss-Jordan elimination procedure can be used to convert $\mathbf{Ax} = \mathbf{b}$ to the canonical form.

However, an easier way is to introduce non-negative auxiliary variables for the "≥ type" and equality constraints, define an auxiliary LP problem, and solve it using the Simplex method. The auxiliary variables are called *artificial variables*. These are variables in addition to the surplus variables for the "≥ type" constraints. They have no physical meaning, but with their addition we obtain an initial basic feasible solution for the auxiliary LP problem by treating the artificial variables as basic along with any slack variables for the "≤ type" constraints. All other variables are treated as nonbasic (i.e., set to zero).

Example 8.12 illustrates the process of adding artificial variables for "≥ type" and equality constraints.

EXAMPLE 8.12 INTRODUCTION OF ARTIFICIAL VARIABLES

Introduce artificial variables and obtain an initial basic feasible solution for the following LP problem:

Maximize

$$z = x_1 + 4x_2 \tag{a}$$

subject to

$$x_1 + 2x_2 \leq 5 \tag{b}$$

$$2x_1 + x_2 = 4 \tag{c}$$

$$-x_1 + x_2 \leq -1 \tag{d}$$

$$x_1 \geq 0; \quad x_2 \geq 0 \tag{e}$$

Solution

To transform the problem into the standard form, we multiply the inequality in Eq. (d) by -1 to make the right side of the constraint non-negative as $x_1 - x_2 \geq 1$. Now we introduce slack and surplus variables for the constraint in Eq. (b) and Eq. (d), respectively, and obtain the problem in the standard form as

Minimize

$$f = -x_1 - 4x_2 \tag{f}$$

subject to

$$x_1 + 2x_2 + x_3 = 5 \tag{g}$$

$$2x_1 + x_2 = 4 \tag{h}$$

$$x_1 - x_2 - x_4 = 1 \tag{i}$$

$$x_i \geq 0; \quad i = 1 \text{ to } 4 \tag{j}$$

where $x_3 \geq 0$ is a slack variable and $x_4 \geq 0$ is a surplus variable.

The linear system $\mathbf{Ax} = \mathbf{b}$ in Eqs. (g) through (i) is not in the canonical form; Therefore, a basic feasible solution is not available to initiate the Simplex method. We introduce non-negative artificial variables $x_5 \geq 0$ and $x_6 \geq 0$ in Eqs (h) and (i) and rewrite the constraints in Eqs. (g) through (i) as

$$x_1 + 2x_2 + x_3 = 5 \tag{k}$$

$$2x_1 + x_2 + x_5 = 4 \tag{l}$$

$$x_1 - x_2 - x_4 + x_6 = 1 \tag{m}$$

The linear system in Eqs. (k) through (m) is in the canonical form as displayed in Table 8.16. x_3, x_5, and x_6 are noted as the identity columns in the system $\mathbf{Ax} = \mathbf{b}$. Therefore, x_3, x_5, and x_6 are identified as the basic variables, and the remaining ones are identified as the nonbasic variables. Association of the basic variables with the rows of the tableau depends on the location of the unit element in the identity columns as seen in Table 8.16. The initial basic feasible solution for the auxiliary problem from Table 8.16 is given as

Basic variables: $x_3 = 5$, $x_5 = 4$, $x_6 = 1$
Nonbasic variables: $x_1 = 0$, $x_2 = 0$, $x_4 = 0$

TABLE 8.16 Initial basic feasible solution for Example 8.12

Basic↓	x_1	x_2	x_3	x_4	x_5	x_6	b
1. x_3	1	2	1	0	0	0	5
2. x_5	2	1	0	0	1	0	4
3. x_6	1	−1	0	−1	0	1	1

8.6.2 Artificial Cost Function

The artificial variable for each equality and "≥ type" constraint is introduced to obtain an initial basic feasible solution for the auxiliary problem. These variables have no physical meaning and need to be eliminated from the problem. To eliminate the artificial variables from the problem, we define an auxiliary cost function called the *artificial cost function* and minimize it subject to the constraints of the problem and the non-negativity of all of the variables. The artificial cost function is simply a sum of all the artificial variables and will be designated as w:

$$w = \sum \text{(all artificial variables)} \tag{8.18}$$

As an example, the artificial cost function in Example 8.12 is given as

$$w = x_5 + x_6 \tag{8.19}$$

8.6.3 Definition of the Phase I Problem

Since the artificial variables are introduced simply to obtain an initial basic feasible solution for the original problem, they need to be eliminated eventually. This elimination is done by defining and solving an LP problem called the Phase I problem. The objective of this problem is to make all the artificial variables nonbasic so they have zero value. In that case, the artificial cost function in Eq. (8.18) will be zero, indicating the end of Phase I.

Thus, the Phase I problem is to minimize the artificial cost function in Eq. (8.18) subject to all of the constraints of the problem. However, it is not yet in a form suitable to initiate the Simplex method. The reason is that the reduced cost coefficients c_j' of the nonbasic variables in the artificial cost function are not yet available to determine the pivot element and perform the pivot step. Currently, the artificial cost function in Eq. (8.18) is in terms of basic variables, such as the one in Eq. (8.19). Therefore, the reduced cost coefficients c_j' cannot be identified. They can be identified only if the artificial cost function w is in terms of nonbasic variables.

To obtain w in terms of nonbasic variables, we use the constraint expressions to eliminate the basic variables from the artificial cost function. For example, we substitute for x_5 and x_6 from Eqs. (l) and (m) in Example 8.12 and obtain the artificial cost function of Eq. (8.19) in terms of the nonbasic variables as

$$w = (4 - 2x_1 - x_2) + (1 - x_1 + x_2 + x_4) = 5 - 3x_1 + x_4 \tag{8.20}$$

If there are also "$\leq$ type" constraints in the original problem, these are cast into the standard LP form by adding slack variables that serve as basic variables in Phase I. Therefore, the number of artificial variables is less than or at the most equal to the total number of constraints.

8.6.4 Phase I Algorithm

The standard Simplex procedure described in Section 8.5 can now be employed to solve the auxiliary optimization problem of Phase I. During this phase, the artificial cost function is used to determine the pivot element. The original cost function is treated as a constraint, and the elimination step is also executed for it. This way, the real cost function is only in terms of the nonbasic variables at the end of Phase I, and the Simplex method can be continued during Phase II. All artificial variables should become nonbasic at the end of Phase I. Since w is the sum of all of the artificial variables, its minimum value is clearly zero. When $w = 0$, an extreme point of the original feasible set is reached; w is then discarded in favor of f and iterations continue in Phase II until the minimum of f is obtained. Example 8.13 illustrates calculations for Phase I of the Simplex method.

Infeasible Problem

Suppose, however, that w cannot be driven to zero at the end of Phase I. This will be apparent when none of the reduced cost coefficients for the artificial cost function are negative and yet w is greater than zero. Clearly, this means that we cannot reach the original feasible set and, therefore, *no feasible solution exists for the original design problem; that is, it is an infeasible problem*. At this point the designer should re-examine the formulation of the problem, which may be over-constrained or improperly formulated.

EXAMPLE 8.13 PHASE I OF THE SIMPLEX METHOD

For Example 8.12, complete Phase I of the Simplex method.

Solution

Using the constraints given in Eqs. (k) through (m) in Example 8.12, the initial tableau for the problem is set up as shown in Table 8.17. The artificial cost function of Eq. (8.20) is written as the last row of the tableau, and the real cost function of Eq. (f) is written as the next-to-last row.

Now the Simplex iterations are carried out using the artificial cost row to determine the pivot element. The Gauss-Jordan elimination procedure, as demonstrated earlier, is used to complete the pivot operations. In the initial tableau, the element $a_{31} = 1$ is identified as the pivot element. The pivot operation is performed in column x_1 using row 3 as the pivot row. With this pivot operation, the artificial variable x_6 is replaced by x_1 in the basic set. Thus the artificial variable x_6 becomes nonbasic and assumes a zero value.

In the second tableau, $a_{22} = 3$ is identified as the pivot element. Thus eliminations are performed in the x_2 column using row 7 as the pivot row. At the end of the second pivot operation, x_5 is replaced by x_2 in the basic set; that is, the artificial variable x_5 assumes a zero value. Since both of the artificial variables have become nonbasic, the artificial cost function has attained a

TABLE 8.17   Phase I of Simplex method for Example 8.13

Initial tableau: x_6 is identified to be replaced with x_1 in the basic set

Basic↓	x_1	x_2	x_3	x_4	x_5	x_6	b	Ratio
1. x_3	1	2	1	0	0	0	5	5/1
2. x_5	2	1	0	0	1	0	4	4/2
3. x_6	1	−1	0	−1	0	1	1	1/1
4. Cost	−1	−4	0	0	0	0	$f-0$	
5. Artificial	−3	0	0	1	0	0	$w-5$	

Second tableau: x_5 is identified to be replaced with x_2 in the basic set

Basic↓	x_1	x_2	x_3	x_4	x_5	x_6	b	Ratio
6. x_3	0	3	1	1	0	−1	4	4/3
7. x_5	0	3	0	2	1	−2	2	2/3
8. x_1	1	−1	0	−1	0	1	1	Negative
9. Cost	0	−5	0	−1	0	1	$f+1$	
10. Artificial	0	−3	0	−2	0	3	$w-2$	

Third tableau: Reduced cost coefficients in nonbasic columns are non-negative; the tableau gives Optimum point. End of Phase I

Basic↓	x_1	x_2	x_3	x_4	x_5	x_6	b
11. x_3	0	0	1	−1	−1	1	2
12. x_2	0	1	0	$\frac{2}{3}$	$\frac{1}{3}$	$-\frac{2}{3}$	$\frac{2}{3}$
13. x_1	1	0	0	$-\frac{1}{3}$	$\frac{1}{3}$	$\frac{1}{3}$	$\frac{5}{3}$
14. Cost	0	0	0	$\frac{7}{3}$	$\frac{5}{3}$	$-\frac{7}{3}$	$f+\frac{13}{3}$
15. Artificial	0	0	0	0	1	1	$w-0$

x_3, slack variable; x_4, surplus variable; x_5, x_6, artificial variables.

zero value. This indicates the end of Phase I of the Simplex method. An examination of the artificial cost row shows that the reduced cost coefficients in the nonbasic columns x_4, x_5, and x_6 are non-negative, which also indicates the end of Phase I.

The basic feasible solution for the original problem is reached as

Basic variables: $x_1 = 5/3$, $x_2 = 2/3$, $x_3 = 2$
Nonbasic variables: $x_4 = x_5 = x_6 = 0$
Cost function: $f = -13/3$

Now we can discard the artificial cost row and the artificial variable columns x_5 and x_6 in Table 8.17 and continue with the Simplex iterations using the real cost row to determine the pivot column.

8.6.5 Phase II Algorithm

In the final tableau from Phase I, the artificial cost row is replaced by the actual cost function equation and the Simplex iterations continue based on the algorithm explained in Section 8.5. The basic variables, however, should not appear in the cost function. Thus, the pivot steps need to be performed on the cost function equation to eliminate the basic variables from it. A convenient way of accomplishing this is to treat the cost function as one of the equations in the Phase I tableau, say the second equation from the bottom. Elimination is performed on this equation along with the others. In this way, the cost function is in the correct form to continue with Phase II.

The foregoing procedure was used in Example 8.13 and Table 8.17. The real cost was written before the artificial cost row, and eliminations were performed in the cost row as well. It took two iterations of the Simplex method to solve the Phase I linear programming problem.

Examining the real cost row in the third tableau in Table 8.17, we observe that the reduced cost coefficient in the nonbasic column x_4 is positive (artificial columns x_5 and x_6 are ignored in Phase II). This indicates the end of Phase II of the algorithm, and the solution obtained from the third tableau is the optimum point. Note that the artificial variable column (x_6) is the negative of the surplus variable column (x_4) for the "$\geq$ type" constraint. This is true in general and can serve as a check for correctness of the Simplex iterations.

The artificial variable columns can also be discarded for Phase II calculations. However, they are kept in the tableau because they provide useful information useful for postoptimality analysis.

Features of the Simplex method
1. **If there is a solution to the LP problem, the method finds it (Example 8.14).**
2. **If the problem is infeasible, the method indicates that (Example 8.15).**
3. **If the problem is unbounded, the method indicates that (Examples 8.11 and 8.16).**
4. **If there are multiple solutions, the method indicates that (Examples 8.7 and 8.10).**

EXAMPLE 8.14 USE OF ARTIFICIAL VARIABLES FOR "$\geq$ TYPE" CONSTRAINTS

Find the optimum solution for the following LP problem using the Simplex method:

Maximize

$$z = y_1 + 2y_2 \tag{a}$$

subject to

$$3y_1 + 2y_2 \leq 12 \tag{b}$$

$$2y_1 + 3y_2 \geq 6 \tag{c}$$

$$y_1 \geq 0, \ y_2 \text{ is unrestricted in sign.} \tag{d}$$

Solution

The graphical solution to the problem is shown in Figure 8.4. It can be seen that the optimum solution is at point B. We will use the two-phase Simplex method to verify the solution. Since y_2 is free in sign, we decompose it as $y_2 = y_2^+ - y_2^-$. To write the problem in the standard form, we redefine the variables as $x_1 = y_1$, $x_2 = y_2^+$, and $x_3 = y_2^-$ and transform the problem:

Minimize
$$f = -x_1 - 2x_2 + 2x_3 \tag{e}$$

subject to
$$3x_1 + 2x_2 - 2x_3 + x_4 = 12 \tag{f}$$

$$2x_1 + 3x_2 - 3x_3 - x_5 = 6 \tag{g}$$

$$x_i \geq 0; \quad i = 1 \text{ to } 5 \tag{h}$$

where x_4 is a slack variable for the first constraint and x_5 is a surplus variable for the second constraint. It is seen that if we select the real variables as nonbasic (i.e., $x_1 = 0$, $x_2 = 0$, $x_3 = 0$), the resulting basic solution is infeasible because $x_5 = -6$. Therefore, we need to use the two-phase algorithm. Accordingly, we introduce an artificial variable x_6 in the second constraint of Eq. (g) as

$$2x_1 + 3x_2 - 3x_3 - x_5 + x_6 = 6 \tag{i}$$

The artificial cost function is defined as $w = x_6$. Since w must be in terms of nonbasic variables (x_6 is basic), we substitute for x_6 from Eq. (i) and obtain w as

$$w = x_6 = 6 - 2x_1 - 3x_2 + 3x_3 + x_5 \tag{j}$$

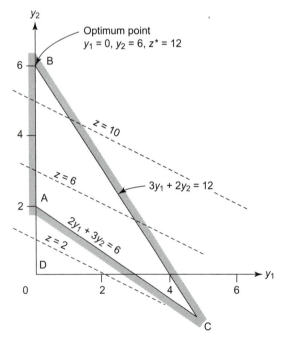

FIGURE 8.4 Graphical solution to the LP problem of Example 8.14.

The initial tableau for Phase I is shown in Table 8.18. The initial basic feasible solution is given as

Basic variables: $x_4 = 12$, $x_6 = 6$
Nonbasic variables: $x_1 = x_2 = x_3 = x_5 = 0$
Cost function: $f = 0$;
Artificial cost function: $w = 6$

This solution corresponds to the infeasible point D in Figure 8.4. According to the Simplex algorithm, the pivot element is identified as $a_{22} = 3$, which implies that x_2 should become basic and x_6 should become nonbasic. Performing the pivot operation in column x_2 using row 2 as the pivot row, we obtain the second tableau given in Table 8.18.

For the second tableau, $x_4 = 8$ and $x_2 = 2$ are the basic variables and all others are nonbasic. This corresponds to the feasible point A in Figure 8.4. Since the reduced cost coefficients of the artificial cost function in the nonbasic columns are non-negative and the artificial cost function is zero, an initial basic feasible solution for the original problem is obtained. Therefore, this is the end of Phase I.

For Phase II, column x_6 should be ignored while determining the pivot column. For the next step, the pivot element is identified as $a_{15} = 2/3$ in the second tableau according to the steps of

TABLE 8.18 Solution by the two-phase Simplex method for Example 8.14

Initial tableau: x_6 is identified to be replaced with x_2 in the basic set

Basic↓	x_1	x_2	x_3	x_4	x_5	x_6	b	Ratio
1. x_4	3	2	−2	1	0	0	12	$\frac{12}{2} = 6$
2. x_6	2	③	−3	0	−1	1	6	$\frac{6}{3} = 2$
3. Cost	−1	−2	2	0	0	0	$f - 0$	
4. Artificial cost	−2	−3	3	0	1	0	$w - 6$	

Second tableau: x_4 is identified to be replaced with x_5 in the basic set. End of Phase I

Basic↓	x_1	x_2	x_3	x_4	x_5	x_6	b	Ratio
5. x_4	$\frac{5}{3}$	0	0	1	$\frac{2}{3}$	$-\frac{2}{3}$	8	$\frac{8}{2/3} = 12$
6. x_2	$\frac{2}{3}$	1	−1	0	$-\frac{1}{3}$	$\frac{1}{3}$	2	Negative
7. Cost	$\frac{1}{3}$	0	0	0	$-\frac{2}{3}$	$\frac{2}{3}$	$f + 4$	
8. Artificial cost	0	0	0	0	0	1	$w - 0$	

Third tableau: Reduced cost coefficients in nonbasic columns are non-negative; the third tableau gives optimum solution. End of Phase II

Basic↓	x_1	x_2	x_3	x_4	x_5	x_6	b
9. x_5	$\frac{5}{2}$	0	0	$\frac{3}{2}$	1	−1	12
10. x_2	$\frac{3}{2}$	1	−1	$\frac{1}{2}$	0	0	6
11. Cost	2	0	0	1	0	0	$f + 12$

the Simplex method. This implies that x_4 should be replaced by x_5 as a basic variable in row 5. After the pivot operation, the third tableau is obtained, as shown in Table 8.18. This tableau yields an optimum solution to the problem since the reduced cost coefficients in the nonbasic columns x_1, x_3, and x_4 are non-negative:

Basic variables: $x_5 = 12$, $x_2 = 6$
Nonbasic variables: $x_1 = x_3 = x_4 = 0$
Cost function: $f = -12$

The solution to the original design problem is then

$$y_1 = x_1 = 0, \quad y_2 = x_2 - x_3 = 6 - 0 = 6, \quad z = -f = 12$$

which agrees with the graphical solution in Figure 8.4.

Note that the artificial variable column (x_6) in the final tableau is the negative of the surplus variable column (x_5). This is true for all "$\geq$ type" constraints.

EXAMPLE 8.15 USE OF ARTIFICIAL VARIABLES FOR EQUALITY CONSTRAINTS (INFEASIBLE PROBLEM)

Solve the LP problem:

Maximize

$$z = x_1 + 4x_2 \tag{a}$$

subject to

$$x_1 + 2x_2 \leq 5 \tag{b}$$

$$2x_1 + x_2 = 4 \tag{c}$$

$$x_1 - x_2 \geq 3 \tag{d}$$

$$x_1, x_2 \geq 0 \tag{e}$$

Solution

The constraints for the problem are plotted in Figure 8.5. It is seen that the problem has no feasible solution. We will solve the problem using the Simplex method to see how we can recognize an infeasible problem. Writing the problem in the standard LP form, we obtain

Minimize

$$f = -x_1 - 4x_2 \tag{f}$$

subject to

$$x_1 + 2x_2 + x_3 = 5 \tag{g}$$

$$2x_1 + x_2 + x_5 = 4 \tag{h}$$

$$x_1 - x_2 - x_4 + x_6 = 3 \tag{i}$$

$$x_i \geq 0; \quad i = 1 \text{ to } 6 \tag{j}$$

Here x_3 is a slack variable, x_4 is a surplus variable, and x_5 and x_6 are artificial variables. Table 8.19 shows the Phase I iterations of the Simplex method. It is seen that after the first pivot step, all the reduced cost coefficients of the artificial cost function for nonbasic variables are positive, indicating the end of Phase I. However, the artificial cost function is not zero ($w = 1$). Therefore, there is no feasible solution to the original problem.

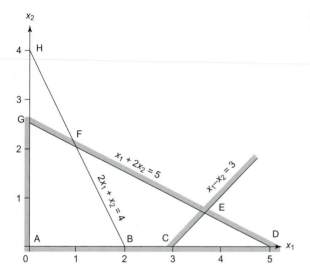

FIGURE 8.5 Graphic of the constraints for the LP problem of Example 8.15: Infeasible problem.

TABLE 8.19 Solution to Example 8.15 (infeasible problem)

Initial tableau: x_5 is identified to be replaced with x_1 in the basic set

Basic↓	x_1	x_2	x_3	x_4	x_5	x_6	b	Ratio
1. x_3	1	2	1	0	0	0	5	$\frac{5}{1} = 5$
2. x_5	[2]	1	0	0	1	0	4	$\frac{4}{2} = 2$
3. x_6	1	−1	0	−1	0	1	3	$\frac{3}{1} = 3$
4. Cost	−1	−4	0	0	0	0	$f - 0$	
5. Artificial cost	$\boxed{-3}$	0	0	1	0	0	$w - 7$	

Second tableau: End of Phase I

Basic↓	x_1	x_2	x_3	x_4	x_5	x_6	b
6. x_3	0	$\frac{3}{2}$	1	0	$-\frac{1}{2}$	0	3
7. x_1	1	$\frac{1}{2}$	0	0	$\frac{1}{2}$	0	2
8. x_6	0	$-\frac{3}{2}$	0	−1	$-\frac{1}{2}$	1	1
9. Cost	0	$-\frac{7}{2}$	0	0	$\frac{1}{2}$	0	$f + 2$
10. Artificial cost	0	$\frac{3}{2}$	0	1	$\frac{3}{2}$	0	$w - 1$

EXAMPLE 8.16 USE OF ARTIFICIAL VARIABLES (UNBOUNDED SOLUTION)

Solve the LP problem:

Maximize

$$z = 3x_1 - 2x_2 \qquad\qquad \text{(a)}$$

subject to

$$x_1 - x_2 \geq 0 \qquad\qquad \text{(b)}$$

$$x_1 + x_2 \geq 2 \qquad\qquad \text{(c)}$$

$$x_1, x_2 \geq 0 \qquad\qquad \text{(d)}$$

Solution

The constraints for the problem are plotted in Figure 8.6. It is seen that the feasible set for the problem is unbounded. We will solve the problem with the Simplex method and see how to recognize *unboundedness*. Transforming the problem to the standard form, we get:

Minimize

$$f = -3x_1 + 2x_2 \qquad\qquad \text{(e)}$$

subject to

$$-x_1 + x_2 + x_3 = 0 \qquad\qquad \text{(f)}$$

$$x_1 + x_2 - x_4 + x_5 = 2 \qquad\qquad \text{(g)}$$

$$x_i \geq 0; \quad i = 1 \text{ to } 5 \qquad\qquad \text{(h)}$$

where x_3 is a slack variable, x_4 is a surplus variable, and x_5 is an artificial variable. Note that the right side of the first constraint is zero, so it can be treated as either "≤ type" or "≥ type." We will treat it as "≤ type." Note also that the second constraint is "≥ type," so we must use an artificial variable and an artificial cost function to find the initial basic feasible solution.

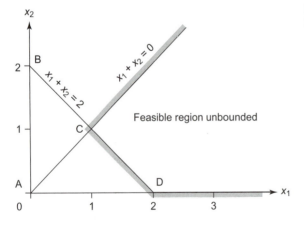

FIGURE 8.6　Constraints for the LP problem of Example 8.16: Unbounded problem.

TABLE 8.20 Solution to Example 8.16 (unbounded problem)

Initial tableau: x_5 is identified to be replaced with x_1 in the basic set

Basic↓	x_1	x_2	x_3	x_4	x_5	b	Ratio
1. x_3	−1	1	1	0	0	0	Negative
2. x_5	1	1	0	−1	1	2	$\frac{2}{1}=2$
3. Cost	−3	2	0	0	0	$f-0$	
4. Artificial cost	−1	−1	0	1	0	$w-2$	

Second tableau: End of Phase I. End of Phase II

Basic↓	x_1	x_2	x_3	x_4	x_5	b	Ratio
5. x_3	0	2	1	−1	1	2	Negative
6. x_1	1	1	0	−1	1	2	Negative
7. Cost	0	5	0	−3	3	$f+6$	
8. Artificial cost	0	0	0	0	1	$w-0$	

The initial set-up of the problem and its solution are given in Table 8.20. For the initial tableau, $x_3 = 0$ and $x_5 = 2$ are basic variables and all others are nonbasic. Note that this is a *degenerate basic feasible solution*. The solution corresponds to point A (the origin) in Figure 8.6. Scanning the artificial cost row, we observe that there are two possibilities for pivot columns x_1 or x_2. If x_2 is selected as the pivot column, then the first row must be the pivot row with $a_{12} = 1$ as the pivot element. This will make x_2 basic and x_3 nonbasic. However, x_2 will remain zero, and the resulting solution will be degenerate, corresponding to point A. One more iteration will be necessary to move from A to D.

If we choose x_1 as the pivot column, then $a_{21} = 1$ will be the pivot element, making x_1 as basic and x_5 as nonbasic. Carrying out the pivot step, we obtain the second tableau as shown in Table 8.20. The basic feasible solution is $x_1 = 2$, $x_3 = 2$, and the other variables are zero. This solution corresponds to point D in Figure 8.6. This is the basic feasible solution to the original problem because the artificial cost function is zero (i.e., $w = 0$). The original cost function has also reduced from 0 to −6. This is the end of Phase I.

Scanning the cost function row in the second tableau, we find that the reduced cost coefficient c_4' is negative, but the pivot element cannot be determined; that is, x_4 cannot be made basic because all the elements in the x_4 column are negative in the second tableau. However, we are not at an optimum point. This indicates the problem to be unbounded.

8.6.6 Degenerate Basic Feasible Solution

It is possible that during iterations of the Simplex method, a basic variable attains a zero value, that is, the basic feasible solution becomes degenerate. What are the implications of this situation? We will discuss them in Example 8.17.

EXAMPLE 8.17 IMPLICATIONS OF THE DEGENERATE BASIC FEASIBLE SOLUTION

Solve the following LP problem by the Simplex method:

Maximize

$$z = x_1 + 4x_2 \tag{a}$$

subject to

$$x_1 + 2x_2 \leq 5 \tag{b}$$

$$2x_1 + x_2 \leq 4 \tag{c}$$

$$2x_1 + x_2 \geq 4 \tag{d}$$

$$x_1 - x_2 \geq 1 \tag{e}$$

$$x_1, x_2 \geq 0. \tag{f}$$

Solution

The problem is transcribed into the standard LP form as follows:

Minimize

$$f = -x_1 - 4x_2 \tag{g}$$

subject to

$$x_1 + 2x_2 + x_3 = 5 \tag{h}$$

$$2x_1 + x_2 + x_4 = 4 \tag{i}$$

$$2x_1 + x_2 - x_5 + x_7 = 4 \tag{j}$$

$$x_1 - x_2 - x_6 + x_8 = 1 \tag{k}$$

$$x_i \geq 0; \quad i = 1 \text{ to } 8 \tag{l}$$

where x_3 and x_4 are slack variables, x_5 and x_6 are surplus variables, and x_7 and x_8 are artificial variables. The two-phase Simplex procedure takes three iterations to reach the optimum point. These iterations are given in Table 8.21.

It is seen that in the third tableau, the basic variable x_4 has a zero value so the *basic feasible solution is degenerate*. At this iteration, it is determined that x_5 should become basic so x_5 is the pivot column. We need to determine the pivot row. We take the ratios of the right sides with the positive elements in the x_5 column. This determines the second row as the pivot row because it has the smallest ratio (zero).

In general, if the element in the pivot column and the row that gives the degenerate basic variable is positive, then that row must always be the pivot row; otherwise, the new solution cannot be feasible. Also, in this case, the new basic feasible solution will be degenerate, as for the final tableau in Table 8.21. The only way the new feasible solution can be nondegenerate is when the element in the pivot column and the degenerate variable row is negative. In that case the new basic feasible solution will be nondegenerate.

It is theoretically possible for the Simplex method to fail by cycling between two degenerate basic feasible solutions. However, in practice this usually does not happen. The final solution to this problem is

Basic variables: $x_1 = 5/3$, $x_2 = 2/3$, $x_3 = 2$, $x_5 = 0$
Nonbasic variables: $x_4 = x_6 = x_7 = x_8 = 0$
Optimum cost function: $f = -13/3$ or $z = 13/3$

TABLE 8.21 Solution to Example 8.17 (degenerate basic feasible solution)

Initial tableau: x_8 is identified to be replaced with x_1 in the basic set

Basic↓	x_1	x_2	x_3	x_4	x_5	x_6	x_7	x_8	b	Ratio
1. x_3	1	2	1	0	0	0	0	0	5	$\frac{5}{1}=5$
2. x_4	[2]	1	0	1	0	0	0	0	4	$\frac{4}{2}=2$
3. x_7	2	1	0	0	−1	0	1	0	4	$\frac{4}{2}=2$
4. x_8	[−1]	−1	0	0	0	−1	0	1	1	$\frac{1}{1}=1$
5. Cost	−1	−4	0	0	0	0	0	0	$f-0$	
6. Artificial	[−3]	0	0	0	1	1	0	0	$w-5$	

Second tableau: x_7 is identified to be replaced with x_2 in the basic set

Basic↓	x_1	x_2	x_3	x_4	x_5	x_6	x_7	x_8	b	Ratio
7. x_3	0	3	1	0	0	1	0	−1	4	$\frac{4}{3}$
8. x_4	0	3	0	1	0	2	0	−2	2	$\frac{2}{3}$
9. x_7	0	[3]	0	0	−1	2	1	−2	2	$\frac{2}{3}$
10. x_1	1	−1	0	0	0	−1	0	1	1	Negative
11. Cost	0	−5	0	0	0	−1	0	1	$f+1$	
12. Artificial	0	[−3]	0	0	1	−2	0	3	$w-2$	

Third tableau: x_4 is identified to be replaced with x_5 in the basic set. End of Phase I

Basic↓	x_1	x_2	x_3	x_4	x_5	x_6	x_7	x_8	b	Ratio
13. x_3	0	0	1	0	1	−1	−1	1	2	$\frac{2}{1}=2$
14. x_4	0	0	0	1	[1]	0	−1	0	0	$\frac{0}{1}=0$
15. x_2	0	1	0	0	$-\frac{1}{3}$	$\frac{2}{3}$	$\frac{1}{3}$	$-\frac{2}{3}$	$\frac{2}{3}$	Negative
16. x_1	1	0	0	0	$-\frac{1}{3}$	$-\frac{1}{3}$	$\frac{1}{3}$	$\frac{1}{3}$	$\frac{5}{3}$	Negative
17. Cost	0	0	0	0	$-\frac{5}{3}$	$\frac{7}{3}$	$\frac{5}{3}$	$-\frac{7}{3}$	$f+\frac{13}{3}$	
18. Artificial	0	0	0	0	0	0	1	0	$w-0$	

Final tableau: End of Phase II

Basic↓	x_1	x_2	x_3	x_4	x_5	x_6	x_7	x_8	b
19. x_3	0	0	1	−1	0	−1	0	1	2
20. x_5	0	0	0	1	1	0	−1	0	0
21. x_2	0	1	0	$\frac{1}{3}$	0	$\frac{2}{3}$	0	$-\frac{2}{3}$	$\frac{2}{3}$
22. x_1	1	0	0	$\frac{1}{3}$	0	$-\frac{1}{3}$	0	$\frac{1}{3}$	$\frac{5}{3}$
23. Cost	0	0	0	$\frac{5}{3}$	0	$\frac{7}{3}$	0	$-\frac{7}{3}$	$f+\frac{13}{3}$

8.7 POSTOPTIMALITY ANALYSIS

The optimum solution to the LP problem depends on the parameters in vectors **c** and **b** and the matrix **A** defined in Eqs. (8.10) through (8.12). These parameters are prone to errors in practical design problems. Thus we are interested not only in the optimum solution but also in how it changes when these parameters change. The changes may be either discrete or continuous.

The study of discrete parameter changes is often called sensitivity analysis, *and that of continuous changes is called* parametric programming. There are five basic parametric changes affecting the optimum solution:

1. Changes in the cost function coefficients, c_j
2. Changes in the resource limits, b_i
3. Changes in the constraint coefficients, a_{ij}
4. The effect of including additional constraints
5. The effect of including additional variables

A thorough discussion of these changes, while not necessarily that difficult, is beyond our scope here. In principle, we can imagine solving a new problem for every change. Fortunately, for a small number of changes there are useful shortcuts. Almost all computer programs for LP problems provide some information about parameter variations. We will study the parametric changes defined in items 1 through 3. *The final tableau for the LP problem contains all of the information needed to study these changes.* We will describe the information contained in the final tableau and its use to study the three parametric changes. For other variations, full-length texts on linear programming may be consulted.

It turns out that the optimum solution to the altered problem can be computed using the optimum solution to the original problem and the information in the final tableau as long as the change in the parameters is within certain limits. This is especially beneficial for problems that take a long time to solve. In the following discussion we use a'_{ij}, c'_j, and b'_i to represent the corresponding values of the parameters a_{ij}, c_j, and b_i in the final tableau.

8.7.1 Changes in Constraint Limits

Recovery of Lagrange Multipliers

First, we study how the optimum value of the cost function for the problem changes if we change the right-side parameters, b_i (also known as *resource limits*), of the constraints. The Constraint Variation Sensitivity Theorem 4.7 can be used to study the effect of these changes. Use of that theorem requires knowledge of the Lagrange multipliers for the constraints. Theorem 8.5 gives a way of recovering the multipliers for the constraints of an LP problem from the *final tableau*. Calculation of the new values of the design variables for the changed problem is explained later.

THEOREM 8.5

Lagrange Multiplier Values Let the standard linear programming problem be solved using the Simplex method:

1. For "≤ type" constraints, the Lagrange multiplier equals the reduced cost coefficient in the slack variable column associated with the constraint.

2. For " = " and "≥ type" constraints, the Lagrange multiplier equals the reduced cost coefficient in the artificial variable column associated with the constraint.

3. The Lagrange multiplier is always ≥ 0 for the "≤ type" constraint, always ≤ 0 for the "≥ type" constraint, and free in sign for the " = type" constraint.

Change in Cost Function

In Section 4.7, the *physical meaning of the Lagrange multipliers* was described. There, the Lagrange multipliers were related to derivatives of the cost function with respect to the right-side parameters. Equality and inequality constraints were treated separately, with v_i and u_i as their Lagrange multipliers, respectively. However, in this section, we use a slightly different notation defined as follows:

e_i the right-side parameter of the ith constraint
y_i the Lagrange multiplier of the ith constraint

Using this notation and Theorem 4.7, we obtain the following derivative of the cost function with respect to the right-side parameters, and the change in the optimum cost function:

$$\frac{\partial f}{\partial e_i} = -y_i; \quad \Delta f = -y_i \Delta e_i = -y_i(e_{inew} - e_{iold}) \tag{8.21}$$

It is noted here that Eq. (8.21) is applicable only for minimization of the cost function. Also, Theorem 8.5 and Eq. (8.21) are applicable only if changes in the right-side parameters are within certain limits; that is, there are upper and lower limits on changes in the resource limits for which Eq. (8.21) is valid. The changes need not be small, as was stipulated for nonlinear problems in Section 4.7. Calculations for the limits are discussed later in this section. Note that the calculation for Δf remains valid for simultaneous changes to multiple constraints; in that case all the changes are added.

It is also noted that Theorem 4.7 and Eq. (8.21) were discussed in Section 4.7 for the general problem written as the minimization of a cost function with "≤ type" and equality constraints. However, Eq. (8.21) is applicable to "≥ type" constraints too, as long as care is exercised in using the appropriate sign for the Lagrange multiplier y_i and the change in right side Δe_i. We will demonstrate use of Theorem 8.5 and Eq. (8.21) with examples.

It is also important to note that if an inequality is inactive at the optimum, then its slack or surplus variable is greater than 0. Therefore, its Lagrange multiplier is 0 to satisfy the switching condition, $y_i s_i = 0$ (except for the abnormal case, where both the Lagrange multiplier and the constraint function have zero value). This observation can help in verifying the correctness of the Lagrange multipliers recovered from the final LP tableau. Example 8.18 describes the recovery of the Lagrange multipliers from the final tableau for the "$\leq$ type" constraints.

The Lagrange multipliers are also called the dual variables (or dual prices) for the problem. The concept of duality in linear programming is described in Chapter 9. Example 8.19 demonstrates recovery of Lagrange multipliers for equality and "$\geq$ type" constraints.

EXAMPLE 8.18 RECOVERY OF LAGRANGE MULTIPLIERS FOR "$\leq$ TYPE" CONSTRAINTS

Consider the problem:

Maximize

$$z = 5x_1 - 2x_2 \tag{a}$$

subject to

$$2x_1 + x_2 \leq 9 \tag{b}$$

$$x_1 - 2x_2 \leq 2 \tag{c}$$

$$-3x_1 + 2x_2 \leq 3 \tag{d}$$

$$x_1, x_2 \geq 0 \tag{e}$$

Solve the problem by the Simplex method. Recover the Lagrange multipliers for the constraints.

Solution

Constraints for the problem and cost function contours are plotted in Figure 8.7. The optimum solution is at point C and is given as $x_1 = 4$, $x_2 = 1$, $z = 18$. The problem is transformed into the standard form as

Minimize

$$f = -5x_1 + 2x_2 \tag{f}$$

subject to

$$2x_1 + x_2 + x_3 = 9 \tag{g}$$

$$x_1 - 2x_2 + x_4 = 2 \tag{h}$$

$$-3x_1 + 2x_2 + x_5 = 3 \tag{i}$$

$$x_i \geq 0; \quad i = 1 \text{ to } 5 \tag{j}$$

where x_3, x_4, and x_5 are the slack variables. Solving the problem using the Simplex method, we obtain the sequence of calculations given in Table 8.22. From the final tableau,

Basic variables: $x_1 = 4$, $x_2 = 1$, $x_5 = 13$
Nonbasic variables: $x_3 = 0$, $x_4 = 0$
Objective function: $z = 18$ ($f = -18$)

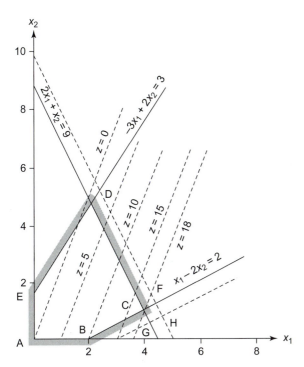

FIGURE 8.7 Graphical solution to the LP problem of Example 8.18.

In the problem formulation, x_3, x_4, and x_5 are the slack variables for the three constraints. Since all constraints are "$\leq$ type," the reduced cost coefficients for the slack variables in the final tableau are the Lagrange multipliers as follows:

1. For $2x_1 + x_2 \leq 9$:

$$y_1 = 1.6 \ (c_3' \text{ in column } x_3) \tag{k}$$

2. For $x_1 - 2x_2 \leq 2$:

$$y_2 = 1.8 \ (c_4' \text{ in column } x_4) \tag{l}$$

3. For $-3x_1 + 2x_2 \leq 3$:

$$y_3 = 0 \ (c_5' \text{ in column } x_5) \tag{m}$$

Therefore, Eq. (8.21) gives partial derivatives of f with respect to e_i as

$$\frac{\partial f}{\partial e_1} = -1.6; \quad \frac{\partial f}{\partial e_2} = -1.8; \quad \frac{\partial f}{\partial e_3} = 0 \tag{n}$$

where $f = -(5x_1 - 2x_2)$. If the right side of the first constraint changes from 9 to 10, the cost function f changes by

$$\Delta f = 1.6(e_{1new} - e_{1old}) = -1.6(10 - 9) = -1.6 \tag{o}$$

That is, the new value of f will be -19.6 ($z = 19.6$). Point F in Figure 8.7 gives the new optimum solution to this case.

If the right side of the second constraint changes from 2 to 3, the cost function f changes by $\Delta f = -1.8(3 - 2) = -1.8$ to -19.8. Point G in Figure 8.7 gives the new optimum solution. Note that any small change in the right side of the third constraint will have no effect on the cost function. *When the right sides of the first and second constraints are changed to 10 and 3 simultaneously,* the net change in the cost function is $-(1.6 + 1.8)$ (i.e., new f will be -21.4). The new solution is at point H in Figure 8.7.

TABLE 8.22   Solution to Example 8.18 by the Simplex method

Initial tableau: x_4 is identified to be replaced with x_1 in the basic set

Basic↓	x_1	x_2	x_3	x_4	x_5	b
1. x_3	2	1	1	0	0	9
2. x_4	1	−2	0	1	0	2
3. x_5	−3	2	0	0	1	3
4. Cost	−5	2	0	0	0	$f - 0$

Second tableau: x_3 is identified to be replaced with x_2 in the basic set

Basic↓	x_1	x_2	x_3	x_4	x_5	b
5. x_3	0	5	1	−2	0	5
6. x_1	1	−2	0	1	0	2
7. x_5	0	−4	0	3	1	9
8. Cost	0	−8	0	5	0	$f + 10$

Third tableau: Reduced cost coefficients in nonbasic columns are non-negative; the tableau gives optimum point

Basic↓	x_1	x_2	x_3	x_4	x_5	b
9. x_2	0	1	0.2	−0.4	0	1
10. x_1	1	0	0.4	0.2	0	4
11. x_5	0	0	0.8	1.4	1	13
12. Cost	0 (c_1')	0 (c_2')	1.6 (c_3')	1.8 (c_4')	0 (c_5')	$f + 18$

x_3, x_4 and x_5 are slack variables.

EXAMPLE 8.19 RECOVERY OF LAGRANGE MULTIPLIERS FOR " = " AND "≥ TYPE" CONSTRAINTS

Solve the following LP problem and recover the proper Lagrange multipliers for the constraints:

Maximize

$$z = x_1 + 4x_2 \tag{a}$$

subject to

$$x_1 + 2x_2 \leq 5 \tag{b}$$

$$2x_1 + x_2 = 4 \tag{c}$$

$$x_1 - x_2 \geq 1 \tag{d}$$

$$x_1, x_2 \geq 0 \tag{e}$$

Solution

Constraints for the problem are plotted in Figure 8.8. It is seen that line E–C is the feasible region for the problem and point E gives the optimum solution. The equality constraint and the "≥ type" constraint in Eq. (d) are active at the optimum point. The optimum point, calculated using the active constraints, is given as (5/3, 2/3).

This problem was solved in Example 8.13 using the two-phase Simplex method. The optimum point was reached after two Simplex iterations, where Phases I and II ended simultaneously. The Simplex iterations are given in Table 8.17. The final tableau for the problem is copied from Table 8.17 into Table 8.23, where x_3 is the slack variable for the "≤ type" constraint in Eq. (b), x_4 is the surplus variable for the "≥ type" constraint in Eq. (d), and x_5 and x_6 are artificial variables for the equality and "≥ type" constraints, respectively. Note that the artificial variable column (x_6) is the negative of the surplus variable column (x_4) for the "≥ type" constraint in Eq. (d). The solution from the final tableau is

Basic variables: $x_1 = 5/3, \ x_2 = 2/3, \ x_3 = 2$ (f)

Nonbasic variables: $x_4 = x_5 = x_6 = 0$ (g)

Cost function: $f = -13/3$ (h)

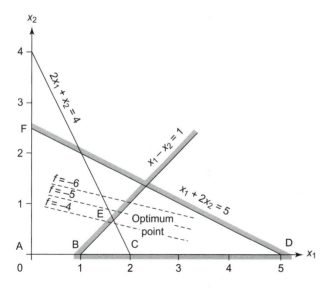

FIGURE 8.8 Constraints for the LP problem from Example 8.19. Feasible region: line E–C.

TABLE 8.23   Final tableau for Example 8.19

Basic↓	x_1	x_2	x_3	x_4	x_5	x_6	b
11. x_3	0	0	1	-1	-1	1	2
12. x_2	0	1	0	$\frac{2}{3}$	$\frac{1}{3}$	$-\frac{2}{3}$	$\frac{2}{3}$
13. x_1	1	0	0	$-\frac{1}{3}$	$\frac{1}{3}$	$\frac{1}{3}$	$\frac{5}{3}$
14. Cost	0 (c_1')	0 (c_2')	0 (c_3')	$\frac{7}{3}(c_4')$	$\frac{5}{3}(c_5')$	$-\frac{7}{3}(c_6')$	$f + \frac{13}{3}$

Note: x_3 = slack variable; x_4 = surplus variable; x_5, x_6 = artificial variables.

Using Theorem 8.5, the Lagrange multipliers for the constraints are

1. For $x_1 + 2x_2 \leq 5$:

$$y_1 = 0 \ (c_3' \text{ in the slack variable column } x_3) \tag{i}$$

2. For $2x_1 + x_2 = 4$:

$$y_2 = \frac{5}{3} \ (c_5' \text{ in the artificial variable column } x_5) \tag{j}$$

3. For $x_1 - x_2 \geq 1$:

$$y_3 = -\frac{7}{3} \ (c_6' \text{ in the artificial variable column } x_6) \tag{k}$$

When the right side of the third constraint is changed to 2 ($x_1 - x_2 \geq 2$), the cost function $f = (-x_1 - 4x_2)$ changes by

$$\Delta f = -y_3 \Delta e_3 = -\left(-\frac{7}{3}\right)(2 - 1) = \frac{7}{3} \tag{l}$$

That is, the cost function will increase by 7/3 from −13/3 to −2 ($z = 2$). This can also be observed in Figure 8.8. We will demonstrate that the same result is obtained when the third constraint is written in the "≤ form" ($-x_1 + x_2 \leq -1$). The Lagrange multiplier for the constraint is 7/3, which is the negative of the preceding value.

Note that it is also c_4' in the surplus variable column x_4. When the right side of the third constraint is changed to 2 (i.e., it becomes $-x_1 + x_2 \leq -2$), the cost function $f = (-x_1 - 4x_2)$ changes by

$$\Delta f = -y_3 \Delta e_3 = -\left(-\frac{7}{3}\right)[-2 - (-1)] = \frac{7}{3} \tag{m}$$

which is the same as before.

When the right side of the equality constraint is changed to 5 from 4, the cost function $f = (-x_1 - 4x_2)$ changes by

$$\Delta f = -y_2 \Delta e_2 = -\frac{5}{3}(5 - 4) = -\frac{5}{3} \tag{n}$$

That is, the cost function will decrease by 5/3, from −13/3 to −6 ($z = 6$).

8.7.2 Ranging Right-Side Parameters

When the right side of a constraint is changed, the constraint boundary moves parallel to itself, changing the feasible region for the problem. However, the isocost lines do not change.

Since the feasible region is changed, the optimum solution, that is, the design variables as well as the cost function, may change. There are, however, certain limits on changes for which the set of active constraints at the optimum point is not altered. In other words, if the changes are within certain limits, the sets of basic and nonbasic variables do not change. In that case, the solution to the altered problem can be obtained from the information contained in the final tableau. Otherwise, Eq. (8.21) cannot be used and more iterations of the Simplex method are needed to obtain a solution to the altered problem. Theorem 8.6 describes determination of the limits and the new right sides when the changes are within the limits.

THEOREM 8.6

Limits on Changes in Resources Let Δ_k be the possible change in the right side b_k of the kth constraint. If Δ_k satisfies the following inequalities, then no more iterations of the Simplex method are required to obtain the solution to the altered problem, and Eq. (8.21) can be used to determine changes to the cost function:

$$\max\{r_i \leq 0, a'_{ij} > 0\} \leq \Delta_k \leq \min\{r_i \geq 0, a'_{ij} < 0\};$$

$$r_i = -\frac{b'_i}{a'_{ij}}, \quad i = 1 \text{ to } m \qquad (8.22)$$

where

b'_i = right-side parameter for the ith constraint in the final tableau.

a'_{ij} = parameters in the jth column of the final tableau; the jth column corresponds to x_j, which is the slack variable for a "$\leq$ type" constraint, or the artificial variable for an equality, or "$\geq$ type" constraint.

r_i = negative of the ratios of the right sides with the parameters in the jth column.

Δ_k = possible change in the right side of the kth constraint; the slack or artificial variable for the kth constraint determines the index j of the column with elements that are used in the inequalities of Eq. (8.22).

To determine the range, we first determine the column index j according to the rules given in Theorem 8.6. Then, using the elements in the jth column, we determine the ratios $r_i = -b'_i/a'_{ij}$ $(a'_{ij} \neq 0)$.

1. The largest negative ratio r_i gives the lower limit on change Δ_k in b_k. If there is no $a'_{ij} > 0$, then the said ratio cannot be found, and so there is no lower bound on Δ_k (i.e., the lower limit is $-\infty$).
2. The smallest positive ratio r_i gives the upper limit on change Δ_k in b_k. If there is no $a'_{ij} < 0$, then the said ratio cannot be found, and there is no upper bound on Δ_k (i.e., the upper limit is ∞).

New Values of the Basic Variables

The new right-side parameters b''_j due to a change of Δ_k in b_k are given as

$$b''_i = b'_i + \Delta_k a'_{ij}; \quad i = 1 \text{ to } m \qquad (8.23)$$

Using Eq. (8.23) and the final tableau, new values for the basic variables in each row can be obtained. Equation (8.23) is applicable only if Δ_k is in the range determined by inequalities of Eq. (8.22).

Example 8.20 demonstrates calculations of the ranges for the right side parameters and new values for the right side (i.e., the basic variables) for a problem with "≤ type" constraints.

EXAMPLE 8.20 RANGES FOR RESOURCE LIMITS—"≤ TYPE" CONSTRAINTS

Find ranges for the right-side parameters of constraints for the problem solved in Example 8.18.

Solution

The graphical solution to the problem is shown in Figure 8.7. The final tableau for the problem is given in Table 8.22. For the first constraint, x_3, is the slack variable, so $j = 3$ in the inequalities of Eq. (8.22) for the calculation of range for Δ_1, which is the change to the constraint's right side. The ratios of the right side parameters with the elements in x_3 column, r_i of Eq. (8.22) are calculated as

$$r_i = -\frac{b_i'}{a_{i3}'} = \left\{ -\frac{1}{0.2}, -\frac{4}{0.4}, -\frac{13}{0.8} \right\} = \{-5.0, -10.0, -16, 25\} \tag{a}$$

Since there is no positive r_i, there is no upper limit on Δ_1. The lower limit is determined as the largest element among the negative ratios according to the inequality of Eq. (8.22):

$$\max\{-5.0, -10.0, -16.25\} \le \Delta_1, \text{ or } -5 \le \Delta_1 \tag{b}$$

Thus, limits for Δ_1 are $-5 \le \Delta_1 \le \infty$ and the range on b_1 is obtained by adding the current value of $b_1 = 9$ to both sides:

$$-5 + 9 \le b_1 \le \infty + 9, \text{ or } 4 \le b_1 \le \infty \tag{c}$$

For the second constraint $(k = 2)$, x_4 is the slack variable. Therefore, we will use elements in the x_4 column of the final tableau $(a_{i4}', j = 4)$ in the inequalities of Eq. (8.22). The ratios of the right-side parameters with the elements in the x_4 column, r_i of Eq. (8.22), are calculated as

$$r_i = -\frac{b_i'}{a_{i4}'} = \left\{ -\frac{1}{-0.4}, -\frac{4}{0.2}, -\frac{13}{1.4} \right\} = \{2.5, -20.0, -9.286\} \tag{d}$$

According to the inequalities in Eq. (8.22), lower and upper limits for Δ_2 are given as

$$\max\{-20, -9.286\} \le \Delta_2 \le \min\{2.5\}, \text{ or } -9.286 \le \Delta_2 \le 2.5 \tag{e}$$

Therefore, the allowed decrease in b_2 is 9.286 and the allowed increase is 2.5. Adding 2 to the above inequality (the current value of b_2), the range of b_2 is given as

$$-7.286 \le b_2 \le 4.5 \tag{f}$$

Similarly, for the third constraint, the ranges for Δ_3 and b_3 are

$$-13 \le \Delta_3 \le \infty, \ -10 \le b_3 \le \infty \tag{g}$$

New Values of Basic Variables

Let us calculate new values for the design variables if the right side of the first constraint is changed from 9 to 10. Note that this change is within the limits determined in the foregoing section. In Eq. (8.23), $k = 1$, so $\Delta_1 = 10 - 9 = 1$. Also, $j = 3$, so we use the third column from Table 8.22 in Eq. (8.23) and obtain new values of the basic variables as

$$x_2 = b_1'' = b_1' + \Delta_1 a_{13}' = 1 + (1)(0.2) = 1.2 \tag{h}$$

$$x_1 = b_2'' = b_2' + \Delta_1 a_{23}' = 4 + (1)(0.4) = 4.4 \tag{i}$$

$$x_5 = b_3'' = b_3' + \Delta_1 a_{33}' = 13 + (1)(0.8) = 13.8 \tag{j}$$

The other variables remain nonbasic, so they have zero values. The new solution corresponds to point F in Figure 8.7.

Similarly, if the right side of the second constraint is changed from 2 to 3, the new values of the variables, using Eq. (8.23) and the x_4 column from Table 8.22, are calculated as $x_2 = 0.6$, $x_1 = 4.2$, and $x_5 = 14.4$. This solution corresponds to point G in Figure 8.7.

When the right sides of two or more constraints are changed simultaneously, we can use Eq. (8.23) to determine the new values of the design variables. However, we have to make sure that the new right sides do not change the basic and nonbasic variable sets, that is, the vertex that gives the optimum solution must not change. In other words, no new constraint becomes active. As an example, let us calculate the new values of design variables using Eq. (8.23) when the right sides of the first and second constraints are changed to 10 and 3, from 9 and 2, respectively:

$$x_2 = b_1'' = b_1' + \Delta_1 a_{13}' + \Delta_2 a_{14}' = 1 + (1)(0.2) + (1)(-0.4) = 0.8 \tag{k}$$

$$x_1 = b_2'' = b_2' + \Delta_1 a_{23}' + \Delta_2 a_{24}' = 4 + (1)(0.4) + (1)(0.2) = 4.6 \tag{l}$$

$$x_5 = b_3'' = b_3' + \Delta_1 a_{33}' + \Delta_2 a_{34}' = 13 + (1)(0.8) + (1)(1.4) = 15.2 \tag{m}$$

It can be verified that the new solution corresponds to point H in Figure 8.7.

Example 8.21 demonstrates calculations of the ranges for the right-side parameters and the new values for the right sides (i.e., the basic variables) for a problem with equality and "≥ type" constraints.

EXAMPLE 8.21 RANGES FOR RESOURCE LIMITS—EQUALITY AND "≥ TYPE" CONSTRAINTS

Find ranges for the right-side parameters of the problem solved in Example 8.19.

Solution

The final tableau for the problem is given earlier in Table 8.23. The graphical solution to the problem is given in Figure 8.8. In the tableau, x_3 is a slack variable for the first constraint, x_4 is a surplus variable for the third constraint, x_5 is an artificial variable for the second constraint, and x_6 is an artificial variable for the third constraint. For the first constraint, x_3 is the slack variable, and therefore, index j for use in the inequalities of Eq. (8.22) is determined as 3. Using the same procedure as for Example 8.20, the ranges for Δ_1 and b_1 are calculated as $-2 \leq \Delta_1 \leq \infty$ and $3 \leq b_1 \leq \infty$.

Since the second constraint is an equality, the index j for use in Eq. (8.22) is determined by the artificial variable x_5 for the constraint, i.e., $j = 5$. Accordingly, ratios r_i in Eq. (8.22) and the range for Δ_2 are calculated as

$$r_i = -\frac{b'_i}{a'_{i5}} = \left\{ -\frac{2}{-1}, \ -\frac{2/3}{1/3}, \ -\frac{5/3}{1/3} \right\} = \{2.0, \ -2.0, \ -5.0\} \tag{a}$$

$$\max\{-2.0, \ -5.0\} \leq \Delta_2 \leq \min\{2.0\}, \text{ or } -2 \leq \Delta_2 \leq 2 \tag{b}$$

The range for b_2 can be found by adding the current value of $b_2 = 4$ to both sides of the above inequality as $2 \leq b_2 \leq 6$.

The third constraint is a "$\geq$ type", so index j for use in the inequalities of Eq. (8.22) is determined by its artificial variable x_6 (i.e., $j = 6$). Accordingly, ratios r_i in Eq. (8.22) and the range for Δ_3 are calculated as

$$r_i = -\frac{b'_i}{a'_{i6}} = \left\{ -\frac{2}{1}, \ -\frac{2/3}{-2/3}, \ -\frac{5/3}{1/3} \right\} = \{-2.0, 1.0, \ -5.0\} \tag{c}$$

$$\max\{-2.0, \ -5.0\} \leq \Delta_3 \leq \min\{1.0\}, \text{ or } -2 \leq \Delta_3 \leq 1 \tag{d}$$

The limits on changes in b_3 are (by adding the current value of $b_3 = 1$ to both sides of the above inequality) $-1 \leq b_3 \leq 2$.

New values of basic variables We can use Eq. (8.23) to calculate the new values of the basic variables for the right-side changes that remain within the previously determined ranges. It can be seen that since the first constraint is not active, it does not affect the optimum solution as long as its right side remains within the range of $3 \leq b_1 \leq \infty$ determined previously.

Let us determine the new solution when the right side of the second constraint is changed to 5 from 4 (the change is within the range determined previously). The second constraint has x_5 as an artificial variable, so we use column 5 ($j = 5$) from Table 6.19 in Eq. (8.23) and obtain the new values of the basic variables as follows:

$$x_3 = b''_1 = b'_1 + \Delta_2 a'_{15} = 2 + (1)(-1) = 1 \tag{e}$$

$$x_2 = b''_2 = b'_2 + \Delta_2 a'_{25} = \tfrac{2}{3} + (1)\left(\tfrac{1}{3}\right) = 1 \tag{f}$$

$$x_1 = b''_3 = b'_3 + \Delta_2 a'_{35} = \tfrac{5}{3} + (1)\left(\tfrac{1}{3}\right) = 2 \tag{g}$$

To determine the new values of the basic variables when the right side of the third constraint is changed from 1 to 2, we use the x_6 column ($j = 6$) from Table 8.23 in Eq. (8.23) and obtain the new solution as

$$x_3 = b''_1 = b'_1 + \Delta_3 a'_{16} = 2 + (1)(1) = 3 \tag{h}$$

$$x_2 = b''_2 = b'_2 + \Delta_3 a'_{26} = \tfrac{2}{3} + (1)\left(-\tfrac{2}{3}\right) = 0 \tag{i}$$

$$x_1 = b''_3 = b'_3 + \Delta_3 a'_{36} = \tfrac{5}{3} + (1)\left(\tfrac{1}{3}\right) = 2 \tag{j}$$

It can easily be seen from Figure 8.8 that the new solution corresponds to point C.

8.7.3 Ranging Cost Coefficients

If a cost coefficient c_k is changed to $c_k + \Delta c_k$, we want to find an admissible range on Δc_k such that the optimum values of the variables are not changed. *Note that when the cost coefficients are changed, the feasible region for the problem does not change.* However, the orientation of the cost function hyperplane and the value of the cost function do change. Limits on the change Δc_k for the coefficient c_k depend on whether x_k is a basic variable at the optimum. Thus, we must consider the two cases separately. Theorems 8.7 and 8.8 give ranges for the cost coefficients for the two cases, respectively.

THEOREM 8.7

Range for Cost Coefficient of the Nonbasic Variables Let c_k be such that x_k^* is not a basic variable. If this c_k is replaced by any $c_k + \Delta c_k$, where $-c_k' \leq \Delta c_k \leq \infty$, then the optimum solution (design variables and cost function) does not change. Here, c_k' is the reduced cost coefficient corresponding to x_k^* in the final tableau.

THEOREM 8.8

Range for the Cost Coefficient of the Basic Variables Let c_k be such that x_k^* is a basic variable, and let $x_k^* = b_r'$ (a superscript * is used to indicate optimum value). Then the range for the change Δc_k in c_k for which the variables' optimum values do not change is given as follows:

$$\max\{d_j < 0\} \leq \Delta c_k \leq \min\{d_j > 0\},$$

$$d_j = \frac{c_j'}{a_{rj}'} \qquad (8.24)$$

where

a_{rj}' = element in the rth row and the jth column of the final tableau. The index r is determined by the row that

determines x_k^*. Index j corresponds to each of the nonbasic columns, excluding artificial columns. (Note that if no $a_{rj}' > 0$, then there is no upper limit; if no $a_{rj}' < 0$, then there is no lower limit.)

c_j' = reduced cost coefficient in the jth nonbasic column, excluding artificial variable columns.

d_j = ratios of reduced cost coefficients, with the elements in the rth row corresponding to nonbasic columns, excluding artificial columns.

When Δc_k satisfies the inequality of Eq. (8.24), the optimum value of the cost function is $f^* + \Delta c_k x_k^*$.

To determine possible changes in the cost coefficient of a basic variable, the first step is to determine the row index r for use in Inequalities (8.24). This represents the row determining the basic variable x_k^*. After r has been determined, we determine ratios of the reduced cost coefficients and elements in the rth row according to the rules given in Theorem 8.8. The lower limit on Δc_k is determined by the maximum of the negative ratios. The upper limit is determined by the minimum of the positive ratios. Example 8.22 demonstrates the procedure for the "$\leq$ type" constraints and Example 8.23 demonstrates it for the equality and "$\geq$ type" constraints.

EXAMPLE 8.22 RANGES FOR COST COEFFICIENTS—"$\leq$ TYPE" CONSTRAINTS

Determine ranges for the cost coefficients of the problem solved in Example 8.18.

Solution

The final tableau for the problem is given earlier in Table 8.22. The problem is solved as a minimization of the cost function $f = -5x_1 + 2x_2$. Therefore, we will find ranges for the cost coefficients $c_1 = -5$ and $= c_2 = 2$. Note that since both x_1 and x_2 are basic variables, Theorem 8.8 will be used.

Since the second row determines the basic variable x_1, $r = 2$ (the row number) for use in the inequalities of Eq. (8.24). Columns 3 and 4 are nonbasic; therefore, $j = 3$, 4 are the column indices for use in Eq. (8.24). After calculating the ratios d_j, the range for Δc_1 is calculated as

$$d_j = \frac{c'_j}{a'_{2j}} = \left\{ \frac{1.6}{0.4}, \frac{1.8}{0.2} \right\} = \{4, 9\}; \quad -\infty \leq \Delta c_1 \leq \min\{4, 9\}; \quad \text{or} \quad -\infty \leq \Delta c_1 \leq 4 \qquad \text{(a)}$$

The range for c_1 is obtained by adding the current value of $c_1 = -5$ to both sides of the above inequality:

$$-\infty - 5 \leq c_1 \leq 4 - 5; \quad -\infty \leq c_1 \leq -1 \qquad \text{(b)}$$

Thus, if c_1 changes from -5 to -4, the new cost function value is given as

$$f^*_{\text{new}} = f^* + \Delta c_1 x^*_1 = -18 + (-4 - (-5))(4) = -14 \qquad \text{(c)}$$

That is, the cost function will increase by 4.

For the second cost coefficient, $r = 1$ (the row number) because the first row determines x_2 as a basic variable. After calculating the ratios d_j, the range for Δc_2 is calculated as

$$d_j = \frac{c'_j}{a'_{1j}} = \left\{ \frac{1.6}{0.2}, \frac{1.8}{-0.4} \right\} = \{8, -4.5\}; \quad \max\{-4.5\} \leq \Delta c_2 \leq \min\{8\}; \quad \text{or} \quad -4.5 \leq \Delta c_2 \leq 8 \qquad \text{(d)}$$

The range for c_2 is obtained by adding the current value of $c_2 = 2$ to both sides of the above inequality:

$$-4.5 + 2 \leq c_2 \leq 8 + 2; \quad -2.5 \leq c_2 \leq 10 \qquad \text{(e)}$$

Thus, if c_2 is changed from 2 to 3, the new cost function value is given as

$$f^*_{\text{new}} = f^* + \Delta c_1 x^*_2 = -18 + (3 - 2)(1) = -17 \qquad \text{(f)}$$

Note that the range for the coefficients of the maximization function ($z = 5x_1 - 2x_2$) can be obtained from Eqs. (b) and (e). To determine these ranges, we multiply Eqs. (b) and (e) by -1. Therefore, the ranges for $d_1 = 5$ and $d_2 = -2$ are given as:

$$1 \leq d_1 \leq \infty; \quad -10 \leq d_2 \leq 2.5 \qquad \text{(g)}$$

EXAMPLE 8.23 RANGES FOR COST COEFFICIENTS—EQUALITY AND "≥ TYPE" CONSTRAINTS

Find ranges for the cost coefficients of the problem solved in Example 8.19.

Solution

The final tableau for the problem is given in Table 8.23. In the tableau, x_3 is a slack variable for the first constraint, x_4 is a surplus variable for the third constraint, and x_5 and x_6 are artificial variables for the second and third constraints, respectively. Since both x_1 and x_2 are basic variables, we will use Theorem 8.8 to find ranges for the cost coefficients $c_1 = -1$ and $c_2 = -4$. Note that the problem is solved as the minimization of the cost function $f = -x_1 - 4x_2$. Columns 4, 5, and 6 are nonbasic. However, since artificial columns 5 and 6 must be excluded, only column 4 can be used in Eq. (8.24).

To find the range for Δc_1, $r = 3$ is used because the third row determines x_1 as a basic variable. Using the inequalities of Eq. (8.24) with $r = 3$ and $j = 4$, we have

$$\max\left\{\frac{7}{3} \bigg/ \left(-\frac{1}{3}\right)\right\} \leq \Delta c_1 \leq \infty; \quad \text{or} \quad -7 \leq \Delta c_1 \leq \infty \tag{a}$$

The range for c_1 is obtained by adding the current value of $c_1 = -1$ to both sides of the inequality:

$$-7 - 1 \leq c_1 \leq \infty - 1; \quad -8 \leq c_1 \leq \infty \tag{b}$$

Thus, if c_1 changes from -1 to -2, the new optimum cost function value is given as

$$f_{new}^* = f^* + \Delta c_1 x_1^* = -\frac{13}{3} + (-2 - (-1))\left(\frac{5}{3}\right) = -6 \tag{c}$$

For the second cost coefficient, $r = 2$ because the second row determines x_2 as a basic variable. Using Eq. (8.24) with $r = 2$ and $j = 4$, the ranges for Δc_2 and for $c_2 = -4$ are given as

$$-\infty \leq \Delta c_2 \leq 3.5; \quad -\infty \leq c_2 \leq -0.5 \tag{d}$$

If c_2 changes from -4 to -3, the new value of the cost function is given as

$$f_{new}^* = f^* + \Delta c_2 x_2^* = -\frac{13}{3} + (-3 - (-4))\left(\frac{2}{3}\right) = -\frac{11}{3} \tag{e}$$

The ranges for coefficients of the *maximization function* $(z = x_1 + 4x_2)$ are obtained by multiplying the preceding ranges by -1:

$$-\infty \leq c_1 \leq 8 \, (-\infty \leq \Delta c_1 \leq 7) \quad \text{and} \quad 0.5 \leq c_2 \leq \infty \, (-3.5 \leq \Delta c_2 \leq \infty) \tag{e}$$

*8.7.4 Changes in the Coefficient Matrix

Any change in the coefficient matrix $\mathbf{A}$ in Eq. (8.11) changes the feasible region for the problem. This may change the optimum solution to the problem depending on whether the change is associated with a basic variable. Let a_{ij} be replaced by $a_{ij} + \Delta a_{ij}$. We will determine limits for Δa_{ij} so that with minor computations the optimum solution to the changed problem can be obtained. We must consider the two cases:

1. when the change is associated with a nonbasic variable
2. when the change is associated with a basic variable

Results for these two cases are summarized in Theorems 8.9 and 8.10, respectively.

THEOREM 8.9

Change That Is Associated with a Nonbasic Variable   Let j in a_{ij} be such that x_j is not a basic variable and let k be the column index for the slack or artificial variable associated with the constraint of the ith row. Define a vector

$$\mathbf{c}_B = [c_{B1} \ c_{B2} \ \ldots \ c_{Bm}]^T \qquad (8.25)$$

where $c_{Bi} = c_j$ if $x_j^* = b_i^*$, $i = 1$ to m (i.e., the index i corresponds to the row that determines the optimum value of variable x_j). Recall that m is the number of constraints. Also define a scalar

$$R = \sum_{r=1}^{m} c_{Br} a'_{rk} \qquad (8.26)$$

With this notation, if Δa_{ij} satisfies one of the following sets of inequalities, then the optimum solution (design variables and cost function) does not change when a_{ij} is replaced by any $a_{ij} + \Delta a_{ij}$:

$$\Delta a_{ij} \geq c'_j / R \text{ when } R > 0, \text{ and}$$
$$\Delta a_{ij} \leq \infty \text{ when } R = 0 \qquad (8.27)$$

or

$$\Delta a_{ij} \leq c'_j / R \text{ when } R < 0, \text{ and}$$
$$\Delta a_{ij} \geq -\infty \text{ when } R = 0 \qquad (8.28)$$

Also, if $R = 0$, the solution does not change for any value of Δa_{ij}.

To use the theorem, a first step is to determine indices j and k. Then we determine the vector $\mathbf{c}_B$ of Eq. (8.25) and the scalar R of Eq. (8.26). Conditions of the inequalities of Eqs. (8.27) and (8.28) then determine whether the given Δa_{ij} will change the optimum solution. If the inequalities are not satisfied, then we have to re-solve the problem to obtain the new solution.

THEOREM 8.10

Change That Is Associated with a Basic Variable   Let j in a_{ij} be such that x_j is a basic variable and let $x_j^* = b_t'$ (i.e., t is the row index that determines the optimum value of x_j).

Let the index k and the scalar R be defined as in Theorem 8.9. Let Δa_{ij} satisfy the following inequalities:

$$\max_{r \neq t} \{b_r' / A_r, A_r < 0\} \leq \Delta a_{ij}$$
$$\leq \min_{r \neq t} \{b_r' / A_r, A_r > 0\} \qquad (8.29)$$

$$A_r = b_t' a'_{rk} - b_r' a'_{tk}, \ r = 1 \text{ to } m; \ r \neq t \qquad (8.30)$$

and

$$\max_q \{-c'_q / B_q, B_q > 0\}$$
$$\leq \Delta a_{ij} \leq \min_q \{-c'_q / B_q, B_q < 0\} \qquad (8.31)$$

$$B_q = (c'_q a'_{tk} + a'_{iq} R) \text{ for all}$$
$$q \text{ not in the basis} \qquad (8.32)$$

and

$$1 + a'_{tk} \Delta a_{ij} > 0 \qquad (8.33)$$

Note that the upper and lower limits on Δa_{ij} do not exist if the corresponding denominators in Eqs. (8.29) and (8.31) do not exist. If Δa_{ij} satisfies the above inequalities, then the optimum solution of the changed problem can be obtained without any further iterations of the Simplex method. If b'_r for $r = 1$ to m is replaced by

$$b''_r = b'_r - \Delta a_{ij} a'_{rk}/(1 + \Delta a_{ij} a'_{tk}), \quad r = 1 \text{ to } m; \quad r \neq t$$
$$b''_t = b'_t/(1 + \Delta a_{ij} a'_{tk})$$

(8.34)

in the final tableau, then the new optimum values for the basic variables can be obtained when a_{ij} is replaced by $a_{ij} + \Delta a_{ij}$. In other words, if $x^*_j = b'_r$, then $x'_j = b''_r$, where x'_j refers to the optimum solution to the changed problem.

To use the theorem, we need to determine indices j, t, and k. Then we determine the constants A_r and B_q from Eqs. (8.30) and (8.32). With these, ranges on Δa_{ij} can be determined from the inequalities of Eqs. (8.29) and (8.31). If these inequalities are satisfied by Δa_{ij}, Eq. (8.34) determines the new solution. If the inequalities are not satisfied, the problem must be re-solved for the new solution.

EXERCISES FOR CHAPTER 8

Section 8.2 Definition of a Standard Linear Programming Problem

8.1 *Answer True or False.*

1. A linear programming problem having maximization of a function cannot be transcribed into the standard LP form.
2. A surplus variable must be added to a "$\leq$ type" constraint in the standard LP formulation.
3. A slack variable for an LP constraint can have a negative value.
4. A surplus variable for an LP constraint must be non-negative.
5. If a "$\leq$ type" constraint is active, its slack variable must be positive.
6. If a "$\geq$ type" constraint is active, its surplus variable must be zero.
7. In the standard LP formulation, the resource limits are free in sign.
8. Only "$\leq$ type" constraints can be transcribed into the standard LP form.
9. Variables that are free in sign can be treated in any LP problem.
10. In the standard LP form, all the cost coefficients must be positive.
11. All variables must be non-negative in the standard LP definition.

Convert the following problems to the standard LP form.

8.2 Minimize $f = 5x_1 + 4x_2 - x_3$
subject to $x_1 + 2x_2 - x_3 \geq 1$
$\quad\quad 2x_1 + x_2 + x_3 \geq 4$
$\quad\quad x_1, x_2 \geq 0; x_3$ is unrestricted in sign

8.3 Maximize $z = x_1 + 2x_2$
subject to $-x_1 + 3x_2 \leq 10$
$\quad\quad x_1 + x_2 \leq 6$
$\quad\quad x_1 - x_2 \leq 2$
$\quad\quad x_1 + 3x_2 \geq 6$
$\quad\quad x_1, x_2 \geq 0$

8.4 Minimize $f = 2x_1 - 3x_2$
subject to $x_1 + x_2 \leq 1$
$$-2x_1 + x_2 \geq 2$$
$$x_1, x_2 \geq 0$$

8.5 Maximize $z = 4x_1 + 2x_2$
subject to $-2x_1 + x_2 \leq 4$
$$x_1 + 2x_2 \geq 2$$
$$x_1, x_2 \geq 0$$

8.6 Maximize $z = x_1 + 4x_2$
subject to $x_1 + 2x_2 \leq 5$
$$x_1 + x_2 = 4$$
$$x_1 - x_2 \geq 3$$
$$x_1, x_2 \geq 0$$

8.7 Maximize $z = x_1 + 4x_2$
subject to $x_1 + 2x_2 \leq 5$
$$2x_1 + x_2 = 4$$
$$x_1 - x_2 \geq 1$$
$$x_1, x_2 \geq 0$$

8.8 Minimize $f = 9x_1 + 2x_2 + 3x_3$
subject to $-2x_1 - x_2 + 3x_3 \leq -5$
$$x_1 - 2x_2 + 2x_3 \geq -2$$
$$x_1, x_2, x_3 \geq 0$$

8.9 Minimize $f = 5x_1 + 4x_2 - x_3$
subject to $x_1 + 2x_2 - x_3 \geq 1$
$$2x_1 + x_2 + x_3 \geq 4$$
$$x_1, x_2 \geq 0; x_3 \text{ is unrestricted in sign}$$

8.10 Maximize $z = -10x_1 - 18x_2$
subject to $x_1 - 3x_2 \leq -3$
$$2x_1 + 2x_2 \geq 5$$
$$x_1, x_2 \geq 0$$

8.11 Minimize $f = 20x_1 - 6x_2$
subject to $3x_1 - x_2 \geq 3$
$$-4x_1 + 3x_2 = -8$$
$$x_1, x_2 \geq 0$$

8.12 Maximize $z = 2x_1 + 5x_2 - 4.5x_3 + 1.5x_4$
subject to $5x_1 + 3x_2 + 1.5x_3 \leq 8$
$$1.8x_1 - 6x_2 + 4x_3 + x_4 \geq 3$$
$$-3.6x_1 + 8.2x_2 + 7.5x_3 + 5x_4 = 15$$
$$x_i \geq 0; i = 1 \text{ to } 4$$

8.13 Minimize $f = 8x_1 - 3x_2 + 15x_3$
subject to $5x_1 - 1.8x_2 - 3.6x_3 \geq 2$
$$3x_1 + 6x_2 + 8.2x_3 \geq 5$$
$$1.5x_1 - 4x_2 + 7.5x_3 \geq -4.5$$
$$-x_2 + 5x_3 \geq 1.5$$
$$x_1, x_2 \geq 0; x_3 \text{ is unrestricted in sign}$$

8.14 Maximize $z = 10x_1 + 6x_2$
subject to $2x_1 + 3x_2 \leq 90$
$4x_1 + 2x_2 \leq 80$
$x_2 \geq 15$
$5x_1 + x_2 = 25$
$x_1, x_2 \geq 0$

8.15 Maximize $z = -2x_1 + 4x_2$
subject to $2x_1 + x_2 \geq 3$
$2x_1 + 10x_2 \leq 18$
$x_1, x_2 \geq 0$

8.16 Maximize $z = x_1 + 4x_2$
subject to $x_1 + 2x_2 \leq 5$
$2x_1 + x_2 = 4$
$x_1 - x_2 \geq 3$
$x_1 \geq 0$, x_2 is unrestricted in sign

8.17 Minimize $f = 3x_1 + 2x_2$
subject to $x_1 - x_2 \geq 0$
$x_1 + x_2 \geq 2$
$x_1, x_2 \geq 0$

8.18 Maximize $z = 3x_1 + 2x_2$
subject to $x_1 - x_2 \geq 0$
$x_1 + x_2 \geq 2$
$2x_1 + x_2 \leq 6$
$x_1, x_2 \geq 0$

8.19 Maximize $z = x_1 + 2x_2$
subject to $3x_1 + 4x_2 \leq 12$
$x_1 + 3x_2 \geq 3$
$x_1 \geq 0$; x_2 is unrestricted in sign

Section 8.3 Basic Concepts Related to Linear Programming Problems, and Section 8.4 Calculation of Basic Solutions

8.20 *Answer True or False.*

1. In the standard LP definition, the number of constraint equations (i.e., rows in the matrix **A**) must be less than the number of variables.

2. In an LP problem, the number of "$\leq$ type" constraints cannot be more than the number of design variables.

3. In an LP problem, the number of "$\geq$ type" constraints cannot be more than the number of design variables.

4. An LP problem has an infinite number of basic solutions.

5. A basic solution must have zero value for some of the variables.

6. A basic solution can have negative values for some of the variables.

7. A degenerate basic solution has exactly m variables with nonzero values, where m is the number of equations.

8. A basic feasible solution has all variables with non-negative values.

9. A basic feasible solution must have m variables with positive values, where m is the number of equations.

II. NUMERICAL METHODS FOR CONTINUOUS VARIABLE OPTIMIZATION

10. The optimum point for an LP problem can be inside the feasible region.
11. The optimum point for an LP problem lies at a vertex of the feasible region.
12. The solution to any LP problem is only a local optimum.
13. The solution to any LP problem is a unique global optimum.

Find all the basic solutions for the following LP problems using the Gauss-Jordan elimination method. Identify basic feasible solutions and show them on graph paper.

8.21 Maximize $z = x_1 + 4x_2$

subject to $x_1 + 2x_2 \leq 5$

$$2x_1 + x_2 = 4$$
$$x_1 - x_2 \geq 1$$
$$x_1, x_2 \geq 0$$

8.22 Maximize $z = -10x_1 - 18x_2$

subject to $x_1 - 3x_2 \leq -3$

$$2x_1 + 2x_2 \geq 5$$
$$x_1, x_2 \geq 0$$

8.23 Maximize $z = x_1 + 2x_2$

subject to $3x_1 + 4x_2 \leq 12$

$$x_1 + 3x_2 \geq 3$$
$$x_1 \geq 0, x_2 \text{ is unrestricted in sign}$$

8.24 Minimize $f = 20x_1 - 6x_2$

subject to $3x_1 - x_2 \geq 3$

$$-4x_1 + 3x_2 = -8$$
$$x_1, x_2 \geq 0$$

8.25 Maximize $z = 5x_1 - 2x_2$

subject to $2x_1 + x_2 \leq 9$

$$x_1 - 2x_2 \leq 2$$
$$-3x_1 + 2x_2 \leq 3$$
$$x_1, x_2 \geq 0$$

8.26 Maximize $z = x_1 + 4x_2$

subject to $x_1 + 2x_2 \leq 5$

$$x_1 + x_2 = 4$$
$$x_1 - x_2 \geq 3$$
$$x_1, x_2 \geq 0$$

8.27 Minimize $f = 5x_1 + 4x_2 - x_3$

subject to $x_1 + 2x_2 - x_3 \geq 1$

$$2x_1 + x_2 + x_3 \geq 4$$
$$x_1, x_3 \geq 0; x_2 \text{ is unrestricted in sign}$$

8.28 Minimize $f = 9x_1 + 2x_2 + 3x_3$

subject to $-2x_1 - x_2 + 3x_3 \leq -5$

$$x_1 - 2x_2 + 2x_3 \geq -2$$
$$x_1, x_2, x_3 \geq 0$$

8.29 Maximize $z = 4x_1 + 2x_2$

    subject to $-2x_1 + x_2 \leq 4$

$$x_1 + 2x_2 \geq 2$$

$$x_1, x_2 \geq 0$$

8.30 Maximize $z = 3x_1 + 2x_2$

    subject to $x_1 - x_2 \geq 0$

$$x_1 + x_2 \geq 2$$

$$x_1, x_2 \geq 0$$

8.31 Maximize $z = 4x_1 + 5x_2$

    subject to $-x_1 + 2x_2 \leq 10$

$$3x_1 + 2x_2 \leq 18$$

$$x_1, x_2 \geq 0$$

Section 8.5 The Simplex Method

Solve the following problems by the Simplex method and verify the solution graphically whenever possible.

8.32 Maximize $z = x_1 + 0.5x_2$

    subject to $6x_1 + 5x_2 \leq 30$

$$3x_1 + x_2 \leq 12$$

$$x_1 + 3x_2 \leq 12$$

$$x_1, x_2 \geq 0$$

8.33 Maximize $z = 3x_1 + 2x_2$

    subject to $3x_1 + 2x_2 \leq 6$

$$-4x_1 + 9x_2 \leq 36$$

$$x_1, x_2 \geq 0$$

8.34 Maximize $z = x_1 + 2x_2$

    subject to $-x_1 + 3x_2 \leq 10$

$$x_1 + x_2 \leq 6$$

$$x_1 - x_2 \leq 2$$

$$x_1, x_2 \geq 0$$

8.35 Maximize $z = 2x_1 + x_2$

    subject to $-x_1 + 2x_2 \leq 10$

$$3x_1 + 2x_2 \leq 18$$

$$x_1, x_2 \geq 0$$

8.36 Maximize $z = 5x_1 - 2x_2$

    subject to $2x_1 + x_2 \leq 9$

$$x_1 - x_2 \leq 2$$

$$-3x_1 + 2x_2 \leq 3$$

$$x_1, x_2 \geq 0$$

8.37 Minimize $f = 2x_1 - x_2$

    subject to $-x_1 + 2x_2 \leq 10$

$$3x_1 + 2x_2 \leq 18$$

$$x_1, x_2 \geq 0$$

8.38 Minimize $f = -x_1 + x_2$
subject to $2x_1 + x_2 \leq 4$
$-x_1 - 2x_2 \geq -4$
$x_1, x_2 \geq 0$

8.39 Maximize $z = 2x_1 - x_2$
subject to $x_1 + 2x_2 \leq 6$
$2 \geq x_1$
$x_1, x_2 \geq 0$

8.40 Maximize $z = x_1 + x_2$
subject to $4x_1 + 3x_2 \leq 12$
$x_1 + 2x_2 \leq 4$
$x_1, x_2 \geq 0$

8.41 Maximize $z = -2x_1 + x_2$
subject to $x_1 \leq 2$
$x_1 + 2x_2 \leq 6$
$x_1, x_2 \geq 0$

8.42 Maximize $z = 2x_1 + x_2$
subject to $4x_1 + 3x_2 \leq 12$
$x_1 + 2x_2 \leq 4$
$x_1, x_2 \geq 0$

8.43 Minimize $f = 9x_1 + 2x_2 + 3x_3$
subject to $2x_1 + x_2 - 3x_3 \geq -5$
$x_1 - 2x_2 + 2x_3 \geq -2$
$x_1, x_2, x_3 \geq 0$

8.44 Maximize $z = x_1 + x_2$
subject to $4x_1 + 3x_2 \leq 9$
$x_1 + 2x_2 \leq 6$
$2x_1 + x_2 \leq 6$
$x_1, x_2 \geq 0$

8.45 Minimize $f = -x_1 - 4x_2$
subject to $x_1 + x_2 \leq 16$
$x_1 + 2x_2 \leq 28$
$24 \geq 2x_1 + x_2$
$x_1, x_2 \geq 0$

8.46 Minimize $f = x_1 - x_2$
subject to $4x_1 + 3x_2 \leq 12$
$x_1 + 2x_2 \leq 4$
$4 \geq 2x_1 + x_2$
$x_1, x_2 \geq 0$

8.47 Maximize $z = 2x_1 + 3x_2$
subject to $x_1 + x_2 \leq 16$
$-x_1 - 2x_2 \geq -28$
$24 \geq 2x_1 + x_2$
$x_1, x_2 \geq 0$

8.48 Maximize $z = x_1 + 2x_2$

subject to $2x_1 - x_2 \geq 0$

$2x_1 + 3x_2 \geq -6$

$x_1, x_2 \geq 0$

8.49 Maximize $z = 2x_1 + 2x_2 + x_3$

subject to $10x_1 + 9x_3 \leq 375$

$x_1 + 3x_2 + x_3 \leq 33$

$2 \geq x_3$

$x_1, x_2, x_3 \geq 0$

8.50 Maximize $z = x_1 + 2x_2$

subject to $-2x_1 - x_2 \geq -5$

$3x_1 + 4x_2 \leq 10$

$x_1 \leq 2$

$x_1, x_2 \geq 0$

8.51 Minimize $f = -2x_1 - x_2$

subject to $-2x_1 - x_2 \geq -5$

$3x_1 + 4x_2 \leq 10$

$x_1 \leq 3$

$x_1, x_2 \geq 0$

8.52 Maximize $z = 12x_1 + 7x_2$

subject to $2x_1 + x_2 \leq 5$

$3x_1 + 4x_2 \leq 10$

$x_1 \leq 2$

$x_2 \leq 3$

$x_1, x_2 \geq 0$

8.53 Maximize $z = 10x_1 + 8x_2 + 5x_3$

subject to $10x_1 + 9x_2 \leq 375$

$5x_1 + 15x_2 + 3x_3 \leq 35$

$3 \geq x_3$

$x_1, x_2, x_3 \geq 0$

Section 8.6 Two-Phase Simplex Method—Artificial Variables

8.54 *Answer True or False.*

1. A pivot step of the Simplex method replaces a current basic variable with a nonbasic variable.
2. The pivot step brings the design point to the interior of the constraint set.
3. The pivot column in the Simplex method is determined by the largest reduced cost coefficient corresponding to a basic variable.
4. The pivot row in the Simplex method is determined by the largest ratio of right-side parameters with the positive coefficients in the pivot column.
5. The criterion for a current basic variable to leave the basic set is to keep the new solution basic and feasible.
6. A move from one basic feasible solution to another corresponds to extreme points of the convex polyhedral set.

II. NUMERICAL METHODS FOR CONTINUOUS VARIABLE OPTIMIZATION

7. A move from one basic feasible solution to another can increase the cost function value in the Simplex method.

8. The right sides in the Simplex tableau can assume negative values.

9. The right sides in the Simplex tableau can become zero.

10. The reduced cost coefficients corresponding to the basic variables must be positive at the optimum.

11. If a reduced cost coefficient corresponding to a nonbasic variable is zero at the optimum point, there may be multiple solutions to the problem.

12. If all elements in the pivot column are negative, the problem is infeasible.

13. The artificial variables must be positive in the final solution.

14. If artificial variables are positive at the final solution, the artificial cost function is also positive.

15. If artificial cost function is positive at the optimum solution, the problem is unbounded.

Solve the following LP problems by the Simplex method and verify the solution graphically, whenever possible.

8.55 Maximize $z = x_1 + 2x_2$
subject to $-x_1 + 3x_2 \leq 10$
$$x_1 + x_2 \leq 6$$
$$x_1 - x_2 \leq 2$$
$$x_1 + 3x_2 \geq 6$$
$$x_1, x_2 \geq 0$$

8.56 Maximize $z = 4x_1 + 2x_2$
subject to $-2x_1 + x_2 \leq 4$
$$x_1 + 2x_2 \geq 2$$
$$x_1, x_2 \geq 0$$

8.57 Maximize $z = x_1 + 4x_2$
subject to $x_1 + 2x_2 \leq 5$
$$x_1 + x_2 = 4$$
$$x_1 - x_2 \geq 3$$
$$x_1, x_2 \geq 0$$

8.58 Maximize $z = x_1 + 4x_2$
subject to $x_1 + 2x_2 \leq 5$
$$2x_1 + x_2 = 4$$
$$x_1 - x_2 \geq 1$$
$$x_1, x_2 \geq 0$$

8.59 Minimize $f = 3x_1 + x_2 + x_3$
subject to $-2x_1 - x_2 + 3x_3 \leq -5$
$$x_1 - 2x_2 + 3x_3 \geq -2$$
$$x_1, x_2, x_3 \geq 0$$

8.60 Minimize $f = 5x_1 + 4x_2 - x_3$
subject to $x_1 + 2x_2 - x_3 \geq 1$
$$2x_1 + x_2 + x_3 \geq 4$$
$$x_1, x_2 \geq 0; x_3 \text{ is unrestricted in sign}$$

8.61 Maximize $z = -10x_1 - 18x_2$

subject to $x_1 - 3x_2 \leq -3$

$2x_1 + 2x_2 \geq 5$

$x_1, x_2 \geq 0$

8.62 Minimize $f = 20x_1 - 6x_2$

subject to $3x_1 - x_2 \geq 3$

$-4x_1 + 3x_2 = -8$

$x_1, x_2 \geq 0$

8.63 Maximize $z = 2x_1 + 5x_2 - 4.5x_3 + 1.5x_4$

subject to $5x_1 + 3x_2 + 1.5x_3 \leq 8$

$1.8x_1 - 6x_2 + 4x_3 + x_4 \geq 3$

$-3.6x_1 + 8.2x_2 + 7.5x_3 + 5x_4 = 15$

$x_i \geq 0; i = 1$ to 4

8.64 Minimize $f = 8x_1 - 3x_2 + 15x_3$

subject to $5x_1 - 1.8x_2 - 3.6x_3 \geq 2$

$3x_1 + 6x_2 + 8.2x_3 \geq 5$

$1.5x_1 - 4x_2 + 7.5x_3 \geq -4.5$

$-x_2 + 5x_3 \geq 1.5$

$x_1, x_2 \geq 0; x_3$ is unrestricted in sign

8.65 Maximize $z = 10x_1 + 6x_2$

subject to $2x_1 + 3x_2 \leq 90$

$4x_1 + 2x_2 \leq 80$

$x_2 \geq 15$

$5x_1 + x_2 = 25$

$x_1, x_2 \geq 0$

8.66 Maximize $z = -2x_1 + 4x_2$

subject to $2x_1 + x_2 \geq 3$

$2x_1 + 10x_2 \leq 18$

$x_1, x_2 \geq 0$

8.67 Maximize $z = x_1 + 4x_2$

subject to $x_1 + 2x_2 \leq 5$

$2x_1 + x_2 = 4$

$x_1 - x_2 \geq 3$

$x_1 \geq 0; x_2$ is unrestricted in sign

8.68 Minimize $f = 3x_1 + 2x_2$

subject to $x_1 - x_2 \geq 0$

$x_1 + x_2 \geq 2$

$x_1, x_2 \geq 0$

8.69 Maximize $z = 3x_1 + 2x_2$

subject to $x_1 - x_2 \geq 0$

$x_1 + x_2 \geq 2$

$2x_1 + x_2 \leq 6$

$x_1, x_2 \geq 0$

II. NUMERICAL METHODS FOR CONTINUOUS VARIABLE OPTIMIZATION

8.70 Maximize $z = x_1 + 2x_2$

subject to $3x_1 + 4x_2 \le 12$

$x_1 + 3x_2 \le 3$

$x_1 \ge 0$; x_2 is unrestricted in sign

8.71 Minimize $f = x_1 + 2x_2$

subject to $-x_1 + 3x_2 \le 20$

$x_1 + x_2 \le 6$

$x_1 - x_2 \le 12$

$x_1 + 3x_2 \ge 6$

$x_1, x_2 \ge 0$

8.72 Maximize $z = 3x_1 + 8x_2$

subject to $3x_1 + 4x_2 \le 20$

$x_1 + 3x_2 \ge 6$

$x_1 \ge 0$; x_2 is unrestricted in sign

8.73 Minimize $f = 2x_1 - 3x_2$

subject to $x_1 + x_2 \le 1$

$-2x_1 + x_2 \ge 2$

$x_1, x_2 \ge 0$

8.74 Minimize $f = 3x_1 - 3x_2$

subject to $-x_1 + x_2 \le 0$

$x_1 + x_2 \ge 2$

$x_1, x_2 \ge 0$

8.75 Minimize $f = 5x_1 + 4x_2 - x_3$

subject to $x_1 + 2x_2 - x_3 \ge 1$

$2x_1 + x_2 + x_3 \ge 4$

$x_1, x_2 \ge 0$; x_3 is unrestricted in sign

8.76 Maximize $z = 4x_1 + 5x_2$

subject to $x_1 - 2x_2 \le -10$

$3x_1 + 2x_2 \le 18$

$x_1, x_2 \ge 0$

8.77 Formulate and solve the optimum design problem of Exercise 2.2. Verify the solution graphically.

8.78 Formulate and solve the optimum design problem of Exercise 2.6. Verify the solution graphically.

8.79 Formulate and solve the optimum design problem of Exercise 2.7. Verify the solution graphically.

8.80 Formulate and solve the optimum design problem of Exercise 2.8. Verify the solution graphically.

*8.81 Formulate and solve the optimum design problem of Exercise 2.18.

*8.82 Formulate and solve the optimum design problem of Exercise 2.20.

8.83 Solve the "saw mill" problem formulated in Section 2.4.

*8.84 Formulate and solve the optimum design problem of Exercise 2.21.

*8.85 Obtain solutions for the three formulations of the "cabinet design" problem given in Section 2.6. Compare solutions for the three formulations.

Section 8.7 Postoptimality Analysis

8.86 Formulate and solve the "crude oil" problem stated in Exercise 2.2. What is the effect on the cost function if the market for lubricating oil suddenly increases to 12,000 barrels? What is the effect on the solution if the price of Crude A drops to $110/bbl? Verify the solutions graphically.

8.87 Formulate and solve the problem stated in Exercise 2.6. What are the effects of the following changes? Verify your solutions graphically.
  1. The supply of material C increases to 120kg.
  2. The supply of material D increases to 100kg.
  3. The market for product A decreases to 60.
  4. The profit for A decreases to $8/kg.

Solve the following problems and determine Lagrange multipliers for the constraints at the optimum point.

8.88 Exercise 8.55	**8.89** Exercise 8.56	**8.90** Exercise 8.57
8.91 Exercise 8.58	**8.92** Exercise 8.59	**8.93** Exercise 8.60
8.94 Exercise 8.61	**8.95** Exercise 8.62	**8.96** Exercise 8.63
8.97 Exercise 8.64	**8.98** Exercise 8.65	**8.99** Exercise 8.66
8.100 Exercise 8.67	**8.101** Exercise 8.68	**8.102** Exercise 8.69
8.103 Exercise 8.70	**8.104** Exercise 8.71	**8.105** Exercise 8.72
8.106 Exercise 8.73	**8.107** Exercise 8.74	**8.108** Exercise 8.75
8.109 Exercise 8.76		

Solve the following problems and determine ranges for the right-side parameters.

8.110 Exercise 8.55	**8.111** Exercise 8.56	**8.112** Exercise 8.57
8.113 Exercise 8.58	**8.114** Exercise 8.59	**8.115** Exercise 8.60
8.116 Exercise 8.61	**8.117** Exercise 8.62	**8.118** Exercise 8.63
8.119 Exercise 8.64	**8.120** Exercise 8.65	**8.121** Exercise 8.66
8.122 Exercise 8.67	**8.123** Exercise 8.68	**8.124** Exercise 8.69
8.125 Exercise 8.70	**8.126** Exercise 8.71	**8.127** Exercise 8.72
8.128 Exercise 8.73	**8.129** Exercise 8.74	**8.130** Exercise 8.75
8.131 Exercise 8.76		

Solve the following problems and determine ranges for the coefficients of the objective function.

8.132 Exercise 8.55	**8.133** Exercise 8.56	**8.134** Exercise 8.57
8.135 Exercise 8.58	**8.136** Exercise 8.59	**8.137** Exercise 8.60
8.138 Exercise 8.61	**8.139** Exercise 8.62	**8.140** Exercise 8.63
8.141 Exercise 8.64	**8.142** Exercise 8.65	**8.143** Exercise 8.66
8.144 Exercise 8.67	**8.145** Exercise 8.68	**8.146** Exercise 8.69
8.147 Exercise 8.70	**8.148** Exercise 8.71	**8.149** Exercise 8.72
8.150 Exercise 8.73	**8.151** Exercise 8.74	**8.152** Exercise 8.75
8.153 Exercise 8.76		

*8.154 Formulate and solve the optimum design problem of Exercise 2.2. Determine Lagrange multipliers for the constraints. Calculate the ranges for the right-side parameters, and the coefficients of the objective function. Verify your results graphically.

*8.155 Formulate and solve the optimum design problem of Exercise 2.6. Determine Lagrange multipliers for the constraints. Calculate the ranges for the parameters of the right side and the coefficients of the objective function. Verify your results graphically.

8.156 Formulate and solve the "diet" problem stated in Exercise 2.7. Investigate the effect on the optimum solution of the following changes:
1. The cost of milk increases to $1.20/kg.
2. The need for vitamin A increases to 6 units.
3. The need for vitamin B decreases to 3 units.
Verify the solution graphically.

8.157 Formulate and solve the problem stated in Exercise 2.8. Investigate the effect on the optimum solution of the following changes:
1. The supply of empty bottles decreases to 750.
2. The profit on a bottle of wine decreases to $0.80.
3. Only 200 bottles of alcohol can be produced.

*8.158 Formulate and solve the problem stated in Exercise 2.18. Investigate the effect on the optimum solution of the following changes:
1. The profit on margarine increases to $0.06/kg.
2. The supply of milk base substances increases to 2500kg.
3. The supply of soybeans decreases to 220,000kg.

8.159 Solve the "saw mill" problem formulated in Section 2.4. Investigate the effect on the optimum solution of the following changes:
1. The transportation cost for the logs increases to $0.16 per kilometer per log.
2. The capacity of Mill A decreases to 200logs/day.
3. The capacity of Mill B decreases to 270logs/day.

*8.160 Formulate and solve the problem stated in Exercise 2.20. Investigate the effect on the optimum solution of the following changes:
1. Due to demand on capital, the available cash decreases to $1.8 million.
2. The initial investment for truck B increases to $65,000.
3. Maintenance capacity decreases to 28 trucks.

*8.161 Formulate and solve the "steel mill" problem stated in Exercise 2.21. Investigate the effect on the optimum solution of the following changes:
1. The capacity of reduction plant 1 increases to 1,300,000.
2. The capacity of reduction plant 2 decreases to 950,000.
3. The capacity of fabricating plant 2 increases to 250,000.
4. The demand for product 2 increases to 130,000.
5. The demand for product 1 decreases to 280,000.

*8.162 Obtain solutions for the three formulations of the "cabinet design" problem given in Section 2.6. Compare the three formulations. Investigate the effect on the optimum solution of the following changes:
1. Bolting capacity is decreased to 5500/day.
2. The cost of riveting the C_1 component increases to $0.70.
3. The company must manufacture only 95 devices per day.

8.163 Given the following problem:

Minimize $f = 2x_1 - 4x_2$
subject to $g_1 = 10x_1 + 5x_2 \leq 15$
$\qquad\quad g_2 = 4x_1 + 10x_2 \leq 36$
$\qquad\quad x_1 \geq 0, x_2 \geq 0$

Slack variables for g_1 and g_2 are x_3 and x_4, respectively. The final tableau for the problem is given in Table E8.163. Using the given tableau:

1. Determine the optimum values of f and $\mathbf{x}$.
2. Determine Lagrange multipliers for g_1 and g_2.
3. Determine the ranges for the right sides of g_1 and g_2.
4. What is the smallest value that f can have, with the current basis, if the right side of g_1 is changed? What is the right side of g_1 for that case?

TABLE E8.163　Final tableau for Exercise 8.163

x_1	x_2	x_3	x_4	b
2	1	$\frac{1}{5}$	0	3
−16	0	−2	1	6
10	0	$\frac{4}{5}$	0	$f + 12$

More on Linear Programming Methods for Optimum Design

Upon completion of this chapter, you will be able to

- Derive the Simplex method and understand the theory behind its steps

- Use an alternate form of the two-phase Simplex method called the Big-M method

- Write a dual problem for the given LP problem

- Recover the solution to the original LP problem from the solution to the dual problem

- Solve quadratic programming (QP) problems using the Simplex method

This chapter presents some additional topics related to linear programming problems. These topics are usually not covered in an undergraduate course on optimum design. They may also be omitted in the first independent reading of the book.

9.1 DERIVATION OF THE SIMPLEX METHOD

In the previous chapter, we presented the basic ideas and concepts of the Simplex method. The steps of the method were described and illustrated in several examples. In this section, we describe the theory that leads to the steps used in the example problems.

9.1.1 General Solution to $Ax = b$

Canonical Form

An $m \times n$ system of simultaneous equations $\mathbf{A}\mathbf{x} = \mathbf{b}$ with rank $(\mathbf{A}) = m$ is said to be in the *canonical form* if each equation has a variable (with unit coefficient) that does not appear in any other equation. A canonical form in general is written as follows:

$$x_i + \sum_{j=m+1}^{n} a_{ij}x_j = b_i; \quad i = 1 \text{ to } m \tag{9.1}$$

Note that variables x_1 to x_m appear in only one of the equations; x_1 appears in the first equation, x_2 in the second equation, and so on. Note also that this sequence of variables x_1 to x_m in Eq. (9.1) is chosen only for convenience. In general, any of the variables x_1 to x_n may be associated with the first equation as long as it does not appear in any other equation. Similarly, the second equation need not be associated with the second variable x_2.

Additionally, it is possible to write the canonical form of Eq. (9.1) as a matrix equation, as was also explained in Section A.4:

$$\mathbf{I}_{(m)}\mathbf{x}_{(m)} + \mathbf{Q}\mathbf{x}_{(n-m)} = \mathbf{b} \tag{9.2}$$

where

$\mathbf{I}_{(m)} = m$-dimensional identity matrix
$\mathbf{x}_{(m)} = [x_1 \, x_2 \, ... \, x_m]^T$; vector of dimension m
$\mathbf{x}_{(n-m)} = [x_{m+1} \, ... \, x_n]^T$; vector of dimension $(n - m)$
$\mathbf{Q} = m \times (n - m)$; matrix consisting of coefficients of the variables x_{m+1} to x_n in Eq. (9.1)
$\mathbf{b} = [b_1 \, b_2 \, ... \, b_m]^T$; vector of dimension m

GENERAL SOLUTION   The canonical form in Eq. (9.1) or Eq. (9.2) gives a *general solution* to $\mathbf{Ax} = \mathbf{b}$ as

$$\mathbf{x}_{(m)} = \mathbf{b} - \mathbf{Q}\mathbf{x}_{(n-m)} \tag{9.3}$$

It is seen that $\mathbf{x}_{(n-m)}$ can be assigned different values and the corresponding values for $\mathbf{x}_{(m)}$ can be calculated from Eq. (9.3). Thus $\mathbf{x}_{(m)}$ are *dependent variables* and $\mathbf{x}_{(n-m)}$ are *independent variables*.

BASIC SOLUTION   A *particular solution* to the equations is obtained if we set the independent variables to zero (i.e., $\mathbf{x}_{(n-m)} = \mathbf{0}$). Then from Eq. (9.3), $\mathbf{x}_{(m)} = \mathbf{b}$. The solution thus obtained is called a *basic solution*.

BASIC FEASIBLE SOLUTION   If the right side parameters b_i are ≥ 0 in Eq. (9.1), then the particular solution (the basic solution) is called a *basic feasible solution*.

NONBASIC VARIABLES   The independent variables that are set to zero in $\mathbf{x}_{(n-m)}$ to obtain a basic solution are called *nonbasic*.

BASIC VARIABLES   The dependent variables $\mathbf{x}_{(m)}$ that are solved from Eq. (9.3) are called *basic*.

THE TABLEAU   It is customary to represent the canonical form in a tableau, as shown in Table 9.1. The leftmost column of the tableau identifies the basic variable associated

TABLE 9.1 Representation of a canonical form in a tableau

No.	Basic↓	x_1	x_2	•	•	•	x_m	x_{m+1}	x_{m+2}	•	•	•	x_n	RS
1	x_1	1	0	•	•	•	0	$a_{1,m+1}$	$a_{1,m+2}$	•	•	•	$a_{1,n}$	b_1
2	x_2	0	1	•	•	•	0	$a_{2,m+1}$	$a_{2,m+2}$	•	•	•	$a_{2,n}$	b_2
3	x_3	0	0	•	•	•	0	$a_{3,m+1}$	$a_{3,m+2}$	•	•	•	$a_{3,n}$	b_3
•	•	•	•	•	•	•	•	•	•	•	•	•	•	•
•	•	•	•	•	•	•	•	•	•	•	•	•	•	•
m	x_m	0	0	•	•	•	1	$a_{m,m+1}$	$a_{m,m+2}$	•	•	•	$a_{m,n}$	b_m

with each row. The rightmost column RS contains the right side of each equation. Each of the remaining columns is associated with a variable of the problem. The tableau contains m identity columns to form the $m \times m$ identity matrix and the remaining columns form the $m \times (n - m)$ matrix **Q**. The tableau gives a basic solution to $\mathbf{Ax} = \mathbf{b}$.

9.1.2 Selection of a Nonbasic Variable that Should Become Basic

If the current basic feasible solution to the problem is not an optimum point, then an improved basic feasible solution needs to be calculated by replacing one of the basic variables with a nonbasic variable. Derivation of the Simplex method is based on answering the two questions posed earlier: (1) which current nonbasic variable should become basic, and (2) which current basic variable should become nonbasic. We will answer the first question in this subsection and the second question in the next subsection.

Cost Function in Terms of Nonbasic Variables

The *main idea* of bringing a nonbasic variable into the basic set is to *improve the design, that is, to reduce the current value of the cost function.* A clue to the desired improvement is obtained if we examine the cost function expression. To do this we *need to transform the cost function to be in terms of the nonbasic variables only.* We substitute the current values of the basic variables from Eq. (9.1) into the cost function to eliminate the basic variables from it. Current values of the basic variables are given in terms of the nonbasic variables from Eq. (9.1) as

$$x_i = b_i - \sum_{j=m+1}^{n} a_{ij} x_j; \quad i = 1 \text{ to } m \tag{9.4}$$

Substituting Eq. (9.4) into the cost function expression in Eq. (8.8) and simplifying, we obtain *an expression for the cost function in terms of the nonbasic variables* ($x_j, j = m + 1$ to n) as

$$f = f_0 + \sum_{j=m+1}^{n} c_j' x_j \tag{9.5}$$

where f_0 is the current value of the cost function given as

$$f_0 = \sum_{i=1}^{m} b_i c_i \tag{9.6}$$

and the parameters c_j' are

$$c_j' = c_j - \sum_{i=1}^{m} a_{ij} c_i; \quad j = (m+1) \text{ to } n \tag{9.7}$$

Since x_j' for $(m+1) \le j \le n$ are nonbasic variables, they have a zero value. Therefore, the current cost function value f is equal to f_0 from Eq. (9.5).

Reduced Cost Coefficients

The cost coefficients c_j' of the nonbasic variables in Eq. (9.7) play a key role in the Simplex method and are called the *reduced or relative cost coefficients*. They are used to identify a nonbasic variable that should become basic to reduce the current value of the cost function. Expressing the cost function in terms of the current nonbasic variables is a key step in the Simplex method. We have seen that this is not difficult to accomplish because the Gauss-Jordan elimination steps can be used routinely on the cost function expression to eliminate basic variables from it. Once this has been done, the reduced cost coefficients c_j' are readily identified.

c_j' **is the reduced cost coefficient associated with the jth nonbasic variable. Since the basic variables do not appear in the cost function, their coefficients have a zero value.**

Optimum Cost Function

In general the reduced cost coefficients c_j' of the nonbasic variables may be positive, negative, or zero. If all c_j' are non-negative, then it is not possible to reduce the cost function any further and the current basic feasible solution is *optimum*. This is revealed by an examination of Eq. (9.5): If any nonbasic variable x_i is made basic (i.e., it attains a positive value) when all c_j' are non-negative, the cost function will either increase or at the most remain the same.

Note that when all c_j' are strictly positive, the optimum solution is unique. If at least one c_j' is zero, then there is a possibility of alternate optima. If the nonbasic variable associated with a zero reduced cost coefficient can be made basic, the extreme point corresponding to an alternate optimum is obtained. Since the reduced cost coefficient is zero, the optimum cost function value will not change, as seen in Eq. (9.5). Any point on the line segment joining the optimum extreme points also corresponds to an optimum. Note that these optima are global as opposed to local, although there is no distinct global optimum. Geometrically, multiple optima for an LP problem imply that the cost function hyperplane is parallel to an active constraint hyperplane.

Selection of a Nonbasic Variable to Become Basic

Let one of c_j' be negative. This identifies the corresponding nonbasic variable x_i to become a basic variable. It is seen from Eq. (9.5) that the current value of f will decrease since x_i will have a positive value (because it will be basic). If more than one negative c_j' is present, a widely used rule of thumb is to choose the nonbasic variable associated with the smallest c_j' (i.e., negative c_j' with the largest absolute value) to become basic. Thus if any c_j' is negative, then it is possible to find a new basic feasible solution (if one exists) that will further reduce the cost function.

Unbounded Problem

Note that if the nonbasic variable associated with the negative reduced cost coefficient c_j' cannot be made basic (e.g., when all a_{ij} in the c_j' column are negative), then the *feasible region is unbounded.*

9.1.3 Selection of a Basic Variable that Should Become Nonbasic

Assume that x_q is a nonbasic variable tapped to become basic. This indicates that the qth nonbasic column should replace some current basic column. After this interchange, there should be all zero elements in the qth column except a positive unit element at one location.

In order to answer the second question posed earlier, which current basic variable should become nonbasic, we need to determine the pivot row for the elimination process. This way the current basic variable associated with that row will become nonbasic after the elimination step. To determine the pivot row, we transfer all of the terms associated with the current nonbasic variable x_q (tapped to become basic) to the right side of the canonical form of Eq. (9.1). The system of equations becomes

$$x_i + \sum_{\substack{j=m+1 \\ j \neq q}}^{n} a_{ij}x_j = b_i - a_{iq}x_q; \quad i = 1 \text{ to } m \tag{9.8}$$

Since the summation term on the left side of Eq. (9.8) is zero, the equation becomes

$$x_i = b_i - a_{iq}x_q; \quad i = 1 \text{ to } m \tag{9.9}$$

Since x_q is to become a basic variable, its value should become non-negative in the new basic feasible solution. The right sides of Eq. (9.9) represent values of the basic variables for the next Simplex iteration once x_q is assigned a value greater than or equal to 0. An examination of these right sides shows that x_q cannot increase arbitrarily. The reason is that if x_q becomes arbitrarily large, then some of the new right-side parameters $(b_i - a_{iq} x_q)$, $i = 1$ to m may become negative. Since right-side parameters are the new values of the basic variables, the new basic solution will not be feasible. Thus for the new solution to be basic and feasible, the following constraint must be satisfied by the right side of Eq. (9.9) in selecting a current basic variable that should become nonbasic (i.e., attain a zero value):

$$b_i - a_{iq}x_q \geq 0; \quad i = 1 \text{ to } m \tag{9.10}$$

Any a_{iq} that are non-positive pose no limit on how much x_q can be increased since Inequality (9.10) remains satisfied; recall that $b_i \geq 0$. For a positive a_{iq}, x_q can be increased from zero until one of the inequalities in Eq. (9.10) becomes active, that is, one of the right sides of Eq. (9.9) becomes zero. A further increase would violate the non-negativity conditions of Eq. (9.10). Thus, the maximum value that the incoming variable x_q can take is given as

$$\frac{b_p}{a_{pq}} = \min_i \left\{ \frac{b_i}{a_{iq}}, \ a_{iq} > 0; \quad i = 1 \text{ to } m \right\} \tag{9.11}$$

where p is the index of the *smallest ratio*, b_t/a_{tq}. Equation (9.11) says that we take ratios of the right-side parameters b_i with the positive elements in the qth column (a_{iq}) and we select the row index p giving the smallest ratio. *In the case of a tie, the choice for the index* p *is arbitrary among the tying indices and in such a case the resulting basic feasible solution will be degenerate.*

Thus, Eq. (9.11) identifies a row with the smallest ratio b_i/a_{iq}. The basic variable x_p associated with this row should become nonbasic. *If all a_{iq} are nonpositive in the qth column, then x_q can be increased indefinitely.* This indicates the LP problem to be *unbounded*. Any practical problem with this situation is not properly constrained, so the problem formulation should be reexamined.

9.1.4 Artificial Cost Function

Artificial Variables

When there are "$\geq$ type" (with positive right side) and equality constraints in the LP problem, an initial basic feasible solution is not readily available. We must use the two-phase Simplex method to solve the problem. To define the Phase I minimization problem, we introduce an artificial variable for each "$\geq$ type" and equality constraint.

For the sake of simplicity of discussion, let us assume that each constraint of the standard LP problem requires an artificial variable in Phase I of the Simplex method. The constraints that do not require an artificial variable can also be treated routinely, as we saw in the example problems in the previous chapter. Recalling that the standard LP problem has n variables and m equality constraints, the constraint equations $\mathbf{Ax} = \mathbf{b}$ augmented with the artificial variables are now given as

$$\sum_{j=1}^{n} a_{ij}x_j + x_{n+i} = b_i; \quad i = 1 \text{ to } m \tag{9.12}$$

where x_{n+i}, $i = 1$ to m are the artificial variables. Thus the initial basic feasible solution to the Phase I problem is given as

Basic variables: $x_{n+i} = b_i$, $i = 1$ to m
Nonbasic variables: $x_j = 0$, $j = 1$ to n

Note that the artificial variables basically augment the convex polyhedron of the original problem. The initial basic feasible solution to the Phase I problem corresponds to an extreme point (vertex) located in the expanded space. The problem now is to traverse the extreme points in the expanded space until an extreme point is reached in the original space. When the original space is reached, all artificial variables will be nonbasic (i.e., they will have zero values) and the artificial cost function will have a zero value. At this point the augmented space is literally removed so that future movements are only among the extreme points of the original space until the optimum point is reached. In short, after creating artificial variables, we eliminate them as quickly as possible.

Artificial Cost Function

To eliminate the artificial variables from the problem, we define an auxiliary function, called the *artificial cost function*, and minimize it subject to the constraints of Eq. (9.12) and the non-negativity of all of the variables. The artificial cost function is simply a sum of all of the artificial variables and will be designated as w:

$$w = x_{n+1} + x_{n+2} + \ldots + x_{n+m} = \sum_{i=1}^{m} x_{n+i} \tag{9.13}$$

The objective of the Phase I problem is to make all of the artificial variables nonbasic so that they have zero value. In that case, the artificial cost function in Eq. (9.13) will be zero, indicating the end of Phase I.

However, the Phase I problem is not yet in a form suitable to initiate the Simplex method. The reason is that the reduced cost coefficients c_j' of the nonbasic variables in the artificial cost function are not yet available to determine the pivot element and perform the pivot step.

Currently, the artificial cost function in Eq. (9.13) is in terms of the basic variables $x_{n+1}, \ldots, x_{n+m}$. Therefore the reduced cost coefficients c_j' cannot be identified. They can be identified only if the artificial cost function w is in terms of the nonbasic variables $x_1, \ldots, x_n$. To obtain w in terms of nonbasic variables, we use the constraint expressions to eliminate the basic variables from the artificial cost function. Calculating $x_{n+1}, \ldots, x_{n+m}$ from Eqs. (9.12) and substituting into Eq. (9.13), we obtain the artificial cost function w in terms of the nonbasic variables as

$$w = \sum_{i=1}^{m} b_i - \sum_{j=1}^{n} \sum_{i=1}^{m} a_{ij} x_j \tag{9.14}$$

The reduced cost coefficients c_j' are identified as the coefficients of the nonbasic variables x_j in Eq. (9.14) as

$$c_j' = - \sum_{i=1}^{m} a_{ij}; \quad j = 1 \text{ to } n \tag{9.15}$$

If there are also "$\leq$ type" constraints in the original problem, these are cast into the standard LP form by adding slack variables that serve as basic variables in Phase I. Therefore, the number of artificial variables is less than m—the total number of constraints. Accordingly, the number of artificial variables required to obtain an initial basic feasible solution is also less than m. This implies that the sums in Eqs. (9.14) and (9.15) are not for all of the m constraints. They are only for the constraints requiring an artificial variable.

9.1.5 The Pivot Step

The pivot step, based on the Guass-Jordan elimination procedure, interchanges a basic variable with a nonbasic variable. Let a basic variable x_p $(1 \leq p \leq m)$ be selected to replace a nonbasic variable x_q for $(n - m) \leq q \leq n$. The pth basic column is to be interchanged with the qth nonbasic column. That is, the qth column will become a column of the identity matrix, and the pth column will no longer be the identity matrix column. This is possible only when the pivot element in the pth column and qth row is nonzero (i.e., $a_{pq} \neq 0$). The current nonbasic variable x_q will be basic if it is eliminated from all of the equations except the pth one. This can be accomplished by performing a *Gauss-Jordan elimination step* on the qth column of the tableau shown earlier in Table 9.1 using the pth row for elimination. This will give $a_{pq} = 1$ and zeros elsewhere in the qth column.

Let a'_{ij} denote the new coefficients in the canonical form of $\mathbf{Ax} = \mathbf{b}$ after the pivot step. Then the *pivot step* for performing elimination in the qth column using the pth row as the pivot row is described by the following general equations.

1. Divide the pivot row (p) by the pivot element a_{pq}:

$$a'_{pj} = a_{pj}/a_{pq} \text{ for } j = 1 \text{ to } n; \quad b'_p = b_p/a_{pq} \tag{9.16}$$

2. Eliminate x_q from all rows except the pth row by performing the Gauss-Jordan elimination step:

$$a'_{ij} = a_{ij} - (a_{pj}/a_{pq})a_{iq}; \quad \begin{cases} i \neq p, \ i = 1 \text{ to } m \\ \quad\quad j = 1 \text{ to } n \end{cases} \tag{9.17}$$

$$b'_i = b_i - (b_p/a_{pq})a_{iq}; \quad i \neq p, \ i = 1 \text{ to } m \tag{9.18}$$

In Eq. (9.16), the pth row of the tableau is simply divided by the pivot element a_{pq}. Equations (9.17) and (9.18)) perform the elimination step in the qth column of the tableau. Elements in the qth column above and below the pth row are reduced to zero by the elimination process, thus eliminating x_q from all of the rows except the pth row. These equations may be coded into a computer program to perform the pivot step. On completion of the pivot step, a new canonical form for the equation $\mathbf{Ax} = \mathbf{b}$ is obtained; that is, a new basic solution to the equations is obtained.

9.1.6 The Simplex Algorithm

The steps of the Simplex method were illustrated in Example 8.7 with only "$\leq$ type" constraints. They are summarized for the general LP problem as follows:

Step 1. *Problem in the standard form.* Transcribe the problem into the standard LP form.

Step 2. *Initial basic feasible solution.* This is readily available if all constraints are "$\leq$ type" because the slack variables are basic and the real variables are nonbasic. If there are equality and/or "$\geq$ type" constraints, then the two-phase Simplex procedure must be used. Introduction of artificial variable for each equality and "≥ type" constraint gives an initial basic feasible solution to the Phase I problem.

Step 3. *Optimality check: The cost function must be in terms of only the nonbasic variables.* This is readily available when there are only "≤ type" constraints. For equality and/or "≥ type" constraints, the artificial cost function for the Phase I problem can also be easily transformed to be in terms of the nonbasic variables.

If all of the reduced cost coefficients for nonbasic variables are non-negative (≥ 0), we have the optimum solution (end of Phase I). Otherwise, there is a possibility of improving the cost function (artificial cost function). We need to select a nonbasic variable that should become basic.

Step 4. *Selection of a nonbasic variable to become basic.* We scan the cost row (the artificial cost row for the Phase I problem) and identify a column having negative reduced cost coefficient because the nonbasic variable associated with this column should become basic to reduce the cost (artificial cost) function from its current value. This is called the *pivot column*.

Step 5. *Selection of a basic variable to become nonbasic.* If all elements in the pivot column are negative, then we have an unbounded problem. If there are positive elements in the pivot column, then we take ratios of the right-side parameters with the positive elements in the pivot column and identify a row with the smallest positive ratio according to Eq. (9.11). In the case of a tie, any row among the tying ratios can be selected. The basic variable associated with this row should become nonbasic (i.e., zero). The selected row is called the *pivot row*, and its intersection with the pivot column identifies the *pivot element*.

Step 6. *Pivot step.* Use the Gauss-Jordan elimination procedure and the pivot row identified in Step 5. *Elimination must also be performed in the cost function (artificial cost) row* so that it is only in terms of nonbasic variables in the next tableau. This step eliminates the nonbasic variable identified in Step 4 from all of the rows except the pivot row; that is, it becomes a basic variable.

Step 7. *Optimum solution.* If the optimum solution is obtained, then read the values of the basic variables and the optimum value of the cost function from the tableau. Otherwise, go to Step 3.

9.2 AN ALTERNATE SIMPLEX METHOD

A slightly different procedure can be used to solve linear programming problems having "≥ type" and equality constraints. The artificial variables are introduced into the problem as before. However, the artificial cost function is not used. Instead, *the original cost function is augmented by adding to it the artificial variables multiplied by large positive constants. The additional terms act as penalties for having artificial variables in the problem.* Since artificial variables are basic, they need to be eliminated from the cost function before the Simplex method can be used to solve the preceding modified problem. This can easily be done using the appropriate equations that contain artificial variables, as in

Eq. (9.14). Once this has been done, the regular Simplex method can be used to solve the problem. We illustrate the procedure, sometimes called the *Big-M Method*, with Example 9.1.

EXAMPLE 9.1   THE BIG-M METHOD FOR EQUALITY AND "≥ TYPE" CONSTRAINTS

Find the numerical solution to the problem given in Example 8.14 using the alternate Simplex procedure:

Maximize

$$z = y_1 + 2y_2 \qquad\qquad\text{(a)}$$

subject to

$$3y_1 + 2y_2 \leq 12 \qquad\qquad\text{(b)}$$

$$2y_1 + 3y_2 \geq 6 \qquad\qquad\text{(c)}$$

$$y_1 \geq 0, \quad y_2 \text{ is unrestricted in sign} \qquad\qquad\text{(d)}$$

Solution

Since y_2 is unrestricted, it has been defined as $y_2 = x_2 - x_3$. Converting the problem to the standard form, and defining $x_1 = y_1$, we obtain:

Minimize

$$f = -x_1 - 2x_2 + 2x_3 \qquad\qquad\text{(e)}$$

subject to

$$3x_1 + 2x_2 - 2x_3 + x_4 = 12 \qquad\qquad\text{(f)}$$

$$2x_1 + 3x_2 - 3x_3 - x_5 + x_6 = 6 \qquad\qquad\text{(g)}$$

$$x_i \geq 0; \quad i = 1 \text{ to } 6 \qquad\qquad\text{(h)}$$

where x_4 is a slack variable, x_5 is a surplus variable, and x_6 is an artificial variable.

Following the alternate Simplex procedure, we add Mx_6 (with, say, $M = 10$) to the cost function and obtain $f = -x_1 - 2x_2 + 2x_3 + 10x_6$. Note that if there is a feasible solution to the problem, then all artificial variables will become nonbasic (i.e., zero), and we will recover the original cost function. Also note that if there are other artificial variables in the problem, they will be multiplied by M and added to the cost function. Now, substituting for x_6 from Eq. (g) into the foregoing cost function, we get the augmented cost function as

$$f = -x_1 - 2x_2 + 2x_3 + 10(6 - 2x_1 - 3x_2 + 3x_3 + x_5) = 60 - 21x_1 - 32x_2 + 32x_3 + 10x_5 \qquad\text{(i)}$$

This is written as $-21x_1 - 32x_2 + 32x_3 + 10x_5 = f - 60$ in the Simplex tableau. With this cost function, iterations of the Simplex method are shown in Table 9.2. It can be seen that the final solution is the same as given in Table 8.18 and Figure 8.4.

TABLE 9.2 Solution to Example 9.1 by alternate Simplex method

Initial tableau: x_6 is identified to be replaced with x_2 in the basic set

Basic↓	x_1	x_2	x_3	x_4	x_5	x_6	b	Ratio
x_4	3	2	−2	1	0	0	12	$\frac{12}{2}=6$
x_6	2	③	−3	0	−1	1	6	$\frac{6}{3}=2$
Cost	−21	−32	32	0	10	0	$f-60$	

Second tableau: x_4 is identified to be replaced with x_5 in the basic set

Basic↓	x_1	x_2	x_3	x_4	x_5	x_6	b	Ratio
x_4	$\frac{5}{3}$	0	0	1	$\frac{2}{3}$	$-\frac{2}{3}$	8	$\frac{8}{2/3}=12$
x_2	$\frac{2}{3}$	1	−1	0	$-\frac{1}{3}$	$\frac{1}{3}$	2	Negative
Cost	$\frac{1}{3}$	0	0	0	$-\frac{2}{3}$	$\frac{32}{3}$	$f+4$	

Third tableau: Reduced cost coefficients in nonbasic columns are non-negative; the tableau gives optimum point

Basic↓	x_1	x_2	x_3	x_4	x_5	x_6	b
x_5	$\frac{5}{2}$	0	0	$\frac{3}{2}$	1	−1	12
x_2	$\frac{3}{2}$	1	−1	$\frac{1}{2}$	0	0	6
Cost	2	0	0	1	0	10	$f+12$

9.3 DUALITY IN LINEAR PROGRAMMING

Associated with every LP problem is another problem called the *dual problem*. The original LP is called the *primal problem*. Some theorems related to dual and primal problems are stated and explained. Dual variables are related to the Lagrange multipliers of the primal constraints. The solution to the dual problem can be recovered from the final primal solution, and vice versa. Therefore, only one of the two problems needs to be solved. This is illustrated with examples.

9.3.1 Standard Primal LP Problem

There are several ways of defining the primal and the corresponding dual problems. We will define a *standard primal problem* as finding $x_1, x_2, \ldots, x_n$ to maximize a primal objective function:

$$z_p = d_1 x_1 + \ldots + d_n x_n = \sum_{i=1}^{n} d_i x_i = \mathbf{d}^T \mathbf{x} \tag{9.19}$$

subject to the constraints

$$a_{11}x_1 + \ldots + a_{1n}x_n \leq e_1$$

$$\ldots \qquad\qquad (\mathbf{Ax} \leq \mathbf{e})$$

$$a_{m1}x_1 + \ldots + a_{mn}x_n \leq e_m$$

$$x_j \geq 0; \quad j = 1 \text{ to } n$$

(9.20)

We will use a subscript p on z to indicate the primal objective function. Also, z is used as the maximization function. It must be understood that in the standard LP problem defined in Eqs. (8.5) through (8.7), all constraints were equalities and right-side parameters b_i were non-negative. However, in the definition of the standard primal problem, all constraints must be "≤ type" and there is no restriction on the sign of the right-side parameters e_i. Therefore, "≥ type" constraints must be multiplied by -1 to convert them to "≤ type." Equalities should be also converted to "≤ type" constraints. This is explained later in this section. Note that to solve the preceding primal LP problem by the Simplex method, we must transform it into the standard Simplex form of Eqs. (8.5) through (8.7).

9.3.2 Dual LP Problem

The dual for the standard primal is defined as follows: Find the dual variables $y_1, y_2, \ldots, y_m$ to minimize a dual objective function:

$$f_d = e_1 y_1 + \ldots + e_m y_m = \sum_{i=1}^{m} e_i y_i = \mathbf{e}^T \mathbf{y} \qquad (9.21)$$

subject to the constraints

$$a_{11}y_1 + \ldots + a_{m1}y_m \geq d_1$$

$$\ldots \qquad\qquad (\mathbf{A}^T\mathbf{y} \geq \mathbf{d})$$

$$a_{1n}y_1 + \ldots + a_{mn}y_m \geq d_n$$

$$y_i \geq 0; \quad i = 1 \text{ to } m$$

(9.22)

We use a subscript d on f to indicate that it is the cost function for the dual problem. Note the following *relations between the primal and dual problems*:

1. The number of dual variables is equal to the number of primal constraints. Each dual variable is associated with a primal constraint. For example, y_i is associated with the ith primal constraint.
2. The number of dual constraints is equal to the number of primal variables. Each primal variable is associated with a dual constraint. For example, x_i is associated with the ith dual constraint.
3. Primal constraints are "≤ type" inequalities, whereas dual constraints are "≥ type."
4. The maximization of the primal objective function is replaced by the minimization of the dual cost function.
5. The coefficients d_i of the primal objective function become the right side of the dual constraints. The right-side parameters e_i of the primal constraints become coefficients for the dual cost function.

6. The coefficient matrix $[a_{ij}]$ of the primal constraints is transposed to $[a_{ji}]$ for the dual constraints.
7. The non-negativity condition applies to both primal and dual variables.

Example 9.2 illustrates how to write the dual problem for a given LP problem.

EXAMPLE 9.2   THE DUAL OF AN LP PROBLEM

Write the dual of the problem

Maximize
$$z_p = 5x_1 - 2x_2 \tag{a}$$

subject to
$$2x_1 + x_2 \le 9 \tag{b}$$
$$x_1 - 2x_2 \le 2 \tag{c}$$
$$-3x_1 + 2x_2 \le 3 \tag{d}$$
$$x_1, x_2 \ge 0 \tag{e}$$

Solution

The problem is already in the standard primal form, and the following associated vectors and matrices can be identified:

$$\mathbf{d} = \begin{bmatrix} 5 \\ -2 \end{bmatrix}, \quad \mathbf{e} = \begin{bmatrix} 9 \\ 2 \\ 3 \end{bmatrix}, \quad \mathbf{A} = \begin{bmatrix} 2 & 1 \\ 1 & -2 \\ -3 & 2 \end{bmatrix} \tag{f}$$

Since there are three primal constraints, there are three dual variables for the problem. Let y_1, y_2, and y_3 be the dual variables associated with the three constraints. Therefore, Eqs. (9.21) and (9.22) give the dual for the problem as

Minimize
$$f_d = 9y_1 + 2y_2 + 3y_3 \tag{g}$$

subject to
$$2y_1 + y_2 - 3y_3 \ge 5 \tag{h}$$
$$y_1 - 2y_2 + 2y_3 \ge -2 \tag{i}$$
$$y_1, y_2, y_3 \ge 0 \tag{j}$$

9.3.3 Treatment of Equality Constraints

Many design problems have equality constraints. Each equality constraint can be replaced by a pair of inequalities. For example, $2x_1 + 3x_2 = 5$ can be replaced by the pair $2x_1 + 3x_2 \ge 5$

and $2x_1 + 3x_2 \leq 5$. We can multiply the "$\geq$ type" inequality by -1 to convert it into the standard primal form. Example 9.3 illustrates treatment of equality and "$\geq$ type" constraints.

EXAMPLE 9.3 THE DUAL OF AN LP PROBLEM WITH EQUALITY AND "$\geq$ TYPE" CONSTRAINTS

Write the dual for the problem

Maximize

$$z_p = x_1 + 4x_2 \tag{a}$$

subject to

$$x_1 + 2x_2 \leq 5 \tag{b}$$

$$2x_1 + x_2 = 4 \tag{c}$$

$$x_1 - x_2 \geq 1 \tag{d}$$

$$x_1, x_2 \geq 0 \tag{e}$$

Solution

The equality constraint $2x_1 + x_2 = 4$ is equivalent to the two inequalities $2x_1 + x_2 \geq 4$ and $2x_1 + x_2 \leq 4$. The "$\geq$ type" constraints are multiplied by -1 to convert them into the "$\leq$" form. Thus, the standard primal for the given problem is

Maximize

$$z_p = x_1 + 4x_2 \tag{f}$$

subject to

$$x_1 + 2x_2 \leq 5 \tag{g}$$

$$2x_1 + x_2 \leq 4 \tag{h}$$

$$-2x_1 - x_2 \leq -4 \tag{i}$$

$$-x_1 + x_2 \leq -1 \tag{j}$$

$$x_1, \ x_2 \geq 0 \tag{k}$$

Using Eqs. (9.21) and (9.22), the dual for the primal is

Minimize

$$f_d = 5y_1 + 4(y_2 - y_3) - y_4 \tag{l}$$

subject to

$$y_1 + 2(y_2 - y_3) - y_4 \geq 1 \tag{m}$$

$$2y_1 + (y_2 - y_3) + y_4 \geq 4 \tag{n}$$

$$y_1, \ y_2, \ y_3, \ y_4 \geq 0 \tag{o}$$

9.3.4 Alternate Treatment of Equality Constraints

We will show that it is not necessary to replace an equality constraint by a pair of inequalities to write the dual problem. Note that there are four dual variables for Example 9.3. The variables y_2 and y_3 correspond to the second and third primal constraints written in the standard form. The second and third constraints are actually equivalent to the original equality constraint. Note also that the term $(y_2 - y_3)$ appears in all of the expressions of the dual problem. We define

$$y_5 = y_2 - y_3 \tag{a}$$

which can be positive, negative, or zero, since it is the difference of two non-negative variables ($y_2 \geq 0$, $y_3 \geq 0$). Substituting for y_5, the dual problem in Example 9.3 is rewritten as

Minimize
$$f_d = 5y_1 + 4y_5 - y_4 \tag{b}$$

subject to
$$y_1 + 2y_5 - y_4 \geq 1 \tag{c}$$

$$2y_1 + y_5 + y_4 \geq 4 \tag{d}$$

$$y_1, \; y_4 \geq 0; \; y_5 = y_2 - y_3 \text{ is unrestricted in sign} \tag{e}$$

The number of dual variables is now only three. Since the number of dual variables is equal to the number of the original primal constraints, the dual variable y_5 must be associated with the equality constraint $2x_1 + x_2 = 4$. Thus, we can draw the following conclusion: *If the ith primal constraint is left as an equality, the ith dual variable is unrestricted in sign.* In a similar manner, we can show that if a primal variable is unrestricted in sign, then the *i*th dual constraint is an equality. This is left as an exercise. Example 9.4 demonstrates recovery of the primal formulation from the dual formulation.

EXAMPLE 9.4 RECOVERY OF PRIMAL FORMULATION FROM DUAL FORMULATION

Note that we can convert a dual problem into the standard primal form and write its dual again. It can be shown that the dual of this problem gives the primal problem back again. To see this, let us convert the preceding dual problem into standard primal form:

Maximize
$$z_p = -5y_1 - 4y_5 + y_4 \tag{a}$$
subject to
$$-y_1 - 2y_5 + y_4 \leq -1 \tag{b}$$

$$-2y_1 - y_5 - y_4 \leq -4 \tag{c}$$

$$y_1, \ y_4 \geq 0; \ y_5 = y_2 - y_3 \text{ is unrestricted in sign} \tag{d}$$

Writing the dual of the above problem, we obtain

Minimize

$$f_d = -x_1 - 4x_2 \tag{e}$$

subject to

$$-x_1 - 2x_2 \geq -5 \tag{f}$$

$$-2x_1 - x_2 = -4 \tag{g}$$

$$x_1 - x_2 \geq 1 \tag{h}$$

$$x_1, \ x_2 \geq 0 \tag{i}$$

which is the same as the original problem (Example 9.3). Note that in the preceding dual problem, the second constraint is an equality because the second primal variable (y_5) is unrestricted in sign. Theorem 9.1 states this result.

THEOREM 9.1

Dual of a Dual Problem   The dual of the dual problem is the primal problem.

9.3.5 Determination of the Primal Solution from the Dual Solution

It remains to be determined how the optimum solution to the primal is obtained from the optimum solution to the dual, or vice versa. First, let us multiply each dual inequality in Eq. (9.22) by $x_1, x_2, \ldots, x_n$ and add them. Since x_i's are restricted to be non-negative, we get the inequality

$$x_1(a_{11}y_1 + \ldots + a_{m1}y_m) + x_2(a_{12}y_1 + \ldots + a_{m2}y_m)$$
$$+ \ldots + x_n(a_{1n}y_1 + \ldots + a_{mn}y_m) \geq d_1x_1 + d_2x_2 + \ldots + d_nx_n \tag{9.23}$$

Or, in matrix form,

$$x^T A^T y \geq x^T d \tag{9.24}$$

Rearranging the equation by collecting terms with $y_1, y_2, \ldots y_m$ (or taking the transpose of the left side as $y^T Ax$), we obtain

$$y_1(a_{11}x_1 + a_{12}x_2 + \ldots + a_{1n}x_n) + y_2(a_{21}x_1 + a_{22}x_2 + \ldots + a_{2n}x_n)$$
$$+ \ldots + y_m(a_{m1}x_1 + a_{m2}x_2 + \ldots + a_{mn}x_n) \geq d_1x_1 + d_2x_2 + \ldots + d_nx_n \tag{9.25}$$

In the matrix form, the preceding inequality can be written as $\mathbf{y}^T A \mathbf{x} \geq \mathbf{x}^T \mathbf{d}$. Each term in parentheses in Eq. (9.25) is less than the corresponding value of e on the right side of Inequalities (9.20). Therefore, replacing these terms with the corresponding e from Inequalities (9.20) preserves the inequality in Eq. (9.25):

$$y_1 e_1 + y_2 e_2 + \ldots + y_m e_m \geq d_1 x_1 + d_2 x_2 + \ldots + d_n x_n, \text{ or } \mathbf{y}^T \mathbf{e} \geq \mathbf{x}^T \mathbf{d} \tag{9.26}$$

Note that in Inequality (9.26) the left side is the dual cost function and the right side is the primal objective function. Therefore, from Inequality (9.26), $f_d \geq z_p$ for all $(x_1, x_2, \ldots, x_n)$ and $(y_1, y_2, \ldots, y_m)$, satisfying Eqs. (9.19) through (9.22). Thus, the vectors $\mathbf{x}$ and $\mathbf{y}$ with $z_p = f_d$ maximize z_p while minimizing f_d. The optimum (minimum) value of the dual cost function is also the optimum (maximum) value of the primal objective function. Theorems 9.2, 9.3, and 9.4 regarding primal and dual problems can be stated as follows.

THEOREM 9.2

The Relationship between Primal and Dual Problems Let $\mathbf{x}$ and $\mathbf{y}$ be in the feasible sets of the primal and dual problems, respectively (as defined in Eqs. (9.19) through (9.22)). Then the following conditions hold:

1. $f_d(\mathbf{y}) \geq z_p(\mathbf{x})$.
2. If $f_d = z_p$, then $\mathbf{x}$ and $\mathbf{y}$ are the solutions to the primal and the dual problems, respectively.

3. If the primal is unbounded, the corresponding dual is infeasible, and vice versa.
4. If the primal is feasible and the dual is infeasible, then the primal is unbounded, and vice versa.

THEOREM 9.3

Primal and Dual Solutions Let both the primal and the dual have feasible points. Then both have optimum solutions in $\mathbf{x}$ and $\mathbf{y}$, respectively, and $f_d(\mathbf{y}) = z_p(\mathbf{x})$.

THEOREM 9.4

Solution to Primal From Dual If the ith dual constraint is a strict inequality at optimum, then the corresponding ith primal variable is nonbasic (i.e., it vanishes). Also, if the ith dual variable is basic, then the ith primal constraint is satisfied at equality.

The conditions of Theorem 9.4 can be written as (if the jth dual constraint is a strict inequality, then the jth primal variable is nonbasic)

$$\text{if } \sum_{i=1}^{m} a_{ij}y_i > d_j, \text{ then } x_j = 0 \tag{9.27}$$

(if the ith dual variable is basic, the ith primal constraint is satisfied at equality that is, active):

$$\text{if } y_i > 0, \text{ then } \sum_{j=1}^{n} a_{ij}x_j = e_i \tag{9.28}$$

These conditions can be used to obtain primal variables using the dual variables. The primal constraints satisfied at equality are identified from the values of the dual variables. The resulting linear equations can be solved simultaneously for the primal variables. However, this is not necessary because the final dual tableau can be used directly to obtain the primal variables. We illustrate the use of these theorems in Example 9.5.

EXAMPLE 9.5 PRIMAL AND DUAL SOLUTIONS

Consider the following problem:

Maximize

$$z_p = 5x_1 - 2x_2 \tag{a}$$

subject to

$$2x_1 + x_2 \leq 9 \tag{b}$$

$$x_1 - 2x_2 \leq 2 \tag{c}$$

$$-3x_1 + 2x_2 \leq 3 \tag{d}$$

$$x_1, \, x_2 \geq 0 \tag{e}$$

Solve the primal and dual problems and study their final tableaux.

Solution

The problem was solved using the Simplex method in Example 8.18 and Table 8.22. The final tableau is reproduced in Table 9.3. From the final primal tableau,

Basic variables:	$x_1 = 4, \quad x_2 = 1, \quad x_5 = 13$	(f)
Nonbasic variables:	$x_3 = 0, \quad x_4 = 0$	(g)
Optimum objective function:	$z_p = 18$ (minimum value is -18)	(h)

Now, let us write the dual for the problem and solve it using the Simplex method. Note that the original problem is already in the standard primal form. There are three primal inequality constraints, so there are three dual variables. There are two primal variables, so there are two dual constraints. Let y_1, y_2, and y_3 be the dual variables. Therefore, the dual of the problem is given as

Minimize

$$f_d = 9y_1 + 2y_2 + 3y_3 \tag{i}$$

subject to

$$2y_1 + y_2 - 3y_3 \geq 5 \tag{j}$$

$$y_1 - 2y_2 + 2y_3 \geq -2 \tag{k}$$

$$y_1, \ y_2, \ y_3 \geq 0 \tag{l}$$

Writing the constraints in the standard Simplex form by introducing slack, surplus, and artificial variables, we obtain

$$2y_1 + y_2 - 3y_3 - y_4 + y_6 = 5 \tag{m}$$

$$-y_1 + 2y_2 - 2y_3 + y_5 = 2 \tag{n}$$

$$y_i \geq 0; \quad i = 1 \text{ to } 6 \tag{o}$$

where y_4 is a surplus variable, y_5 is a slack variable, and y_6 is an artificial variable. The two-phase Simplex procedure is used to solve the problem, as displayed in Table 9.4. From the final dual tableau, we obtain the following solution:

Basic variables:	$y_1 = 1.6, \ y_2 = 1.8$
Nonbasic variables:	$y_3 = 0, \ y_4 = 0, \ y_5 = 0$
Optimum value of dual function:	$f_d = 18$

Note that at the optimum $f_d = z_p$, which satisfies the conditions of Theorems 9.2 and 9.3. Using Theorem 9.4, we see that the first and second primal constraints must be satisfied at equality since the dual variables y_1 and y_2 associated with the constraints are positive (basic) in Table 9.4. Therefore, primal variables x_1 and x_2 are obtained as a solution to the first two primal constraints satisfied at equality: $2x_1 + x_2 = 9$, $x_1 - 2x_2 = 2$. The solution to these equations is given as $x_1 = 4$, $x_2 = 1$, which is the same as obtained from the final primal tableau.

9.3.6 Use of the Dual Tableau to Recover the Primal Solution

It turns out that we do not need to follow the preceding procedure (Theorem 9.4) to recover the primal variables. The final dual tableau contains all of the information to recover the primal solution. Similarly, the final primal tableau contains all of the information to recover the dual solution. Looking at the final tableau in Table 9.4 for Example 9.5, we observe that the elements in the last row of the dual tableau match the elements in the last column of the primal tableau in Table 9.3. Similarly, the reduced cost coefficients in the final primal tableau match the dual variables. To recover the primal variables from the final dual tableau, we use reduced cost coefficients in the columns corresponding to the slack or surplus variables. We note that the reduced cost coefficient in column y_4 is precisely x_1 and the reduced cost coefficient in column y_5 is precisely x_2.

TABLE 9.3 Final tableau for Example 9.5 by Simplex method (primal solution)

Basic↓	x_1	x_2	x_3	x_4	x_5	b
x_2	0	1	0.2	−0.4	0	1
x_1	1	0	0.4	0.2	0	4
x_5	0	0	0.8	1.4	1	13
Cost	0	0	1.6	1.8	0	$f_p + 18$

TABLE 9.4 Solution to dual of the problem in Example 9.5

Initial tableau: y_6 is identified to be replaced with y_1 in the basic set

Basic↓	y_1	y_2	y_3	y_4	y_5	y_6	b
y_6	2	1	−3	−1	0	1	5
y_5	−1	2	−2	0	1	0	2
Cost	9	2	3	0	0	0	$f_d - 0$
Artificial cost	−2	−1	3	1	0	0	$w - 5$

Second tableau: End of Phase I. y_5 is identified to be replaced with y_2 in the basic set

Basic↓	y_1	y_2	y_3	y_4	y_5	y_6	b
y_1	1	0.5	−1.5	−0.5	0	0.5	2.5
y_5	0	2.5	−3.5	−0.5	1	0.5	4.5
Cost	0	−2.5	16.5	4.5	0	−4.5	$f_d - 22.5$
Artificial cost	0	0	0	0	0	1	$w - 0$

Third tableau: Reduced cost coefficients in nonbasic columns are non-negative; the tableau gives optimum point. End of Phase II

Basic↓	y_1	y_2	y_3	y_4	y_5	y_6	b
y_1	1	0	−0.8	−0.4	−0.2	0.4	1.6
y_2	0	1	−1.4	−0.2	0.4	0.2	1.8
Cost	0	0	13.0	4.0	1.0	−4.0	$f_d - 18$

Reduced cost coefficients corresponding to the slack and surplus variables in the final dual tableau give the values of the primal variables.

Similarly, if we solve the primal problem, we can recover the dual solution from the final primal tableau. Theorem 9.5 summarizes this result.

THEOREM 9.5

Recovery of the Primal Solution From the Dual Tableau Let the dual of the standard primal defined earlier in Eqs. (9.19) and (9.20) (i.e., maximize $\mathbf{d}^T\mathbf{x}$ subject to $\mathbf{Ax} \le \mathbf{e}, \mathbf{x} \ge \mathbf{0}$) be solved by the standard Simplex method. Then the value of the ith primal variable equals the reduced cost coefficient of the slack or surplus variable associated with the ith dual constraint in the final dual tableau. In addition, if a dual variable is nonbasic, then its reduced cost coefficient equals the value of the slack or surplus variable for the corresponding primal constraint.

Note that if a dual variable is nonbasic (i.e., has a zero value), then its reduced cost coefficient equals the value of the slack or surplus variable for the corresponding primal constraint. In Example 9.5, y_3, the dual variable corresponding to the third primal constraint, is nonbasic. The reduced cost coefficient in the y_3 column is 13. Therefore, the slack variable for the third primal constraint has a value of 13; that is, the constraint is inactive. This is the same as obtained from the final primal tableau. We also note that the dual solution can be obtained from the final primal tableau using Theorem 9.5 as $y_1 = 1.6$, $y_2 = 1.8$, $y_3 = 0$, which is the same solution as before. While using Theorem 9.5, the following additional points should be noted:

1. When the final primal tableau is used to recover the dual solution, the dual variables correspond to the primal constraints expressed in the "$\le$" form only. However, the primal constraints must be converted to standard Simplex form while solving the problem. Recall that all of the right sides of the constraints must be non-negative for the Simplex method. The dual variables are non-negative only for the constraints written in the "$\le$" form.

2. When a primal constraint is an equality, it is treated in the Simplex method by adding an artificial variable in Phase I. There is no slack or surplus variable associated with an equality. We also know from the previous discussion that the dual variable associated with the equality constraint is unrestricted in sign. Then the question becomes how to recover the dual variable for the equality constraint from the final primal tableau. There are a couple of ways of doing this.

    The first procedure is to convert the equality constraint into a pair of inequalities, as noted previously. For example, the constraint $2x_1 + x_2 = 4$ is written as the pair of inequalities $2x_1 + x_2 \le 4$, $-2x_1 - x_2 \le -4$. The two inequalities are treated in a standard way in the Simplex method. The corresponding dual variables are recovered from the final primal tableau using Theorem 9.5. Let $y_2 \ge 0$ and $y_3 \ge 0$ be the dual variables associated with the two inequality constraints, respectively, and y_1 be the dual variable associated with the original equality constraint. Then $y_1 = y_2 - y_3$. Accordingly, y_1 is unrestricted in sign and its value is known using y_2 and y_3.

    The second way of recovering the dual variable for the equality constraint is to carry along its artificial variable column in Phase II of the Simplex method. Then the dual variable for the equality constraint is the reduced cost coefficient in the artificial variable column in the final primal tableau. We illustrate these procedures in Example 9.6.

EXAMPLE 9.6   USE OF THE FINAL PRIMAL TABLEAU TO RECOVER DUAL SOLUTIONS

Solve the following LP problem and recover its dual solution from the final primal tableau:

Maximize

$$z_p = x_1 + 4x_2 \tag{a}$$

subject to

$$x_1 + 2x_2 \le 5, \tag{b}$$

$$2x_1 + x_2 = 4, \tag{c}$$

$$x_1 - x_2 \ge 1, \tag{d}$$

$$x_1, \quad x_2 \ge 0 \tag{e}$$

Solution

When the equality constraint is converted into a pair of inequalities—that is, $2x_1 + x_2 \le 4$, $-2x_1 - x_2 \le -4$—the problem is the same as the one solved in Example 8.17. The final tableau for the problem was given in Table 8.21. Using Theorem 9.5, the dual variables for the preceding four constraints are

1. $x_1 + 2x_2 \le 5$: $y_1 = 0$, reduced cost coefficient of x_3, the slack variable
2. $2x_1 + x_2 \le 4$: $y_2 = 5/3$, reduced cost coefficient of x_4, the slack variable
3. $-2x_1 - x_2 \le -4$: $y_3 = 0$, reduced cost coefficient of x_5, the surplus variable
4. $-x_1 + x_2 \le -1$: $y_4 = 7/3$, reduced cost coefficient of x_6, the surplus variable

Thus, from the above discussion, the dual variable for the equality constraint $2x_1 + x_2 = 4$ is $y_2 - y_3 = 5/3$. Note also that $y_4 = 7/3$ is the dual variable for the fourth constraint, written as $-x_1 + x_2 \le -1$, and not for the constraint $x_1 - x_2 \ge 1$.

Now let us re-solve the same problem with the equality constraint as it is. The problem is the same as the one solved in Example 8.19. The final tableau for the problem is given in Table 8.23. Using Theorem 9.5 and the preceding discussion, the dual variables for the given three constraints are

1. $x_1 + 2x_2 \le 5$: $y_1 = 0$, reduced cost coefficient of x_3, the slack variable
2. $2x_1 + x_2 = 4$: $y_2 = 5/3$, reduced cost coefficient of x_5, the artificial variable
3. $-x_1 + x_2 \le -1$: $y_3 = 7/3$, reduced cost coefficient of x_4, the surplus variable

We see that the foregoing two solutions are the same. Therefore, we do not have to replace an equality constraint by two inequalities in the standard Simplex method. The reduced cost coefficient corresponding to the artificial variable associated with the equality constraint gives the value of the dual variable for the constraint.

9.3.7 Dual Variables as Lagrange Multipliers

Section 8.7 describes how the optimum value of the cost function for the problem changes if we change the right-side parameters of constraints b_i, the resource limits. The

Constraint Variation Sensitivity Theorem 4.7 (Chapter 4) is used to study this effect. Use of that theorem requires knowledge of the Lagrange multipliers for the constraints that must be determined. It turns out that the dual variables of the problem are related to the Lagrange multipliers. Theorem 9.6 gives this relationship.

THEOREM 9.6

Dual Variables as Lagrange Multipliers
Let $\mathbf{x}$ and $\mathbf{y}$ be optimal solutions for the primal and dual problems stated in Eqs. (9.19) through (9.22), respectively. Then the dual variables $\mathbf{y}$ are also the Lagrange multipliers for the primal constraints of Eq. (9.20).

Proof The theorem can be proved by writing the Karush-Kuhn-Tucker (KKT) necessary conditions of Theorem 4.6 for the primal problem defined in Eqs. (9.19) and (9.20). To write these conditions, convert the primal problem to a minimization problem and then define a Lagrange function of Eq. (4.46) as

$$L = -\sum_{j=1}^{n} d_j x_j + \sum_{i=1}^{m} y_i \left(\sum_{j=1}^{n} a_{ij} x_j - e_i \right) - \sum_{j=1}^{n} v_j x_j \quad (a)$$
$$= -\mathbf{d}^T \mathbf{x} + \mathbf{y}^T (\mathbf{A}\mathbf{x} - \mathbf{e}) - \mathbf{v}^T \mathbf{x}$$

where y_i is the Lagrange multiplier for the ith primal constraint of Eq. (9.20) and v_j is the Lagrange multiplier for the jth non-negativity constraint for the variable x_j. Write the KKT necessary conditions of Theorem 4.6 as

$$-d_j + \sum_{i=1}^{m} y_i a_{ij} - v_j = 0; \quad (b)$$
$$j = 1 \text{ to } n \quad (\partial L / \partial x_j = 0)$$

$$y_i \left(\sum_{j=1}^{n} a_{ij} x_j - e_i \right) = 0, \quad i = 1 \text{ to } m \quad (c)$$

$$v_i x_i = 0, \quad x_i \geq 0, \quad i = 1 \text{ to } n \quad (d)$$

$$y_i \geq 0, \quad i = 1 \text{ to } m \quad (e)$$

$$v_i \geq 0, \quad i = 1 \text{ to } n \quad (f)$$

Rewrite Eq. (b) as

$$-d_j + \sum_{i=1}^{m} a_{ij} y_i = v_j; \quad j = 1 \text{ to } n \quad (-\mathbf{d} + \mathbf{A}^T \mathbf{y} = \mathbf{v})$$

Using conditions (f) in the preceding equation, we conclude that

$$\sum_{i=1}^{m} a_{ij} y_i \geq d_j; \quad j = 1 \text{ to } n \quad (\mathbf{A}^T \mathbf{y} \geq \mathbf{d}) \quad (g)$$

Thus y_i are feasible solutions for the dual constraints of Eq. (9.22).

Now let x_i represent the optimum solution to the primal problem. Then m of the x_i are positive (barring degeneracy), and the corresponding v_i are equal to zero from Eq. (d). The remaining x_i are zero and the corresponding v_i are greater than zero. Therefore, from Eq. (g), we obtain

$$(i) \; v_j > 0, \; x_j = 0, \quad \sum_{i=1}^{m} a_{ij} y_i > d_j \quad (h)$$

$$(ii) \; v_j = 0, \; x_j > 0, \quad \sum_{i=1}^{m} a_{ij} y_i = d_j \quad (i)$$

Adding the m rows given in Eq. (c), interchanging the sums on the left side, and rearranging, we obtain

$$\sum_{j=1}^{n} x_j \sum_{i=1}^{m} a_{ij} y_i = \sum_{i=1}^{m} y_i e_i \quad (\mathbf{x}^T \mathbf{A}^T \mathbf{y} = \mathbf{y}^T \mathbf{e}) \quad (j)$$

Using Eqs. (h) and (i), Eq. (j) can be written as

$$\sum_{j=1}^{n} d_j x_j = \sum_{i=1}^{m} y_i e_i \quad (\mathbf{d}^T \mathbf{x} = \mathbf{y}^T \mathbf{e}) \quad (k)$$

Equation (k) also states that

$$z_p = \sum_{i=1}^{m} y_i e_i = \mathbf{y}^T \mathbf{e} \quad (l)$$

The right side of Eq. (l) represents the dual cost function. According to Theorem 9.2, if the primal and dual functions have the same values and if $\mathbf{x}$ and $\mathbf{y}$ are feasible points for the primal and dual problems, then they are optimum solutions for the respective problems. Thus, the dual variables y_i, $i = 1$ to m that solve the dual problem defined in Eqs. (9.21) and (9.22) are also the Lagrange multipliers for the primal constraints.

9.4 KKT CONDITIONS FOR THE LP PROBLEM

Now we write the KKT optimality conditions for the linear programming (LP) problem and show that the Simplex method essentially solves these conditions in a systematic way. The LP problem is defined in the standard form as

$$f(\mathbf{x}) = \mathbf{c}^T \mathbf{x} \tag{9.29}$$

$$\mathbf{A}\mathbf{x} = \mathbf{b}; \quad \mathbf{b} \geq \mathbf{0} \tag{9.30}$$

$$\mathbf{x} \geq \mathbf{0} \tag{9.31}$$

where the dimensions of various vectors and matrices are $\mathbf{c}_{(n)}$, $\mathbf{x}_{(n)}$, $\mathbf{b}_{(m)}$, and $\mathbf{A}_{(m \times n)}$. It is assumed that the rank of $\mathbf{A}$ is m; that is., all of the equations are linearly independent in Eq. (9.30).

9.4.1. KKT Optimality Conditions

The Lagrangian for the problem is defined as

$$L = \mathbf{c}^T \mathbf{x} + \mathbf{y}^T (\mathbf{A}\mathbf{x} - \mathbf{b}) + \mathbf{v}^T (-\mathbf{x}) \tag{9.32}$$

where $\mathbf{y}_{(m)}$ and $\mathbf{v}_{(n)}$ are the Lagrange multipliers. The KKT necessary conditions give

$$\frac{\partial L}{\partial \mathbf{x}} = \mathbf{c} + \mathbf{A}^T \mathbf{y} - \mathbf{v} = \mathbf{0} \tag{9.33}$$

$$\mathbf{A}\mathbf{x} = \mathbf{b} \tag{9.34}$$

$$v_i x_i = 0, \quad v_i \geq 0; \quad i = 1 \text{ to } n \tag{9.35}$$

9.4.2. Solution to the KKT Conditions

Equations (9.33) through (9.35) can be solved for $\mathbf{x}$, $\mathbf{y}$, and $\mathbf{v}$. Partition various vectors and matrices into basic and nonbasic parts as

$$\mathbf{x} = \begin{bmatrix} \mathbf{x}_B \\ \mathbf{x}_N \end{bmatrix}; \quad \mathbf{x}_{B(m)}, \mathbf{x}_{N(n-m)} \tag{9.36}$$

$$\mathbf{v} = \begin{bmatrix} \mathbf{v}_B \\ \mathbf{v}_N \end{bmatrix}; \quad \mathbf{v}_{B(m)}, \mathbf{v}_{N(n-m)} \tag{9.37}$$

$$\mathbf{c} = \begin{bmatrix} \mathbf{c}_B \\ \mathbf{c}_N \end{bmatrix}; \quad \mathbf{c}_{B(m)}, \mathbf{c}_{N(n-m)} \tag{9.38}$$

$$\mathbf{A} = \begin{bmatrix} \mathbf{B} & \mathbf{N} \end{bmatrix}; \quad \mathbf{B}_{(m \times m)}, \quad \mathbf{N}_{(m \times n-m)} \tag{9.39}$$

Equation (9.34) can be now partitioned as

$$\mathbf{B}\mathbf{x}_B + \mathbf{N}\mathbf{x}_N = \mathbf{b}, \text{ or } \mathbf{x}_B = \mathbf{B}^{-1}(-\mathbf{N}\mathbf{x}_N + \mathbf{b}) \tag{9.40}$$

It is assumed that $\mathbf{x}_B$ and $\mathbf{x}_N$ are selected in such a way that $\mathbf{B}^{-1}$ exists. This is always possible since $\mathbf{A}$ has full row rank. Equation (9.40) expresses the basic variables $\mathbf{x}_B$ in terms of the nonbasic variables $\mathbf{x}_N$. This is called a general solution to $\mathbf{A}\mathbf{x} = \mathbf{b}$. Any specification of $\mathbf{x}_N$ gives $\mathbf{x}_B$. In particular, if $\mathbf{x}_N$ is set to zero, $\mathbf{x}_B$ is given as

$$\mathbf{x}_B = \mathbf{B}^{-1}\mathbf{b} \tag{9.41}$$

This is called a *basic solution to* $\mathbf{A}\mathbf{x} = \mathbf{b}$. If all of the basic variables are also non-negative, then the solution is called a *basic feasible solution*.

Equation (9.33) can be written in the partitioned form as

$$\begin{bmatrix} \mathbf{c}_B \\ \mathbf{c}_N \end{bmatrix} + \begin{bmatrix} \mathbf{B}^T \\ \mathbf{N}^T \end{bmatrix} \mathbf{y} - \begin{bmatrix} \mathbf{v}_B \\ \mathbf{v}_N \end{bmatrix} = \begin{bmatrix} \mathbf{0} \\ \mathbf{0} \end{bmatrix} \tag{9.42}$$

Since $\mathbf{x}_B \neq \mathbf{0}$, $\mathbf{v}_B = \mathbf{0}$ to satisfy Eq. (9.35). Therefore, from the first row of Eq. (9.42), we obtain

$$\mathbf{y} = -\mathbf{B}^{-T}\mathbf{c}_B \tag{9.43}$$

From the second row of Eq. (9.42), we obtain

$$\mathbf{v}_N = \mathbf{c}_N + \mathbf{N}^T\mathbf{y} \tag{9.44}$$

Substituting for $\mathbf{y}$ from Eq. (9.43) into Eq. (9.44), we obtain

$$\mathbf{v}_N = \mathbf{c}_N - \mathbf{N}^T\mathbf{B}^{-T}\mathbf{c}_B \tag{9.45}$$

The cost function is written in terms of the basic and nonbasic variables as

$$f = \mathbf{c}_B^T\mathbf{x}_B + \mathbf{c}_N^T\mathbf{x}_N \tag{9.46}$$

Substituting for $\mathbf{x}_B$ from Eq. (9.40) into Eq. (9.46) and collecting terms, we obtain

$$f = \mathbf{c}_B^T(\mathbf{B}^{-1}\mathbf{b}) + (\mathbf{c}_N - \mathbf{N}^T\mathbf{B}^{-T}\mathbf{c}_B)^T\mathbf{x}_N \tag{9.47}$$

Substituting Eq. (9.45) into Eq. (9.47), we obtain

$$f = \mathbf{c}_B^T(\mathbf{B}^{-1}\mathbf{b}) + \mathbf{v}_N^T\mathbf{x}_N \tag{9.48}$$

This equation expresses the cost function in terms of only the nonbasic variables. The coefficients of the nonbasic variables are the reduced cost coefficients. Since these are also the Lagrange multipliers, they are required to be non-negative at the optimum point; that is,

$$\mathbf{v}_N \geq \mathbf{0} \tag{9.49}$$

This was precisely the condition that was used to recognize the optimum solution in the Simplex method. Also, since $\mathbf{x}_N$ are nonbasic variables, they have a zero value. Therefore, the optimum cost function is given from Eq. (9.48) as

$$f^* = \mathbf{c}_B^T(\mathbf{B}^{-1}\mathbf{b}) \tag{9.50}$$

The optimization procedure is to determine the basic variables $\mathbf{x}_B \geq 0$ and nonbasic variables $\mathbf{x}_N$ such that the optimality condition $\mathbf{v}_N \geq 0$ is satisfied for a feasible point. This is achieved by the Simplex method of linear programming, as seen earlier.

The Simplex tableau is set up using Eqs. (9.40) and (9.48) as follows:

$$\begin{array}{ccc} \mathbf{x}_B & \mathbf{x}_N & RS \\ \begin{bmatrix} \mathbf{I} & \mathbf{B}^{-1}\mathbf{N} & \mathbf{B}^{-1}\mathbf{b} \\ \mathbf{0} & \mathbf{u}_N & -\mathbf{c}_B^T(\mathbf{B}^{-1}\mathbf{b}) \end{bmatrix} \end{array} \tag{9.51}$$

The first row is Eq. (9.40) and the second row is the cost function in Eq. (9.48). The first column belongs to $\mathbf{x}_B$ and the second column belongs to $\mathbf{x}_N$. The third column is the right side of Eqs. (9.40) and (9.48). A Simplex tableau can be set up to carry out the pivot operations until the solution is obtained.

9.5 QUADRATIC PROGRAMMING PROBLEMS

A quadratic programming (QP) problem has a quadratic cost function and linear constraints. Such problems are encountered in many real-world applications. In addition, many general nonlinear programming algorithms require solution to a quadratic programming subproblem at each iteration. The QP subproblem is obtained when a nonlinear problem is linearized and a quadratic step size constraint is imposed (Chapter 12).

It is important to solve the QP subproblem efficiently so that large-scale problems can be treated. Thus, it is not surprising that substantial research effort has been expended in developing and evaluating many algorithms for solving QP problems (Gill et al., 1981; Luenberger, 1984; Nocedal and Wright, 2006). Also, several commercially available software packages are available for solving QP problems, for example, MATLAB, QPSOL (Gill et al., 1984), VE06A (Hopper, 1981), and E04NAF (NAG, 1984). Some of the available LP codes also have an option for solving QP problems (Schrage, 1991).

To give a flavor of the calculations needed to solve QP problems, we will describe a method that is an extension of the *Simplex method*. Many other methods are available to solve QP problems.

9.5.1. Definition of a QP Problem

Let us define a general QP problem as follows:

Minimize

$$q(\mathbf{x}) = \mathbf{c}^T\mathbf{x} + \frac{1}{2}\mathbf{x}^T\mathbf{H}\mathbf{x} \tag{9.52}$$

subject to the linear equality and inequality constraints

$$N^T x = e \tag{9.53}$$

$$A^T x \le b \tag{9.54}$$

and the non-negativity of the variables

$$x \ge 0 \tag{9.55}$$

where

c = n-dimensional constant vector
x = n-dimensional vector of unknowns
b = m-dimensional constant vector
e = p-dimensional constant vector
H = $n \times n$ constant Hessian matrix
A = $n \times m$ constant matrix
N = $n \times p$ constant matrix

Note that all of the linear inequality constraints are expressed in the "$\le$ form." This is needed because we will use the KKT necessary conditions of Section 4.6, which require this form. Note also that if the matrix H is positive semidefinite, the QP problem is convex, so any solution (if one exists) represents a global minimum point (which need not be unique). Further, if the matrix H is positive definite, the problem is strictly convex. Therefore, the problem has a unique global solution (if one exists). We will assume that the matrix H is at least positive semidefinite. This is not an unreasonable assumption in practice, as many applications satisfy it.

Note also that the variables x are required to be non-negative in Eq. (9.55). Variables that are free in sign can be easily treated by the method described in Section 8.2.

9.5.2. KKT Necessary Conditions for the QP Problem

A procedure for solving the QP problem of Eqs. (9.52) through (9.55) is to first write the KKT necessary conditions of Section 4.6 and then transform them into a form that can be treated by Phase I of the Simplex method of Section 8.5. To write the necessary conditions, we introduce slack variables s for Inequalities (9.54) and transform them into equalities as

$$A^T x + s = b; \quad \text{with} \quad s \ge 0 \tag{9.56}$$

The slack variable for the jth inequality in Eq. (9.54) can be expressed using Eq. (9.56) as

$$s_j = b_j - \sum_{i=1}^{n} a_{ij} x_i \quad (s = b - A^T x) \tag{9.57}$$

Note the non-negativity constraints of Eq. (9.55) (when expressed in the standard form $-x \le 0$) do not need slack variables because $x_i \ge 0$ itself is a slack variable.

Let us now define the Lagrange function of Eq. (4.46) for the QP problem as

$$L = \mathbf{c}^T\mathbf{x} + 0.5\mathbf{x}^T\mathbf{H}\mathbf{x} + \mathbf{u}^T(\mathbf{A}^T\mathbf{x} + \mathbf{s} - \mathbf{b}) - \boldsymbol{\xi}^T\mathbf{x} + \mathbf{v}^T(\mathbf{N}^T\mathbf{x} - \mathbf{e}) \tag{9.58}$$

where $\mathbf{u}$, $\mathbf{v}$, and $\boldsymbol{\xi}$ are the Lagrange multiplier vectors for the inequality constraints of Eq. (9.56), the equality constraints of Eq. (9.53), and the non-negativity constraints $(-\mathbf{x} \le \mathbf{0})$, respectively. The KKT necessary conditions give

$$\frac{\partial L}{\partial \mathbf{x}} = \mathbf{c} + \mathbf{H}\mathbf{x} + \mathbf{A}\mathbf{u} - \boldsymbol{\xi} + \mathbf{N}\mathbf{v} = \mathbf{0} \tag{9.59}$$

$$\mathbf{A}^T\mathbf{x} + \mathbf{s} - \mathbf{b} = \mathbf{0} \tag{9.60}$$

$$\mathbf{N}^T\mathbf{x} - \mathbf{e} = \mathbf{0} \tag{9.61}$$

$$u_i s_i = 0; \quad i = 1 \text{ to } m \tag{9.62}$$

$$\xi_i x_i = 0; \quad i = 1 \text{ to } n \tag{9.63}$$

$$s_i, u_i \ge 0 \text{ for } i = 1 \text{ to } m; \quad x_i, \xi_i \ge 0 \text{ for } i = 1 \text{ to } n \tag{9.64}$$

These conditions need to be solved for $\mathbf{x}$, $\mathbf{u}$, $\mathbf{v}$, $\mathbf{s}$, and $\boldsymbol{\xi}$.

9.5.3. Transformation of KKT Conditions

Before discussing the method for solving the KKT conditions, we will transform them into a more compact form in this subsection. Since the Lagrange multipliers $\mathbf{v}$ for the equality constraints are free in sign, we may decompose them as

$$\mathbf{v} = \mathbf{y} - \mathbf{z} \text{ with } \mathbf{y}, \mathbf{z} \ge \mathbf{0} \tag{9.65}$$

Now, writing Eqs. (9.59) through (9.61) into a matrix form, we get

$$\begin{bmatrix} \mathbf{H} & \mathbf{A} & -\mathbf{I}_{(n)} & \mathbf{0}_{(n \times m)} & \mathbf{N} & -\mathbf{N} \\ \mathbf{A}^T & \mathbf{0}_{(m \times m)} & \mathbf{0}_{(m \times n)} & \mathbf{I}_{(m)} & \mathbf{0}_{(m \times p)} & \mathbf{0}_{(m \times p)} \\ \mathbf{N}^T & \mathbf{0}_{(p \times m)} & \mathbf{0}_{(p \times n)} & \mathbf{0}_{(p \times m)} & \mathbf{0}_{(p \times p)} & \mathbf{0}_{(p \times p)} \end{bmatrix} \begin{bmatrix} \mathbf{x} \\ \mathbf{u} \\ \boldsymbol{\xi} \\ \mathbf{s} \\ \mathbf{y} \\ \mathbf{z} \end{bmatrix} = \begin{bmatrix} -\mathbf{c} \\ \mathbf{b} \\ \mathbf{e} \end{bmatrix} \tag{9.66}$$

where $\mathbf{I}_{(n)}$ and $\mathbf{I}_{(m)}$ are $n \times n$ and $m \times m$ identity matrices, respectively, and $\mathbf{0}$ are zero matrices of the indicated order. In a compact matrix notation, Eq. (9.66) becomes

$$\mathbf{B}\mathbf{X} = \mathbf{D} \tag{9.67}$$

where matrix $\mathbf{B}$ and vectors $\mathbf{X}$ and $\mathbf{D}$ are identified from Eq. (9.66) as

$$\mathbf{B} = \begin{bmatrix} \mathbf{H} & \mathbf{A} & -\mathbf{I}_{(n)} & \mathbf{0}_{(n \times m)} & \mathbf{N} & -\mathbf{N} \\ \mathbf{A}^T & \mathbf{0}_{(m \times m)} & \mathbf{0}_{(m \times n)} & \mathbf{I}_{(m)} & \mathbf{0}_{(m \times p)} & \mathbf{0}_{(m \times p)} \\ \mathbf{N}^T & \mathbf{0}_{(p \times m)} & \mathbf{0}_{(p \times n)} & \mathbf{0}_{(p \times m)} & \mathbf{0}_{(p \times p)} & \mathbf{0}_{(p \times p)} \end{bmatrix}_{[(n+m+p) \times (2n+2m+2p)]} \tag{9.68}$$

$$X = \begin{bmatrix} \mathbf{x} \\ \mathbf{u} \\ \boldsymbol{\xi} \\ \mathbf{s} \\ \mathbf{y} \\ \mathbf{z} \end{bmatrix}_{(2n+2m+2p)} \qquad D = \begin{bmatrix} -\mathbf{c} \\ \mathbf{b} \\ \mathbf{e} \end{bmatrix}_{(n+m+p)} \tag{9.69}$$

The KKT conditions are now reduced to finding X as a solution to the linear system in Eq. (9.67) subject to the constraints of Eqs. (9.62) through (9.64). In the new variables X_i, the complementary slackness conditions of Eqs. (9.62) and (9.63), reduce to

$$X_i X_{n+m+i} = 0; \quad i = 1 \text{ to } (n + m) \tag{9.70}$$

and the non-negativity conditions of Eq. (9.64) reduce to

$$X_i \geq 0; \quad i = 1 \text{ to } (2n + 2m + 2p) \tag{9.71}$$

9.5.4. The Simplex Method for Solving QP Problem

A solution to the linear system in Eq. (9.67) that satisfies the complementary slackness condition of Eq. (9.70) and the non-negativity condition of Eq. (9.71) is a solution to the original QP problem. Note that the complementary slackness condition of Eq. (9.70) is nonlinear in the variables X_i. Therefore, it may appear that the Simplex method for LP problems cannot be used to solve Eq. (9.67). However, a procedure developed by Wolfe (1959) and refined by Hadley (1964) can be used instead. The procedure converges to a solution in the finite number of steps provided that the matrix H in Eq. (9.52) is positive definite. It can be further shown (Kunzi and Krelle, 1966, p. 123) that the method converges even when H is positive semidefinite provided that the vector c in Eq. (9.52) is zero.

The method is based on Phase I of the Simplex procedure from Chapter 8, where we introduced an artificial variable for each equality constraint, defined an artificial cost function, and used it to determine an initial basic feasible solution. Following that procedure, we introduce an artificial variable Y_i for each of the Eqs. (9.67) as

$$BX + Y = D \tag{9.72}$$

where Y is an $(n + m + p)$-dimensional vector. In this way, we initially choose all X_i as nonbasic variables and all Y_j as basic variables. Note that all of the elements in D must be non-negative for the initial basic solution to be feasible. If any of the elements in D are negative, the corresponding equation in Eq. (9.67) must be multiplied by -1 to have a non-negative element on the right side.

The artificial cost function for the problem is defined as the summation of all of the artificial variables:

$$w = \sum_{i=1}^{n+m+p} Y_i \tag{9.73}$$

To use the Simplex procedure, we need to express the artificial cost function w in terms of the nonbasic variables $\mathbf{X}$ only. We eliminate basic variables Y_i from Eq. (9.73) by substituting Eq. (9.72) into it as

$$w = \sum_{i=1}^{n+m+p} D_i - \sum_{j=1}^{2(n+m+p)} \sum_{i=1}^{n+m+p} B_{ij} X_j = w_0 + \sum_{j=1}^{2(n+m+p)} C_j X_j \tag{9.74}$$

$$C_j = - \sum_{i=1}^{n+m+p} B_{ij} \quad \text{and} \quad w_0 = \sum_{i=1}^{n+m+p} D_i \tag{9.75}$$

Thus w_0 is the initial value of the artificial cost function and C_j is the initial relative cost coefficient obtained by adding the elements of the jth column of the matrix $\mathbf{B}$ and changing its sign.

Before we can use Phase I of the Simplex method, we need to develop a procedure to impose the complementary slackness condition of Eq. (9.70). The condition is satisfied if both X_i and X_{n+m+i} are not simultaneously basic variables. Or, if they are, then one of them must have a zero value (degenerate basic feasible solution). These conditions can be easily checked while determining the pivot element in the Simplex method.

It is useful to note here that a slightly different procedure to solve the KKT necessary conditions for the QP problem has been developed by Lemke (1965). It is known as the *complementary pivot method*. Numerical experiments have shown that method to be computationally more attractive than many other methods for solving QP problems when matrix $\mathbf{H}$ is positive semidefinite (Ravindran and Lee, 1981).

Example 9.7 illustrates the use of the Simplex method to solve a QP problem.

EXAMPLE 9.7   THE SOLUTION TO A QP PROBLEM

Minimize

$$f(\mathbf{x}) = (x_1 - 3)^2 + (x_2 - 3)^2 \tag{a}$$

subject to

$$x_1 + x_2 \le 4 \tag{b}$$

$$x_1 - 3x_2 = 1 \tag{c}$$

$$x_1, x_2 \ge 0 \tag{d}$$

Solution

The cost function for the problem can be expanded as $f(\mathbf{x}) = x_1^2 - 6x_1 + x_2^2 - 6x_2 + 18$. We will ignore the constant 18 in the cost function and minimize the following quadratic function expressed in the form of Eq. (9.52):

$$q(\mathbf{x}) = [-6 \quad -6] \begin{bmatrix} x_1 \\ x_2 \end{bmatrix} + 0.5[x_1 \quad x_2] \begin{bmatrix} 2 & 0 \\ 0 & 2 \end{bmatrix} \begin{bmatrix} x_1 \\ x_2 \end{bmatrix} \tag{e}$$

From the foregoing equations, the following quantities can be identified:

$$\mathbf{c} = \begin{bmatrix} -6 \\ -6 \end{bmatrix}; \quad \mathbf{H} = \begin{bmatrix} 2 & 0 \\ 0 & 2 \end{bmatrix}; \quad \mathbf{A} = \begin{bmatrix} 1 \\ 1 \end{bmatrix}; \quad \mathbf{b} = [4]; \quad \mathbf{N} = \begin{bmatrix} 1 \\ -3 \end{bmatrix}; \quad \mathbf{e} = [1] \tag{f}$$

Using these quantities, matrix $\mathbf{B}$ and vectors $\mathbf{D}$ and $\mathbf{X}$ of Eqs. (9.68) and (9.69) are identified as

$$\mathbf{B} = \begin{bmatrix} 2 & 0 & 1 & -1 & 0 & 0 & 1 & -1 \\ 0 & 2 & 1 & 0 & -1 & 0 & -3 & 3 \\ 1 & 1 & 0 & 0 & 0 & 1 & 0 & 0 \\ 1 & -3 & 0 & 0 & 0 & 0 & 0 & 0 \end{bmatrix} \quad \mathbf{D} = \begin{bmatrix} 6 & 6 & | & 4 & | & 1 \end{bmatrix}^T$$

$$\mathbf{X} = \begin{bmatrix} x_1 & x_2 & | & u_1 & | & \xi_1 & \xi_2 & | & s_1 & | & y_1 & z_1 \end{bmatrix}^T \tag{g}$$

Table 9.5 shows the initial Simplex tableau as well as the four iterations to reach the optimum solution. Note that the relative cost coefficient C_j in the initial tableau is obtained by adding all of the elements in the jth column and changing the sign of the sum. Also, the complementary slackness condition of Eq. (9.70) requires $X_1 X_4 = 0$, $X_2 X_5 = 0$, and $X_3 X_6 = 0$, implying that X_1 and X_4, X_2 and X_5, and X_3 and X_6 cannot be basic variables simultaneously. We impose these conditions while determining the pivots in Phase I of the Simplex procedure.

After four iterations of the Simplex method, all of the artificial variables are nonbasic and the artificial cost function is zero. Therefore, the optimum solution is given as

$$X_1 = \frac{13}{4}, \quad X_2 = \frac{3}{4}, \quad X_3 = \frac{3}{4}, \quad X_s = \frac{3}{4} \tag{h}$$
$$X_4 = 0, \quad X_5 = 0, \quad X_6 = 0, \quad X_7 = 0$$

Using these values, the optimum solution to the original QP problem is recovered as

$$x_1 = \frac{13}{4}, \quad x_2 = \frac{3}{4}, \quad u_1 = \frac{3}{4}, \quad \xi_1 = 0, \quad \xi_2 = 0$$

$$s_1 = 0, \quad y_1 = 0, \quad z_1 = \frac{5}{4}, \quad v_1 = y_1 - z_1 = -\frac{5}{4} \tag{i}$$

$$f\left(\frac{13}{4}, \frac{3}{4}\right) = \frac{41}{8}$$

It can be verified that the solution satisfies all of the KKT optimality conditions for the problem.

TABLE 9.5 Simplex solution procedure for QP problem of Example 9.7

	X_1	X_2	X_3	X_4	X_5	X_6	X_7	X_8	Y_1	Y_2	Y_3	Y_4	D
Initial													
Y_1	2	0	1	−1	0	0	1	−1	1	0	0	0	6
Y_2	0	2	1	0	−1	0	−3	3	0	1	0	0	6
Y_3	1	1	0	0	0	1	0	0	0	0	1	0	4
Y_4	$\underline{1}$	−3	0	0	0	0	0	0	0	0	0	1	1
	$\underline{-4}$	0	−2	1	1	−1	2	−2	0	0	0	0	$w-17$
1st iteration													
Y_1	0	$\underline{6}$	1	−1	0	0	1	−1	1	0	0	−2	4
Y_2	0	2	1	0	−1	0	−3	3	0	1	0	0	6
Y_3	0	4	0	0	0	1	0	0	0	0	1	−1	3
X_1	1	−3	0	0	0	0	0	0	0	0	0	1	1
	0	$\underline{-12}$	2	1	1	−1	2	−2	0	0	0	4	$w-13$
2nd iteration													
X_2	0	1	$\frac{1}{6}$	$-\frac{1}{6}$	0	0	$\frac{1}{6}$	$-\frac{1}{6}$	$\frac{1}{6}$	0	0	$-\frac{1}{3}$	$\frac{2}{3}$
Y_2	0	0	$\frac{2}{3}$	$\frac{1}{3}$	−1	0	$-\frac{10}{3}$	$\frac{10}{3}$	$-\frac{1}{3}$	1	0	$\frac{2}{3}$	$\frac{14}{3}$
Y_3	0	0	$-\frac{2}{3}$	$\frac{2}{3}$	0	1	$-\frac{2}{3}$	$\frac{2}{3}$	$-\frac{2}{3}$	0	1	$\frac{1}{3}$	$\frac{1}{3}$
X_1	1	0	$\frac{1}{2}$	$-\frac{1}{2}$	0	0	$\frac{1}{2}$	$-\frac{1}{2}$	$\frac{1}{2}$	0	0	0	3
	0	0	0	−1	1	−1	4	$\underline{-4}$	2	0	0	0	$w-5$
3rd iteration													
X_2	0	1	0	0	0	$\frac{1}{4}$	0	0	0	0	$\frac{1}{4}$	$-\frac{1}{4}$	$\frac{3}{4}$
Y_2	0	0	$\underline{4}$	−3	−1	−5	0	0	3	1	−5	−1	3
X_8	0	0	−1	1	0	$\frac{3}{2}$	−1	1	−1	0	$\frac{3}{2}$	1	$\frac{1}{2}$
X_1	1	0	0	0	0	$\frac{3}{4}$	0	0	0	0	$\frac{3}{4}$	$\frac{1}{4}$	$1\frac{3}{4}$
	0	0	$\underline{-4}$	3	1	5	0	0	−2	0	6	2	$w-3$
4th iteration													
X_2	0	1	0	0	0	$\frac{1}{4}$	0	0	0	0	$\frac{1}{4}$	$-\frac{1}{4}$	$\frac{3}{4}$
X_3	0	0	1	$-\frac{3}{4}$	$-\frac{1}{4}$	$-\frac{5}{4}$	0	0	$\frac{3}{4}$	$\frac{1}{4}$	$-\frac{5}{4}$	$-\frac{1}{4}$	$\frac{3}{4}$
X_8	1	0	0	$\frac{1}{4}$	$-\frac{1}{4}$	$\frac{1}{4}$	−1	1	$-\frac{1}{4}$	$\frac{1}{4}$	$\frac{1}{4}$	$\frac{1}{4}$	$\frac{5}{4}$
X_1	0	0	0	0	0	$\frac{3}{4}$	0	0	0	0	$\frac{3}{4}$	$\frac{1}{4}$	$\frac{13}{4}$
	0	0	0	0	0	0	0	0	1	1	1	1	$w-0$

EXERCISES FOR CHAPTER 9

Write dual problems for the following problems; solve the dual problem and recover the values of the primal variables from the final dual tableau; verify the solution graphically whenever possible.

9.1 Exercise 8.55	**9.2** Exercise 8.56	**9.3** Exercise 8.57
9.4 Exercise 8.58	**9.5** Exercise 8.59	**9.6** Exercise 8.60
9.7 Exercise 8.61	**9.8** Exercise 8.62	**9.9** Exercise 8.63
9.10 Exercise 8.64	**9.11** Exercise 8.65	**9.12** Exercise 8.66
9.13 Exercise 8.67	**9.14** Exercise 8.68	**9.15** Exercise 8.69
9.16 Exercise 8.70	**9.17** Exercise 8.71	**9.18** Exercise 8.72
9.19 Exercise 8.73	**9.20** Exercise 8.74	**9.21** Exercise 8.75
9.22 Exercise 8.76		

10

Numerical Methods for Unconstrained Optimum Design

Upon completion of this chapter, you will be able to

- Explain gradient-based search and direct search methods (derivative-free) for design optimization

- Explain the concept of iterative numerical search methods for smooth optimum design problems

- Explain two basic calculations in gradient-based search methods for optimum design: (1) calculation of a search direction, and (2) calculation of a step size in the search direction

- Explain the basic concept of a descent direction for unconstrained optimization

- Verify the descent condition for a given search direction for unconstrained optimization

- Calculate the search direction for the steepest-descent and conjugate gradient methods

- Calculate a step size along the search direction using an interval-reducing method

When some or all of the functions of the problem (cost function and/or constraint functions) are nonlinear for an optimization problem, it is called a *nonlinear programming* (NLP) problem. This chapter and the next concentrate on the concepts and description of methods for unconstrained nonlinear optimization problems. Chapters 12 and 13 treat constrained problems. Before presenting the numerical methods for solving optimization problems, we discuss the two classes of methods for unconstrained and constrained problems.

10.1 GRADIENT-BASED AND DIRECT SEARCH METHODS

Gradient-Based Search Methods

These methods, as the name implies, use gradients of the problem functions to perform the search for the optimum point. Therefore, all of the problem functions are assumed to be smooth and at least twice continuously differentiable everywhere in the feasible design space. Also, the design variables are assumed to be continuous that can have any value in their allowable ranges.

The gradient-based methods have been developed extensively since the 1950s, and many good ones are available to solve smooth nonlinear optimization problems. Since these methods use only local information (functions and their gradients at a point) in their search process, they converge only to a local minimum point for the cost function. However, based on these methods strategies have been developed to search for global minimum points for the cost function. Such methods are discussed in Chapter 18.

Direct Search Methods

The term "direct search methods" refers to a class of methods that do not calculate, use, or approximate derivatives of the problem functions. Only the function values are used in the search process. The methods were developed in 1960s and 1970s. They have been employed quite regularly since then because of their simplicity and ease of use. Recently, convergence properties of the methods have been studied, and it has been shown that under certain conditions the methods are convergent to the minimum point for the function (Lewis et al., 2000; Kolda et al., 2003). We will discuss two prominent methods in this class: Hooke-Jeeves in Chapter 11, and Nelder-Mead in Chapter 18.

Nature-Inspired Search Methods

Nature-inspired methods also use only the function values in their search process. Problem functions need not be differentiable or even continuous. The methods, developed since 1980s, use stochastic ideas in their search. Many methods have been developed and evaluated. It turns out that they tend to converge to a global minimum point for the cost function as opposed to a local minimum as with gradient-based methods. Another good feature of the methods is that they are more general than gradient-based methods because they can be used for smooth and nonsmooth problems as well as problems with discrete, integer, and binary variables. Their drawbacks are that (1) they do not have a good stopping criterion (since no optimality conditions are used) and (2) they are slower than gradient-based methods. We will present some of these methods in Chapters 16 and 19.

In this chapter, we describe the basic concepts of *gradient-based methods* for smooth unconstrained optimization problems.

10.2 GENERAL CONCEPTS: GRADIENT-BASED METHODS

In this section, we describe some basic concepts that are applicable to numerical optimization methods. The idea of iterative numerical algorithms is introduced to search for optimum solutions for the design problem. The algorithms are initiated with an estimate for the optimum solution that is improved iteratively if it does not satisfy the optimality conditions.

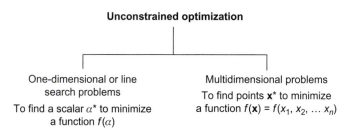

Unconstrained optimization

One-dimensional or line search problems
To find a scalar α^* to minimize a function $f(\alpha)$

Multidimensional problems
To find points $\mathbf{x}^*$ to minimize a function $f(\mathbf{x}) = f(x_1, x_2, \dots x_n)$

FIGURE 10.1   Graphic of the classification of unconstrained optimization problems.

10.2.1 General Concepts

Gradient-based search methods are iterative where the same calculations are repeated in every iteration. In such approaches, we estimate an initial design and improve it until optimality conditions are satisfied. Many *numerical methods* have been developed for NLP problems. Detailed derivations and theories of the various methods are beyond the scope of the present text. However, it is important to understand a few basic concepts, ideas, and procedures that are used in most algorithms for unconstrained and constrained optimization. Therefore, *the approach followed in this text is to stress these underlying concepts with example problems.*

Some details of the numerical algorithms are also presented to give the student a flavor of the calculations performed while searching for the optimum solution. These algorithms are rarely done "by hand"; they require a computer program for their effective use. *Many computer programs for optimization are available for general use such as MATLAB, Excel, and others. Therefore, coding of the algorithms should be attempted only as a last resort.* It is, however, important to understand the underlying ideas to be able to use the programs and the methods properly.

The unconstrained optimization problems are classified as one-dimensional and multidimensional problems, as shown in Figure 10.1. Numerical methods for solving unconstrained problems have been developed over the last several decades. Substantial work, however, was done during the 1950s and 1960s because it was shown that constrained optimization problems could be transformed into a sequence of unconstrained problems (these procedures are presented in Chapter 11). Consequently, the methods have gained considerable importance, and substantial effort has been expended in developing efficient algorithms and computer programs for unconstrained optimization problems.

10.2.2 A General Iterative Algorithm

Many gradient-based optimization methods are described by the following iterative prescription:

Vector form:

$$\mathbf{x}^{(k+1)} = \mathbf{x}^{(k)} + \Delta\mathbf{x}^{(k)}; \quad k = 0, 1, 2, \dots \tag{10.1}$$

Component form:

$$x_i^{(k+1)} = x_i^{(k)} + \Delta x_i^{(k)}; \quad i = 1 \text{ to } n; \quad k = 0, 1, 2, \dots \tag{10.2}$$

where

k = superscript representing the iteration number
i = subscript denoting the design variable number
$\mathbf{x}^{(0)}$ = starting point
$\Delta \mathbf{x}^{(k)}$ = change in the current point

The iterative scheme described in Eq. (10.1) or (10.2) is continued until the optimality conditions are satisfied or a termination criterion is satisfied. This iterative scheme is applicable to constrained as well as unconstrained problems. For unconstrained problems, calculations for $\Delta \mathbf{x}^{(k)}$ depend on the cost function and its derivatives at the current design point. For constrained problems, the constraints must also be considered while computing the change in design $\Delta \mathbf{x}^{(k)}$. Therefore, in addition to the cost function and its derivatives, the constraint functions and their derivatives play a role in determining $\Delta \mathbf{x}^{(k)}$. There are several methods for calculating $\Delta \mathbf{x}^{(k)}$ for unconstrained and constrained problems. This chapter focuses on methods for unconstrained optimization problems.

In most methods, the change in design $\Delta \mathbf{x}^{(k)}$ is further decomposed into two parts as

$$\Delta \mathbf{x}^{(k)} = \alpha_k \mathbf{d}^{(k)} \tag{10.3}$$

where

$\mathbf{d}^{(k)}$ = "desirable" search direction in the design space
α_k = positive scalar called the step size in the search direction

If the direction $\mathbf{d}^{(k)}$ is any "good," then the step size must be greater than 0; this will become clearer when we relate the search direction to a descent direction for the cost function. Thus, the process of computing $\Delta \mathbf{x}^{(k)}$ involves solving two separate subproblems:

1. The direction-finding subproblem
2. The step length determination subproblem (scaling along the direction)

The process of moving from one design point to the next is illustrated in Figure 10.2. In the figure, B is the current design point $\mathbf{x}^{(k)}$, $\mathbf{d}^{(k)}$ is the search direction, and α_k is a

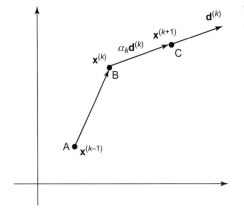

FIGURE 10.2   Conceptual diagram for iterative steps of an optimization method.

step length. Therefore, when $\alpha_k \mathbf{d}^{(k)}$ is added to the current design $\mathbf{x}^{(k)}$, we reach a new point C in the design space, $\mathbf{x}^{(k+1)}$. The entire process is repeated from point C. There are many procedures for calculating the step size α_k and the search direction $\mathbf{d}^{(k)}$. Various combinations of these procedures are used to develop different optimization algorithms.

A General Algorithm

The iterative process just described represents an organized search through the design space for points that represent local minima. The process is summarized as a *general algorithm* that is applicable to both constrained and unconstrained problems:

Step 1. Estimate a reasonable starting design $\mathbf{x}^{(0)}$. Set the iteration counter $k = 0$.
Step 2. Compute a search direction $\mathbf{d}^{(k)}$ at the point $\mathbf{x}^{(k)}$ in the design space. This calculation generally requires a cost function value and its gradient for unconstrained problems and, in addition, constraint functions and their gradients for constrained problems.
Step 3. Check for convergence of the algorithm. If it has converged, stop; otherwise, continue.
Step 4. Calculate a positive step size α_k in the direction $\mathbf{d}^{(k)}$.
Step 5. Update the design as follows, set $k = k + 1$, and go to Step 2:

$$\mathbf{x}^{(k+1)} = \mathbf{x}^{(k)} + \alpha_k \mathbf{d}^{(k)} \tag{10.4}$$

In the remaining sections of this chapter, we present some basic methods for calculating the step size α_k and the search direction $\mathbf{d}^{(k)}$ for unconstrained optimization problems to implement the above algorithm.

10.3 DESCENT DIRECTION AND CONVERGENCE OF ALGORITHMS

The unconstrained minimization problem is defined as finding $\mathbf{x}$ to

Minimize

$$f(\mathbf{x}) \tag{10.5}$$

Since we want to minimize the cost function, the idea of a *descent step* is introduced, which simply means that *changes in the design at every search step must reduce the cost function value.* Convergence of an algorithm and its rate of convergence are also briefly described.

10.3.1 Descent Direction and Descent Step

We have referred to $\mathbf{d}^{(k)}$ as a desirable direction of design change in the iterative process. Now we discuss what we mean by a *desirable direction*. The objective of the iterative optimization process is to reach a minimum point for the cost function $f(\mathbf{x})$.

Let us assume that we are in the kth iteration and have determined that $\mathbf{x}^{(k)}$ is not a minimum point; that is, the optimality conditions of Theorem 4.4 are not satisfied. If $\mathbf{x}^{(k)}$ is not a minimum point, then we should be able to find another point $\mathbf{x}^{(k+1)}$ with a smaller cost function value than the one at $\mathbf{x}^{(k)}$. This statement can be expressed mathematically as

$$f(\mathbf{x}^{(k+1)}) < f(\mathbf{x}^{(k)}) \tag{10.6}$$

Substitute $\mathbf{x}^{(k+1)}$ from Eq. (10.4) into the preceding inequality to obtain

$$f(\mathbf{x}^{(k)} + \alpha_k \mathbf{d}^{(k)}) < f(\mathbf{x}^{(k)}) \tag{10.7}$$

Approximating the left side of Eq. (10.7) by the linear Taylor's expansion at the point $\mathbf{x}^{(k)}$, we get

$$f(\mathbf{x}^{(k)}) + \alpha_k (\mathbf{c}^{(k)} \cdot \mathbf{d}^{(k)}) < f(\mathbf{x}^{(k)}) \tag{10.8}$$

where

$\mathbf{c}^{(k)} = \nabla f(\mathbf{x}^{(k)})$ is the gradient of $f(\mathbf{x})$ at the point $\mathbf{x}^{(k)}$
$(\mathbf{a} \cdot \mathbf{b}) = $ dot product of the vectors $\mathbf{a}$ and $\mathbf{b}$

Subtracting $f(\mathbf{x}^{(k)})$ from both sides of Inequality (10.8), we get $\alpha_k(\mathbf{c}^{(k)} \cdot \mathbf{d}^{(k)}) < 0$. Since $\alpha_k > 0$, it may be dropped without affecting the inequality. Therefore, we get the condition

$$(\mathbf{c}^{(k)} \cdot \mathbf{d}^{(k)}) < 0 \tag{10.9}$$

Since $\mathbf{c}^{(k)}$ is a known vector (the gradient of the cost function), the search direction $\mathbf{d}^{(k)}$ must satisfy Inequality (10.9). Any small move in such a direction will decrease the cost function. Geometrically, using the definition of the dot product of two vectors, the inequality shows that the angle between the vectors $\mathbf{c}^{(k)}$ and $\mathbf{d}^{(k)}$ must be between $90°$ and $270°$.

We can now define a *desirable direction of change* as any vector $\mathbf{d}^{(k)}$ satisfying Inequality (10.9). Such vectors are also called *directions of descent* for the cost function, and Inequality (10.9) is called the *descent condition*. A step of the iterative optimization method based on these directions is called a *descent step*. There can be several directions of descent at a design point and each optimization algorithm computes it differently. A method based on the idea of a *descent step* is called a *descent method*. Clearly, such a method will not converge to a local maximum point for the function.

The descent direction is also sometimes called the "downhill" direction. The problem of minimizing $f(\mathbf{x})$ can be considered as a problem of trying to reach the bottom of a hill from a high point. From the top, we find a downhill direction and travel along it to the lowest point. From the lowest point in the direction, we repeat the process until the bottom of the hill is reached.

The concepts of *descent directions* and descent step are used in most gradient-based optimization methods. Therefore, they should be clearly understood. Example 10.1 illustrates the concept of a descent direction.

EXAMPLE 10.1 CHECK FOR THE DESCENT CONDITION

For the function

$$f(\mathbf{x}) = x_1^2 - x_1 x_2 + 2x_2^2 - 2x_1 + e^{(x_1 + x_2)} \tag{a}$$

check if the direction $\mathbf{d} = (1,2)$ at point $(0, 0)$ is a descent direction for the function f.

Solution

If $\mathbf{d} = (1,2)$ is a descent direction, then it must satisfy Inequality (10.9). To verify this, we calculate the gradient $\mathbf{c}$ of the function $f(\mathbf{x})$ at $(0, 0)$ and evaluate $(\mathbf{c} \cdot \mathbf{d})$ as

$$\mathbf{c} = (2x_1 - x_2 - 2 + e^{(x_1 + x_2)}, \quad -x_1 + 4x_2 + e^{(x_1 + x_2)}) = (-1, 1) \tag{b}$$

$$(\mathbf{c} \cdot \mathbf{d}) = (-1, 1)\begin{bmatrix} 1 \\ 2 \end{bmatrix} = -1 + 2 = 1 > 0 \tag{c}$$

Inequality (10.9) is violated, and thus the given $\mathbf{d}$ is not a descent direction for the function $f(\mathbf{x})$.

10.3.2 Convergence of Algorithms

The central idea behind numerical methods of optimization is to search for the optimum point in an iterative manner, generating a sequence of designs. It is important to note that the success of a method depends on the guarantee of convergence of the sequence to the optimum point. The property of convergence to a local optimum point irrespective of the starting point is called *global convergence* of the numerical method. It is desirable to employ such convergent numerical methods because they are more reliable. For unconstrained problems, a convergent algorithm must reduce the cost function at each iteration until a minimum point is reached. It is important to note that the algorithms converge to a local minimum point only, as opposed to a global minimum, since they use only local information about the cost function and its derivatives in the search process. Methods to search for global minima are described in Chapter 18.

10.3.3 Rate of Convergence

In practice, a numerical method may take a large number of iterations to reach the optimum point. Therefore, it is important to employ methods having a convergence rate that is faster. An algorithm's rate of convergence is usually measured by the number of iterations and function evaluations that is needed to obtain an acceptable solution. *Rate of convergence is a measure of how fast the difference between the solution point and its estimates goes to zero.* Faster algorithms usually use second-order information about the problem functions when calculating the search direction. They are known as *Newton methods*. Many algorithms also approximate second-order information using only first-order information. They are known as *quasi-Newton methods* and they are described in Chapter 11.

10.4 STEP SIZE DETERMINATION: BASIC IDEAS

Unconstrained numerical optimization methods are based on the iterative formula that is given in Eq. (10.1). As discussed earlier, the problem of obtaining the design change $\Delta \mathbf{x}$ is usually decomposed into two subproblems, as expressed in Eq. (10.3):

1. Direction-finding subproblem
2. Step size determination subproblem

We need to discuss numerical methods for solving both subproblems. In the following paragraphs, we first discuss basic ideas related to the problem of *step size determination*. This is often called *one-dimensional* (or *line*) search. Such problems are simpler to solve. This is one reason for discussing them first. After presenting one-dimensional minimization numerical methods in Section 10.5, two methods are described in Sections 10.6 and 10.7 for computing a *descent direction* $\mathbf{d}$ in the design space.

10.4.1 Definition of the Step Size Determination Subproblem

Reduction to a Function of One Variable

For an optimization problem with several variables, the direction-finding subproblem must be solved first. Then, a step size must be determined by searching for the minimum of the cost function along the search direction. This is always a one-dimensional minimization problem, also called a *line search* problem. To see how the line search will be used in multidimensional problems, let us assume for the moment that a search direction $\mathbf{d}^{(k)}$ has been computed.

Then, in Eqs. (10.1) and (10.3), scalar α_k is the only unknown. Since the best step size α_k is yet unknown, we replace it by α in Eq. (10.3), which is then treated as an unknown in the step size calculation subproblem. Then, using Eqs. (10.1) and (10.3), the cost function f(x) at the new point $\mathbf{x}^{(k+1)}$ is given as $f(\mathbf{x}^{(k+1)}) = f(\mathbf{x}^{(k)} + \alpha \mathbf{d}^{(k)})$. Now, since $\mathbf{d}^{(k)}$ is known, the right side becomes a function of the scalar parameter α only. This process is summarized in the following equations:

Design update:
$$\mathbf{x}^{(k+1)} = \mathbf{x}^{(k)} + \alpha \mathbf{d}^{(k)} \tag{10.10}$$

Cost function evaluation:
$$f(\mathbf{x}^{(k+1)}) = f(\mathbf{x}^{(k)} + \alpha \mathbf{d}^{(k)}) = \bar{f}(\alpha) \tag{10.11}$$

where $\bar{f}(\alpha)$ is the new function, with α as the only independent variable (in the sequel, we will drop the over bar for functions of a single variable). Note that at $\alpha = 0$, $f(0) = f(\mathbf{x}^{(k)})$ from Eq. (10.11), which is the current value of the cost function. It is important to understand this *reduction of a function of n variables to a function of only one variable* since this procedure is used in almost all gradient-based optimization methods. It is also important to understand the geometric significance of Eq. (10.11). We elaborate on these ideas in the following paragraphs.

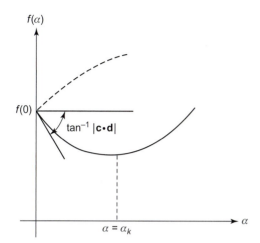

$f(\alpha)$

$f(0)$

$\tan^{-1}|\mathbf{c}\cdot\mathbf{d}|$

$\alpha = \alpha_k$

α

FIGURE 10.3   Graph of $f(\alpha)$ versus α.

One-Dimensional Minimization Problem

If $\mathbf{x}^{(k)}$ is not a minimum point, then it is possible to find a descent direction $\mathbf{d}^{(k)}$ at the point and reduce the cost function further. *Recall that a small move along $\mathbf{d}^{(k)}$ reduces the cost function.* Therefore, using Eqs. (10.6) and (10.11), the descent condition for the cost function can be expressed as the inequality:

$$f(\alpha) < f(0) \tag{10.12}$$

Since $f(\alpha)$ is a function of single variable (also called the *line search function*), we can plot $f(\alpha)$ versus α. To satisfy Inequality (10.12), the curve $f(\alpha)$ versus α must have a negative slope at the point $\alpha = 0$. Such a function is shown by the solid curve in Figure 10.3. It must be understood that if the search direction is that of descent, the graph of $f(\alpha)$ versus α cannot be the one shown by the dashed curve because a small positive α would cause the function $f(\alpha)$ to increase, violating Inequality (10.12). This is also a contradiction as $\mathbf{d}^{(k)}$ is a direction of descent for the cost function.

Therefore, the graph of $f(\alpha)$ versus α must be the solid curve in Figure 10.3 for all problems. In fact, the slope of the curve $f(\alpha)$ at $\alpha = 0$ is calculated by differentiating Eq. (10.11) as $f'(0) = (\mathbf{c}^{(k)} \cdot \mathbf{d}^{(k)})$, which is negative, as seen in Eq. (10.9). This discussion shows that if $\mathbf{d}^{(k)}$ is a descent direction, then α *must always be a positive scalar* in Eq. (10.3). Thus, the step size determination subproblem is to find α to

Minimize

$$f(\alpha) \tag{10.13}$$

Solving this problem gives the step size $\alpha_k = \alpha^*$ for use in Eq. (10.3).

10.4.2 Analytical Method to Compute Step Size

If $f(\alpha)$ is a simple function, then we can use the necessary and sufficient conditions of Section 4.3 to determine α_k. The necessary condition is $df(\alpha_k)/d\alpha = 0$, and the sufficient condition is $d^2f(\alpha_k)/d\alpha^2 > 0$. We will illustrate the analytical line search procedure with Example 10.2. Note that differentiation of $f(\mathbf{x}^{(k+1)})$ in Eq. (10.11) with respect to α, using the chain rule of differentiation and setting it to zero, gives

$$\frac{df(\mathbf{x}^{(k+1)})}{d\alpha} = \frac{\partial f^T(\mathbf{x}^{(k+1)})}{\partial \mathbf{x}} \frac{d(\mathbf{x}^{(k+1)})}{d\alpha} = \nabla f(\mathbf{x}^{(k+1)}) \cdot \mathbf{d}^{(k)} = \mathbf{c}^{(k+1)} \cdot \mathbf{d}^{(k)} = 0 \qquad (10.14)$$

Since the dot product of two vectors is zero in Eq. (10.14), the gradient of the cost function at the new point is orthogonal to the search direction at the kth iteration; that is, $\mathbf{c}^{(k+1)}$ is normal to $\mathbf{d}^{(k)}$. The condition in Eq. (10.14) is important for two reasons:

1. It can be used directly to obtain an equation in terms of α whose smallest root gives the exact step size, α_k.
2. It can be used to check the accuracy of the step size in a numerical procedure to calculate α; thus it is called the *line search termination criterion*.

Many times numerical line search methods will give an approximate or inexact value of the step size along the search direction. The line search termination criterion is useful for determining the accuracy of the step size (i.e., by checking the condition $(\mathbf{c}^{(k+1)} \cdot \mathbf{d}^{(k)}) = 0$).

EXAMPLE 10.2 ANALYTICAL STEP SIZE DETERMINATION

Let a direction of change for the function

$$f(\mathbf{x}) = 3x_1^2 + 2x_1x_2 + 2x_2^2 + 7 \qquad (a)$$

at the point (1,2) be given as $(-1,-1)$. Compute the step size α_k to minimize $f(\mathbf{x})$ in the given direction.

Solution

For the given point $\mathbf{x}^{(k)} = (1,2)$, $f(\mathbf{x}^{(k)}) = 22$, and $\mathbf{d}^{(k)} = (-1,-1)$. We first check to see if $\mathbf{d}^{(k)}$ is a direction of descent using Inequality (10.9). To do that, we need the gradient of the cost function, which is given as

$$\mathbf{c} = \begin{bmatrix} 6x_1 + 2x_2 \\ 2x_1 + 4x_2 \end{bmatrix} \qquad (b)$$

Thus substituting the current point (1,2) into Eq. (b), the gradient of the function at (1,2) is given as $\mathbf{c}^{(k)} = (10,10)$ and $(\mathbf{c}^{(k)} \cdot \mathbf{d}^{(k)}) = 10(-1) + 10(-1) = -20 < 0$. Therefore, $(-1,-1)$ is a direction of descent.

The new point $\mathbf{x}^{(k+1)}$ using Eq. (10.10) is given in terms of α as

$$\begin{bmatrix} x_1 \\ x_2 \end{bmatrix}^{(k+1)} = \begin{bmatrix} 1 \\ 2 \end{bmatrix} + \alpha \begin{bmatrix} -1 \\ -1 \end{bmatrix}, \text{ or } x_1^{(k+1)} = 1 - \alpha; \ x_2^{(k+1)} = 2 - \alpha \qquad (c)$$

Substituting these equations into the cost function of Eq. (a), we get

$$f(\mathbf{x}^{(k+1)}) = 3(1 - \alpha)^2 + 2(1 - \alpha)(2 - \alpha) + 2(2 - \alpha)^2 + 7 = 7\alpha^2 - 20\alpha + 22 = f(\alpha) \qquad (d)$$

Therefore, along the given direction $(-1, -1)$, $f(\mathbf{x})$ becomes a function of the single variable α. Note from Eq. (d) that $f(0) = 22$, which is the cost function value at the current point, and that $f'(0) = -20 < 0$, which is the slope of $f(\alpha)$ at $\alpha = 0$ (also recall that $f'(0) = \mathbf{c}^{(k)} \cdot \mathbf{d}^{(k)}$, which is -20).

Now, using the necessary and sufficient conditions of optimality for $f(\alpha)$ in Eq. (d), we obtain

$$\frac{df}{d\alpha} = 14\alpha_k - 20 = 0; \quad \alpha_k = \frac{10}{7}; \quad \frac{d^2f}{d\alpha^2} = 14 > 0 \tag{e}$$

Therefore, $\alpha_k = 10/7$ minimizes $f(\mathbf{x})$ in the direction $(-1,-1)$. The new point is obtained by substituting the step size in Eq. (c) as

$$\begin{bmatrix} x_1 \\ x_2 \end{bmatrix}^{(k+1)} = \begin{bmatrix} 1 \\ 2 \end{bmatrix} + \left(\frac{10}{7}\right)\begin{bmatrix} -1 \\ -1 \end{bmatrix} = \begin{bmatrix} -\dfrac{3}{7} \\ \dfrac{4}{7} \end{bmatrix} \tag{f}$$

Substituting the new design $(-3/7, 4/7)$ into the cost function $f(\mathbf{x})$, we find the new value of the cost function as 54/7. This is a substantial reduction from the cost function value of 22 at the previous point.

Note that Eq. (e) for calculation of step size α can also be obtained by directly using the condition given in Eq. (10.14). Using Eq. (c), the gradient of f at the new design point in terms of α is

$$\mathbf{c}^{(k+1)} = (6x_1 + 2x_2, 2x_1 + 4x_2) = (10 - 8\alpha, 10 - 6\alpha) \tag{g}$$

Using the condition of Eq. (10.14), we get $14\alpha - 20 = 0$, which is same as Eq. (e).

10.5 NUMERICAL METHODS TO COMPUTE STEP SIZE

10.5.1 General Concepts

In Example 10.2, it *was possible to simplify expressions* and obtain an explicit form for the function $f(\alpha)$. Also, the functional form of $f(\alpha)$ was quite simple. Therefore, it was possible to use the necessary and sufficient conditions of optimality to find the minimum of $f(\alpha)$ and analytically calculate the step size α_k. For many problems, it is not possible to obtain an explicit expression for $f(\alpha)$. Moreover, even if the functional form of $f(\alpha)$ is known, it may be too complicated to lend itself to analytical solution. Therefore, a *numerical method must be used to find* α_k to minimize $f(\mathbf{x})$ in the known direction $\mathbf{d}^{(k)}$.

Unimodal Functions

The numerical line search process is itself iterative, requiring several iterations before a minimum point for $f(\alpha)$ is reached. Many line search techniques are based on comparing function values at several points along the search direction. Usually, we must make some assumptions on the form of the line search function $f(\alpha)$ to compute step size by numerical methods. For example, it must be assumed that a minimum exists and that it is unique in some interval of interest. A function with this property is called the *unimodal function*.

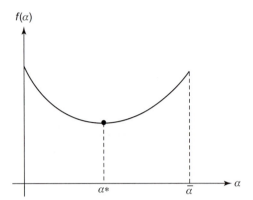

FIGURE 10.4 Unimodal function $f(\alpha)$.

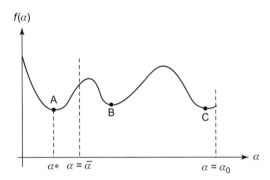

FIGURE 10.5 Nonunimodal function $f(\alpha)$ for $0 \leq \alpha \leq \alpha_0$ (unimodal for $0 \leq \alpha \leq \alpha$).

Figure 10.4 shows the graph of such a function that decreases continuously until the minimum point is reached. Comparing Figures 10.3 and 10.4, we observe that $f(\alpha)$ is a unimodal function in some interval. Therefore, it has a unique minimum point.

Most one-dimensional search methods assume the line search function to be a unimodal function in some interval. This may appear to be a severe restriction on the methods; however, it is not. For functions that are not unimodal, we can think of locating only a local minimum point that is closest to the starting point (i.e., closest to $\alpha = 0$). This is illustrated in Figure 10.5, where the function $f(\alpha)$ is not unimodal for $0 \leq \alpha \leq \alpha_0$. Points A, B, and C are all local minima. If we restrict α to lie between 0 and $\overline{\alpha}$, however, there is only one local minimum point A because the function $f(\alpha)$ is unimodal for $0 \leq \alpha \leq \overline{\alpha}$. Thus, the assumption of unimodality is not as restrictive as it may appear.

Interval-Reducing Methods

The line search problem, then, is to find α in an interval $0 \leq \alpha \leq \overline{\alpha}$ at which the function $f(\alpha)$ has a global minimum. This statement of the problem, however, requires some modification. Since we are dealing with numerical methods, it is not possible to locate the exact minimum point α^*. In fact, *what we determine is the interval in which the minimum lies*—some lower and upper limits α_l and α_u for α^*. The interval (α_l, α_u) is called the *interval of uncertainty* and is designated as $I = \alpha_u - \alpha_l$. Most numerical methods iteratively reduce the interval of uncertainty until it satisfies a specified tolerance ε, (i.e., $I < \varepsilon$). Once this stopping criterion is satisfied, α^* is taken as $0.5(\alpha_l + \alpha_u)$.

Methods based on the preceding philosophy are called *interval-reducing methods*. In this chapter, we will only present line search methods that are based on this idea. The *basic procedure for these methods* can be divided into *two phases*. In Phase I, the location of the minimum point is bracketed and the initial interval of uncertainty is established. In Phase II, the interval of uncertainty is reduced by eliminating regions that cannot contain the minimum. This is done by computing and comparing function values in the interval of uncertainty. We will describe the two phases for these methods in more detail in the following subsections.

It is important to note that the performance of most optimization methods depends heavily on the step size calculation procedure. Therefore, it is not surprising that numerous procedures have been developed and evaluated for step size calculation. In the following paragraphs, we describe two rudimentary methods to give students a flavor of the calculations needed to evaluate a step size. In Chapter 11, more advanced methods based on the concept of an inexact line search are described and discussed.

10.5.2 Equal-Interval Search

Initial Bracketing of Minimum—Phase I

As mentioned earlier, the basic idea of any interval-reducing method is to successively reduce the interval of uncertainty to a small acceptable value. To clearly describe the idea, we start with a very simple approach called *equal-interval search*. The idea is quite elementary, as illustrated in Figure 10.6. In the interval $0 \leq \alpha \leq \bar{\alpha}$, the function $f(\alpha)$ is evaluated at several points using a uniform grid on the α axis in Phase I. To do this, we select a small number δ and evaluate the function at the α values of δ, 2δ, $3\delta, \ldots, q\delta$, $(q + 1)\,\delta$, and so on, as can be seen in Figure 10.6(a). We compare the values of the function at two successive points, say q and $(q + 1)$. Then, if the function at the point $q\delta$ is larger than that at the next point $(q + 1)\delta$—that is, $f(q\delta) > f((q + 1)\delta)$—the minimum point has not been surpassed yet.

However, if the function has started to increase, that is,

$$f(q\delta) < f((q + 1)\delta) \tag{10.15}$$

then the minimum has been surpassed. Note that once the condition in Eq. (10.15) is satisfied for points $q\delta$ and $(q + 1)\delta$, the minimum can be between either the points $(q-1)\delta$ and $q\delta$ or the points $q\delta$ and $(q + 1)\delta$. To account for both possibilities, we take the minimum to lie between the points $(q - 1)\delta$ and $(q + 1)\delta$. Thus, lower and upper limits for the interval of uncertainty are established as

$$\alpha_l = (q - 1)\delta, \quad \alpha_u = (q + 1)\delta, \quad I = \alpha_u - \alpha_l = 2\delta \tag{10.16}$$

Reducing the Interval of Uncertainty—Phase II

Establishment of the lower and upper limits on the minimum value of α indicates the end of Phase I. In Phase II, we restart the search process from the lower end of the interval

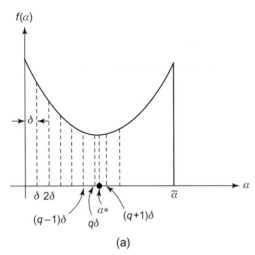

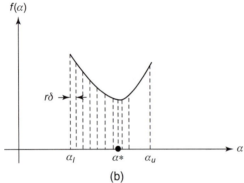

FIGURE 10.6 Equal-interval search process. (a) Phase I: initial bracketing of minimum. (b) Phase II: reducing the interval of uncertainty.

of uncertainty $\alpha = \alpha_l$ with some reduced value for the increment δ, say $r\delta$, where $r \ll 1$. Then the preceding process of Phase I is repeated from $\alpha = \alpha_l$ with the reduced δ, and the minimum is again bracketed. Now the interval of uncertainty I is reduced to $2r\delta$. This is illustrated in Figure 10.6(b). The value of the increment is further reduced to, say, $r^2\delta$, and the process is repeated until the interval of uncertainty is reduced to an acceptable value ε. Note that the method is *convergent* for unimodal functions and can be easily coded into a computer program.

The efficiency of a method, such as equal-interval search, depends on the number of function evaluations needed to achieve the desired accuracy. Clearly, this is dependent on the initial choice for the value of δ. If δ is very small, the process may take many function evaluations to initially bracket the minimum. An advantage of using a smaller δ, however, is that the interval of uncertainty at the end of the Phase I is fairly small. Subsequent improvements to the interval of uncertainty require fewer function evaluations. It is usually advantageous to start with a larger value of δ and quickly bracket the minimum point first. Then Phase II calculations are continued until the accuracy requirement is met.

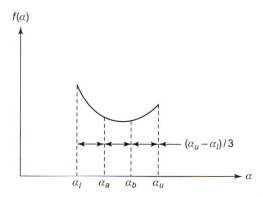

$f(\alpha)$

$(\alpha_u - \alpha_l)/3$

α

$\alpha_l \quad \alpha_a \quad \alpha_b \quad \alpha_u$

FIGURE 10.7 Graphic of an alternate equal-interval solution process.

10.5.3 Alternate Equal-Interval Search

A slightly different computational procedure can be followed to reduce the interval of uncertainty in Phase II once the minimum has been bracketed in Phase I. This procedure is a precursor to the more efficient golden sections search presented in the next subsection. The procedure is to evaluate the function at two new points, say α_a and α_b in the interval of uncertainty. The points α_a and α_b are located at a distance of $I/3$ and $2I/3$ from the lower limit α_l, respectively, where $I = \alpha_u - \alpha_l$. That is,

$$\alpha_a = \alpha_l + \frac{1}{3}I; \quad \alpha_b = \alpha_l + \frac{2}{3}I = \alpha_u - \frac{1}{3}I \qquad (10.17)$$

This is shown in Figure 10.7.

Next the function is evaluated at the two new points α_a and α_b. Let these be designated as $f(\alpha_a)$ and $f(\alpha_b)$. The following two conditions must now be checked:

1. If $f(\alpha_a) < f(\alpha_b)$, then the minimum lies between α_l and α_b. The right one-third interval between α_b and α_u is discarded. New limits for the interval of uncertainty are $\alpha'_l = \alpha_l$ and $\alpha'_u = \alpha_b$ (the prime on α is used to indicate revised limits for the interval of uncertainty). Therefore, the reduced interval of uncertainty is $I' = \alpha'_u - \alpha'_l = \alpha_b - \alpha_l$. The procedure is repeated with the new limits.
2. If $f(\alpha_a) > f(\alpha_b)$, then the minimum lies between α_a and α_u. The interval between α_l and α_a is discarded. The procedure is repeated with $\alpha'_l = \alpha_a$ and $\alpha'_u = \alpha_u(I' = \alpha'_u - \alpha'_l)$.

With the preceding calculations, the interval of uncertainty is reduced to $I' = 2I/3$ after every set of two function evaluations. The entire process is continued until the interval of uncertainty is reduced to an acceptable value.

10.5.4 Golden Section Search

Golden section search is an improvement over the alternate equal-interval search and is one of the better methods in the class of interval-reducing methods. The basic idea of the method is still the same: Evaluate the function at predetermined points, compare them to bracket the minimum in Phase I, and then converge on the minimum point in Phase II by systematically reducing the interval of uncertainty. The method uses fewer function evaluations to reach the minimum point compared with other similar methods. The number of function evaluations is reduced during both the phases, the initial bracketing phase as well as the interval-reducing phase.

Initial Bracketing of Minimum—Phase I

In the equal-interval methods, the selected increment δ is kept fixed to bracket the minimum initially. This can be an inefficient process if δ happens to be a small number. An alternate procedure is to vary the increment at each step, that is, multiply it by a constant $r > 1$. This way initial bracketing of the minimum is rapid; however, the length of the initial interval of uncertainty is increased. The *golden section* search procedure is such a *variable-interval search method*. In it the value of r is not selected arbitrarily. It is selected as the *golden ratio*, which can be derived as 1.618 in several different ways. One derivation is based on the *Fibonacci sequence*, defined as

$$F_0 = 1; \quad F_1 = 1; \quad F_n = F_{n-1} + F_{n-2}, \quad n = 2, 3, \ldots \qquad (10.18)$$

Any number of the Fibonacci sequence for $n > 1$ is obtained by adding the previous two numbers, so the sequence is given as 1, 1, 2, 3, 5, 8, 13, 21, 34, 55, 89.... The sequence has the property

$$\frac{F_n}{F_{n-1}} \to 1.618 \text{ as } n \to \infty \qquad (10.19)$$

That is, as n becomes large, the ratio between two successive numbers F_n and F_{n-1} in the Fibonacci sequence reaches a constant value of 1.618 or $(\sqrt{5} + 1)/2$. This golden ratio has many other interesting properties that will be exploited in the one-dimensional search procedure. One property is that $1/1.618 = 0.618$.

Figure 10.8 illustrates the process of initially bracketing the minimum using a sequence of larger increments based on the golden ratio. In the figure, starting at $q = 0$, we evaluate $f(\alpha)$ at $\alpha = \delta$, where $\delta > 0$ is a small number. We check to see if the value $f(\delta)$ is smaller than the value $f(0)$. If it is, we then take an increment of 1.618δ in the step size (i.e., the

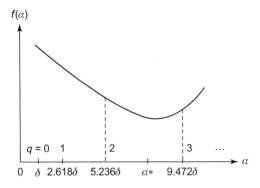

FIGURE 10.8   Initial bracketing of the minimum point in the golden section method.

increment is 1.618 times the previous increment δ). This way we evaluate the function at the following points and compare them:

$$q = 0; \quad \alpha_0 = \delta$$

$$q = 1; \quad \alpha_1 = \delta + 1.618\delta = 2.618\delta = \sum_{j=0}^{1} \delta(1.618)^j$$

$$q = 2; \quad \alpha_2 = 2.618\delta + 1.618(1.618\delta) = 5.236\delta = \sum_{j=0}^{2} \delta(1.618)^j \qquad (10.20)$$

$$q = 3; \quad \alpha_2 = 5.236\delta + 1.618^3\delta = 9.472\delta = \sum_{j=0}^{3} \delta(1.618)^j$$

$$\cdots$$

$$\cdots$$

In general, we continue to evaluate the function at the points

$$\alpha_q = \sum_{j=0}^{q} \delta(1.618)^j; \quad q = 0, 1, 2, \ldots \qquad (10.21)$$

Let us assume that the function at α_{q-1} is smaller than that at the previous point α_{q-2} and the next point α_q; that is,

$$f(\alpha_{q-1}) < f(\alpha_{q-2}) \text{ and } f(\alpha_{q-1}) < f(\alpha_q) \qquad (10.22)$$

Therefore, the minimum point has been surpassed. Actually, the minimum point lies between the previous two intervals, that is, between α_{q-2} and α_q, as in equal-interval search. Therefore, *upper and lower limits on the interval of uncertainty* are

$$\alpha_u = \alpha_q = \sum_{j=0}^{q} \delta(1.618)^j; \quad \alpha_l = \alpha_{q-2} = \sum_{j=0}^{q-2} \delta(1.618)^j \qquad (10.23)$$

Thus, the initial interval of uncertainty is calculated as

$$I = \alpha_u - \alpha_l = \sum_{j=0}^{q} \delta(1.618)^j - \sum_{j=0}^{q-2} \delta(1.618)^j = \delta(1.618)^{q-1} + \delta(1.618)^q$$

$$= \delta(1.618)^{q-1}(1 + 1.618) = 2.618(1.618)^{q-1}\delta \qquad (10.24)$$

Reducing the Interval of Uncertainty—Phase II

The next task is to start reducing the interval of uncertainty by evaluating and comparing functions at some points in the established interval of uncertainty I. *The method uses two function values* within the interval I, just as in the alternate equal-interval search shown in Figure 10.7. However, the points α_a and α_b are not located at $I/3$ from either end of the

(a)

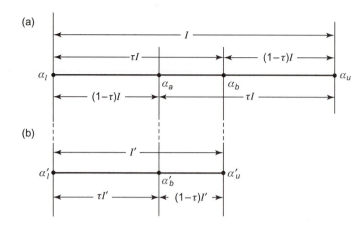

FIGURE 10.9   Graphic of a golden section partition.

(b)

interval of uncertainty. Instead, they are located at a distance of $0.382I$ (or $0.618I$) from either end. The factor 0.382 is related to the golden ratio, as we will see in the following.

To see how the factor 0.618 is determined, consider two points symmetrically located a distance from either end, as shown in Figure 10.9(a)—points α_a and α_b are located at distance τI from either end of the interval. Comparing function values at α_a and α_b, either the left (α_l, α_a) or the right (α_b, α_u) portion of the interval is discarded because the minimum cannot lie there. Let us assume that the right portion is discarded, as shown in Figure 10.9(b), so α'_l and α'_u are the new lower and upper bounds on the minimum. The new interval of uncertainty is $I' = \tau I$.

There is one point in the new interval at which the function value is known. It is required that this point be located at a distance of $\tau I'$ from the left end; therefore, $\tau I' = (1 - \tau)I$. Since $I' = \tau I$, this gives the equation $\tau^2 + \tau - 1 = 0$. The positive root of this equation is $\tau = (-1 + \sqrt{5})/2 = 0.618$. Thus the two points are located at a distance of $0.618I$ or $0.382I$ from either end of the interval.

The golden section search can be initiated once the initial interval of uncertainty is known. If the initial bracketing is done using the variable step increment (with a factor of 1.618, which is $1/0.618$), then the function value at one of the points α_{q-1} is already known. It turns out that α_{q-1} is automatically the point α_a. This can be seen by multiplying the initial interval I in Eq. (10.24) by 0.382. If the preceding procedure is not used to initially bracket the minimum, then the points α_a and α_b will have to be located at a distance of $0.382I$ from the lower and upper limits for the interval of uncertainty.

Algorithm for One-Dimensional Search by Golden Sections

Find α to minimize $f(\alpha)$.

Step 1. *Phase I*: For a chosen small number δ, calculate $f(0)$, $f(\alpha_0)$, $f(\alpha_1)$, ..., where α_i are given by Eq. (10.21). Let q be the smallest integer to satisfy Eqs. (10.22), where α_q, α_{q-1}, and α_{q-2} are calculated from Eq. (10.21). The upper and lower bounds $(\alpha_l$ and $\alpha_u)$ on α^* (optimum value for α) are given by Eq. (10.23). The interval of uncertainty is given as $I = \alpha_u - \alpha_l$.

Step 2. *Phase II*: Compute $f(\alpha_b)$, where $\alpha_b = \alpha_l + 0.618I$. Note that, at the first iteration, $\alpha_a = \alpha_l + 0.382I = \alpha_{q-1}$, and so $f(\alpha_a)$ is already known.

Step 3. Compare $f(\alpha_a)$ and $f(\alpha_b)$, and go to (i), (ii), or (iii).

   (i) If $f(\alpha_a) < f(\alpha_b)$, then minimum point α^* lies between α_l and α_b, that is, $\alpha_l \leq \alpha^* \leq \alpha_b$. The new limits for the reduced interval of uncertainty are $\alpha'_l = \alpha_l$ and $\alpha'_u = \alpha_b$. Also, $\alpha'_b = \alpha_a$. Compute $f(\alpha'_a)$, where $\alpha'_a = \alpha'_l + 0.382(\alpha'_u - \alpha'_l)$ and go to Step 4.

  (ii) If $f(\alpha_a) > f(\alpha_b)$, then minimum point α^* lies between α_a and α_u, that is, $\alpha_a \leq \alpha^* \leq \alpha_u$. Similar to the procedure in Step 3(i), let $\alpha'_l = \alpha_a$ and $\alpha'_u = \alpha_u$, so that $\alpha'_a = \alpha_b$. Compute $f(\alpha'_b)$, where $\alpha'_b = \alpha'_l + 0.618(\alpha'_u - \alpha'_l)$ and go to Step 4.

 (iii) If $f(\alpha_a) = f(\alpha_b)$, let $\alpha_l = \alpha_a$ and $\alpha_u = \alpha_b$ and return to Step 2.

Step 4. If the new interval of uncertainty $I' = \alpha'_u - \alpha'_l$ is small enough to satisfy a stopping criterion (i.e., $I' < \varepsilon$), let $\alpha^* = (\alpha'_u + \alpha'_l)/2$ and stop. Otherwise, delete the primes on α'_l, α'_a, and α'_b and return to Step 3.

Example 10.3 illustrates the golden sections method for step size calculation.

EXAMPLE 10.3 MINIMIZATION OF A FUNCTION BY GOLDEN SECTION SEARCH

Consider the function $f(\alpha) = 2 - 4\alpha + e^{\alpha}$. Use golden section search to find the minimum within an accuracy of $\varepsilon = 0.001$. Use $\delta = 0.5$.

Solution

Analytically, the solution is $\alpha^* = 1.3863$, $f(\alpha^*) = 0.4548$. In the golden section search, we need to first bracket the minimum point (Phase I) and then iteratively reduce the interval of uncertainty (Phase II). Table 10.1 shows various iterations of the method. In Phase I, we use Eq. (10.21) to calculate the trial step sizes α_q and the function values at these points. It is seen that the minimum point is bracketed in only four iterations, as seen in the first part of Table 10.1 because $f(2.618034) > f(1.309017)$. The initial interval of uncertainty is calculated as $I = (\alpha_u - \alpha_l) = 2.618034 - 0.5 = 2.118034$. Note that this interval is larger than the one obtained using equal-interval search.

Now, to reduce the interval of uncertainty in Phase II, let us calculate α_b as $(\alpha_l + 0.618I)$ or $\alpha_b = \alpha_u - 0.382I$ (each iteration of Phase II is shown in the second part of Table 10.1). Note that α_a and $f(\alpha_a)$ are already known and need no further calculation. This is the main advantage of the golden section search; only one additional function evaluation is needed in the interval of uncertainty in each iteration, compared with the two function evaluations needed for the alternate equal-interval search. We describe Iteration 1 of Phase II in Table 10.1 as follows.

We calculate $\alpha_b = 1.809017$ and $f(\alpha_b) = 0.868376$. Note that the new calculation of the function is shown in boldface for each iteration. Since $f(\alpha_a) < f(\alpha_b)$, we are at Step 3(i) of the algorithm, and the new limits for the reduced interval of uncertainty are $\alpha'_l = 0.5$ and $\alpha'_u = 1.809017$. Also, $\alpha'_b = 1.309017$, at which the function value is already known. We need to compute only $f(\alpha'_a)$, where $\alpha'_a = \alpha'_l + 0.382(\alpha'_u - \alpha'_l) = 1.000$. The known values for α and the function are transferred to row 2 of Table 10.1 in Phase II, as indicated by the arrows. The cell for which new α and function values are evaluated is shaded.

Further refinement of the interval of uncertainty is repetitive and can be accomplished by writing a computer program.

TABLE 10.1   Golden section search for $f(\alpha) = 2 - 4\alpha + e^{\alpha}$ *from Example 10.3*

Phase I: Initial Bracketing of Minimum

Number q	Trial step α	Function value, $f(\alpha)$
1. $\alpha = 0$	0.000000	3.000000
2. $q = 0$	$\alpha_0 = \delta\ 0.500000 \leftarrow \alpha_l$	1.648721
3. $q = 1$	$\alpha_1 = \sum_{j=0}^{1} \delta(1.618)^j = 1.309017$	0.466464
4. $q = 2$	$\alpha_2 = \sum_{j=0}^{2} \delta(1.618)^j = 2.618034 \leftarrow \alpha_u$	5.236610

Phase II: Reducing Interval of Uncertainty

Iteration no.	$\alpha_l; [f(\alpha_l)]$	$\alpha_a; [f(\alpha_a)]$	$\alpha_b; [f(\alpha_b)]$	$\alpha_u; [f(\alpha_u)]$	$I = \alpha_u - \alpha_l$
1	0.500000 [1.648721] ↓	1.309017 [0.466 464] ↘	**1.809017** **[0.868376]** ↘	2.618034 [5.236610]	2.118034
2	0.500000 [1.648721]	**1.000000** ↙ **[0.718282]**	1.309017 ↙ [0.466464]	1.809017 [0.868376] ↓	1.309017
3	1.000000 [0.718282]	1.309017 [0.466 464]	**1.500000** **[0.481689]**	1.809017 [0.868376]	0.809017
—	—	—	—	—	—
—	—	—	—	—	—
16	1.385438 [0.454824]	1.386031 [0.454823]	**1.386398** **[0.454823]**	1.386991 [0.454824]	0.001553
17	1.386031 [0.454823]	1.386398 [0.454823]	**1.386624** **[0.454823]**	1.386991 [0.454823]	0.000960

$\alpha^* = 0.5(1.386398 + 1.386624) = 1.386511$; $f(\alpha) = 0.454823$

Note: *New calculation for each iteration is shown as boldfaced and shaded; the arrows indicate direction of data transfer to the next row/iteration.*

A subroutine, GOLD, implementing the golden section search procedure is given in Appendix B. The minimum for the function f is obtained at $\alpha^* = 1.386511$ with $f(\alpha^*) = 0.454823$ in 22 function evaluations, as shown in Table 10.1. The number of function evaluations is a measure of an algorithm's efficiency. The problem was also solved using equal-interval search, and 37 function evaluations were needed to obtain the same solution. This verifies our earlier observation that golden section search is a better method for a specified accuracy and initial step length.

It may appear that if the initial step length δ is too large in the equal-interval or golden section method, the line search fails, that is, $f(\delta) > f(0)$. Actually, it indicates that initial δ is not proper and needs to be reduced until $f(\delta) < f(0)$. With this procedure, convergence of the method can be numerically enforced. This numerical procedure is implemented in the GOLD subroutine given in Appendix B.

10.6 SEARCH DIRECTION DETERMINATION: THE STEEPEST-DESCENT METHOD

Thus far we have assumed that a search direction in the design space is known and we have tackled the problem of step size determination. In this section, and the next, we will address the question of how to determine the search direction **d**. The *basic requirement for* **d** *is that the cost function be reduced* if we take a small step along **d**; that is, the descent condition of Eq. (10.9) must be satisfied. This will be called the *descent direction*.

Several methods are available for determining a descent direction for unconstrained optimization problems. The *steepest-descent method* is the simplest, the oldest, and probably the best known numerical method for unconstrained optimization. The philosophy of the method, introduced by Cauchy in 1847, is to find the direction **d**, at the current iteration, in which the cost function $f(\mathbf{x})$ decreases most rapidly, at least locally. Because of this philosophy, the method is called the *steepest-descent* search technique. Also, properties of the gradient of the cost function are used in the iterative process, which is the reason for its alternate name: the *gradient method*. The steepest-descent method is a *first-order method* since only the gradient of the cost function is calculated and used to evaluate the search direction. In the next chapter, we will discuss *second-order methods* in which the Hessian of the function will also be used in determining the search direction.

The gradient of a scalar function $f(x_1, x_2, \ldots, x_n)$ was defined in Chapter 4 as the column vector:

$$\mathbf{c} = \nabla f = \left[\frac{\partial f}{\partial x_1} \frac{\partial f}{\partial x_2} \cdots \frac{\partial f}{\partial x_n} \right]^T \tag{10.25}$$

To simplify the notation, we will use vector **c** to represent the gradient of the cost function $f(\mathbf{x})$; that is, $c_i = \partial f / \partial x_i$. We will use a superscript to denote the point at which this vector is calculated:

$$\mathbf{c}^{(k)} = \mathbf{c}(\mathbf{x}^{(k)}) = \left[\frac{\partial f(\mathbf{x}^{(k)})}{\partial x_i} \right]^T \tag{10.26}$$

The *gradient vector has several properties* that are used in the steepest-descent method. These will be discussed in the next chapter in more detail. The most important property is that the *gradient at a point* **x** *points in the direction of maximum increase in the cost function*. Thus the direction of maximum decrease is opposite to that, that is, negative of the gradient vector. Any small move in the negative gradient direction will result in the maximum local rate of decrease in the cost function. The negative gradient vector thus represents a *direction of steepest descent* for the cost function and is written as

$$\mathbf{d} = -\mathbf{c}, \text{ or } d_i = -c_i = -\frac{\partial f}{\partial x_i}; \quad i = 1 \text{ to } n \tag{10.27}$$

Note that since $\mathbf{d} = -\mathbf{c}$, the descent condition of inequality (10.9) is always satisfied as

$$(\mathbf{c} \cdot \mathbf{d}) = -||\mathbf{c}||^2 < 0 \tag{10.28}$$

Steepest-Descent Algorithm

Equation (10.27) gives a direction of change in the design space for use in Eq. (10.4). Based on the preceding discussion, the *steepest-descent algorithm* is stated as follows:

Step 1. Estimate a starting design $x^{(0)}$ and set the iteration counter $k = 0$. Select a convergence parameter $\varepsilon > 0$.
Step 2. Calculate the gradient of $f(x)$ at the current point $x^{(k)}$ as $c^{(k)} = \nabla f(x^{(k)})$.
Step 3. Calculate the length of $c^{(k)}$ as $||c^{(k)}||$. If $||c^{(k)}|| < \varepsilon$, then stop the iterative process because $x^* = x^{(k)}$ is a local minimum point. Otherwise, continue.
Step 4. Let the search direction at the current point $x^{(k)}$ be $d^{(k)} = -c^{(k)}$.
Step 5. Calculate a step size α_k that minimizes $f(\alpha) = f(x^{(k)} + \alpha d^{(k)})$ in the direction $d^{(k)}$. Any one-dimensional search algorithm may be used to determine α_k.
Step 6. Update the design using Eq. (10.4) as $x^{(k+1)} = x^{(k)} + \alpha_k d^{(k)}$. Set $k = k + 1$, and go to Step 2.

The basic idea of the steepest-descent method is quite simple. We start with an initial estimate for the minimum design. The direction of steepest descent is computed at that point. If the direction is nonzero, we move as far as possible along it to reduce the cost function. At the new design point, we calculate the steepest-descent direction again and repeat the entire process. Examples 10.4 and 10.5 illustrate the calculations involved in the steepest-descent method.

EXAMPLE 10.4 USE OF THE STEEPEST-DESCENT ALGORITHM

Minimize

$$f(x_1, x_2) = x_1^2 + x_2^2 - 2x_1 x_2 \tag{a}$$

using the steepest-descent method starting from point (1, 0).

Solution

To solve the problem, we follow the steps of the steepest-descent algorithm:

1. The starting design is given as $x^{(0)} = (1, 0)$. Set $k = 0$ and $\varepsilon = 0.0001$.
2. $c^{(0)} = (2x_1 - 2x_2, 2x_2 - 2x_1) = (2, -2)$.
3. $||c^{(0)}|| = \sqrt{2^2 + 2^2} = 2\sqrt{2} > \varepsilon$; continue.
4. Set $d^{(0)} = -c^{(0)} = (-2, 2)$.
5. Calculate α to minimize $f(\alpha) = f(x^{(0)} + \alpha d^{(0)})$, where $x^{(0)} + \alpha d^{(0)} = (1 - 2\alpha, 2\alpha)$:

$$f(x^{(0)} + \alpha d^{(0)}) = (1 - 2\alpha)^2 + (2\alpha)^2 + (2\alpha)^2 - 2(1 - 2\alpha)(2\alpha) = 16\alpha^2 - 8\alpha + 1 = f(\alpha) \tag{b}$$

Since this is a simple function of α, we can use necessary and sufficient conditions to solve for the optimum step length. In general, a numerical one-dimensional search will have to be used to calculate α. Using the analytic approach to solve for optimum α, we get

$$\frac{df(\alpha)}{d\alpha} = 0; \quad 32\alpha - 8 = 0 \quad \text{or} \quad \alpha_0 = 0.25 \tag{c}$$

$$\frac{d^2 f(\alpha)}{d\alpha^2} = 32 > 0 \tag{d}$$

Therefore, the sufficiency condition for a minimum for $f(\alpha)$ is satisfied.

6. Update the design $(\mathbf{x}^{(0)} + \alpha_0 \mathbf{d}^{(0)})$: $x_1^{(1)} = 1 - 0.25(2) = 0.5$, $x_2^{(1)} = 0 + 0.25(2) = 0.5$. Solving for $\mathbf{c}^{(1)}$ from the expression in Step 2, we see that $\mathbf{c}^{(1)} = (0, 0)$, which satisfies the stopping criterion. Therefore, (0.5, 0.5) is a minimum point for $f(\mathbf{x})$ where $f^* = f(\mathbf{x}^*) = 0$.

The preceding problem is quite simple and an optimum point is obtained in only one iteration of the steepest-descent method. This is because the condition number of the Hessian of the cost function is 1 (the condition number is a scalar associated with a square matrix; refer to Section A.7). In such a case, the steepest-descent method converges in just one iteration with any starting point. In general, the algorithm will require several iterations before an acceptable optimum is reached.

EXAMPLE 10.5 USE OF THE STEEPEST-DESCENT ALGORITHM

Minimize

$$f(x_1, x_2, x_3) = x_1^2 + 2x_2^2 + 2x_3^2 + 2x_1 x_2 + 2x_2 x_3 \tag{a}$$

using the steepest-descent method with a starting design of (2, 4, 10). Select 0.005 as the convergence parameter ε. Perform a line search by golden section search with an initial step length $\delta = 0.05$ and an accuracy of 0.0001.

Solution

1. The starting point is set as $\mathbf{x}^{(0)} = (2, 4, 10)$. Set $k = 0$ and $\varepsilon = 0.005$.
2. $\mathbf{c} = \nabla f = (2x_1 + 2x_2, 4x_2 + 2x_1 + 2x_3, 4x_3 + 2x_2)$; $\mathbf{c}^{(0)} = (12, 40, 48)$.
3. $\|\mathbf{c}^{(0)}\| = \sqrt{12^2 + 40^2 + 48^2} = \sqrt{4048} = 63.6 > \varepsilon$ (continue).
4. $\mathbf{d}^{(0)} = -\mathbf{c}^{(0)} = (-12, -40, -48)$.
5. Calculate α_0 by golden section search to minimize $f(\alpha) = f(\mathbf{x}^{(0)} + \alpha \mathbf{d}^{(0)})$; $\alpha_0 = 0.1587$.
6. Update the design as $\mathbf{x}^{(1)} = \mathbf{x}^{(0)} + \alpha_0 \mathbf{d}^{(0)} = (0.0956, -2.348, 2.381)$. At the new design, $\mathbf{c}^{(1)} = (-4.5, -4.438, 4.828)$, $\|\mathbf{c}^{(1)}\| = 7.952 > \varepsilon$.

Note that $(\mathbf{c}^{(1)} \cdot \mathbf{d}^{(0)}) = 0$, which verifies the exact line search termination criterion given in Eq. (10.14). The steps in steepest-descent algorithm should be repeated until the convergence criterion is satisfied. Appendix B contains the computer program and user-supplied subroutines FUNCT and GRAD to implement the steps of the steepest-descent algorithm. The optimum results for the problem from the computer program are given in Table 10.2. The true optimum cost function value is 0.0 and the optimum point is $\mathbf{x}^* = (0,0,0)$. Note that large numbers of iterations and function evaluations are needed to reach the optimum for this problem.

TABLE 10.2   Optimum solution for Example 10.5 with the steepest-descent method

$f(x_1, x_2, x_3) = x_1^2 + 2x_2^2 + 2x_3^2 + 2x_1x_2 + 2x_2x_3$	
Starting values of design variables	2, 4, 10
Optimum design variables	8.04787E-03, −6.81319E-03, 3.42174E-03
Optimum cost function value	2.473 47E-05
Norm of gradient of the cost function at optimum	4.970 71E-03
Number of iterations	40
Total number of function evaluations	753

Although the method of steepest descent is quite simple and robust (it is convergent), it has some drawbacks:

1. Even if convergence of the steepest-descent method is guaranteed, a large number of iterations may be required to reach the minimum point.
2. Each iteration of the method is started independently of others, which can be inefficient. Information calculated at the previous iterations is not used.
3. Only first-order information about the function is used at each iteration to determine the search direction. This is one reason that convergence of the method is slow. The rate of convergence of the steepest-descent method depends on the condition number of the Hessian of the cost function at the optimum point. If the condition number is large, the rate of convergence of the method is slow.
4. Practical experience with the steepest-descent method has shown that a substantial decrease in the cost function is achieved in the initial few iterations and then this decreases slows considerably in later iterations.

10.7 SEARCH DIRECTION DETERMINATION: THE CONJUGATE GRADIENT METHOD

Many optimization methods are based on the concept of conjugate gradients; however, we will describe a method attributed to Fletcher and Reeves (1964) in this section. The conjugate gradient method is a very simple and effective modification of the steepest-descent method. It is shown in the next chapter that the steepest-descent directions at two consecutive steps are orthogonal to each other. This tends to slow down the steepest-descent method, although it is guaranteed to converge to a local minimum point. The conjugate gradient directions are not orthogonal to each other. Rather, these directions tend to cut diagonally through the orthogonal steepest-descent directions. Therefore, they improve the rate of convergence of the steepest-descent method considerably. Actually, the *conjugate gradient directions* $\mathbf{d}^{(i)}$ are orthogonal with respect to a symmetric and positive definite matrix $\mathbf{A}$, that is,

$$\mathbf{d}^{(i)T}\mathbf{A}\mathbf{d}^{(j)} = 0 \text{ for all } i \text{ and } j, \ i \neq j \qquad (10.29)$$

Conjugate Gradient Algorithm

Step 1. Estimate a starting design as $\mathbf{x}^{(0)}$. Set the iteration counter $k = 0$. Select the convergence parameter ε. Calculate

$$\mathbf{d}^{(0)} = -\mathbf{c}^{(0)} = -\nabla f(\mathbf{x}^{(0)}) \tag{10.30}$$

Check the stopping criterion. If $\|\mathbf{c}^{(0)}\| < \varepsilon$, then stop. Otherwise, go to Step 5 (note that the first iteration of the conjugate gradient and steepest-descent methods is the same).

Step 2. Compute the gradient of the cost function as $\mathbf{c}^{(k)} = \nabla f(\mathbf{x}^{(k)})$.

Step 3. Calculate $\|\mathbf{c}^{(k)}\|$. If $\|\mathbf{c}^{(k)}\| < \varepsilon$, then stop; otherwise continue.

Step 4. Calculate the new conjugate direction as

$$\mathbf{d}^{(k)} = -\mathbf{c}^{(k)} + \beta_k \mathbf{d}^{(k-1)} \tag{10.31}$$

$$\beta_k = \left(\frac{\|\mathbf{c}^{(k)}\|}{\|\mathbf{c}^{(k-1)}\|} \right)^2 \tag{10.32}$$

Step 5. Compute a step size $\alpha_k = \alpha$ to minimize $f(\alpha) = f(\mathbf{x}^{(k)} + \alpha \mathbf{d}^{(k)})$.

Step 6. Change the design as follows: set $k = k + 1$ and go to Step 2.

$$\mathbf{x}^{(k+1)} = \mathbf{x}^{(k)} + \alpha_k \mathbf{d}^{(k)} \tag{10.33}$$

Note that the conjugate direction shown in Eq. (10.31) satisfies the descent condition of Inequality (10.9). This can be shown by substituting $\mathbf{d}^{(k)}$ from Eq. (10.31) into Inequality (10.9) and using the step size determination condition given in Eq. (10.14). The first iteration of the conjugate gradient method is just the steepest-descent iteration. The only difference between the conjugate gradient and steepest-descent methods is in Eq. (10.31). In this equation, the current steepest-descent direction is modified by adding a scaled direction that was used in the previous iteration.

The scale factor is determined by using lengths of the gradient vector at the two iterations, as shown in Eq. (10.32). Thus, the conjugate direction is nothing but a deflected steepest-descent direction. This is a simple modification that requires little additional calculation. It is, however, very effective in substantially improving the rate of convergence of the steepest-descent method. Therefore, *the conjugate gradient method should always be preferred over the steepest-descent method*. In the next chapter, an example is discussed that compares the rate of convergence of the steepest-descent, conjugate gradient, and Newton's methods. We will see there that the conjugate gradient method performs quite well compared with the other two.

Convergence of the Conjugate Gradient Method

The conjugate gradient algorithm finds the minimum in n iterations for positive definite quadratic functions having n design variables. For general functions, if the minimum has not been found by then, the iterative process needs to be restarted every $(n + 1)$ iterations for computational stability. That is, set $\mathbf{x}^{(0)} = \mathbf{x}^{(n+1)}$ and restart the process from Step 1 of the algorithm. The algorithm is very simple to program and works very well for general unconstrained minimization problems. Example 10.6 illustrates the calculations involved in the conjugate gradient method.

EXAMPLE 10.6 USE OF THE CONJUGATE GRADIENT ALGORITHM

Consider the problem solved in Example 10.5:

Minimize

$$f(x_1, x_2, x_3) = x_1^2 + 2x_2^2 + 2x_3^2 + 2x_1x_2 + 2x_2x_3 \tag{a}$$

Carry out two iterations of the conjugate gradient method starting from the design (2, 4, 10).

Solution

The first iteration of the conjugate gradient method is the same as the steepest descent given in Example 10.5:

$$\mathbf{c}^{(0)} = (12, 40, 48); \quad \|\mathbf{c}^{(0)}\| = 63.6, \; f(\mathbf{x}^{(0)}) = 332.0 \tag{b}$$

$$\mathbf{x}^{(1)} = (0.0956, \; -2.348, \; 2.381) \tag{c}$$

The second iteration starts from Step 2 of the conjugate gradient algorithm:

2. Calculate the gradient and the cost function at $\mathbf{x}^{(1)}$

$$\mathbf{c}^{(1)} = (2x_1 + 2x_2, 2x_1 + 4x_2 + 2x_3, 2x_2 + 4x_3) = (-4.5, \; -4.438, \; 4.828), \; f(\mathbf{x}^{(1)}) = 10.75 \tag{d}$$

3. $\|\mathbf{c}^{(1)}\| = \sqrt{(-4.5)^2 + (-4.438)^2 + (4.828)^2} = 7.952 > \varepsilon$, so continue.
4.
$$\beta_1 = [\|\mathbf{c}^{(1)}/\mathbf{c}^{(0)}\|]^2 = (7.952/63.3)^2 = 0.015633 \tag{e}$$

$$\mathbf{d}^{(1)} = -\mathbf{c}^{(1)} + \beta_1\mathbf{d}^{(0)} = -\begin{bmatrix} -4.500 \\ -4.438 \\ 4.828 \end{bmatrix} + (0.015633)\begin{bmatrix} -12 \\ -40 \\ -48 \end{bmatrix} = \begin{bmatrix} 4.31241 \\ 3.81268 \\ -5.57838 \end{bmatrix} \tag{f}$$

5. The step size in the direction $\mathbf{d}^{(1)}$ is calculated as $\alpha = 0.3156$.
6. The design is updated as

$$\mathbf{x}^{(2)} = \begin{bmatrix} 0.0956 \\ -2.348 \\ 2.381 \end{bmatrix} + \alpha \begin{bmatrix} 4.31241 \\ 3.81268 \\ -5.57838 \end{bmatrix} = \begin{bmatrix} 1.4566 \\ -1.1447 \\ 0.6205 \end{bmatrix} \tag{g}$$

Calculating the gradient at this point, we get $\mathbf{c}^{(2)} = (0.6238, -0.4246, 0.1926)$. $\|\mathbf{c}^{(2)}\| = 0.7788 > \varepsilon$, so we need to continue the iterations. It can be verified that $(\mathbf{c}^{(2)} \cdot \mathbf{d}^{(1)}) = 0$; that is, the line search termination criterion is satisfied for the step size of $\alpha = 0.3156$.

The problem is solved using the conjugate gradient method available in Excel Solver with $\varepsilon = 0.0001$. Table 10.3 summarizes the performance results for the method. It can be seen that a very precise optimum is obtained in only four iterations. Comparing these with the steepest-descent method results given in Table 10.2, we conclude that the conjugate gradient method is superior for this example.

TABLE 10.3 Optimum solution for Example 10.6 with the conjugate gradient method

$f(x_1, x_2, x_3) = x_1^2 + 2x_2^2 + 2x_3^2 + 2x_1x_2 + 2x_2x_3$	
Starting values of design variables	2, 4, 10
Optimum design variables	1.01E-07, −1.70E-07, 1.04E-09
Optimum cost function value	−4.0E-14
Norm of gradient at optimum	5.20E-07
Number of iterations	4

10.8 OTHER CONJUGATE GRADIENT METHODS

Several other formulas for β in Eq. (10.32) have been derived, giving different conjugate gradient methods. In this section we briefly describe these methods. First we define the difference in the gradients of the cost function at two successive iterations as

$$\mathbf{y}^{(k)} = \mathbf{c}^{(k)} - \mathbf{c}^{(k-1)} \tag{10.34}$$

Different formulas for β are given as

Hestenes-Stiefel (1952):

$$\beta_k = \frac{(\mathbf{c}^{(k)} \cdot \mathbf{y}^{(k)})}{(\mathbf{d}^{(k-1)} \cdot \mathbf{y}^{(k)})} \tag{10.35}$$

Fletcher-Reeves (1964):

$$\beta_k = \frac{(\mathbf{c}^{(k)} \cdot \mathbf{c}^{(k)})}{(\mathbf{c}^{(k-1)} \cdot \mathbf{c}^{(k-1)})} \tag{10.36}$$

Polak-Ribiére (1969):

$$\beta_k = \frac{(\mathbf{c}^{(k)} \cdot \mathbf{y}^{(k)})}{(\mathbf{c}^{(k-1)} \cdot \mathbf{c}^{(k-1)})} \tag{10.37}$$

The Fletcher-Reeves formula given in Eq. (10.36) is the same as given in Eq. (10.32). All of the three formulas given are equivalent for the quadratic function with a positive definite Hessian matrix when exact line search is used. However, for a general function they can have quite different values. The Fletcher-Reeves and Polak-Ribiére formulas have shown good numerical performance. Based on numerical experiments, the following procedure is recommended for selection of a β value:

$$\beta_k = \begin{cases} \beta_k^{pr}, & \text{if } 0 \le \beta_k^{pr} \le \beta_k^{fr} \\ \beta_k^{fr}, & \text{if } \beta_k^{pr} > \beta_k^{fr} \\ 0, & \text{if } \beta_k^{pr} < 0 \end{cases} \tag{10.38}$$

where β_k^{pr} is the value obtained using the Polak-Ribiére formula in Eq. (10.37) and β_k^{fr} is the value obtained using the Fletcher-Reeves formula in Eq. (10.36).

EXERCISES FOR CHAPTER 10

Section 10.3 Descent Direction and Convergence of Algorithms

10.1 *Answer True or False.*

1. All optimum design algorithms require a starting point to initiate the iterative process.
2. A vector of design changes must be computed at each iteration of the iterative process.
3. The design change calculation can be divided into step size determination and direction finding subproblems.
4. The search direction requires evaluation of the gradient of the cost function.
5. Step size along the search direction is always negative.
6. Step size along the search direction can be zero.
7. In unconstrained optimization, the cost function can increase for an arbitrary small step along the descent direction.
8. A descent direction always exists if the current point is not a local minimum.
9. In unconstrained optimization, a direction of descent can be found at a point where the gradient of the cost function is zero.
10. The descent direction makes an angle of $0-90°$ with the gradient of the cost function.

Determine whether the given direction at the point is that of descent for the following functions (show all of the calculations).

10.2 $f(\mathbf{x}) = 3x_1^2 + 2x_1 + 2x_2^2 + 7$; $\mathbf{d} = (-1, 1)$ at $\mathbf{x} = (2, 1)$
10.3 $f(\mathbf{x}) = x_1^2 + x_2^2 - 2x_1 - 2x_2 + 4$; $\mathbf{d} = (2, 1)$ at $\mathbf{x} = (1, 1)$
10.4 $f(\mathbf{x}) = x_1^2 + 2x_2^2 + 2x_3^2 + 2x_1x_2 + 2x_2x_3$; $\mathbf{d} = (-3, 10, -12)$ at $\mathbf{x} = (1, 2, 3)$
10.5 $f(\mathbf{x}) = 0.1x_1^2 + x_2^2 - 10$; $\mathbf{d} = (1, 2)$ at $\mathbf{x} = (4, 1)$
10.6 $f(\mathbf{x}) = (x_1 - 2)^2 + (x_2 - 1)^2$; $\mathbf{d} = (2, 3)$ at $\mathbf{x} = (4, 3)$
10.7 $f(\mathbf{x}) = 10(x_2 - x_1^2)^2 + (1 - x_1)^2$; $\mathbf{d} = (162, -40)$ at $\mathbf{x} = (2, 2)$
10.8 $f(\mathbf{x}) = (x_1 - 2)^2 + x_2^2$; $\mathbf{d} = (-2, 2)$ at $\mathbf{x} = (1, 1)$
10.9 $f(\mathbf{x}) = 0.5x_1^2 + x_2^2 - x_1x_2 - 7x_1 - 7x_2$; $\mathbf{d} = (7, 6)$ at $\mathbf{x} = (1, 1)$
10.10 $f(\mathbf{x}) = (x_1 + x_2)^2 + (x_2 + x_3)^2$; $\mathbf{d} = (4, 8, 4,)$ at $\mathbf{x} = (1, 1, 1)$
10.11 $f(\mathbf{x}) = x_1^2 + x_2^2 + x_3^2$; $\mathbf{d} = (2, 4, -2)$ at $\mathbf{x} = (1, 2, -1)$
10.12 $f(\mathbf{x}) = (x_1 + 3x_2 + x_3)^2 + 4(x_1 - x_2)^2$; $\mathbf{d} = (-2, -6, -2)$ at $\mathbf{x} = (-1, -1, -1)$
10.13 $f(\mathbf{x}) = 9 - 8x_1 - 6x_2 - 4x_3 - 2x_1^2 + 2x_2^2 + x_3^2 + 2x_1x_2 + 2x_2x_3$; $\mathbf{d} = (-2, 2, 0)$ at $\mathbf{x} = (1, 1, 1)$
10.14 $f(\mathbf{x}) = (x_1 - 1)^2 + (x_2 - 2)^2 + (x_3 - 3)^2 + (x_4 - 4)^2$; $\mathbf{d} = (2, -2, 2, -2)$ at $\mathbf{x} = (2, 1, 4, 3)$

Section 10.5 Numerical Methods to Compute Step Size

10.15 *Answer True or False.*

1. Step size determination is always a one-dimensional problem.
2. In unconstrained optimization, the slope of the cost function along the descent direction at zero step size is always positive.
3. The optimum step lies outside the interval of uncertainty.
4. After initial bracketing, the golden section search requires two function evaluations to reduce the interval of uncertainty.

10.16 Find the minimum of the function $f(\alpha) = 7\alpha^2 - 20\alpha + 22$ using the equal-interval search method within an accuracy of 0.001. Use $\delta = 0.05$.

10.17 For the function $f(\alpha) = 7\alpha^2 - 20\alpha + 22$, use the golden section method to find the minimum with an accuracy of 0.005 (final interval of uncertainty should be less than 0.005). Use $\delta = 0.05$.

10.18 Write a computer program to implement the alternate equal-interval search process shown in Figure 10.7 for any given function $f(\alpha)$. For the function $f(\alpha) = 2 - 4\alpha + e^{\alpha}$, use your program to find the minimum within an accuracy of 0.001. Use $\delta = 0.50$.

10.19 Consider the function $f(x_1, x_2, x_3) = x_1^2 + 2x_2^2 + 2x_3^2 + 2_1x_2 + 2x_2x_3$. Verify whether the vector $\mathbf{d} = (-12, -40, -48)$ at the point (2, 4, 10) is a descent direction for f. What is the slope of the function at the given point? Find an optimum step size along $\mathbf{d}$ by any numerical method.

10.20 Consider the function $f(\mathbf{x}) = x_1^2 + x_2^2 - 2x_1 - 2x_2 + 4$. At the point (1, 1), let a search direction be defined as $\mathbf{d} = (1, 2)$. Express f as a function of one variable at the given point along $\mathbf{d}$. Find an optimum step size along $\mathbf{d}$ analytically.

For the following functions, direction of change at a point is given. Derive the function of one variable (line search function) that can be used to determine optimum step size (show all calculations).

10.21 $f(\mathbf{x}) = 0.1x_1^2 + x_2^2 - 10$; $\mathbf{d} = (-1, -2)$ at $\mathbf{x} = (5, 1)$
10.22 $f(\mathbf{x}) = (x_1 - 2)^2 + (x_2 - 1)^2$; $\mathbf{d} = (-4, -6)$ at $\mathbf{x} = (4, 4)$
10.23 $f(\mathbf{x}) = 10(x_2 - x_1^2)^2 + (1 - x_1)^2$; $\mathbf{d} = (-162, 40)$ at $\mathbf{x} = (2, 2)$
10.24 $f(\mathbf{x}) = (x_1 - 2)^2 + x_2^2$; $\mathbf{d} = (2, -2)$ at $\mathbf{x} = (1, 1)$
10.25 $f(\mathbf{x}) = 0.5x_1^2 + x_2^2 - x_1x_2 - 7x_1 - 7x_2$; $\mathbf{d} = (7, 6)$ at $\mathbf{x} = (1, 1)$
10.26 $f(\mathbf{x}) = (x_1 + x_2)^2 + (x_2 + x_3)^2$; $\mathbf{d} = (-4, -8, -4)$ at $\mathbf{x} = (1, 1, 1)$
10.27 $f(\mathbf{x}) = x_1^2 + x_2^2 + x_3^2$; $\mathbf{d} = (-2, -4, 2)$ at $\mathbf{x} = (1, 2, -1)$
10.28 $f(\mathbf{x}) = (x_1 + 3x_2 + x_3)^2 + 4(x_1 - x_2)^2$; $\mathbf{d} = (1, 3, 1)$ at $\mathbf{x} = (-1, -1, -1)$
10.29 $f(\mathbf{x}) = 9 - 8x_1 - 6x_2 - 4x_3 + 2x_1^2 + 2x_2^2 + x_3^2 + 2x_1x_2 + 2x_2x_3$; $\mathbf{d} = (2, -2, 0)$ at $\mathbf{x} = (1, 1, 1)$
10.30 $f(\mathbf{x}) = (x_1 - 1)^2 + (x_2 - 2)^2 + (x_3 - 3)^2 + (x_4 - 4)^2$; $\mathbf{d} = (-2, 2, -2, 2)$ at $\mathbf{x} = (2, 1, 4, 3)$

For the following problems, calculate the initial interval of uncertainty for the equal-interval search with $\delta = 0.05$ at the given point and the search direction.

10.31 Exercise 10.21	**10.32** Exercise 10.22	**10.33** Exercise 10.23
10.34 Exercise 10.24	**10.35** Exercise 10.25	**10.36** Exercise 10.26
10.37 Exercise 10.27	**10.38** Exercise 10.28	**10.39** Exercise 10.29
10.40 Exercise 10.30		

For the following problems, calculate the initial interval of uncertainty for the golden section search with $\delta = 0.05$ at the given point and the search direction; then complete two iterations of the Phase II of the method.

10.41 Exercise 10.21	**10.42** Exercise 10.22	**10.43** Exercise 10.23
10.44 Exercise 10.24	**10.45** Exercise 10.25	**10.46** Exercise 10.26
10.47 Exercise 10.27	**10.48** Exercise 10.28	**10.49** Exercise 10.29
10.50 Exercise 10.30		

Section 10.6 Search Direction Determination: Steepest-Descent Method

10.51 *Answer True or False.*

1. The steepest-descent method is convergent.
2. The steepest-descent method can converge to a local maximum point starting from a point where the gradient of the function is nonzero.
3. Steepest-descent directions are orthogonal to each other.
4. Steepest-descent direction is orthogonal to the cost surface.

For the following problems, complete two iterations of the steepest-descent method starting from the given design point.

10.52 $f(x_1, x_2) = x_1^2 + 2x_2^2 - 4x_1 - 2x_1x_2$; starting design (1, 1)

10.53 $f(x_1, x_2) = 12.096x_1^2 + 21.504x_2^2 - 1.7321x_1 - x_2$; starting design (1, 1)

10.54 $f(x_1, x_2) = 6.983x_1^2 + 12.415x_2^2 - x_1$; starting design (2, 1)

10.55 $f(x_1, x_2) = 12.096x_1^2 + 21.504x_2^2 - x_2$; starting design (1, 2)

10.56 $f(x_1, x_2) = 25x_1^2 + 20x_2^2 - 2x_1 - x_2$; starting design (3, 1)

10.57 $f(x_1, x_2, x_3) = x_1^2 + 2x_2^2 + 2x_3^2 + 2x_1x_2 + 2x_2x_3$; starting design (1, 1, 1)

10.58 $f(x_1, x_2) = 8x_1^2 + 8x_2^2 - 80\sqrt{x_1^2 + x_2^2 - 20x_2 + 100} + 80\sqrt{x_1^2 + x_2^2 + 20x_2 + 100} - 5x_1 - 5x_2$
Starting design (4, 6); the step size may be approximated or calculated using a computer program.

10.59 $f(x_1, x_2) = 9x_1^2 + 9x_2^2 - 100\sqrt{x_1^2 + x_2^2 - 20x_2 + 100} - 64\sqrt{x_1^2 + x_2^2 + 16x_2 + 64} - 5x_1 - 41x_2$
Starting design (5, 2); the step size may be approximated or calculated using a computer program.

10.60 $f(x_1, x_2) = 100(x_2 - x_1^2)^2 + (1 - x_1)^2$; starting design (5,2)

10.61 $f(x_1, x_2, x_3, x_4) = (x_1 - 10x_2)^2 + 5(x_3 - x_4)^2 + (x_2 - 2x_3)^4 + 10(x_1 - x_4)^4$
Let the starting design be (1, 2, 3, 4).

10.62 Solve Exercises 10.52 to 10.61 using the computer program given in Appendix B for the steepest-descent method.

10.63 Consider the following three functions:

$$f_1 = x_1^2 + x_2^2 + x_3^2; \quad f_2 = x_1^2 + 10x_2^2 + 100x_3^2; \quad f_3 = 100x_1^2 + x_2^2 + 0.1x_3^2$$

Minimize f_1, f_2, and f_3 using the program for the steepest-descent method given in Appendix B. Choose the starting design to be (1, 1, 2) for all functions. What do you conclude from observing the performance of the method on the foregoing functions?

10.64 Calculate the gradient of the following functions at the given points by the forward, backward, and central difference approaches with a 1 percent change in the point and compare them with the exact gradient:

1. $f(\mathbf{x}) = 12.096x_1^2 + 21.504x_2^2 - 1.7321x_1 - x_2$ at (5, 6)
2. $f(\mathbf{x}) = 50(x_2 - x_1^2)^2 + (2 - x_1)^2$ at (1, 2)
3. $f(\mathbf{x}) = x_1^2 + 2x_2^2 + 2x_3^2 + 2x_1x_2 + 2x_2x_3$ at (1, 2, 3)

10.65 Consider the following optimization problem

$$\text{maximize} \sum_{i=1}^{n} u_i \frac{\partial f}{\partial x_i} = (\mathbf{c} \cdot \mathbf{u})$$

$$\text{subject to the constraint } \sum_{i=1}^{n} u_i^2 = 1$$

Here $\mathbf{u} = (u_1, u_2, \ldots, u_n)$ are components of a unit vector. Solve this optimization problem and show that the $\mathbf{u}$ that maximizes the preceding objective function is indeed in the direction of the gradient $\mathbf{c}$.

Section 10.7 Search Direction Determination: Conjugate Gradient Method

10.66 *Answer True or False.*

1. The conjugate gradient method usually converges faster than the steepest-descent method.
2. Conjugate directions are computed from gradients of the cost function.
3. Conjugate directions are normal to each other.
4. The conjugate direction at the kth point is orthogonal to the gradient of the cost function at the $(k+1)$th point when an exact step size is calculated.
5. The conjugate direction at the kth point is orthogonal to the gradient of the cost function at the $(k-1)$th point.

For the following problems, complete two iterations of the conjugate gradient method.

10.67 Exercise 10.52	**10.68** Exercise 10.53	**10.69** Exercise 10.54
10.70 Exercise 10.55	**10.71** Exercise 10.56	**10.72** Exercise 10.57
10.73 Exercise 10.58	**10.74** Exercise 10.59	**10.75** Exercise 10.60
10.76 Exercise 10.61		

10.77 Write a computer program to implement the conjugate gradient method (or, modify the steepest-descent program given in Appendix B). Solve Exercises 10.52 to 10.61 using your program.

For the following problems, write an Excel worksheet and solve the problems using Solver.

10.78 Exercise 10.52	**10.79** Exercise 10.53	**10.80** Exercise 10.54
10.81 Exercise 10.55	**10.82** Exercise 10.56	**10.83** Exercise 10.57
10.84 Exercise 10.58	**10.85** Exercise 10.59	**10.86** Exercise 10.60
10.87 Exercise 10.61		

11

More on Numerical Methods for Unconstrained Optimum Design

Upon completion of this chapter, you will be able to

- Use some alternate procedures for step size calculation
- Explain the properties of the gradient vector used in the steepest-descent method
- Use scaling of design variables to improve the performance of optimization methods
- Use the second-order methods for unconstrained optimization, such as the Newton method, and understand their limitations
- Use approximate second-order methods for unconstrained optimization, called the quasi-Newton methods
- Transform constrained problems into unconstrained problems and use unconstrained optimization methods to solve them
- Explain the rate of convergence of algorithms
- Explain and use direct search methods

The material in this chapter builds on the basic concepts and numerical methods for unconstrained problems presented in the previous chapter. Topics covered include polynomial interpolation for step size calculation, inexact line search, properties of the gradient vector, a Newton method that uses the Hessian of the cost function in numerical optimization, scaling of design variables, approximate second-order methods (quasi-Newton methods), and transformation methods that transform a constrained problem into a problem that is unconstrained so that unconstrained optimization methods can be used to solve constrained problems. These topics may be omitted in an undergraduate course on optimum design or in a first independent reading of the text.

Recall that the unconstrained minimization problem is to find an n-dimensional vector $\mathbf{x}$ to minimize the function $f(\mathbf{x})$ without any constraints.

11.1 MORE ON STEP SIZE DETERMINATION

The interval-reducing methods described in Chapter 10 can require too many function evaluations during line search to determine an appropriate step size. In realistic engineering design problems, function evaluation requires a significant amount of computational effort. Therefore, methods such as golden section search are inefficient for many practical applications. In this section, we present some other line search methods such as polynomial interpolation and inexact line search.

Recall that the step size calculation problem is to find α to

Minimize

$$f(\alpha) = f(\mathbf{x}^{(k)} + \alpha \mathbf{d}^{(k)}) \tag{11.1}$$

It is assumed that the search direction $\mathbf{d}^{(k)}$ is that of descent at the current point $\mathbf{x}^{(k)}$, that is,

$$\mathbf{c}^{(k)} \cdot \mathbf{d}^{(k)} < 0 \tag{11.2}$$

Also, differentiating $f(\alpha)$ with respect to α and using the chain rule of differentiation, we get

$$f'(\alpha) = (\mathbf{c}(\mathbf{x}^{(k)} + \alpha \mathbf{d}^{(k)}) \cdot \mathbf{d}^{(k)}); \quad \mathbf{c}(\mathbf{x}^{(k)} + \alpha \mathbf{d}^{(k)}) = \nabla f(\mathbf{x}^{(k)} + \alpha \mathbf{d}^{(k)}) \tag{11.3}$$

where "prime" indicates the first derivative of $f(\alpha)$. Evaluating Eq. (11.3) at $\alpha = 0$, we get

$$f'(0) = (\mathbf{c}^{(k)} \cdot \mathbf{d}^{(k)}) < 0 \tag{11.4}$$

Thus the slope of the curve $f(\alpha)$ versus α is negative at $\alpha = 0$, as can be observed in Figure 10.3. If an exact step size is determined as α_k, then $f'(\alpha_k) = 0$, which gives the following condition from Eq. (11.3), called the *line search termination criterion*:

$$\mathbf{c}^{(k+1)} \cdot \mathbf{d}^{(k)} = 0 \tag{11.5}$$

11.1.1 Polynomial Interpolation

Instead of evaluating the function at numerous trial points, we can pass a curve through a limited number of points and use the analytical procedure to calculate the step size. Any continuous function on a given interval can be approximated as closely as desired by passing a higher-order polynomial through its data points and then calculating its minimum explicitly. The minimum point of the approximating polynomial is often a good estimate of the exact minimum of the line search function $f(\alpha)$. Thus, polynomial interpolation can be an efficient technique for one-dimensional search. Whereas many polynomial interpolation schemes can be devised, we will present two procedures based on *quadratic interpolation*.

Quadratic Curve Fitting

Many times it is sufficient to approximate the function $f(\alpha)$ on an interval of uncertainty by a quadratic function. To replace a function in an interval with a quadratic function, we need to know the function value at three distinct points to determine the three coefficients of the

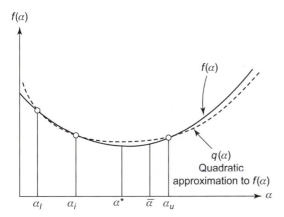

$f(\alpha)$

$f(\alpha)$

$q(\alpha)$
Quadratic
approximation to $f(\alpha)$

α

$\alpha_l \quad \alpha_i \quad \alpha^* \quad \bar{\alpha} \quad \alpha_u$

FIGURE 11.1 Graphic of a quadratic approximation for function $f(\alpha)$.

quadratic polynomial. It must also be assumed that the function $f(\alpha)$ is sufficiently smooth and unimodal, and that the initial interval of uncertainty (α_l, α_u) is known. Let α_i be any intermediate point in the interval (α_l, α_u), and let $f(\alpha_l)$, $f(\alpha_i)$, and $f(\alpha_u)$ be the function values at the respective points. Figure 11.1 shows the function $f(\alpha)$ and the quadratic function $q(\alpha)$ as its approximation in the interval (α_l, α_u). $\bar{\alpha}$ is the minimum point of the quadratic function $q(\alpha)$, whereas α^* is the exact minimum point of $f(\alpha)$. Iteration can be used to improve the estimate $\bar{\alpha}$ for α^*.

Any quadratic function $q(\alpha)$ can be expressed in the general form as

$$q(\alpha) = a_0 + a_1\alpha + a_2\alpha^2 \tag{11.6}$$

where a_0, a_1, and a_2 are the unknown coefficients. Since the function $q(\alpha)$ must have the same value as the function $f(\alpha)$ at the points α_l, α_i, and α_u, we get three equations in three unknowns a_0, a_1, and a_2 as follows:

$$a_0 + a_1\alpha_l + a_2\alpha_l^2 = f(\alpha_l) \tag{11.7}$$

$$a_0 + a_1\alpha_i + a_2\alpha_i^2 = f(\alpha_i) \tag{11.8}$$

$$a_0 + a_1\alpha_u + a_2\alpha_u^2 = f(\alpha_u) \tag{11.9}$$

Solving the system of linear simultaneous equations for a_0, a_1, and a_2, we get

$$a_2 = \frac{1}{(\alpha_u - \alpha_i)}\left[\frac{f(\alpha_u) - f(\alpha_l)}{(\alpha_u - \alpha_l)} - \frac{f(\alpha_i) - f(\alpha_l)}{(\alpha_i - \alpha_l)}\right] \tag{11.10}$$

$$a_1 = \frac{f(\alpha_i) - f(\alpha_l)}{(\alpha_i - \alpha_l)} - a_2(\alpha_l + \alpha_i) \tag{11.11}$$

$$a_0 = f(\alpha_l) - a_1\alpha_l - a_2\alpha_l^2 \tag{11.12}$$

The minimum point $\bar{\alpha}$ of the quadratic function $q(\alpha)$ in Eq. (11.6) is calculated by solving the necessary condition $dq/d\alpha = 0$ and verifying the sufficiency condition $d^2q/d\alpha^2 > 0$:

$$\bar{\alpha} = -\frac{1}{2a_2}a_1; \quad \text{if} \quad \frac{d^2q}{d\alpha^2} = 2a_2 > 0 \tag{11.13}$$

Thus, if $a_2 > 0$, $\overline{\alpha}$ is a minimum of $q(\alpha)$. Additional iterations may be used to further refine the interval of uncertainty. The quadratic curve-fitting technique may now be given in the form of a computational algorithm:

Step 1. Select a small number δ, and locate the initial interval of uncertainty (α_l, α_u). Any method discussed in Chapter 10 may be used.

Step 2. Let α_i be an intermediate point in the interval (α_l, α_u) and $f(\alpha_i)$ be the value of $f(\alpha)$ at α_i.

Step 3. Compute the coefficients a_0, a_1, and a_2 from Eqs. (11.10) through (11.12), $\overline{\alpha}$ from Eq. (11.13), and $f(\overline{\alpha})$.

Step 4. Compare α_i and $\overline{\alpha}$. If $\alpha_i < \overline{\alpha}$, continue with this step. Otherwise, go to Step 5.

  (a) If $f(\alpha_i) < f(\overline{\alpha})$, then $\alpha_l \le \alpha^* \le \overline{\alpha}$. The new limits of the reduced interval of uncertainty are $\alpha_l' = \alpha_l, \alpha_u' = \overline{\alpha}$, and $\alpha_i' = \alpha_i$. Go to Step 6 (a prime on α indicates its updated value).

  (b) If $f(\alpha_i) > f(\overline{\alpha})$, then $\alpha_i \le \alpha^* \le \alpha_u$. The new limits of the reduced interval of uncertainty are $\alpha_l' = \alpha_i$, $\alpha_u' = \alpha_u$, and $\alpha_i' = \overline{\alpha}$. Go to Step 6.

Step 5.

  (a) If $f(\alpha_i) < f(\overline{\alpha})$, then $\overline{\alpha} \le \alpha^* \le \alpha_u$. The new limits of the reduced interval of uncertainty are $\alpha_l' = \overline{\alpha}$, $\alpha_u' = \alpha_u$, and $\alpha_i' = \alpha_i$. Go to Step 6.

  (b) If $f(\alpha_i) > f(\overline{\alpha})$, then $\alpha_l \le \alpha^* \le \alpha_i$. The new limits of the reduced interval of uncertainty are $\alpha_l' = \alpha_l$, $\alpha_u' = \alpha_i$, and $\alpha_i' = \overline{\alpha}$. Go to Step 6.

Step 6. If the two successive estimates of the minimum point of $f(\alpha)$ are sufficiently close, then stop. Otherwise, delete the primes on α_l', α_i', and α_u' and return to Step 3.

Example 11.1 illustrates the evaluation of step size using quadratic interpolation.

EXAMPLE 11.1 ONE-DIMENSIONAL MINIMIZATION WITH QUADRATIC INTERPOLATION

Find the minimum point of

$$f(\alpha) = 2 - 4\alpha + e^{\alpha} \tag{a}$$

from Example 10.3 by polynomial interpolation. Use the golden section search with $\delta = 0.5$ to bracket the minimum point initially.

Solution

Iteration 1 From Example 10.3 the following information is known.

$$\alpha_l = 0.50, \quad \alpha_i = 1.309017, \quad \alpha_u = 2.618034 \tag{b}$$

$$f(\alpha_l) = 1.648721, \quad f(\alpha_i) = 0.466464, \quad f(\alpha_u) = 5.236610 \tag{c}$$

The coefficients a_0, a_1, and a_2 are calculated from Eqs. (11.10) through (11.12) as

$$a_2 = \frac{1}{1.30902}\left(\frac{3.5879}{2.1180} - \frac{-1.1823}{0.80902}\right) = 2.410 \tag{d}$$

$$a_1 = \frac{-1.1823}{0.80902} - (2.41)(1.80902) = -5.821 \tag{e}$$

$$a_0 = 1.648271 - (-5.821)(0.50) - 2.41(0.25) = 3.957 \tag{f}$$

Therefore, $\overline{\alpha} = 1.2077$ from Eq. (11.13), and $f(\overline{\alpha}) = 0.5149$. Note that $\alpha_i > \overline{\alpha}$ and $f(\alpha_i) < f(\overline{\alpha})$; therefore, Step 5(a) of the above algorithm should be used. The new limits of the reduced interval of uncertainty are $\alpha'_l = \overline{\alpha} = 1.2077$, $\alpha'_u = \alpha_u = 2.618034$, and $\alpha'_i = \alpha_i = 1.309017$.

Iteration 2 We have the new limits for the interval of uncertainty, the intermediate point, and the respective values as

$$\alpha_l = 1.2077, \quad \alpha_i = 1.309017, \quad \alpha_u = 2.618034 \tag{g}$$

$$f(\alpha_l) = 0.5149, \quad f(\alpha_i) = 0.466464, \quad f(\alpha_u) = 5.23661 \tag{h}$$

In Step 3 of the algorithm, coefficients a_0, a_1, and a_2 are calculated as before: $a_0 - 5.7129$, $a_1 = -7.8339$, and $a_2 = 2.9228$. Thus, $\overline{\alpha} = 1.34014$ and $f(\overline{\alpha}) = 0.4590$.

Comparing these results with the optimum solution given in Table 10.1, we observe that $\overline{\alpha}$ and $f(\overline{\alpha})$ are quite close to the final solution. One more iteration can give a very good approximation of the optimum step size. Note that only five function evaluations are used to obtain a fairly accurate optimum step size for the function $f(\alpha)$. Therefore, the polynomial interpolation approach can be quite efficient for one-dimensional minimization.

Alternate Quadratic Interpolation

In this approach, we use known information about the function at $\alpha = 0$ to perform quadratic interpolation; that is, we can use $f(0)$ and $f'(0)$ in the interpolation process. Example 11.2 illustrates this alternate quadratic interpolation procedure.

EXAMPLE 11.2 ONE-DIMENSIONAL MINIMIZATION WITH ALTERNATE QUADRATIC INTERPOLATION

Find the minimum point of

$$f(\alpha) = 2 - 4\alpha + e^{\alpha} \tag{a}$$

using $f(0)$, $f'(0)$, and $f(\alpha_u)$ to fit a quadratic curve, where α_u is an upper bound on the minimum point of $f(\alpha)$.

Solution

Let the general equation for a quadratic curve be $a_0 + a_1\alpha + a_2\alpha^2$, where a_0, a_1, and a_2 are the unknown coefficients. Let us select the upper bound on α^* to be 2.618034 (α_u) from the golden section search. Using the given function $f(\alpha)$, we have $f(0) = 3$, $f(2.618034) = 5.23661$, and $f'(0) = -3$. Now, as before, we get the following three equations to solve for the unknown coefficients a_0, a_1, and a_2:

$$a_0 = f(0) = 3 \tag{b}$$

$$f(2.618034) = a_0 + 2.618034a_1 + 6.854a_2 = 5.23661 \tag{c}$$

$$a_1 = f'(0) = -3 \tag{d}$$

Solving the three equations simultaneously, we get $a_0 = 3$, $a_1 = -3$, and $a_2 = 1.4722$. The minimum point of the parabolic curve using Eq. (11.13) is given as $\bar{\alpha} = 1.0189$ and $f(\bar{\alpha}) = 0.69443$. This estimate can be improved using an iteration, as demonstrated in Example 11.1.

Note that an estimate of the minimum point of the function $f(\alpha)$ is found in only two function evaluations $f(0)$ and $f(2.618034)$. However, in an optimization algorithm only one function evaluation is needed since $f(0)$ is the current value of the cost function, which is already available. Also, the slope $f'(0) = \mathbf{c}^{(k)} \cdot \mathbf{d}^{(k)}$ is already known.

11.1.2 Inexact Line Search: Armijo's Rule

Exact line search during unconstrained or constrained minimization can be quite time-consuming. Therefore, inexact line search procedures that also satisfy global convergence requirements are usually employed in most computer implementations. The *basic concept of inexact line search* is that the step size should not be too large or too small, and there should be a sufficient decrease in the cost function value along the search direction. Using these requirements, several inexact line search procedures have been developed and used. Here, we discuss some basic concepts and present a procedure for inexact line search.

Recall that a step size $\alpha_k > 0$ exists if $\mathbf{d}^{(k)}$ satisfies the *descent condition* $(\mathbf{c}^{(k)} \cdot \mathbf{d}^{(k)}) < 0$. Generally, an iterative method, such as quadratic interpolation, is used during line search, and the process is terminated when the step size is sufficiently accurate; that is, the line search termination criterion $(\mathbf{c}^{(k+1)} \cdot \mathbf{d}^{(k)}) = 0$ of Eq. (11.5) is satisfied with sufficient accuracy. However, note that to check this condition, we need to calculate the gradient of the cost function at each trial step size, which can be quite expensive. Therefore, some other simple strategies have been developed that do not require this calculation. One such strategy is called *Armijo's rule*.

The essential idea of Armijo's rule is first to guarantee that the selected step size α is not too large; that is, the current step is not far beyond the optimum step size. Next, the step size should not be too small such that there is little progress toward the minimum point (i.e., there is very little reduction in the cost function).

Let the line search function be defined as $f(\alpha) = f(\mathbf{x}^{(k)} + \alpha \mathbf{d}^{(k)})$, as in Eq. (10.11). Armijo's rule uses a linear function of α as

$$q(\alpha) = f(0) + \alpha[\rho f'(0)] \tag{11.14}$$

where ρ is a fixed number between 0 and 1; $0 < \rho < 1$. This function is shown as the dashed line in Figure 11.2. A value of α is considered *not too large* if the corresponding function value $f(\alpha)$ lies below the dashed line; that is,

$$f(\alpha) \leq q(\alpha) \tag{11.15}$$

In other words, the step size α is to the left of point C in Figure 11.2. This is also called the *sufficient-decrease condition*.

To ensure that α is *not too small*, a number $\eta > 1$ is selected. Then α is considered not too small if it satisfies the following inequality:

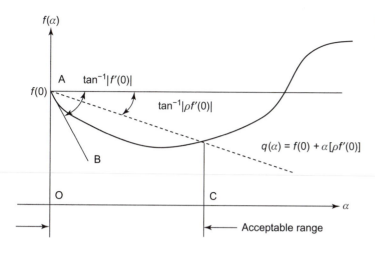

$f(\alpha)$

A $\tan^{-1}|f'(0)|$

$f(0)$

$\tan^{-1}|\rho f'(0)|$

$q(\alpha) = f(0) + \alpha[\rho f'(0)]$

B

O

C

α

Acceptable range

FIGURE 11.2 This is a graphic of an inexact line search that uses Armijo's rule.

$$f(\eta\alpha) > q(\eta\alpha) \tag{11.16}$$

This means that if α is increased by a factor η, it will not meet the test given in Eq. (11.15); that is, $f(\eta\alpha)$ is above the dashed line in Figure 11.2 and the point $\eta\alpha$ is to the right of point C.

Algorithm for Armijo's Rule

Armijo's rule can be used to *determine* the step size as follows: We start with an arbitrary α. If it satisfies Eq. (11.15), it is repeatedly increased by η ($\eta = 2$ and $\rho = 0.2$ are often used) until Eq. (11.15) is violated. The largest α satisfying Eq. (11.15) is selected as the step size. If, on the other hand, the starting value of α does not satisfy Eq. (11.15), it is repeatedly divided by η until Inequality (11.15) is satisfied.

Another procedure, known as the backtracking algorithm, is to start with a larger step size, say $\alpha = 1$. The condition of Eq. (11.15) is checked and, if it is violated, the step size is divided by η. This process is continued until the condition of Eq. (11.15) is satisfied.

It is noted that once $f(\alpha)$ is known at several points, an interpolation scheme (quadratic or cubic) can always be used to obtain a better estimate for the step size α.

Use of a procedure similar to Armijo's rule is demonstrated in a numerical algorithm for constrained problems in Chapter 13.

11.1.3 Inexact Line Search: Wolfe Conditions

The sufficient-decrease condition of Eq. (11.15) is not enough by itself to ensure that the algorithm is making reasonable progress, because it can be satisfied by small values for α. To overcome this drawback, Wolfe (Nocedal and Wright, 2006) introduced another condition for the step size, known as the *curvature condition*, which requires α to satisfy

$$f'(\alpha) \geq \beta f'(0) \tag{11.17}$$

for some constant β, $\rho < \beta < 1$ (also, $0 < \rho < 0.5$). This condition says that at the acceptable step size the slope of $f(\alpha)$ is greater than that at $\alpha = 0$ by the factor β (recall that the slope at $\alpha = 0$ is negative). This is because, if $f'(\alpha)$ is strongly negative, we can further reduce the function $f(\alpha)$. The sufficient-decrease condition of Eq. (11.15) and the curvature condition of Eq. (11.17) are known as *Wolfe conditions*.

Note that the curvature condition of Eq. (11.17) is satisfied even when $f'(\alpha)$ is a large positive number. This implies that the acceptable step size can be far away from the true minimum for $f(\alpha)$ where $f'(\alpha) = 0$. To overcome this, the curvature condition is modified using absolute values for the slopes as

$$|f'(\alpha)| \leq \beta |f'(0)| \tag{11.18}$$

Generally $\beta = 0.1$ to 0.9 and $\rho = 10^{-4}$ to $\rho = 10^{-4}$ to 10^{-3} are taken. Note that a smaller β gives a more accurate step size. For Newton's and quasi-Newton methods, β is selected typically as 0.9 and for the conjugate gradient method as 0.1 (since the conjugate gradient method requires more accurate step size). Conditions in Eqs. (11.15) and (11.18) are called *strong Wolfe conditions* (Nocedal and Wright, 2006).

The criterion of Eq. (11.18) requires evaluation of the gradient of the cost function at each trial step size. If the gradient evaluation is too expensive, a finite-difference approximation may be used for it and Eq. (11.18) may be replaced by

$$\frac{|f(\alpha) - f(\nu)|}{\alpha - \nu} \leq \beta |f'(0)| \tag{11.19}$$

where ν is any scalar such that $0 \leq \nu < \alpha$.

11.1.4 Inexact Line Search: Goldstein Test

The Goldstein test is somewhat similar to Armijo's rule. A value of α is considered *not too large* if it satisfies Eq. (11.15), with ρ given as $0 < \rho < 0.5$. A value of α is considered *not too small* in the Goldstein test if

$$f(\alpha) \geq f(0) + \alpha[(1 - \rho)f'(0)] \tag{11.20}$$

That is, $f(\alpha)$ must lie above the lower dashed line shown in Figure 11.3. In terms of the original function, an acceptable value of α satisfies

$$\rho \leq \frac{f(\alpha) - f(0)}{\alpha f'(0)} \leq (1 - \rho) \tag{11.21}$$

Goldstein conditions are often used in Newton-type methods but are not well suited to quasi-Newton methods (Nocedal and Wright 2006). The Goldstein condition in Eq. (11.20) can be easily checked in the Armijo procedure for the step size calculation described earlier. Note that unless ρ is assigned a proper value, the Goldstein tests in Eqs. (11.15) and (11.20) can omit the true minimum point of $f(\alpha)$.

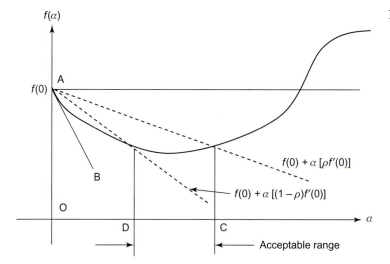

FIGURE 11.3 Goldstein Test.

11.2 MORE ON THE STEEPEST-DESCENT METHOD

In this section we will study the properties of the gradient vector that is used in the steepest-descent method. Proofs of the properties are given since they are quite instructive. We will also show that the steepest-descent directions at successive iterations are orthogonal to each other.

11.2.1 Properties of the Gradient Vector

Property 1

The gradient vector c of a function $f(x_1, x_2, \ldots, x_n)$ at the given point $x^* = (x_1^*, x_2^*, \ldots, x_n^*)$ is orthogonal (normal) to the tangent hyperplane for the surface $f(x_1, x_2, \ldots, x_n) = $ constant.

Proof This is an important property of the gradient vector and is shown graphically in Figure 11.4. The figure shows the surface $f(x) = $ constant; x^* is a point on the surface; C is any curve on the surface through the point x^*; T is a vector tangent to C at point x^*; u is any unit vector; and c is the gradient vector at x^*. According to the above property, vectors c and T are normal to each other; that is, their dot product is zero, $c \cdot T = 0$.

To prove this property, we take any curve C on the surface $f(x_1, x_2, \ldots, x_n) = $ constant, as was shown in Figure 11.4. Let the curve pass through the point $x^* = (x_1^*, x_2^*, \ldots, x_n^*)$. Also, let s be a parameter along C. Then a unit tangent vector T along C at point x^* is given as

$$T = \left[\frac{\partial x_1}{\partial s} \frac{\partial x_2}{\partial s} \cdots \frac{\partial x_n}{\partial s} \right]^T \tag{a}$$

Since $f(x) = $ constant, the derivative of f along curve C is zero; that is, $df/ds = 0$ (the directional derivative of f in the direction s). Or, using the chain rule of differentiation, we get

$$\frac{df}{ds} = \frac{\partial f}{\partial x_1}\frac{\partial x_1}{\partial s} + \cdots + \frac{\partial f}{\partial x_n}\frac{\partial x_n}{\partial s} = 0 \tag{b}$$

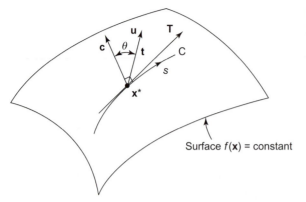

Surface $f(\mathbf{x}) =$ constant

Writing Eq. (b) in the vector form after identifying $\partial f/\partial x_i$ and $\partial x_i/\partial s$ (from Eq. (a)) as components of the gradient and the unit tangent vectors, we obtain $\mathbf{c} \cdot \mathbf{T} = 0$, or $\mathbf{c}^T\mathbf{T} = 0$. Since the dot product of the gradient vector $\mathbf{c}$ with the tangential vector $\mathbf{T}$ is zero, the vectors are normal to each other. But $\mathbf{T}$ is any tangent vector at $\mathbf{x}^*$, and so $\mathbf{c}$ is orthogonal to the tangent hyperplane for the surface $f(\mathbf{x}) =$ constant at point $\mathbf{x}^*$.

Property 2

The second property is that the gradient represents a direction of maximum rate of increase for the function $f(\mathbf{x})$ at the given point $\mathbf{x}^*$.

Proof To show this, let $\mathbf{u}$ be a unit vector in any direction that is not tangent to the surface. This is shown in Figure 11.4. Let t be a parameter along $\mathbf{u}$. The derivative of $f(\mathbf{x})$ in the direction $\mathbf{u}$ at the point $\mathbf{x}^*$ (i.e., the directional derivative of f) is given as

$$\frac{df}{dt} = \lim_{\varepsilon \to 0}\frac{f(\mathbf{x} + \varepsilon\mathbf{u}) - f(\mathbf{x})}{\varepsilon} \tag{c}$$

where ε is a small number and t is a parameter along $\mathbf{u}$. Using Taylor's expansion, we have

$$f(\mathbf{x} + \varepsilon\mathbf{u}) = f(\mathbf{x}) + \varepsilon\left[u_1\frac{\partial f}{\partial x_1} + u_2\frac{\partial f}{\partial x_2} + \cdots + u_n\frac{\partial f}{\partial x_n}\right] + o(\varepsilon^2) \tag{d}$$

where u_i are components of the unit vector $\mathbf{u}$ and $o(\varepsilon^2)$ are terms of order ε^2. Rewriting the foregoing equation,

$$f(\mathbf{x} + \varepsilon\mathbf{u}) - f(\mathbf{x}) = \varepsilon\sum_{i=1}^{n} u_i\frac{\partial f}{\partial x_i} + o(\varepsilon^2) \tag{e}$$

Substituting Eq. (e) into Eq. (c) and taking the indicated limit, we get

$$\frac{df}{dt} = \sum_{i=1}^{n} u_i\frac{\partial f}{\partial x_i} = \mathbf{c} \cdot \mathbf{u} = \mathbf{c}^T\mathbf{u} \tag{f}$$

Using the definition of the dot product in Eq. (e), we get

$$\frac{df}{dt} = ||\mathbf{c}||\,||\mathbf{u}||\,\cos\theta \tag{g}$$

where θ is the angle between the $\mathbf{c}$ and $\mathbf{u}$ vectors. The right side of Eq. (g) will have extreme values when $\theta = 0°$, or $180°$. When $\theta = 0°$, vector $\mathbf{u}$ is along $\mathbf{c}$ and $\cos\theta = 1$.

Therefore, from Eq. (g), df/dt represents the maximum rate of increase for $f(\mathbf{x})$ when $\theta = 0°$. Similarly, when $\theta = 180°$, vector $\mathbf{u}$ points in the negative $\mathbf{c}$ direction. From Eq. (g), then, df/dt represents the maximum rate of decrease for $f(\mathbf{x})$ when $\theta = 180°$.

According to the foregoing property of the gradient vector, if we need to move away from the surface $f(\mathbf{x}) = $ constant, the function increases most rapidly along the gradient vector compared with a move in any other direction. In Figure 11.4, a small move along the direction $\mathbf{c}$ will result in a larger increase in the function, compared with a similar move along the direction $\mathbf{u}$. Of course, any small move along the direction $\mathbf{T}$ results in no change in the function since $\mathbf{T}$ is tangent to the surface.

Property 3

The maximum rate of change in $f(\mathbf{x})$ at any point $\mathbf{x}^*$ is the magnitude of the gradient vector.

Proof Since $\mathbf{u}$ is a unit vector, the maximum value of df/dt from Eq. (g) is given as

$$max \left| \frac{df}{dt} \right| = ||\mathbf{c}|| \tag{h}$$

since the maximum value of $\cos\theta$ is 1 when $\theta = 0°$. However, for $\theta = 0°$, $\mathbf{u}$ is in the direction of the gradient vector. Therefore, the magnitude of the gradient represents the maximum rate of change for the function $f(\mathbf{x})$.

These properties show that the gradient vector at any point $\mathbf{x}^*$ represents a direction of maximum increase in the function $f(\mathbf{x})$ and the rate of increase is the magnitude of the vector. The gradient is therefore called a direction of *steepest ascent* for the function $f(\mathbf{x})$ and the negative of the gradient is called the *direction of steepest descent*. Example 11.3 verifies the properties of the gradient vector.

EXAMPLE 11.3 VERIFICATION OF THE PROPERTIES OF THE GRADIENT VECTOR

Verify the properties of the gradient vector for the following function when it is at the point $\mathbf{x}^{(0)} = (0.6, 4)$.

$$f(\mathbf{x}) = 25x_1^2 + x_2^2 \tag{a}$$

Solution

Figure 11.5 shows in the $x_1 - x_2$ plane the contours of values 25 and 100 for the function f. The value of the function at $(0.6, 4)$ is $f(0.6, 4) = 25$. The gradient of the function at $(0.6, 4)$ is given as

$$\mathbf{c} = \nabla f(0.6, 4) = (\partial f/\partial x_1, \partial f/\partial x_2) = (50x_1, 2x_2) = (30, 8) \tag{b}$$

$$||\mathbf{c}|| = \sqrt{30 \times 30 + 8 \times 8} = 31.04835 \tag{c}$$

Therefore, a unit vector along the gradient is given as

$$\mathbf{C} = \mathbf{c}/||\mathbf{c}|| = (0.966235, 0.257663) \tag{d}$$

Using the given function, a vector tangent to the curve at the point $(0.6, 4)$ is given as

$$\mathbf{t} = (-4, 15) \tag{e}$$

This vector is obtained by differentiating the equation for the following curve at the point (0.6, 4) with respect to the parameter s along the curve:

$$25x_1^2 + x_2^2 = 25 \tag{f}$$

Differentiating this equation with respect to s at the point (0.6, 4) gives

$$25 \times 2x_1 \frac{\partial x_1}{\partial s} + 2x_2 \frac{\partial x_2}{\partial s} = 0, \quad or \quad \partial x_1/\partial s = -(4/15)\partial x_2/\partial s \tag{g}$$

Then the vector $\mathbf{t}$ tangent to the curve is obtained as $(\partial x_1/\partial s, \partial x_2/\partial s)$. The unit tangent vector is calculated as

$$\mathbf{T} = \mathbf{t}/||\mathbf{t}|| = (-0.257663, 0.966235) \tag{h}$$

Property 1

If the gradient is normal to the tangent, then $\mathbf{C} \cdot \mathbf{T} = 0$. This is indeed true for the preceding data. We can also use the condition that if two lines are orthogonal, then $m_1 m_2 = -1$, where m_1 and m_2 are the slopes of the two lines (this result can be proved using the rotational transformation of coordinates through 90 degrees). To calculate the slope of the tangent, we use the equation for the curve $25x_1^2 + x_2^2 = 25$, or $x_2 = 5\sqrt{1 - x_1^2}$. Therefore, the slope of the tangent at the point (0.6, 4) is given as

$$m_1 = \frac{dx_2}{dx_1} = -5x_1/\sqrt{1 - x_1^2} = -\frac{15}{4} \tag{i}$$

This slope is also obtained directly from the tangent vector $\mathbf{t} = (-4, 15)$. The slope of the gradient vector $\mathbf{c} = (30, 8)$ is $m_2 = \frac{8}{30} = \frac{4}{15}$. Thus, $m_1 m_2$ is indeed -1, and the two lines are normal to each other.

Property 2

Consider any arbitrary direction $\mathbf{d} = (0.501034, 0.865430)$ at the point (0.6, 4), as shown in Figure 11.5. If $\mathbf{C}$ is the direction of steepest ascent, then the function should increase more rapidly along $\mathbf{C}$ than along $\mathbf{d}$. Let us choose a step size $\alpha = 0.1$ and calculate two points, one along $\mathbf{C}$ and the other along $\mathbf{d}$:

$$\mathbf{x}^{(1)} = \mathbf{x}^{(0)} + \alpha \mathbf{C} = \begin{bmatrix} 0.6 \\ 4.0 \end{bmatrix} + 0.1 \begin{bmatrix} 0.966235 \\ 0.257633 \end{bmatrix} = \begin{bmatrix} 0.6966235 \\ 4.0257663 \end{bmatrix} \tag{j}$$

$$\mathbf{x}^{(2)} = \mathbf{x}^{(0)} + \alpha \mathbf{d} = \begin{bmatrix} 0.6 \\ 4.0 \end{bmatrix} + 0.1 \begin{bmatrix} 0.501034 \\ 0.865430 \end{bmatrix} = \begin{bmatrix} 0.6501034 \\ 4.0865430 \end{bmatrix} \tag{k}$$

Now we calculate the function at these points and compare their values: $f(\mathbf{x}^{(1)}) = 28.3389$, $f(\mathbf{x}^{(2)}) = 27.2657$. Since $f(\mathbf{x}^{(1)}) > f(\mathbf{x}^{(2)})$, the function increases more rapidly along $\mathbf{C}$ than along $\mathbf{d}$.

Property 3

If the magnitude of the gradient vector represents the maximum rate of change in $f(\mathbf{x})$, then $(\mathbf{c} \cdot \mathbf{c}) > (\mathbf{c} \cdot \mathbf{d})$, $(\mathbf{c} \cdot \mathbf{c}) = 964.0$, and $(\mathbf{c} \cdot \mathbf{d}) = 21.9545$. Therefore, the gradient vector satisfies this property also.

Note that the last two properties are valid only in a local sense—that is, only in a small neighborhood of the point at which the gradient is evaluated.

11.2.2 Orthogonality of Steepest-Descent Directions

It is interesting that the successive directions of steepest descent are normal to one another; that is, $(\mathbf{c}^{(k)} \cdot \mathbf{c}^{(k+1)}) = 0$. This can be shown quite easily using necessary conditions

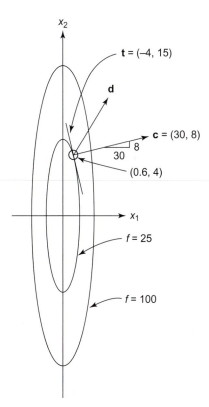

FIGURE 11.5 Contours of the function $f = 25x_1^2 + x_2^2$ for $f = 25$ and 100.

to determine the optimum step size. The step size determination problem is to compute α_k that minimizes $f(\mathbf{x}^{(k)} + \alpha \mathbf{d}^{(k)})$. The necessary condition for this is $df/d\alpha = 0$. Using the chain rule of differentiation, we get

$$\frac{df(\mathbf{x}^{(k+1)})}{d\alpha} = \left[\frac{\partial f(\mathbf{x}^{(k+1)})}{\partial \mathbf{x}}\right]^T \frac{\partial \mathbf{x}^{(k+1)}}{\partial \alpha} = 0 \tag{11.22}$$

which gives (since $\mathbf{d}^{(k+1)} = -\mathbf{c}^{(k+1)}$, the steepest-descent direction)

$$(\mathbf{c}^{(k+1)} \cdot \mathbf{d}^{(k)}) = 0 \quad \text{or} \quad (\mathbf{c}^{(k+1)} \cdot \mathbf{c}^{(k)}) = 0 \tag{11.23}$$

$$\mathbf{c}^{(k+1)} = \frac{\partial f(\mathbf{x}^{(k+1)})}{\partial \mathbf{x}} \quad \text{and} \quad \frac{\partial \mathbf{x}^{(k+1)}}{\partial \alpha} = \frac{\partial}{\partial \alpha}(\mathbf{x}^{(k)} + \alpha \mathbf{d}^{(k)}) = \mathbf{d}^{(k)} \tag{11.24}$$

In the two-dimensional case, $\mathbf{x} = (x_1, x_2)$. Figure 11.6 is an illustration of the design variable space. The closed curves in the figure are contours of the cost function $f(\mathbf{x})$. The figure shows several steepest-descent directions that are orthogonal to each other.

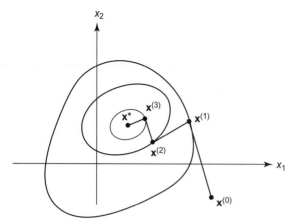

FIGURE 11.6   The graphic shows the orthogonal steepest-descent paths.

11.3 SCALING OF DESIGN VARIABLES

The rate of convergence of the steepest-descent method is at best linear even for a quadratic cost function. It is possible to accelerate this rate of convergence of the steepest-descent method if the condition number of the Hessian of the cost function can be reduced by scaling the design variables. For a quadratic cost function, it is possible to scale the design variables such that the condition number of the Hessian matrix, with respect to the new design variables, is unity (the *condition number* of a matrix is calculated as the ratio of the largest to the smallest eigenvalues of the matrix).

The steepest-descent method converges in only one iteration for a positive definite quadratic function with a unit condition number. To obtain the optimum point with the original design variables, we can then unscale the transformed design variables. The main objective of scaling the design variables, then, is to define transformations such that the condition number of the Hessian with respect to the transformed variables is 1. We will demonstrate the advantage of scaling the design variables with Examples 11.4 and 11.5.

EXAMPLE 11.4 EFFECT OF SCALING THE DESIGN VARIABLES

Minimize

$$f(x_1, x_2) = 25x_1^2 + x_2^2 \tag{a}$$

starting from (1,1) using the steepest-descent method. How would we scale the design variables to accelerate the rate of convergence?

Solution

Let us solve the problem using the computer program for the steepest-descent method given in Appendix B. The results are summarized in Table 11.1. Note the inefficiency of the method with such a simple quadratic cost function; the method takes 5 iterations and 111 function evaluations. Figure 11.7 shows the contours of the cost function and the progress of the method from the initial design.

The Hessian of $f(x_1, x_2)$ is a diagonal matrix, given as

$$\mathbf{H} = \begin{bmatrix} 50 & 0 \\ 0 & 2 \end{bmatrix} \tag{b}$$

The condition number of the Hessian is $50/2 = 25$ since its eigenvalues are 50 and 2. Now let us introduce new design variables y_1 and y_2 such that

$$\mathbf{x} = \mathbf{Dy} \quad \text{where } \mathbf{D} = \begin{bmatrix} \dfrac{1}{\sqrt{50}} & 0 \\ 0 & \dfrac{1}{\sqrt{2}} \end{bmatrix} \tag{c}$$

Note that in general we may use $D_{ii} = 1/\sqrt{H_{ii}}$ for $i = 1$ to n if the Hessian is a diagonal matrix (the diagonal elements are the eigenvalues of $\mathbf{H}$). The transformation in Eq. (c) gives

$$x_1 = y_1/\sqrt{50} \quad \text{and} \quad x_2 = y_2/\sqrt{2} \quad \text{and} \quad f(y_1, y_2) = \frac{1}{2}(y_1^2 + y_2^2). \tag{d}$$

The minimum point of $f(y_1, y_2)$ is found in just one iteration by the steepest-descent method, compared with the five iterations for the original function, since the condition number of the transformed Hessian is 1.

The optimum point is $(0, 0)$ in the new design variable space. To obtain the minimum point in the original design space, we have to unscale the transformed design variables as $x_1^* = y_1/\sqrt{50} = 0$ and $x_2^* = y_2/\sqrt{2} = 0$. Therefore, for this example, the use of design variable scaling is quite beneficial.

TABLE 11.1 Optimum solution to Example 11.4 with the steepest-descent method

$f(x) = 25x_1^2 + x_2^2$	
Starting values of design variables	1, 1
Optimum design variables	−2.35450E−06, 1.37529E−03
Optimum cost function value	1.89157E−06
Norm of gradient at optimum	2.75310E−03
Number of iterations	5
Number of function evaluations	111

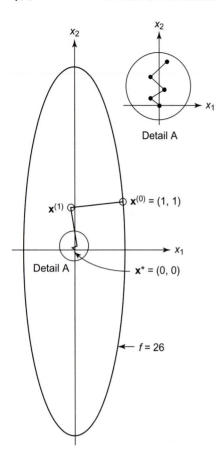

FIGURE 11.7 Iteration history for Example 11.4 with the steepest-descent method.

EXAMPLE 11.5 EFFECT OF SCALING THE DESIGN VARIABLES

Minimize

$$f(x_1, x_2) = 6x_1^2 - 6x_1x_2 + 2x_2^2 - 5x_1 + 4x_2 + 2 \tag{a}$$

starting from $(-1, -2)$ using the steepest-descent method. Scale the design variables to have a condition number of unity for the Hessian matrix of the function with respect to the new design variables.

Solution

Note that, unlike the previous example, the function f in this problem contains the cross-product term x_1x_2. Therefore, the Hessian matrix is not diagonal, and we need to compute its eigenvalues and eigenvectors to find a suitable scaling or transformation of the design variables. The Hessian **H** of the function f is given as

$$\mathbf{H} = \begin{bmatrix} 12 & -6 \\ -6 & 4 \end{bmatrix} \tag{b}$$

The eigenvalues of the Hessian are calculated as 0.7889 and 15.211 (therefore, the condition number = 15.211/0.7889 = 19.3). The corresponding eigenvectors are (0.4718, 0.8817) and (−0.8817, 0.4718). Now let us define new variables y_1 and y_2 by the following transformation:

$$x = Qy \quad \text{where } Q = \begin{bmatrix} 0.4718 & -0.8817 \\ 0.8817 & 0.4718 \end{bmatrix} \tag{c}$$

Note that the columns of Q are the eigenvectors of the Hessian matrix H. The transformation of variables defined by Eq. (c) gives the function in terms of y_1 and y_2 as

$$f(y_1, y_2) = 0.5(0.7889y_1^2 + 15.211y_2^2) + 1.678y_1 + 6.2957y_2 + 2 \tag{d}$$

The condition number of the Hessian matrix in the new design variables y_1 and y_2 is still not unity. To achieve the condition number equal to unity for the Hessian, we must define another transformation of y_1 and y_2 using the eigenvalues of the Hessian matrix as

$$y = Dz, \quad \text{where } D = \begin{bmatrix} \dfrac{1}{\sqrt{0.7889}} & 0 \\ 0 & \dfrac{1}{\sqrt{15.211}} \end{bmatrix} \tag{e}$$

where z_1 and z_2 are the new design variables, which can be calculated from the equations

$$y_1 = \frac{z_1}{\sqrt{0.7889}} \quad \text{and} \quad y_2 = \frac{z_2}{\sqrt{15.211}} \tag{f}$$

The transformed objective function is given as

$$f(z_1, z_2) = 0.5(z_1^2 + z_2^2) + 1.3148z_1 + 1.6142z_2 \tag{g}$$

Since the condition number of the Hessian of $f(z_1, z_2)$ is 1, the steepest-descent method converges to the solution to $f(z_1, z_2)$ in just one iteration as $(-1.3158, -1.6142)$. The minimum point in the original design space is found by defining the inverse transformation as $x = QDz$. This gives the minimum point in the original design space as $(-\frac{1}{3}, -\frac{3}{2})$.

It is important to note that the Hessian matrix for Examples 11.4 and 11.5 is a constant matrix. Therefore, the transformation matrix for the variables is quite easily obtained. In general, the Hessian matrix depends on the design variables. Therefore, the transformation matrix depends on the design variables and will keep changing from iteration to iteration. Actually, we need the Hessian of the function at the minimum point that we are trying to find. Therefore, some approximations must be used to develop the transformation matrices.

11.4 SEARCH DIRECTION DETERMINATION: NEWTON'S METHOD

With the steepest-descent method, only first-order derivative information is used to determine the search direction. If second-order derivatives are available, we can use them to represent the cost surface more accurately, and a better search direction can be found. With the inclusion of second-order information, we can expect a better rate of convergence as well. For example, Newton's method, which uses the Hessian of the function in calculating the search direction, has a *quadratic* rate of convergence (meaning that it converges very rapidly when the design point is within certain radius of the minimum point). For any positive definite quadratic function, the method converges in just one iteration with a step size of one.

11.4.1 Classical Newton's Method

The *basic idea of the classical Newton's method* is to use a second-order Taylor's expansion of the function around the current design point. This gives a quadratic expression for the change in design Δx. The necessary condition for minimization of this function then gives an explicit calculation for design change. In the following, we will omit the argument $x^{(k)}$ from all functions because the derivation applies to any design iteration. Using a second-order Taylor's expansion for the function $f(x)$, we obtain

$$f(x + \Delta x) = f(x) + c^T \Delta x + 0.5\Delta x^T H \Delta x \qquad (11.25)$$

where Δx is a small change in design and H is the Hessian of f at the point x (sometimes denoted $\nabla^2 f$). Equation (11.25) is a quadratic function in terms of Δx. The theory of convex programming problems (Chapter 4) guarantees that if H is positive semidefinite, then there is a Δx that gives a global minimum for the function of Eq. (11.25). In addition, if H is positive definite, then the minimum for Eq. (11.25) is unique.

Writing optimality conditions $(\partial f / \partial(\Delta x) = 0)$ for the function of Eq. (11.25),

$$c + H\Delta x = 0 \qquad (11.26)$$

Assuming H to be nonsingular, we get an expression for Δx as

$$\Delta x = -H^{-1}c \qquad (11.27)$$

Using this value for Δx, the design is updated as

$$x^{(1)} = x^{(0)} + \Delta x \qquad (11.28)$$

Since Eq. (11.25) is just an approximation for f at the point $x^{(0)}$, $x^{(1)}$ will probably not be the precise minimum point of $f(x)$. Therefore, the process will have to be repeated to obtain improved estimates until the minimum is reached.

Each iteration of Newton's method requires computation of the Hessian of the cost function. Since it is a symmetric matrix, it needs computation of $n(n + 1)/2$ second-order derivatives of $f(x)$ (recall that n is the number of design variables). This can require considerable computational effort.

11.4.2 Modified Newton's Method

Note that the classical Newton's method does not have a step size associated with the calculation of design change $\Delta\mathbf{x}$ in Eq. (11.27); that is, the step size is taken as one (a step of length one is called an *ideal step size* or a *Newton's step*). Therefore, there is no way to ensure that the cost function will be reduced at each iteration (i.e., to ensure that $f(\mathbf{x}^{(k+1)}) < f(\mathbf{x}^{(k)})$). Thus, the method is not guaranteed to converge to a local minimum point even with the use of second-order information that requires large calculations.

This situation can be corrected if we incorporate the use of a step size in the calculation of the design change $\Delta\mathbf{x}$. In other words, we treat the solution of Eq. (11.27) as the search direction and use any of the one-dimensional search methods to calculate the step size in the search direction. This is called the *modified Newton's method* and is stated as a step-by-step algorithm:

Step 1. Make an engineering estimate for a starting design $\mathbf{x}^{(0)}$. Set the iteration counter $k = 0$. Select a tolerance ε for the stopping criterion.

Step 2. Calculate $c_i^{(k)} = \partial f(\mathbf{x}^{(k)})/\partial x_i$ for $i = 1$ to n. If $||\mathbf{c}^{(k)}|| < \varepsilon$, stop the iterative process. Otherwise, continue.

Step 3. Calculate the Hessian matrix $\mathbf{H}^{(k)}$ at the current point $\mathbf{x}^{(k)}$.

Step 4. Calculate the search by solving Eq. (11.27) as

$$\mathbf{d}^{(k)} = -[\mathbf{H}^{(k)}]^{-1}\mathbf{c}^{(k)} \tag{11.29}$$

Note that the calculation of $\mathbf{d}^{(k)}$ is symbolic. For computational efficiency, the linear equation $\mathbf{H}^{(k)}\mathbf{d}^{(k)} = -\mathbf{c}^{(k)}$ is solved directly instead of evaluating the inverse of the Hessian matrix.

Step 5. Update the design as $\mathbf{x}^{(k+1)} = \mathbf{x}^{(k)} + \alpha_k\mathbf{d}^{(k)}$, where α_k is calculated to minimize $f(\mathbf{x}^{(k)} + \alpha\mathbf{d}^{(k)})$. Any one-dimensional search procedure may be used to calculate α.

Step 6. Set $k = k + 1$ and go to Step 2.

It is important to note here that unless $\mathbf{H}$ is positive definite, the direction $\mathbf{d}^{(k)}$ determined from Eq. (11.29) may not be that of descent for the cost function. To see this, we substitute $\mathbf{d}^{(k)}$ from Eq. (11.29) into the descent condition of Eq. (11.2) to obtain

$$-\mathbf{c}^{(k)T}\mathbf{H}^{-1}\mathbf{c}^{(k)} < 0 \tag{11.30}$$

The foregoing condition will always be satisfied if $\mathbf{H}$ is positive definite. If $\mathbf{H}$ is negative definite or negative semidefinite, the condition is always violated. With $\mathbf{H}$ as indefinite or positive semidefinite, the condition may or may not be satisfied, so we must check for it. If the direction obtained in Step 4 is not that of descent for the cost function, then we should stop there because a positive step size cannot be determined. Based on the foregoing discussion, it is suggested that the descent condition of Eq. (11.2) should be checked for Newton's search direction at each iteration before calculating the step size. Later we will present methods known as quasi-Newton methods; these use an approximation for the Hessian matrix that is kept positive definite. Because of that the search direction is always that of descent.

Examples 11.6 and 11.7 demonstrate use of the modified Newton's method.

EXAMPLE 11.6 USE OF THE MODIFIED NEWTON'S METHOD

Minimize

$$f(\mathbf{x}) = 3x_1^2 + 2x_1x_2 + 2x_2^2 + 7 \tag{a}$$

using the modified Newton's algorithm starting from the point (5, 10). Use $\varepsilon = 0.0001$ as the stopping criterion.

Solution

We will follow the steps of the modified Newton's method.

Step 1. $\mathbf{x}^{(0)}$ is given as (5, 10).

Step 2. The gradient vector $\mathbf{c}^{(0)}$ at the point (5, 10) is given as

$$\mathbf{c}^{(0)} = (6x_1 + 2x_2, 2x_1 + 4x_2) = (50, 50) \tag{b}$$

$$\|\mathbf{c}^{(0)}\| = \sqrt{50^2 + 50^2} = 50\sqrt{2} > \varepsilon \tag{c}$$

Therefore, the convergence criterion is not satisfied.

Step 3. The Hessian matrix at the point (5, 10) is given as

$$\mathbf{H}^{(0)} = \begin{bmatrix} 6 & 2 \\ 2 & 4 \end{bmatrix} \tag{d}$$

Note that the Hessian does not depend on the design variables and is positive definite (since its eigenvalues are 7.24 and 2.76). Therefore, Newton's direction satisfies the descent condition at each iteration.

Step 4. The direction of design change is

$$\mathbf{d}^{(0)} = -\mathbf{H}^{-1}\mathbf{c}^{(0)} = \frac{-1}{20}\begin{bmatrix} 4 & -2 \\ -2 & 6 \end{bmatrix}\begin{bmatrix} 50 \\ 50 \end{bmatrix} = \begin{bmatrix} -5 \\ -10 \end{bmatrix} \tag{e}$$

Step 5. Calculate the Step size α to minimize $f(\mathbf{x}^{(0)} + \alpha\mathbf{d}^{(0)})$:

$$\mathbf{x}^{(1)} = \mathbf{x}^{(0)} + \alpha\mathbf{d}^{(0)} = \begin{bmatrix} 5 \\ 10 \end{bmatrix} + \alpha\begin{bmatrix} -5 \\ -10 \end{bmatrix} = \begin{bmatrix} 5 - 5\alpha \\ 10 - 10\alpha \end{bmatrix} \tag{f}$$

$$\frac{df}{d\alpha} = 0; \quad \text{or} \quad \nabla f(\mathbf{x}^{(1)}) \cdot \mathbf{d}^{(0)} = 0 \tag{g}$$

where the chain rule of differentiation shown in Eq. (11.22) has been used in Eq. (g). Using the Step 2 calculations, calculate $\nabla f(\mathbf{x}^{(1)})$ and the dot product $\nabla f(\mathbf{x}^{(1)}) \cdot \mathbf{d}^{(0)}$:

$$\nabla f(\mathbf{x}^{(1)}) = \begin{bmatrix} 6(5 - 5\alpha) + 2(10 - 10\alpha) \\ 2(5 - 5\alpha) + 4(10 - 10\alpha) \end{bmatrix} = \begin{bmatrix} 50 - 50\alpha \\ 50 - 50\alpha \end{bmatrix} \tag{h}$$

$$\nabla f(\mathbf{x}^{(1)}) \cdot \mathbf{d}^{(0)} = (50 - 50\alpha, 50 - 50\alpha)\begin{bmatrix} -5 \\ -10 \end{bmatrix} = 0 \tag{i}$$

$$\text{Or} \quad -5(50 - 50\alpha) - 10(50 - 50\alpha) = 0 \tag{j}$$

Solving the preceding equation, we get $\alpha = 1$. Note that the golden section search also gives $\alpha = 1$. Therefore,

$$\mathbf{x}^{(1)} = \begin{bmatrix} 5 - 5\alpha \\ 10 - 10\alpha \end{bmatrix} = \begin{bmatrix} 0 \\ 0 \end{bmatrix} \tag{k}$$

The gradient of the cost function at $\mathbf{x}^{(1)}$ is calculated as

$$\mathbf{c}^{(1)} = \begin{bmatrix} 50 - 50\alpha \\ 50 - 50\alpha \end{bmatrix} = \begin{bmatrix} 0 \\ 0 \end{bmatrix} \tag{l}$$

Since $\|\mathbf{c}^{(k)}\| < \varepsilon$, Newton's method has given the solution in just one iteration. This is because the function is a positive definite quadratic form (the Hessian of f is positive definite everywhere). Note that the condition number of the Hessian is not 1; therefore the steepest-descent method will not converge in one iteration, as was the case in Examples 11.4 and 11.5.

A computer program based on the modified Newton's method is given in Appendix B. It needs three user-supplied subroutines FUNCT, GRAD, and HASN. These subroutines evaluate the cost function, the gradient, and the Hessian matrix of the cost function, respectively. The program is used to solve the problem in Example 11.7.

EXAMPLE 11.7 USE OF THE MODIFIED NEWTON'S METHOD

Minimize

$$f(\mathbf{x}) = 10x_1^4 - 20x_1^2 x_2 + 10x_2^2 + x_1^2 - 2x_1 + 5 \tag{a}$$

using the computer program for the modified Newton's method given in Appendix B from the point $(-1, 3)$. Golden section search may be used for step size determination with $\delta = 0.05$ and line search accuracy equal to 0.0001. For the stopping criterion, use $\varepsilon = 0.005$.

Solution

Note that $f(\mathbf{x})$ is not a quadratic function in terms of the design variables. Thus, we cannot expect Newton's method to converge in one iteration. The gradient of $f(\mathbf{x})$ is given as

$$\mathbf{c} = \nabla f(\mathbf{x}) = (40x_1^3 - 40x_1 x_2 + 2x_1 - 2, \quad -20x_1^2 + 20x_2) \tag{b}$$

and the Hessian matrix of $f(\mathbf{x})$ is

$$\mathbf{H} = \nabla^2 f(\mathbf{x}) = \begin{bmatrix} 120x_1^2 - 40x_2 + 2 & -40x_1 \\ -40x_1 & 20 \end{bmatrix} \tag{c}$$

Results with the modified Newton's method for the problem are given in Table 11.2. The optimum point is $(1, 1)$ and the optimum value of $f(\mathbf{x})$ is 4.0. Newton's method has converged to the optimum solution in eight iterations. Figure 11.8 shows the contours for the function and the progress of the method from the starting design $(-1, 3)$. It is noted that the step size was approximately equal to one in the last phase of the iterative process. This is because the function resembles a quadratic function sufficiently close to the optimum point and the step size is equal to unity for a quadratic function.

TABLE 11.2   Optimum solution to Example 11.7 with the modified Newton's method

$f(x) = 10x_1^4 - 20x_1^2x_2 + 10x_2^2 + x_1^2 - 2x_1 + 5$	
Starting point	−1, 3
Optimum design variables	9.99880E−01, 9.99681E−01
Optimum cost function value	4.0
Norm of gradient at optimum	3.26883E−03
Number of iterations	8
Number of function evaluations	198

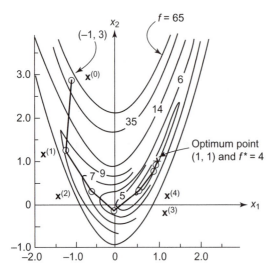

FIGURE 11.8   Iteration history for Example 11.7 with Newton's method.

The *drawbacks* of the modified Newton method for general applications include the following:

1. It requires calculations of second-order derivatives at each iteration, which is usually quite time-consuming. In some applications it may not even be possible to calculate such derivatives. Also, a linear system of equations in Eq. (11.29) needs to be solved. Therefore, each iteration of the method requires substantially more calculations compared with the steepest-descent or conjugate gradient method.
2. The Hessian of the cost function may be singular at some iterations. Thus, Eq. (11.29) cannot be used to compute the search direction. Also, unless the Hessian is positive definite, the search direction cannot be guaranteed to be that of descent for the cost function, as discussed earlier.
3. The method is not convergent unless the Hessian remains positive definite and a step size is calculated along the search direction to update the design. However, the method

has a quadratic rate of convergence when it works. For a strictly convex quadratic function, the method converges in just one iteration from any starting design.

A comparison of the steepest-descent, conjugate gradient, and modified Newton's methods is presented in Example 11.8.

EXAMPLE 11.8 COMPARISON OF STEEPEST-DESCENT, CONJUGATE GRADIENT, AND MODIFIED NEWTON METHODS

Minimize

$$f(\mathbf{x}) = 50(x_2 - x_1^2)^2 + (2 - x_1)^2 \qquad \text{(a)}$$

starting from the point $(5, -5)$. Use the steepest-descent, Newton, and conjugate gradient methods, and compare their performance.

Solution

The minimum point for the function is $(2, 4)$ with $f(2,4) = 0$. We use exact gradient expressions and $\varepsilon = 0.005$ to solve the problem using the steepest-descent and Newton's method programs given in Appendix B, and the conjugate gradient method available in IDESIGN. Table 11.3 summarizes the final results with the three methods.

For the steepest-descent method, $\delta_0 = 0.05$ and a line search termination criterion of 0.00001 are used. For the modified Newton's method, they are 0.05 and 0.0001, respectively. Golden section search is used with both methods. It can be observed again that for the present example the steepest-descent method is the most inefficient and the conjugate gradient is the most efficient. Therefore, the conjugate gradient method is recommended for general applications.

TABLE 11.3 Comparative evaluation of three methods for Example 11.8

$f(x) = 50(x_2 - x_1^2)^2 + (2 - x_1)^2$

	Steepest-descent	Conjugate gradient	Modified Newton's
x_1	1.9941E+00	2.0000E+00	2.0000E+00
x_2	3.9765E+00	3.9998E+00	3.9999E+00
f	3.4564E−05	1.0239E−08	2.5054E−10
$\|\mathbf{c}\|$	3.3236E−03	1.2860E−04	9.0357E−04
Number of function evaluations	138236	65	349
Number of iterations	9670	22	13

11.4.3 Marquardt Modification

As noted before, the modified Newton's method has several drawbacks that can cause numerical difficulties. For example, if the Hessian $\mathbf{H}$ of the cost function is not positive definite, the direction found from Eq. (11.29) may not be that of descent for the cost function. In that case, a step cannot be executed along the direction. Marquardt (1963) suggested a modification to the direction-finding process that has the desirable features of the steepest-descent and Newton's methods. It turns out that far away from the solution point, the method behaves like the steepest-descent method, which is quite good there. Near the solution point, it behaves like the Newton's method, which is very effective there.

In Marquardt's method, the Hessian is modified as $(\mathbf{H} + \lambda\mathbf{I})$, where λ is a positive constant. λ is initially selected as a large number that is reduced as iterations progress. The search direction is computed from Eq. (11.29) as

$$\mathbf{d}^{(k)} = -[\mathbf{H}^{(k)} + \lambda_k\mathbf{I}]^{-1}\mathbf{c}^{(k)} \tag{11.31}$$

Note that when λ is large, the effect of $\mathbf{H}$ is essentially neglected and $\mathbf{d}^{(k)}$ is essentially $-(1/\lambda)$ $\mathbf{c}^{(k)}$, which is the steepest-descent direction with $1/\lambda$ as the step size.

As the algorithm proceeds, λ is reduced (i.e., the step size is increased). When λ becomes sufficiently small, then the effect of $\lambda\mathbf{I}$ is essentially neglected and the Newton direction is obtained from Eq. (11.31). If the direction $\mathbf{d}^{(k)}$ of Eq. (11.31) does not reduce the cost function, then λ is increased (the step size is reduced) and the search direction is recomputed. *Marquardt's algorithm* is summarized in the following steps.

Step 1. Make an engineering estimate for the starting design $\mathbf{x}^{(0)}$. Set the iteration counter at $k = 0$. Select a tolerance ε as the stopping criterion and λ_0 as a large constant (say 1000).
Step 2. Calculate $c_i^{(k)} = \partial f(\mathbf{x}^{(k)})/\partial x_i$ for $i = 1$ to n. If $||\mathbf{c}^{(k)}|| < \varepsilon$, stop. Otherwise, continue.
Step 3. Calculate the Hessian matrix $\mathbf{H}(\mathbf{x}^{(k)})$.
Step 4. Calculate the search direction by solving Eq. (11.31).
Step 5. If $f(\mathbf{x}^{(k)} + \mathbf{d}^{(k)}) < f(\mathbf{x}^{(k)})$, then continue. Otherwise, increase λ_k (say to $2\lambda_k$), and go to Step 4.
Step 6. Reduce λ_k, say to $\lambda_{k+1} = 0.5\lambda_k$. Set $k = k + 1$ and go to Step 2.

11.5 SEARCH DIRECTION DETERMINATION: QUASI-NEWTON METHODS

In Section 10.6 the steepest-descent method was described. Some of the drawbacks of that method were pointed out. It was noted that the method has a poor rate of convergence because only first-order information is used. This flaw is corrected with Newton's method, where second-order derivatives are used. Newton's method has very good convergence properties. However, it can be inefficient because it requires calculation of $n(n + 1)/2$ second-order derivatives to generate the Hessian matrix (recall that n is the number of design variables). For most engineering design problems, calculation of second-order derivatives may be tedious or even impossible. Also, Newton's method runs into difficulties if the Hessian of the function is singular at any iteration.

The methods presented in this section overcome the drawbacks of Newton's method by generating an approximation for the Hessian matrix or its inverse at each iteration. Only the first derivatives of the function are used to generate these approximations. Therefore the methods have desirable features of both the steepest-descent and the Newton's methods. They are called *quasi-Newton methods.*

The quasi-Newton methods were initially developed for positive definite quadratic functions. For such functions they converge to the exact optimum in at most n iterations. However, this ideal behavior does not carry over to general cost functions, and the methods usually need to be restarted at every $(n + 1)$th iteration, just as with the conjugate gradient method.

There are several ways to approximate the Hessian or its inverse. The basic idea is to update the current approximation of the Hessian using two pieces of information: changes in design and the gradient vectors between two successive iterations. While updating, the properties of symmetry and positive definiteness are preserved. Positive definiteness is essential because without it the search direction may not be a descent direction for the cost function.

The derivation of the updating procedures is based on the so-called quasi-Newton condition, also called the *secant equation* (Gill et al., 1981; Nocedal and Wright, 2006). This condition is derived by requiring the curvature of the cost function in the search direction $\mathbf{d}^{(k)}$ to be the same at two consecutive points $\mathbf{x}^{(k)}$ and $\mathbf{x}^{(k+1)}$. The enforcement of this condition gives the updating formulas for the Hessian of the cost function or its inverse. For a strictly convex quadratic function, the updating procedure converges to the exact Hessian in n iterations. We will describe two of the most popular methods in the class of quasi-Newton methods.

11.5.1 Inverse Hessian Updating: The DFP Method

The DFP method, initially proposed by Davidon (1959), was modified by Fletcher and Powell (1963); their version is presented here. DFP is one of the most powerful methods for the minimization of a general function $f(\mathbf{x})$. The method builds an *approximate inverse of the Hessian* of $f(\mathbf{x})$ using only the first derivatives. It is often called the DFP (Davidon-Fletcher-Powell) method:

Step 1. Estimate an initial design $\mathbf{x}^{(0)}$. Choose a symmetric positive definite $n \times n$ matrix $\mathbf{A}^{(0)}$ as an estimate for the inverse of the Hessian of the cost function. In the absence of more information, $\mathbf{A}^{(0)} = \mathbf{I}$ may be chosen. Also, specify a convergence parameter ε. Set $k = 0$. Compute the gradient vector as $\mathbf{c}^{(0)} = \nabla f(\mathbf{x}^{(0)})$.

Step 2. Calculate the norm of the gradient vector as $||\mathbf{c}^{(k)}||$. If $||\mathbf{c}^{(k)}|| < \varepsilon$, then stop the iterative process. Otherwise, continue.

Step 3. Calculate the search direction as

$$\mathbf{d}^{(k)} = -\mathbf{A}^{(k)}\mathbf{c}^{(k)} \tag{11.32}$$

Step 4. Compute the optimum step size:

$$\alpha_k = \alpha \text{ to minimize } f(\mathbf{x}^{(k)} + \alpha \mathbf{d}^{(k)}) \tag{11.33}$$

Step 5. Update the design as

$$\mathbf{x}^{(k+1)} = \mathbf{x}^{(k)} + \alpha_k \mathbf{d}^{(k)} \tag{11.34}$$

Step 6. Update the matrix $\mathbf{A}^{(k)}$—the approximation for the inverse of the Hessian of the cost function—as

$$\mathbf{A}^{(k+1)} = \mathbf{A}^{(k)} + \mathbf{B}^{(k)} + \mathbf{C}^{(k)} \tag{11.35}$$

where the correction matrices $\mathbf{B}^{(k)}$ and $\mathbf{C}^{(k)}$ are calculated using the quasi-Newton condition mentioned earlier:

$$\mathbf{B}^{(k)} = \frac{\mathbf{s}^{(k)}\mathbf{s}^{(k)T}}{(\mathbf{s}^{(k)} \cdot \mathbf{y}^{(k)})}; \quad \mathbf{C}^{(k)} = \frac{-\mathbf{z}^{(k)}\mathbf{z}^{(k)T}}{(\mathbf{y}^{(k)} \cdot \mathbf{z}^{(k)})} \tag{11.36}$$

$$\mathbf{s}^{(k)} = \alpha_k\mathbf{d}^{(k)}(\text{change in design}); \quad \mathbf{y}^{(k)} = \mathbf{c}^{(k+1)} - \mathbf{c}^{(k)}(\text{change in gradient}) \tag{11.37}$$

$$\mathbf{c}^{(k+1)} = \nabla f(\mathbf{x}^{(k+1)}); \quad \mathbf{z}^{(k)} = \mathbf{A}^{(k)}\mathbf{y}^{(k)} \tag{11.38}$$

Step 7. Set $k = k + 1$ and go to Step 2.

Note that the first iteration of the method is the same as that for the steepest-descent method. Fletcher and Powell (1963) prove the following properties of the algorithm:

1. The matrix $\mathbf{A}^{(k)}$ is positive definite for all k. This implies that the method will always converge to a local minimum point, since

$$\frac{d}{d\alpha}f(\mathbf{x}^{(k)} + \alpha\mathbf{d}^{(k)})\big|_{a=0} = -\mathbf{c}^{(k)T}\mathbf{A}^{(k)}\mathbf{c}^{(k)} < 0 \tag{11.39}$$

as long as $\mathbf{c}^{(k)} \neq \mathbf{0}$. This means that $f(\mathbf{x}^{(k)})$ may be decreased by choosing $\alpha > 0$ if $\mathbf{c}^{(k)} \neq \mathbf{0}$ (i.e., $\mathbf{d}^{(k)}$ is a direction of descent).
2. When this method is applied to a positive definite quadratic form, $\mathbf{A}^{(k)}$ converges to the inverse of the Hessian of the quadratic form.

Example 11.9 illustrates calculations for two iterations of the DFP method.

EXAMPLE 11.9 APPLICATION OF THE DFP METHOD

Execute two iterations of the DFP method for the problem starting from the point (1,2):

Minimize

$$f(x) = 5x_1^2 + 2x_1x_2 + x_2^2 + 7 \tag{a}$$

Solution

We will follow steps of the algorithm.

Iteration 1 $(k = 0)$

1. $\mathbf{x}^{(0)} = (1, 2)$; $\mathbf{A}^{(0)} = \mathbf{I}$, $k = 0$, $\varepsilon = 0.001$
   $\mathbf{c}^{(0)} = (10x_1 + 2x_2, 2x_1 + 2x_2) = (14, 6)$
2. $\|\mathbf{c}^{(0)}\| = \sqrt{14^2 + 6^2} = 15.232 > \varepsilon$, so continue
3. $\mathbf{d}^{(0)} = -\mathbf{c}^{(0)} = (-14, -6)$
4. $\mathbf{x}^{(1)} = \mathbf{x}^{(0)} + \alpha\mathbf{d}^{(0)} = (1 - 14\alpha, 2 - 6\alpha)$

$$f(\mathbf{x}^{(1)}) = f(\alpha) = 5(1 - 14\alpha)^2 + 2(1 - 14\alpha)(2 - 6\alpha) + (2 - 6\alpha)^2 + 7 \tag{b}$$

$$\frac{df}{d\alpha} = 5(2)(-14)(1 - 14\alpha) + 2(-14)(2 - 6\alpha) + 2(-6)(1 - 14\alpha) + 2(-6)(2 - 6\alpha) = 0 \tag{c}$$

$$\alpha_0 = 0.099$$

$$\frac{d^2f}{d\alpha^2} = 2348 > 0$$

Therefore, a step size of $\alpha = 0.099$ is acceptable.

5. $\mathbf{x}^{(1)} = \mathbf{x}^{(0)} + \alpha_0\mathbf{d}^{(0)} = \begin{bmatrix} 1 \\ 2 \end{bmatrix} + 0.099\begin{bmatrix} -14 \\ -6 \end{bmatrix} = \begin{bmatrix} -0.386 \\ 1.407 \end{bmatrix}$

6.
$$\mathbf{s}^{(0)} = \alpha_0\mathbf{d}^{(0)} = (-1.386, \ -0.593); \quad \mathbf{c}^{(1)} = (-1.046, \ 2.042) \tag{d}$$

$$\mathbf{y}^{(0)} = \mathbf{c}^{(1)} - \mathbf{c}^{(0)} = (-15.046, \ -3.958); \quad \mathbf{z}^{(0)} = \mathbf{y}^{(0)} = (-15.046, \ -3.958) \tag{e}$$

$$\mathbf{s}^{(0)} \cdot \mathbf{y}^{(0)} = 23.20; \quad \mathbf{y}^{(0)} \cdot \mathbf{z}^{(0)} = 242.05 \tag{f}$$

$$\mathbf{s}^{(0)}\mathbf{s}^{(0)T} = \begin{bmatrix} 1.921 & 0.822 \\ 0.822 & 0.352 \end{bmatrix}; \quad \mathbf{B}^{(0)} = \frac{\mathbf{s}^{(0)}\mathbf{s}^{(0)T}}{\mathbf{s}^{(0)} \cdot \mathbf{y}^{(0)}} = \begin{bmatrix} 0.0828 & 0.0354 \\ 0.0354 & 0.0152 \end{bmatrix} \tag{g}$$

$$\mathbf{z}^{(0)}\mathbf{z}^{(0)T} = \begin{bmatrix} 226.40 & 59.55 \\ 59.55 & 15.67 \end{bmatrix}; \quad \mathbf{C}^{(0)} = -\frac{\mathbf{z}^{(0)}\mathbf{z}^{(0)T}}{\mathbf{y}^{(0)} \cdot \mathbf{z}^{(0)}} = \begin{bmatrix} -0.935 & -0.246 \\ -0.246 & -0.065 \end{bmatrix} \tag{h}$$

$$\mathbf{A}^{(1)} = \mathbf{A}^{(0)} + \mathbf{B}^{(0)} + \mathbf{C}^{(0)} = \begin{bmatrix} 0.148 & -0.211 \\ -0.211 & 0.950 \end{bmatrix} \tag{i}$$

Iteration 2 (k = 1)

2. $\|\mathbf{c}^{(1)}\| = 2.29 > \varepsilon$, so continue
3. $\mathbf{d}^{(1)} = -\mathbf{A}^{(1)}\mathbf{c}^{(1)} = (0.586, -1.719)$; compare this to the steepest-descent direction, $\mathbf{d}^{(1)} = -\mathbf{c}^{(1)} = (1.046, -2.042)$
4. Step size determination:

   Minimize $f(\mathbf{x}^{(1)} + \alpha\mathbf{d}^{(1)})$; $\alpha_1 = 0.776$
5. $\mathbf{x}^{(2)} = \mathbf{x}^{(1)} + \alpha_1\mathbf{d}^{(1)} = (-0.386, 1.407) + 0.776(0.586, -1.719) = (0.069, 0.073)$
6. $\mathbf{s}^{(1)} = \alpha_1\mathbf{d}^{(1)} = (0.455, \ -1.334)$

$$\mathbf{c}^{(2)} = (0.836, 0.284); \quad \mathbf{y}^{(1)} = \mathbf{c}^{(2)} - \mathbf{c}^{(1)} = (1.882, \ -1.758) \tag{j}$$

$$\mathbf{z}^{(1)} = \mathbf{A}^{(1)}\mathbf{y}^{(1)} = (0.649, -2.067); \quad \mathbf{s}^{(1)} \cdot \mathbf{y}^{(1)} = 3.201; \quad \mathbf{y}^{(1)} \cdot \mathbf{z}^{(1)} = 4.855 \tag{k}$$

$$\mathbf{s}^{(1)}\mathbf{s}^{(1)T} = \begin{bmatrix} 0.207 & -0.607 \\ -0.607 & 1.780 \end{bmatrix}; \quad \mathbf{B}^{(1)} = \frac{\mathbf{s}^{(1)}\mathbf{s}^{(1)T}}{\mathbf{s}^{(1)} \cdot \mathbf{y}^{(1)}} = \begin{bmatrix} 0.0647 & -0.19 \\ -0.19 & 0.556 \end{bmatrix} \tag{l}$$

$$\mathbf{z}^{(1)}\mathbf{z}^{(1)T} = \begin{bmatrix} 0.421 & -1.341 \\ -1.341 & 4.272 \end{bmatrix}; \quad \mathbf{C}^{(1)} = -\frac{\mathbf{z}^{(1)}\mathbf{z}^{(1)T}}{\mathbf{y}^{(1)} \cdot \mathbf{z}^{(1)}} = \begin{bmatrix} -0.0867 & 0.276 \\ 0.276 & -0.880 \end{bmatrix} \tag{m}$$

$$\mathbf{A}^{(2)} = \mathbf{A}^{(1)} + \mathbf{B}^{(1)} + \mathbf{C}^{(1)} = \begin{bmatrix} 0.126 & -0.125 \\ -0.125 & 0.626 \end{bmatrix} \tag{o}$$

It can be verified that the matrix $\mathbf{A}^{(2)}$ is quite close to the inverse of the Hessian of the cost function. One more iteration of the DFP method will yield the optimum solution of (0, 0).

11.5.2 Direct Hessian Updating: The BFGS Method

It is possible to update the Hessian rather than its inverse at every iteration. Several such updating methods are available; however, we will present a popular method that has proven to be most effective in applications. Detailed derivation of the method is given in works by Gill and coworkers (1981) and Nocedal and Wright (2006). It is known as the *Broyden-Fletcher-Goldfarb-Shanno* (BFGS) *method* and is summarized in the following algorithm:

Step 1. Estimate an initial design $\mathbf{x}^{(0)}$. Choose a symmetric positive definite $n \times n$ matrix $\mathbf{H}^{(0)}$ as an estimate for the Hessian of the cost function. In the absence of more information, let $\mathbf{H}^{(0)} = \mathbf{I}$. Choose a convergence parameter ε. Set $k = 0$, and compute the gradient vector as $\mathbf{c}^{(0)} = \nabla f(\mathbf{x}^{(0)})$.

Step 2. Calculate the norm of the gradient vector as $||\mathbf{c}^{(k)}||$. If $||\mathbf{c}^{(k)}|| < \varepsilon$, stop the iterative process; otherwise, continue.

Step 3. Solve the following linear system of equations to obtain the search direction:

$$\mathbf{H}^{(k)}\mathbf{d}^{(k)} = -\mathbf{c}^{(k)} \tag{11.40}$$

Step 4. Compute the optimum step size:

$$\alpha_k = \alpha \text{ to minimize } f(\mathbf{x}^{(k)} + \alpha\mathbf{d}^{(k)}) \tag{11.41}$$

Step 5. Update the design as

$$\mathbf{x}^{(k+1)} = \mathbf{x}^{(k)} + \alpha_k\mathbf{d}^{(k)} \tag{11.42}$$

Step 6. Update the Hessian approximation for the cost function as

$$\mathbf{H}^{(k+1)} = \mathbf{H}^{(k)} + \mathbf{D}^{(k)} + \mathbf{E}^{(k)} \tag{11.43}$$

where the correction matrices $\mathbf{D}^{(k)}$ and $\mathbf{E}^{(k)}$ are given as

$$\mathbf{D}^{(k)} = \frac{\mathbf{y}^{(k)}\mathbf{y}^{(k)T}}{(\mathbf{y}^{(k)} \cdot \mathbf{s}^{(k)})}; \quad \mathbf{E}^{(k)} = \frac{\mathbf{c}^{(k)}\mathbf{c}^{(k)T}}{(\mathbf{c}^{(k)} \cdot \mathbf{d}^{(k)})} \tag{11.44}$$

$$\mathbf{s}^{(k)} = a_k\mathbf{d}^{(k)}(\text{change in design}); \quad \mathbf{y}^{(k)} = \mathbf{c}^{(k+1)} - \mathbf{c}^{(k)} \text{ (change in gradient)}; \quad \mathbf{c}^{(k+1)} = \nabla f(\mathbf{x}^{(k+1)}) \tag{11.45}$$

Step 7. Set $k = k + 1$ and go to Step 2.

Note again that the first iteration of the method is the same as that for the steepest-descent method when $\mathbf{H}^{(0)} = \mathbf{I}$. It can be shown that the BFGS update formula keeps the Hessian approximation positive definite if an accurate line search is used. This is important to know because the search direction is guaranteed to be that of descent for the cost function if $\mathbf{H}^{(k)}$ is positive definite. In numerical calculations, difficulties can arise because the Hessian can become singular or indefinite as a result of inexact line search and round-off and truncation errors. Therefore, some safeguards against numerical difficulties must be incorporated into computer programs for stable and convergent calculations.

Another numerical procedure that is extremely useful is to update decomposed factors (Cholesky factors) of the Hessian rather than the Hessian itself. With that procedure, the matrix can numerically be guaranteed to remain positive definite, and the linear equation $\mathbf{H}^{(k)}\mathbf{d}^{(k)} = -\mathbf{c}^{(k)}$ can be solved more efficiently.

Example 11.10 illustrates calculations for two iterations of the BFGS method.

EXAMPLE 11.10 APPLICATION OF THE BFGS METHOD

Execute two iterations of the BFGS method for the problem starting from the point (1, 2):

Minimize
$$f(\mathbf{x}) = 5x_1^2 + 2x_1x_2 + x_2^2 + 7.$$

Solution

We will follow the steps of the algorithm. Note that the first iteration gives the steepest-descent step for the cost function.

Iteration 1 ($k = 0$)

1. $\mathbf{x}^{(0)} = (1, 2)$, $\mathbf{H}^{(0)} = \mathbf{I}$, $\varepsilon = 0.001$, $k = 0$

$$\mathbf{c}^{(0)} = (10x_1 + 2x_2,\ 2x_1 + 2x_2) = (14, 6) \tag{a}$$

2. $\left\| \mathbf{c}^{(0)} \right\| = \sqrt{14^2 + 6^2} = 15.232 > \varepsilon$, so continue

3. $\mathbf{d}^{(0)} = -\mathbf{c}^{(0)} = (-14, -6)$; since $\mathbf{H}^{(0)} = \mathbf{I}$

4. Step size determination (same as Example 11.9): $\alpha_0 = 0.099$

5. $\mathbf{x}^{(1)} = \mathbf{x}^{(0)} + \alpha_0\mathbf{d}^{(0)} = (-0.386, 1.407)$

6. $\mathbf{s}^{(0)} = \alpha_0\mathbf{d}^{(0)} = (-1.386, -0.593)$; $\mathbf{c}^{(1)} = (-1.046, 2.042)$

$$\mathbf{y}^{(0)} = \mathbf{c}^{(1)} - \mathbf{c}^{(0)} = (-15.046, -3.958); \quad \mathbf{y}^{(0)} \cdot \mathbf{s}^{(0)} = 23.20; \quad \mathbf{c}^{(0)} \cdot \mathbf{d}^{(0)} = -232.0 \tag{b}$$

$$\mathbf{y}^{(0)}\mathbf{y}^{(0)T} = \begin{bmatrix} 226.40 & 59.55 \\ 59.55 & 15.67 \end{bmatrix}; \quad \mathbf{D}^{(0)} = \frac{\mathbf{y}^{(0)}\mathbf{y}^{(0)T}}{\mathbf{y}^{(0)} \cdot \mathbf{s}^{(0)}} = \begin{bmatrix} 9.760 & 2.567 \\ 2.567 & 0.675 \end{bmatrix} \tag{c}$$

$$\mathbf{c}^{(0)}\mathbf{c}^{(0)T} = \begin{bmatrix} 196 & 84 \\ 84 & 36 \end{bmatrix}; \quad \mathbf{E}^{(0)} = \frac{\mathbf{c}^{(0)}\mathbf{c}^{(0)T}}{\mathbf{c}^{(0)} \cdot \mathbf{d}^{(0)}} = \begin{bmatrix} -0.845 & -0.362 \\ -0.362 & -0.155 \end{bmatrix} \tag{d}$$

$$\mathbf{H}^{(1)} = \mathbf{H}^{(0)} + \mathbf{D}^{(0)} + \mathbf{E}^{(0)} = \begin{bmatrix} 9.915 & 2.205 \\ 2.205 & 1.520 \end{bmatrix} \tag{e}$$

Iteration 2 ($k = 1$)

2. $\|\mathbf{c}^{(1)}\| = 2.29 > \varepsilon$, so continue

3. $\mathbf{H}^{(1)}\mathbf{d}^{(1)} = -\mathbf{c}^{(1)}$; or, $\mathbf{d}^{(1)} = (0.597, -2.209)$

4. Step size determination: $\alpha_1 = 0.638$

5. $\mathbf{x}^{(2)} = \mathbf{x}^{(1)} + \alpha_1\mathbf{d}^{(1)} = (-0.005, -0.002)$

6. $\mathbf{s}^{(1)} = \alpha_1\mathbf{d}^{(1)} = (0.381, -1.409)$; $\mathbf{c}^{(2)} = (-0.054, -0.014)$

$$\mathbf{y}^{(1)} = \mathbf{c}^{(2)} - \mathbf{c}^{(1)} = (0.992, -2.056); \quad \mathbf{y}^{(1)} \cdot \mathbf{s}^{(1)} = 3.275; \quad \mathbf{c}^{(1)} \cdot \mathbf{d}^{(1)} = -5.135 \tag{f}$$

$$\mathbf{y}^{(1)}\mathbf{y}^{(1)T} = \begin{bmatrix} 0.984 & -2.04 \\ -2.04 & 4.227 \end{bmatrix}; \quad \mathbf{D}^{(1)} = \frac{\mathbf{y}^{(1)}\ \mathbf{y}^{(1)T}}{\mathbf{y}^{(1)} \cdot \mathbf{s}^{(1)}} = \begin{bmatrix} 0.30 & -0.623 \\ -0.623 & 1.291 \end{bmatrix} \tag{g}$$

$$\mathbf{c}^{(1)}\mathbf{c}^{(1)T} = \begin{bmatrix} 1.094 & -2.136 \\ -2.136 & 4.170 \end{bmatrix}; \quad \mathbf{E}^{(1)} = \frac{\mathbf{c}^{(1)}\ \mathbf{c}^{(1)T}}{\mathbf{c}^{(1)} \cdot \mathbf{d}^{(1)}} = \begin{bmatrix} -0.213 & 0.416 \\ 0.416 & -0.812 \end{bmatrix} \tag{h}$$

$$\mathbf{H}^{(2)} = \mathbf{H}^{(1)} + \mathbf{D}^{(1)} + \mathbf{E}^{(1)} = \begin{bmatrix} 10.002 & 1.998 \\ 1.998 & 1.999 \end{bmatrix} \tag{i}$$

It can be verified that $\mathbf{H}^{(2)}$ is quite close to the Hessian of the given cost function. One more iteration of the BFGS method will yield the optimum solution of (0, 0).

11.6 ENGINEERING APPLICATIONS OF UNCONSTRAINED METHODS

There are several engineering applications where unconstrained optimization methods can be used. For example, linear as well as nonlinear equations can be solved with unconstrained optimization methods. Such equations arise while calculating the response of structural and mechanical systems. The procedures have been incorporated into some commercial software packages as well, such as finite element analysis programs.

11.6.1 Data Interpolation

Another very common application of unconstrained optimization techniques is the interpolation of discrete numerical data. Here we want to develop an analytical representation for the discrete numerical data. These data may be collected from experiments or some other observations. For example, we may have discrete data (x_i, y_i), $i = 1$ to n that relate two variables x and y. The data need to be represented as a function $y = q(x)$. The function $q(x)$ may be linear (straight line), polynomial (curve), exponential, logarithmic, or any other function. Similarly, we may have data involving three or more variables. In that case a function of several variables needs to be developed.

The problem of data interpolation is called *regression analysis*. The problem can be formulated as an unconstrained optimization problem where the error between the available data and its analytical representation is minimized. The parameters that characterize the interpolation function are treated as the design variables for the optimization problem. Different error functions may be defined. The most common error function is the sum of squares of the errors at each discrete point, defined as

$$f(q) = \sum_{i=1}^{n} [y_i - q(x_i)]^2 \tag{11.46}$$

Thus the unconstrained optimization problem is to minimize

$$f(q) = \sum_{i=1}^{n} [y_i - q(x_i)]^2 \tag{11.47}$$

This is known as the least squares minimization problem.

If a linear relationship is desired between the variables x and y, then $q(x)$ is represented as

$$q(x) = ax + b \tag{11.48}$$

where a and b are the unknown parameters. Substituting Eq. (11.48) in Eq. (11.47), we obtain the *linear least squares problem* as

Minimize

$$f(a,b) = \sum_{i=1}^{n} [y_i - (ax_i + b)]^2 \tag{11.49}$$

The problem, then, is to determine a and b to minimize the error function of Eq. (11.49). This problem can be solved in a closed form by writing the optimality condition and solving the resulting system of two linear equations.

Depending on the variability in the available data, many other functions may be used for $q(x)$ in Eq. (11.48), such as higher-order polynomials, logarithmic functions, exponential functions, and the like.

11.6.2 Minimization of Total Potential Energy

The equilibrium states of structural and mechanical systems are characterized by the stationary points of the total potential energy of the system. This is known as the *principle of stationary potential energy*. If at a stationary point the potential energy actually has a minimum value, the equilibrium state is called stable. In structural mechanics, these principles are of fundamental importance and form the basis for numerical methods of structural analysis.

To demonstrate the principle, we consider the symmetric two-bar truss shown in Figure 11.9. The structure is subjected to a load W at node C. Under the action of this load, node C moves to a point C′. The problem is to compute the displacements x_1 and x_2 of

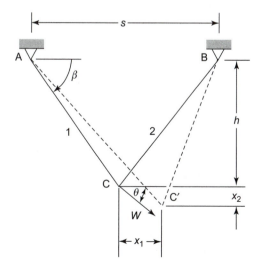

FIGURE 11.9   Two-bar truss.

node C. This can be done by writing the total potential energy of the structure in terms of x_1 and x_2 and then minimizing it. Once the displacements x_1 and x_2 are known, member forces and stresses can be calculated using them. Let

E = modulus of elasticity, N/m^2 (this is the property of a material that relates stresses in the material to strains)

s = span of the truss, m

h = height of the truss, m

A_1 = cross-sectional area of member 1, m^2

A_2 = cross-sectional area of member 2, m^2

θ = angle at which load W is applied, degrees

L = length of the members; $L = \sqrt{h^2 + 0.25s^2}$, m

W = load, N

x_1 = horizontal displacement, m

x_2 = vertical displacement, m

The total potential energy of the system, assuming small displacements, is given as

$$P(x_1, x_2) = \frac{EA_1}{2L}(x_1 \ \cos\beta + x_2 \ \sin\beta)^2 + \frac{EA_2}{2L}(-x_1 \ \cos\beta + x_2 \ \sin\beta)^2$$

$$- Wx_1 \ \cos\theta - Wx_2 \ \sin\theta, \quad \text{N·m} \tag{11.50}$$

where the angle β is shown in Figure 11.9. Minimization of P with respect to x_1 and x_2 gives the displacements x_1 and x_2 for the equilibrium state of the two-bar structure. Example 11.11 demonstrates this calculation.

EXAMPLE 11.11 MINIMIZATION OF THE TOTAL POTENTIAL ENERGY OF A TWO-BAR TRUSS

For the two-bar truss problem, use the following numerical data:

$A_1 = A_2 = 10^{-5}$ m^2

$h = 1.0$ m, $s = 1.5$ m

$W = 10$ kN

$\theta = 30°$

$E = 207$ GPa

Minimize the total potential energy given in Eq. (11.50) by (1) the graphical method, (2) the analytical method, and (3) the conjugate gradient method.

Solution

Substituting these data into Eq. (11.50) and simplifying, we get (note that $\cos\beta = s/2L$ and $\sin\beta = h/L$)

$$P(x_1, \ x_2) = \frac{EA}{L}\left(\frac{s}{2L}\right)^2 x_1^2 + \frac{EA}{L}\left(\frac{h}{2L}\right)^2 x_2^2 - Wx_1 \ \cos\theta - Wx_2 \ \sin\theta \tag{a}$$

$$= (5.962 \times 10^6)x_1^2 + (1.0598 \times 10^6)x_2^2 - 8660x_1 - 5000x_2, \ \text{N·m}$$

Contours for the function are shown in Figure 11.10. The optimum solution from the graph is calculated as $x_1 = (7.2634 \times 10^{-2})$ m; $x_2 = (2.3359 \times 10^{-2})$ m; $P = -37.348$ N $\cdot$ m.

Using the necessary conditions of optimality ($\nabla P = 0$), we get

$$2(5.962 \times 10^6)x_1 - 8660 = 0, \quad x_1 = (7.2629 \times 10^{-3}), \text{m} \tag{b}$$

$$2(1.0598 \times 10^6)x_2 - 5000 = 0, \quad x_2 = (2.3589 \times 10^{-3}), \text{m} \tag{c}$$

The conjugate gradient method given in IDESIGN (Arora and Tseng, 1987a,b) also converges to the same solution.

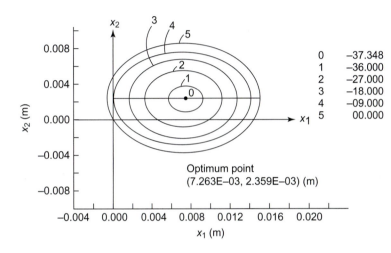

0	−37.348
1	−36.000
2	−27.000
3	−18.000
4	−09.000
5	00.000

Optimum point
(7.263E–03, 2.359E–03) (m)

FIGURE 11.10 Contours of the potential energy function $P(x_1, x_2)$ for a two-bar truss ($P = 0, -9.0, -18.0, -27.0, -36.0,$ and -37.348 N $\cdot$ m).

11.6.3 Solutions to Nonlinear Equations

Unconstrained optimization methods can be used to find the roots of a nonlinear system of equations. To demonstrate this, we consider the following 2×2 system:

$$F_1(x_1, x_2) = 0; \quad F_2(x_1, x_2) = 0 \tag{11.51}$$

We define a function that is the sum of the squares of the functions F_1 and F_2 as

$$f(x_1, x_2) = F_1^2(x_1, x_2) + F_2^2(x_1, x_2) \tag{11.52}$$

Note that if x_1 and x_2 are the roots of Eq. (11.51), then $f = 0$ in Eq. (11.52). If x_1 and x_2 are not the roots, then the function $f > 0$ represents the sum of the squares of the errors in the equations $F_1 = 0$ and $F_2 = 0$. Thus, the optimization problem is to find x_1 and x_2 to minimize the function $f(x_1, x_2)$ of Eq. (11.52). We need to show that the necessary conditions for minimization of $f(\mathbf{x})$ give the roots for the nonlinear system of equations. The necessary conditions of optimality give

$$\frac{\partial f}{\partial x_1} = 2F_1\frac{\partial F_1}{\partial x_1} + 2F_2\frac{\partial F_2}{\partial x_1} = 0 \tag{11.53}$$

$$\frac{\partial f}{\partial x_2} = 2F_1 \frac{\partial F_1}{\partial x_2} + 2F_2 \frac{\partial F_2}{\partial x_2} = 0 \tag{11.54}$$

Note that the necessary conditions are satisfied if $F_1 = F_2 = 0$, x_1 and x_2 are the roots of the equations $F_1 = 0$, and $F_2 = 0$. At this point $f = 0$. Note also that the necessary conditions can be satisfied if $\partial F_i / \partial x_j = 0$ for i, $j = 1$, 2. If $\partial F_i / \partial x_j = 0$, x_1 and x_2 are stationary points for the functions F_1 and F_2. For most problems it is unlikely that the stationary points for F_1 and F_2 will also be the roots of $F_1 = 0$ and $F_2 = 0$, so we may exclude these cases. In any case, if x_1 and x_2 are the roots of the equations, then f must have a zero value. Also if the optimum value of f is different from zero ($f \neq 0$), then x_1 and x_2 cannot be the roots of the nonlinear system. Thus, if the optimization algorithm converges with $f \neq 0$, then the optimum point for the problem of minimization of f is not a root of the nonlinear system. The algorithm should be restarted from a different initial point. Example 11.12 illustrates this root-finding process.

EXAMPLE 11.12 FINDING ROOTS OF NONLINEAR EQUATIONS BY UNCONSTRAINED MINIMIZATION

Find the roots of the equations

$$F_1(\mathbf{x}) = 3x_1^2 + 12x_2^2 + 10x_1 = 0; \quad F_2(\mathbf{x}) = 24x_1 x_2 + 4x_2 + 3 = 0 \tag{a}$$

Solution

We define the error function $f(\mathbf{x})$ as

$$f(\mathbf{x}) = F_1^2 + F_2^2 = (3x_1^2 + 12x_2^2 + 10x_1)^2 + (24x_1 x_2 + 4x_2 + 3)^2 \tag{b}$$

To minimize this function, we can use any of the methods discussed previously. Table 11.4 shows the iteration history with the conjugate gradient method available in IDESIGN (Arora and Tseng, 1987a,b).

One root of the equations is $x_1 = -0.3980$, $x_2 = 0.5404$ starting from the point $(-1, 1)$. Starting from a different point $(-50, 50)$, another root is found as $(-3.331, 0.03948)$. However, starting from another point $(2, 3)$, the program converges to $(0.02063, -0.2812)$ with $f = 4.351$. Since $f \neq 0$, this point is not a root of the given system of equations. When this happens, we start from a different point and re-solve the problem.

TABLE 11.4   Root of the nonlinear equations in Example 11.12

Number	x_1	x_2	F_1	F_2	f
0	−1.0000	1.0000	5.0000	−17.0000	314.0000
1	−0.5487	0.4649	−1.9900	−1.2626	5.5530
2	−0.4147	0.5658	0.1932	−0.3993	0.1968
3	−0.3993	0.5393	−0.0245	−0.0110	7.242E−4
4	−0.3979	0.5403	−9.377E−4	−1.550E−3	2.759E−6
5	−0.3980	0.5404	−4.021E−4	−3.008E−4	1.173E−8

Note that the preceding procedure can be generalized to a system of n equations in n unknowns. In this case, the error function $f(\mathbf{x})$ will be defined as

$$f(\mathbf{x}) = \sum_{i=1}^{n} [F_i(\mathbf{x})]^2 \tag{11.55}$$

11.7 SOLUTIONS TO CONSTRAINED PROBLEMS USING UNCONSTRAINED OPTIMIZATION METHODS

It turns out that unconstrained optimization methods can also be used to solve constrained design problems. This section briefly describes such methods that transform the constrained problem to a sequence of unconstrained problems. The basic idea is to construct a composite function using the cost and constraint functions. The composite function also contains certain parameters—called penalty parameters—that penalize the composite function for violation of constraints. The larger the violation, the larger the penalty. Once the composite function is defined for a set of penalty parameters, it is minimized using any of the unconstrained optimization techniques. The penalty parameters are then adjusted based on certain conditions, and the composite function is redefined and minimized again. The process is continued until there is no significant improvement in the estimate for the optimum point.

Methods based on the foregoing philosophy have generally been called sequential unconstrained minimization techniques, or SUMTs (Fiacco and McCormick, 1968). It can be seen that the basic idea of SUMTs is quite straightforward. Because of their simplicity, the methods have been extensively developed and tested for engineering design problems. A very brief discussion of the basic concepts and philosophy of the methods is included here to give students a flavor for the techniques. For more detailed presentations, texts by Gill, Murray, and Wright (1981), Nocedal and Wright (2006), and others should be consulted.

The term "transformation method" is used to describe any method that solves the constrained optimization problem by transforming it into one or more unconstrained problems. Such methods include the so-called penalty and barrier function methods (exterior and interior penalty methods, respectively) as well as the *multiplier methods* (also called *augmented Lagrangian methods*). To remind the reader of the original constrained problem that we are trying to solve, we restate it as follows: Find an n-vector $\mathbf{x} = (x_1, x_2, \ldots, x_n)$ to

Minimize

$$f = f(\mathbf{x}) \tag{11.56}$$

subject to

$$h_i(\mathbf{x}) = 0; \quad i = 1 \text{ to } p \tag{11.57}$$

$$g_i(\mathbf{x}) \leq 0; \quad i = 1 \text{ to } m \tag{11.58}$$

All transformation methods convert this constrained optimization problem into an unconstrained problem using a *transformation function* of the form:

$$\Phi(\mathbf{x}, \mathbf{r}) = f(\mathbf{x}) + P(\mathbf{h}(\mathbf{x}), \mathbf{g}(\mathbf{x}), \mathbf{r}) \tag{11.59}$$

where **r** is a vector of penalty parameters and P is a real-valued function whose action of imposing the penalty on the cost function is controlled by **r**. The form of penalty function P depends on the method used.

The basic procedure is to choose an initial design estimate $\mathbf{x}^{(0)}$ and define the function Φ of Eq. (11.59). The penalty parameters **r** are also initially selected. The function Φ is minimized for **x**, keeping **r** fixed. Then the parameters **r** are adjusted and the procedure is repeated until no further improvement is possible.

11.7.1 Sequential Unconstrained Minimization Techniques

Sequential unconstrained minimization techniques consist of two different types of penalty functions. The first one is called the *penalty function* method and the second is called the *barrier function* method.

Penalty Function Method

The basic idea of the penalty function approach is to define the function P in Eq. (11.48) in such a way that if there are constraint violations, the cost function $f(\mathbf{x})$ is penalized by addition of a positive value. Several penalty functions can be defined. The most popular one is called the *quadratic loss function*, defined as

$$P(\mathbf{h}(\mathbf{x}), \mathbf{g}(\mathbf{x}), r) = r\left\{\sum_{i=1}^{p}[h_i(\mathbf{x})]^2 + \sum_{i=1}^{m}[g_i^+(\mathbf{x})]^2\right\}; \quad g_i^+(\mathbf{x}) = \max\,(0, g_i(\mathbf{x})) \qquad (11.60)$$

where $r > 0$ is a scalar penalty parameter. Note that $g_i^+(\mathbf{x}) \geq 0$; it is zero if the inequality is active or inactive $(g_i(\mathbf{x}) \leq 0)$ and it is positive if the inequality is violated. It can be seen that if the equality constraint is not satisfied $(h_i(\mathbf{x}) \neq 0)$ or the inequality is violated $(g_i(\mathbf{x}) > 0)$, then Eq. (11.60) gives a positive value to the function P, and the cost function is penalized, as seen in Eq. (11.59). The starting point for the method can be arbitrary. *The methods based on the philosophy of penalty functions are sometimes called* the *exterior methods because they iterate through the infeasible region.*

The advantages and disadvantages of the penalty function method are as follows:

1. It is applicable to general constrained problems with equality and inequality constraints.
2. The starting point can be arbitrary.
3. The method iterates through the infeasible region where the cost and/or constraint functions may be undefined.
4. If the iterative process terminates prematurely, the final point may not be feasible and hence not usable.

Barrier Function Methods

The following methods are applicable only to the inequality-constrained problems. Popular barrier functions are

1. *Inverse barrier function:* $\qquad P(\mathbf{g}(\mathbf{x}), r) = \dfrac{1}{r}\sum_{i=1}^{m}\dfrac{-1}{g_i(\mathbf{x})}$ $\qquad\qquad$ (11.61)

2. *Log barrier function:*
$$P(\mathbf{g}(\mathbf{x}), r) = \frac{1}{r} \sum_{i=1}^{m} log(-g_i(\mathbf{x})) \tag{11.62}$$

These are called the barrier function methods because a large barrier is constructed around the feasible region. In fact, the function P becomes infinite if any of the inequalities is active. Thus, when the iterative process is started from a feasible point, it is not possilbe for it to go into the infeasible region because it cannot cross the huge barrier on the boundary of the feasible set.

For both penalty function and barrier function methods, it can be shown that as $r \to \infty$, $\mathbf{x}(r) \to \mathbf{x}^*$, where $\mathbf{x}(r)$ is a point that minimizes the transformed function $\Phi(\mathbf{x}, r)$ of Eq. (11.59) and $\mathbf{x}^*$ is a solution of the original constrained optimization problem.

The advantages and disadvantages of the barrier function methods are as follows:

1. The methods are applicable to inequality-constrained problems only.
2. The starting point must be feasible. It turns out, however, that the penalty function defined in Eq. (11.60) can be minimized to determine a feasible starting point (Haug and Arora, 1979).
3. The method always iterates through the feasible region, so if it terminates prematurely, the final point is feasible and hence usable.

The sequential unconstrained minimization techniques have certain weaknesses that are most serious when r is large. The penalty and barrier functions tend to be ill-behaved near the boundary of the feasible set, where the optimum points usually lie. There is also a problem of selecting the sequence $r^{(k)}$. The choice of $r^{(0)}$ and the rate at which $r^{(k)}$ tends to infinity can seriously affect the computational effort to find a solution. Furthermore, the Hessian matrix of the unconstrained function becomes ill-conditioned as $r \to \infty$.

11.7.2 Augmented Lagrangian (Multiplier) Methods

To alleviate some of the difficulties of the methods presented in the previous subsection, a different class of transformation methods has been developed in the literature. These are called the *multiplier or augmented Lagrangian methods*. In these methods, there is no need for the penalty parameters r to go to infinity. As a result the transformation function Φ has good conditioning with no singularities. The multiplier methods are convergent, as are the SUMTs. That is, they converge to a local minimum starting from any point. It has been proven that they possess a faster rate of convergence than the two methods of the previous subsection.

The augmented Lagrangian function can be defined in several different ways (Arora, Chahande, and Paeng, 1991). As the name implies, this transformed function adds a penalty term to the Lagrangian function for the problem. There are different forms of the penalty function for the augmented Lagrangian methods. A form that uses a penalty parameter and a multiplier for each constraint separately in the penalty function is defined as

$$P(\mathbf{h}(\mathbf{x}), \mathbf{g}(\mathbf{x}), \mathbf{r}, \mathbf{\theta}) = \frac{1}{2} \sum_{i=1}^{p} r_i'(h_i + \theta_i')^2 + \frac{1}{2} \sum_{i=1}^{m} r_i[(g_i + \theta_i)^+]^2 \tag{11.63}$$

where $\theta_i > 0$, $r_i > 0$, and θ_i', $r_i' > 0$ are parameters associated with the ith inequality and equality constraints, respectively, and $(g_i + \theta_i)^+ = \max(0, g_i + \theta_i)$.

If $\theta_i = \theta_i' = 0$ and $r_i = r_i' = r$, then Eq. (11.63) reduces to the well-known quadratic loss function given in Eq. (11.49), where convergence is enforced by letting $r \to \infty$. However, the objective of the multiplier methods is to keep each r_i and r_i' finite and still achieve convergence of the numerical algorithm. The idea of multiplier methods is to start with some r_i, r_i', θ_i', and θ_i and to minimize the transformation function of Eq. (11.59). The parameters r_i, r_i', θ_i', and θ_i are then adjusted using some procedures and the entire process is repeated until all of the optimality conditions are satisfied. This form of the augmented Lagrangian function has been implemented and applied to several engineering design applications (Belegundu and Arora, 1984; Arora et al., 1991), especially for dynamic response optimization problems (Paeng and Arora, 1989; Chahande and Arora, 1993, 1994).

Another common form for the augmented Lagrangian uses the Lagrange multipliers for the constraints directly and only one penalty parameter for all of the constraints (Gill, Murray, and Wright, 1991; Nocedal and Wright, 2006). Before we define this augmented Lagrangian for the general optimization problem, let us first define the augmented Lagrangian function for the equality-constrained problem as

$$\Phi_E(\mathbf{x}, \mathbf{h}(\mathbf{x}), r) = f(\mathbf{x}) + \sum_{i=1}^{p} \left[v_i h_i(\mathbf{x}) + \frac{1}{2} r h_i^2(\mathbf{x}) \right] \tag{11.64}$$

where $r > 0$ is a penalty parameter, and v_i is the Lagrange multiplier for the ith equality constraint. Now the augmented function for the equality-inequality−constrained problem is defined as

$$\Phi(\mathbf{x}, \mathbf{h}(\mathbf{x}), \mathbf{g}(\mathbf{x}), r) = \Phi_E(\mathbf{x}, \mathbf{h}(\mathbf{x}), r) + \sum_{j=1}^{m} \begin{cases} u_j g_j(\mathbf{x}) + \frac{1}{2} r g_j^2(\mathbf{x}), & \text{if } g_j + \frac{u_j}{r} \geq 0 \\ -\frac{1}{2r} u_j^2, & \text{if } g_j + \frac{u_j}{r} < 0 \end{cases} \tag{11.65}$$

where $u_j \geq 0$ is the Lagrange multiplier for the jth inequality constraint.

Augmented Lagrangian Algorithm

The steps of the augmented Lagrangian algorithm are as follows:

Step 1. Set the iteration counter at $k = 0$, $K = \infty$ (a large number); estimate vectors $\mathbf{x}^{(0)}$, $\mathbf{v}^{(0)}$, $\mathbf{u}^{(0)} \geq 0$, $r > 0$ and scalars $\alpha > 1$, $\beta > 1$, $\varepsilon > 0$, where ε is the desired accuracy; α is used to enforce a sufficient decrease in the constraint violations, and β is used to increase the penalty parameter.
Step 2. Set $k = k + 1$
Step 3. Minimize $\Phi(\mathbf{x}, \mathbf{h}(\mathbf{x}), \mathbf{g}(\mathbf{x}), r_k)$ of Eq. (11.65) with respect to $\mathbf{x}$, starting from the point $\mathbf{x}^{(k-1)}$. Let $\mathbf{x}^{(k)}$ be the best point obtained in this step.

Step 4. Evaluate the constraint functions $h_i(\mathbf{x}^{(k)})$, $i = 1$ *to* p, and $g_i(\mathbf{x}^{(k)})$, $j = 1$ to m. Calculate the maximum constraint violation parameter $\overline{K}$ as follows:

$$\overline{K} = \max\{|h_i|, i = 1 \text{ to } p; \ |\max(g_j, -u_j/r_k)|, \ j = 1 \ \text{ to } \ m\} \tag{11.66}$$

Check for convergence of the algorithm; if the termination criteria are satisfied, stop. Otherwise, continue to Step 5.

Step 5. If $\overline{K} \geq K$ (i.e., the constraint violation did not improve), set $r_{k+1} = \beta r_k$ and go to Step 2. That is, increase the penalty parameter by the factor β while keeping the Lagrange multipliers unchanged. If $\overline{K} < K$, continue to Step 6.

Step 6. Update the multipliers (this step is executed only when the constraint violations have improved) by setting:

$$v_i^{(k+1)} = v_i^k + r_k h_i(\mathbf{x}^{(k)}); \quad i = 1 \text{ to } p \tag{11.67}$$

$$u_j^{(k+1)} = u_j^{(k)} + r_k \max \ [g_j(\mathbf{x}^{(k)}), -u_j^{(k)}/r_k]; \quad j = 1 \text{ to } m \tag{11.68}$$

If $\overline{K} \leq K/\alpha$ (the constraint violation has improved by the factor α), set $K = \overline{K}$ and go to Step 2; otherwise, continue to Step 7.

Step 7. Set $r_{k+1} = \beta r_k$ (note that this step is executed only when the constraint violations do not improve by the factor α). Set $K = \overline{K}$ and go to Step 2.

In Step 4 of the algorithm, the following termination criteria are used:

$$\overline{K} \leq \varepsilon_1 \tag{11.69}$$

$$||\nabla\Phi(\mathbf{x}^{(k)})|| \ \leq \varepsilon_2\{\max (1, ||\mathbf{x}^{(k)}||)\}; \quad \text{Or}, \quad ||\nabla\Phi(\mathbf{x}^{(k)})|| \leq \varepsilon_2\{\max(1, |\Phi(\mathbf{x}^{(k)})|)\} \tag{11.70}$$

where ε_1 and ε_2 are the user-specified parameters.

11.8 RATE OF CONVERGENCE OF ALGORITHMS

In this section, we briefly discuss concepts related to the rate of convergence of iterative sequences—the focus is on the sequences generated by optimization algorithms (Luenberger, 1984).

11.8.1 Definitions

We now consider unconstrained optimization algorithms that have *global convergence properties*. These algorithms generate a sequence of vectors that converges to a solution point $\mathbf{x}^*$ starting from any point $\mathbf{x}^{(0)}$, that is, a local minimum point for the function $f(\mathbf{x})$. In the following, we assume that the algorithm generates a sequence $\{\mathbf{x}^{(k)}\}$ that converges to $\mathbf{x}^*$. The most effective way to measure the rate of convergence is to compare improvement toward the solution point at two successive iterations of the algorithm, that is, the closeness of $\mathbf{x}^{(k+1)}$ to $\mathbf{x}^*$ relative to the closeness of $\mathbf{x}^{(k)}$ to $\mathbf{x}^*$.

ORDER OF CONVERGENCE   A sequence $\{x^{(k)}\}$ is said to converge to x^* with order p, where p is the largest number, such that

$$0 \leq \lim_{k \to \infty} \frac{||x^{(k+1)} - x^*||}{||x^{(k)} - x^*||^p} < \infty \tag{11.70}$$

Note that this definition is applicable to the *tail end* of the sequence. Also, the condition needs to be checked in the limit as $k \to \infty$. p is usually referred as the *rate of convergence*.

CONVERGENCE RATIO   If the sequence $\{x^{(k)}\}$ converges to x^* with an order of convergence p, then the following limit is called the convergence ratio:

$$\beta = \lim_{k \to \infty} \frac{||x^{(k+1)} - x^*||}{||x^{(k)} - x^*||^p} \tag{11.71}$$

β is sometimes called the *asymptotic error constant*. Note that the ratio on the right side of Eq. (11.71) can also be viewed as the ratio of errors from the solution point at the $(k+1)$th and kth iterations when $p = 1$. In terms of β, Eq. (11.70) is written as $0 \leq \beta < \infty$, which implies that the convergence ratio β is bounded (remains finite). The comparison of algorithms is based on their convergence ratio: The smaller the ratio the faster the rate.

LINEAR CONVERGENCE   If $p = 1$ in Eq. (11.70), the sequence is said to display linear convergence; in this case β must be less than one for the sequence to converge.

QUADRATIC CONVERGENCE   If $p = 2$ in Eq. (11.70), the sequence is said to have quadratic convergence.

SUPERLINEAR CONVERGENCE   If $\beta = 0$ when p is taken as unity in Eq. (11.70), the associated convergence is called superlinear. Note that an order of convergence greater than unity implies superlinear convergence.

11.8.2 Steepest-Descent Method

We here discuss the convergence properties of the steepest-descent method.

QUADRATIC FUNCTION   We first consider a quadratic function:

$$q(x) = \frac{1}{2} x^T Q x - b^T x \tag{11.72}$$

where Q is a symmetric, constant, and positive definite matrix. Further define an error function $E(x)$ as

$$E(x) = \tfrac{1}{2}(x - x^*)^T Q(x - x^*) \tag{11.73}$$

Expanding the term on the right side of Eq. (11.73) and using the condition $Qx^* = b$ at the minimum point of quadratic function $q(x)$, we obtain $E(x) = q(x) + \tfrac{1}{2} x^{*T} Q x^*$. This shows that $E(x)$ differs from $q(x)$ by a constant.

THEOREM 11.1

For any starting point $\mathbf{x}(0)$, the method of steepest descent converges to the unique minimum point $\mathbf{x}^*$ of $q(\mathbf{x})$. Furthermore, with $E(\mathbf{x})$ defined in Eq. (11.73), there holds at every step k

$$E(\mathbf{x}^{(k+1)}) \le \left(\frac{r-1}{r+1}\right)^2 E(\mathbf{x}^{(k)}) \qquad (11.74)$$

where $r = \lambda_{max}/\lambda_{min}$ is the *condition number* of $\mathbf{Q}$ and λ_{max} and λ_{min} are the largest and smallest eigenvalues of $\mathbf{Q}$.

Thus the method converges *linearly* with a convergence ratio no greater than $[(r-1)/(r+1)]^2$. It should be clear that as r becomes large, the number $[(r-1)/(r+1)]^2$ approaches unity and the convergence is slowed; that is, the error in Eq. (11.73) reduces quite slowly from one iteration to the next. The method converges in one iteration if $r = 1$.

NONQUADRATIC CASE The above result is generalized to the nonquadratic case. We replace $\mathbf{Q}$ by the Hessian $\mathbf{H}(\mathbf{x}^*)$ at the minimum point.

THEOREM 11.2

Let $f(\mathbf{x})$ have a local minimum at $\mathbf{x}^*$. Let $\mathbf{H}(\mathbf{x}^*)$ be positive definite with λ_{max} and λ_{min} as its largest and its smallest eigenvalues, and r as its condition number. If the sequence $\{\mathbf{x}^{(k)}\}$ generated by the steepest-descent method converges to $\mathbf{x}^*$, then the sequence of cost function values $\{f(\mathbf{x}^{(k)})\}$ converges to $f(\mathbf{x}^*)$ *linearly* with a convergence ratio no greater than β:

$$\beta = \left(\frac{r-1}{r+1}\right)^2 \qquad (11.75)$$

11.8.3 Newton's Method

The following theorem defines the order of convergence of the Newton's method.

THEOREM 11.3

Let $f(\mathbf{x})$ be continuously differentiable three times, and the Hessian $\mathbf{H}(\mathbf{x}^*)$ be positive definite. Then the sequence of points generated by Newton's method converges to $\mathbf{x}^*$. The order of convergence is *at least two*.

11.8.4 Conjugate Gradient Method

For a *positive definite quadratic function*, the conjugate gradient method converges in n iterations. For the *nonquadratic* case, assume that $\mathbf{H}(\mathbf{x}^*)$ is positive definite. We expect the asymptotic convergence rate per step to be at least as good as the steepest-descent method, since this is true in the quadratic case. In addition to this bound on the single-step rate, we expect the method to have quadratic convergence with respect to each complete cycle of n steps; that is,

$$\left\| \mathbf{x}^{(k+n)} - \mathbf{x}^* \right\| \leq c \left\| \mathbf{x}^{(k)} - \mathbf{x}^* \right\|^2 \tag{11.76}$$

for some c and $k = 0, n, 2n, \ldots$

11.8.5 Quasi-Newton Methods

DFP METHOD: QUADRATIC CASE   To study the rate of convergence of quasi-Newton methods, consider the quadratic function defined in Eq. (11.72). Using the DFP quasi-Newton method, the search direction is determined as

$$\mathbf{d}^{(k)} = -\mathbf{A}^{(k)} \, \mathbf{c}^{(k)} \tag{11.77}$$

where $\mathbf{c}^{(k)}$ is the gradient of the cost function and $\mathbf{A}^{(k)}$ is the quasi-Newton approximation for the Hessian inverse at the kth iteration. It can be shown that the directions generated by Eq. (11.77) are $\mathbf{Q}$ conjugate. Therefore, the method is a conjugate gradient method where the minimum point of $q(\mathbf{x})$ is obtained in n iterations; moreover, $\mathbf{A}^{(n)} = \mathbf{Q}^{-1}$. The rate of convergence of the DFP method is then determined by the eigenvalue structure of the matrix $(\mathbf{A}^{(k)}\mathbf{Q})$.

THEOREM 11.4

Let $\mathbf{x}^*$ be the unique minimum point of $q(\mathbf{x})$, and define the error function $E(\mathbf{x})$ as in Eq. (11.73). Then, for the DFP quasi-Newton algorithm, there holds at every step k

where r_k is the *condition number* for the matrix $\mathbf{A}^{(k)}\mathbf{Q}$.

$$E(\mathbf{x}^{(k+1)}) \leq \left(\frac{r_k - 1}{r_k + 1} \right)^2 E(\mathbf{x}^{(k)}) \qquad (11.78)$$

The preceding theorem shows that the order of convergence is one. However, if $\mathbf{A}^{(k)}$ is close to $\mathbf{Q}^{-1}$, then the condition number of $(\mathbf{A}^{(k)}\mathbf{Q})$ is close to unity and the convergence ratio in Eq. (11.78) is close to zero.

DFP METHOD: NONQUADRATIC CASE   The method is globally convergent. It needs to be restarted after every n iterations, just as in the conjugate gradient method. Each cycle, if at least n steps in duration, will then contain one complete cycle of an approximation to the conjugate gradient method. Asymptotically, at the tail of the generated sequence, this approximation becomes arbitrarily accurate, and hence we may conclude (as for any method approaching the conjugate gradient method asymptotically) that the method converges superlinearly (at least if viewed at the end of each cycle).

BFGS QUASI-NEWTON METHOD Under the assumptions of continuity and boundedness of the Hessian matrix, the method can be shown to be convergent to a minimum point x^* starting from any point $x^{(0)}$. The rate of convergence is superlinear (Nocedal and Wright, 2006).

11.9 DIRECT SEARCH METHODS

In this section, we discuss methods commonly known as *direct search*. The term was introduced by Hooke and Jeeves (1961) and refers to methods that do not require derivatives of the functions in their search strategy. This means that the methods can be used for problems where the derivatives are expensive to calculate or are unavailable due to lack of differentiability of functions. However, convergence of the methods can be proved if functions are assumed to be continuous and differentiable. A more detailed discussion of the state of the art of direct search methods is presented by Lewis et al. (2000) and Kolda et al. (2003).

There are two prominent methods in the direct search class: the Nelder-Mead simplex method (not to be confused with the Simplex method of linear programming), described in Chapter 18 as part of a global optimization algorithm; and the Hooke-Jeeves method, described in this section. Before describing Hooke-Jeeves, we present another direct search method, called univariate search.

11.9.1 Univariate Search

This method is based on a simple idea to minimize the function with respect to one variable at a time, keeping all other variables fixed. In other words, minimize $f(x)$ with respect to x_1, keeping all other variables fixed at their current value, then minimize with respect to x_2, and so on. The method is described by the following iterative equation:

$$x_i^{(k+1)} = x_i^{(k)} + \alpha_i; \quad i = 1 \text{ to } n \tag{11.79}$$

where superscript k refers to the iteration number, $x^{(0)}$ is an initial point, and the increment α_i is calculated to minimize the function in the coordinate direction x_i:

$$\underset{\alpha}{\text{minimize}} \, f(x_i^{(k)} + \alpha) \tag{11.80}$$

Any one-dimensional minimization technique that uses only the function values may be used to solve this one-dimensional problem. If the problem in Eq. (11.80) fails, then an increment of $-\alpha$ is tried in Eqs. (11.79) and (11.80). If that also fails, then $x_i^{(k)}$ is unchanged and the search moves to the next design variable (i.e., the next coordinate direction).

This one-variable-at-a-time approach can result in very small step sizes, which can be quite inefficient. It has been shown that cycling can occur, resulting in the failure of the method. Pattern search methods, such as Hooke-Jeeves, have been developed to overcome the drawbacks of univariate search and improve its efficiency.

11.9.2 Hooke-Jeeves Method

Hooke-Jeeves falls into a class of direct search methods known as *pattern search methods*. The univariate search discussed in the previous subsection is always performed along the fixed directions (i.e., the coordinate directions). In the pattern search method, the search direction is not always fixed. At the end of one cycle of complete univariate search, the search direction is calculated as the difference between the two previous design points, and the design is incremented in that direction. From there the univariate search is resumed. The first part of this method is exploratory search and the second part is pattern search. These are described as follows.

Exploratory Search

Here univariate search is performed with a fixed step size in each coordinate direction (or a search for a minimum can be performed in each coordinate direction, as described in the previous section). The exploration starts from an initial point. The point is incremented in a coordinate direction by the specified step in that direction. The cost function is evaluated; if it does not increase, the move is accepted. If it increases, the move is made in the opposite direction, and the cost function is evaluated again. If the cost function does not increase, the move is accepted; otherwise, it is rejected. When all of the n coordinate directions have been explored, the exploratory search step is complete. If the search is successful, the new design point is called the *base point*. On the other hand, if the search fails, the step sizes are reduced by a factor and the search is repeated.

Pattern Search

Pattern search consists of a single step in the direction determined by the difference in the two most recent base points. The design is updated as

$$x_p^{(k+1)} = x^{(k)} + d^{(k)}; \quad d^{(k)} = (x^{(k)} - x^{(k-1)}) \tag{11.81}$$

where $d^{(k)}$ is the search direction and $x_p^{(k+1)}$ is the temporary new base point. To accept this point, we perform exploratory search from there. If this search is successful, that is, if the cost function value reduces, then the temporary base point is accepted as the new base point; otherwise, it is rejected and a new exploratory search is performed from the current base point $x^{(k)}$. This procedure is continued until the exploratory search fails. Then the step sizes are reduced by a factor and the exploratory search is repeated. Eventually the entire search process stops when the step sizes become sufficiently small.

Hooke-Jeeves Algorithm

To state the foregoing procedure in a step-by-step algorithm, we introduce the following notation:

$x^{(k)}$ = current base point
$x^{(k-1)}$ = previous base point

$x_p^{(k+1)}$ = temporary base point
$x^{(k+1)}$ = new base point

Step 0. Select a starting point $x^{(0)}$; step sizes α_i, $i = 1$ to n; step reduction parameter $\eta > 1$; and termination parameter ε. Set the iteration counter to $k = 1$.
Step 1. Perform the exploratory search from the current point. Let the result of this search be $x^{(k)}$. If the exploratory search is successful, go to Step 4; otherwise, continue.
Step 2. Check the termination criterion: If $||\alpha|| < \varepsilon$, stop; otherwise, continue.
Step 3. Reduce the step sizes: $\alpha_i = \frac{\alpha_i}{\eta}, i = 1$ to n. Go to Step 1.
Step 4. Calculate the temporary base point $x_p^{(k+1)}$ using Eq. (11.81).
Step 5. Perform the exploratory search from $x_p^{(k+1)}$, resulting in a new base point $x^{(k+1)}$. If this exploratory search is not successful, go to Step 3. Otherwise, set $k = k + 1$, and go to Step 4.

The algorithm given above can be implemented in several different ways. For example, once the search direction $d^{(k)}$ has been determined in Eq. (11.81), a step size can be calculated to minimize $f(x)$ in that direction.

EXERCISES FOR CHAPTER 11*

Section 11.1 More on Step Size Determination

11.1 Write a computer program to implement the polynomial interpolation with a quadratic curve fitting. Choose a function $f(\alpha) = 7\alpha^2 - 20\alpha + 22$. Use the golden section method to initially bracket the minimum point of $f(\alpha)$ with $\delta = 0.05$. Use your program to find the minimum point of $f(\alpha)$. Comment on the accuracy of the solution.

11.2 For the function $f(\alpha) = 7\alpha^2 - 20\alpha + 22$, use two function values, $f(0)$ and $f(\alpha_u)$, and the slope of f at $\alpha = 0$ to fit a quadratic curve. Here α_u is any upper bound on the minimum point of $f(\alpha)$. What is the estimate of the minimum point from the preceding quadratic curve? How many iterations will be required to find α^*? Why?

11.3 Under what situation can the polynomial interpolation approach not be used for one-dimensional minimization?

11.4 Given

$$f(x) = 10 - x_1 + x_1 x_2 + x_2^2$$

$$x^{(0)} = (2, 4); \quad d^{(0)} = (-1, -1)$$

For the one-dimensional search, three values of α, $\alpha_l = 0$, $\alpha_i = 2$, and $\alpha_u = 4$ are tried. Using quadratic polynomial interpolation, determine

1. At what value of α is the function a minimum? Prove that this is a minimum point and not a maximum.
2. At what values of α is $f(\alpha) = 15$?

Section 11.2 More on the Steepest-Descent Method

Verify the properties of the gradient vector for the following functions at the given point.

11.5 $f(x) = 6x_1^2 - 6x_1 x_2 + 2x_2^2 - 5x_1 + 4x_2 + 2$; $x^{(0)} = (-1, -2)$
11.6 $f(x) = 3x_1^2 + 2x_1 x_2 + 2x_2^2 + 7$; $x^{(0)} = (5, 10)$
11.7 $f(x) = 10(x_1^2 - x_2) + x_1^2 - 2x_1 + 5$; $x^{(0)} = (-1, 3)$

Section 11.3 Scaling of Design Variables

11.8 Consider the following three functions:

$$f_1 = x_1^2 + x_2^2 + x_3^2; \quad f_2 = x_1^2 + 10x_2^2 + 100x_3^2; \quad f_3 = 100x_1^2 + x_2^2 + 0.1x_3^2$$

Minimize f_1, f_2, and f_3 using the program for the steepest-descent method given in Appendix B. Choose the starting design to be (1, 1, 2) for all functions. What do you conclude from observing the performance of the method on the foregoing functions? How would you scale the design variables for the functions f_2 and f_3 to improve the rate of convergence of the method?

Section 11.4 Search Direction Determination: Newton's Method

11.9 *Answer True or False.*

1. In Newton's method, it is always possible to calculate a search direction at any point.
2. The Newton direction is always that of descent for the cost function.
3. Newton's method is convergent starting from any point with a step size of 1.
4. Newton's method needs only gradient information at any point.

For the following problems, complete one iteration of the modified Newton's method; also check the descent condition for the search direction.

11.10 Exercise 10.52	**11.11** Exercise 10.53	**11.12** Exercise 10.54
11.13 Exercise 10.55	**11.14** Exercise 10.56	**11.15** Exercise 10.57
11.16 Exercise 10.58	**11.17** Exercise 10.59	**11.18** Exercise 10.60
11.19 Exercise 10.61		

11.20 Write a computer program to implement the modified Newton's algorithm. Use equal interval search for line search. Solve Exercises 10.52 to 10.61 using the program.

Section 11.5 Search Direction Determination: Quasi-Newton Methods

11.21 *Answer True or False for unconstrained problems.*

1. The DFP method generates an approximation to the inverse of the Hessian.
2. The DFP method generates a positive definite approximation to the inverse of the Hessian.
3. The DFP method always gives a direction of descent for the cost function.
4. The BFGS method generates a positive definite approximation to the Hessian of the cost function.
5. The BFGS method always gives a direction of descent for the cost function.
6. The BFGS method always converges to the Hessian of the cost function.

For the following problems, complete two iterations of the Davidon-Fletcher-Powell method.

11.22 Exercise 10.52	**11.23** Exercise 10.53	**11.24** Exercise 10.54
11.25 Exercise 10.55	**11.26** Exercise 10.56	**11.27** Exercise 10.57
11.28 Exercise 10.58	**11.29** Exercise 10.59	**11.30** Exercise 10.60
11.31 Exercise 10.61		

11.32 Write a computer program to implement the Davidon-Fletcher-Powell method. Solve Exercises 10.52 to 10.61 using the program.

For the following problems, complete two iterations of the BFGS method.

11.33 Exercise 10.52   **11.34** Exercise 10.53   **11.35** Exercise 10.54

11.36 Exercise 10.55   **11.37** Exercise 10.56   **11.38** Exercise 10.57

11.39 Exercise 10.58   **11.40** Exercise 10.59   **11.41** Exercise 10.60

11.42 Exercise 10.61

11.43 Write a computer program to implement the BFGS method. Solve Exercises 10.52 to 10.61 using the program.

Section 11.6 Engineering Applications of Unconstrained Methods

Find the equilibrium configuration for the two-bar structure of Figure 11.9 using the following numerical data.

11.44 $A_1 = 1.5$ cm^2, $A_2 = 2.0$ cm^2, $h = 100$ cm, $s = 150$ cm, $W = 100{,}000$ N, $\theta = 45°$, $E = 21$ MN/cm^2

11.45 $A_1 = 100$ mm^2, $A_2 = 200$ mm^2, $h = 1000$ mm, $s = 1500$ mm, $W = 50{,}000$ N, $\theta = 60°$, $E = 210{,}000$ N/mm^2

Find the roots of the following nonlinear equations using the conjugate gradient method.

11.46 $F(x) = 3x - e^x = 0$

11.47 $F(x) = \sin x = 0$

11.48 $F(x) = \cos x = 0$

11.49 $F(x) = \frac{2x}{3} - \sin x = 0$

11.50 $F_1(x) = 1 - \frac{10}{x_1^2 x_2} = 0, \quad F_2(x) = 1 - \frac{2}{x_1 x_2^2} = 0$

11.51 $F_1(x) = 5 - \frac{1}{8} x_1 x_2 - \frac{1}{4x_1^2} x_2^2 = 0, \quad F_2(x) = -\frac{1}{16} x_1^2 + \frac{1}{2x_1} x_2 = 0$

12

Numerical Methods for Constrained Optimum Design

Upon completion of this chapter, you will be able to

- Explain the basic steps of a numerical algorithm for solving smooth constrained nonlinear optimization problems

- Explain the concepts of descent direction and descent step for smooth constrained nonlinear optimization problems

- Linearize a constrained nonlinear optimization problem and define a linear programming subproblem

- Use a sequential linear programming algorithm to solve constrained nonlinear optimization problems

- Define a quadratic programming subproblem with a solution that gives a search direction for the constrained nonlinear optimization problem

- Use an optimization algorithm to solve nonlinear constrained optimization problems

In the previous chapter, the constrained nonlinear programming problem was transformed into a sequence of unconstrained problems for its solution. In this chapter, we describe numerical methods to directly solve the original constrained problem. For convenience of reference, the problem defined in Section 2.11 is restated as follows: Find $\mathbf{x} = (x_1, \ldots, x_n)$, a design variable vector of dimension n, to minimize $f = f(\mathbf{x})$ subject to

$$h_i(\mathbf{x}) = 0, \ \ i = 1 \text{ to } p; \ \ \ g_i(\mathbf{x}) \leq 0, \ \ i = 1 \text{ to } m \tag{12.1}$$

and the explicit bounds on design variables $x_{iL} \leq x_i \leq x_{iU}$; $i = 1$ to n, where x_{iL} and x_{iU} are, respectively, the smallest and largest allowed values for the ith design variable x_i.

These simple bound constraints are easy to treat in actual numerical implementations. However, in the discussion and illustration of the numerical methods, we will assume that they are included in the inequality constraints in Eq. (12.1). Note also that we will present only the methods that can treat the general constrained problem with the equality and inequality constraints defined in Eq. (12.1). That is, the methods that treat only equalities or only inequalities are not presented.

Just as for unconstrained problems, several methods have been developed and evaluated for the general constrained optimization problems in Eq. (12.1). Most methods follow the two-phase approach as for the unconstrained problems: the *search direction* and *step size* determination phases. The approach followed here is to describe the underlying *ideas* and *concepts* of the methods. Comprehensive coverage of all of the methods giving their advantages and disadvantages is avoided. Only a few simple and generally applicable methods are described and illustrated with examples.

In Section 10.6 we described the steepest-descent method for solving unconstrained optimization problems. That method is quite straightforward. It is, however, not directly applicable to constrained problems. One reason is that we must consider constraints while computing the search direction and the step size. In this chapter, we will describe a *constrained steepest-descent method* that computes the direction of design change considering the local behavior of cost and constraint functions.

The method (and most others) is based on linearization of the problem about the current estimate of the optimum design. Therefore, linearization of the problem is quite important and is discussed in detail. Once the problem has been linearized, it is natural to ask if it can be solved using linear programming methods. The answer is yes, and we first describe a method that is a simple extension of the Simplex method of linear programming. Then we describe the constrained steepest-descent method.

12.1 BASIC CONCEPTS RELATED TO NUMERICAL METHODS

This section contains basic concepts, ideas, and definitions of the terms used in numerical methods for constrained optimization. The status of a constraint at a design point is defined, along with active, inactive, violated, and ε-active constraints. Normalization of constraints and its advantages are explained with examples. The ideas of a descent function and convergence of algorithms are explained.

12.1.1 Basic Concepts Related to Algorithms for Constrained Problems

In the *numerical search methods*, we select a design to initiate the iterative process, as for the unconstrained methods described in Chapters 10 and 11. The iterative process is continued until no further moves are possible and the optimality conditions are satisfied. Most of the general concepts of iterative numerical algorithms discussed in Section 10.2 also apply to methods for constrained optimization problems. Therefore, those concepts should be thoroughly reviewed.

Iterative Process

All numerical methods discussed in this chapter are based on the following iterative prescription, as also given in Eqs. (10.1) and (10.2) for unconstrained problems:

Vector form: $\qquad$ $\mathbf{x}^{(k+1)} = \mathbf{x}^{(k)} + \Delta\mathbf{x}^{(k)}; \quad k = 0, 1, 2, \ldots$ $\qquad$ (12.2)

Component form: $\quad x_i^{(k+1)} = x_i^{(k)} + \Delta x_i^{(k)}; \quad k = 0, 1, 2, \ldots; \quad i = 1 \text{ to } n$ $\qquad$ (12.3)

The superscript k represents the iteration or design cycle number, the subscript i refers to the ith design variable, $\mathbf{x}^{(0)}$ is the starting design estimate, and $\Delta\mathbf{x}^{(k)}$ represents a change in the current design.

As in the unconstrained numerical methods, the change in design $\Delta\mathbf{x}^{(k)}$ is decomposed as

$$\Delta\mathbf{x}^{(k)} = \alpha_k \mathbf{d}^{(k)} \qquad (12.4)$$

where α_k is a step size in the search direction $\mathbf{d}^{(k)}$. Thus, design improvement involves solving the search direction and step size determination subproblems. Solution of both subproblems can involve values of cost and constraint functions as well as their gradients at the current design point.

IMPLEMENTATION OF ITERATIONS

Conceptually, algorithms for unconstrained and constrained optimization problems are based on the same iterative philosophy. There is one important difference, however: Constraints must be considered while determining the search direction as well as the step size. A different procedure for determining either one can give a different optimization algorithm. We will describe, in general terms, a couple of ways in which the algorithms may proceed in the design space. All algorithms need a design estimate to initiate the iterative process. The starting design can be feasible or infeasible. If it is inside the feasible set as is point A in Figure 12.1, then there are two possibilities:

1. The gradient of the cost function vanishes at the point, so it is an unconstrained stationary point. We need to check the second-order conditions for optimality of the point.
2. If the current point is not stationary, then we can reduce the cost function by moving along a descent direction, say, the steepest-descent direction ($-\mathbf{c}$), as shown in Figure 12.1. We continue such iterations until either a constraint is encountered or an unconstrained minimum point is reached.

For the remaining discussion, we assume that the optimum point is on the boundary of the feasible set; that is, some constraints are active (inequality is satisfied at equality). Once the constraint boundary is encountered at point B, one strategy is to travel along a tangent to the boundary such as the direction B–C in Figure 12.1. This results in an infeasible point from where the constraints are corrected in order to again reach the

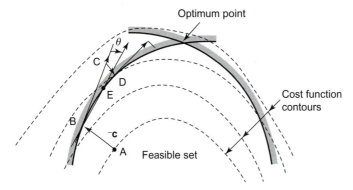

FIGURE 12.1 The conceptual steps of the constrained optimization algorithms initiated from a feasible point.

feasible point D. From there the preceding steps are repeated until the optimum point is reached.

Another strategy is to deflect the tangential direction B–C toward the feasible region by a certain angle θ when there are no equality constraints. Then a line search is performed through the feasible region to reach the boundary point E, as shown in Figure 12.1. The procedure is then repeated from there.

When the starting point is infeasible, like point A in Figure 12.2, one strategy is to correct constraints to reach the constraint boundary at point B. From there, the strategies described in the preceding paragraph can be followed to reach the optimum point. This is shown in path 1 in Figure 12.2. The second strategy is to iterate through the infeasible region by computing directions that take successive design points closer to the optimum point, shown as path 2 in Figure 12.2.

Several algorithms based on the strategies described in the foregoing have been developed and evaluated. Some algorithms are better for a certain class of problems than others. A few algorithms work well if the problem has only inequality constraints, whereas others can treat both equality and inequality constraints simultaneously. In this text, we will concentrate mostly on general algorithms that have no restriction on the form of the functions or the constraints.

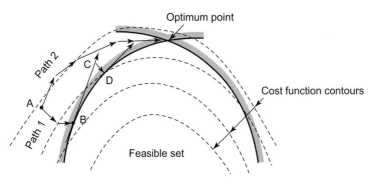

FIGURE 12.2 The conceptual steps of the constrained optimization algorithms initiated from a point that is infeasible.

Most of the algorithms that we will describe in this chapter and the next can treat feasible or infeasible initial designs. They are based on the following *four basic steps* of a numerical algorithm to solve constrained optimization problems.

1. Linearization of cost and constraint functions about the current design point.
2. *Definition of a search direction determination subproblem* using the linearized functions.
3. *Solution of the subproblem* that gives a search direction in the design space.
4. Calculation of a *step size* to minimize a descent function in the search direction.

12.1.2 Constraint Status at a Design Point

An inequality constraint can be either active, ε-active, violated, or inactive at a design point. On the other hand, an equality constraint is either active or violated at a design point. The precise definitions of the status of a constraint at a design point are needed in the development and discussion of numerical methods.

Active constraint: An inequality constraint $g_i(x) \leq 0$ is said to be *active* (or *tight*) at a design point $x^{(k)}$ if it is satisfied as an equality at that point (i.e., $g_i(x^{(k)}) = 0$).
Inactive constraint: An inequality constraint $g_i(x) \leq 0$ is said to be inactive at a design point $x^{(k)}$ if it has negative value at that point (i.e., $g_i(x^{(k)}) < 0$).
Violated constraint: An *inequality* constraint $g_i(x) \leq 0$ is said to be *violated* at a design point $x^{(k)}$ if it has a positive value there (i.e., $g_i(x^{(k)}) > 0$). An *equality* constraint $h_i(x^{(k)}) = 0$ is *violated* at a design point $x^{(k)}$ if it has a nonzero value there (i.e., $h_i(x^{(k)}) \neq 0$). Note that by these definitions, an equality constraint is always either active or violated at a design point.
ε-*Active inequality constraint:* Any inequality constraint $g_i(x^{(k)}) \leq 0$ is said to be ε-active at the point $x^{(k)}$ if $g_i(x^{(k)}) < 0$ but $g_i(x^{(k)}) + \varepsilon \geq 0$, where $\varepsilon > 0$ is a small number. This means that the point is close to the constraint boundary on the feasible side (within an ε-band, as shown in Figure 12.3). That is, the constraint is strictly inactive but it is close to becoming active. Note that the concept of an ε-active constraint applies only to inequality constraints.

To understand the idea of the status of a constraint, refer to Figure 12.3. Consider the *i*th inequality constraint $g_i(x) \leq 0$. The constraint boundary (the surface in the *n*-dimensional space), $g_i(x) = 0$, is plotted, and feasible and infeasible sides for the constraint are identified. An artificial boundary at a distance of ε from the boundary $g_i(x) = 0$ and inside the feasible region is also plotted. We consider four design points A, B, C, and D, as shown in

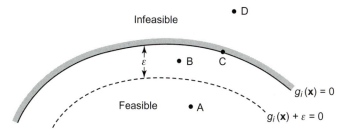

FIGURE 12.3 Status of a constraint at design points A, B, C, and D.

Figure 12.3. For design point A, the constraint $g_i(\mathbf{x})$ is negative and even $g_i(\mathbf{x}) + \varepsilon < 0$. Thus, the constraint is *inactive* for design point A. For design point B, $g_i(\mathbf{x})$ is strictly less than zero, so it is inactive. However, $g_i(\mathbf{x}) + \varepsilon > 0$, so the constraint is *$\varepsilon$-active* for design point B. For design point C, $g_i(\mathbf{x}) = 0$, so it is *active* there. For design point D, $g_i(\mathbf{x})$ is greater than zero, so the constraint is *violated*.

If $g_i(\mathbf{x})$ were an equality constraint, it would be active at point C and violated at points A, B and D in Figure 12.3.

12.1.3 Constraint Normalization

In numerical calculations, it is desirable to normalize all of the constraint functions with respect to their limit values. As noted earlier, active and violated constraints are used in computing a desirable direction of design change. In addition, feasibility of a constraint needs to be checked at the optimum point. Usually one value for ε (say 0.01) is used in checking the status of all of the constraints to check for the ε-active constraint condition. Since different constraints involve different orders of magnitude, it is not proper to use the same ε for all of the constraints unless they are normalized. For example, consider a stress constraint as

$$\sigma \le \sigma_a, \quad \text{or} \quad \sigma - \sigma_a \le 0 \tag{12.5}$$

and a displacement constraint as

$$\delta \le \delta_a, \quad \text{or} \quad \delta - \delta_a \le 0 \tag{12.6}$$

where

σ = calculated stress at a point
σ_a = allowable stress
δ = calculated deflection at a point
δ_a = allowable deflection

Note that the units for the two constraints are different. The constraint of Eq. (12.5) involves stress, which has units of Pascals (Pa, N/m^2). For example, the allowable stress for steel is 250 MPa. The constraint in Eq. (12.6) involves deflections of the structure, which may be only a few centimeters. The allowable deflection δ_a may be, say, only 2 cm. Thus, the values of the two constraints are of widely differing orders of magnitude. If the constraints are violated, it is difficult to judge the severity of their violation. We can, however, normalize the constraints by dividing them by their respective allowable values to obtain the normalized constraint as

$$R - 1.0 \le 0 \tag{12.7}$$

where $R = \sigma/\sigma_a$ for the stress constraint and $R = \delta/\delta_a$ for the deflection constraint. Here, both σ_a and δ_a are assumed to be positive; otherwise, the sense of the inequality changes. For normalized constraints, it is easy to check for ε-active constraints using the same value of ε for both of them.

There are other constraints that must be written in the form

$$1.0 - R \le 0 \tag{12.8}$$

when normalized with respect to their nominal value. For example, the fundamental vibration frequency ω of a structure or a structural element must be above a given threshold value of ω_a (i.e., $\omega \geq \omega_a$). When the constraint is normalized and converted to the standard "less than" form, it is given as in Eq. (12.8) with $R = \omega/\omega_a$. In subsequent discussions, it is assumed that all equality as well as inequality constraints have been converted to the normalized form.

There are some constraints that cannot be normalized. For these the allowable values may be zero. One example is the lower bound on some design variables. Such constraints cannot be normalized with respect to lower bounds. They may be kept in their original form, or they may be divided by 100 to transform them into a percent value.

Example 12.1 illustrates the constraint normalization process and checking of the constraint status.

EXAMPLE 12.1 CONSTRAINT NORMALIZATION AND STATUS AT A POINT

Consider the two constraints

$$\bar{h} = x_1^2 + \frac{1}{2}x_2 = 18 \tag{a}$$

$$\bar{g} = 500x_1 - 30{,}000x_2 \leq 0 \tag{b}$$

At the design points $(1, 1)$ and $(-4.5, -4.5)$, investigate whether the constraints are active, violated, ε-active, or inactive. Use $\varepsilon = 0.1$ to check the ε-active constraints.

Solution

Let us normalize the equality constraint in Eq. (a) and express it in the standard form as

$$h = \frac{1}{18}x_1^2 + \frac{1}{36}x_2 - 1.0 = 0 \tag{c}$$

Evaluating the constraint at the two given points $(1, 1)$ and $(-4.5, -4.5)$, we get

$$h(1,1) = \frac{1}{18}(1)^2 + \frac{1}{36}(1) - 1 = -0.9166 \neq 0 \tag{d}$$

$$h(-4.5,-4.5) = \frac{1}{18}(-4.5)^2 + \frac{1}{36}(-4.5) - 1 = 0 \tag{e}$$

Therefore, the equality constraint is violated at $(1, 1)$ and satisfied (active) at $(-4.5, -4.5)$.

The inequality constraint cannot be normalized by dividing it by $500x_1$ or $30{,}000x_2$ because x_1 and x_2 can have negative values that will change the sense of the inequality. We must normalize the constraint functions using only positive constants or positive variables. To treat this situation, we may divide the constraint by $30{,}000|x_2|$ and obtain a normalized constraint as $\dfrac{x_1}{60|x_2|} - \dfrac{x_2}{|x_2|} \leq 0$. However, this type of normalization is not desirable since it changes the nature of the constraint from linear to nonlinear. Linear constraints are more efficient to treat than the nonlinear constraints in numerical calculations. Therefore, care and judgment need to be exercised while normalizing some of the constraints. If a normalization procedure changes the nature of the constraint, then another procedure should be tried.

In some cases, it may be better to use the constraints in their original form, especially the equality constraints. Thus, in numerical calculations some experimentation with normalization of constraints may be needed for some constraint forms. For the present constraint in Eq. (b), we normalize it with respect to the constant 500 and then divide by 100 to obtain it in the percent form (or we use some typical value for the constraint):

$$g = \frac{1}{100}(x_1 - 60x_2) \le 0 \tag{f}$$

At the points $(1, 1)$ and $(-4.5, -4.5)$, the constraint is evaluated as

$$g(1,1) = \frac{1}{100}(1 - 60 \times 1) = -0.59 < 0 \tag{g}$$

$$g(-4.5,-4.5) = \frac{1}{100}(-4.5 - 60(-4.5)) = 2.655 > 0 \tag{h}$$

Thus, the constraint is inactive at the point $(1, 1)$ and violated at the point $(-4.5, -4.5)$. At the point $(1, 1)$, the constraint is not even ε-active since

$$g(1,1) + \varepsilon = -0.59 + 0.10 = -0.49 < 0 \tag{i}$$

12.1.4 The Descent Function

For unconstrained optimization, each algorithm in Chapters 10 and 11 required reduction in the cost function at every design iteration. With that requirement, a descent toward the minimum point was maintained. *A function used to monitor progress toward the minimum is called the* descent, *or merit,* function. The cost function is used as the descent function in unconstrained optimization problems.

The idea of a descent function is very important in constrained optimization as well. With some constrained optimization methods, the cost function can be used as the descent function. However, with many modern numerical methods, it cannot be so used. Therefore, many other descent functions have been proposed and used. We will discuss one such function later in this chapter.

At this point, the purpose of the descent function should be well understood. The basic idea is to compute a search direction $\mathbf{d}^{(k)}$ and then a step size along it such that the descent function is reduced. With this requirement, proper progress toward the minimum point is maintained. The descent function also has the property that its minimum value is the same as that of the original cost function.

12.1.5 Convergence of an Algorithm

The idea of convergence of an algorithm is very important in constrained optimization problems. We first define and then discuss its importance and how to achieve it. *An algorithm is said to be convergent if it reaches a local minimum point starting from an arbitrary point.* An algorithm that has been proven to converge starting from an arbitrary point is called a *robust* method. In practical applications of optimization, such *reliable algorithms* are highly desirable.

Many engineering design problems require considerable numerical effort to evaluate functions and their gradients. Failure of the algorithm in such applications can have disastrous effects with respect to wastage of valuable resources as well as the morale of designers. For this reason, it is important to use convergent algorithms for practical applications.

A convergent algorithm satisfies the following requirements:
1. **There is a descent function for the algorithm. The idea is that the descent function must decrease at each iteration. This way, progress toward the minimum point can be monitored.**
2. **The direction of design change d$^{(k)}$ is a continuous function of the design variables. This is also an important requirement. It implies that a proper direction can be found such that descent toward the minimum point can be maintained. This requirement also avoids "oscillations" or "zigzagging" in the descent function.**
3. **The feasible set must be closed and bounded.**

The algorithm may or may not converge if these conditions are not satisfied. Note that the feasible set is *closed* if all of the boundary points are included in the set; that is, *there are no strict inequalities in the problem formulation*. A *bounded set* implies that there are upper and lower bounds on the elements of the set. These two requirements are satisfied if all functions of the problem are continuous. The preceding requirements are not unreasonable for many engineering applications.

12.2 LINEARIZATION OF THE CONSTRAINED PROBLEM

At each iteration, most numerical methods for constrained optimization compute design change by solving a subproblem that is obtained by writing linear Taylor's expansions for the cost and constraint functions. This idea of approximate or linearized subproblems is central to the development of many numerical optimization methods and should be thoroughly understood.

All search methods start with a design estimate and iteratively improve it, as seen in Eq. (12.2) or Eq. (12.3). Let $\mathbf{x}^{(k)}$ be the design estimate at the kth iteration and $\Delta\mathbf{x}^{(k)}$ be the change in design. Writing Taylor's expansion of the cost and constraint functions about the point $\mathbf{x}^{(k)}$, we obtain the linearized subproblem as

Minimize

$$f(\mathbf{x}^{(k)} + \Delta\mathbf{x}^{(k)}) \cong f(\mathbf{x}^{(k)}) + \nabla f^T(\mathbf{x}^{(k)})\Delta\mathbf{x}^{(k)} \tag{12.9}$$

subject to the linearized equality constraints

$$h_j(\mathbf{x}^{(k)} + \Delta\mathbf{x}^{(k)}) \cong h_j(\mathbf{x}^{(k)}) + \nabla h_j^T(\mathbf{x}^{(k)})\Delta\mathbf{x}^{(k)} = 0; \quad j = 1 \text{ to } p \tag{12.10}$$

and the linearized inequality constraints

$$g_j(\mathbf{x}^{(k)} + \Delta\mathbf{x}^{(k)}) \cong g_j(\mathbf{x}^{(k)}) + \nabla g_j^T(\mathbf{x}^{(k)})\Delta\mathbf{x}^{(k)} \leq 0; \quad j = 1 \text{ to } m \tag{12.11}$$

where ∇f, ∇h_j, and ∇g_j are the gradients of the cost function, the jth equality constraint, and the jth inequality constraint, respectively, and "$\cong$" implies approximate equality. All of the functions and gradients are evaluated at the current point $\mathbf{x}^{(k)}$.

NOTATION FOR THE LINEARIZED SUBPROBLEM In the following presentation, we introduce some simplified notations for the current design $\mathbf{x}^{(k)}$ as follows:

Cost function value:

$$f_k = f(\mathbf{x}^{(k)}) \tag{12.12}$$

Negative of the jth equality constraint function value:

$$e_j = -h_j(\mathbf{x}^{(k)}) \tag{12.13}$$

Negative of the jth inequality constraint function value:

$$b_j = -g_j(\mathbf{x}^{(k)}) \tag{12.14}$$

Derivative of the cost function with respect to x_i:

$$c_i = \partial f(\mathbf{x}^{(k)})/\partial x_i \tag{12.15}$$

Derivative of h_j with respect to x_i:

$$n_{ij} = \partial h_j(\mathbf{x}^{(k)})/\partial x_i \tag{12.16}$$

Derivative of g_j with respect to x_i:

$$a_{ij} = \partial g_j(\mathbf{x}^{(k)})/\partial x_i \tag{12.17}$$

Design change:

$$d_i = \Delta x_i^{(k)} \tag{12.18}$$

Note also that problem linearization is done at any design iteration, so the argument $\mathbf{x}^{(k)}$, as well as the superscript k indicating the iteration number, will be omitted for some quantities.

DEFINITION OF THE LINEARIZED SUBPROBLEM Using these notations and dropping f_k in the linearized cost function, the approximate subproblem given in Eqs. (12.9) through (12.11) is defined as follows:

Minimize

$$\bar{f} = \sum_{i=1}^{n} c_i d_i \quad (\bar{f} = \mathbf{c}^T \mathbf{d}) \tag{12.19}$$

subject to the linearized equality constraints

$$\sum_{i=1}^{n} n_{ij} d_i = e_j; \quad j = 1 \text{ to } p \quad (\mathbf{N}^T \mathbf{d} = \mathbf{e}) \tag{12.20}$$

and the linearized inequality constraints

$$\sum_{i=1}^{n} a_{ij} d_i \le b_j; \quad j = 1 \text{ to } m \quad (\mathbf{A}^T \mathbf{d} \le \mathbf{b}) \tag{12.21}$$

where columns of the matrix $\mathbf{N}$ $(n \times p)$ are the gradients of the equality constraints and the columns of the matrix $\mathbf{A}$ $(n \times m)$ are the gradients of the inequality constraints.

Note that f_k is a *constant*, which does not affect solution of the linearized subproblem; it is *dropped* from Eq. (12.19). Therefore, $\bar{f}$ represents the linearized change in the original cost function. Let $\mathbf{n}^{(j)}$ and $\mathbf{a}^{(j)}$ represent the gradients of the jth equality and the jth inequality constraints, respectively. Therefore, they are given as the column vectors:

$$\mathbf{n}^{(j)} = \left(\frac{\partial h_j}{\partial x_1} \quad \frac{\partial h_j}{\partial x_2} \quad \cdots \quad \frac{\partial h_j}{\partial x_n} \right)^T \tag{12.22}$$

$$\mathbf{a}^{(j)} = \left(\frac{\partial g_j}{\partial x_1} \quad \frac{\partial g_j}{\partial x_2} \quad \cdots \quad \frac{\partial g_j}{\partial x_n} \right)^T \tag{12.23}$$

The matrices $\mathbf{N}$ and $\mathbf{A}$ are formed using gradients of the constraints as their columns:

$$\mathbf{N} = [\mathbf{n}^{(j)}]_{(n \times p)} \tag{12.24}$$

$$\mathbf{A} = [\mathbf{a}^{(j)}]_{(n \times m)} \tag{12.25}$$

Examples 12.2 and 12.3 illustrate the linearization process for nonlinear optimization problems.

EXAMPLE 12.2 DEFINITION OF A LINEARIZED SUBPROBLEM

Consider the optimization problem of Example 4.31:

Minimize

$$f(\mathbf{x}) = x_1^2 + x_2^2 - 3x_1x_2 \tag{a}$$

subject to

$$g_1(\mathbf{x}) = \frac{1}{6}x_1^2 + \frac{1}{6}x_2^2 - 1.0 \le 0 \tag{b}$$

$$g_2(\mathbf{x}) = -x_1 \le 0, \quad g_3(\mathbf{x}) = -x_2 \le 0 \tag{c}$$

Linearize the cost and constraint functions about the point $\mathbf{x}^{(0)} = (1,1)$ and write the approximate problem given by Eqs. (12.19) through (12.21).

Solution

Note that the constraint in Eq. (b) has been normalized using the constant 6. The graphical solution to the problem is shown in Figure 12.4. It is seen that the optimum solution is at the

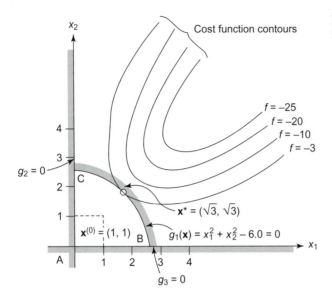

FIGURE 12.4 Graphical representation of the cost and constraints in Example 12.2.

point $\left(\sqrt{-3},\ \sqrt{-3}\right)$ with the cost function as -3. The given point $(1,\ 1)$ is inside the feasible region.

Function values Evaluating the cost and constraint functions at the point $(1, 1)$, we get

$$f(1,1) = (1)^2 + (1)^2 - 3(1)(1) = -1 \tag{d}$$

$$g_1(1,1) = \left(\frac{1}{6}(1)^2 + \frac{1}{6}(1)^2 - 1\right) = -\frac{2}{3} < 0\ (\textit{inactive}) \tag{e}$$

$$g_2(1,1) = -1 < 0\ (\textit{inactive}) \tag{f}$$

$$g_3(1,1) = -1 < 0\ (\textit{inactive}) \tag{g}$$

Thus the given point $(-1, -1)$ is feasible, as seen in Figure 12.4.

Function gradients The gradients of the cost and constraint functions at $(1, 1)$ are calculated as

$$c^{(0)} = \nabla f(1,1) = (2x_1 - 3x_2,\ -3x_1 + 2x_2) = (2 \times 1 - 3 \times 1,\ -3 \times 1 + 2 \times 1) = (-1,\ -1) \tag{h}$$

$$\nabla g_1(1,1) = \left(\frac{2}{6}x_1,\ \frac{2}{6}x_2\right) = \left(\frac{1}{3},\ \frac{1}{3}\right) \tag{i}$$

$$\nabla g_2(1,1) = (-1, 0) \tag{j}$$

$$\nabla g_3(1,1) = (0, -1) \tag{k}$$

Linearized subproblem Using the Taylor's expansion of Eq. (12.9), the linearized cost function at the point (1,1) is given as

$$\bar{f} = f(\mathbf{x}^{(0)}) + \nabla f(\mathbf{x}^{(0)}) \cdot \mathbf{d} = -1 + \begin{bmatrix} -1 & -1 \end{bmatrix} \begin{bmatrix} d_1 \\ d_2 \end{bmatrix} = -1 - d_1 - d_2 \tag{l}$$

Similarly, linearizing the constraint functions using Eq. (12.11), we obtain

$$\frac{1}{3}d_1 + \frac{1}{3}d_2 \le \frac{2}{3} \tag{m}$$

$$-d_1 \le 1, -d_2 \le 1 \tag{n}$$

Thus the linearized subproblem is defined as minimizing the cost function of Eq. (l) subject to the constraints of Eqs. (m) and (n).

Note that the matrix $\mathbf{A}$ of Eq. (12.17), vector $\mathbf{b}$ of Eq. (12.14), and vector $\mathbf{c}$ of Eq. (12.15) are identified from Eqs. (l) to (n) as

$$\mathbf{A} = \begin{bmatrix} \frac{1}{3} & -1 & 0 \\ \frac{1}{3} & 0 & -1 \end{bmatrix}, \quad \mathbf{b} = \begin{bmatrix} \frac{2}{3} \\ 1 \\ 1 \end{bmatrix}, \quad \mathbf{c} = \begin{bmatrix} -1 \\ -1 \end{bmatrix} \tag{o}$$

Linearization in Terms of Original Variables Note also that the linearized subproblem is in terms of the design changes d_1 and d_2. *We may also write the subproblem in terms of the original variables x_1 and x_2.* To do this we substitute $\mathbf{d} = (\mathbf{x} - \mathbf{x}^{(0)})$ in all of the foregoing expressions or in the linear Taylor's expansion, and obtain:

$$\bar{f}(x_1, x_2) = f(\mathbf{x}^{(0)}) + \nabla f \cdot (\mathbf{x} - \mathbf{x}^{(0)}) = -1 + \begin{bmatrix} -1 & -1 \end{bmatrix} \begin{bmatrix} (x_1 - 1) \\ (x_2 - 1) \end{bmatrix} = -x_1 - x_2 + 1 \tag{j}$$

$$\bar{g}_1(x_1, x_2) = g_1(\mathbf{x}^{(0)}) + \nabla g \cdot (\mathbf{x} - \mathbf{x}^{(0)}) = -\frac{2}{3} + \begin{bmatrix} \frac{1}{3} & \frac{1}{3} \end{bmatrix} \begin{bmatrix} (x_1 - 1) \\ (x_2 - 1) \end{bmatrix}$$

$$= \frac{1}{3}(x_1 + x_2 - 4) \le 0 \tag{k}$$

$$\bar{g}_2 = -x_1 \le 0; \quad \bar{g}_3 = -x_2 \le 0 \tag{l}$$

In the foregoing expressions, the overbar for a function indicates its linearized approximation. The feasible regions for the linearized subproblem at the point (1, 1) and the original problem are shown in Figure 12.5. Since the linearized cost function is parallel to the linearized first constraint $\bar{g}_1$, the optimum solution for the linearized subproblem is any point on the line D−E in Figure 12.5.

It is important to note that the linear approximations for the functions of the problem change from point to point. Therefore, the feasible region for the linearized subproblem will change with the point at which the linearization is performed.

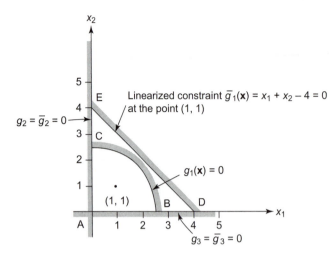

FIGURE 12.5 Graphical representation of the linearized feasible region of Example 12.2.

EXAMPLE 12.3 LINEARIZATION OF RECTANGULAR BEAM DESIGN PROBLEM

Linearize the rectangular beam design problem formulated that is in Section 3.8 at the point (50,200) mm.

Solution

The normalized problem is defined as follows: Find width b and depth d to

Minimize

$$f(b,d) = bd \tag{a}$$

subject to

$$g_1 = \frac{(2.40 \times 10^7)}{bd^2} - 1.0 \leq 0 \tag{b}$$

$$g_2 = \frac{(1.125 \times 10^5)}{bd} - 1.0 \leq 0 \tag{c}$$

$$g_3 = \frac{1}{100}(-2b + d) \leq 0 \tag{d}$$

$$g_4 = -b \leq 0; \quad g_5 = -d \leq 0 \tag{e}$$

Note that constraint g_3 in Eq. (d) has been normalized with respect to 100.

Evaluation of problem functions At the given point, the problem functions are evaluated as

$$f(50,200) = 50 \times 200 = 10,000 \tag{f}$$

$$g_1(50,200) = \frac{2.40 \times 10^7}{50 \times 200^2} - 1 = 11 > 0 \text{ (violation)} \tag{g}$$

$$g_2(50,200) = \frac{1.125 \times 10^5}{50 \times 200} - 1 = 10.25 > 0 \text{ (violation)} \tag{h}$$

$$g_3(50,200) = \frac{1}{100}(-2 \times 50 + 200) = 1 > 0 \text{ (violation)} \tag{i}$$

$$g_4(50,200) = -50 < 0 \text{ (inactive)} \tag{j}$$

$$g_5(50,200) = -200 < 0 \text{ (inactive)} \tag{k}$$

Evaluation of gradients In the following calculations, we will ignore constraints g_4 and g_5, assuming that they will remain satisfied; that is, the design will remain in the first quadrant. The gradients of the functions are evaluated as

$$\nabla f(50,200) = (d, b) = (200,50) \tag{l}$$

$$\nabla g_1(50,200) = \left(\frac{-(2.40 \times 10^7)}{b^2 d^2}, \frac{-(2.40 \times 10^7)}{bd^3} \right)$$
$$= \left(\frac{-(2.40 \times 10^7)}{50^2 \times 200^2}, \frac{-(2.40 \times 10^7)}{50 \times 200^3} \right) = (-0.24, -0.12) \tag{m}$$

$$\nabla g_2(50,200) = \left(\frac{-(1.125 \times 10^7)}{b^2 d}, \frac{-(1.125 \times 10^7)}{bd^2} \right)$$
$$= \left(\frac{-(1.125 \times 10^7)}{50^2 \times 200}, \frac{-(1.125 \times 10^7)}{50 \times 200^2} \right) = (-0.225, -0.05625) \tag{n}$$

$$\nabla g_3(50,200) = \left(\frac{-2}{100}, \frac{1}{100} \right) = (-0.02, 0.01) \tag{o}$$

Linearized subproblem Using the function values and their gradients, the linear Taylor's expansions given in Eqs. (12.11) through (12.13) give the linearized subproblem at point $(50, 200)$ in terms of the original variables as

Minimize

$$\bar{f}(b, d) = 10,000 + 200(b - 50) + 50(d - 200)$$
$$= 200b + 50d - 10,000 \tag{p}$$

subject to

$$\bar{g}_1(b, d) = 11 - 0.24(b - 50) - 0.12(d - 200)$$
$$= -0.24b - 0.12d + 47 \le 0 \tag{q}$$

$$\bar{g}_2(b, d) = 10.25 - 0.225(d - 50) - 0.05625(b - 200)$$
$$= -0.225b - 0.05625d + 32.75 \le 0 \tag{r}$$

$$\bar{g}_3(b, d) = 1 - 0.02(b - 50) + 0.01(d - 200)$$
$$= -0.02b + 0.01d \le 0 \tag{s}$$

Note that the linearized constraint g_3 in Eq. (s) has the same form as the original constraint in Eq. (d), as expected. The linearized constraint functions are plotted in Figure 12.6 and their feasible region is identified. The feasible region for the original constraints is also identified. It can be observed that the two regions are quite different. Also observe that the linearized cost function is parallel to constraint $\bar{g}_2$. The optimum solution lies at the point H, which is at the intersection of constraints $\bar{g}_1$ and $\bar{g}_3$ and is given as

$$b = 97.9 \text{ mm}, \quad d = 195.8 \text{ mm}, \quad \bar{f} = 19{,}370 \text{ mm}^2 \tag{t}$$

For this point, the original constraints g_1 and g_2 are still violated. Apparently, for nonlinear constraints, iterations are needed to correct constraint violations and reach the feasible set.

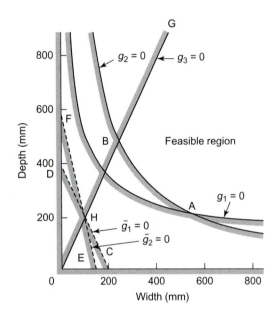

FIGURE 12.6   Feasible region for the original and the linearized constraints of the rectangular beam design problem in Example 12.3.

12.3 THE SEQUENTIAL LINEAR PROGRAMMING ALGORITHM

Note that all of the functions in Eqs. (12.19) through (12.21) are linear in the variables d_i. Therefore, linear programming methods can be used to solve for d_i. Such procedures where linear programming is used to compute design change are referred to as *sequential linear programming*, or SLP for short. In this section, we will briefly describe such a procedure and discuss its advantages and drawbacks. The idea of move limits and their needs are explained and illustrated.

12.3.1 Move Limits in SLP

To solve the LP by the standard Simplex method, the right-side parameters e_i and b_j in Eqs. (12.13) and (12.14) must be nonnegative. If any b_j is negative, we must multiply the

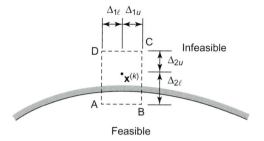

FIGURE 12.7   Linear move limits on design changes.

corresponding constraint by -1 to make the right side nonnegative. This will change the sense of the inequality in Eq. (12.21); that is, it will become a "$\geq$ type" constraint.

It must be noted that the problem defined in Eqs. (12.19) through (12.21) may not have a bounded solution, or the changes in design may become too large, thus invalidating the linear approximations. Therefore, limits must be imposed on changes in design. Such constraints are usually called *move limits*, expressed as

$$-\Delta_{il}^{(k)} \leq d_i^{(k)} \leq \Delta_{iu}^{(k)} \qquad i = 1 \text{ to } n \tag{12.26}$$

where $\Delta_{il}^{(k)}$ and $\Delta_{iu}^{(k)}$ are the maximum allowed decrease and increase in the ith design variable, respectively, at the kth iteration. The problem is still linear in terms of d_i, so LP methods can be used to solve it. Note that the iteration counter k is used to specify $\Delta_{il}^{(k)}$ and $\Delta_{iu}^{(k)}$. That is, the move limits may change at every iteration. Figure 12.7 shows the effect of imposing the move limits on changes in the design $\mathbf{x}^{(k)}$; the new design estimate is required to stay in the rectangular area ABCD for a two-dimensional problem.

The move limits in Eq. (12.26) achieve two important objectives in the linearized subproblem:

1. They make the linearized subproblem bounded.
2. They give the design changes directly without performing the line search for a step size.

Selection of Proper Move Limits

Selecting proper move limits is of critical importance because it can mean success or failure of the SLP algorithm. Their specification, however, requires some experience with the method as well as knowledge of the problem being solved. Therefore, the user should not hesitate to try different move limits if one specification leads to failure or improper design. Many times lower and upper bounds are specified on the real design variables x_i. Therefore, move limits must be selected to remain within the specified bounds.

Also, since linear approximations of the functions are used, the design changes should not be very large and the move limits should not be excessively large. Usually $\Delta_{il}^{(k)}$ and $\Delta_{iu}^{(k)}$ are *selected as some fraction of the current design variable values* (this may vary from 1 to 100 percent). If the resulting LP problem turns out to be infeasible, the move limits will need to be relaxed (i.e., larger changes in the design must be allowed) and the subproblem solved again. Usually, a certain amount of experience with the problem is necessary in order to select proper move limits and adjust them at every iteration to solve the problem successfully.

Positive/Negative Design Changes

Another point must be noted before an SLP algorithm can be stated. This concerns the sign of the variables d_i (or Δx_i), which can be positive or negative. In other words, the current values of the design variables can increase or decrease. To allow for such a change, we must treat the LP variables d_i as free in sign. This can be done as explained in Section 8.1. Each free variable d_i is replaced as $d_i = d_i^+ - d_i^-$ in all of the expressions. The LP subproblem defined in Eqs. (12.19) through (12.21) is then transformed to the standard form for the Simplex method.

12.3.2 An SLP Algorithm

We must define some stopping criteria before stating the algorithm:

1. All constraints must be satisfied. This can be expressed as $g_i \leq \varepsilon_1$; $i = 1$ to m and $|h_i| \leq \varepsilon_1$; $i = 1$ to p, where $\varepsilon_1 > 0$ is a specified small number defining tolerance for constraint violations.
2. The changes in design should be almost zero; that is, $||\mathbf{d}|| \leq \varepsilon_2$, where $\varepsilon_2 > 0$ is a specified small number.

The *sequential linear programming* algorithm is now stated as follows:

Step 1. Estimate a starting design as $\mathbf{x}^{(0)}$. Set $k = 0$. Specify two small positive numbers, ε_1 and ε_2.
Step 2. Evaluate the cost and constraint functions at the current design $\mathbf{x}^{(k)}$; that is, calculate f_k, b_j; $j = 1$ to m, and e_j; $j = 1$ to p, as defined in Eqs. (12.12) through (12.14). Also, evaluate the cost and constraint function gradients at the current design $\mathbf{x}^{(k)}$.
Step 3. Select move limits $\Delta_{il}^{(k)}$ and $\Delta_{iu}^{(k)}$ as some fraction of the current design. Define the LP subproblem of Eqs. (12.19) through (12.21).
Step 4. If needed, convert the LP subproblem to the standard Simplex form (refer to Section 8.2), and solve it for $\mathbf{d}^{(k)}$.
Step 5. Check for convergence. If $g_i \leq \varepsilon_1$; $i = 1$ to m; $|h_i| \leq \varepsilon_1$; $i = 1$ to p; and $||\mathbf{d}^{(k)}|| \leq \varepsilon_2$, then stop. Otherwise, continue.
Step 6. Update the design as $\mathbf{x}^{(k+1)} = \mathbf{x}^{(k)} + \mathbf{d}^{(k)}$. Set $k = k + 1$ and go to Step 2.

It is interesting to note here that the LP problem defined in Eqs. (12.19) through (12.21) can be *transformed to be in the original variables* by substituting $d_i = x_i - x_i^{(k)}$. This was demonstrated in Examples 12.2 and 12.3. The move limits on d_i of Eq. (12.26) can also be transformed to be in the original variables. This way the solution of the LP problem directly gives the estimate for the next design point.

Examples 12.4 and 12.5 illustrate the use of sequential linear programming algorithm.

EXAMPLE 12.4 STUDY OF THE SEQUENTIAL LINEAR PROGRAMMING ALGORITHM

Consider the problem given in Example 12.2. Define the linearized subproblem at the point (3, 3) and discuss its solution after imposing the proper move limits.

Solution

To define the linearized subproblem, the problem functions and their gradients are calculated at the given point (3, 3):

$$f(3,3) = 3^2 + 3^2 - 3 \times 3 \times 3 = -9 \tag{a}$$

$$g_1(3,3) = \frac{1}{6}(3^2) + \frac{1}{6}(3^2) - 1 = 2 > 0 \text{ (violation)} \tag{b}$$

$$g_2(3,3) = -x_1 = -3 < 0 \text{ (inactive)} \tag{c}$$

$$g_3(3,3) = -x_2 = -3 < 0 \text{ (inactive)} \tag{d}$$

$$c(3,3) = \nabla f = (2x_1 - 3x_2,\ 2x_2 - 3x_1) = (2 \times 3 - 3 \times 3,\ 2 \times 3 - 3 \times 3) = (-3,\ -3) \tag{e}$$

$$\nabla g_1(3,3) = \left(\frac{2x_1}{6},\ \frac{2x_2}{6}\right) = \left(\frac{2 \times 3}{6},\ \frac{2 \times 3}{6}\right) = (1,\ 1) \tag{f}$$

$$\nabla g_2(3,3) = (-1, 0), \quad \nabla g_3(3,3) = (0, -1)$$

The given point is in the infeasible region, as the first constraint is violated. The linearized subproblem is defined according to Eqs. (12.19) through (12.21) as

Minimize

$$\bar{f} = [-3 \ -3]\begin{bmatrix} d_1 \\ d_2 \end{bmatrix} \tag{g}$$

subject to the linearized constraints

$$\begin{bmatrix} 1 & 1 \\ -1 & 0 \\ 0 & -1 \end{bmatrix}\begin{bmatrix} d_1 \\ d_2 \end{bmatrix} \leq \begin{bmatrix} -2 \\ 3 \\ 3 \end{bmatrix} \tag{h}$$

The subproblem has only two variables, so it can be solved using the graphical solution procedure shown in Figure 12.8. This figure, when superimposed on Figure 12.4, represents a linearized approximation of the original problem at the point (3, 3). The feasible solution for the linearized subproblem must lie in the region ABC in Figure 12.8. The cost function is parallel to the line B–C; thus any point on the line minimizes the function. We may choose $d_1 = -1$ and $d_2 = -1$ as the solution that satisfies all of the linearized constraints (note that the linearized change in cost is 6). If 100 percent move limits are selected (i.e., $-3 \leq d_1 \leq 3$ and $-3 \leq d_2 \leq 3$), then the solution to the LP subproblem must lie in the region ADEF. If the move limits are set as 20 percent of the current value of design variables, the solution must satisfy $-0.6 \leq d_1 \leq 0.6$ and $-0.6 \leq d_2 \leq 0.6$.

In this case, the solution must lie in the region $A_1D_1E_1F_1$. It can be seen that there is no feasible solution to this linearized subproblem because region $A_1D_1E_1F_1$ does not intersect the line B–C. We must enlarge this region by increasing the move limits. Thus, we note that if the move limits are too restrictive, the linearized subproblem may not have any solution.

If we choose $d_1 = -1$ and $d_2 = -1$, then the improved design is given as (2, 2). This is still an infeasible point, as can be seen in Figure 12.4. Therefore, although the linearized constraint is satisfied with $d_1 = -1$ and $d_2 = -1$, the original nonlinear constraint g_1 is still violated.

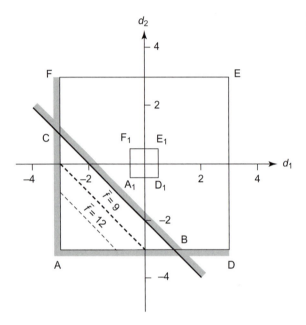

FIGURE 12.8   Graphical solution for the linearized subproblem in Example 12.4.

EXAMPLE 12.5   USE OF SEQUENTIAL LINEAR PROGRAMMING

Consider the problem given in Example 12.2. Perform one iteration of the SLP algorithm. Use $\varepsilon_1 = \varepsilon_2 = 0.001$ and choose move limits such that a 15 percent design change is permissible. Let $\mathbf{x}^{(0)} = (1,1)$ be the starting design.

Solution

The given point represents a feasible solution for the problem, as may be seen in Figure 12.4. The linearized subproblem with 15 percent move limits on design changes d_1 and d_2 at the point $\mathbf{x}^{(0)}$ is obtained in Example 12.2 as

Minimize

$$\bar{f} = -d_1 - d_2 \tag{a}$$

subject to

$$\frac{1}{3}d_1 + \frac{1}{3}d_2 \le \frac{2}{3} \tag{b}$$

$$-(1+d_1) \le 0, \ -(1+d_2) \le 0 \tag{c}$$

$$-0.15 \le d_1 \le 0.15, \ -0.15 \le d_2 \le 0.15 \tag{d}$$

The graphical solution to the linearized subproblem is given in Figure 12.9. Move limits of 15 percent define the solution region as DEFG. The optimum solution for the problem is at point F where $d_1 = 0.15$ and $d_2 = 0.15$. It is seen that much larger move limits are possible in the present case.

We will solve the problem using the Simplex method as well. Note that in the linearized subproblem, the design changes d_1 and d_2 are free in sign. If we wish to solve the problem by the Simplex method, we must define new variables, A, B, C, and D such that $d_1 = A - B$, $d_2 = C - D$, with A, B, C, and $D \ge 0$. Variables A, B, C, D may conflict with symbols in the figure. Therefore, substituting these decompositions into the foregoing equations, we get the following problem, written in standard form:

Minimize

$$\bar{f} = -A + B - C + D \tag{e}$$

subject to

$$\frac{1}{3}(A - B + C - D) \le \frac{2}{3} \tag{f}$$

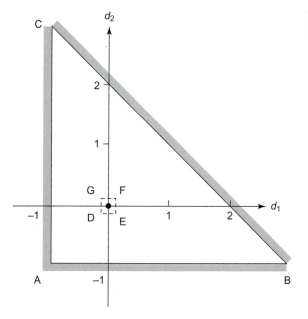

FIGURE 12.9   Graphical solution to the linearized subproblem in Example 12.5.

$$-A + B \le 1.0, \quad -C + D \le 1.0 \tag{g}$$

$$A - B \le 0.15, \quad B - A \le 0.15 \tag{h}$$

$$C - D \le 0.15, \quad D - C \le 0.15 \tag{i}$$

$$A, B, C, D \ge 0 \tag{j}$$

The solution to the foregoing LP problem with the Simplex method is obtained as: $A = 0.15$, $B = 0$, $C = 0.15$, and $D = 0$. Therefore, $d_1 = A - B = 0.15$ and $d_2 = C - D = 0.15$. This gives the updated design as $\mathbf{x}^{(1)} = \mathbf{x}^{(0)} + \mathbf{d}^{(0)} = (1.15, 1.15)$. At the new design $(1.15, 1.15)$, we have $f(\mathbf{x}^{(1)}) = -1.3225$ and $g_1(\mathbf{x}^{(1)}) = -0.5592$. Note that the cost function has decreased for the new design $\mathbf{x}^{(1)}$ without violating the constraint. This indicates that the new design is an improvement over the previous one. Since the norm of the design change, $\|\mathbf{d}\| = 0.212$, is larger than the permissible tolerance (0.001), we need to go through more iterations to satisfy the stopping criterion.

Linearization in terms of original variables It should also be noted that the linearized subproblem at the point $(1, 1)$ can be written in the original variables. This was done in Example 12.2, and the linearized subproblem was obtained as

Minimize

$$\bar{f} = -x_1 - x_2 + 1 \tag{k}$$

subject to

$$\bar{g}_1 = \frac{1}{3}(x_1 + x_2 - 4) \le 0, \quad \bar{g}_2 = -x_1 \le 0, \quad \bar{g}_3 = -x_2 \le 0 \tag{l}$$

The 15 percent move limits can also be transformed to be in the original variables using $-\Delta_{il} \le x_i - x_i^{(0)} \le \Delta_{iu}$:

$$-0.15 \le (x_1 - 1) \le 0.15 \quad \text{or} \quad 0.85 \le x_1 \le 1.15 \tag{m}$$

$$-0.15 \le (x_2 - 1) \le 0.15 \quad \text{or} \quad 0.85 \le x_2 \le 1.15 \tag{n}$$

Solving the subproblem, we obtain the same solution $(1.15, 1.15)$ as before.

Observe that when the problem is transformed in terms of the original variables, there is no need to split the variables into their positive and negative parts since the original variables are required to be nonnegative anyway.

12.3.3 The SLP Algorithm: Some Observations

The sequential linear programming algorithm is a simple and straightforward approach to solving constrained optimization problems. It can be applied to engineering design problems, especially those having a large number of design variables. The following observations highlight some features and limitations of the SLP method.

1. *The method should not be used as a black box approach for engineering design problems.* The selection of move limits is one of trial and error and can be best achieved in an interactive mode.
2. *The method may not converge to the precise minimum* since no descent function is defined, and line search is not performed along the search direction to compute a step size.

3. *The method can cycle between two points* if the optimum solution is not a vertex of the feasible set.

4. The method is quite simple conceptually as well as numerically. Although it may not be possible to reach the precise local minimum point with it, it may be used to obtain improved designs in practice.

12.4 SEQUENTIAL QUADRATIC PROGRAMMING

As observed in the previous section, SLP is a simple algorithm to obtain improved designs for general constrained optimization problems. However, the method has some limitations, the major one being its lack of robustness. To overcome SLP's drawbacks, several other derivative-based methods have been developed to solve smooth nonlinear programming problems. These include the gradient projection method, the feasible directions method, and the generalized reduced gradient method. The basic concepts of these methods are described in Chapter 13. Some methods have good performance for equality-constrained problems only; others have good performance for inequality-constrained problems only.

In this text, we primarily focus on general methods that can treat equality as well as inequality constraints. Sequential quadratic programming (SQP) methods are relatively new and have become quite popular as a result of their generality, robustness, and efficiency. Also, they can incorporate second-order information about the problem functions relatively easily. This is explained in the next chapter.

Here, we describe the basic concepts and steps associated with SQP methods. These methods basically implement the iterative concepts in Eqs. (12.2) through (12.4). That is, they implements the following two steps:

Step 1. A search direction in the design space is calculated by utilizing the values and the gradients of the problem functions; a quadratic programming subproblem is defined and solved.

Step 2. A step size along the search direction is calculated to minimize a descent function; a step size calculation subproblem is defined and solved.

It can be imagined that the two subproblems can be defined in several ways and solved using different numerical methods, giving different SQP methods. We will discuss these in the Sections 12.5 and 12.6 and in Chapter 13.

In most methods, the direction-finding subproblem still uses linearized approximations of Eqs. (12.19) through (12.21) for the nonlinear cost and constraint functions. However, the linear move limits of Eq. (12.26) are abandoned in favor of a step size calculation procedure. In the SQP method the linearized cost function is modified by adding a second-order term so that it becomes a quadratic function. Thus the direction-finding subproblem becomes a *quadratic programming* (QP) subproblem, and it becomes a bounded problem as well. There are a number of ways to define the modified linearized cost function. In the next section we define a quadratic programming subproblem with a search direction that can be interpreted as the constrained steepest-descent direction or the steepest-descent direction projected onto the constraint hyperplane. This subproblem uses only the first-order information about the problem functions. In Chapter 13, we define another QP subproblem that uses second-order information about the problem functions.

12.5 SEARCH DIRECTION CALCULATION: THE QP SUBPROBLEM

In this section, we define a QP subproblem to determine the search direction. We also discuss a method for solving the problem. Note that in the SLP method of the previous section, the vector **d** represents design change at the current point. In this section, the vector **d** represents a direction of design change (search direction). A step along this direction then gives the design change.

12.5.1 Definition of the QP Subproblem

The move limits of Eq. (12.26) in SLP play two roles in the solution process: (1) they make the linearized subproblem bounded, and (2) they give the design change without performing line search. It turns out that these two roles of the move limits of Eq. (12.26) can be achieved by defining and solving a slightly different subproblem to determine the search direction and then performing a line search for the step size along the search direction to calculate the design change.

The linearized subproblem can be bounded if we require minimization of the length of the search direction $\|\mathbf{d}\|$ in addition to minimization of the linearized cost function, $(\mathbf{c} \cdot \mathbf{d})$, in Eq. (12.19). This can be accomplished by combining these two objectives. Since this combined objective is a quadratic function in terms of the search direction **d**, the resulting subproblem is called a QP subproblem. It is defined as

Minimize

$$\bar{f} = \mathbf{c}^T \mathbf{d} + \frac{1}{2} \mathbf{d}^T \mathbf{d} \tag{12.27}$$

subject to the linearized equality and inequality constraints of Eqs. (12.20) and (12.21):

$$\mathbf{N}^T \mathbf{d} = \mathbf{e} \tag{12.28}$$

$$\mathbf{A}^T \mathbf{d} \le \mathbf{b} \tag{12.29}$$

The factor of $1/2$ with the second term in Eq. (12.27) is introduced to eliminate the factor of 2 during differentiations. Also, square of the length of **d** is used instead of its length. The following two observations about the QP subproblem are noteworthy:

1. The QP subproblem is strictly convex and therefore its minimum (if one exists) is global and unique.
2. The cost function of Eq. (12.27) represents an equation of a hypersphere with its center at $-\mathbf{c}$ (circle in two dimensions, sphere in three dimensions).

Example 12.6 demonstrates how to define a quadratic programming subproblem at a given point.

EXAMPLE 12.6 DEFINITION OF A QP SUBPROBLEM

Consider the constrained optimization problem:

Minimize

$$f(\mathbf{x}) = 2x_1^3 + 15x_2^2 - 8x_1x_2 - 4x_1 \tag{a}$$

subject to equality and inequality constraints as

$$h(\mathbf{x}) = x_1^2 + x_1x_2 + 1.0 = 0 \tag{b}$$

$$g(\mathbf{x}) = x_1 - \frac{1}{4}x_2^2 - 1.0 \le 0 \tag{c}$$

Linearize the cost and constraint functions about a point (1, 1) and define the QP subproblem of Eqs. (12.27) to (12.29).

Solution

Note that the constraints for the problem are already written in the normalized form. Figure 12.10 is a graphical representation of the problem. The equality constraint has two branches that are shown as $h = 0$; the boundary of the inequality constraint is $g = 0$. The feasible region for the inequality constraint is identified and several cost function contours are shown.

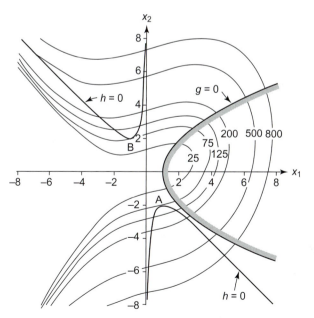

FIGURE 12.10 Graphical representation of Example 12.6.

Since the equality constraint must be satisfied, the optimum point must lie on the two curves $h = 0$. The inequality constraint plays no role in locating the optimum points; its boundary does not intersect the equality constraint curves. Two local minimum points are identified as

Point A:
$$\mathbf{x}^* = (1, -2), \ f(\mathbf{x}^*) = 74 \tag{d}$$

Point B:
$$\mathbf{x}^* = (-1, 2), \ f(\mathbf{x}^*) = 78 \tag{e}$$

Evaluation of functions The cost and the constraint functions are evaluated at the point $(1, 1)$ as
$$f(1,1) = 2(1)^3 + 15(1)^2 - 8 \times 1 \times 1 - 4 \times 1 = 5 \tag{f}$$

$$h(1,1) = (1)^2 + 1 \times 1 + 1 = 3 \neq 0 \tag{g}$$

$$g(1,1) = 1 - \frac{1}{4}(1)^2 - 1 = -0.25 < 0 \tag{h}$$

Gradient evaluation The gradients of the cost and constraint functions are evaluated as
$$\mathbf{c}(1,1) = \nabla f(1,1) = (6x_1^2 - 8x_2 - 4, \ 30x_2 - 8x_1) = (-6, 22) \tag{i}$$
$$\nabla h(1,1) = (2x_1 + x_2, \ x_1) = (3, 1) \tag{j}$$

$$\nabla g(1,1) = \left(1, -\frac{1}{2}x_2\right) = (1, -0.5) \tag{k}$$

Linearized subproblem Substituting Eqs. (f) and (i) into Eq. (12.9), the linearized cost function is
$$\bar{f} = 5 + [-6 \ 22]\begin{bmatrix} d_1 \\ d_2 \end{bmatrix} = 5 - 6d_1 + 22d_2 \tag{l}$$

Similarly the linearized forms of the constraint functions can be written and the linearized subproblem defined as

Minimize
$$\bar{f} = -6d_1 + 22d_2 \tag{m}$$

subject to
$$3d_1 + d_2 = -3 \tag{n}$$
$$d_1 - 0.5d_2 \leq 0.25 \tag{o}$$

Note that the constant 5 has been dropped from the linearized cost function in Eq. (m) because it does not affect the solution to the subproblem. Also, the constants 3 and -0.25 in the linearized constraints have been transferred to the right side in Eqs. (n) and (o).

If 50% move limits are specified for the current point $(1, 1)$ in the SLP algorithm, then each design variable can change by ± 0.5. This gives the move limit constraints as
$$-0.5 \leq d_1 \leq 0.5, \quad -0.5 \leq d_2 \leq 0.5 \tag{p}$$

QP subproblem The QP subproblem of Eqs. (12.27) and (12.28) is defined as

Minimize

$$\bar{f} = (-6d_1 + 22d_2) + \frac{1}{2}(d_1^2 + d_2^2) \tag{q}$$

subject to the linearized constraints of Eqs. (n) and (o).

Note that the move limits of Eqs. (p) are not needed for the QP subproblem.

Solving LP and QP subproblems To compare the solutions, the preceding LP and QP subproblems are plotted in Figures 12.11 and 12.12, respectively. In these figures, the solution must satisfy the linearized equality constraint, so it must lie on the line C–D. The feasible region for the linearized inequality constraint is also shown. Therefore, the solution to the subproblem must lie on the line G–C. It can be seen in Figure 12.11 that with 50 percent move limits, the linearized subproblem is infeasible. The move limits require the changes to lie in the square HIJK, which does not intersect the line G–C. If we relax the move limits to 100 percent, then point L gives the optimum solution:

$$d_1 = -\frac{2}{3}, \quad d_2 = -1.0, \quad \bar{f} = -18 \tag{r}$$

Thus, we again see that the design change with the linearized subproblem is affected by the move limits.

With the QP subproblem, the constraint set remains the same but there is no need for the move limits as seen in Figure 12.12. The cost function $\bar{f}$ is quadratic in the variables. Actually,

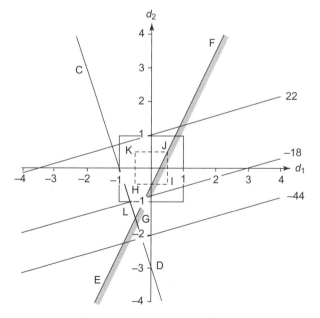

FIGURE 12.11 Solution to linearized subproblem in Example 12.6 at the point (1, 1).

the cost function in Eq. (q) is an equation of a circle with its center at $-\mathbf{c}$ (i.e., at $(6, -22)$). The optimum solution is at point G:

$$d_1 = -0.5, \quad d_2 = -1.5, \quad \bar{f} = -28.75 \tag{s}$$

Note that the direction determined by the QP subproblem is unique, but it depends on the move limits with the LP subproblem. The two directions determined by LP and QP subproblems are in general different.

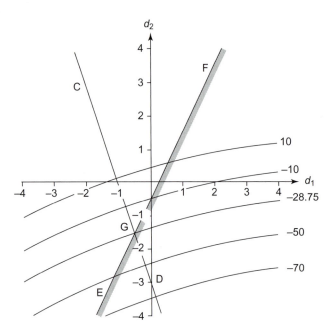

FIGURE 12.12 Solution to the quadratic programming subproblem in Example 12.6 at the point $(1, 1)$.

12.5.2 Solving the QP Subproblem

As noted earlier, many general nonlinear programming algorithms require solving a quadratic programming subproblem at each design cycle. In addition, QP problems are encountered in many real-world applications. Therefore, it is important to solve a quadratic programming subproblem efficiently, so it is not surprising that substantial research effort has been expended in developing and evaluating many numerical methods for solving QP problems (Gill et al., 1981; Luenberger, 1984; Nocedal and Wright, 2006). Many good programs have been developed to solve such problems.

Solution to the QP subproblem gives the search direction **d**. In addition, it gives values for the Lagrange multipliers for the constraints. These multipliers will be needed to calculate the descent function, as we will see later. In Chapter 9, we described a method for solving general QP problems that is a simple extension of the Simplex method of linear programming. If the problem is simple, we can solve it using the Karush-Kuhn-Tucker (KKT) conditions of optimality given in Theorem 4.6. To aid the KKT solution process, we can use a graphical representation of two variable problems to identify the possible solution case and solve that case only. We present such a procedure in Example 12.7.

EXAMPLE 12.7 SOLUTION TO THE QP SUBPROBLEM

Consider the problem of Example 12.2 linearized as

Minimize

$$\bar{f} = -d_1 - d_2 \tag{a}$$

subject to

$$\frac{1}{3}d_1 + \frac{1}{3}d_2 \le \frac{2}{3}; \quad -d_1 \le 1, -d_2 \le 1 \tag{b}$$

Define the quadratic programming subproblem and solve it.

Solution

The linearized cost function in Eq. (a) is modified to a quadratic function as follows:

$$\bar{f} = (-d_1 - d_2) + 0.5(d_1^2 + d_2^2) \tag{c}$$

The quadratic cost function in Eq. (c) corresponds to an equation of a circle with its center at $(-c_1, -c_2)$ where c_i are components of the gradient of the cost function (i.e., at (1, 1)). The graphical solution to the problem is shown in Figure 12.13, where the triangle ABC represents the feasible set for the QP subproblem. Cost function contours are circles of different radii. The optimum

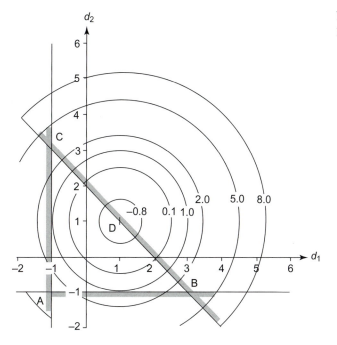

FIGURE 12.13 Solution to the QP subproblem in Example 12.7 at the point (1, 1).

solution is at point D, where $d_1 = 1$ and $d_2 = 1$. Note that the QP subproblem is strictly convex and thus has a unique solution.

A numerical method must generally be used to solve the subproblem. However, since the present problem is quite simple, it can be solved by writing the KKT necessary conditions of Theorem 4.6 as follows:

$$L = (-d_1 - d_2) + 0.5(d_1^2 + d_2^2) + u_1\left(\frac{1}{3}(d_1 + d_2 - 2) + s_1^2\right)$$

(d)

$$+ u_2(-d_1 - 1 + s_2^2) + u_3(-d_2 - 1 + s_3^2)$$

$$\frac{\partial L}{\partial d_1} = -1 + d_1 + \frac{1}{3}u_1 - u_2 = 0, \qquad \frac{\partial L}{\partial d_2} = -1 + d_2 + \frac{1}{3}u_1 - u_3 = 0$$

(e)

$$\frac{1}{3}(d_1 + d_2 - 2) + s_1^2 = 0$$

(f)

$$(-d_1 - 1) + s_2^2 = 0; \quad (-d_2 - 1) + s_3^2 = 0$$

(g)

$$u_i s_i = 0, \quad u_i \geq 0, \quad s_i^2 \geq 0, \quad i = 1, 2, 3$$

(h)

where u_1, u_2, and u_3 are the Lagrange multipliers for the three constraints and s_1^2, s_2^2, and s_3^2 are the corresponding slack variables.

The switching conditions $u_i s_i = 0$ in Eqs. (h) give eight solution cases. However, only one case can give the optimum solution. The graphical solution shows that only the first inequality is active at the optimum, giving the case as $s_1 = 0$, $u_2 = 0$, $u_3 = 0$. Solving this case from Eqs. (e) and (f), we get the solution as

Direction vector: $\mathbf{d} = (1, 1)$;   Lagrange multiplier vector: $\mathbf{u} = (0, 0, 0)$ (i)

12.6 THE STEP SIZE CALCULATION SUBPROBLEM

In this section, we address the problem of step size calculation along the search direction. A descent function that needs to be minimized along the search direction is defined. A class of methods called the interval-reducing methods to minimize the descent function is described. Some other methods to determine the step size are described in Chapter 13.

12.6.1 The Descent Function

Recall that in unconstrained optimization methods the cost function is used as the descent function to monitor the progress of the algorithms toward the optimum point. Although the cost function can be used as a descent function with some constrained optimization methods, it cannot be used for general SQP-type methods. For most methods, the descent function is constructed by adding a *penalty* for constraint violations to the current value of the cost function. Based on this idea, many descent functions can be formulated. In this section, we will describe one of them and show its use.

One of the properties of a descent function is that its value at the optimum point for the optimization problem must be the same as that for the original cost function. Also, it should have the property that a unit step size is admissible in the neighborhood of the optimum point. We will introduce *Pshenichny's descent function* (also called the *exact penalty function*) because of its simplicity and success in solving a large number of engineering design problems (Pshenichny and Danilin, 1982; Belegundu and Arora, 1984a,b). Other descent functions will be discussed in Chapter 13.

Pshenichny's descent function Φ at any point $\mathbf{x}$ is defined as

$$\Phi(\mathbf{x}) = f(\mathbf{x}) + RV(\mathbf{x}) \tag{12.30}$$

where $R > 0$ is a strictly positive number called the *penalty parameter* (initially specified by the user), $V(\mathbf{x}) \geq 0$ is either the *maximum constraint violation* among all of the constraints or zero, and $f(\mathbf{x})$ is the cost function value at $\mathbf{x}$. As an example, the descent function at the point $\mathbf{x}^{(k)}$ during the kth iteration is calculated as

$$\Phi_k = f_k + RV_k \tag{12.31}$$

where Φ_k and V_k are the values of $\Phi(\mathbf{x})$ and $V(\mathbf{x})$ at $\mathbf{x}^{(k)}$:

$$\Phi_k = \Phi(\mathbf{x}^{(k)}); \quad V_k = V(\mathbf{x}^{(k)}) \tag{12.32}$$

and R is the most current value of the penalty parameter. As explained later with examples, the penalty parameter may change during the iterative process. Actually, it must be ensured that it is greater than or equal to the sum of all of the Lagrange multipliers of the QP subproblem at the point $\mathbf{x}^{(k)}$. This is a necessary condition given as

$$R \geq r_k \tag{12.33}$$

where r_k is the sum of all of the Lagrange multipliers at the kth iteration:

$$r_k = \sum_{i=1}^{p} \left| v_i^{(k)} \right| + \sum_{i=1}^{m} u_i^{(k)} \tag{12.34}$$

Since the Lagrange multiplier $v_i^{(k)}$ for an equality constraint is free in sign, its absolute value is used in Eq. (12.34). $u_i^{(k)}$ is the multiplier for the ith inequality constraint. Thus if R_k is the current value of the penalty parameter, the necessary condition of Eq. (12.33) is satisfied if R is selected as

$$R = \max(R_k, r_k) \tag{12.35}$$

The *parameter* $V_k \geq 0$ *related to the maximum constraint violation* at the kth iteration is determined using the calculated values of the constraint functions at the design point $\mathbf{x}^{(k)}$ as

$$V_k = \max\{0; \ |h_1|, |h_2|, \ \ldots, \ |h_p|; \ g_1, g_2, \ldots, g_m\} \tag{12.36}$$

Since the equality constraint is violated if it is different from zero, the absolute value is used with each h_i in Eq. (12.36). Note that V_k is always nonnegative; that is, $V_k \geq 0$. If all constraints are satisfied at $\mathbf{x}^{(k)}$, then $V_k = 0$.

Thus to determine the step size, we *minimize the descent function* of Eq. (12.30). This implies that we must be able to calculate its values at different points along the search direction. Example 12.8 illustrates the calculations for the descent function.

EXAMPLE 12.8 CALCULATION OF DESCENT FUNCTION

A design problem is formulated as follows:

Minimize

$$f(\mathbf{x}) = x_1^2 + 320x_1x_2 \tag{a}$$

subject to four inequalities:

$$\frac{1}{100}(x_1 - 60x_2) \le 0 \tag{b}$$

$$1 - \frac{x_1(x_1 - x_2)}{3600} \le 0 \tag{c}$$

$$-x_1 \le 0, \ -x_2 \le 0 \tag{d}$$

Taking the penalty parameter R as 10,000, calculate the value of the descent function at the point $\mathbf{x}^{(0)} = (40, 0.5)$.

Solution

The cost and constraint functions at the given point $\mathbf{x}^{(0)} = (40, 0.5)$ are evaluated as

$$f_0 = f(40, 0.5) = (40)^2 + 320(40)(0.5) = 8000 \tag{e}$$

$$g_1 = \frac{1}{100}(40 - 60 \times 0.5) = 0.1 > 0 \text{ (violation)} \tag{f}$$

$$g_2 = 1 - \frac{40(40 - 0.5)}{3600} = 0.5611 > 0 \text{ (violation)} \tag{g}$$

$$g_3 = -40 < 0 \text{ (inactive)} \tag{h}$$

$$g_4 = -0.5 < 0 \text{ (inactive)} \tag{i}$$

Thus, the maximum constraint violation is determined using Eq. (12.36) as

$$V_0 = \max\{0; 0.1, 0.5611, -40, -0.5\} = 0.5611 \tag{j}$$

Using Eq. (12.31), the descent function is calculated as

$$\Phi_0 = f_0 + RV_0 = 8000 + (10,000)(0.5611) = 13,611 \tag{k}$$

12.6.2 Step Size Calculation: Line Search

Once the search direction $\mathbf{d}^{(k)}$ is determined at the current point $\mathbf{x}^{(k)}$, the updated design in Eqs. (12.2) through (12.4) becomes a function of the step size α as

$$\mathbf{x}^{(k+1)} = \mathbf{x}^{(k)} + \alpha\mathbf{d}^{(k)} \tag{12.37}$$

Substituting this updated design into Eq. (12.30), the descent function becomes a function of the step size α as

$$\Phi(\alpha) = \Phi(\mathbf{x}^{(k)} + \alpha \mathbf{d}^{(k)}) \tag{12.38}$$

Thus the step size calculation subproblem becomes finding α to

Minimize

$$\Phi(\alpha) = \Phi(\mathbf{x}^{(k)} + \alpha \mathbf{d}^{(k)}) \tag{12.39}$$

Before the constrained steepest-descent algorithm can be stated, a step size determination procedure is needed. The step size determination problem is to calculate an α_k for use in Eq. (12.4) that minimizes the descent function Φ of Eq. (12.30). In most practical implementations of the algorithm, an *inexact line search* that has worked fairly well is used to determine the step size. We will describe that procedure and illustrate its use with examples in Chapter 13.

In this section we assume that a step size along the search direction can be calculated using the golden section method described in Chapter 10. However, it is realized that the method can be inefficient; therefore, inexact line search should be preferred in most constrained optimization methods.

In performing line search for the minimum value of the descent function Φ, we need a notation to represent the trial design points and values of the descent, cost, and constraint functions. The following notation will be used at iteration k:

α_j = jth trial step size
$x_i^{(k,j)}$ = ith design variable value at the jth trial step size
$f_{k,j}$ = cost function value at the jth trial point
$\Phi_{k,j}$ = descent function value at the jth trial point
$V_{k,j}$ = absolute maximum constraint function value at the jth trial point
R_k = penalty parameter value that is kept fixed during line search as long as the necessary condition of Eq. (12.33) is satisfied.

Thus the descent function of Eq. (12.30) is evaluated at the trial step size α_j using the following equation:

$$\Phi_{k,j} = f_{k,j} + R_k V_{k,j} \tag{12.40}$$

Example 12.9 illustrates the calculations for the descent function during golden section search.

EXAMPLE 12.9 CALCULATION OF THE DESCENT FUNCTION FOR GOLDEN SECTION SEARCH

For the design problem in Example 12.8, the QP subproblem has been defined and solved at the starting point $\mathbf{x}^{(0)} = (40, 0.5)$. The search direction is determined as $\mathbf{d}^{(0)} = (25.60, 0.45)$ and the Lagrange multipliers for the constraints are determined as $\mathbf{u} = (16,300, 19,400, 0, 0)$. Let the initial value of the penalty parameter be given as $R_0 = 1$. Calculate the descent function value at the two points during initial bracketing of the step size in the golden section search using $\delta = 0.1$. Compare the descent function values.

Solution

Since we are evaluating the step size at the initial design point, $k = 0$ and j will be taken as 0, 1, and 2 in Eq. (12.40).

Descent function value at the initial point, $\alpha = 0$ Using the calculations given in Example 12.8 at the initial design point, we get

$$f_{0,0} = 8000, \quad V_{0,0} = 0.5611 \tag{a}$$

To check the necessary condition of Eq. (12.33) for the penalty parameter, we need to evaluate r_0 using Eq. (12.34) as follows:

$$r_0 = \sum_{i=1}^{m} u_i^{(0)} = 16{,}300 + 19{,}400 + 0 + 0 = 35{,}700 \tag{b}$$

The necessary condition of Eq. (12.33) is satisfied if we select the penalty parameter R as

$$R = \max(R_0, \ r_0) = \max(1, \ 35{,}700) = 35{,}700 \tag{c}$$

Thus, the descent function value at the starting point is given as

$$\Phi_{0,0} = f_{0,0} + RV_{0,0} = 8000 + (35{,}700)(0.5611) = 28{,}031 \tag{d}$$

Descent function value at the first trial point Now let us calculate the descent function at the first trial step size $\delta = 0.1$ (i.e., $\alpha_0 = 0.1$). Updating the current design point in the search direction, we get

$$\mathbf{x}^{(0,1)} = \begin{bmatrix} 40 \\ 0.5 \end{bmatrix} + (0.1) \begin{bmatrix} 25.6 \\ 0.45 \end{bmatrix} = \begin{bmatrix} 42.56 \\ 0.545 \end{bmatrix} \tag{e}$$

Various functions for the problem given in Eqs. (a) through (d) are calculated at $\mathbf{x}^{(0,1)}$ as

$$f_{0,1}(42.56, 0.545) = (42.56)^2 + 320(42.56)(0.545) = 9233.8 \tag{f}$$

$$g_1(42.56, 0.545) = \frac{1}{100}(42.56 - 60 \times 0.545) = 0.0986 > 0 \ (\text{violation}) \tag{g}$$

$$g_2(42.56, 0.545) = 1 - \frac{42.56(42.56 - 0.545)}{3600} = 0.5033 > 0 \ (\text{violation}) \tag{h}$$

$$g_3(42.56, 0.545) = -42.56 < 0 \ (\text{inactive}) \tag{i}$$

$$g_4(42.56, 0.545) = -0.545 < 0 \ (\text{inactive}) \tag{j}$$

Thus, the maximum constraint violation is determined using Eq. (12.36) as

$$V_{0,1} = \max\{0; 0.0986, 0.5033, -42.56, -0.545\} = 0.5033 \tag{k}$$

Now the descent function at the trial step size of $\alpha_0 = 0.1$ is given (note that the value of the penalty parameter R is not changed during step size calculation):

$$\Phi_{0,1} = f_{0,1} + RV_{0,1} = 9233.8 + (35{,}700)(0.5033) = 27{,}202 \tag{l}$$

Since $\Phi_{0,1} < \Phi_{0,0}$ (27,202 < 28,031), we need to continue the process of initial bracketing of the optimum step size.

Descent function value at the second trial point In the golden section search procedure, the next trial step size has an increment of (1.618 × the previous increment) and is given as (Section 10.5.4)

$$\alpha_1 = \delta + 1.618\delta = 2.618\delta = 2.618(0.1) = 0.2618 \tag{m}$$

The next trial design point using Eq. (12.2) is obtained as

$$x^{(0,2)} = \begin{bmatrix} 40 \\ 0.5 \end{bmatrix} + (0.2618)\begin{bmatrix} 25.6 \\ 0.45 \end{bmatrix} = \begin{bmatrix} 46.70 \\ 0.618 \end{bmatrix} \tag{n}$$

Following the foregoing procedure, various quantities and the descent function are calculated at the point (46.70, 0.618) as

$$f_{0,2} = 11,416.3; \quad g_1 = 0.0962; \quad g_2 = 0.4022; \quad g_3 = -46.70; \quad g_4 = -0.618 \tag{o}$$

$$V_{0,2} = \max\{0; \quad 0.0962, \, 0.4022, \, -46.70, \, -0.618\} = 0.4022 \tag{p}$$

$$\Phi_{0,2} = f_{0,2} + RV_{0,2} = 11,416.3 + (35,700)(0.4022) = 25,775 \tag{q}$$

Since $\Phi_{0,2} < \Phi_{0,1}(25,775 < 27,202)$, the minimum for the descent function has not been surpassed yet. Therefore we need to continue the initial bracketing process. The next trial step size with an increment of (1.618 × the previous increment) is given as

$$\alpha_2 = 2.618\delta + 1.61'8(1.618\delta) = 0.5236 \tag{r}$$

Following the foregoing procedure, $\Phi_{0,3}$ can be calculated and compared with $\Phi_{0,2}$.

Note that the value of the penalty parameter R is calculated at the beginning of the line search along the search direction and then kept fixed during all subsequent calculations for step size determination.

12.7 THE CONSTRAINED STEEPEST-DESCENT METHOD

In this section, we summarize a general method, called the constrained steepest-descent (CSD) method, that can treat equality as well as inequality constraints in its computational steps. It also requires inclusion of only a few of the critical constraints in the calculation of the search direction at each iteration; that is, the QP subproblem of Eqs. (12.27) and (12.28) may be defined using only the active and violated constraints. This can lead to efficient calculations for larger-scale engineering design problems, as explained in Chapter 13.

The CSD method has been proved to be convergent to a local minimum point starting from any point. This is considered a model algorithm that illustrates how most optimization algorithms work.

The method can be extended for more efficient calculations by including second-order information about the problem, as explained in Chapter 13. Here, a step-by-step procedure is given to show the kind of calculations needed to implement the method for numerical calculations. It is important to understand these steps and calculations to effectively use optimization software and to diagnose errors when something goes wrong with an application.

Note that when there are no constraints, or no active ones, minimization of the quadratic function of Eq. (12.27) using the necessary condition $\partial \bar{f} / \partial \mathbf{d} = \mathbf{0}$ gives

$$\mathbf{d} = -\mathbf{c} \qquad\qquad (12.41)$$

This is just the steepest-descent direction of Section 10.6 for unconstrained problems. When there are constraints, their effect must be included in calculating the search direction. The search direction must satisfy all of the linearized constraints. Since the search direction is a modification of the steepest-descent direction to satisfy constraints, it is called the *constrained steepest-descent direction*. It is actually a direction obtained by projecting the steepest-descent direction on to the constraint hyperplane.

It is important to note that the CSD method presented in this section is the most introductory and simple interpretation of more powerful *sequential quadratic programming* (SQP) methods. Not all features of the algorithms are discussed here in order to keep the presentation of the key ideas simple. It is noted, however, that the methods work equally well when initiated from feasible or infeasible points.

12.7.1 The CSD Algorithm

We are now ready to state the constrained steepest-descent algorithm in a step-by-step form. It has been proved that the solution point of the sequence $\mathbf{x}^{(k)}$ generated by the algorithm is a KKT point for the general constrained optimization problem (Pshenichny and Danilin, 1982). The stopping criterion for the algorithm is that $||\mathbf{d}|| \leq \varepsilon$ for a feasible point. Here ε is a small positive number and $\mathbf{d}$ is the search direction that is obtained as a solution to the QP subproblem. The CSD method is now summarized in the form of a *computational algorithm*.

Step 1. Set $k = 0$. Estimate initial values for the design variables as $\mathbf{x}^{(0)}$. Select an initial value for the penalty parameter R_0, and two small numbers $\varepsilon_1 > 0$ and $\varepsilon_2 > 0$ that define the permissible constraint violation and convergence parameter values, respectively. $R_0 = 1$ is a reasonable selection.

Step 2. At $\mathbf{x}^{(k)}$, compute the cost and constraint functions and their gradients. Calculate the maximum constraint violation V_k, as defined in Eq. (12.36).

Step 3. Using the cost and constraint function values and their gradients, define the QP subproblem given in Eqs. (12.27) and (12.28). Solve the QP subproblem to obtain the search direction $\mathbf{d}^{(k)}$ and the Lagrange multipliers vectors $\mathbf{v}^{(k)}$ and $\mathbf{u}^{(k)}$.

Step 4. Check for the stopping criteria $||\mathbf{d}^{(k)}|| \leq \varepsilon_2$ and the maximum constraint violation $V_k \leq \varepsilon_1$. If these criteria are satisfied then stop. Otherwise, continue.

Step 5. To check the necessary condition of Eq. (12.33) for the penalty parameter R, calculate the sum r_k of the Lagrange multipliers defined in Eq. (12.34). Set $R = \max \{R_k, r_k\}$. This will always satisfy the necessary condition of Eq. (12.33).

Step 6. Set $\mathbf{x}^{(k+1)} = \mathbf{x}^{(k)} + \alpha_k \mathbf{d}^{(k)}$, where $\alpha = \alpha_k$ is a proper step size. As for the unconstrained problems, the step size can be obtained by minimizing the descent function of Eq. (12.30) along the search direction $\mathbf{d}^{(k)}$. Any of the procedures, such as golden section search, can be used to determine a step size.

Step 7. Save the current penalty parameter as $R_{k+1} = R$. Update the iteration counter as $k = k + 1$, and go to Step 2.

The CSD algorithm, along with the foregoing step size determination procedure, is convergent, provided that second derivatives of all of the functions are piece-wise continuous (this is the so-called Lipschitz condition) that and the set of design points $\mathbf{x}^{(k)}$ are bounded as follows:

$$\Phi(\mathbf{x}^{(k)}) \leq \Phi(\mathbf{x}^{(0)}); \quad k = 1, 2, 3, \ldots \tag{12.42}$$

12.7.2 The CSD Algorithm: Some Observations

These observations can be made about the constrained steepest-descent algorithm:

1. The CSD algorithm is a *first-order* method that can treat equality and inequality constraints. The algorithm converges to a local minimum point starting from an arbitrary point, feasible or infeasible.
2. The *potential constraint strategy* discussed in the next chapter is not introduced in the algorithm for the sake of simplicity of presentation. This strategy is useful for engineering applications and can be easily incorporated into the algorithm (Belegundu and Arora, 1984).
3. Golden section search can be inefficient and is generally not recommended for engineering applications. Inexact line search, described in Chapter 13, works quite well and is recommended.
4. The rate of convergence of the CSD algorithm can be improved by including second-order information in the QP subproblem. This is discussed in Chapter 13.
5. The starting point can affect performance of the algorithm. For example, at some points the QP subproblem may not have any solution. This need not mean that the original problem is infeasible. The original problem may be highly nonlinear, and the linearized constraints may be inconsistent, giving an infeasible QP subproblem. This situation can be handled by either temporarily deleting the inconsistent linearized constraints or starting from another point. For more discussion on the implementation of the algorithm, Tseng and Arora (1988) may be consulted.

EXERCISES FOR CHAPTER 12

Section 12.1 Basic Concepts Related to Numerical Methods

12.1 *Answer True or False.*

1. The basic numerical iterative philosophy for solving constrained and unconstrained problems is the same.
2. Step size determination is a one-dimensional problem for unconstrained problems.
3. Step size determination is a multidimensional problem for constrained problems.
4. An inequality constraint $g_i(\mathbf{x}) \leq 0$ is violated at $\mathbf{x}^{(k)}$ if $g_i(\mathbf{x}^{(k)}) > 0$.

5. An inequality constraint $g_i(\mathbf{x}) \leq 0$ is active at $\mathbf{x}^{(k)}$ if $g_i(\mathbf{x}^{(k)}) > 0$.
6. An equality constraint $h_i(\mathbf{x}) = 0$ is violated at $\mathbf{x}^{(k)}$ if $h_i(\mathbf{x}^{(k)}) < 0$.
7. An equality constraint is always active at the optimum.
8. In constrained optimization problems, search direction is found using the cost gradient only.
9. In constrained optimization problems, search direction is found using the constraint gradients only.
10. In constrained problems, the descent function is used to calculate the search direction.
11. In constrained problems, the descent function is used to calculate a feasible point.
12. Cost function can be used as a descent function in unconstrained problems.
13. One-dimensional search on a descent function is needed for convergence of algorithms.
14. A robust algorithm guarantees convergence.
15. A feasible set must be closed and bounded to guarantee convergence of algorithms.
16. A constraint $x_1 + x_2 \leq -2$ can be normalized as $(x_1 + x_2)/(-2) \leq 1.0$.
17. A constraint $x_1^2 + x_2^2 \leq 9$ is active at $x_1 = 3$ and $x_2 = 3$.

Section 12.2 Linearization of the Constrained Problem

12.2 *Answer True or False.*

1. Linearization of cost and constraint functions is a basic step for solving nonlinear optimization problems.
2. General constrained problems cannot be solved by solving a sequence of linear programming subproblems.
3. In general, the linearized subproblem without move limits may be unbounded.
4. The sequential linear programming method for general constrained problems is guaranteed to converge.
5. Move limits are essential in the sequential linear programming procedure.
6. Equality constraints can be treated in the sequential linear programming algorithm.

Formulate the following design problems, transcribe them into the standard form, create a linear approximation at the given point, and plot the linearized subproblem and the original problem on the same graph.

12.3 Beam design problem formulated in Section 3.8 at the point $(b, d) = (250, 300)$ mm.
12.4 Tubular column design problem formulated in Section 2.7 at the point $(R, t) = (12, 4)$ cm. Let $P = 50$ kN, $E = 210$ GPa, $l = 500$ cm, $\sigma_a = 250$ MPa, and $\rho = 7850$ kg/m^3.
12.5 Wall bracket problem formulated in Section 4.9.1 at the point $(A_1, A_2) = (150, 150)$ cm^2.
12.6 Exercise 2.1 at the point $h = 12$ m, $A = 4000$ m^2.
12.7 Exercise 2.3 at the point $(R, H) = (6, 15)$ cm.
12.8 Exercise 2.4 at the point $R = 2$ cm, $N = 100$.
12.9 Exercise 2.5 at the point $(W, D) = (100, 100)$ m.
12.10 Exercise 2.9 at the point $(r, h) = (6, 16)$ cm.
12.11 Exercise 2.10 at the point $(b, h) = (5, 10)$ m.
12.12 Exercise 2.11 at the point, width $= 5$ m, depth $= 5$ m, and height $= 5$ m.
12.13 Exercise 2.12 at the point $D = 4$ m and $H = 8$ m.
12.14 Exercise 2.13 at the point $w = 10$ m, $d = 10$ m, $h = 4$ m.

12.15 Exercise 2.14 at the point $P_1 = 2$ and $P_2 = 1$.

Section 12.3 The Sequential Linear Programming Algorithm

Complete one iteration of the sequential linear programming algorithm for the following problems (try 50 percent move limits and adjust them if necessary).

12.16 Beam design problem formulated in Section 3.8 at the point $(b, d) = (250, 300)$ mm.

12.17 Tubular column design problem formulated in Section 2.7 at the point $(R, t) = (12, 4)$ cm. Let $P = 50$ kN, $E = 210$ GPa, $l = 500$ cm, $\sigma_a = 250$ MPa, and $\sigma = 7850$ kg/m^3.

12.18 Wall bracket problem formulated in Section 4.9.1 at the point $(A_1, A_2) = (150, 150)$ cm^2.

12.19 Exercise 2.1 at the point $h = 12$ m, $A = 4000$ m^2.

12.20 Exercise 2.3 at the point $(R, H) = (6, 15)$ cm.

12.21 Exercise 2.4 at the point $R = 2$ cm, $N = 100$.

12.22 Exercise 2.5 at the point $(W, D) = (100, 100)$ m.

12.23 Exercise 2.9 at the point $(r, h) = (6, 16)$ cm.

12.24 Exercise 2.10 at the point $(b, h) = (5, 10)$ m.

12.25 Exercise 2.11 at the point, width $= 5$ m, depth $= 5$ m, and height $= 5$ m.

12.26 Exercise 2.12 at the point $D = 4$ m and $H = 8$ m.

12.27 Exercise 2.13 at the point $w = 10$ m, $d = 10$ m, $h = 4$ m.

12.28 Exercise 2.14 at the point $P_1 = 2$ and $P_2 = 1$.

Section 12.5 Search Direction Calculation: The QP Subproblem

Solve the following QP problems using KKT optimality conditions.

12.29 Minimize $f(\mathbf{x}) = (x_1 - 3)^2 + (x_2 - 3)^2$
subject to $x_1 + x_2 \leq 5$
$\qquad x_1, x_2 \geq 0$

12.30 Minimize $f(\mathbf{x}) = (x_1 - 1)^2 + (x_2 - 1)^2$
subject to $x_1 + 2x_2 \leq 6$
$\qquad x_1, x_2 \geq 0$

12.31 Minimize $f(\mathbf{x}) = (x_1 - 1)^2 + (x_2 - 1)^2$
subject to $x_1 + 2x_2 \leq 2$
$\qquad x_1, x_2 \geq 0$

12.32 Minimize $f(\mathbf{x}) = x_1^2 + x_2^2 - x_1 x_2 - 3x_1$
subject to $x_1 + x_2 \leq 3$
$\qquad x_1, x_2 \geq 0$

12.33 Minimize $f(\mathbf{x}) = (x_1 - 1)^2 + (x_2 - 1)^2 - 2x_2 + 2$
subject to $x_1 + x_2 \leq 4$
$\qquad x_1, x_2 \geq 0$

12.34 Minimize $f(\mathbf{x}) = 4x_1^2 + 3x_2^2 - 5x_1 x_2 - 8x_1$
subject to $x_1 + x_2 = 4$
$\qquad x_1, x_2 \geq 0$

12.35 Minimize $f(\mathbf{x}) = x_1^2 + x_2^2 - 2x_1 - 2x_2$
subject to $x_1 + x_2 - 4 = 0$
$\qquad x_1 - x_2 - 2 = 0$
$\qquad x_1, x_2 \geq 0$

12.36 Minimize $f(\mathbf{x}) = 4x_1^2 + 3x_2^2 - 5x_1x_2 - 8x_1$

subject to $x_1 + x_2 \leq 4$

$x_1, x_2 \geq 0$

12.37 Minimize $f(\mathbf{x}) = x_1^2 + x_2^2 - 4x_1 - 2x_2$

subject to $x_1 + x_2 \geq 4$

$x_1, x_2 \geq 0$

12.38 Minimize $f(\mathbf{x}) = 2x_1^2 + 6x_1x_2 + 9x_2^2 - 18x_1 + 9x_2$

subject to $x_1 - 2x_2 \leq 10$

$4x_1 - 3x_2 \leq 20$

$x_1, x_2 \geq 0$

12.39 Minimize $f(\mathbf{x}) = x_1^2 + x_2^2 - 2x_1 - 2x_2$

subject to $x_1 + x_2 - 4 \leq 0$

$2 - x_1 \leq 0$

$x_1, x_2 \geq 0$

12.40 Minimize $f(\mathbf{x}) = 2x_1^2 + 2x_2^2 + x_3^2 + 2x_1x_2 - x_1x_3 - 0.8x_2x_3$

subject to $1.3x_1 + 1.2x_2 + 1.1x_3 \geq 1.15$

$x_1 + x_2 + x_3 = 1$

$x_1 \leq 0.7$

$x_2 \leq 0.7$

$x_3 \leq 0.7$

$x_1, x_2, x_3 \geq 0$

For the following problems, obtain the quadratic programming subproblem, plot it on a graph, obtain the search direction for the subproblem, and show the search direction on the graphical representation of the original problem.

12.41 Beam design problem formulated in Section 3.8 at the point $(b, d) = (250, 300)$ mm.

12.42 Tubular column design problem formulated in Section 2.7 at the point $(R, t) = (12, 4)$ cm. Let $P = 50$ kN, $E = 210$ GPa, $l = 500$ cm, $\sigma_a = 250$ MPa, and $\rho = 7850$ kg/m^3.

12.43 Wall bracket problem formulated in Section 4.9.1 at the point $(A_1, A_2) = (150, 150)$ cm^2.

12.44 Exercise 2.1 at the point $h = 12$ m, $A = 4000$ m^2.

12.45 Exercise 2.3 at the point $(R, H) = (6, 15)$ cm.

12.46 Exercise 2.4 at the point $R = 2$ cm, $N = 100$.

12.47 Exercise 2.5 at the point $(W, D) = (100, 100)$ m.

12.48 Exercise 2.9 at the point $(r, h) = (6, 16)$ cm.

12.49 Exercise 2.10 at the point $(b, h) = (5, 10)$ m.

12.50 Exercise 2.11 at the point, width $= 5$ m, depth $= 5$ m, and height $= 5$ m.

12.51 Exercise 2.12 at the point $D = 4$ m and $H = 8$ m.

12.52 Exercise 2.13 at the point $w = 10$ m, $d = 10$ m, $h = 4$ m.

12.53 Exercise 2.14 at the point $P_1 = 2$ and $P_2 = 1$.

Section 12.7 The Constrained Steepest-descent Method

12.54 *Answer True or False.*

  1. The constrained steepest-descent (CSD) method, when there are active constraints, is based on using the cost function gradient as the search direction.

  2. The constrained steepest-descent method solves two subproblems: the search direction and step size determination.

3. The cost function is used as the descent function in the CSD method.
4. The QP subproblem in the CSD method is strictly convex.
5. The search direction, if one exists, is unique for the QP subproblem in the CSD method.
6. Constraint violations play no role in step size determination in the CSD method.
7. Lagrange multipliers of the subproblem play a role in step size determination in the CSD method.
8. Constraints must be evaluated during line search in the CSD method.

For the following problems, calculate the descent function values Φ_0, Φ_1, and Φ_2 at the trial step sizes $\alpha = 0$, δ and 2.618δ (let $R_0 = 1$, and $\delta = 0.1$).

12.55 Beam design problem formulated in Section 3.8 at the point $(b, d) = (250, 300)$ mm.
12.56 Tubular column design problem formulated in Section 2.7 at the point $(R, t) = (12, 4)$ cm. Let $P = 50$ kN, $E = 210$ GPa, $l = 500$ cm, $\sigma_a = 250$ MPa, and $\rho = 7850$ kg/m^3.
12.57 Wall bracket problem formulated in Section 4.9.1 at the point $(A_1, A_2) = (150, 150)$ cm^2.
12.58 Exercise 2.1 at the point $h = 12$ m, $A = 4000$ m^2.
12.59 Exercise 2.3 at the point $(R, H) = (6, 15)$ cm.
12.60 Exercise 2.4 at the point $R = 2$ cm, $N = 100$.
12.61 Exercise 2.5 at the point $(W, D) = (100, 100)$ m.
12.62 Exercise 2.9 at the point $(r, h) = (6, 16)$ cm.
12.63 Exercise 2.10 at the point $(b, h) = (5, 10)$ m.
12.64 Exercise 2.11 at the point, width $= 5$ m, depth $= 5$ m, and height $= 5$ m.
12.65 Exercise 2.12 at the point $D = 4$ m and $H = 8$ m.
12.66 Exercise 2.13 at the point $w = 10$ m, $d = 10$ m, $h = 4$ m.
12.67 Exercise 2.14 at the point $P_1 = 2$ and $P_2 = 1$.

More on Numerical Methods
for Constrained Optimum Design

Upon completion of this chapter, you will be able to

- Use potential constraint strategy in numerical optimization algorithms for constrained problems

- Use inexact step size calculation for constrained optimization methods

- Explain the bound-constrained optimization algorithm

- Use quasi-Newton methods to solve constrained nonlinear optimization problems

- Explain the basic concepts associated with quadratic programming

- Explain the basic ideas behind the feasible directions, gradient projection, and generalized reduced gradient methods

In Chapter 12, basic concepts and steps related to constrained optimization methods were presented and illustrated. In this chapter, we build upon those basic ideas and describe some concepts and methods that are more appropriate for practical applications. Topics such as inexact line search, constrained quasi-Newton methods, and potential constraint strategy to define the quadratic programming subproblem are discussed and illustrated. The bound-constrained optimization problem is defined and an algorithm to solve the problem is presented. Methods to solve the quadratic programming problem for determining the search direction are discussed. These topics usually are not covered in an undergraduate course on optimum design or on a first independent reading of the text.

For convenience of reference, the general constrained optimization problem treated in the previous chapter is restated as: find $\mathbf{x} = (x_1, \ldots, x_n)$, a design variable vector of dimension n, to

$$\underset{\mathbf{x} \in S}{\text{minimize}} \, f(\mathbf{x}); \quad S = \left\{ \mathbf{x} | h_i(\mathbf{x}) = 0, \; i = 1 \text{ to } p; \; g_i(\mathbf{x}) \leq 0, \; i = 1 \text{ to } m \right\} \tag{13.1}$$

13.1 POTENTIAL CONSTRAINT STRATEGY

It is important to note that for most problems, only a subset of the inequality constraints is active at the minimum point. However, this subset of active constraints is not known a priori and must be determined as part of the solution to the problem. Here we introduce the concept of constraints that could be *potentially active* at the minimum point. The concept can be incorporated into the numerical algorithms for constrained optimization to effect efficiency of calculations, especially for large-scale problems.

To evaluate the search direction in numerical methods for constrained optimization, we need to know the cost and constraint functions and their gradients. The numerical algorithms for constrained optimization can be classified based on whether the gradients of all of the constraints or only a subset of them are required to define the search direction determination subproblem. The numerical algorithms that use gradients of only a subset of the constraints in the definition of this subproblem are said to use *potential constraint strategy*. To implement this strategy, a potential constraint index set needs to be defined, which is composed of active, ε-active, and violated constraints at the current design $\mathbf{x}^{(k)}$. At the kth iteration, we define this *potential constraint index set I_k* as follows:

$$I_k = \left[\{j|j = 1 \text{ to } p \text{ for equalities}\} \quad \text{and} \quad \{i|g_i(\mathbf{x}^{(k)}) + \varepsilon \geq 0, \ i = 1 \text{ to } m\}\right] \qquad (13.2)$$

where $\varepsilon > 0$ is a small number. Note that the set I_k contains a list of constraints that satisfy the criteria given in Eq. (13.2); all of the *equality constraints are always included in I_k by definition*. The inequalities that do not meet the criterion given in Eq. (13.2) are ignored at the current iteration in defining the subproblem for calculation of the search direction.

The main effect of using the potential constraint strategy in an algorithm is on the efficiency of the entire iterative process. This is particularly true for large and complex applications where the evaluation of gradients of constraints is expensive. With the potential set strategy, gradients of only the constraints in the set I_k are calculated and used in defining the search direction determination subproblem. The original problem may have hundreds of constraints, but only a few of them may be in the potential set. Thus, with this strategy, not only is the number of gradient evaluations reduced; the dimension of the subproblem for the search direction is substantially reduced as well. This can result in additional savings in computational effort. Therefore, *the potential set strategy is beneficial and should be used in practical applications of optimization*. Before using software to solve a problem, the designer should inquire whether the program uses the potential constraint strategy.

Example 13.1 illustrates determination of a potential constraint set for an optimization problem.

EXAMPLE 13.1   DETERMINATION OF A POTENTIAL CONSTRAINT SET

Consider the following six constraints:

$$2x_1^2 + x_2 \leq 36 \qquad (a)$$

$$x_1 \geq 60x_2 \qquad (b)$$

$$x_2 \leq 10 \tag{c}$$

$$x_2 + 2 \geq 0 \tag{d}$$

$$x_1 \leq 10; \quad x_1 \geq 0 \tag{e}$$

Let $\mathbf{x}^{(k)} = (-4.5, -4.5)$ and $\varepsilon = 0.1$. Form the potential constraint index set I_k of Eq. (13.2).

Solution

After normalization and conversion to the standard form, the constraints are given as shown in the following equations:

$$g_1 = \frac{1}{18}x_1^2 + \frac{1}{36}x_2 - 1 \leq 0, \quad g_2 = \frac{1}{100}(-x_1 + 60x_2) \leq 0 \tag{f}$$

$$g_3 = \frac{1}{10}x_2 - 1 \leq 0, \quad g_4 = -\frac{1}{2}x_2 - 1 \leq 0 \tag{g}$$

$$g_5 = \frac{1}{10}x_1 - 1 \leq 0, \quad g_6 = -x_1 \leq 0 \tag{h}$$

Since the second constraint does not have a constant in its expression, it is divided by 100 to get a percent value of it. Evaluating the constraints at the given point $(-4.5, -4.5)$, and checking for ε-active constraints, we obtain

$$g_1 + \varepsilon = \frac{1}{18}(-4.5)^2 + \frac{1}{36}(-4.5) - 1.0 + 0.1 = 0.10 > 0 \quad (\varepsilon\text{-active}) \tag{i}$$

$$g_2 + \varepsilon = \frac{1}{100}[-(-4.5) + 60(-4.5)] + 0.10 = -2.555 < 0 \quad (\text{inactive}) \tag{j}$$

$$g_3 + \varepsilon = \frac{-4.5}{10} - 1.0 + 0.10 = -1.35 < 0 \quad (\text{inactive}) \tag{k}$$

$$g_4 + \varepsilon = -\frac{1}{2}(-4.5) - 1.0 + 0.10 = 1.35 > 0 \quad (\text{violated}) \tag{l}$$

$$g_5 + \varepsilon = \frac{1}{10}(-4.5) - 1.0 + 0.10 = -1.35 < 0 \quad (\text{inactive}) \tag{m}$$

$$g_6 + \varepsilon = -(-4.5) + 0.10 = 4.6 > 0 \quad (\text{violated}) \tag{n}$$

Therefore, we see that g_1 is active (also ε-active); g_4 and g_6 are violated; and g_2, g_3, and g_5 are inactive. The potential constraint index set is thus given as

$$I_k = \{1, 4, 6\} \tag{o}$$

Note that the elements of the index set depend on the value of ε used in Eq. (13.2). Also, the search direction with different index sets can be different, giving a different path to the optimum point.

It is important to note that a numerical algorithm using the potential constraint strategy must be proved to be convergent. The potential set strategy has been incorporated into the CSD algorithm of Chapter 12, which has been proved to be convergent to a local minimum point starting from any point (Pshenichny and Danilin, 1982).

Example 13.2 calculates the search directions with and without the potential set strategy and shows that they are different.

EXAMPLE 13.2 SEARCH DIRECTION WITH AND WITHOUT POTENTIAL CONSTRAINT STRATEGY

Consider the design optimization problem:

Minimize

$$f(\mathbf{x}) = x_1^2 - 3x_1x_2 + 4.5x_2^2 - 10x_1 - 6x_2 \tag{a}$$

subject to

$$x_1 - x_2 \leq 3 \tag{b}$$

$$x_1 + 2x_2 \leq 12 \tag{c}$$

$$x_1, \ x_2 \geq 0 \tag{d}$$

At the point (4, 4), calculate the search directions with and without the potential set strategy. Use $\varepsilon = 0.1$.

Solution

Writing constraints in the standard normalized form, we get

$$g_1 = \frac{1}{3}(x_1 - x_2) - 1 \leq 0 \tag{e}$$

$$g_2 = \frac{1}{12}(x_1 + 2x_2) - 1 \leq 0 \tag{f}$$

$$g_3 = -x_1 \leq 0; \quad g_4 = -x_2 \leq 0 \tag{g}$$

At the point (4, 4), functions and their gradients are calculated as

$$f(4,4) = -24, \quad \mathbf{c} = \nabla f = (2x_1 - 3x_2 - 10, -3x_1 + 9x_2 - 6) = (-14, \ 18) \tag{h}$$

$$g_1(4,4) = -1 < 0 \ (\text{inactive}), \quad \mathbf{a}^{(1)} = \nabla g_1 = \left(\frac{1}{3}, -\frac{1}{3}\right) \tag{i}$$

$$g_2(4,4) = 0 \ (\text{active}), \quad \mathbf{a}^{(2)} = \nabla g_2 = \left(\frac{1}{12}, \frac{1}{6}\right) \tag{j}$$

$$g_3(4,4) = -4 < 0 \ (\text{inactive}), \quad \mathbf{a}^{(3)} = \nabla g_3 = (-1, \ 0) \tag{k}$$

$$g_4(4,4) = -4 < 0 \ (\text{inactive}), \quad \mathbf{a}^{(4)} = \nabla g_4 = (0, \ -1) \tag{l}$$

Note that checking for ε-active constraints ($g_i + \varepsilon \geq 0$) gives g_2 as ε-active and all others as inactive constraints.

When the potential constraint strategy is not used, the QP subproblem of Eqs. (12.27) through (12.29) is defined as

Minimize

$$\bar{f} = -14d_1 + 18d_2 + \frac{1}{2}(d_1^2 + d_2^2) \tag{m}$$

subject to

$$
\begin{bmatrix}
\dfrac{1}{3} & -\dfrac{1}{3} \\[2mm]
\dfrac{1}{12} & \dfrac{1}{6} \\[2mm]
-1 & 0 \\[2mm]
0 & -1
\end{bmatrix}
\begin{bmatrix}
d_1 \\ d_2
\end{bmatrix}
\le
\begin{bmatrix}
1 \\ 0 \\ 4 \\ 4
\end{bmatrix}
\tag{n}
$$

The problem's solution, using the Karush-Kuhn-Tucker (KKT) necessary conditions of Theorem 4.6, is given as

$$
\mathbf{d} = (-0.5, -3.5), \quad \mathbf{u} = (43.5, 0, 0, 0) \tag{o}
$$

If we use the potential constraint strategy, the index set I_k is defined as $I_k = \{2\}$; that is only the second constraint needs to be considered in defining the QP subproblem. With this strategy, the QP subproblem is defined as

Minimize

$$
\bar{f} = -14d_1 + 18d_2 + \frac{1}{2}(d_1^2 + d_2^2) \tag{p}
$$

subject to

$$
\frac{1}{12}d_1 + \frac{1}{6}d_2 \le 0 \tag{q}
$$

The solution to this problem using the KKT necessary conditions is given as

$$
\mathbf{d} = (14, -18), \quad u = 0 \tag{r}
$$

Thus it is seen that the search directions determined by the two subproblems are quite different. The path to the optimum solution and the computational effort will also be different.

13.2 INEXACT STEP SIZE CALCULATION

13.2.1 Basic Concept

In Chapter 12, the constrained steepest-descent (CSD) algorithm was presented. There it was proposed to calculate the step size using golden section search. Although that method is quite good among the interval-reducing methods, it is inefficient for many complex engineering applications. The method can require too many function evaluations, which for many engineering problems require solving a complex analysis problem. Therefore, in most practical implementations of algorithms, an *inexact line search* that has worked fairly well is used to determine an approximate step size. We will describe the procedure and illustrate its use in an example.

The philosophy of inexact line search that we will present is quite similar to Armijo's procedure, which was presented for unconstrained problems in Chapter 11; other procedures and checks can also be incorporated. The cost function was used to determine the approximate step size in Chapter 11; here the descent function Φ_k defined in Eq. (12.31) will be used:

$$\Phi_k = f_k + RV_k \qquad (13.3)$$

where f_k is the cost function value, $R > 0$ is the penalty parameter, and $V_k \geq 0$ is the maximum constraint violation as defined in Eq. (12.36) as follows:

$$V_k = \max\{0; \quad |h_1|, |h_2|, \ldots, |h_p|; \quad g_1, g_2, \ldots, g_m\} \qquad (13.4)$$

The basic idea of the approach is to try different step sizes until the condition of sufficient reduction in the descent function is satisfied. Note that all constraints are included in the calculation of V_k in Eq. (13.4).

To determine an acceptable step size, define a sequence of trial step sizes t_j:

$$t_j = (\mu)^j; \quad j = 0, 1, 2, 3, 4, \ldots \qquad (13.5)$$

Typically $\mu = 0.5$ has been used. In this case, an acceptable step size is one of the numbers in the sequence $\{1, \frac{1}{2}, \frac{1}{4}, \frac{1}{8}, \frac{1}{16}, \ldots\}$ of trial step sizes. Basically, we start with the trial step size as $t_0 = 1$. If a certain descent condition (defined in the following paragraph) is not satisfied, the trial step is taken as half of the previous trial (i.e., $t_1 = \frac{1}{2}$). If the descent condition is still not satisfied, the trial step size is bisected again. The procedure is continued until the descent condition is satisfied. $\mu = 0.6$ has also been used; however, in the following examples we will use $\mu = 0.5$.

13.2.2 Descent Condition

In the following development, we will use *a second subscript or superscript to indicate the values of certain quantities at the trial step sizes*. For example, let t_j be the trial step size at the kth optimization iteration. Then the trial design point for which the descent condition is checked is calculated as

$$x^{(k+1,j)} = x^{(k)} + t_j d^{(k)} \qquad (13.6)$$

where $d^{(k)}$ is the search direction at the current design point $x^{(k)}$ that is calculated by solving a quadratic programming (QP) subproblem as defined in Chapter 12. At the kth iteration, we determine an acceptable step size as $\alpha_k = t_j$, with j as the smallest integer (or the largest number in the sequence $1, \frac{1}{2}, \frac{1}{4}, \ldots$) to satisfy the *descent condition*

$$\Phi_{k+1,j} \leq \Phi_k - t_j \beta_k \qquad (13.7)$$

where $\Phi_{k+1,j}$ is the descent function of Eq. (13.3) evaluated at the trial step size t_j and the corresponding design point $x^{(k+1,j)}$:

$$\Phi_{k+1,j} = \Phi(x^{(k+1,j)}) = f_{k+1,j} + RV_{k+1,j} \qquad (13.8)$$

with $f_{k+1,j} = f(x^{(k+1,j)})$ and $V_{k+1,j} \geq 0$ as the maximum constraint violation at the trial design point calculated using Eq. (13.4). Note that in evaluating $\Phi_{k+1,j}$ and Φ_k in Eq. (13.7), the most recent value of the penalty parameter R is used. The constant β_k in Eq. (13.7) is determined using the search direction $d^{(k)}$:

$$\beta_k = \gamma \| d^{(k)} \|^2 \qquad (13.9)$$

where γ is a specified constant between 0 and 1. We will later study the effect of γ on the step size determination process.

Note that in the kth iteration, β_k defined in Eq. (13.9) is a constant. As a matter of fact t_j is the only variable on the right side of Inequality (13.7). However, when t_j is changed, the design point is changed, affecting the cost and constraint function values. This way the descent function value on the left side of Inequality (13.7) is changed.

Inequality (13.7) is called the *descent condition*. It is an important condition that must be satisfied at each iteration to obtain a convergent algorithm. Basically, the condition of Eq. (13.7) requires that the descent function must be reduced by a certain amount at each iteration of the optimization algorithm.

To understand the meaning of the descent condition (13.7), consider Figure 13.1, where various quantities are plotted as functions of t. For example, the horizontal line A−B represents the constant Φ_k, which is the value of the descent function at the current design point $\mathbf{x}^{(k)}$; line A−C represents the function $(\Phi_k - t\beta_k)$, the right side of Inequality (13.7); and the curve AHGD represents the descent function Φ plotted as a function of the step size parameter t and originating from point A. The line A−C and the curve AHGD intersect at point J which corresponds to the point $t = \bar{t}$ on the t-axis. For the descent condition of Inequality (13.7) to be satisfied, the curve AHGD must be below the line A−C. This gives only the portion AHJ of the curve AHGD.

Thus we see from the figure that a step size larger than $\bar{t}$ does not satisfy the descent condition of Inequality (13.7). To verify this, consider points D and E on the line $t_0 = 1$. Point D represents $\Phi_{k+1,0} = \Phi(\mathbf{x}^{(k+1,0)})$ and point E represents $(\Phi_k - t_0\beta_k)$. Thus point D represents the left side (LS) and point E represents the right side (RS) of Inequality (13.7). Since point D is higher than point E, Inequality (13.7) is violated. Similarly, points G and F on the line $t_1 = \frac{1}{2}$ violate the descent condition. Points I and H on the line $t_2 = \frac{1}{4}$ satisfy the descent condition, so the step size α_k at the kth iteration is taken as $\frac{1}{4}$ for the example of Figure 13.1.

It is important to understand the effect of γ on step size determination. γ is selected as a positive number between 0 and 1. Let us select $\gamma_1 > 0$ and $\gamma_2 > 0$ with $\gamma_2 > \gamma_1$. A larger γ gives a larger value for the constant β_k in Eq. (13.9). Since β_k is the slope of the line $t\beta_k$, we

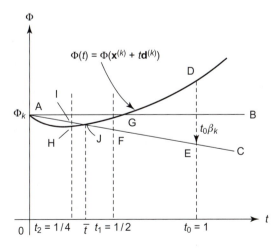

FIGURE 13.1 Geometrical interpretation of the descent condition for determining of step size in the constrained steepest-descent algorithm.

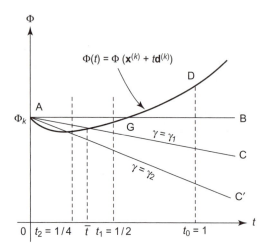

FIGURE 13.2   The effect of parameter γ on step-size determination.

designate line A–C as $\gamma = \gamma_1$ and A–C' as $\gamma = \gamma_2$ in Figure 13.2. We observe from the figure, then, that a larger γ tends to reduce the acceptable range for the step size in order to satisfy the descent condition of Inequality (13.7). Note that with a larger γ the true minimum of the descent function Φ may be outside the acceptable range for the step size.

For the purpose of checking the descent condition in actual calculations, it may be more convenient to write the inequality of Eq. (13.7) as

$$\Phi_{k+1,j} + t_j \beta_k \leq \Phi_k; \quad j = 0, 1, 2 \dots \tag{13.10}$$

We illustrate the procedure for calculating step size in Example 13.3.

EXAMPLE 13.3   CALCULATIONS FOR STEP SIZE IN THE CONSTRAINED STEEPEST-DESCENT METHOD

An engineering design problem is formulated as

Minimize

$$f(\mathbf{x}) = x_1^2 + 320 x_1 x_2 \tag{a}$$

subject to

$$g_1(\mathbf{x}) = \frac{1}{100}(x_1 - 60 x_2) \leq 0 \tag{b}$$

$$g_2(\mathbf{x}) = 1 - \frac{x_1(x_1 - x_2)}{3600} \leq 0 \tag{c}$$

$$g_3(\mathbf{x}) = -x_1 \leq 0, \quad g_4(\mathbf{x}) = -x_2 \leq 0 \tag{d}$$

At a design point $\mathbf{x}^{(0)} = (40, 0.5)$, the search direction is calculated as $\mathbf{d}^{(0)} = (25.6, 0.45)$. The Lagrange multiplier vector for the constraint is calculated as $\mathbf{u} = [16{,}300 \ \ 19{,}400 \ \ 0 \ \ 0]^T$. Choose $\gamma = 0.5$ and calculate the step size for design change using the inexact line search procedure.

Solution

Since the Lagrange multipliers for the constraints are given, the initial value of the penalty parameter is calculated as

$$R = \sum_{i=1}^{4} u_i = 16{,}300 + 19{,}400 = 35{,}700 \tag{e}$$

It is important to note that same value of R is to be used on both sides of the descent condition in Eq. (13.7) or Eq. (13.10). The constant β_0 of Eq. (13.9) is calculated as

$$\beta_0 = 0.5(25.6^2 + 0.45^2) = 328 \tag{f}$$

Calculation of Φ_0 The cost and constraint functions at the starting point $\mathbf{x}^{(0)} = (40, 0.5)$ are calculated as

$$f_0 = f(40, 0.5) = 40^2 + 320(40)(0.5) = 8000 \tag{g}$$

$$g_1(40, 0.5) = \frac{1}{100}(40 - 60 \times 0.5) = 0.10 > 0 \ \ \text{(violation)} \tag{h}$$

$$g_2(40, 0.5) = 1 - \frac{40(40 - 0.5)}{3600} = 0.5611 > 0 \ \ \text{(violation)} \tag{i}$$

$$g_3(40, 0.5) = -40 < 0 \ \ \text{(inactive)}; \ \ g_4(40, 0.5) = -0.5 < 0 \ \ \text{(inactive)} \tag{j}$$

The maximum constraint violation using Eq. (12.36) is given as

$$V_0 = \max\{0; \ \ 0.10, 0.5611, -40, -0.5\} = 0.5611 \tag{k}$$

Using Eq. (13.3), the current descent function is evaluated as

$$\Phi_0 = f_0 + RV_0 = 8000 + (35{,}700)(0.5611) = 28{,}031 \tag{l}$$

Trial step size $t_0 = 1$ Let $j = 0$ in Eq. (13.5), so the trial step size is $t_0 = 1$. The trial design point in the search direction is calculated from Eq. (13.6) as

$$x_1^{(1,0)} = x_1^{(0)} + t_0 d_1^{(0)} = 40 + (1.0)(25.6) = 65.6$$
$$x_2^{(1,0)} = x_2^{(0)} + t_0 d_2^{(0)} = 0.5 + (1.0)(0.45) = 0.95 \tag{m}$$

The cost and constraint functions at the trial design point are calculated as

$$f_{1,0} = f(65.6, 0.95) = (65.6)^2 + 320(65.6)(0.95) = 24{,}246$$

$$g_1(65.6, 0.95) = \frac{1}{100}(65.6 - 60 \times 0.95) = 0.086 > 0 \ \ \text{(violation)} \tag{n}$$

$$g_2(65.6, 0.95) = 1 - \frac{65.6(65.6 - 0.95)}{3600} = -0.1781 < 0 \ \ \text{(inactive)}$$
$$g_3(65.6, 0.95) = -65.6 < 0 \ \ \text{(inactive)} \tag{o}$$
$$g_4(65.6, 0.95) = -0.95 < 0 \ \ \text{(inactive)}$$

The maximum constraint violation using Eq. (12.36) is given as

$$V_{1,0} = \max\{0; \ \ 0.086, -0.1781, -65.6, -0.95\} = 0.086 \tag{p}$$

The descent function at the first trial point is calculated using Eq. (13.8) as

$$\Phi_{1,0} = f_{1,0} + RV_{1,0} = 24{,}246 + 35{,}700(0.086) = 27{,}316 \tag{q}$$

For the descent condition of Eq. (13.10), we get

$$\Phi_{1,0} + t_0\beta_0 = 27{,}316 + 1(328) = 27{,}644 < \Phi_0 = 28{,}031 \tag{r}$$

Therefore, Inequality (13.10) is satisfied and the step size of $t_0 = 1$ is acceptable. If Inequality (13.10) had been violated, the step size of $t_1 = 0.5$ would have been tried and the foregoing steps would have been repeated.

13.2.3 CSD Algorithm with Inexact Step Size

Example 13.4 illustrates calculation of the approximate step size in the CSD algorithm.

EXAMPLE 13.4   USE OF THE CSD ALGORITHM

Consider the problem of Example 12.2:

Minimize

$$f(\mathbf{x}) = x_1^2 + x_2^2 - 3x_1x_2 \tag{a}$$

subject to

$$g_1(\mathbf{x}) = \frac{1}{6}x_1^2 + \frac{1}{6}x_2^2 - 1.0 \le 0 \tag{b}$$

$$g_2(\mathbf{x}) = -x_1 \le 0, \quad g_3(\mathbf{x}) = -x_2 \le 0 \tag{c}$$

Let $\mathbf{x}^{(0)} = (1,1)$ be the starting design. Use $R_0 = 10$, $\gamma = 0.5$, and $\varepsilon_1 = \varepsilon_2 = 0.001$ in the CSD method. Perform only two iterations.

Solution

The functions of the problem are plotted in Figure 12.4. The optimum solution to the problem is obtained as $\mathbf{x} = (\sqrt{3}, \sqrt{3})$, $\mathbf{u} = (3,0,0)$, $f = -3$.

Iteration 1 ($k = 0$)   For the CSD method the following steps are implemented.

Step 1. The initial data are specified as $\mathbf{x}^{(0)} = (1,1)$; $R_0 = 10$; $\gamma = 0.5$ $(0 < \gamma < 1)$; $\varepsilon_1 = \varepsilon_2 = 0.001$.
Step 2. To define the QP subproblem for calculating the search direction, the cost and constraint function values and their gradients must be evaluated at the initial design point $\mathbf{x}^{(0)}$:

$$f(1,1) = -1, \qquad\qquad \nabla f(1,1) = (-1, -1)$$

$$g_1(1,1) = -\frac{2}{3} < 0 \text{ (inactive)} \quad \nabla g_1(1,1) = \left(\frac{1}{3}, \frac{1}{3}\right)$$

$$g_2(1,1) = -1 < 0 \text{ (inactive)} \quad \nabla g_2(1,1) = (-1, 0)$$

$$g_3(1,1) = -1 < 0 \text{ (inactive)} \quad \nabla g_3(1,1) = (0, -1)$$

(d)

Note that all constraints are inactive at the starting point, so $V_0 = 0$ is calculated from Eq. (13.4) as $V_0 = \max\{0;\ -2/3,\ -1,\ -1\}$. The linearized constraints are plotted in Figure 12.5. **Step 3**. Using the preceding values, the QP subproblem of Eqs. (12.27) through (12.29) at (1, 1) is defined as

Minimize

$$\bar{f} = (-d_1 - d_2) + 0.5(d_1^2 + d_2^2) \tag{e}$$

subject to

$$\frac{1}{3}d_1 + \frac{1}{3}d_2 \leq \frac{2}{3}; \quad -d_1 \leq 1; \quad -d_2 \leq 1 \tag{f}$$

Note that the QP subproblem is strictly convex and thus has a unique solution. A numerical method must generally be used to solve the subproblem. However, since the present problem is quite simple, it can be solved by writing the KKT necessary conditions of Theorem 4.6 as follows:

$$L = (-d_1 - d_2) + 0.5(d_1^2 + d_2^2) + u_1 \left[\frac{1}{3}(d_1 + d_2 - 2) + s_1^2 \right] + u_2(-d_1 - 1 - s_2^2) \tag{g}$$

$$+ u_3(-d_2 - 1 + s_3^2)$$

$$\frac{\partial L}{\partial d_1} = -1 + d_1 + \frac{1}{3}u_1 - u_2 = 0$$

$$\frac{\partial L}{\partial d_2} = -1 + d_2 + \frac{1}{3}u_1 - u_3 = 0 \tag{h}$$

$$\frac{1}{3}(d_1 + d_2 - 2) + s_1^2 = 0 \tag{i}$$

$$(-d_1 - 1) + s_2^2 = 0, \quad (-d_2 - 1) + s_3^2 = 0$$

$$u_i s_i = 0; \quad \text{and} \quad s_i^2, u_i \geq 0; \quad i = 1, 2, 3 \tag{j}$$

where u_1, u_2, and u_3 are the Lagrange multipliers for the three constraints and s_1^2, s_2^2, and s_3^2 are the corresponding slack variables. Solving the foregoing KKT conditions, we get the direction vector $\mathbf{d}^{(0)} = (1,1)$, with $\bar{f} = -1$ and $\mathbf{u}^{(0)} = (0,0,0)$. This solution agrees with the graphical solution given in Figure 12.13. The feasible region for the subproblem is the triangle ABC, and the optimum solution is at Point D.

Step 4. Because $||\mathbf{d}^{(0)}|| = \sqrt{2} > \varepsilon_2$, the convergence criterion is not satisfied.

Step 5. Calculate $r_0 = \sum_{i=1}^{m} u_i^{(0)} = 0$, as was defined in Eq. (12.34). To satisfy the necessary condition of Inequality (12.31), let $R = \max\{R_0, r_0\} = \max\{10, 0\} = 10$. It is important to note that $R = 10$ is to be used throughout the first iteration to satisfy the descent condition of Eq. (13.7) or Eq. (13.10). **Step 6.** For step size determination, we use inexact line search, which is described earlier in this section. The current value of the descent function Φ_0 of Eq. (13.3) and the constant β_0 of Eq. (13.9) are calculated as

$$\Phi_0 = f_0 + R V_0 = -1 + (10)(0) = -1 \tag{k}$$

$$\beta_0 = \gamma \|\mathbf{d}^{(0)}\|^2 = 0.5(1 + 1) = 1 \tag{l}$$

Let the trial step size be $t_0 = 1$ and evaluate the new value of the descent function to check the descent condition of Eq. (13.7):

$$\mathbf{x}^{(1,0)} = \mathbf{x}^{(0)} + t_0 \mathbf{d}^{(0)} = (2, 2) \tag{m}$$

At the trial design point, evaluate the cost and constraint functions, and then evaluate the maximum constraint violation to calculate the descent function:

$$f_{1,0} = f(2,2) = -4$$

$$V_{1,0} = V(2,2) = \max\left\{0; \frac{1}{3}, -2, -2\right\} = \frac{1}{3} \tag{n}$$

$$\Phi_{1,0} = f_{1,0} + RV_{1,0} = -4 + (10)\frac{1}{3} = -\frac{2}{3}$$

$$\Phi_0 - t_0\beta_0 = -1 - 1 = -2$$

Since $\Phi_{1,0} > \Phi_0 - t_0\beta_0$, the descent condition of Inequality (13.7) is not satisfied.

Let us try $j = 1$ in Eq. (13.5) (i.e., bisect the step size to $t_1 = 0.5$), and evaluate the new value of the descent function to check the descent condition of Eq. (13.7). The design is updated as

$$\mathbf{x}^{(1,1)} = \mathbf{x}^{(0)} + t_1 \mathbf{d}^{(0)} = (1.5, 1.5) \tag{o}$$

At the new trial design point, evaluate the cost and constraint functions, and then evaluate the maximum constraint violation to calculate the descent function:

$$f_{1,1} = f(1.5,1.5) = -2.25$$

$$V_{1,1} = V(1.5,1.5) = \max\left\{0; -\frac{1}{4}, -1.5, -1.5\right\} = 0 \tag{p}$$

$$\Phi_{1,1} = f_{1,1} + RV_{1,1} = -2.25 + (10)0 = -2.25$$

$$\Phi_0 - t_1\beta_0 = -1 - 0.5 = -1.5$$

Now the descent condition of Inequality (13.7) is satisfied (i.e., $\Phi_{1,1} < \Phi_0 - t_1\beta_0$), and thus $\alpha_0 = 0.5$ is acceptable and $\mathbf{x}^{(1)} = (1.5, 1.5)$.

Step 7. Set $R_{0+1} = R_0 = 10$, $k = 1$ and go to Step 2.

Iteration 2 ($k = 1$) For the second iteration, Steps 3 through 7 of the CSD algorithm are repeated as follows:

Step 3. The QP subproblem of Eqs. (12.27) through (12.29) at $\mathbf{x}^{(1)} = (1.5, 1.5)$ is defined as follows:

Minimize

$$\bar{f} = (-1.5d_1 - 1.5d_2) + 0.5(d_1^2 + d_2^2)$$

subject to

$$0.5d_1 + 0.5d_2 \le 0.25 \quad \text{and} \quad -d_1 \le 1.5, -d_2 \le 1.5 \tag{q}$$

Since all constraints are inactive, the maximum violation $V_1 = 0$ from Eq. (12.36). The new cost function is given as $f_1 = -2.25$. The solution to the preceding quadratic programming subproblem is $\mathbf{d}^{(1)} = (0.25, 0.25)$ and $\mathbf{u}^{(1)} = (2.5, 0, 0)$.

Step 4. Because $||\mathbf{d}^{(1)}|| = 0.3535 > \varepsilon_2$, the convergence criterion is not satisfied.

Step 5. Evaluate $r_1 = \sum_{i=1}^{m} u_i^{(1)} = 2.5$. Therefore, $R = \max\{R_1, r_1\} = \max\{10, 2.5\} = 10$.

Step 6. For line search, try $j = 0$ in Inequality (13.7) (i.e., $t_0 = 1$):

$$\Phi_1 = f_1 + RV_1 = -2.25 + (10)0 = -2.25$$

$$\beta_1 = \gamma \|\mathbf{d}^{(1)}\|^2 = 0.5(0.125) = 0.0625 \tag{r}$$

Let the trial step size be $t_0 = 1$ and evaluate the new value of the descent function to check the descent condition of Eq. (13.7):

$$\mathbf{x}^{(2,0)} = \mathbf{x}^{(1)} + t_0 \mathbf{d}^{(1)} = (1.75, 1.75)$$

$$f_{2,0} = f(1.75, 1.75) = -3.0625 \tag{s}$$

$$V_{2,0} = V(1.75, 1.75) = \max\{0; 0.0208, -1.75, -1.75\} = 0.0208$$

$$\Phi_{2,0} = f_{2,0} + RV_{2,0} = -3.0625 + (10)0.0208 = -2.8541 \tag{t}$$

$$\Phi_1 - t_0 \beta_1 = -2.25 - (1)(0.0625) = -2.3125$$

Because the descent condition of Inequality (13.7) is satisfied, $\alpha_1 = 1.0$ is acceptable and $\mathbf{x}^{(2)} = (1.75, 1.75)$.

Step 7. Set $R_2 = R = 10$, $k = 2$, and go to Step 2.

The maximum constraint violation at the new design $\mathbf{x}^{(2)} = (1.75, 1.75)$ is 0.0208, which is greater than the permissible constraint violation. Therefore, we need to go through more iterations of the CSD algorithm to reach the optimum point and the feasible set. Note, however, that since the optimum point is (1.732, 1.732), the current point is quite close to the solution with $f_2 = -3.0625$. Also, it is observed that the algorithm iterates through the infeasible region for the present problem.

Example 13.5 examines the effect of γ (for use in Eq. (13.9)) on the step size determination in the CSD method.

EXAMPLE 13.5 EFFECT OF γ ON THE PERFORMANCE OF THE CSD ALGORITHM

For the optimum design problem of Example 13.4, study the effect of variations in the parameter γ on the performance of the CSD algorithm.

Solution

In Example 13.4, $\gamma = 0.5$ is used. Let us see what happens if a very small value of γ (say 0.01) is used. All calculations up to Step 6 of Iteration 1 are unchanged. In Step 6, the value of β_0 is changed to $\beta_0 = \gamma \|\mathbf{d}^{(0)}\|^2 = 0.01(2) = 0.02$. Therefore,

$$\Phi_0 - t_0 \beta_0 = -1 - 1(0.02) = -1.02 \tag{a}$$

which is smaller than $\Phi_{1,0}$, so the descent condition of Inequality (13.7) is violated. Thus, the step size in Iteration 1 will be 0.5 as before.

The calculations in Iteration 2 are unchanged until Step 6, where $\beta_1 = \gamma||\mathbf{d}^{(1)}||^2 = 0.01(0.125) = 0.00125$. Therefore,

$$\Phi_1 - t_0\beta_1 = -2.25 - (1)(0.00125) = -2.25125 \tag{b}$$

The descent condition of Inequality (13.7) is satisfied. A smaller value of γ thus has no effect on the first two iterations.

Let us see what happens if a larger value for γ (say 0.9) is chosen. It can be verified that in Iteration 1, there is no difference in the calculations. In Iteration 2, the step size is reduced to 0.5. Therefore, the new design point is $\mathbf{x}^{(2)} = (1.625, 1.625)$. At this point $f_2 = -2.641$, $g_1 = -0.1198$, and $V_1 = 0$, and so, a larger γ results in a smaller step size and the new design point remains strictly feasible.

Example 13.6 examines the effect of the initial value of the penalty parameter R on the step size calculation in the CSD method.

EXAMPLE 13.6   EFFECT OF PENALTY PARAMETER R ON CSD ALGORITHM

For the optimum design problem of Example 13.4, study the effect of variations in the parameter R on the performance of the CSD algorithm.

Solution

In Example 13.4, the initial R is selected as 10. Let us see what happens if R is selected as 1.0. There is no change in the calculations up to Step 5 in Iteration 1. In Step 6,

$$\Phi_{1,0} = -4 + (1)\left(\frac{1}{3}\right) = -\frac{11}{3} \tag{a}$$

$$\Phi_0 - t_0\beta_0 = -1 + (1)(0) = -1$$

Therefore, $\alpha_0 = 1$ satisfies the descent condition of Inequality (13.7) and the new design is given as $\mathbf{x}^{(1)} = (2,2)$. This is different from what was obtained in Example 13.4.

Iteration 2 Since the acceptable step size in Iteration 1 has changed compared with that in Example 13.4, the calculations for Iteration 2 need to be performed again.

Step 3. The QP subproblem of Eqs. (12.27) and (12.29) at $\mathbf{x}^{(1)} = (2,2)$ is defined as follows:

Minimize
$$\bar{f} = (-2d_1 - 2d_2) + 0.5(d_1^2 + d_2^2) \tag{b}$$

subject to
$$\frac{2}{3}d_1 + \frac{2}{3}d_2 \leq -\frac{1}{3}, \quad -d_1 \leq 2, \quad -d_2 \leq 2 \tag{c}$$

At the point $(2, 2)$, $V_1 = 1/3$ and $f_1 = -4$. The solution to the QP subproblem is given as

$$\mathbf{d}^{(1)} = (-0.25, -0.25) \text{ and } \mathbf{u}^{(1)} = \left(\frac{27}{8}, 0, 0\right) \tag{d}$$

Step 4. As $||\mathbf{d}^{(1)}|| = 0.3535 > \varepsilon_2$, the convergence criterion is not satisfied.

Step 5. Evaluate $r_1 = \sum\limits_{i=1}^{3} u_i^{(1)} = 27/8$. Therefore,

$$R = \max\{R_1, r_1\} = \max\left\{1, \frac{27}{8}\right\} = \frac{27}{8} \tag{e}$$

Step 6. For line search, try $j = 0$ in Inequality (13.7), that is, $t_0 = 1$:

$$\Phi_1 = f_1 + RV_1 = -4 + \left(\frac{27}{8}\right)\left(\frac{1}{3}\right) = -2.875$$

$$\Phi_{2,0} = f_{2,0} + RV_{2,0} = -3.0625 + \left(\frac{27}{8}\right)(0.0208) = -2.9923 \tag{f}$$

$$\beta_1 = \gamma\|\mathbf{d}^{(1)}\|^2 = 0.5(0.125) = 0.0625$$

$$\Phi_1 - t_0\beta_1 = -2.875 - (1)(0.0652) = -2.9375$$

As the descent condition of Eq. (13.7) is satisfied, $\alpha_1 = 1.0$ is acceptable and $\mathbf{x}^{(2)} = (1.75, 1.75)$.
Step 7. Set $R_2 = R_1 = 27/8$, $k = 2$, and go to Step 2.

The design at the end of the second iteration is the same as in Example 13.4. This is just a coincidence. We observe that a smaller R gave a larger step size in the first iteration. In general, this can change the history of the iterative process.

Example 13.7 illustrates use of the CSD method for an engineering design problem.

EXAMPLE 13.7   MINIMUM AREA DESIGN OF A RECTANGULAR BEAM

For the minimum area beam design problem of Section 3.8, find the optimum solution using the CSD algorithm starting from the points (50, 200) mm and (1000, 1000) mm.

Solution

The problem was formulated and solved graphically in Section 3.8. After normalizing the constraints, we define the problem as follows: Find width b and depth d to minimize the cross-sectional area subject to various constraints:

$$f(b, d) = bd \tag{a}$$

Bending stress constraint:
$$\frac{(2.40 \times 10^7)}{bd^2} - 1.0 \leq 0 \tag{b}$$

Shear stress constraint:
$$\frac{(1.125 \times 10^5)}{bd} - 1.0 \leq 0 \tag{c}$$

Depth constraint:
$$\frac{1}{100}(d - 2b) \leq 0 \tag{d}$$

Explicit bound constraint: $\quad 10 \leq b \leq 1000, \quad 10 \leq d \leq 1000 \tag{e}$

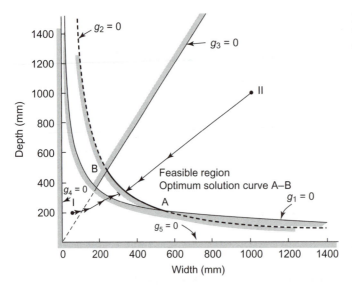

FIGURE 13.3　History of the iterative process for the rectangular beam design problem.

The graphical solution to the problem is given in Figure 13.3; any point on the curve AB gives an optimum solution.

The problem is solved starting from the given points with the CSD algorithm that is available in the IDESIGN software package (Arora and Tseng, 1987a,b). The algorithm has been implemented using the potential constraint strategy. The constraint violation tolerance and the convergence parameter are set as 0.0001. Iteration histories with the two starting points I and II are shown in Figure 13.3. Results of the optimization process are summarized in Table 13.1.

The starting point I is infeasible with a maximum constraint violation of 1100 percent. The program finds the optimum solution in eight iterations. The algorithm iterates through the infeasible region to reach the optimum solution, which agrees with the one obtained analytically in Section 3.8.

The starting point II is feasible and takes six iterations to converge to the optimum. Although the first starting point takes more iterations (eight) to converge to the optimum point compared with

TABLE 13.1　Results of the optimum design process for the rectangular beam design problem

	Starting Point I (50, 200) mm	Starting Point II (1000, 1000) mm
Optimum point	(315.2, 356.9)	(335.4, 335.4)
Optimum area	1.125×10^5	1.125×10^5
Number of iterations to reach optimum	8	6
Number of calls for function evaluations	8	12
Total number of constraint gradients evaluated	14	3
Active constraints at optimum	Shear stress	Shear stress
Lagrange multipliers for constraints	1.125×10^5	1.125×10^5

the second point (six), the number of calls for function evaluations is smaller for the first point. The total numbers of constraint gradient evaluations with the two points are 14 and 3, respectively.

Note that if the potential constraint strategy had not been used, the total number of gradient evaluations would have been 24 and 18, respectively, for the two points. These are substantially higher than the actual number of gradient evaluations with the potential set strategy. It is clear that for large-scale applications, the potential set strategy can have a substantial impact on the efficiency of calculations for an optimization algorithm.

13.3 BOUND-CONSTRAINED OPTIMIZATION

The bound-constrained optimization problem is defined as

Minimize

$$f(\mathbf{x}) \tag{13.11}$$

subject to

$$x_{iL} \leq x_i \leq x_{iU}; \quad i = 1 \text{ to } n \tag{13.12}$$

where x_{iL} and x_{iU} are the lower and upper bounds on the ith variable. Such problems are encountered in some practical applications—for example, optimal trajectory determination. In addition, a constrained optimization problem can be transformed into a sequence of unconstrained problems by using the penalty or the augmented Lagrangian approach, as discussed in Chapter 11. In such cases it is more efficient to treat the bound constraints on the variables explicitly in the numerical algorithm. Therefore, it is useful to develop special algorithms to solve the problem defined in Eqs. (13.11) and (13.12). In this section, we present one such algorithm.

13.3.1 Optimality Conditions

First we study the KKT optimality conditions for the problem defined in Eqs. (13.11) and (13.12). We define the Lagrangian function for the problem as

$$L = f(\mathbf{x}) + \sum_{i=1}^{n} V_i(x_{iL} - x_i) + \sum_{i=1}^{n} U_i(x_i - x_{iU}) \tag{13.13}$$

where $V_i \geq 0$ and $U_i \geq 0$ are the Lagrange multipliers for the lower- and upper-bound constraints, respectively. The optimality conditions give

$$c_i - V_i + U_i = 0; \quad i = 1 \text{ to } n \text{ where } c_i = \frac{\partial f}{\partial x_i} \tag{13.14}$$

$$V_i(x_{iL} - x_i) = 0 \tag{13.15}$$

$$U_i(x_i - x_{iU}) = 0 \tag{13.16}$$

Note that these conditions lead to the following conclusions:

If $x_i = x_{iL}$, then $V_i = c_i \geq 0$ and $U_i = 0$ by Eqs. (13.14) and (13.16) $\qquad$ (13.17)

If $x_i = x_{iU}$, then $U_i = -c_i \geq 0$ and $V_i = 0$ by Eqs. (13.14) and (13.15)   (13.18)

If $x_{iL} < x_i < x_{iU}$, then Eqs. (13.14) to (13.16) give $V_i = 0$ and $U_i = 0 \Rightarrow c_i = 0$ (13.19)

In the numerical algorithm to solve the problem defined in Eqs. (13.11) and (13.12), the set of active constraints at the optimum point is not known and must be determined as a part of the solution. The following steps can be used to determine the set of active constraints and the solution point:

If

$$x_i < x_{iL}, \quad \text{then set } x_i = x_{iL}$$ (13.20)

If

$$x_i > x_{iU}, \quad \text{then set } x_i = x_{iU}$$ (13.21)

If

$x_i = x_{iL}$ and $c_i > 0$ (Eq. 13.17), then keep $x_i = x_{iL}$;   otherwise, release it.   (13.22)

If

$x_i = x_{iU}$ and $c_i < 0$ (Eq. 13.18), then keep $x_i = x_{iU}$;   otherwise, release it.   (13.23)

Once a variable is released, its value can go away from the bound. Thus at any iteration of the numerical algorithm, the sign of the derivative of the cost function with respect to a variable (i.e., the Lagrange multiplier) determines whether that variable is to remain on its bound or not.

13.3.2 Projection Methods

Equations (13.20) and (13.21) constitute a projection of the design variable vector onto the constraint hyperplane. Methods that incorporate these equations are called projection methods (projected conjugate gradient method, projected BFGS method, etc.).

The basic idea of the projection method is to identify variables that are on their bounds at each iteration and keep them fixed at their bound values as long as the conditions in Eq. (13.20) or (13.21)—whichever are applicable—are satisfied. The problem is then reduced to an unconstrained problem in the remaining variables. Any method, such as the conjugate gradient method or a quasi-Newton method, can be used to compute the search direction in terms of the free variables. Then the line search is performed to update the design variables. At the new point the entire process is repeated by identifying the set of active design variables again. This process can identify the final set of active variables quite rapidly.

To state the numerical algorithm (Schwarz and Polak, 1997), let us define

Projection Operator:

$$P_i(z) = \begin{cases} x_{iL}, & \text{if } z \leq x_{iL} \\ z, & \text{if } x_{iL} < z < x_{iU} \\ x_{iU}, & \text{if } z \geq x_{iU} \end{cases}$$ (13.24)

The projection operator basically ensures that all of the design variables remain within or on their bounds:

Feasible Set:

$$S = \{ \mathbf{x} \in R^n | x_{iL} \leq x_i \leq x_{iU}, \quad i = 1 \text{ to } n \} \tag{13.25}$$

Set of Active Variables:

$$A_k = A(\mathbf{x}^{(k)})$$

$$A_k = \left\{ i | x_{iL} \leq x_i^{(k)} \leq x_{iL} + \varepsilon \quad \text{and} \quad c_i^{(k)} > 0 \right.$$
$$\left. \text{or} \quad x_{iU} - \varepsilon \leq x_i^{(k)} \leq x_{iU} \quad \text{and} \quad c_i^{(k)} < 0 \right\} \tag{13.26}$$

where $\varepsilon = 0$ is a small number. The active set A_k is created to satisfy the conditions given in Eqs. (13.17) and (13.18), and contains a list of variables that are on or close to their boundaries (within the ε band). If a design variable is close to its limit, it is included in the active set.

Set of Inactive Variables:

$$I_k = I(\mathbf{x}^{(k)})$$

I_k is the complement of A_k in $\{1, 2, \ldots n\}$ \hspace{2cm} (13.27)

Step-by-Step Algorithm

For the data, select

$$\rho, \beta \in (0, 1), \quad \sigma_1 \in (0, 1), \quad \sigma_2 \in (1, \infty), \quad \varepsilon \in (0, \infty), \quad \mathbf{x}^{(0)} \in S \tag{13.28}$$

Step 0: Set $k = 0$.
Step 1: *Active/Inactive Variables.* Compute $\mathbf{c}^{(k)} = \nabla f(\mathbf{x}^{(k)})$ and define the active and inactive sets of variables using Eqs. (13.26) and (13.27):

$$A_k = A(\mathbf{x}^{(k)}), \quad I_k = I(\mathbf{x}^{(k)}) \tag{13.29}$$

If the following conditions are satisfied, stop; otherwise, continue:

$$\left\| \mathbf{c}^{(k)} \right\|_{I_k} = 0 \quad \text{and} \quad x_i^{(k)} = x_{iL} \quad \text{or} \quad x_{iU} \quad \text{for} \quad i \in A_k; U_i, V_i \geq 0 \quad \text{for} \quad i = 1 \text{ to } n \tag{13.30}$$

Step 2: *Search Direction Definition.* Calculate a search direction $\mathbf{d}^{(k)}$ satisfying the following conditions:

$$d_i^{(k)} = -c_i^{(k)} \quad \text{for all} \quad i \in A_k \tag{13.31}$$

$$(\mathbf{d}^{(k)} \cdot \mathbf{c}^{(k)})_{I_k} \leq -\sigma_1 \| \mathbf{c}^{(k)} \|_{I_k}^2 \tag{13.32}$$

$$\left\| \mathbf{d}^{(k)} \right\|_{I_k} \leq \sigma_2 \left\| \mathbf{c}^{(k)} \right\|_{I_k} \tag{13.33}$$

The search direction for the variables in the active set is the steepest-descent direction as seen in Eq. (13.31). The search direction for inactive variables $(\mathbf{d}^{(k)})_{I_k}$, used in Eq. (13.32), can be calculated using any unconstrained optimization method, such as steepest descent, conjugate gradient or quasi-Newton.

Step 3: *Design Update.* During step size calculation, the design needs to be updated for the trial step size α, which is updated using the projection operator defined in Eq. (13.24) as

$$\mathbf{x}^{(k)}(\alpha, \mathbf{d}^{(k)}) = \mathbf{P}(\mathbf{x}^{(k)} + \alpha\mathbf{d}^{(k)}) \tag{13.34}$$

Then, if α_k is the step size (calculation is discussed later) at $\mathbf{x}^{(k)}$, the design is updated as

$$\mathbf{x}^{(k+1)} = \mathbf{x}^{(k)}(\alpha_k, \mathbf{d}^{(k)}) = \mathbf{P}(\mathbf{x}^{(k)} + \alpha_k\mathbf{d}^{(k)}) \tag{13.35}$$

Step 4. Set $k = k + 1$ and go to Step 1.

The algorithm has been proved to identify the correct active set in a finite number of iterations. In addition, it has been proved to converge to a local minimum point starting from any point (Schwarz and Polak, 1997).

The search direction in Step 2 can be computed by any method. Schwarz and Polak (1997) have implemented steepest-descent, conjugate gradient, and a limited-memory BFGS quasi-Newton methods for search direction calculation. Whereas all of the methods worked well, the BFGS method worked the best, taking the smallest CPU time. The conjugate gradient method also performed comparably well. For the conjugate gradient method, the algorithmic parameters were selected as

$$\rho = \frac{1}{2}, \quad \sigma_1 = 0.2, \quad \sigma_2 = 10, \quad \beta = \frac{3}{5}, \quad \varepsilon = 0.2 \tag{13.36}$$

For the BFGS method, the algorithmic parameters were selected as

$$\rho = \frac{1}{3}, \quad \sigma_1 = 0.0002, \quad \sigma_2 = \sqrt{1000} \times 10^3, \quad \beta = \frac{3}{5}, \quad \varepsilon = 0.2 \tag{13.37}$$

After a satisfactory step size was obtained using the Armijo-like rule, the quadratic interpolation was used to further refine the step size. The limit on parameter m in the step size calculation was set at 20.

13.3.3 Step Size Calculation

The step size for the bound-constrained algorithm can be calculated using any of the procedures discussed earlier. However, we will discuss an Armijo-like procedure (refer to Chapter 11 for Armijo's rule) presented by Schwarz and Polak (1997). We will first give the step size determination criterion and then discuss its implementation.

An acceptable step size is $\alpha_k = \beta^m$, where $0 < \beta < 1$ and m is the smallest integer to satisfy the Armijo-like rule:

$$f(\alpha_k) \leq f(0) + \rho\left[\alpha_k(\mathbf{c}^{(k)} \cdot \mathbf{d}^{(k)})_{I_k} + (\mathbf{c}^{(k)} \cdot \{\mathbf{x}^{(k+1)} - \mathbf{x}^{(k)}\})_{A_k}\right] \tag{13.38}$$

$$f(\alpha_k) = f(\mathbf{x}^{(k)}(\alpha_k, \mathbf{d}^{(k)})); \quad f(0) = f(\mathbf{x}^{(k)}) \tag{13.39}$$

where the search direction $\mathbf{d}^{(k)}$ is calculated as explained in Step 2 of the algorithm. During line search, the design is updated using Eq. (13.34).

Note that if the set A_k is empty (i.e., there is no active or nearly active variable), then the terms $(\mathbf{c}^{(k)} \cdot \{\mathbf{x}^{(k+1)} - \mathbf{x}^{(k)}\})_{A_k}$ on the right side of Eq. (13.38) vanish and the condition reduces to the Armijo's rule given in Eq. (11.15). If the set A_k contains all of the variables that are on their bounds, then $(\mathbf{x}^{(k+1)} - \mathbf{x}^{(k)})_{A_k} = \mathbf{0}$ and the foregoing term again vanishes and the condition of Eq. (13.38) reduces to the Armijo's rule in Eq. (11.15). In other words, we find the step size to minimize $f(\alpha)$ considering only the inactive variables and keeping the active variables fixed at their bounds. The cases that we need to consider in more detail are the ones when a variable is close to its bound and is in the active set A_k.

Variable Close to Its Lower Bound

Let the ith variable be in the active set A_k and close to its lower bound. Then, according to Eq. (13.26), $x_i^{(k)}$ is within the ε-band from the lower bound with $c_i^{(k)} > 0$. Note that if $c_i^{(k)} < 0$, then the variable is not in the set A_k. Since $d_i^{(k)} = -c_i^{(k)}$, the ith variable can only move closer to its lower bound; that is, $x_i^{(k+1)} - x_i^{(k)} < 0$. The term $c_i^{(k)}(x_i^{(k+1)} - x_i^{(k)}) < 0$ on the right side of Eq. (13.38) which satisfies the descent condition, as for this inactive variable. Thus the step size calculation criterion in Eq. (13.38) allows the variable to move closer to its lower bound.

Variable Close to Its Upper Bound

Let the ith variable be in the active set A_k and close to its upper bound. Then, according to Eq. (13.26), $x_i^{(k)}$ is within the ε-band from the upper bound with $c_i^{(k)} < 0$. Note that if $c_i^{(k)} > 0$, then the variable is not in the set A_k. Since $d_i^{(k)} = -c_i^{(k)}$, the ith variable can move only closer to its upper bound; that is $x_i^{(k+1)} - x_i^{(k)} > 0$. The term $c_i^{(k)}(x_i^{(k+1)} - x_i^{(k)}) < 0$ on the right side of Eq. (13.38) which satisfies the descent condition, as for the inactive variables. Thus the step size calculation criterion in Eq. (13.38) allows the variable to move closer to its upper bound.

13.4 SEQUENTIAL QUADRATIC PROGRAMMING: SQP METHODS

Thus far we have used only linear approximation for the cost and constraint functions in defining the search direction determination subproblem for the nonlinear programming (NLP) problem. The rate of convergence of algorithms based on such subproblems can be slow. This rate can be improved if second-order information about the problem functions is incorporated into the solution process. It turns out that the QP subproblem defined in Section 12.5 can be modified slightly to introduce curvature information for the Lagrange function into the quadratic cost function of Eq. (12.27) (Wilson, 1963). Since second-order derivatives of the Lagrange function are quite tedious and difficult to calculate, they are approximated using only the first-order information (Han, 1976, 1977; Powell, 1978a,b).

The basic idea is the same as for the unconstrained quasi-Newton methods described in Section 11.5. Therefore it is important to review that material at this point. There we used gradients of the cost function at two points for generating the approximate Hessian of the cost function. Here we use the gradient of the Lagrange function at the two points to update approximation to the Hessian of the Lagrange function.

These are generally called *sequential quadratic programming methods.* **In the literature, they have also been called** *constrained quasi-Newton, constrained variable metric,* **or** *recursive quadratic* **programming methods.**

Several variations of the SQP methods can be generated. However, we will extend the constrained steepest-descent algorithm to include the Hessian of the Lagrange function in the definition of the QP subproblem. Derivation of the subproblem is given, and the procedure for updating the approximate Hessian is explained. Sequential quadratic programming methods are quite simple and straightforward, but very effective in their numerical performance. The methods are illustrated with examples, and numerical aspects are discussed.

13.4.1 Derivation of the Quadratic Programming Subproblem

There are several ways to derive the quadratic programming (QP) subproblem that must be solved for the search direction at each optimization iteration. An understanding of the detailed derivation of the QP subproblem is not necessary in using the SQP methods. Therefore, the reader who is not interested in the derivation can skip this subsection.

It is customary to derive the QP subproblem by considering only the equality-constrained design optimization problem as

Minimize

$$f(\mathbf{x})$$

subject to

$$h_i(\mathbf{x}) = 0; \quad i = 1 \text{ to } p \tag{13.40}$$

Later on, the inequality constraints will be easily incorporated into the subproblem.

The procedure for derivation of the QP subproblem is to write KKT necessary conditions of Theorem 4.6 for the problem defined in Eq. (13.40), and then solve the resulting nonlinear equations by the Newton-Raphson method. Each iteration of this method can then be interpreted as being equivalent to the a QP subproblem solution. In the following derivations, we assume that all functions are twice continuously differentiable and that the gradients of all constraints are linearly independent.

To write the KKT necessary conditions for the optimization problem defined in Eq. (13.40), we write the Lagrange function of Eq. (4.46) as

$$L(\mathbf{x},\mathbf{v}) = f(\mathbf{x}) + \sum_{i=1}^{p} v_i h_i(\mathbf{x}) = f(\mathbf{x}) + \mathbf{v} \cdot \mathbf{h}(\mathbf{x}) \tag{13.41}$$

where v_i is the Lagrange multiplier for the ith equality constraint $h_i(x) = 0$. Note that there is no restriction on the sign of v_i. The KKT necessary conditions give

$$\nabla L(x,v) = 0, \quad \text{or} \quad \nabla f(x) + \sum_{i=1}^{p} v_i \nabla h_i(x) = 0 \tag{13.42}$$

$$h_i(x) = 0; \quad i = 1 \text{ to } p \tag{13.43}$$

Note that Eq. (13.42) actually represents n equations because the dimension of the design variable vector is n. These equations along with the p equality constraints in Eq. (13.43) give $(n + p)$ equations in $(n + p)$ unknowns (n design variables in x and p Lagrange multipliers in v). These are nonlinear equations, so the Newton-Raphson method can be used to solve them.

Let us write Eqs. (13.42) and (13.43) in a compact notation as

$$F(y) = 0 \tag{13.44}$$

where F and y are identified as

$$F = \begin{bmatrix} \nabla L \\ h \end{bmatrix}_{(n+p)} \quad \text{and} \quad y = \begin{bmatrix} x \\ v \end{bmatrix}_{(n+p)} \tag{13.45}$$

Now, using the iterative procedure of the Newton-Raphson method, we assume that $y^{(k)}$ at the kth iteration is known and a change $\Delta y^{(k)}$ is desired. Using linear Taylor's expansion for Eq. (13.44), $\Delta y^{(k)}$ is given as a solution to the linear system:

$$\nabla F^T(y^{(k)}) \Delta y^{(k)} = -F(y^{(k)}) \tag{13.46}$$

where ∇F is an $(n + p) \times (n + p)$ Jacobian matrix for the nonlinear equations whose ith column is the gradient of the function $F_i(y)$ with respect to the vector y. Substituting definitions of F and y from Eq. (13.45) into Eq. (13.46), we obtain

$$\begin{bmatrix} \nabla^2 L & N \\ N^T & 0 \end{bmatrix}^{(k)} \begin{bmatrix} \Delta x \\ \Delta v \end{bmatrix}^{(k)} = -\begin{bmatrix} \nabla L \\ h \end{bmatrix}^{(k)} \tag{13.47}$$

where the superscript k indicates that the quantities are calculated at the kth iteration, $\nabla^2 L$ is an $n \times n$ Hessian matrix of the Lagrange function, N is an $n \times p$ matrix defined in Eq. (12.24) with an ith column that is the gradient of the equality constraint h_i, $\Delta x^{(k)} = x^{(k+1)} - x^{(k)}$, and $\Delta v^{(k)} = v^{(k+1)} - v^{(k)}$.

Equation (13.47) can be converted to a slightly different form by writing the first row as

$$\nabla^2 L^{(k)} \Delta x^{(k)} + N^{(k)} \Delta v^{(k)} = -\nabla L^{(k)} \tag{13.48}$$

Substituting $\Delta v^{(k)} = v^{(k+1)} - v^{(k)}$ and ∇L from Eq. (13.42) into Eq. (13.48), we obtain

$$\nabla^2 L^{(k)} \Delta x^{(k)} + N^{(k)}(v^{(k+1)} - v^{(k)}) = -\nabla f(x^{(k)}) - N^{(k)} v^{(k)} \tag{13.49}$$

Or the equation is simplified to

$$\nabla^2 L^{(k)} \Delta x^{(k)} + N^{(k)} v^{(k+1)} = -\nabla f(x^{(k)}) \tag{13.50}$$

Combining Eq. (13.50) with the second row of Eq. (13.47), we obtain

$$\begin{bmatrix} \nabla^2 L & \mathbf{N} \\ \mathbf{N}^T & 0 \end{bmatrix}^{(k)} \begin{bmatrix} \Delta \mathbf{x}^{(k)} \\ \mathbf{v}^{(k+1)} \end{bmatrix} = - \begin{bmatrix} \nabla f \\ \mathbf{h} \end{bmatrix}^{(k)} \tag{13.51}$$

Solving Eq. (13.51) gives a change in the design $\Delta \mathbf{x}^{(k)}$ and a new value for the Lagrange multiplier vector $\mathbf{v}^{(k+1)}$. The foregoing Newton-Raphson iterative procedure to solve the KKT necessary conditions is continued until a stopping criterion is satisfied.

It is now shown that Eq. (13.51) is also the solution to a certain QP problem defined at the kth iteration as (note that the superscript k is omitted for simplicity of presentation):

Minimize

$$\nabla f^T \Delta \mathbf{x} + 0.5 \Delta \mathbf{x}^T \nabla^2 L \Delta \mathbf{x} \tag{13.52}$$

subject to the linearized equality constraints

$$h_i + \mathbf{n}^{(i)T} \Delta \mathbf{x} = 0; \quad i = 1 \text{ to } p \tag{13.53}$$

where $\mathbf{n}^{(i)}$ is the gradient of the function h_i. The Lagrange function of Eq. (4.46) for the problem defined in Eqs. (13.52) and (13.53) is written as $\bar{L}$:

$$\bar{L} = \nabla f^T \Delta \mathbf{x} + 0.5 \Delta \mathbf{x}^T \nabla^2 L \Delta \mathbf{x} + \sum_{i=1}^{p} v_i (h_i + \mathbf{n}^{(i)T} \Delta \mathbf{x}) \tag{13.54}$$

The KKT necessary conditions of Theorem 4.6 treating $\Delta \mathbf{x}$ as the unknown variable give

$$\frac{\partial \bar{L}}{\partial (\Delta \mathbf{x})} = 0; \quad \text{or} \quad \nabla f + \nabla^2 L \Delta \mathbf{x} + \mathbf{N} \mathbf{v} = 0 \tag{13.55}$$

$$h_i + \mathbf{n}^{(i)T} \Delta \mathbf{x} = 0; \quad i = 1 \text{ to } p \tag{13.56}$$

It can be seen that if we combine Eqs. (13.55) and (13.56) and write them in a matrix form, we get Eq. (13.51). Thus, the problem of minimizing $f(\mathbf{x})$ subject to $h_i(\mathbf{x}) = 0$; $i = 1$ to p can be solved by iteratively solving the QP subproblem defined in Eqs. (13.52) and (13.53).

Just as in Newton's method for unconstrained problems, the solution $\Delta \mathbf{x}$ is treated as a search direction and step size is determined by minimizing an appropriate descent function to obtain a convergent algorithm. Defining the search direction $\mathbf{d} = \Delta \mathbf{x}$ and *including inequality constraints*, the QP subproblem for the general constrained optimization problem is defined as

Minimize

$$\bar{f} = \mathbf{c}^T \mathbf{d} + 0.5 \mathbf{d}^T \mathbf{H} \mathbf{d} \tag{13.57}$$

subject to the constraints of Eqs. (13.4) and (13.5) as

$$\mathbf{n}^{(i)T} \mathbf{d} = e_i; \quad i = 1 \text{ to } p \tag{13.58}$$

$$\mathbf{a}^{(i)T} \mathbf{d} \leq b_i; \quad i = 1 \text{ to } m \tag{13.59}$$

where the notation defined in Section 12.2 is used, $\mathbf{c}$ is the gradient of the cost function, and $\mathbf{H}$ is the Hessian matrix $\nabla^2 L$ or its approximation.

Usually, the potential constraint strategy is used in reducing the number of inequalities in Eq. (13.59), as discussed in Section 13.1. We will further elaborate on this point later.

13.4.2 Quasi-Newton Hessian Approximation

Just as for the quasi-Newton methods of Section 11.5 for unconstrained problems, we can approximate the Hessian of the Lagrangian function in Eq. (13.57) for the constrained problems. We assume that the approximate Hessian $\mathbf{H}^{(k)}$ at the kth iteration is available and we want to update it to $\mathbf{H}^{(k+1)}$. The BFGS formula shown in Section 11.5 for direct updating of the Hessian can be used. It is important to note that the updated Hessian should be kept positive definite because, with this property, the QP subproblem defined in Eqs. (13.57) through (13.59) remains strictly convex. Thus, a unique search direction is obtained as the solution for the problem.

It turns out that the standard BFGS updating formula can lead to a singular or indefinite Hessian. To overcome this difficulty, Powell (1978a) suggested a modification to the standard BFGS formula. Although the modification is based on intuition, it has worked well in most applications. We will give the modified BFGS formula.

Several intermediate scalars and vectors must be calculated before the final formula can be given. We define these as follows:

Design change vector ($\alpha_k = step\ size$):
$$\mathbf{s}^{(k)} = \alpha_k \mathbf{d}^{(k)} \tag{13.60}$$

Vector:
$$\mathbf{z}^{(k)} = \mathbf{H}^{(k)} \mathbf{s}^{(k)} \tag{13.61}$$

Difference in the gradients of the Lagrange function at two points:
$$\mathbf{y}^{(k)} = \nabla L(\mathbf{x}^{(k+1)}, \mathbf{u}^{(k)}, \mathbf{v}^{(k)}) - \nabla L(\mathbf{x}^{(k)}, \mathbf{u}^{(k)}, \mathbf{v}^{(k)}) \tag{13.62}$$

Scalar:
$$\xi_1 = \mathbf{s}^{(k)} \cdot \mathbf{y}^{(k)} \tag{13.63}$$

Scalar:
$$\xi_2 = \mathbf{s}^{(k)} \cdot \mathbf{z}^{(k)} \tag{13.64}$$

Scalar:
$$\theta = 1 \text{ if } \xi_1 \geq 0.2\ \xi_2; \quad \text{otherwise,} \quad \theta = 0.8\ \xi_2/(\xi_2 - \xi_1) \tag{13.65}$$

Vector:
$$\mathbf{w}^{(k)} = \theta \mathbf{y}^{(k)} + (1 - \theta)\mathbf{z}^{(k)} \tag{13.66}$$

Scalar:
$$\xi_3 = \mathbf{s}^{(k)} \cdot \mathbf{w}^{(k)} \tag{13.67}$$

An n × n correction matrix:
$$\mathbf{D}^{(k)} = (1/\xi_3)\mathbf{w}^{(k)}\mathbf{w}^{(k)^T} \tag{13.68}$$

An n × n correction matrix:
$$\mathbf{E}^{(k)} = (1/\xi_2)\mathbf{z}^{(k)}\mathbf{z}^{(k)^T} \tag{13.69}$$

With the preceding definition of matrices $\mathbf{D}^{(k)}$ and $\mathbf{E}^{(k)}$, the Hessian is updated as

$$\mathbf{H}^{(k+1)} = \mathbf{H}^{(k)} + \mathbf{D}^{(k)} - \mathbf{E}^{(k)} \tag{13.70}$$

It turns out that if the scalar ξ_1 in Eq. (13.63) is negative, the original BFGS formula can lead to an indefinite Hessian. The use of the modified vector $\mathbf{w}^{(k)}$ given in Eq. (13.66) tends to alleviate this difficulty. Because of the usefulness of incorporating a Hessian into an optimization algorithm, several updating procedures have been developed in the literature (Gill et al., 1981; Nocedal and Wright, 2006). For example, the Cholesky factors of the Hessian can be directly updated. In numerical implementations, it is useful to incorporate such procedures because numerical stability can be guaranteed.

13.4.3 SQP Algorithm

The CSD algorithm of Section 12.7 has been extended to include Hessian updating and potential set strategy (Belegundu and Arora, 1984a; Lim and Arora, 1986; Thanedar et al., 1986; Huang and Arora, 1996). The original algorithm did not use the potential set strategy (Han, 1976, 1977; Powell, 1978a,b,c). The new algorithm has been extensively investigated numerically, and several computational enhancements have been incorporated into it to make it robust as well as efficient. In the following, we describe a very basic algorithm as a simple extension of the CSD algorithm and refer to it as the *SQP method*:

Step 1. The same as Step 1 in the CSD algorithm of Section 12.7, except also set the initial estimate or the approximate Hessian as identity(i.e., $\mathbf{H}^{(0)} = \mathbf{I}$).
Step 2. Calculate the cost and constraint functions at $\mathbf{x}^{(k)}$ and calculate the gradients of the cost and constraint functions. Calculate the maximum constraint violation V_k as defined in Eq. (12.36). If $k > 0$, update the Hessian of the Lagrange function using Eqs. (13.60) to (13.70). If $k = 0$, skip updating and go to Step 3.
Step 3. Define the QP subproblem of Eqs. (13.57) through (13.59) and solve it for the search direction $\mathbf{d}^{(k)}$ and the Lagrange multipliers $\mathbf{v}^{(k)}$ and $\mathbf{u}^{(k)}$.
Steps 4–7. Same as for the CSD algorithm of Section 12.7.

Thus we see that the only difference between the two algorithms is in Steps 2 and 3. We demonstrate the use of the SQP algorithm with Example 13.9.

EXAMPLE 13.9　USE OF SQP METHOD

Complete two iterations of the SQP algorithm for Example 13.5:

Minimize

$$f(\mathbf{x}) = x_1^2 + x_2^2 - 3x_1 x_2 \tag{a}$$

subject to

$$g_1(\mathbf{x}) = \frac{1}{6}x_1^2 + \frac{1}{6}x_2^2 - 1.0 \le 0 \tag{b}$$

$$g_2(\mathbf{x}) = -x_1 \le 0, \quad g_3(\mathbf{x}) = -x_2 \le 0. \tag{c}$$

The starting point is $(1, 1)$, $R_0 = 10$, $\gamma = 0.5$, $\varepsilon_1 = \varepsilon_2 = 0.001$.

Solution

The first iteration of the SQP algorithm is the same as in the CSD algorithm. From Example 13.5, the results of the first iteration are

$$\mathbf{d}^{(0)} = (1,1); \quad \alpha = 0.5, \quad \mathbf{x}^{(1)} = (1.5,1.5)$$
$$\mathbf{u}^{(0)} = (0,0,0); \quad R_1 = 10, \quad \mathbf{H}^{(0)} = \mathbf{I}.$$

(d)

Iteration 2 At the point $\mathbf{x}^{(1)} = (1.5,1.5)$, the cost and constraint functions and their gradients are evaluated as

$$f = -6.75; \quad \nabla f = (-1.5, -1.5)$$
$$g_1 = -0.25; \quad \nabla g_1 = (0.5, 0.5)$$
$$g_2 = -1.5; \quad \nabla g_2 = (-1, 0)$$
$$g_3 = -1.5; \quad \nabla g_3 = (0, -1)$$

(e)

To update the Hessian matrix, we define the vectors in Eqs. (13.60) and (13.61) as

$$\mathbf{s}^{(0)} = \alpha_0 \mathbf{d}^{(0)} = (0.5, 0.5), \quad \mathbf{z}^{(0)} = \mathbf{H}^{(0)} \mathbf{s}^{(0)} = (0.5, 0.5)$$

(f)

Since the Lagrange multiplier vector $\mathbf{u}^{(0)} = (0,0,0)$, the gradient of the Lagrangian ∇L is simply the gradient of the cost function ∇f. Therefore, vector $\mathbf{y}^{(0)}$ of Eq. (13.62) is calculated as

$$\mathbf{y}^{(0)} = \nabla f(\mathbf{x}^{(1)}) - \nabla f(\mathbf{x}^{(0)}) = (-0.5, -0.5)$$

(g)

Also, the scalars in Eqs. (13.63) and (13.64) are calculated as

$$\xi_1 = \mathbf{s}^{(0)} \cdot \mathbf{y}^{(0)} = -0.5, \quad \xi_2 = \mathbf{s}^{(0)} \cdot \mathbf{z}^{(0)} = 0.5$$

(h)

Since $\xi_1 < 0.2 \, \xi_2$, θ in Eq. (13.65) is calculated as

$$\theta = 0.8(0.5)/(0.5 + 0.5) = 0.4$$

(i)

The vector $\mathbf{w}^{(0)}$ in Eq. (13.66) is calculated as

$$\mathbf{w}^{(0)} = 0.4 \begin{bmatrix} -0.5 \\ -0.5 \end{bmatrix} + (1 - 0.4) \begin{bmatrix} 0.5 \\ 0.5 \end{bmatrix} = \begin{bmatrix} 0.1 \\ 0.1 \end{bmatrix}$$

(j)

The scalar ξ_3 in Eq. (13.67) is calculated as

$$\xi_3 = (0.5, 0.5) \cdot (0.1, 0.1) = 0.1$$

(k)

The two correction matrices in Eqs. (13.68) and (13.69) are calculated as

$$\mathbf{D}^{(0)} = \begin{bmatrix} 0.1 & 0.1 \\ 0.1 & 0.1 \end{bmatrix}; \quad \mathbf{E}^{(0)} = \begin{bmatrix} 0.5 & 0.5 \\ 0.5 & 0.5 \end{bmatrix}$$

(l)

Finally, from Eq. (13.70), the updated Hessian is given as

$$\mathbf{H}^{(1)} = \begin{bmatrix} 1 & 0 \\ 0 & 1 \end{bmatrix} + \begin{bmatrix} 0.1 & 0.1 \\ 0.1 & 0.1 \end{bmatrix} - \begin{bmatrix} 0.5 & 0.5 \\ 0.5 & 0.5 \end{bmatrix} = \begin{bmatrix} 0.6 & -0.4 \\ -0.4 & 0.6 \end{bmatrix}$$

(m)

Step 3. With the updated Hessian and other data previously calculated, the QP subproblem of Eqs. (13.57) through (13.59) is defined as:

Minimize

$$\bar{f} = -1.5 d_1 - 1.5 d_2 + 0.5(0.6 d_1^2 - 0.8 d_1 d_2 + 0.6 d_2^2)$$

(n)

subject to

$$0.5d_1 + 0.5d_2 \leq 0.25, \quad -d_1 \leq 1.5, \quad -d_2 \leq 1.5 \tag{o}$$

The QP subproblem is strictly convex and thus has a unique solution. Using the KKT conditions, the solution is obtained as

$$\mathbf{d}^{(1)} = (0.25, 0.25), \quad \mathbf{u}^{(1)} = (2.9, 0, 0) \tag{p}$$

This solution is the same as in Example 13.4. Therefore, the rest of the steps have the same calculations. It is seen that in this example, inclusion of the approximate Hessian does not actually change the search direction at the second iteration. In general, it gives different directions and better convergence.

EXAMPLE 13.10 SOLUTION TO SPRING DESIGN PROBLEM USING THE SQP METHOD

Solve the spring design problem (Shigley and Mischke, 2001) formulated in Section 2.9 using the SQP method with the data given there.

Solution

The problem was also solved in Section 6.5 using the Excel Solver. Here we solve the problem using the SQP method available in the IDESIGN program. The problem is restated in the normalized form as: find d, D, and N to

Minimize

$$f = (N + 2)Dd^2 \tag{a}$$

subject to the deflection constraint

$$g_1 = 1.0 - \frac{D^3 N}{(71875d^4)} \leq 0 \tag{b}$$

the shear stress constraint

$$g_2 = \frac{D(4D - d)}{12566d^3(D - d)} + \frac{2.46}{12566d^2} - 1.0 \leq 0 \tag{c}$$

the surge wave frequency constraint

$$g_3 = 1.0 - \frac{140.54d}{D^2 N} \leq 0 \tag{d}$$

the outer diameter constraint

$$g_4 = \frac{D + d}{1.5} - 1.0 \leq 0 \tag{e}$$

The lower and upper bounds on the design variables are selected as follows:

$$0.05 \leq d \leq 0.20 \text{ in}$$
$$0.25 \leq D \leq 1.30 \text{ in} \tag{f}$$
$$2 \leq N \leq 15$$

Note that the constant $\pi^2 \rho / 4$ in the cost function of Eq. (a) has been neglected. This simply scales the cost function value without affecting the final optimum solution. The problem has three design variables and 10 inequality constraints in Eqs. (b) through (f). If we attempt to solve the problem analytically using the KKT conditions of Section 4.6, we will have to consider 2^{10} cases, which is tedious and time-consuming.

The history of the iterative design process with the SQP algorithm is shown in Table 13.2. The table shows iteration number (Iter.), maximum constraint violation (Max. vio.), convergence parameter (Conv. parm.), cost function (Cost), and design variable values at each iteration. It also gives the constraint activity at the optimum point, indicating whether a constraint is active or not, constraint function values, and their Lagrange multipliers. Design variable activity is shown at the optimum point, and the final cost function value and the number of calls to the user-supplied subroutines are given.

The following stopping criteria are used for the present problem:

1. The maximum constraint violation should be less than ε_1, that is, $V \leq \varepsilon_1$ in Step 4 of the algorithm. ε_1 is taken as $1.00E-04$.
2. The length of the direction vector (convergence parameter) should be less than ε_2, that is, $\|\mathbf{d}\| \leq \varepsilon_2$ in Step 4 of the algorithm. ε_2 is taken as $1.00E-03$.

The starting design estimate is (0.2, 1.3, 2.0), where the maximum constraint violation is 96.2 percent and the cost function value is 0.208. At the sixth iteration, a feasible design (the maximum constraint violation is $1.97E-05$) is obtained at a cost function value of ($1.76475E-02$).

Note that in this example, the constraint correction is accompanied by a substantial reduction in the cost function (by a factor of 10). However, the constraint correction will most often result in an increase in cost. The program takes another 12 iterations to reach the optimum design. At the optimum point, the deflection and shear stress constraints of Eqs. (c) and (d) are active. The Lagrange multiplier values are ($1.077E-02$) and ($2.4405E-02$). Design variable 1 (wire diameter) is close to its lower bound.

13.4.4 Observations on SQP Methods

The quasi-Newton methods are considered to be most efficient, reliable, and generally applicable. Schittkowski and coworkers (1980, 1981, 1987) extensively analyzed the methods and evaluated them against several other methods using a set of nonlinear programming test problems. Their conclusion was that the quasi-Newton methods are far superior to others. Lim and Arora (1986), Thanedar et al. (1986), Thanedar, Arora, et al. (1987), and Arora and Tseng (1987b) evaluated the methods for a class of engineering design problems. Gabrielle and Beltracchi (1987) discussed several enhancements of Pshenichny's constrained steepest-descent (CSD) algorithm, including incorporation of quasi-Newton updates of the Hessian of the Lagrangian. In general, these investigations showed the quasi-Newton methods to be superior. Therefore, they are recommended for general engineering design applications.

Numerical implementation of an algorithm is an art. Considerable care, judgment, safeguards, and user-friendly features must be designed and incorporated into the software. Numerical calculations must be robustly implemented. Each step of the algorithm must be analyzed and proper numerical procedures developed in order to implement the intent of the step. The software must be properly evaluated for performance by solving many different problems.

Many aspects of numerical implementation of algorithms are discussed by Gill et al. (1981). The steps of the SQP algorithm have been analyzed (Tseng and Arora, 1988).

TABLE 13.2 History of the iterative optimization process for the spring design problem

Iteration no.	Maximum violation	Convergence parameter	Cost	d	D	N
1	9.61791E−01	1.00000E+00	2.08000E−01	2.0000E−01	1.3000E+00	2.0000E+00
2	2.48814E+00	1.00000E+00	1.30122E−02	5.0000E−02	1.3000E+00	2.0038E+00
3	6.89874E−01	1.00000E+00	1.22613E−02	5.7491E−02	9.2743E−01	2.0000E+00
4	1.60301E−01	1.42246E−01	1.20798E−02	6.2522E−02	7.7256E−01	2.0000E+00
5	1.23963E−02	8.92216E−03	1.72814E−02	6.8435E−02	9.1481E−01	2.0336E+00
6	1.97357E−05	6.47793E−03	1.76475E−02	6.8770E−02	9.2373E−01	2.0396E+00
7	9.25486E−06	3.21448E−02	1.76248E−02	6.8732E−02	9.2208E−01	2.0460E+00
8	2.27139E−04	7.68889E−02	1.75088E−02	6.8542E−02	9.1385E−01	2.0782E−00
9	5.14338E−03	8.80280E−02	1.69469E−02	6.7635E−02	8.7486E−01	2.2346E+00
10	8.79064E−02	8.87076E−02	1.44839E−02	6.3848E−02	7.1706E−01	2.9549E+00
11	9.07017E−02	6.66881E−02	1.31958E−02	6.0328E−02	5.9653E−01	4.0781E+00
12	7.20705E−02	7.90647E−02	1.26517E−02	5.7519E−02	5.1028E−01	5.4942E+00
13	6.74501E−02	6.86892E−02	1.22889E−02	5.4977E−02	4.3814E−01	7.2798E+00
14	2.81792E−02	4.50482E−02	1.24815E−02	5.3497E−02	4.0092E−01	8.8781E+00
15	1.57825E−02	1.94256E−02	1.25465E−02	5.2424E−02	3.7413E−01	1.0202E+01
16	5.85935E−03	4.93063E−03	1.26254E−02	5.1790E−02	3.5896E−01	1.1113E+01
17	1.49687E−04	2.69244E−05	1.26772E−02	5.1698E−02	3.5692E−01	1.1289E+01
18	0.00000E+00	9.76924E−08	1.26787E−02	5.1699E−02	3.5695E−01	1.1289E+01

Constraint activity

Iteration no.	Active	Value	Lagrange multiplier
1	Yes	−4.66382E−09	1.07717E−02
2	Yes	−2.46286E−09	2.44046E−02
3	No	−4.04792E+00	0.00000E+00
4	No	−7.27568E−01	0.00000E+00

Design variable activity

Iteration no.	Active	Design	Lower	Upper	Lagrange multiplier
1	Lower	5.16987E−02	5.00000E−02	2.00000E−01	0.00000E+00
2	Lower	3.56950E−01	2.50000E−01	1.30000E+00	0.00000E+00
3	No	1.12895E+01	2.00000E+00	1.50000E+01	0.00000E+00

Note: *Number of calls for cost function evaluation = 18; number of calls for evaluation of cost function gradients = 18; number of calls for constraint function evaluation = 18; number of calls for evaluation of constraint function gradients = 18; number of total gradient evaluations = 34.*

Various potential constraint strategies have been incorporated and evaluated. Several descent functions have been investigated. Procedures to resolve inconsistencies in the QP subproblem have been developed and evaluated. As a result of these enhancements and evaluations, a very powerful, robust, and general algorithm for engineering design applications has become available.

13.4.5 Descent Functions

Descent functions play an important role in SQP methods, so we will discuss them briefly. Some of the descent functions are nondifferentiable, while others are differentiable. For example, the descent function of Eq. (12.30) is nondifferentiable. Another nondifferentiable descent function has been proposed by Han (1977) and Powell (1978c). We will denote this as Φ_H and define it as follows at the kth iteration:

$$\Phi_H = f(\mathbf{x}^{(k)}) + \sum_{i=1}^{p} r_i^{(k)}|h_i| + \sum_{i=1}^{m} \mu_i^{(k)}\max\{0, g_i\} \tag{13.71}$$

where $r_i^{(k)} \geq |v_i^{(k)}|$ are the penalty parameters for equality constraints and $\mu_i^{(k)} \geq u_i^{(k)}$ are the penalty parameters for inequality constraints. Because the penalty parameters sometimes become very large, Powell (1978c) suggested a procedure to adjust them as follows:

First iteration:

$$r_i^{(0)} = \left|v_i^{(0)}\right|; \quad \mu_i^{(0)} = u_i^{(0)} \tag{13.72}$$

Subsequent iterations:

$$r_i^{(k)} = \max\left\{|v_i^{(k)}|, \frac{1}{2}(r_i^{(k-1)} + |v_i^{(k)}|)\right\}$$
$$\mu_i^{(k)} = \max\left\{u_i^{(k)}, \frac{1}{2}(\mu_i^{(k-1)} + u_i^{(k)})\right\} \tag{13.73}$$

Schittkowski (1981) has suggested using the following augmented Lagrangian function Φ_A as the descent function:

$$\Phi_A = f(\mathbf{x}) + P_1(\mathbf{v},\mathbf{h}) + P_2(\mathbf{u},\mathbf{g}) \tag{13.74}$$

$$P_1(\mathbf{v},\mathbf{h}) = \sum_{i=1}^{p}\left(v_i h_i + \frac{1}{2}r_i h_i^2\right) \tag{13.75}$$

$$P_2(\mathbf{u},\mathbf{g}) = \sum_{i=1}^{m}\begin{cases}\left(u_i g_i + \frac{1}{2}\mu_i g_i^2\right), & \text{if } (g_i + u_i/\mu_i) \geq 0 \\[2mm] \frac{1}{2}u_i^2/\mu_i, & \text{otherwise}\end{cases} \tag{13.76}$$

where the penalty parameters r_i and μ_i were defined previously in Eqs. (13.72) and (13.73). One good feature of Φ_A is that the function and its gradient are continuous.

13.5 OTHER NUMERICAL OPTIMIZATION METHODS

Many other methods and their variations for constrained optimization have been developed and evaluated in the literature. For more details about this, Gill et al. (1981), Luenberger (1984), and Ravindran et al. (2006) should be consulted. This section briefly discusses the basic ideas of three methods—feasible directions, gradient projection, and generalized reduced gradient—that have been used quite successfully for engineering design problems.

13.5.1 Method of Feasible Directions

The method of feasible directions is one of the earliest for solving constrained optimization problems. The *basic idea of the method is to move from one feasible point to an improved feasible point in the design space*. Thus, given a feasible design $\mathbf{x}^{(k)}$, an "improving feasible direction" $\mathbf{d}^{(k)}$ is determined such that for a sufficiently small step size $\alpha > 0$, the following two properties are satisfied:

1. The new design, $\mathbf{x}^{(k+1)} = \mathbf{x}^{(k)} + \alpha \mathbf{d}^{(k)}$ is feasible.
2. The new cost function is smaller than the current one (i.e., $f(\mathbf{x}^{(k+1)}) < f(\mathbf{x}^{(k)})$).

Once $\mathbf{d}^{(k)}$ is determined, a line search is performed to determine how far to proceed along $\mathbf{d}^{(k)}$. This leads to a new feasible design $\mathbf{x}^{(k+1)}$, and the process is repeated from there.

The method is based on the general algorithm that was described in Section 12.1.1, where the design change determination is decomposed into search direction and step size determination subproblems. The direction is determined by defining a linearized subproblem at the current feasible point, and the step size is determined to reduce the cost function as well as maintain feasibility of design. Since linear approximations are used, it is difficult to maintain feasibility with respect to the equality constraints.

Therefore, the method has been developed for and applied mostly to inequality-constrained problems. Some procedures have been developed to treat equality constraints in these methods. However, we will describe the method for problems with only inequality constraints.

Now we define a subproblem that yields an improving feasible direction at the current design point. An *improving feasible direction* is defined as the one that reduces the cost function while remaining strictly feasible for a small step size. Thus, it is a direction of descent for the cost function and it points toward the inside of the feasible region. The improving feasible direction $\mathbf{d}$ satisfies the conditions

$$\mathbf{c}^T \mathbf{d} < 0 \quad \text{and} \quad \mathbf{a}^{(i)T} \mathbf{d} < 0 \quad \text{for} \quad i \in I_k \tag{13.77}$$

where I_k is a potential constraint set at the current point as defined in Eq. (13.2). Such a direction is obtained by solving the following min-max optimization problem:

Minimize
$$\left\{ \text{maximum} \left(\mathbf{c}^T \mathbf{d}; \quad \mathbf{a}^{(i)T} \mathbf{d} \quad \text{for} \quad i \in I_k \right) \right\} \tag{13.78}$$

To solve this problem, we transform it into a minimization problem only. Denoting the maximum of the terms within the regular brackets by β, the *direction-finding subproblem* is transformed as

Minimize
$$\beta \tag{13.79}$$

subject to
$$\mathbf{c}^T\mathbf{d} \le \beta \tag{13.80}$$

$$\mathbf{a}^{(i)T}\mathbf{d} \le \beta \quad \text{for} \quad i \in I_k \tag{13.81}$$

$$-1 \le d_j \le 1; \quad j = 1 \text{ to } n \tag{13.82}$$

The normalization constraints of Eq. (13.82) have been introduced to obtain a bounded solution. Other forms of normalization constraints can also be used.

Let $(\beta, \mathbf{d})$ be an optimum solution for the problem defined in Eqs. (13.79) through (13.82). Note that this problem is one of linear programming. Therefore, any method to solve it can be used. If, at the solution to this problem, $\beta < 0$, then $\mathbf{d}$ is an improving feasible direction. If $\beta = 0$, then the current design point satisfies the KKT necessary conditions and the optimization process is terminated. Otherwise, an improving feasible direction at the current design iteration is obtained. To compute the improved design in this direction, a step size is needed. Any of the several step size determination methods can be used here.

To determine a better feasible direction $\mathbf{d}^{(k)}$, the constraints of Eq. (13.81) can be expressed as

$$\mathbf{a}^{(i)T}\mathbf{d} \le \theta_i\beta \tag{13.83}$$

where $\theta_i > 0$ are called the "push-off" factors. The greater the value of θ_i, the more the direction vector $\mathbf{d}$ is pushed into the feasible region. The reason for introducing θ_i is to prevent the iterations from repeatedly hitting the constraint boundary and slowing down the convergence.

Figure 13.4 shows the physical significance of θ_i in the direction-finding subproblem. It depicts a two-variable design space with one active constraint. If θ_i is taken as zero, then the right side of Eq. (13.81) ($\theta_i\beta$) becomes zero. The direction $\mathbf{d}$ in this case tends to follow the active constraint; that is, it is tangent to the constraint surface. On the other hand, if θ_i is very large, the direction $\mathbf{d}$ tends to follow the cost function contour. Thus, a small value of θ_i will result in a direction that rapidly reduces the cost function. It may, however, rapidly encounter the same constraint surface due to nonlinearities. Larger values of θ_i will reduce the risk of re-encountering the same constraint, but will not reduce the cost function as fast. A value of $\theta_i = 1$ yields acceptable results for most problems.

The *disadvantages* of the method are these:

1. A feasible starting point is needed—special procedures must be used to obtain such a point if it is not known.
2. Equality constraints are difficult to impose and require special procedures for their implementation.

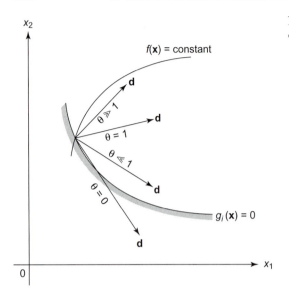

13.5.2 Gradient Projection Method

The gradient projection method was developed by Rosen in 1961. Just as in the feasible directions method, this one uses first-order information about the problem at the current point. The feasible directions method requires solving an LP problem at each iteration to find the search direction. In some applications, this can be an expensive calculation. Thus, Rosen was motivated to develop a method that does not require solving an LP problem. His idea was to develop a procedure in which the direction vector could be calculated easily, although it may not be as good as the one obtained from the feasible directions approach. Thus, he derived an explicit expression for the search direction.

In this method, if the initial point is inside the feasible set, the steepest-descent direction for the cost function is used until a constraint boundary is encountered. If the starting point is infeasible, then the constraint correction step is used to reach the feasible set. When the point is on the boundary, a direction that is tangent to the constraint surface is calculated and used to change the design. This direction is computed by projecting the steepest-descent direction for the cost function onto the tangent hyperplane. This was termed the constrained steepest-descent (CSD) direction in Section 12.7. A step is executed in the negative projected gradient direction. Since the direction is tangent to the constraint surface, the new point will be infeasible. Therefore, a series of correction steps needs to be executed to reach the feasible set.

The iterative process of the gradient projection method is illustrated in Figure 13.5. At the point $\mathbf{x}^{(k)}$, $-\mathbf{c}^{(k)}$ is the steepest-descent direction and $\mathbf{d}^{(k)}$ is the negative projected gradient (constrained steepest-descent) direction. An arbitrary step takes the point $\mathbf{x}^{(k)}$ to $\mathbf{x}^{(k,1)}$, from which point constraint correction steps are executed to reach the feasible point $\mathbf{x}^{(k+1)}$. Comparing the gradient projection method and the CSD method of Section 12.7, we observe that at a feasible point where some constraints are active, the two methods have identical

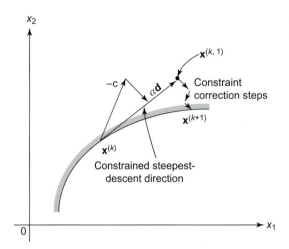

FIGURE 13.5 Graphic of the steps in the gradient projection method.

directions. The only difference is in step size determination. Therefore, the CSD method is preferred over the gradient projection method because it has been proved to converge to a local minimum point starting from an arbitrary point.

Philosophically, the idea of the gradient projection method is quite good, that is, the search direction is easily computable, although it may not be as good as the feasible direction. However, numerically the method has considerable uncertainty. The step size specification is arbitrary; the constraint correction process is quite tedious. A serious drawback is that convergence of the algorithm is tedious to enforce. For example, during the constraint correction steps, it must be ensured that $f(\mathbf{x}^{(k+1)}) < f(\mathbf{x}^{(k)})$. If this condition cannot be satisfied or constraints cannot be corrected, then the step size must be reduced and the entire process must be repeated from the previously updated point. This can be tedious to implement, resulting in additional calculations. Despite these drawbacks, the method has been applied quite successfully to some engineering design problems (Haug and Arora, 1979). In addition, many variations of the method have been investigated in the literature (Gill et al., 1981; Luenberger, 1984; Belegundu and Arora, 1985).

13.5.3 Generalized Reduced Gradient Method

In 1967, Wolfe developed the reduced gradient method based on a simple variable elimination technique for equality-constrained problems (Abadie, 1970). The generalized reduced gradient (GRG) method is an extension of the reduced gradient method to accommodate nonlinear inequality constraints. In this method, a search direction is found such that for any small move, the current active constraints remain precisely active. If some active constraints are not precisely satisfied because of nonlinearity of the constraint functions, the Newton-Raphson method is used to return to the constraint boundary. Thus, the GRG method can be considered somewhat similar to the gradient projection method.

Since inequality constraints can always be converted to equalities by adding slack variables, we can form an equality-constrained NLP model. Also, we can employ the potential constraint strategy and treat all of the constraints in the subproblem as equalities. The direction-finding subproblem in the GRG method can be defined in the following way (Abadie and Carpenter, 1969): Let us partition the design variable vector $\mathbf{x}$ as $[\mathbf{y}^T, \mathbf{z}^T]^T$, where $\mathbf{y}_{(n-p)}$ and $\mathbf{z}_{(p)}$ are vectors of independent and dependent design variables, respectively. First-order changes in the cost and constraint functions (treated as equalities) are given as

$$\Delta f = \frac{\partial f^T}{\partial \mathbf{y}} \Delta \mathbf{y} + \frac{\partial f^T}{\partial \mathbf{z}} \Delta \mathbf{z} \tag{13.84}$$

$$\Delta h_i = \frac{\partial h_i^T}{\partial \mathbf{y}} \Delta \mathbf{y} + \frac{\partial h_i^T}{\partial \mathbf{z}} \Delta \mathbf{z} \tag{13.85}$$

Since we started with a feasible design, any change in the variables must keep the current equalities satisfied at least to first order (i.e., $\Delta h_i = 0$). Therefore, using Eq. (13.85), this requirement is written in matrix form as

$$\mathbf{A}^T \Delta \mathbf{y} + \mathbf{B}^T \Delta \mathbf{z} = 0, \quad \text{or} \quad \Delta \mathbf{z} = -(\mathbf{B}^{-T} \mathbf{A}^T) \Delta \mathbf{y} \tag{13.86}$$

where columns of matrices $\mathbf{A}_{((n-p) \times p)}$ and $\mathbf{B}_{(p \times p)}$ contain gradients of equality constraints with respect to $\mathbf{y}$ and $\mathbf{z}$, respectively. Equation (13.86) can be viewed as the one that determines $\Delta \mathbf{z}$ (change in dependent variable) when $\Delta \mathbf{y}$ (change in independent variable) is specified. Substituting $\Delta \mathbf{z}$ from Eq. (13.86) into Eq. (13.84), we can calculate Δf and identify $df/d\mathbf{y}$ as

$$\Delta f = \left(\frac{\partial f^T}{\partial \mathbf{y}} - \frac{\partial f^T}{\partial \mathbf{z}} \mathbf{B}^{-T} \mathbf{A}^T \right) \Delta \mathbf{y}; \quad \frac{df}{d\mathbf{y}} = \frac{\partial f}{\partial \mathbf{y}} - \mathbf{A} \mathbf{B}^{-1} \frac{\partial f}{\partial \mathbf{z}} \tag{13.87}$$

$df/d\mathbf{y}$ is commonly known as the *reduced gradient*.

In line search, the cost function is treated as the descent function. For a trial value of α, the design variables are updated using $\Delta \mathbf{y} = -\alpha \, df/d\mathbf{y}$, and $\Delta \mathbf{z}$ from Eq. (13.86). If the trial design is not feasible, then independent design variables are considered to be fixed and dependent variables are changed iteratively by applying the Newton-Raphson method (Eq. (13.86) until we get a feasible design point. If the new feasible design satisfies the descent condition, line search is terminated; otherwise, the previous trial step size is discarded and the procedure is repeated with a reduced step size. It can be observed that when $df/d\mathbf{y} = 0$ in Eq. (13.87), the KKT conditions of optimality are satisfied for the original NLP problem.

The main computational burden associated with the GRG algorithm arises from the Newton-Raphson iterations during line search. Strictly speaking, the gradients of the constraints need to be recalculated and the Jacobian matrix $\mathbf{B}$ needs to be inverted at every iteration during line search. This can be expensive. Toward this end, many efficient numerical schemes have been suggested—for example, the use of a quasi-Newton formula to update $\mathbf{B}^{-1}$ without recomputing the gradients but requiring only constraint function values. This can cause problems if the set of independent variables changes during iterations. Another difficulty is to select a feasible starting point. Different procedures must be used to handle arbitrary starting points, as in the feasible directions method.

In the literature, the reduced gradient method and the gradient projection method are considered essentially the same (Sargeant, 1974). There can be some differences between them in implementations depending on how the inequality constraints are treated. It turns out that if a potential constraint strategy is used to treat inequalities, the reduced gradient method becomes essentially the same as the gradient projection method (Belegundu and Arora, 1985). On the other hand, if inequalities are converted to equalities by adding slack variables, it behaves quite differently from the gradient projection method.

The generalized reduced gradient method has been implemented in the Excel Solver program that was described and used in Chapter 6. The program has been used success-fully for numerous applications.

13.6 SOLUTION TO THE QUADRATIC PROGRAMMING SUBPROBLEM

It is seen that the quadratic programming (QP) subproblem needs to be solved for the search direction in many numerical optimization methods for constrained nonlinear opti-mization problems. In this section we discuss a couple of the methods to solve the QP sub-problem. Many methods have been developed for solution to such problems; Nocedal and Wright (2006) should be consulted for more detailed discussion of the algorithms.

The QP subproblem discussed in this section is defined as

Minimize

$$q(\mathbf{d}) = \mathbf{c}^T \mathbf{d} + 0.5 \mathbf{d}^T \mathbf{H} \mathbf{d} \tag{13.88}$$

subject to

$$g_j = \sum_{i=1}^{n} n_{ij} d_i - e_j = 0; \quad j = 1 \text{ to } p \quad (\mathbf{N}_E^T \mathbf{d} = \mathbf{e}_E) \tag{13.89}$$

$$g_j = \sum_{i=1}^{n} n_{ij} d_i - e_j \leq 0; \quad j = p + 1 \text{ to } m \quad (\mathbf{N}_I^T \mathbf{d} \leq \mathbf{e}_I) \tag{13.90}$$

where the dimensions of various vectors and matrices are

$$\mathbf{c}_{n \times 1}, \mathbf{d}_{n \times 1}, \mathbf{e}_{Ep \times 1}, \mathbf{e}_{I(p-m) \times 1}, \mathbf{H}_{n \times n}, \mathbf{N}_{En \times p}, \text{ and } \mathbf{N}_{In \times (m-p)}.$$

It is assumed that the columns of matrices $\mathbf{N}_E$ and $\mathbf{N}_I$ are linearly independent. It is also assumed that the Hessian $\mathbf{H}$ of $q(\mathbf{d})$ is a constant and positive definite matrix. Therefore, the QP subproblem is strictly convex, and if a solution exists, it is a global minimum for the objective function. Note that a slightly different notation is used to define the con-straints in this section to define the QP subproblem in order to present the numerical algo-rithms in a more compact form; p is the number of equality constraints and m is the total number of constraints.

It is noted that $\mathbf{H}$ is an approximation of the Hessian of the Lagrangian function for the constrained nonlinear programming problem. Also, an explicit $\mathbf{H}$ may not be available because a limited-memory updating procedure may be used to update $\mathbf{H}$ and to calculate the product of $\mathbf{H}$ with a vector or the product of the inverse of $\mathbf{H}$ with a vector.

There are two basic approaches to solving the QP subproblem: The first one is to write the KKT necessary conditions and solve them for the minimum point; the second one is to use a search method to solve for the minimum point directly. Both approaches are briefly discussed.

13.6.1 Solving the KKT Necessary Conditions

Since the QP subproblem is strictly convex, solving the KKT necessary conditions, if a solution exists, gives a global minimum point for the function. One method for solving such a system was discussed in Section 9.5, where the KKT necessary conditions for the problem were transformed for solution by the Simplex method of linear programming. Here we discuss some other approaches for solving the KKT necessary conditions.

To write the KKT conditions, we define the Lagrangian for the problem in Eqs. (13.88) through (13.90) as

$$L = c^T d + 0.5 d^T H d + v^T (N^T d - e) \tag{13.91}$$

where $v_{m \times 1}$ is a vector of Lagrange multipliers for the constraints, and the matrix $N_{n \times m}$, and the vector $e_{m \times 1}$ are defined as

$$N_{n \times m} = \begin{bmatrix} N_E & N_I \end{bmatrix} \text{ and } e_{m \times 1} = \begin{bmatrix} e_E \\ e_I \end{bmatrix} \tag{13.92}$$

The KKT optimality conditions give

$$c + Hd + Nv = 0 \tag{13.93}$$

$$N_E^T d = e_E; \quad N_I^T d \le e_I \tag{13.94}$$

$$v_j \ge 0; \quad v_j g_j = 0; \quad j = p+1 \text{ to } m \tag{13.95}$$

Solving for d from Eq. (13.93), we get

$$d = -H^{-1}(Nv + c) \tag{13.96}$$

Now, substituting d from Eq. (13.96) into Eqs. (13.94) and assuming for the moment all inequalities to be active, we obtain

$$N^T d = e, \quad \text{or} \quad N^T[-H^{-1}(Nv + c)] = e \tag{13.97}$$

Simplifying this equation, we obtain

$$(N^T H^{-1} N)v = -(e + N^T H^{-1} c) \tag{13.98}$$

This linear system in v is difficult to solve because the constraints $v_j \ge 0$ for $j = p+1$ to m must be imposed. However, we can transcribe the problem to a bound-constrained optimization problem as follows: find v to

Minimize

$$q(v) = 0.5 v^T (N^T H^{-1} N)v + v^T (e + N^T H^{-1} c) \tag{13.99}$$

subject to

$$v_j \ge 0 \quad \text{for} \quad j = p+1 \text{ to } m \tag{13.100}$$

This bound-constrained optimization problem can be solved using the algorithm presented in Section 13.3.

The gradient of the function in Eq. (13.99) is needed in the minimization procedure for the solution to the problem in Eqs. (13.99) and (13.100). It can be calculated quite efficiently using the limited-memory Hessian updating procedure:

$$\nabla q(v) = (N^T H^{-1} N)v + (e + N^T H^{-1} c) = N^T H^{-1}(Nv + c) + e \tag{13.101}$$

At the kth iteration of the algorithm to solve the bound-constrained optimization problem, the current value of vector $\mathbf{v}^{(k)}$ is known, and $\mathbf{d}^{(k)}$ can be calculated from Eq. (13.96) as

$$\mathbf{d}^{(k)} = -\mathbf{H}^{-1}(\mathbf{Nv}^{(k)} + \mathbf{c}) \tag{13.102}$$

Therefore, substituting $\mathbf{d}^{(k)}$ into Eq. (13.101), we obtain the gradient of $q(\mathbf{v})$ as

$$\nabla q(\mathbf{v}^{(k)}) = -\mathbf{N}^T \mathbf{d}^{(k)} + \mathbf{e} \tag{13.103}$$

If an inverse Hessian is available, then Eq. (13.102) can be used directly to calculate the gradient of the function. However, using the limited-memory updating procedure, $\mathbf{H}^{-1}(\mathbf{Nv}^{(k)} + \mathbf{c})$ can be calculated quite efficiently in Eq. (13.102) (Huang and Arora, 1996; Nocedal, 1980; Luenberger, 1984; Liu and Nocedal, 1989).

A similar procedure can be used to calculate the quadratic cost function of Eq. (13.99) during the line search:

$$q(\mathbf{v}) = \mathbf{v}^T \mathbf{N}^T \mathbf{H}^{-1}(0.5\mathbf{Nv} + \mathbf{c}) + \mathbf{v}^T \mathbf{e} \tag{13.104}$$

Thus the Lagrange multipliers $\mathbf{v}$ can be calculated by minimizing the function in Eq. (13.99) subject to the simple bounds of Eq. (13.100), and the vector $\mathbf{d}^{(k)}$ is readily recovered from Eq. (13.102).

It is important to note that the problem defined in Eqs. (13.99) and (13.100) can also be derived using the duality theory presented in Chapter 5, Section 5.5. To derive the dual problem defined in Eq. (5.38), we need to derive an expression for the dual function first. It is obtained by substituting the expression for $\mathbf{d}$ in Eq. (13.96) into the Lagrange function in Eq. (13.91). The dual function is then maximized subject to the nonnegativity of the Lagrange multipliers for the inequality constraints. This gives the same problem for $\mathbf{v}$ as in Eqs. (13.99) and (13.100), except for an irrelevant constant $0.5\mathbf{c}^T\mathbf{H}^{-1}\mathbf{c}$ in the cost function in Eq. (13.99).

One drawback of the foregoing dual approach is that the simple bound constraints on the design variables must be treated as any other inequality constraint. This may be inefficient unless special care is exercised in numerical implementation of the procedure.

13.6.2 Direct Solution to the QP Subproblem

We can use a search method to directly solve the QP subproblem defined in Eqs. (13.88) through (13.90). For this purpose, we can use the augmented Lagrangian procedure of Section 11.7. The augmented Lagrangian is defined as

$$\Phi = q(\mathbf{d}) + \Phi_E + \Phi_I \tag{13.105}$$

where Φ_E and Φ_I are the terms that are associated with the equality and inequality constraints, given as

$$\Phi_E = \sum_{j=1}^{p}(v_j g_j + 0.5 r g_j^2) \tag{13.106}$$

$$\Phi_I = \sum_{j=p+1}^{m} \begin{cases} v_j g_j + 0.5 r g_j^2, & \text{if } r g_j + v_j \geq 0 \\ -\dfrac{v_j^2}{2r}, & \text{if } r g_j + v_j < 0 \end{cases} \tag{13.107}$$

where $r > 0$ is a penalty parameter. Thus the optimization problem becomes

Minimize
$$\Phi \tag{13.108}$$

subject to
$$d_{iL} \leq d_i \leq d_{iU} \tag{13.109}$$

where d_{iL} and d_{iU} are the lower and upper bounds on d_i. Here the bound-constrained optimization algorithm given in Section 13.3 can be used to solve the problem defined in Eqs. (13.108) and (13.109).

In the iterative solution process, the gradient of the augmented function Φ is needed. Differentiating Φ in Eq. (13.105), we obtain

$$\nabla\Phi = \nabla q(\mathbf{d}) + \nabla\Phi_E + \nabla\Phi_I \tag{13.110}$$

where

$$\nabla q(\mathbf{d}) = \mathbf{Hd} + \mathbf{c} \tag{13.111}$$

$$\nabla\Phi_E = \sum_{j=1}^{p}(v_j\nabla g_j + rg_j\nabla g_j) = \sum_{j=1}^{p}(v_j + rg_j)\nabla g_j \tag{13.112}$$

$$\nabla\Phi_I = \sum_{j=p+1}^{m}\begin{cases}(v_j + rg_j)\nabla g_j, & \text{if } rg_j + v_j \geq 0 \\ 0, & \text{if } rg_j + v_j < 0\end{cases} \tag{13.113}$$

Note that the product $\mathbf{Hd}$ is needed in Eq. (13.111). This can be calculated directly if $\mathbf{H}$ is available. Alternatively, the product $\mathbf{Hd}$ can be calculated quite efficiently using the limited-memory BFGS updating procedure as mentioned earlier.

EXERCISES FOR CHAPTER 13

Section 13.3 Approximate Step Size Determination

For the following problems, complete one iteration of the constrained steepest descent method for the given starting point (let $R_0 = 1$ and $\gamma = 0.5$, use the approximate step size determination procedure).

13.1 Beam design problem formulated in Section 3.8 at the point $(b, d) = (250, 300)$ mm.

13.2 Tubular column design problem formulated in Section 2.7 at the point $(R, t) = (12, 4)$ cm. Let $P = 50$ kN, $E = 210$ GPa, $l = 500$ cm, $\sigma_a = 250$ MPa, and $\rho = 7850$ kg/m^3.

13.3 Wall bracket problem formulated in Section 4.7.1 at the point $(A_1, A_2) = (150, 150)$ cm^2.

13.4 Exercise 2.1 at the point $h = 12$ m, $A = 4000$ m^2.

13.5 Exercise 2.3 at the point $(R, H) = (6, 15)$ cm.

13.6 Exercise 2.4 at the point $R = 2$ cm, $N = 100$.

13.7 Exercise 2.5 at the point $(W, D) = (100, 100)$ m.

13.8 Exercise 2.9 at the point $(r, h) = (6, 16)$ cm.

13.9 Exercise 2.10 at the point $(b, h) = (5, 10)$ m.

13.10 Exercise 2.11 at the point, width $= 5$ m, depth $= 5$ m, and height $= 5$ m.

13.11 Exercise 2.12 at the point $D = 4$ m and $H = 8$ m.

13.12 Exercise 2.13 at the point $w = 10$ m, $d = 10$ m, $h = 4$ m.

13.13 Exercise 2.14 at the point $P_1 = 2$ and $P_2 = 1$.

Section 13.4 Constrained Quasi-Newton Methods

Complete two iterations of the constrained quasi-Newton method and compare the search directions with the ones obtained with the CSD algorithm (note that the first iteration is the same for both methods; let $R_0 = 1$, $\gamma = 0.5$).

13.14 Beam design problem formulated in Section 3.8 at the point $(b, d) = (250, 300)$ mm.

13.15 Tubular column design problem formulated in Section 2.7 at the point $(R, t) = (12, 4)$ cm.
Let $P = 50$ kN, $E = 210$ GPa, $l = 500$ cm, $\sigma_a = 250$ MPa, and $\rho = 7850$ kg/m^3.

13.16 Wall bracket problem formulated in Section 4.7.1 at the point $(A_1, A_2) = (150, 150)$ cm^2.

13.17 Exercise 2.1 at the point $h = 12$ m, $A = 4000$ m^2.

13.18 Exercise 2.3 at the point $(R, H) = (6, 15)$ cm.

13.19 Exercise 2.4 at the point $R = 2$ cm, $N = 100$.

13.20 Exercise 2.5 at the point $(W, D) = (100, 100)$ m.

13.21 Exercise 2.9 at the point $(r, h) = (6, 16)$ cm.

13.22 Exercise 2.10 at the point $(b, h) = (5, 10)$ m.

13.23 Exercise 2.11 at the point, width $= 5$ m, depth $= 5$ m, and height $= 5$ m.

13.24 Exercise 2.12 at the point $D = 4$ m and $H = 8$ m.

13.25 Exercise 2.13 at the point $w = 10$ m, $d = 10$ m, $h = 4$ m.

13.26 Exercise 2.14 at the point $P_1 = 2$ and $P_2 = 1$.

Formulate and solve the following problems using Excel Solver or other software.

*13.27 Exercise 3.34	*13.28 Exercise 3.35	*13.29 Exercise 3.36
*13.30 Exercise 3.50	*13.31 Exercise 3.51	*13.32 Exercise 3.52
*13.33 Exercise 3.53	*13.34 Exercise 3.54	

Practical Applications
of Optimization

- Explain what is meant by practical applications that have implicit functions

- Explain how to evaluate derivatives of implicit functions for your problem

- Determine which software components need to be integrated to solve problems with implicit functions

- Formulate practical design optimization problems

- Understand alternative formulations of the practical design optimization problems

Thus far we have considered simpler engineering design problems to describe optimization concepts and computational methods. For such problems, explicit expressions for all the functions of the problem in terms of the design variables could be derived. Whereas some practical problems can be formulated with explicit functions, there are numerous other applications for which explicit dependence of the problem functions on design variables is not known; that is, explicit expressions in terms of the design variables cannot be derived.

In addition, complex systems require large and more sophisticated analysis models. The number of design variables and constraints can be quite large. A check for convexity of the problem is almost impossible. The existence of even feasible designs is not guaranteed, much less the optimum solution. The calculation of problem functions can require large computational effort. In many cases large special-purpose software must be used to compute the problem functions.

Although we will discuss some methods in Chapters 15 and 16 that do not require gradients of functions, computational algorithms for problems with smooth and continuous variables require gradients of cost and constraint functions. When an explicit form of the problem functions in terms of the design variables is not known, gradient evaluation requires special procedures that must be developed and implemented into proper software. Finally, various software components must also be integrated properly to create an optimum design capability for a particular class of design problems.

In this chapter, issues of the optimum design of complex practical engineering systems are addressed. Formulation of the problem, gradient evaluation, and other practical issues, such as algorithm and software selection, are discussed. Some alternative formulations of the problem are also discussed. These formulations do not have any implicit functions and therefore do not require any special treatment. The important problem of interfacing a particular application with design optimization software is discussed, and several engineering design applications are described.

Although most of the applications discussed in this chapter are related to mechanical and structural systems, the issues here are also relevant to other areas. Therefore, the methodologies presented and illustrated can serve as guidelines for other application areas.

14.1 FORMULATION OF PRACTICAL DESIGN OPTIMIZATION PROBLEMS

14.1.1 General Guidelines

The problem formulation of a design task is an important step that must define a realistic model for the engineering system under consideration. The mathematics of optimization methods can easily give rise to situations that are absurd or that violate the laws of physics. Therefore, to transcribe a design task correctly into a mathematical model, the designers must use intuition, skill, and experience. The following points can serve as guiding principles to generate a mathematical model that is faithful to the real-world design task.

1. In an initial formulation of the problem, all of the possible parameters should be viewed as potential design variables. That is, considerable flexibility and freedom must be allowed for analyzing different possibilities. As we gain more knowledge about the problem, redundant design variables can be fixed or eliminated from the model.

2. The *existence of an optimum solution* to a design optimization model depends on its formulation. If the constraints are too restrictive, there may not be any feasible solution to the problem. In such a case, the constraints must be relaxed by allowing larger resource limits for inequality constraints.

3. The problem of optimizing more than one objective function simultaneously (*multi-objective problems*) can be transformed into the standard problem by assigning weighting factors to different objective functions to combine them into a single objective function (Chapter 17). Or the most important criterion can be treated as the cost function and the remaining ones as constraints.

4. The *potential cost functions* for many structural, mechanical, automotive, and aerospace systems are weight, volume, mass, fundamental frequency, stress at a point, performance, and system reliability, among others.

5. It is important to have *continuous and differentiable cost and constraint functions* for derivative-based methods. In certain instances, it may be possible to replace a

nondifferentiable function, such as $|x|$ with a smooth function x^2 without changing the problem definition drastically.

6. In general, *it is desirable to normalize all of the constraints with respect to their limit values*, as was discussed in Section 12.1.3. In numerical computations, this procedure leads to more stable behavior.

7. It is sometimes desirable to establish a feasible design to determine feasibility of the problem formulation.

   How to Determine Feasible Points: **Ignore the real cost function and solve the optimization problem with the cost function as a constant, as described in Chapter 6.**

14.1.2 Example of a Practical Design Optimization Problem

Optimum design formulation of complex engineering systems requires more general tools and procedures than the ones discussed previously. We will demonstrate this by considering a class of problems that has a wide range of applications in automotive, aerospace, mechanical, and structural engineering. This important application area is chosen to demonstrate the procedure of problem formulation and explain the treatment of implicit functions of the problem. Evaluation of functions and their gradients will be explained. Readers unfamiliar with this application area should use the material as a guideline for their area of interest because similar analyses and procedures will need to be used in other practical applications.

The application area that we have chosen to investigate is the optimum design of systems modeled by the finite-element technique. It is common practice to analyze complex structural and mechanical systems using this technique, which is available in many commercial software packages. Displacements, stresses, and strains at various points, vibration frequencies, and buckling loads for the system can be computed and constraints imposed on them. We will describe an optimum design formulation for this application area.

Let $\mathbf{x}$ represent an n-component vector containing design variables for the system. This may contain thicknesses of members, cross-sectional areas, parameters describing the shape of the system, and stiffness and material properties of elements. Once $\mathbf{x}$ is specified, a design of the system is known. To analyze the system (i.e., calculate stresses, strains and frequencies, buckling load, and displacements), the procedure is to first calculate displacements at some key points—called the grid, or nodal, points—of the finite-element model. From these displacements, strains (relative displacement of the material particles), and stresses at various points of the system can be calculated (Cook, 1981; Huebner and Thornton, 1982; Grandin, 1986; Chandrupatla and Belegundu, 1997; Bhatti, 2005).

Let $\mathbf{U}$ be a vector having l components representing generalized displacements at the system's key points. The basic equation that determines the displacement vector $\mathbf{U}$ for a linear elastic system—which is called the equilibrium equation in terms of displacements—is given as

$$\mathbf{K(x)U = F(x)} \tag{14.1}$$

where $\mathbf{K(x)}$ is an $l \times l$ matrix called the stiffness matrix and $\mathbf{F(x)}$ is an effective load vector having l components. The stiffness matrix $\mathbf{K(x)}$ is a property of the structural system that depends explicitly on the design variables, material properties, and geometry of the system.

Systematic procedures have been developed to automatically calculate the matrix with different types of structures at the given design $\mathbf{x}$. The load vector $\mathbf{F(x)}$, in general, can also depend on design variables. We will not discuss procedures to calculate $\mathbf{K(x)}$ because that is beyond the scope of the present text. Our objective is to demonstrate how the design can be optimized once a finite-element model for the problem (meaning Eq. (14.1)) has been developed. We will pursue that objective assuming that the finite-element model for the system has been developed.

It is seen that once the design $\mathbf{x}$ is specified, the displacements $\mathbf{U}$ can be calculated by solving the linear system of Eq. (14.1). Note that a different $\mathbf{x}$ will give, in general, different values for the displacements $\mathbf{U}$. Thus $\mathbf{U}$ is a function of $\mathbf{x}$ (i.e., $\mathbf{U} = \mathbf{U(x)}$). However, an explicit expression for $\mathbf{U(x)}$ in terms of the design variables $\mathbf{x}$ cannot be written. That is, $\mathbf{U}$ is an implicit function of the design variables $\mathbf{x}$. The stress σ_i at the ith point is calculated using the displacements and is an explicit function of $\mathbf{U}$ and $\mathbf{x}$ as $\sigma_i(\mathbf{U}, \mathbf{x})$. However, since $\mathbf{U}$ is an implicit function of $\mathbf{x}$, σ_i becomes an implicit function of the design variables $\mathbf{x}$ as well. The stress and displacement constraints can be written in a functional form as

$$g_i\,(\mathbf{x}, \mathbf{U}) \leq 0 \qquad\qquad (14.2)$$

In many automotive, aerospace, mechanical, and structural engineering applications, the amount of material used must be minimized for efficient and cost-effective systems. Thus, the usual cost function for this class of applications is the weight, mass, or material volume of the system, which is usually an explicit function of the design variables $\mathbf{x}$. Implicit cost functions, such as stress, displacement, vibration frequencies, and so forth, can also be treated by introducing artificial design variables (Haug and Arora, 1979).

In summary, a general formulation for the design problem involving explicit and implicit functions of design variables is defined as: Find an n-dimensional vector $\mathbf{x}$ of design variables to minimize a cost function $f(\mathbf{x})$ satisfying the implicit design constraints of Eq. (14.2), with $\mathbf{U}$ satisfying the system of Eq. (14.1). Note that equality constraints, if present, can be routinely included in the formulation as in the previous chapters. We illustrate the procedure of problem formulation in Example 14.1.

EXAMPLE 14.1   DESIGN OF A TWO-MEMBER FRAME

Consider the design of a two-member frame subjected to out-of-plane loads, as shown in Figure 14.1. Such frames are encountered in numerous automotive, aerospace, mechanical, and structural engineering applications. We want to formulate the problem of minimizing the volume of the frame subject to stress and size limitations (Bartel, 1969).

Solution

Since the optimum structure will be symmetric, the two members of the frame are identical. Also, it has been determined that hollow rectangular sections will be used as members with three design variables defined as

d = width of the member, in

h = height of the member, in
t = wall thickness, in

Thus, the design variable vector is $\mathbf{x} = (d, h, t)$.

The volume for the structure is taken as the cost function, which is an explicit function of the design variables given as

$$f(\mathbf{x}) = 2L(2dt + 2ht - 4t^2) \tag{14.3}$$

To calculate stresses, we need to solve the analysis problem. The members are subjected to both bending and torsional stresses, and the combined stress constraint needs to be imposed at points 1 and 2.

Let σ and τ be the maximum bending and torsional shear stresses in the member, respectively. The failure criterion for the member is based on a combined stress theory, known as the von Mises (or maximum distortion energy) yield condition (Crandall et al., 1978). With this criterion, the effective stress σ_e is given as $\sqrt{\sigma^2 + 3\tau^2}$ and the *stress constraint* is written in a normalized form as

$$\frac{1}{\sigma_a^2}(\sigma^2 + 3\tau^2) - 1.0 \le 0 \tag{14.4}$$

where σ_a is the allowable design stress.

The stresses are calculated from the member-end moments and torques, which are calculated using the finite-element procedure. The three generalized nodal displacements (deflections and rotations) for the finite-element model shown in Figure 14.1 are defined as

U_1 = vertical displacement at node 2
U_2 = rotation about line 3–2
U_3 = rotation about line 1–2

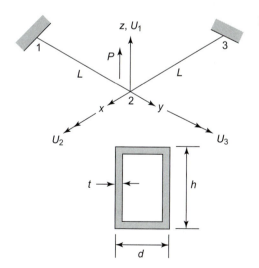

FIGURE 14.1 Graphic of a two-member frame.

Using these, the equilibrium equation (Eq. (14.1)) for the finite-element model that determines the displacements U_1, U_2, and U_3, is given as (for details of the procedure to obtain the equation using individual member equilibrium equations, refer to texts by Cook, 1981; Haug and Arora, 1979; Huebner and Thornton, 1982; Grandin, 1986; Chandrupatla and Belegundu, 1997; Bhatti, 2005):

$$\frac{EI}{L^3}\begin{bmatrix} 24 & -6L & 6L \\ -6L & \left(4L^2 + \frac{GJ}{EI}L^2\right) & 0 \\ 6L & 0 & \left(4L^2 + \frac{GJ}{EI}L^2\right) \end{bmatrix}\begin{bmatrix} U_1 \\ U_2 \\ U_3 \end{bmatrix} = \begin{bmatrix} P \\ 0 \\ 0 \end{bmatrix} \tag{14.5}$$

where

E = modulus of elasticity, 3×10^7 psi

L = member length, 100 in

G = shear modulus, 1.154×10^7 psi

P = load at node 2, $-10{,}000$ lb

$$I = \text{moment of inertia} = \tfrac{1}{12}\left[dh^3 - (d - 2t)(h - 2t)^3\right], \ \ \text{in}^4 \tag{14.6}$$

$$J = \text{polar moment of inertia} = \frac{2t(d - t)^2(h - t)^2}{(d + h - 2t)}, \ \ \text{in}^4 \tag{14.7}$$

$$A = \text{area for calculation of torsional shear stress} = (d - t)(h - t), \ \ \text{in}^2 \tag{14.8}$$

From Eq. (14.5), the stiffness matrix $\mathbf{K}(\mathbf{x})$ and the load vector $\mathbf{F}(\mathbf{x})$ of Eq. (14.1) can be identified. Note that in the present example, the load vector $\mathbf{F}$ does not depend on the design variables.

As can be seen from Eq. (14.5), $\mathbf{U}$ is an implicit function of $\mathbf{x}$. If $\mathbf{K}$ can be inverted explicitly in terms of the design variables $\mathbf{x}$, then $\mathbf{U}$ can be written as an explicit function of $\mathbf{x}$. This is possible in the present example; however, we will deal with the implicit form to illustrate the procedures for evaluating constraints and their gradients.

For a given design, once the displacements U_1, U_2, and U_3 have been calculated from Eq. (14.5), the torque and bending moment at points 1 and 2 for member 1–2 are calculated as

$$T = -\frac{GJ}{L}U_3, \ \ \text{lb-in} \tag{14.9}$$

$$M_1 = \frac{2EI}{L^2}(-3U_1 + U_2 L), \ \ \text{lb-in (moment at end 1)} \tag{14.10}$$

$$M_2 = \frac{2EI}{L^2}(-3U_1 + 2U_2 L), \ \ \text{lb-in (moment at end 2)} \tag{14.11}$$

Using these moments, the torsional shear and bending stresses are calculated as

$$\tau = \frac{T}{2At}, \ \ \text{psi} \tag{14.12}$$

$$\sigma_1 = \frac{1}{2I}M_1 h, \ \ \text{psi (bending stress at end 1)} \tag{14.13}$$

$$\sigma_2 = \frac{1}{2I}M_2 h, \ \ \text{psi (bending stress at end 2)} \tag{14.14}$$

Thus the stress constraints of Eq. (14.4) at points 1 and 2 are given as

$$g_1(\mathbf{x}, \mathbf{U}) = \frac{1}{\sigma_a^2}(\sigma_1^2 + 3\tau^2) - 1.0 \leq 0 \tag{14.15}$$

$$g_2(\mathbf{x}, \mathbf{U}) = \frac{1}{\sigma_a^2}(\sigma_2^2 + 3\tau^2) - 1.0 \leq 0 \tag{14.16}$$

We observe that since moments T, M_1, and M_2 are implicit functions of the design variables, the stresses are also implicit functions. They are also explicit functions of the design variables, as seen in Eqs. (14.13) and (14.14). Therefore, the stress constraints of Eqs. (14.15) and (14.16) are implicit as well as explicit functions of the design variables. This observation is important because gradient evaluation for implicit constraint functions requires special procedures, which are explained in the next section.

In addition to the two stress constraints, the following upper- and lower-bound constraints on design variables are imposed:

$$2.5 \leq d \leq 10.0$$
$$2.5 \leq h \leq 10.0 \tag{14.17}$$
$$0.1 \leq t \leq 1.0$$

As can easily be observed, the explicit forms of the constraint functions g_1 and g_2 in terms of the design variables d, h, and t are quite difficult to obtain even for this simple problem. We will need an explicit form for the displacements U_1, U_2, and U_3 in Eqs. (14.9) through (14.11) to have an explicit form for the stress τ, σ_1, and σ_2. To have an explicit form for U_1, U_2, and U_3, we will have to explicitly invert the coefficient matrix for the equilibrium equation (14.5). Although this is not impossible for the present example, it is quite impossible to do in general. Thus we observe that the constraints are implicit functions of the design variables.

To illustrate the procedure, we select a design point as (2.5, 2.5, 0.1) and calculate the displacements and stresses. Using the given data, we calculate the following quantities that are needed in further calculations:

$$I = \frac{1}{12}\left[2.5^4 - 2.3^4\right] = 0.9232 \text{ in}^4$$

$$J = \frac{1}{4.8}\left[2(0.1)(2.4)^2(2.4)^2\right] = 1.3824 \text{ in}^4$$

$$A = (2.4)(2.4) = 5.76 \text{ in}^2$$

$$GJ = (1.154 \times 10^7)(1.3824) = (1.5953 \times 10^7) \tag{14.18}$$

$$EI = (3.0 \times 10^7)(0.9232) = (2.7696 \times 10^7)$$

$$4L^2 + \frac{GJ}{EI}L^2 = \left(4 + \frac{1.5953}{2.7696}\right)100^2 = (4.576 \times 10^4)$$

Using the foregoing data, the equilibrium equation (Eq. 14.5), is given as

$$27.696\begin{bmatrix} 24 & -600 & 600 \\ -600 & 45,760 & 0 \\ 600 & 0 & 45,760 \end{bmatrix}\begin{bmatrix} U_1 \\ U_2 \\ U_3 \end{bmatrix} = \begin{bmatrix} -10,000 \\ 0 \\ 0 \end{bmatrix} \tag{14.19}$$

Solving the preceding equation, the three generalized displacements of node 2 are given as

$$U_1 = -43.68190 \text{ in}$$
$$U_2 = -0.57275 \tag{14.20}$$
$$U_3 = 0.57275$$

Using Eqs. (14.9) through (14.11), the torque in the member and the bending moments at points 1 and 2 are

$$T = -\frac{1.5953 \times 10^7}{100}(0.57275) = -(9.1371 \times 10^4) \text{ lb-in}$$

$$M_1 = \frac{2(2.7696 \times 10^7)}{(100)(100)}[-3(-43.68190) - 0.57275(100)]$$
$$= (4.0861 \times 10^5) \text{ lb-in}$$

$$\tag{14.21}$$

$$M_2 = \frac{2(2.7696 \times 10^7)}{(100)(100)}[-3(-43.6819) - 2(0.57275)(100)]$$
$$= (9.1373 \times 10^4) \text{ lb-in}$$

Since $M_1 > M_2$, σ_1 will be larger than σ_2, as observed from Eqs. (14.13) and (14.14). Therefore, only the g_1 constraint of Eq. (14.15) needs to be imposed.

Torsional shear and bending stresses at point 1 are calculated from Eqs. (14.12) and (14.13) as

$$\tau = \frac{-(9.13731 \times 10^4)}{2(5.76)(0.1)} = -(7.9317 \times 10^4) \text{ psi}$$

$$\sigma_1 = \frac{(4.08631 \times 10^5)(2.5)}{2(0.9232)} = (5.53281 \times 10^5) \text{ psi} \tag{14.22}$$

Taking the allowable stress σ_a as 40,000 psi, the effective stress constraint of Eq. (14.15) is given as

$$g_1 = \frac{1}{(4.0\text{E} + 04)^2}[(5.53281 \times 10^5)^2 + 3(-7.9317 \times 10^4)^2] - 1 = 202.12 > 0 \tag{14.23}$$

Therefore, the constraint is very severely violated at the given design.

14.2 GRADIENT EVALUATION OF IMPLICIT FUNCTIONS

To use a derivative-based optimization method, we need to evaluate gradients of constraint functions with respect to the design variables. When the constraint functions are implicit in the design variables, we need to develop and utilize special procedures for gradient evaluation. We will develop a procedure using the finite-element application of Section 14.1.

Let us consider the constraint function $g_i(\mathbf{x}, \mathbf{U})$ of Eq. (14.2). Using the chain rule of differentiation, the total derivative of g_i with respect to the jth design variable is given as

$$\frac{dg_i}{dx_j} = \frac{\partial g_i}{\partial x_j} + \frac{\partial g_i^T}{\partial \mathbf{U}}\frac{d\mathbf{U}}{dx_j} \tag{14.24}$$

where

$$\frac{\partial g_i}{\partial \mathbf{U}} = \left[\frac{\partial g_i}{\partial U_1} \frac{\partial g_i}{\partial U_2} \cdots \frac{\partial g_i}{\partial U_l}\right]^T \tag{14.25}$$

and

$$\frac{d\mathbf{U}}{dx_j} = \left[\frac{\partial U_1}{\partial x_j} \frac{\partial U_2}{\partial x_j} \cdots \frac{\partial U_l}{\partial x_j}\right]^T \tag{14.26}$$

Therefore, to calculate the gradient of a constraint, we need to calculate the partial derivatives $\partial g_i/\partial x_j$ and $\partial g_i/\partial \mathbf{U}$, and the total derivatives $d\mathbf{U}/dx_j$. The partial derivatives $\partial g_i/\partial x_j$ and $\partial g_i/\partial \mathbf{U}$ are quite easy to calculate using the form of the function $g_i(\mathbf{x}, \mathbf{U})$. To calculate $d\mathbf{U}/dx_j$, we differentiate the equilibrium Eq. (14.1) to obtain

$$\frac{\partial \mathbf{K(x)}}{\partial x_j}\mathbf{U} + \mathbf{K(x)}\frac{d\mathbf{U}}{dx_j} = \frac{\partial \mathbf{F}}{\partial x_j} \tag{14.27}$$

This equation can be rearranged as

$$\mathbf{K(x)}\frac{d\mathbf{U}}{dx_j} = \frac{\partial \mathbf{F}}{\partial x_j} - \frac{\partial \mathbf{K(x)}}{\partial x_j}\mathbf{U} \tag{14.28}$$

The equation can be used to calculate $d\mathbf{U}/dx_j$. The derivative of the stiffness matrix $\partial \mathbf{K(x)}/\partial x_j$ can be easily calculated if the explicit dependence of $\mathbf{K}$ on $\mathbf{x}$ is known. Note that Eq. (14.28) needs to be solved for each design variable. Once $d\mathbf{U}/dx_j$ are known, the gradient of the constraint is calculated from Eq. (14.24). The derivative vector in Eq. (14.24) is often called the design gradient. We will illustrate the procedure with an example problem.

It should be noted that substantial work has been done in developing and implementing efficient procedures for calculating derivatives of implicit functions with respect to the design variables (Arora and Haug, 1979; Adelman and Haftka, 1986; Arora, 1995). The subject is generally known as design sensitivity analysis. For efficiency considerations and proper numerical implementations, the foregoing literature should be consulted. The procedures have been programmed into general-purpose software for automatic computation of design gradients.

EXAMPLE 14.2 GRADIENT EVALUATION FOR A TWO-MEMBER FRAME

Calculate the gradient of the stress constraint $g_1(\mathbf{x}, \mathbf{U})$ for the two-member frame of Example 14.1 at the design point (2.5, 2.5, 0.1).

Solution

The problem was formulated in Example 14.1. The finite-element model was defined there, and nodal displacements and member stresses were calculated. We will use Eqs. (14.24) and (14.28) to evaluate the gradient of the stress constraint of Eq. (14.15).

The partial derivatives of the constraint of Eq. (14.15) with respect to x and U are given as

$$\frac{\partial g_1}{\partial \mathbf{x}} = \frac{1}{\sigma_a^2}\left[2\sigma_1 \frac{\partial \sigma_1}{\partial \mathbf{x}} + 6\tau \frac{\partial \tau}{\partial \mathbf{x}}\right] \tag{14.29}$$

$$\frac{\partial g_1}{\partial \mathbf{U}} = \frac{1}{\sigma_a^2}\left[2\sigma_1 \frac{\partial \sigma_1}{\partial \mathbf{U}} + 6\tau \frac{\partial \tau}{\partial \mathbf{U}}\right] \tag{14.30}$$

Using Eqs. (14.9) through (14.14), the partial derivatives of τ and σ_1 with respect to x and U are calculated as follows.

Partial derivatives of shear stress Differentiating the expression for shear stress in Eq. (14.12) with respect to the design variables x, we get

$$\frac{\partial \tau}{\partial \mathbf{x}} = \frac{1}{2At}\frac{\partial T}{\partial \mathbf{x}} - \frac{T}{2A^2 t}\frac{\partial A}{\partial \mathbf{x}} - \frac{T}{2At^2}\frac{\partial t}{\partial \mathbf{x}} \tag{14.31}$$

where the partial derivatives of the torque T with respect to the design variable x are given as

$$\frac{\partial T}{\partial \mathbf{x}} = -\frac{GU_3}{L}\frac{\partial J}{\partial \mathbf{x}} \tag{14.32}$$

with $\partial J/\partial \mathbf{x}$ calculated as

$$\frac{\partial J}{\partial d} = \frac{4t(d-t)(h-t)^2(d+h-2t) - 2t(d-t)^2(h-t)^2}{(d+h-2t)^2} = 0.864 \tag{14.33}$$

$$\frac{\partial J}{\partial h} = \frac{4t(d-t)^2(h-t)(d+h-2t) - 2t(d-t)^2(h-t)^2}{(d+h-2t)^2} = 0.864 \tag{14.34}$$

$$\frac{\partial J}{\partial t} = \frac{2(d-t)^2(h-t)^2 - 4t(d-t)(h-t)^2 - 4t(d-t)^2(h-t)}{(d+h-2t)}$$

$$- \frac{2t(d-t)^2(h-t)^2(-2)}{(d+h-2t)^2} = 12.096 \tag{14.35}$$

Therefore, $\partial J/\partial \mathbf{x}$ is assembled as

$$\frac{\partial J}{\partial \mathbf{x}} = \begin{bmatrix} 0.864 \\ 0.864 \\ 12.096 \end{bmatrix} \tag{14.36}$$

and Eq. (14.32) gives $\partial T/\partial \mathbf{x}$ as

$$\frac{\partial T}{\partial \mathbf{x}} = -\frac{(1.154 \times 10^7)}{100}(0.57275)\begin{bmatrix} 0.864 \\ 0.864 \\ 12.096 \end{bmatrix}$$

$$= -(6.610 \times 10^4)\begin{bmatrix} 0.864 \\ 0.864 \\ 12.096 \end{bmatrix} \tag{14.37}$$

Other quantities needed to complete the calculations in Eq. (14.31) are $\partial A/\partial \mathbf{x}$ and $\partial t/\partial \mathbf{x}$, which are calculated as

$$\frac{\partial A}{\partial \mathbf{x}} = \begin{bmatrix} (h-t) \\ (d-t) \\ -(h-t)-(d-t) \end{bmatrix} = \begin{bmatrix} 2.4 \\ 2.4 \\ -4.6 \end{bmatrix}$$

(14.38)

$$\frac{\partial t}{\partial \mathbf{x}} = \begin{bmatrix} 0 \\ 0 \\ 1 \end{bmatrix}$$

Substituting various quantities into Eq. (14.31), we get the partial derivative of τ with respect to $\mathbf{x}$ as

$$\frac{\partial \tau}{\partial \mathbf{x}} = \frac{1}{2At} \left[\frac{\partial T}{\partial \mathbf{x}} - \frac{T}{A}\frac{\partial A}{\partial \mathbf{x}} - \frac{T}{t}\frac{\partial t}{\partial \mathbf{x}} \right] = \begin{bmatrix} -1.653 \times 10^4 \\ -1.653 \times 10^4 \\ 3.580 \times 10^4 \end{bmatrix}$$

(14.39)

Differentiating the expression for the shear stress τ in Eq. (14.12) with respect to the generalized displacements $\mathbf{U}$, we get

$$\frac{\partial \tau}{\partial \mathbf{U}} = \frac{1}{2At}\frac{\partial T}{\partial \mathbf{U}}$$

(14.40)

where Eq. (14.9) gives

$$\frac{\partial T}{\partial \mathbf{U}} = \begin{bmatrix} 0 \\ 0 \\ -GJ/L \end{bmatrix} = \begin{bmatrix} 0 \\ 0 \\ -1.5953 \times 10^5 \end{bmatrix}$$

(14.41)

Therefore, $\partial \tau / \partial \mathbf{U}$ is given as

$$\frac{\partial \tau}{\partial \mathbf{U}} = \begin{bmatrix} 0 \\ 0 \\ -1.3848 \times 10^5 \end{bmatrix}$$

(14.42)

Partial derivatives of bending stress Differentiating the expression for σ_1 given in Eq. (14.13) with respect to the design variables $\mathbf{x}$, we get

$$\frac{\partial \sigma_1}{\partial \mathbf{x}} = \frac{h}{2I}\frac{\partial M_1}{\partial \mathbf{x}} + \frac{M_1}{2I}\frac{\partial h}{\partial \mathbf{x}} - \frac{M_1 h}{2I^2}\frac{\partial I}{\partial \mathbf{x}}$$

(14.43)

where $\partial M_1/\partial \mathbf{x}$, $\partial I/\partial \mathbf{x}$, and $\partial h/\partial \mathbf{x}$ are given as

$$\frac{\partial M_1}{\partial \mathbf{x}} = \frac{2E}{L^2}(-3U_1 + U_2 L)\frac{\partial I}{\partial \mathbf{x}}$$

$$\frac{\partial I}{\partial d} = \frac{1}{12}\left[h^3 - (h-2t)^3\right] = 0.288167$$

$$\frac{\partial I}{\partial h} = \frac{1}{4}\left[dh^2 - (d-2t)(h-2t)^2\right] = 0.8645$$

(14.44)

$$\frac{\partial I}{\partial t} = \frac{(h-2t)^3}{6} + \frac{(h-2t)^2(d-2t)}{2} = 8.11133$$

$$\frac{\partial h}{\partial \mathbf{x}} = \begin{bmatrix} 0 \\ 1 \\ 0 \end{bmatrix}$$

Substituting various quantities into Eq. (14.43),

$$\frac{\partial \sigma_1}{\partial \mathbf{x}} = \begin{bmatrix} 0 \\ 2.2131 \times 10^5 \\ 0 \end{bmatrix} \tag{14.45}$$

Differentiating the expression for σ_1 in Eq. (14.13) with respect to generalized displacements $\mathbf{U}$, we get

$$\frac{\partial \sigma_1}{\partial \mathbf{U}} = \frac{h}{2I} \frac{\partial M_1}{\partial \mathbf{U}} \tag{14.46}$$

where $\partial M_1/\partial \mathbf{U}$ is given from Eq. (14.10) as

$$\frac{\partial M_1}{\partial \mathbf{U}} = \frac{2EI}{L^2} \begin{bmatrix} -3 \\ L \\ 0 \end{bmatrix} \tag{14.47}$$

Therefore, $\partial \sigma_1/\partial \mathbf{U}$ is given as

$$\frac{\partial \sigma_1}{\partial \mathbf{U}} = \frac{Eh}{L^2} \begin{bmatrix} -3 \\ L \\ 0 \end{bmatrix} = \begin{bmatrix} -2.25 \times 10^5 \\ 7.50 \times 10^5 \\ 0 \end{bmatrix} \tag{14.48}$$

Substituting various quantities into Eqs. (14.29) and (14.30), we obtain the partial derivatives of constraints as

$$\frac{\partial g_1}{\partial \mathbf{x}} = \begin{bmatrix} 4.917 \\ 157.973 \\ -10.648 \end{bmatrix}$$

$$\frac{\partial g_1}{\partial \mathbf{U}} = \begin{bmatrix} -15.561 \\ 518.700 \\ 41.190 \end{bmatrix} \tag{14.49}$$

Derivatives of the displacements To calculate the derivatives of the displacements, we use Eq. (14.28). Since the load vector does not depend on the design variables, $\partial \mathbf{F}/\partial x_j = 0$ for $j = 1, 2, 3$ in Eq. (14.28). To calculate $(\partial \mathbf{K}(\mathbf{x})/\partial x_j)\mathbf{U}$ on the right side of Eq. (14.28), we differentiate Eq. (14.5) with respect to the design variables. For example, differentiation of Eq. (14.5) with respect to d gives the following vector:

$$\frac{\partial \mathbf{K}(\mathbf{x})}{\partial d} \mathbf{U} = \begin{bmatrix} -3.1214 \times 10^3 \\ -2.8585 \times 10^4 \\ 2.8585 \times 10^4 \end{bmatrix} \tag{14.50}$$

Similarly, by differentiating with respect to h and t, we obtain

$$\frac{\partial \mathbf{K}(\mathbf{x})}{\partial \mathbf{x}} \mathbf{U} = (1.0E+03) \begin{bmatrix} -3.1214 & -9.3642 & -87.861 \\ -28.5850 & 28.4540 & 3.300 \\ 28.5850 & -28.4540 & -3.300 \end{bmatrix} \tag{14.51}$$

Since $\mathbf{K}(\mathbf{x})$ is already known in Example 14.1, we use Eq. (14.28) to calculate $d\mathbf{U}/d\mathbf{x}$ as

$$\frac{d\mathbf{U}}{d\mathbf{x}} = \begin{bmatrix} 16.9090 & 37.6450 & 383.4200 \\ 0.2443 & 0.4711 & 5.0247 \\ -0.2443 & -0.4711 & -5.0247 \end{bmatrix} \tag{14.52}$$

Finally, substituting all of the quantities in Eq. (14.24), we obtain the design gradient for the effective stress constraint of Eq. (14.15) as

$$\frac{dg_1}{d\mathbf{x}} = \begin{bmatrix} -141.55 \\ -202.87 \\ -3577.30 \end{bmatrix} \tag{14.53}$$

As noted in Example 14.1, the stress constraint of Eq. (14.15) is severely violated at the given design. The signs of the foregoing design derivatives indicate that all of the variables will have to be increased to reduce the constraint violation at the point (2.5, 2.5, 0.1).

14.3 ISSUES IN PRACTICAL DESIGN OPTIMIZATION

Several issues need to be considered for practical design optimization. For example, careful consideration needs to be given to the selection of an algorithm and the associated software. Improper choice of either one can mean failure of the optimum design process. In this section, we will discuss some of the issues that can have a significant impact on practical applications of the optimization methodology. This material augments the related discussion presented in Chapter 6.

14.3.1 Selection of an Algorithm

Many algorithms have been developed and evaluated for practical design optimization. We need to consider several aspects when selecting an algorithm for practical applications, such as robustness, efficiency, generality, and ease of use.

Robustness

Characteristics of a robust algorithm were discussed in Section 12.1.5. For practical applications, it is important to use a method that is theoretically guaranteed to converge. *A method having such a guarantee starting from any initial design estimate is called robust* (so called *globally convergent* to a local minimum point). Robust algorithms usually require a few more calculations during each iteration compared with algorithms that have no proof of convergence. However, they save the designer's time in the long run and remove uncertainty about the optimum solution.

Potential Constraint Strategy

To evaluate the search direction in numerical methods for constrained optimization, we need to know the cost and constraint functions and their gradients. *The numerical algorithms can be classified into two categories based on whether gradients of all of the constraints or only a subset of them are required during design iteration.* The numerical algorithms that need the gradients of only a subset of the constraints are said to use *potential constraint strategy*. The potential constraint set, in general, is composed of active, nearly active, and violated constraints at the current iteration. For further discussion on the topic of potential set strategy, refer to Section 13.1.

14.3.2 Attributes of a Good Optimization Algorithm

Based on the preceding discussion, the attributes of a good algorithm for practical design applications are defined as follows:

Reliability. The algorithm must be reliable for general design applications because such algorithms converge to a minimum point starting from any initial design estimate. Reliability of an algorithm is guaranteed if it is theoretically proven to converge.

Generality. The algorithm must be general, which implies that it should be able to treat equality as well as inequality constraints. In addition, it should not impose any restrictions on the form of the problem functions.

Ease of use. The algorithm must be easy to use by the experienced as well as the inexperienced designer. From a practical standpoint, this is an important requirement because an algorithm requiring selection of tuning parameters is difficult to use. The proper specification of the parameters usually requires not only a complete knowledge and understanding of the algorithm's mathematical structure but also experimentation with each problem. Such an algorithm is unsuitable for practical design applications.

Efficiency. The algorithm must be efficient for general engineering applications. An efficient algorithm has (1) a faster rate of convergence to the minimum point, and (2) the fewest number of calculations within one design iteration. The *rate of convergence* can be accelerated by incorporating second-order information about the problem into the algorithm. Incorporation of second-order information, however, requires additional calculations during an iteration. Therefore, there is a trade-off between the efficiency of calculation within an iteration and the rate of convergence. Some existing algorithms use second-order information whereas others do not.

As regards the last item, efficiency within an iteration implies a minimum number of calculations for search direction and step size. One way to achieve efficiency is to use a *potential constraint strategy* in calculating the search direction. Some algorithms use this strategy in their calculations while others do not. When the potential constraint strategy is used, the direction-finding subproblem needs the gradients of only the potentially active constraints. Otherwise, the gradients of all constraints are needed, which is inefficient in most practical applications.

Another consideration for improving efficiency within an iteration is to keep the number of function evaluations for step size determination to a minimum. This can be achieved by using step size determination procedures requiring fewer calls for function evaluations (e.g., inexact line search and polynomial interpolation).

The designer needs to ask the following questions (all answers should be yes) before selecting an optimization algorithm for practical applications:

1. Does the algorithm have proof of convergence? That is, is it theoretically guaranteed to converge to an optimum point starting from any initial design estimate?
2. Can the starting design be infeasible (i.e., arbitrary)?
3. Can the algorithm solve a general optimization problem without any restrictions on the form of the constraint functions?
4. Can the algorithm treat equality as well as inequality constraints?
5. Is the algorithm easy to use? (In other words, it does not require tuning for each problem.)

14.4 USE OF GENERAL-PURPOSE SOFTWARE

As we saw in previous sections, practical systems require considerable computer analysis before optimum solutions are obtained. For a particular application, problem functions and gradient evaluation software, as well as optimization software, must be integrated to create an optimum design capability. Depending on the application, each of the software components can be very large. Therefore, to create a design optimization capability, the most sophisticated and modern computer facilities need to be used to integrate the software components.

For the example of structures modeled by the finite elements that were discussed in Section 14.1.2, large analysis packages must be used to analyze the structure. From the calculated response, constraint functions must be evaluated and programs must be developed to calculate gradients. All of the software components must be integrated to create the optimum design capability for structures modeled by finite elements.

In this section we will discuss the issues involved in selecting a general-purpose optimization software. Interfacing of the software with a particular application will also be discussed.

14.4.1 Software Selection

Several issues need to be explored before general-purpose optimization software is selected for integration with other application-dependent software. The most important question pertains to the optimization algorithm and how well it is implemented.

The attributes of a good algorithm were given in Section 14.3.2. The software must contain at least one algorithm that satisfies all of the requirements stated there. The algorithm should also be robustly implemented because a good algorithm when badly implemented is not very useful. The proof of convergence of most algorithms is based on certain assumptions. These need to be strictly adhered to while implementing the algorithm. In addition, most algorithms have some numerical uncertainties in their steps that need to be recognized, and proper procedures need to be developed for their numerical implementation. It is also important that the software be well tested on a range of applications of varying difficulty.

Several other user-friendly features are desirable. For example, how good is the user's manual? What sample problems are available with the program and how well are they documented? How easy is it to install the program on different computer systems? Does the program require the user to select algorithm-related parameters? All of these questions should be investigated before selecting the software.

14.4.2 Integration of an Application into General-Purpose Software

Each general-purpose program for optimization requires that the particular design application be integrated into the software. Ease of integration of the software components for various applications can influence program selection. Also, the amount of data preparation needed to use the program is important.

Some general-purpose libraries are available that contain various subprograms implementing different algorithms. The user is required to write a main program, defining various data before invoking the subprogram. Subprograms for function and gradient evaluation must also be written. These are then called by the optimization program during calculation of the search direction. Also, the function evaluation subprograms are called during the line search to determine the step size.

The other approach is to develop a computer program with options for various optimization algorithms. Each application is implanted in the program through a standard interface that consists of "subprogram calls." The user prepares a few subprograms to describe the design problem only. All of the data between the program and the subprograms flow through the subprogram arguments. For example, design variable data are sent to the subprograms and the expected output is the constraint function values and their gradients. Interactive capabilities, graphics, and other user-friendly features may be available in the program.

Both procedures described in the foregoing paragraphs have been successfully used for many practical applications of optimization. In most cases, the choice of the procedure has been dictated by the availability of the software. We will use the program IDESIGN, which is based on the second approach, to solve several design optimization problems. The program is combined with user-supplied subprograms to create an executable module. The user-supplied subprograms can be quite simple for problems having explicit functions and complex for problems having implicit functions. External programs may have to be called upon to generate the function values and their gradients needed by IDESIGN. This has been done, and several complex problems have been solved and reported in the literature (Tseng and Arora, 1987, 1988).

14.5 OPTIMUM DESIGN OF A TWO-MEMBER FRAME WITH OUT-OF-PLANE LOADS

Figure 14.1 shows the two-member frame subjected to out-of-plane loads. The members of the frame are subjected to torsional, bending, and shearing loads. The objective of the problem is to design a structure having minimum volume without material failure due to applied loads. The problem was formulated in Section 14.1.2 using the finite-element approach. In defining the stress constraint, the von Mises yield criterion is used and the shear stress due to the transverse load is neglected.

The formulation and equations given in Sections 14.1.2 and 14.2 are used to develop appropriate subprograms for IDESIGN. The data given there are used to optimize the problem. Also, as noted, only the constraint g_1 needs to be imposed. Two widely separated starting designs, (2.5, 2.5, 0.1) and (10, 10, 1), are tried to observe their effect on the convergence rate.

For the first starting point, all of the variables are at their lower bounds; for the second point, they are all at their upper bounds. Both starting points converge to the same optimum solution with almost the same values for the design variables and Lagrange multipliers for active constraints, as shown in Tables 14.1 and 14.2. However, the number of iterations and the number of calls for function and gradient evaluations are quite different. For the first starting point, the stress constraint is severely violated (by 20,212 percent). Several iterations are expended to bring the design close to the feasible region. For the second starting point, the stress constraint is satisfied and the program takes only six iterations

TABLE 14.1 History of the iterative process and optimum solution for a two-member frame, starting point (2.5, 2.5, 0.10)

Iteration no.	Maximum violation	Convergence parameter	Cost	d	h	t
1	2.02119E+02	1.00000E+00	1.92000E+02	2.5000E+00	2.5000E+00	1.0000E−01
2	8.57897E+01	1.00000E+00	2.31857E+02	2.5000E+00	3.4964E+00	1.0000E−01
3	3.58717E+01	1.00000E+00	2.85419E+02	2.5000E+00	4.8355E+00	1.0000E−01
:	:	:	:	:	:	:
17	6.78824E−01	1.00000E+00	6.14456E+02	5.5614E+00	1.0000E+01	1.0000E−01
18	1.58921E−01	6.22270E−01	6.76220E+02	7.1055E+00	1.0000E+01	1.0000E−01
19	1.47260E−02	7.01249E−02	7.01111E+02	7.7278E+00	1.0000E+01	1.0000E−01
20	1.56097E−04	7.59355E−04	7.03916E+02	7.7979E+00	1.0000E+01	1.0000E−01

Constraint activity

Iteration no.	Active	Value	Lagrange multiplier
1	Yes	1.56097E−04	1.94631E+02

Design variable activity

Iteration no.	Active	Design	Lower	Upper	Lagrange multiplier
1	No	7.79791E+00	2.50000E+00	1.00000E+01	0.00000E+00
2	Upper	1.00000E+01	2.50000E+00	1.00000E+01	7.89773E+01
3	Lower	1.00000E−01	1.00000E−01	1.00000E+00	3.19090E+02

Cost function at optimum = 703.92; number of calls for cost function evaluation = 20; number of calls for cost function gradient evaluation = 20; number of calls for constraint function evaluation = 20; number of calls for constraint function gradient evaluation = 20; number of total gradient evaluations = 20.

to find the optimum solution. Note that both solutions reported in Tables 14.1 and 14.2 are obtained using the *Sequential Quadratic Programming* option (SQP) available in IDESIGN. Also, a very severe stopping criterion is used to obtain the precise optimum point.

The notation used in the tables is defined as follows:

- Max violation (maximum violation of among constraints)
- Convergence parameter
- Cost (cost function value)
- Lagrange multiplier (for a constraint)

The preceding discussion shows that the starting design estimate for the iterative process can have a substantial impact on the rate of convergence of an algorithm. In many practical applications, a good starting design is available or can be obtained after some preliminary analyses. Such a starting design for the optimization algorithm is desirable because the optimal solution can be obtained rather quickly.

TABLE 14.2 History of the iterative process and optimum solution for a two-member frame, starting point (10, 10, 1)

Iteration no.	Maximum violation	Convergence parameter	Cost	d	h	t
1	0.00000E+00	6.40000E+03	7.20000E+03	1.0000E+01	1.0000E+01	1.0000E+00
2	0.00000E+00	2.27873E+01	7.87500E+02	9.9438E+00	9.9438E+00	1.0000E−01
3	1.25020E−02	1.31993E+00	7.13063E+02	9.0133E+00	9.0133E+00	1.0000E−01
4	2.19948E−02	1.03643E−01	6.99734E+02	7.6933E+00	1.0000E+01	1.0000E−01
5	3.44115E−04	1.67349E−03	7.03880E+02	7.7970E+00	1.0000E+01	1.0000E−01
6	9.40469E−08	4.30513E−07	7.03947E+02	7.7987E+00	1.0000E+01	1.0000E−01

Constraint activity

Iteration no.	Active	Value	Lagrange multiplier		
1	Yes	9.40469E−08	1.94630E+02		

Design variable activity

Iteration no.	Active	Design	Lower	Upper	Lagrange multiplier
1	No.	7.79867E+00	2.50000E+00	1.00000E+01	0.00000E+00
2	Upper	1.00000E+01	2.50000E+00	1.00000E+01	7.89767E+01
3	Lower	1.00000E−01	1.00000E−00	1.00000E+00	3.19090E+02

Cost function at optimum = 703.95; number of calls for cost function evaluation = 9; number of calls for cost function gradient evaluation = 6; number of calls for constraint function evaluation = 9; number of calls for constraint function gradient evaluation = 4; number of total gradient evaluations = 4.

14.6 OPTIMUM DESIGN OF A THREE-BAR STRUCTURE FOR MULTIPLE PERFORMANCE REQUIREMENTS

In the previous section, we discussed design of a structural system for one performance requirement—the material must not fail under the applied loads. In this section, we discuss a similar application where the system must perform safely under several operating environments. The problem that we have chosen is the three-bar structure that was formulated in Section 2.10. The structure was shown in Figure 2.6. The design requirement is to minimize the weight of the structure and satisfy the constraints of member stress, deflection at node 4, buckling of members, vibration frequency, and explicit bounds on the design variables. We will optimize the symmetric and asymmetric structures and compare the solutions. A very strict stopping criterion will be used to obtain the precise optimum designs.

14.6.1 Symmetric Three-Bar Structure

A detailed formulation for the symmetric structure where members 1 and 3 are similar was discussed in Section 2.10. In the present application, the structure is designed to

TABLE 14.3   Design data for a three-bar structure

Structure	Data		
Allowable stress	Members 1 and 3, $\sigma_{1a} = \sigma_{3a} = 5000$ psi Member 2, $\sigma_{2a} = 20{,}000$ psi		
Height	$l = 10$ in		
Allowable displacements	$u_a = 0.005$ in $v_a = 0.005$ in		
Modulus of elasticity	$E = 10^7$ psi		
Weight density	$\gamma = 0.10$ lb/in^3		
Constant	$\beta = 1.0$		
Lower limit on design	$(0.1, 0.1, 0.1)$ in^2		
Upper limit on design	$(100, 100, 100)$ in^2		
Starting design	$(1, 1, 1)$ in^2		
Lower limit on frequency	2500 Hz		
Loading conditions	3		
Angle, θ (degrees)	45	90	135
Load, P (lb)	40,000	30,000	20,000

withstand three loading conditions and the foregoing constraints. Table 14.3 contains all of the data used for designing the structure. All of the expressions programmed for IDESIGN were given in Section 2.10. The constraint functions are appropriately normalized and expressed in the standard form. The cost function is taken as the total weight of the truss, which is given as volume × weight density.

To study the effect of imposing more performance requirements, the following three design cases are defined (note that explicit design variable–bound constraints are included in all of the cases):

Case 1. Stress constraints only (total constraints = 13).
Case 2. Stress and displacement constraints (total constraints = 19).
Case 3. All constraints—stress, displacement, member buckling, and frequency (total constraints = 29).

Tables 14.4 through 14.6 that follow contain the history of the iterative process with the SQP method in IDESIGN. The constraint-numbering scheme for the problem is shown in Table 14.6. The active constraints at the optimum point and their Lagrange multipliers (for normalized constraints) are

Case 1. Stress in member 1 under loading condition 1, 21.11.

TABLE 14.4 History of the iterative process and final solution for a symmetric three-bar structure, Case 1—Stress constraints

Iteration no.	Maximum violation	Convergence parameter	Cost	$A_1 = A_3$	A_2
1	4.65680E+00	1.00000E+00	3.82843E+00	1.0000E+00	1.0000E+00
2	2.14531E+00	1.00000E+00	6.72082E+00	1.9528E+00	1.1973E+00
:	:	:	:	:	:
8	2.20483E−04	3.97259E−03	2.11068E+01	6.3140E+00	3.2482E+00
9	1.58618E−06	5.34172E−05	2.11114E+01	6.3094E+00	3.2657E+00

Cost function at optimum = 21.11; number of calls for cost function evaluation = 9; number of calls for cost function gradient evaluation = 9; number of calls for constraint function evaluation = 9; number of calls for constraint function gradient evaluation = 9; number of total gradient evaluations = 19.

TABLE 14.5 History of the iterative process and final solution for a symmetric three-bar structure, Case 2—stress and displacement constraints

Iteration no.	Maximum violation	Convergence parameter	Cost	$A_1 = A_3$	A_2
1	6.99992E+00	1.00000E+00	3.82843E+00	1.0000E+00	1.0000E+00
2	3.26663E+00	1.00000E+00	6.90598E+00	1.8750E+00	1.6027E+00
:	:	:	:	:	:
8	1.50650E−04	3.05485E−04	2.29695E+01	7.9999E+00	3.4230E−01
9	2.26886E−08	4.53876E−08	2.29704E+01	7.9999E+00	3.4320E−01

Cost function at optimum = 22.97; number of calls for cost function evaluation = 9; number of calls for cost function gradient evaluation = 9; number of calls for constraint function evaluation = 9; number of calls for constraint function gradient evaluation = 9; number of total gradient evaluations = 48.

Case 2. Stress in member 1 under loading condition 1, 0.0; horizontal displacement under loading condition 1, 16.97; horizontal displacement under loading condition 1, 6.00; vertical displacement under loading condition 2, 0.0.
Case 3. Same as for Case 2.

Note that the cost function value at the optimum point increases for Case 2 as compared with Case 1. This is consistent with the hypothesis that more constraints for the system imply a smaller feasible region, thus giving a higher value for the optimum cost function. There is no difference between the solutions for Cases 2 and 3 because none of the additional constraints for Case 3 is active.

14.6.2 Asymmetric Three-Bar Structure

When the symmetry condition for the structure (i.e., member 1 same as member 3) is relaxed, we get three design variables for the problem compared with only two for the

TABLE 14.6 History of the iterative process and final solution for a symmetric three-bar structure, Case 3—All constraints

Iteration no.	Maximum violation	Convergence parameter	Cost	$A_1 = A_3$	A_2
1	6.99992E+00	1.00000E+00	3.82843E+00	1.0000E+00	1.0000E+00
2	2.88848E+00	1.00000E+00	7.29279E+00	2.0573E+00	1.4738E+00
:	:	:	:	:	:
7	7.38741E−05	3.18776E−04	2.29691E+01	7.9993E+00	3.4362E−01
8	5.45657E−09	2.31529E−08	2.29704E+01	7.9999E+00	3.4320E−01

Constraint activity

Iteration no.	Active	Value	Lagrange multiplier	Notes
1	No	−1.56967E−01	0.00000E+00	(Frequency)
2	Yes	−2.86000E−02	0.00000E+00	(σ_1, Loading Condition 1)
3	No	−9.71400E−01	0.00000E+00	(σ_2, Loading Condition 1)
4	No	−7.64300E−01	0.00000E+00	(σ_3, Loading Condition 1)
5	Yes	5.45657E−09	1.69704E+01	(u, Loading Condition 1)
6	Yes	−5.72000E−02	0.00000E+00	(v, Loading Condition 1)
7	No	−5.00000E−01	0.00000E+00	(σ_1, Loading Condition 2)
8	No	−5.00000E−01	0.00000E+00	(σ_2, Loading Condition 2)
9	No	−7.50000E−01	0.00000E+00	(σ_3, Loading Condition 2)
10	No	−1.00000E+00	0.00000E+00	(u, Loading Condition 2)
11	Yes	0.00000E+00	6.00000E+00	(v, Loading Condition 2)
12	No	−9.85700E−01	0.00000E+00	(σ_1, Loading Condition 3)
13	No	−5.14300E−01	0.00000E+00	(σ_2, Loading Condition 3)
14	No	−8.82150E−01	0.00000E+00	(σ_3, Loading Condition 3)
15	No	−5.00000E−01	0.00000E+00	(u, Loading Condition 3)
16	No	−5.28600E−01	0.00000E+00	(v, Loading Condition 3)

Design variable activity

Iteration no.	Active	Design	Lower	Upper	Lagrange multiplier
1	No	7.99992E+00	1.00000E−01	1.00000E+02	0.00000E+00
2	No	3.43200E−01	1.00000E−01	1.00000E+02	0.00000E+00

Cost function at optimum = 22.97; number of calls for cost function evaluation = 8; number of calls for cost function gradient evaluation = 8; number of calls for constraint function evaluation = 8; number of calls for constraint function gradient evaluation = 8; number of total gradient evaluations = 50.

symmetric case (i.e., areas A_1, A_2, and A_3 for members 1, 2, and 3, respectively). With this, the design space becomes expanded so we can expect better optimum designs compared with the previous cases.

The data used for the problem are the same as given in Table 14.3. The weight of the structure is minimized for the following three cases (note that the explicit design variable–bound constraints are included in all cases):

Case 4. Stress constraints only (total constraints = 15)
Case 5. Stress and displacement constraints (total constraints = 21)
Case 6. All constraints—stress, displacement, buckling, and frequency (total constraints = 31)

The structure can be analyzed by considering either the equilibrium of node 4 or the general finite-element procedures. By following the general procedures, the following expressions for displacements, member stresses, and fundamental vibration frequency are obtained (note that the notations are defined in Section 2.10):

Displacements:

$$u = \frac{l}{E}\left[\frac{(A_1 + 2\sqrt{2}A_2 + A_3)P_u + (A_1 - A_3)P_v}{A_1A_2 + \sqrt{2}A_1A_3 + A_2A_3}\right], \text{ in}$$

(14.54)

$$v = \frac{l}{E}\left[\frac{-(A_1 - A_3)P_u + P_v(A_1 + A_3)}{A_1A_2 + \sqrt{2}A_1A_3 + A_2A_3}\right], \text{ in}$$

Member stresses:

$$\sigma_1 = \frac{(\sqrt{2}A_2 + A_3)P_u + A_3P_v}{A_1A_2 + \sqrt{2}A_1A_3 + A_2A_3}, \text{ psi}$$

$$\sigma_2 = \frac{-(A_1 - A_3)P_u + (A_1 + A_3)P_v}{A_1A_2 + \sqrt{2}A_1A_3 + A_2A_3}, \text{ psi}$$

(14.55)

$$\sigma_3 = \frac{-(A_1 + \sqrt{2}A_2)P_u + A_1P_v}{A_1A_2 + \sqrt{2}A_1A_3 + A_2A_3}, \text{ psi}$$

Lowest eigenvalue:

$$\zeta = \frac{3E}{2\sqrt{2}\rho l^2}\left[\frac{A_1 + \sqrt{2}A_2 + A_3 - [(A_1 - A_3)^2 + 2A_2^2]^{1/2}}{\sqrt{2}(A_1 + A_3) + A_2}\right]$$

(14.56)

Fundamental frequency:

$$\omega = \frac{1}{2\pi}\sqrt{\zeta}, \text{ Hz}$$

(14.57)

Tables 14.7 through 14.9 contain the history of the iterative process with the SQP method in IDESIGN. The active constraints at the optimum point and their Lagrange multipliers (for normalized constraints) are as follows.

Case 4. Stress in member 1 under loading condition 1, 11.00; stress in member 3 under loading condition 3, 4.97.

TABLE 14.7 History of the iterative process and final solution for an asymmetric three-bar structure, Case 4—Stress constraints

Iteration no.	Maximum violation	Convergence parameter	Cost	A_1	A_2	A_3
1	4.65680E+00	1.00000E+00	3.82843E+00	1.0000E+00	1.0000E+00	1.0000E+00
2	2.10635E+00	1.00000E+00	6.51495E+00	1.9491E+00	1.4289E+00	1.6473E+00
:	:	:	:	:	:	:
8	4.03139E−04	2.52483E−03	1.59620E+01	7.0220E+00	2.1322E+00	2.7572E+00
9	4.80986E−07	6.27073E−05	1.59684E+01	7.0236E+00	2.1383E+00	2.7558E+00

Cost function at optimum = 15.97; number of calls for cost function evaluation = 9; number of calls for cost function gradient evaluation = 9; number of calls for constraint function evaluation = 9; number of calls for constraint function gradient evaluation = 9; number of total gradient evaluations = 26.

TABLE 14.8 History of the iterative process and final solution for an asymmetric three-bar structure, Case 5—Stress and displacement constraints

Iteration no.	Maximum violation	Convergence parameter	Cost	A_1	A_2	A_3
1	6.99992E+00	1.00000E+00	3.82843E+00	1.0000E+00	1.0000E+00	1.0000E+00
2	3.26589E+00	1.00000E+00	6.77340E+00	1.9634E+00	1.6469E+00	1.6616E+00
:	:	:	:	:	:	:
9	2.18702E−05	3.83028E−04	2.05432E+01	8.9108E+00	1.9299E+00	4.2508E+00
10	6.72142E−09	1.42507E−06	2.05436E+01	8.9106E+00	1.9295E+00	4.2516E+00

Cost function at optimum = 20.54; number of calls for cost function evaluation = 10; number of calls for cost function gradient evaluation = 10; number of calls for constraint function evaluation = 10; number of calls for constraint function gradient evaluation = 10; number of total gradient evaluations = 43.

TABLE 14.9 History of the iterative process and final solution for an asymmetric three-bar structure, Case 6—All constraints

Iteration no.	Maximum violation	Convergence parameter	Cost	A_1	A_2	A_3
1	6.99992E+00	1.00000E+00	3.82843E+00	1.0000E+00	1.0000E+00	1.0000E+00
2	2.88848E+00	1.00000E+00	7.29279E+00	2.0573E+00	1.4738E+00	2.0573E+00
:	:	:	:	:	:	:
7	6.75406E−05	2.25516E−04	2.10482E+01	8.2901E+00	1.2017E+00	6.7435E+00
8	6.46697E−09	1.88151E−08	2.10494E+01	8.2905E+00	1.2013E+00	5.7442E+00

Cost function at optimum = 21.05; number of calls for cost function evaluation = 8; number of calls for cost function gradient evaluation = 8; number of calls for constraint function evaluation = 8; number of calls for constraint function gradient evaluation = 8; number of total gradient evaluations = 48.

TABLE 14.10   Comparison of optimum costs for six cases of a three-bar structure

	Symmetric structure			Asymmetric structure		
	Case 1	Case 2	Case 3	Case 4	Case 5	Case 6
Optimum weight (lb)	21.11	22.97	22.97	15.97	20.54	21.05
NIT[a]	9	9	8	9	10	8
NCF[a]	9	9	8	9	10	8
NGE[a]	19	48	50	26	43	48

[a]NIT, *number of iterations*; NCF, *number of calls for function evaluation*; NGE, *total number of gradients evaluations*.

Case 5. Horizontal displacement under loading condition 1, 11.96; vertical displacement under loading condition 2, 8.58.

Case 6. Frequency constraint, 6.73; horizontal displacement under loading condition 1, 13.28; vertical displacement under loading condition 2, 7.77.

Note that the optimum weight for Case 5 is higher than that for Case 4, and for Case 6 it is higher than that for Case 5. This is consistent with the previous observation; the number of constraints for Case 5 is larger than that for Case 4, and for Case 6 it is larger than that for Case 5.

14.6.3 Comparison of Solutions

Table 14.10 contains a comparison of solutions for all six cases. Since an asymmetric structure has a larger design space, the optimum solutions should be better than those for the symmetric case, and they are; Case 4 is better than Case 1, Case 5 is better than Case 2, and Case 6 is better than Case 3.

These results show that for better practical solutions, more flexibility should be allowed in the design process by defining more design variables—that is, by allowing more design degrees of freedom.

14.7 OPTIMAL CONTROL OF SYSTEMS BY NONLINEAR PROGRAMMING

14.7.1 A Prototype Optimal Control Problem

Optimal control problems are dynamic in nature. A brief discussion of the differences between optimal control and optimum design problems is given here. It turns out that some optimal control problems can be formulated and solved by the nonlinear programming methods described in Chapters 12 and 13. In this section, we consider a simple optimal control problem that has numerous practical applications. Various formulations of the problem are described and optimal solutions are obtained and discussed.

The application area that we have chosen to demonstrate the use of nonlinear programming methods is the vibration control of systems. This is an important class of problems that is encountered in numerous real-world applications. Examples include the control of structures under earthquake and wind loads, vibration control of sensitive instruments to blast loading or shock input, control of large-space structures, and precision control of machines, among others. To treat these problems we will consider a simple model of the system to demonstrate the basic formulation and solution procedures. Using the demonstrated procedures, more complex models can be treated to simulate real-world systems more accurately.

To treat optimal control problems, dynamic response analysis capability must be available. In the present text, we will assume that students have some background in vibration analysis of systems. In particular, we will model systems as single-degree-of-freedom linear spring-mass systems. This leads to a second-order linear differential equation whose closed-form solution is available (Clough and Penzien, 1975; Chopra, 2007). It may be worthwhile for the students to briefly review the material on the solution of linear differential equations in appropriate textbooks.

To demonstrate the formulation and the solution process, we consider the cantilever structure shown in Figure 14.2. The data for the problem and various notations used in the figure are defined in Table 14.11. The structure is a highly idealized model of many systems that are used in practice. The length of the structure is L and its cross-section is rectangular with its width as b and its depth as h. The system is at rest initially at time $t = 0$. It experiences a sudden load due to a shock wave or other similar phenomenon.

The problem is to control the vibrations of the system such that the displacements are not too large and the system comes to rest in a controlled manner. The system has proper sensors and actuators that generate the desired force to suppress the vibrations and bring the system to rest. The control force may also be generated by properly designed dampers or viscoelastic support pads along the length of the structure. We will not discuss the

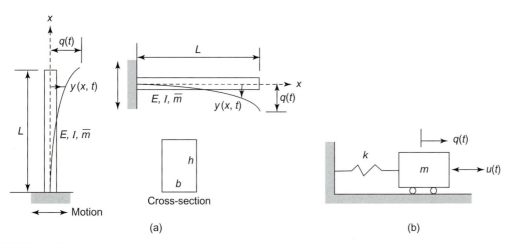

FIGURE 14.2 Model of a system subjected to shock input. (a) Cantilever structure subjected to shock input at support. (b) Equivalent single degree of freedom model.

TABLE 14.11   Data for the optimal control problem

Data	Specifics
Length of the structure	$L = 1.0$ m
Width of cross section	$b = 0.01$ m
Depth of cross section	$h = 0.02$ m
Modulus of elasticity	$E = 200$ GPa
Mass density	$\rho = 7800$ kg/m^3
Moment of inertia	$I = (6.667 \times 10^{-9})$ m^4
Mass per unit length	$\overline{m} = 1.56$ kg/m
Control function	$u(t) =$ to be determined
Limit on the control function	$u_a = 30$ N
Initial velocity	$v_0 = 1.5$ m/s

detailed design of the control force—generating mechanisms, but we will discuss the problem of determining the optimum shape of the control force.

The governing equation that describes the motion of the system is a second-order partial differential equation. To simplify the analysis, we use separation of variables and express the deflection function $y(x,t)$ as

$$y(x,t) = \psi(x)q(t) \tag{14.58}$$

where $\psi(x)$ is a known function called the shape function, and $q(t)$ is the displacement at the tip of the cantilever, as shown in Figure 14.2. Several shape functions can be used; however, we will use the following one:

$$\psi(x) = \frac{1}{2}(3\xi^2 - \xi^3); \quad \xi = \frac{x}{L} \tag{14.59}$$

Using kinetic and potential energies for the system, $\psi(x)$ of Eq. (14.59), and the data in Table 14.11, the mass and spring constants for an equivalent single-degree-of-freedom system shown in Figure 14.2 are calculated as follows (Clough and Penzien, 1975; Chopra, 2007):

Mass:

$$\text{kinetic energy} = \frac{1}{2}\int_0^L \overline{m}\dot{y}^2(t)dx = \frac{1}{2}\left[\int_0^L \overline{m}\psi^2(x)dx\right]\dot{q}^2(t) = \frac{1}{2}m\dot{q}^2(t) \tag{14.60}$$

where the mass m is identified as

$$m = \int_0^L \overline{m}\psi^2(x)dx = \frac{33}{140}\overline{m}L = \frac{33}{140}(1.56)(1.0) = 0.3677 \text{ kg} \tag{14.61}$$

Spring constant:

$$\text{strain energy} = \frac{1}{2}\int_0^L EI[y''(x)]^2 dx = \frac{1}{2}\left[\int_0^L EI(\psi''(x))^2 dx\right]q^2(t) = \frac{1}{2}kq^2(t) \tag{14.62}$$

where the spring constant k is identified as

$$k = \int_0^L EI(\psi''(x))^2 dx = \frac{3EI}{L^3} = 3(2.0 \times 10^{11})(6.667 \times 10^{-9})/(1.0)^3 = 4000\,\text{N/m} \qquad (14.63)$$

In the foregoing, a dot over a variable indicates derivatives with respect to time; a prime indicates derivatives with respect to the coordinate x.

The equation of motion for the single-degree-of-freedom system and the initial conditions (initial displacement q_0, initial velocity v_0) are given as

$$m\ddot{q}(t) + kq(t) = u(t) \qquad (14.64)$$

$$q(0) = q_0, \quad \dot{q}(0) = v_0 \qquad (14.65)$$

where $u(t)$ is the control force needed to suppress vibrations due to the initial velocity v_0 (*shock loading* for the system is transformed to an equivalent initial velocity calculated as the impulse of the force divided by the mass). Note that the material damping for the system is neglected. Therefore, if no control force $u(t)$ is used, the system will continue to oscillate. Figure 14.3 shows the displacement response of the system for the initial 0.10 s when $u(t) = 0$ (i.e., when no control mechanism is used). The velocity also keeps oscillating between 1.5 and -1.5 m/s.

The control problem is to determine the forcing function $u(t)$ such that the system comes to rest in a specified time. We can pose the problem as follows: Determine the control force to minimize the time to bring the system to rest. We will investigate several of formulations in the following paragraphs.

We note here that for the preceding simple problem, solution procedures other than the nonlinear programming methods are available (Meirovitch, 1985). Those procedures may be

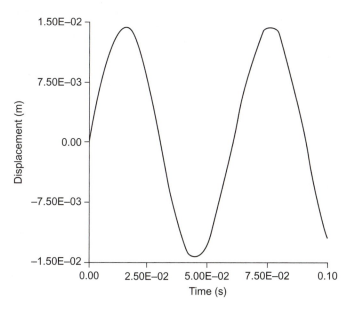

FIGURE 14.3 Graphic of the displacement response of the equivalent single-degree-of-freedom system to shock loading with no control force.

better for other applications and real-time control problems. However, we will use nonlinear programming formulations to solve the problem to demonstrate generality of the method.

14.7.2 Minimization of Error in the State Variable

As a first formulation, we define the performance index (cost function) as minimization of error in the state variable (response) in the time interval 0 to T as

$$f_1 = \int_0^T q^2(t)dt \tag{14.66}$$

Constraints are imposed on the terminal response, the displacement response, and the control force as follows:

Displacement constraint:

$$|q(t)| \le q_a \text{ in the time interval 0 to } T \tag{14.67}$$

Terminal displacement constraint:

$$q(T) = q_T \tag{14.68}$$

Terminal velocity constraint:

$$\dot{q}(T) = v_T \tag{14.69}$$

Control force constraint:

$$|u(t)| \le u_a \text{ in the interval 0 to } T \tag{14.70}$$

q_a is the maximum allowed displacement of the system, q_T and v_T are small specified constants, and u_a is the limit on the control force. Thus, the design problem is to compute the control function $u(t)$ in the time interval 0 to T to minimize the performance index of Eq. (14.66) subject to the constraints of Eqs. (14.67) through (14.70) and satisfaction of the equations of motion (14.64) and the initial conditions in Eq. (14.65). Note that the constraints of Eqs. (14.67) and (14.70) are dynamic in nature and need to be satisfied over the entire time interval 0 to T.

Another performance index can be defined as the sum of the squares of the displacement and the velocity:

$$f_2 = \int_0^T [q^2(t) + \dot{q}^2(t)]\, dt \tag{14.71}$$

Formulation for a Numerical Solution

To obtain numerical results, the following data are used:

Allowable time to suppress motion: $T = 0.10$ s
Initial velocity: $\qquad\qquad v_0 = 1.5$ m/s
Initial displacement: $\qquad\quad q_0 = 0.0$ m
Allowable displacement: $\qquad q_a = 0.01$ m

Terminal velocity: $\qquad$ $v_T = 0.0 \text{ m/s}$
Terminal displacement: $\qquad$ $q_T = 0.0 \text{ m}$
Limit on the control force: $\qquad$ $u_a = 30.0 \text{ N}$

For the present example, the equation of motion is quite simple, and its analytical solution can be written using Duhamel's integral (Chopra, 2007) as follows:

$$q(t) = \frac{1}{\omega}v_0 \ \sin\omega t + q_0 \ \cos\omega t + \frac{1}{m\omega}\int_0^t u(\eta) \ \sin\omega(t-\eta)d\eta \qquad (14.72)$$

$$\dot{q}(t) = v_0 \ \cos\omega t - q_0\omega \ \sin\omega t + \frac{1}{m}\int_0^t u(\eta) \ \cos\omega(t-\eta)d\eta \qquad (14.73)$$

In more complex applications, the equations of motion will have to be integrated using numerical methods (Shampine, 1994; Hsieh and Arora, 1984).

Since explicit forms for the displacement and velocity in terms of the design variable u (t) are known, we can calculate their derivatives by differentiating Eqs. (14.72) and (14.73) with respect to $u(\eta)$, where η is a point between 0 and T:

$$\frac{dq(t)}{du(\eta)} = \frac{1}{m\omega}\sin\omega(t-\eta) \quad \text{for} \quad t \geq \eta$$
$$= 0 \qquad\qquad\qquad \text{for} \quad t < \eta \qquad (14.74)$$

$$\frac{d\dot{q}(t)}{du(\eta)} = \frac{1}{m}\cos\omega(t-\eta) \quad \text{for} \quad t \geq \eta$$
$$= 0 \qquad\qquad\qquad \text{for} \quad t < \eta \qquad (14.75)$$

In the foregoing expressions, $du(t)/du(\eta) = \delta(t-\eta)$ has been used, where $\delta(t-\eta)$ is the Dirac delta function. The derivative expressions can easily be programmed to impose constraints on the problem. For more general applications, derivatives must be evaluated using numerical computational procedures. Several such procedures developed and evaluated by Hsieh and Arora (1984) and Tseng and Arora (1987) can be used for more complex applications.

Equations (14.72) through (14.75) are used to develop the user-supplied subroutines for the IDESIGN program. Several procedures are needed to solve the problem numerically. First of all, a grid must be used to discretize the time at which displacement, velocity, and control force are evaluated. Interpolation methods, such as cubic splines, B-splines (De Boor, 1978), and the like, can be used to evaluate the functions at points other than the grid points.

Another difficulty concerns the dynamic displacement constraint of Eq. (14.67). The constraint must be imposed during the entire time interval 0 to T. Several treatments for such constraints have been investigated (Hsieh and Arora, 1984; Tseng and Arora, 1987). For example, the constraint can be replaced by several constraints imposed at the local maximum points for the function $q(t)$; it may be replaced by an integral constraint; or it may be imposed at each grid point.

In addition to the foregoing numerical procedures, a numerical integration scheme, such as simple summation, trapezoidal rule, Simpson's rule, Gaussian quadrature, and so on, must be selected for evaluating the integrals in Eqs. (14.66), (14.72), and (14.73). Based on some preliminary investigations, the following numerical procedures are selected for their simplicity to solve the present problem:

- *Numerical integration:*   Simpson's method
- *Dynamic constraint:*   imposed at each grid point
- *Design variable (control force):* value at each grid point

Numerical Results

Using the foregoing procedures and the numerical data, the problem is solved using the SQP method of Section 13.4, available in the IDESIGN software package. The number of grid points is selected as 41, so there are 41 design variables. The displacement constraint of Eq. (14.67) is imposed at the grid points with its limit set as $q_a = 0.01$ m. As an initial estimate, $u(t)$ is set to zero, so the constraints of Eqs. (14.67) through (14.69) are violated.

The algorithm finds a feasible design in just three iterations. During these iterations, the cost function of Eq. (14.66) is also reduced. The algorithm reaches near to the optimum point at the 11th iteration. As a result of the strict stopping criteria, it takes another 27 iterations to satisfy the specified criteria. The cost function history is plotted in Figure 14.4. For all practical purposes, the optimum solution is obtained somewhere between the 15th and 20th iterations.

The final displacement response and the control force history are shown in Figures 14.5 and 14.6. It is noted that the displacement and velocity both go to zero at about 0.05 s, so the system comes to rest at that point. The control force also has a zero value after that

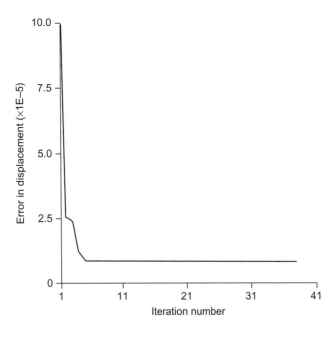

FIGURE 14.4   Cost function history for the optimal control problem of minimizing the error in the state variable (cost function f_1).

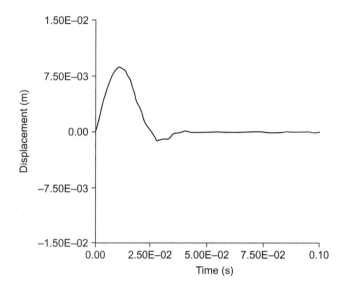

FIGURE 14.5 The displacement response at optimum with minimization of error in the state variable as performance index (cost function f_1).

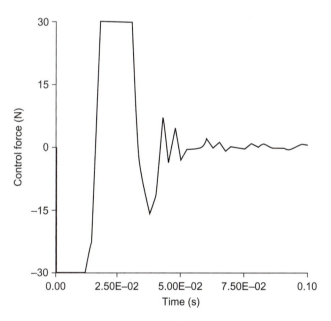

FIGURE 14.6 Optimum control force to minimize error in the state variable due to shock input (cost function f_1).

point and reaches its limit value at several grid points during that interval. The final cost function value is ($8.536E-07$).

Effect of Problem Normalization

It turns out that for the present application it is advantageous to normalize the problem and optimize it with normalized variables. We will briefly discuss these normalizations, which can be useful in other applications. Without normalization of the present problem, the cost function and its gradient, as well as the constraint functions and their gradients, have quite small values. The algorithm requires a very small value for the convergence parameter (1.0E−09) to converge to the same optimum solution as with the normalized problem. In addition, the rate of convergence without normalization is slow. This numerical behavior of the problem is overcome by the normalization procedure that is described next.

The independent variable transformation for the time is defined as

$$t = \tau T \quad \text{or} \quad \tau = \frac{t}{T} \tag{14.76}$$

where τ is the normalized independent variable. With this transformation, when t varies between 0 and T, τ varies between 0 and 1. The displacement is normalized as

$$q(t) = Tq_{max}\bar{q}(\tau) \quad \text{or} \quad \bar{q}(\tau) = \frac{q(t)}{Tq_{max}} \tag{14.77}$$

where $\bar{q}(\tau)$ is the normalized displacement and q_{max} is taken as 0.015. Derivatives of the displacement with respect to time are transformed as

$$\dot{q}(t) = q_{max}\dot{\bar{q}}(\tau) \tag{14.78}$$

$$\dot{q}(0) = q_{max}\dot{\bar{q}}(0) \quad \text{or} \quad \dot{\bar{q}}(0) = \frac{v_0}{q_{max}} \tag{14.79}$$

$$\ddot{q}(t) = \frac{1}{T}q_{max}\ddot{\bar{q}}(\tau) \tag{14.80}$$

The control force is normalized as

$$u(t) = u_{max}\bar{u}(\tau), \quad \text{or} \quad \bar{u}(\tau) = \frac{u(t)}{u_{max}} \tag{14.81}$$

With this normalization, $\bar{q}(\tau)$ varies between −1 and 1 as $u(t)$ varies between $-u_{max}$ and u_{max}.

Substituting the preceding transformations into Eqs. (14.64) and (14.65), we get

$$\bar{m}\ddot{\bar{q}}(\tau) + \bar{k}\bar{q}(\tau) = \bar{u}(\tau) \tag{14.82}$$

$$\bar{q}(0) = \frac{q_0}{Tq_{max}}, \quad \dot{\bar{q}}(0) = \frac{v_0}{q_{max}}$$

$$\bar{k} = \frac{kT}{u_{max}}q_{max}, \quad \bar{m} = \frac{m}{Tu_{max}}q_{max} \tag{14.83}$$

The constraints of Eqs. (14.67) to (14.70) are also normalized as

Displacement constraint:

$$|\bar{q}(\tau)| \leq \frac{q_a}{Tq_{max}} \quad \text{in the interval } 0 \leq \tau \leq 1 \tag{14.84}$$

Terminal displacement constraint:

$$\bar{q}(1) = \frac{1}{Tq_{max}}q_T \tag{14.85}$$

Terminal velocity constraint:

$$\dot{\bar{q}}(1) = \frac{1}{q_{max}} v_T \tag{14.86}$$

Control force constraint:

$$|\bar{u}(\tau)| \le 1 \text{ in the interval } 0 \le \tau \le 1 \tag{14.87}$$

With the foregoing normalizations, the numerical algorithm behaved considerably better and the convergence to the optimum solution as reported earlier was quite rapid. Therefore, for general usage, normalization of the problem is recommended whenever possible. We will use the foregoing normalizations in the two additional formulations of the problem discussed in Sections 14.7.3 and 14.7.4.

Discussion of Results

The final solution to the problem can be affected by the number of grid points and the stopping criterion. The solution reported previously was obtained using 41 grid points and a convergence criterion of (10^{-3}). A stricter convergence criterion of (10^{-6}) also gave the same solution, using a few more iterations.

The number of grid points can also affect the accuracy of the final solution. The use of 21 grid points also gave approximately the same solution. The shape of the final control force was slightly different. The final cost function value was slightly higher than that with 41 grid points, as expected.

It is also important to note that the problem can become infeasible if the limit q_a on the displacement in Eq. (14.67) is too restrictive. For example, when q_a was set to 0.008 m, the problem was infeasible with 41 grid points. However, with 21 grid points a solution was obtained. This also shows that when the number of grid points is smaller, the displacement constraint may actually be violated between the grid points, although it is satisfied at the grid points. Therefore, the number of grid points should be selected judiciously.

The foregoing discussion shows that to impose the time-dependent constraints more precisely, the exact local max-points should be located and the constraint imposed there. To locate the exact max-points, interpolation procedures may be used, or bisection of the interval in which the max-point lies can be used (Hsieh and Arora, 1984). The gradient of the constraint must be evaluated at the max-points. For the present problem, the preceding procedure is not too difficult to implement because the analytical form of the response is known. For more general applications, the computational and programming efforts can increase substantially to implement the foregoing procedure.

It is worthwhile to note that several other starting points for the control force, such as $u(t) = -30 \text{ N}$ or $u(t) = 30 \text{ N}$, converged to the same solution as given in Figures 14.5 and 14.6. The computational effort varied somewhat. The CPU time with 21 grid points was about 20 percent of that with 41 grid points when $u(t) = 0$ was used as the starting point.

It is interesting that, at the optimum, the dynamic constraint of Eq. (14.67) is not active at any time grid point. It is violated at many intermediate iterations. Also, the terminal response constraints of Eqs. (14.68) and (14.69) are satisfied at the optimum with normalized Lagrange multipliers as $(-7.97E-04)$ and $(5.51E-05)$. Since the multipliers are almost zero, the constraints can be somewhat relaxed without affecting the optimal solution. This can be observed from the final displacement response shown in Figure 14.5. Since the system is essentially at rest after $t = 0.05$ s, there is no effect of imposing the terminal constraints of Eqs. (14.68) and (14.69).

The control force is at its limit value ($u_a = 30$ N) at several grid points; for example, it is at its lower limit at the first six grid points and at its upper limit at the next six grid points. The Lagrange multiplier for the constraint has its largest value initially and gradually decreases to zero after the 13th grid point. According to the Constraint Variation Sensitivity Theorem 4.7, the optimum cost function can be reduced substantially if the limit on the control force is relaxed for a small duration after the system is disturbed.

14.7.3 Minimum Control Effort Problem

Another formulation for the problem is possible where we minimize the total control effort, calculated as

$$f_3 = \int_0^T u^2(t)dt \tag{14.88}$$

The constraints are the same as defined in Eqs. (14.67) through (14.70) and Eqs. (14.64) and (14.65). The numerical procedures for obtaining an optimum solution to the problem are the same as described in Section 14.7.2.

This formulation of the problem is quite well behaved. The same optimum solution is obtained quite rapidly (9–27 iterations) with many different starting points. Figures 14.7 and 14.8 give the displacement response and the control force at the optimum solution, which is obtained starting from $u(t) = 0$ and 41 grid points. This solution was obtained in 13 iterations of the SQP method. The final control effort of 7.481 is much smaller than that for the first case, where it was 28.74. However, the system comes to rest at 0.10 s

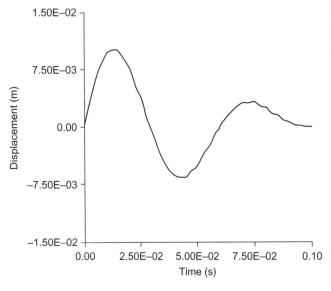

FIGURE 14.7 Graphic of the displacement response at optimum with minimization of control effort as performance index (cost function f_3).

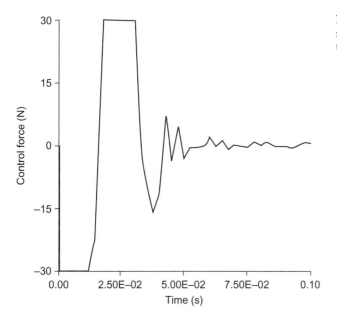

FIGURE 14.8 Optimum control force to minimize the control effort to bring the system to rest after shock input (cost function f_3).

compared with 0.05 s in the previous case. The solution with 21 grid points resulted in a slightly smaller control effort as a result of the numerical procedures used, as explained earlier.

Note that the displacement constraint of Eq. (14.67) is active at the eighth grid point with the normalized Lagrange multiplier as (2.429E−02). The constraints on the terminal displacement and velocity of Eqs. (14.68) and (14.69) are also active with the normalized Lagrange multipliers as (−1.040E−02) and (−3.753E−04). In addition, the control force is at its lower limit at the first grid point with the Lagrange multiplier as (7.153E−04).

14.7.4 Minimum Time Control Problem

The idea of this formulation is to minimize the time required to suppress the motion of the system subject to various constraints. In the previous formulations, the desired time to bring the system to rest was specified. In the present formulation, however, we try to minimize the time T. Therefore, the cost function is

$$f_4 = T \qquad (14.89)$$

The constraints on the system are the same as defined in Eqs. (14.64), (14.65), and (14.67) through (14.70). Note that compared with the previous formulations, gradients of constraints with respect to T are also needed. They can be computed quite easily, since analytical expressions for the functions are known.

The same optimum solution is obtained by starting from several points, such as $T = 0.1, 0.04, 0.02$, and $u(t) = 0, 30, −30$. Figures 14.9 through 14.11 show the displacement and velocity responses and the control force at the optimum with 41 grid points, $T = 0.04$,

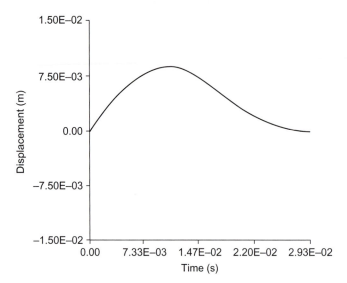

FIGURE 14.9 Displacement response at optimum with minimization of time as performance index (cost function f_4).

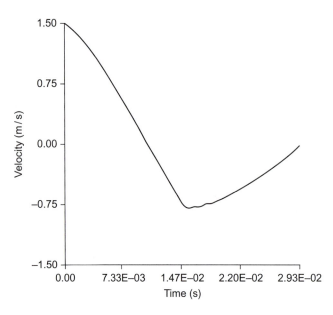

FIGURE 14.10 Velocity response at optimum with minimization of time as performance index (cost function f_4).

and $u(t) = 0$ as the starting point. It takes 0.02933 s to bring the system to rest, requiring 21 iterations of the SQP method. Depending on the starting point, the number of iterations to converge to the final solution varies between 20 and 56.

Constraints on the terminal displacement and velocity of Eqs. (14.68) and (14.69) are active with the normalized Lagrange multipliers as (6.395E−02) and (−1.771E−01), respectively. The control force is at its lower limit for the first 22 grid points and at its upper limit at the remaining points.

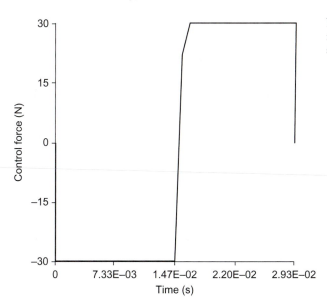

FIGURE 14.11 Optimum control force to minimize time to bring the system to rest after shock input (cost function f_4).

TABLE 14.12 Summary of optimum solutions for three formulations of the problem of optimal control of motion of a system subjected to shock input

	Formulation 1: Minimization of error in state variable	Formulation 2: Minimization of control effort	Formulation 3: Minimization of terminal time
f_1	8.53607E−07	2.32008E−06	8.64466E−07
f_2	1.68241E−02	2.73540E−02	1.45966E−02
f_3	2.87398E+01	7.48104	2.59761E+01
f_4	0.10	0.10	2.9336E−02
NIT	38	13	20
NCF	38	13	20
NGE	100	68	64

NIT, *number of iterations;* NCF, *number of calls for function evaluation;* NGE, *total number of gradients evaluated.*

14.7.5 Comparison of Three Formulations for the Optimal Control of System Motion

It is interesting to compare the three formulations for the optimal control of motion of the system shown in Figure 14.2. Table 14.12 is a summary of the optimum solutions with the three formulations. All of the solutions are obtained with 41 grid points and $u(t) = 0$ as the starting point. For the third formulation, $T = 0.04$ s is used as the starting point.

The results in Table 14.12 show that the control effort is the greatest with the first formulation and the least with the second one. The second formulation turns out to be the most efficient as well as convenient to implement. By varying the total time T, this formulation can be used to generate results for Formulation 3. For example, using $T = 0.05$ and 0.02933 s, solutions with Formulation 2 were obtained. With $T = 0.02933$ s, the same results as with Formulation 3 were obtained. Also, when $T = 0.025$ s was used, Formulation 2 resulted in an infeasible problem.

14.8 ALTERNATIVE FORMULATIONS FOR STRUCTURAL OPTIMIZATION PROBLEMS

We have seen that problems with implicit functions require solving an analysis problem to evaluate the constraints. In addition, special procedures must be used to evaluate the derivatives of the functions. This can be time-consuming in terms of calculations as well as numerical implementations. However, there are alternative formulations that treat the analysis problem and design problem simultaneously in the optimization formulation (Arora and Wang, 2005). In such formulations, the analysis variables are also treated as design variables and the equilibrium equations are treated as equality constraints. In this way all of the functions of the optimization problem are explicit in terms of the optimization variables.

In the optimization process with this formulation, the functions and their derivatives can be evaluated quite easily. Note, however, that the size of the optimization problem is increased substantially in terms of the number of variables and the number of constraints. Therefore, algorithms that are particularly suitable for large-scale optimization problems must be used.

Alternate Formulation for Design of Two-member Frame

As an alternate formulation for this problem, the nodal displacement U_1, U_2, U_3 are treated as design variables in addition to the cross-sectional dimensions. The optimization problem becomes: Find the variables d, h, t, U_1, U_2, U_3 to minimize the cost function in Eq. (14.3):

$$f = 2L(2dt + 2ht - 4t^2) \tag{14.90}$$

subject to the equality constraints (equilibrium equations):

$$h_1: P - \frac{EI}{L^3}[24U_1 - 6LU_2 + 6LU_3] = 0 \tag{14.91}$$

$$h_2: \frac{EI}{L^3}\left[-6LU_1 + \left(4L^2 + \frac{GJ}{EI}L^2\right)U_2\right] = 0 \tag{14.92}$$

$$h_3: \frac{EI}{L^3}\left[6LU_1 + \left(4L^2 + \frac{GJ}{EI}L^2\right)U_3\right] = 0 \tag{14.93}$$

and the inequality constraints:

$$g_1 = \frac{1}{\sigma_a^2}(\sigma_1^2 + 3\tau^2) - 1.0 \le 0 \tag{14.94}$$

$$g_2 = \frac{1}{\sigma_a^2}(\sigma_2^2 + 3\tau^2) - 1.0 \le 0 \tag{14.95}$$

The stresses are calculated in terms of the design variables and the displacements using Eqs. (14.12) through (14.14) as

$$\tau = \frac{T}{2At} = -\frac{GJU_3}{2LAt} \tag{14.96}$$

$$\sigma_1 = \frac{M_1 h}{2I} = \frac{Eh}{L^2}(-3U_1 + U_2 L) \tag{14.97}$$

$$\sigma_2 = \frac{M_2 h}{2I} = \frac{Eh}{L^2}(-3U_1 + 2U_2 L) \tag{14.98}$$

In addition, the explicit constraints on the design variables given in Eq. (14.17) must be imposed. The optimum solution obtained with this formulation was the same as reported in Section 14.5.

Alternative formulations have been developed for truss structures (Wang and Arora, 2005b) and framed structures (Wang and Arora, 2006). These formulations have worked quite well, and their advantages and disadvantages are discussed in the work of Wang and Arora. It is important to note that most of the problem functions in the alternative formulations depend on only a few design variables. Therefore, the Jacobian matrix for the constraints (matrix of partial derivatives) is quite sparsely populated. For large-scale problems, then, optimization algorithms that take advantage of this sparsity must be used for efficiency of calculations (Arora and Wang, 2005; Arora, 2007; Wang and Arora, 2007).

14.9 ALTERNATIVE FORMULATIONS FOR TIME-DEPENDENT PROBLEMS

Time-dependent optimization problems involve integration of linear or nonlinear differential-algebraic equations (DAEs) or just differential equations (DEs) to determine the response of the system to external inputs. Then, using the response variables, cost and constraint functions for the problem are formulated. These constraints are implicit functions of the design variables, and they are time-dependent, which adds to the complexity of the problem.

The most common approach to optimization of such problems has been one in which only the design variables are treated as optimization variables (Arora, 1999). All of the response variables, such as displacements, velocities, and accelerations, are treated as implicit functions of the design variables. Therefore, in the optimization process a system of DAEs is integrated to obtain the response (state) variables and to calculate the values of various functions of the optimization problem. An optimization algorithm is then used to update the design. This nested process of DAE solution and design update, also called the *conventional approach*, is repeated until a stopping criterion is satisfied.

This optimization process is difficult to use in practice. The main difficulty is that the response-related quantities are implicit functions of the design variables, which requires special methods for their gradient evaluation. These methods also require integration of additional DAEs.

It is useful to develop alternative formulations for time-dependent problems that do not require explicit solution of DAEs at each iteration and there is no need for special procedures to calculate derivatives of problem functions with respect to the design variables. By formulating the optimization problem in a mixed space of design and state variables, these two objectives can be met. A complete discussion of all of the alternative formulations is beyond the scope of this text. Therefore we present an overview of recent developments of alternative formulations in two applications areas.

Mechanical and Structural Design Problems

In these applications, various state variables, such as the displacements, velocities and accelerations, are treated as independent variables in addition to the design variables. The equations of motion become equality constraints in the formulations. Since the state variables are functions of time, they need to be parameterized for numerical calculations. Several possible ways to parameterize these variables have been investigated (Wang and Arora, 2005, 2009).

With alternative formulations, all constraints of the problem are expressed explicitly in terms of the optimization variables. Therefore, their gradient evaluations become quite simple. Although the resulting optimization problem is large, it is quite sparse which can be solved using sparse nonlinear programming algorithms. The formulations have been applied to some sample problems to evaluate them and study their advantages and disadvantages.

Digital Human Modeling

Another application area that has seen intense activity recently is the digital human modeling and simulation. The problem here is to simulate various human tasks using a skeletal model of the human body. The problem is to determine all of the joint angle profiles (time-dependent functions) to accomplish a given task. The motion of the model is governed by nonlinear differential equations. These equations are difficult to integrate in order to generate the motion of the skeletal model. Therefore alternative formulations have been developed where an objective function for the problem is defined and the joint angle profiles are treated as optimization variables. In the numerical solution process, the joint angle profiles are parameterized using the B-spline basis functions. This optimization-based approach for simulation of human motion has been called *predictive dynamics* (Xiang, Chung, et al., 2010). The approach has been used successfully to simulate many human activities, such as normal and abnormal walking (Xiang et al., 2009, 2011), lifting of objects (Xiang, Arora, et al., 2010), and throwing of objects (Kim, Xiang, et al., 2010).

EXERCISES FOR CHAPTER 14*

Formulate and solve the following design problems using a nonlinear programming algorithm starting with a reasonable design estimate. Also solve the problems graphically whenever possible and trace the history of the iterative process on the graph of the problem.

14.1 Exercise 3.34	**14.2** Exercise 3.35	**14.3** Exercise 3.36
14.4 Exercise 3.50	**14.5** Exercise 3.51	**14.6** Exercise 3.52
14.7 Exercise 3.53	**14.8** Exercise 3.54	**14.9** Exercise 7.9
14.10 Exercise 7.10	**14.11** Exercise 7.11	**14.12** Exercise 7.12

14.13 *Design of a tapered flag pole.* Formulate the flag pole design problem of Exercise 3.52 for the data given there. Use a hollow tapered circular tube with constant thickness as the structural member. The mass of the pole is to be minimized subject to various constraints. Use a numerical optimization method to obtain the final solution and compare it with the optimum solution for the uniform flag pole.

14.14 *Design of a sign support.* Formulate the sign support column design problem described in Exercise 3.53 for the data given there. Use a hollow tapered circular tube with constant thickness as the structural member. The mass of the pole is to be minimized subject to various constraints. Use a numerical optimization method to obtain the final solution and compare it with the optimum solution for the uniform column.

14.15 Repeat the problem of Exercise 14.13 for a hollow square tapered column of uniform thickness.

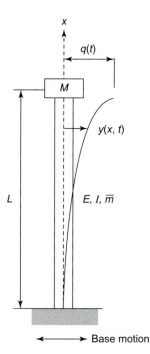

FIGURE E14.20 Cantilever structure with mass at the tip. (See exercise on next page.)

14.16 Repeat the problem of Exercise 14.14 for a hollow square tapered column of uniform thickness.

14.17 For the optimal control problem of minimization of error in the state variable formulated and solved in Section 14.7.2, study the effect of changing the limit on the control force (u_a) to 25 N and then to 35 N.

14.18 For the minimum control effort problem formulated and solved in Section 14.7.3, study the effect of changing the limit on the control force (u_a) to 25 N and then to 35 N.

14.19 For the minimum time control problem formulated and solved in Section 14.7.4, study the effect of changing the limit on the control force (u_a) to 25 N and then to 35 N.

14.20 For the optimal control problem of minimization of error in the state variable formulated and solved in Section 14.7.2, study the effect of having an additional lumped mass M at the tip of the beam ($M = 0.05$ kg) as shown in Figure E14.20.

14.21 For the minimum control effort problem formulated and solved in Section 14.7.3, study the effect of having an additional mass M at the tip of the beam ($M = 0.05$ kg).

14.22 For the minimum time control problem formulated and solved in Section 14.7.4, study the effect of having an additional lumped mass M at the tip of the beam ($M = 0.05$ kg).

14.23 For Exercise 14.20, what will be the optimum solution if the tip mass M is treated as a design variable with limits on it as $0 \leq M \leq 0.10$ kg?

14.24 For Exercise 14.21, what will be the optimum solution if the tip mass M is treated as a design variable with limits on it as $0 \leq M \leq 0.10$ kg?

14.25 For Exercise 14.22, what will be the optimum solution if the tip mass M is treated as a design variable with limits on it as $0 \leq M \leq 0.10$ kg?

14.26 For the optimal control problem of minimization of error in the state variable formulated and solved in Section 14.7.2 study the effect of including a 1 percent critical damping in the formulation.

14.27 For the minimum control effort problem formulated and solved in Section 14.7.3, study the effect of including a 1 percent critical damping in the formulation.

14.28 For the minimum time control problem formulated and solved in Section 14.7.4, study the effect of including a 1 percent critical damping in the formulation.

14.29 For the spring-mass-damper system shown in Figure E14.29, formulate and solve the problem of determining the spring constant and damping coefficient to minimize the maximum acceleration of the system over a period of 10 s when it is subjected to an initial velocity of 5 m/s. The mass is specified as 5 kg.

　　The displacement of the mass should not exceed 5 cm for the entire time interval of 10 s. The spring constant and the damping coefficient must also remain within the limits

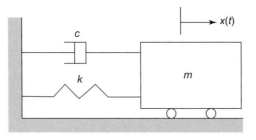

FIGURE E14.29 Graphic of a damped single-degree-of-freedom system.

$1000 \leq k \leq 3000$ N/m; $0 \leq c \leq 300$ N $\cdot$ s/m. (*Hint*: The objective of minimizing the maximum acceleration is a min–max problem, which can be converted to a nonlinear programming problem by introducing an artificial design variable. Let $a(t)$ be the acceleration and A be the artificial variable. Then the objective can be to minimize A subject to an additional constraint $|a(t)| \leq A$ for $0 \leq t \leq 10$).

14.30 Formulate the problem of optimum design of steel transmission poles described in Kocer and Arora (1996b). Solve the problem as a continuous variable optimization problem.

Discrete Variable Optimum Design Concepts and Methods

Upon completion of this chapter, you will be able to

- Formulate mixed continuous-discrete variable optimum design problems

- Use the terminology associated with mixed continuous-discrete variable optimization problems

- Explain concepts associated with various types of mixed continuous-discrete variable optimum design problems and methods

- Determine an appropriate method to solve a mixed continuous-discrete variable optimization problem

In many practical applications, discrete and integer design variables occur naturally in the problem formulation. For example: plate thickness must be selected from the available ones; number of bolts must be an integer; material properties must correspond to the available materials; number of teeth in a gear must be an integer; number of reinforcing bars in a concrete member must be an integer; diameter of reinforcing bars must be selected from the available ones; number of strands in a prestressed member must be an integer; structural members must be selected from commercially available ones, and many more. Types of discrete variables and cost and constraint functions can dictate the method used to solve such problems.

Discrete Variable. A variable is called discrete if its value must be assigned from a given set of values.

Integer Variable. A variable that can have only integer values is called an integer variable. Note that the integer variables are just a special class of discrete variables.

Linked Discrete Variable. If assignment of a value to a variable specifies the values for a group of parameters, then it is called a linked discrete variable.

Binary Variable. A discrete variable that can have a value of 0 or 1 is called a binary variable.

For the sake of brevity, we will refer to these problems as mixed variable (discrete, continuous, integer) optimization problems, or, in short, MV-OPT. In this chapter, we will describe various types of MV-OPT problems, and concepts and terminologies associated with their solution. Various methods for solution of different types of problems will be described. The approach taken is to stress the basic concepts of the methods and point out their advantages and disadvantages.

Because of the importance of this class of problems for practical applications, considerable interest has been shown in the literature to study and develop appropriate methods for their solution. Material for the present chapter is introductory in nature and describes various solution strategies in the most basic form. The material is derived from several publications of the author and his coworkers, and numerous other references cited there (Arora et al., 1994; Arora and Huang, 1996; Huang and Arora, 1995, 1997a,b; Huang et al., 1997; Arora, 1997, 2002; Kocer and Arora 1996a,b, 1997, 1999, 2002). These references contain numerous examples of various classes of discrete variable optimization problems. Only a few of these examples are covered in this chapter.

15.1 BASIC CONCEPTS AND DEFINITIONS

15.1.1 Definition of a Mixed Variable Optimum Design Problem: MV-OPT

The standard design optimization model defined and treated in earlier chapters with equality and inequality constraints can be extended by defining some of the variables as continuous and others as discrete, as follows:

Minimize

$$f(\mathbf{x})$$

subject to

$$h_i = 0, \ i = 1 \text{ to } p, \ g_j \leq 0, \ j = 1 \text{ to } m \tag{15.1}$$

$$x_i \in D_i, \ D_i = (d_{i1}, d_{i2}, \ldots, d_{iq_i}), \ i = 1 \text{ to } n_d, \ x_{iL} \leq x_i \leq x_{iU}, \ i = (n_d + 1) \text{ to } n$$

where f, h_i, and g_j are cost and constraint functions, respectively; x_{iL} and x_{iU} are the lower and upper bounds for the continuous design variable x_i; p, m, and n are the number of equality constraints, inequality constraints, and design variables, respectively; n_d is the number of *discrete design variables*; D_i is the set of discrete values for the ith variable; q_i is the number of allowable discrete values; and d_{ik} is the kth possible discrete value for the ith variable.

Note that the foregoing problem definition includes *integer variable* as well as 0-1 *variable* problems. The formulation in Eq. (15.1) can also be used to solve design problems with linked discrete variables (Arora and Huang 1996; Huang and Arora, 1997a). There are many design applications where such linked discrete variables are encountered. We will describe some of them in a later section.

15.1.2 Classification of Mixed Variable Optimum Design Problems

Depending on the type of design variables, and the cost and constraint functions, mixed continuous-discrete variable problems can be classified into five different categories as described in the following paragraphs. Depending on the type of problem, one discrete variable optimization method may be more effective than another to solve it. In the following, we assume that the continuous variables in the problem can be treated with an appropriate continuous variable optimization method. Or, if appropriate, a continuous variable is transformed into a discrete variable by defining a grid for it. Thus we focus only on the discrete variables.

MV-OPT 1: MIXED DESIGN VARIABLES   Problem functions are twice continuously differentiable. Discrete variables can have nondiscrete values during the solution process (i.e., functions can be evaluated at nondiscrete points). Several solution strategies are available for this class of problem. There are numerous examples of it—such as plate thickness from specified values and member radii from the ones available in the market.

MV-OPT 2: MIXED DESIGN VARIABLES   Problem functions are nondifferentiable at least at some points in the feasible set; however, discrete variables can have nondiscrete values during the solution process. An example of this class of problem includes design problems where constraints from a design code are imposed. Many times these constraints are based on experiments and experience, and are not differentiable everywhere in the feasible set. One example is given in Huang and Arora (1997a,b).

MV-OPT 3: MIXED DESIGN VARIABLES   Problem functions may or may not be differentiable; some of the discrete variables must have only discrete values in the solution process; some of the problem functions can be evaluated only at discrete design variable values during the solution process. Examples of such variables are the number of strands in a prestressed beam or column, the number of teeth in a gear, and the number of bolts for a joint. A problem is not classified as MV-OPT 3 if the effects of the nondiscrete design points can be "simulated" somehow. For instance, a coil spring must have an integer number of coils. However, during the solution process, having a noninteger number of coils is acceptable (it may or may not have any physical meaning) as long as function evaluations are possible.

MV-OPT 4: MIXED DESIGN VARIABLES   Problem functions may or may not be differentiable; some of the discrete variables are linked to others; assignment of a value to one variable specifies values for others. This type of a problem covers many practical applications, such as structural design with members selected from a catalog, material selection, and engine-type selection for automotive and other applications.

MV-OPT 5: COMBINATORIAL PROBLEMS   These are purely discrete nondifferentiable problems. A classic example of this class is the traveling salesman problem. The total distance traveled to visit a number of cities needs to be minimized. A set of integers (cities)

TABLE 15.1   Characteristics of design variables and functions for problem types

MV-OPT	Variable types	Functions differentiable?	Functions defined at nondiscrete points?	Nondiscrete variables allowed for discrete variables?	Variables linked?
1	Mixed	Yes	Yes	Yes	No
2	Mixed	No	Yes	Yes	No
3	Mixed	Yes/No	No	No	No
4	Mixed	Yes/No	No	No	Yes
5	Discrete	No	No	No	Yes/No

can be arranged in different orders to specify a travel schedule (a design). A particular integer can appear only once in a sequence. Examples of this type of engineering design problem include design of a bolt insertion sequence, a welding sequence, and a member placement sequence between a given set of nodes (Huang et al., 1997).

As will be seen later, some of the discrete variable methods assume that the functions and their derivatives can be evaluated at nondiscrete points. Such methods are not applicable to some of the problem types defined above. Various characteristics of the five problem types are summarized in Table 15.1.

15.1.3 Overview of Solution Concepts

Enumerating on allowable discrete values for each of the design variables can always solve discrete variable optimization problems. The number of combinations N_c to be evaluated in such a calculation is given as

$$N_c = \prod_{i=1}^{n_d} q_i \tag{15.2}$$

The number of combinations that are to be analyzed, however, increases rapidly with an increase in n_d, the number of design variables, and q_i, the number of allowable discrete values for each variable. For this reason, full enumeration can lead to an extremely large computational effort to solve the problem. Thus many discrete variable optimization methods try to reduce the search to only a partial list of possible combinations using various strategies and heuristic rules. This is sometimes called *implicit enumeration*.

Most of the methods guarantee an optimum solution to only a very restricted class of problems (linear or convex). For more general nonlinear problems, good usable solutions can be obtained depending on how much computation is allowed. Note that at a discrete optimum point, none of the inequalities may be active unless the discrete point happens to be exactly on the boundary of the feasible set. Also, the final solution is affected by how widely separated the allowable discrete values are in the sets D_i in Eq. (15.1).

It is important to note that if the problem is an MV-OPT 1 type, it is useful to solve it first using a continuous variable optimization method. *The optimum cost function value for the continuous solution represents a lower bound for the value corresponding to a discrete solution.*

The requirement of discreteness of design variables represents additional constraints on the problem. Therefore, the optimum cost function with discrete design variables will have higher value compared with that for the continuous solution. In this way the penalty paid for a discrete solution can be analyzed.

There are two basic classes of method for MV-OPT: enumerative and stochastic. In the enumerative category full enumeration is a possibility; however, partial enumeration is most common, based on branch-and-bound methods. In the stochastic category, the most common methods are simulated annealing, genetic algorithms, and other such algorithms (presented in Chapter 19). Simulated annealing will be discussed later in this chapter; genetic algorithms will be discussed in Chapter 16.

15.2 BRANCH-AND-BOUND METHODS

The branch and bound (BBM) method was originally developed for discrete variable linear programming (LP) problems for which a global optimum solution is obtained. It is sometimes called an *implicit enumeration* method because it reduces the full enumeration in a systematic manner. It is one of the earliest and the best-known methods for discrete variable problems and has also been used to solve MV-OPT problems. The concepts of *branching, bounding, and fathoming* are used to perform the search, as is explained later. The following definitions are useful for description of the method, especially when applied to continuous variable problems.

Half-Bandwidth. When r allowable discrete values are taken below and $(r-1)$ values are taken above a given discrete value for a variable, giving $2r$ allowable values, the parameter r is called the half-bandwidth. It is used to limit the number of allowable values for a discrete variable, for example, based on the rounded-off continuous solution.
Completion. Assignment of discrete values from the allowable ones to all of the variables is called a completion.
Feasible Completion. It is a completion that satisfies all of the constraints.
Partial Solution. It is an assignment of discrete values to some, but not all, of the variables for a continuous discrete problem.
Fathoming. A partial solution for a continuous problem, or a discrete intermediate solution for a discrete problem (*node of the solution tree*), is said to be fathomed if it is determined that no feasible completion of smaller cost than the one previously known can be determined from the current point. It implies that all possible completions have been *implicitly enumerated* from this node.

15.2.1 Basic BBM

The first use of the branch and bound method for linear problems is attributed to Land and Doig (1960). Dakin (1965) later modified the algorithm that has been subsequently used for many applications. There are two basic implementations of the BBM. In the first one, nondiscrete values for the discrete variables are not allowed (or they are not possible) during the solution process. This implementation is quite straightforward; the concepts of *branching*, *bounding*, and *fathoming* are used directly to obtain the final solution. No subproblems are defined or solved; only the problem functions are evaluated for different combinations of design variables.

In the second implementation, nondiscrete values for the design variables are allowed. Forcing a variable to have a discrete value generates a node of the solution tree. This is done by defining additional constraints to force out a discrete value for the variable. The subproblem is solved using either LP or NLP methods depending on the problem type. Example 15.1 demonstrates use of the BBM when only discrete values for the variables are allowed.

EXAMPLE 15.1 BBM WITH ONLY DISCRETE VALUES ALLOWED

Solve the following LP problem:

Minimize

$$f = -20x_1 - 10x_2 \tag{a}$$

subject to

$$g_1 = -20x_1 - 10x_2 + 75 \leq 0 \tag{b}$$

$$g_2 = 12x_1 + 7x_2 - 55 \leq 0 \tag{c}$$

$$g_3 = 25x_1 + 10x_2 - 90 \leq 0 \tag{d}$$

$$x_1 \in \{0, 1, 2, 3\}, \text{ and } x_2 \in \{0, 1, 2, 3, 4, 5, 6\} \tag{e}$$

Solution

In this implementation of the BBM, variables x_1 and x_2 can have only discrete values from the given four and seven values, respectively. Full enumeration would require evaluation of problem functions for 28 combinations; however, the BBM can find the final solution in fewer evaluations. For the problem, the derivatives of f with respect to x_1 and x_2 are always negative. This information can be used to advantage in the BBM. We can enumerate the discrete points in descending order of x_1 and x_2 to ensure that the cost function is always increased when one of the variables is perturbed to the next lower discrete value.

The BBM for the problem is illustrated in Figure 15.1. For each point (called a *node*), the cost and constraint function values are shown. From each node, assigning the next smaller value to each of the variables generates two more nodes. This is called *branching*. At each node, all of the problem functions are evaluated again. If there is any constraint violation at a node, further branching is necessary from that node. Once a feasible completion is obtained, the node requires no further branching since no point with a lower cost is possible from there. Such nodes are said to have *fathomed*; that is, they have reached their lowest point on the branch and no further branching will produce a solution with lower cost. Nodes 6 and 7 are fathomed this way where the cost function has a value of −80.

For the remaining nodes, this value becomes an upper bound for the cost function. This is called *bounding*. Later, any node having a cost function value higher than the current bound is also fathomed. Nodes 9, 10, and 11 are fathomed because the designs are infeasible with the cost function value larger than or equal to the current bound of −80. Since no further branching is possible, the global solution for the problem is found at Nodes 6 and 7 in 11 function evaluations.

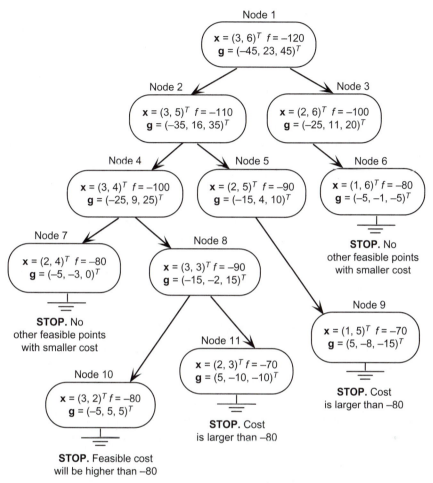

FIGURE 15.1 Basic branch and bound method without solving continuous subproblems (Huang and Arora, 1997; Arora, 2002).

15.2.2 BBM with Local Minimization

For optimization problems where the discrete variables can have nondiscrete values during the solution process and all of the functions are differentiable, we can take advantage of the local minimization procedures to reduce the number of nodes in the solution tree. In such a BBM procedure, initially an optimum point is obtained by treating all of the discrete variables as continuous. If the solution is discrete, an optimum point is obtained and the process is terminated. If one of the variables does not have a discrete value, then its value lies between two discrete values; for example, $d_{ij} < x_i < d_{ij+1}$. Now two subproblems are defined, one with the constraint $x_i \le d_{ij}$ and the other with $x_i \ge d_{ij+1}$. This process is also called *branching*, which is slightly different from the one explained in Example 15.1 for purely discrete problems. It basically eliminates some portion of the continuous feasible region that is not feasible for the discrete problem. However, none of the discrete feasible solutions is eliminated.

The two subproblems are solved again, and the optimum solutions are stored as nodes of the tree containing optimum values for all of the variables, the cost function, and the appropriate bounds on the variables. This process of branching and solving continuous problems is continued until a feasible discrete solution is obtained. Once this has been achieved, the cost function corresponding to the discrete feasible solution becomes an *upper bound* on the cost function for the remaining subproblems (nodes) to be solved later. The solutions that have cost values higher than the upper bound are eliminated from further consideration (i.e., they are fathomed).

The foregoing process of branching and fathoming is repeated from each of the unfathomed nodes. The search for the optimum solution terminates when all of the nodes have been fathomed as a result of one of the following reasons: (1) a discrete optimum solution is found, (2) no feasible continuous solution can be found, or (3) a feasible solution is found but the cost function value is higher than the established upper bound. Example 15.2 illustrates use of the BBM where nondiscrete values for the variables are allowed during the solution process.

EXAMPLE 15.2   BBM WITH LOCAL MINIMIZATIONS

Re-solve the problem of Example 15.1 treating the variables as continuous during the branching and bounding process.

Solution

Figure 15.2 shows implementation of the BBM where requirements of discreteness and non-differentiability of the problem functions are relaxed during the solution process. Here we start with a continuous solution for the problem. From that solution two subproblems are defined by imposing an additional constraint requiring that x_1 not be between 1 and 2. Subproblem 1 imposes the constraint $x_1 \leq 1$ and Subproblem 2, $x_1 \geq 2$. Subproblem 1 is solved using the continuous variable algorithm that gives a discrete value for x_1 but a nondiscrete value for x_2. Therefore, further branching is needed from this node. Subproblem 2 is also solved using the continuous variable algorithm that gives discrete values for the variables with the cost function of -80. This gives an upper bound for the cost function, and no further branching is needed from this node.

Using the solution of Subproblem 1, two subproblems are defined by requiring that x_2 not be between 6 and 7. Subproblem 3 imposes the constraint $x_2 \leq 6$ and Subproblem 4, $x_2 \geq 7$. Subproblem 3 has a discrete solution with $f = -80$, which is the same as the current upper bound. Since the solution is discrete, there is no need to branch further from there by defining more subproblems. Subproblem 4 does not lead to a discrete solution with $f = -80$. Since further branching from this node cannot lead to a discrete solution with the cost function value smaller than the current upper bound of -80, the node is fathomed. Thus, Subproblems 2 and 3 give the two optimum discrete solutions for the problem, as before.

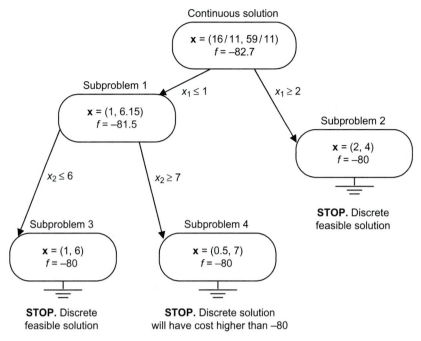

FIGURE 15.2   Branch and bound method with solution of continuous subproblems (Arora, 2002).

Since the foregoing problem has only two design variables, it is fairly straightforward to decide how to create various nodes of the solution process. When there are more design variables, node creation and the branching processes are not unique. These aspects are discussed further for nonlinear problems.

15.2.3 BBM for General MV-OPT

In most practical applications for nonlinear discrete problems, the latter version of the BBM has been used most often, where functions are assumed to be differentiable and the design variables can have nondiscrete values during the solution process. Different methods have been used to solve nonlinear optimization subproblems to generate the nodes. The branch and bound method has been used successfully to deal with discrete design variable problems and has proved to be quite robust. However, for problems with a large number of discrete design variables, the number of subproblems (nodes) becomes large.

Therefore, considerable effort has been spent in investigating strategies to reduce the number of nodes by trying different fathoming and branching rules. For example, the variable that is used for branching to its upper and lower values for the two subproblems is fixed to the assigned value, thus eliminating it from further consideration. This reduces dimensionality of the subproblem that can result in efficiency. As the iterative process progresses, more and more variables are fixed and the size of the optimization problem keeps on decreasing.

Many other variations of the BBM for nonlinear continuous problems have been investigated to improve its efficiency. Since an early establishment of a good upper bound on the cost is important, it may be possible to accomplish this by choosing an appropriate variable for branching. More nodes or subproblems may be fathomed early if a smaller upper bound is established. Several ideas have been investigated in this regard. For example, the distance of a continuous variable from its nearest discrete value, and the cost function value when a variable is assigned a discrete value, can be used to decide the variable to be branched.

It is important to note that the BBM is guaranteed to find the global optimum only if the problem is linear or convex. In the case of general nonlinear nonconvex problems, there is no such guarantee. It is possible that a node is fathomed too early and one of its branches actually contains the true global solution.

15.3 INTEGER PROGRAMMING

Optimization problems where the variables are required to take on integer values are called integer programming (IP) problems. If some of the variables are continuous, we get a mixed variable problem. With all functions as linear, an integer linear programming (ILP) problem is obtained; otherwise, it is nonlinear. The ILP problem can be converted to a 0-1 programming problem. Linear problems with discrete variables can also be converted to 0-1 programming problems. Several algorithms are available to solve such problems (Sysko et al., 1983; Schrijver, 1986), such as BBM, discussed earlier. Nonlinear discrete problems can also be solved by sequential linearization procedures if the problem functions are continuous and differentiable, as discussed later.

In this section, we show how to transform an ILP into a 0-1 programming problem. To do that, let us consider an ILP as follows:

Minimize

$$f = \mathbf{c}^T \mathbf{x}$$

subject to

$$\mathbf{Ax} \le \mathbf{b}$$
$$x_i \ge 0 \text{ integer}; \quad i = 1 \text{ to } n_d; \quad x_{iL} \le x_i \le x_{iU}, \quad i = n_d + 1 \text{ to } n \tag{15.3}$$

Define z_{ij} as the 0-1 variables ($z_{ij} = 0$ or 1 for all i and j). Then the *ith* integer variable is expressed as

$$x_i = \sum_{j=1}^{q_i} z_{ij} d_{ij}; \quad \sum_j z_{ij} = 1, \quad i = 1 \text{ to } n_d \tag{15.4}$$

where q_i and d_{ij} are defined in Eq. (15.1). Substituting this into the foregoing mixed ILP problem, it is converted to a 0-1 programming problem in terms of z_{ij}, as

Minimize

$$f = \sum_{i=1}^{n_d} c_i \left[\sum_{j=1}^{q_i} z_{ij} d_{ij} \right] + \sum_{k=n_d+1}^{n} c_k x_k$$

subject to

$$\sum_{j=1}^{n_d} a_{ij} \left[\sum_{m=1}^{q_j} z_{jm} d_{jm} \right] + \sum_{k=n_d+1}^{n} a_{ik} x_k \leq b_i \tag{15.5}$$

$$\sum_{j=1}^{q_i} z_{ij} = 1, \quad i = 1 \text{ to } n_d; \quad z_{ij} = 0 \text{ or } 1 \text{ for all } i \text{ and } j; \quad x_{iL} \leq x_i \leq x_{iU}; \quad i = n_d + 1 \text{ to } n$$

It is important to note that many modern computer programs for linear programming have an option to solve discrete variable LP problems; for example, LINDO (Schrage, 1991).

15.4 SEQUENTIAL LINEARIZATION METHODS

If the problem functions are differentiable, a reasonable approach to solving the MV-OPT is to linearize the nonlinear problem at each iteration. Then discrete-integer linear programming (LP) methods can be used to solve the linearized subproblem. There are several ways in which a linearized subproblem can be defined and solved. For example, it can be converted to a 0-1 variable problem. This way the number of variables is increased considerably. However, several methods are available to solve integer linear programming problems. Therefore, MV-OPT can be solved using the sequential LP approach and existing codes.

A modification of this approach is to obtain a continuous optimum point first, if possible, and then linearize and use integer programming methods. This process can reduce the number of integer LP subproblems to be solved. Restricting the number of discrete values to those in the neighborhood of the continuous solution (a small value for r, the half-bandwidth) can also reduce the size of the ILP problem. It is noted here that once a continuous solution has been obtained, any discrete variable optimization method can be used with a reduced set of discrete values for the variables.

Another possible approach to solving an MV-OPT problem is to optimize for discrete and continuous variables in a sequence. The problem is first linearized in terms of the discrete variables, but keeping the continuous variables fixed at their current values. The linearized discrete subproblem is solved using a discrete variable optimization method. Then the discrete variables are fixed at their current values, and the continuous subproblem is solved using a nonlinear programming method. The process is repeated a few times to obtain the final solution.

15.5 SIMULATED ANNEALING

Simulated annealing (SA) is a stochastic approach for locating a good approximation to the global minimum of a function. The name of the approach comes from the annealing process in metallurgy. This process involves heating and controlled cooling of a material to increase the size of its crystals and reduce their defects. At high temperatures, the atoms become loose from their initial configuration and move randomly, perhaps through states of higher internal energy, to reach a configuration having absolute minimum energy. The cooling process should be slow, and enough time needs to be spent at each temperature, giving more chance for the atoms to find configurations of lower internal energy. If the temperature is not lowered slowly and enough time is not spent at each temperature, the process can get trapped in a state of local minimum for the internal energy. The resulting crystal may have many defects or the material may even become glass with no crystalline order.

The simulated annealing method for optimization of systems emulates this process. Given a long enough time to run, an algorithm based on this concept finds global minima for continuous-discrete-integer variable nonlinear programming problems.

The basic procedure for implementation of this analogy to the annealing process is to generate random points in the neighborhood of the current best point and evaluate the problem functions there. If the cost function *(penalty function for constrained problems)* value is smaller than its current best value, the point is accepted and the best function value is updated. If the function value is higher than the best value known thus far, the point is sometimes accepted and sometimes rejected. The point's acceptance is based on the value of the probability density function of the Bolzman-Gibbs distribution. If this probability density function has a value greater than a random number, then the trial point is accepted as the best solution even if its function value is higher than the known best value.

In computing the probability density function, a parameter called the *temperature* is used. For the optimization problem, this temperature can be a target value for the optimum value of the cost function. Initially, a larger target value is selected. As the trials progress, the target value (the temperature) is reduced (this is called the *cooling schedule*), and the process is terminated after a large number of trials. The acceptance probability steadily decreases to zero as the temperature is reduced. Thus, in the initial stages, the method sometimes accepts worse designs, while in the final stages the worse designs are almost always rejected. This strategy avoids getting trapped at a local minimum point.

It is seen that the SA method requires evaluation of cost and constraint functions only. Continuity and differentiability of functions are not required. Thus the method can be useful for nondifferentiable problems, and problems for which gradients cannot be calculated or are too expensive to calculate. It is also possible to implement the algorithm on parallel computers to speed up the calculations. The deficiencies of the method are the unknown rate for reduction of the target level for the global minimum, and the uncertainty in the total number of trials and the point at which the target level needs to be reduced.

Simulated Annealing Algorithm

It is seen that the algorithm is quite simple and easy to program. The following steps illustrate the basic ideas of the algorithm.

Step 1. Choose an initial temperature T_0 (expected global minimum for the cost function) and a feasible trial point $\mathbf{x}^{(0)}$. Compute $f(\mathbf{x}^{(0)})$. Select an integer L (e.g., a limit on the number of trials to reach the expected minimum value), and a parameter $r < 1$. Initialize the iteration counter as $K = 0$ and another counter $k = 1$.

Step 2. Generate a new point $\mathbf{x}^{(k)}$ randomly in a neighborhood of the current point. If the point is infeasible, generate another random point until feasibility is satisfied (a variation of this step is explained later). Compute $f(\mathbf{x}^{(k)})$ and $\Delta f = f(\mathbf{x}^{(k)}) - f(\mathbf{x}^{(0)})$.

Step 3. If $\Delta f < 0$, then accept $\mathbf{x}^{(k)}$ as the new best point $\mathbf{x}^{(0)}$, set $f(\mathbf{x}^{(0)}) = f(\mathbf{x}^{(k)})$ and go to Step 4. Otherwise, calculate the probability density function:

$$p(\Delta f) = exp\left(\frac{-\Delta f}{T_K}\right) \tag{15.6}$$

Generate a random number z uniformly distributed in [0,1]. If $z < p(\Delta f)$, then accept $\mathbf{x}^{(k)}$ as the new best point $\mathbf{x}^{(0)}$ and go to Step 4. Otherwise go to Step 2.

Step 4. If $k < L$, then set $k = k + 1$ and go to Step 2. If $k > L$ and any of the stopping criteria is satisfied, then stop. Otherwise, go to Step 5.

Step 5. Set $K = K + 1$, $k = 1$; set $T_K = rT_{K-1}$; go to Step 2.

The following points are noted for implementation of the algorithm:

1. In Step 2 only one point is generated at a time within a certain neighborhood of the current point. Thus, although SA randomly generates design points without the need for function or gradient information, it is not a pure random search within the entire design space. At the early stage, a new point can be located far away from the current point to speed up the search process and to avoid getting trapped at a local minimum point. Once the temperature gets low, the new point is usually created nearby in order to focus on the local area. This can be controlled by defining a step size procedure.

2. In Step 2, the newly generated point is required to be feasible. If it is not, another point is generated until feasibility is attained. Another method for treating constraints is to use the penalty function approach; that is, the constrained problem is converted to an unconstrained one, as discussed in Chapter 11. The cost function is replaced by the penalty function in the algorithm. Therefore, the feasibility requirements are not imposed explicitly in Step 2.

3. The following stopping criteria are suggested in Step 4:
   (a) The algorithm stops if change in the best function value is less than some specified value for the last J consecutive iterations.
   (b) The *program* stops if $I/L < \delta$, where L is a limit on the number of trials (or the number of feasible points generated) within one iteration, and I is the number of trials that satisfy $\Delta f < 0$ (see Step 3).
   (c) The algorithm stops if K reaches a preset value.

The foregoing ideas from statistical mechanics can also be used to develop methods for global optimization of continuous variable problems. For such problems, simulated annealing may be combined with a local minimization procedure. However, the temperature T is slowly and continuously decreased so that the effect is similar to annealing. Using the probability density function given in Eq. (15.6), a criterion can be used to decide whether to start a local search from a particular point.

15.6 DYNAMIC ROUNDING-OFF METHOD

A simple approach for MV-OPT 1 type problems is to first obtain an optimum solution using a continuous approach. Then, using heuristics, the variables are rounded off to their nearest available discrete values to obtain a discrete solution. Rounding-off is a simple idea that has been used often, but it can result in infeasible designs for problems having a large number of variables. The main concern of the rounding-off approach is the selection of variables to be increased and the variables to be decreased. The strategy may not converge, especially in case of high nonlinearity and widely separated allowable discrete values. In that case, the discrete minimum point need not be in a neighborhood of the continuous solution.

Dynamic Rounding-off Algorithm

The dynamic rounding-off algorithm is a simple extension of the usual rounding-off procedure. The basic idea is to round off variables in a sequence rather than all of them at the same time. After a continuous variable optimum solution is obtained, one or a few variables are selected for discrete assignment. This assignment can be based on the penalty that needs to be paid for the increase in the cost function or the Lagrangian function. These variables are then eliminated from the problem and the continuous variable optimization problem is solved again. This idea is quite simple because an existing optimization program can be used to solve discrete variable problems of type MV-OPT 1. The process can be carried out in an interactive mode, or it can be implemented manually. Whereas the dynamic rounding-off strategy can be implemented in many different ways, the following algorithm illustrates one simple procedure:

Step 1. Assume that all of the design variables to be continuous and solve the NLP problem.
Step 2. If the solution is discrete, stop. Otherwise, continue.
Step 3. FOR $k = 1$ to n
Calculate the Lagrangian function value for each k, with the kth variable perturbed to its discrete neighbors.
END FOR
Step 4. Choose a design variable that minimizes the Lagrangian in Step 3 and remove that variable from the design variable set. This variable is assigned the selected discrete value. Set $n = n - 1$ and if $n = 1$, stop; otherwise, go to Step 2.

The number of additional continuous problems that needs to be solved by the preceding method is $(n - 1)$. However, the number of design variables is reduced by one for each subsequent continuous problem. In addition, more variables may be assigned discrete values each time, thus reducing the number of continuous problems to be solved. The dynamic rounding-off strategy has been used successfully to solve several optimization problems (Al-Saadoun and Arora, 1989; Huang and Arora, 1997a,b).

15.7 NEIGHBORHOOD SEARCH METHOD

When the number of discrete variables is small and each discrete variable has only a few choices, the simplest way to find the solution to a mixed variable problem may be just to explicitly enumerate all of the possibilities. With all of the discrete variables fixed at their chosen values, the problem is then optimized for the continuous variables. This approach has some advantages over the BBM: It can be implemented easily with an existing optimization software, the problem to be solved is smaller, and the gradient information with respect to the discrete variables is not needed. However, the approach is far less efficient than an implicit enumeration method, such as the BBM, as the number of discrete variables and the size of the discrete set of values become large.

When the number of discrete variables is large and the number of discrete values for each variable is large, then a simple extension of the above approach is to solve the optimization problem first by treating all of the variables as continuous. Based on that solution, a reduced set of allowable discrete values for each variable is then selected. Now the neighborhood search approach is used to solve the MV-OPT 1 problem. A drawback is that the search for a discrete solution is restricted to only a small neighborhood of the continuous solution.

15.8 METHODS FOR LINKED DISCRETE VARIABLES

Linked discrete variables occur in many applications. For example, in the design of a coil spring problem formulated in Chapter 2, we may have choice of three materials as shown in Table 15.2. Once a material type is specified, all of the properties associated with it must be selected and used in all calculations. The optimum design problem is to determine the material type and other variables to optimize an objective function and satisfy all of the constraints. The problem has been solved in Huang and Arora (1997a,b).

TABLE 15.2   Material data for the spring design problem

Material type	G, lb/in^2	ρ, lb-s^2/in^4	τ_a, lb/in^2	U_p
1	11.5×10^6	7.38342×10^{-4}	80,000	1.0
2	12.6×10^6	8.51211×10^{-4}	86,000	1.1
3	13.7×10^6	9.71362×10^{-4}	87,000	1.5

G = shear modulus; ρ = mass density; τ_a = shear stress; U_p = relative unit price.

Another practical example where linked discrete variables are encountered is the optimum design of framed structural systems. Here the structural members must be selected from the ones available in the manufacturer's catalog. Table 15.3 shows some of the standard wide-flange sections (W-shapes) available in the catalog. The optimum design problem is to find the best possible sections for members of a structural frame to minimize a cost function and satisfy all of the performance constraints. The section number, section area, moment of inertia, or any other section property can be designated as a linked discrete design variable for the frame member. Once a value for such a discrete variable is specified from the table, each of its linked variables (properties) must also be assigned the unique value and used in the optimization process.

These properties affect values of the cost and constraint functions for the problem. A certain value for a particular property can only be used when appropriate values for other properties are also assigned. Relationships among such variables and their linked properties cannot be expressed analytically, and so a gradient-based optimization method may be applicable only after some approximations. It is not possible to use one of the properties as the only continuous design variable because other section properties cannot be calculated using just that property. Also, were each property to be treated as an independent design variable, the final solution would generally be unacceptable since the variables would have values that cannot co-exist in the table. Solutions for such problems are presented in Huang and Arora (1997a,b).

It is seen that problems with linked variables are discrete and the problem functions are not differentiable with respect to them. Therefore, they must be treated by a discrete variable optimization algorithm that does not require gradients of functions. There are several algorithms for such problems, such as simulated annealing and genetic algorithms. Simulated annealing was discussed earlier, and genetic algorithms are presented in Chapter 16.

It is noted that for each class of problems having linked discrete variables, it is possible to develop strategies to treat the problem more efficiently by exploiting the structure of

TABLE 15.3   Some wide flange standard sections

Section	A	d	t_w	b	t_f	I_x	S_x	r_x	I_y	S_y	r_y
W36 × 300	88.30	36.74	0.945	16.655	1.680	20300	1110	15.20	1300	156	3.830
W36 × 280	82.40	36.52	0.885	16.595	1.570	18900	1030	15.10	1200	144	3.810
W36 × 260	76.50	36.26	0.840	16.550	1.440	17300	953	15.00	1090	132	3.780
W36 × 245	72.10	36.08	0.800	16.510	1.350	16100	895	15.00	1010	123	3.750
W36 × 230	67.60	35.90	0.760	16.470	1.260	15000	837	14.90	940	114	3.730
W36 × 210	61.80	36.69	0.830	12.180	1.360	13200	719	14.60	411	67.5	2.580
W36 × 194	57.00	36.49	0.765	12.115	1.260	12100	664	14.60	375	61.9	2.560

Wn × m, n = nominal depth of the section, in, m = weight/ft; A = cross-sectional area, in²; I_x = moment of inertia about the x−x axis, in⁴; d = depth, in; S_x = elastic section modulus about the x−x axis, in³; t_w = web thickness, in; r_x = radius of gyration with respect to the x−x axis, in; b = flange width, in; I_y = moment of inertia about the y−y axis, in⁴; t_f = flange thickness, in; S_y = elastic section modulus about the y−y axis, in³; r_y = radius of gyration with respect to the y−y axis, in.

the problem and knowledge of the problem functions. Two or more algorithms may be combined to develop strategies that are more effective than the use of a purely discrete algorithm. For the structural design problem, several such strategies have been developed (Arora, 2002).

15.9 SELECTION OF A METHOD

Selection of a method to solve a particular mixed variable optimization problem depends on the nature of the problem functions. Features of the methods and their suitability for various types of MV-OPT problems are summarized in Table 15.4. It is seen that branch and bound, simulated annealing, and genetic algorithms (discussed in Chapter 16) are the most general methods. They can be used to solve all the problem types. However, these are also the most expensive ones in terms of computational effort.

If the problem functions are differentiable and discrete variables can be assigned non-discrete values during the iterative solution process, then there are numerous strategies for their solution that are more efficient than the three methods just discussed. Most of these involve a combination of two or more algorithms.

Huang and Arora (1997a,b) have evaluated the discrete variable optimization methods presented in this chapter using 15 types of test problems. Applications involving linked discrete variables are described in Huang and Arora (1997), Arora and Huang (1996), and Arora (2002). Applications of discrete variable optimization methods to electrical transmission line structures are described in Kocer and Arora (1996, 1997, 1999, 2002). Discrete variable optimum solutions for the plate girder design problem formulated and solved in Section 6.6 are described and discussed in Arora et al. (1997).

TABLE 15.4 Characteristics of discrete variable optimization methods

Method	MV-OPT problem type solved	Can find feasible discrete solution?	Can find global minimum for convex problem?	Need gradients?
Branch and bound	1–5	Yes	Yes	No/Yes
Simulated annealing	1–5	Yes	Yes	No
Genetic algorithm	1–5	Yes	Yes	No
Sequential linearization	1	Yes	Yes	Yes
Dynamic round-off	1	Yes	No guarantee	Yes
Neighborhood search	1	Yes	Yes	Yes

15.10 ADAPTIVE NUMERICAL METHOD FOR DISCRETE VARIABLE OPTIMIZATION

In Section 2.11.8, we described a simple adaptive procedure for discrete variable optimization. In this section, we demonstrate that procedure for a simple design problem. The basic idea of the procedure is to obtain an optimum solution of the problem with continuous variables, if that is possible. Then the variables that are close to their discrete values are assigned that value. They are then held fixed and the problem is re-optimized. The procedure is continued until all of the variables have been assigned discrete values.

The application area that we have chosen is the optimum design of aerospace, automotive, mechanical, and structural systems, by employing finite-element models. The problem is to design a minimum-weight system with constraints on various performance specifications. As a sample application, we will consider the 10-bar cantilever structure shown in Figure 15.3. The loading and other design data for the problem are given in Table 15.5. The set of discrete values given there is taken from the American Institute of Steel Construction (AISC) *Manual*. The final design for the structure must be selected from this set.

The cross-sectional area of each member is treated as a design variable giving a total of 10 variables. Constraints are imposed on member stress (10), nodal displacement (8), member buckling (10), vibration frequency (1), and explicit bounds on the design variables (20). This gives a total of 49 constraints. In imposing the member buckling constraint, the moment of inertia is taken as $I = \beta A^2$, where β is a constant and A is the member cross-sectional area.

The formulation of the problem is quite similar to that for the three-bar structure discussed in Section 2.10. The only difference is that the explicit form of the constraint function is not known. Therefore, we must use the finite-element procedures described in Sections 14.1 and 14.2 for structural analysis and the gradient evaluation of constraints.

15.10.1 Continuous Variable Optimization

To compare solutions, the continuous variable optimization problem is solved first. To use the program IDESIGN, user subroutines are developed using the material of Sections 14.1 and 14.2 to evaluate the functions and their gradients. These subroutines contain analysis of trusses and evaluation of gradients of constraints. The optimum solution, using a very strict stopping criterion and a uniform starting design of 1.62 in^2, is obtained as

Design variables: 28.28, 1.62, 27.262, 13.737, 1.62, 4.0026, 13.595, 17.544, 19.13, 1.62
Optimum cost function: 5396.5 lb

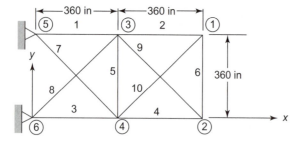

FIGURE 15.3   Ten-bar cantilever truss.

TABLE 15.5 Design data for a 10-bar structure

Element	Data
Modulus of elasticity	$E = (10^7)$ psi
Material weight density	$\gamma = (0.1)$ lb/in^3
Displacement limit	± 2.0 in
Stress limit	25,000 psi
Frequency limit	22 Hz
Lower limit on design variables	1.62 in^2
Upper limit on design variables	None
Constant β $(I = \beta A^2)$	1.0

LOADING DATA	
Node no.	**Load in y-direction (lb)**
1	50,000
2	−150,000
3	50,000
4	−150,000
Available member areas (in^2)	1.62, 1.80, 1.99, 2.13, 2.38, 2.62, 2.63, 2.88, 2.93, 3.09, 3.13, 3.38, 3.47, 3.55, 3.63, 3.84, 3.87, 3.88, 4.18, 4.22, 4.49, 4.59, 4.80, 4.97, 5.12, 5.74, 7.22, 7.97, 11.50, 13.50, 13.90, 14.20, 15.50, 16.00, 16.90, 18.80, 19.90, 22.00, 22.90, 26.50, 30.00, 33.50

Number of iterations: 19
Number of analyses: 21
Maximum constraint violation at optimum: (8.024E−10)
Convergence parameter at optimum: (2.660E−05)
Active constraints at optimum and their Lagrange multipliers:
    Frequency: 392.4
    Stress in member 2: 38.06
    Displacement at node 2 in the y direction: 4967
    Lower bound for member 2: 7.829
    Lower bound for member 5: 205.1
    Lower bound for member 10: 140.5

15.10.2 Discrete Variable Optimization

We use the adaptive numerical optimization procedure that was described in Section 2.11.8 to obtain a discrete variable solution. The procedure is to use the IDESIGN program in the interactive mode. Design conditions are monitored and decisions made to

TABLE 15.6   Interactive solution for a 10-Member structure with discrete variables

Iteration no.	Maximum violation (%)	Cost function	Algorithm used	Variables fixed to value shown in parentheses
1	1.274E+04	679.83	CC	All free
2	4.556E+03	1019.74	CC	All free
3	1.268E+03	1529.61	CC	All free
4	4.623E+02	2294.42	CC	All free
5	1.144E+02	3441.63	CC	All free
6	2.020E+01	4722.73	CC	5(1.62), 10(1.62)
7	2.418E+00	5389.28	CCC	2(1.80)
11	1.223E−01	5402.62	SQP	1(30.0), 6(3.84), 7(13.5)
13	5.204E−04	5411.13	SQP	3(26.5), 9(19.9)
14	1.388E+00	5424.69	—	4(13.5), 8(16.9)

CC = constraint correction algorithm; CCC = constraint correction at constant cost; SQP = sequential quadratic programming.

fix design variables that are not changing. The interactive facilities used include design variable histories, maximum constraint violation, and the cost function.

Table 15.6 contains a snapshot of the design conditions at various iterations and the decisions made. It can be seen that for the first five iterations the constraint violations are very large, so the constraint correction (CC) algorithm is used to correct the constraints. At the 6th iteration, it is determined that design variables 5 and 10 are not changing, so they are fixed to their current value. Similarly, at other iterations, variables are assigned values from the available set. At the 14th iteration, all variables have discrete values, the constraint violation is about 1.4 percent, and the structural weight is 5424.69, which is an increase of less than 1 percent from the optimum with continuous variables. This is a reasonable final solution.

It should be noted, that with the discrete variables, several solutions near the true optimum point are possible. A different sequence of fixing variables can give a different solution. For example, starting from the optimum solution with continuous variables, the following acceptable discrete solutions are obtained interactively:

1. 30.0, 1.62, 26.5, 13.9, 1.62, 4.18, 13.5, 18.8, 18.8, 1.62; cost = 5485.6, maximum violation = 4.167 percent for stress in member 2.
2. Same as (1) except the 8th design variable is 16.9; cost = 5388.9, and maximum violation = 0.58 percent.
3. Same as (1) except design variables 2 and 6 are 2.38 and 2.62; cost = 5456.8, maximum violation = 3.74 percent for stress in member 2
4. Same as (3) except design variable 2 is 2.62; cost = 5465.4; all constraints are satisfied.

EXERCISES FOR CHAPTER 15*

15.1 Solve Example 15.1 with the available discrete values for the variables as $x_1 \in \{0, 1, 2, 3\}$, and $x_2 \in \{0, 1, 2, 3, 4, 5, 6\}$. Assume that the functions of the problem are not differentiable.

15.2 Solve Example 15.1 with the available discrete values for the variables as $x_1 \in \{0, 1, 2, 3\}$, and $x_2 \in \{0, 1, 2, 3, 4, 5, 6\}$. Assuming that the functions of the problem are differentiable, use a continuous variable optimization procedure to solve for discrete variables.

15.3 Formulate and solve Exercise 3.34 using the outside diameter d_o and the inside diameter d_i as design variables. The outside diameter and thickness must be selected from the following available sets:

$$d_o \in \{0.020, 0.022, 0.024, \ldots, 0.48, 0.50\} \text{ m}; \quad t \in \{5, 7, 9, \ldots, 23, 25\} \text{ mm}$$

Check your solution using the graphical method of Chapter 3. Compare continuous and discrete solutions.

15.4 Consider the minimum mass tubular column problem formulated in Section 2.7. Find the optimum solution for the problem using the following data: $P = 100$ kN, length, $l = 5$ m, Young's modulus, $E = 210$ GPa, allowable stress, $\sigma_a = 250$ MPa, mass density, $\rho = 7850$ kg/m³, $R \leq 0.4$ m, $t \leq 0.05$ m, and $R, t \geq 0$. The design variables must be selected from the following sets:

$$R \in \{0.01, 0.012, 0.014, \ldots, 0.38, 0.40\} \text{ m}; \quad t \in \{4, 6, 8, \ldots, 48, 50\} \text{ mm}$$

Check your solution using the graphical method of Chapter 3. Compare continuous and discrete solutions.

15.5 Consider the plate girder design problem described and formulated in Section 6.6. The design variables for the problem must be selected from the following sets

$$h, b \in \{0.30, 0.31, 0.32, \ldots, 2.49, 2.50\} \text{ m}; \quad t_w, t_f \in \{10, 12, 14, \ldots, 98, 100\} \text{ mm}$$

Assume that the functions of the problem are differentiable and a continuous variable optimization program can be used to solve subproblems, if needed. Solve the discrete variable optimization problem. Compare the continuous and discrete solutions.

15.6 Consider the plate girder design problem described and formulated in Section 6.6. The design variables for the problem must be selected from the following sets

$$h, b \in \{0.30, 0.31, 0.32, \ldots, 2.49, 2.50\} \text{ m}; \quad t_w, t_f \in \{10, 12, 14, \ldots, 98, 100\} \text{ mm}$$

Assume functions of the problem to be nondifferentiable. Solve the discrete variable optimization problem. Compare the continuous and discrete solutions.

15.7 Consider the plate girder design problem described and formulated in Section 6.6. The design variables for the problem must be selected from the following sets

$$h, b \in \{0.30, 0.31, 0.32, \ldots, 2.48, 2.50\} \text{ m}; \quad t_w, t_f \in \{10, 14, 16, \ldots, 96, 100\} \text{ mm}$$

Assume that the functions of the problem are differentiable and a continuous variable optimization program can be used to solve subproblems, if needed. Solve the discrete variable optimization problem. Compare the continuous and discrete solutions.

15.8 Consider the plate girder design problem described and formulated in Section 6.6. The design variables for the problem must be selected from the following sets

$$h, b \in \{0.30, 0.32, 0.34, \ldots, 2.48, 2.50\} \text{ m}; \quad t_w, t_f \in \{10, 14, 16, \ldots, 96, 100\} \text{ mm}$$

Assume functions of the problem to be nondifferentiable. Solve the discrete variable optimization problem. Compare the continuous and discrete solutions.

15.9 Solve the problems of Exercises 15.3 and 15.5. Compare the two solutions, commenting on the effect of the size of the discreteness of variables on the optimum solution. Also, compare the continuous and discrete solutions.

15.10 Consider the spring design problem formulated in Section 2.9 and solved in Section 6.5. Assume that the wire diameters are available in increments of 0.01 in, the coils can be fabricated in increments of 1/16th of an inch, and the number of coils must be an integer. Assume functions of the problem to be differentiable. Compare the continuous and discrete solutions.

15.11 Consider the spring design problem formulated in Section 2.9 and solved in Section 6.5. Assume that the wire diameters are available in increments of 0.01 in, the coils can be fabricated in increments of 1/16th of an inch, and the number of coils must be an integer. Assume the functions of the problem to be nondifferentiable. Compare the continuous
and discrete solutions.

15.12 Consider the spring design problem formulated in Section 2.9 and solved in Section 6.5. Assume that the wire diameters are available in increments of 0.015 in, the coils can be fabricated in increments of 1/8th of an inch, and the number of coils must be an integer. Assume functions of the problem to be differentiable. Compare the continuous and discrete solutions.

15.13 Consider the spring design problem formulated in Section 2.9 and solved in Section 6.5. Assume that the wire diameters are available in increments of 0.015 in, the coils can be fabricated in increments of 1/8th of an inch, and the number of coils must be an integer. Assume the functions of the problem to be nondifferentiable. Compare the continuous and discrete solutions.

15.14 Solve problems of Exercises 15.8 and 15.10. Compare the two solutions, commenting on the effect of the size of the discreteness of variables on the optimum solution. Also, compare the continuous and discrete solutions.

15.15 Formulate the problem of optimum design of prestressed concrete transmission poles described in Kocer and Arora (1996a). Use a mixed variable optimization procedure to solve the problem. Compare the solution to that given in the reference.

15.16 Formulate the problem of optimum design of steel transmission poles described in Kocer and Arora (1996b). Solve the problem as a continuous variable optimization problem.

15.17 Formulate the problem of optimum design of steel transmission poles described in Kocer and Arora (1996b). Assume that the diameters can vary in increments of 0.5 in and the thicknesses can vary in increments of 0.05 in. Solve the problem as a discrete variable optimization problem.

15.18 Formulate the problem of optimum design of steel transmission poles using standard sections described in Kocer and Arora (1997). Compare your solution to the solution given there.

15.19 Solve the following mixed variable optimization problem (Hock and Schittkowski, 1981):

Minimize

$$f = (x_1 - 10)^2 + 5(x_2 - 12)^2 + 3(x_4 - 11)^2 + 10x_5^6$$
$$+ 7x_6^2 + x_7^4 - 4x_6x_7 - 10x_6 - 8x_7$$

subject to

$$g_1 = 2x_1^2 + 3x_2^4 + x_3 + 4x_4^2 + 5x_5 \le 127$$

$$g_2 = 7x_1 + 3x_2 + 10x_3^2 + x_4 - x_5 \le 282$$

$$g_3 = 23x_1 + x_2^2 + 6x_6^2 - 8x_7 \le 196$$

$$g_4 = 4x_1^2 + x_2^2 - 3x_1x_2 + 2x_3^2 + 5x_6 - 11x_7 \le 0$$

The first three design variables must be selected from the following sets

$$x_1 \in \{1, 2, 3, 4, 5\}; \quad x_2, x_3 \in \{1, 2, 3, 4, 5\}$$

15.20 Formulate and solve the three-bar truss of Exercise 3.50 as a discrete variable problem where the cross-sectional areas must be selected from the following discrete set:

$$A_i \in \{50, 100, 150, \ldots, 4950, 5000\} \, \text{mm}^2$$

Check your solution using the graphical method of Chapter 3. Compare continuous and discrete solutions.

16

Genetic Algorithms for Optimum Design

Genetic algorithms (GA) belong to the class of *stochastic search optimization methods*, such as the simulated annealing method described in Chapter 15. These algorithms also belong to a class of methods known as *evolutionary methods* or *nature-inspired methods*, presented in Chapter 19. As you get to know basics of the algorithms, you will see that decisions made in most computational steps of the algorithms are based on random number generation. Therefore, executed at different times, the algorithm can lead to a different sequence of designs and a different problem solution even with the same initial conditions.

The genetic algorithms use only the function values in the search process to make progress toward a solution without regard to how the functions are evaluated. Continuity or differentiability of the problem functions is neither required nor used in calculations of the algorithms. Therefore, the algorithms are very general and can be applied to all kinds of problems—discrete, continuous, and nondifferentiable. In addition, the methods determine global optimum solutions as opposed to the local solutions determined by a *derivative-based optimization algorithm*. The methods are easy to use and program since they do not require the use of gradients of cost or constraint functions.

The *drawbacks* of genetic algorithms are as follows:

1. They require a large amount of calculation for even reasonably sized problems or for problems where evaluation of functions itself requires massive calculation.
2. There is no absolute guarantee that a global solution has been obtained.

The first drawback can be overcome to some extent by the use of massively parallel computers. The second drawback can be overcome to some extent by executing the algorithm several times and allowing it to run longer.

Recall that the optimization problem considered is defined as

Minimize

$$f(\mathbf{x}) \quad \text{for } \mathbf{x} \in S \tag{16.1}$$

where S is the set of feasible designs defined by equality and inequality constraints. For unconstrained problems, S is the entire design space. Note that to use a genetic algorithm, the constrained problem is converted to an unconstrained problem using the penalty approach discussed in Chapter 11.

In the remaining sections of this chapter, concepts and terminology associated with genetic algorithms are defined and explained for the optimization problem. Fundamentals of genetic algorithms are presented and explained. Although the algorithm can be used for continuous problems, our focus will be on discrete variable optimization problems. Various steps of a genetic algorithm are described that can be implemented in different ways.

Most of the material for this chapter is derived from the work of the author and his coworkers and is introductory in nature (Arora et al., 1994; Huang and Arora, 1997; Huang et al., 1997; Arora, 2002). Numerous other good references on the subject are available (e.g., Holland, 1975; Goldberg, 1989; Mitchell, 1996; Gen and Cheng, 1997; Coello-Coello et al., 2002; Osyczka, 2002; Pezeshk and Camp, 2002).

16.1 BASIC CONCEPTS AND DEFINITIONS

Genetic algorithms loosely parallel *biological evolution* and are based on Darwin's theory of natural selection. The specific mechanics of the algorithm uses the language of microbiology, and its implementation mimics genetic operations. We will explain this in subsequent paragraphs and sections. *The basic idea of the approach is to start with a set of designs,* randomly generated using the allowable values for each design variable. Each design is also assigned a fitness value, usually using the cost function for unconstrained problems or the penalty function for constrained problems. From the current set of designs, a subset is selected randomly with a bias allocated to more fit members of the set. Random processes are used to generate new designs using the selected subset of designs.

The size of the design set is kept fixed. Since more fit members of the set are used to create new designs, the successive sets of designs have a higher probability of having designs with better fitness values. The process is continued until a stopping criterion is met. In the following paragraphs, some details of implementing these basic steps are presented and explained. First we will define and explain various terms associated with the algorithm.

Population. The set of design points at the current iteration is called a population. It represents a group of designs as potential solution points. N_p = number of designs in a population; this is also called the population size.
Generation. An iteration of the genetic algorithm is called a generation. A generation has a population of size N_p that is manipulated in a genetic algorithm.

Chromosome. This term is used to represent a design point. Thus a chromosome represents a design of the system, whether feasible or infeasible. It contains values for all the design variables of the system.

Gene. This term is used for a scalar component of the design vector; that is, it represents the value of a particular design variable.

Design Representation

A method is needed to represent design variable values in their allowable sets and to represent design points so that they can be used and manipulated in the algorithm. This is called a *schema*, and it needs to be encoded (i.e., defined). Although binary encoding is the most common approach, real-number coding and integer encoding are also possible. Binary encoding implies a string of 0s and 1s. Binary strings are also useful because it is easier to explain the operations of the genetic algorithm with them.

A binary string of 0s and 1s can represent a design variable (a gene). Also, an L-digit string with a 0 or 1 for each digit, where L is the total number of binary digits, can be used to specify a design point (a chromosome). Elements of a binary string are called *bits*; a bit can have a value of 0 or 1. We will use the *term "V-string" for a binary string that represents the value of a variable*; that is, the component of a design vector (a gene). Also, we will use the *term "D-string" for a binary string that represents a design of the system*—that is, a particular combination of n V-strings, where n is the number of design variables. This is also called a *genetic string* (or a chromosome).

An m-digit binary string has 2^m possible 0–1 combinations implying that 2^m discrete values can be represented. The following method can be used to transform a V-string consisting of a combination of m 0's and 1's to its corresponding discrete value of a variable having N_c allowable discrete values: let m be the smallest integer satisfying $2^m > N_c$; calculate the integer j:

$$j = \sum_{i=1}^{m} ICH(i)2^{(i-1)} + 1 \tag{16.2}$$

where $ICH(i)$ is the value of the ith digit (either 0 or 1). Thus the jth allowable discrete value corresponds to this 0–1 combination; that is, the jth discrete value corresponds to this V-string. Note that when $j > N_c$ in Eq. (16.2), the following procedure can be used to adjust j such that $j \le N_c$:

$$j = INT\left(\frac{N_c}{2^m - N_c}\right)(j - N_c) \tag{16.3}$$

where INT(x) is the integer part of x. As an example, consider a problem with three design variables each having $N_c = 10$ possible discrete values. Thus, we will need a 4-digit binary string to represent discrete values for each design variable; that is, $m = 4$ implying that 16 possible discrete values can be represented. Let a design point $\mathbf{x} = (x_1, x_2, x_3)$ be encoded as the following D-string (genetic string):

$$\begin{bmatrix} x_1 & x_2 & x_3 \\ |\,0110\,| & |\,1111\,| & |\,1101\,| \end{bmatrix} \tag{16.4}$$

Using Eq. (16.2), the j values for the three V-strings are calculated as 7, 16, and 14. Since the last two numbers are larger than $N_c = 10$, they are adjusted by using Eq. (16.3) as 6 and 4, respectively. Thus the foregoing D-string (genetic string) represents a design point where the seventh, sixth, and fourth allowable discrete values are assigned to the design variables x_1, x_2, and x_3, respectively.

Initial Generation/Starting Design Set

With a method to represent a design point defined, the first population consisting of N_p designs needs to be created. This means that N_p D-strings need to be created. In some cases, the designer already knows some good usable designs for the system. These can be used as *seed designs* to generate the required number of designs for the population using some random process. Otherwise, the initial population can be generated randomly via the use of a random number generator. Several methods can be used for this purpose. The following procedure shows a way to produce a 32-digit D-string:

1. Generate two random numbers between 0 and 1 as "0.3468 0254 7932 7612 and 0.6757 2163 5862 3845."
2. Create a string by combining the two numbers as "3468 0254 7932 7612 6757 2163 5862 3845."
3. The 32 digits of the above string are converted to 0's and 1's by using a rule in which "0" is used for any value between 0 and 4 and "1" for any value between 5 and 9, as "0011 0010 1100 1100 1111 0010 1110 0101."

Fitness Function

The fitness function defines the relative importance of a design. A higher fitness value implies a better design. The fitness function may be defined in several different ways; it can be defined using the cost function value as follows:

$$F_i = (1 + \varepsilon)f_{max} - f_i, \tag{16.5}$$

where f_i is the cost function (penalty function value for a constrained problems) for the ith design, f_{max} is the largest recorded cost (penalty) function value, and ε is a small value (e.g., 2×10^{-7}) to prevent numerical difficulties when F_i becomes 0.

16.2 FUNDAMENTALS OF GENETIC ALGORITHMS

The basic idea of a genetic algorithm is to generate a new set of designs (population) from the current set such that the average fitness of the population is improved. The process is continued until a stopping criterion is satisfied or the number of iterations exceeds a specified limit. Three genetic operators are used to accomplish this task: reproduction, crossover, and mutation.

Reproduction is an operator where an old design (D-string) is copied into the new population according to the design's fitness. There are many different strategies to implement this reproduction operator. This is also called the *selection process*.

Crossover corresponds to allowing selected members of the new population to exchange characteristics of their designs among themselves. Crossover entails selection of starting and ending positions on a pair of randomly selected strings (called *mating strings*), and simply exchanging the string of 0s and 1s between these positions. *Mutation* is the third step that safeguards the process from a complete premature loss of valuable genetic material during reproduction and crossover. In terms of a binary string, this step corresponds to selection of a few members of the population, determining a location on the strings at random, and switching the 0 to 1 or vice versa.

The foregoing three steps are repeated for successive generations of the population until no further improvement in fitness is attainable. The member in this generation with the highest level of fitness is taken as the optimum design. Some details of the GA implemented by Huang and Arora (1997a) are described in the sequel.

Reproduction Procedure

Reproduction is a process of selecting a set of designs (D-strings) from the current population and carrying them into the next generation. The *selection process* is biased toward more fit members of the current design set (population). Using the fitness value F_i for each design in the set, its probability of selection is calculated as

$$P_i = \frac{F_i}{Q}; \qquad Q = \sum_{j=1}^{N_p} F_j \qquad (16.6)$$

It is seen that the members with higher fitness value have larger probability of selection. To explain the process of selection, let us consider a roulette wheel with a handle shown in Figure 16.1. The wheel has N_p segments to cover the entire population, with the size of the ith segment proportional to the probability P_i. Now a random number w is generated between 0 and 1. The wheel is then rotated clockwise, with the rotation proportional to the

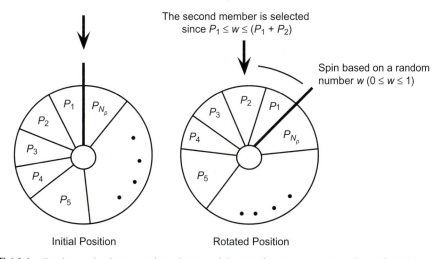

FIGURE 16.1   Roulette wheel process for selection of designs for new generation (reproduction).

random number w. After spinning the wheel, the member pointed to by the arrow at the starting location is selected for inclusion in the next generation. In the example shown in Figure 16.1, member 2 is carried into the next generation. Since the segments on the wheel are sized according to the probabilities P_i, the selection process is biased toward the more fit members of the current population.

Note that a member copied to the mating pool remains in the current population for further selection. Thus, the new population may contain identical members and may not contain some of the members found in the current population. This way, the average fitness of the new population is increased.

Crossover

Once a new set of designs is determined, *crossover* is conducted as a means to introduce variation into a population. Crossover is the process of combining or mixing two different designs (chromosomes) into the population. Although there are many methods for performing crossover, the most common ones are the *one-cut-point* and *two-cut-point methods*. A cut point is a position on the D-string (genetic string). In the one-cut method a position on the string is randomly selected that marks the point at which two parent designs (chromosomes) split. The resulting four halves are then exchanged to produce new designs (children).

The process is illustrated in Figure 16.2, where the cut point is determined as four digits from the right end. The new designs produced $x^{1'}$ and $x^{2'}$ and replace the old designs (parents). Similarly, the two-cut-point method is illustrated in Figure 16.3. Selecting how many or what percentage of chromosomes cross over, and at what points the crossover operation occurs, is part of the heuristic nature of genetic algorithms. There are many different approaches, and most are based on random selections.

Mutation

Mutation is the next operation on the members of the new design set (population). The idea of mutation is to safeguard the process from a complete premature loss of valuable genetic material during the reproduction and crossover steps. In terms of a genetic string,

$\mathbf{x}^1 = 101110\,1001$ $\mathbf{x}^2 = 010100\,1011$ FIGURE 16.2   Crossover operation with one-cut point.

(a)

$\mathbf{x}^{1'} = 101110\,1011$ $\mathbf{x}^{2'} = 010100\,1001$

(b)

$\mathbf{x}^1 = 101\,1101\,001$ $\mathbf{x}^2 = 010\,1001\,011$ FIGURE 16.3   Crossover operation with two-cut point.

(a)

$\mathbf{x}^{1'} = 101\,1001\,001$ $\mathbf{x}^{2'} = 010\,1101\,011$

(b)

this step corresponds to selecting a few members of the population, determining a location on each string randomly, and switching 0 to 1 or vice versa. The number of members selected for mutation is based on heuristics, and the selection of location on the string for mutation is based on a random process. Let us select a design as "10 1110 1001" and select location 7 from the right end of its D-string. The mutation operation involves replacing the current value of 1 at the seventh location with 0 as "10 1010 1001."

Number of Crossovers and Mutations

For each generation (iteration), three operators—reproduction or selection, crossover, and mutation—are performed. While the number of the reproduction operations is always equal to the size of the population, the number of crossovers and mutation can be adjusted to fine-tune the performance of the algorithm.

To show the type of operations needed to implement the mutation and crossover at each generation, we present a possible procedure as follows.

1. Set I_{max} as an integer that controls the amount of crossover. Calculate I_m, which controls the amount of mutation as $I_m = \text{INT}(P_m N_p)$, where P_m represents a fraction of the population that is selected for mutation, and N_p is the size of the population. Too many crossovers can result in a poorer performance of the algorithm since it may produce designs that are far away from the mating designs. Therefore, I_{max} should be set to a small number. The mutation, however, changes designs in the neighborhood of the current design; therefore, a larger amount of mutation may be allowed. Note also that the population size N_p needs to be set to a reasonable number for each problem. It may be heuristically related to the number of design variables and the number of all possible designs determined by the number of allowable discrete values for each variable.

2. Let f_K^* denote the best cost (or penalty) function value for the population at the Kth iteration. If the improvement in f_K^* is less than some small positive number ε' for the last two consecutive iterations, then I_{max} is doubled temporarily. This "doubling" strategy continues at the subsequent iterations and returns to the original value as soon as f_K^* is reduced. The concept behind this is that we do not want too much crossover or mutation to ruin the good designs in D-strings as long as they keep producing better offspring. On the other hand, we need more crossover and mutation to trigger changes when progress stops.

3. If improvement in f_K^* is less than ε' for the last I_g consecutive iterations, P_m is doubled.

4. The crossover and mutation may be performed as follows:

```
FOR i = 1, I_max
   Generate a random number z uniformly distributed in [0, 1]
   If z > 0.5, perform crossover.
   If z ≤ 0.5, skip crossover.
   FOR j = 1, I_m
   Generate a random number z uniformly distributed in [0, 1]
   If z > 0.5, perform mutation.
   If z ≤ 0.5, skip to next j.
  ENDFOR
ENDFOR
```

Leader of the Population

At each generation, the member having the lowest cost function value among all of the designs is defined as the "leader" of the population. If several members have the same lowest cost, only one of them is chosen as the leader. The leader is replaced if another member with lower cost appears. In this way, it is safeguarded from extinction (as a result of reproduction, crossover, or mutation). In addition, the leader is guaranteed a higher probability of selection for reproduction. One benefit of using a leader is that the best cost (penalty) function value of the population can never increase from one iteration to another, and some of the best design variable values (V-strings or genes) will be able to always survive.

Stopping Criteria

If the improvement for the best cost (penalty) function value is less than ε' for the last I consecutive iterations, or if the number of iterations exceeds a specified value, then the algorithm terminates.

Genetic Algorithm

Based on the ideas presented here, a sample genetic algorithm is stated.

Step 1. Define a schema to represent different design points. Randomly generate N_p genetic strings (members of the population) according to the schema, where N_p is the population size. Or use the seed designs to generate the initial population. For *constrained problems*, only the feasible strings are accepted when the penalty function approach is not used. Set the iteration counter $K = 0$. Define a fitness function for the problem, as in Eq. (16.5).

Step 2. Calculate the fitness values for all the designs in the population. Set $K = K + 1$, and the counter for the number of crossovers $I_c = 1$.

Step 3. *Reproduction*: Select designs from the current population according to the roulette wheel selection process described earlier for the mating pool (next generation) from which members for crossover and mutation are selected.

Step 4. *Crossover*: Select two designs from the mating pool. Randomly choose two sites on the genetic strings and swap strings of 0's and 1's between the two chosen sites. Set $I_c = I_c + 1$.

Step 5. *Mutation*: Choose a fraction (P_m) of the members from the mating pool and switch a 0 to 1 or vice versa at a randomly selected site on each chosen string. If, for the past I_g consecutive generations, the member with the lowest cost remains the same, the mutation fraction P_m is doubled. I_g is an integer defined by the user.

Step 6. If the member with the lowest cost remains the same for the past two consecutive generations, then increase I_{max}. If $I_c < I_{max}$, go to Step 4. Otherwise, continue.

Step 7. *Stopping criterion*: If after the mutation fraction P_m is doubled, the best value of the fitness is not updated for the past I_g consecutive generations, then stop. Otherwise, go to Step 2.

Immigration

It may be useful to introduce completely new designs into the population in an effort to increase diversity. This is called immigration, which may be done at a few iterations during the solution process when progress toward the solution point is slow.

Multiple Runs for a Problem

It is seen that the genetic algorithms make decisions at several places based on random number generation. Therefore, when the same problem is run at different times, it may give different final designs. It is suggested that the problem be run a few times to ensure that the best possible solution has been obtained.

16.3 GENETIC ALGORITHM FOR SEQUENCING-TYPE PROBLEMS

There are many applications in engineering where the sequence of operations needs to be determined. To introduce the type of problems being treated, let us consider the design of a metal plate that is to have 10 bolts at the locations shown in Figure 16.4. The bolts are to be inserted into predrilled holes by a computer-controlled robotic arm. The objective is to minimize the movement of the robot arm while it passes over and inserts a bolt into each hole. This class of problems is generally known as *traveling salesperson problem*, which is defined as follows: Given a list of N cities and a means to calculate the traveling distance between the two cities, we must plan a salesperson's route that passes through each city once (with the option of returning to the starting point) while minimizing the total distance.

For such problems, a feasible design is a string of numbers (a sequence of the cities to be visited) that do not repeat themselves (e.g., "1 3 4 2" is feasible and "3 1 3 4" is not). Typical operators used in genetic algorithms, such as crossover and mutation, are not applicable to these types of problems since they usually create infeasible designs with repeated numbers. Therefore, other operators need to be used to solve such problems. We will describe some such operators in the following paragraphs.

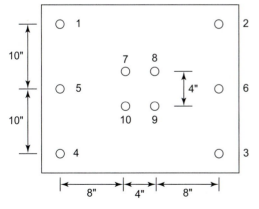

FIGURE 16.4   Bolt insertion sequence determination at 10 locations (Huang et al., 1997).

Permutation Type 1. Let n_1 be a fraction for selection of the mating pool members for carrying out Type 1 permutation. Choose Nn_1 members from the mating pool at random, and reverse the sequence between two randomly selected sites on each chosen string. For example, a chosen member with a string of "345216" and two randomly selected sites of "4" and "1," is changed to "312546."

Permutation Type 2. Let n_2 be a fraction for selection of the mating pool members for carrying out the Type 2 permutation. Choose Nn_2 members from the mating pool at random, and exchange the numbers of two randomly selected sites on each chosen string. For example, a chosen member with a string of "345216" and two randomly selected sites of "4" and "1," is changed to "315246."

Permutation Type 3. Let n_3 be a fraction for selection of the mating pool members for carrying out the Type 3 permutation. Choose Nn_3 members from the mating pool at random, and exchange the numbers of one randomly selected site and the site next to it on each chosen string. For example, a chosen member with a string of "345216" and a randomly selected site of "4", is changed to "354216".

Relocation

Let n_r be a fraction for selection of the mating pool members for carrying out relocation. Choose Nn_r members from the mating pool at random, remove the number of a randomly selected site, and insert it in front of another randomly selected site on each chosen string. For example, a chosen member with a string of "345216" and two randomly selected sites of "4" and "1", is changed to "352416."

A computer program based on the previously mentioned operators is developed and used to solve the bolt insertion sequence problem in Example 16.1.

EXAMPLE 16.1 BOLT INSERTION SEQUENCE DETERMINATION AT 10 LOCATIONS

Solve the problem shown in Figure 16.4 using the genetic algorithm to minimize the total distance travelled by the robotic arm.

Solution

The problem is solved by using a genetic algorithm (Huang and Arora, 1997). The population size N_p is set to 150, and I_g (the number of consecutive iterations for which the best cost function did not improve by at least ε') is set to 10. No seed designs are used for the problem. The optimum bolting sequence is not unique to the problem. With hole 1 as the starting point, the optimum sequence is determined as (1, 5, 4, 10, 7, 8, 9, 3, 6, 2) and the cost function value is 74.63 in. The number of function evaluations is 1445, which is much smaller than the total number of possibilities ($10! = 3,628,800$).

Two other problems are solved in Huang and Arora (1997). The first problem concerns determining the bolting sequence for 16 locations. The optimum sequence is not unique for this example either. The solution is obtained in 3358 function evaluations compared with the total number of possibilities, $16! \cong 2.092 \times 10^{13}$.

The second example concerns the *A-pillar subassembly welding sequence* determination for a passenger vehicle. There are 14 welding locations. The objective is to determine the best welding sequence that minimizes the deformation at some critical points of the structure. Cases where one and two welding guns are used are also considered. This is equivalent to having two salesmen traveling between N cities for the traveling salesman problem. The optimum sequences are obtained with 3341 and 3048 function evaluations for the two cases, which are much smaller than those for the full enumeration.

16.4 APPLICATIONS

Numerous applications of genetic algorithms for different classes of problems have been presented in the literature. There are specialty conferences focusing on developments in genetic and other evolutionary algorithms and their applications. The literature in this area is substantial. Therefore, a survey of all the applications is not attempted here. For mechanical and structural design, some of the applications are covered in Arora (2002), Pezeshk and Camp (2002), Arora and Huang (1996), and Chen and Rajan (2000). Applications of the genetic algorithms for optimum design of electric transmission line structures are given in Kocer and Arora (1996, 1997, 1999, 2002).

EXERCISES FOR CHAPTER 16*

Solve the following problems using a genetic algorithm.

16.1 Example 15.1 with the available discrete values for the variables as $x_1 \in \{0, 1, 2, 3\}$, and $x_2 \in \{0, 1, 2, 3, 4, 5, 6\}$. Compare the solution with that obtained with the branch and bound method.

16.2 Exercise 3.34 using the outside diameter d_0 and the inside diameter d_i as design variables. The outside diameter and thickness must be selected from the following available sets:

$$d_0 \in \{0.020, 0.022, 0.024, \ldots, 0.48, 0.50\} \text{ m}; \quad t \in \{5, 7, 9, \ldots, 23, 25\} \text{ mm}$$

Check your solution using the graphical method of Chapter 3. Compare continuous and discrete solutions. Study the effect of reducing the number of elements in the available discrete sets.

16.3 Formulate the minimum mass tubular column problem described in Section 2.7 using the following data: $P = 100$ kN, length, $l = 5$ m, Young's modulus, $E = 210$ GPa, allowable stress, $\sigma_a = 250$ MPa, mass density, $\rho = 7850$ kg/m^3, $R \leq 0.4$ m, $t \leq 0.05$ m, and $R, t \geq 0$. The design variables must be selected from the following sets:

$$R \in \{0.01, 0.012, 0.014, \ldots, 0.38, 0.40\} \text{ m}; \quad t \in \{4, 6, 8, \ldots, 48, 50\} \text{ mm}$$

Check your solution using the graphical method of Chapter 3. Compare continuous and discrete solutions. Study the effect of reducing the number of elements in the available discrete sets.

16.4 Consider the plate girder design problem described and formulated in Section 6.6. The design variables for the problem must be selected from the following sets

$$h, b \in \{0.30, 0.31, 0.32, \ldots, 2.49, 2.50\} \text{ m}; \quad t_w, t_f \in \{10, 12, 14, \ldots, 98, 100\} \text{ mm}$$

Compare the continuous and discrete solutions. Study the effect of reducing the number of elements in the available discrete sets.

16.5 Consider the plate girder design problem described and formulated in Section 6.6. The design variables for the problem must be selected from the following sets

$$h, b \in \{0.30, 0.32, 0.34, \ldots, 2.48, 2.50\} \text{ m}; \quad t_w, t_f \in \{10, 14, 16, \ldots, 96, 100\} \text{ mm}$$

Compare the continuous and discrete solutions. Study the effect of reducing the number of elements in the available discrete sets.

16.6 Solve problems of Exercises 16.4 and 16.5. Compare the two solutions, commenting on the effect of the size of the discreteness of variables on the optimum solution. Also, compare the continuous and discrete solutions.

16.7 Formulate the spring design problem described in Section 2.9 and solved in Section 6.5. Assume that the wire diameters are available in increments of 0.01 in., the coils can be fabricated in increments of $^1/_{16}$ in, and the number of coils must be an integer. Compare the continuous and discrete solutions. Study the effect of reducing the number of elements in the available discrete sets.

16.8 Formulate the spring design problem described in Section 2.9 and solved in Section 6.5. Assume that the wire diameters are available in increments of 0.015 in, the coils can be fabricated in increments of $^1/_8$ in, and the number of coils must be an integer. Compare the continuous and discrete solutions. Study the effect of reducing the number of elements in the available discrete sets.

16.9 Solve problems of Exercises 16.7 and 16.8. Compare the two solutions, commenting on the effect of the size of the discreteness of variables on the optimum solution. Also, compare the continuous and discrete solutions.

16.10 Formulate the problem of optimum design of prestressed concrete transmission poles described in Kocer and Arora (1996a). Compare your solution to that given in the reference.

16.11 Formulate the problem of optimum design of steel transmission poles described in Kocer and Arora (1996b). Solve the problem as a continuous variable optimization problem.

16.12 Formulate the problem of optimum design of steel transmission poles described in Kocer and Arora (1996b). Assume that the diameters can vary in increments of 0.5 in and the thicknesses can vary in increments of 0.05 in. Compare your solution to that given in the reference.

16.13 Formulate the problem of optimum design of steel transmission poles using standard sections described in Kocer and Arora (1997). Compare your solution to the solution given in the reference.

16.14 Formulate and solve three-bar truss of Exercise 3.50 as a discrete variable problem where the cross-sectional areas must be selected from the following discrete set:

$$A_i \in \{50, 100, 150, \ldots, 4950, 5000\} \text{ mm}^2$$

Check your solution using the graphical method of Chapter 3. Compare continuous and discrete solutions. Study the effect of reducing the number of elements in the available discrete sets.

16.15 Solve Example 16.1 of bolt insertion sequence at 10 locations. Compare your solution to the one given in the example.

TABLE E16.17   Material data for the spring design problem

Material Type	G, lb/in^2	ρ, lb-s^2/in^4	τ_a, lb/in^2	U_p
1	11.5×10^6	7.38342×10^{-4}	80,000	1.0
2	12.6×10^6	8.51211×10^{-4}	86,000	1.1
3	13.7×10^6	9.71362×10^{-4}	87,000	1.5

G = shear modulus, ρ = mass density, τ_a = shear stress, U_p = relative unit price.

16.16 Solve the 16-bolt insertion sequence determination problem described in Huang and coworkers (1997). Compare your solution to the one given in the reference.

16.17 The material for the spring in Exercise 16.7 must be selected from one of three possible materials given in Table E16.17 (refer to Section 15.8 for more discussion of the problem) (Huang and Arora, 1997). Obtain a solution to the problem.

16.18 The material for the spring in Exercise 16.8 must be selected from one of three possible materials given in Table E16.17 (refer to Section 15.8 for more discussion of the problem) (Huang and Arora, 1997). Obtain a solution to the problem.

Multi-objective Optimum Design Concepts and Methods

Upon completion of this chapter, you will be able to

- Explain basic terminology and concepts related to multi-objective optimization problems

- Explain the concepts of Pareto optimality and Pareto optimal set

- Solve your multi-objective optimization problem using a suitable formulation

Thus far, we have considered problems in which only one objective function needed to be optimized. However, there are many practical applications where the designer may want to optimize two or more objective functions simultaneously. These are called *multi-objective, multicriteria,* or *vector optimization* problems; we refer to them as *multi-objective optimization problems.*

In this chapter, we describe basic terminology, concepts, and solution methods for such problems. The material is introductory in nature and is derived from Marler and Arora (2004, 2009) and many other references cited there (e.g., Ehrgott and Grandibleaux, 2000).[1]

---

[1]The original draft of this chapter was provided by R. T. Marler. The contribution to this book is very much appreciated.

17.1 PROBLEM DEFINITION

The general design optimization model defined in Chapter 2 is modified to treat multi-objective optimization problems as follows:

Minimize

$$\mathbf{f}(\mathbf{x}) = (f_1(\mathbf{x}), f_2(\mathbf{x}), \ldots, f_k(\mathbf{x})) \tag{17.1}$$

subject to

$$h_i(\mathbf{x}) = 0; \quad i = 1 \text{ to } p \tag{17.2}$$

$$g_j(\mathbf{x}) \le 0; \quad j = 1 \text{ to } m \tag{17.3}$$

where k is the number of objective functions, p is the number of equality constraints, and m is the number of inequality constraints. $\mathbf{f}(\mathbf{x})$ is a k-dimensional vector of objective functions. Recall that the *feasible set S* (also called the *feasible design space*) is defined as a collection of all of the feasible design points:

$$S = \left\{ \mathbf{x} | h_i(\mathbf{x}) \le 0; \quad i = 1 \text{ to } p; \quad \text{and} \quad g_j(\mathbf{x}) \le 0; \quad j = 1 \text{ to } m \right\} \tag{17.4}$$

The problem shown in Eqs. (17.1) through (17.3) usually does not have a unique solution, and this idea is illustrated by contrasting single-objective and multi-objective problems. *Note that we will use the terms "cost function" and "objective function" interchangeably in this chapter.* Examples 17.1 and 17.2 illustrate the basic difference between single-objective and multi-objective optimization problems.

EXAMPLE 17.1 SINGLE-OBJECTIVE OPTIMIZATION PROBLEM

Minimize

$$f_1(\mathbf{x}) = (x_1 - 2)^2 + (x_2 - 5)^2 \tag{a}$$

subject to

$$g_1 = -x_1 - x_2 + 10 \le 0 \tag{b}$$

$$g_2 = -2x_1 + 3x_2 - 10 \le 0 \tag{c}$$

Solution

Figure 17.1 is a graphical representation of the problem. The feasible set S is convex, as shown in the figure. A few objective function contours are also shown. It is seen that the problem has a distinct minimum at the point A(4, 6) with the objective function value of $f_1(4,6) = 5$. At the minimum point, both constraints are active. Note that since the objective function is also strictly convex, point A represents the unique global minimum for the problem.

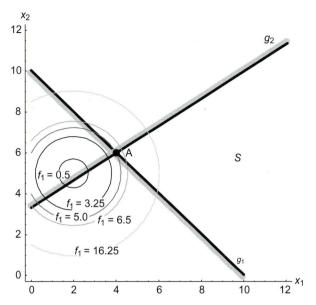

FIGURE 17.1 Graphical representation of a single-objective optimization problem.

EXAMPLE 17.2 TWO-OBJECTIVE OPTIMIZATION PROBLEM

A second objective function is added to Example 17.1 to obtain the following two-objective problem:

Minimize

$$f_1(\mathbf{x}) = (x_1 - 2)^2 + (x_2 - 5)^2 \tag{a}$$

$$f_2(\mathbf{x}) = (x_1 - 4.5)^2 + (x_2 - 8.5)^2 \tag{b}$$

subject to same constraints as for Example 17.1.

Solution

Figure 17.2 is a modification of Figure 17.1, where the contours of the second objective function are also shown. The minimum value of f_2 is 3.25 at point B(5.5, 7.0). Note that f_2 is also a strictly convex function, and so point B is a unique global minimum point for f_2. The minimum points for the two objective functions are different. Therefore, if one wishes to minimize f_1 and f_2 simultaneously, pinpointing a single optimum point is not possible. In fact, there are *infinitely many possible solution points, called the Pareto optimal set*, which is explained later. The challenge is to find a solution that suits requirements of the designer. This dilemma requires the description of additional terminology and solution concepts.

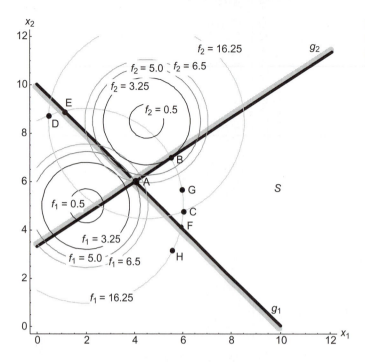

FIGURE 17.2   Graphical representation of a two-objective optimization problem.

17.2 TERMINOLOGY AND BASIC CONCEPTS

17.2.1 Criterion Space and Design Space

Example 17.2 is depicted in the *design space* in Figure 17.2. That is, the constraints g_1 and g_2 and the objective function contours are plotted as functions of the design variables x_1 and x_2. Alternatively, a multi-objective optimization problem may also be depicted in the *criterion space* (also called the *cost space*), where the axes represent different objective functions. For the present problem, f_1 and f_2 are the axes in the criterion space, as shown in Figures 17.3 and 17.4. q_1 represents the g_1 boundary, and q_2 represents the g_2 boundary.

In general, a curve in the design space in the form $g_j(\mathbf{x}) = 0$ is translated into a curve q_j in the criterion space simply by evaluating the values of the objective functions at different points on the constraint curve in the design space. The feasible criterion space Z is defined simply as the set of objective function values corresponding to the feasible points in the design space; that is,

$$Z = \left\{ f(\mathbf{x}) | \mathbf{x} \text{ in the feasible set } S \right\} \tag{17.5}$$

The feasible points in the design space map onto a set of points in the criterion space. Note that although q_j represents g_j in the criterion space, it may not necessarily represent the boundaries of the *feasible criterion space*. This is seen in Figure 17.3, where the feasible criterion space for the problem in Example 17.2 is displayed. All portions of the curves q_1 and q_2 do not form boundaries of the feasible criterion space. This concept of feasible criterion space is important and used frequently, so we will discuss it further.

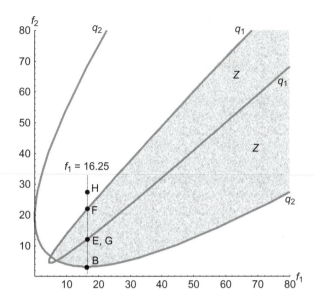

FIGURE 17.3 Graphical representation of a two-objective optimization problem in the criterion space.

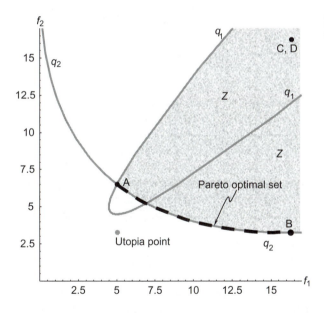

FIGURE 17.4 Illustration of Pareto optimal set and utopia point in the criterion space.

Let us first consider the single-objective function problem depicted in Figure 17.1. The feasible criterion space for the problem is the line f_1 that starts at 5, the minimum value for the function, and goes to infinity. Note that each feasible design point corresponds only to one objective function value; it maps onto only one point on the feasible criterion line. However, for one objective function value, there may be many different feasible design points in the feasible design space S. For instance, in Figure 17.1, there are infinitely many design points that result in $f_1 = 16.25$, as seen for the contour $f_1 = 16.25$. Note also that the contour $f_1 = 16.25$ passes through the infeasible region as well. Thus, for a given objective function value (a given point in the feasible criterion space), there can be feasible or infeasible points in the design space. Note also that the objective function values for design points on the constraint boundaries for g_1 and g_2 fall on the line f_1 in the criterion space.

Now let us consider the problem with two objective functions and study the relationship between constraint boundaries in the design space and the corresponding curves in the criterion space. Let us consider two feasible points E and F in the criterion space, as shown in Figure 17.3. Both points are on the curve q_1 and have a value of 16.25 for f_1. Since both points are on the curve q_1, they must also lie on the constraint boundary g_1 in the design space. They indeed are on the g_1 line in the design space at E(1.076, 8.924) and F(5.924, 4.076), as shown in Figure 17.2. Whereas both points satisfy the constraint g_1, the point E violates the constraint g_2 and thus is not in the feasible set S. The question is, then, how can point E be in the feasible criterion space?

It turns out that there is another point G on the $f_1 = 16.25$ contour in Figure 17.2 that is feasible and has the same value for f_2 as for the point E. Therefore, point G also maps onto point E in the criterion space, as shown in Figure 17.3. Thus, a feasible point in the criterion space can map onto multiple points in the design space, some of which may violate constraints. Note that the feasible point C and the infeasible point D in Figure 17.2 both map onto a feasible point C in the criterion space, as can be seen in Figure 17.4. An infeasible point H in the criterion space in Figure 17.3 having $f_1 = 16.25$ maps onto an infeasible point H in the design space, as seen in Figure 17.2. Thus, the feasible criterion space consists of all the points obtained using the feasible points in the design space.

A concept that is related to the feasibility of design points is that of *attainability*. Feasibility of a design implies that no constraint is violated in the design space. Attainability implies that a point in the criterion space can be related to a point in the feasible design space. Whereas each point in the feasible design space translates to a point in the criterion space, the reverse may not be true; that is, every point in the criterion space does not necessarily correspond to a single point $\mathbf{x}$ in the feasible design space S, as we saw in the foregoing example. Thus, even with an unconstrained problem, only certain points in the criterion space are *attainable*. We will use the symbol Z to indicate points in the criterion space that are attainable and correspond to a feasible point in the set S. The set Z is also referred to as the *attainable set*.

For the relatively simple problem in Example 17.2, it is possible to depict Z, as illustrated in Figures 17.3 and 17.4, but generally it is not possible to depict the feasible criterion space directly.

As noted earlier in Example 17.1, the feasible criterion space is the real line starting at 5 and going up to infinity. Therefore, only the objective function values of 5 and higher are *attainable* and constitute the feasible criterion space Z for that problem.

17.2.2 Solution Concepts

From a classical standpoint, optimizing a single function simply entails determining a set of stationary points, identifying a local maximum or minimum, and possibly finding the global optimum, such as point A in Figure 17.1. In contrast, the process of determining a solution for a multi-objective optimization problem is slightly more complex and less definite than that for a single-objective problem. As seen in Example 17.2 and depicted in Figure 17.2, point A is the minimum for f_1 and point B is the minimum for f_2. But which design point minimizes both f_1 and f_2 simultaneously? This is not clear even for this simple problem. Therefore, it is not clear what is meant by the minimum of multiple functions that may have opposing characteristics since what decreases the value of one function may increase the value of another. Therefore, in this section we describe some solution concepts related to multi-objective optimization problems.

Pareto Optimality

The predominant solution concept in defining solutions for multi-objective optimization problems is that of *Pareto optimality* (Pareto, 1906). A point x^* in the feasible design space S is called Pareto optimal if there is no other point x in the set S that reduces at least one objective function without increasing another one. This is defined more precisely as follows:

> A point x^* in the feasible design space S is *Pareto optimal* if and only if there does not exist another point x in the set S such that $\mathbf{f}(x) \leq \mathbf{f}(x^*)$ with at least one $f_i(x) < f_i(x^*)$.

Note that inequalities between vectors apply to every component of each vector; for example, $\mathbf{f}(x) \leq \mathbf{f}(x^*)$ implies $f_1 \leq f_1^*$, $f_2 \leq f_2^*$, and so on. The set of all Pareto optimal points is called the *Pareto optimal set*, and this term can refer to points in the design space, or to points in the criterion space. The preceding definition means that for x^* to be called the Pareto optimal point, no other point exists in the feasible design space S that improves at least one objective function while keeping others unchanged.

As an example of a Pareto optimal point, consider point A in Figures 17.2 and 17.4. It is not possible to move from this point and simultaneously decrease the value of f_1 and f_2 without violating a constraint, that is, without moving into the infeasible region. Therefore, point A is a Pareto optimal point. However, it is possible to move from point C and simultaneously reduce the values of both f_1 and f_2. This can be seen most clearly in Figure 17.4. Thus, point C is not Pareto optimal. In Figure 17.2, the points on the line between A and B along g_2 represent the Pareto optimal set. In Figure 17.4, the Pareto optimal set is shown as the curve between points A and B along q_2. In fact, *the Pareto optimal set is always on the boundary of the feasible criterion space Z*. When there are just two objective functions, as with this example, then the minimum points of each objective function define the endpoints of the Pareto optimal curve, assuming the minima to be unique.

Note that although the Pareto optimal set is always on the boundary of Z, it is not necessarily defined by the constraints. As noted earlier, Z exists even for unconstrained problems. In such a case, the Pareto optimal set is defined by the relationship between the gradients of the objective functions. In the simple case *when there are just two objective functions, the gradients of the functions point in opposite directions at all Pareto optimal points*. An exception to this rule is the individual minimum points for the functions at which the gradient is zero.

For Example 17.2 without the constraints g_1 and g_2, the Pareto optimal set is along the line connecting the centers of the circles for the two objective functions—that is, points (2, 5) and (4.5, 8.5). This line maps onto a curve in the criterion space in Figure 17.3.

Weak Pareto Optimality

A concept closely related to Pareto optimality is that of weak Pareto optimality. At the weakly Pareto optimal points, it is possible to improve some objective functions without penalizing others. A *weakly Pareto optimal* point is defined as follows:

> A point $\mathbf{x}^*$ in the feasible design space S is weakly Pareto optimal if and only if there does not exist another point $\mathbf{x}$ in the set S such that $\mathbf{f}(\mathbf{x}) < \mathbf{f}(\mathbf{x}^*)$. That is, there is no point that improves all of the objective functions simultaneously; however, there may be points that improve some of the objectives while keeping others unchanged.

In contrast to weakly Pareto optimal points, no objective function can be improved from a Pareto optimal point without detriment to another objective function. It will be seen later that there are numerical algorithms for multi-objective optimization that may converge to weakly Pareto optimal solutions as opposed to always giving Pareto optimal solutions.

Efficiency and Dominance

Efficiency is another primary concept in multi-objective optimization and is defined as follows:

> A point $\mathbf{x}^*$ in the feasible design space S is efficient if and only if there does not exist another point $\mathbf{x}$ in the set S such that $\mathbf{f}(\mathbf{x}) \leq \mathbf{f}(\mathbf{x}^*)$ with at least one $f_i(\mathbf{x}) < f_i(\mathbf{x}^*)$. Otherwise, $\mathbf{x}^*$ is inefficient. The set of all efficient points is called the *efficient frontier*.

Another common concept is that of *nondominated* and *dominated* points, which is defined as follows:

> A vector of objective functions $\mathbf{f}^* = \mathbf{f}(\mathbf{x}^*)$ in the feasible criterion space Z is nondominated if and only if there does not exist another vector $\mathbf{f}$ in the set Z such that $\mathbf{f} \leq \mathbf{f}^*$, with at least one $f_i < f_i^*$. Otherwise, $\mathbf{f}^*$ is dominated.

Note that the definitions of Pareto optimal and efficient points are the same. Also, the definitions of efficient and nondominated points are similar. The only distinction is that efficiency refers to points in the design space and nondominance refers to the points in the criterion space. Pareto optimality, however, generally refers to both the design and the criterion spaces. In numerical algorithms, the idea of nondomination in the criterion space is often used for a subset of points; one point may be nondominated compared with other points in the subset. Pareto optimality, on the other hand, implies a condition in terms of the complete feasible design or criterion space.

Genetic algorithms and some random search methods for multi-objective optimization update and store a discrete set of potential solution points in each iteration. Each new addition to this set is compared with all of the objective function values of potential solution points to determine if the new point is dominated. If it is nondominated, then it is kept in the set of potential solution points; note, however, that this point may not be Pareto optimal.

Utopia Point

This is a unique point in the criterion space that is defined as follows:

A point $\mathbf{f}^\circ$ in the criterion space is called the utopia point if $f_i^\circ = \min\{f_i(\mathbf{x}) \mid$ for all $\mathbf{x}$ in the set $S\}$, $i = 1$ to k. It is also called the *ideal point*.

The utopia point is obtained by minimizing each objective function without regard for other objective functions. Each minimization yields a design point in the design space and the corresponding value for the objective function. It is rare that each minimization will end up at the same point in the design space. That is, one design point cannot simultaneously minimize all of the objective functions. Thus, *the utopia point exists only in the criterion space and, in general, it is not attainable.*

Figure 17.4 shows the Pareto optimal set and the utopia point for the problem in Example 17.2. The Pareto optimal set is on the boundary of Z and coincides with the curve q_2. The utopia point is located at the point (5, 3.25), as calculated before. Note that the utopia point is not in Z, and is therefore unattainable.

Compromise Solution

The next best thing to a utopia point is a solution that is as close as possible to the utopia point. Such a solution is called a *compromise solution*. The term *closeness* can be defined in several different ways. Usually, it implies that one minimizes the Euclidean distance $D(\mathbf{x})$ from the utopia point in the criterion space, which is defined as follows:

$$D(\mathbf{x}) = ||\mathbf{f}(\mathbf{x}) - \mathbf{f}^\circ|| = \left\{ \sum_{i=1}^k [f_i(\mathbf{x}) - f_i^\circ]^2 \right\}^{1/2} \tag{17.6}$$

where f_i° represents a component of the utopia point in the criterion space. *Compromise solutions are Pareto optimal.*

17.2.3 Preferences and Utility Functions

Because mathematically there are infinitely many Pareto optimal solutions, we often have to make decisions concerning which solution is preferred. Fundamentally, this specification of preferences is based on opinions concerning points in the criterion space. Ideally, a multi-objective optimization method should reflect the user's preferences, if known; that is, it should incorporate how the user feels about different solution points. However, having a mathematical model or algorithm to represent one's preferences perfectly is usually impossible. Nonetheless, different methods for multi-objective optimization try to incorporate preferences in different ways.

This idea of accurately incorporating and reflecting preferences is a common and significant issue for multi-objective optimization methods. Consequently, some work has been done to develop methods that effectively incorporate preferences. These methods typically try to capture knowledge about the problem functions and incorporate it into mathematical expressions that are then used in multi-objective optimization methods. One such recent method that captures this knowledge quite accurately is called *physical programming*. The method has been used successfully in several applications (Messac, 1996; Chen et al., 2000, Messac et al., 2001; Messac and Mattson, 2002).

Essentially, there are three approaches to expressing preferences about different objective functions. Preferences can be declared before solving the multi-objective optimization problem. For instance, one may specify weights associated with each objective, indicating their relative importance. Alternatively, preferences can be indicated by interacting with the optimization routine and making choices based on intermediate optimization results. For engineering applications, such approaches can be awkward, especially with problems that require a significant amount of time to evaluate the problem functions.

Finally, it is possible to calculate the complete Pareto optimal set (or its approximation) and then select a single solution point after the problem has been solved. This, however, is not practical for more than three objective functions (although selected subsets of two or three objectives functions may be displayed). In some instances, the decision maker may not be able to concretely define preferences. Thus, as a special case, one may choose not to declare preferences at all.

A *utility function* is a mathematical expression that attempts to model the decision maker's preferences. It is most relevant to methods that indicate preferences before the problem is solved. In this context, utility, which is modeled with a utility function, represents an individual's degree of contentment. Utility emphasizes a decision maker's satisfaction, which is slightly different from the usual meaning of usefulness or worth. The utility function is a scalar function incorporating various objective functions.

17.2.4 Vector Methods and Scalarization Methods

Two predominant classifications for multi-objective approaches are *scalarization* and *vector optimization methods*. With scalarization methods, the components of objective function vectors are combined to form a scalar objective function. Then we can use standard single-objective methods to optimize the result. Alternatively, the term "vector optimization" implies that each objective function is treated independently. We will describe examples of both approaches.

17.2.5 Generation of Pareto Optimal Set

A key characteristic of multi-objective optimization methods is the nature of the solutions that they provide. Some methods always yield Pareto optimal solutions but may skip certain points in the Pareto optimal set; that is, they may not be able to yield all of the Pareto optimal points. Other methods are able to capture all of the points in the Pareto optimal set (when the problem is solved by changing the method parameters), but may also provide non-Pareto optimal points. The former quality is beneficial when we are interested in using a method to obtain just one solution point. The latter quality is useful when the complete

Pareto optimal set needs to be generated. We will note this feature of each method when it is described in later sections. The tendency of a particular method to result in non-Pareto optimal points, and the ability of a method to capture all of the Pareto optimal points, depend not only on the method itself but also on the nature of the problem being solved.

17.2.6 Normalization of Objective Functions

Many multi-objective optimization methods involve comparing and making decisions about different objective functions. However, values of different functions may have different units and/or significantly different orders of magnitude, making comparisons difficult. Thus, it is usually necessary to transform the objective functions such that they all have similar orders of magnitude. Although there are many approaches, the most robust is to normalize the objective functions as follows:

$$f_i^{\text{norm}} = \frac{f_i(\mathbf{x}) - f_i^{\circ}}{f_i^{\text{max}} - f_i^{\circ}} \tag{17.7}$$

where f_i° is the utopia point. $f_i^{\text{norm}}(\mathbf{x})$ generally has values between 0 and 1, depending on the accuracy and method with which $f_i^{\text{max}}(\mathbf{x})$ and $f_i^{\circ}(\mathbf{x})$ are determined.

There are two approaches for determining $f_i^{\text{max}}(\mathbf{x})$. One can define it such that $f_i^{\text{max}}(\mathbf{x}) = \max_{1 \leq j \leq k} f_i(\mathbf{x}_j^*)$, where $\mathbf{x}_j^*$ is the point that minimizes the jth objective function. This implies that each objective $f_j(\mathbf{x})$ needs to be minimized to determine $\mathbf{x}_j^*$. Then all objective functions need to be evaluated at $\mathbf{x}_j^*$. The maximum of all of the f_i values is $f_i^{\text{max}}(\mathbf{x})$. This process also determines the utopia point $f_i^{\circ}(\mathbf{x})$. It is noted that this normalization process may not be practical in some cases. Therefore, instead of $f_i^{\text{max}}(\mathbf{x})$, we can use the absolute maximum value of $f_i(\mathbf{x})$, or its approximation based on engineering intuition. Similarly, the utopia point may be replaced with a reasonable estimate (called the *aspiration point, target value,* or *goal*). We will assume that the objective functions have been normalized. Note, however, that if all of the objective functions have similar values, normalization may not be needed.

17.2.7 Optimization Engine

Most approaches for solving multi-objective optimization problems actually entail formulating the multiple objective functions into a single-objective problem or a series of problems. Then a standard single-objective optimization routine is used to solve the consequent formulation. We call this routine the *optimization engine*. The performance of most multi-objective methods depends on which optimization engine is used.

17.3 MULTI-OBJECTIVE GENETIC ALGORITHMS

Genetic algorithms (GAs) for single-objective optimization can be extended to provide an effective approach for solving multi-objective optimization problems as well. Since GAs for multi-objective optimization build on the GAs for single-objective optimization, the concepts and procedures described previously in Chapter 16 should be reviewed.

Because genetic algorithms do not require gradient information, they can be effective regardless of the nature of the problem functions. They combine the use of random numbers and information from previous iterations to evaluate and improve a population of points (a group of potential solutions) rather than a single point at a time. Another appeal of genetic algorithms is their *ability to converge to the Pareto optimal set* rather than a single Pareto optimal point (Osyczka, 2002).

Although the algorithms in this section are intended for application to engineering problems, much of the literature uses terminology from the fields of biology and genetics. Thus, for the sake of clarity, the basic definitions from Chapter 16 are reviewed here and some new terms are introduced.

A *population* represents a set of design points in the design space. A *subpopulation* is a subset of points in a generation. *Generation* refers to a computational iteration. To say that a *point survives* into the next generation means that the point is *selected* for use in the next iteration. A *niche* is a group of points that are close together (typically in terms of distance in the criterion space).

Multi-objective Genetic Algorithms

The primary questions when developing genetic algorithms for multi-objective problems are how to evaluate fitness, how to incorporate the idea of Pareto optimality, and how to determine which potential solution points should be selected (will survive) for the next iteration (generation). Note that the fitness of a design point (determined usually by a fitness function) is used in the selection process, that is, to decide whether to include the design in the next generation.

However, in some multi-objective genetic algorithms, the fitness of a design is neither defined nor used; instead, some selection strategy is used directly to select the designs for the next iteration (generation). The approaches that are described in this section collectively address these two issues. Different selection techniques are discussed that serve as potential ingredients in a genetic multi-objective optimization algorithm. Once a set of designs is selected for the next generation, the crossover and mutation operators, described in Chapter 16, are used to create a new set of designs that are subjected to the selection process again; thus, the iterations continue this way.

In general, genetic algorithms for multi-objective optimization continue to evolve. We will describe some basic ideas and techniques that can be combined, modified, and used in different ways in a specific genetic algorithm for selection of designs for the next generation.

Vector-Evaluated Genetic Algorithm

One of the first treatments of multi-objective genetic algorithms was presented by Schaffer (1985); it has provided a foundation for later developments. The general idea behind the approach, called the *vector-evaluated genetic algorithm* (VEGA), involves producing smaller subsets (subpopulations) of the current designs (population) in a given iteration (generation). One subset is created by evaluating one objective function at a time rather than aggregating all of the functions. The *selection process* is composed of a series of computational loops, and during each loop the fitness of each member of the current set of designs is evaluated using a single-objective function. Then certain members of the population are selected and passed on to the next generation using the stochastic processes discussed in Chapter 16. This process is repeated for each objective function. Consequently, for a problem with k objectives, k subsets are created, each with N_p/k members, where N_p is the size of the entire set (population size). The resulting subsets are then combined to yield a new population.

The selection process is based on the idea that the minimum of a single-objective function is a Pareto optimal point (assuming that the minimum is unique). Such minima generally define the vertices of the Pareto optimal set. Consequently, Schaffer's method does not yield an even distribution of Pareto optimal points. Solutions in a given generation tend to cluster around individual function minima. This is analogous to the evolution of species, where a *species* is a class of organisms with common attributes.

Ranking

A class of alternatives to VEGA, when it comes to evaluating fitness and selecting designs for the next generation, involves giving each design a rank based on whether it is dominated in the criterion space (Goldberg, 1989; Srinivas and Deb, 1995; Cheng and Li, 1998). Fitness is then based on a design's rank within a population. The means of determining rank and of assigning fitness values associated with it may vary from method to method, but a general approach is common, as described in the following discussion.

For a given set of designs, the objective functions are evaluated at each point. All non-dominated points receive a rank of 1. Determining whether a point is dominated (a *nondominated check*) entails comparing the vector of objective function values at that point with the vector at all other points. Then the points with a rank of 1 are temporarily removed from consideration, and the points that are nondominated relative to the remaining group are given a rank of 2. This process is repeated until all points are ranked. The points with the lowest rank have the highest fitness value. That is, fitness is determined such that it is inversely proportional to rank.

Pareto Fitness Function

The fitness function for a problem with multiple objectives can be defined in many different ways (Balling et al., 1999, 2000; Balling, 2000). The following function, called the *maximin fitness function*, has been used successfully in some applications (Balling, 2003):

$$F(\mathbf{x}_i) = \max_{j \neq i;\ j \in P} \left[\min_{1 \leq s \leq k} \{ f_s(\mathbf{x}_i) - f_s(\mathbf{x}_j) \} \right] \tag{17.8}$$

Here, $F(\mathbf{x}_i)$ is the fitness of the ith design, and P is the set of nondominated points in the current population; it is assumed that each objective function has been scaled by dividing it by an appropriate positive constant. Thus, with each iteration one must first determine all of the nondominated points before evaluating the fitness of the designs. Note that the nondominated points have negative fitness values. This fitness function automatically penalizes clustering of the nondominated points. Thus, compared with other selection approaches, this one is relatively simple and effective.

Pareto-Set Filter

It is possible to have a Pareto optimal point in a particular iteration that does not appear in subsequent iterations; that is, it may be dropped from further consideration during the selection process. To guard against this situation, a *Pareto-set filter* can be used. Regardless of how fitness is determined, most genetic multi-objective optimization methods incorporate some type of Pareto-set filter to avoid losing potential Pareto optimal solutions. One type is described as follows (Cheng and Li, 1997). Basically, the algorithm stores two sets of solutions: the current population and the *filter* (another set of potential solutions). The filter is called an *approximate Pareto set* and provides an approximation of the Pareto optimal set.

With each iteration, points with a rank of 1 are saved in the filter. When new points from subsequent iterations are added to the filter, they are subjected to a nondominated check within the filter; the dominated points are discarded. The capacity of the filter is typically set to the size of the population. When the filter is full, points at a minimum distance from other points are discarded in order to maintain an even distribution of Pareto optimal points. The filter eventually converges on the true Pareto optimal set.

Elitist Strategy

Although the elitist strategy is similar to the Pareto-set filter approach, it provides an alternative means for ensuring that Pareto optimal solutions are not lost (Ishibuchi and Murata, 1996; Murata et al., 1996). It functions independently of the ranking scheme. As with the Pareto-set filter, two sets of solutions are stored: a current population and a *tentative set of nondominated solutions*, which is an approximate Pareto optimal set. With each iteration, all points in the current population that are not dominated by any points in the tentative set are added to the tentative set. Then the dominated points in the tentative set are discarded. After crossover and mutation operations are applied, a user-specified number of points from the tentative set are reintroduced into the current population. These are called *elite points*. In addition, the k solutions with the best values for each objective function can be regarded as elite points and preserved for the next generation.

Tournament Selection

Tournament selection is another technique for choosing designs that are used in subsequent iterations. Although it concerns the selection process, it circumvents the idea of fitness. It is an alternative to the ranking approach previously described.

Tournament selection proceeds as follows (Horn et al., 1994; Srinivas and Deb, 1995). Two points, called *candidate points*, are randomly selected from the current population. These candidate points essentially compete for survival in the next generation. A separate set of points, called a *tournament* (or *comparison*) set also is randomly compiled. The candidate points are then compared with each member of the tournament set. If both points are dominated by the points in the tournament set, another pair is selected. If there is only one candidate that is nondominated relative to the tournament set, that candidate is selected for use in the next iteration.

However, if there is no preference between candidates, or if there is a tie, *fitness sharing*, which is explained later, is used to select a candidate. The size of the tournament set is prespecified as a percentage of the total population. It imposes the degree of difficulty in surviving, which is called the *domination pressure*. An insufficient number of Pareto optimal points will be found if the tournament size is too small, and premature convergence may result if the tournament size is too large.

Niche Techniques

A *niche* in genetic algorithms is a group of points that are close to each other, typically in the criterion space. *Niche techniques* (also called *niche schemes* or *niche-formation methods*) are methods for ensuring that a set of designs does not converge to a niche (i.e., a limited number of Pareto optimal points). Thus, these techniques foster an even spread of points (in the criterion space). Genetic multi-objective algorithms tend to create a limited number of niches; they converge to or cluster around a limited set of Pareto optimal points. This phenomenon is known as *genetic* or (*population*) drift, and niche techniques force the development of multiple niches while limiting the growth of any single niche.

Fitness sharing is a common niche technique, the basic idea of which is to penalize the fitness of points in crowded areas, thus reducing the probability of their survival for the next iteration (Deb, 1989; Fonseca and Fleming, 1993; Horn et al., 1994; Srinivas and Deb, 1995; Narayana and Azarm, 1999). The fitness of a given point is divided by a constant that is proportional to the number of other points within a specified distance in the criterion space. In this way the fitness of all points in a niche is shared in some sense—thus the term "fitness sharing."

In reference to tournament selection, when two candidates are both either nondominated or dominated, the most fit candidate is the one with the least number of individuals surrounding it (within a specified distance in the criterion space). This is called *equivalence class sharing*.

17.4 WEIGHTED SUM METHOD

The most common approach to multi-objective optimization is the *weighted sum* method:

$$U = \sum_{i=1}^{k} w_i f_i(\mathbf{x}) \tag{17.9}$$

Here $\mathbf{w}$ is a vector of weights typically set by the decision maker such that $\sum_{i=1}^{k} w_i = 1$ and $\mathbf{w} > \mathbf{0}$. If objectives are not normalized, w_i need not add to 1.

As with most methods that involve objective function weights, setting one or more of the weights to 0 can result in weakly Pareto optimal points. The relative value of the weights generally reflects the relative importance of the objectives. This is another common characteristic of weighted methods. If all of the weights are omitted or are set to 1, then all objectives are treated equally.

The weights can be used in two ways. The user may either set $\mathbf{w}$ to reflect preferences before the problem is solved or systematically alter $\mathbf{w}$ to yield different Pareto optimal points (to generate the Pareto optimal set). In fact, *most methods that involve weights can be used in both of these capacities*—to generate a single solution or multiple solutions.

This method is easy to use, and if all of the weights are positive, the minimum of Eq. (17.9) is always Pareto optimal. However, there are a few recognized difficulties with the weighted sum method (Koski, 1985; Das and Dennis, 1997). First, even with some of the methods discussed in the literature for determining weights, a satisfactory a priori weight selection does not necessarily guarantee that the final solution will be acceptable; one may have to re-solve the problem with new weights. In fact, this is true of most weighted methods.

The second problem is that it is impossible to obtain points on nonconvex portions of the Pareto optimal set in the criterion space (Marler and Arora, 2010). Although nonconvex Pareto optimal sets are relatively uncommon, some examples are noted in the literature (Koski, 1985; Stadler and Dauer, 1992; Stadler, 1995). The final difficulty with the weighted sum method is that varying the weights consistently and continuously may not necessarily result in an even distribution of Pareto optimal points and an accurate, complete representation of the Pareto optimal set.

17.5 WEIGHTED MIN-MAX METHOD

The *weighted min-max* method (also called the *weighted Tchebycheff* method) is formulated to minimize U, which is given as follows:

$$U = \max_i \{w_i[f_i(\mathbf{x}) - f_i^\circ]\} \tag{17.10}$$

A common approach to treatment of Eq. (17.10) is to introduce an additional unknown parameter λ as follows:

Minimize λ

subject to additional constraints

$$w_i[f_i(\mathbf{x}) - f_i^\circ] - \lambda \leq 0; \quad i = 1 \text{ to } k \tag{17.11}$$

Whereas the weighted sum method discussed in Section 17.4 always yields Pareto optimal points, but may miss certain points when the weights are varied, this method can provide all of the Pareto optimal points (the complete Pareto optimal set). However, it may provide non-Pareto optimal points as well. Nonetheless, the solution using the min-max approach is always weakly Pareto optimal, and if the solution is unique, then it is Pareto optimal.

The *advantages* of the method are as follows:

1. It provides a clear interpretation of minimizing the largest difference between $f_i(\mathbf{x})$ and f_i°.
2. It can provide all of the Pareto optimal points.
3. It always provides a weakly Pareto optimal solution.
4. It is relatively well suited for generating the complete Pareto optimal set (with variation in the weights).

The *disadvantages* are as follows:

1. It requires the minimization of each objective when using the utopia point, which can be computationally expensive.
2. It requires that additional constraints be included.
3. It is not clear exactly how to set the weights when only one solution point is desired.

17.6 WEIGHTED GLOBAL CRITERION METHOD

This is a scalarization method that combines all objective functions to form a single function which is then minimized. Although the term "global criterion" can refer to any scalarized function, it has been used in the literature primarily for formulations similar to the ones presented in this section. Although a global criterion may be a mathematical function with no correlation to preferences, *a weighted* global criterion is a type of utility function in which the parameters are used to model preferences. The most common weighted global criterion is defined as follows:

$$
U = \left\{ \sum_{i=1}^{k} \left[w_i \left(f_i(\mathbf{x}) - f_i^{\circ} \right) \right]^p \right\}^{1/p}
\tag{17.12}
$$

Solutions using the global criterion formulation depend on the values of both $\mathbf{w}$ and p. Generally, p is proportional to the amount of emphasis placed on minimizing the function with the largest difference between $f_i(\mathbf{x})$ and f_i°. The root $1/p$ may be omitted because the formulations with and without it theoretically provide the same solution. Typically, p and $\mathbf{w}$ are not varied or determined in unison. Rather, a fixed value for p is selected, and then $\mathbf{w}$ is either selected to reflect preferences before the problem is solved or systematically altered to yield different Pareto optimal points.

Depending on how p is set, the global criteria can be reduced to other common methods. For instance, when $p = 1$, Eq. (17.12) is similar to a weighted sum with the objective functions adjusted with the utopia point. When $p = 2$ and weights are equal to 1, Eq. (17.12) represents the distance from the utopia point, and the solution is usually considered a compromise, as discussed earlier. When $p = \infty$, Eq. (17.12) reduces to Eq. (17.10).

With the weighted global criterion method, increasing the value of p can increase its effectiveness in providing the complete Pareto optimal set (Athan and Papalambros, 1996; Messac et al., 2000a,b). This explains why the weighted min-max approach can provide the complete Pareto optimal set with variation in the weights; as shown in Eq. (17.10) it is the limit of Eq. (17.12) as $p \rightarrow \infty$.

For computational efficiency, or in cases where a function's independent minimum may be difficult to determine, *one may approximate the utopia point* with $\mathbf{z}$, which is called an *aspiration point, reference point, goal,* or *target point*. When this is done, U is called an *achievement function*. The user thus has three different parameters that can be used to specify different types of preferences: $\mathbf{w}$, p, and $\mathbf{z}$. Assuming that $\mathbf{w}$ is fixed, every Pareto optimal point may be captured using a different aspiration point $\mathbf{z}$, as long as the aspiration point is not in the feasible criterion space Z.

However, this is not a practical approach for generating the complete Pareto optimal set. Often it is not possible to determine whether **z** is in the feasible criterion space Z before solving the problem. In addition, if the aspiration point is in the feasible criterion space, the method may provide non-Pareto optimal solutions. Thus, it is recommended that the utopia point be used whenever possible. In addition, the aspiration point should not be varied as a parameter of the method but only as an approximation of the utopia point.

Equation (17.12) always yields a Pareto optimal solution as long as **w** > 0 and as long as the utopia point is used. However, it may skip certain Pareto optimal points, depending on the nature of the objective functions and the value of p that is used. Generally, using a higher value for p enables us to better capture all Pareto optimal points (with variation in **w**).

We can view the arguments of the summation in Eq. (17.12) in two ways: as transformations of the original objective functions or as components of a distance function that minimizes the distance between the solution point and the utopia point in the criterion space. Consequently, global criterion methods are often called *utopia point*, or *compromise programming*, methods, as the decision maker usually has to compromise between the final solution and the utopia point.

The *advantages* of the global criterion method are as follows:

1. It gives a clear interpretation of minimizing the distance from the utopia point (or the aspiration point).
2. It gives a general formulation that reduces to many other approaches.
3. It allows multiple parameters to be set to reflect preferences.
4. It always provides a Pareto optimal solution when the utopia point is used.

The *disadvantages* are as follows:

1. The use of the utopia point requires minimization of each objective function, which can be computationally expensive.
2. The use of an aspiration point requires that it be infeasible in order to yield a Pareto optimal solution.
3. The setting of parameters is not intuitively clear when only one solution point is desired.

17.7 LEXICOGRAPHIC METHOD

With the *lexicographic method*, preferences are imposed by ordering the objective functions according to their importance or significance, rather than by assigning weights. In this way, the following optimization problems are solved one at a time:

Minimize

$$f_i(\mathbf{x})$$

subject to

$$f_j(\mathbf{x}) \le f_j(\mathbf{x}_j^*); \quad j = 1 \text{ to } (i-1); \quad i > 1; \quad i = 1 \text{ to } k \tag{17.13}$$

Here i represents a function's position in the preferred sequence, and $f_j(\mathbf{x}_j^*)$ represents the minimum value for the jth objective function, found in the jth optimization problem. Note that after the first iteration ($j > 1$), $f_j(\mathbf{x}_j^*)$ is not necessarily the same as the independent minimum of $f_j(\mathbf{x})$ because new constraints are introduced for each problem. The algorithm terminates once a unique optimum is determined. Generally, this is indicated when two consecutive optimization problems yield the same solution point. However, determining if a solution is unique (within the feasible design space S) can be difficult, especially with local gradient-based optimization engines.

For this reason, often with continuous problems, this approach terminates after simply finding the optimum of the first objective $f_1(\mathbf{x})$. Thus, it is best to use a global optimization engine with this approach. In any case, the solution is, theoretically, always Pareto optimal. Note that this method is classified as a vector multi-objective optimization method because each objective is treated independently.

The *advantages* of the method are as follows:

1. It is a unique approach to specifying preferences.
2. It does not require that the objective functions be normalized.
3. It always provides a Pareto optimal solution.

The *disadvantages* are as follows:

1. It can require the solution of many single-objective problems to obtain just one solution point.
2. It requires that additional constraints to be imposed.
3. It is most effective when used with a global optimization engine, which can be expensive.

17.8 BOUNDED OBJECTIVE FUNCTION METHOD

The *bounded objective function method* minimizes the single most important objective function $f_s(\mathbf{x})$ with other objective functions treated as constraints: $l_i \leq f_i(\mathbf{x}) \leq \varepsilon_i$; $i = 1$ to k; $i \neq s$. l_i and ε_i are the lower and upper bounds for $f_i(\mathbf{x})$, respectively. In this way, the user imposes preferences by setting limits on the objectives. l_i is obsolete unless the intent is to achieve a goal or fall within a range of values for $f_i(\mathbf{x})$.

The ε-*constraint* approach (also referred to as *e-constraint* or *trade-off*) is a variation of the bounded objective function method in which l_i is excluded. In this case, a systematic variation of ε_i yields a set of Pareto optimal solutions. However, improper selection of the ε-vector can result in a formulation with no feasible solution. Guidelines for selecting ε-values that reflect preferences are discussed in the literature (Cohon, 1978; Stadler, 1988). A general mathematical guideline for selecting ε_i is provided as follows (Carmichael, 1980):

$$f_i(\mathbf{x}_i^*) \leq \varepsilon_i \leq f_s(\mathbf{x}_i^*) \tag{17.14}$$

A solution to the ε-constraint formulation, if it exists, is weakly Pareto optimal. If the solution is unique, then it is Pareto optimal. Of course, uniqueness can be difficult to verify, although if the problem is convex and if $f_s(\mathbf{x})$ is strictly convex, then the solution is necessarily unique. Solutions with active e-constraints (and nonzero Lagrange multipliers) are necessarily Pareto optimal (Carmichael, 1980).

The *advantages* of the method are as follows:

1. It focuses on a single objective with limits on others.
2. It always provides a weakly Pareto optimal point, assuming that the formulation gives a solution.
3. It is not necessary to normalize the objective functions.
4. It gives Pareto optimal solution if one exists and is unique.

The only *disadvantage* is that the optimization problem may be infeasible if the bounds on the objective functions are not appropriate.

17.9 GOAL PROGRAMMING

With *goal programming*, goals b_j are specified for each objective function $f_j(\mathbf{x})$. Then the total deviation from the goals $\sum_{i=1}^{k}|d_j|$ is minimized, where d_j is the deviation from the goal b_j for the jth objective function. To model the absolute values, d_j is split into positive and negative parts such that $d_j = d_j^+ - d_j^-$ with $d_j^+ \geq 0$, $d_j^- \geq 0$, and $d_j^+ d_j^- = 0$. Consequently, $|d_j| = (d_j^+ + d_j^-)$. d_j^+ and d_j^- represent underachievement and overachievement, respectively, where achievement implies that a goal has been reached. The optimization problem is formulated as follows:

Minimize

$$\sum_{i=1}^{k}(d_i^+ + d_i^-) \tag{17.15}$$

subject to

$$f_j(\mathbf{x}) + d_j^+ - d_j^- = b_j; \quad d_j^+, d_j^- \geq 0; \quad d_j^+ d_j^- = 0; \quad i = 1 \text{ to } k \tag{17.16}$$

In the absence of any other information, goals may be set to the utopia point: $b_j = f_j^\circ$. In this case Eq. (17.15) can be considered a type of global criterion method. Lee and Olson (1999) provide an extensive review of applications for goal programming. However, despite its popularity, there is no guarantee that this method provides a Pareto optimal solution. Also, Eq. (17.15) has additional variables and nonlinear equality constraints, both of which can be troublesome with larger problems.

The *advantages* of the method are as follows:

1. It is easy to assess whether the predetermined goals have been reached.
2. It is easy to tailor the method to a variety of problems.

The *disadvantages* are as follows:

1. There is no guarantee that the solution is even weakly Pareto optimal.
2. There is an increase in the number of variables.
3. There is an increase in the number of constraints.

17.10 SELECTION OF METHODS

Deciding which multi-objective optimization method is the most appropriate or the most effective can be difficult. It depends on the user's preferences and what types of solutions might be acceptable (Floudas et al., 1990). Knowledge of the problem functions can aid in the selection process. The following key characteristics of the methods discussed in this chapter are helpful in selecting the most appropriate one for a particular application:

1. Always provides a Pareto optimal solution.
2. Can provide all of the Pareto optimal solutions.
3. Involves weights to express preferences.
4. Depends on the continuity of the problem functions.
5. Uses the utopia point or its approximation.

Table 17.1 summarizes these characteristics.

TABLE 17.1   Characteristics of multi-objective optimization methods

Method	Always yields Pareto optimal point?	Can yield all Pareto optimal points?	Involves weights?	Depends on function continuity?	Uses utopia point?
Genetic	Yes	Yes	No	No	No
Weighted sum	Yes	No	Yes	Problem type and optimization engine determines this	Utopia point or its approximation is needed for function normalization or in the formulation of the method
Weighted min-max	Yes[1]	Yes	Yes	Same as above	Same as above
Weighted global criterion	Yes	No	Yes	Same as above	Same as above
Lexicographic	Yes[2]	No	No	Same as above	No
Bounded objective func.	Yes[3]	No	No	Same as above	No
Goal programming	No	No	No[4]	Same as above	No

[1]*Sometimes solution is only weakly Pareto optimal.*
[2]*Lexicographic method always provides Pareto optimal solution only if global optimization engine is used or if solution point is unique.*
[3]*Always weak Pareto optimal if it exists; Pareto optimal if solution is unique.*
[4]*Weights may be incorporated into objective function to represent relative significance of deviation from particular goal.*

EXERCISES FOR CHAPTER 17*

17.1 In the design space, plot the objective function contours for the following unconstrained problem and sketch the Pareto optimal set, which should turn out to be a curve:

Minimize

$$f_1 = (x_1 - 0.75)^2 + (x_2 - 2)^2$$

$$f_2 = (x_1 - 2.5)^2 + (x_2 - 1.5)^2$$

Draw the gradients of each function at any point on the Pareto optimal curve. Comment on the relationship between the two gradients.

17.2 Sketch the Pareto optimal set for Exercise 17.1 in the criterion space.

17.3 In the design space, plot the following constrained problem and sketch the Pareto optimal set:

Minimize

$$f_1 = (x_1 - 3)^2 + (x_2 - 7)^2$$

$$f_2 = (x_1 - 9)^2 + (x_2 - 8)^2$$

subject to

$$g_1 = 70 - 4x_2 - 8x_1 \leq 0$$

$$g_2 = -2.5x_2 + 3x_1 \leq 0$$

$$g_3 = -6.8 + x_1 \leq 0$$

17.4 Identify the weakly Pareto optimal points in the plot in Figure E17.4.

17.5 Plot the following global criterion contour in the criterion space, using p-values of 1, 2, 5, and 20 (plot one contour line for each p-value):

$$U = (f_1^p + f_2^p)^{1/p} = 1.0$$

Comment on the difference between the shapes of the different contours. Which case represents the weighted sum utility function (with all weights equal to 1)?

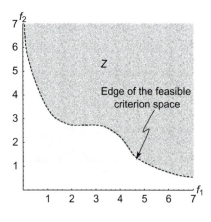

FIGURE E17-4   Identification of weakly Pareto optimal points.

17.6 Plot contours for the following min-max utility function in the criterion space:

$$U = \max[f_1, f_2]$$

Compare the shape of these contours with those determined in Exercise 17.5.

17.7 Solve the following problem using the Karush-Kuhn-Tucker (KKT) optimality conditions of Chapter 4 with the weighted sum method:

Minimize

$$f_1 = (x_1 - 3)^2 + (x_2 - 7)^2$$

$$f_2 = (x_1 - 9)^2 + (x_2 - 8)^2$$

Write your solution for the design variables in terms of the two weights w_1 and w_2. Comment on how the different weights affect the final solution.

17.8 Solve the following problem using KKT optimality conditions of Chapter 4 with the weighted sum method:

Minimize

$$f_1 = 20(x_1 - 0.75)^2 + (2x_2 - 2)^2$$

$$f_2 = 5(x_1 - 1.6)^2 + 2x_2$$

subject to

$$g_1 = -x_2 \le 0$$

First, use $w_1 = 0.1$ and $w_2 = 0.9$. Then, resolve the problem using $w_1 = 0.9$ and $w_2 = 0.1$. Comment on the constraint activity in each case.

17.9 Formulate the following problem (Stadler and Dauer, 1992) and solve it using Excel with the weighted sum method.

Determine the optimal height and radius of a closed-end cylinder necessary to simultaneously maximize the volume and minimize the surface area. Assume that the cylinder has negligible thickness. The height must be at least 0.1 m, and the radius must be at least half the height. Neither the height nor the radius should be greater than 2.0 m.

Use a starting point of $x^{(0)} = (1, 1)$. Use the following vectors of weights and comment on the solution that each yields: $w = (1, 0)$; $w = (0.75, 0.25)$; $w = (0.5, 0.5)$; $w = (0.25, 0.75)$; $w = (0, 1)$.

17.10 Solve Exercise 17.9 using Excel with a weighted global criterion. Use $x^{(0)} = (1, 1)$, $w = (0.5, 0.5)$, and $p = 2.0$. Compare the solution with those determined in Exercise 17.9.

17.11 Plot the objective functions contours for the following problem (on the same graph) and solve the problem using the lexicographic method:

Minimize

$$f_1 = (x - 1)^2(x - 4)^2$$

$$f_2 = 4(x - 2)^2$$

$$f_3 = 8(x - 3)^2$$

Indicate the final solution point on the graph. Assume the functions are prioritized in the following order: f_1, f_2, f_3, with f_1 being the most important.

18

Global Optimization Concepts and Methods

Upon completion of this chapter, you will be able to

- Explain the basic concepts associated with finding a global solution to a design problem

- Explain the basic ideas, procedures, and limitations of *deterministic* and *stochastic methods* for global optimization

- Use an appropriate method for solving a global optimization problem

The standard design optimization model treated in this text is minimizing $f(\mathbf{x})$ for $\mathbf{x}$ in the feasible set S defined as

$$S = \left\{ \mathbf{x} | h_i(\mathbf{x}) = 0, \ i = 1 \text{ to } p; \ g_j(\mathbf{x}) \leq 0, \ j = 1 \text{ to } m \right\} \tag{18.1}$$

The discrete variables in the problem are treated as explained in Chapter 15. Thus far in this text, we have addressed mainly the problem of finding a local minimum for the cost function in the feasible set. In this chapter, we focus on presentation and discussion of concepts and methods for the global optimum solutions because, in some practical applications, it is important to find such solutions as opposed to local ones.

The material for the chapter is introductory in nature and is derived from the work of the author and his coworkers (Arora et al., 1995; Elwakeil and Arora, 1996a,b). Numerous other references are cited in these articles that contain more exposition on the subject (Dixon and Szego, 1978; Evtushenko, 1985; Pardalos and Rosen, 1987; Rinnooy and Timmer, 1987a,b; Törn and Žilinskas, 1989; Pardalos et al., 2002).

18.1 BASIC CONCEPTS OF SOLUTION METHODS

18.1.1 Basic Solution Concepts

Most of the methods presented in this chapter assume continuous variables and functions. For discrete and nondifferentiable problems, the simulated annealing and genetic algorithms are appropriate for global optimization and may be used as described in Chapters 15 and 16. It is important to note that many methods for global optimization presented in the literature consider the unconstrained problem only. *It is assumed that the constraints can be treated implicitly using the penalty or augmented Lagrangian methods that are discussed in Chapter 11.* A possible disadvantage of this approach is that some methods can terminate at infeasible points. Many, however, can treat, or may even require, explicit bound constraints on the design variables. To discuss such methods, let us define a set S_b of feasible points with respect to the explicit bound constraints as

$$S_b = \{x_i | x_{iL} \le x_i \le x_{iU}; \quad i = 1 \text{ to } n\} \tag{18.2}$$

Recall that n is the number of design variables, and x_{iL} and x_{iu} are the lower and upper bounds on the ith variable.

Also, note that many global optimization methods repeatedly search for local minima in their algorithm. These methods are relatively easy to implement and use for solving global optimization problems. It is important to use robust and efficient software to search for local minima. We will assume that such an optimization engine is available for use with these global optimization methods.

> *Definition of Global Minimum:* **Before describing the methods for finding global minima, let us first recall definitions of local and global minima from Chapter 4. A point x* is called a *local minimum* for the problem if $f(x^*) \le f(x)$ for all x in a *small feasible neighborhood* of the point x*. A point x_G^* is defined as a *global minimum* for the problem if $f(x_G^*) \le f(x)$ for all x in the feasible set S.**

Characterization of a Global Minimum

A problem can have multiple global minimum points that must have the same cost function value. If the feasible set S is closed and bounded and the cost function is continuous on it, the Weierstrass Theorem 4.1 in Chapter 4 guarantees the *existence of a global minimum* point. However, finding it is a different matter altogether. At a local optimum point, the Karush-Kuhn-Tucker (KKT) necessary conditions apply (as was described in Chapters 4 and 5). Although a global minimum point must also be a local minimum point, *there are no mathematical conditions that characterize a global minimum point*, except when the problem can be shown to be *convex*. In most practical applications, however, it is difficult to check for convexity of the problem. Therefore, the problem is generally assumed to be *nonconvex*.

> **There are no mathematical conditions that characterize a global minimum point, except when the problem can be shown to be *convex*.**

Have We Found a Global Minimum?

An important issue, then, is how we know that a numerical search process has terminated at a global minimum point. The answer is that, in general, we do not know. Because of this, it is difficult to define a precise stopping criterion for a computational algorithm for global optimization. Usually, the best solution obtained by an algorithm after it is allowed to run for a long time is accepted as the global solution to the problem. The quality of the solution usually depends on how long the algorithm is allowed to run. It is important to note that the computational effort to solve a global optimization problem is substantial, and it increases enormously as the number of design variables increase. Thus, solving global optimization problems remains a challenge from a mathematical as well as a computational viewpoint. It is noted, however, that some algorithms can be implemented on parallel processors, which can reduce the "wall clock" time to solve the problem.

When to Stop Searching for a Global Minimum

It is seen that because of the lack of global optimality conditions for general problems, a global solution to the problem can be obtained only by an *exhaustive search of the design space* (the feasible set S). The procedure for such a search is to specify some sample points in the set S_b and evaluate the cost function at them. The point where the function has the smallest value is taken as the global minimum point. We see that the location and value of the global minimum depend on the sample size. An exact solution to the problem requires an infinite number of calculations. Generally, this infinite calculation is avoided by accepting the best solution as a global minimum point obtained by an algorithm after it is allowed to run for a sufficiently long time. When a point within a distance ε from x_G^* is sought, *many strategies* exist that only require a *finite number* of function evaluations. These strategies, however, are of limited practical use since ε cannot be specified because x_G^* is not known. Thus, either a *further restriction* on the class of cost functions or a *further relaxation* of what is required of an algorithm is necessary.

18.1.2 Overview of Methods

Deterministic and Stochastic Methods

Global optimization methods can be divided into two major categories: *deterministic* and *stochastic*. This classification is mainly based on whether the method incorporates any stochastic elements to solve the global optimization problem. Deterministic methods find the global minimum by an exhaustive search over the set S_b. The success of the method can be guaranteed for only the functions that satisfy certain conditions. We will describe four deterministic methods: covering, zooming, generalized descent, and tunneling.

Several *stochastic methods* have been developed as variations of pure random search. Some are useful for only discrete optimization problems while others can be used for problems that are both discrete and continuous. All of the stochastic methods involve random elements to determine the global minimum point, each one trying to reduce the computational burden of pure random search. At the outset, a random sample of points in the set S_b is picked. Then each method manipulates the sample points in a different manner. In some cases the two operations are simultaneous; that is, a random point is picked and manipulated or used before the next one is chosen. We will briefly describe some of these methods: multistart, clustering, control random search, acceptance-rejection, stochastic integration, stochastic zooming, and domain elimination.

In the remaining sections of this chapter, we describe the basic concepts and ideas underlying various methods for global optimization. Algorithms for some of the methods are described and discussed to give the student a flavor of the type of calculations needed to find a global solution to a design problem. Some of the methods describe calculations for the global minimum of the cost function without reference to constraints. It is assumed in these methods that the constraints are used to define a penalty function, which is then minimized. Some of the algorithms have been implemented on the computer to evaluate their performance on mathematical programming test problems as well as on structural design problems. These numerical experiments are described, and the performance results are discussed.

18.2 OVERVIEW OF DETERMINISTIC METHODS

Deterministic methods find the global minimum by an exhaustive search over the set S_b. If an absolute guarantee of success is desired for such a method, additional assumptions about the cost function are needed to avoid huge calculations. The most popular approach is to assume the *Lipschitz continuity condition* for the function: There exists a *Lipschitz constant L* such that for all $\mathbf{x}$, $\mathbf{y}$ in the set S_b, $|f(\mathbf{x}) - f(\mathbf{y})| \le L||\mathbf{x} - \mathbf{y}||$; that is, *the rate of change in the function is bounded.* The upper bound on the rate of change of $f(\mathbf{x})$ (first derivatives of the function) implied by the Lipschitz constant can be used in various ways to perform an exhaustive search over the set S_b (Evtushenko, 1985). Unfortunately, it is difficult in practice to verify whether a function satisfies such a condition for all points in the set S_b.

Deterministic methods for global optimization are further classified as *finite exact* and *heuristic. Finite exact methods* provide an *absolute guarantee* that the global minimum will be found in a *finite* number of steps. Generally, the number of steps is very large, so the methods require large computational effort, especially when the number of design variables is more than two. However, for some problems, it is essential to find the global minimum with an absolute guarantee, irrespective of the computational effort needed. Since no other method gives an absolute guarantee of finding the global minimum in a finite number of steps, these methods become important. *Heuristic methods,* on the other hand, offer only an *empirical guarantee* of finding the global optimum.

18.2.1 Covering Methods

As the name implies, the basic idea of covering methods is to "cover" the set S_b by evaluating the cost function at all of the points in searching for the global minimum. This is, of course, an infinite calculation and is therefore impossible to implement and use. All of the covering methods therefore devise procedures to evaluate the functions at selected points but still cover the entire set S_b implicitly.

Some covering methods take advantage of certain properties of the cost function to define a mesh of points that may or may not be uniform for evaluating functions at these points. Some of the covering methods are relatively efficient but can treat only simple cost functions (which occur only in standard test problems). In these methods, upper and lower bounds on the cost function over a subset of S_b are computed by *interval arithmetic*. Different means to exclude inferior intervals are then used. The *branch-and-bound* methods discussed in Chapter 15 are based on such ideas. Other methods successively form closer approximations (of given functions) that can be separated into convex and concave terms.

The covering method of Evtushenko (1985) uses a nonuniform mesh to cover the set S_b. An approximation of the solution point x_G^* is obtained for a given positive tolerance ε such that it belongs to the set A_ε of points with cost function values less than $f(x_G^*) + \varepsilon$; that is, A_ε is defined as

$$A_\varepsilon = \{x \in S | (f(x) - \varepsilon) \le f(x_G^*)\} \tag{18.3}$$

The set A_ε can never be constructed since x_G^* is not known. The solution, however, is guaranteed to belong to it; that is, it is within ε of the global minimum value.

In some covering methods, the mesh density is determined using the Lipschitz constant L. The upper bound on the rate of change in $f(x)$ implied by the Lipschitz constant is used in various ways to sequentially generate a mesh and perform an exhaustive search over the set S_b. Unfortunately, it is hard, in practice, to verify whether a function satisfies such a condition for all points in the set S_b. Also, the computational effort required to compute L is substantial, so only an approximation to L can be used.

In Evtushenko's method, the mesh points are generated as centers of hyperspheres. The union of these spheres has to completely cover S_b for the approximate solution to be valid. The covering is done sequentially; one sphere after another is constructed until the entire set is covered. Therefore, the total number of mesh points is not known until the covering is complete. In multidimensional problems, covering by hyperspheres is difficult and inefficient, as the hyperspheres must overlap to cover the entire set S_b.

For this reason, hypercubes inscribed in the hyperspheres are used instead. In two dimensions the design space is filled with squares; in three dimensions it is filled with cubes, and so on. The resulting mesh is nonuniform in the first variable and uniform in the rest of the variables. Since finding the true value of the Lipschitz constant L is a difficult task, a smaller approximation for L and a larger value for the tolerance ε are used initially. Then the approximation for L is increased and that of ε is decreased, and the entire covering procedure is repeated. The repetition is continued until the difference between two consecutive solutions is less than ε. In some methods, departing from purely deterministic procedures, the Lipschitz constant is estimated using some statistical models of the cost function. An advantage to Evtushenko's method is that it yields a guaranteed estimate of the global minimum for any upper-bound approximation of the Lipschitz constant.

It is seen that the covering methods are generally not practical for problems having more than two variables. However, any two-variable problem can be solved more efficiently by the graphical optimization method of Chapter 3.

18.2.2 Zooming Method

The *zooming method* was designed especially for problems with general constraints. It strives to achieve a specified target value for the global minimum of the cost function.

Once the target is achieved, it is reduced further to "zoom in" on the global minimum. The method combines a local minimization method with successive truncation of the feasible set S to eliminate regions of local minima to *zoom in* on the global solution. The basic idea is to initiate the search for a constrained local minimum from any point—feasible or infeasible. Once a local minimum point has been found, the problem is redefined in such a way that the current solution is eliminated from any further search by adding the following constraint to the problem:

$$f(\mathbf{x}) \leq \gamma f(\mathbf{x}^*) \tag{18.4}$$

where $f(\mathbf{x}^*)$ is the cost function value at the current local minimum point and $0 < \gamma < 1$ if $f(\mathbf{x}^*) > 0$, and $\gamma > 1$ if $f(\mathbf{x}^*) < 0$. The redefined problem is solved again and the process continued until no more minimum points can be found.

The zooming method appears to be a good alternative to stochastic (discussed later in this chapter) and other methods for constrained global optimization problems. It is quite simple to use: The formulation is modified slightly by adding the zooming constraint of Eq. (18.4), and existing local minimization software is used.

However, there are certain *limitations* of the method. As the target level for the global minimum of the cost function is lowered, the feasible set for the problem keeps on shrinking. It may also result in a disjointed feasible set. Therefore, as the global minimum is approached, finding even a feasible point for the redefined problem becomes quite difficult. Several different trial starting points need to be tried before declaring the redefined problem to be infeasible and accepting the previous local minimum as the global minimum. The only stopping criterion is the limit on the number of trials allowed to search for a feasible point for the reduced feasible set. An improvement in the method, described later, introduces some stochastic elements into the computational procedure.

18.2.3 Methods of Generalized Descent

Generalized descent methods are classified as *heuristic deterministic*. They are a generalization of the descent methods, described in Chapters 10 and 11, in which finite-descent steps are taken along straight lines—that is, the search directions. For nonquadratic problems, it is sometimes difficult to find a suitable step size along the search direction. Therefore, it may be more effective if we deliberately follow a *curvilinear path* in the design space (also called a *trajectory*). Before describing the generalized descent methods for global optimization, we will describe the basic ideas of *trajectory methods* that generate curvilinear paths in the design space in search of minimum points.

A trajectory can be considered a design history of the cost function from the starting point $\mathbf{x}^{(0)}$ to a local minimum point $\mathbf{x}^*$. Let the design vector $\mathbf{x}$ be dependent on the parameter t that monotonically increases along the solution curve $\mathbf{x}(t)$ and is zero at $\mathbf{x}^{(0)}$. The simplest path from an arbitrary initial point $\mathbf{x}^{(0)}$ to $\mathbf{x}^*$ is a continuous steepest-descent trajectory given as solution of the vector differential equation:

$$\dot{\mathbf{x}}(t) = -\nabla f(\mathbf{x}), \quad \text{with} \quad \mathbf{x}(0) = \mathbf{x}^{(0)} \tag{18.5}$$

where an "over-dot" represents the derivative with respect to t. We can also use a continuous Newton's trajectory by changing the right side of Eq. (18.5) to $-[\mathbf{H}(\mathbf{x})]^{-1}\nabla f(\mathbf{x})$, where $\mathbf{H}(\mathbf{x})$ is the Hessian of the cost function that is assumed to be nonsingular for all $\mathbf{x}$. It is noted that good software is available to solve the first-order differential equation (18.5).

The generalized descent methods for global optimization are extensions of the foregoing trajectory methods. In these methods, the trajectories are solutions for certain second-order differential equations rather than the first-order Eq. (18.5). The search for the global minimum is based on solution properties of these differential equations. The most important property is that their trajectories pass through the majority of the stationary points of the cost function (or in their neighborhood). Certain conditions determine if the trajectory will pass through all of the local minima of the function. In that case, the global minimum is guaranteed to be found. The differential equations use the function values and function gradients along the trajectories.

There are two types of generalized descent methods: (1) *trajectory methods*, which modify the differential equation describing the local descent trajectory so as to make it converge to a global rather than a local minimum, and (2) *penalty methods*, which apply the standard local algorithm repeatedly to a modified cost function so as to prevent the descent trajectory from converging to local minima previously found. Examples of penalty methods are: algebraic functions, filled function, and tunneling. In this section we will describe the trajectory methods only. The tunneling methods will be described in the next section.

Alternation of Descents and Ascents

Trajectory methods have been implemented in two ways: alternation of descents and ascents and the so-called golf methods. The first method consists of three subalgorithms that are modifications of the local descent algorithms. These are: *descent* to a local minimum, *ascent* from a local minimum, and *pass through* a saddle point. First, to descend to a local minimum from a starting point, we use a combination of steepest-descent and Newton's methods based on whether or not the Hessian matrix of the cost function is positive definite at a point along the trajectory (this is the *modified Newton's method* for local minimization).

Second, to get from the local minimum point to a saddle point, we use the eigenvector corresponding to the maximum eigenvalue of the Hessian as the search direction. Third, to pass through the saddle point, we use Newton's method. To descend to the next local minimum, we use the direction of the last step of the passing operation as a starting direction for the descent operation. The three operations are repeated until some stopping criterion is satisfied. Note that all local minimum points are recorded so that if the trajectory retraces itself, a new initial point is chosen to restart the algorithm.

The disadvantages of this method are the large number of function evaluations wasted in ascending from a local minimum, and difficulties with solving problems of dimensions larger than two. It is also difficult to apply the method if we do not have an expression for the cost function gradient.

Golf Methods

Golf methods are analogous to the mechanics of inertial motion of a particle of mass m moving in a force field. The resulting trajectory is analogous to that of the optimization problem. Mathematically, the assignment of mass means introducing a second-order term into the particle's equation of motion. Taking the mass as a function of time $m(t)$, the particle is thus moving in a force field defined by the cost function $f(\mathbf{x})$ and subjected to a dissipating or nonconservative resistance force (e.g., air resistance) given by $-n(t)\dot{\mathbf{x}}(t)$, where $n(t)$ is the resistance function. The force field of $f(\mathbf{x})$ is given as $-\nabla f(\mathbf{x})$. Thus, the motion of the particle is described by the system of differential equations:

$$m(t)\ddot{\mathbf{x}}(t) - n(t)\dot{\mathbf{x}}(t) = -\nabla f(\mathbf{x}), \quad m(t) \geq 0 \quad n(t) \leq 0 \tag{18.6}$$

where $\dot{\mathbf{x}}(t)$ and $\ddot{\mathbf{x}}(t)$ are the velocity and acceleration vectors of the particle, respectively.

Under some conditions, the trajectory, which is a solution for the system of equations, converges to a local minimum point of $f(\mathbf{x})$. Moreover, the trajectory leaves some local minima that are not deep enough—hence the name, golf methods. It is obvious that algorithm efficiency is a function of the mass and resistance functions $m(t)$ and $n(t)$. For some functions, the differential equation is simplified by assuming the mass of the particle as 1 and a frictionless force field (i.e., $m(t) = 1$ and $n(t) = 0$). In this case the differential equation is simplified to $\ddot{\mathbf{x}}(t) = -\nabla f(\mathbf{x})$. Such a class of functions is encountered if $f(\mathbf{x})$ is an interpolation of noisy data from some experiments.

18.2.4 Tunneling Method

The *tunneling method* falls into the class of heuristic generalized descent penalty methods. It was initially developed for unconstrained problems and then extended for constrained problems (Levy and Gomez, 1985). The basic idea is to execute the following two phases successively until some stopping criterion is satisfied: the *local minimization* phase and the *tunneling* phase. The first phase consists of finding a local minimum $\mathbf{x}^*$ for the problem using any reliable and efficient method. The tunneling phase determines a starting point that is different from $\mathbf{x}^*$ but has a cost function value smaller than or equal to the known minimum value. However, finding a suitable point in the tunneling phase is also a global problem that is as hard as the original problem. When a rough estimate of the global minimum is required, the tunneling or zooming method is justified instead of one that can guarantee a global minimum at the expense of a large computational effort.

The basic idea of the tunneling method is depicted graphically in Figure 18.1 for a one-dimensional problem. Starting from the initial minimum point $\mathbf{x}^{*(1)}$, the method *tunnels* under many other local minima and locates a new starting point $\mathbf{x}^{0(2)}$. From there, a new local minimum is found, and the process is repeated. The tunneling phase is accomplished by finding a root of the nonlinear *tunneling function*, $T(\mathbf{x})$. This function is defined in such a way that it avoids previously determined local minima and the starting points. The new point found during the tunneling phase should not be a local minimum either because the local minimization procedure cannot be started from such a point.

Therefore, if the tunneling phase yields a local minimum point, a new root of the tunneling function is sought. Once a suitable point is obtained through the tunneling

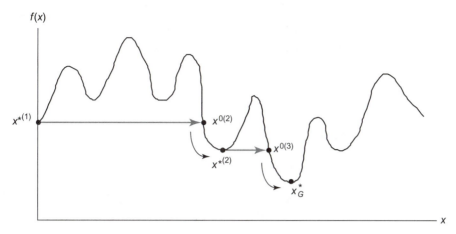

FIGURE 18.1 Basic concept of tunneling method. The method tunnels below irrelevant minima to approach the global minimum.

phase, local minimization is started to obtain a new local minimum point. The two phases are repeated until no suitable roots of the tunneling function can be found, which is realized numerically when $T(\mathbf{x}) \geq 0$ for all $\mathbf{x}$. We note here that such a criterion is very expensive to satisfy in terms of the number of function evaluations needed. The first few tunneling phases are relatively efficient in that they require little computing effort. As the tunneling level approaches the global minimum, the number of computations increases because there are fewer roots of the tunneling function. This difficulty is similar to the one noted for the zooming method.

The tunneling method has the *global descent property*: The local minima, obtained by the minimization phase approaches the global minimum in an orderly fashion. Tunneling takes place below irrelevant local minima regardless of their number or location. Because of this property, the tunneling method can be efficient relative to other methods, especially for problems with a large number of local minima. It has the advantage that a point with a smaller cost function value is reached at each iteration. Therefore, a point with a relatively small cost function value is obtained very quickly, as with the zooming method. Such a solution is acceptable for some engineering applications. In these cases, the tunneling or the zooming method is justified instead of a method that can guarantee a global minimum at the expense of a large computational effort.

18.3 OVERVIEW OF STOCHASTIC METHODS

Let S_* be the *set of all local minima* of the optimization problem. The aim of many stochastic methods is to determine this set S_*. The best point of the set is then claimed as the global minimum point. Generally, far better results have been obtained using *stochastic* methods than deterministic methods.

Stochastic methods usually have *two phases*: global and local. In the *global phase*, the function is evaluated at a number of randomly sampled points. In the *local phase*, the sample points are manipulated, for example, by local searches, to yield candidate global minima.

The global phase is necessary because there is no local improvement strategy that, starting from an arbitrary point, can be guaranteed to converge to the global minimum. The global phase locates a candidate global minimum point in every subset of the feasible set S_b to ensure *reliability* of the method. Local search techniques are efficient tools for finding a point with a relatively small function value. Therefore, the local phase is incorporated into stochastic methods to improve their *efficiency*. A challenge for global optimization algorithms is to increase their efficiency while maintaining reliability.

There are many stochastic methods for global optimization, such as random search, multistart, clustering, controlled random search, simulated annealing, acceptance-rejection, stochastic integration, genetic, and tabu search. We will describe only the ideas that lay the foundation for computations of these methods. More details can be found in Arora and et al. (1995) and other references cited therein.

Most *stochastic methods* are based on some variation of pure random search. They are used in two ways: (1) to develop stopping criteria, and (2) to develop techniques to approximate the *region of attraction* for a local minimum point, which is defined as follows:

> When the search for the local minimum started from a point within a certain region around the minimum converges to the same minimum point, the region is called the region of attraction for that local minimum.

The goal of many stochastic methods is to develop good approximations of the regions of attraction for local minima so that the search for a local minimum is performed only once.

Usually, most of the stochastic algorithms use *uniform distribution* of sampling over the set S_b. However, mechanisms for modifying the sampling distribution based on the information obtained in previous iterations may be more appropriate. A stochastic approximation that is of the type used in sampling can be employed to determine a sampling distribution that peaks in unexplored regions of attraction to discover new local minima. Even though the stochastic methods do not offer an *absolute guarantee* of success, *the probability that a point within a distance ε of $\mathbf{x}_G^*$ will be found approaches 1 as the sample size increases.*

Note that since some stochastic methods use random processes, such as simulated annealing and genetic algorithms, an algorithm run at different times can generate different design histories and local minima. Therefore, a particular problem needs to be run several times before the solution is accepted as the global optimum.

18.3.1 Pure Random Search Method

Pure random search is the *simplest stochastic method* for global optimization, and most other stochastic methods are variations of it. Though very inefficient, it is described here to introduce a basis for those other methods. Pure random search consists only of a global phase: Evaluate $f(\mathbf{x})$ at N sample points drawn from a random uniform distribution over the set S_b. The smallest function value found is the candidate global minimum for $f(\mathbf{x})$.

Pure random search is asymptotically guaranteed to converge, in a probabilistic sense, to the global minimum point. It is quite inefficient because of the large number of function evaluations required to provide such a guarantee. A simple extension of the method is so-called *single start*. In this, a single local search is performed (if the problem is continuous) starting from the best point in the sample set at the end of pure random search.

18.3.2 Multistart Method

The multistart method is one of several extensions of pure random search where a local phase is added to the global phase to improve efficiency. In multistart, in contrast to single start, each sample point is used as a starting point for the local minimization procedure. The best local minimum point found is a *candidate* for the global minimum $\mathbf{x}_G^*$. The method is reliable, but it is not efficient since many sample points will lead to the same local minimum. The algorithm consists of three simple steps:

Step 1. Take a random point $\mathbf{x}^{(0)}$ from a *uniform distribution* over the set S_b.
Step 2. Start a local minimization procedure from $\mathbf{x}^{(0)}$.
Step 3. Return to Step 1 unless a *stopping criterion* is satisfied.

Once the stopping criterion is satisfied, the local minimum with the smallest function value is taken as the global minimum $\mathbf{x}_G^*$. It can be seen that a particular local minimum may be *reached several times* starting from different points. Strategies to eliminate this inefficiency in the algorithm have been developed and are discussed in the following sections.

Stopping Criterion

Several ideas for terminating an algorithm have been proposed; however, most of them are not practical. Here we describe a criterion that has been used most often. Since the starting points of the multistart method are uniformly distributed over S_b, a local minimum has a fixed *probability* of being found in each trial. In a *Bayesian approach*, in which the unknowns are themselves assumed to be random variables with a *uniform prior distribution*, the following result can be proved: Given that M distinct local minima have been found in L searches, the *optimal Bayesian estimate* of the unknown number of local minima K is given by

$$K = \text{integer}\left[\frac{M(L-1)}{L-M-2}\right] \quad \text{provided } L \geq M+3 \tag{18.7}$$

The multistart method can be stopped when $M = K$. It can be shown that this stopping rule is appropriate for other methods as well.

18.3.3 Clustering Methods

Clustering methods remove the inefficiency of the multistart method by trying to use the local search procedure only once for each local minimum point. To do this, random sample points are linked into groups to form clusters. Each cluster is considered to represent one region of attraction for a local minimum point. Each local minimum point has a *region of attraction* such that a search initiated from any point in the region converges to the same local minimum point. Four clustering methods have been used for development of the regions of attraction: density clustering, single linkage, mode analysis, and vector quantization multistart. They differ in the way in which these regions of attraction are constructed. A major disadvantage of all clustering methods is that their performance depends heavily on the dimension of the problem, that is, the number of design variables.

Reduced Sample Points

Let A_N be a set of random points drawn from a uniform distribution over the set S_b (details of how to define A_N are given later). In the clustering methods, a preprocessing of the sample is performed to produce regions that are likely to contain local minima. This can be done in two ways: reduction and concentration. In *reduction*, a set A_q of sample points having cost function values smaller than or equal to some value f_q is constructed:

$$A_q = \{\mathbf{x} \in A_N \mid f(\mathbf{x}) \le f_q\} \tag{18.8}$$

This is called an f_q-*level set* of $f(\mathbf{x})$, or simply the reduced set, and points $\mathbf{x}$ in the set A_q are called the *reduced sample points*. The set A_q may be composed of a number of components that are disjointed. Each component will contain at least one local minimum point.

Figure 18.2(a) is an example of a uniform sample in a set, and Figure 18.2(b) shows the reduced sample points—the set A_q. The set consists of three components, each containing one local minimum point. A component of A_q is called a cluster, which is taken as an approximation of the region of attraction for a local minimum point. Note that depending on the value of f_q, a component may contain more than one local minimum point. Furthermore, the local minimum points $\mathbf{x}^*$ having function values higher than f_q will not belong to A_q and therefore may not be found.

In the second preprocessing procedure, called *concentration*, a few *steepest-descent* steps are applied to every sample point. However, in this case, unlike in reduction, the transformed points are not uniformly distributed. Usually, a uniform distribution is assumed in clustering methods; therefore, the former method of transformation is preferred.

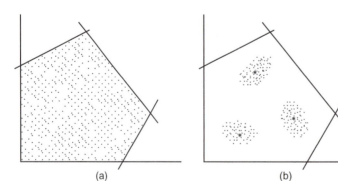

FIGURE 18.2 An example of random and reduced sample points. (a) A random sample of points. (b) Reduced sample points. The * indicated a local minimum point.

(a)   (b)

Individual Clustering Methods

Four clustering methods are available in the literature: *density, single linkage, mode analysis*, and *vector quantization multistart*. In these methods it is assumed that

1. All local minima *of f*(**x**) lie in the interior of S_b.
2. Stationary points are isolated.
3. Local search is always that of descent.

The methods execute a basic algorithm a number of times. In every iteration, a set of A_N consisting of κN sample points from a uniform random distribution is used, where κ is an integer containing the number of times the algorithm has been executed. That is, the sample size keeps on increasing with each execution of the algorithm.

Before clustering, the sample is reduced to produce the set A_q defined in Eq. (18.8). The clustering algorithm is then applied to A_q. The iterations are continued until a stopping criterion is satisfied. The stopping criterion used for multistart can be applied to all of the clustering methods. In applying these rules, we have to assume that the way A_q changes with different samples does not affect analysis. More important, we have to assume that each local minimum with a function value smaller than f_q is actually found.

DENSITY CLUSTERING   In density clustering, clusters are identified based on the density of the reduced sample points defined in Eq. (18.8); hence the name "density clustering." Regions of attraction are approximated by hyperspheres or ellipsoids with centers at local minimum points. Reduced sample points are added to clusters based on their distance from the centers (also called the *seed points*)—that is, all points within a critical radius of a center belong to the cluster. A cluster is started with a local minimum point as a seed and then expanded in stages by increasing its critical radius. All points within the new radius are added to the cluster and so on. At the end of a stage, the best unclustered point is used in a local search procedure to find a local minimum point. If the local minimum found is new, it is taken as the seed for a new cluster; otherwise that minimum point is the seed for an already existing cluster that needs to be expanded. This is continued until all of the reduced sample points are clustered.

SINGLE LINKAGE CLUSTERING   In single linkage clustering, a better approximation to the clusters is achieved by not enforcing a particular shape. Points are linked to others in their proximity as opposed to being linked to the clusters' centers or seeds. A point is assigned to a cluster if it falls within a critical distance r_κ from any point that already belongs to that cluster.

MODE ANALYSIS CLUSTERING   The density clustering and the single linkage clustering methods use information at only two points at a time. In the mode analysis method, on the other hand, clusters are formed using more information. Here the set S_b is partitioned into nonoverlapping, small hypercubic *cells* that cover S_b entirely. The cell is said to be *full* if it contains at least G reduced sample points; otherwise, it is empty.

VECTOR QUANTIZATION   In the *vector quantization method, the theories of lattices and vector quantization* are used to form clusters. The basic idea is to cluster cells rather than sample points, as in mode analysis. Thus the entire space S_b is divided into a finite number of cells and a *code point* is associated with each one. The code point is used to represent all of the points in that cell during the clustering process. The point with the smallest function value of a cell is the most suitable code point. Further, code points need not be sample points; they can be generated independently. They may also be centroids of the cells. Identification of a cluster is done using *vector quantization* of the reduced sample points. The aim is to approximate the clusters in a more efficient way than with the previous three methods.

18.3.4 Controlled Random Search: Nelder-Mead Method

The *basic idea of controlled random search* (CRS), which is another variation of pure random search, is to use the sample points in a way so as to move toward the global minimum point (Price, 1987). The method does not use gradients of the cost function, and so continuity and differentiability of the functions are not required. It uses the idea of a *simplex*, which is a geometric figure formed by a set of $n + 1$ points in the n-dimensional space (recall that n is the number of design variables). When the points are equidistant, the simplex is said to be *regular*. In two dimensions, the simplex is just a triangle; in three dimensions, it is a tetrahedron (see Figure 18.3), and so on. The method has global and local phases.

GLOBAL PHASE   The notation used in the *global phase* of the algorithm that follows is defined as

$$\mathbf{x}^W, f^W = \text{worst point and the corresponding cost function value (largest)}$$
$$\mathbf{x}^L, f^L = \text{best point and the corresponding cost function value (smallest)}$$
$$\mathbf{x}^C, f^C = \text{centroid of } n \text{ points, and corresponding cost function value} \left(\mathbf{x}^C = \frac{1}{n} \sum_{k=1}^{n} \mathbf{x}^{(k)} \right)$$
$$\mathbf{x}^P, f^P = \text{trial point and corresponding cost function value}$$

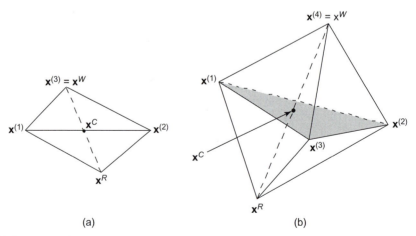

FIGURE 18.3   Reflection operations. (a) Two-dimensional Simplex. (b) Three-dimensional Simplex. The points $\mathbf{x}^W, \mathbf{x}^C$, and $\mathbf{x}^R$ lie on a straight line.

Step 1. Generate N random points uniformly distributed over S_b. Evaluate the cost function at the points.

Step 2. Find the worst point $\mathbf{x}^W$ with the function value f^W (largest) and the best point $\mathbf{x}^L$ with the function value f^L (smallest).

Step 3. Let $\mathbf{x}^{(1)} = \mathbf{x}^L$. Randomly choose n distinct points $\mathbf{x}^{(2)}, \ldots, \mathbf{x}^{(n+1)}$ from the remaining $N-1$ sample points. Determine the centroid $\mathbf{x}^C$ of the points $\mathbf{x}^{(1)}, \ldots, \mathbf{x}^{(n)}$. Compute a new trial point $\mathbf{x}^P = 2\mathbf{x}^C - \mathbf{x}^{(n+1)}$.

Step 4. If $\mathbf{x}^P$ is feasible, evaluate f^P, and go to Step 5. Otherwise, go to Step 3.

Step 5. If $f^P < f^W$, replace $\mathbf{x}^W$ by $\mathbf{x}^P$ and go to Step 6. Otherwise, go to Step 3.

Step 6. If a stopping criterion is satisfied, stop. Otherwise, go to Step 2.

As the algorithm proceeds, the current set of n points tends to cluster around the minimum point. Note that the point $\mathbf{x}^{(n+1)}$ used in the calculation of the new trial point $\mathbf{x}^P$ in Step 3 is arbitrarily chosen. This point is called the *vertex* of the simplex. Once the global phase has terminated, the local phase starts.

LOCAL PHASE The basic idea of the local phase is to compare cost function values at the $n+1$ vertices of the simplex and move this simplex gradually toward the minimum point. The movement of the simplex is achieved using three operations, known as *reflection*, *expansion*, and *contraction*. The following additional notation is used in describing these operations:

$\mathbf{x}^S, f^S$ = second worst point and corresponding cost function value
$\mathbf{x}^R, f^R$ = reflected point and corresponding cost function value
$\mathbf{x}^E, f^E$ = expansion point and corresponding cost function value
$\mathbf{x}^Q, f^Q$ = contraction point and corresponding cost function value

REFLECTION Let $\mathbf{x}^{(1)}, \ldots, \mathbf{x}^{(n+1)}$ be the $n+1$ points that define the simplex and let the worst point $(\mathbf{x}^W)$ be the vertex with the largest cost function value. It can be expected that the point $\mathbf{x}^R$ obtained by reflecting $\mathbf{x}^W$ in the opposite face of the simplex will have a smaller cost function value. If this is the case, then a new simplex can be constructed by rejecting the point $\mathbf{x}^W$ from the simplex and including the new point $\mathbf{x}^R$. In Figure 18.3(b), the original simplex is given by points $\mathbf{x}^{(1)}, \mathbf{x}^{(2)}, \mathbf{x}^{(3)}$, and $\mathbf{x}^{(4)} = \mathbf{x}^W$, and the new simplex is given by $\mathbf{x}^{(1)}, \mathbf{x}^{(2)}, \mathbf{x}^{(3)}$, and $\mathbf{x}^R$. The point $\mathbf{x}^C$ is the centroid of the n points of the original simplex excluding $\mathbf{x}^W$. It is seen that the direction of movement of the simplex is always away from the worst point. Mathematically, the reflection point $\mathbf{x}^R$ is given by

$$\mathbf{x}^R = (1 + \alpha_R)\mathbf{x}^C - \alpha_R \mathbf{x}^W, \text{ with } 0 < \alpha_R \leq 1 \qquad (18.9)$$

EXPANSION If the reflection procedure produces a better point, one can generally expect to reduce the function value further by moving along the direction $\mathbf{x}^C$ to $\mathbf{x}^R$. An expansion point $\mathbf{x}^E$ along this direction is calculated(its use is explained later the local phase algorithm):

$$\mathbf{x}^E = (1 + \alpha_E)\mathbf{x}^C - \alpha_E \mathbf{x}^W, \text{ with } \alpha_E > 1 \qquad (18.10)$$

CONTRACTION If the point obtained by reflection is not satisfactory, a contraction point $\mathbf{x}^Q$ along the direction $\mathbf{x}^C$ to $\mathbf{x}^R$ can be calculated (its use is explained later in the local phase algorithm):

$$\mathbf{x}^Q = (1 + \alpha_Q)\mathbf{x}^C - \alpha_Q \mathbf{x}^W, \text{ with } -1 < \alpha_Q < 0 \tag{18.11}$$

The *local search algorithm* given in the following uses the relations in Eqs. (18.9) through (18.11) with $\alpha_R = 1$, $\alpha_E = 3$, and $\alpha_Q = -1/2$, respectively. The $n + 1$ best points of the random sample of the global phase constitute a simplex in the n-dimensional space for this algorithm. The algorithm does not use the gradient of the cost function $f(\mathbf{x})$. Therefore, a compatible local search procedure that also does not use gradients is needed. An algorithm that is an adaptation of basic *Nelder-Mead simplex algorithm* (1965) for general constrained optimization problems may be used (this simplex method should not be confused with the Simplex method of linear programming).

NELDER-MEAD SIMPLEX ALGORITHM Nelder-Mead is a direct search algorithm that can be used to find local minimum point of a function. One of the better methods in the direct search class, it can be implemented in several different ways (Lagarias et al., 1998; Price et al., 2002; Singer and Singer, 2004). The method is also available in MATLAB as function `fminsearch`, whose use was demonstrated in Chapter 7. The following steps represent one possible implementation:

Step 1. For the simplex formed of $n + 1$ points, let $\mathbf{x}^W$ be the worst point and $\mathbf{x}^C$ be the centroid of the other n points. Let $\mathbf{x}^S$ be the second worst point of the simplex with the function value f^S. Compute three trial points based on reflection, expansion, and contraction as $\mathbf{x}^R = 2\mathbf{x}^C - \mathbf{x}^W$, $\mathbf{x}^E = 4\mathbf{x}^C - 3\mathbf{x}^W$, and $\mathbf{x}^Q = (\mathbf{x}^C + \mathbf{x}^W)/2$.
Step 2. If $\mathbf{x}^R$ is not in S_b, go to Step 4. Otherwise, evaluate the function value f^R at $\mathbf{x}^R$. If $f^R < f^S$, go to Step 3. Otherwise, go to Step 4.
Step 3. *Expansion:* If $\mathbf{x}^E$ is not in S_b, accept $\mathbf{x}^R$ as the replacement point and go to Step 5. Otherwise, calculate the function value f^E at $\mathbf{x}^E$. If $f^E < f^S$, accept $\mathbf{x}^E$ as the replacement point and go to Step 5. Otherwise, accept $\mathbf{x}^R$ as the replacement point and go to Step 5.
Step 4. If $\mathbf{x}^Q$ is infeasible, then stop; no further improvement is possible. Otherwise evaluate the function value f^Q at $\mathbf{x}^Q$. If $f^Q < f^S$, then accept $\mathbf{x}^Q$ as the replacement point and go to Step 5. Otherwise, stop.
Step 5. Update the simplex by replacing $\mathbf{x}^W$ with the replacement point. Return to Step 1.

The global and local phases of the method described in the foregoing are combined as follows: Execute the global phase and generate a new trial point $\mathbf{x}^P$ in Step 3. Let the N sample points be sorted in descending order of their cost function values. If $\mathbf{x}^P$ is feasible and falls within the bottom $n + 1$ points of the sample, then execute the local phase starting with those $n + 1$ points as a simplex. Continue execution of the two phases until the global phase stops in Step 6.

The following features of the method should be noted. The local phase operates only on the best $n + 1$ points in the database of sample points. Thus, it has minimal effect on the performance of the global phase. In Step 2 of the composite algorithm, the local phase may improve the best point in the database. Thus, it tends to speed up convergence because the global phase always uses the best point. However, this may reduce, to a small degree, the global search capability. If desired, it is easy to counter this effect by requiring that the global phase not include the best point in some iterations. In other words, in Step 3 of the algorithm, choose all of the $n + 1$ distinct points $\mathbf{x}^{(1)}$ to $\mathbf{x}^{(n+1)}$ randomly from the N sample points.

Taking the sample size $N = 10(n + 1)$ gives satisfactory results. In Step 1 of the composite algorithm, if $f^W/f^L < 1 + \varepsilon$ ($\varepsilon > 0$ is a small number), then the global phase may be terminated. Any other stopping criterion may also be used.

18.3.5 Acceptance-Rejection Methods

The *acceptance-rejection* (A-R) methods are *modifications of the multistart algorithm* to improve its efficiency by using ideas from statistical mechanics. In the multistart method, a local minimization is started from each randomly generated point. Thus, the number of local minimizations is very large and many of them converge to the same local minimum point. A strategy to improve this situation is to start the local minimization procedure only when the randomly generated point has a smaller cost function value than that of the local minimum that was previously obtained. This forces the algorithm to tunnel below irrelevant local minima. This modification, however, has been shown to be inefficient. As a result, the tunneling process has been pursued only by means of the deterministic algorithms explained earlier.

The acceptance–rejection methods modify this tunneling procedure. The basic idea of this is to sometimes start local minimization from a randomly generated point even if it has a higher cost function value than that at a previously obtained local minimum. This is called the *acceptance phase*, which involves calculation of certain probabilities. If the local minimization procedure started from an accepted point produces a local minimum that has a higher cost function value than a previously obtained minimum, then the new minimum point is rejected (*rejection phase*). The procedure just described is sometimes called *random tunneling*.

A possible formulation of the *acceptance criterion* to start the local minimization is suggested by the statistical mechanics approach described in Chapter 15, simulated annealing. The acceptance-rejection methods thus resemble the simulated annealing approach. The local minimization is started from a point $\mathbf{x}$ only if it has the probability given by

$$p(\mathbf{x}) = \exp\left(\frac{[f(\mathbf{x}) - \bar{f}]^+}{-F}\right) \tag{18.12}$$

where $\bar{f}$ is an estimate of the upper bound of the global minimum, F is a target value for the global minimum, and $[h]^+ = \max(0, h)$.

The initial value of F is usually provided by the user, or it may be estimated using a few random points. In this algorithm, unlike simulated annealing, the choice of schedule for reduction of the target level does not prevent convergence. Nevertheless, the schedule is critical for performance of the algorithm. $\bar{f}$ is adjusted at each iteration as the best approximation to the global minimum value. At the start, it may be taken as the smallest cost function value among some randomly generated points, or it can be supplied by the user if a better value is known.

18.3.6 Stochastic Integration

In *stochastic integration*, a suitable stochastic perturbation of the system of equations for the trajectory methods (described in Section 18.2.3) is introduced in order to force the trajectory to a global minimum point. This is achieved by monitoring the cost function value along the trajectories. By changing some coefficients in the differential equations, we get different solution processes starting from the same initial point. This idea is similar to simulated annealing, but here the differential equation parameter is decreased continuously. We describe the stochastic integration global minimization method using the steepest-descent trajectory.

In this, a stochastic perturbation is introduced in Eq. (18.5) in order to increase the chance of the trajectory reaching the global minimum point. The resulting system of stochastic differential equations is given as

$$dx(t) = -\nabla f(\mathbf{x})dt + \varepsilon(t)d\mathbf{w}(t), \quad \text{with } \mathbf{x}(0) = \mathbf{x}^{(0)} \tag{18.13}$$

where $\mathbf{w}(t)$ is an n-dimensional standard *Wiener process* and $\varepsilon(t)$ is a real function called the *noise coefficient*. In actual implementation, a standard Gaussian distribution is usually used instead of the Wiener process.

Let $\mathbf{x}(t)$ be the solution to Eq. (18.13) starting from $\mathbf{x}^{(0)}$ with a constant noise coefficient $\varepsilon(t) = \varepsilon_0$. Then, as is well known in the field of statistical mechanics, the probability density function of $\mathbf{x}(t)$ approaches the limit density $Z\exp[-2f(\mathbf{x})/\varepsilon_0^2]$, as $t \to \infty$, where Z is a normalization constant. The limit density is independent of $\mathbf{x}^{(0)}$ and peaks around the global minima of $f(\mathbf{x})$. The peaks become narrower with a smaller ε_0; that is, ε_0 is equivalent to the target level F that decreases in the simulated annealing method. In this method, an attempt is made to obtain the global minima by looking at the asymptotic (as $t \to \infty$) values of a numerically computed sample trajectory of Eq. (18.13), where the noise function $\varepsilon(t)$ is continuous and suitably tends to zero as $t \to \infty$. In other words, unlike simulated annealing, the target level is lowered continuously. The computational effort in this method can be reduced by observing that a correct numerical computation of the gradient in Eq. (18.13) is not really needed since a stochastic term is added to it. An approximate finite difference gradient may be used instead.

Computing a single trajectory of Eq. (18.13) by decreasing $\varepsilon(t)$ (for $t > 0$) and following the trajectory for a long time to obtain a global solution may not be very efficient. Therefore, in actual implementation an alternative strategy can be used where several trajectories are generated simultaneously. The cost function values along all of the trajectories are monitored and compared with each other. A point corresponding to the smallest cost function value on any of the trajectories at any trial is stored. If some trajectories are not progressing satisfactorily, they may be discarded and new ones initiated. As with other stochastic methods, the procedure is executed several times before accepting the best point as the global optimum point.

18.4 TWO LOCAL-GLOBAL STOCHASTIC METHODS

In this section, we describe two stochastic global optimization methods that have both local and global phases. The algorithms have been designed to treat general constraints in the problem explicitly. They can be viewed as a modification of the multistart procedure but with the ability to learn as the search progresses.

18.4.1 Conceptual Local-Global Algorithm

As explained in the last section, most stochastic methods have local and global phases. In this subsection, we describe a conceptual algorithm having both of these phases that forms the basis for the two algorithms described in the next two sections.

Step 1. Generate a random point $\mathbf{x}^{(0)}$ in the set S_b.

Step 2. Check some rejection criteria (discussed later) based on the proximity of $\mathbf{x}^{(0)}$ to one of the previous starting points, local minimum points, or rejected points. If a rejection criterion is satisfied, add $\mathbf{x}^{(0)}$ to the set of rejected points and go to Step 1. Otherwise, execute the local phase by continuing with Step 3.

Step 3. Add $\mathbf{x}^{(0)}$ to the set of starting points and find a local minimum $\mathbf{x}^*$ in the feasible set S for the problem.

Step 4. Check if $\mathbf{x}^*$ is a new local minimum; if so, add it to the set of local minima, otherwise add $\mathbf{x}^{(0)}$ to the set of rejected points. Go to Step 1.

Steps 1 and 2 constitute the global phase and Steps 3 and 4 constitute the local phase of the algorithm. The basic idea is to explore the entire feasible domain in a systematic way for the global minimum.

Bearing in mind that generation and evaluation of a random point are much cheaper than one local minimization, which may require many function and gradient evaluations, more emphasis is placed on the algorithm's global phase. The algorithm avoids searching near any local minimum point and all of the points leading to it, thus increasing the chance of finding a new local minimum in the unexplored region. To do this, several sets that contain certain types of points are constructed. For example, a set is reserved for all of the local minimum points found, and another contains all of the starting points for local searches. The following sets, in addition to S and S_b defined earlier, are used in the two algorithms:

$S_* = $ set of local minima

$S_0 = $ set of starting points $\mathbf{x}^{(0)}$

$S_r = $ set of rejected points

A uniformly distributed random point generation scheme over S_b is used so that the entire feasible domain is explored with a uniform probability of finding the global minimum. In Step 1 of the algorithm, the point $\mathbf{x}^{(0)}$ in S_b is accepted because finding a point in S can be a difficult problem (Elwakeil and Arora, 1995). The use of a uniform distribution enables the application of some well-known stopping rules.

The next two sections present the domain elimination and stochastic zooming methods. In both, a random point is generated in S_b. Other constraints are ignored at this stage because a local minimization procedure that does not require a feasible starting point can be used. Also, both methods require very little programming effort since the local phase can use existing software. In the algorithms, a modified local phase is used instead of the one given in Step 3: The local search is performed using many subsearches, each one consisting of a few iterations (two or three). Certain criteria, explained in the following sections, are checked after each subsearch to determine if the next subsearch should be started or the local phase should be terminated.

18.4.2 Domain Elimination Method

The basic idea of this algorithm is to systematically explore the entire feasible domain for the problem to find the global minimum. To accomplish this, each local search is attempted from a point that is likely to lead to a new local minimum point. Figure 18.4 shows a conceptual flow diagram for the major steps of the algorithm, which starts with

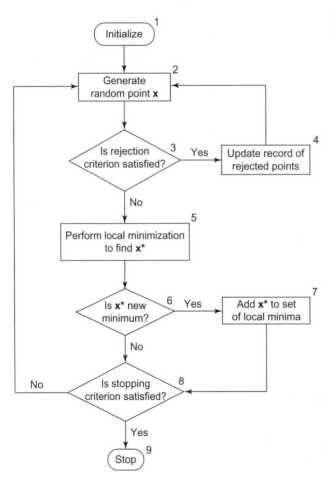

FIGURE 18.4   Flow diagram for domain elimination method.

the selection of a random point from a uniform distribution over the set (block 2). The point is rejected or accepted based on certain criteria (block 3); if the point is accepted, then a search for a local minimum is initiated from there (block 5). If it is rejected, then it is added to the set of rejected points (block 4) and a new random point is selected.

To accept or reject a random point, records for the following three types of points are kept: previous starting points for local minimization, local minimum points, and rejected points. A random point is rejected if it is within a critical distance of one of the foregoing sets. The distance between the random point and the points in the sets can be calculated in different ways; the infinity norm, being the simplest, is suggested. If the point is accepted, a search is initiated for a new local minimum point. The local search process is also monitored, and if the search is going toward a known local minimum point, it is terminated. This is done by checking the closeness of the design points generated during the local search to stored trajectories between previous starting points and the corresponding local minimum points.

Thus, it is seen that the domain elimination method *eliminates domains* around the known local minimum points and previously rejected points. At the beginning of the search for a global minimum, most of the random points are accepted as starting points for local minimization. However, near the tail end of the search process, fewer local minimizations are performed.

The following *counters* are used in the algorithm given next:

c_1 = number of elements in S_r, the *set of rejected points*
c_2 = number of rejected points that were near S_r
c_3 = number of elements in S_0, the *set of starting points*
c_4 = number of elements in S_*, the *set of local minimum points*

Step 1. Initialize the sets S_*, S_0, and S_r. Select a value for the parameter M that specifies the number of local search iterations to be performed to determine an intermediate point x^M for checking the rejection criteria. Set global iteration counter $i = 0$. Set four counters c_1, c_2, c_3, and c_4 to 0; these are used to keep track of the number of elements in various sets defined earlier.

Step 2. If one of the stopping criteria is met, then stop. Otherwise, generate a random point x^R drawn from a uniform distribution over S_b.

Step 3. If the current point x^R is in or near S_* or S_0 (determined using a procedure to be presented in Section 18.4.4), then add x^R to S_r and set $c_1 = c_1 + 1$; else, if it is in or near S_r, set $c_2 = c_2 + 1$; else, if x^R is near a trajectory (a mapping $S_0 \rightarrow S_*$), add x^R to S_r and set $c_1 = c_1 + 1$. If any of the four conditions is true, go to Step 2 (this avoids starting the local phase, which has a low probability of finding a new minimum). Otherwise, add x^R to S_0, and set $x^{(0)} = x^R$ and $c_3 = c_3 + 1$.

Step 4. Execute the local minimization for M iterations to yield an intermediate point x^M. If x^M is a local minimum point, then set $x^* = x^M$ and go to Step 6. Otherwise, continue.

Step 5. If the current point x^M is in or near S_0, then add x^M to S_r and set $c_1 = c_1 + 1$; else, if x^M is in or near S_r, set $c_2 = c_2 + 1$; else, if x^M is near a trajectory, add x^M to S_r and set $c_1 = c_1 + 1$. If any of the three conditions is true, then store the trajectory from $x^{(0)}$ to x^M and go to Step 2. Otherwise, go to Step 4.

Step 6. If x^* is a new local minimum, then add it to S_* and set $c_4 = c_4 + 1$; otherwise, increment the associated indicator of the number of times x^* was found.

Step 7. Set $i = i + 1$; if i is larger than a specified limit, then stop. Otherwise, go to Step 2.

Stopping Criteria

Several stopping criteria can be used, and two of them are discussed next. In Step 2 of the algorithm, the procedure is stopped if any of the sets S_0, S_*, and S_r become full based on prespecified limits on set sizes. The maximum size of a set is determined based on the number of design variables and a confidence level parameter. Since the random points are drawn from a uniform distribution, the number of points generated before stopping the search is proportional to n. The size of set S_* of local minimum points is usually much smaller than the other two sets (since there can be only one minimum for each starting point). In the numerical evaluation presented later, the sizes chosen for the current implementation are $10n$ for S_* and $40n$ for S_0 and S_r.

Another criterion is to stop the algorithm if the number of local minima found exceeds the Bayesian estimate for the number of local minima, as was explained in Section 18.3.2, or if a specified limit on iterations is exceeded.

18.4.3 Stochastic Zooming Method

This method is an extension of the zooming method described in Section 18.2.2. Recall that as the target level for the cost function grows closer to the global optimum, it becomes difficult to find a feasible point in the set S. Eventually, the modified problem needs to be declared infeasible to stop the algorithm. To overcome this difficulty, the algorithm is modified by adding a global phase to it to ensure that the set S is reasonably well searched before declaring the modified problem to be infeasible and accepting the previous local minimum as the global minimum.

The major difference between stochastic zooming and domain elimination is the addition of the zooming constraint of Eq. (18.4). Therefore, the domain elimination algorithm can be used with some minor modification to keep track of the number of local searches that did not terminate at a feasible minimum point. To do this, the number of iterations to search for a local minimum in Steps 4 and 5 is monitored. If this number exceeds a specified limit, then the local search process is declared to have failed. The number of such failures is also monitored, and if it exceeds a specified limit, the algorithm is terminated and the best local minimum is taken as the global minimum point for the problem. In addition to this stopping criterion, if the cost function value reaches a target value F specified by the user, $(f(x) \leq F)$, the algorithm is terminated.

18.4.4 Operations Analysis of Methods

It is seen that the domain elimination and stochastic zooming methods differ primarily in the inclusion of the constraint from Eq. (18.4) in the latter. This is an important difference, however, because it changes the behavior of the basic algorithm considerably. This section presents an analysis of the operations and choice of design variable bounds used in both methods. The analysis includes numerical requirements and performance with available implementation alternatives for the algorithms' rejection criteria. The following calculations are needed in the algorithms:

1. Distance between a point and a set of points
2. Approximation of the trajectory between a starting point and a local minimum
3. Distance between a point and a trajectory

Each can be accomplished with several different procedures. The operations count for each option is necessary for choosing the most efficient procedure.

Checking the Proximity of a Point to a Set

After generating a random point, we need to determine if it will yield a new local minimum. For this purpose, the random point x^R is compared with the points in the three sets S_*, S_0, and S_r. If it is within a certain critical distance D_{cr} from any point in these sets, it is discarded and a new random point is generated. The same procedure is used for the intermediate point x^M (Step 4 of the algorithm). Two methods are presented for checking the proximity of a point to a set.

Let x^S be a point belonging to one of the sets mentioned previously and x^M be either a random point or an intermediate point. The first method is to construct a hypersphere of either a constant or a variable radius around x^S. The proposed point (x^R or x^M) is rejected if it lies inside the hypersphere. This involves calculation of the distance between the two points $D = ||x^S - x^M||$, and the point is rejected if $D \leq D_{cr}$, where D_{cr} is specified as $\alpha ||x||$ with $x = x^R$ or x^M and $0.01 \leq \alpha \leq 0.20$.

The second method is to construct a hyperprism around x^S rather than a hypersphere. If the proposed point lies inside the hyperprism, it is rejected. In this case, the distance between the two points is not required. Each of the design variables is compared in turn with the corresponding value for the prism's center. If the difference is larger than the corresponding critical value, then the rest of the variables need not be compared and the point is accepted. This can be represented by the following pseudocode (let $D_{cr(i)} = \alpha |x_i|$ be a vector with $x_i = (x^R \text{ or } x^M)_i$):

```
for i = 1 to n do
    if (xˢ - xᴹ)ᵢ ≥ Dcr(i) then accept xᴹ
end do
reject xᴹ
```

Based on the operations count, it is seen that the second approach is less expensive.

Trajectory Approximation

The random points x^R selected for starting a local search, as well as the intermediate points x^M during local minimization, are examined for proximity to each stored trajectory. A trajectory is the design history from a starting point to the corresponding local minimum point. There can be many trajectories meeting at one local minimum point. The selected point is rejected if it is near any trajectory. This is done to prevent unnecessary minimization steps that would otherwise lead to already known local minima. The trajectory can be approximated using several techniques. The simplest approximation is a straight line connecting $x^{(0)}$ and the corresponding x^*. Experiments have shown that actual trajectories usually do not follow straight lines, especially at the beginning of the search and for nonlinear problems. Other alternatives to approximate the trajectory include

1. Passing a least squares straight line through several points along the trajectory.
2. Passing straight line segments through selected points along the trajectory.
3. Passing a quadratic curve through three points.
4. Passing quadratic segments through groups of three points.
5. Constructing higher-order polynomial or spline approximations.

Several issues affect the choice of the technique to use: the number of points needed (which have to be stored), the number of operations, and the accuracy of the approximation. Any technique other than straight line approximation requires more intermediate points to be saved and more calculations. Therefore, use of a straight line approximation is suggested.

Distance between a Point and a Trajectory

With the linear approximation of the trajectory, the decision of whether a point x lies near the trajectory can be made in more than one way. The first procedure is to calculate the internal angle $x^{(0)} - x - x^*$ of the triangle formed by the three points. The point x is rejected if the angle is larger than a threshold value. The second approach is to calculate an offset distance by generating $\bar{x}$ as a projection of x on the line $x^{(0)} - x^*$. If $\bar{x}$ lies outside the line segment $x^{(0)} - x^*$, then it is accepted; otherwise, the offset length $x - \bar{x}$ is calculated. If it is larger than a critical value, then x is considered far from the trajectory.

Geometrical representation of the triangle method indicates construction of an ellipsoidal body around the line segment $x^{(0)} - x^*$. The offset method, on the other hand, can be represented by a cylinder constructed with $x^{(0)} - x^*$ as its axis. A point x is considered close to the trajectory if it lies inside the ellipsoidal body or the cylinder, respectively. The offset method features a uniform critical distance from the linear trajectory, whereas with the triangle method the distance is made smaller near the endpoints. In other words, the critical offset distance in the triangle method is related to the trajectory's length.

This makes physical sense because the trajectory's length can be related to the size of the region of attraction for the local minimum. The same effect can be achieved for the cylinder method by requiring the critical offset to be proportional to the length of the line segment $x^{(0)} - x^*$ (i.e., $\beta \|x^{(0)} - x^*\|$ for some $\beta > 0$). The proportional offset has the advantage of accounting for the problem's scale and thus maintains accuracy. Another advantage is that large subdomains are eliminated from larger regions of attraction and vice versa.

A third approach is to construct a truncated cone with the larger base at x^* and the smaller base at $x^{(0)}$, thus allowing a better identification of close regions of attraction. The cone can be constructed by requiring that the critical offset distance be proportional to the distance $\|\bar{x} - x^{(0)}\|$. This option requires more calculations than the simple cylinder method.

Based on the operations count, the triangle method is chosen for the implementation that is described later since it uses fewer multiplications. The critical angle in the range $150° \leq \theta \leq 175°$ has shown good performance ($\theta = 170°$ is used in the implementation).

Design Variable Constraints

In local minimization algorithms, any simple bounds on the design variables can be treated efficiently. Specification of appropriate bounds on the design variables is more important in stochastic global optimization methods than in others. The further apart these bounds are, the larger the number of random points generated in the set S_b. Consequently, the number of local searches performed is increased, which reduces efficiency. For this reason, the *design variable bounds for a global optimization problem must be chosen carefully* to reflect the nature of the problem. A simple numerical experiment where the allowable range for one of the design variables out of a total of 40 was doubled, showed that the numerical effort to obtain the same global minimum point increased by 50 percent (Elwakeil and Arora, 1996a). This clearly shows the importance of selecting appropriate bounds on the design variables in global optimization.

18.5 NUMERICAL PERFORMANCE OF METHODS

Various concepts and aspects of global optimization methods have been described in this chapter. Details of some of the algorithms have been presented to give a flavor of the computations that are needed to solve a problem. It is seen that solving a global optimization problem is a computational challenge, especially when a true global minimum is required. The main reason is that *even if the global minimum point has been reached during the search process, it is not possible to recognize this fact.* That is, there is no definite stopping criterion. Therefore, the search process needs to be continued and the algorithm needs to be executed repeatedly to ensure that the global minimum point has not been missed. In other words, the entire feasible set needs to be thoroughly searched, implicitly or explicitly.

For practical applications, however, improved local minima or improved feasible designs are acceptable. In this case, reasonably efficient and effective computational algorithms are available or can be devised to achieve this objective. In addition, *many algorithms can be implemented on parallel processors to reduce the "wall clock" time* needed for practical applications.

In this section, we summarize features of the methods described earlier. The numerical performance of some of them is described using a limited set of test problems to gain insights into the type of computation they need and their behavior (Elwakeil and Arora, 1996a,b). Several structural design problems have also been devised and solved to study that class of problems for global optimization.

18.5.1 Summary of Features of Methods

It is *difficult to recommend a single global optimization method for all applications.* Selection depends on the characteristics of the problem and what is desired. For example, if all of the local minima are desired, then tunneling or zooming is not suitable. If the problem has discrete variables and the functions are not differentiable, a method that requires and uses gradients will not do. If an absolute guarantee of a global solution is desired, then certain methods that do not guarantee this are not appropriate. It is suggested that the problem characteristics and requirements be analyzed before selecting an algorithm for global optimization.

TABLE 18.1   Characteristics of global optimization methods

Method	Can solve discrete problems?	General constraints?	Tries to find all x^*?	Phases	Needs gradients?
Covering (D)	No	No	Yes	G	1
Zooming (D)	Yes[1]	Yes	No	L	1
Generalized descent (D)	No	No	No	G	Yes
Tunneling (D)	No	Yes	No	L + G	1
Multistart (S)	Yes[1]	Yes	Yes	L + G	1
Clustering (S)	Yes[1]	Yes	Yes	L + G	1
Controlled random search (S)	Yes	No	No	L + G	No
Acceptance-rejection (S)	Yes[1]	Yes	No	G	No
Stochastic integration (S)	No	No	No	G	No
Genetic (S)	Yes	No	No	G	No
Stochastic zooming (S)	Yes[1]	Yes	No	L + G	1
Domain elimination (S)	Yes[1]	Yes	Yes	L + G	1

Note: D: *deterministic methods*; S: *stochastic methods*; G: *global phase*; L: *local phase*.
[1]*Depends on the local minimization procedure used.*

We must be aware that no matter which algorithm is selected, the computational effort to reach a solution point is substantial. Therefore, we must be willing and able to bear the enormous cost of finding an estimate of a global solution to our problem. Table 18.1 summarizes the following characteristics of various global optimization algorithms:

1. Classification of method: deterministic (D) or stochastic (S).
2. Ability of the method to solve discrete problems; it is desirable that the method be able to do so.
3. Ability to treat general constraints explicitly; this is a desirable feature.
4. Ability to find all of the local minima; this depends on the desire of the user.
5. Use of local and global phases; methods using both are generally more reliable and efficient.
6. Need for gradients; if a method definitely needs function gradients, its applicability is limited to continuous problems.

18.5.2 Performance of Some Methods with Unconstrained Problems

For a first numerical performance study, the following four methods were implemented (Elwakeil and Arora, 1996a): covering, acceptance-rejection (A-R), controlled random search (CRS), and simulated annealing (SA). The numerical tests were performed on 29 unconstrained problems available in the literature. These problems had one to six design variables and only explicit bounds. Global solutions for the problems were known.

Based on the results, it was concluded that the covering methods were not practical because of their inefficiency for problems with $n > 2$. They required very large computational effort. Also, it was difficult to generate a good estimate for the Lipschitz constant that is needed in the algorithm. Both A-R and CRS methods performed better than simulated annealing and the covering method. However, the fact that A-R does not include any stopping criterion makes it undesirable for practical applications. It worked efficiently on test problems because it was stopped upon finding the known global optimum point.

The CRS method contains a stopping criterion and is more efficient compared with other methods. An attempt to treat general constraints explicitly with it was not successful because constraint violations could not be corrected with reasonable computational effort.

18.5.3 Performance of Stochastic Zooming and Domain Elimination Methods

In another study, the stochastic zooming method (ZOOM) and the domain elimination (DE) method were also implemented (in addition to CRS and SA), and their performance was evaluated using ten mathematical programming test problems (Elwakeil and Arora, 1996a). The test problems included were constrained as well as unconstrained. The reason was that, even though most engineering application problems are constrained, it is beneficial to test performance of the algorithms on unconstrained problems as well.

The CRS method could be used only for unconstrained problems. It is noted, however, that the problems classified as unconstrained included simple bounds on the design variables. The sequential quadratic programming (SQP) method was used in all local searches performed using ZOOM and DE. For ZOOM, the percent reduction required from one local minimum to the next was set arbitrarily to 15 percent (i.e., $\gamma = 0.85$ in Eq. (18.4)) for all of the test problems.

The ten test problems used in the study had the following characteristics:

- Four problems had no constraints.
- The number of design variables varied from 2 to 15.
- The total number of general constraints varied from 2 to 29.
- Two problems had equality constraints.
- All problems had 2 or more local minima.
- Two problems had 2 global minima and one had 4.
- One problem had a global minimum of 0.
- Four problems had negative global minimum values.

To compare the performance of different algorithms, each of the test problems was solved five times and averages for the following evaluation criteria were recorded:

- Number of random starting points
- Number of local searches performed
- Number of iterations used during the local search
- Number of local minima found by the method

- Cost function value of the best local minimum (the global minimum)
- Total number of calls for function evaluations
- CPU time used

Because a random point generator with a random seed was used, the performance of the algorithms changed each time they were executed. The seed was automatically chosen based on wall clock time. The results differed in the number of local minima found as well as in the other evaluation criteria.

DE found the global solution to 9 out of 10 problems, whereas ZOOM found a global minimum for 7 out of 10. In general, DE found more local minima than ZOOM did. This is attributed to the latter requiring a reduction in the cost function value after each local minimum was found. As noted earlier, ZOOM is designed to "tunnel" under some minima with relatively close cost function values.

In terms of the number of function evaluations and CPU time, DE was cheaper than ZOOM. This was because the latter performed more local iterations for a particular search without finding a feasible solution. On the other hand, the number of iterations during a local search performed in DE was smaller since it could find a solution in most cases.

The CPU time needed by CRS was considerably smaller than that for other methods, even with a larger number of function evaluations. This was due to the use of a local search procedure that did not require gradients or line search. However, the method is applicable to only unconstrained problems.

Simulated annealing failed to locate the global minimum for six problems. For the successful problems, the CPU time required was three to four times that for DE. The tests also showed that there was a drastic increase in the computational effort it required as the number of design variables increased. This implementation of SA was thus considered inefficient and unreliable compared with those of both DE and ZOOM. It is noted that SA may be more suitable for problems with discrete variables only.

18.5.4 Global Optimization of Structural Design Problems

The DE and ZOOM methods were used to find global solutions to structural design problems by Elwakeil and Arora (1996b). In this section, we summarize and discuss the results of that study, which used the following six structures:

- 10-bar cantilever truss
- 200-bar truss
- 1-bay, 2-story frame
- 2-bay, 6-story frame
- 10-member cantilever frame
- 200-member frame

These structures had been used previously to test various algorithms for local minimization (Haug and Arora, 1979). A variety of constraints were imposed on the structures. They included constraints and other requirements given in the *Specification of the American Institute of Steel Construction* (AISC, 1989) and the *Aluminum Association Specifications*

(AA, 1986), as well as displacement constraints and constraints on the natural frequency of the structure.

Some of the structures were subjected to multiple loading cases. For all problems, the weight of the structure was minimized. Using the six structures, 28 test problems were devised by varying the cross-sectional shape of members to hollow circular tubes or I-sections, and changing the material from steel to aluminum. The number of design variables varied from 4 to 116, the number of stress constraints varied from 10 to 600, the number of deflection constraints varied from 8 to 675, and the number of local buckling constraints for the members varied from 0 to 72. The total number of general inequality constraints varied from 19 to 1276. These test problems can be considered to be large compared with the ones used in the previous section.

Detailed results using DE and ZOOM can be found in Elwakeil (1995). Each problem was solved five times with a different seed for the random number generator. The five runs were then combined and all of the optimum solutions found were stored.

It was observed that all six structures tested possessed many local minima. ZOOM found only one local minimum for all but two problems. For most of the problems, the global minimum was found with the first random starting point. Therefore, other local minima were not found since they had a higher cost function value. DE found many local minima for all problems except for one, which turned out to be infeasible. The method did not find all of the local minima in one run because of the imposed limit on the number of random starting designs.

From the recorded CPU times, it was difficult to draw a general conclusion about the relative efficiency of the two methods because for some problems one method was more efficient and for the remaining the second method was more efficient. However, both can be useful depending on the requirements. If only the global minimum is sought, then ZOOM can be used. If all or most of the local minima are wanted, then DE should be used. The zooming method can be used to determine lower-cost practical designs by appropriately selecting the parameter γ in Eq. (18.4).

Some problems showed only a small difference between weights for the best and the worst local minima. This indicates a flat feasible domain, perhaps with small variations in the weight which results in multiple global minima. One of the problems was infeasible because of an unreasonable requirement for the natural frequency to be no less than 22 Hz. However, when the constraint was gradually relaxed, a solution was found at a value of 17 Hz.

It is clear that the designer's experience with, and knowledge of, the problems to be solved, as well as the design requirements, can affect the performance of global optimization algorithms. For example, by setting a correct limit on the number of local minima desired, the computational effort required by the domain elimination method can be reduced substantially. For the zooming method, the computational effort is reduced if the parameter γ in Eq. (18.4) is selected judiciously. In this regard, it may be possible to develop a strategy to automatically adjust the value of γ dynamically during local searches so as to avoid the infeasible problems that constitute a major computational effort in the zooming method. A realistic value for F, the target value for the global minimum cost function, would also improve the method's efficiency.

EXERCISES FOR CHAPTER 18*

Calculate a global minimum point for the following problems.

18.1 (See Branin and Hoo, 1972)

Minimize

$$f(\mathbf{x}) = \left(4 - 2.1x_1^2 + \frac{1}{3}x_1^4\right)x_1^2 + x_1x_2 + (-4 + 4x_2^2)x_2^2$$

subject to

$$-3 \le x_1 \le 3$$
$$-2 \le x_2 \le 2$$

18.2 (See Lucidi and Piccioni, 1989)

Minimize

$$f(\mathbf{x}) = \frac{\pi}{n}\left\{10\sin^2(\pi x_1) + \sum_{i=1}^{n-1}\left[(x_i - 1)^2(1 + 10\sin^2(\pi x_{i+1}))\right] + (x_n - 1)^2\right\}$$

subject to

$$-10 \le x_i \le 10; \quad i = 1 \text{ to } 5$$

18.3 (See Walster et al., 1984)

Minimize

$$f(\mathbf{x}) = \sum_{i=1}^{11}\left[a_i - x_1\frac{b_i^2 - b_ix_2}{b_i^2 + b_ix_3 + x_4}\right]$$

subject to

$$-2 \le x_i \le 2; \quad i = 1 \text{ to } 4$$

where the coefficients (a_i, b_i) $(i = 1 \text{ to } 11)$ are given as follows: (0.1975, 4), (0.1947, 2), (0.1735, 1), (0.16, 0.5), (0.0844, 0.25), (0.0627, 0.1667), (0.0456, 0.125), (0.0342, 0.1), (0.0323, 0.0833), (0.0235, 0.0714), (0.0246, 0.0625).

18.4 (See Evtushenko, 1974)

Minimize

$$f(\mathbf{x}) = -\left[\sum_{i=1}^{6}\frac{1}{6}\sin 2\pi\left(x_i + \frac{i}{5}\right)\right]^2$$

subject to

$$0 \le x_i \le 1; \quad i = 1 \text{ to } 6$$

18.5 Minimize

$$f(\mathbf{x}) = 2x_1 + 3x_2 - x_1^3 - 2x_2^2$$

subject to

$$\frac{1}{6}x_1 + \frac{1}{2}x_2 - 1.0 \le 0$$

$$\frac{1}{2}x_1 + \frac{1}{5}x_2 - 1.0 \le 0$$

$$x_1, \ x_2 \ge 0$$

18.6 (See Hock and Schittkowski, 1981)

Minimize

$$f(\mathbf{x}) = \sum_{i=1}^{99} f_i^2(\mathbf{x})$$

$$f_i(\mathbf{x}) = -\frac{i}{100} + \exp\left(-\frac{1}{x_1}(u_i - x_2)^{x_3}\right)$$

$$u_i = 25 + \left[-50\ln(0.01i)\right]^{2/3}; \quad i = 1 \text{ to } 99$$

subject to

$$0.1 \le x_1 \le 100, \quad 0.0 \le x_2 \le 25.6, \quad 0.0 \le x_3 \le 5$$

18.7 (See Hock and Schittkowski, 1981)

Minimize

$$f(\mathbf{x}) = (x_1^2 - x_2^2) + (x_2^2 - x_3^2) + (x_3^2 - x_4^2) + (x_4^2 - x_5^2)$$

subject to

$$x_1 + x_2^2 + x_3^3 - 3 = 0$$

$$x_1 - x_3^2 + x_4 - 1 = 0$$

$$x_1 x_5 - 1 = 0$$

18.8 (See Hock and Schittkowski, 1981)

Minimize

$$f(\mathbf{x}) = -75.196 + b_1 x_1 + b_2 x_1^3 - b_3 x_1^4 + b_4 x_2 - b_5 x_1 x_2 + b_6 x_2 x_1^2 + b_7 x_1^4 x_2 - b_8 x_2^2 + c_1 x_2^3 - c_2 x_2^4$$

$$+ 28.106/(x_2 + 1) + c_3 x_1^2 x_2^2 + c_4 x_1^3 x_2^2 - c_5 x_1^3 x_2^3 - c_6 x_1 x_2^2 + c_7 x_1 x_2^3$$

$$+ 2.8673 \, \exp\left(\frac{x_1 x_2}{2000}\right) - c_8 x_1^3 x_2$$

subject to

$$x_1 x_2 - 700 \ge 0$$

$$x_2 - x_1^2/125 \ge 0$$

$$(x_2 - 50)^2 - 5(x_1 - 55) \ge 0$$

$$0 \le x_1 \le 75, \quad 0 \le x_2 \le 65$$

where the parameters (b_i, c_i) $(i = 1$ to $8)$ are given as
(3.8112E+00, 3.4604E−03), (2.0567E−03, 1.3514E−05),
(1.0345E−05, 5.2375E−06), (6.8306E+00, 6.3000E−08),
(3.0234E−02, 7.0000E−10), (1.2814E−03, 3.4050E−04),
(2.2660E−07, 1.6638E−06), (2.5645E−01, 3.5256E−05).

18.9 (See Hock and Schittkowski, 1981)

Minimize

$$f(\mathbf{x}) = x_1 x_4 (x_1 + x_2 + x_3) + x_3$$

subject to

$$x_1 x_2 x_3 x_4 - 25 \geq 0$$

$$x_1^2 + x_2^2 + x_3^2 + x_4^2 - 40 = 0$$

$$1 \leq x_i \leq 5; \quad i = 1 \text{ to } 4$$

18.10 (See Hock and Schittkowski, 1981)

Minimize

$$f(\mathbf{x}) = \sum_{k=0}^{4} \left(2.3 x_{3k+1} - (1.0E{-}4) x_{3k+1}^3 + 1.7 x_{3k+2} \right.$$

$$\left. + (1.0E{-}4) x_{3k+2}^2 + 2.2 x_{3k+3} + (1.5E{-}4) x_{3k+3}^2 \right)$$

subject to

$$0 \leq x_{3j+1} - x_{3j-2} + 7 \leq 13; \quad j = 1 \text{ to } 4$$

$$0 \leq x_{3j+3} - x_{3j-2} + 7 \leq 14; \quad j = 1 \text{ to } 4$$

$$0 \leq x_{3j+3} - x_{3j} + 7 \leq 13; \quad j = 1 \text{ to } 4$$

$$x_1 + x_2 + x_3 - 60 \geq 0$$

$$x_4 + x_5 + x_6 - 50 \geq 0$$

$$x_7 + x_8 + x_9 - 70 \geq 0$$

$$x_{10} + x_{11} + x_{12} - 85 \geq 0$$

$$x_{13} + x_{14} + x_{15} - 105 \geq 0$$

and the bounds are ($k = 1$ to 4):

$$8.0 \leq x_1 \leq 21.0$$

$$43.0 \leq x_2 \leq 57.0$$

$$3.0 \leq x_3 \leq 16.0$$

$$0.0 \leq x_{3k+1} \leq 90.0$$

$$0.0 \leq x_{3k+2} \leq 120.0$$

$$0.0 \leq x_{3k+3} \leq 60.0$$

Find all of the local minimum points for the following problems and determine a global minimum point.

18.11 Exercise 18.1   **18.12** Exercise 18.2   **18.13** Exercise 18.3

18.14 Exercise 18.4   **18.15** Exercise 18.5   **18.16** Exercise 18.6

18.17 Exercise 18.7   **18.18** Exercise 18.8   **18.19** Exercise 18.9

18.20 Exercise 18.10

19

Nature-Inspired Search Methods

Upon completion of this chapter, you will be able to

- Explain and use the differential evolution algorithm
- Explain and use the particle swarm optimization algorithm
- Explain and use the ant colony optimization algorithm

In this chapter, optimization algorithms inspired by natural phenomena are described. These fall into the general class of *direct search methods* described in Chapter 11. However, in contrast to some direct search methods, they do not require the continuity or differentiability of problem functions. The only requirement is that we be able to evaluate functions at any point within the allowable ranges for the design variables. Nature-inspired methods use stochastic ideas and random numbers in their calculations to search for the optimum point. They tend to converge to a global minimum point for the function, but there is no guarantee of convergence or global optimality of the final solution.

Nature-inspired approaches have been called stochastic programming, evolutionary algorithms, genetic programming, swarm intelligence, and evolutionary computation. They are also called nature-inspired metaheuristics methods, as they make no assumptions about the optimization problem and can search very large spaces for candidate solutions.

Nature-inspired algorithms can overcome some of the challenges that are due to multiple objectives, mixed design variables, irregular/noisy problem functions, implicit problem functions, expensive and/or unreliable function gradients, and uncertainty in the model and the environment. For this reason, there has been considerable interest in their development and in their application to a wide variety of practical problems. Several books on various methods have been published; a few examples are Price et al. (2005), Qing (2009), Glover and Kochenberger (2002), Corne et al. (1999), and Kennedy et al. (2001).

There have also been conferences and workshops on various nature-inspired methods such as the IEEE Congress on Evolutionary Computation, Soft Computing, GECCO (Genetic and Evolutionary Computation Conference), PPSN (International Conference on Parallel Problem Solving from Nature), ANTS (Ant Colony Optimization and Swarm Intelligence), the Evolutionary Programming Conference, and others. Journals devoted to research on nature-inspired methods include these: *IEEE Transactions on Evolutionary Computation, Applied Intelligence, Neural Network World, Artificial Intelligence Review, Applied Soft Computing, Physics of Life Reviews, AI Communications, Evolutionary Computing, Journal of Artificial Intelligence Research, Journal of Heuristics*, and *Artificial Life*.

The methods usually start with a collection of design points called the *population*. Using certain stochastic processes, they try to come up with a better design point for each *generation* (iteration of the algorithm). In Chapter 16, concepts and details of genetic algorithms were presented. Those methods also fall into the nature-inspired class. Some of the terminology and concepts used there are used here as well.

To give a flavor of nature-inspired methods, we will describe three methods in this chapter that are relatively popular. (Other methods in this class are noted in Das and Suganthan (2011).) Each one uses specific terminology from the corresponding biological phenomenon or other natural phenomena that may be unfamiliar to engineers, so we will describe such terminology wherever used.

The methods presented here treat the following optimization problem:

Minimize

$$f(\mathbf{x}) \text{ for } \mathbf{x} \in S \tag{19.1}$$

where S is the feasible set of designs and $\mathbf{x}$ is the n-dimensional design variable vector. If the problem is unconstrained, the set S is the entire design space, and if it is constrained, S is determined by the constraints. The methods presented in this chapter are generally used for unconstrained problems.

Constrained optimization problems, however, can be addressed using the penalty function approach described in Chapter 11 or the exact penalty function defined in Chapter 12. In the following presentation, the terms *design vector*, *design point*, and *design* are used interchangeably. They all refer to the n-dimensional design variable vector $\mathbf{x}$.

19.1 DIFFERENTIAL EVOLUTION ALGORITHM

The differential evolution (DE) algorithm works with a population of designs. At each iteration, called a generation, a new design is generated using some current designs and random operations. If the new design is better than a preselected parent design, then it replaces that design in the population; otherwise, the old design is retained and the process is repeated. In this section, the steps of a basic DE algorithm are described. The material is derived from a recent article by Das and Suganthan (2011).

Compared to genetic algorithms (GAs), DE algorithms are easier to implement on the computer. Unlike GAs, they do not require binary number coding and encoding, as seen later (although GAs have been implemented with real number coding as well). Therefore,

TABLE 19.1   Notation and terminology for the DE algorithm

Notation	Terminology
Cr	Crossover rate; an algorithm parameter
F	Scale factor, usually in the interval [0.4, 1.0]; an algorithm parameter
k	kth generation of the iterative process
k_{max}	Limit on the number of generations
n	Number of design variables
N_p	Number of design points in the population; population size
r_{ij}	Random number uniformly distributed between 0 and 1 for the ith design and its jth component
x_j	jth component of the design variable vector $\mathbf{x}$
$U^{(p,k)}$	Trial design vector at the kth generation/iteration associated with the parent design p
$V^{(p,k)}$	Donor design vector at the kth generation/iteration associated with the parent design p
$\mathbf{x}^{(i,k)}$	ith design point of the population at the kth generation/iteration
$x^{(p,k)}$	Parent design (also called the target design) of the population at the kth generation/iteration
$\mathbf{x}_L$	Vector containing the lower limits on the design variables
$\mathbf{x}_U$	Vector containing the upper limits on the design variables

they are quite popular for numerous practical applications (Das and Suganthan, 2011). There are four steps in executing the basic DE algorithm:

Step 1. Generation of the initial population of designs.
Step 2. Mutation with difference of vectors to generate a so-called *donor design vector*.
Step 3. Crossover/recombination to generate a so-called *trial design vector*.
Step 4. Selection, that is, acceptance or rejection of the trial design vector using the *fitness function*, which is usually the cost function.

These steps are described in the following subsections. The notation and terminology listed in Table 19.1 are used.

19.1.1 Generation of an Initial Population

A first step in DE is to generate an initial population of N_p design points; N_p is usually selected as a large number, say, between $5n$ and $10n$. Each design point/vector is also called a *chromosome*. Initial designs can be generated by any procedure that tries to cover the entire design space in a uniformly distributed random manner. If some designs for the system are known, they can be included in the initial population. One way to generate the initial set of designs is to use the lower and upper limits on the design variables and uniformly distributed random numbers. For example, the ith member (design) of the population may be generated as follows:

$$x_j^{(i,0)} = x_{jL} + r_{ij}(x_{jU} - x_{jL}); \quad j = 1 \text{ to } n \tag{19.2}$$

where r_{ij} is a uniformly distributed random number between 0 and 1 that is generated for each component of the design point. Each member of the population is a potential solution/optimum point.

19.1.2 Generation of a Donor Design

In this subsection, we describe the idea of a donor design and its generation. A donor design is generated using mutation of a selected design with the difference of two other distinct designs in the population. Biologically, mutation means a change in the *gene* (a component of the design vector) characteristics of a *chromosome* (the complete design vector). The donor design point is created by changing a design point of the current population. This change is accomplished by combining the design vector with the difference of two other vectors of the population, all selected randomly. The design vector thus generated is called the donor design/vector. In the context of donor design, then, mutation implies changing all components of a design vector.

To generate the donor design vector, we randomly select three distinct design points from the current population in the generation k: $\mathbf{x}^{(r_1,k)}$, $\mathbf{x}^{(r_2,k)}$ and $\mathbf{x}^{(r_3,k)}$, where the superscripts r_1, r_2, and r_3 refer to three different designs. In addition, we select a fourth point $\mathbf{x}^{(p,k)}$, called the *parent/target* design point; its use in the crossover operation is explained later (the superscript p refers to the parent design). We then form a difference vector using two design points, say r_2 and r_3, as $(\mathbf{x}^{(r_2,k)} - \mathbf{x}^{(r_3,k)})$. This difference vector is scaled and added to the third vector to form the donor design vector $\mathbf{V}^{(p,k)}$:

$$\mathbf{V}^{(p,k)} = \mathbf{x}^{(r_1,k)} + F \times (\mathbf{x}^{(r_2,k)} - \mathbf{x}^{(r_3,k)}) \tag{19.3}$$

where F is a scale factor, typically selected between 0.4 and 1. Note that any procedure can be used to randomly select the foregoing four members of the current population; one example is the roulette wheel procedure described in Chapter 16.

19.1.3 Crossover Operation to Generate the Trial Design

A crossover operation is performed after generating the donor design through mutation. In it, the donor design vector $\mathbf{V}^{(p,k)}$ exchanges some of its components with the parent design vector to form the trial design vector $x_j^{(p,k)}$. The crossover operation is described in the following equation:

$$U_j^{(p,k)} = \begin{cases} V_j^{(p,k)}, & \text{if } r_{pj} \leq Cr \text{ or } j = j_r \\ x_j^{(p,k)}, & \text{otherwise} \end{cases} ; \quad j = 1 \text{ to } n \tag{19.4}$$

where r_{pj} is a uniformly distributed random number between 0 and 1 and j_r is a randomly generated index between 1 and n that ensures that $\mathbf{U}^{(p,\,k)}$ receives at least one component from $\mathbf{V}^{(p,k)}$.

The crossover operation in Eq. (19.4) says that when the random number r_{pj} for each component of the design vector does not exceed the Cr value, or if $j = j_r$, set the trial design component $U_j^{(p,k)}$ to the donor design component $V_j^{(p,k)}$; otherwise, replace it with the parent design component $x_j^{(p,k)}$. With this approach, the number of components inherited from the donor design vector has a (nearly) binomial distribution. Therefore, this operation is called *binomial crossover*.

19.1.4 Acceptance/Rejection of the Trial Design

The next step of the algorithm is to check if the trial design $U^{(p,k)}$ is better than the parent design $x^{(p,k)}$; if it is, it replaces the parent design in the population to keep the population size constant (as a variation, both vectors may be retained sometimes increasing the size of the population by one every time). Usually called the *selection* step, this is described in the following equation:

$$x^{(p,k+1)} = \begin{cases} U^{(p,k)}, & \text{if } f(U^{(p,k)}) \leq f(x^{(p,k)}) \\ x^{(p,k)}, & \text{otherwise} \end{cases} \tag{19.5}$$

Accordingly, if the cost function value for the trial design point does not exceed that for the parent design, it replaces the parent design point in the next generation; otherwise, the parent design is retained. Thus the population either gets better or remains the same in fitness status, but it never deteriorates. Note that in Eq. (19.5) the parent design is replaced by the trial design even if both yield the same value for the cost function. This allows the design vectors to move over the flat fitness landscape.

19.1.5 DE Algorithm

The basic DE algorithm is quite straightforward to implement. It requires specification of only three parameters: N_p, F, and Cr. A flow diagram describing the basic steps of the DE algorithm is shown in Figure 19.1.

The termination criteria for the algorithm are defined as follows:

1. A specified limit $k_{\max}$ on the number of generations is reached.
2. The best fitness/cost function value of the population does not change appreciably for several generations.
3. A prespecified value for the cost function is reached.

Because of its simplicity, the DE algorithm has been quite popular in many application fields since its inception in the mid-1990s. It was inspired by the Nelder-Mead (1965) direct search

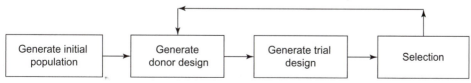

FIGURE 19.1 Main steps of the differential evolution algorithm.

method, which also uses the difference of vectors, as described in Chapter 18. Numerous variations on the algorithm have been studied and evaluated. It has been used to solve continuous variable, mixed-discrete-continuous variable, and multi-objective optimization problems, and it has also been evaluated against many other nature-inspired algorithms. A detailed review is beyond the scope of the present text. An excellent recent survey paper by Das and Suganthan (2011) and numerous references cited there should be consulted.

19.2 ANT COLONY OPTIMIZATION

Ant colony optimization (ACO), another nature-inspired approach, emulates the food-searching behavior of ants. It was developed by Dorigo (1992) to search for an optimal path for a problem represented by a graph based on the behavior of ants seeking the shortest path between their colony and a food source. ACO falls into the metaheuristics and swarm intelligence methods class. It can be viewed as a stochastic technique for solving computational problems that can be reduced to finding optimal paths through graphs.

Ants are social insects that live in colonies. From the colony, they go out to search for food and, surprisingly, find the shortest path from the colony to the food source. In this section, the process that ants use is described and translated into a computational algorithm for design optimization. The algorithm was developed originally for discrete variable combinatorial optimization problems, although it has been applied to continuous variable and other problems as well. Some of the material in this section is derived from Blum (2005) and associated references.

ACO uses the following terminology:

Pheromone. The word is derived from the Greek words *pherin* (to transport) and *hormone* (to stimulate). It refers to a secreted or excreted chemical factor that triggers a social response in members of the same species. Pheromones are capable of acting outside the body of the secreting individual in order to impact the behavior of the receiving individual. This is also called a chemical messenger.

Pheromone Trail. Ants deposit pheromones wherever they go. This is called the pheromone trail. Other ants can smell the pheromones and are likely to follow an existing trail.

Pheromone Density. When ants travel on the same path again and again, they continuously deposit pheromones on it. In this way the amount of pheromones increases and is called the pheromone density. The ants are likely to follow paths having higher pheromone densities.

Pheromone Evaporation. Pheromones have the property of evaporation over time. Therefore, if a path is not being traveled by the ants, the pheromones evaporate and the path disappears over time.

19.2.1 Ant Behavior

A first step in developing the ACO algorithm is to understand the behavior of ants, which is described in this subsection. Initially ants move from their nest randomly to search for food. Upon finding it, they return to their colony following the path they took to it while laying down pheromone trails. If other ants find such a path, they are likely to follow it instead of moving randomly. The path is thus reinforced, since ants deposit more pheromone on it. However, the pheromone evaporates over time; the longer the path, the more time there is available for it to evaporate. For a shorter path, pheromone reinforcement is quicker as more and more ants travel this route. Therefore, the pheromone density is higher on shorter paths than on the longer ones. Pheromone provides a positive feedback mechanism for ants, so eventually all the ants follow the shortest path.

The basic idea of an ant colony algorithm is to emulate this behavior with "artificial ants," which means that we need to model the pheromone deposit, measure its density, and model its evaporation. The following notation and terminology are used in this section:

Q = positive constant; an algorithm parameter
ρ = pheromone evaporation rate, $\rho \in (0, 1]$; an algorithm parameter
N_a = number of ants
τ_i = pheromone value for the ith path

A Simple Model/Algorithm

To transcribe the ants' food-searching behavior into a computational algorithm, we consider a simplified model consisting of two paths from the ant colony to the food source and six ants, as shown in Figure 19.2(a). This is a highly idealized model, introduced to explain the transcription of ant behavior into a computational algorithm. The model can be represented in a graph $G = (N, L)$, where N consists of two nodes (n_c, representing the ant colony, and n_f, representing the food source; in general a graph has many nodes as seen later), and L consists of two links, L_1 and L_2, between n_c and n_f.

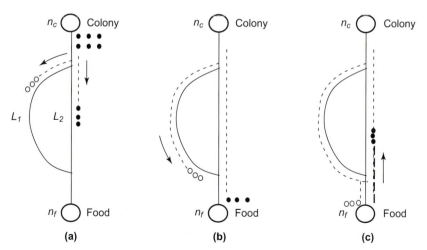

FIGURE 19.2   A simple set up showing shortest path finding capability of ants. (a) Movement of ants from colony to food source. (b) Ants taking the shorter route have reached the food source. (c) Ants taking the longer route have reached the food source while ants that took the shorter route are already returning to their colony.

Let L_1 have a length of d_1, and L_2 a length of d_2, with $d_1 > d_2$, implying that L_2 is a shorter path between n_c and n_f. Figure 19.2 is a graph that shows various stages of ant movement, which are explained as follows:

(a) Six ants start from their colony in search for food. Randomly, three ants (shown as solid circles) take the shorter route and three (shown as open circles) take the longer route.
(b) The three ants that took the shorter route have reached their destination, while the ants on the longer route are still traveling. Initially, the pheromone concentration is the same for the two routes, as shown by the dashed lines.
(c) The ants that took the shorter route are on their return journey to the colony while the ants taking the longer route are just arriving at their destination. Pheromone concentration on the shorter route is higher, as shown by the heavier dashed line.

The ants deposit pheromone while traveling on a route. The pheromone trails are modeled by introducing an artificial pheromone value τ_i for each of the two routes, $i = 1, 2$ (initially both values may be set as one). Such a value indicates the strength of the pheromone trail on the corresponding route.

Each ant behaves as follows: Starting from the node n_c (i.e., the colony), the ant chooses between route L_1 and route L_2 to reach n_f with the probability

$$p_i = \frac{\tau_i}{\tau_1 + \tau_2}, \quad i = 1, 2 \tag{19.6}$$

If $\tau_2 > \tau_1$, the probability of choosing L_2 is higher, and vice versa. The selection of a path by an ant is based on some selection scheme that uses probabilities from Eq. (19.6) and a random number, such as the roulette wheel selection procedure described in Chapter 16. While returning from the node n_f to the node n_c, the ant uses the same route it chose to reach n_f. It deposits additional artificial pheromone on the route to increase its density (this is also called pheromone reinforcement) as follows:

$$\tau_i \leftarrow \tau_i + \frac{Q}{d_i} \tag{19.7}$$

where the positive constant Q is a parameter of the model. Equation (19.7) models the higher amount of artificial pheromone deposit for a shorter path and a smaller amount for a longer path.

In the iterative process, all ants start from the node n_c at the beginning of each iteration. Each ant moves from that node n_c to node n_f depositing pheromone on the chosen route. However, with time the pheromone is subject to evaporation. This evaporation process in the artificial model is simulated as follows:

$$\tau_i \leftarrow (1 - \rho)\tau_i \tag{19.8}$$

where $\rho \in (0, 1]$ is a parameter of the model that regulates evaporation. After reaching the food source, the ants return to their colony, reinforcing the chosen path by depositing more pheromone on it.

19.2.2 ACO Algorithm for the Traveling Salesman Problem

The procedure described in the previous subsection to simulate the food-searching behavior of ants cannot be used directly for combinatorial optimization problems. The reason is that we assume the solution to the problem to be known and the pheromone values to be associated with the solution, as in Eq. (19.7). In general this is not the case because we are trying to find the optimum solution and the associated path with the minimum distance. Therefore, for combinatorial optimization problems, the pheromone values are associated with the solution components. Solution components are the units from which the entire solution to the problem can be constructed. This will become clearer later, when we describe the ACO algorithm for combinatorial optimization problems.

In this subsection, we describe an ant colony algorithm for discrete variable, or *traveling salesperson* (TS), problems. The TS problem is a classical *combinatorial optimization* problem. In it a traveling salesperson is required to visit a specified number of cities (called a *tour*). The goal is to visit a city only once while minimizing the total distance traveled. Many practical problems can be modeled as TS problems; one example is the welding sequence problem described and solved in Chapter 16.

The following assumptions are made in deriving the algorithm:

1. While a real ant can take a return path to the colony that is different from the original path depending on the pheromone values, an artificial ant takes the return path that is the same as the original path.
2. The artificial ant always finds a feasible solution and deposits pheromone only on its way back to the nest.
3. While real ants evaluate a solution based on the length of the path from their nest to the food source, artificial ants evaluate their solution based on a cost function value.

To describe the ACO algorithm for the TS problem, we consider a simple problem of touring four cities by the traveling salesperson. The situation is depicted in Figure 19.3, where the cities are represented as the nodes c_1 through c_4 of the graph, with distances between the cities known. From each city, there are links to other cities; that is, the salesperson can travel to any other city, but travel to the already visited cities (i.e.,

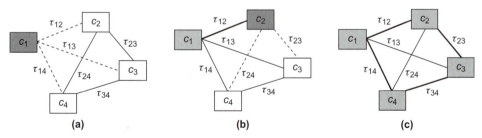

(a) **(b)** **(c)**

FIGURE 19.3 Traveling salesperson problem for four cities. (a) Start of tour at c_1; feasible links from the current city are shown by dashed lines; current city is displayed with darker shading. (b) The link already traveled is shown with a thicker line; city already visited is displayed with lighter shading. (c) A feasible solution is shown.

backtracking) is not allowed. Thus, a feasible solution to the problem consists of a sequence of cities visited on a tour—for example, $c_1 c_3 c_2 c_4 c_1$. The distance traveled on a tour is the cost function $f(\mathbf{x})$, which depends on the links used.

The definition of the task for the artificial ant changes from "finding a path from the nest to the food source" to "finding a feasible solution to the TS problem."

The TS tour must start from a city that can be randomly selected. We will call it c_1; the remaining cities are numbered randomly. To complete a four-city tour, four links need to be selected. The following notation and terminology are used in this subsection:

Qa = positive constant; an algorithm parameter
ρ = pheromone evaporation rate, $\rho \in (0, 1]$; an algorithm parameter
n = number of design variables; four for the example
N_a = number of artificial ants used in the algorithm
τ_{ij} = pheromone value for the link ij
x_j = jth component of the design variable vector $\mathbf{x}$; represents the link selected from the jth city
x_{ij} = link between the ith city and the jth city; also represents the distance between them
D_i = the list of integers corresponding to the cities that can be visited from the ith city

For the example in Figure 19.3, x_{12}, x_{13} and x_{14} are the links from city c_1 to cities c_2, c_3 and c_4, respectively; $D_1 = \{2, 3, 4\}$ for city c_1, with the associated feasible links given as $\{x_{12}, x_{13}, x_{14}\}$. The design variable vector is given as $\mathbf{x} = [x_1 \; x_2 \; x_3 \; x_4]^T$. A feasible solution to the problem is given as $\mathbf{x} = [x_{12} \; x_{24} \; x_{43} \; x_{31}]^T$.

Now let us begin the tour. From each city, selection of the next city to visit by the artificial ant is based on certain probabilities. For the ACO algorithm, the probabilities are calculated using the pheromone values τ_{ij} for each of the links from the current city; initially all τ_{ij} can be selected as 1 for all links. Also, the number of artificial ants N_a is selected as reasonable depending on the number of design variables (say $5n$ to $10n$). Individual ants can start randomly from any city. Their task is to construct a feasible solution (i.e., a feasible tour) for the TS problem, one component at a time; that is, from each city visited, a link to the next feasible city is determined in a sequence.

Each ant constructs a feasible solution (tour) for the problem, starting from a randomly selected city and moving from one city to another one that has not been visited. At each step, the traveled link is added to the solution under construction by a specific ant. In this way the ACO algorithm constructs a solution one component at a time: For example, x_1 and then x_2, and so on. Different ants pursue feasible solutions concurrently, although different ants may find the same one. When no unvisited city is left for a specific ant, that ant moves to the starting city to complete the tour. This solution process implies that an ant has memory M to store already visited cities. Using this memory, we can construct an index set D_i of feasible cities to visit from the current city i.

The ACO algorithm constructs a feasible solution, one component (i.e., one design variable) at a time.

Figure 19.3(a) shows the starting city for an artificial ant as c_1; the starting city is identified by darker shading. The feasible links from the city are shown with dashed lines:

$D_1 = \{2, 3, 4\}$, and the associated link list is $\{x_{12}, x_{13}, x_{14}\}$. The probability of taking a feasible route from the ith city is calculated as

$$p_{ij} = \frac{\tau_{ij}}{\sum_{k \in D_i}(\tau_{ik})}; \quad \text{for all } j \in D_i \tag{19.9}$$

where D_i is the list of feasible cities that can be visited from city i. For Figure 19.3(a), the probabilities for the cities that can be visited from city c_1 are calculated as

$$p_{1j} = \frac{\tau_{1j}}{\tau_{12} + \tau_{13} + \tau_{14}}; \quad j = 2, 3, 4 \tag{19.10}$$

Once these probabilities are calculated, a selection process is used for the route and the city to visit next. The roulette wheel selection process described in Chapter 16, or any other procedure, can be used for this. That process requires calculation of a random number between 0 and 1. Based on it, let the next city to visit be c_2. Thus the link x_{12} is used here and the design variable x_1 is set as x_{12}. This is shown by a darker line in Figure 19.3(b). From c_2, city c_3 or c_4 can be visited. This is shown by the dashed lines in Figure 19.3(b). The cities that have already been visited are shown by lighter shading. Therefore, $D_2 = \{3, 4\}$ and the associated link list is $\{x_{23}, x_{24}\}$. The probabilities of visiting cities c_3 and c_4 from city c_2 are given as

$$p_{2j} = \frac{\tau_{2j}}{\tau_{23} + \tau_{24}}; \quad j = 3, 4 \tag{19.11}$$

Using the foregoing procedure, the artificial ant completes the tour as follows:

$$c_1 \rightarrow c_2 \rightarrow c_3 \rightarrow c_4 \rightarrow c_1 \tag{19.12}$$

This gives the design variable values as

$$\mathbf{x} = [x_{12}\ x_{23}\ x_{34}\ x_{41}]^T \tag{19.13}$$

Using this design, the cost function $f(\mathbf{x})$, which is the total distance traveled on this tour, can be calculated.

Once all artificial ants have constructed their solution, pheromone evaporation (i.e., a reduction in the pheromone density for each link) is performed as follows:

$$\tau_{ij} \leftarrow (1 - \rho)\tau_{ij} \quad \text{for all } i \text{ and } j \tag{19.14}$$

Now the artificial ants start their return journey, depositing pheromone on the path that was used to reach the destination. This is equivalent to increasing the pheromone level for the links belonging to each ant's solution. For the kth ant, pheromone deposit is performed as follows:

$$\tau_{ij} \leftarrow \tau_{ij} + \frac{Q}{f(\mathbf{x}^{(k)})} \quad \text{for all } i, j \text{ belonging to } k\text{th ant's solution} \tag{19.15}$$

where Q is a positive constant and $f(\mathbf{x}^{(k)})$ is the cost function value for the kth ant's solution $\mathbf{x}^{(k)}$. The process of pheromone deposit in Eq. (19.15) is repeated for the solution of each of the N_a ants. Note that a tour (solution) that has a smaller cost function value deposits a larger pheromone value. Also, a link that is traveled in multiple solutions receives a pheromone deposit multiple times.

The foregoing process represents one iteration of the ACO algorithm. It is repeated several times until a stopping criterion is satisfied—that is, all ants follow the same route or the limit on the number of iterations or on CPU time is reached.

19.2.3 ACO Algorithm for Design Optimization

Problem Definition

In this subsection, we discuss the ACO algorithm for the following unconstrained discrete variable design optimization problem:

$$\text{Minimize} \qquad\qquad\qquad f(\mathbf{x}) \qquad\qquad\qquad (19.16)$$

$$x_i \in D_i; \quad D_i = (d_{i1},\ d_{i2}, \ldots, d_{iq_i}), \quad i = 1 \text{ to } n \qquad\qquad (19.17)$$

where D_i is the set of discrete values and q_i is the number of discrete values allowed for the ith design variable. This type of design problem is encountered quite frequently in practical applications, as was discussed in Chapter 15. For example, the thickness of members must be selected from an available set, structural members must be selected from the members available in the catalog, concrete reinforcing bars must be selected from the available bars on the market, and so forth.

The problem described in Eqs. (19.16) and (19.17) is quite similar to the TS problem described and discussed in the previous subsection. One major difference is that the set of available values for a design variable is predefined, whereas for the TS problem it must be determined once a city is reached (i.e., once a component of the design variable vector has been determined). The procedure described in the previous subsection can be adapted to solve this discrete variable optimization problem.

Example Problem

To describe the solution algorithm, we consider a simpler problem having three design variables, with each variable having four allowable discrete values. Therefore, $n = 3$, and $q_i = 4$, $i = 1$ to 4 in Eqs. (19.16) and (19.17). The problem can be displayed in a multilayered graph as shown in Figure 19.4. The graph shows the starting node 00 as the nest and the destination node as the food source. The starting point is called Layer 0. Layer 1 represents the allowable values for the design variable x_1 in the set D_1; each allowable value is represented as a node, such as node d_{12}. There are links from the nest to each of these nodes. Level 2 represents the allowable values for the design variable x_2 as nodes. For example, from d_{13} there are links to d_{21}, d_{22}, d_{23}, and d_{24}. Similarly, from d_{11} there are links to d_{21}, d_{22}, d_{23}, and d_{24}, and so on.

The ACO algorithm proceeds as follows: An ant starts from the nest and chooses a link to travel to a node at Layer 1 based on probabilities such as the link to node d_{13}; that is, design variable x_1 is assigned the value d_{13} From this node, the probabilities are calculated again for all links to the next layer on the graph, and the ant moves to, say, node d_{22}. This procedure is repeated for the next layer, and the ant moves to node d_{34}. Since there are no further layers, this ant has reached its destination. Its feasible solution is obtained as

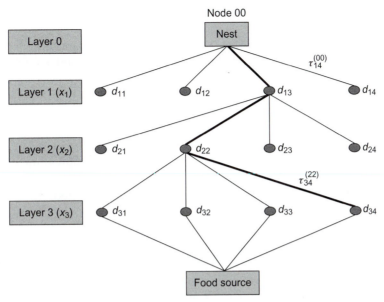

FIGURE 19.4 A multilayered graphical representation of a discrete variable problem with 3 design variables each one having 4 allowable values; the links chosen by the ant are shown using the darker lines. (Note that all possible links are not shown.)

$\mathbf{x} = (d_{13}, d_{22}, d_{34})$, with the cost function value as $f(\mathbf{x})$. The route for this ant is shown by the darker lines in Figure 19.4.

Once all the ants have found feasible solutions, pheromone evaporation is performed for all of the links using Eq. (19.14) or one similar to it. Then each ant traces its path back to the nest, depositing pheromone using Eq. (19.15) or one similar to it on each link that it previously traveled. This is equivalent to updating (increasing) the pheromone values for the links traveled by the ants. The entire process is then repeated until a stopping criterion is satisfied.

Finding Feasible Solutions

The foregoing procedure can be generalized to the case of n design variables (an n-layered graph), each having q_i discrete values. The following notation is used:

$\tau_{ij}^{(rs)}$ = pheromone value for the link from node rs to node ij; note that since the procedure moves from one layer to the next, $i = r + 1$ (e.g., $\tau_{34}^{(22)}$ between nodes d_{22} and d_{34} in Figure 19.4). Thus, the superscript r represents the layer number (design variable number), the superscript s represents the allowable value number for the design variable number r, subscript i represents the next layer (next design variable), and subscript j represents the allowable design variable number for the ith design variable.

$p_{ij}^{(rs)}$ = probability of selection of the link from node rs to node ij.

To find a feasible solution for an artificial ant k, the following steps are suggested.

STEP 1. SELECTION OF AN INITIAL LINK   Ant k starts from the nest (i.e., node 00 of Layer 0). Calculate probabilities for the links from node 00 to all nodes for Layer 1 (design variable x_1) as follows:

$$p_{1j}^{(00)} = \frac{\tau_{1j}^{(00)}}{\sum_{r=1}^{q_1} \tau_{1r}^{(00)}}; \quad j=1 \ to \ q_1 \tag{19.18}$$

Using these probabilities and a selection process, choose a link to a node at Layer 1 and go to that node. Let this node be k_1; the design variable x_1 is thus assigned the value d_{kl}.

STEP 2. SELECTION OF A LINK FROM LAYER R   Let ant k be at node rs. Calculate probabilities of the links from node rs to all nodes at the next layer:

$$p_{ij}^{(rs)} = \frac{\tau_{ij}^{(rs)}}{\sum_{l=1}^{q_i} \tau_{il}^{(rs)}}; \quad j=1 \ to \ q_i \tag{19.19}$$

Note that $i = r + 1$. Using these probabilities and a selection procedure, select a link to the next layer and the corresponding node for ant k to travel. Repeat this step until the nth layer is reached, at which point ant k has reached its destination and a feasible solution has been obtained.

STEP 3. OBTAINING FEASIBLE SOLUTIONS FOR ALL ANTS   Repeat Steps 1 and 2 for each artificial ant to obtain all N_a feasible solutions. Let the solutions and the corresponding cost function values be represented as

$$\mathbf{x}^{(k)}, \ f(\mathbf{x}^{(k)}); \quad k=1 \ to \ N_a \tag{19.20}$$

Pheromone Evaporation

Once all of the ants have reached their destination (all of them have found solutions), pheromone evaporation (i.e., reduction in the pheromone level) is performed for all links as follows:

$$\tau_{ij}^{(rs)} \leftarrow (1 - \rho)\tau_{ij}^{(rs)} \ \text{for all} \ r, \ s, \ i \ \text{and} \ j \tag{19.21}$$

Pheromone Deposit

After pheromone evaporation, the ants start their journey back to their nest, which means that they will deposit pheromone on the return trail. This involves increasing the pheromone density of the links that they have traveled. For the kth ant, the pheromone deposit is performed as follows:

$$\tau_{ij}^{(rs)} \leftarrow \tau_{ij}^{(rs)} + \frac{Q}{f(\mathbf{x}^{(k)})} \ \text{for all} \ r, s, i, j \ \text{belonging to} \ k\text{th ant's solution} \tag{19.22}$$

The operation in Eq. (19.22) is performed for all solutions obtained by the ants. It is seen that the solutions that have a smaller cost function value receive more pheromone deposit.

Also, the links that are traveled multiple times receive more reinforcement of pheromone. A larger value of the pheromone for a link gives a larger probability value from Eq. (19.18), which then favors it selection for travel by the artificial ants in subsequent iterations of the ACO algorithm.

We see that the ACO algorithm is quite simple to implement, requiring specification of only three parameters, N_a, ρ, and Q. N_a can be given a reasonable value of, say, $5n$ to $10n$; $\rho \in (0, 1)$, a value of, say, 0.4 to 0.8; Q may be selected as a typical value for the cost function $f(\mathbf{x})$.

19.3 PARTICLE SWARM OPTIMIZATION

Particle swarm optimization (PSO), another nature-inspired method, mimics the social behavior of bird flocking or fish schooling. It falls into the metaheuristics and swarm intelligence methods class. It is also a population-based stochastic optimization technique, introduced by Kennedy and Eberhart in 1995. PSO shares many similarities with evolutionary computation techniques such as GA and DE. Just like those approaches, PSO starts with a randomly generated set of solutions called the initial population. An optimum solution is then searched by updating generations.

An attractive feature of PSO is that it has fewer algorithmic parameters to specify compared to GAs. It does not use any of the GAs' evolutionary operators such as crossover and mutation. Also, unlike GAs, the algorithm does not require binary number encoding or decoding and thus is easier to implement on the computer. PSO has been successfully applied to many classes of problems, such as mechanical and structural optimization and multi-objective optimization, artificial neural network training, and fuzzy system control.

In this section, we present the basic ideas of PSO and a simple particle swarm optimization algorithm. Many variations on the method arc available in the literature, and research on the subject continues to develop better algorithms and expand the range of their application (Kennedy et al., 2001).

19.3.1 Swarm Behavior and Terminology

The PSO computational algorithm tries to emulate the social behavior of a swarm of animals, such as a flock of birds or a school of fish (moving in search for food). In a swarm, an individual behaves according to its limited intelligence as well as to the intelligence of the group. Each individual observes the behavior of its neighbors and adjusts its own behavior accordingly. If an individual member discovers a good path to food, other members follow this path no matter where they are situated in the swarm.

PSO uses the following terminology:

Particle. This term is used to identify an individual in the swarm (e.g., a bird in the flock or a fish in the school). *Agent* is also used in some circles. Each particle has a location in the swarm. In the optimization algorithm, each particle location represents a design point that is a potential solution to the problem.

Particle Position. This term refers to the coordinates of the particle. In the optimization algorithm, it refers to a design point (a vector of design variables).

Particle Velocity. The term refers to the rate at which the particles are moving in space. In the optimization algorithm, it refers to the design change.

Swarm Leader. This is the particle having the best position. For the optimization algorithm, the term refers to a design point having the smallest value for the cost function.

19.3.2 Particle Swarm Optimization Algorithm

The PSO translates the social behavior of the swarm described above into a computational algorithm. The notation shown in Table 19.2 is used in the subsequent step-by-step algorithm.

Each particle in the swarm keeps track of its own current position and its best position (solution) achieved during the running of the algorithm. This implies that each point stores not only its current value but also its best value achieved thus far. The best position for the ith particle (design point) is denoted $x_P^{(i,k)}$. Another "best" value that is tracked by the particle swarm optimizer is the best position for the entire swarm, denoted $x_G^{(k)}$. The PSO algorithm consists of changing, at each time step (iteration), the velocity of each particle toward its own best position as well as the swarm's best position (also sometimes referred to as accelerating the particle toward the best known position).

TABLE 19.2   Notation and terminology for the step-by-step algorithm

Notation	Terminology
c_1	Algorithm parameter (i.e., cognitive parameter); taken between 0 and 4, usually set to 2
c_2	Algorithm parameter (i.e., social parameter); taken between 0 and 4, usually set to 2
r_1, r_2	Random numbers between 0 and 1
k	Iteration counter
k_{max}	Limit on the number of iterations
n	Number of design variables
N_p	Number of particles (design points) in the swarm; *swarm size* (usually $5n$ to $10n$)
x_j	jth component of the design variable vector x
$v^{(i,k)}$	Velocity of the ith particle (design point) of the swarm at the kth generation/iteration
$x^{(i,k)}$	Location of the ith particle (design point) of the swarm at the kth generation/iteration
$x_P^{(i,k)}$	Best position of the ith particle based on its travel history at the kth generation/iteration
$x_G^{(k)}$	Best solution for the swarm at the kth generation; considered the leader of the swarm
x_L	Vector containing lower limits on the design variables
x_U	Vector containing upper limits on the design variables

The step-by-step PCO algorithm is stated as follows.

Step 0. *Initialization.* Select N_p, c_1, c_2, and k_{max} as the maximum number of iterations. Set the initial velocity of the particle $\mathbf{v}^{(i,0)}$ to 0. Set the iteration counter at $k = 1$.

Step 1. *Initial Generation.* Using a random procedure, generate N_p particles $\mathbf{x}^{(i,0)}$. The procedure described in Eq. (19.2) can be used to generate these points within their allowable ranges. Evaluate the cost function for each of these points $f(\mathbf{x}^{(i,0)})$. Determine the best solution among all particles as $\mathbf{x}_G^{(k)}$—that is, a point having the smallest cost function value.

Step 2. *Calculate Velocities.* Calculate the velocity of each particle as

$$\mathbf{v}^{(i,k+1)} = \mathbf{v}^{(i,k)} + c_1 r_1 (\mathbf{x}_P^{(i,k)} - \mathbf{x}^{(i,k)}) + c_2 r_2 (\mathbf{x}_G^{(k)} - \mathbf{x}^{(i,k)}); \quad i = 1 \text{ to } N_p \tag{19.23}$$

Update the positions of the particles as

$$\mathbf{x}^{(i,k+1)} = \mathbf{x}^{(i,k)} + \mathbf{v}^{(i,k+1)}; \quad i = 1 \text{ to } N_p \tag{19.24}$$

Check and enforce bounds on the particle positions:

$$\mathbf{x}_L \leq \mathbf{x}^{(i,k+1)} \leq \mathbf{x}_U \tag{19.25}$$

Step 3. *Update the Best Solution.* Calculate the cost function at all new points $f(\mathbf{x}^{(i,k+1)})$. For each particle, perform the following check:

$$\begin{aligned} &\text{If } f(\mathbf{x}^{(i,k+1)}) \leq f(\mathbf{x}_P^{(i,k)}), \text{ then } \mathbf{x}_P^{(i,k+1)} = \mathbf{x}^{(i,k+1)}; \\ &\text{otherwise } \mathbf{x}_P^{(i,k+1)} = \mathbf{x}_P^{(i,k)} \text{ for each } i = 1 \text{ to } N_p \end{aligned} \tag{19.26}$$

$$\text{If } f(\mathbf{x}_P^{(i,k+1)}) \leq f(\mathbf{x}_G), \text{ then } \mathbf{x}_G = \mathbf{x}_P^{(i,k+1)}, \quad i = 1 \text{ to } N_p \tag{19.27}$$

Step 4. *Stopping Criterion.* Check for convergence of the iterative process. If a stopping criterion is satisfied (i.e., $k = k_{max}$ or if all of the particles have converged to the best swarm solution), stop. Otherwise, set $k = k + 1$ and go to Step 2.

EXERCISES FOR CHAPTER 19*

19.1 Implement the DE algorithm into a computer program. Solve the Example 16.1 of bolt insertion sequence determination using your program. Compare performance of the DE and GA algorithms.

19.2 Implement the ACO algorithm into a computer program. Solve the Example 16.1 of bolt insertion sequence determination using your program. Compare performance of the ACO and GA algorithms.

19.3 Implement the PSO algorithm into a computer program. Solve the Example 16.1 of bolt insertion sequence determination using your program. Compare performance of the PSO and GA algorithms.

20

Additional Topics on Optimum Design

Upon completion of this chapter, you will be able to:

- Understand and use the concept of a meta-model for design optimization

- Understand and use the concept of design of experiments for selection of sample points

- Understand and use the robust design approach for practical engineering problems

- Understand and use the reliability-based design optimization approach

In this chapter, various optimization topics of practical interest are presented and discussed. These include generation of meta-models for practical design optimization problems, design of experiments for response surface generation, robust design, and reliability-based design optimization, or design under uncertainty. These topics may not be covered in an undergraduate course or in a first independent reading of the text. Material for this chapter is derived from several sources, such as Park et al. (2006), Park (2007), Beyer and Sandhoff (2007), and Choi et al. (2007).[1]

20.1 META-MODELS FOR DESIGN OPTIMIZATION

20.1.1 Meta-Model

Many practical applications require a detailed model of the system to accurately predict system response to various inputs. These models can be quite large and complex, requiring enormous calculation. Optimization of such systems is difficult, if not impossible,

[1]The original draft of this chapter was provided by G. J. Park. The contribution to this book is very much appreciated.

because evaluation of cost and constraint functions requires large numbers of calculations. In addition, calculation of the gradients of functions requires special procedures since the cost and/or the constraint functions for such problems are implicit in terms of the design variables. This calculation is also quite tedious and time-consuming.

It was seen in Chapters 10 through 14 that the functions and their derivatives need to be repeatedly calculated in the optimization process. For a large analysis and design model of the system, this calculation process is quite time-consuming. Therefore it is useful to develop simplified functions for design optimization that have explicit forms in terms of the design variables. The explicit function is a model of the model and is called a meta-model. A meta-model can be generated by conducting experimental observations and/or numerical simulations.

Suppose that we have a mathematical model of the form

$$f = f(\mathbf{x}) \tag{20.1}$$

where $f(\mathbf{x})$ does not have an explicit expression in terms of the design variables $\mathbf{x}$. The function $f(\mathbf{x})$ can be approximated by a simplified explicit function (meta-model) using the information at some sample points $\mathbf{x}_i$. To make the meta-model, f is evaluated at k points as follows:

$$f_i = f(\mathbf{x}_i), \quad i = 1 \text{ to } k \tag{20.2}$$

The meta-model is constructed using the f_i that may be obtained by experiments or numerical simulations. Examples of meta-models are illustrated in Figure 20.1. Here, we have one variable x and $f(x)$ is the original model for which we may not have an explicit expression in terms of x. Three points (x_1, x_2, x_3) are selected and f is evaluated at these points. $\hat{f}_j(x)$, $j = 1, 2, 3$ are the three meta-models constructed using some method.

Once a meta-model has been developed, we can use it instead of the original model in the optimization process. Generally, the meta-model has errors due to mathematical approximations, experimental errors, or computational approximations. The error $\varepsilon(\mathbf{x})$ in the meta-model $\hat{f}(\mathbf{x})$ is expressed as

$$\varepsilon(\mathbf{x}) = f(\mathbf{x}) - \hat{f}(\mathbf{x}) \tag{20.3}$$

Usually the meta-model is derived to minimize this error $\varepsilon(\mathbf{x})$.

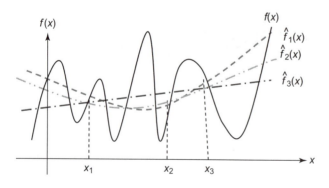

FIGURE 20.1   Examples of meta-models. *Source: Adapted from Park, 2007.*

20.1.2 Response Surface Method

The Response Surface Method (RSM) is a representative method for generating meta-models. The original model is evaluated at multiple sample points and the meta-model is constructed usually as a linear or a quadratic function. The coefficients of the meta-model function are determined by minimizing the error in Eq. (20.3). In other words, the response surface of a model is approximated by an explicit polynomial function using the *least squares method* to minimize the error in Eq. (20.3). An example of a response surface meta-model is illustrated in Figure 20.2.

In statistics, the RSM has been utilized to generate various statistical data since the 1950s. In design, the approximated functions are used in the optimization process. The following aspects should be considered for optimization with RSM:

- Selection of the sample points to generate the response surface
- Generation of the response surface using the function value at the sample points

Quadratic Response Surface Generation

Several types of functions can be used to generate the response surface approximation, such as linear, quadratic, cubic, and some other special functions. Here we demonstrate the response surface generation using a quadratic function approximation. The approach is to evaluate the function to be approximated at several sample points. Then we interpolate a quadratic function through them. The coefficients of the quadratic function are calculated by minimizing the error between the original function and the approximated function.

First, the sample design points are selected where the function is evaluated. In many cases, the sample points are arbitrarily selected. Sometimes they are selected using a method such as orthogonal arrays, explained in the next section. Let a design point be represented as an n-vector $x_1, x_2, \ldots, x_n$. A quadratic approximation for the function $f(\mathbf{x})$ in terms of the design variables is defined as

$$
\begin{aligned}
f = a_{00} + a_{10}x_1 + a_{20}x_2 + \ldots + a_{n0}x_n + a_{11}x_1^2 + \ldots + a_{nn}x_n^2 \\
+ a_{12}x_1x_2 + \ldots + a_{n-1,n}x_{n-1}x_n + \varepsilon
\end{aligned}
\tag{20.4}
$$

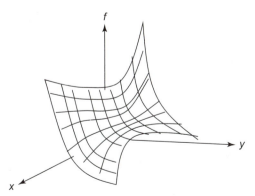

FIGURE 20.2 Example of a response surface $f(x, y)$ in two variables. *Source: Adapted from Park, 2007.*

where ε is the error in the approximation process, and a_{ij}; $i, j = 0$ to n are the unknown coefficients.

Let k be the number of sample design points at which the function $f(\mathbf{x})$ has been evaluated. We shall use double subscripts for design variables to represent their values at the sample points as

$$x_{ij}; \quad i = 1 \text{ to } k; \quad j = 1 \text{ to } n \tag{20.5}$$

where $k = $ number of sample points; $n = $ number of design variables.

Thus, x_{ij} represents value of the jth design variable for the ith sample point; i.e., that is, the first subscript corresponds to the sample point number and the second one corresponds to the design variable number. For example, x_{23} represents the value of the third design variable for the second sample point. The sample point data can be represented in a table where the columns represent the design variables and the rows represent the sample design points. This table format is used later in example problems.

The function value at the ith sample point is now written using Eq. (20.4) and the double-subscripted variables of Eq. (20.5) as

$$f_i = a_{00} + a_{10}x_{i1} + a_{20}x_{i2} + \ldots + a_{n0}x_{in} + a_{11}x_{i1}^2 + \ldots + a_{nn}x_{in}^2$$
$$+ a_{12}x_{i1}x_{i2} + \ldots + a_{n-1,n}x_{in-1}x_{in} + \varepsilon \tag{20.6}$$

In this equation, the only unknowns are the coefficients a_{ij}. Therefore, the linear and quadratic terms in Eq. (20.6) can be treated in a similar manner.

We introduce the following simplifying notation to develop the response surface methodology:

$$a_{00} \to d_0, a_{10} \to d_1, \ldots, a_{11} \to d_{n+1}, \ldots, a_{n-1,n} \to d_l \tag{20.7}$$

$$x_1 \to \xi_1, \ldots, x_1^2 \to \xi_{n+1}, \ldots, x_{n-1}x_n \to \xi_l \tag{20.8}$$

where l is the total number of linear and quadratic terms given as

$$l = \frac{1}{2}n(n + 1) + n \tag{20.9}$$

Thus the function in Eq. (20.4) is written in terms of the new variables as

$$f = d_0 + d_1\xi_1 + d_2\xi_2 + \ldots + d_l\xi_l + \varepsilon = d_0 + \sum_{i=1}^{l} d_i\xi_i + \varepsilon \tag{20.10}$$

The approximated function $\hat{f}$ is defined by dropping the error ε in Eq. (20.10):

$$\hat{f} = d_0 + \sum_{i=1}^{l} d_i\xi_i \tag{20.11}$$

To represent the value of the function in Eq. (20.11) at different sample points, we introduce the double-subscript notation for the variables ξ_j:

$$\xi_{ij}; \quad i = 1 \text{ to } k, \quad j = 1 \text{ to } l \tag{20.12}$$

Thus ξ_{ij} represents the value of the variable ξ_j at the ith sample point. For the ith sample point, then, the function value in Eq. (20.11) is now written as

$$\hat{f}_i = d_0 + d_1\xi_{i1} + d_2\xi_{i2} + \ldots + d_l\xi_{il} = d_0 + \sum_{j=1}^{l} d_j\xi_{ij}; \quad i = 1 \text{ to } k \tag{20.13}$$

To determine the coefficients d_i in Eq. (20.13), an error function E is defined, which is the sum of the square of the error ε_i at each sample point, as

$$E = \sum_{i=1}^{k} \varepsilon_i^2 = \sum_{i=1}^{k} (f_i - \hat{f}_i)^2 = \sum_{i=1}^{k} \left[f_i - \left(d_0 + \sum_{j=1}^{l} d_j\xi_{ij} \right) \right]^2 \tag{20.14}$$

where f_i is the ith observed value of f. The coefficients $d_0, d_1, \ldots, d_1$ are determined to minimize E using the following optimality condition:

$$\frac{\partial E}{\partial d_i} = 0, \quad i = 0 \text{ to } l \tag{20.15}$$

These conditions result in a linear system of equations for the unknowns d_i as (recall that k is the number of sample points at which the function has been evaluated):

$$\begin{bmatrix} k & \sum_{i=1}^{k}\xi_{i1} & \sum_{i=1}^{k}\xi_{i2} & \cdots & \sum_{i=1}^{k}\xi_{il} \\ \sum_{i=1}^{k}\xi_{i1} & \sum_{i=1}^{k}\xi_{i1}^2 & \sum_{i=1}^{k}\xi_{i1}\xi_{i2} & \cdots & \sum_{i=1}^{k}\xi_{i1}\xi_{il} \\ \vdots & \vdots & \vdots & \vdots & \vdots \\ \sum_{i=1}^{k}\xi_{il} & \sum_{i=1}^{k}\xi_{il}\xi_{i1} & \sum_{i=1}^{k}\xi_{il}\xi_{i2} & & \sum_{i=1}^{k}\xi_{il}^2 \end{bmatrix} \begin{bmatrix} d_0 \\ d_1 \\ \vdots \\ d_l \end{bmatrix} = \begin{bmatrix} \sum_{i=1}^{k}f_i \\ \sum_{i=1}^{k}\xi_{i1}f_i \\ \vdots \\ \sum_{i=1}^{k}\xi_{il}f_i \end{bmatrix} \tag{20.16}$$

The coefficient vector $\mathbf{d} = (d_0, d_1, \ldots d_l)$ is obtained by solving Eq. (20.16). This process is exactly the same as the least squares method. The number of calculations depends on how many terms are included in Eq. (20.4). If $l = n$, only the linear terms in Eq. (20.4) are included in the approximation. If $l = 2n$, the linear and the perfect square terms are included. If $l = n(n + 1)/2 + n$, both the linear and quadratic terms are included. Generally, when l is large more terms are considered and the cost of generating the meta-model increases. Therefore, the designer should select l in accordance with desired accuracy and computational cost. It is noted that the number of sample points k should be equal to or greater than the number of unknown coefficients ($l + 1$). Otherwise, the matrix shown in Eq. (20.16) is singular.

EXAMPLE 20.1 GENERATION OF QUADRATIC RESPONSE SURFACE

Suppose f is a function of two design variables x_1 and x_2 as follows:

$$f = f(x_1, x_2) \tag{a}$$

We have experimental values for f at nine sample points ($k = 9$), as shown in Table 20.1 (the numbers in the columns x_1 and x_2 represent the x_{ij} values). Generate a response surface with all of the linear and quadratic terms using the least squares method.

Solution

The quadratic approximation for the function f using Eqs. (20.4), (20.7), (20.8) and (20.11) is given as

$$\hat{f} = a_{00} + a_{10}x_1 + a_{20}x_2 + a_{11}x_1^2 + a_{22}x_2^2 + a_{12}x_1x_2 = d_0 + d_1\xi_1 + d_2\xi_2 + d_3\xi_3 + d_4\xi_4 + d_5\xi_5 \quad \text{(b)}$$

Using the design variable data given in Table 20.1, and Eqs. (20.8) and (20.12), we obtain the values of ξ_{ij} shown in Table 20.2.

Using the function data from Table 20.1 and the ξ_{ij} data from Table 20.2 in Eq. (20.16), we obtain the following system of linear equations:

$$\begin{bmatrix} 9 & 11.25 & 0 & 59.4375 & 54 & 0 \\ 11.25 & 59.4375 & 0 & 187.734 & 67.5 & 0 \\ 0 & 0 & 54 & 0 & 0 & 67.5 \\ 59.4375 & 187.734 & 0 & 790.512 & 356.625 & 0 \\ 54 & 67.5 & 0 & 356.625 & 486 & 0 \\ 0 & 0 & 67.5 & 0 & 0 & 356.625 \end{bmatrix} \begin{bmatrix} d_0 \\ d_1 \\ d_2 \\ d_3 \\ d_4 \\ d_5 \end{bmatrix} = \begin{bmatrix} 106.352 \\ 177.663 \\ -0.228 \\ 937.577 \\ 809.73 \\ -546.46 \end{bmatrix} \quad \text{(c)}$$

Solving Eq. (c), and substituting the values of d_i into Eq. (b), we obtain the quadratic function representation for $f(\mathbf{x})$ in terms of the variables ξ_i as

$$\hat{f} = 0.475 - 1.712\xi_1 + 2.504\xi_2 + 1.079\xi_3 + 1.0594\xi_4 - 2.006\xi_5 \quad \text{(d)}$$

Substituting the definitions of ξ_i from Eq. (b) or Eq. (20.8) into Eq. (d), the response surface model for the function $f(\mathbf{x})$ is given as

$$\hat{f} = 0.475 - 1.712x_1 + 2.504x_2 + 1.079x_1^2 + 1.0594x_2^2 - 2.006x_1x_2 \quad \text{(e)}$$

TABLE 20.1　Nine sample design points ($k = 9$) for Example 20.1

Sample point	x_1	x_2	f
1	−1.5	−3	−1.022
2	−1.5	0	4.503
3	−1.5	3	31.997
4	1.25	−3	8.704
5	1.25	0	1.636
6	1.25	3	8.793
7	4	−3	37.341
8	4	0	10.243
9	4	3	4.157

TABLE 20.2 Values for variables ξ_{ij} at nine sample points ($k = 9$) for Example 20.1

Sample point	ξ_1	ξ_2	ξ_3	ξ_4	ξ_5
1	−1.5	−3	2.25	9	4.5
2	−1.5	0	2.25	0	0
3	−1.5	3	2.25	9	−4.5
4	1.25	−3	1.563	9	−3.75
5	1.25	0	1.563	0	0
6	1.25	3	1.563	9	3.75
7	4	−3	16	9	−12
8	4	0	16	0	0
9	4	3	16	9	12
Maximum	4	3	16	9	12
Minimum	−1.5	−3	1.563	0	−12

20.1.3 Normalization of Variables

In numerical calculations, it is useful to normalize the variables of the response surface generation problem. This results in the well-conditioned system of linear equations in Eq. (20.16), thus avoiding numerical instability. Several normalization procedures can be used. Here we demonstrate two procedures.

Normalization of Variables: Procedure 1

The following transformation of variables is defined such that the transformed variables receive values between −1 and 1:

$$w_j = \frac{\xi_j - \left[\overset{\max}{m}(\xi_{mj}) + \overset{\min}{m}(\xi_{mj}) \right]/2}{\left[\overset{\max}{m}(\xi_{mj}) - \overset{\min}{m}(\xi_{mj}) \right]/2} \tag{20.17}$$

where $\overset{\max}{m}(\xi_{mj})$ and $\overset{\min}{m}(\xi_{mj})$ represent the maximum and minimum values, respectively, in the jth column of the ξ_{mj} table, as can be seen in Table 20.2. Thus, in terms of the normalized variables w_j, the approximated function in Eq. (20.11) is written as

$$\hat{f} = c_0 + c_1 w_1 + c_2 w_2 + \ldots + c_l w_l \tag{20.18}$$

where c_j are the coefficients of the normalized variables w_j. For the ith sample point, the function value is written using Eq. (20.18) as

$$\hat{f}_i = c_0 + c_1 w_{i1} + c_2 w_{i2} + \ldots + c_l w_{il}, \quad i = 1 \text{ to } k \tag{20.19}$$

where w_{ij} is the value of w_j at the ith sample point.

In terms of the normalized variables, minimization of the error function in Eq. (20.14) results in the linear system in Eq. (20.16), where ξ_{ij} are replaced by w_{ij} and d_i are replaced by c_i. Example 20.2 illustrates development of the response surface using this normalization procedure.

EXAMPLE 20.2 RESPONSE SURFACE USING THE NORMALIZATION PROCEDURE 1

Re-solve the problem in Example 20.1 using the normalization procedure given in Eqs. (20.17) and (20.18).

Solution

The quadratic approximation for the function f in terms of the normalized variables is given as follows:

$$\hat{f} = a_{00} + a_{10}x_1 + a_{20}x_2 + a_{11}x_1^2 + a_{22}x_2^2 + a_{12}x_1x_2 = c_0 + c_1w_1 + c_2w_2 + c_3w_3 + c_4w_4 + c_5w_5 \quad \text{(a)}$$

where w_j are the normalized variables ξ_j. Using the data given in Table 20.2 for ξ_{ij} and the $\overset{max}{\underset{m}{}}(\xi_{mj})$ values, the transformation of variables is defined using Eq. (20.17) as

$$w_1 = \frac{\xi_1 - (4 - 1.5)/2}{(4 + 1.5)/2} = \frac{x_1 - 1.25}{2.75} \quad \text{(b)}$$

$$w_2 = \frac{\xi_2 - (3 - 3)/2}{(3 + 3)/2} = \frac{x_2}{3} \quad \text{(c)}$$

$$w_3 = \frac{\xi_3 - (16 + 1.5625)/2}{(16 - 1.5625)/2} = \frac{x_1^2 - 8.782}{7.219} \quad \text{(d)}$$

$$w_4 = \frac{\xi_4 - (9 + 0)/2}{(9 - 0)/2} = \frac{x_2^2 - 4.5}{4.5} \quad \text{(e)}$$

$$w_5 = \frac{\xi_5 - (12 - 12)/2}{(12 + 12)/2} = \frac{x_1x_2}{12} \quad \text{(f)}$$

Table 20.3 contains the normalized values w_{ij} obtained using Eqs. (b) through (f) for the ξ_{ij} data given in Table 20.2. For example, w_{53} is calculated using Eq. (d) as

$$w_{53} = \frac{\xi_{53} - 8.782}{7.219} = \frac{1.563 - 8.782}{7.219} = -1 \quad \text{(g)}$$

Using the w_{ij} data given in Table 20.3 for the nine sample points ($k = 9$), and the function data given in Table 20.1 in Eq. (20.16), we obtain the following linear system of equations:

$$\begin{bmatrix} 9 & 0 & 0 & -2.715 & 3 & 0 \\ 0 & 6 & 0 & 5.715 & 0 & 0 \\ 0 & 0 & 6 & 0 & 0 & 1.875 \\ -2.715 & 5.715 & 0 & 8.457 & -0.905 & 0 \\ 3 & 0 & 0 & -0.905 & 9 & 0 \\ 0 & 0 & 1.875 & 0 & 0 & 2.477 \end{bmatrix} \begin{bmatrix} c_0 \\ c_1 \\ c_2 \\ c_3 \\ c_4 \\ c_5 \end{bmatrix} = \begin{bmatrix} 106.352 \\ 16.263 \\ -0.076 \\ 0.500 \\ 73.588 \\ -45.538 \end{bmatrix} \quad \text{(h)}$$

It is seen that the coefficient matrix in Eq. (h) is diagonally dominant due to the normalization process, whereas the coefficient matrix in Eq. (c) of Example 20.1 is not diagonally dominant. Solving Eq. (h) for c_i, and substituting these values into Eq. (20.18), we obtain the function representation in terms of the normalized variables w_i as

$$\hat{f} = 12.577 - 4.708w_1 + 7.510w_2 + 7.789w_3 + 4.767w_4 - 24.074w_5 \tag{i}$$

Using the transformations given in Eqs. (b) to (f), the response surface model for the function $f(\mathbf{x})$ is given as

$$\hat{f}(\mathbf{x}) = 12.577 - 4.708\left(\frac{x_1 - 1.25}{2.75}\right) + 7.510\left(\frac{x_2}{3}\right) + 7.789\left(\frac{x_1^2 - 8.782}{7.219}\right) + 4.767\left(\frac{x_2^2 - 4.5}{4.5}\right)$$
$$- 24.074\left(\frac{x_1 x_2}{12}\right) \tag{j}$$
$$= 0.475 - 1.712x_1 + 2.504x_2 + 1.079x_1^2 + 1.0594x_2^2 - 2.006x_1 x_2$$

This equation is the same as obtained earlier in Example 20.1.

TABLE 20.3   Values for variables w_{ij} for Example 20.2 for nine sample points ($k = 9$)

Sample point	w_1	w_2	w_3	w_4	w_5
1	−1	−1	−0.905	1	0.375
2	−1	0	−0.905	−1	0
3	−1	1	−0.905	1	−0.375
4	0	−1	−1	1	−0.3125
5	0	0	−1	−1	0
6	0	1	−1	1	0.3125
7	1	−1	1	1	−1
8	1	0	1	−1	0
9	1	1	1	1	1

Normalization of Variables: Procedure 2

Another procedure is to normalize the variables x_i using the data of the sample points in Eq. (20.17). In this way, the transformed variables receive values between −1 and 1. Using these normalized variables, the normalized data w_{ij} can be calculated for use in Eq. (20.16). The procedure is demonstrated in Example 20.3.

EXAMPLE 20.3   RESPONSE SURFACE USING NORMALIZATION PROCEDURE 2

Re-solve the problem in Example 20.1 using the normalization procedure given in Eqs. (20.17) and (20.18).

Solution

The quadratic approximation for the function $f(\mathbf{x})$ is given as

$$\hat{f} = d_0 + d_1 x_1 + d_2 x_2 + d_3 x_1^2 + d_4 x_2^2 + d_5 x_1 x_2 \tag{a}$$

Using the data given in the first two columns of Table 20.2 (or the data given in Table 20.1 for the variables x_1 and x_2) and the corresponding maximum and minimum values, the transformation of variables is defined using Eq. (20.17) as

$$w_1 = \frac{x_1 - (4 - 1.5)/2}{(4 + 1.5)/2} = \frac{x_1 - 1.25}{2.75} \tag{b}$$

$$w_2 = \frac{x_2 - (3 - 3)/2}{(3 + 3)/2} = \frac{x_2}{3} \tag{c}$$

Using these transformed variables, the quadratic approximation of Eq. (a) for the function f is obtained as

$$\hat{f} = c_0 + c_1 w_1 + c_2 w_2 + c_3 w_1^2 + c_4 w_2^2 + c_5 w_1 w_2 \tag{d}$$

Now we define the variables w_3, w_4, and w_5 as follows:

$$w_3 = w_1^2 = \left(\frac{x_1 - 1.25}{2.75}\right)^2 \tag{e}$$

$$w_4 = w_2^2 = \left(\frac{x_2}{3}\right)^2 \tag{f}$$

$$w_5 = w_1 w_2 = \left(\frac{x_1 - 1.25}{2.75}\right)\left(\frac{x_2}{3}\right) \tag{g}$$

Thus the quadratic approximation of Eq. (d) is given as

$$\hat{f} = c_0 + c_1 w_1 + c_2 w_2 + c_3 w_3 + c_4 w_4 + c_5 w_5 \tag{h}$$

Using the design variable data given in Table 20.1, and the transformation Eqs (b) and (c), the w_{ij} values are calculated as shown in Table 20.4. Using these values for w_{ij} in Eq. (20.16), we obtain the following linear system of equations:

$$\begin{bmatrix} 9 & 0 & 0 & 6 & 6 & 0 \\ 0 & 6 & 0 & 0 & 0 & 0 \\ 0 & 0 & 6 & 0 & 0 & 0 \\ 6 & 0 & 0 & 6 & 4 & 0 \\ 6 & 0 & 0 & 4 & 6 & 0 \\ 0 & 0 & 0 & 0 & 0 & 4 \end{bmatrix} \begin{bmatrix} c_0 \\ c_1 \\ c_2 \\ c_3 \\ c_4 \\ c_5 \end{bmatrix} = \begin{bmatrix} 106.352 \\ 16.264 \\ -0.076 \\ 87.219 \\ 89.970 \\ -66.204 \end{bmatrix} \tag{i}$$

Again, we observe that the coefficient matrix in Eq. (i) is diagonally dominant. Solving Eq. (i) for c_i, and substituting them into Eq. (h), we obtain the function representation in terms of the normalized variables w_i as

$$\hat{f} = 0.0214 + 2.711w_1 - 0.01267w_2 + 8.159w_3 + 9.534w_4 - 16.551w_5 \qquad \text{(j)}$$

Substituting $w_3 = w_1^2$, $w_4 = w_2^2$ and $w_5 = w_1 w_2$ into Eq. (j), we obtain

$$\hat{f} = 0.0214 + 2.711w_1 - 0.01267w_2 + 8.159w_1^2 + 9.534w_2^2 - 16.551w_1 w_2 \qquad \text{(k)}$$

Now, substituting the transformation of variables given in Eqs. (b) and (c) into Eq. (k), we obtain

$$\hat{f}(\mathbf{x}) = 0.0214 + 2.711\left(\frac{x_1 - 1.25}{2.75}\right) - 0.01267\left(\frac{x_2}{3}\right) + 8.159\left(\frac{x_1 - 1.25}{2.75}\right)^2 + 9.534\left(\frac{x_2}{3}\right)^2$$
$$- 16.551\left(\frac{x_1 - 1.25}{2.75}\right)\left(\frac{x_2}{3}\right) \qquad \text{(l)}$$
$$= 0.475 - 1.712x_1 + 2.504x_2 + 1.079x_1^2 + 1.0593x_2^2 - 2.006x_1 x_2$$

Equation (l) is the same as Eq. (e), obtained in Example 20.1.

TABLE 20.4 Values for the normalized variables w_{ij} in Example 20.3 for nine sample points ($k = 9$)

Sample point	w_1	w_2	w_3	w_4	w_5
1	−1	−1	1	1	1
2	−1	0	1	0	0
3	−1	1	1	1	−1
4	0	−1	0	1	0
5	0	0	0	0	0
6	0	1	0	1	0
7	1	−1	1	1	−1
8	1	0	1	0	0
9	1	1	1	1	1

20.2 DESIGN OF EXPERIMENTS FOR RESPONSE SURFACE GENERATION

As noted in the preceding section, the first step of the RSM is selection of the sample points. They can be selected arbitrarily or, if possible, based on designer's intuition and experience. Where intuition and experience cannot be used to select sample points, we can use a selection method capable of covering the entire design range. The orthogonal arrays method, which offers a good approach for this selection, is explained here.

An orthogonal array is represented by a two-dimensional matrix. There are a few methods to develop orthogonal arrays. We will use Taguchi's (1987) approach and convention to express them as

$$
L_N(\prod_{i=1}^{n} s_i^{k_i})
\tag{20.20}
$$

where N is the number of rows in the orthogonal array (the number of sample points; also called the *number of experiments*), n is the number of design variable groups with a specific level of values, s_i is the number of levels, and k_i is the number of design variables with s_i levels. *Levels* refers to the number of different values for a design variable or parameter. For example, three levels for a design variable imply three different values for it for specification of sample points.

Table 20.5 shows the $L_{18}(2^1 3^7)$ orthogonal array having 18 rows, one design variable with two levels, and seven design variables with three levels. Each level for a design

TABLE 20.5 The $L_{18}(2^1 3^7)$ orthogonal array showing design variable levels for generating sample points

Sample point	Column							
	x_1	x_2	x_3	x_4	x_5	x_6	x_7	x_8
1	1	1	1	1	1	1	1	1
2	1	1	2	2	2	2	2	2
3	1	1	3	3	3	3	3	3
4	1	2	1	1	2	2	3	3
5	1	2	2	2	3	3	1	1
6	1	2	3	3	1	1	2	2
7	1	3	1	2	1	3	2	3
8	1	3	2	3	2	1	3	1
9	1	3	3	1	3	2	1	2
10	2	1	1	3	3	2	2	1
11	2	1	2	1	1	3	3	2
12	2	1	3	2	2	1	1	3
13	2	2	1	2	3	1	3	2
14	2	2	2	3	1	2	1	3
15	2	2	3	1	2	3	2	1
16	2	3	1	3	2	3	1	2
17	2	3	2	1	3	1	2	3
18	2	3	3	2	1	2	3	1

variable actually refers to a numerical value for it. For example, in the table design variable 1 has two levels, implying that two values for this variable are selected within its acceptable range; levels are presented by the numbers 1 and 2 in column x_1. Similarly, design variables 2 through 8 have three levels; therefore, three values are selected for each of these variables within their acceptable ranges. These are represented by the numbers 1, 2, and 3 in columns x_2 through x_8.

Once the number of levels is decided, each level is assigned a numerical value. Thus, each row of the orthogonal array represents a design point (sample point) at which the problem functions are evaluated. The numbers in each column of the orthogonal array represent the design variable level used for each sample point.

The orthogonal array is named as such because the columns are orthogonal to each other. If the integers 1 and 2 are replaced by the integers -1 and 1 in the column x_1 in Table 20.5, and integers 1, 2, and 3 are replaced by -1, 0, and 1 in columns x_2 through x_8, the dot product of any two columns is 0 (i.e., they are orthogonal). There are no specific rules to generate an orthogonal array. Various orthogonal arrays are being designed by researchers. The designer can choose one from the database of orthogonal arrays that can cover the number of design variables and their desired levels. The rule for choosing the smallest orthogonal array is given in Taguchi (1987) and Park (2007). It should be remembered that the number of rows is the number of sample points and must be equal to or greater than the number of unknowns $(l + 1)$ in the meta-model.

Suppose that we want to approximate a function f by RSM and we have four design variables, each having three specified levels. The design variable values are shown in Table 20.6, where x_{ij} represents the value of the ith design variable at the jth level. An $L_9(3^4)$ orthogonal array, shown in Table 20.7, can be used for selection of sample points for this example, having four design variables each having three levels. The integers in the columns refer to the level number; the number of experiments (in this case 9) is the number of sample points. The integers 1, 2, and 3 refer to the design variable levels 1, 2, and 3, respectively. These numbers can be normalized to be between -1 and 1 using a normalization procedure such as the one in Eq. (20.17). Using that procedure, the integers 1, 2, and 3 are transformed to -1, 0, and 1 as shown in brackets in Table 20.7. Thus the three levels for each design variable can be represented by the integers -1, 0, and 1.

When -1, 0, 1 are used as levels, all of the columns are orthogonal to each other, thus the name orthogonal array. The right most column of Table 20.7 represents the value of

TABLE 20.6 Four design variables, each having three levels

Design variable	Level 1	Level 2	Level 3
x_1	x_{11}	x_{12}	x_{13}
x_2	x_{21}	x_{22}	x_{23}
x_3	x_{31}	x_{32}	x_{33}
x_4	x_{41}	x_{42}	x_{43}

TABLE 20.7 The $L_9(3^4)$ orthogonal array

Experiment no./ Sample point	Design variables and levels				Function value
	x_1	x_2	x_3	x_4	
1	1(−1)	1(−1)	1(−1)	1(−1)	f_1
2	1(−1)	2(0)	2(0)	2(0)	f_2
3	1(−1)	3(1)	3(1)	3(1)	f_3
4	2(0)	1(−1)	2(0)	3(1)	f_4
5	2(0)	2(0)	3(1)	1(−1)	f_5
6	2(0)	3(1)	1(−1)	2(0)	f_6
7	3(1)	1(−1)	3(1)	2(0)	f_7
8	3(1)	2(0)	1(−1)	3(1)	f_8
9	3(1)	3(1)	2(0)	1(−1)	f_9

Note: *Integers 1, 2, and 3 refer to levels 1, 2, and 3 for design variables, respectively. Also normalized integers* −1, 0, *and* 1 *refer to the three levels 1, 2, and 3, respectively.*

the function f for each sample point. We have all of the information in Table 20.7 needed for response surface generation. Now the approximated function $\hat{f}$ in Eq. (20.11) is defined according to the process described in Section 20.1.2. It is noted that alternate methods for selection of sample points can replace the orthogonal arrays method.

EXAMPLE 20.4 GENERATION OF A RESPONSE SURFACE USING AN ORTHOGONAL ARRAY

Generate a response surface for a function with four design variables as follows:

$$f = f(x_1, x_2, x_3, x_4) \tag{a}$$

The sample points are selected using the orthogonal array $L_9(3^4)$ shown in Table 20.7. The levels of the design variables and their numerical values are shown in Table 20.8. The value of function f for each row of the orthogonal array is shown in Table 20.9 on page 760. Generate a quadratic response surface without the cross-product terms.

Solution

From Table 20.9, the maximum and the minimum values for each variable are shown in Table 20.10. Since the cross-product terms are ignored, $n = 4$ and $l = 2n = 8$. Thus, to approximate the function we need at least nine sample points. The approximate function is represented as

$$\hat{f} = c_0 + c_1 w_1 + c_2 w_2 + c_3 w_3 + c_4 w_4 + c_5 w_5 + c_6 w_6 + c_7 w_7 + c_8 w_8 \tag{b}$$

Using Eq. (20.17) and the data given in Table 20.10, the normalized variables w_i are defined as

$$w_1 = \frac{x_1 - 0}{1.5}, \quad w_2 = \frac{x_2 - 0}{3}, \quad w_3 = \frac{x_3 - (-3)}{3}, \quad w_4 = \frac{x_4 - 2.4}{1.2} \tag{c}$$

Using normalization procedure 2, illustrated in Example 20.3, Eq. (20.16) gives

$$
\begin{bmatrix}
9 & 0 & 0 & 0 & 0 & 6 & 6 & 6 & 6 \\
0 & 6 & 0 & 0 & 0 & 0 & 0 & 0 & 0 \\
0 & 0 & 6 & 0 & 0 & 0 & 0 & 0 & 0 \\
0 & 0 & 0 & 6 & 0 & 0 & 0 & 0 & 0 \\
0 & 0 & 0 & 0 & 6 & 0 & 0 & 0 & 0 \\
6 & 0 & 0 & 0 & 0 & 6 & 4 & 4 & 4 \\
6 & 0 & 0 & 0 & 0 & 4 & 6 & 4 & 4 \\
6 & 0 & 0 & 0 & 0 & 4 & 4 & 6 & 4 \\
6 & 0 & 0 & 0 & 0 & 4 & 4 & 4 & 6
\end{bmatrix}
\begin{bmatrix}
c_0 \\ c_1 \\ c_2 \\ c_3 \\ c_4 \\ c_5 \\ c_6 \\ c_7 \\ c_8
\end{bmatrix}
=
\begin{bmatrix}
-245.538 \\
-21.6 \\
-32.216 \\
155.52 \\
-89.584 \\
-160.452 \\
-140.477 \\
-202.572 \\
-166.86
\end{bmatrix}
\tag{d}
$$

The solution for Eq. (d) is

$$
\begin{aligned}
& \begin{bmatrix} c_0 & c_1 & c_2 & c_3 & c_4 & c_5 & c_6 & c_7 & c_8 \end{bmatrix}^T \\
& = \begin{bmatrix} -22.085 & -3.6 & -5.369 & 25.92 & -14.931 & 1.62 & 11.608 & -19.44 & -1.584 \end{bmatrix}^T
\end{aligned}
\tag{e}
$$

Using $w_5 = w_1^2$, $w_6 = w_2^2$, $w_7 = w_3^2$, and $w_8 = w_4^2$ in Eq. (b), the approximated function is

$$
\begin{aligned}
\hat{f} &= -22.085 - 3.6w_1 - 5.369w_2 + 25.92w_3 - 14.931w_4 + 1.62w_5 + 11.608w_6 - 19.44w_7 - 1.584w_8 \\
&= -22.085 - 3.6\left(\frac{x_1 - 0}{1.5}\right) - 5.369\left(\frac{x_2 - 0}{3}\right) + 25.92\left(\frac{x_3 - (-3)}{3}\right) - 14.931\left(\frac{x_4 - 2.4}{1.2}\right) \\
&\quad + 1.62\left(\frac{x_1 - 0}{1.5}\right)^2 + 11.608\left(\frac{x_2 - 0}{3}\right)^2 - 19.44\left(\frac{x_3 - (-3)}{3}\right)^2 - 1.584\left(\frac{x_4 - 2.4}{1.2}\right)^2 \\
&= -7.921 - 2.4x_1 - 1.79x_2 - 4.32x_3 - 7.163x_4 + 0.72x_1^2 + 1.29x_2^2 - 2.16x_3^2 - 1.1x_4^2
\end{aligned}
\tag{f}
$$

We can use the results of the RSM in the optimization process. The cost and constraint functions for the problem are also approximated and optimization is performed with the approximated functions. The following example demonstrates this process.

TABLE 20.8 Four design variables, each having three levels

Design variable	Level		
	1(−1)	2(0)	3(1)
x_1	−1.5	0	1.5
x_2	−3.0	0	3.0
x_3	−6.0	−3.0	0
x_4	1.2	2.4	3.6

TABLE 20.9 Sample points using the $L_9(3^4)$ orthogonal array and function values for Example 20.4

Experiment no.	Design variables and levels				Function value
	x_1	x_2	x_3	x_4	
1	−1.5(−1)	−3(−1)	−6(−1)	1.2(−1)	−31.901
2	−1.5(−1)	0(0)	−3(0)	2.4(0)	−16.865
3	−1.5(−1)	3(1)	0(1)	3.6(1)	−20.661
4	0(0)	−3(−1)	−3(0)	3.6(1)	−21.622
5	0(0)	0(0)	0(1)	1.2(−1)	−2.258
6	0(0)	3(1)	−6(−1)	2.4(0)	−61.206
7	1.5(1)	−3(−1)	0(1)	2.4(0)	−0.608
8	1.5(1)	0(0)	−6(−1)	3.6(1)	−85.939
9	1.5(1)	3(1)	−3(0)	1.2(−1)	−4.479

TABLE 20.10 Data for normalization of variables for Example 20.4

	$x_1(w_1)$	$x_2(w_2)$	$x_3(w_3)$	$x_4(w_4)$	$x_1{}^2(w_5)$	$x_2{}^2(w_6)$	$x_3{}^2(w_7)$	$x_4{}^2(w_8)$
max	1.5	3	0	3.6	2.25	9	36	12.96
min	−1.5	−3	−6	1.2	0	0	0	1.44
$\frac{max + min}{2}$	0	0	−3	2.4	1.125	4.5	18	7.2
$\frac{max - min}{2}$	1.5	3	3	1.2	1.125	4.5	18	5.76

EXAMPLE 20.5   OPTIMIZATION USING RSM

Solve the optimization problem in Example 13.7 by the Response Surface Method with all quadratic terms included, and compare the solution with that in Example 13.7. There are nine sample points using the orthogonal array $L_9(3^4)$, as shown in Table 20.11.

Solution

Using the process of RSM and the data in Table 20.11, the functions are approximated as follows:

Cost function:

$$\hat{f} = c_0 + c_1 w_1 + c_2 w_2 + c_3 w_3 + c_4 w_4 + c_5 w_5$$
$$= 126{,}000 + 18{,}000 w_1 + 10{,}500 w_2 + 1{,}500 w_5$$
$$= 126{,}000 + 18{,}000 \left(\frac{b - 350}{50}\right) + 10{,}500 \left(\frac{d - 360}{30}\right) + 1{,}500 \left(\frac{b - 350}{50}\right)\left(\frac{d - 360}{30}\right) \tag{a}$$
$$= bd$$

Bending stress constraint:

$$\hat{g}_1 = c_{10} + c_{11}w_{11} + c_{12}w_{12} + c_{13}w_{13} + c_{14}w_{14} + c_{15}w_{15}$$
$$= -0.4710 - 0.0782w_{11} - 0.0907w_{12} + 0.0112w_{13} + 0.0113w_{14} + 0.0130w_{15}$$
$$= -0.4710 - 0.0782\left(\frac{b-350}{50}\right) - 0.0907\left(\frac{d-360}{30}\right) + 0.0112\left(\frac{b-350}{50}\right)^2$$
$$+ 0.0113\left(\frac{d-360}{30}\right)^2 + 0.0130\left(\frac{b-350}{50}\right)\left(\frac{d-360}{30}\right)$$
$$= 4.433 - 0.00782b - 0.0151d + 4.48 \times 10^{-6}b^2 + 1.26 \times 10^{-5}d^2 + 8.67 \times 10^{-6}bd \tag{b}$$

Shear stress constraint:

$$\hat{g}_2 = c_{20} + c_{21}w_{21} + c_{22}w_{22} + c_{23}w_{23} + c_{24}w_{24} + c_{25}w_{25}$$
$$= -0.1072 - 0.1308w_{21} - 0.0760w_{22} + 0.0187w_{23} + 0.0063w_{24} + 0.0109w_{25}$$
$$= -0.1072 - 0.1308\left(\frac{b-350}{50}\right) - 0.0760\left(\frac{d-360}{30}\right) + 0.0187\left(\frac{b-350}{50}\right)^2$$
$$+ 0.0063\left(\frac{d-360}{30}\right)^2 + 0.0109\left(\frac{b-350}{50}\right)\left(\frac{d-360}{30}\right)$$
$$= 4.46 - 0.0105b - 0.0101d + 7.48 \times 10^{-6}b^2 + 7.0 \times 10^{-6}d^2 + 7.27 \times 10^{-6}bd \tag{c}$$

Depth constraint:

$$\hat{g}_3 = c_{30} + c_{31}w_{31} + c_{32}w_{32} + c_{33}w_{33} + c_{34}w_{34} + c_{35}w_{35}$$
$$= -3.4 - w_{31} + 0.3w_{32}$$
$$= -3.4 - \left(\frac{b-350}{50}\right) + 0.3\left(\frac{d-360}{30}\right) \tag{d}$$
$$= -0.02b + 0.01d$$

The optimum design problem with functions approximated by RSM is now defined as

Cost function:

$$f = bd \tag{e}$$

Bending stress constraint:

$$\hat{g}_1 = 4.433 - 0.00782b - 0.0151d + 4.48 \times 10^{-6}b^2 + 1.26 \times 10^{-5}d^2 + 8.67 \times 10^{-6}bd \le 0 \tag{f}$$

Shear stress constraint:

$$\hat{g}_2 = 4.46 - 0.0105b - 0.0101d + 7.48 \times 10^{-6}b^2 + 7.0 \times 10^{-6}d^2 + 7.27 \times 10^{-6}bd \le 0 \tag{g}$$

Depth constraint:

$$\hat{g}_3 = -0.02b + 0.01d \le 0 \tag{h}$$

Since the bound constraints on the design variables are linear functions, the functions in Eq. (d) of Example 13.7 are directly utilized. Optimization is carried out with the approximated functions in Eqs. (e) through (h) and the bound constraints. The optimum values from RSM are shown and compared with those of Example 13.7 in Table 20.12. The values in that table are those from the approximated functions. The constraints are satisfied when the approximated functions are used in the optimization process. However, they can be violated when the original constraint functions are evaluated. We should be careful on this aspect when RSM is utilized in the optimization process. The original functions given in Example 13.7 with the optimum solution from RSM in Table 20.12 are calculated as follows:

$$\hat{f} = 1.109 \times 10^5, \quad g_1 = -0.076 \text{ (satisfied)}, \quad g_2 = 0.014 \text{ (violated)}, \quad g_3 = -7.143 \text{ (satisfied)} \qquad \text{(i)}$$

When RSM gives a better objective function value compared to the true optimum as shown in Eq. (i), the original constraints are usually violated. For the present example, the constraint violation is acceptable (1.4%). When the violations are larger, the optimization process with RSM should be repeated by adding more sample points to get better approximate model, modification of the design bounds, and so forth.

As shown in Table 20.12, when RSM is used, the number of calls for function evaluations is the same as the number of sample points, and we do not need the information for the gradients of the original functions. Therefore, RSM can be utilized when the gradient information for the original functions is not available or is expensive to evaluate. The optimum point may not satisfy the original constraints because it is obtained using the approximated functions.

When the ranges for the design variables are large, we may not be able to obtain a precise optimum. In that case, more experiments are needed to obtain better approximations. Generally, there is a limit on the number of design variables when using the Response Surface Method. While not mathematically proved, the approximation deteriorates when the number of design variables is larger than 10.

TABLE 20.11 Sample points and function values for Example 20.5

Sample point	$b(w_1)$	$d(w_2)$	Empty col.	Empty col.	Cost f	Bending g_1	Shear g_2	Depth g_3
1	300(−1)	330(−1)	−	−	99,000	−0.2654	0.1364	−0.45
2	300(−1)	360(0)	−	−	108,000	−0.3827	0.0417	−0.4
3	300(−1)	390(1)	−	−	117,000	−0.4740	−0.0385	−0.35
4	350(0)	330(−1)	−	−	115,500	−0.3703	−0.0260	−0.5286
5	350(0)	360(0)	−	−	126,000	−0.4709	−0.1071	−0.4857
6	350(0)	390(1)	−	−	136,500	−0.5492	−0.1758	−0.4429
7	400(1)	330(−1)	−	−	132,000	−0.4490	−0.1477	−0.5875
8	400(1)	360(0)	−	−	144,000	−0.5370	−0.2188	−0.55
9	400(1)	390(1)	−	−	156,000	−0.6055	−0.2788	−0.5125

TABLE 20.12 Optimum solution for Example 20.5

	RSM solution	Results of Example 13.7
Optimum point	(474.1, 234.0)	(335.4, 335.4)
Optimum area	1.109×10^5	1.125×10^5
Number of calls for function evaluations	9	6
Total number of constraint gradients evaluated		12

20.3 DISCRETE DESIGN WITH ORTHOGONAL ARRAYS

In the previous section, orthogonal arrays were utilized to select sample points for generating response surfaces to be used in the optimization process. Orthogonal arrays can be used directly for determination of design variables in the discrete design space, i.e., design variables that must have discrete values. In this section, we discuss how to use orthogonal arrays to determine a discrete design from the specified discrete values.

We consider the unconstrained optimization problem of minimizing $f(\mathbf{x})$. The constrained problem can be treated using the penalty function approach, as was explained in Chapter 11. To demonstrate the procedure, consider a problem having four design variables. Let the possible discrete values for the variables be as shown earlier in Table 20.6; that is, we have four design variables, each having three levels (three discrete values). We can use each combination of discrete design variable values and evaluate the cost function there. Then we select the point with the smallest cost function value as the optimum point. However, this full enumeration requires evaluation of the cost function at 3^4 design points. Thus the process is usually quite expensive because, as the number of design variables and/or the number of discrete values for each design variable increases, the enumeration grows very rapidly.

It turns out that we can use an orthogonal array and the function value at the sample points to obtain the discrete optimum point. This means that only a partial enumeration is needed with the use of orthogonal arrays. For the case of four design variables, we can use the $L_9(3^4)$ orthogonal array as shown earlier in Table 20.7. There the cost function is evaluated at nine sample points. With this procedure, then, only nine combinations of design variables are used in determining the discrete solution, compared with the full enumeration of $3^4 = 81$ combinations.

For the purpose of demonstrating this procedure, let the numerical values of the cost function for the nine sample points in Table 20.7 be given as shown in Table 20.13 (Park, 2007). The means of various combinations of these nine cost function values are used to determine the discrete solution point. This process is accordingly called "analysis of means" (ANOM). It is a statistical technique used in illustrating variations among groups of data. It compares the mean of each group to the overall mean to detect statistically significant differences. The approach has been applied commonly in quality control of products and processes.

TABLE 20.13 Function value for each experiment (sample point)

Experiment number	f_1	f_2	f_3	f_4	f_5	f_6	f_7	f_8	f_9
Function value	20	50	30	25	45	30	45	65	70

The procedure for discrete variable design, developed by Taguchi (1987), uses the additive model for approximating the function. It is assumed that the function values beyond what are available in the orthogonal array cannot be calculated because the calculation is too tedious or expensive, or new experiments cannot be performed. Therefore, the values need to be approximated somehow. The additive model assumes that the function values can be approximated by adding to the mean of all function values the deviation from the mean caused by setting a design variable to a particular level. These deviations can be calculated as the means of different groups of the known cost function values. In the additive model, cross-product terms involving two or more design variables are not allowed. More details on the additive model can be found in Taguchi (1987) and Phadke (1989).

First, the mean of the cost function for the nine values is calculated as

$$\mu = \frac{1}{9}\sum_{i=1}^{9} f_i = \frac{1}{9}(20 + 50 + 30 + 25 + 45 + 30 + 45 + 65 + 70) = 42.2 \qquad (20.21)$$

This mean is used to evaluate the effect of various design variables on the cost function. To investigate the effect of the various levels (discrete values) of each design variable on the cost function value, we calculate several cost function value means. To do this, we define the following notation for various mean values:

μ_{ij} = mean of cost functions calculated using sample design points containing the ith design variable and its jth level $\qquad (20.22)$

For example, μ_{13} is the mean of the cost function values calculated using sample points containing design variable x_1 and its level 3 numerical value x_{13}. Referring to Tables 20.7 and 20.13, μ_{13} is calculated as

$$\mu_{13} = \frac{1}{3}(f_7 + f_8 + f_9) \qquad (20.23)$$

Therefore, the effect of level 3 of variable x_1 on the cost function is $(\mu_{13} - \mu)$. This is called the deviation from the mean μ caused by setting design variable x_1 to the level 3 value. In a similar manner, μ_{11} and μ_{12} are calculated using the data in Tables 20.7 and 20.13, as

$$\mu_{11} = \frac{1}{3}(f_1 + f_2 + f_3) \qquad (20.24)$$

$$\mu_{12} = \frac{1}{3}(f_4 + f_5 + f_6) \qquad (20.25)$$

The effects of levels 1 and 2 of design variable x_1 on the cost function are $(\mu_{13} - \mu)$, $(\mu_{11} - \mu)$, and $(\mu_{12} - \mu)$, respectively. The deviations $(\mu_{11} - \mu)$, $(\mu_{12} - \mu)$, and $(\mu_{13} - \mu)$ add to 0; that is, for each design variable, the effects of all of its levels satisfy the following equation:

$$(\mu_{i1} - \mu) + (\mu_{i2} - \mu) + (\mu_{i3} - \mu) = \sum_{j=1}^{3} ((\mu_{ij} - \mu)) = 0 \tag{20.26}$$

This is a property of the *additive model*.

The means of various levels (discrete values) of design variables x_2, x_3, and x_4 are calculated in a similar manner to that in Eqs. (20.23) through (20.25). Results derived using the data given in Tables 20.7 and 20.13 are shown in Table 20.14, which is called a one-way table. It can be verified that the one-way table satisfies the property of the additive model of Eq. (20.26):

$$i = 1: \quad (33.3 - 42.2) + (33.3 - 42.2) + (60 - 42.2) = 0 \tag{20.27}$$

$$i = 2: \quad (30 - 42.2) + (53.3 - 42.2) + (43.3 - 42.2) = 0 \tag{20.28}$$

$$i = 3: \quad (38.3 - 42.2) + (48.3 - 42.2) + (40 - 42.2) = 0 \tag{20.29}$$

$$i = 4: \quad (45 - 42.2) + (41.6 - 42.2) + (40 - 42.2) = 0 \tag{20.30}$$

For each design variable, the level that gives the least value for the mean of f in the one-way table is considered as its final value because of the additive model assumption.

For the cost function values shown in Table 20.13, the results from the calculations in Table 20.14 are plotted in Figure 20.3. The figure shows, along the vertical axis, the means of the cost functions for each design variable and its three levels. From the figure, we note the level that gives the least value for the mean for each design variable. For x_1, x_{11}, or x_{12} gives the lowest mean; for x_2, it is x_{21}; for x_3 it is x_{31}; and for x_4 it is x_{43}. Therefore, the optimum design variable values are $x_{11}x_{21}x_{31}x_{43}$ or $x_{12}x_{21}x_{31}x_{43}$.

The ANOM process is based on the additive model for the function f which has some inherent errors. Therefore, a confirmation experiment is required with the optimum design variable values, which were obtained with the above procedure to verify the final solution. That is, the cost function is evaluated at the optimum levels of the design variables. The result of this confirmation experiment is compared with the function values listed earlier in Table 20.13, which are calculated for the cases in the orthogonal array in Table 20.7. If we use the $L_9(3^4)$ orthogonal array, we have 10 cases from which the best solution can be selected.

TABLE 20.14 Example of the one-way table for an orthogonal array

Design variable	Level 1	2	3
x_1	$\mu_{11} = \frac{f_1 + f_2 + f_3}{3} = 33.3$	$\mu_{12} = \frac{f_4 + f_5 + f_6}{3} = 33.3$	$\mu_{13} = \frac{f_7 + f_8 + f_9}{3} = 60$
x_2	$\mu_{21} = \frac{f_1 + f_4 + f_7}{3} = 30$	$\mu_{22} = \frac{f_2 + f_5 + f_8}{3} = 53.3$	$\mu_{23} = \frac{f_3 + f_6 + f_9}{3} = 43.3$
x_3	$\mu_{31} = \frac{f_1 + f_6 + f_8}{3} = 38.3$	$\mu_{32} = \frac{f_2 + f_4 + f_9}{3} = 48.3$	$\mu_{33} = \frac{f_3 + f_5 + f_7}{3} = 40$
x_4	$\mu_{41} = \frac{f_1 + f_5 + f_9}{3} = 45$	$\mu_{42} = \frac{f_2 + f_6 + f_7}{3} = 41.6$	$\mu_{43} = \frac{f_3 + f_4 + f_8}{3} = 40$

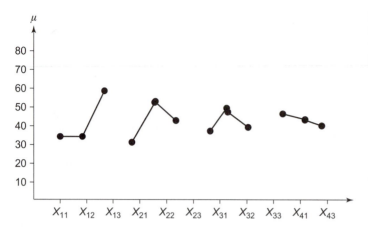

FIGURE 20.3 Graphical representation of the one-way table of mean values in Table 20.14. *Source: Adapted from Park, 2007.*

The ANOM method is illustrated for an unconstrained optimization problem in the foregoing example. However, most optimization problems have constraints on design variables. To use the method for such problems, we adopt the penalty function approach to convert the constrained problem to an unconstrained problem, and then use the foregoing procedure. To transform the constrained problem to an unconstrained one, we define an augmented function $\Phi(\mathbf{x})$ as

$$\Phi(\mathbf{x}) = f(\mathbf{x}) + P(\mathbf{h}(\mathbf{x}),\ \mathbf{g}(\mathbf{x}),\ R) \tag{20.31}$$

where $P(\mathbf{h}(\mathbf{x}), \mathbf{g}(\mathbf{x}), R)$ is a penalty function that depends on the equality and inequality constraint functions $\mathbf{h}(\mathbf{x})$ and $\mathbf{g}(\mathbf{x})$ and a penalty parameter $R > 0$.

The penalty function $P(\mathbf{h}(\mathbf{x}), \mathbf{g}(\mathbf{x}), R)$ adds a penalty for violation of constraints to the cost function. It can be defined in several different ways, for example, as in Eq. (11.60) and Eq. (12.30). Using Eq. (12.30), we define the exact penalty function as

$$P(\mathbf{h}(\mathbf{x}), \mathbf{g}(\mathbf{x}),\ R) = RV(\mathbf{x}) \tag{20.32}$$

where $V(\mathbf{x})$ is the maximum constraint violation defined as

$$V(\mathbf{x}) = \max\big\{0;\ |h_i|,\ \ i = 1\ \ \text{to}\ p;\ \ g_j,\ \ j = 1\ \text{to}\ m\big\} \tag{20.33}$$

The confirmation experiment should be carried out for constrained problems as well. The final solution is the best feasible solution among the cases of the orthogonal array and the confirmation experiment.

EXAMPLE 20.6　DISCRETE DESIGN WITH AN ORTHOGONAL ARRAY

Solve the optimization problem in Example 13.7 in a discrete space using the $L_9(3^4)$ orthogonal array. Table 20.11 shows nine experimental results using the orthogonal array $L_9(3^4)$. Find the best discrete design out of all of the cases, including the confirmation experiment, and compare the results with those of Example 13.7. Let the penalty parameter R be 1,000,000.

Solution

For each row of the orthogonal array, the problem functions are evaluated that are shown in Table 20.15 (since this example has only two design variables, columns 3 and 4 in the table are empty). The one-way table using the data from Table 20.15 is shown in Table 20.16. For example, μ_{b1} is calculated as $\dfrac{235,364 + 149,667 + 117,000}{3} = 167,344$. From the one-way table, the levels that have the minimum mean value are level 2 for b and level 3 for d. Thus, the solution is $b = 350$, $d = 390$, and $f = 136,500$; all constraints are satisfied.

This solution is compared to all cases of the orthogonal array in Table 20.15. It is seen that the fourth case in Table 20.15 gives a better solution. Therefore, the final solution is $b = 350$, $d = 330$, and $f = 115,500$. We note that this solution is not as good as the one in Example 13.7. This is because the current solution is obtained in a discrete design space while Example 13.7 solves the problem in a continuous design space.

TABLE 20.15 Sample points and function values for Example 20.6

Experiment no.	$b(w_1)$	$d(w_2)$	Empty col.	Empty col.	Cost f	Maximum violation of constraints	P	Φ
1	300(−1)	330(−1)	−	−	99,000	0.1364	136,364	235,364
2	300(−1)	360(0)	−	−	108,000	0.0417	41,667	149,667
3	300(−1)	390(1)	−	−	117,000	0	0	117,000
4	350(0)	330(−1)	−	−	115,500	0	0	115,500
5	350 (0)	360(0)	−	−	126,000	0	0	126,000
6	350 (0)	390(1)	−	−	136,500	0	0	136,500
7	400(1)	330(−1)	−	−	132,000	0	0	132,000
8	400(1)	360(0)	−	−	144,000	0	0	144,000
9	400(1)	390(1)	−	−	156,000	0	0	156,000

TABLE 20.16 One-way table for Example 20.6

Design variable	Level		
	1	2	3
b	$\mu_{b1} = 167,344$	$\mu_{b2} = 126,000$	$\mu_{b3} = 144,000$
d	$\mu_{d1} = 160,955$	$\mu_{d2} = 139,889$	$\mu_{d3} = 136,500$

20.4 ROBUST DESIGN APPROACH

What is a robust design?

Taguchi (1987), a pioneer of the robust design approach, defined *robustness* as follows: "robustness is the state where the technology, product, or process performance is minimally sensitive to factors causing variability (either in manufacturing or in the user's environment) and aging at the lowest unit manufacturing cost." This concept of *robustness* has been developed to improve product quality and reliability, and manufacturing processes in industrial engineering. It can account for noise factors, such as environmental variation during a product's use, manufacturing variation, and product deterioration. It has also been extended and applied to all kinds of design situations. This section presents an introduction to the robust design approach. Detailed reviews of the approach have been presented by Park et al. (2006) and Beyer and Sandhoff (2007).

The design process always has some uncertainties in the design variables and/or the problem parameters. The problem parameters are the ones that are considered constants in the design process. Examples of problem parameters in structural design are external forces, material properties, temperature, length of members, dimensions of parts, member support conditions, and so forth. Uncertainties in the final design are introduced from tolerances on the design variables and noise in the problem parameters.

These uncertainties may be known or unknown, and the question is how to treat them in the design process. Designers always want to have steady performance for their design even though the uncertainties noted above exist. In other words, the performance of the designed product should be robust (insensitive) with respect to the uncertainties. The robust design approach attempts to accomplish this objective.

A robust design is relatively insensitive to variations in the problem-related parameters and variables. The procedure to find designs that are insensitive to parameter variations is called the robust design method.

When the robust design approach is related to the optimization process, it is called robust design optimization. The optimization methods and the problem formulations discussed in Chapters 8 through 13 are called deterministic approaches; that is, uncertainties of any parameters are not considered. In robust optimization, the effect of uncertainties in the problem parameters is incorporated into the formulation of the problem using the mean and variance of various data and functions.

Robustness is usually defined for the cost function, although robustness of the constraints can be considered as well. Two main approaches for robust design are presented in this section: (1) robust optimization, which uses conventional optimization algorithms, and (2) the Taguchi method.

20.4.1 Robust Optimization

Mean

Mean is defined as the simple average that is the sum of all values divided by the number of values. It is also called the *average value* or the *expected value*. The standard symbol for the mean is μ. Let a function f be observed (evaluated) at l points as $f_1, f_2, \ldots, f_l$.

The *mean* μ_f of these function values is the average calculated as the sum of all of the observations divided by the number of observations:

$$\mu_f = \frac{\sum\limits_{i=1}^{l} f_i}{l} \tag{20.34}$$

Note that the mean can be positive, negative, or zero.

Variance

Another important statistical property is *variance*, which is defined as the average of the squared difference from the mean (i.e., $(f_i - \mu_f)^2$), usually written as σ^2. Squaring the differences makes them all positive numbers. It also makes the larger ones stand out. For the above example of l observations of function f, σ_f^2 for the function values is calculated as

$$\sigma_f^2 = \frac{\sum\limits_{i=1}^{l} (f_i - \mu_f)^2}{l} \tag{20.35}$$

Variance represents dispersion of the data from its mean value. When data are extracted as samples from a large population, the degree of freedom becomes $(l-1)$. In that case, l in Eq. (20.35) is replaced by $(l-1)$.

Standard Deviation

The standard deviation is defined as the square root of the variance; that is, σ_f. Thus, it is also a measure of the variability or dispersion of the data from the average (mean or expected value). Its standard symbol is σ. Standard deviation is a more meaningful measure of the dispersion of the data and is most commonly used as such. Along with the mean, it gives us a standard way of knowing what is normal—that is, any data falling within the standard deviation of the mean. A smaller standard deviation indicates that the data points tend to be very close to the mean, whereas a larger standard deviation indicates that the data are spread out over a larger band around it.

If X is a random variable with mean value μ, then it is expressed as

$$E[X] = \mu \tag{20.36}$$

where the operator $E[X]$ denotes the average or expected value of X. The standard deviation of X is given as

$$\sigma = \sqrt{E[(X - \mu)^2]} \tag{20.37}$$

The standard deviation is the square root of the variance of X; that is, it is the square root of the average value or expected value of $(X - \mu)^2$.

Probability Density Function

A continuous random variable X takes on various values x within the range $-\infty < x < \infty$. *A random variable is usually expressed by an uppercase letter, while its particular value is denoted by a lowercase letter.* A mathematical function that describes the distribution of a continuous random variable is called the *probability density function* (PDF) and is designated as $f_X(x)$. That is, it is a function that describes the relative likelihood (probability) for this random variable to occur at a given point x. The probability density function is nonnegative everywhere, and its integral over the entire space is equal to one:

$$f_X(\mathbf{x}) \geq 0 \tag{20.38}$$

$$\int_{-\infty}^{\infty} f_X(x)dx = 1 \tag{20.39}$$

Problem Definition

Since most engineering design problems do not usually involve equality constraints, we consider the inequality-constrained optimization problem:

Minimize

$$f(\mathbf{x}) \tag{20.40}$$

subject to

$$g_j(\mathbf{x}) \leq 0; \quad j = 1 \text{ to } m \tag{20.41}$$

Design variables are changed during the optimization process while any problem parameters are kept fixed. Since there can be uncertainties in the design variables and problem parameters, the problem definition needs to be modified. The cost and constraint functions in Eqs. (20.40) and (20.41) are redefined to include uncertainties:

$$f(\mathbf{x}, \mathbf{y}) \rightarrow f(\mathbf{x} + \mathbf{z}^\mathbf{x}, \mathbf{y} + \mathbf{z}^\mathbf{y}) \tag{20.42}$$

$$g(\mathbf{x}, \mathbf{y}) \rightarrow g(\mathbf{x} + \mathbf{z}^\mathbf{x}, \mathbf{y} + \mathbf{z}^\mathbf{y}) \tag{20.43}$$

where $\mathbf{y} = (y_1, y_2, \ldots, y_r)$ is the problem parameter vector, and $\mathbf{z}^\mathbf{x} = (z^{x_1}, z^{x_2}, \ldots, z^{x_n})$ and $\mathbf{z}^\mathbf{y} = (z^{y_1}, z^{y_2}, \ldots, z^{y_r})$ are uncertainties in the design variable vector and the problem parameter vector, respectively.

The uncertainties can be interpreted as perturbations or noise in the variables and are treated as random variables. In robust optimization, the optimization problem in Eqs. (20.40) and (20.41) is changed to incorporate the perturbations as follows: Find an n-vector $\mathbf{x} = (x_1, x_2, \ldots, x_n)$ of design variables to

minimize a cost function

$$F(\mathbf{x}, \mathbf{y}, \mathbf{z}^\mathbf{x}, \mathbf{z}^\mathbf{y}) = F(f(\mathbf{x} + \mathbf{z}^\mathbf{x}, \mathbf{y} + \mathbf{z}^\mathbf{y}))$$
$$= F(f(x_1 + z^{x_1}, x_2 + z^{x_2}, \ldots, x_n + z^{x_n}; \ y_1 + z^{y_1}, y_2 + z^{y_2}, \ldots, y_r + z^{y_r})) \tag{20.44}$$

subject to the m inequality constraints

$$G_j(\mathbf{x}, \mathbf{y}, \mathbf{z}^\mathbf{x}, \mathbf{z}^\mathbf{y}) = G_j(g_j(\mathbf{x} + \mathbf{z}^\mathbf{x}, \mathbf{y} + \mathbf{z}^\mathbf{y}))$$
$$= G_j(g_j(x_1 + z^{x_1}, x_2 + z^{x_2}, \ldots, x_n + z^{x_n}; \tag{20.45}$$
$$y_1 + z^{y_1}, y_2 + z^{y_2}, \ldots, y_r + z^{y_r})) \leq 0$$

where functions F and G_j are derived by considering noise in the functions f and g_j, respectively. Usually functions F and G_j are defined using the mean and variance of the functions f and g_j, respectively.

When the uncertainties in the variables are given by the probability density functions, the mean μ_f and the variance σ_f^2 of f given in Eq. (20.42) are calculated (Phadke, 1989) as

$$\mu_f = E[f(\mathbf{x}, \mathbf{y})] = \int \int \cdots \int f(\mathbf{x} + \mathbf{z}^{\mathbf{x}}, \mathbf{y} + \mathbf{z}^{\mathbf{y}})$$
$$\times u_1(z^{x_1}) \cdots u_n(z^{x_n}) v_1(z^{y_1}) \cdots v_r(z^{y_r}) dz^{x_1} \cdots dz^{x_n} dz^{y_1} \cdots dz^{y_r} \tag{20.46}$$

$$\sigma_f^2 = E[(f(\mathbf{x}, \mathbf{y}) - \mu_f)^2] = \int \int \cdots \int [f(\mathbf{x} + \mathbf{z}^{\mathbf{x}}, \mathbf{y} + \mathbf{z}^{\mathbf{y}}) - \mu_f]^2$$
$$\times u_1(z^{x_1}) \cdots u_n(z^{x_n}) v_1(z^{y_1}) \cdots v_r(z^{y_r}) dz^{x_1} \cdots dz^{x_n} dz^{y_1} \cdots dz^{y_r} \tag{20.47}$$

where $E[b]$ is the expected value of b and $u_i(z^{x_i})$ and $v_i(z^{y_i})$ are the probability density functions of the uncertainties z^{x_i} and z^{y_i}, respectively. In Eqs. (20.46) and (20.47), the uncertainties are assumed to be statistically independent. If they follow the Gaussian (normal) distribution, the probability density function $u_i(z^{x_i})$ is given as

$$u_i(z^{x_i}) = \frac{1}{\sigma_{x_i}\sqrt{2\pi}} \exp\left[\frac{-(x_i - \mu_{x_i})^2}{2\sigma_{x_i}^2}\right] \tag{20.48}$$

where μ_{x_i} and σ_{x_i} are the mean and the standard deviation of the ith design variable x_i. The probability density function $v_i(z^{y_i})$ is defined in the same manner as Eq. (20.48).

The robust optimization problem in Eqs. (20.44) and (20.45) is defined using the mean and the variance in Eqs. (20.46) and (20.47). Robust optimization tries to reduce the dispersion of the cost function with respect to the uncertainties because the dispersion is equivalent to the sensitivity. This implies that standard deviation of the cost function should be minimized. Since the mean of the cost function should be simultaneously minimized, this becomes a two-objective optimization problem. Using a weighted sum method (refer to Section 17.4), the cost function of Eq. (20.44) for robust design optimization is defined as

$$F = w_1\mu_f + w_2\sigma_f \tag{20.49}$$

where w_1 and w_2 are the weighting coefficients. If w_1 is larger, minimization of the cost function is emphasized more than obtaining a robust design and vice versa. If a different method for multi-objective optimization is used, Eq. (20.49) is modified according to that method.

The constraint in Eq. (20.45) should be defined so that the original constraint is satisfied even though uncertainties exist. To sufficiently satisfy the constraint, the constraint in Eq. (20.45) is defined as

$$G_j \equiv \mu_{g_j} + k\sigma_{g_j}^2 \leq 0 \tag{20.50}$$

where $k > 0$ is a user-defined constant depending on the design purpose and $\sigma_{g_j}^2$ denotes the dispersion of data for the constraint g_j. If the uncertainties are bounded by some upper and lower limits, the worst case of g_j can be considered as follows:

$$G_j = \mu_{g_j} + k^{x_j} \sum_{i=1}^{n} \left| \frac{\partial g_j}{\partial x_i} \right| |z^{x_i}| + k^{y_j} \sum_{i=1}^{r} \left| \frac{\partial g_j}{\partial y_i} \right| |z^{y_i}| \tag{20.51}$$

where $|z^{x_i}|$ and $|z^{y_i}|$ denote the maximum values of the uncertainty (tolerance) ranges, and $k^{x_j} > 0$ and $k^{y_j} > 0$ are the user-defined constants. Equation (20.51) is obtained by writing a linear Taylor expansion of G_j and using the absolute values for the quantities in the second and third terms to obtain the worst case.

If we have distribution for the design variables, then we have distribution for the cost function f as well, so we can calculate the mean of f by integration in Eq. (20.46). This calculation, however, is quite costly. Therefore, the mean and the variance of the cost function are approximated as

$$\mu_f \cong f(\mu_x, \mu_y) \tag{20.52}$$

$$\sigma_f^2 \cong \sum_{i=1}^{n} \left(\frac{\partial f}{\partial x_i} \right)^2 \sigma_{x_i}^2 + \sum_{i=1}^{r} \left(\frac{\partial f}{\partial y_i} \right)^2 \sigma_{y_i}^2 \tag{20.53}$$

where μ_x is a vector of means of the design variable vector x and μ_y is a vector of means for the problem parameter vector y. These are calculated using their corresponding probability density functions. The mean and the variance for the constraint functions are defined in a similar manner.

Equation (20.53) is derived using a Taylor series for f. The first-order Taylor series expansion of the cost function $f(x, y)$ at the points μ_x and μ_y is defined as follows:

$$f(x, y) \cong f(\mu_x, \mu_y) + \sum_{i=1}^{n} \left(\frac{\partial f}{\partial x_i} \right)(x_i - \mu_{x_i}) + \sum_{i=1}^{r} \left(\frac{\partial f}{\partial y_i} \right)(y_i - \mu_{y_i}) \tag{20.54}$$

If all random variables and parameters are statistically uncorrelated, the variance of the cost function can be approximated as follows:

$$Var[f(x, y)] = \sigma_f^2 \cong Var\left[f(\mu_x, \mu_y) + \sum_{i=1}^{n} \left(\frac{\partial f}{\partial x_i} \right)(x_i - \mu_{x_i}) + \sum_{i=1}^{r} \left(\frac{\partial f}{\partial y_i} \right)(y_i - \mu_{y_i}) \right]$$

$$= Var\left[f(\mu_x, \mu_y) \right] + Var\left[\sum_{i=1}^{n} \left(\frac{\partial f}{\partial x_i} \right)(x_i - \mu_{x_i}) \right] + Var\left[\sum_{i=1}^{r} \left(\frac{\partial f}{\partial y_i} \right)(y_i - \mu_{y_i}) \right]$$

$$= 0 + \sum_{i=1}^{n} \left(\frac{\partial f}{\partial x_i} \right)^2 Var[x_i] + \sum_{i=1}^{r} \left(\frac{\partial f}{\partial y_i} \right)^2 Var[y_i]$$

$$= \sum_{i=1}^{n} \left(\frac{\partial f}{\partial x_i} \right)^2 \sigma_{x_i}^2 + \sum_{i=1}^{r} \left(\frac{\partial f}{\partial y_i} \right)^2 \sigma_{y_i}^2 \tag{20.55}$$

In robust optimization, first-order derivatives of the cost function f are included in the variances; therefore, we need second-order derivatives of f in the optimization process if a gradient-based method is used. The calculation of the second-order derivatives can be quite expensive, especially for a large-scale problem. Therefore, other methods have been used to avoid this calculation, such as the direct search methods discussed in Chapter 11 or the nature-inspired methods discussed in Chapter 19.

Equation (20.44) or (20.49) is called the robustness index. The robustness index can be defined to have other forms. Also, Eq. (20.45) or (20.50) can be defined differently according to the design purpose.

To transform an optimization problem into a robust optimization problem, probability density functions of uncertainties must be known or assumed. The robust optimization problem is then defined in terms of the mean and variance of the cost and constraint functions.

EXAMPLE 20.7 ROBUST OPTIMIZATION

Solve a robust optimization problem formulated as

Minimize

$$f = x_1 x_2 \, \cos x_1 + x_1^2 - \frac{1}{4} x_2^2 - e^{x_2} \tag{a}$$

subject to the constraints

$$g_1(\mathbf{x}) = (x_1 - 1)^2 + x_2^2 - x_1 - 6 \le 0 \tag{b}$$

$$g_2(\mathbf{x}) = \tfrac{3}{7} x_1^2 - \frac{1}{10} x_2 + (x_2 - 1)^2 - 5 \le 0 \tag{c}$$

$$-2.0 \le x_1 \le 2.0; \quad -2.0 \le x_2 \le 2.0 \tag{d}$$

The design variables x_1 and x_2 have the normal distribution with $\sigma_{x_1} = \sigma_{x_2} = 0.1$. The maximum tolerances $|z^{x_1}| = |z^{x_2}| = 0.3$. Use Eqs. (20.49) and (20.51) with $w_1 = w_2 = 0.5$ and $k^{x_1} = k^{x_2} = 0.5$. Use (0.4, 0.4) for the initial design.

Solution

For this problem, the standard deviations for design variables x_1 and x_2 are given and the uncertainties are bounded as ± 0.3. Using Eqs. (20.52) and (20.53), the mean μ_f and the standard deviation σ_f of f are calculated as follows:

$$\mu_f = x_1 x_2 \, \cos x_1 + x_1^2 - \frac{1}{4} x_2^2 - e^{x_2} \tag{e}$$

$$\sigma_f = \sqrt{\left(\frac{\partial f}{\partial x_1}\right)^2 \sigma_{x_1}^2 + \left(\frac{\partial f}{\partial x_2}\right)^2 \sigma_{x_2}^2}$$
$$= \sqrt{(-x_1 x_2 \, \sin x_1 + x_2 \, \cos x_1 + 2x_1)^2 \sigma_{x_1}^2 + (x_1 \, \cos x_1 - 0.5 x_2 - e^{x_2})^2 \sigma_{x_2}^2} \tag{f}$$

Due to the approximation in Eq. (20.52), the expression for the mean of the cost function is the same as that for the original cost function. A normalization process is used to define the cost function in Eq. (20.49) because μ_f and σ_f have different orders of magnitude. Let μ_f^* and σ_f^* represent μ_f and σ_f at the initial point (0.4, 0.4), respectively. Therefore, from Eqs. (e) and (f), we have

$$\mu_f^* = (0.4)(0.4)\cos 0.4 + 0.4^2 - 0.25(0.4^2) - e^{0.4} = -1.224 \tag{g}$$

$$\sigma_f^* = \sqrt{\{(-0.4)(0.4)\sin 0.4 + 0.4\cos 0.4 + (2)(0.4)\}^2(0.1^2) + \{0.4\cos 0.4 - (0.5)(0.4) - e^{0.4}\}^2(0.1^2)} \tag{h}$$
$$= 0.172$$

Thus the multi-objective function $F = w_1\mu_f + w_2\sigma_f$ is normalized as follows:

$$F = w_1\frac{\mu_f}{|\mu_f^*|} + w_2\frac{\sigma_f}{\sigma_f^*} = (0.5)\frac{x_1x_2\cos x_1 + x_1^2 - \frac{1}{4}x_2^2 - e^{x_2}}{|-1.224|}$$
$$+ (0.5)\frac{\sqrt{(-x_1x_2\sin x_1 + x_2\cos x_1 + 2x_1)^2\sigma_{x_1}^2 + (x_1\cos x_1 - 0.5x_2 - e^{x_2})^2\sigma_{x_2}^2}}{0.172} \tag{i}$$

The constraints considering robustness are given using Eq. (20.51):

$$G_1 = \mu_{g_1} + k^{x_j}\sum_{i=1}^{2}\left|\frac{\partial g_1}{\partial x_i}\right||z^{x_i}| = (x_1 - 1)^2 + x_2^2 - x_1 - 6 + (0.5)\left\{\left|2(x_1 - 1) - 1\right|(0.3) + |2x_2|0.3\right\} \tag{j}$$

$$G_2 = \mu_{g_2} + k^{x_j}\sum_{i=1}^{2}\left|\frac{\partial g_2}{\partial x_i}\right||z^{x_i}| = \frac{3}{7}x_1^2 - \frac{1}{10}x_2 + (x_2 - 1)^2 - 5$$
$$+ (0.5)\left\{\left|\frac{6}{7}x_1\right|(0.3) + \left|2(x_2 - 1) - \frac{1}{10}\right|(0.3)\right\} \tag{k}$$

Equations (i) through (k) are used in the optimization process. The optimization results are shown in Table 20.17. We see that the deterministic optimum has a better μ_f while the robust optimum shows a significantly reduced σ_f.

TABLE 20.17   Robust optimization solution for Example 20.7

	Initial point	Deterministic optimum	Robust optimum
(x_1, x_2)	(0.4,0.4)	(−0.303, 2.0)	(0.265, −0.593)
μ_f	−1.224	−8.875	−0.722
σ_f	0.172	0.875	3.01E-06
f	−1.224	−8.875	−0.722
g_1	−5.88	8.1E-04	−5.373
g_2	−4.611	−4.16	−2.373

20.4.2 The Taguchi method

The Taguchi method (1987) was developed for quality improvement of products and processes. Initially, it was applied to process rather than product design, and it did not use formal optimization methods. Also, robustness of only the cost function was considered. Recently, the method has been extended to product design problems as well. In this section, the Taguchi method is explained from the viewpoint of robust design because quality improvement can be considered equivalent to a robust design.

Taguchi introduced a *quadratic loss function* to represent robustness (insensitiveness) as

$$L(f) = k(f - m_f)^2 \tag{20.56}$$

where L is the loss function, m_f is the target value for the cost function f, and $k > 0$ is a constant. As shown in Eq. (20.42), the cost function is a function of the design variables and problem parameters. Generally, constraints are ignored in the Taguchi method and only the robustness of the cost function is considered. The loss function means the loss when the design does not meet the target value for the cost function due to uncertainty (noise). The higher the loss, the farther away the solution from the target cost function value. Thus, *if the loss function is reduced, quality is enhanced*. This is the objective of the Taguchi method: Minimize the loss function in Eq. (20.56).

When we have various cases due to different noises, the expected value of the loss function in Eq. (20.56) is derived as follows:

$$E[L(f)] = E[k(f^2 - 2m_f f + m_f^2)] = kE[f^2 - 2m_f f + m_f^2] = k\left\{ E[f^2] - 2m_f E[f] + E[m_f^2] \right\} \tag{20.57}$$

The variance of f is calculated using the following identity:

$$\text{Var}[f] = E[f^2] - (E[f])^2; \quad \text{or} \quad E[f^2] = \text{Var}[f] + (E[f])^2 = \sigma_f^2 + \mu_f^2 \tag{20.58}$$

Thus, substituting Eq. (20.58) into Eq. (20.57), the expected value of the loss function is given from Eq. (20.57) as

$$E[L(f)] = k(\sigma_f^2 + \mu_f^2 - 2m_f\mu_f + m_f^2) = k(\sigma_f^2 + (\mu_f - m_f)^2) \equiv Q \tag{20.59}$$

where σ_f and μ_f are the standard deviation and the mean value of the cost function f, respectively. It is noted that the loss function Q in Eq. (20.59) is similar to the cost function of robust optimization in Eq. (20.49). This loss function is minimized to obtain a robust design.

Sometimes, the loss function is modified by a scale factor. Suppose we have a scale factor $s = m_f/\mu_f$, which can adjust the current mean μ_f to the target value m_f for the cost function. Since this factor scales σ_f and μ_f as $\frac{m_f}{\mu_f}\sigma_f$ and $\frac{m_f}{\mu_f}\mu_f$, a new loss function Q_a is obtained from Eq. (20.59):

$$Q_a = k\left(\left(\frac{m_f}{\mu_f}\sigma_f \right)^2 + \left(\mu_f \frac{m_f}{\mu_f} - m_f \right)^2 \right) = km_f^2 \frac{\sigma_f^2}{\mu_f^2} \tag{20.60}$$

The new loss function is the predicted amount of the loss when the current design is changed to the target value. It is noted that σ_f and μ_f are calculated at the current design.

To enhance the additive effect of the design variables, Eq. (20.60) is transformed (see Taguchi, 1987; Phadke, 1989) into:

$$\eta = 10 \, log_{10} \frac{\mu_f^2}{\sigma_f^2} \tag{20.61}$$

This equation is obtained using Q_a as $log_{10}(1/Q_a)$, ignoring the constant km_f^2 and then multiplying the result by the factor 10. Use of a logarithm enhances the additive effect of the design variables. Equation (20.61) is the ratio of the power of the signal μ_f to the power of the noise σ_f. It is called the signal-to-noise (S/N) ratio.

The power of a signal refers to the property the designer wants to improve. In this case, the designer wants to meet the target value m_f. The power of noise is the amount of uncertainty (variance). We try to find the design parameters so that the influence of the noise is minimized; that is, we maximize the S/N ratio. This is equivalent to minimizing the loss function in Eq. (20.60). In other words, maximizing S/N ratio η in Eq. (20.61) results in a robust design.

In the Taguchi method, we try to find values of the design parameters (variables) that minimize the loss function or maximize the S/N ratio.

The response having f at the target value of m_f is referred as "nominal-the-best"; a response with a target value of 0 is referred as "smaller-the-better"; and a response with a target value of infinity is referred as "larger-the-better." These examples of S/N ratios are summarized in Table 20.18, where c is the number of sample points (repetitions). For the "smaller-the-better" case in the table, the S/N ratio is derived as follows.

Since the target value m_f is zero for this case, the loss function in Eq. (20.56) is given as

$$L(f) = kf^2 \tag{20.62}$$

The expected value of this loss function is

$$Q = E\left[L(f)\right] = E\left[kf^2\right] = k\left(\frac{1}{c}\sum_{i=1}^{c} f_i^2\right) \tag{20.63}$$

where c is the total number of sample points (experiments) f_i. Ignoring the factor k, taking the logarithm of $1/Q$, and multiplying it by a factor of 10, we get

$$\eta = -10 \, log\left(\frac{1}{c}\sum_{i=1}^{c} f_i^2\right) \tag{20.64}$$

TABLE 20.18 Example requirements for the S/N ratio

Characteristic	S/N ratio
Nominal-the-best	$\eta = 10 \, log \frac{\mu_f^2}{\sigma_f^2}$
Smaller-the-better	$\eta = -10 \, log\left[\frac{1}{c}\sum_{i=1}^{c} f_i^2\right]$
Larger-the-better	$\eta = -10 \, log\left[\frac{1}{c}\sum_{i=1}^{c} \frac{1}{f_i^2}\right]$

A similar procedure can be used to derive the S/N ratio for the third case of "larger-the-better" in Table 20.18.

We need the mean and variance to calculate the loss function or the S/N ratio, and repeated experiments (function evaluation) are required to calculate these quantities. The S/N ratio is usually employed in process design. The loss function is usually directly used in product design although the S/N ratio can also be used (i.e., the one shown in the second row of Table 20.18).

In the Taguchi method, an orthogonal array is used for discrete design, as described in the previous section. Suppose a cost function is to be minimized and there are four design variables with three levels, as shown in Table 20.6. Then we can use the $L_9(3^4)$ orthogonal array shown in Table 20.19. The S/N ratio for each row of the orthogonal array is calculated as shown in the rightmost column of the table. For each row (design point), the experiments (function evaluations) are repeatedly carried out to calculate the S/N ratio. Although the levels of design variables are fixed for each row, the response f can be different because the unknown uncertainties (noises) are included in each design variable. The following loss function is frequently used as well:

$$Q = \frac{1}{c}\sum_{i=1}^{c} f_i^2 \tag{20.65}$$

For the problem of minimizing f, the S/N ratio in Table 20.19 is maximized or the loss function in Eq. (20.65) is minimized.

TABLE 20.19 The $L_9(3^4)$ orthogonal array

Experiment no.	Design variables and levels				Signal-to-noise ratio
	x_1	x_2	x_3	x_4	
1	1	1	1	1	$\eta_1 = -10\log\left[\frac{1}{c}\sum_{i=1}^{c} f_i^2\right]$
2	1	2	2	2	$\eta_2 = -10\log\left[\frac{1}{c}\sum_{i=1}^{c} f_i^2\right]$
3	1	3	3	3	$\eta_3 = -10\log\left[\frac{1}{c}\sum_{i=1}^{c} f_i^2\right]$
4	2	1	2	3	$\eta_4 = -10\log\left[\frac{1}{c}\sum_{i=1}^{c} f_i^2\right]$
5	2	2	3	1	$\eta_5 = -10\log\left[\frac{1}{c}\sum_{i=1}^{c} f_i^2\right]$
6	2	3	1	2	$\eta_6 = -10\log\left[\frac{1}{c}\sum_{i=1}^{c} f_i^2\right]$
7	3	1	3	2	$\eta_7 = -10\log\left[\frac{1}{c}\sum_{i=1}^{c} f_i^2\right]$
8	3	2	1	3	$\eta_8 = -10\log\left[\frac{1}{c}\sum_{i=1}^{c} f_i^2\right]$
9	3	3	2	1	$\eta_9 = -10\log\left[\frac{1}{c}\sum_{i=1}^{c} f_i^2\right]$

When physical experiments are performed for each design point in the orthogonal array, the results are different because the experiments automatically include noises. Therefore, we can calculate the variance of function f using the experimental results. If we conduct a numerical simulation for function evaluation instead of experiments, the same results are always obtained for each row of the orthogonal array (each design point). Thus, artificial noise needs to be introduced into the design variable values to obtain different simulation results. The artificial noise is constructed by perturbing the design variables or the problem parameters. The perturbation can be arbitrarily determined by the user. Alternatively, an outer array is constructed for each row of the orthogonal array, such as the one in Table 20.19.

The *outer array* is constructed as follows: First the noise levels (the values for the noise) are defined. For each row in Table 20.19, each design variable value is then perturbed systematically by different noise levels to generate perturbed design points at which the function is evaluated. The concept and procedure of orthogonal arrays as described previously is used here as well. For example, if three noise levels are selected for each of the four design variables, then each design point (each row of the orthogonal array) will generate nine perturbed points. These nine points define another table called the outer array. This will become clearer in the example problems presented below. The original orthogonal array of design points in Table 20.19 is called the *inner array*.

EXAMPLE 20.8   APPLICATION OF THE TAGUCHI METHOD

Solve a robust optimization problem formulated as

Minimize

$$f = x_1 x_2 \cos x_1 + x_1^2 - \frac{1}{4} x_2^2 - e^{x_2} + 5 \tag{a}$$

Each design variable has the levels -1.0, 0.0, and 1.0, and the disturbance for each design variable is given as $-0.1 \leq z_i \leq 0.1$. For the repetition of numerical experiments, use an outer array.

Solution

The orthogonal array $L_9(3^4)$ is used as the inner array for the problem. To generate the outer array, three levels are selected for the disturbance z_i in the ith design variable as -0.1, 0.0, and 0.1. Therefore, for each row of the inner array, nine perturbed design points are generated. The inner array and the outer array for the first row of the inner array are shown in Table 20.20 (note that since there are only two design variables, columns 3 an 4 are ignored in the orthogonal array). The outer array is generated by systematically perturbing each design variable by three levels of disturbance -0.1, 0.0, and 0.1. Therefore, we have nine cases for each row of the inner array. In the outer array, we calculate μ_f and σ_f for each row of the inner array, as shown at the top of Table 20.20.

The S/N ratio for the character "smaller-the-better" in Table 20.18 is used for this example because the problem is to minimize f. For the first row of the inner array, the S/N ratio for the "smaller-the-better" problem is calculated from the outer array as

$$\text{S/N ratio} = -10 \, log\left[\frac{1}{c}\sum_{i=1}^{c} f_i^2\right] = -10 \, log\left[\frac{1}{9}(6.12^2 + 6.09^2 + \cdots + 5.75^2 + 5.70^2)\right] = -15.45 \quad \text{(b)}$$

In this way, the S/N ratio for each row of the inner array is calculated. The results are shown in Table 20.21. The one-way table described in the previous section is constructed using the data in Table 20.21, as shown in Table 20.22. From this table, the levels (2, 3) for **x** yield a best solution as **x** = (0.0, 1.0) (since we are maximizing the S/N ratio, the levels corresponding to the largest

TABLE 20.20   The inner and outer arrays for the first row of Example 20.8

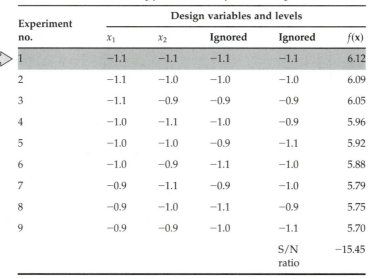

	Inner Array			
Sample point	Design variables and levels			
	x_1	x_2	Ignored	Ignored
1	−1.0	−1.0	−1.0	−1.0
2	−1.0	0.0	0.0	0.0
3	−1.0	1.0	1.0	1.0
4	0.0	−1.0	0.0	1.0
5	0.0	0.0	1.0	−1.0
6	0.0	1.0	−1.0	0.0
7	1.0	−1.0	1.0	0.0
8	1.0	0.0	−1.0	1.0
9	1.0	1.0	0.0	−1.0

	Outer Array for First Row of Inner Arrary				
Experiment no.	Design variables and levels				
	x_1	x_2	Ignored	Ignored	$f(\mathbf{x})$
1	−1.1	−1.1	−1.1	−1.1	6.12
2	−1.1	−1.0	−1.0	−1.0	6.09
3	−1.1	−0.9	−0.9	−0.9	6.05
4	−1.0	−1.1	−1.0	−0.9	5.96
5	−1.0	−1.0	−0.9	−1.1	5.92
6	−1.0	−0.9	−1.1	−1.0	5.88
7	−0.9	−1.1	−0.9	−1.0	5.79
8	−0.9	−1.0	−1.1	−0.9	5.75
9	−0.9	−0.9	−1.0	−1.1	5.70
				S/N ratio	−15.45

values for each design variable in the one-way table are selected). This solution is compared with the best solution in the rows of the inner array. In Table 20.21, the solution that maximizes the S/N ratio corresponds to the sixth row; this solution is the same as the one from the one-way table. Therefore, the solution $x = (0.0, 1.0)$ is selected as the final solution.

TABLE 20.21   S/N ratios for Example 20.8

| Sample point | Design variables and levels | | | | S/N ratio |
	x_1	x_2	Ignored	Ignored	
1	−1.0	−1.0	−1.0	−1.0	−15.45
2	−1.0	0.0	0.0	0.0	−13.99
3	−1.0	1.0	1.0	1.0	−8.03
4	0.0	−1.0	0.0	1.0	−12.84
5	0.0	0.0	1.0	−1.0	−12.05
6	0.0	1.0	−1.0	0.0	−6.22
7	1.0	−1.0	1.0	0.0	−13.73
8	1.0	0.0	−1.0	1.0	−13.99
9	1.0	1.0	0.0	−1.0	−11.05

TABLE 20.22   The one-way table for Example 20.8

| Design variable | Level | | |
	1	2	3
x_1	−12.49	−10.37	−12.92
x_2	−14.01	−13.34	−8.43
Ignored	−11.88	−12.63	−11.27
Ignored	−12.85	−11.31	−11.62

EXAMPLE 20.9   APPLICATION OF THE TAGUCHI METHOD

Solve the constrained optimization problem of Example 20.7 using the Taguchi method. Each design variable has three levels −1.0, 0.0, and 1.0, and the disturbance for each design variable is given as $-0.1 \leq z_i \leq 0.1$. For repetition of numerical experiments, use an outer array.

Solution

The orthogonal array $L_9(3^4)$ is used as the inner array for the problem. To generate the outer array, three levels are selected for the disturbance z_i in the ith design variable as −0.1, 0.0, and 0.1.

Therefore, for each row of the inner array, nine perturbed design points are generated. The inner array and the outer array for the first row of the inner array are shown in Table 20.23 on the next page (note that since there are only two design variables, columns 3 and 4 are ignored in the orthogonal array). The outer array is generated by systematically perturbing each design variable by three levels of disturbance -0.1, 0.0, and 0.1, so we have nine cases for each row of the inner array. In the outer array, we calculate μ_f and σ_f for each row of the inner array as shown in Table 20.23.

To use the loss function in Eq. (20.59), we need a target; the target of this example is set as $-\infty$. Thus, we cannot use the function in Eq. (20.59). The characteristic "smaller-the-better" in Table 20.18 can be used for a minimization problem; however, the cost function should be positive to use this index. In this case a new index is defined. Since this is a constrained minimization problem, the robustness index F is defined as

$$F = w_1\mu_f + w_2\sigma_f + P(\mathbf{x}) \tag{a}$$

where w_1 and w_2 are weighting factors, and $P(\mathbf{x})$ is a penalty function as defined in Eq. (20.32). That is, when constraints are violated, F is increased.

For each row of the inner array we calculate F using an outer array. The values of μ_f and σ_f are calculated as shown in Table 20.23, and F is evaluated for each row of the inner array. The results for each row of the inner array are shown in Table 20.24. In this problem, the factors are defined as $w_1 = w_2 = 0.5$ and the penalty parameter $R = 100$ for use in Eq. (20.32). The one-way table described in the previous section is constructed using the data in Table 20.24, as shown in Table 20.25. From this table, levels (2, 3) for $\mathbf{x}$ yield the solution as $\mathbf{x} = (0.0, 1.0)$ (since we are minimizing F, the levels corresponding to the smallest values in the one-way table are selected). This solution is compared with the best solution in the rows of the inner array in Table 20.24. The best solution corresponds to the sixth row and it is the same as the one from the one-way table. Therefore, the solution $\mathbf{x} = (0.0, 1.0)$ is selected as the final solution.

20.5 RELIABILITY-BASED DESIGN OPTIMIZATION—DESIGN UNDER UNCERTAINTY

A reliable design is one that satisfies the design criteria even with some uncertainties in the design variables or the problem parameters. Reliability is measured by the probability of satisfying a design criterion. An optimization procedure that incorporates reliability requirements in its calculations is called *reliability-based design optimization* (RBDO). In an RBDO formulation of the problem, a reliability constraint is defined so that the probability of violating the original constraint is less than a specified value. Therefore, reliability is imposed on constraints in RBDO. This is in contrast to the robust design approach (discussed in the previous section), where robustness is imposed on the cost function.

This section presents an introduction to the topic of RBDO. It is noted that considerable work has been done on this subject over the last 30 years. Consult Nikolaidis et al. (2005) and Choi et al. (2007) for more detailed discussion on the subject.

TABLE 20.23   The inner and outer arrays for the first row of Example 20.9

	Inner Array			
	Design variables and levels			
Experiment no.	x_1	x_2	Ignored	Ignored
1	−1.0	−1.0	−1.0	−1.0
2	−1.0	0.0	0.0	0.0
3	−1.0	1.0	1.0	1.0
4	0.0	−1.0	0.0	1.0
5	0.0	0.0	1.0	−1.0
6	0.0	1.0	−1.0	0.0
7	1.0	−1.0	1.0	0.0
8	1.0	0.0	−1.0	1.0
9	1.0	1.0	0.0	−1.0

	Outer Array for First Row of Inner Array				
	Design variables and levels				
Experiment no.	x_1	x_2	Ignored	Ignored	$f(x)$
1	−1.1	−1.1	−1.1	−1.1	1.12
2	−1.1	−1.0	−1.0	−1.0	1.09
3	−1.1	−0.9	−0.9	−0.9	1.05
4	−1.0	−1.1	−1.0	−0.9	0.96
5	−1.0	−1.0	−0.9	−1.1	0.92
6	−1.0	−0.9	−1.1	−1.0	0.88
7	−0.9	−1.1	−0.9	−1.0	0.79
8	−0.9	−1.0	−1.1	−0.9	0.75
9	−0.9	−0.9	−1.0	−1.1	0.70
				μ_f	0.92
				σ_f	0.15

20.5.1 Review of Background Material for RBDO

The basic idea of RBDO is to transform the constraints of the optimization problem into reliability-based constraints. This transformation process uses probability and statistics concepts and procedures, some of which are reviewed in this section.

TABLE 20.24   Results of Example 20.9

Experiment no.	Design variables and levels				Constraints				
	x_1	x_2	Ignored	Ignored	g_1	g_2	μ_f	σ_f	F
1	−1.0	−1.0	−1.0	−1.0	0.00	−0.47	0.92	0.15	0.53
2	−1.0	0.0	0.0	0.0	−1.00	−3.57	0.00	0.22	0.11
3	−1.0	1.0	1.0	1.0	0.00	−4.67	−2.51	0.38	−1.06
4	0.0	−1.0	0.0	1.0	−4.00	−0.90	−0.61	0.09	−0.26
5	0.0	0.0	1.0	−1.0	−5.00	−4.00	−1.00	0.09	−0.46
6	0.0	1.0	−1.0	0.0	−4.00	−5.10	−2.97	0.29	−1.34
7	1.0	−1.0	1.0	0.0	−6.00	−0.47	−0.15	0.21	0.03
8	1.0	0.0	−1.0	1.0	−7.00	−3.57	0.00	0.18	0.09
9	1.0	1.0	0.0	−1.0	−6.00	−4.67	−1.44	0.28	−0.58

TABLE 20.25   The one-way table for Example 20.9

Design variable	Level		
	1	2	3
x_1	−0.14	−0.69	−0.15
x_2	0.10	−0.09	−0.99
Ignored	−0.24	−0.25	−0.50
Ignored	−0.17	−0.40	−0.41

Probability Density Function

A continuous random variable X takes on various values x that are within the range $-\infty < x < \infty$. A mathematical function that describes the distribution of a continuous random variable is called the *probability density function* (PDF) and is designated $f_X(x)$. That is, PDF is a function that describes the relative likelihood (probability) of this random variable X occurring at a given point x.

Notation. **A random variable is expressed by an uppercase letter, while its particular value is denoted by a lowercase letter: for example, random variable X and its value x.**

The *probability density function* (also called the *probability distribution function* or *probability mass function*) of one variable is nonnegative everywhere and its integral over the entire space is equal to one:

$$f_X(x) \geq 0, \quad \int_{-\infty}^{\infty} f_X(x)dx = 1 \qquad (20.66)$$

The probability P that the random variable will fall within a particular region is given by the integral of this variable's probability density over the region. For example, the probability that X will lie within a differential interval dx between x and $x + dx$ is given as

$$P[x \leq X \leq x + dx] = f_X(x)dx \qquad (20.67)$$

Thus the probability that X will fall between a and b, written as $P[a \leq X \leq b]$, is given as the integral of Eq. (20.67):

$$P[a \leq X \leq b] = \int_a^b f_X(x)dx \qquad (20.68)$$

The probability density function having a normal (Gaussian) distribution is shown in Figure 20.4(a).

Cumulative Distribution Function

The *cumulative distribution function* (CDF) $F_X(x)$ describes the probability that a random variable X with a given probability distribution will be found at a value less than or equal to x. This function is given as

$$F_X(x) = P[X \leq x] = \int_{-\infty}^{x} f_X(u)du \qquad (20.69)$$

That is, for a given value x, $F_X(x)$ is the probability that the observed value of X is less than or equal to x. If f_X is continuous at x, then the probability density function is the derivative of the cumulative distribution function:

$$f_X(x) = \frac{dF_X(x)}{dx} \qquad (20.70)$$

The CDF also has the following properties:

$$\lim_{x \to -\infty} F(x) = 0; \quad \lim_{x \to \infty} F(x) = 1 \qquad (20.71)$$

The cumulative distribution function is illustrated in Figure 20.4(b). It shows that the probability of X being less than or equal to x_l is $F_X(x_l)$. This is a point on the $F_X(x)$ versus x curve in Figure 20.4(b) and it is the shaded area in Figure 20.4(a).

Probability of Failure

A reliability-based constraint for the jth inequality constraint $G_j(\mathbf{X}) \geq 0$ is defined as

$$P_f = P[G_j(\mathbf{x} + \mathbf{z}^x, \mathbf{y} + \mathbf{z}^y) \leq 0] \leq P_{j,0}, \quad j = 1, \ldots, m \qquad (20.72)$$

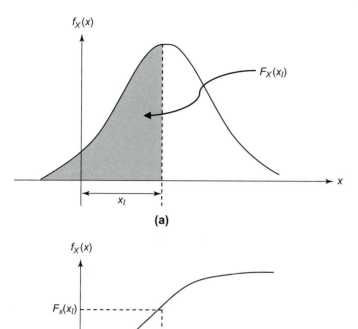

FIGURE 20.4   Graphic of (a) probability density function and (b) cumulative distribution function.

where in the preceding equation P_f is the probability of failure, $P[b]$ denotes the probability of b, $\mathbf{x}$ is the n-dimensional design variable vector, $\mathbf{y}$ is an r-dimensional vector of problem parameters, and $\mathbf{z}^x$ and $\mathbf{z}^y$ are the n-dimensional and r-dimensional vectors containing uncertainties in the design variables and problem parameters, respectively. $P_{j,0}$ represents the limit on the probability of failure for the jth constraint. If the probabilistic distributions of uncertainties are known, the probability of failure in Eq. (20.72) is given as

$$P_f = P[G_j(\mathbf{x} + \mathbf{z}^x, \mathbf{y} + \mathbf{z}^y) \leq 0] = \int_{G_j(\mathbf{x}+\mathbf{z}^x,\mathbf{y}+\mathbf{z}^y) \leq 0} d(\mathbf{z}^x, \mathbf{z}^y) d\mathbf{z}^x d\mathbf{z}^y, \quad j = 1, \ldots, m \quad (20.73)$$

where $d(\mathbf{z}^x, \mathbf{z}^y)$ is the joint probability density function of the probabilistic variables $\mathbf{z}^x$ and $\mathbf{z}^y$ and

$$d\mathbf{z}^x d\mathbf{z}^y = (dz^{x_1} dz^{x_2} \ldots dz^{x_n})(dz^{y_1} dz^{y_2} \ldots dz^{y_r}) \quad (20.74)$$

Because the joint probability density function is a density function distributed by multiple variables, it must be known for the random variables in order to calculate the probability of failure.

Expected Value

The *expected value* (or *expectation, mean,* or *first moment*) of a random variable is the weighted average of all possible values that this variable can have. The weights used in computing this average correspond to the probabilities in the case of a discrete random variable, or the densities in the case of a continuous random variable. The expected value is the integral of the random variable with respect to its probability measure. It is the sample mean as the sample size goes to infinity.

Let the random variable X take value x_i with probability p_i for $i = 1$ to k. X's expected value $E(X)$ is then defined as

$$E[X] = \sum_{i=1}^{k} x_i p_i \tag{20.75}$$

Since all of the probabilities p_i add to one ($\sum_{i=1}^{k} p_i = 1$), the expected value can be viewed as the weighted average of $x_i's$, with $p_i's$ being the weights:

$$E[X] = \frac{\sum_{i=1}^{k} x_i p_i}{\sum_{i=1}^{k} p_i} = \sum_{i=1}^{k} x_i p_i \tag{20.76}$$

If the probability distribution of X admits a probability density function $f_X(x)$, then the expected value is computed as

$$E[X] = \int_{-\infty}^{\infty} x f_X(x) dx \tag{20.77}$$

It is seen that this is the first moment of X; hence X is also called the first moment.

The expected value of $G(X)$, a function of random variable X, with respect to the probability density function $f_X(x)$ is given as

$$E[G(X)] = \int_{-\infty}^{\infty} g(x) f_X(x) dx \tag{20.78}$$

The expected value of $G(X) = X^m$ is called the *m*th moment of X and is given as

$$E[X^m] = \int_{-\infty}^{\infty} x^m f_X(x) dx \tag{20.79}$$

MEAN AND VARIANCE   The mean and variance of the random variable X are the first and second moments of X calculated as follows:

$$\mu_X = E[X] = \int_{-\infty}^{\infty} x f_X(x) dx \tag{20.80}$$

$$
\begin{aligned}
\mathrm{Var}[X] = E[(X - \mu_X)^2] &= \sigma_X^2 \\
&= \int_{-\infty}^{\infty} (x - \mu_X)^2 f_X(x)dx = \int_{-\infty}^{\infty} (x^2 - 2x\mu_X + \mu_X^2)f_X(x)dx \\
&= \int_{-\infty}^{\infty} x^2 f_X(x)dx - 2\mu_X \int_{-\infty}^{\infty} x f_X(x)dx + \mu_X^2 \int_{-\infty}^{\infty} f_X(x)dx \\
&= E[X^2] - 2\mu_X^2 + \mu_X^2 = E[X^2] - \mu_X^2
\end{aligned}
\tag{20.81}
$$

STANDARD DEVIATION   The standard deviation σ_X of X is given as

$$
\sigma_X = \sqrt{\mathrm{Var}[X]}
\tag{20.82}
$$

COEFFICIENT OF VARIATION   The *coefficient of variation* δ_X indicates the relative amount of uncertainty, defined as the ratio of the standard deviation of X to the mean of X:

$$
\delta_X = \frac{\sigma_X}{\mu_X}
\tag{20.83}
$$

RELIABILITY INDEX   The *reliability index* β is defined as the reciprocal of the coefficient of variation δ_X; that is, it is the ratio of the mean of X to the standard deviation of X:

$$
\beta = \frac{\mu_X}{\sigma_X}
\tag{20.84}
$$

COVARIANCE   If two random variables X and Y are correlated, the correlation is represented by the covariance σ_{XY}, calculated as follows:

$$
\sigma_{XY} = \mathrm{Cov}(X, Y) = E[(X - \mu_X)(Y - \mu_Y)] = \int_{-\infty}^{\infty} \int_{-\infty}^{\infty} (x - \mu_X)(y - \mu_Y)d(x, y)dxdy
\tag{20.85}
$$

where $d(x, y)$ is the joint probability density function of X and Y.

CORRELATION COEFFICIENT   The *correlation coefficient* is a nondimensional measure of the correlation, defined as

$$
\rho_{XY} = \frac{\sigma_{XY}}{\sigma_X \sigma_Y}
\tag{20.86}
$$

GAUSSIAN (NORMAL) DISTRIBUTION   The Gaussian (normal) distribution is used in many engineering and science fields and is defined using the mean and standard deviation of X as the probability density function:

$$
f_X(x) = \frac{1}{\sigma_X \sqrt{2\pi}} \exp\left[-\frac{1}{2}\left(\frac{x - \mu_X}{\sigma_X}\right)^2 \right], \quad -\infty < x < \infty
\tag{20.87}
$$

It is represented as $N(\mu_X, \sigma_X)$. The Gaussian distribution can be normalized using a transformation of variable X as

$$U = (X - \mu_X)/\sigma_X \tag{20.88}$$

This yields the standard normal distribution $N(0,1)$, and the corresponding probability density function becomes

$$f_U(u) = \frac{1}{\sqrt{2\pi}}\exp\left(\frac{-u^2}{2}\right), \quad -\infty < u < \infty \tag{20.89}$$

Cumulative distribution with respect to u is then obtained as the cumulative distribution function (CDF) $\Phi(u)$:

$$\Phi(u) = F_U(u) = \int_{-\infty}^{u} \frac{1}{\sqrt{2\pi}}\exp\left(\frac{-\xi^2}{2}\right)d\xi \tag{20.90}$$

Numerical values of $\Phi(u)$ can be found in statistics texts. Since the normal distribution in Eq. (20.90) is symmetric with respect to $x = 0$,

$$\Phi(-u) = 1 - \Phi(u) \tag{20.91}$$

INVERSE If the CDF is strictly increasing and continuous, then $\Phi^{-1}(p)$, $p \in [0,1]$ is the unique number u_p such that $\Phi(u_p) = p$; that is, $u_p = \Phi^{-1}(p)$. Also,

$$u_p = \Phi^{-1}(p) = -\Phi^{-1}(1-p) \tag{20.92}$$

where u_p is the standard normalized variable, p is the corresponding cumulative probability, and Φ^{-1} is the inverse of the CDF.

20.5.2 Calculation of the Reliability Index

In this subsection, calculation of the reliability index that is used in the optimization process is explained. Knowing the reliability index, the probability of failure can be calculated, or the index can be used directly in the optimization process.

Limit State Equation

In structural design, the limit state indicates the margin of safety between structural resistance and the structural load. The limit state function ($G(\cdot)$) and the probability of failure (P_f) are defined as

$$G(X) = R(X) - S(X) \tag{20.93}$$

$$P_f = P[G(\mathbf{X}) \leq 0] \tag{20.94}$$

where R is the structural resistance and S is the loading. $G(X) < 0$, $G(X) = 0$, and $G(X) > 0$ indicate the failure region, the failure surface, and the safe region, respectively. They are illustrated in Figure 20.5 (Choi et al., 2007).

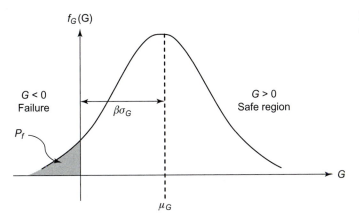

FIGURE 20.5   Probability density function for the limit state $G(X)$.

Using Eq. (20.93), the mean and standard deviation of $G(X)$ are calculated as

$$\mu_G = \mu_R - \mu_S \tag{20.95}$$

$$\sigma_G = \sqrt{\sigma_R^2 + \sigma_S^2 - 2\rho_{RS}\sigma_R\sigma_S} \tag{20.96}$$

where μ_R, μ_S, and ρ_{RS} are the mean of R, the mean of S, and the correlation coefficient between R and S, respectively. The variance of $G(X)$ is calculated as

$$Var[G(X)] = Var[R(X) - S(X)] = Var[R(X)] + Var[S(X)] - 2Cov[R(X), S(X)]$$
$$= \sigma_R^2 + \sigma_S^2 - 2\sigma_{RS} = \sigma_R^2 + \sigma_S^2 - 2\rho_{RS}\sigma_R\sigma_S \tag{20.97}$$
$$Var[G(X)] = \sigma_G^2 = \sigma_R^2 + \sigma_S^2 - 2\rho_{RS}\sigma_R\sigma_S$$

Thus the standard deviation is given as in Eq. (20.96).

The reliability index for G is given using Eqs. (20.84), (20.95), and (20.96) as

$$\beta = \frac{\mu_G}{\sigma_G} = \frac{\mu_R - \mu_S}{\sqrt{\sigma_R^2 + \sigma_S^2 - 2\rho_{RS}\sigma_R\sigma_S}} \tag{20.98}$$

Suppose that R and S are normally distributed; then the probability density function of the limit state function is

$$f_G(g) = \frac{1}{\sigma_G\sqrt{2\pi}} \exp\left[-\frac{1}{2}\left(\frac{g - \mu_G}{\sigma_G}\right)^2\right] \tag{20.99}$$

The probability of failure using Eq. (20.69) is given as

$$P_f = P[G(X) \le 0] = \int_{-\infty}^{0} f_G(g)dg = \int_{-\infty}^{0} \frac{1}{\sigma_G\sqrt{2\pi}} \exp\left[-\frac{1}{2}\left(\frac{g - \mu_G}{\sigma_G}\right)^2\right] dg \tag{20.100}$$

After introducing the following normalizing transformation for the random variable G, the probability of failure is obtained from Eq. (20.100) as

$$U = (G - \mu_G)/\sigma_G \tag{20.101}$$

$$P_f = \int_{-\infty}^{-\beta} \frac{1}{\sqrt{2\pi}} \exp\left(-\frac{u^2}{2}\right) du \tag{20.102}$$

where $\beta = \mu_G/\sigma_G = (\mu_R - \mu_S)/\sqrt{(\sigma_R^2 + \sigma_S^2)}$, which is obtained by assuming that the correlation coefficient in Eq. (20.98) is 0. Using the definition of cumulative distribution function and its property in Eqs. (20.90) and (20.91), we get

$$P_f = \Phi(-\beta) = 1 - \Phi(\beta) \tag{20.103}$$

Linear Limit State Equation

If we suppose that a limit state equation is a linear combination of random variables X_i, $i = 1, \ldots, n$ with normal distributions, the limit state function is given as follows:

$$G = a_0 + \sum_{i=1}^{n} a_i X_i \tag{20.104}$$

Then, assuming a normal distribution for G, the mean and variance of G are calculated as

$$\mu_G = a_0 + \sum_{i=1}^{n} a_i \mu_i \tag{20.105}$$

$$\sigma_G^2 = \sum_{i=1}^{n} \sum_{j=1}^{n} a_i a_j \text{Cov}[X_i, X_j] = \sum_{i=1}^{n} \sum_{j=1}^{n} a_i a_j \rho_{ij} \sigma_i \sigma_j \tag{20.106}$$

where the mean and variance of X_i are μ_i and σ_i, respectively. The probability of failure for G is

$$P_f = P[G \leq 0] = \Phi\left(-\frac{\mu_G}{\sigma_G}\right) = \Phi(-\beta) \tag{20.107}$$

Nonlinear Limit State Equation

When the limit state equation is a nonlinear function of the random variable vector $\mathbf{X} = (X_1, X_2, \ldots, X_n)$, it is linearized by the Taylor series around the mean $\mu_X = (\mu_1, \mu_2, \ldots, \mu_n)$ of $\mathbf{X}$ as

$$G(\mathbf{X}) = G(\mu_X) + \sum_{i=1}^{n} \frac{\partial G}{\partial X_i}\Big|_{\mu_X}(X_i - \mu_i) \tag{20.108}$$

The mean and variance of this linearized equation are

$$\mu_G = G(\boldsymbol{\mu}_X) \tag{20.109}$$

$$\sigma_G = \sum_{i=1}^{n}\sum_{j=1}^{n} \frac{\partial G}{\partial X_i}\bigg|_{\mu_X} \frac{\partial G}{\partial X_j}\bigg|_{\mu_X} Cov\left[X_i, X_j\right] = \sum_{i=1}^{n}\sum_{j=1}^{n} \frac{\partial G}{\partial X_i}\bigg|_{\mu_X} \frac{\partial G}{\partial X_j}\bigg|_{\mu_X} \rho_{ij}\sigma_i\sigma_j \tag{20.110}$$

It is noted that $\partial G/\partial X_i|_{\mu_X}$ and $\partial G/\partial X_j|_{\mu_X}$ correspond to a_i and a_j in Eq. (20.106). Therefore, the reliability index can be calculated using Eq. (20.98). This is called the mean value first-order second-moment method (MVFOSM). MVFOSM has a drawback. Since the limit state equation is linearized around the mean point, the value of the reliability index depends on the equation's form. When the form of the limit-equation is changed by a scale, the reliability index is also changed. That is, the approach lacks the invariance of the reliability index.

Advanced First-Order Second Moment Method

To overcome the lack of invariance, Hasofer and Lind (1974) proposed the advanced first-order second-moment method (AFOSM). First, a random variable for the standard normal distribution $N(0, 1)$ is defined as

$$U_i = \frac{X_i - \mu_i}{\sigma_i}, \quad i = 1 \text{ to } n \tag{20.111}$$

Substituting for X_i from Eq. (20.111), the limit state equation in Eq. (20.104) is transformed as

$$G(\mathbf{U}) = a_0 + \sum_{i=1}^{n} a_i(\mu_i + \sigma_i U_i) \tag{20.112}$$

The mean value μ_G of $G(\mathbf{U})$ in Eq. (20.112) is calculated as

$$\mu_G = E[G(\mathbf{U})] = E\left[a_0 + \sum_{i=1}^{n} a_i(\mu_i + \sigma_i U_i)\right]$$

$$= a_0 + \sum_{i=1}^{n} E[a_i(\mu_i + \sigma_i U_i)]$$

$$= a_0 + \sum_{i=1}^{n} a_i E[\mu_i + \sigma_i U_i]$$

$$= a_0 + \sum_{i=1}^{n} a_i(\mu_i + \sigma_i E[U_i])$$

$$= a_0 + \sum_{i=1}^{n} a_i\mu_i, \quad \text{since } E[U_i] = 0 \tag{20.113}$$

Therefore, μ_G can be written with all $U_i = 0$ in Eq. (20.112) as

$$\mu_G = a_0 + \sum_{i=1}^{n} a_i\mu_i = |G(\text{all } U_i = 0)| \tag{20.114}$$

The variance of $G(U)$ in Eq. (20.112) is derived as follows:

$$\sigma_G^2 = \text{Var}[G(\mathbf{U})] = \text{Var}\left[a_0 + \sum_{i=1}^{n} a_i(\mu_i + \sigma_i U_i)\right]$$

$$= \text{Var}[a_0] + \text{Var}\left[\sum_{i=1}^{n} a_i\mu_i\right] + \text{Var}\left[\sum_{i=1}^{n} a_i\sigma_i U_i\right]$$

$$= 0 + 0 + \sum_{i=1}^{n} a_i^2\sigma_i^2\text{Var}[U_i]$$

$$= \sum_{i=1}^{n} a_i^2\sigma_i^2, \quad \text{since} \quad \text{Var}[U_i] = 1 \tag{20.115}$$

Therefore, the standard deviation of $G(U)$ is given as

$$\sigma_G = \sqrt{\sum_{i=1}^{n}(a_i\sigma_i)^2} = \sqrt{\sum_{i=1}^{n}\left(\frac{\partial G}{\partial U_i}\right)^2} \tag{20.116}$$

So the reliability index β is given from Eq. (20.84) as

$$\beta = \frac{\mu_G}{\sigma_G} = \frac{\left|G(\text{all } U_i = 0)\right|}{\sqrt{\sum_{i=1}^{n}\left(\frac{\partial G}{\partial U_i}\right)^2}} \tag{20.117}$$

For the two-variable case, Eq. (20.112) gives

$$G(\mathbf{U}) = a_0 + a_1(\mu_1 + \sigma_1 U_1) + a_2(\mu_2 + \sigma_2 U_2) \tag{20.118}$$

This equation is plotted as a straight line $G(\mathbf{U}) = 0$ in Figure 20.6, designated as line AB. The shortest distance from the origin to this line is given as

$$\frac{\left|a_0 + a_1\mu_1 + a_2\mu_2\right|}{\sqrt{(a_1\sigma_1)^2 + (a_2\sigma_2)^2}} \tag{20.119}$$

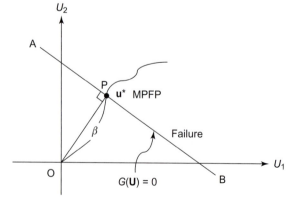

FIGURE 20.6 Geometric representation of the reliability index.

Equation (20.119) is derived by first developing the equation for a line normal to the line AB that also passes through the origin O. Then the coordinates of the point P are determined as the intersection of lines AB and the line normal to it. Knowing the coordinates of the point P, the distance OP can be calculated.

The formula for the shortest distance in Eq. (20.119) is the same as the reliability index in Eq. (20.117). Thus, the geometric meaning of the reliability index is that it is the shortest distance from the origin to the limit state equation. The point on the limit state surface that is closest to the origin is called the most probable failure point (MPFP). In the literature, this has also been called the most probable point (MPP). The MPFP is denoted u^* in Figure 20.6.

In the X space, the coordinates of the MPFP using Eq. (20.111) are

$$x_i^* = \mu_i + u_i^* \sigma_i, \quad i = 1, \ldots, n \tag{20.120}$$

where n is the number of design variables. A nonlinear limit state equation is linearized around the MPFP $\mathbf{x}^*$ as

$$G(\mathbf{X}) \approx \sum_{i=1}^{n} \left. \frac{\partial G}{\partial X_i} \right|_{\mathbf{X}=\mathbf{x}^*} (X_i - x_i^*) \tag{20.121}$$

where $G(\mathbf{x}^*) = 0$ is used.

Using Eqs. (20.109) and (20.110), the mean and variance of $G(\mathbf{X}) = 0$ are

$$\mu_G = \sum_{i=1}^{n} \left. \frac{\partial G}{\partial X_i} \right|_{\mathbf{X}=\mathbf{x}^*} (\mu_i - x_i^*) \tag{20.122}$$

$$\sigma_G^2 = \sum_{i=1}^{n} \left(\left. \frac{\partial G}{\partial X_i} \right|_{\mathbf{X}=\mathbf{x}^*} \right)^2 \sigma_i^2 \tag{20.123}$$

By the chain rule, we get

$$\frac{\partial G}{\partial X_i} = \frac{\partial G}{\partial U_i} \frac{\partial U_i}{\partial X_i} = \frac{1}{\sigma_i} \frac{\partial G}{\partial U_i} \tag{20.124}$$

Substituting Eq. (20.124) into Eqs. (20.122) and (20.123), and using Eq. (20.111), the mean and variance become

$$\mu_G = -\sum_{i=1}^{n} \left. \frac{\partial G}{\partial U_i} \right|_{\mathbf{u}^*} u_i^* \tag{20.125}$$

$$\sigma_G^2 = \sum_{i=1}^{n} \left(\left. \frac{\partial G}{\partial U_i} \right|_{\mathbf{u}^*} \right)^2 \tag{20.126}$$

where U_i is the normalized standard variable vector and u_i^* is the MPFP.

The linearized equation in Eq. (20.121) is in terms of the normalized variables and is given as

$$G(\mathbf{U}) = \sum_{i=1}^{n} \left. \frac{\partial G}{\partial U_i} \right|_{\mathbf{u}^*} (U_i - u_i^*) \tag{20.127}$$

Therefore, if a nonlinear state equation is linearized around the MPFP, the lack of invariance is avoided because the reliability index in Eq. (20.117) is expressed by the normalized variable U_i (Hasofer and Lind, 1974).

In AFOSM, the MPFP is obtained from a nonlinear limit state equation and the limit state equation is linearized around the MPFP. Because the MPFP is the closest point from the origin to the limit state equation, it is obtained typically by solving the following optimization problem:

Minimize

$$\beta = \sqrt{\mathbf{U}^T\mathbf{U}}, \quad \text{subject to } G(\mathbf{U}) = 0 \tag{20.128}$$

where β is the reliability index.

Using an iterative method (Hasofer and Lind, 1974), the iterative equation for solution to the problem in Eq. (20.128) is given as

$$\mathbf{U}^{(k+1)} = \frac{G_{\mathbf{U}}^{(k)T}\mathbf{U}^{(k)} - G(\mathbf{U}^{(k)})}{G_{\mathbf{U}}^{(k)T}G_{\mathbf{U}}^{(k)}} G_{\mathbf{U}}^{(k)} \tag{20.129}$$

where $G_{\mathbf{U}}^{(k)} = \frac{\partial G}{\partial \mathbf{U}}$ at $\mathbf{U} = \mathbf{U}^{(k)}$ is an n-dimensional vector.

To derive the iterative Eq. (20.129), we proceed as follows: At the kth iteration, we want to update the vector $\mathbf{U}^{(k)}$ to $\mathbf{U}^{(k+1)}$, which is expressed as

$$\mathbf{U}^{(k+1)} = \mathbf{U}^{(k)} + \Delta\mathbf{U} \tag{20.130}$$

At $\mathbf{U}^{(k)}$, we write linear the Taylor expansion of the constraint $G(\mathbf{U}^{(k+1)}) = 0$, as

$$G(\mathbf{U}^{(k)}) + G_{\mathbf{U}}^{(k)T}\Delta\mathbf{U} = 0 \tag{20.131}$$

The question now is how to determine $\Delta\mathbf{U}$. We define a minimization problem for $\Delta\mathbf{U}$ using the original problem defined in Eq. (20.128) as

Minimize
$$(\mathbf{U}^{(k)} + \Delta\mathbf{U})^T(\mathbf{U}^{(k)} + \Delta\mathbf{U}) \quad \text{subject to } G(\mathbf{U}^{(k)}) + G_{\mathbf{U}}^{(k)T}\Delta\mathbf{U} = 0 \tag{20.132}$$

Note that the cost function in Eq. (20.128) has been replaced by its square, which does not affect the solution to the problem. The problem defined in Eq. (20.132) can be solved by writing the optimality conditions given in Chapter 4 for the equality-constrained problem. For that, we define the Lagrangian function and differentiate it with respect to $\Delta\mathbf{U}$ as

$$L = (\mathbf{U}^{(k)} + \Delta\mathbf{U})^T(\mathbf{U}^{(k)} + \Delta\mathbf{U}) + \lambda(G(\mathbf{U}^{(k)}) + G_{\mathbf{U}}^{(k)T}\Delta\mathbf{U}) \tag{20.133}$$

$$\frac{\partial L}{\partial(\Delta\mathbf{U})} = 2(\mathbf{U}^{(k)} + \Delta\mathbf{U}) + \lambda(G_{\mathbf{U}}^{(k)}) = 0 \tag{20.134}$$

where λ is the Lagrange multiplier for the equality constraint. The equality constraint in Eq. (20.131) and the optimality conditions in Eq. (20.134) provide just the right number of equations to solve for λ and $\Delta\mathbf{U}$.

There are a number of ways to solve for the Lagrange multiplier λ. One is to pre-multiply Eq. (20.134) by $G_U^{(k)T}$, and solve for λ as

$$\lambda = -\frac{2}{G_U^{(k)T} G_U^{(k)}} G_U^{(k)T}(\mathbf{U}^{(k)} + \Delta \mathbf{U}) \tag{20.135}$$

Substituting for λ from Eq. (20.135) into Eq. (20.134) and taking the second term to the right side, we get

$$2(\mathbf{U}^{(k)} + \Delta \mathbf{U}) = \left[\frac{2}{G_U^{(k)T} G_U^{(k)}} G_U^{(k)T}(\mathbf{U}^{(k)} + \Delta \mathbf{U})\right](G_U^{(k)}) \tag{20.136}$$

Now, replacing the left side of Eq. (20.136) with Eq. (20.130), and substituting the equality constraint from Eq. (20.132) into the right side of Eq. (20.136), we get

$$\mathbf{U}^{(k+1)} = \left[\frac{\{G_U^{(k)T} \mathbf{U}^{(k)} - G(\mathbf{U}^{(k)})\}}{G_U^{(k)T} G_U^{(k)}}\right] G_U^{(k)} \tag{20.137}$$

which is same as Eq. (20.129).

Once the MPFP is determined, the reliability index is obtained as

$$\beta = \sqrt{\mathbf{u}^{*T} \mathbf{u}^*} = -\frac{G_U^{*T} \mathbf{u}^*}{\sqrt{G_U^{*T} G_U^*}} \tag{20.138}$$

In Eq. (20.138), the following vector α is a measure of the sensitivity of the reliability index to each random variable:

$$\alpha = -\frac{G_U^*}{\sqrt{G_U^{*T} G_U^*}} \tag{20.139}$$

MVFOSM and AFOSM are attributed to the first-order reliability method (FORM). There are many other methods to calculate the reliability index. For details, the reader is referred to Choi et al. (2007).

The preceding method is a mathematical way to calculate the probability of failure of a constraint. Sampling methods are also utilized for this purpose and the most commonly used method in this class is Monte Carlo simulation (MCS). In MCS, many trials are conducted. If N trials are conducted for G_i, the probability of failure is approximately given by

$$P_f = \frac{N_f}{N} \tag{20.140}$$

where N_f is the number of trials for which G_i is violated out of the N trials conducted. For a large-scale problem, the MCS method needs a large number of computations. Thus, MCS is utilized only for small-scale problems or as a reference method for a new approach to calculate the reliability index.

EXAMPLE 20.10 CALCULATION OF THE RELIABILITY INDEX

Evaluate the reliability index β for the function $g_1(x)$ of Example 20.7. The initial design is $(-0.8, 0.8)$ and $\sigma_1 = \sigma_2 = 0.1$.

Solution

The function $G_1(X)$ of Example 20.7 is given as follows:

$$G_1(\mathbf{X}) = -(X_1 - 1)^2 - X_2^2 + X_1 + 6 \geq 0 \tag{a}$$

Since the initial design is $(-0.8, 0.8)$, $(\mu_1, \mu_2) = (-0.8, 0.8)$.

The iterative method is used for calculation of β. The termination criterion for the iterative process is defined as

$$\varepsilon = \left| \frac{\beta^{(k)} - \beta^{(k+1)}}{\beta^{(k)}} \right| \leq 0.001 \tag{b}$$

Iteration 1 (k = 0) Using Eq. (20.111), $\mathbf{u}^{(0)}$ is obtained at $(x_1^{(0)}, x_2^{(0)}) = (-0.8, 0.8)$ as follows:

$$u_1^{(0)} = \frac{x_1^{(0)} - \mu_1}{\sigma_1} = 0, \quad u_2^{(0)} = \frac{x_2^{(0)} - \mu_2}{\sigma_2} = 0 \tag{c}$$

$$G_1(\mathbf{x}^{(0)}) = 1.32 \tag{d}$$

G_U is calculated using Eq. (20.124) as

$$G_U = \frac{\partial G}{\partial U_i} = \frac{\partial G}{\partial X_i} \sigma_i \tag{e}$$

where

$$\frac{\partial G}{\partial X_1} = -2X_1 + 3, \quad \frac{\partial G}{\partial X_2} = -2X_2 \tag{f}$$

Therefore,

$$G_U^{(0)} = \begin{bmatrix} (-2x_1^{(0)} + 3)\sigma_1 \\ (-2x_2^{(0)})\sigma_2 \end{bmatrix} = \begin{bmatrix} 0.46 \\ -0.16 \end{bmatrix} \tag{g}$$

Using Eq. (20.129)

$$\mathbf{u}^{(1)} = \frac{G_U^{(0)T} \mathbf{u}^{(0)} - G(\mathbf{u}^{(0)})}{G_U^{(0)T} G_U^{(0)}} G_U^{(0)} = \frac{\begin{bmatrix} 0.46 & -0.16 \end{bmatrix} \begin{bmatrix} 0 \\ 0 \end{bmatrix} - 1.32}{\begin{bmatrix} 0.46 & -0.16 \end{bmatrix} \begin{bmatrix} 0.46 \\ -0.16 \end{bmatrix}} \begin{bmatrix} 0.46 \\ -0.16 \end{bmatrix} = \begin{bmatrix} -2.560 \\ 0.891 \end{bmatrix} \tag{h}$$

From Eq. (20.138), the reliability index is

$$\beta^{(1)} = \sqrt{\mathbf{u}^{(1)T} \mathbf{u}^{(1)}} = 2.710 \tag{i}$$

Iteration 2 (k = 1)

$$\begin{aligned} x_1^{(1)} &= \mu_1 + \sigma_1 u_1^{(1)} = (-0.8) + (0.1)(-2.560) = -1.0560 \\ x_2^{(1)} &= \mu_2 + \sigma_2 u_2^{(1)} = 0.8 + (0.1)(0.891) = 0.889 \end{aligned} \tag{j}$$

$$G_1(\mathbf{x}^{(1)}) = -0.0736 \tag{k}$$

$$G_U^{(1)} = \begin{bmatrix} (-2x_1^{(1)} + 3)\sigma_1 \\ (-2x_2^{(1)})\sigma_2 \end{bmatrix} = \begin{bmatrix} 0.511 \\ -0.178 \end{bmatrix} \tag{l}$$

$$\mathbf{u}^{(2)} = \frac{G_U^{(1)T}\mathbf{u}^{(1)} - G(\mathbf{u}^{(1)})}{G_U^{(1)T}G_U^{(1)}} G_U^{(1)} = \frac{\begin{bmatrix} 0.511 & -0.178 \end{bmatrix} \begin{bmatrix} -2.560 \\ 0.891 \end{bmatrix} + 0.0736}{\begin{bmatrix} 0.511 & -0.178 \end{bmatrix} \begin{bmatrix} 0.511 \\ -0.178 \end{bmatrix}} \begin{bmatrix} 0.511 \\ -0.178 \end{bmatrix} = \begin{bmatrix} -2.431 \\ 0.846 \end{bmatrix} \tag{m}$$

$$\beta^{(2)} = \sqrt{\mathbf{u}^{(2)T}\mathbf{u}^{(2)}} = 2.575 \tag{n}$$

$$\varepsilon = \left| \frac{2.710 - 2.575}{2.710} \right| = 0.05 \geq 0.001 \tag{o}$$

The convergence criterion is not satisfied.

Iteration 3 (k = 2)

$$x_1^{(2)} = \mu_1 + \sigma_1 u_1^{(2)} = (-0.8) + (0.1)(-2.431) = -1.043$$
$$x_2^{(2)} = \mu_2 + \sigma_2 u_2^{(2)} = 0.8 + (0.1)(0.846) = 0.885 \tag{p}$$

$$G_1(\mathbf{x}^{(2)}) = -0.0001754 \tag{q}$$

$$G_U^{(2)} = \begin{bmatrix} (-2x_1^{(2)} + 3)\sigma_2 \\ (-2x_2^{(2)})\sigma_2 \end{bmatrix} = \begin{bmatrix} 0.509 \\ -0.177 \end{bmatrix} \tag{r}$$

$$\mathbf{u}^{(3)} = \frac{G_U^{(2)T}\mathbf{u}^{(2)} - G(\mathbf{u}^{(2)})}{G_U^{(2)T}G_U^{(2)}} G_U^{(2)} = \frac{\begin{bmatrix} 0.509 & -0.177 \end{bmatrix} \begin{bmatrix} -2.431 \\ 0.846 \end{bmatrix} + 0.0001754}{\begin{bmatrix} 0.509 & -0.177 \end{bmatrix} \begin{bmatrix} 0.509 \\ -0.177 \end{bmatrix}} \begin{bmatrix} 0.509 \\ -0.177 \end{bmatrix} = \begin{bmatrix} -2.431 \\ 0.846 \end{bmatrix} \tag{s}$$

$$\beta^{(3)} = \sqrt{\mathbf{u}^{(3)T}\mathbf{u}^{(3)}} = 2.574 \tag{t}$$

$$\varepsilon = \left| \frac{2.575 - 2.574}{2.575} \right| = 0.000128 \leq 0.001 \tag{u}$$

The convergence criterion is satisfied.

$$x_1^{(3)} = \mu_1 + \sigma_1 u_1^{(3)} = (-0.8) + (0.1)(-2.431) = -1.043$$
$$x_2^{(3)} = \mu_2 + \sigma_2 u_2^{(3)} = 0.8 + (0.1)(0.846) = 0.885 \tag{v}$$

Thus the location of MPFP is $(-1.043, 0.885)$. The reliability index is $\beta = 2.574$, and the reliability is $\Phi(\beta) = 0.9947$ from Eq. (20.90). The value of $\Phi(\beta)$ can be read from the table for the normal distribution.

20.5.3 Formulation of Reliability-Based Design Optimization

We now present the formulation for reliability-based design optimization (RBDO). The design variable vector is $\mathbf{X} = (X_1, X_2, \ldots, X_n)$, and the cost function to be minimized is $F(\mathbf{X})$. The reliability constraints are defined as

$$P_f = P\left[G_j(\mathbf{X}) \leq 0\right] = \int_{G_j(X) \leq 0} f_G(g_j)dg_j = \Phi\left(-\frac{\mu_{G_j}}{\sigma_{G_j}}\right) = \Phi(-\beta) \leq P_{j,0}, \quad j = 1, \ldots, m \quad (20.141)$$

Each constraint in Eq. (20.141) is an inner optimization problem in itself. These are the inner optimization problems in the outer reliability-based design optimization problem. Therefore, we need the sensitivity information of the inner problem. This sensitivity is called *optimum sensitivity* and is explained in detail in Choi et al. (2007) and Park (2007).

EXAMPLE 20.11　RELIABILITY-BASED DESIGN OPTIMIZATION

Perform RBDO with Example 20.7. The initial design is $(-0.8, 0.8)$ and $\sigma_1 = \sigma_2 = 0.1$. Compare the solution with that of the one obtained in Example 20.7.

Solution

The optimization formulation for reliability-based design optimization is

Design variables:
$$\mathbf{X} \qquad\qquad\qquad (a)$$

Cost function:
$$\text{Minimize } F(\mathbf{X}) \qquad\qquad (b)$$

Constraints:
$$\beta_i(\mathbf{X}) \geq \beta_{i,target} \quad (i = 1, 2) \qquad (c)$$

The target reliability index is set as $\beta_{target} = 3.0$ (it is specified by the user). In solving this optimization problem, the inner optimization problems are solved using the procedure described in Example 20.10. The outer problem has inner optimization problems as constraints. Therefore, their derivatives are needed in the outer problem. These are calculated by the sensitivity analysis of the inner problems at their optimum points (Choi et al., 2007; Park, 2007).

The optimization results are shown in Table 20.26. The results of the deterministic optimization approach are also shown. Since RBDO has some reliability margins on the constraints, the optimum cost of RBDO is slightly higher than the cost of the deterministic approach.

TABLE 20.26　Optimum solutions for Example 20.11

		Initial point	Deterministic optimum	RBDO solution
Design variables	(x_1, x_2)	$(-0.8, 0.8)$	$(-0.304, 2.0)$	$(-0.148, 2.0)$
Objective function	f	-2.191	-8.876	-8.660
Constraints	g_1	-1.320	-0.003	-0.534
	g_2	-4.766	-4.161	-4.191

Vector and Matrix Algebra

Matrix and vector notation is compact and useful in describing many numerical methods and derivations. Matrix and vector algebra is a necessary tool in developing methods for the optimum design of systems. The solution to linear optimization problems (linear programming) involves an understanding of the solution process for a system of linear equations. Therefore, it is important to understand operations of vector and matrix algebra and to be comfortable with their notation. *The subject is often referred to as linear algebra* and has been well developed. It has become a standard tool in almost all engineering and scientific applications.

In this appendix, some fundamental properties of vectors and matrices are reviewed. For more comprehensive treatment of the subject, several excellent textbooks are available and should be consulted (Hohn, 1964; Franklin, 1968; Cooper and Steinberg, 1970; Stewart, 1973; Bell, 1975; Strang, 1976; Jennings, 1977; Deif, 1982; Gere and Weaver, 1983). In addition, most software libraries have subroutines for linear algebra operations which should be directly utilized.

After reviewing the basic vector and matrix notations, special matrices, determinants, and matrix rank, the solution to a simultaneous system of linear equations is discussed: first an $n \times n$ system and then a rectangular $m \times n$ system. A section on linear independence of vectors is also included. Finally, the eigenvalue problem encountered in many fields of engineering is discussed. Such problems play a prominent role in convex programming problems and sufficiency conditions for optimization.

A.1 DEFINITION OF MATRICES

A matrix is defined as a rectangular array of quantities that can be real numbers, complex numbers, or functions of several variables. The entries in the rectangular array are also called the elements of the matrix. Since the solution to simultaneous linear equations is the most common application of matrices, we use them to develop the notion of matrices.

Consider this system of two simultaneous linear equations in three unknowns:

$$x_1 + 2x_2 + 3x_3 = 6$$
$$-x_1 + 6x_2 - 2x_3 = 3 \tag{A.1}$$

The symbols x_1, x_2, x_3 represent the *solution variables for the system of equations*. Note that the variables x_1, x_2, and x_3 can be replaced by any other variables, say w_1, w_2, and w_3, without affecting the solution. Therefore, they are sometimes called *dummy variables*.

Since they are dummy variables, they can be omitted while writing the equations in a matrix form. For example, Eqs. (A.1) can be written in a rectangular array as

$$\begin{bmatrix} 1 & 2 & 3 & | & 6 \\ -1 & 6 & -2 & | & 3 \end{bmatrix} \tag{A.2}$$

The entries to the left of the vertical line are coefficients of the variables x_1, x_2, and x_3, and to the right of the vertical line are the numbers on the right side of the equations. It is customary to enclose the array with square brackets as shown. Thus, we see that the system of equations in Eqs. (A.1) can be represented by a matrix having two rows and four columns.

An array with m rows and n columns is called a matrix of order "m by n," written as (m, n) or as $m \times n$. To distinguish between matrices and scalars, we will boldface the variables that represent matrices. In addition, uppercase letters will be used to represent matrices. For example, a general matrix $\mathbf{A}$ of order $m \times n$ can be represented as

$$\mathbf{A} = \begin{bmatrix} a_{11} & a_{12} & \cdots & a_{1n} \\ a_{21} & a_{22} & \cdots & a_{2n} \\ \vdots & \vdots & & \vdots \\ a_{m1} & a_{m2} & \cdots & a_{mn} \end{bmatrix} \tag{A.3}$$

The coefficients a_{ij} are called elements of the matrix $\mathbf{A}$; the subscripts i and j indicate the row and column numbers for the element a_{ij} (e.g., a_{32} represents the element in the third row and second column). Although the elements can be real numbers, complex numbers, or functions, we will not deal with complex matrices in the present text. We will encounter matrices having elements as functions of several variables, for example, the Hessian matrix of a function discussed in Chapter 4.

It is useful to employ a more compact notation for matrices. For example, a matrix $\mathbf{A}$ of order $m \times n$ with a_{ij} as its elements is written compactly as

$$\mathbf{A} = [a_{ij}]_{(m \times n)} \tag{A.4}$$

Often, the size of the matrix is not shown, and $\mathbf{A}$ is written as $[a_{ij}]$. If a matrix has the same number of rows and columns, it is called a *square matrix*. In Eq. (A.3) or Eq. (A.4), if $m = n$, $\mathbf{A}$ is a square matrix. It is called a matrix of order n.

It is important to understand the matrix notation for a set of linear equations because we will encounter such equations quite often in this text. For example, Eqs. (A.1) can be written as

$$\begin{array}{cccc} x_1 & x_2 & x_3 & \mathbf{b} \end{array}$$
$$\begin{bmatrix} 1 & 2 & 3 & | & 6 \\ -1 & 6 & -2 & | & 3 \end{bmatrix} \tag{A.5}$$

The array in Eq. (A.5) containing coefficients of the equations and the right-side parameters is called the *augmented matrix*.

Note that *each column of the matrix is identified with a variable*; the first column is associated with the variable x_1 because it contains coefficients of x_1 for all equations; the second is associated with x_2; the third, with x_3; and the last column, with the right-side vector, which we call **b**. This interpretation is important while solving linear equations (discussed later) or linear programming problems (discussed in Chapter 8).

A.2 TYPES OF MATRICES AND THEIR OPERATIONS

A.2.1 Null Matrix

A matrix having all zero elements is called a *null (zero) matrix*, denoted by a boldfaced zero, **0**. Any *zero matrix* of proper order when premultiplied or postmultiplied by any other matrix (or scalar) results in a zero matrix.

A.2.2 Vector

A matrix of order $1 \times n$ is called a *row matrix*, or simply a *row vector*. Similarly, a matrix of order $n \times 1$ is called a *column matrix*, or simply a *column vector*. A vector with n elements is called an n-component vector, or simply an *n-vector*. In this text, all vectors are considered to be column vectors and denoted by a lowercase letter in boldface.

A.2.3 Addition of Matrices

If **A** and **B** are two matrices of the order $m \times n$, then their *sum* is also an $m \times n$ matrix defined as

$$C_{(m \times n)} = A + B; \quad c_{ij} = a_{ij} + b_{ij} \text{ for all } i \text{ and } j \tag{A.6}$$

Matrix addition satisfies the following properties

$$A + B = B + A \text{ (commutative)} \tag{A.7}$$

If **A**, **B**, and **C** are three matrices of the same order, then

$$A + (B + C) = (A + B) + C \text{ (associative)} \tag{A.8}$$

If **A**, **B**, and **C** are of same order, then

$$A + C = B + C \text{ implies } A = B \tag{A.9}$$

where **A** = **B** implies that the matrices are equal. Two matrices **A** and **B** of order $m \times n$ are equal if $a_{ij} = b_{ij}$ for $i = 1$ to m and $j = 1$ to n.

A.2.4 Multiplication of Matrices

Multiplication of a matrix $\mathbf{A}$ of order $m \times n$ by a scalar k is defined as

$$k\mathbf{A} = [ka_{ij}]_{(m \times n)} \qquad (A.10)$$

The multiplication (product) $\mathbf{AB}$ of two matrices $\mathbf{A}$ and $\mathbf{B}$ is defined only if $\mathbf{A}$ and $\mathbf{B}$ are of proper order. The number of columns of $\mathbf{A}$ must be equal to the number of rows of $\mathbf{B}$ for the product $\mathbf{AB}$ to be defined. In that case, the matrices are said to be *conformable* for multiplication. If $\mathbf{A}$ is $m \times n$ and $\mathbf{B}$ is $r \times p$, then the multiplication $\mathbf{AB}$ is defined only when $n = r$, and multiplication $\mathbf{BA}$ is defined only when $m = p$. Multiplication of two matrices of proper order results in a third matrix. If $\mathbf{A}$ and $\mathbf{B}$ are of order $m \times n$ and $n \times p$ respectively, then

$$\mathbf{AB} = \mathbf{C} \qquad (A.11)$$

where $\mathbf{C}$ is a matrix of order $m \times p$. Elements of the matrix $\mathbf{C}$ are determined by multiplying the elements of a row of $\mathbf{A}$ with the elements of a column of $\mathbf{B}$ and adding all the multiplications. Thus

$$\begin{bmatrix} a_{11} & a_{12} & \cdots & a_{1n} \\ a_{21} & a_{22} & \cdots & a_{2n} \\ \vdots & \vdots & & \vdots \\ a_{m1} & a_{m2} & \cdots & a_{mn} \end{bmatrix} \begin{bmatrix} b_{11} & b_{12} & \cdots & b_{1p} \\ b_{21} & b_{22} & \cdots & b_{2p} \\ \vdots & \vdots & & \vdots \\ b_{n1} & b_{n2} & \cdots & b_{np} \end{bmatrix} = \begin{bmatrix} c_{11} & c_{12} & \cdots & c_{1p} \\ c_{21} & c_{22} & \cdots & c_{2p} \\ \vdots & \vdots & & \vdots \\ c_{m1} & c_{m2} & \cdots & c_{mp} \end{bmatrix} \qquad (A.12)$$

where elements c_{ij} are calculated as

$$c_{ij} = a_{i1}b_{1j} + a_{i2}b_{2j} + \ldots + a_{in}b_{nj} = \sum_{k=1}^{n} a_{ik}b_{kj} \qquad (A.13)$$

Note that if $\mathbf{B}$ is an $n \times 1$ matrix (i.e., a vector), then $\mathbf{C}$ is an $m \times 1$ matrix. We will encounter this type of matrix multiplication quite often in this text; for example, a system of linear equations is represented as $\mathbf{Ax} = \mathbf{b}$, where $\mathbf{x}$ contains the solution variables and $\mathbf{b}$ the right-side parameters. Equation (A.1) can be written in this form.

In the product $\mathbf{AB}$, the matrix $\mathbf{A}$ is said to be *postmultiplied* by $\mathbf{B}$ or $\mathbf{B}$ is said to be *premultiplied* by $\mathbf{A}$. Whereas the matrix addition satisfies the commutative law, matrix multiplication, in general, does not; that is, $\mathbf{AB} \neq \mathbf{BA}$. Also, even if $\mathbf{AB}$ is well defined, $\mathbf{BA}$ may not be defined.

EXAMPLE A.1   MULTIPLICATION OF MATRICES

$$\mathbf{A} = \begin{bmatrix} 2 & 3 & 1 \\ 6 & 3 & 2 \\ 4 & 2 & 0 \\ 0 & 3 & 5 \end{bmatrix}_{(4 \times 3)} \qquad \mathbf{B} = \begin{bmatrix} 2 & -1 \\ 1 & 0 \\ 3 & -2 \end{bmatrix}_{(3 \times 2)} \qquad \text{(a)}$$

$$AB = \begin{bmatrix} 2 & 3 & 1 \\ 6 & 3 & 2 \\ 4 & 2 & 0 \\ 0 & 3 & 5 \end{bmatrix} \begin{bmatrix} 2 & -1 \\ 1 & 0 \\ 3 & -2 \end{bmatrix} = \begin{bmatrix} (2\times2+3\times1+1\times3) & (-2\times1+3\times0-1\times2) \\ (6\times2+3\times1+2\times3) & (-6\times1+3\times0-2\times2) \\ (4\times2+2\times1+0\times3) & (-4\times1+2\times0-0\times2) \\ (0\times2+3\times1+5\times3) & (-0\times1+3\times0-5\times2) \end{bmatrix} = \begin{bmatrix} 10 & -4 \\ 21 & -10 \\ 10 & -4 \\ 18 & -10 \end{bmatrix}_{(4\times2)} \quad (b)$$

Note that the product **BA** is not defined because the number of columns in **B** is not equal to the number of rows in **A**.

EXAMPLE A.2 MULTIPLICATION OF MATRICES

$$A = \begin{bmatrix} 4 & 1 \\ -2 & 0 \\ 8 & 3 \end{bmatrix} \quad B = \begin{bmatrix} -3 & 8 & 6 \\ 8 & 3 & -1 \end{bmatrix} \quad (a)$$

$$AB = \begin{bmatrix} 4 & 1 \\ -2 & 0 \\ 8 & 3 \end{bmatrix} \begin{bmatrix} -3 & 8 & 6 \\ 8 & 3 & -1 \end{bmatrix} = \begin{bmatrix} -4 & 35 & 23 \\ 6 & -16 & -12 \\ 0 & 73 & 45 \end{bmatrix} \quad (b)$$

$$BA = \begin{bmatrix} -3 & 8 & 6 \\ 8 & 3 & -1 \end{bmatrix} \begin{bmatrix} 4 & 1 \\ -2 & 0 \\ 8 & 3 \end{bmatrix} = \begin{bmatrix} 20 & 15 \\ 18 & 5 \end{bmatrix} \quad (c)$$

Note that for the products **AB** and **BA** to be matrices of the same order, **A** and **B** must be square matrices.

Note also that even if matrices **A**, **B**, and **C** are properly defined, **AB** = **AC** does not imply **B** = **C**. Also, **AB** = **0** does not imply either **B** = **0** or **A** = **0**. The matrix multiplication, however, satisfies two important laws: associative and distributive. Let matrices **A**, **B**, **C**, **D**, and **F** be of proper dimension. Then

Associative law:

$$(AB)\,C = A\,(BC) \tag{A.14}$$

Distributive law:

$$B\,(C + D) = BC + BD \tag{A.15}$$

$$(C + D)\,F = CF + DF \tag{A.16}$$

$$(A + B)(C + D) = AC + AD + BC + BD \tag{A.17}$$

A.2.5 Transpose of a Matrix

We can write rows of a matrix as columns and obtain another matrix. Such an operation is called the *transpose* of a matrix. If $\mathbf{A} = [a_{ij}]$ is an $m \times n$ matrix, then its transpose, denoted as $\mathbf{A}^T$, is an $n \times m$ matrix. It is obtained from $\mathbf{A}$ by changing its rows to columns (or columns to rows). The first column of $\mathbf{A}$ is the first row of $\mathbf{A}^T$; the second column of $\mathbf{A}$ is the second row of $\mathbf{A}^T$; and so on. Thus, if $\mathbf{A} = [a_{ij}]$, then $\mathbf{A}^T = [a_{ji}]$. The operation of transposing a matrix is illustrated by the following 2×3 matrix:

$$\mathbf{A} = \begin{bmatrix} a_{11} & a_{12} & a_{13} \\ a_{21} & a_{22} & a_{23} \end{bmatrix}; \quad \mathbf{A}^T = \begin{bmatrix} a_{11} & a_{21} \\ a_{12} & a_{22} \\ a_{13} & a_{23} \end{bmatrix} \tag{A.18}$$

Some properties of the transpose are

1. $(\mathbf{A}^T)^T = \mathbf{A}$ $\tag{A.19}$

2. $(\mathbf{A} + \mathbf{B})^T = \mathbf{A}^T + \mathbf{B}^T$ $\tag{A.20}$

3. $(\alpha\mathbf{A})^T = \alpha\mathbf{A}^T, \quad \alpha = \text{scalar}$ $\tag{A.21}$

4. $(\mathbf{AB})^T = \mathbf{B}^T\mathbf{A}^T$ $\tag{A.22}$

A.2.6 Elementary Row–Column Operations

There are three simple but extremely useful operations for rows or columns of a matrix. They are used in later discussions, so we state them here:

1. Interchange any two rows (columns).
2. Multiply any row (column) by a nonzero scalar.
3. Add to any row (column) a scalar multiple of another row (column).

A.2.7 Equivalence of Matrices

A matrix $\mathbf{A}$ is said to be *equivalent* to another matrix $\mathbf{B}$, written as $\mathbf{A} \sim \mathbf{B}$, if $\mathbf{A}$ can be transformed into $\mathbf{B}$ by means of one or more elementary row and/or column operations. If only row (column) operations are used, we say $\mathbf{A}$ is *row* (*column*) equivalent to $\mathbf{B}$.

A.2.8 Scalar Product—Dot Product of Vectors

A special case of matrix multiplication of particular interest is the multiplication of a row vector by a column vector. If $\mathbf{x}$ and $\mathbf{y}$ are two n-component vectors, then

$$\mathbf{x}^T\mathbf{y} = \sum_{j=1}^{n} x_j y_j \tag{A.23}$$

where

$$\mathbf{x}^T = [x_1 \quad x_2 \quad \ldots x_n] \text{ and } y = [y_1 \quad y_2 \quad \ldots y_n]^T \tag{A.24}$$

The product in Eq. (A.23) is called the *scalar* or *dot product* of **x** and **y**. It is also denoted as $(\mathbf{x} \cdot \mathbf{y})$. Note that since the dot product of two vectors is a scalar, $\mathbf{x}^T\mathbf{y} = \mathbf{y}^T\mathbf{x}$.

Associated with any vector **x** is a scalar called the *norm or length* of the vector, defined as

$$(\mathbf{x}^T\mathbf{x})^{1/2} = \left(\sum_{i=1}^{n} x_i^2 \right)^{1/2} \tag{A.25}$$

Often the norm of **x** is designated as $||\mathbf{x}||$.

A.2.9 Square Matrices

A matrix having the same number of rows and columns is called a *square matrix*; otherwise, it is called a *rectangular matrix*. The elements a_{ii}, $i = 1$ to n are called the *main diagonal elements* and others are called the *off-diagonal elements*. A square matrix having zero entries at all off-diagonal locations is called a *diagonal matrix*. If all main diagonal elements of a diagonal matrix are equal, it is called a *scalar matrix*.

A square matrix **A** is *symmetric* if $\mathbf{A}^T = \mathbf{A}$ and *asymmetric* or *unsymmetric* otherwise. It is *antisymmetric* if $\mathbf{A}^T = -\mathbf{A}$. If all elements below the main diagonal of a square matrix are zero ($a_{ij} = 0$ for $i > j$), it is an *upper triangular matrix*. Similarly, a *lower triangular matrix* has all zero elements above the main diagonal ($a_{ij} = 0$ for $i < j$). A matrix that has all zero entries except in a band around the main diagonal is a *banded matrix*.

A square matrix having unit elements on the main diagonal and zeros elsewhere is an *identity matrix*. An identity matrix of order n is denoted as $\mathbf{I}_{(n)}$. Identity matrices are useful because their pre- or postmultiplication with another matrix does not change them. For example, let **A** be any $m \times n$ matrix; then

$$\mathbf{I}_{(m)}\mathbf{A} = \mathbf{A} = \mathbf{A}\mathbf{I}_{(n)} \tag{A.26}$$

A scalar matrix $\mathbf{S}_{(n)}$ having diagonal elements as α can be written as

$$\mathbf{S}_{(n)} = \alpha\mathbf{I}_{(n)} \tag{A.27}$$

Note that premultiplying or postmultiplying any matrix by a scalar matrix of proper order results in multiplying the original matrix by the scalar. This can be proved for any $m \times n$ matrix **A** as follows:

$$\mathbf{S}_{(m)}\mathbf{A} = \alpha\mathbf{I}_{(m)}\mathbf{A} = \alpha\mathbf{A} = \alpha(\mathbf{A}\mathbf{I}_{(n)}) = \mathbf{A}(\alpha\mathbf{I}_{(n)}) = \mathbf{A}\mathbf{S}_{(n)} \tag{A.28}$$

A.2.10 Partitioning of Matrices

It is often useful to divide vectors and matrices into a smaller group of elements. This can be done by partitioning a matrix into smaller rectangular arrays, called *submatrices*, and a vector into *subvectors*. For example, consider a matrix $\mathbf{A}$ as

$$\mathbf{A} = \begin{bmatrix} 2 & 1 & -6 & 4 & 3 \\ 2 & 3 & 8 & -1 & -3 \\ 1 & -6 & 2 & 3 & 8 \\ -3 & 0 & 5 & -2 & 7 \end{bmatrix}_{(4\times5)} \tag{A.29}$$

A possible partitioning of $\mathbf{A}$ is

$$\mathbf{A} = \left[\begin{array}{ccc|cc} 2 & 1 & -6 & 4 & 3 \\ 2 & 3 & 8 & -1 & -3 \\ \hline 1 & -6 & 2 & 3 & 8 \\ -3 & 0 & 5 & -2 & 7 \end{array} \right]_{(4\times5)} \tag{A.30}$$

Therefore, the submatrices of $\mathbf{A}$ are

$$\mathbf{A}_{11} = \begin{bmatrix} 2 & 1 & -6 \\ 2 & 3 & 8 \end{bmatrix}_{(2\times3)} \qquad \mathbf{A}_{12} = \begin{bmatrix} 4 & 3 \\ -1 & -3 \end{bmatrix}_{(2\times2)} \tag{A.31}$$

$$\mathbf{A}_{21} = \begin{bmatrix} 1 & -6 & 2 \\ -3 & 0 & 5 \end{bmatrix}_{(2\times3)} \qquad \mathbf{A}_{22} = \begin{bmatrix} 3 & 8 \\ -2 & 7 \end{bmatrix}_{(2\times2)} \tag{A.32}$$

where $\mathbf{A}_{ij}$ are matrices of proper order. Thus, $\mathbf{A}$ can be written in terms of submatrices as

$$\mathbf{A} = \left[\begin{array}{c|c} \mathbf{A}_{11} & \mathbf{A}_{12} \\ \hline \mathbf{A}_{21} & \mathbf{A}_{22} \end{array} \right] \tag{A.33}$$

Note that partitioning of vectors and matrices must be proper so that the operations of addition or multiplication remain defined. To see how two partitioned matrices are multiplied, consider $\mathbf{A}$ an $m \times n$ matrix and $\mathbf{B}$ an $n \times p$ matrix. Let these be partitioned as

$$\mathbf{A} = \left[\begin{array}{c|c} \mathbf{A}_{11} & \mathbf{A}_{12} \\ \hline \mathbf{A}_{21} & \mathbf{A}_{22} \end{array} \right]_{(m \times n)} \; ; \quad \mathbf{B} = \left[\begin{array}{c|c} \mathbf{B}_{11} & \mathbf{B}_{12} \\ \hline \mathbf{B}_{21} & \mathbf{B}_{22} \end{array} \right]_{(n \times p)} \tag{A.34}$$

Then the product $\mathbf{AB}$ can be written as

$$\mathbf{AB} = \left[\begin{array}{c|c} \mathbf{A}_{11} & \mathbf{A}_{12} \\ \hline \mathbf{A}_{21} & \mathbf{A}_{22} \end{array} \right] \left[\begin{array}{c|c} \mathbf{B}_{11} & \mathbf{B}_{12} \\ \hline \mathbf{B}_{21} & \mathbf{B}_{22} \end{array} \right] = \left[\begin{array}{c|c} (\mathbf{A}_{11}\mathbf{B}_{11} + \mathbf{A}_{12}\mathbf{B}_{21}) & (\mathbf{A}_{11}\mathbf{B}_{12} + \mathbf{A}_{12}\mathbf{B}_{22}) \\ \hline (\mathbf{A}_{21}\mathbf{B}_{11} + \mathbf{A}_{22}\mathbf{B}_{21}) & (\mathbf{A}_{21}\mathbf{B}_{12} + \mathbf{A}_{22}\mathbf{B}_{22}) \end{array} \right]_{(m \times p)} \tag{A.35}$$

Note that the partitioning of matrices $\mathbf{A}$ and $\mathbf{B}$ must be such that the matrix products $\mathbf{A}_{11}\mathbf{B}_{11}$, $\mathbf{A}_{12}\mathbf{B}_{21}$, $\mathbf{A}_{11}\mathbf{B}_{12}$, $\mathbf{A}_{12}\mathbf{B}_{22}$, and so on, are proper. In addition, the pairs of matrices $\mathbf{A}_{11}\mathbf{B}_{11}$ and $\mathbf{A}_{12}\mathbf{B}_{21}$, $\mathbf{A}_{11}\mathbf{B}_{12}$, and $\mathbf{A}_{12}\mathbf{B}_{22}$, and so on, must be of the same order.

A.3 SOLVING n LINEAR EQUATIONS IN n UNKNOWNS

A.3.1 Linear Systems

Linear equations are encountered in numerous engineering and scientific applications. Therefore, substantial work has been done to devise numerical solution procedures for them. It is important to understand the basic ideas and concepts related to linear equations because we use them quite often in this text. In this section we will describe a basic procedure, known as *Gaussian elimination*, for solving an $n \times n$ *(square) system of linear equations*. More general methods for solving a *rectangular $m \times n$ system* are discussed in the next section.

It turns out that the idea of determinants is closely related to solving a set of linear equations, so first we discuss determinants and their properties. It also turns out that the solution to a square system can be found by inverting the matrix associated with the system, so we describe methods for inverting matrices.

Let us consider the following system of *n* equations in *n* unknowns:

$$\mathbf{A}\mathbf{x} = \mathbf{b} \tag{A.36}$$

where $\mathbf{A}$ is an $n \times n$ matrix of specified constants, $\mathbf{x}$ is an *n*-vector of solution variables, and $\mathbf{b}$ is an *n*-vector of specified constants known as the right-side vector. $\mathbf{A}$ is called the *coefficient matrix* and when the vector $\mathbf{b}$ is added as the $(n + 1)$th column of $\mathbf{A}$ as $[\mathbf{A}|\mathbf{b}]$, the result is called the *augmented matrix* for the given system of equations. If the right-side vector $\mathbf{b}$ is zero, Eq. (A.36) is called a *homogeneous* system; otherwise, it is called a *nonhomogeneous system* of equations.

The equation $\mathbf{A}\mathbf{x} = \mathbf{b}$ can also be written in the following summation form:

$$\sum_{j=1}^{n} a_{ij}x_j = b_i; \quad i = 1 \text{ to } n \tag{A.37}$$

If each row of the matrix $\mathbf{A}$ is interpreted as an *n*-dimensional row vector, $\overline{\mathbf{a}}^{(i)}$, the left side of Eq. (A.37) can be interpreted as the dot product of two vectors:

$$(\overline{\mathbf{a}}^{(i)} \cdot \mathbf{x}) = b_i; \quad i = 1 \text{ to } n \tag{A.38}$$

If each column of $\mathbf{A}$ is interpreted as an *n*-dimensional column vector, $\mathbf{a}^{(i)}$, the left side of Eq. (A.36) can be interpreted as the summation of the scaled columns of the matrix $\mathbf{A}$:

$$\sum_{i=1}^{n} \mathbf{a}^{(i)}x_i = \mathbf{b} \tag{A.39}$$

These interpretations can be useful in devising solution strategies for the system $\mathbf{A}\mathbf{x} = \mathbf{b}$ and in their implementation. For example, Eq. (A.39) shows that the solution variable x_i is simply a scale factor for the *i*th column of $\mathbf{A}$; that is, variable x_i is associated with the *i*th column.

A.3.2 Determinants

To develop the solution strategies for the linear system $\mathbf{A}\mathbf{x} = \mathbf{b}$, we begin by introducing the concept of determinants and studying their properties. The methods for calculating determinants are intimately related to the procedures for solving linear equations, so we will also discuss them.

Every square matrix has a scalar associated with it, called the determinant, calculated from its elements. To introduce the idea of determinants, we set $n = 2$ in Eq. (A.36) and consider the following 2×2 system of simultaneous equations:

$$\begin{bmatrix} a_{11} & a_{12} \\ a_{21} & a_{22} \end{bmatrix} \begin{bmatrix} x_1 \\ x_2 \end{bmatrix} = \begin{bmatrix} b_1 \\ b_2 \end{bmatrix} \tag{A.40}$$

The number $(a_{11}a_{22} - a_{21}a_{12})$ calculated using elements of the coefficient matrix is called its determinant. To see how this number arises, we will solve the system in Eq. (A.40) by the *elimination process.*

Multiplying the first row by a_{22} and the second by a_{12} in Eq. (A.40), we get

$$\begin{bmatrix} a_{11}a_{22} & a_{12}a_{22} \\ a_{12}a_{21} & a_{12}a_{22} \end{bmatrix} \begin{bmatrix} x_1 \\ x_2 \end{bmatrix} = \begin{bmatrix} a_{22}b_1 \\ a_{12}b_2 \end{bmatrix} \tag{A.41}$$

Subtracting the second row from the first one in Eq. (A.41), we eliminate x_2 from the first equation and obtain

$$(a_{11}a_{22} - a_{12}a_{21})x_1 = a_{22}b_1 - a_{12}b_2 \tag{A.42}$$

Now, repeating the foregoing process to eliminate x_1 from the second row in Eq. (A.40) by multiplying the first equation by a_{21} and the second by a_{11} and subtracting, we obtain

$$(a_{11}a_{22} - a_{12}a_{21})x_2 = a_{11}b_2 - a_{21}b_1 \tag{A.43}$$

The coefficient of x_1 and x_2 in Eqs. (A.42) and (A.43) must be nonzero for a unique solution to the linear system, that is, $(a_{11}a_{22} - a_{12}a_{21}) \neq 0$, and the values of x_1 and x_2 are calculated as

$$x_1 = \frac{a_{22}b_1 - a_{12}b_2}{a_{11}a_{22} - a_{12}a_{21}}, \quad x_2 = \frac{a_{11}b_2 - a_{21}b_1}{a_{11}a_{22} - a_{12}a_{21}} \tag{A.44}$$

The denominator $(a_{11}a_{22} - a_{12}a_{21})$ in Eqs. (A.44) is identified as the *determinant* of the matrix $\mathbf{A}$ of Eq. (A.40). It is denoted by $\det(\mathbf{A})$, or $|\mathbf{A}|$. Thus, for any 2×2 matrix $\mathbf{A}$,

$$|\mathbf{A}| = a_{11}a_{22} - a_{12}a_{21} \tag{A.45}$$

Using the definition of Eq. (A.45), we can rewrite Eq. (A.44) as

$$x_1 = \frac{|\mathbf{B}_1|}{|\mathbf{A}|}, \quad x_2 = \frac{|\mathbf{B}_2|}{|\mathbf{A}|} \tag{A.46}$$

where $\mathbf{B}_1$ is obtained by replacing the first column of $\mathbf{A}$ with the right side, and $\mathbf{B}_2$ is obtained by replacing the second column of $\mathbf{A}$ with the right side:

$$\mathbf{B}_1 = \begin{bmatrix} b_1 & a_{12} \\ b_2 & a_{22} \end{bmatrix}, \quad \mathbf{B}_2 = \begin{bmatrix} a_{11} & b_1 \\ a_{21} & b_2 \end{bmatrix} \tag{A.47}$$

Equation (A.46) is known as *Cramer's rule.* According to this rule, we need to compute only the three determinants—$|\mathbf{A}|$, $|\mathbf{B}_1|$, and $|\mathbf{B}_2|$—to determine the solution to any 2×2 system of linear equations. If $|\mathbf{A}| = 0$, *there is no unique solution to Eq. (A.40). There may be an infinite number of solutions or no solution at all.* These cases are investigated in the next section.

The preceding concept of a determinant can be generalized to $n \times n$ matrices. For such a system, there are n equations in Eq. (A.46) based on Cramer's rule. *For every square matrix* **A** *of any order, we can associate a unique scalar, called the determinant of* **A**. There are many ways of calculating the determinant of a matrix. These procedures are closely related to those used to solve the linear system of equations that we will discuss later in this section.

Properties of Determinants

The determinants have several properties that are useful in devising procedures for their calculation. Therefore, these should be clearly understood.

1. The determinant of any square matrix **A** is also equal to the determinant of the transpose of the matrix (i.e., $|\mathbf{A}| = |\mathbf{A}^T|$).
2. If a square matrix **A** has two identical columns (or rows), then its determinant is zero (i.e., $|\mathbf{A}| = 0$).
3. If a new matrix is formed by interchanging any two columns (or rows) of a given matrix **A** (elementary row–column operation 1), the determinant of the resulting matrix is the negative of the determinant of the original matrix.
4. If a new matrix is formed by adding any multiple of one column (row) to a different column (row) of a given matrix (elementary row–column operation 3), the determinant of the resulting matrix is equal to the determinant of the original matrix.
5. If a square matrix **B** is identical to a matrix **A**, except some column (or row) is a scalar multiple c of the corresponding column (or row) of **A** (elementary row–column operation 2), then $|\mathbf{B}| = c|\mathbf{A}|$.
6. If elements of a column (or row) of a square matrix **A** are zero, then $|\mathbf{A}| = 0$.
7. If a square matrix **A** is lower or upper triangular, then the determinant of **A** is equal to the product of the diagonal elements:

$$|\mathbf{A}| = a_{11}a_{22} \cdot \ldots \cdot a_{nn} \tag{A.48}$$

8. If **A** and **B** are any two square matrices of the same order, then $|\mathbf{AB}| = |\mathbf{A}\ \mathbf{B}|$.
9. Let $|\mathbf{A}_{ij}|$ denote the determinant of a matrix obtained by deleting the ith row and the jth column of **A** (yielding a square matrix of order $n-1$); the scalar $|\mathbf{A}_{ij}|$ is called the *minor* of the element a_{ij} of matrix **A**.
10. The *cofactor* of a_{ij} is defined as

$$\mathrm{cofac}(a_{ij}) = (-1)^{i+j}|\mathbf{A}_{ij}| \tag{A.49}$$

The determinant of **A** is calculated in terms of the cofactors as

$$|\mathbf{A}| = \sum_{j=1}^{n} a_{ij}\ \mathrm{cofac}(a_{ij}), \quad \text{for any } i \tag{A.50}$$

or

$$|\mathbf{A}| = \sum_{i=1}^{n} a_{ij}\ \mathrm{cofac}(a_{ij}), \quad \text{for any } j \tag{A.51}$$

Note that the cofac(a_{ij}) is also a scalar obtained from the minor $|A_{ij}|$, but having a positive or negative sign determined by the indices i and j as $(-1)^{i+j}$. Equation (A.50) is called the cofactor expansion for $|\mathbf{A}|$ by the ith row; Eq. (A.51) is called the cofactor expansion for $|\mathbf{A}|$ by the jth column. Equations (A.50) and (A.51) can be used to prove Properties 2, 5, 6, and 7 directly.

It is important to note that it is difficult to use Eq. (A.50) or Eq. (A.51) to calculate the determinant of $\mathbf{A}$. These equations require calculation of the cofactors of the elements a_{ij}, which are determinants in themselves. However, using the *elementary row and column operations*, a square matrix can be converted to either lower or upper triangular form. The determinant is then computed using Eq. (A.48). This is illustrated in an example later in this section.

Singular Matrix

A matrix having a zero determinant is called a singular matrix; a matrix with a nonzero determinant is called nonsingular. A nonhomogeneous n $\times$ n *system of equations has a unique solution if and only if the matrix of coefficients is nonsingular.* These properties are discussed and used subsequently to develop methods for solving a system of linear equations.

Leading Principal Minor

Every $n \times n$ square matrix $\mathbf{A}$ has certain scalars associated with it, called the leading principal minors. They are obtained as determinants of certain submatrices of $\mathbf{A}$, and are useful in determining the "form" of a matrix that is needed to check sufficiency conditions for optimality as well as the convexity of functions, as discussed in Chapter 4. Therefore, we examine leading principal minors here.

Let M_k, $k = 1$ to n be called the leading principal minors of $\mathbf{A}$. Then each M_k is defined as the determinant of the following submatrix:

$$M_k = |\mathbf{A}_{kk}|, \quad k = 1 \text{ to } n \tag{A.52}$$

where $\mathbf{A}_{kk}$ is a $k \times k$ submatrix of $\mathbf{A}$ obtained by deleting the last $(n - k)$ columns and the corresponding rows. For example, $M_1 = a_{11}$, $M_2 = $ determinant of a 2×2 matrix obtained by deleting all rows and columns of $\mathbf{A}$ except the first two, and so on, are the leading principal minors of the matrix $\mathbf{A}$.

A.3.3 Gaussian Elimination Procedure

The elimination process described in Section A.3.1 for solving a 2×2 system of equations can be generalized to solve any $n \times n$ system of equations. The entire process can be organized and explained using matrix notation. The procedure, which can also be used to calculate the determinant of any matrix, is known as *Gaussian elimination*. We will describe it in detail in the following.

Using the three elementary row–column operations defined in Section A.2, the system $\mathbf{Ax} = \mathbf{b}$ of Eq. (A.36) can be transformed into the following form:

$$\begin{bmatrix} 1 & \bar{a}_{12} & \bar{a}_{13} & \cdots & \bar{a}_{1n} \\ 0 & 1 & \bar{a}_{23} & \cdots & \bar{a}_{2n} \\ \vdots & \vdots & \vdots & & \vdots \\ 0 & 0 & 0 & \cdots & 1 \end{bmatrix} \begin{bmatrix} x_1 \\ x_2 \\ \vdots \\ x_n \end{bmatrix} = \begin{bmatrix} \bar{b}_1 \\ \bar{b}_2 \\ \vdots \\ \bar{b}_n \end{bmatrix} \qquad (A.53)$$

Or, in expanded form, Eq. (A.53) becomes

$$\begin{aligned} x_1 + \bar{a}_{12}x_2 + \bar{a}_{13}x_3 + \; \cdots \; + \bar{a}_{1n}x_n &= \bar{b}_1 \\ x_2 + \bar{a}_{23}x_3 + \; \cdots \; + \bar{a}_{2n}x_n &= \bar{b}_2 \\ x_3 + \; \cdots \; + \bar{a}_{3n}x_n &= \bar{b}_3 \\ \vdots \qquad \vdots \\ x_n &= \bar{b}_n \end{aligned} \qquad (A.54)$$

Note that we use $\bar{a}_{ij}$ and $\bar{b}_i$ to represent modified elements a_{ij} and b_j of the original system. From the nth equation of the system (A.54), we have $x_n = \bar{b}_n$. If we substitute this value into the $(n-1)$th equation of (A.54), we can solve for x_{n-1}:

$$x_{n-1} = \bar{b}_{n-1} - \bar{a}_{n-1,n}x_n = \bar{b}_{n-1} - \bar{a}_{n-1,n}\bar{b}_n \qquad (A.55)$$

Equation (A.55) can now be substituted into the $(n-2)$th equation of (A.54) and x_{n-2} can be determined. Continuing in this manner, each of the unknowns can be solved in reverse order: $x_n, x_{n-1}, x_{n-2}, \ldots, x_2, x_1$. The procedure of reducing a system of n equations in n unknowns and then solving successively for $x_n, x_{n-1}, x_{n-2}, \ldots, x_2, x_1$ is called *Gaussian elimination* or *Gauss reduction*. The latter part of the method (solving successively for x_n, $x_{n-1}, x_{n-2}, \ldots, x_2, x_1$) is called *backward substitution, or backward pass*.

Elimination Procedure

The Gaussian elimination procedure uses elementary row–column operations to convert the main diagonal elements of the given coefficient matrix to 1 and the elements below the main diagonal to zero. To carry out these operations we use the following steps:

1. We start with the first row and the first column of the given matrix augmented with the right side of the system of equations.
2. To make the diagonal element 1, the first row is divided by the diagonal element.
3. To convert the elements in the first column below the main diagonal to zero, we multiply the first row by the element $\bar{a}_{i1}$ in the ith row ($i = 2$ to n). The resulting elements of the first row are subtracted from the ith row. This makes the element $\bar{a}_{i1}$ zero in the ith row.
4. The operations in Step 3 are carried out for each row using the first row for elimination each time.
5. Once all of the elements below the main diagonal are zero in the first column, the procedure is repeated for the second column using the second row for elimination, and so on.

The row used to obtain zero elements in a column is called the *pivot row,* and the column in which elimination is performed is called the *pivot column.* We will illustrate this procedure in an example later.

The foregoing operations of converting elements below the main diagonal to zero can be explained in another way. When we make the elements below the main diagonal in the first column zero, we are eliminating the variable x_1 from all of the equations except the first one (x_1 is associated with the first column). For this elimination step we use the first equation. In general, when we reduce the elements below the main diagonal in the ith column to zero, we use the ith row as the pivot row. Thus, we eliminate the ith variable from all equations below the ith row. This explanation is quite straightforward once we realize that each column of the coefficient matrix has a variable associated with it, as noted before.

EXAMPLE A.3 SOLUTIONS TO EQUATIONS BY GAUSSIAN ELIMINATION

Solve the following 3×3 system of equations:

$$
\begin{aligned}
x_1 - x_2 + x_3 &= 0 \\
x_1 - x_2 + 2x_3 &= 1 \\
x_1 + x_2 + 2x_3 &= 5
\end{aligned}
\tag{a}
$$

Solution

We will illustrate the Gaussian elimination procedure in a step-by-step manner using the augmented matrix idea. The augmented matrix for Eq. (a) is defined using the coefficients of the variables in each equation and the right-side parameters:

$$
\mathbf{B} =
\begin{array}{cccc}
x_1 & x_2 & x_3 & \mathbf{b} \\
\end{array}
\left[
\begin{array}{cccc}
1 & -1 & 1 & 0 \\
1 & -1 & 2 & 1 \\
1 & 1 & 2 & 5
\end{array}
\right]
\tag{b}
$$

To convert the preceding system to the form of Eq. (A.53), we use the elementary row–column operations as follows:

1. Add $-1 \times$ row 1 to row 2 and $-1 \times$ row 1 to row 3 (eliminating x_1 from the second and third equations; elementary row operation 3):

$$
\mathbf{B} \sim
\begin{array}{cccc}
x_1 & x_2 & x_3 & \mathbf{b} \\
\end{array}
\left[
\begin{array}{ccc|c}
1 & -1 & 1 & 0 \\
0 & 0 & 1 & 1 \\
0 & 2 & 1 & 5
\end{array}
\right]
\left(
\begin{array}{l}
\text{recall that symbol "}\sim\text{"} \\
\text{means equivalence} \\
\text{between matrices}
\end{array}
\right)
\tag{c}
$$

2. Since the element at location (2, 2) is zero, interchange rows 2 and 3 to bring a nonzero element to that location (elementary row operation 1). Then dividing the new second row by 2 gives

$$
\mathbf{B} \sim
\begin{array}{cccc}
x_1 & x_2 & x_3 & \mathbf{b} \\
\end{array}
\left[
\begin{array}{ccc|c}
1 & -1 & 1 & 0 \\
0 & 1 & 0.5 & 2.5 \\
0 & 0 & 1 & 1
\end{array}
\right]
\tag{d}
$$

3. Since the element at the location (3, 3) is one and all elements below the main diagonal are zero, the foregoing matrix puts the system of Eqs. (a) into the form of Eq. (A.53):

$$\begin{bmatrix} 1 & -1 & 1 \\ 0 & 1 & 0.5 \\ 0 & 0 & 1 \end{bmatrix} \begin{bmatrix} x_1 \\ x_2 \\ x_3 \end{bmatrix} = \begin{bmatrix} 0 \\ 2.5 \\ 1 \end{bmatrix} \qquad \text{(e)}$$

Performing backward substitution, we obtain

$$\begin{aligned} x_3 &= 1 \text{(from third row)} \\ x_2 &= 2.5 - 0.5x_3 = 2 \text{(from second row)} \\ x_1 &= 0 - x_3 + x_2 = 1 \text{(from first row)} \end{aligned} \qquad \text{(f)}$$

Therefore, the solution to Eq. (a) is

$$x_1 = 1, \quad x_2 = 2, \quad x_3 = 1 \qquad \text{(g)}$$

The Gaussian elimination method can easily be transcribed into a general-purpose computer program that can handle any given system of linear equations. However, certain modifications must be made to the procedure because numerical calculations on a machine with a finite number of digits introduce round-off errors. These errors can become significantly large if certain precautions are not taken. The modifications primarily involve a reordering of the rows or columns of the augmented matrix in such a way that possible round-off effects tend to be minimized.

This reordering must be performed at each step of the elimination process so that the diagonal element of the pivot row is the absolute largest among the elements of the remaining matrix on the lower right side. This is known as the *total pivoting* procedure. When only the rows are interchanged to bring the absolute largest element from a column to the diagonal location, the procedure is known as *partial pivoting*. Note that many programs are available to solve a system of linear equations. Thus, before attempting to write a program for Gaussian elimination, availability of existing programs should be explored.

EXAMPLE A.4 DETERMINANT OF A MATRIX BY GAUSSIAN ELIMINATION

The Gaussian elimination procedure can also be used to calculate the determinant of a matrix. We illustrate the procedure for the following 3×3 matrix:

$$\mathbf{A} = \begin{bmatrix} 2 & 3 & 0 \\ 1 & 2 & 1 \\ 0 & 3 & 4 \end{bmatrix} \qquad \text{(a)}$$

Solution

Using the Gaussian elimination procedure, we make the elements below the main diagonal zero, but this time the diagonal elements are not converted to unity. Once the matrix is converted to that form, the determinant is obtained using Eq. (A.48).

$$\begin{bmatrix} 2 & 3 & 0 \\ 1 & 2 & 1 \\ 0 & 3 & 4 \end{bmatrix} \sim \begin{bmatrix} 2 & 3 & 0 \\ 0 & 0.5 & 1 \\ 0 & 3 & 4 \end{bmatrix} \quad \text{(elimination in the first column)}$$

$$\sim \begin{bmatrix} 2 & 3 & 0 \\ 0 & 0.5 & 1 \\ 0 & 0 & -2 \end{bmatrix} \quad \text{(elimination in the second column)}$$

(b)

The preceding system is in the upper-triangular form, and $|\mathbf{A}|$ is given simply by multiplication of all the diagonal elements:

$$|\mathbf{A}| = (2)(0.5)(-2) = -2 \qquad\qquad (c)$$

A.3.4 Inverse of a Matrix: Gauss-Jordan Elimination

If the multiplication of two square matrices results in an identity matrix, they are called the inverse of each other. Let $\mathbf{A}$ and $\mathbf{B}$ be two square matrices of order n. Then $\mathbf{B}$ is called the inverse of $\mathbf{A}$ if

$$\mathbf{AB} = \mathbf{BA} = \mathbf{I}_{(n)} \qquad\qquad (A.56)$$

The inverse of $\mathbf{A}$ is usually denoted $\mathbf{A}^{-1}$. We will later describe methods to calculate the inverse of a matrix.

Not every square matrix may have an inverse. A matrix having no inverse is called a *singular matrix*. If the coefficient matrix of an $n \times n$ system of equations has an inverse, then the system can be solved for the unknown variables. Consider the $n \times n$ system of equations $\mathbf{Ax} = \mathbf{b}$, where $\mathbf{A}$ is the coefficient matrix and $\mathbf{b}$ is the right-side vector. Premultiplying both sides of the equation by $\mathbf{A}^{-1}$, we get

$$\mathbf{A}^{-1}\mathbf{Ax} = \mathbf{A}^{-1}\mathbf{b} \qquad\qquad (A.57)$$

Since $\mathbf{A}^{-1}\mathbf{A} = \mathbf{I}$, the equation reduces to

$$\mathbf{x} = \mathbf{A}^{-1}\mathbf{b} \qquad\qquad (A.58)$$

Thus, if we know the inverse of matrix $\mathbf{A}$, the preceding equation can be used to solve for the unknown vector $\mathbf{x}$.

Inverse by Cofactors

There are a couple of ways to calculate the inverse of a nonsingular matrix. The first is based on use of the cofactors of $\mathbf{A}$ and its determinant. If $\mathbf{B}$ is the inverse of $\mathbf{A}$, its elements are given as (called *inverse using cofactors*)

$$b_{ji} = \frac{\text{cofac}(a_{ij})}{|\mathbf{A}|}; \quad i = 1 \text{ to } n; \quad j = 1 \text{ to } n \qquad\qquad (A.59)$$

Note that indices on the left side of the equation are ji and on the right side they are ij. Thus, the cofactors of the row of matrix $\mathbf{A}$ generate the corresponding column of the inverse matrix $\mathbf{B}$.

The procedure just described is reasonable for smaller matrices up to, say, 3×3. For larger matrices, it becomes cumbersome and inefficient.

Inverse by Gaussian Elimination

A clue to the second procedure for calculating the inverse is provided by Eq. (A.56). In that equation, elements of $\mathbf{B}$ can be considered as unknowns for the system of linear equations

$$\mathbf{AB} = \mathbf{I} \tag{A.60}$$

Thus, the system can be solved using the Gaussian elimination procedure to obtain the inverse of $\mathbf{A}$. We illustrate the procedure with an example.

EXAMPLE A.5 INVERSE OF A MATRIX BY COFACTORS AND GAUSS-JORDAN REDUCTION

Compute the inverse of the following 3×3 matrix:

$$\mathbf{A} = \begin{bmatrix} 1 & 3 & 0 \\ 1 & 2 & 0 \\ 0 & 3 & 1 \end{bmatrix} \tag{a}$$

Solution

Inverse by cofactors Let $\mathbf{B}$ be a 3×3 matrix that is the inverse of matrix $\mathbf{A}$. In order to use the *cofactors approach* given in Eq. (A.59), we first calculate the determinant of $\mathbf{A}$ as $|\mathbf{A}| = -1$. Using Eq. (A.49), the cofactors of the first row of $\mathbf{A}$ are

$$\text{cofac}(a_{11}) = (-1)^{1+1} \begin{vmatrix} 2 & 0 \\ 3 & 1 \end{vmatrix} = 2 \tag{b}$$

$$\text{cofac}(a_{12}) = (-1)^{1+2} \begin{vmatrix} 1 & 0 \\ 0 & 1 \end{vmatrix} = -1 \tag{c}$$

$$\text{cofac}(a_{13}) = (-1)^{1+3} \begin{vmatrix} 1 & 2 \\ 0 & 3 \end{vmatrix} = 3 \tag{d}$$

Similarly, the cofactors of the second and third rows are

$$-3, \ 1, \ -3; \ 0, \ 0, \ -1 \tag{e}$$

Thus, Eq. (A.59) gives the inverse of $\mathbf{A}$ as

$$\mathbf{B} = \begin{bmatrix} -2 & 3 & 0 \\ 1 & -1 & 0 \\ -3 & 3 & 1 \end{bmatrix} \tag{f}$$

Inverse by Gaussian elimination We will first demonstrate the Gaussian elimination procedure before presenting the Gauss-Jordan procedure. Since **B** is the inverse of **A**, $\mathbf{AB} = \mathbf{I}$. Or, writing this in the expanded form,

$$\begin{bmatrix} 1 & 3 & 0 \\ 1 & 2 & 0 \\ 0 & 3 & 1 \end{bmatrix} \begin{bmatrix} b_{11} & b_{12} & b_{13} \\ b_{21} & b_{22} & b_{23} \\ b_{31} & b_{32} & b_{33} \end{bmatrix} = \begin{bmatrix} 1 & 0 & 0 \\ 0 & 1 & 0 \\ 0 & 0 & 1 \end{bmatrix} \tag{g}$$

where b_{ij} are the elements of **A**. The foregoing equation can be considered a system of simultaneous equations having three different right-side vectors. We can solve for each unknown column on the left side corresponding to each right-side vector by using the Gaussian elimination procedure. For example, considering the first column of **B** only, we obtain

$$\begin{bmatrix} 1 & 3 & 0 \\ 1 & 2 & 0 \\ 0 & 3 & 1 \end{bmatrix} \begin{bmatrix} b_{11} \\ b_{21} \\ b_{31} \end{bmatrix} = \begin{bmatrix} 1 \\ 0 \\ 0 \end{bmatrix} \tag{h}$$

Using the elimination procedure in the augmented matrix form, we obtain

$$\begin{bmatrix} 1 & 3 & 0 & | & 1 \\ 1 & 2 & 0 & | & 0 \\ 0 & 3 & 1 & | & 0 \end{bmatrix} \sim \begin{bmatrix} 1 & 3 & 0 & | & 1 \\ 0 & -1 & 0 & | & -1 \\ 0 & 3 & 1 & | & 0 \end{bmatrix} \text{ (elimination in the first column)}$$

$$\sim \begin{bmatrix} 1 & 3 & 0 & | & 1 \\ 0 & 1 & 0 & | & 1 \\ 0 & 0 & 1 & | & -3 \end{bmatrix} \text{ (elimination in the second column)} \tag{i}$$

Using back substitution, we obtain the first column of **B** as $b_{31} = -3$, $b_{21} = 1$, $b_{11} = -2$. Similarly, we find $b_{12} = 3$, $b_{22} = -1$, $b_{32} = 3$, $b_{13} = 0$, $b_{23} = 0$, and $b_{33} = 1$. Therefore, the inverse of **A** is given as

$$\mathbf{B} = \begin{bmatrix} -2 & 3 & 0 \\ 1 & -1 & 0 \\ -3 & 3 & 1 \end{bmatrix} \tag{j}$$

Inverse by Gauss-Jordan elimination We can organize the procedure for calculating the inverse of a matrix slightly differently. The augmented matrix can be defined with all three columns of the right side. The Gaussian elimination process can be carried out below as well as above the main diagonal. With this procedure, the left 3×3 matrix is converted to an identity matrix; the right 3×3 matrix then contains the inverse of the matrix. When elimination is performed below as well as above the main diagonal, the procedure is called *Gauss-Jordan elimination*. The process proceeds as follows for calculating the inverse of **A**:

$$\begin{bmatrix} 1 & 3 & 0 & | & 1 & 0 & 0 \\ 1 & 2 & 0 & | & 0 & 1 & 0 \\ 0 & 3 & 1 & | & 0 & 0 & 1 \end{bmatrix} \text{ (augmented martrix)}$$

$$\sim \begin{bmatrix} 1 & 3 & 0 & | & 1 & 0 & 0 \\ 0 & -1 & 0 & | & -1 & 1 & 0 \\ 0 & 3 & 1 & | & 0 & 0 & 1 \end{bmatrix} \quad \text{(elimination in the first column)}$$

(k)

$$\sim \begin{bmatrix} 1 & 0 & 0 & | & -2 & 3 & 0 \\ 0 & 1 & 0 & | & 1 & -1 & 0 \\ 0 & 0 & 1 & | & -3 & 3 & 1 \end{bmatrix} \quad \text{(elimination in the second column)}$$

There is no need to perform elimination on the third column since $\bar{a}_{13} = \bar{a}_{23} = 0$ and $\bar{a}_{33} = 1$. We observe from the above matrix that the last three columns give precisely the matrix **B**, which is the inverse of **A**.

The Gauss-Jordan procedure for computing the inverse of a 3×3 matrix can be generalized to any nonsingular $n \times n$ matrix. It can also be coded systematically into a general-purpose computer program to compute the inverse of a matrix.

A.4 SOLUTION TO m LINEAR EQUATIONS IN n UNKNOWNS

In the last section, the concept of determinants was used to ascertain the existence of a unique solution for any $n \times n$ system of equations. There are many instances in engineering applications where the number of equations is not equal to the number of variables—that is, rectangular systems. In a system of m equations in n unknowns ($m \neq n$), the matrix of coefficients is not square. Therefore, no determinant can be associated with it. Thus, to treat such systems, a more general concept than determinants is needed. We introduce such a concept in this section.

A.4.1 Rank of a Matrix

The general concept needed to develop the solution procedure for a general $m \times n$ system of equations is known as the *rank of the matrix, defined as the order of the largest nonsingular square submatrix of the given matrix.* Using the idea of rank of a matrix, we can develop a general theory for the solution of a system of linear equations.

Let r be the rank of an $m \times n$ matrix **A**. Then r satisfies the following conditions:

1. For $m < n$, $r \leq m < n$ (if $r = m$, the matrix is said to have full row rank).
2. For $n < m$, $r \leq n < m$ (if $r = n$, the matrix is said to have full column rank).
3. For $n = m$, $r \leq n$ (if $r = n$, the square matrix is called nonsingular).

To determine the rank of a matrix, we need to check the determinants of all of the submatrices. This is a cumbersome and time-consuming process. However, it turns out that the Gauss-Jordan elimination process can be used to solve the linear system as well as determine the rank of the matrix.

Using the Gauss-Jordan elimination procedure, we can transform any $m \times n$ matrix $\mathbf{A}$ into the following equivalent form (for $m < n$):

$$\mathbf{A} \sim \begin{bmatrix} \mathbf{I}_{(r)} & \mathbf{0}_{(r \times n - r)} \\ \mathbf{0}_{(m - r \times r)} & \mathbf{0}_{(m - r \times n - r)} \end{bmatrix} \qquad (A.61)$$

where $\mathbf{I}_{(r)}$ is the $r \times r$ identity matrix. Then r is the matrix's rank, where r satisfies one of the preceding three conditions. Note that the identity matrix $\mathbf{I}_{(r)}$ is unique for any given matrix.

EXAMPLE A.6   RANK DETERMINATION BY ELEMENTARY OPERATIONS

Determine rank of the following matrix:

$$\mathbf{A} = \begin{bmatrix} 2 & 6 & 2 & 4 \\ -2 & -4 & 2 & 2 \\ 1 & 2 & -1 & -1 \end{bmatrix} \qquad (a)$$

Solution

The elementary operations lead to the following matrices:

$$\mathbf{A} \sim \begin{bmatrix} 1 & 3 & 1 & 2 \\ -2 & -4 & 2 & 2 \\ 1 & 2 & -1 & -1 \end{bmatrix} \text{ (obtained by multiplying row 1 by } \tfrac{1}{2} \text{ in Eq. (a))} \qquad (b)$$

$$\mathbf{A} \sim \begin{bmatrix} 1 & 3 & 1 & 2 \\ 0 & 2 & 4 & 6 \\ 0 & -1 & -2 & -3 \end{bmatrix} \begin{pmatrix} \text{obtained by adding 2 times row 1 to row 2} \\ \text{and } -1 \text{ times row 1 to row 3 in Eq. (b)} \end{pmatrix} \qquad (c)$$

$$\mathbf{A} \sim \begin{bmatrix} 1 & 0 & 0 & 0 \\ 0 & 2 & 4 & 6 \\ 0 & -1 & -2 & -3 \end{bmatrix} \begin{pmatrix} \text{obtained by adding } -3 \text{ times column 1 to column 2,} \\ -1 \text{ times column 1 to column 3,} \\ -2 \text{ times column 1 to column 4 in Eq. (c)} \end{pmatrix} \qquad (d)$$

$$\mathbf{A} \sim \begin{bmatrix} 1 & 0 & 0 & 0 \\ 0 & 1 & 2 & 3 \\ 0 & 0 & 0 & 0 \end{bmatrix} \begin{pmatrix} \text{obtained by multiplying row 2 by } \tfrac{1}{2} \\ \text{and adding it to row 3 in Eq. (d)} \end{pmatrix} \qquad (e)$$

$$\mathbf{A} \sim \begin{bmatrix} 1 & 0 & 0 & 0 \\ 0 & 1 & 0 & 0 \\ 0 & 0 & 0 & 0 \end{bmatrix} \begin{pmatrix} \text{obtained by adding } -2 \text{ times column 2 to column 3} \\ \text{and } -3 \text{ times column 2 to column 4 in Eq. (e)} \end{pmatrix} \qquad (f)$$

The matrix in Eq. (e) is in the form of Eq. (A.61). The rank of $\mathbf{A}$ is 2, since a 2×2 identity matrix is obtained at the upper left corner.

A.4.2 General Solution of $m \times n$ Linear Equations

Let us now consider solving a system of m simultaneous equations in n unknowns. The existence of a solution for such a system depends on the rank of the system's coefficient matrix and the augmented matrix. Let the system be represented as

$$\mathbf{Ax} = \mathbf{b} \tag{A.62}$$

where $\mathbf{A}$ is an $m \times n$ matrix, $\mathbf{b}$ is an m-vector, and $\mathbf{x}$ is an n-vector of the unknowns. Note that m may be larger than n; that is, there may be more equations than unknowns. In that case, either the system is *inconsistent* (has no solution) or some of the equations are redundant and may be deleted. The solution process described in the following can treat these cases.

Note that if an equation is multiplied by a constant, the solution to the system is unchanged. If c times one equation is added to another, the solution to the resulting system is the same as for the original system. Also, if two columns of the coefficient matrix are interchanged (for example, columns i and j), the resulting set of equations is equivalent to the original system; however, the solution variables x_i and x_j are interchanged in the vector $\mathbf{x}$ as follows:

$$\mathbf{x} = [x_1 \quad x_2 \ldots x_{i-1} \quad \overset{\downarrow}{x_j} \quad x_{i+1} \ldots x_{j-1} \quad \overset{\downarrow}{x_i} \quad x_{j+1} \ldots x_n]^T \tag{A.63}$$

This indicates that each column of the coefficient matrix has a variable associated with it, as also noted earlier, for example, x_i and x_j for the ith and the jth columns, respectively.

Using the elementary row–column operations, it is always possible to convert a system of m equations in n unknowns in Eq. (A.62) into an equivalent system of the form shown in the Eq. (A.64). In the equation, a bar over each element indicates its new value, obtained by performing row–column operations on the augmented matrix of the original system. The value of the subscript r in Eq. (A.64) is the rank of the *coefficient matrix.*

$$\begin{bmatrix} 1 & \bar{a}_{12} & \bar{a}_{13} & \bar{a}_{14} & \cdots & \bar{a}_{1r} & \cdots & \bar{a}_{1n} \\ 0 & 1 & \bar{a}_{23} & \bar{a}_{24} & \cdots & \bar{a}_{2r} & \cdots & \bar{a}_{2n} \\ 0 & 0 & 1 & \bar{a}_{34} & \cdots & \bar{a}_{3r} & \cdots & \bar{a}_{3n} \\ & & & 1 & & & & \\ \cdot & \cdot & \cdot & \cdot & & \cdot & & \cdot \\ \cdot & \cdot & \cdot & \cdot & & \cdot & & \cdot \\ \cdot & \cdot & \cdot & \cdot & & \cdot & & \cdot \\ 0 & 0 & 0 & 0 & \cdots & 1 & \cdots & \bar{a}_{rn} \\ 0 & 0 & 0 & 0 & \cdots & 0 & \cdots & 0 \\ \cdot & & & & & \cdot & & \cdot \\ \cdot & & & & & \cdot & & \cdot \\ \cdot & & & & \cdots & \cdot & & \cdot \\ 0 & 0 & 0 & 0 & \cdots & 0 & & 0 \end{bmatrix} \begin{bmatrix} x_1 \\ x_2 \\ x_3 \\ \cdot \\ \cdot \\ \cdot \\ \cdot \\ \cdot \\ \cdot \\ x_n \end{bmatrix} = \begin{bmatrix} \bar{b}_1 \\ \bar{b}_2 \\ \bar{b}_3 \\ \cdot \\ \cdot \\ \cdot \\ \bar{b}_r \\ \bar{b}_{r+1} \\ \cdot \\ \cdot \\ \bar{b}_m \end{bmatrix} \tag{A.64}$$

Note that if $\bar{b}_{r+1} = \bar{b}_{r+2} = \ldots = \bar{b}_m = 0$ in Eq. (A.64), then the last $(m - r)$ equations become

$$0x_1 + 0x_2 + \cdots + 0x_n = 0 \tag{A.65}$$

These rows can be eliminated from further consideration. However, if any of the last $(m - r)$ components of vector $\bar{\mathbf{b}}$ is not 0, then at least one of the last $(m - r)$ equations is *inconsistent* and the system has no solution. Note also that the rank of the coefficient matrix equals the rank of the augmented matrix if and only if $\bar{b}_i = 0$, $i = (r + 1)$ to m. Thus, a *system of m equations in n unknowns is consistent (i.e., possesses solutions) if and only if the rank of the coefficient matrix equals the rank of the augmented matrix.*

If elementary operations are performed below as well as above the main diagonal to eliminate off-diagonal elements (Gauss-Jordan elimination procedure), an equivalent system of the following form is obtained:

$$\left[\begin{array}{c|c} \mathbf{I}_{(r)} & \mathbf{Q}_{(r\times n-r)} \\ \hline \mathbf{0}_{(m-r\times r)} & \mathbf{0}_{(m-r\times n-r)} \end{array}\right] \left[\begin{array}{c} \mathbf{x}_{(r)} \\ \mathbf{x}_{(n-r)} \end{array}\right] = \left[\begin{array}{c} \mathbf{q}_{(r+1)} \\ \mathbf{p}_{(m-r\times 1)} \end{array}\right] \tag{A.66}$$

Here $\mathbf{I}_{(r)}$ is an $r \times r$ identity matrix, and $\mathbf{x}_{(r)}$ and $\mathbf{x}_{(n-r)}$ are the r-component and $(n-r)$-component subvectors of vector $\mathbf{x}$.

Note that, depending on the values of r, n, and m, the equation can have several different forms. For example, if $r = n$, the matrices $\mathbf{Q}_{(r\times n-r)}$, $\mathbf{0}_{(m-r\times n-r)}$, and the vector $\mathbf{x}_{(n-r)}$ disappear; similarly, if $r = m$, matrices $\mathbf{0}_{(m-r\times r)}$, $\mathbf{0}_{(m-r\times n-r)}$, and the vector $\mathbf{p}_{(m-r\times 1)}$ disappear. The system of equations (A.62) is consistent only if vector $\mathbf{p} = \mathbf{0}$ in Eq. (A.66). *It must be remembered that for every interchange of columns necessary to produce Eq. (A.66), the corresponding components of $\mathbf{x}$ must be interchanged.*

When the system is consistent, the first line of Eq. (A.66) gives

$$\mathbf{I}_{(r)}\mathbf{x}_{(r)} + \mathbf{Q}\mathbf{x}_{(n-r)} = \mathbf{q} \tag{A.67}$$

or

$$\mathbf{x}_{(r)} = \mathbf{q} - \mathbf{Q}\mathbf{x}_{(n-r)} \tag{A.68}$$

Equation (A.68) gives r components of $\mathbf{x}$ in terms of the remaining $(n-r)$ components. If the system is consistent, Eq. (A.68) represents the *general solution* to the system of equations $\mathbf{A}\mathbf{x} = \mathbf{b}$. The last $(n-r)$ components of $\mathbf{x}$ can be assigned arbitrary values; any assignment to $x_{r+1}, \ldots, x_n$ yields a solution. Thus, the system of equations has infinitely many solutions. If $r = n$, the solution is unique. Equation (A.66) is known as the *canonical representation* of the system of equations $\mathbf{A}\mathbf{x} = \mathbf{b}$. This form representation is very useful in solving the linear programming problems given in Chapter 8.

The following examples illustrate the Gauss-Jordan elimination procedure.

EXAMPLE A.7   GENERAL SOLUTION BY GAUSS-JORDAN REDUCTION

Find a general solution to the set of equations

$$\begin{aligned} x_1 + x_2 + x_3 + 5x_4 &= 6 \\ x_1 + x_2 - 2x_3 - x_4 &= 0 \\ x_1 + x_2 - x_3 + x_4 &= 2 \end{aligned} \tag{a}$$

Solution

The augmented matrix for the set of equations is given as

$$\mathbf{A} \sim \begin{array}{c} \begin{array}{ccccc} x_1 & x_2 & x_3 & x_4 & \mathbf{b} \end{array} \\ \left[\begin{array}{cccc|c} 1 & 1 & 1 & 5 & 6 \\ 1 & 1 & -2 & -1 & 0 \\ 1 & 1 & -1 & 1 & 2 \end{array}\right] \end{array} \quad \text{and} \quad \mathbf{x} = \left[\begin{array}{c} x_1 \\ x_2 \\ x_3 \\ x_4 \end{array}\right] \tag{b}$$

The following elimination steps are used to transform the system into canonical form:

1. Subtracting row 1 from rows 2 and 3, we convert elements below the main diagonal in the first column (a_{21} and a_{31}) to 0; that is, we eliminate x_1 from equations 2 and 3, and obtain

$$
\mathbf{A} \sim
\begin{array}{cccc}
x_1 & x_2 & x_3 & x_4 & \mathbf{b}
\end{array}
\left[
\begin{array}{cccc|c}
1 & 1 & 1 & 5 & 6 \\
0 & 0 & -3 & -6 & -6 \\
0 & 0 & -2 & -4 & -4
\end{array}
\right]
\tag{c}
$$

2. Now, since a_{22} is zero we cannot proceed any further with the elimination process. We must interchange rows and/or columns to bring a nonzero element to location a_{22}. We can interchange either column 3 or column 4 with column 2 to bring a nonzero element into position a_{22}. (Note that the last column can never be interchanged with any other column; it is the right side of the system $\mathbf{Ax} = \mathbf{b}$, so it does not correspond to a variable.) Interchanging column 2 with column 3 (elementary column operation 1), we obtain

$$
\mathbf{A} \sim
\begin{array}{cccc}
x_1 & x_3 & x_2 & x_4 & \mathbf{b}
\end{array}
\left[
\begin{array}{ccc|c|c}
1 & 1 & 1 & 5 & 6 \\
0 & -3 & 0 & -6 & -6 \\
0 & -2 & 0 & -4 & -4
\end{array}
\right]
\quad \text{and} \quad
\mathbf{x} =
\begin{bmatrix}
x_1 \\ x_3 \\ x_2 \\ x_4
\end{bmatrix}
\tag{d}
$$

Note that the positions of the variables x_2 and x_3 are also interchanged in the vector $\mathbf{x}$.

3. Now, dividing row 2 by -3, multiplying it by 2, and adding to row 3 gives

$$
\mathbf{A} \sim
\begin{array}{ccccc}
x_1 & x_3 & x_2 & x_4 & \mathbf{b}
\end{array}
\left[
\begin{array}{cc|cc|c}
1 & 1 & 1 & 5 & 6 \\
0 & 1 & 0 & 2 & 2 \\
0 & 0 & 0 & 0 & 0
\end{array}
\right]
\tag{e}
$$

Thus elements below the main diagonal in Eq. (e) are zero and the Gaussian elimination process is complete.

4. To put the equations in the canonical form of Eq. (A.66), we need to perform elimination above the main diagonal also (Gauss-Jordan elimination). Subtracting row 2 from row 1, we obtain

$$
\mathbf{A} \sim
\begin{array}{ccccc}
x_1 & x_3 & x_2 & x_4 & \mathbf{b}
\end{array}
\left[
\begin{array}{cccc|c}
1 & 0 & 1 & 3 & 4 \\
0 & 1 & 0 & 2 & 2 \\
0 & 0 & 0 & 0 & 0
\end{array}
\right]
\tag{f}
$$

Note that the third row in both Eq. (e) and Eq. (f) has all zeros. This implies that the third equation in Eq. (a) is linearly dependent on the others. Since the right side of this equation is also zero in Eq. (f), the linear system in Eq. (a) is consistent. The rank of the coefficient matrix and the augmented matrix is 2.

5. Using the matrix of Eq. (f), the given system of equations is transformed into the canonical form of Eq. (A.66) as follows:

$$
\begin{bmatrix}
1 & 0 & 1 & 3 \\
0 & 1 & 0 & 2 \\
0 & 0 & 0 & 0
\end{bmatrix}
\begin{bmatrix}
x_1 \\ x_3 \\ x_2 \\ x_4
\end{bmatrix}
=
\begin{bmatrix}
4 \\ 2 \\ 0
\end{bmatrix}
\tag{g}
$$

or

$$\left[\begin{array}{c|c} \mathbf{I}_{(2)} & \mathbf{Q}_{(2\times2)} \\ \hline \mathbf{0}_{(1\times2)} & \mathbf{0}_{(1\times2)} \end{array}\right] \begin{bmatrix} x_1 \\ x_3 \\ x_2 \\ x_4 \end{bmatrix} = \begin{bmatrix} \mathbf{q}_{(2\times1)} \\ \mathbf{p}_{(1\times1)} \end{bmatrix} \tag{h}$$

where

$$\mathbf{Q} = \begin{bmatrix} 1 & 3 \\ 0 & 2 \end{bmatrix}; \quad \mathbf{q} = \begin{bmatrix} 4 \\ 2 \end{bmatrix}; \quad \mathbf{p} = \mathbf{0} \tag{i}$$

$$\mathbf{x}_{(r)} = (x_1, \ x_3), \quad \mathbf{x}_{(n-r)} = (x_2, \ x_4)$$

6. Since $\mathbf{p} = \mathbf{0}$, the given system of equations is consistent (i.e., it has solutions). Its general solution, in the form of Eq. (A.68), is

$$\begin{bmatrix} x_1 \\ x_3 \end{bmatrix} = \begin{bmatrix} 4 \\ 2 \end{bmatrix} - \begin{bmatrix} 1 & 3 \\ 0 & 2 \end{bmatrix} \begin{bmatrix} x_2 \\ x_4 \end{bmatrix} \tag{j}$$

Or, in the expanded notation, the general solution is

$$x_1 = 4 - x_2 - 3x_4$$
$$x_3 = 2 - 2x_4 \tag{k}$$

7. It can be seen that the general solution to Eq. (k) gives x_1 and x_3 in terms of x_2 and x_4; that is, x_2 and x_4 are *independent variables* and x_1 and x_3 are *dependent* on them. The system has infinite solutions because any specification for x_2 and x_4 gives a solution.

Basic Solutions

In the preceding general solution in Eq. (k), we see that x_2 and x_4 can be given arbitrary values and that the corresponding x_1 and x_3 can be calculated. *Therefore the system has an infinite number of solutions.* A *particular solution* of much interest in linear programming (LP) is obtained by setting $\mathbf{x}_{(n-r)} = \mathbf{0}$ in the general solution of Eq. (A.68). Such a solution is called the *basic solution* for the linear equations $\mathbf{Ax} = \mathbf{b}$. For the present example, a basic solution is $x_1 = 4$, $x_2 = 0$, $x_3 = 2$, and $x_4 = 0$, which is obtained from Eq. (k) by setting $x_2 = x_4 = 0$.

Note that although Eq. (k) gives an infinite number of solutions to the system of equations, the number of basic solutions is finite. For example, another basic solution can be obtained by setting $x_2 = x_3 = 0$ and solving for x_1 and x_4. It can be verified that this basic solution is $x_1 = 1$, $x_2 = 0$, $x_3 = 0$, and $x_4 = 1$. The fact that the number of basic solutions is finite is very important for the linear programming problems discussed in Chapter 8. The reason is that the *optimum solution for an LP problem is one of the basic solutions.*

EXAMPLE A.8 GAUSS-JORDAN REDUCTION PROCESS IN TABULAR FORM

Find a general solution to the following set of equations using a tabular form of the Gauss-Jordan elimination process:

$$
\begin{aligned}
-x_1 + 2x_2 - 3x_3 + x_4 &= -1 \\
2x_1 + x_2 + x_3 - 2x_4 &= 2 \\
x_1 - x_2 + 2x_3 + x_4 &= 3 \\
x_1 + 3x_2 - 2x_3 - x_4 &= 1
\end{aligned}
\tag{a}
$$

Solution

The iterations of the Gauss-Jordan reduction process for the linear system are explained in Table A.1. Three iterations are needed to reduce the given system to the canonical form of Eq. (A.66). Note that at the second step, the element a_{33} is zero so it cannot be used as a pivot element. Therefore, we use element a_{34} as the pivot element and perform elimination in the x_4 column. This effectively means that we interchange column x_3 with column x_4 (similar to what was done in Example A.7).

Rewriting the results from the third step of Table A.1 in the form of Eq. (A.66), we get

$$
\begin{matrix}
x_1 & x_2 & x_4 & & x_3 \\
\end{matrix}
\begin{bmatrix}
1 & 0 & 0 & 1 \\
0 & 1 & 0 & -1 \\
0 & 0 & 1 & 0 \\
0 & 0 & 0 & 0
\end{bmatrix}
\begin{bmatrix}
x_1 \\ x_2 \\ x_4 \\ x_3
\end{bmatrix}
=
\begin{bmatrix}
2 \\ 0 \\ 1 \\ 0
\end{bmatrix}
\tag{b}
$$

Since the last equation essentially gives $0 = 0$, the given system of equations is consistent (i.e., it has solutions). Also, since the rank of the coefficient matrix is 3, which is less than the number of equations, there are an infinite number of solutions for the linear system.

From Eq. (b), the general solution is given as

$$
\begin{aligned}
x_1 &= 2 - x_3 \\
x_2 &= 0 + x_3 \\
x_4 &= 1
\end{aligned}
\tag{c}
$$

A basic solution is obtained by setting x_3 to zero as $x_1 = 2$, $x_2 = 0$, $x_3 = 0$, and $x_4 = 1$.

To summarize the results of this section, we note that

1. The $m \times n$ system of equations (A.62) is *consistent* if the rank of the coefficient matrix is the same as the rank of the augmented matrix. A consistent system implies that it has a solution.
2. If the number of equations is less than the number of variables ($m < n$) and the system is consistent, having a rank less than or equal to m ($r \leq m$), then it has infinitely many solutions.
3. If $m = n = r$, then the system in Eq. (A.62) has a unique solution.

TABLE A.1   General solution for the linear system of equations in Example A.8 by Gauss-Jordan elimination

Step	x_1	x_2	x_3	x_4	b	
	−1	2	−3	1	−1	Divide row 1 by −1 and use it to perform elimination in column x_1, e.g., multiply new row 1 by 2 and subtract it from row 2, etc.
Initial	2	1	1	−2	2	
	1	−1	2	1	3	
	1	3	−2	−1	1	
	1	−2	3	−1	1	
First iteration	0	5	−5	0	0	Divide row 2 by 5 and perform elimination in column x_2
	0	1	−1	2	2	
	0	5	−5	0	0	
	1	0	1	−1	1	
Second iteration	0	1	−1	0	0	
	0	0	0	2	2	Divide row 3 by 2 and perform elimination in column x_4
	0	0	0	0	0	
	1	0	1	0	2	Canonical form with columns x_1, x_2, and x_4 containing the identity matrix
Third iteration	0	1	−1	0	0	
	0	0	0	1	1	
	0	0	0	0	0	

A.5 CONCEPTS RELATED TO A SET OF VECTORS

In several applications, we come across a set of vectors. It is useful to discuss some concepts related to these sets, such as the linear independence of vectors and vector spaces. In this section, we briefly discuss these concepts and describe a procedure for checking the linear independence of a set of vectors.

A.5.1 Linear Independence of a Set of Vectors

Consider a set of k vectors, each of dimension n:

$$A = \left\{ \mathbf{a}^{(1)}, \ \mathbf{a}^{(2)}, \ \ldots, \ \mathbf{a}^{(k)} \right\} \tag{A.69}$$

where a superscript (i) represents the ith vector. A *linear combination* of vectors in the set A is another vector obtained by scaling each vector in A and adding all the resulting vectors. That is, if $\mathbf{b}$ is a linear combination of a vector in A, it is defined as

$$b = x_1\mathbf{a}^{(1)} + x_2\mathbf{a}^{(2)} + \ldots + x_k\mathbf{a}^{(k)} = \sum_{i=1}^{k} x_i\mathbf{a}^{(i)} \qquad \text{(A.70)}$$

where $x_1, x_2, \ldots, x_k$ are some scalars. The preceding equation can be written compactly in matrix form as

$$\mathbf{b} = \mathbf{Ax} \qquad \text{(A.71)}$$

where $\mathbf{x}$ is a k-component vector and $\mathbf{A}$ is an $n \times k$ matrix with vectors $\mathbf{a}^{(i)}$ as its columns.

To determine whether the set of vectors is *linearly independent* or *dependent*, we set the linear combination of Eq. (A.70) to zero:

$$x_1\mathbf{a}^{(1)} + x_2\mathbf{a}^{(2)} + \ldots + x_k\mathbf{a}^{(k)} = 0; \quad \text{or} \quad \mathbf{Ax} = 0 \qquad \text{(A.72)}$$

This gives a homogeneous system of equations with x_i as unknowns. There are n equations in k unknowns. Note that $\mathbf{x} = 0$ satisfies Eq. (A.72).

If $\mathbf{x} = 0$ is the only solution, then the set of vectors is linearly independent. In this case, rank r of the matrix $\mathbf{A}$ must be equal to k (the number of vectors in the set). If there exists a set of scalars x_i not all zero that satisfies Eq. (A.72), then the given set A of vectors $\mathbf{a}^{(1)}, \mathbf{a}^{(2)}, \ldots, \mathbf{a}^{(k)}$ is said to be linearly dependent. In this case rank r of $\mathbf{A}$ is less than k.

If a set of vectors is linearly dependent, then one or more vectors are parallel to each other, or there is at least one vector that can be expressed as a linear combination of the rest. That is, at least one of the scalars $x_1, x_2, \ldots, x_k$ is nonzero. If we assume x_j to be nonzero, then Eq. (A.72) can be written as follows:

$$-x_j\mathbf{a}^{(j)} = x_1\mathbf{a}^{(1)} + x_2\mathbf{a}^{(2)} + \ldots + x_{j-1}\mathbf{a}^{(j-1)} + x_{j+1}\mathbf{a}^{(j+1)} + \ldots + x_k\mathbf{a}^{(k)} = \sum_{i=1}^{k} x_i\mathbf{a}^{(i)}; \quad i \neq j \qquad \text{(A.73)}$$

Or, since $x_j \neq 0$, we can divide both sides by it to obtain

$$\mathbf{a}^{(j)} = -\sum_{i=1}^{k}(x_i/x_j)\,\mathbf{a}^{(i)}; \quad i \neq j \qquad \text{(A.74)}$$

In Eq. (A.74) we have expressed $\mathbf{a}^{(j)}$ as a *linear combination* of the vectors $\mathbf{a}^{(1)}, \mathbf{a}^{(2)}, \ldots, \mathbf{a}^{(j-1)}, \mathbf{a}^{(j+1)}, \ldots, \mathbf{a}^{(k)}$. In general, we see that if a set of vectors is linearly dependent, then at least one of them can be expressed as a linear combination of the rest.

EXAMPLE A.9 CHECK FOR LINEAR INDEPENDENCE OF VECTORS

Check the linear independence of the following set of vectors:

(i)

$$\mathbf{a}^{(1)} = \begin{bmatrix} 2 \\ 5 \\ 2 \\ -1 \end{bmatrix}, \quad \mathbf{a}^{(2)} = \begin{bmatrix} 3 \\ 2 \\ 1 \\ 0 \end{bmatrix}, \quad \mathbf{a}^{(3)} = \begin{bmatrix} 8 \\ 9 \\ 4 \\ -1 \end{bmatrix}$$

(ii)
$$\mathbf{a}^{(1)} = \begin{bmatrix} 2 \\ 6 \\ 2 \\ -2 \end{bmatrix}, \quad \mathbf{a}^{(2)} = \begin{bmatrix} 4 \\ 3 \\ 2 \\ 0 \end{bmatrix}, \quad \mathbf{a}^{(3)} = \begin{bmatrix} 6 \\ 9 \\ 4 \\ 1 \end{bmatrix}$$

Solution

To check for linear independence, we form the linear combination of Eq. (A.70) and set it to zero as in Eq. (A.72). The resulting homogeneous system of equations is solved for the scalars x_i. If all of the scalars are zero, then the given set of vectors is linearly independent; otherwise, it is dependent.

The vectors in set (i) are linearly dependent, since $x_1 = 1$, $x_2 = 2$, and $x_3 = -1$ give the linear combination of Eq. (A.72) a zero value; that is,

$$\mathbf{a}^{(1)} + 2\mathbf{a}^{(2)} - \mathbf{a}^{(3)} = 0 \tag{a}$$

It may also be checked that the rank of the following matrix, whose columns are the given vectors, is only 2; thus the set of vectors is linearly dependent:

$$\mathbf{A} = \begin{bmatrix} 2 & 3 & 8 \\ 5 & 2 & 9 \\ 2 & 1 & 4 \\ -1 & 0 & -1 \end{bmatrix} \tag{b}$$

For set (ii), let us form a linear combination of the given vectors and set it to zero:

$$x_1 \mathbf{a}^{(1)} + x_2 \mathbf{a}^{(2)} + x_3 \mathbf{a}^{(3)} = 0 \tag{c}$$

This is a vector equation that gives the following system when written in the expanded form:

$$2x_1 + 4x_2 + 6x_3 = 0 \tag{d}$$

$$6x_1 + 3x_2 + 9x_3 = 0 \tag{e}$$

$$2x_1 + 2x_2 + 4x_3 = 0 \tag{f}$$

$$-2x_1 + x_3 = 0 \tag{g}$$

We solve the preceding system of equations by the elimination process.

From Eq. (g), we find $x_3 = 2x_1$. Equations (d) through (f) then become

$$14x_1 + 4x_2 = 0 \tag{h}$$

$$24x_1 + 3x_2 = 0 \tag{i}$$

$$10x_1 + 2x_2 = 0 \tag{j}$$

From Eq. (j), we find $x_2 = -5x_1$. Substituting this result into Eqs. (h) and (i) gives

$$14x_1 + 4(-5x_1) = -6x_1 = 0 \tag{k}$$

$$24x_1 + 3(-5x_1) = 9x_1 = 0 \tag{l}$$

Equations (k) and (l) imply that $x_1 = 0$; therefore, $x_2 = -5x_1 = 0$, $x_3 = 2x_1 = 0$. Thus the only solution to Eq. (c) is the trivial one, $x_1 = x_2 = x_3 = 0$. The vectors $\mathbf{a}^{(1)}$, $\mathbf{a}^{(2)}$, and $\mathbf{a}^{(3)}$ are therefore linearly independent.

Equation (A.72) may be considered a set of n simultaneous equations in k unknowns. To see this, define the k vectors as

$$\mathbf{a}^{(1)} = \begin{bmatrix} a_{11} \\ a_{21} \\ a_{31} \\ . \\ . \\ . \\ a_{n1} \end{bmatrix}, \quad \mathbf{a}^{(2)} = \begin{bmatrix} a_{12} \\ a_{22} \\ a_{32} \\ . \\ . \\ . \\ a_{n2} \end{bmatrix} \quad \ldots, \quad \mathbf{a}^{(k)} = \begin{bmatrix} a_{1k} \\ a_{2k} \\ a_{3k} \\ . \\ . \\ . \\ a_{nk} \end{bmatrix}, \quad \mathbf{x} = \begin{bmatrix} x_1 \\ x_2 \\ . \\ . \\ . \\ x_k \end{bmatrix} \qquad (A.75)$$

Also, let $\mathbf{A}_{(n \times k)} = [\mathbf{a}^{(1)}, \mathbf{a}^{(2)}, \ldots, \mathbf{a}^{(k)}]$; that is, $\mathbf{A}$ is a matrix with an ith column that is the ith vector $\mathbf{a}^{(i)}$. Then Eq. (A.72) can be written as

$$\mathbf{A}\mathbf{x} = 0 \qquad (A.76)$$

The results of Section A.4 show that there is a unique solution to Eq. (A.76) if and only if the rank r of $\mathbf{A}$ is equal to k ($r = k < n$), the number of columns of $\mathbf{A}$. In that case, the unique solution is $\mathbf{x} = \mathbf{0}$. Therefore, *the vectors $\mathbf{a}^{(1)}, \mathbf{a}^{(2)}, \ldots, \mathbf{a}^{(k)}$ are linearly independent if and only if the rank of the matrix $\mathbf{A}$ is k* (the number of vectors in the set).

Note that if $k > n$, then the rank of $\mathbf{A}$ cannot exceed n. Therefore, $\mathbf{a}^{(1)}, \mathbf{a}^{(2)}, \ldots, \mathbf{a}^{(k)}$ *will always be linearly dependent if $k > n$*. The maximum number of linearly dependent n-component vectors is thus n. Any set of $(n + 1)$ vectors is always linearly dependent.

Given any set of n linearly independent (n-component) vectors, $\mathbf{a}^{(1)}, \mathbf{a}^{(2)}, \ldots, \mathbf{a}^{(n)}$, any other ($n$-component) vector $\mathbf{b}$ can be expressed as a unique linear combination of these vectors. The problem is to choose a set of scalars $x_1, x_2, \ldots, x_n$ such that

$$x_1 \mathbf{a}^{(1)} + x_2 \mathbf{a}^{(2)} + \ldots + x_n \mathbf{a}^{(n)} = \mathbf{b}; \quad \text{or} \quad \mathbf{A}\mathbf{x} = \mathbf{b} \qquad (A.77)$$

We wish to show that a solution exists for Eq. (A.77) and that it is unique. Note that $\mathbf{a}^{(1)}, \mathbf{a}^{(2)}, \ldots, \mathbf{a}^{(n)}$ are linearly independent. Therefore, the rank of the coefficient matrix $\mathbf{A}$ is n, and the rank of the augmented matrix $[\mathbf{A}, \mathbf{b}]$ is also n. It cannot be $(n + 1)$ because the matrix has only n rows. Thus, Eq. (A.77) always possesses a solution for any given $\mathbf{A}$. Moreover, $\mathbf{A}$ is nonsingular; hence, the solution is unique.

In summary, we state the following points for a k set of vectors, each having n components:

1. If $k > n$, the set of vectors is always linearly dependent—for example, three vectors each having two components. In other words, the number of linearly independent vectors is always less than or equal to n (e.g., for two-component vectors, there are at most two linearly independent vectors).
2. If there are n linearly independent vectors, each of dimension n, then any other n-component vector can be expressed as a unique linear combination of them. For example, given two linearly independent vectors $\mathbf{a}^{(1)} = (1, 0)$ and $\mathbf{a}^{(2)} = (0, 1)$ of dimension 2, any other vector, such as $\mathbf{b} = (b_1, b_2)$, can be expressed as a unique linear combination of $\mathbf{a}^{(1)}$ and $\mathbf{a}^{(2)}$.
3. The linear independence of the given set of vectors can be determined in two ways:
   a. Form the matrix $\mathbf{A}$ of dimension $n \times k$ whose columns are the given vectors. Then, if rank r is equal to k ($r = k$), the given set is linearly independent; otherwise, it is dependent.

b. Set the linear combination of the given vectors to zero, as $\mathbf{Ax} = \mathbf{0}$. If $\mathbf{x} = \mathbf{0}$ is the only solution to the resulting system, then the set is independent; otherwise, it is dependent.

A.5.2 Vector Spaces

Before giving a definition for vector space, let us define closure under addition and scalar multiplication:

Closure under addition. A set of vectors is said to be closed under addition if the sum of any two vectors in the set is also in the set.

Closure under scalar multiplication. A set of vectors is said to be closed under scalar multiplication if the product of any vector in the set by a scalar gives a vector in the set.

Vector space. A nonempty set S of elements (vectors) $\mathbf{x}, \mathbf{y}, \mathbf{z}, \ldots$ is called a vector space if the two algebraic operations on them (vector addition and multiplication by a real scalar) satisfy the following properties:

1. *Closure under addition*: if $\mathbf{x} \in S$ and $\mathbf{y} \in S$, then $\mathbf{x} + \mathbf{y} \in S$.
2. *Commutative in addition*: $\mathbf{x} + \mathbf{y} = \mathbf{y} + \mathbf{x}$.
3. *Associative in addition*: $(\mathbf{x} + \mathbf{y}) + \mathbf{z} = \mathbf{x} + (\mathbf{y} + \mathbf{z})$.
4. *Identity for addition*: There exists a zero vector $\mathbf{0}$ in the set S such that $\mathbf{x} + \mathbf{0} = \mathbf{x}$ for all $\mathbf{x}$.
5. *Inverse for addition*: There exists a $-\mathbf{x}$ in the set S such that $\mathbf{x} + (-\mathbf{x}) = \mathbf{0}$ for all $\mathbf{x}$.
6. *Closure under scalar multiplication*: For real scalars $\alpha, \beta, \ldots$, if $\mathbf{x} \in S$, then $\alpha\mathbf{x} \in S$.
7. *Distributive*: $(\alpha + \beta)\mathbf{x} = \alpha\mathbf{x} + \beta\mathbf{x}$.
8. *Distributive*: $\alpha(\mathbf{x} + \mathbf{y}) = \alpha\mathbf{x} + \alpha\mathbf{y}$.
9. *Associative in scalar multiplication*: $(\alpha\beta)\mathbf{x} = \alpha(\beta\mathbf{x})$.
10. *Identity in scalar multiplication*: $1\mathbf{x} = \mathbf{x}$.

In Section A.5.1, it was noted that the maximum number of linearly independent vectors in the set of all n-component vectors is n. Thus, for every subset of this set, there exists some maximum number of linearly independent vectors. In particular, every vector space has a maximum number of linearly independent vectors. This number is called the *dimension of the vector space*. If a vector space has dimension k, then any set of k linearly independent vectors in the vector space is called *a basis* for the vector space. Any other vector in the vector space can be expressed as a unique linear combination of the given set of basis vectors.

EXAMPLE A.10 CHECK FOR VECTOR SPACE

Check if the set of vectors $S = \{(x_1, x_2, x_3) \mid x_1 = 0\}$ is a vector space.

Solution

To see this, consider any two vectors in S as

$$\mathbf{x} = \begin{bmatrix} 0 \\ a \\ b \end{bmatrix} \quad \text{and} \quad \mathbf{y} = \begin{bmatrix} 0 \\ c \\ d \end{bmatrix} \tag{a}$$

where scalars a, b, c, and d are completely arbitrary. Then

$$\mathbf{x} + \mathbf{y} = \begin{bmatrix} 0 \\ a+c \\ b+d \end{bmatrix} \tag{b}$$

Therefore, $\mathbf{x} + \mathbf{y}$ is in the set S. Also, for any scalar α,

$$\alpha \mathbf{x} = \begin{bmatrix} 0 \\ \alpha a \\ \alpha b \end{bmatrix} \tag{c}$$

Therefore, $\alpha \mathbf{x}$ is in the set S, and so S is closed under addition and scalar multiplication. All other properties for the definition of a vector space can be easily proved. To show the property (2), we have

$$\mathbf{x} + \mathbf{y} = \begin{bmatrix} 0 \\ a+c \\ b+d \end{bmatrix} = \begin{bmatrix} 0 \\ c+a \\ d+b \end{bmatrix} = \begin{bmatrix} 0 \\ c \\ d \end{bmatrix} + \begin{bmatrix} 0 \\ a \\ b \end{bmatrix} = \mathbf{y} + \mathbf{x} \tag{d}$$

"Associative in addition" is shown as

$$(\mathbf{x} + \mathbf{y}) + \mathbf{z} = \begin{bmatrix} 0 \\ a+c \\ b+d \end{bmatrix} + \begin{bmatrix} 0 \\ e \\ f \end{bmatrix} = \begin{bmatrix} 0 \\ a+c+e \\ b+d+f \end{bmatrix} = \begin{bmatrix} 0 \\ a \\ b \end{bmatrix} + \begin{bmatrix} 0 \\ c+e \\ d+f \end{bmatrix} = \mathbf{x} + (\mathbf{y} + \mathbf{z}) \tag{e}$$

For identity in addition, we have a zero vector in the set S as

$$\mathbf{0} = \begin{bmatrix} 0 \\ 0 \\ 0 \end{bmatrix} \tag{f}$$

such that

$$\mathbf{x} + \mathbf{0} = \begin{bmatrix} 0 \\ a+0 \\ b+0 \end{bmatrix} = \begin{bmatrix} 0 \\ a \\ b \end{bmatrix} = \mathbf{x} \tag{g}$$

The inverse in addition exists if we define $-\mathbf{x}$ as

$$-\mathbf{x} = - \begin{bmatrix} 0 \\ a \\ b \end{bmatrix} = \begin{bmatrix} 0 \\ -a \\ -b \end{bmatrix} \tag{h}$$

such that

$$\mathbf{x} + (-\mathbf{x}) = \begin{bmatrix} 0 \\ a+(-a) \\ b+(-b) \end{bmatrix} = \begin{bmatrix} 0 \\ 0 \\ 0 \end{bmatrix} = \mathbf{0} \tag{i}$$

In a similar way, properties (7) through (10) can easily be shown. Therefore, the set S is a vector space. Note that the set $V = \{(x_1, x_2, x_3) | x_1 = 1\}$ is not a vector space.

Let us now determine the dimension of S. Note that if $\mathbf{A}$ is a matrix whose columns are vectors in S, it has three rows, with the first row containing only zeros. Thus, the rank of $\mathbf{A}$ must be less than or equal to 2, and the dimension of S is either 1 or 2. To show that it is in fact 2, we need only find two linearly independent vectors. The following are three such sets of two linearly independent vectors from the set S:

(i)
$$\mathbf{a}^{(1)} = \begin{bmatrix} 0 \\ 1 \\ 0 \end{bmatrix}, \quad \mathbf{a}^{(2)} = \begin{bmatrix} 0 \\ 0 \\ 1 \end{bmatrix} \tag{j}$$

(ii)
$$\mathbf{a}^{(3)} = \begin{bmatrix} 0 \\ 2 \\ 1 \end{bmatrix}, \quad \mathbf{a}^{(4)} = \begin{bmatrix} 0 \\ 0 \\ 1 \end{bmatrix} \tag{k}$$

(iii)
$$\mathbf{a}^{(5)} = \begin{bmatrix} 0 \\ 1 \\ 1 \end{bmatrix}, \quad \mathbf{a}^{(6)} = \begin{bmatrix} 0 \\ 1 \\ -1 \end{bmatrix} \tag{l}$$

Each of these three sets is a basis for S. Any vector in S can be expressed as a linear combination of each one. If $\mathbf{x} = (0, c, d)$ is any element of S, then

(i)
$$\mathbf{x} = c\mathbf{a}^{(1)} + d\mathbf{a}^{(2)}$$
$$\begin{bmatrix} 0 \\ c \\ d \end{bmatrix} = c \begin{bmatrix} 0 \\ 1 \\ 0 \end{bmatrix} + d \begin{bmatrix} 0 \\ 0 \\ 1 \end{bmatrix} \tag{m}$$

(ii)
$$\mathbf{x} = \frac{c}{2}\mathbf{a}^{(3)} + \left(c - \frac{d}{2}\right)\mathbf{a}^{(4)}$$
$$\begin{bmatrix} 0 \\ c \\ d \end{bmatrix} = \frac{c}{2} \begin{bmatrix} 0 \\ 2 \\ 1 \end{bmatrix} + \left(c - \frac{d}{2}\right) \begin{bmatrix} 0 \\ 0 \\ 1 \end{bmatrix} \tag{n}$$

(iii)
$$\mathbf{x} = \left(\frac{c+d}{2}\right)\mathbf{a}^{(5)} + \left(\frac{c-d}{2}\right)\mathbf{a}^{(6)}$$
$$\begin{bmatrix} 0 \\ c \\ d \end{bmatrix} = \frac{c+d}{2} \begin{bmatrix} 0 \\ 1 \\ 1 \end{bmatrix} + \frac{c-d}{2} \begin{bmatrix} 0 \\ 1 \\ -1 \end{bmatrix} \tag{o}$$

A.6 EIGENVALUES AND EIGENVECTORS

Given an $n \times n$ matrix $\mathbf{A}$, any nonzero vector $\mathbf{x}$ satisfying

$$\mathbf{A}\mathbf{x} = \lambda\mathbf{x} \tag{A.78}$$

where λ is a scale factor, is called an *eigenvector* (proper or characteristic vector). The scalar λ is called the *eigenvalue* (proper or characteristic value). Since $\mathbf{x} \neq \mathbf{0}$, from Eq. (A.78) we see that λ is given as the roots of the characteristic equation

$$|\mathbf{A} - \lambda \mathbf{I}| = 0 \tag{A.79}$$

Equation (A.79) gives an nth degree polynomial in λ. The roots of this polynomial are the required eigenvalues. After the eigenvalues have been determined, the eigenvectors can be determined from Eq. (A.78).

The coefficient matrix $\mathbf{A}$ may be symmetric or asymmetric. For many applications, $\mathbf{A}$ is a symmetric matrix, so we consider this case in the text. Two properties of eigenvalues and eigenvectors are as follows:

1. Eigenvalues and eigenvectors of a real symmetric matrix are real. They may be complex for real nonsymmetric matrices.
2. Eigenvectors corresponding to distinct eigenvalues of real symmetric matrices are orthogonal to each other (that is, their dot product vanishes).

EXAMPLE A.11 CALCULATION OF EIGENVALUES AND EIGENVECTORS

Find the eigenvalues and eigenvectors of the matrix

$$\mathbf{A} = \begin{bmatrix} 2 & 1 \\ 1 & 2 \end{bmatrix} \tag{a}$$

Solution

The eigenvalue problem is defined as

$$\begin{bmatrix} 2 & 1 \\ 1 & 2 \end{bmatrix} \begin{bmatrix} x_1 \\ x_2 \end{bmatrix} = \lambda \begin{bmatrix} x_1 \\ x_2 \end{bmatrix} \tag{b}$$

The characteristic polynomial is given by $|\mathbf{A} - \lambda \mathbf{I}| = 0$:

$$\begin{vmatrix} 2 - \lambda & 1 \\ 1 & 2 - \lambda \end{vmatrix} = 0 \tag{c}$$

or

$$\lambda^2 - 4\lambda + 3 = 0 \tag{d}$$

The roots of this polynomial are

$$\lambda_1 = 3, \quad \lambda_2 = 1 \tag{e}$$

Therefore, the eigenvalues are 3 and 1.

The eigenvectors are determined from Eq. (A.78). For $\lambda_1 = 3$, Eq. (A.78) is

$$\begin{bmatrix} (2-3) & 1 \\ 1 & (2-3) \end{bmatrix} \begin{bmatrix} x_1 \\ x_2 \end{bmatrix} = \begin{bmatrix} 0 \\ 0 \end{bmatrix} \tag{f}$$

or $x_1 = x_2$. Therefore, a solution to the above equation is (1, 1). After normalization (dividing by its length), the first eigenvector becomes

$$\mathbf{x}^{(1)} = \frac{1}{\sqrt{2}} \begin{bmatrix} 1 \\ 1 \end{bmatrix} \tag{g}$$

For $\lambda_2 = 1$, Eq. (A.78) is

$$\begin{bmatrix} (2-1) & 1 \\ 1 & (2-1) \end{bmatrix} \begin{bmatrix} x_1 \\ x_2 \end{bmatrix} = \begin{bmatrix} 0 \\ 0 \end{bmatrix} \tag{h}$$

or $x_1 = -x_2$. Therefore, a solution to the above equation is (1, −1). After normalization, the second eigenvector is

$$\mathbf{x}^{(2)} = \frac{1}{\sqrt{2}} \begin{bmatrix} 1 \\ -1 \end{bmatrix} \tag{i}$$

It may be verified that $\mathbf{x}^{(1)} \cdot \mathbf{x}^{(2)}$ is zero; that is, $\mathbf{x}^{(1)}$ and $\mathbf{x}^{(2)}$ are orthogonal to each other.

*A.7 NORM AND CONDITION NUMBER OF A MATRIX

A.7.1 Norm of Vectors and Matrices

Every n-dimensional vector $\mathbf{x}$ has a scalar-valued function associated with it, denoted as $||\mathbf{x}||$. It is called a norm of $\mathbf{x}$ if it satisfies the following three conditions:

1. $||\mathbf{x}|| > 0$ for $\mathbf{x} \neq \mathbf{0}$, and $||\mathbf{x}|| = 0$ only when $\mathbf{x} = \mathbf{0}$.
2. $||\mathbf{x} + \mathbf{y}|| \leq ||\mathbf{x}|| + ||\mathbf{y}||$ (triangle inequality).
3. $||a\mathbf{x}|| = |a| \, ||\mathbf{x}||$ where a is a scalar.

The ordinary length of a vector for $n \leq 3$ satisfies the foregoing three conditions. The concept of norm is therefore a generalization of the ordinary length of a vector in one-, two-, or three-dimensional Euclidean space. For example, it can be verified that the Euclidean distance in the n-dimensional space

$$\|\mathbf{x}\| = \sqrt{\mathbf{x}^T \mathbf{x}} = \sqrt{\mathbf{x} \cdot \mathbf{x}} \tag{A.80}$$

satisfies the three norm conditions and hence is a norm.

Every $n \times n$ matrix $\mathbf{A}$ has a *scalar function* associated with it called its norm. It is denoted $||\mathbf{A}||$ and is calculated as

$$\|\mathbf{A}\| = \max_{\mathbf{x} \neq 0} \frac{\|\mathbf{A}\mathbf{x}\|}{\|\mathbf{x}\|} \tag{A.81}$$

Note that since $\mathbf{A}\mathbf{x}$ is a vector, Eq. (A.81) says that the norm of $\mathbf{A}$ is determined by the vector $\mathbf{x}$ that maximizes the ratio $||\mathbf{A}\mathbf{x}||/||\mathbf{x}||$.

The three conditions of the norm can be verified easily for Eq. (A.81) as follows:

1. $||\mathbf{A}|| > 0$ unless it is a null matrix in which case it is zero.
2. $||\mathbf{A} + \mathbf{B}|| \leq ||\mathbf{A}|| + ||\mathbf{B}||$.
3. $||a\mathbf{A}|| = |a| \, ||\mathbf{A}||$ where a is a scalar.

Other vector norms can also be defined. For example, the summation norm and the max-norm (called the "∞-norm") are defined as

$$\|\mathbf{x}\| = \sum_{i=1}^{n} |x_i|, \quad \text{or} \quad \|\mathbf{x}\| = \max_{1 \leq i \leq n} |x_i| \tag{A.82}$$

They also satisfy the three conditions of the norm of the vector $\mathbf{x}$.

If λ_1^2 is the largest eigenvalue of $\mathbf{A}^T\mathbf{A}$, then it can be shown, using Eq. (A.81), that the norm of $\mathbf{A}$ is also defined as

$$\|\mathbf{A}\| = \lambda_1 > 0 \tag{A.83}$$

Similarly, if λ_n^2 is the smallest eigenvalue of $\mathbf{A}^T\mathbf{A}$, then the norm of $\mathbf{A}^{-1}$ is defined as

$$\|\mathbf{A}^{-1}\| = \lambda_n > 0 \tag{A.84}$$

A.7.2 Condition Number of a Matrix

The condition number is another scalar associated with an $n \times n$ matrix. It is useful when solving a linear system of equations $\mathbf{A}\mathbf{x} = \mathbf{b}$. Often there is uncertainty in elements of the coefficient matrix $\mathbf{A}$ or the right side of vector $\mathbf{b}$. The question, then, is how the solution vector $\mathbf{x}$ changes for small perturbations in $\mathbf{A}$ and $\mathbf{b}$. The answer to this question is in the condition number of the matrix $\mathbf{A}$.

It can be shown that the condition number of an $n \times n$ matrix $\mathbf{A}$, denoted cond($\mathbf{A}$), is given as

$$\text{cond}(\mathbf{A}) = \lambda_1/\lambda_n \geq 0 \tag{A.85}$$

where λ_1^2 and λ_n^2 are the largest and the smallest eigenvalues of $\mathbf{A}^T\mathbf{A}$.

It turns out that a larger condition number indicates that the solution $\mathbf{x}$ is very sensitive to variations in the elements of $\mathbf{A}$ and $\mathbf{b}$. That is, small changes in $\mathbf{A}$ and $\mathbf{b}$ give large changes in $\mathbf{x}$. A very large condition number for the matrix $\mathbf{A}$ indicates that it is nearly singular. The corresponding system of equations $\mathbf{A}\mathbf{x} = \mathbf{b}$ is called ill-conditioned.

EXERCISES FOR APPENDIX A

Evaluate the following determinants.

A.1

$$\begin{vmatrix} 2 & 1 & 3 \\ 1 & 2 & 1 \\ 3 & 1 & 5 \end{vmatrix}$$

A.2
$$\begin{vmatrix} 0 & 2 & 3 & 2 \\ 0 & 4 & 5 & 4 \\ 1 & -2 & -2 & 1 \\ 3 & -1 & 2 & 1 \end{vmatrix}$$

A.3
$$\begin{vmatrix} 0 & 0 & 0 & -2 \\ 0 & 0 & 5 & 3 \\ 0 & 1 & -1 & 1 \\ 2 & 3 & -3 & 2 \end{vmatrix}$$

For the following determinants, calculate the values of the scalar λ for which the determinants vanish.

A.4
$$\begin{vmatrix} 2-\lambda & 1 & 0 \\ 1 & 3-\lambda & 0 \\ 0 & 3 & 2-\lambda \end{vmatrix}$$

A.5
$$\begin{vmatrix} 2-\lambda & 2 & 0 \\ 1 & 2-\lambda & 0 \\ 0 & 0 & 2-\lambda \end{vmatrix}$$

Determine the rank of the following matrices.

A.6
$$\begin{bmatrix} 3 & 0 & 1 & 3 \\ 2 & 0 & 3 & 2 \\ 0 & 2 & -8 & 1 \\ -2 & -1 & 2 & -1 \end{bmatrix}$$

A.7
$$\begin{bmatrix} 1 & 2 & 2 & 2 & 4 \\ 1 & 6 & 3 & 0 & 3 \\ 2 & 2 & 3 & 3 & 2 \\ 1 & 3 & 2 & 5 & 1 \end{bmatrix}$$

A.8
$$\begin{bmatrix} 1 & 2 & 3 & 4 \\ 0 & 0 & 0 & 1 \\ 3 & 2 & 3 & 0 \\ 2 & 3 & 1 & 4 \\ 2 & 0 & 6 & 0 \\ 1 & 2 & 1 & 4 \end{bmatrix}$$

Obtain the solutions to the following equations using the Gaussian elimination procedure.

A.9 $2x_1 + 2x_2 + x_3 = 5$
$\quad\quad x_1 - 2x_2 + 2x_3 = 1$
$\quad\quad x_2 + 2x_3 = 3$

A.10 $x_2 - x_3 = 0$
$\quad\quad x_1 + x_2 + x_3 = 3$
$\quad\quad x_1 - 3x_2 = -2$

A.11 $2x_1 + x_2 + x_3 = 7$
$\quad\quad 4x_2 - 5x_3 = -7$
$\quad\quad x_1 - 2x_2 + 4x_3 = 9$

A.12 $2x_1 + x_2 - 3x_3 + x_4 = 1$
$x_1 + 2x_2 + 5x_3 - x_4 = 7$
$-x_1 + x_2 + x_3 + 4x_4 = 5$
$2x_1 - 3x_2 + 2x_3 - 5x_4 = -4$

A.13 $3x_1 + x_2 + x_3 = 8$
$2x_1 - x_2 - x_3 = -3$
$x_1 + 2x_2 - x_3 = 2$

A.14 $x_1 + x_2 - x_3 = 2$
$2x_1 - x_2 + x_3 = 4$
$-x_1 + 2x_2 + 3x_3 = 3$

A.15 $-x_1 + x_2 - x_3 = -2$
$-2x_1 + x_2 + 2x_3 = 6$
$x_1 + x_2 + x_3 = 6$

A.16 $-x_1 + 2x_2 + 3x_3 = 4$
$2x_1 - x_2 - 2x_3 = -1$
$x_1 - 3x_2 + 4x_3 = 2$

A.17 $x_1 + x_2 + x_3 + x_4 = 2$
$2x_1 + x_2 - x_3 + x_4 = 2$
$-x_1 + 2x_2 + 3x_3 + x_4 = 1$
$3x_1 + 2x_2 - 2x_3 - x_4 = 8$

A.18 $x_1 + x_2 + x_3 + x_4 = -1$
$2x_1 - x_2 + x_3 - 2x_4 = 8$
$3x_1 + 2x_2 + 2x_3 + 2x_4 = 4$
$-x_1 - x_2 + 2x_3 - x_4 = -2$

Check if the following systems of equations are consistent. If they are, calculate their general solutions.

A.19 $3x_1 + x_2 + 5x_3 + 2x_4 = 2$
$2x_1 - 2x_2 + 4x_3 = 2$
$2x_1 + 2x_2 + 3x_3 + 2x_4 = 1$
$x_1 + 3x_2 + x_3 + 2x_4 = 0$

A.20 $x_1 + x_2 + x_3 + x_4 = 10$
$-x_1 + x_2 - x_3 + x_4 = 2$
$2x_1 - 3x_2 + 2x_3 - 2x_4 = -6$

A.21 $x_2 + 2x_3 + x_4 = -2$
$x_1 - 2x_2 - x_3 - x_4 = 1$
$x_1 - 2x_2 - 3x_3 + x_4 = 1$

A.22 $x_1 + x_2 + x_3 + x_4 = 0$
$2x_1 + x_2 - 2x_3 - x_4 = 6$
$3x_1 + 2x_2 + x_3 + 2x_4 = 2$

A.23 $x_1 + x_2 + x_3 + 3x_4 - x_5 = 5$
$2x_1 - x_2 + x_3 - x_4 + 3x_5 = 4$
$-x_1 + 2x_2 - x_3 + 3x_4 - 2x_5 = 1$

A.24 $2x_1 - x_2 + x_3 + x_4 - x_5 = 2$
$-x_1 + x_2 - x_3 - x_4 + x_5 = -1$
$4x_1 + 2x_2 + 3x_3 + 2x_4 - x_5 = 20$

A.25 $3x_1 + 3x_2 + 2x_3 + x_4 = 19$
$2x_1 - x_2 + x_3 + x_4 - x_5 = 2$
$4x_1 + 2x_2 + 3x_3 + 2x_4 - x_5 = 20$

A.26 $x_1 + x_2 + 2x_4 - x_5 = 5$
$x_1 + x_2 + x_3 + 3x_4 - x_5 = 5$
$2x_1 - x_2 + x_3 - x_4 + 3x_5 = 4$
$-x_1 + 2x_2 - x_3 + 3x_4 - 2x_5 = 1$

A.27 $x_2 + 2x_3 + x_4 + 3x_5 + 2x_6 = 9$
$-x_1 + 5x_2 + 2x_3 + x_4 + 2x_5 + x_7 = 10$
$5x_1 - 3x_2 + 8x_3 + 6x_4 + 3x_5 - 2x_8 = 17$
$2x_1 - x_2 + x_4 + 5x_5 - 2x_8 = 5$

Check the linear independence of the following set of vectors.

A.28

$$\mathbf{a}^{(1)} = \begin{bmatrix} 3 \\ 2 \\ 1 \end{bmatrix}, \quad \mathbf{a}^{(2)} = \begin{bmatrix} -3 \\ -4 \\ 1 \end{bmatrix}, \quad \mathbf{a}^{(3)} = \begin{bmatrix} 2 \\ 3 \\ 0 \end{bmatrix}, \quad \mathbf{a}^{(4)} = \begin{bmatrix} 4 \\ 0 \\ 1 \end{bmatrix}$$

A.29

$$\mathbf{a}^{(1)} = \begin{bmatrix} 1 \\ 2 \\ 3 \\ 4 \\ 5 \end{bmatrix}, \quad \mathbf{a}^{(2)} = \begin{bmatrix} -2 \\ 1 \\ 0 \\ 1 \\ -1 \end{bmatrix}, \quad \mathbf{a}^{(3)} = \begin{bmatrix} 4 \\ 0 \\ -3 \\ 2 \\ 1 \end{bmatrix}$$

Find eigenvalues for the following matrices.

A.30

$$\begin{bmatrix} 1 & 2 \\ 2 & 5 \end{bmatrix}$$

A.31

$$\begin{bmatrix} 2 & 2 \\ 2 & 4 \end{bmatrix}$$

A.32

$$\begin{bmatrix} 1 & 1 & 0 \\ 1 & 4 & 0 \\ 0 & 0 & 5 \end{bmatrix}$$

A.33

$$\begin{bmatrix} 1 & 0 & 0 \\ 0 & 0 & 1 \\ 0 & 1 & 2 \end{bmatrix}$$

A.34

$$\begin{bmatrix} 0 & 0 & 0 \\ 0 & 1 & 1 \\ 0 & 1 & 5 \end{bmatrix}$$

Sample Computer Programs

This appendix contains some computer programs based on the algorithms for numerical methods of unconstrained optimization given in Chapters 10 and 11. The objective is to educate the student on how to transform a step-by-step numerical algorithm into a program. Note that the programs we discuss are not claimed to be the most efficient. The key idea is to highlight the essential numerical aspects of the algorithms in a simple and straightforward way. A beginner in the numerical techniques of optimization is expected to experiment with these programs to get a feel for various methods by solving some numerical examples. For this reason, a black box approach is discouraged.

B.1 EQUAL INTERVAL SEARCH

As discussed in Chapter 10, equal interval search is the simplest method of one-dimensional minimization. A computer program based on it is shown in Figure B.1. It is assumed that the one-dimensional function is unimodal and continuous and that it has a negative slope in the interval of interest. The initial step length (δ) and line search accuracy (ε) must be specified in the main program.

The subroutine EQUAL is called from the main program to perform line search. The three major tasks to be accomplished in it are (1) to establish the initial step length δ such that $f(0) > f(\delta)$; (2) to establish the initial interval of uncertainty, (α_l, α_u); and (3) to reduce the interval of uncertainty such that $(\alpha_u - \alpha_l) \le \varepsilon$.

The subroutine EQUAL calls the subroutine FUNCT to determine the value of the one-dimensional function at various trial steps. FUNCT is supplied by the user. To illustrate, $f(\alpha) = 2 - 4\alpha + e^{\alpha}$ is chosen as the one-dimensional minimization function. The figure's program listing is self-explanatory. In the EQUAL subroutine, the following notation is used:

AL = lower limit on α, α_l
AU = upper limit on α, α_u
FL = function value at α_l, $f(\alpha_l)$
FU = function value at α_u, $f(\alpha_u)$
AA = intermediate point α_a
FA = function value at α_a, $f(\alpha_a)$

```
C       MAIN PROGRAM FOR EQUAL INTERVAL SEARCH

        IMPLICIT DOUBLE PRECISION (A-H,O-Z)

        DELTA  = 5.0D-2
        EPSLON = 1.0D-3
        NCOUNT = 0
        F      = 0.0D0
        ALFA   = 0.0D0
C
C       TO PERFORM LINE SEARCH CALL SUBROUTINE EQUAL
C
        CALL EQUAL(ALFA,DELTA,EPSLON,F,NCOUNT)
        WRITE(*,10) ' MINIMUM =', ALFA
        WRITE(*,10) ' MINIMUM FUNCTION VALUE =', F
        WRITE(*,*) 'NO. OF FUNCTION EVALUATIONS =', NCOUNT
10      FORMAT(A,1PE14.5)

        STOP
        END



        SUBROUTINE EQUAL(ALFA,DELTA,EPSLON,F,NCOUNT)
C       ---------------------------------------------------------
C       THIS SUBROUTINE IMPLEMENTS EQUAL INTERVAL SEARCH
C       ALFA   = OPTIMUN VALUE ON RETURN
C       DELTA  = INITIAL STEP LENGTH
C       EPSLON = CONVERGENCE PARAMETER
C       F      = OPTIMUM VALUE OF THE FUNCTION ON RETURN
C       NCOUNT = NUMBER OF FUNCTION EVALUATIONS ON RETURN
C       ---------------------------------------------------------

        IMPLICIT DOUBLE PRECISION (A-H,O-Z)
C
C       ESTABLISH INITIAL DELTA
C
        AL = 0.0D0
        CALL FUNCT(AL,FL,NCOUNT)
10      CONTINUE
        AA = DELTA
        CALL FUNCT(AA,FA,NCOUNT)
        IF (FA .GT. FL) THEN
           DELTA = DELTA * 0.1D0
           GO TO 10
        END IF
```

FIGURE B.1   Program for Equal Interval Search.

```
C
C       ESTABLISH INITIAL INTERVAL OF UNCERTAINTY
C
20      CONTINUE
        AU = AA + DELTA
        CALL FUNCT(AU,FU,NCOUNT)
        IF (FA .GT. FU) THEN
           AL = AA
           AA = AU
           FL = FA
           FA = FU
           GO TO 20
        END IF
C
C       REFINE THE INTERVAL OF UNCERTAINTY FURTHER
C
30      CONTINUE
        IF ((AU - AL) .LE. EPSLON) GO TO 50
        DELTA = DELTA * 0.1D0
        AA = AL
        FA = FL
40      CONTINUE
        AU = AA + DELTA
        CALL FUNCT(AU,FU,NCOUNT)
        IF (FA .GT. FU) THEN
           AL = AA
           AA = AU
           FL = FA
           FA = FU
           GO TO 40
        END IF
        GO TO 30
C
C       MINIMUM IS FOUND
C
50      ALFA = (AU + AL) * 0.5D0
        CALL FUNCT(ALFA,F,NCOUNT)

        RETURN
        END
```

FIGURE B.1   (*Continued*)

```
      SUBROUTINE FUNCT(AL,F,NCOUNT)
C     ---------------------------------------------------------
C     CALCULATES THE FUNCTION VALUE
C     AL    = VALUE OF ALPHA, INPUT
C     F     = FUNCTION VALUE ON RETURN
C     NCOUNT = NUMBER OF CALLS FOR FUNCTION EVALUATION
C     ---------------------------------------------------------

      IMPLICIT DOUBLE PRECISION (A-H,O-Z)

      NCOUNT = NCOUNT + 1
C      F = 1.0D0 - 3.0D0 * AL + DEXP(2.0D0 * AL)
       F = 18.5D0*AL**2-85.0D0*AL-13.5D0


      RETURN
      END
```

FIGURE B.1   (*Continued*)

B.2 GOLDEN SECTION SEARCH

Golden section search is considered one of the more efficient methods, requiring only function values. The subroutine GOLD, shown in Figure B.2, implements the golden section search algorithm given in Chapter 10, and is called from the main program shown in Figure B.1; the call to EQUAL is replaced by a call to GOLD.

The initial step length and initial interval of uncertainty are established in GOLD, as in EQUAL. The interval of uncertainty is reduced further to satisfy line search accuracy by implementing Step 3 of the algorithm given in Chapter 10. The subroutine FUNCT is used to evaluate the function value at a trial step.

The following notation is used in the subroutine GOLD:

$AA = \alpha_a$
$AB = \alpha_b$
$AL = \alpha_l$
$AU = \alpha_u$
$FA = f(\alpha_a)$
$FB = f(\alpha_b)$
$FL = f(\alpha_l)$
$FU = f(\alpha_u)$
$GR =$ golden ratio $(\sqrt{5} + 1)/2$.

```
      SUBROUTINE GOLD(ALFA,DELTA,EPSLON,F,NCOUNT)
C     --------------------------------------------------------
C     THIS SUBROUTINE IMPLEMENTS GOLDEN SECTION SEARCH
C     ALFA   = OPTIMUM VALUE OF ALPHA ON RETURN
C     DELTA  = INITIAL STEP LENGTH
C     EPSLON = CONVERGENCE PARAMETER
C     F      = OPTIMUM VALUE OF THE FUNCTION ON RETURN
C     NCOUNT = NUMBER OF FUNCTION EVALUATIONS ON RETURN
C     --------------------------------------------------------

      IMPLICIT DOUBLE PRECISION(A-H,O-Z)

      GR = 0.5D0 * SQRT(5.0D0) + 0.5D0
C
C     ESTABLISH INITIAL DELTA
C
      AL = 0.0D0
      CALL FUNCT(AL,FL,NCOUNT)
10    CONTINUE
      AA = DELTA
      CALL FUNCT(AA,FA,NCOUNT)
      IF (FA .GT. FL) THEN
         DELTA = DELTA * 0.1D0
         GO TO 10
      END IF
C
C     ESTABLISH INITIAL INTERVAL OF UNCERTAINTY
C
      J = 0
20    CONTINUE
      J = J + 1
      AU = AA + DELTA * (GR ** J)
      CALL FUNCT(AU,FU,NCOUNT)
      IF (FA .GT. FU) THEN
         AL = AA
         AA = AU
         FL = FA
         FA = FU
         GO TO 20
      END IF
C
C     REFINE THE INTERVAL OF UNCERTAINTY FURTHER
C
      AB = AL + (AU - AL) / GR
      CALL FUNCT(AB,FB,NCOUNT)
30    CONTINUE
```

FIGURE B.2 Subroutine GOLD for Golden Section Search.

```fortran
      IF ((AU - AL) .LE. EPSLON) GO TO 80
C
C   IMPLEMENT STEPS 4, 5 OR 6 OF THE ALGORITHM
C
      IF (FA - FB) 40, 60, 50
C
C   FA IS LESS THAN FB (STEP 4)

40    AU = AB
      FU = FB
      AB = AA
      FB = FA
      AA = AL + (AU - AL) * (1.0D0 - 1.0D0 / GR)
      CALL FUNCT(AA,FA,NCOUNT)
      GO TO 30
C
C   FA IS GREATER THAN FB (STEP 5)
C
50    AL = AA
      FL = FA
      AA = AB
      FA = FB
      AB = AL + (AU - AL) / GR
      CALL FUNCT(AB,FB,NCOUNT)
      GO TO 30
C
C   FA IS EQUAL TO FB (STEP 6)
C
60    AL = AA
      FL = FA
      AU = AB
      FU = FB
      AA = AL + (1.0D0 - 1.0D0 / GR) * (AU - AL)
      CALL FUNCT(AA,FA,NCOUNT)
      AB = AL + (AU - AL) / GR
      CALL FUNCT(AB,FB,NCOUNT)
      GO TO 30
C
C   MINIMUM IS FOUND
C
80    ALFA = (AU + AL) * 0.5D0
      CALL FUNCT(ALFA,F,NCOUNT)
      RETURN
      END
```

FIGURE B.2   (*Continued*)

B.3 STEEPEST-DESCENT METHOD

The steepest-descent method is the simplest of the gradient-based methods for unconstrained optimization. A computer program for it is shown in Figure B.3. The basic steps in the algorithm, which the main program essentially follows, are (1) to evaluate the gradient of the cost function at the current point; (2) to evaluate an optimum step size along the negative gradient direction; and (3) to update the design, check the convergence criterion, and, if necessary, repeat the preceding steps.

The arrays declared in the main program must have the dimensions of the design variable vector. Also, the initial data and starting point are user-provided. The cost function and its gradient must be provided in FUNCT and GRAD, respectively. Line search is performed in the subroutine GOLDM by golden section search for a multivariate problem. For example, $f(x) = x_1^2 + 2x_2^2 + 2x_3^2 + 2x_1x_2 + 2x_2x_3$ is chosen as the cost function.

B.4 MODIFIED NEWTON'S METHOD

The modified Newton's Method evaluates the gradient as well as the Hessian for the function and thus has a quadratic rate of convergence. Note that, even though this method has a superior rate of convergence, it may fail to converge because of the singularity or indefiniteness of the Hessian matrix of the cost function. A modified Newton's program is shown in Figure B.4. The cost function, the gradient vector and the Hessian matrix are calculated in the subroutines FUNCT, GRAD, and HASN, respectively. As an example, $f(\mathbf{x}) = x_1^2 + 2x_2^2 + 2x_3^2 + 2x_1x_2 + 2x_2x_3$ is chosen as the cost function.

The Newton direction is obtained by solving a system of linear equations in the subroutine SYSEQ. It is likely that the Newton direction may not be one of descent, in which case the line search will fail to evaluate an appropriate step size. The iterative loop is stopped in this case and an appropriate message is printed. The main program for the modified Newton's method and related subroutines is shown in Figure B.4.

```
C       THE MAIN PROGRAM FOR STEEPEST-DESCENT METHOD
C       --------------------------------------------------------------
C       DELTA  = INITIAL STEP LENGTH FOR LINE SEARCH
C       EPSLON = LINE SEARCH ACCURACY
C       EPSL   = STOPPING CRITERION FOR STEEPEST-DESCENT METHOD
C       NCOUNT = NO. OF FUNCTION EVALUATIONS
C       NDV    = NO. OF DESIGN VARIABLES
C       NOC    = NO. OF CYCLES OF THE METHOD
C       X      = DESIGN VARIABLE VECTOR
C       D      = DIRECTION VECTOR
C       G      = GRADIENT VECTOR
C       WK     = WORK ARRAY USED FOR TEMPORARY STORAGE
C       --------------------------------------------------------------

        IMPLICIT DOUBLE PRECISION (A-H, O-Z)
        DIMENSION X(4), D(4), G(4), WK(4)
C
C       DEFINE INITIAL DATA
C
        DELTA  = 5.0D-2
        EPSLON = 1.0D-4
        EPSL   = 5.0D-3
        NCOUNT = 0
        NDV    = 3
        NOC    = 100
C
C       STARTING VALUES OF THE DESIGN VARIABLES
C
        X(1)=2.0D0
        X(2)=4.0D0
        X(3)=10.0D0

        CALL GRAD(X,G,NDV)
        WRITE(*,10)
10      FORMAT(' NO.      COST FUNCT       STEP SIZE',
     &         '    NORM OF GRAD  ')
        DO 20 K = 1, NOC
            CALL SCALE (G,D,-1.0D0,NDV)
            CALL GOLDM(X,D,WK,ALFA,DELTA,EPSLON,F,NCOUNT,NDV)
            CALL SCALE(D,D,ALFA,NDV)
            CALL PRINT(K,X,ALFA,G,F,NDV)
            CALL ADD(X,D,X,NDV)
            CALL GRAD(X,G,NDV)
            IF(TNORM(G,NDV) .LE. EPSL) GO TO 30
20      CONTINUE
```

FIGURE B.3   Computer program for steepest-descent method.

```
        WRITE(*,*)
        WRITE(*,*)' LIMIT ON NO. OF CYCLES HAS EXCEEDED'
        WRITE(*,*)' THE CURRENT DESIGN VARIABLES ARE:'
        WRITE(*,*) X
        CALL EXIT

30      WRITE(*,*)
        WRITE(*,*) 'THE OPTIMAL DESIGN VARIABLES ARE:'
        WRITE(*,40) X
40      FORMAT (3F15.6)


        CALL FUNCT(X,F,NCOUNT,NDV)
        WRITE(*,50)' THE OPTIMUM COST FUNCTION VALUE IS :', F
50      FORMAT(A, F13.6)
        WRITE(*,*)'TOTAL NO. OF FUNCTION EVALUATIONS ARE', NCOUNT

        STOP
        END




        SUBROUTINE GRAD(X,G,NDV)
C
C       CALCULATES THE GRADIENT OF F(X) IN VECTOR G
C
        IMPLICIT DOUBLE PRECISION (A-H, O-Z)
        DIMENSION X(NDV),G(NDV)

        G(1) = 2.0D0 * X(1) + 2.0D0 * X(2)
        G(2) = 2.0D0 * X(1) + 4.0D0 * X(2) + 2.0D0 * X(3)
        G(3) = 2.0D0 * X(2) + 4.0D0 * X(3)

        RETURN
        END




        SUBROUTINE SCALE(A,X,S,M)
C
C       MULTIPLIES VECTOR A(M) BY SCALAR S AND STORES IN X(M)
C
        IMPLICIT DOUBLE PRECISION (A-H, O-Z)
        DIMENSION A(M),X(M)
```

FIGURE B.3   (*Continued*)

```fortran
      DO 10 I = 1, M
         X(I) = S * A(I)
10    CONTINUE

      RETURN
      END




      REAL*8 FUNCTION TNORM(X,N)
C
C     CALCULATES NORM OF VECTOR X(N)
C
      IMPLICIT DOUBLE PRECISION (A-H, O-Z)
      DIMENSION X(N)

      SUM = 0.0D0
      DO 10 I = 1, N
         SUM = SUM + X(I) * X(I)
10    CONTINUE
      TNORM = DSQRT(SUM)

      RETURN
      END




      SUBROUTINE ADD(A,X,C,M)
C
C     ADDS VECTORS A(M) AND X(M) AND STORES IN C(M)
C
      IMPLICIT DOUBLE PRECISION (A-H, O-Z)
      DIMENSION A(M), X(M), C(M)

      DO 10 I = 1, M
         C(I) = A(I) + X(I)
10    CONTINUE

      RETURN
      END
```

FIGURE B.3   (*Continued*)

```fortran
      SUBROUTINE PRINT(I,X,ALFA,G,F,M)
C
C     PRINTS THE OUTPUT
C
      IMPLICIT DOUBLE PRECISION (A-H, O-Z)
      DIMENSION X(M),G(M)

      WRITE(*,10) I, F, ALFA, TNORM(G,M)
10    FORMAT(I4, 3F15.6)

      RETURN
      END




      SUBROUTINE FUNCT(X,F,NCOUNT,NDV)
C
C     CALCULATES THE FUNCTION VALUE
C
      IMPLICIT DOUBLE PRECISION (A-H, O-Z)
      DIMENSION X(NDV)

      NCOUNT = NCOUNT + 1
      F = X(1) ** 2 + 2.D0 * (X(2) **2) + 2.D0 * (X(3) ** 2)
     &    + 2.0D0 * X(1) * X(2) + 2.D0 * X(2) * X(3)

      RETURN
      END

      SUBROUTINE UPDATE (XN,X,D,AL,NDV)
C
C     UPDATES THE DESIGN VARIABLE VECTOR
C
      IMPLICIT DOUBLE PRECISION (A-H, O-Z)
      DIMENSION XN(NDV), X(NDV), D(NDV)

      DO 10 I = 1, NDV
         XN(I) = X(I) + AL * D(I)
10    CONTINUE

      RETURN
      END
```

FIGURE B.3   (*Continued*)

```
      SUBROUTINE GOLDM(X,D,XN,ALFA,DELTA,EPSLON,F,NCOUNT,NDV)
C     ----------------------------------------------------------
C     IMPLEMENTS GOLDEN SECTION SEARCH FOR MULTIVARIATE PROBLEMS
C     X      = CURRENT DESIGN POINT
C     D      = DIRECTION VECTOR
C     XN     = CURRENT DESIGN + TRIAL STEP * SEARCH DIRECTION
C     ALFA   = OPTIMUM VALUE OF ALPHA ON RETURN
C     DELTA  = INITIAL STEP LENGTH
C     EPSLON = CONVERGENCE PARAMETER
C     F      = OPTIMUM VALUE OF THE FUNCTION
C     NCOUNT = NUMBER OF FUNCTION EVALUATIONS ON RETURN
C     ----------------------------------------------------------

      IMPLICIT DOUBLE PRECISION (A-H, O-Z)
      DIMENSION X(NDV), D(NDV), XN(NDV)

      GR = 0.5D0 * DSQRT(5.0D0) + 0.5D0
      DELTA1 = DELTA
C
C     ESTABLISH INITIAL DELTA
C
      AL = 0.0D0
      CALL UPDATE(XN,X,D,AL,NDV)
      CALL FUNCT(XN,FL,NCOUNT,NDV)
      F = FL
10    CONTINUE
      AA = DELTA1
      CALL UPDATE(XN,X,D,AA,NDV)
      CALL FUNCT(XN,FA,NCOUNT,NDV)
      IF (FA .GT. FL) THEN
         DELTA1 = DELTA1 * 0.1D0
         GO TO 10
      END IF
C
C     ESTABLISH INITIAL INTERVAL OF UNCERTAINTY
C
      J = 0
20    CONTINUE
      J = J + 1
      AU = AA + DELTA1 * (GR ** J)

      CALL UPDATE(XN,X,D,AU,NDV)
      CALL FUNCT(XN,FU,NCOUNT,NDV)
      IF (FA .GT. FU) THEN
         AL = AA
```

FIGURE B.3 (*Continued*)

```
              AA = AU
              FL = FA
              FA = FU
              GO TO 20
          END IF
C
C         REFINE THE INTERVAL OF UNCERTAINTY FURTHER
C
          AB = AL + (AU - AL) / GR
          CALL UPDATE(XN,X,D,AB,NDV)
          CALL FUNCT(XN,FB,NCOUNT,NDV)
30        CONTINUE
          IF((AU-AL) .LE. EPSLON) GO TO 80
C
C         IMPLEMENT STEPS 4 ,5 OR 6 OF THE ALGORITHM
C
          IF (FA-FB) 40, 60, 50
C
C         FA IS LESS THAN FB (STEP 4)
C
40        AU = AB
          FU = FB
          AB = AA
          FB = FA
          AA = AL + (1.0D0 - 1.0D0 / GR) * (AU - AL)
          CALL UPDATE(XN,X,D,AA,NDV)
          CALL FUNCT(XN,FA,NCOUNT,NDV)
          GO TO 30
C
C         FA IS GREATER THAN FB (STEP 5)
C
50        AL = AA
          FL = FA
          AA = AB
          FA = FB
          AB = AL + (AU - AL) / GR
          CALL UPDATE(XN,X,D,AB,NDV)
          CALL FUNCT(XN,FB,NCOUNT,NDV)
          GO TO 30
C
C         FA IS EQUAL TO FB (STEP 6)
C
```

FIGURE B.3 (*Continued*)

```
60     AL = AA
       FL = FA
       AU = AB
       FU = FB
       AA = AL + (1.0D0 - 1.0D0 / GR) * (AU - AL)
       CALL UPDATE(XN,X,D,AA,NDV)
       CALL FUNCT(XN,FA,NCOUNT,NDV)
       AB = AL + (AU - AL) / GR
       CALL UPDATE(XN,X,D,AB,NDV)
       CALL FUNCT(XN,FB,NCOUNT,NDV)
       GO TO 30
C
C      MINIMUM IS FOUND
C
80     ALFA = (AU + AL) * 0.5D0

       RETURN
       END
```

FIGURE B.3   (*Continued*)

```
C      THE MAIN PROGRAM FOR MODIFIED NEWTON'S METHOD
C      -------------------------------------------------------------
C      DELTA  = INITIAL STEP LENGTH FOR LINE SEARCH
C      EPSLON = LINE SEARCH ACCURACY
C      EPSL   = STOPPING CRITERION FOR MODIFIED NEWTON'S METHOD
C      NCOUNT = NO. OF FUNCTION EVALUATIONS
C      NDV    = NO. OF DESIGN VARIABLES
C      NOC    = NO. OF CYCLES OF THE METHOD
C      X      = DESIGN VARIABLE VECTOR
C      D      = DIRECTION VECTOR
C      G      = GRADIENT VECTOR
C      H      = HESSIAN MATRIX
C      WK     = WORK ARRAY USED FOR TEMPORARY STORAGE
C      -------------------------------------------------------------

       IMPLICIT DOUBLE PRECISION (A-H,O-Z)
       DIMENSION X(3), D(3), G(3), H(3,3), WK(3)
C
C      DEFINE INITIAL DATA
```

FIGURE B.4   A program for Newton's method.

```
C
      DELTA  = 5.0D-2
      EPSLON = 1.0D-4
      EPSL   = 5.0D-3
      NCOUNT = 0
      NDV    = 3
      NOC    = 100
C
C     STARTING VALUES OF THE DESIGN VARIABLES
C
      X(1) = 2.0D0
      X(2) = 4.0D0
      X(3) = 10.0D0

      CALL GRAD(X,G,NDV)
      WRITE(*,10)
10    FORMAT(' NO.     COST FUNCT     STEP SIZE',
     &       '   NORM OF GRAD  ')
      DO 20 K = 1, NOC
         CALL HASN(X,H,NDV)
         CALL SCALE (G,D,-1.0D0,NDV)
         CALL SYSEQ(H,NDV,D)
         IF (DOT(G,D,NDV) .GE. 1.0E-8) GO TO 60
         CALL GOLDM(X,D,WK,ALFA,DELTA,EPSLON,F,NCOUNT,NDV)
         CALL SCALE(D,D,ALFA,NDV)
         CALL PRINT(K,X,ALFA,G,F,NDV)
         CALL ADD(X,D,X,NDV)
         CALL GRAD(X,G,NDV)
         IF(TNORM(G,NDV) .LE. EPSL) GO TO 30
20    CONTINUE

      WRITE(*,*)
      WRITE(*,*)' LIMIT ON NO. OF CYCLES HAS EXCEEDED'
      WRITE(*,*)' THE CURRENT DESIGN VARIABLES ARE:'
      WRITE(*,*) X
      CALL EXIT
30    WRITE(*,*)
      WRITE(*,*) 'THE OPTIMAL DESIGN VARIABLES ARE  :'
      WRITE(*,40) X
40    FORMAT(4X,3F15.6)
      CALL FUNCT(X,F,NCOUNT,NDV)
      WRITE(*,50) ' OPTIMUM COST FUNCTION VALUE IS    :', F
50    FORMAT(A, F13.6)
      WRITE(*,*) 'NO. OF FUNCTION EVALUATIONS ARE   :   ', NCOUNT
      CALL EXIT
```

FIGURE B.4 (*Continued*)

```
60      WRITE(*,*)
        WRITE(*,*)' DESCENT DIRECTION CANNOT BE FOUND'
        WRITE(*,*)' THE CURRENT DESIGN VARIABLES ARE:'
        WRITE(*,40) X

        STOP
        END


        DOUBLE PRECISION FUNCTION DOT(X,Y,N)
C
C       CALCULATES DOT PRODUCT OF VECTORS X AND Y
C
        IMPLICIT DOUBLE PRECISION (A-H,O-Z)
        DIMENSION X(N),Y(N)

        SUM = 0.0D0
        DO 10 I = 1, N
        SUM = SUM + X(I) * Y(I)
10      CONTINUE
        DOT = SUM

        RETURN
        END
        SUBROUTINE HASN(X,H,N)
C
C       CALCULATES THE HESSIAN MATRIX H AT X
C
        IMPLICIT DOUBLE PRECISION (A-H,O-Z)
        DIMENSION X(N),H(N,N)

        H(1,1) = 2.0D0
        H(2,2) = 4.0D0
        H(3,3) = 4.0D0
        H(1,2) = 2.0D0
        H(1,3) = 0.0D0
        H(2,3) = 2.0D0
        H(2,1) = H(1,2)
        H(3,1) = H(1,3)
        H(3,2) = H(2,3)
        RETURN
        END


        SUBROUTINE SYSEQ(A,N,B)
C
```

FIGURE B.4   (*Continued*)

```
C       SOLVES AN N X N SYMMETRIC SYSTEM OF LINEAR EQUATIONS
        AX = B
C       A IS THE COEFFICIENT MATRIX; B IS THE RIGHT HAND SIDE;
C       THESE ARE INPUT

C       B CONTAINS SOLUTION ON RETURN
C
        IMPLICIT DOUBLE PRECISION (A-H,O-Z)
        DIMENSION A(N,N), B(N)
C
C       REDUCTION OF EQUATIONS
C
        M = 0
50      M = M + 1
        MM = M + 1
        B(M) = B(M) / A(M,M)
        IF (M - N) 70, 130, 70
70      DO 80 J = MM, N
           A(M,J) = A(M,J) / A(M,M)
80      CONTINUE
C
C       SUBSTITUTION INTO REMAINING EQUATIONS
C
        DO 120 I = MM, N
           IF(A(I,M)) 90, 120, 90
90         DO 100 J = I, N
              A(I,J) = A(I,J) - A(I,M) * A(M,J)
              A(J,I) = A(I,J)
100        CONTINUE
           B(I) = B(I) - A(I,M) * B(M)
120     CONTINUE
        GO TO 50
C
C       BACK SUBSTITUTION
C
130     M = M - 1
        IF(M .EQ. 0) GO TO 150
        MM = M + 1
        DO 140 J = MM, N
           B(M) = B(M) - A(M,J) * B(J)
140     CONTINUE
        GO TO 130

150     RETURN
        END
```

FIGURE B.4   (*Continued*)

Bibliography

AA (1986). *Construction manual series, Section 1, No. 30*. Washington, DC: Aluminum Association.

AASHTO (1992). *Standard specifications for highway bridges* (15th ed.). Washington, DC: American Association of State Highway and Transportation Officials.

Abadie, J. (Ed.), (1970). *Nonlinear programming*. Amsterdam: North Holland.

Abadie, J., & Carpenter, J. (1969). Generalization of the Wolfe reduced gradient method to the case of nonlinear constraints. In R. Fletcher (Ed.), *Optimization* (pp. 37–47). New York: Academic Press.

ABNT (Associação Brasileira de Normas Técnicas), (1988). Forças Devidas ao Vento em Edificações, NBR-6123 [in Portuguese].

Ackoff, R. L., & Sasieni, M. W. (1968). *Fundamentals of operations research*. New York: John Wiley.

Adelman, H., & Haftka, R. T. (1986). Sensitivity analysis of discrete structural systems. *AIAA Journal*, 24(5), 823–832.

AISC (2005). *Manual of steel construction* (13th ed.). Chicago: American Institute of Steel Construction.

Al-Saadoun, S. S., & Arora, J. S. (1989). Interactive design optimization of framed structures. *Journal of Computing in Civil Engineering, ASCE*, 3(1), 60–74.

Antoniou, A., & Lu, W-S. (2007). *Practical optimization: Algorithms and engineering applications*. Norwell, MA: Springer.

Aoki, M. (1971). *Introduction to optimization techniques*. New York: Macmillan.

Arora, J. S. (1984). An algorithm for optimum structural design without line search. In E. Atrek, R. H. Gallagher, K. M. Ragsdell, & O. C. Zienkiewicz (Eds.), *New directions in optimum structural design* (pp. 429–441). New York: John Wiley.

Arora, J. S. (1990a). Computational design optimization: A review and future directions. *Structural Safety, 7*, 131–148.

Arora, J. S., (1990b). Global optimization methods for engineering design. *Proceedings of the 31st AIAA/ASME/ASCE/AHS/ASC structures, structural dynamics and materials conference* (pp. 123–135), Long Beach, CA. Reston, VA: American Institute of Aeronautics and Astronautics.

Arora, J. S. (1995). Structural design sensitivity analysis: Continuum and discrete approaches. In J. Herskovits (Ed.), *Advances in structural optimization* (pp. 47–70). Boston: Kluwer Academic.

Arora, J. S. (Ed.), (1997). *Guide to structural optimization*. ASCE Manuals and Reports on Engineering Practice, No. 90. Reston, VA: American Society of Civil Engineering.

Arora, J. S. (1999). Optimization of structures subjected to dynamic loads. In C. T. Leondes (Ed.), *Structural dynamic systems: Computational techniques and optimization* (Vol. 7, pp. 1–73). Newark, NJ: Gordon & Breech.

Arora, J. S. (2002). Methods for discrete variable structural optimization. In S. Burns (Ed.), *Recent advances in optimal structural design* (pp. 1–40). Reston, VA: Structural Engineering Institute.

Arora, J. S. (2007). *Optimization of structural and mechanical systems*. Singapore: World Scientific Publishing.

Arora, J. S., & Haug, E. J. (1979). Methods of design sensitivity analysis in structural optimization. *AIAA Journal*, 17(9), 970–974.

Arora, J. S., & Baenziger, G. (1986). Uses of artificial intelligence in design optimization. *Computer Methods in Applied Mechanics and Engineering, 54*, 303–323.

Arora, J. S., & Thanedar, P. B. (1986). Computational methods for optimum design of large complex systems. *Computational Mechanics, 1*(2), 221–242.

Arora, J. S., & Baenziger, G., (1987). A nonlinear optimization expert system. In D. R. Jenkins, (Ed.), *Proceedings of the ASCE Structures Congress, Computer Applications in Structural Engineering* (pp. 113–125). Reston, VA: American Society of Civil Engineers.

Arora, J. S., & Tseng, C. H. (1987a). *User's manual for IDESIGN: Version 3.5*. Optimal Design Laboratory, College of Engineering, University of Iowa.

Arora, J. S., & Tseng, C. H. (1987b). An investigation of Pshenichnyi's recursive quadratic programming method for engineering optimization—A discussion. *Journal of Mechanisms, Transmissions and Automation in Design, Transactions of the ASME, 109*(6), 254–256.

Arora, J. S., & Tseng, C. H. (1988). Interactive design optimization. *Engineering Optimization, 13,* 173–188.

Arora, J. S., & Huang, M. W. (1996). Discrete structural optimization with commercially available sections: A review. *Journal of Structural and Earthquake Engineering, JSCE, 13*(2), 93–110.

Arora, J. S., & Wang, Q. (2005). Review of formulations for structural and mechanical system optimization. *Structural and Multidisciplinary Optimization, 30,* 251–272.

Arora, J. S., Chahande, A. I., & Paeng, J. K. (1991). Multiplier methods for engineering optimization. *International Journal for Numerical Methods in Engineering, 32,* 1485–1525.

Arora, J. S., Huang, M. W., & Hsieh, C. C. (1994). Methods for optimization of nonlinear problems with discrete variables: A review. *Structural Optimization, 8*(2/3), 69–85.

Arora, J. S., Elwakeil, O. A., Chahande, A. I., & Hsieh, C. C. (1995). Global optimization methods for engineering applications: A review. *Structural Optimization, 9,* 137–159.

Arora, J. S., Burns, S., & Huang, M. W. (1997). What is optimization? In J. S. Arora (Ed.), *Guide to structural optimization, ASCE manual on engineering practice, 90,* 1–23. Reston, VA: American Society of Civil Engineers.

ASCE (2005). *Minimum design loads for buildings and other structures.* Reston, VA: American Society of Civil Engineers.

Athan, T. W., & Papalambros, P. Y. (1996). A note on weighted criteria methods for compromise solutions in multi-objective optimization. *Engineering Optimization, 27,* 155–176.

Atkinson, K. E. (1978). *An introduction to numerical analysis.* New York: John Wiley.

Balling, R. J. (2000). Pareto sets in decision-based design. *Journal of Engineering Valuation and Cost Analysis, 3*(2), 189–198.

Balling, R. J., (2003). The maximum fitness function: Multi-objective city and regional planning. In C. M. Fonseca, P. J. Fleming, E. Zitzler, K. Deb, & L. Thiele (Eds.), *Second international conference on evolutionary multi-criterion optimization* (pp. 1–15), Faro, Portugal, April 8–11, Berlin: Springer.

Balling, R. J., Taber, J. T., Brown, M. R., & Day, K. (1999). Multiobjective urban planning using a genetic algorithm. *Journal of Urban Planning and Development, 125*(2), 86–99.

Balling, R. J., Taber, J. T., Day, K., & Wilson, S. (2000). Land use and transportation planning for twin cities using a genetic algorithm. *Transportation Research Record, 1722,* 67–74.

Bartel, D. L. (1969). *Optimum design of spatial structures.* Doctoral dissertation, College of Engineering, University of Iowa.

Bazarra, M. S., Sherali, H. D., & Shetty, C. M. (2006). *Nonlinear programming: Theory and applications* (3rd ed.). Hoboken, NJ: Wiley-Interscience.

Belegundu, A. D., & Arora, J. S. (1984a). A recursive quadratic programming algorithm with active set strategy for optimal design. *International Journal for Numerical Methods in Engineering, 20*(5), 803–816.

Belegundu, A. D., & Arora, J. S. (1984b). A computational study of transformation methods for optimal design. *AIAA Journal, 22*(4), 535–542.

Belegundu, A. D., & Arora, J. S. (1985). A study of mathematical programming methods for structural optimization. *International Journal for Numerical Methods in Engineering, 21*(9), 1583–1624.

Belegundu, A. D., & Chandrupatla, T. R. (2011). *Optimization concepts and applications in engineering* (2nd ed.). New York: Cambridge University Press.

Belegundu, A. D., & Arora, J. S. (1984). A computational study of transformation methods for optimal design. *AIAA Journal, 22*(4), 535–542.

Bell, W. W. (1975). *Matrices for scientists and engineers.* New York: Van Nostrand Reinhold.

Bertsekas, D. P. (1995). *Nonlinear programming.* Belmont, MA: Athena Scientific.

Beyer, H-G., & Sandhoff, B. (2007). Robust optimization—A comprehensive survey. *Computer Methods in Applied Mechanics and Engineering, 196*(33–34), 3190–3218.

Bhatti, M. A. (2000). *Practical optimization methods with mathematica Applications.* New York: Springer Telos.

Bhatti, M. A. (2005). *Fundamental finite element analysis and applications with mathematica and MATLAB computations.* New York: John Wiley.

Blank, L., & Tarquin, A. (1983). *Engineering economy* (2nd ed.). New York: McGraw-Hill.

Blum, C. (2005). Ant colony optimization: Introduction and recent trends. *Physics of Life Reviews, 2,* 353–373.

Box, G. E. P., & Wilson, K. B. (1951). On the experimental attainment of optimum conditions. *Journal of the Royal Statistical Society, Series B, XIII,* 1–45.

Branin, F. H., & Hoo, S. K. (1972). A method for finding multiple extrema of a function of *n* variables. In F. A. Lootsma (Ed.), *Numerical methods of nonlinear optimization.* London: Academic Press.

Carmichael, D. G. (1980). Computation of pareto optima in structural design. *International Journal for Numerical Methods in Engineering*, 15, 925–952.

Cauchy, A. (1847). Method generale pour la resolution des systemes d'equations simultanees. *Comptes Rendus. de Academie Scientifique*, 25, 536–538.

Chahande, A. I., & Arora, J. S. (1993). Development of a multiplier method for dynamic response optimization problems. *Structural Optimization*, 6(2), 69–78.

Chahande, A. I., & Arora, J. S. (1994). Optimization of large structures subjected to dynamic loads with the multiplier method. *International Journal for Numerical Methods in Engineering*, 37(3), 413–430.

Chandrupatla, T. R., & Belegundu, A. D. (1997). *Introduction to finite elements in engineering* (2nd ed.). Upper Saddle River, NJ: Prentice-Hall.

Chen, S. Y., & Rajan, S. D. (2000). A robust genetic algorithm for structural optimization. *Structural Engineering and Mechanics*, 10, 313–336.

Chen, W., Sahai, A., Messac, A., & Sundararaj, G. (2000). Exploration of the effectiveness of physical programming in robust design. *Journal of Mechanical Design*, 122, 155–163.

Cheng, F. Y., & Li., D. (1997). Multiobjective optimization design with Pareto genetic algorithm. *Journal of Structural Engineering*, 123, 1252–1261.

Cheng, F. Y., & Li., D. (1998). Genetic algorithm development for multiobjective optimization of structures. *AIAA Journal*, 36, 1105–1112.

Choi, S. K., Grandhi, R. V., & Canfield, R. A. (2007). *Reliability-based structural design*. Berlin: Springer-Verlag.

Chong, K. P., & Zak, S. H. (2008). *An introduction to optimization* (3rd ed.). New York: John Wiley.

Chopra, A. K. (2007). *Dynamics of structures: Theory and applications to earthquake engineering* (3rd ed.). Upper Saddle River, NJ: Prentice-Hall.

Clough, R. W., & Penzien, J. (1975). *Dynamics of structures*. New York: McGraw-Hill.

Coello-Coello, C. A., Van Veldhuizen, D. A., & Lamont, G. B. (2002). *Evolutionary algorithms for solving multi-objective problems*. New York: Kluwer Academic.

Cohon, J. L. (1978). *Multiobjective programming and planning*. New York: Academic Press.

Cook, R. D. (1981). *Concepts and applications of finite element analysis*. New York: John Wiley.

Cooper, L., & Steinberg, D. (1970). *Introduction to methods of optimization*. Philadelphia: W. B. Saunders.

Corcoran, P. J. (1970). Configuration optimization of structures. *International Journal of Mechanical Sciences*, 12, 459–462.

Corne, D., Dorigo, M., & Glover, F. (Eds.), (1999). *New ideas in optimization* New York: McGraw-Hill.

Crandall, S. H., Dahl, H. C., & Lardner, T. J. (1999). *Introduction to mechanics of solids* (2nd ed.). New York: McGraw-Hill.

Dakin, R. J. (1965). A tree-search algorithm for mixed integer programming problems. *Computer Journal*, 8, 250–255.

Dano, S. (1974). *Linear programming in industry* (4th ed.). New York: Springer-Verlag.

Dantzig, G. B., & Thapa, M. N. (1997). *Linear programming, 1: Introduction*. New York: Springer-Verlag.

Das, I., & Dennis, J. E. (1997). A closer look at drawbacks of minimizing weighted sums of objectives for pareto set generation in multicriteria optimization problems. *Structural Optimization*, 14, 63–69.

Das, S., & Suganthan, N. (2011). Differential evolution: a survey of the state-of-the-art. *IEEE Transactions on Evolutionary Computation*, 15(1), 4–31.

Davidon, W. C. (1959). *Variable metric method for minimization, research and development report ANL-5990*. Argonne, IL: Argonne National Laboratory.

Day, H. J., & Dolbear, F., (1965). Regional water quality management. *Proceedings of the 1st annual meeting of the American Water Resources Association* (pp. 283–309). Chicago: University of Chicago Press.

De Boor, C. (1978). *A practical guide to splines: Applied mathematical sciences* (Vol. 27). New York: Springer-Verlag.

Deb, K., (1989). *Genetic algorithms in multimodal function optimization*. Master's thesis (TCGA report No. 89002). University of Alabama, Tuscaloosa.

Deb, K. (2001). *Multi-objective optimization using evolutionary algorithms*. Chichester, UK: John Wiley.

Deif, A. S. (1982). *Advanced matrix theory for scientists and engineers*. New York: Halsted Press.

Deininger, R. A., (1975). Water quality management—The planning of economically optimal pollution control systems. *Proceedings of the 1st annual meeting of the American Water Resources Association* (pp. 254–282). Chicago: University of Chicago Press.

Dixon, L. C. W., & Szego, G. P. (Eds.), (1978). *Towards global optimization 2*. Amsterdam: North-Holland.

Dorigo, M., (1992). *Optimization, learning and natural algorithms*. Ph.D. thesis, Politecnico di Milano, Italy.

Drew, D. (1968). *Traffic flow theory and control*. New York: McGraw-Hill.

Ehrgott, M., & Gandibleux, X. (Eds.), (2002). *Multiple criteria optimization: State of the art annotated bibliographic surveys* Boston: Kluwer Academic.

Elwakeil, O. A. (1995). *Algorithms for global optimization and their application to structural optimization problems.* Doctoral dissertation, University of Iowa.

Elwakeil, O. A., & Arora, J. S. (1995). Methods for finding feasible points in constrained optimization. *AIAA Journal, 33*(9), 1715–1719.

Elwakeil, O. A., & Arora, J. S. (1996a). Two algorithms for global optimization of general NLP problems. *International Journal for Numerical Methods in Engineering, 39*, 3305–3325.

Elwakeil, O. A., & Arora, J. S. (1996b). Global optimization of structural systems using two new methods. *Structural Optimization, 12*, 1–12.

Evtushenko, Yu. G. (1974). Methods of search for the global extremum. *Operations Research, Computing Center of the U.S.S.R Akad. of Sci., 4*, 39–68.

Evtushenko, Yu. G. (1985). *Numerical optimization techniques*. New York: Optimization Software.

Fang, S. C., & Puthenpura, S. (1993). *Linear optimization and extensions: Theory and algorithms*. Englewood Cliffs, NJ: Prentice-Hall.

Fiacco, A. V., & McCormick, G. P. (1968). *Nonlinear programming: Sequential unconstrained minimization techniques.* Philadelphia: Society for Industrial and Applied Mathematics.

Fletcher, R., & Powell, M. J. D. (1963). A rapidly convergent descent method for minimization. *The Computer Journal, 6*, 163–180.

Fletcher, R., & Reeves, R. M. (1964). Function minimization by conjugate gradients. *The Computer Journal, 7*, 149–160.

Floudas, C. A., et al. (1990). *Handbook of test problems in local and global optimization*. Norwell, MA: Kluwer Academic.

Fonseca, C. M., & Fleming, P. J., (1993). Genetic algorithms for multiobjective optimization: Formulation, discussion, and generalization. *Fifth international conference on genetic algorithms* (pp. 416–423). Urbana-Champaign, IL, San Mateo, CA: Morgan Kaufmann.

Forsythe, G. E., & Moler, C. B. (1967). *Computer solution of linear algebraic systems*. Englewood Cliffs, NJ: Prentice-Hall.

Franklin, J. N. (1968). *Matrix theory*. Englewood Cliffs, NJ: Prentice-Hall.

Gabrielle, G. A., & Beltracchi, T. J. (1987). An investigation of Pschenichnyi's recursive quadratic programming method for engineering optimization. *Journal of Mechanisms, Transmissions and Automation in Design, Transactions of the ASME, 109*(6), 248–253.

Gen, M., & Cheng, R. (1997). *Genetic algorithms and engineering design*. New York: John Wiley.

Gere, J. M., & Weaver, W. (1983). *Matrix algebra for engineers*. Monterey, CA: Brooks/Cole Engineering Division.

Gill, P. E., Murray, W., & Wright, M. H. (1981). *Practical optimization*. New York: Academic Press.

Gill, P. E., Murray, W., & Wright, M. H. (1991). *Numerical linear algebra and optimization* (Vol. 1). New York: Addison-Wesley.

Gill, P. E., Murray, W., Saunders, M. A., & Wright, M. H. (1984). *User's guide for QPSOL: Version 3.2*. Stanford, CA: Systems Optimization Laboratory, Department of Operations Research, Stanford University.

Glover, F., & Kochenberger, G. (Eds.), (2002). *Handbook on metaheuristics* Norwell, MA: Kluwer Academic.

Goldberg, D. E. (1989). *Genetic algorithms in search, optimization and machine learning*. Reading, MA: Addison-Wesley.

Grandin, H. (1986). *Fundamentals of the finite element method*. New York: Macmillan.

Grant, E. L., Ireson, W. G., & Leavenworth, R. S. (1982). *Principles of engineering economy* (7th ed.). New York: John Wiley.

Hadley, G. (1961). *Linear programming*. Reading, MA: Addison-Wesley.

Hadley, G. (1964). *Nonlinear and dynamic programming*. Reading, MA: Addison-Wesley.

Haftka, R. T., & Gurdal, Z. (1992). *Elements of structural optimization*. Norwell, MA: Kluwer Academic.

Han, S. P. (1976). Superlinearly convergent variable metric algorithms for general nonlinear programming. *Mathematical Programming, 11*, 263–282.

Han, S. P. (1977). A globally convergent method for nonlinear programming. *Journal of Optimization Theory and Applications, 22*, 297–309.

Hasofer, A. M., & Lind, N. C. (1974). Exact and invariant second-moment code format. *Journal of the Engineering Mechanics Division, ASCE, 100*, 111–121.

Haug, E. J., & Arora, J. S. (1979). *Applied optimal design.* New York: Wiley-Interscience.

Hestenes, M. R., & Stiefel, E. (1952). Methods of conjugate gradients for solving linear systems. *Journal of Research of the National Bureau of Standards, 49*, 409–436.

Hibbeler, R. C. (2007). *Mechanics of materials* (7th ed.). Upper Saddle River, NJ: Prentice-Hall.

Hock, W., & Schittkowski, K. (1981). *Test examples for nonlinear programming codes, Lecture notes in economics and mathematical systems, 187.* New York: Springer-Verlag.

Hock, W., & Schittkowski, K. (1983). A comparative performance evaluation of 27 nonlinear programming codes. *Computing, 30*, 335–358.

Hohn, F. E. (1964). *Elementary matrix algebra.* New York: Macmillan.

Holland, J. H. (1975). *Adaptation in natural and artificial systems.* Ann Arbor: University of Michigan Press.

Hooke, R., & Jeeves, T. A. (1961). Direct search solution of numerical and statistical problems. *Journal of the Association of Computing Machinery, 8*(2), 212–229.

Hopper, M. J. (1981). *Harwell subroutine library.* Oxfordshire, UK: Computer Science and Systems Division, AERE Harwell.

Horn, J., Nafpliotis, N., & Goldberg, D. E., (1994). A niched pareto genetic algorithm for multiobjective optimization. *First IEEE conference on evolutionary computation* (pp. 82–87), Orlando, FL. Piscataway, NJ: IEEE Neural Networks Council.

Hsieh, C. C., & Arora, J. S. (1984). Design sensitivity analysis and optimization of dynamic response. *Computer Methods in Applied Mechanics and Engineering, 43*, 195–219.

Huang, M. W., & Arora, J. S., (1995). Engineering optimization with discrete variables. *Proceedings of the 36th AIAA SDM conference* (pp. 1475–1485), New Orleans, April 10–12.

Huang, M. W., & Arora, J. S. (1996). A self-scaling implicit SQP method for large scale structural optimization. *International Journal for Numerical Methods in Engineering, 39*, 1933–1953.

Huang, M. W., & Arora, J. S. (1997a). Optimal design with discrete variables: Some numerical experiments. *International Journal for Numerical Methods in Engineering, 40*, 165–188.

Huang, M. W., & Arora, J. S. (1997b). Optimal design of steel structures using standard sections. *Structural and Multidisciplinary Optimization, 14*, 24–35.

Huang, M. W., Hsieh, C. C., & Arora, J. S. (1997). A genetic algorithm for sequencing type problems in engineering design. *International Journal for Numerical Methods in Engineering, 40*, 3105–3115.

Huang, M. -W., & Arora, J. S. (1996). A self-scaling implicit SQP method for large scale structural optimization. *International Journal for Numerical Methods in Engineering, 39*, 1933–1953.

Huebner, K. H., & Thornton, E. A. (1982). *The finite element method for engineers.* New York: John Wiley.

Hyman, B. (2003). *Fundamentals of engineering design* (2nd ed.). Upper Saddle River, NJ: Prentice-Hall.

IEEE/ASTM (1997). *Standard for use of the international system of units (SI): The modern metric system.* New York: The Institute of Electrical and Electronics Engineers/American Society for Testing of Materials.

Ishibuchi, H., & Murata, T., (1996). Multiobjective genetic local search algorithm. *IEEE international conference on evolutionary computation* (pp. 119–124), Nagoya, Japan. Piscataway, NJ: Institute of Electrical and Electronics Engineers.

Iyengar, N. G. R., & Gupta, S. K. (1980). *Programming Methods in Structural Design.* New York: John Wiley.

Javonovic, V., & Kazerounian, K. (2000). Optimal design using chaotic descent method. *Journal of Mechanical Design, ASME, 122*(3), 137–152.

Jennings, A. (1977). *Matrix computations for engineers.* New York: John Wiley.

Karush, W. (1939). *Minima of functions of several variables with inequalities as side constraints.* Master's thesis, Chicago: Department of Mathematics, University of Chicago.

Kennedy, J., & Eberhart, R. C., (1995). Particle swarm optimization. *Proceedings of IEEE international conference on neural network, IV*, 1942–1948. IEEE Service Center, Piscataway, NJ.

Kennedy, J., Eberhart, R. C., & Shi, Y. (2001). *Swarm intelligence.* San Francisco: Morgan Kaufmann.

Kim, C. H., & Arora, J. S. (2003). Development of simplified dynamic models using optimization: Application to crushed tubes. *Computer Methods in Applied Mechanics and Engineering, 192*(16–18), 2073–2097.

Kim, J. H., Xiang, Y., Yang, J., Arora, J. S., & Abdel-Malek, K. (2010). Dynamic motion planning of overarm throw for a biped human multibody system. *Multibody System Dynamics, 24*(1), 1–24.

Kirsch, U. (1993). *Structural optimization.* New York: Springer-Verlag.

Kirsch, U. (1981). *Optimum structural design.* New York: McGraw-Hill.

Kocer, F. Y., & Arora, J. S. (1996a). Design of prestressed concrete poles: An optimization approach. *Journal of Structural Engineering, ASCE, 122*(7), 804–814.

Kocer, F. Y., & Arora, J. S. (1996b). Optimal design of steel transmission poles. *Journal of Structural Engineering, ASCE, 122*(11), 1347–1356.

Kocer, F. Y., & Arora, J. S. (1997). Standardization of transmission pole design using discrete optimization methods. *Journal of Structural Engineering, ASCE, 123*(3), 345–349.

Kocer, F. Y., & Arora, J. S. (1999). Optimal design of H-frame transmission poles subjected to earthquake loading. *Journal of Structural Engineering, ASCE, 125*(11), 1299–1308.

Kocer, F. Y., & Arora, J. S. (2002). Optimal design of latticed towers subjected to earthquake loading. *Journal of Structural Engineering, ASCE, 128*(2), 197–204.

Kolda, T. G., Lewis, R. M., & Torczon, V. (2003). Optimization by direct search: New perspective on some classical and modern methods. *SIAM Review, 45*(3), 385–482.

Koski, J. (1985). Defectiveness of weighting method in multicriterion optimization of structures. *Communications in Applied Numerical Methods, 1*, 333–337.

Kunzi, H. P., & Krelle, W. (1966). *Nonlinear programming.* Waltham, MA: Blaisdell.

Lagarias, J. C., Reeds, J. A., Wright, M. H., & Wright, P. E. (1998). Convergence properties of the Nelder-Mead Simplex method in low dimensions. *SIAM Journal of Optimization, 9*, 112–147.

Land, A. M., & Doig, A. G. (1960). An automatic method of solving discrete programming problems. *Econometrica, 28*, 497–520.

Lee, S. M., & Olson, D. L. (1999). Goal programming. In T. Gal, T. J. Stewart, & T. Hanne (Eds.), *Multicriteria decision making: Advances in MCDM models, algorithms, theory, and applications.* Boston: Kluwer Academic.

Lemke, C. E. (1965). Bimatrix equilibrium points and mathematical programming. *Management Science, 11*, 681–689.

Levy, A. V., & Gomez, S. (1985). The tunneling method applied to global optimization. In P. T. Boggs, R. H. Byrd, & R. B. Schnabel (Eds.), *Numerical optimization 1984.* Philadelphia: Society for Industrial and Applied Mathematics.

Lewis, R. M., Torczon, V., & Trosset, M. W. (2000). Direct search methods: Then and now. *Journal of Computational and Applied Mathematics., 124*, 191–207.

Lim, O. K., & Arora, J. S (1986). An active set RQP algorithm for optimal design. *Computer Methods in Applied Mechanics and Engineering, 57*, 51–65.

Lim, O. K., & Arora, J. S (1987). Dynamic response optimization using an active set RQP algorithm. *International Journal for Numerical Methods in Engineering, 24*(10), 1827–1840.

Liu, D. C., & Nocedal, J. (1989). On the limited memory BFGS method for large scale optimization. *Mathematical Programming, 45*, 503–528.

Lucidi, S., & Piccioni, M. (1989). Random tunneling by means of acceptance-rejection sampling for global optimization. *Journal of Optimization Theory and Applications, 62*(2), 255–277.

Luenberger, D. G. (1984). *Linear and nonlinear programming.* Reading, MA: Addison-Wesley.

Marler, T. R., & Arora, J. S. (2004). Survey of multiobjective optimization methods for engineering. *Structural and Multidisciplinary Optimization, 26*(6), 369–395.

Marler, R. T., & Arora, J. S. (2009). *Multi-objective optimization: Concepts and methods for engineering.* Saarbrucken, Germany: VDM Verlag.

Marler, R. T., & Arora, J. S. (2010). The weighted sum method for multi-objective optimization: New insights. *Structural and Multidisciplinary Optimization, 41*(6), 453–462.

Marquardt, D. W. (1963). An algorithm for least squares estimation of nonlinear parameters. *SIAM Journal, 11*, 431–441.

MathWorks (2001). *Optimization toolbox for use with MATLAB, User's guide, Ver. 2.* Natick, MA: The MathWorks, Inc.

McCormick, G. P. (1967). Second-order conditions for constrained optima. *SIAM Journal Applied Mathematics, 15*, 641–652.

Meirovitch, L. (1985). *Introduction to Dynamics and Controls.* New York: John Wiley.

Messac, A. (1996). Physical programming: Effective optimization for computational design. *AIAA Journal*, *34*(1), 149–158.

Messac, A., & Mattson, C. A. (2002). Generating well-distributed sets of Pareto points for engineering design using physical programming. *Optimization and Engineering*, *3*, 431–450.

Messac, A., Puemi-Sukam, C., & Melachrinoudis, E. (2000a). Aggregate objective functions and pareto frontiers: Required relationships and practical implications. *Optimization and Engineering*, *1*, 171–188.

Messac, A., Sundararaj, G. J., Tappeta, R. V., & Renaud, J. E. (2000b). Ability of objective functions to generate points on nonconvex pareto frontiers. *AIAA Journal*, *38*(6), 1084–1091.

Messac, A., Puemi-Sukam, C., & Melachrinoudis, E. (2001). Mathematical and pragmatic perspectives of physical programming. *AIAA Journal*, *39*(5), 885–893.

Metropolis, N., Rosenbluth, A. W., Rosenbluth, M. N., Teller, A. H., & Teller, E. (1953). Equations of state calculations by fast computing machines. *Journal of Chemical Physics*, *21*, 1087–1092.

Microsoft. *Microsoft EXCEL, Version 11.0*, Redmond, WA: Microsoft.

Minoux, M. (1986). *Mathematical programming theory and algorithms*. New York: John Wiley.

Mitchell, M. (1996). *An introduction to genetic algorithms*. Cambridge, MA: MIT Press.

Moré, J. J., & Wright, S. J. (1993). *Optimization software guide*. Philadelphia: Society for Industrial and Applied Mathematics.

Murata, T., Ishibuchi, H., & Tanaka, H. (1996). Multiobjective genetic algorithm and its applications to flowshop scheduling. *Computers and Industrial Engineering*, *30*, 957–968.

NAG (1984). *FORTRAN library manual*. Downers Grove, IL: Numerical Algorithms Group.

Narayana, S., & Azarm, S. (1999). On improving multiobjective genetic algorithms for design optimization. *Structural Optimization*, *18*, 146–155.

Nash, S. G., & Sofer, A. (1996). *Linear and nonlinear programming*. New York: McGraw-Hill.

Nelder, J. A., & Mead, R. A. (1965). A Simplex method for function minimization. *Computer Journal*, *7*, 308–313.

Nemhauser, G. L., & Wolsey, S. J. (1988). *Integer and combinatorial optimization*. New York: John Wiley.

Nikolaidis, E., Ghiocel, D. M., & Singhal, S. (2005). *Engineering design reliability handbook*. Boca Raton, FL: CRC Press.

Nocedal, J. (1980). Updating quasi-Newton matrices with limited storage. *Mathematics of Computation*, *35*(151), 773–782.

Nocedal, J., & Wright, S. J. (2006). *Numerical optimization* (2nd ed.). New York: Springer Science.

Norton, R. L. (2000). *Machine design: An integrated approach* (2nd ed.). Upper Saddle River, NJ: Prentice-Hall.

Onwubiko, C. (2000). *Introduction to engineering design optimization*. Upper Saddle River, NJ: Prentice-Hall.

Osman, M. O. M., Sankar, S., & Dukkipati, R. V. (1978). Design synthesis of a multi-speed machine tool gear transmission using multiparameter optimization. *Journal of Mechanical Design, Transactions of ASME*, *100*, 303–310.

Osyczka, A. (2002). *Evolutionary algorithms for single and multicriteria design optimization*. Berlin: Physica Verlag.

Paeng, J. K., & Arora, J. S. (1989). Dynamic response optimization of mechanical systems with multiplier methods. *Journal of Mechanisms, Transmissions, and Automation in Design, Transactions of the ASME*, *111*(1), 73–80.

Papalambros, P. Y., & Wilde, D. J. (2000). *Principles of Optimal Design: Modeling and Computation* (2nd ed.). New York: Cambridge University Press.

Pardalos, P. M., & Rosen, J. B. (1987). Constrained global optimization: Algorithms and applications. In G. Goos, & J. Hartmanis (Eds.), *Lecture notes in computer science*. New York: Springer-Verlag.

Pardalos, P. M., Migdalas, A., & Burkard, R. (2002). *Combinatorial and global optimization, Series on Applied Mathematics* (Vol. 14). River Edge, NJ: World Scientific Publishing.

Pardalos, P. M., Romeijn, H. E., & Tuy, H. (2000). Recent developments and trends in global optimization. *Journal of Computational and Applied Mathematics*, *124*, 209–228.

Pareto, V., (1971). Manuale di economica politica, societa editrice libraria. A. S. Schwier, A. N., Page, & A. M. Kelley (Eds., Trans.). New York: Augustus M. Kelley. Originally published in 1906.

Park, G. J. (2007). *Analytical methods in design practice*. Berlin: Springer-Verlag.

Park, G-J., Lee, T-H, Lee, K. H., & Hwang, K-H. (2006). Robust design: An overview. *AIAA Journal*, *44*(1), 181–191.

Pederson, D. R., Brand, R. A., Cheng, C., & Arora, J. S. (1987). Direct comparison of muscle force predictions using linear and nonlinear programming. *Journal of Biomechanical Engineering, Transactions of the ASME*, *109*(3), 192–199.

Pezeshk, S., & Camp, C. V. (2002). State-of-the-art on use of genetic algorithms in design of steel structures. In S. Burns (Ed.), *Recent advances in optimal structural design*. Reston, VA: Structural Engineering Institute, ASCE.

Phadke, M. S. (1989). *Quality engineering using robust design*. Englewood Cliff, NJ: Prentice-Hall.

Polak, E., & Ribiére, G. (1969). Note sur la convergence de méthods de directions conjuguées. *Revue Française d'Informatique et de Recherche Opérationnelle, 16*, 35–43.

Powell, M. J. D. (1978a). A fast algorithm for nonlinearly constrained optimization calculations. In G. A. Watson, et al. (Eds.), *Lecture notes in mathematics*. Berlin: Springer-Verlag. Also published in *Numerical Analysis, Proceedings of the Biennial Conference*, Dundee, Scotland, June 1977.

Powell, M. J. D. (1978b). The convergence of variable metric methods for nonlinearity constrained optimization calculations. In O. L. Mangasarian, R. R. Meyer, & S. M. Robinson (Eds.), *Nonlinear programming 3*. New York: Academic Press.

Powell, M. J. D. (1978c). Algorithms for nonlinear functions that use Lagrange functions. *Mathematical Programming, 14*, 224–248.

Price, C. J., Coope, I. D., & Byatt, D. (2002). A Convergent Variant of the Nelder-Mead Algorithm. *Journal of Optimization Theory and Applications, 113*(1), 5–19.

Price, K., Storn, R., & Lampinen, J. (2005). *Differential evolution—A practical approach to global optimization*. Berlin: Springer.

Price, W. L. (1987). Global optimization algorithms for a CAD workstation. *Journal of Optimization Theory and Applications, 55*, 133–146.

Pshenichny, B. N. (1978). Algorithms for the general problem of mathematical programming. *Kibernetica, 5*, 120–125.

Pshenichny, B. N., & Danilin, Y. M. (1982). *Numerical methods in extremal problems* (2nd ed.). Moscow: Mir Publishers.

Qing, A. (2009). *Differential evolution—fundamentals and applications in electrical engineering*. New York: Wiley-Interscience.

Randolph, P. H., & Meeks, H. D. (1978). *Applied linear optimization*. Columbus, OH: GRID.

Rao, S. S. (2009). *Engineering optimization: Theory and practice*. Hoboken, NJ: John Wiley.

Ravindran, A., & Lee, H. (1981). Computer experiments on quadratic programming algorithms. *European Journal of Operations Research, 8*(2), 166–174.

Ravindran, A., Ragsdell, K. M., & Reklaitis, G. V. (2006). *Engineering optimization: Methods and applications*. New York: John Wiley.

Rinnooy, A. H. G., & Timmer, G. T. (1987a). Stochastic global optimization methods. Part I: Clustering methods. *Mathematical Programming, 39*, 27–56.

Rinnooy, A. H. G., & Timmer, G. T. (1987b). Stochastic global optimization methods. Part II: Multilevel methods. *Mathematical Programming, 39*, 57–78.

Roark, R. J., & Young, W. C. (1975). *Formulas for stress and strain* (5th ed.). New York: McGraw-Hill.

Rosen, J. B. (1961). The gradient projection method for nonlinear programming. *Journal of the Society for Industrial and Applied Mathematics, 9*, 514–532.

Rubinstein, M. F., & Karagozian, J. (1966). Building design under linear programming. *Proceedings of the ASCE, 92* (ST6), 223–245.

Salkin, H. M. (1975). *Integer programming*. Reading, MA: Addison-Wesley.

Sargeant, R. W. H. (1974). Reduced-gradient and projection methods for nonlinear programming. In P. E. Gill, & W. Murray (Eds.), *Numerical methods for constrained optimization* (pp. 149–174). New York: Academic Press.

Sasieni, M., Yaspan, A., & Friedman, L. (1960). *Operations—methods and problems*. New York: John Wiley.

Schaffer, J. D., (1985). Multiple objective optimization with vector evaluated GENETIC algorithms. *First international conference on genetic algorithms and their applications* (pp. 93–100), Pittsburgh. Hillsdale, NJ: Erlbaum.

Schittkowski, K. (1981). The nonlinear programming method of Wilson, Han and Powell with an augmented Lagrangian type line search function, Part 1: Convergence analysis, Part 2: An efficient implementation with linear least squares subproblems. *Numerische Mathematik, 38*, 83–127.

Schittkowski, K. (1987). *More test examples for nonlinear programming codes*. New York: Springer-Verlag.

Schmit, L. A., (1960). Structural design by systematic synthesis. *Proceedings of the second ASCE conference on electronic computations* (pp. 105–122), Pittsburgh. Reston, VA: American Society of Civil Engineers.

Schrage, L. (1991). *LINDO: Text and software*. Palo Alto, CA: Scientific Press.

Schrijver, A. (1986). *Theory of linear and integer programming*. New York: John Wiley.

Shampine, L. F., & Gordon, M. K. (1975). *Computer simulation of ordinary differential equations: The initial value problem*. San Francisco: W. H. Freeman.

Shampine, L. F. (1994). *Numerical solution of ordinary differential equations*. New York: Chapman & Hall.

Shigley, J. E., Mischke, C. R., & Budynas, R. (2004). *Mechanical engineering design* (7th ed.). New York: McGraw-Hill.

Siddall, J. N. (1972). *Analytical decision-making in engineering design*. Englewood Cliffs, NJ: Prentice-Hall.

Singer, S., & Singer, S. (2004). Efficient implementation of the Nelder-Mead search algorithm. *Applied Numerical Analysis and Computational Mathematics, 1*(3), 524–534.

Spotts, M. F. (1953). *Design of Machine Elements* (2nd ed.). Englewood Cliffs, NJ: Prentice-Hall.

Srinivas, N., & Deb, K. (1995). Multiobjective optimization using nondominated sorting in general algorithms. *Evolutionary Computations, 2*, 221–248.

Stadler, W. (1977). Natural structural shapes of shallow arches. *Journal of Applied Mechanics, 44*, 291–298.

Stadler, W. (1988). Fundamentals of multicriteria optimization. In W. Stadler (Ed.), *Multicriteria Optimization in Engineering and in the Sciences* (pp. 1–25). New York: Plenum.

Stadler, W. (1995). Caveats and boons of multicriteria optimization. *Microcomputers in Civil Engineering, 10*, 291–299.

Stadler, W., & Dauer, J. P. (1992). Multicriteria optimization in engineering: A tutorial and survey. In M. P. Kamat (Ed.), *Structural optimization: Status and promise* (pp. 211–249). Washington, DC: American Institute of Aeronautics and Astronautics.

Stark, R. M., & Nicholls, R. L. (1972). *Mathematical foundations for design: Civil engineering systems*. New York: McGraw-Hill.

Stewart, G. (1973). *Introduction to matrix computations*. New York: Academic Press.

Stoecker, W. F. (1971). *Design of thermal systems*. New York: McGraw-Hill.

Strang, G. (1976). *Linear algebra and its applications*. New York: Academic Press.

Sun, P. F., Arora, J. S., & Haug, E. J. (1975). *Fail-safe optimal design of structures*. Technical report No. 19. Department of Civil and Environmental Engineering? University of Iowa.

Syslo, M. M., Deo, N., & Kowalik, J. S. (1983). *Discrete optimization algorithms*. Englewood Cliffs, NJ: Prentice-Hall.

Taguchi, G. (1987). *Systems of experimental design* (Vols. I, II). New York: Kraus International.

Thanedar, P. B., Arora, J. S., & Tseng, C. H. (1986). A hybrid optimization method and its role in computer aided design. *Computers and Structures, 23*(3), 305–314.

Thanedar, P. B., Arora, J. S., Tseng, C. H., Lim, O. K., & Park, G. J. (1987). Performance of some SQP algorithms on structural design problems. *International Journal for Numerical Methods in Engineering, 23*(12), 2187–2203.

Törn, A., & Zilinskas, A. (1989). Global optimization. In G. Goos, & J. Hartmanis (Eds.), *Lecture notes in computer science*. New York: Springer-Verlag.

Tseng, C. H., & Arora, J. S. (1987). *Optimal design for dynamics and control using a sequential quadratic programming algorithm*. Technical report No. ODL-87.10, Optimal Design Laboratory, College of Engineering, University of Iowa.

Tseng, C. H., & Arora, J. S. (1988). On implementation of computational algorithms for optimal design 1: Preliminary investigation; 2: Extensive numerical investigation. *International Journal for Numerical Methods in Engineering, 26*(6), 1365–1402.

Vanderplaats, G. N. (1984). *Numerical optimization techniques for engineering design with applications*. New York: McGraw-Hill.

Vanderplaats, G. N., & Yoshida, N. (1985). Efficient calculation of optimum design sensitivity. *AIAA Journal, 23*(11), 1798–1803.

Venkataraman, P. (2002). *Applied optimization with MATLAB programming*. New York: John Wiley.

Wahl, A. M. (1963). *Mechanical springs* (2nd ed.). New York: McGraw-Hill.

Walster, G. W., Hansen, E. R., & Sengupta, S. (1984). Test results for a global optimization algorithm. In T. Boggs, et al. (Eds.),*Numerical optimization* (pp. 280–283). Philadelphia: SIAM.

Wang, Q., & Arora, J. S. (2005a). Alternative formulations for transient dynamic response optimization. *AIAA Journal, 43*(10), 2188–2195.

Wang, Q., & Arora, J. S. (2005b). Alternative formulations for structural optimization: An evaluation using trusses. *AIAA Journal, 43*(10), 2202–2209.

Wang, Q., & Arora, J. S. (2006). Alternative formulations for structural optimization: An evaluation using frames. *Journal of Structural Engineering, 132*(12), 1880–1889.

Wang, Q., & Arora, J. S. (2007). Optimization of large scale structural systems using sparse SAND formulations. *International Journal for Numerical Methods in Engineering, 69*(2), 390–407.

Wang, Q., & Arora, J. S. (2009). Several alternative formulations for transient dynamic response optimization: An evaluation. *International Journal for Numerical Methods for Engineering, 80*, 631–650.

Wilson, R. B. (1963). *A simplicial algorithm for concave programming*. Doctoral dissertation, School of Business Administration, Harvard University.

Wolfe, P. (1959). The Simplex method for quadratic programming. *Econometica, 27*(3), 382–398.

Wu, N., & Coppins, R. (1981). *Linear programming and extensions*. New York: McGraw-Hill.

Xiang, Y., Arora, J. S., & Abdel-Malek, K. (2011). Optimization-based prediction of asymmetric human gait. *Journal of Biomechanics, 44*(4), 683–693.

Xiang, Y, Arora, J. S., Rahamatalla, S., & Abdel-Malek, K. (2009). Optimization-based dynamic human walking prediction: One step formulation. *International Journal for Numerical Methods in Engineering, 79*(6), 667–695.

Xiang, Y., Arora, J. S., Rahmatalla, S., Marler, T., Bhatt, R., & Abdel-Malek, K. (2010). Human lifting simulation using a multi-objective optimization approach. *Multibody System Dynamics, 23*(4), 431–451.

Xiang, Y., Chung, H. J., Kim, J., Bhatt, R., Marler, T., & Rahmatalla, S., et al. (2010). Predictive dynamics: An optimization-based novel approach for human motion simulation. *Structural and Multidisciplinary Optimization, 41*(3), 465–480.

Yang, X-E. (2010). *Engineering optimization: An introduction to metaheuristic applications*. Hoboken, NJ: John Wiley.

Zhou, C. S., & Chen, T. L. (1997). Chaotic annealing and optimization. *Physical Review E, 55*(3), 2580–2587.

Zoutendijk, G. (1960). *Methods of feasible directions*. Amsterdam: Elsevier.

Answers to Selected Exercises

Chapter 3 Graphical Optimization

3.1 $x^* = (2, 2)$, $f^* = 2$.   **3.2** $x^* = (0, 4)$, $F^* = 8$.   **3.3** $x^* = (8, 10)$, $f^* = 38$.   **3.4** $x^* = (4, 3.333, 2)$, $F^* = 11.33$.   **3.5** $x^* = (10, 10)$, $F^* = 400$.   **3.6** $x^* = (0, 0)$, $f^* = 0$.   **3.7** $x^* = (0, 0)$, $f^* = 0$.   **3.8** $x^* = (2, 3)$, $f^* = -22$.   **3.9** $x^* = (-2.5, 1.58)$, $f^* = -3.95$.   **3.10** $x^* = (-0.5, 0.167)$, $f^* = -0.5$.   **3.11** Global minimum: $x^* = (0.71, 0.71)$, $f^* = -3.04$; Global maximum: $x^* = (-0.71, -0.71)$, $f^* = 4.04$.   **3.12** Global minimum: $x^* = (2.17, 1.83)$, $f^* = -8.33$; No local maxima.   **3.13** Global minimum: $x^* = (2.59, -2.02)$, $f^* = 15.3$; Local minimum: $x^* = (-3.73, 3.09)$, $f^* = 37.88$; Global maximum: $x^* = (-3.63, -3.18)$, $f^* = 453.2$; Local maximum: $x^* = (1.51, 3.27)$, $f^* = 244.53$.   **3.14** Global minimum: $x^* = (2.0)$, $f^* = -4$; Local minimum: $x^* = (0, 0)$, $f^* = 0$; Local minimum: $x^* = (0, 2)$, $f^* = -2$; Local minimum: $x^* = (1.39, 1.54)$, $f^* = 0$; Global maximum: $x^* = (0.82, 0.75)$, $f^* = 2.21$.   **3.15** Global minimum: $x^* = (7, 5)$, $f^* = 10$; Global maximum: $x^* = (0, 0)$, $f^* = 128$; Local maximum: $x^* = (12, 0)$, $f^* = 80$.   **3.16** Global minimum: $x^* = (2, 1)$, $f^* = -25$; Global maximum: $x^* = (-2.31, 0.33)$, $f^* = 24.97$.   **3.17** Global minimum: $x^* = (2.59, -2.01)$, $f^* = 15.25$; Local minimum: $x^* = (-3.73, 3.09)$, $f^* = 37.87$; No local maxima.   **3.18** Global minimum: $x^* = (4, 4)$, $f^* = 0$; Global maximum: $x^* = (0, 10)$, $f^* = 52$; Local maximum: $x^* = (0, 0)$, $f^* = 32$; Local maximum: $x^* = (5, 0)$, $f^* = 17$.   **3.19** No local minima; Global maximum: $x^* = (28, 18)$, $f^* = 8$.   **3.20** Global minimum: $x^* = (3, 2)$, $f^* = 1$; Global maximum: $x^* = (0, 5)$, $f^* = 25$; Local maximum: $x^* = (0, 0)$, $f^* = 20$.   **3.21** $b^* = 24.66$ cm, $d^* = 49.32$ cm, $f^* = 1216$ cm^3.   **3.22** $R_o^* = 20$ cm, $R_i^* = 19.84$ cm, $f^* = 79.1$ kg.   **3.23** $R^* = 53.6$ mm, $t^* = 5.0$ mm, $f^* = 66$ kg.   **3.24** $R_o^* = 56$ mm, $R_i^* = 51$ mm, $f^* = 66$ kg.   **3.25** $w^* = 93$ mm, $t^* = 5$ mm, $f^* = 70$ kg.   **3.26** Infinite optimum points, $f^* = 0.812$ kg.   **3.27** $A^* = 5000$, $h^* = 14$, $f^* = \$13.4$ million.   **3.28** $R^* \cong 1.0$ m, $t^* = 0.0167$ m, $f^* \cong 8070$ kg.   **3.29** $A_1^* = 6.1$ cm^2, $A_2^* = 2.0$ cm^2, $f^* = 5.39$ kg.   **3.31** $t^* = 8.45$, $f^* = 1.91 \times 10^5$.   **3.32** $R^* = 7.8$ m, $H^* = 15.6$ m, $f^* = \$1.75 \times 10^6$.   **3.33** Infinite optimum points; one point: $R^* = 0.4$ m, $t^* = 1.59 \times 10^{-3}$ m, $f^* = 15.7$ kg.   **3.34** For $l = 0.5$ m, $T_o = 10$ kN $\cdot$ m, $T_{max} = 20$ kN $\cdot$ m, $x_1^* = 103$ mm, $x_2^* = 0.955$, $f^* = 2.9$ kg.   **3.35** For $l = 0.5$, $T_o = 10$ kN $\cdot$ m, $T_{max} = 20$ kN $\cdot$ m, $d_o^* = 103$ mm, $d_i^* = 98.36$ mm, $f^* = 2.9$ kg.   **3.36** $R^* = 50.3$ mm, $t^* = 2.35$ mm, $f^* = 2.9$ kg.   **3.37** $R^* = 20$ cm, $H^* = 7.2$ cm, $f^* = -9000$ cm^3.   **3.38** $R^* = 0.5$ cm, $N^* = 2550$, $f^* = -8000$ ($l = 10$).   **3.39** $R^* = 33.7$ mm, $t^* = 5.0$ mm, $f^* = 41$ kg.   **3.40** $R^* = 21.5$ mm, $t^* = 5.0$ mm, $f^* = 26$ kg.   **3.41** $R^* = 27$, $t^* = 5$ mm, $f^* = 33$ kg.   **3.42** $R_o^* = 36$ mm, $R_i^* = 31$ mm, $f^* = 41$ kg.   **3.43** $R_o^* = 24.0$ mm, $R_i^* = 19.0$ mm, $f^* = 26$ kg.   **3.44** $R_o^* = 29.5$ mm, $R_i^* = 24.5$ mm, $f^* = 33$ kg.   **3.45** $D^* = 8.0$ cm, $H^* = 8.0$ cm, $f^* = 301.6$ cm^2.   **3.46** $A_1^* = 413.68$ mm, $A_2^* = 163.7$ mm, $f^* = 5.7$ kg.   **3.47** Infinite optimum points; one point: $R^* = 20$ mm, $t^* = 3.3$ mm, $f^* = 8.1$ kg.   **3.48** $A^* = 390$ mm^2, $h^* = 500$ mm, $f^* = 5.5$ kg.   **3.49** $A^* = 410$ mm^2, $s^* = 1500$ mm, $f^* = 8$ kg.   **3.50** $A_1^* = 300$ mm^2, $A_2^* = 50$ mm^2, $f^* = 7$ kg.   **3.51** $R^* = 130$ cm, $t^* = 2.86$ cm, $f^* = 57{,}000$ kg.   **3.52** $d_o^* = 41.56$ cm, $d_i^* = 40.19$ cm, $f^* = 680$ kg.   **3.53** $d_o^* = 1310$ mm, $t^* = 14.2$ mm, $f^* = 92{,}500$ N.   **3.54** $H^* = 50.0$ cm, $D^* = 3.42$ cm, $f^* = 6.6$ kg.

Chapter 4 Optimum Design Concepts: Optimality Conditions

4.2 $\cos x = 1.044 - 0.15175x - 0.35355x^2$ at $x = \pi/4$.   **4.3** $\cos x = 1.1327 - 0.34243x - 0.25x^2$ at $x = \pi/3$.   **4.4** $\sin x = -0.02199 + 1.12783x - 0.25x^2$ at $x = \frac{\pi}{6}$.   **4.5** $\sin x = 0.06634 + 1.2625x - 0.35355x^2$ at $x = \frac{\pi}{4}$.   **4.6** $e^x = 1 + x + 0.5x^2$ at $x = 0$.   **4.7** $e^x = 7.389 - 7.389x + 3.6945x^2$ at $x = 2$.   **4.8** $\bar{f}(x) = 41x_1^2 - 42x_1 - 40x_1x_2 + 20x^2 + 10x_2^2 + 15$; $\bar{f}(1.2, 0.8) = 7.64$, $f(1.2, 0.8) = 8.136$, Error $= f - \bar{f} = 0.496$.   **4.9** Indefinite.   **4.10** Indefinite.   **4.11** Indefinite.   **4.12** Positive definite.   **4.13** Indefinite.   **4.14** Indefinite.   **4.15** Positive definite.   **4.16** Indefinite.   **4.17** Indefinite.   **4.18** Positive definite.   **4.19** Positive definite.   **4.20** Indefinite.   **4.22** $\mathbf{x} = (0, 0)$ − local minimum, $f = 7$.   **4.23** $\mathbf{x}^* = (0, 0)$ − inflection point.   **4.24** $\mathbf{x}_1^* = (-3.332, 0.0395)$ − local maximum, $f = 18.58$; $\mathbf{x}_2^* = (-0.398, 0.5404)$ − inflection point.   **4.25** $\mathbf{x}_1^* = (4, 8)$ − inflection point; $\mathbf{x}_2^* = (-4, -8)$ − inflection point.   **4.26** $x^* = (2n + 1)\pi$, $n = 0, \pm1, \pm2, \ldots$ local minima, $f^* = -1$; $x^* = 2n\pi$, $n = 0, \pm1, \pm2, \ldots$ local maxima, $f^* = 1$.   **4.27** $\mathbf{x}^* = (0, 0)$ − local minimum, $f^* = 0$.   **4.28** $x^* = 0$ − local minimum, $f^* = 0$; $x^* = 2$ − local maximum, $f^* = 0.541$.   **4.29** $\mathbf{x}^* = (3.684, 0.7368)$ − local minimum, $f^* = 11.0521$.   **4.30** $\mathbf{x}^* = (1, 1)$ − local minimum, $f^* = 1$.   **4.31** $\mathbf{x}^* = (-2/7, -6/7)$ − local minimum, $f^* = -24/7$.   **4.32** $\mathbf{x}_1^* = (241.7643, 0.03099542)$ − local minimum, $U^* = 483{,}528.6$; $\mathbf{x}_2^* = (-241.7643, -0.03099542)$ − local maximum.   **4.43** $\mathbf{x}^* = (13/6, 11/6)$, $v^* = -1/6$, $f^* = -25/3$.   **4.44** $\mathbf{x}^* = (13/6, 11/6)$, $v^* = 1/6$, $F^* = -25/3$.   **4.45** $\mathbf{x}^* = (32/13, -4/13)$, $v^* = -6/13$, $f^* = 9/13$.   **4.46** $\mathbf{x}^* = (-0.4, 2.6/3)$, $v^* = 7.2$, $f^* = 27.2$.   **4.47** $\mathbf{x}^* = (1.717, -0.811, 1.547)$, $v_1^* = -0.943$, $v_2^* = 0.453$, $f^* = 2.132$.   **4.48** $\mathbf{x}_1^* = (1.5088, 3.272)$, $v^* = -17.1503$, $f^* = 244.528$; $\mathbf{x}_2^* = (2.5945, -2.0198)$, $v^* = -1.4390$, $f^* = 15.291$; $\mathbf{x}_3^* = (-3.630, -3.1754)$, $v^* = -23.2885$, $f^* = 453.154$; $\mathbf{x}_4^* = (-3.7322, 3.0879)$, $v^* = -2.122$, $f^* = 37.877$.   **4.49** $\mathbf{x}^* = (2, 2)$, $v^* = -2$, $f^* = 2$.   **4.50** (i) No, (ii) Solution of equalities, $\mathbf{x}^* = (3, 1)$, $f^* = 4$.   **4.51** $\mathbf{x}^* = (11/6, 13/6)$, $v^* = -23/6$, $f^* = -1/3$.   **4.52** $\mathbf{x}^* = (11/6, 13/6)$, $v^* = 23/6$, $F^* = -1/3$.   **4.54** $\mathbf{x}_1^* = (0, 0)$, $F^* = -8$; $\mathbf{x}_1^* = (11/6, 13/6)$, $F^* = -1/3$.   **4.55** $\mathbf{x}^* = (0, 0)$, $f^* = -8$.   **4.56** $\mathbf{x}_1^* = (48/23, 40/23)$, $F^* = -192/13$; $\mathbf{x}_2^* = (13/6, 11/6)$, $u^* = 1/6$, $F^* = -25/3$.   **4.57** $\mathbf{x}^* = (3, 1)$, $v^* = -2$, $u^* = 2$, $f^* = 4$.   **4.58** $\mathbf{x}^* = (3, 1)$, $v^* = -2$, $u^* = 2$, $f^* = 4$.   **4.59** $\mathbf{x}^* = (3, 1)$, $u_1^* = 2$, $u_2^* = 2$, $f^* = 4$.   **4.60** $\mathbf{x}^* = (6, 6)$, $\mathbf{u}^* = (0, 4, 0)$, $f^* = 4$.   **4.61** $\mathbf{x}_1^* = (0.816, 0.75)$, $\mathbf{u}^* = (0, 0, 0, 0)$, $f^* = 2.214$; $\mathbf{x}_2^* = (0.816, 0)$, $\mathbf{u}^* = (0, 0, 0, 3)$, $f^* = 1.0887$; $\mathbf{x}_3^* = (0, 0.75)$, $\mathbf{u}^* = (0, 0, 2, 0)$, $f^* = 1.125$; $\mathbf{x}_4^* = (1.5073, 1.2317)$, $\mathbf{u}^* = (0, 0.9632, 0, 0)$, $f^* = 0.251$; $\mathbf{x}_5^* = (1.0339, 1.655)$, $\mathbf{u}^* = (1.2067, 0, 0, 0)$, $f^* = 0.4496$; $\mathbf{x}_6^* = (0, 0)$, $\mathbf{u}^* = (0, 0, 2, 3)$, $f^* = 0$; $\mathbf{x}_7^* = (2, 0)$, $\mathbf{u}^* = (0, 2, 0, 7)$, $f^* = -4$; $\mathbf{x}_8^* = (0, 2)$, $\mathbf{u}^* = (5/3, 0, 11/3, 0)$, $f^* = -2$; $\mathbf{x}_9^* = (1.386, 1.538)$, $\mathbf{u}^* = (0.633, 0.626, 0, 0)$, $f^* = -0.007388$.   **4.62** $\mathbf{x}^* = (48/23, 40/23)$, $u^* = 0$, $f^* = -192/23$.   **4.63** $\mathbf{x}^* = (2.5, 1.5)$, $u^* = 1$, $f^* = 1.5$.   **4.64** $\mathbf{x}^* = (6.3, 1.733)$, $u^* = (0, 0.8, 0, 0)$, $f^* = -56.901$.   **4.65** $\mathbf{x}^* = (1, 1)$, $u^* = 0$, $f^* = 0$.   **4.66** $\mathbf{x}^* = (1, 1)$, $\mathbf{u}^* = (0, 0)$, $f^* = 0$.   **4.67** $\mathbf{x}^* = (2, 1)$, $\mathbf{u}^* = (0, 2)$, $f^* = 1$.   **4.68** $\mathbf{x}_1^* = (2.5945, 2.0198)$, $u_1^* = 1.439$, $f^* = 15.291$; $\mathbf{x}_2^* = (-3.63, 3.1754)$, $u_1^* = 23.2885$, $f^* = 453.154$; $\mathbf{x}_3^* = (1.5088, -3.2720)$, $u_1^* = 17.1503$, $f^* = 244.53$; $\mathbf{x}_4^* = (-3.7322, -3.0879)$, $u_1^* = 2.1222$, $f^* = 37.877$.   **4.69** $\mathbf{x}^* = (3.25, 0.75)$, $v^* = -1.25$, $u^* = 0.75$, $f^* = 5.125$.   **4.70** $\mathbf{x}_1^* = (4/\sqrt{3}, 1/3)$, $u^* = 0$, $f^* = -24.3$; $\mathbf{x}_2^* = (-4/\sqrt{3}, 1/3)$, $u^* = 0$, $f^* = 24.967$; $\mathbf{x}_3^* = (0, 3)$, $u^* = 16$, $f^* = -21$; $\mathbf{x}_4^* = (2, 1)$, $u^* = 4$, $f^* = -25$.   **4.71** $\mathbf{x}^* = (-2/7, -6/7)$, $u^* = 0$, $f^* = -24/7$.   **4.72** $x^* = 4$, $y^* = 6$, $\mathbf{u}^* = (0, 0, 0, 0)$, $f^* = 0$.   **4.74** Three local maxima: $x^* = 0$, $y^* = 0$, $\mathbf{u}^* = (0, 0, -18, -12)$, $F^* = 52$; $x^* = 6$, $y^* = 0$, $\mathbf{u}^* = (0, -4, 0, -12)$, $F^* = 40$; $x^* = 0$, $y^* = 12$, $\mathbf{u}^* = (-12, 0, -4, 0)$, $F^* = 52$; One stationary point: $x^* = 5$, $y^* = 7$, $\mathbf{u}^* = (-2, 0, 0, 0)$, $F^* = 2$.   **4.79** $D^* = 7.98$ cm, $H^* = 8$ cm, $\mathbf{u}^* = (0.5, 0, 0, 0.063, 0)$, $f^* = 300.6$ cm^2.   **4.80** $R^* = 7.871686 \times 10^{-2}$, $t^* = 1.574337 \times 10^{-3}$, $\mathbf{u}^* = (0, 3.056 \times 10^{-4}, 0.3038, 0, 0)$, $f^* = 30.56$ kg.   **4.81** $R_o^* = 7.950204 \times 10^{-2}$, $R_i^* = 7.792774 \times 10^{-2}$, $\mathbf{u}^* = (0, 3.056 \times 10^{-4}, 0.3055, 0, 0)$, $f^* = 30.56$ kg.   **4.82** $x_1^* = 60.50634$,

$x_2^* = 1.008439$, $u_1^* = 19{,}918$, $u_2^* = 23{,}186$, $u_3^* = u_4^* = 0$, $f = 23{,}186.4$. **4.83** $h^* = 14$ m, $A^* = 5000$ m^2, $u_1^* = 5.9 \times 10^{-4}$, $u_2^* = 6.8 \times 10^{-4}$, $u_3^* = u_4^* = u_5^* = 0$, $f^* = \$13.4$ million. **4.84** $A^* = 20{,}000$, $B^* = 10{,}000$, $u_1^* = 35$, $u_3^* = 27$ (or, $u_1^* = 8$, $u_2^* = 108$), $f^* = -\$1{,}240{,}000$. **4.85** $R^* = 20$ cm, $H^* = 7.161973$ cm, $u_1^* = 10$, $u_3^* = 450$, $f^* = -9000$ cm^3. **4.86** $R^* = 0.5$ cm, $N = 2546.5$, $u_1^* = 16{,}000$, $u_2^* = 4$, $f = -8000$ cm^2. **4.87** $W^* = 70.7107$ m, $D^* = 141.4214$ m, $u_3^* = 1.41421$, $u_4^* = 0$, $f^* = \$28{,}284.28$. **4.88** $A^* = 70$ kg, $B = 76$ kg, $u_1^* = 0.4$, $u_4^* = 16$, $f^* = -\$1308$. **4.89** $B^* = 0$, $M^* = 2.5$ kg, $u_1^* = 0.5$, $u_3^* = 1.5$, $f^* = \$2.5$. **4.90** $x_1^* = 316.667$, $x_2^* = 483.33$, $u_1^* = 2/3$, $u_2^* = 10/3$, $f^* = -\$1283.333$. **4.91** $r^* = 4.57078$ cm, $h^* = 9.14156$ cm, $v_1^* = -0.364365$, $u_1^* = 43.7562$, $f^* = 328.17$ cm^2. **4.92** $b^* = 10$ m, $h^* = 18$ m, $u_1^* = 0.04267$, $u_2^* = 0.00658$, $f^* = 0.545185$. **4.94** $D^* = 5.758823$ m, $H^* = 5.758823$ m, $n_1^* = -277.834$, $f^* = \$62{,}512.75$. **4.96** $P_1^* = 30.4$, $P_2^* = 29.6$, $u_1^* = 59.8$, $f^* = \$1789.68$. **4.134** (i) $\pi \le x \le 2\pi$ (ii) $\frac{\pi}{2} \le x \le 3\pi/2$. **4.135** Convex everywhere. **4.136** Not convex. **4.137** $S = \left\{ x \mid x_1 \ge -5/3, \ (x_1 + 11/12)^2 - 4x_2^2 - 9/16 \ge 0 \right\}$. **4.138** Not convex. **4.139** Convex everywhere. **4.140** Convex if $C \ge 0$. **4.141** Fails convexity check. **4.142** Fails convexity check. **4.143** Fails convexity check. **4.144** Fails convexity check. **4.145** Fails convexity check. **4.146** Fails convexity check. **4.147** Convex. **4.148** Fails convexity check. **4.149** Convex. **4.150** Convex. **4.151** $18.43° \le \theta \le 71.57°$. **4.152** $\theta \ge 71.57°$. **4.153** No solution. **4.154** $\theta \le 18.43°$.

Chapter 5 More on Optimum Design Concepts: Optimality Conditions

5.4 $x_1^* = 2.1667$, $x_2^* = 1.8333$, $v^* = -0.1667$; isolated minimum. **5.9** (1.5088, 3.2720), $v^* = -17.15$; not a minimum point; (2.5945, −2.0198), $v^* = -1.439$; isolated local minimum; (−3.6300, −3.1754), $v^* = -23.288$; not a minimum point; (−3.7322, 3.0879), $v^* = -2.122$; isolated local minimum. **5.20** (0.816, 0.75), $\mathbf{u}^* = (0, 0, 0, 0)$; not a minimum point; (0.816, 0), $\mathbf{u}^* = (0, 0, 0, 3)$; not a minimum point; (0, 0.75), $\mathbf{u}^* = (0, 0, 2, 0)$; not a minimum point; (1.5073, 1.2317), $\mathbf{u}^* = (0, 0.9632, 0, 0)$; not a minimum point; (1.0339, 1.6550), $\mathbf{u}^* = (1.2067, 0, 0, 0)$; not a minimum point; (0, 0), $\mathbf{u}^* = (0, 0, 2, 3)$; isolated local minimum; (2, 0), $\mathbf{u}^* = (2, 0, 0, 7)$; isolated local minimum; (0, 2), $\mathbf{u}^* = (1.667, 0, 3.667, 0)$; isolated local minimum; (1.386, 1.538), $\mathbf{u}^* = (0.633, 0.626, 0, 0)$; isolated local minimum. **5.21** (2.0870, 1.7391), $u^* = 0$; isolated global minimum. **5.22** $\mathbf{x}^* = (2.5, 1.5)$, $u^* = 1$, $f^* = 1.5$. **5.23** $\mathbf{x}^* = (6.3, 1.733)$, $\mathbf{u}^* = (0, 0.8, 0, 0)$, $f^* = -56.901$. **5.24** $\mathbf{x}^* = (1, 1)$, $u^* = 0$, $f^* = 0$. **5.25** $\mathbf{x}^* = (1, 1)$, $\mathbf{u}^* = (0, 0)$, $f^* = 0$. **5.26** $\mathbf{x}^* = (2, 1)$, $\mathbf{u}^* = (0, 2)$, $f^* = 1$. **5.27** (2.5945, 2.0198), $u^* = 1.4390$; isolated local minimum; (−3.6300, 3.1754), $u^* = 23.288$; not a minimum; (1.5088, −3.2720), $u^* = 17.150$; not a minimum; (−3.7322, −3.0879), $u^* = 2.122$; isolated local minimum. **5.28** (3.25, 0.75), $u^* = 0.75$, $n^* = -1.25$; isolated global minimum. **5.29** (2.3094, 0.3333), $u^* = 0$; not a minimum; (−2.3094, 0.3333), $u^* = 0$; not a minimum; (0, 3), $u^* = 16$; not a minimum; (2, 1), $u^* = 4$; isolated local minimum. **5.30** (−0.2857, −0.8571), $u^* = 0$; isolated local minimum. **5.38** $R_o^* = 20$ cm, $R_i^* = 19.84$ cm, $f^* = 79.1$ kg, $\mathbf{u}^* = (3.56 \times 10^{-3}, 0, 5.29, 0, 0, 0)$. **5.39** Multiple optima between (31.83, 1.0) and (25.23, 1.26)mm, $f^* = 45.9$ kg. **5.40** $R^* = 1.0077$ m, $t^* = 0.0168$ m, $f^* = 8182.8$ kg, $\mathbf{u}^* = (0.0417, 0.00408, 0, 0, 0)$. **5.41** $R^* = 0.0787$ m, $t^* = 0.00157$ m, $f^* = 30.56$ kg. **5.42** $R_o^* = 0.0795$ m, $R_i^* = 0.0779$ m, $f^* = 30.56$ kg. **5.43** $H^* = 8$ cm, $D^* = 7.98$ cm, $f^* = 300.6$ cm^2. **5.44** $A^* = 5000$ m^2, $h^* = 14$ m, $f^* = \$13.4$ million. **5.45** $x_1^* = 102.98$ mm, $x_2^* = 0.9546$, $f^* = 2.9$ kg, $\mathbf{u}^* = (4.568 \times 10^{-3}, 0, 3.332 \times 10^{-8}, 0, 0, 0)$. **5.46** $d_o^* = 103$ mm, $d_i^* = 98.36$ mm, $f^* = 2.9$ kg, $\mathbf{u}^* = (4.657 \times 10^{-3}, 0, 3.281 \times 10^{-8}, 0, 0, 0)$. **5.47** $R^* = 50.3$ mm, $t^* = 2.34$ mm, $f^* = 2.9$ kg, $\mathbf{u}^* = (4.643 \times 10^{-3}, 0, 3.240 \times 10^{-8}, 0, 0, 0)$. **5.48** $H^* = 50$ cm, $D^* = 3.42$ cm, $f^* = 6.6$ kg, $\mathbf{u}^* = (0, 9.68 \times 10^{-5}, 0, 4.68 \times 10^{-2}, 0, 0)$. **5.50** Not a convex

programming problem; $D^* = 10$ m, $H^* = 10$ m, $f^* = 60,000\pi$ m^3; $\Delta f = -800\pi$ m^3. **5.51** Convex; $A_1^* = 2.937 \times 10^{-4}$ m^2, $A_2^* = 6.556 \times 10^{-5}$ m^2, $f^* = 7.0$ kg. **5.52** $h^* = 14$ m, $A^* = 5000$ m^2, $u_1^* = 5.9 \times 10^{-4}$, $u_2^* = 6.8 \times 10^{-4}$, $u_3^* = u_4^* = u_5^* = 0$, $f^* = \$13.4$ million. **5.53** $R^* = 20$, $H^* = 7.16$, $u_1^* = 10$, $u_3^* = 450$, $f^* = -9000$ cm$_3$. **5.54** $R^* = 0.5$ cm, $N^* = 2546.5$, $u_1^* = 16,022$, $u_2^* = 4$, $f^* = 8000$ cm^2. **5.55** $W^* = 70.7107$, $D^* = 141.4214$, $u_3^* = 1.41421$, $u_4^* = 0$, $f^* = \$28,284.28$.
5.56 $r^* = 4.57078$ cm, $h^* = 9.14156$ cm, $v_1^* = -0.364365$, $u_1^* = 43.7562$, $f^* = 328.17$ cm^2.
5.57 $b^* = 10$ m, $h^* = 18$ m, $u_1^* = 0.04267$, $u_2^* = 0.00658$, $f^* = 0.545185$. **5.58** $D^* = 5.758823$ m, $H^* = 5.758823$ m, $v_1^* = -277.834$, $f^* = \$62,512.75$. **5.59** $P_1^* = 30.4$, $P_2^* = 29.6$, $u_1^* = 59.8$, $f^* = \$1789.68$. **5.60** $R_o^* = 20$ cm, $R_i^* = 19.84$ cm, $f^* = 79.1$ kg. **5.61** Multiple optima between $(31.83, 1.0)$ and $(25.23, 1.26)$mm, $f^* = 45.9$ kg. **5.62** $R^* = 0.0787$ m, $t^* = 0.00157$ m, $\mathbf{u}^* = (0, 3.056 \times 10^{-4}, 0.3038, 0, 0)$, $f^* = 30.56$ kg. **5.63** $R_o^* = 0.0795$ m, $R_i^* = 0.0779$ m, $\mathbf{u}^* = (0, 3.056 \times 10^{-4}, 0.3055, 0, 0)$, $f^* = 30.56$ kg. **5.64** $D^* = 7.98$ cm, $H^* = 8$ cm, $\mathbf{u}^* = (0.5, 0, 0, 0.063, 0)$, $f^* = 300.6$ cm^2. **5.65** $R^* = 1.0077$ m, $t^* = 0.0168$ m, $f^* = 8182.8$ kg, $\mathbf{u}^* = (0.0417, 0.00408, 0, 0, 0, 0)$. **5.66** $x_1^* = 102.98$ mm, $x_2^* = 0.9546$, $f^* = 2.9$ kg, $\mathbf{u}^* = (4.568 \times 10^{-3}, 0, 3.332 \times 10^{-8}, 0, 0, 0)$. **5.67** $d_o^* = 103$ mm, $d_i^* = 98.36$ mm, $f^* = 2.9$ kg, $\mathbf{u}^* = (4.657 \times 10^{-3}, 0, 3.281 \times 10^{-8}, 0, 0, 0, 0)$. **5.68** $R^* = 50.3$ mm, $t^* = 2.34$, $f^* = 2.9$ kg, $\mathbf{u}^* = (4.643 \times 10^{-3}, 0, 3.240 \times 10^{-8}, 0, 0, 0, 0)$. **5.69** $R^* = 33.7$ mm, $t^* = 5.0$ mm, $f^* = 41.6$ kg, $\mathbf{u}^* = (0, 2.779 \times 10^{-4}, 0, 0, 0, 5.54, 0)$. **5.70** $R^* = 21.3$ mm, $t^* = 5.0$ mm, $f^* = 26.0$ kg, $\mathbf{u}^* = (0, 1.739 \times 10^{-4}, 0, 0, 0, 3.491, 0)$. **5.71** $R^* = 27.0$ mm, $t^* = 5.0$ mm, $f^* = 33.0$ kg, $\mathbf{u}^* = (0, 2.165 \times 10^{-4}, 0, 0, 0, 4.439, 0)$.
5.72 $A_1^* = 413.68$ mm^2, $A_2^* = 163.7$ mm^2, $f^* = 5.7$ kg, $\mathbf{u}^* = (0, 1.624 \times 10^{-2}, 0, 6.425 \times 10^{-3}, 0)$.
5.73 Multiple solutions $R^* = 20.0$ mm, $t^* = 3.3$ mm, $f^* = 8.1$ kg, $\mathbf{u}^* = (0.0326, 0, 0, 0, 0, 0, 0)$.
5.74 $A^* = 390$ mm^2, $h^* = 500$ mm, $f^* = 5.5$ kg, $\mathbf{u}^* = (2.216 \times 10^{-2}, 0, 0, 0, 0, 0, 1.67 \times 10^{-3})$.
5.75 $A^* = 415$ mm^2, $s^* = 1480$ mm, $f^* = 8.1$ kg, $u_1^* = 0.0325$, all others are zero. **5.76** $A_1^* = 300$ mm^2, $A_2^* = 50$ mm^2, $f^* = 7.04$ kg, $\mathbf{u}^* = (0.0473, 0, 0, 0, 0, 0, 0, 0)$. **5.77** $R^* = 130$ cm, $t^* = 2.86$ cm, $f^* = 57,000$ kg, $\mathbf{u}^* = (28170, 0, 294, 0, 0, 0, 0, 0)$. **5.78** $d_o^* = 41.6$ cm, $d_i^* = 40.2$ cm, $f^* = 680$ kg, $\mathbf{u}^* = (0, 0, 35.7, 6.1, 0, ...)$. **5.79** $d_o^* = 1310$ mm, $t^* = 14.2$ mm, $f^* = 92,500$ N, $\mathbf{u}^* = (0, 508, 462, 0, ...)$.
5.80 $H^* = 50.0$ cm, $D^* = 3.42$ cm, $f^* = 6.6$ kg, $\mathbf{u}^* = (0, 9.68 \times 10^{-5}, 0, 4.68 \times 10^{-2}, 0, 0)$.

Chapter 6 Optimum Design with Excel Solver

6.1 $\mathbf{x}^* = (241.8, 0.0310)$, $f^* = 483,528.61$. **6.2** $\mathbf{x}^* = (4.15, 0.362)$, $f^* = -1616.2$. **6.3** $\mathbf{x}^* = (3.73, 0.341)$, $f^* = -1526.6$. **6.4** $\mathbf{x}^* = (1.216, 1.462)$, $f^* = 0.0752$. **6.5** $\mathbf{x}^* = (0.246, 0.0257, 0.1808, 0.205)$, $f^* = 0.01578$. **6.6** $\mathbf{x}^* = (2, 4)$, $f^* = 10$. **6.7** $\mathbf{x}^* = (3.67, 0.667)$, $f^* = 6.33$. **6.8** $\mathbf{x}^* = (0, 1.67, 2.33)$, $f^* = 4.33$. **6.9** $\mathbf{x}^* = (1.34, 0.441, 0, 3.24)$, $f^* = 9.73$. **6.10** $\mathbf{x}^* = (0.654, 0.0756, 0.315)$, $f^* = 9.73$. **6.11** $\mathbf{x}^* = (0, 25)$, $f^* = 150$. **6.12** $\mathbf{x}^* = (103.0, 98.3)$, $f^* = 2.90$. **6.13** $\mathbf{x}^* = (294, 65.8)$, $f^* = 7.04$. **6.14** $\mathbf{x}^* = (2.84, 129.0)$, $f^* = 562$. **6.15** $\mathbf{x}^* = (50, 3.42)$, $f^* = 6.61$. **6.16** $\mathbf{x}^* = (0.0705, 0.444, 10.16)$, $f^* = 0.0268$. **6.17** $\mathbf{x}^* = (0.05, 0.282)$, $f^* = 0.0155$. **6.18** $\mathbf{x}^* = (0.0601, 0.334, 8.74)$, $f^* = 0.0130$. **6.19** $\mathbf{x}^* = (2.5, 0.3, 0.045, 0.013)$, $f^* = 2.067$. **6.20** $\mathbf{x}^* = (2.11, 0.403, 0.0156, 0.0115)$, $f^* = 0.921$. **6.21** $\mathbf{x}^* = (0.4503, 0.0675)$, $f^* = 2.87$. **6.22** $\mathbf{x}^* = (13.7, 5, 0.335, 0.23)$, $f^* = 21.6$, select W14x22. **6.23** $\mathbf{x}^* = (11.9, 4.85, 0.322, 0.269)$, $f^* = 20.9$, select W12x22. **6.24** $\mathbf{x}^* = (8.083, 5.46, 0.539, 0.331)$, $f^* = 27.9$, select W8x28. **6.25** $\mathbf{x}^* = (9.73, 4.59, 0.516, 0.398)$, $f^* = 27.9$, select W10x30. **6.26** $\mathbf{x}^* = (16.4, 16, 1.663, 0.970)$, $f^* = 224$, select W14x233. **6.27** $\mathbf{x}^* = (14.7, 12.8, 1.441, 0.876)$, $f^* = 160.4$, select W12x170. **6.28** $\mathbf{x}^* = (17.7, 11.7, 1.376, 1.108)$, $f^* = 165.6$, select W18x175. **6.29** $\mathbf{x}^* = (13.7, 8.44, 0.335, 0.966)$, $f^* = 62.0$, select W14x68. **6.30** $\mathbf{x}^* = (13.2, 8.28, 0.333, 0.926)$, $f^* = 58.0$, select W14x61. **6.31** $\mathbf{x}^* = (16.4, 16, 0.944, 0.23)$, $f^* = 113.9$, select W14x145. **6.32** $\mathbf{x}^* = (19.69, 11.7, 1.254, 0.3)$, $f^* = 117.2$, select W18x143. **6.33** $\mathbf{x}^* = (21.5, 6.13, 0.335, 0.23)$, $f^* = 30.2$, select W21x44.

6.34 $x^* = (27.8, 17.44, 0.566, 0.1944)$, $f^* = 84.7$, select W18x143.   **6.35** $x^* = (16.4, 15.46, 1.023, 0.23)$, $f^* = 118.6$, select W14x145.   **6.36** $x^* = (25.5, 5, 1.56, 0.548)$, $f^* = 94.5$, select W21x111.   **6.37** $x^* = (25.5, 12.79, 0.699, 0.23)$, $f^* = 79.5$, select W21x111.   **6.38** $x^* = (25.5, 7.71, 0.360, 0.23)$, $f^* = 38.2$, select W21x57.   **6.39** $x^* = (0.3, 0.0032, 0.01874)$, $f^* = 33030.7$.   **6.40** $x^* = (0.3, 0.0032, 0.01874)$, $f^* = 33030.7$.   **6.41** $x^* = (0.3, 0.00619, 0.01173)$, $f^* = 46614$.

Chapter 7 Optimum Design with MATLAB

7.1 For $l = 0.5$ m, $T_o = 10$ kN $\cdot$ m, $T_{max} = 20$ kN $\cdot$ m, $x_1^* = 103$ mm, $x_2^* = 0.955$, $f^* = 2.9$ kg.   **7.2** For $l = 0.5$, $T_o = 10$ kN $\cdot$ m, $T_{max} = 20$ kN $\cdot$ m, $d_o^* = 103$ mm, $d_i^* = 98.36$ mm, $f^* = 2.9$ kg.   **7.3** $R^* = 50.3$ mm, $t^* = 2.35$ mm, $f^* = 2.9$ kg.   **7.4** $A_1^* = 300$ mm^2, $A_2^* = 50$ mm^2, $f^* = 7$ kg.   **7.5** $R^* = 130$ cm, $t^* = 2.86$ cm, $f^* - 57,000$ kg.   **7.6** $d_o^* = 41.56$ cm, $d_i^* = 40.19$ cm, $f^* - 680$ kg.   **7.7** $d_o^* = 1310$ mm, $t^* = 14.2$ mm, $f^* = 92,500$ N.   **7.8** $H^* = 50.0$ cm, $D^* = 3.42$ cm, $f^* = 6.6$ kg.   **7.9** $b^* = 0.5$ in, $h^* = 0.28107$ in, $f^* = 0.140536$ in^2; Active constraints: fundamental vibration frequency and lower limit on b.
7.10 $b^* = 50.4437$ cm, $h^* = 15.0$ cm, $t_1^* = 1.0$ cm, $t_2^* = 0.5218$ cm, $f^* = 16,307.2$ cm^3; Active constraints: axial stress, shear stress, upper limit on t_1 and upper limit on h.   **7.11** $A_1^* = 1.4187$ in^2, $A_2^* = 2.0458$ in^2, $A_3^* = 2.9271$ in^2, $x_1^* = -4.6716$ in, $x_2^* = 8.9181$ in, $x_3^* = 4.6716$ in, $f^* = 75.3782$ in^3; Active stress constraints: member 1—loading condition 3, member 2—loading condition 1, member 3—loading conditions 1 and 3.   **7.12** For $\phi = \sqrt{2}$: $x_1^* = 2.4138$, $x_2^* = 3.4138$, $x_3^* = 3.4141$, $f^* = 1.2877 \times 10^{-7}$; For $\phi = 2^{1/3}$: $x_1^* = 2.2606$, $x_2^* = 2.8481$, $x_3^* = 2.8472$, $f^* = 8.03 \times 10^{-7}$.

Chapter 8 Linear Programming Methods for Optimum Design

8.21 $(0, 4, -3, -5)$; $(2, 0, 3, 1)$; $(1, 2, 0, -2)$; $(5/3, 2/3, 2, 0)$.   **8.22** $(0, 0, -3, -5)$; $(0, 1, 0, -3)$; $(0, 2.5, 4.5, 0)$; $(-3, 0, 0, -11)$; $(2.5, 0, -5.5, 0)$; $(9/8, 11/8, 0, 0)$.   **8.23** Decompose x_2 into two variables; $(0, 0, 0, 12, -3)$; $(0, 0, -3, 0, 6)$; $(0, 0, -1, 8, 0)$; $(0, 3, 0, 0, 6)$; $(0, 1, 0, 8, 0)$; $(4, 0, 0, 0, 1)$; $(3, 0, 0, 3, 0)$; $(4.8, 0, 0.6, 0, 0)$; $(4.8, -0.6, 0, 0, 0)$.   **8.24** $(0, -8/3, -1/3)$; $(2, 0, 3)$; $(0.2, -2.4, 0)$.   **8.25** $(0, 0, 9, 2, 3)$; $(0, 9, 0, 20, -15)$; $(0, -1, 10, 0, 5)$; $(0, 1.5, 7.5, 5, 0)$; $(4.5, 0, 0, -2.5, 16.5)$; $(2, 0, 5, 0, 9)$; $(-1, 0, 11, 3, 0)$; $(4, 1, 0, 0, 13)$; $(15/7, 33/7, 0, 65/7, 0)$; $(-2.5, -2.25, 16.25, 0, 0)$.   **8.26** $(0, 4, -3, -7)$; $(4, 0, 1, 1)$; $(3, 1, 0, -1)$; $(3.5, 0.5, 0.5, 0)$.   **8.27** Decompose x_2 into two variables; 15 basic solutions; basic feasible solutions are $(0, 4, 0, 0, 7, 0)$; $(0, 5/3, 0, 7/3, 0, 0)$; $(2, 0, 0, 0, 1, 0)$; $(5/3, 0, 0, 2/3, 0, 0)$; $(7/3, 0, 2/3, 0, 0, 0)$.   **8.28** Ten basic solutions; basic feasible solutions are $(2.5, 0, 0, 0, 4.5)$; $(1.6, 1.8, 0, 0, 0)$.   **8.29** $(0, 0, 4, -2)$; $(0, 4, 0, 6)$; $(0, 1, 3, 0)$; $(-2, 0, 0, -4)$; $(2, 0, 8, 0)$; $(-1.2, 1.6, 0, 0)$.   **8.30** $(0, 0, 0, -2)$; $(0, 2, -2, 0)$; $(0, 0, 0, -2)$; $(2, 0, 2, 0)$; $(0, 0, 0, -2)$; $(1, 1, 0, 0)$.
8.31 $(0, 0, 10, 18)$; $(0, 5, 0, 8)$; $(0, 9, -8, 0)$; $(-10, 0, 0, 48)$; $(6, 0, 16, 0)$; $(2, 6, 0, 0)$.   **8.32** $x^* = (10/3, 2)$; $f^* = -13/3$.   **8.33** Infinite solutions between $x^* = (0, 3)$ and $x^* = (2, 0)$; $f^* = 6$.
8.34 $x^* = (2, 4)$; $z^* = 10$.   **8.35** $x^* = (6, 0)$; $z^* = 12$.   **8.36** $x^* = (3.667, 1.667)$; $z^* = 15$.   **8.37** $x^* = (0, 5)$; $f^* = -5$.   **8.38** $x^* = (2, 0)$; $f^* = -2$.   **8.39** $x^* = (2, 0)$; $z^* = 4$.   **8.40** $x^* = (2.4, 0.8)$; $z^* = 3.2$.   **8.41** $x^* = (0, 3)$; $z^* = 3$.   **8.42** $x^* = (0, 4)$; $z^* = 22/3$.   **8.43** $x^* = (0, 0)$; $z^* = 0$.   **8.44** $x^* = (0, 3)$; $z^* = 3$.
8.45 $x^* = (0, 14)$; $f^* = -56$.   **8.46** $x^* = (0, 2)$; $f^* = -2$.   **8.47** $x^* = (0, 14)$; $z^* = 42$.   **8.48** $x^* = (0, 0)$; $z^* = 0$.   **8.49** $x^* = (33, 0, 0)$; $z^* = 66$.   **8.50** $x^* = (0, 2.5)$; $z^* = 5$.   **8.51** $x^* = (2, 1)$; $f^* = -5$.   **8.52** $x^* = (2, 1)$; $z^* = 31$.   **8.53** $x^* = (7, 0, 0)$; $z^* = 70$.   **8.55** $x^* = (2, 4)$; $z^* = 10$.   **8.56** Unbounded.   **8.57** $x^* = (3.5, 0.5)$; $z^* = 5.5$.   **8.58** $x^* = (1.667, 0.667)$; $z^* = 4.333$.   **8.60** $x^* = (0, 1.667, 2.333)$; $f^* = 4.333$.
8.61 $x^* = (1.125, 1.375)$; $f^* = 36$.   **8.62** $x^* = (2, 0)$; $f^* = 40$.   **8.63** $x^* = (1.3357, 0.4406, 0, 3.2392)$; $z^* = 9.7329$.   **8.64** $x^* = (0.6541, 0.0756, 0.3151)$; $f^* = 9.7329$.   **8.65** $x^* = (0, 25)$; $z^* = 150$.
8.66 $x^* = (2/3, 5/3)$; $z^* = 16/3$.   **8.67** $x^* = (7/3, -2/3)$; $z^* = -1/3$.   **8.68** $x^* = (1, 1)$; $f^* = 5$.
8.69 $x^* = (2, 2)$; $f^* = 10$.   **8.70** $x^* = (4.8, -0.6)$; $z^* = 3.6$.   **8.71** $x^* = (0, 2)$; $f^* = 4$.   **8.72** $x^* = (0, 5)$; $z^* = 40$.

8.73 Infeasible problem. **8.74** Infinite solutions; $f^* = 0$. **8.75** $\mathbf{x}^* = (0, 5/3, 7/3)$; $f^* = -13/3$.
8.76 $\mathbf{x}^* = (0, 9)$; $f^* = 4.5$. **8.77** $A^* = 20{,}000$, $B^* = 10{,}000$, Profit = \$4,600,000 (Irregular
optimum point). **8.78** $A^* = 70$, $B^* = 76$, Profit = \$1308. **8.79** Bread = 0, Milk = 2.5 kg; Cost =
\$2.5. **8.80** Bottles of wine = 316.67, Bottles of whiskey = 483.33; Profit = \$1283.3. **8.81** Shortening
produced = 149,499.5 kg, Salad oil produced = 50,000 kg, Margarine produced = 10,000 kg; Profit =
\$19,499.2. **8.82** $A^* = 10$, $B^* = 0$, $C^* = 20$; Capacity = 477,000. **8.83** $x_1^* = 0$, $x_2^* = 0$, $x_3^* = 200$, $x_4^* = 100$;
$f^* = 786$. **8.84** $f^* = 1{,}333{,}679$ ton. **8.85** $\mathbf{x}^* = (0, 800, 0, 500, 1500, 0)$; $f^* = 7500$; $\mathbf{x}^* = (0, 0, 4500, 4000,$
$3000, 0)$; $f^* = 7500$; $\mathbf{x}^* = (0, 8, 0, 5, 15, 0)$, $f^* = 7500$. **8.86** Irregular optimum point; Lagrange
multipliers are not unique: 20, 420, 0, 0. (a) No effect (b) Cost decreases by 200,000 (Profit increases
by 200,000). **8.87** 1. No effect; 2. Out of range, re-solve the problem; $A^* = 70$, $B^* = 110$; Profit =
\$1580; 3. Profit reduces by \$4; 4. Out of range, re-solve the problem; $A^* = 41.667$, $B^* = 110$;
Profit = \$1213.33. **8.88** $y_1 = 0.25$, $y_2 = 1.25$, $y_3 = 0$, $y_4 = 0$. **8.89** Unbounded. **8.90** $y_1 = 0$, $y_2 = 2.5$,
$y_3 = -1.5$. **8.91** $y_1 = 0$, $y_2 = 5/3$, $y_3 = -7/3$. **8.92** $y_1 = 4$, $y_2 = -1$. **8.93** $y_1 = -5/3$,
$y_2 = -2/3$. **8.94** $y_1 = 2$, $y_2 = -6$. **8.95** $y_1 = 0$, $y_2 = 5$. **8.96** $y_1 = 0.654$, $y_2 = -0.076$, $y_3 = 0.315$.
8.97 $y_1 = -1.336$, $y_2 = -0.441$, $y_3 = 0$, $y_4 = -3.239$. **8.98** $y_1 = 0$, $y_2 = 0$, $y_3 = 0$, $y_4 = 6$.
8.99 $y_1 = -1.556$, $y_2 = 0.556$. **8.100** $y_1 = 0$, $y_2 = 5/3$, $y_3 = -7/3$. **8.101** $y_1 = -0.5$, $y_2 = -2.5$.
8.102 $y_1 = -1/3$, $y_2 = 0$, $y_3 = 5/3$. **8.103** $y_1 = 0.2$, $y_2 = 0.4$. **8.104** $y_1 = 0$, $y_2 = 0$, $y_3 = 0$,
$y_4 = -2/3$. **8.105** $y_1 = 2$, $y_2 = 0$. **8.106** Infeasible problem. **8.107** $y_1 = 3$, $y_2 = 0$. **8.110** For
$b_1 = 10$: $-8 \le \Delta_1 \le 8$; for $b_2 = 6$: $-2.667 \le \Delta_2 \le 8$; for $b_3 = 2$: $-4 \le \Delta_3 \le \infty$; for
$b_4 = 6$: $-\infty \le \Delta_4 \le 8$. **8.111** Unbounded problem. **8.112** For $b_1 = 5$: $-0.5 \le \Delta_1 \le \infty$; for
$b_2 = 4$: $-1 \le \Delta_2 \le 0.333$; for $b_3 = 3$: $-1 \le \Delta_3 \le 1$. **8.113** For $b_1 = 5$: $-2 \le \Delta_1 \le \infty$; for
$b_2 = 4$: $-2 \le \Delta_2 \le 2$; for $b_3 = 1$: $-2 \le \Delta_3 \le 1$. **8.114** For $b_1 = -5$: $-\infty \le \Delta_1 \le 4$; for $b_2 = -2$:
$-8 \le \Delta_2 \le 4.5$. **8.115** $b_1 = 1$: $-5 \le \Delta_1 \le 7$; for $b_2 = 4$: $-4 \le \Delta_2 \le \infty$. **8.116** For $b_1 = -3$:
$-4.5 \le \Delta_1 \le 5.5$; for $b_2 = 5$: $-3 \le \Delta_2 \le \infty$. **8.117** For $b_1 = 3$: $-\infty \le \Delta_1 \le 3$; for
$b_2 = -8$: $-\infty \le \Delta_2 \le 4$. **8.118** For $b_1 = 8$: $-8 \le \Delta_1 \le \infty$; for $b_2 = 3$: $-14.307 \le \Delta_2 \le 4.032$;
for $b_3 = 15$: $-20.16 \le \Delta_3 \le 101.867$. **8.119** For $b_1 = 2$: $-3.9178 \le \Delta_1 \le 1.1533$; for $b_2 = 5$:
$-0.692 \le \Delta_2 \le 39.579$; for $b_3 = -4.5$: $-\infty \le \Delta_3 \le 7.542$; for $b_4 = 1.5$: $-2.0367 \le \Delta_4 \le 0.334$. **8.120**
For $b_1 = 90$: $-15 \le \Delta_1 \le \infty$; for $b_2 = 80$: $-30 \le \Delta_2 \le \infty$; for $b_3 = 15$: $-\infty \le \Delta_3 \le 10$; for $b_4 = 25$:
$-10 \le \Delta_4 \le 5$. **8.121** For $b_1 = 3$: $-1.2 \le \Delta_1 \le 15$; for $b_2 = 18$: $-15 \le \Delta_2 \le 12$. **8.122** For $b_1 = 5$:
$-4 \le \Delta_1 \le \infty$; for $b_2 = 4$: $-7 \le \Delta_2 \le 2$; for $b_3 = 3$: $-1 \le \Delta_3 \le \infty$. **8.123** For $b_1 = 0$: $-2 \le \Delta_1 \le 2$; for
$b_2 = 2$: $-2 \le \Delta_2 \le \infty$. **8.124** For $b_1 = 0$: $-6 \le \Delta_1 \le 3$; for $b_2 = 2$: $-\infty \le \Delta_2 \le 2$; for $b_3 = 6$:
$-3 \le \Delta_3 \le \infty$. **8.125** For $b_1 = 12$: $-3 \le \Delta_1 \le \infty$; for $b_2 = 3$: $-\infty \le \Delta_2 \le 1$. **8.126** For $b_1 = 10$:
$-8 \le \Delta_1 \le 8$; for $b_2 = 6$: $-2.667 \le \Delta_2 \le 8$; for $b_3 = 2$: $-4 \le \Delta_3 \le \infty$; for $b_4 = 6$: $-\infty \le \Delta_4 \le 8$.
8.127 For $b_1 = 20$: $-12 \le \Delta_1 \le \infty$; for $b_2 = 6$: $-\infty \le \Delta_2 \le 9$. **8.128** Infeasible problem. **8.129** For
$b_1 = 0$: $-2 \le \Delta_1 \le 2$; for $b_2 = 2$: $-2 \le \Delta_2 \le \infty$. **8.132** For $c_1 = -1$: $-1 \le \Delta c_1 \le 1.667$; for
$c_2 = -2$: $-\infty \le \Delta c_2 \le 1$. **8.133** Unbounded problem. **8.134** For $c_1 = 1$: $-\infty \le \Delta c_1 \le 3$; for $c_2 = 4$:
$-3 \le \Delta c_2 \le \infty$. **8.135** For $c_1 = 1$: $-\infty \le \Delta c_1 \le 7$; for $c_2 = 4$: $-3.5 \le \Delta c_2 \le \infty$. **8.136** For $c_1 = 9$:
$-5 \le \Delta c_1 \le \infty$; for $c_2 = 2$: $-9.286 \le \Delta c_2 \le 2.5$; for $c_3 = 3$: $-13 \le \Delta c_3 \le \infty$. **8.137** For $c_1 = 5$:
$-2 \le \Delta c_1 \le \infty$; for $c_2 = 4$: $-2 \le \Delta c_2 \le 2$; for $c_3 = -1$: $0 \le \Delta c_3 \le 2$; for $c_4 = 1$: $0 \le \Delta c_4 \le \infty$.
8.138 For $c_1 = -10$: $-8 \le \Delta c_1 \le 16$; for $c_2 = -18$: $-\infty \le \Delta c_2 \le 8$. **8.139** For $c_1 = 20$: $-12 \le \Delta c_1 \le \infty$;
for $c_2 = -6$: $-9 \le \Delta c_2 \le \infty$. **8.140** For $c_1 = 2$: $-3.918 \le \Delta c_1 \le 1.153$; for $c_2 = 5$: $-0.692 \le \Delta c_2$
≤ 39.579; for $c_3 = -4.5$: $-\infty \le \Delta c_3 \le 7.542$; for $c_4 = 1.5$: $-3.573 \le \Delta c_4 \le 0.334$. **8.141** $c_1 = 8$:
$-8 \le \Delta c_1 \le \infty$; for $c_2 = -3$: $-4.032 \le \Delta c_2 \le 14.307$; for $c_3 = 15$: $0 \le \Delta c_3 \le 101.8667$;

for $c_4 = -15$: $0 \le \Delta c_4 \le \infty$. **8.142** For $c_1 = 10$: $-\infty \le \Delta c_1 \le 20$; for $c_2 = 6$: $-4 \le c_2 \le \infty$. **8.143** For $c_1 = -2$: $-\infty \le \Delta c_1 \le 2.8$; for $c_2 = 4$: $-5 \le \Delta c_2 \le \infty$. **8.144** For $c_1 = 1$: $-\infty \le \Delta c_1 \le 7$; for $c_2 = 4$: $-\infty \le \Delta c_2 \le 0$; for $c_3 = -4$: $-\infty \le \Delta c_3 \le 0$. **8.145** For $c_1 = 3$: $-1 \le \Delta c_1 \le \infty$; for $c_2 = 2$: $-5 \le \Delta c_2 \le 1$. **8.146** For $c_1 = 3$: $-5 \le \Delta c_1 \le 1$; for $c_2 = 2$: $-0.5 \le \Delta c_2 \le \infty$. **8.147** For $c_1 = 1$: $-0.3333 \le \Delta c_1 \le 0.5$; for $c_2 = 2$: $-\infty \le \Delta c_2 \le 0$; for $c_3 = -2$: $-1 \le \Delta c_3 \le 0$. **8.148** For $c_1 = 1$: $-1.667 \le \Delta c_1 \le 1$; for $c_2 = 2$: $-1 \le \Delta c_2 \le \infty$. **8.149** For $c_1 = 3$: $-\infty \le \Delta c_1 \le 3$; for $c_2 = 8$: $-4 \le \Delta c_2 \le 0$; for $c_3 = -8$: $-\infty \le \Delta c_3 \le 0$. **8.150** Infeasible problem. **8.151** For $c_1 = 3$: $0 \le \Delta c_1 \le \infty$; for $c_2 = -3$: $0 \le \Delta c_2 \le 6$. **8.154** $20{,}000 \le b_1 \le 30{,}000$; $5000 \le b_2 \le 10{,}000$; $20{,}000 \le b_3 \le \infty$; $10{,}000 \le b_4 \le \infty$. For $c_1 = -180$: $-20 \le \Delta c_1 \le 105$; for $c_2 = -100$: $-140 \le \Delta c_2 \le 10$. **8.155** For $c_1 = -10$: $-\infty \le \Delta c_1 \le 0.4$; for $c_2 = -8$: $-0.3333 \le \Delta c_2 \le 8$. **8.156** 1. $\Delta f = 0.5$; 2. $\Delta f = 0.5$ (Bread = 0, Milk = 3, $f^* = 3$); 3. $\Delta f = 0$. **8.157** 1. $\Delta f = 33.33$ (Wine bottles = 250, Whiskey bottles = 500, Profit = 1250); 2. $\Delta f = 63.33$. 3. $\Delta f = 83.33$ (Wine bottles = 400, Whiskey bottles = 400, Profit = 1200). **8.158** 1. Re-solve; 2. $\Delta f = 0$; 3. No change. **8.159** 1. Cost function increases by \$52.40; 2. No change; 3. Cost function increases by \$11.25, $x_1^* = 0$, $x_2^* = 30$, $x_3^* = 200$, $x_4^* = 70$. **8.160** 1. $\Delta f = 0$; 2. No change; 3. $\Delta f = 1800$ ($A^* = 6$, $B^* = 0$, $C^* = 22$, $f^* = -475{,}200$). **8.161** 1. $\Delta f = 0$; 2. $\Delta f = 2{,}485.65$; 3. $\Delta f = 0$; 4. $\Delta f = 14{,}033.59$; 5. $\Delta f = -162{,}232.3$. **8.162** 1. $\Delta f = 0$; 2. $\Delta f = 400$; 3. $\Delta f = -375$. **8.163** 1. $x_1^* = 0$, $x_2^* = 3$, $f^* = -12$; 2. $y_1 = 4/5$, $y_2 = 0$; 3. $-15 \le \Delta_1 \le 3$, $-6 \le \Delta_2 \le \infty$; 4. $f^* = -14.4$, $b_1 = 18$.

Chapter 9 More on Linear Programming Methods for Optimum Design

9.1 $y_1^* = 1/4$, $y_2^* = 5/4$, $y_3^* = 0$, $y_4^* = 0$, $f_d^* = 10$. **9.2** Dual problem is infeasible. **9.3** $y_1^* = 0$, $y_2^* = 2.5$, $y_3^* = 1.5$, $f_d^* = 5.5$. **9.4** $y_1^* = 0$, $y_2^* = 1.6667$, $y_3^* = 2.3333$, $f_d^* = 4.3333$. **9.5** $y_1^* = 1.4$, $y_2^* = 0.2$, $f_d^* = -6.6$. **9.6** $y_1^* = 1.6667$, $y_2^* = 0.6667$, $f_d^* = -4.3333$. **9.7** $y_1^* = 2$, $y_2^* = 6$, $f_d^* = -36$. **9.8** $y_1^* = 0$, $y_2^* = 5$, $f_d^* = -40$. **9.9** $y_1^* = 0.65411$, $y_2^* = 0.075612$, $y_3^* = 0.315122$, $f_d^* = 9.732867$. **9.10** $y_1^* = 1.33566$, $y_2^* = 0.44056$, $y_3^* = 0$, $y_4^* = 3.2392$, $f_d^* = -9.732867$. **9.11** $y_1^* = 0$, $y_2^* = 0$, $y_3^* = 0$, $y_4^* = 6$, $f_d^* = 150$. **9.12** $y_1^* = 14/9$, $y_2^* = 5/9$, $f_d^* = 16/3$. **9.13** $y_1^* = 0$, $y_2^* = 5/3$, $y_3^* = 7/3$, $f_d^* = -1/3$. **9.14** $y_1^* = 0.5$, $y_2^* = 2.5$, $f_d^* = -5$. **9.15** $y_1^* = 1/3$, $y_2^* = 0$, $y_3^* = 5/3$, $f_d^* = 10$. **9.16** $y_1^* = 0.2$, $y_2^* = 0.4$, $f_d^* = 3.6$. **9.17** $y_1^* = 0$, $y_2^* = 0$, $y_3^* = 0$, $y_4^* = 2/3$, $f_d^* = -4$. **9.18** $y_1^* = 2$, $y_2^* = 0$, $f_d^* = 40$. **9.19** Unbounded dual problem. **9.20** $y_1^* = 0$, $y_2^* = 3$, $f_d^* = 0$. **9.21** $y_1^* = 5/3$, $y_2^* = 2/3$, $f_d^* = -13/3$. **9.22** $y_1^* = 0$, $y_2^* = 2.5$, $f_d^* = 45$.

Chapter 10 Numerical Methods for Unconstrained Optimum Design

10.2 Yes. **10.3** No. **10.4** Yes. **10.5** No. **10.6** No. **10.7** No. **10.8** No. **10.9** Yes. **10.10** No. **10.11** No. **10.12** No. **10.13** No. **10.14** No. **10.16** $\alpha^* = 1.42850$, $f^* = 7.71429$. **10.17** $\alpha^* = 1.42758$, $f^* = 7.71429$. **10.18** $\alpha^* = 1.38629$, $f^* = 0.454823$. **10.19** d is descent direction; slope $= -4048$; $\alpha^* = 0.15872$. **10.20** $\alpha^* = 0$. **10.21** $f(\alpha) = 4.1\alpha^2 - 5\alpha - 6.5$. **10.22** $f(\alpha) = 52\alpha^2 - 52\alpha + 13$. **10.23** $f(\alpha) = 6.88747 \times 10^9 \alpha^4 - 3.6111744 \times 10^8 \alpha^3 + 5.809444 \times 10^6 \alpha^2 - 27844\alpha + 41$. **10.24** $f(\alpha) = 8\alpha^2 - 8\alpha + 2$. **10.25** $f(\alpha) = 18.5\alpha^2 - 85\alpha - 13.5$. **10.26** $f(\alpha) = 288\alpha^2 - 96\alpha + 8$. **10.27** $f(\alpha) = 24\alpha^2 - 24\alpha + 6$. **10.28** $f(\alpha) = 137\alpha^2 - 110\alpha + 25$. **10.29** $f(\alpha) = 8\alpha^2 - 8\alpha$. **10.30** $f(\alpha) = 16\alpha^2 - 16\alpha + 4$. **10.31** $\alpha^* = 0.61$. **10.32** $\alpha^* = 0.5$. **10.33** $\alpha^* = 3.35\text{E-}03$. **10.34** $\alpha^* = 0.5$. **10.35** $\alpha^* = 2.2973$. **10.36** $\alpha^* = 0.16665$. **10.37** $\alpha^* = 0.5$. **10.38** $\alpha^* = 0.40145$. **10.39** $\alpha^* = 0.5$. **10.40** $\alpha^* = 0.5$. **10.41** $\alpha^* = 0.6097$. **10.42** $\alpha^* = 0.5$. **10.43** $\alpha^* = 3.45492\text{E-}03$. **10.44** $\alpha^* = 0.5$. **10.45** $\alpha^* = 2.2974$. **10.46** $\alpha^* = 0.1667$. **10.47** $\alpha^* = 0.5$. **10.48** $\alpha^* = 0.4016$. **10.49** $\alpha^* = 0.5$. **10.50** $\alpha^* = 0.5$. **10.52** $x^{(2)} = (5/2, 3/2)$. **10.53** $x^{(2)} = (0.1231, 0.0775)$. **10.54** $x^{(2)} = (0.222, 0.0778)$. **10.55** $x^{(2)} = (0.0230, 0.0688)$.

10.56 $\mathbf{x}^{(2)} = (0.0490, 0.0280)$. **10.57** $\mathbf{x}^{(2)} = (0.259, -0.225, 0.145)$. **10.58** $\mathbf{x}^{(2)} = (4.2680, 0.2244)$.
10.59 $\mathbf{x}^{(2)} = (3.8415, 0.48087)$. **10.60** $\mathbf{x}^{(2)} = (-1.590, 2.592)$. **10.61** $\mathbf{x}^{(2)} = (2.93529, 0.33976, 1.42879,$
$2.29679)$. **10.62** (10.52) $\mathbf{x}^* = (3.996096, 1.997073)$, $f^* = -7.99999$; (10.53) $\mathbf{x}^* = (0.071659, 0.023233)$,
$f^* = -0.073633$; (10.54) $\mathbf{x}^* = (0.071844, -0.000147)$, $f^* = -0.035801$; (10.55) $\mathbf{x}^* = (0.000011, 0.023273)$,
$f^* = -0.011626$; (10.56) $\mathbf{x}^* = (0.040028, 0.02501)$, $f^* = -0.0525$; (10.57) $\mathbf{x}^* = (0.006044, -0.005348, 0.002467)$,
$f^* = 0.000015$; (10.58) $\mathbf{x}^* = (4.1453, 0.361605)$, $f^* = -1616.183529$; (10.59) $\mathbf{x}^* = (3.733563, 0.341142)$,
$f^* = -1526.556493$; (10.60) $\mathbf{x}^* = (0.9087422, 0.8256927)$, $f^* = 0.008348$, 1000 iterations; (10.61) $\mathbf{x}^* = (0.13189,$
$0.013188, 0.070738, 0.072022)$, $f^* = 0.000409$, 1000 iterations. **10.63** $\mathbf{x}^* = (0.000023, 0.000023, 0.000045)$,
$f_1^* = 0$, 1 iteration; $\mathbf{x}^* = (0.002353, 0.0, 0.000007)$, $f_2^* = 0.000006$, 99 iterations; $\mathbf{x}^* = (0.000003, 0.0, 0.023598)$,
$f_3^* = 0.000056$, 135 iterations. **10.64** Exact gradients are: 1. $\nabla f = (119.2, 258.0)$, 2. $\nabla f = (-202, 100)$,
3. $\nabla f = (6, 16, 16)$. **10.65** $\mathbf{u} = \mathbf{c}/2v$, $v =$ Lagrange multiplier for the equality constraints.
10.67 $\mathbf{x}^{(2)} = (4.2)$. **10.68** $\mathbf{x}^{(2)} = (0.07175, 0.02318)$. **10.69** $\mathbf{x}^{(2)} = (0.072, 0.0)$. **10.70** $\mathbf{x}^{(2)} = (0.0,$
$0.0233)$. **10.71** $\mathbf{x}^{(2)} = (0.040, 0.025)$. **10.72** $\mathbf{x}^{(2)} = (0.257, -0.229, 0.143)$. **10.73** $\mathbf{x}^{(2)} = (4.3682,$
$0.1742)$. **10.74** $\mathbf{x}^{(2)} = (3.7365, 0.2865)$. **10.75** $\mathbf{x}^{(2)} = (-1.592, 2.592)$. **10.76** $\mathbf{x}^{(2)} = (3.1134, 0.32224, 1.34991,$
$2.12286)$.

Chapter 11 More on Numerical Methods for Unconstrained Optimum Design
11.1 $\alpha^* = 1.42857$, $f^* = 7.71429$. **11.2** $a^* = 10/7$, $f^* = 7.71429$, one iteration. **11.4** 1. $a^* = 13/4$
2. $\alpha = 1.81386$ or 4.68614. **11.10** $\mathbf{x}^{(1)} = (4, 2)$. **11.11** $\mathbf{x}^{(1)} = (0.071598, 0.023251)$. **11.12**
$\mathbf{x}^{(1)} = (0.071604, 0.0)$. **11.13** $\mathbf{x}^{(1)} = (0.0, 0.0232515)$. **11.14** $\mathbf{x}^{(1)} = (0.04, 0.025)$. **11.15** $\mathbf{x}^{(1)} = (0, 0, 0)$.
11.16 $\mathbf{x}^{(1)} = (-2.7068, 0.88168)$. **11.17** $\mathbf{x}^{(1)} = (3.771567, 0.335589)$. **11.18** $\mathbf{x}^{(1)} = (4.99913,$
$24.99085)$. **11.19** $\mathbf{x}^{(1)} = (-1.26859, -0.75973, 0.73141, 0.39833)$. **11.22** $\mathbf{x}^{(2)} = (4, 2)$. **11.23** $\mathbf{x}^{(2)} =$
$(0.0716, 0.02325)$. **11.24** $\mathbf{x}^{(2)} = (0.0716, 0.0)$. **11.25** $\mathbf{x}^{(2)} = (0.0, 0.02325)$. **11.26** $\mathbf{x}^{(2)} = (0.04,$
$0.025)$. **11.27** DFP: $\mathbf{x}^{(2)} = (0.2571, -0.2286, 0.1428)$; BFGS: $\mathbf{x}^{(2)} = (0.2571, -0.2286, 0.1429)$.
11.28 DFP: $\mathbf{x}^{(2)} = (4.37045, 0.173575)$; BFGS: $\mathbf{x}^{(2)} = (4.37046, 0.173574)$. **11.29** $\mathbf{x}^{(2)} = (3.73707,$
$0.28550)$. **11.30** $\mathbf{x}^{(2)} = (-1.9103, -1.9078)$. **11.31** DFP: $\mathbf{x}^{(2)} = (3.11339, 0.32226, 1.34991, 2.12286)$;
BFGS: $\mathbf{x}^{(2)} = (3.11339, 0.32224, 1.34991, 2.12286)$. **11.44** $x_1 = 3.7754$ mm, $x_2 = 2.2835$ mm.
11.45 $x_1 = 2.2213$ mm, $x_2 = 1.8978$ mm. **11.46** $x^* = 0.619084$. **11.47** $x^* = 9.424753$.
11.48 $x^* = 1.570807$. **11.49** $x^* = 1.496045$. **11.50** $\mathbf{x}^* = (3.667328, 0.739571)$. **11.51** $\mathbf{x}^* = (4.000142,$
$7.999771)$.

Chapter 12 Numerical Methods for Constrained Optimum Design
12.29 $\mathbf{x}^* = (5/2, 5/2)$, $u^* = 1$, $f^* = 0.5$. **12.30** $\mathbf{x}^* = (1, 1)$, $u^* = 0$, $f^* = 0$. **12.31** $\mathbf{x}^* =$
$(4/5, 3/5)$, $u^* = 2/5$, $f^* = 1/5$. **12.32** $\mathbf{x}^* = (2, 1)$, $u^* = 0$, $f^* = -3$. **12.33** $\mathbf{x}^* = (1, 2)$, $u^* = 0$,
$f^* = -1$. **12.34** $\mathbf{x}^* = (13/6, 11/6)$, $v^* = -1/6$, $f^* = -25/3$. **12.35** $\mathbf{x}^* = (3, 1)$, $v_1^* = -2$,
$v_2^* = -2$, $f^* = 2$. **12.36** $\mathbf{x}^* = (48/23, 40/23)$, $u^* = 0$, $f^* = -192/23$. **12.37** $\mathbf{x}^* = (5/2, 3/2)$,
$u^* = 1$, $f^* = -9/2$. **12.38** $\mathbf{x}^* = (63/10, 26/15)$, $u_1^* = 0$, $u_2^* = 4/5$, $f^* = -3547/50$.
12.39 $\mathbf{x}^* = (2, 1)$, $u_1^* = 0$, $u_2^* = 2$, $f^* = -1$. **12.40** $\mathbf{x}^* = (0.241507, 0.184076, 0.574317)$; $\mathbf{u}^*$
$= (0, 0, 0, 0)$, $v_1^* = -0.7599$, $f^* = 0.3799$.

Chapter 14 Design Optimization Applications with Implicit Functions
14.1 For $l = 500$ mm, $d_o^* = 102.985$ mm, $d_o^*/d_i^* = 0.954614$, $f^* = 2.900453$ kg; Active constraints:
shear stress and critical torque. **14.2** For $l = 500$ mm, $d_o^* = 102.974$ mm, $d_i^* = 98.2999$ mm,
$f^* = 2.90017$ kg; Active constraints: shear stress and critical torque. **14.3** For $l = 500$ mm,
$R^* = 50.3202$ mm, $t^* = 2.33723$ mm, $f^* = 2.90044$ kg; Active constraints: shear stress and critical

torque. **14.5** $R^* = 129.184$ cm, $t^* = 2.83921$ cm, $f^* = 56,380.61$ kg; Active constraints: combined stress and diameter/thickness ratio. **14.6** $d_o^* = 41.5442$ cm, $d_i^* = 40.1821$ cm, $f^* = 681.957$ kg; Active constraints: deflection and diameter/thickness ratio. **14.7** $d_o^* = 1308.36$ mm, $t^* = 14.2213$ mm, $f^* = 92,510.7$ N; Active constraints: diameter/thickness ratio and deflection. **14.8** $H^* = 50$ cm, $D^* = 3.4228$ cm, $f^* = 6.603738$ kg; Active constraints: buckling load and minimum height. **14.9** $b^* = 0.5$ in, $h^* = 0.28107$ in, $f^* = 0.140536$ in^2; Active constraints: fundamental vibration frequency and lower limit on b. **14.10** $b^* = 50.4437$ cm, $h^* = 15.0$ cm, $t_1^* = 1.0$ cm, $t_2^* = 0.5218$ cm, $f^* = 16,307.2$ cm^3; Active constraints: axial stress, shear stress, upper limit on t_1 and upper limit on h. **14.11** $A_1^* = 1.4187$ in^2, $A_2^* = 2.0458$ in^2, $A_3^* = 2.9271$ in^2, $x_1^* = -4.6716$ in, $x_2^* = 8.9181$ in, $x_3^* = 4.6716$ in, $f^* = 75.3782$ in^3; Active stress constraints: member 1—loading condition 3, member 2—loading condition 1, member 3—loading conditions 1 and 3. **14.12** For $\phi = \sqrt{2}$: $x_1^* = 2.4138$, $x_2^* = 3.4138$, $x_3^* = 3.4141$, $f^* = 1.2877 \times 10^{-7}$; For $\phi = 2^{1/3}$: $x_1^* = 2.2606$, $x_2^* = 2.8481$, $x_3^* = 2.8472$, $f^* = 8.03 \times 10^{-7}$. **14.13** d_o^* at base $= 48.6727$ cm, d_o^* at top $= 16.7117$ cm, $t^* = 0.797914$ cm, $f^* = 623.611$ kg. **14.14** d_o^* at base $= 1419$ mm, d_o^* at top $= 956.5$ mm, $t^* = 15.42$ mm, $f^* = 90,894$ kg. **14.15** Outer dimension at base $= 42.6407$ cm, outer dimension at top $= 14.6403$ cm, $t^* = 0.699028$ cm, $f^* = 609.396$ kg. **14.16** Outer dimension at base $= 1243.2$ mm, outer dimension at top $= 837.97$ mm, $t^* = 13.513$ mm, $f^* = 88,822.2$ kg. **14.17** $u_a = 25$: $f_1 = 1.07301E{-}06$, $f_2 = 1.83359E{-}02$, $f_3 = 24.9977$; $u_a = 35$: $f_1 = 6.88503E{-}07$, $f_2 = 1.55413E{-}02$, $f_3 = 37.8253$. **14.18** $u_a = 25$: $f_1 = 2.31697E{-}06$, $f_2 = 2.74712E{-}02$, $f_3 = 7.54602$; $u_a = 35$: $f_1 = 2.31097E{-}06$, $f_2 = 2.72567E{-}02$, $f_3 = 7.48359$. **14.19** $u_a = 25$: $f_1 = 1.11707E{-}06$, $f_2 = 1.52134E{-}02$, $f_3 = 19.815$, $f_4 = 3.3052E{-}02$; $u_a = 3.5$: $f_1 = 6.90972E{-}07$, $f_2 = 1.36872E{-}0_2$, $f_3 = 31.479$, $f_4 = 2.3974E{-}02$. **14.20** $f_1 = 1.12618E{-}06$, $f_2 = 1.798E{-}02$, $f_3 = 33.5871$, $f_4 = 0.10$. **14.21** $f_1 = 2.34615E{-}06$, $f_2 = 2.60131E{-}02$, $f_3 = 10.6663$, $f_4 = 0.10$. **14.22** $f_1 = 1.15097E{-}06$, $f_2 = 1.56229E{-}02$, $f_3 = 28.7509$, $f_4 = 3.2547E{-}02$. **14.23** $f_1 = 8.53536E{-}07$, $f_2 = 1.68835E{-}02$, $f_3 = 31.7081$, $f_4 = 0.10$. **14.24** $f_1 = 2.32229E{-}06$, $f_2 = 2.73706E{-}02$, $f_3 = 7.48085$, $f_4 = 0.10$. **14.25** $f_1 = 8.65157E{-}07$, $f_2 = 1.4556E{-}02$, $f_3 = 25.9761$, $f_4 = 2.9336E{-}02$. **14.26** $f_1 = 8.27815E{-}07$, $f_2 = 1.65336E{-}02$, $f_3 = 28.2732$, $f_4 = 0.10$. **14.27** $f_1 = 2.313E{-}06$, $f_2 = 2.723E{-}02$, $f_3 = 6.86705$, $f_4 = 0.10$. **14.28** $f_1 = 8.39032E{-}07$, $f_2 = 1.43298E{-}2$, $f_3 = 25.5695$, $f_4 = 2.9073E{-}02$. **14.29** $k^* = 2084.08$, $c^* = 300$ (upper limit), $f^* = 1.64153$.

Chapter 18 Global Optimization Concepts and Methods for Optimum Design

18.1 Six local minima, two global minima: $(0.0898, -0.7126)$, $(-0.0898, 0.7126)$, $f_G^* = -1.0316258$. **18.2** 10^n local minima; global minimum: $x_i^* = 1$, $f_G^* = 0$. **18.3** Many local minima; global minimum: $\mathbf{x}^* = (0.195, -0.179, 0.130, 0.130)$, $f_G^* = 3.13019 \times 10^{-4}$. **18.4** Many local minima; two global minima: $\mathbf{x}^* = (0.05, 0.85, 0.65, 0.45, 0.25, 0.05)$, $\mathbf{x}^* = (0.55, 0.35, 0.15, 0.95, 0.75, 0.55)$, $f_G^* = -1$. **18.5** Local minima: $\mathbf{x}^* = (0, 0)$, $f^* = 0$; $\mathbf{x}^* = (0, 2)$, $f^* = -2$; $\mathbf{x}^* = (1.38, 1.54)$, $f^* = -0.04$; Global minimum: $\mathbf{x}^* = (2, 0)$, $f^* = -4$.

Appendix B Vector and Matrix Algebra

A.1 $|\mathbf{A}| = 1$. **A.2** $|\mathbf{A}| = 14$. **A.3** $|\mathbf{A}| = -20$. **A.4** $\lambda_1 = (5 - \sqrt{5})/2$, $\lambda_2 = 2$, $\lambda_3 = (5 - \sqrt{5})/2$. **A.5** $\lambda_1 = (2 - \sqrt{2})$, $\lambda_2 = 2$, $\lambda_3 = (2 + \sqrt{2})$. **A.6** $r = 4$. **A.7** $r = 4$. **A.8** $r = 4$. **A.9** $x_1 = 1$, $x_2 = 1$, $x_3 = 1$. **A.10** $x_1 = 1$, $x_2 = 1$, $x_3 = 1$. **A.11** $x_1 = 1$, $x_2 = 2$, $x_3 = 3$. **A.12** $x_1 = 1$, $x_2 = 1$, $x_3 = 1$, $x_4 = 1$. **A.13** $x_1 = 1$, $x_2 = 2$, $x_3 = 3$. **A.14** $x_1 = 2$, $x_2 = 1$, $x_3 = 1$. **A.15** $x_1 = 1$, $x_2 = 2$, $x_3 = 3$. **A.16** $x_1 = 1$, $x_2 = 1$, $x_3 = 1$. **A.17** $x_1 = 2$, $x_2 = 1$, $x_3 = 1$, $x_4 = -2$. **A.18** $x_1 = 6$, $x_2 = -15$, $x_3 = -1$, $x_4 = 9$. **A.19** $x_1 = (3 - 7x_3 - 2x_4)/4$, $x_2 = (-1 + x_3 - 2x_4)/4$. **A.20** $x_1 = (4 - x_3)$, $x_2 = 2$, $x_4 = 4$.

A.21 $x_1 = (-3 - 4x_4)$, $x_2 = (-2 - 3x_4)$, $x_3 = x_4$.   **A.22** $x_1 = -x_4$, $x_2 = (2 + x_4)$, $x_3 = (-2 - x_4)$.
A.27 $x_1 = 4 + (2x_2 - 8x_3 - 5x_4 + 2x_5)/3$; $x_6 = (9 - x_2 - 2x_3 - x_4 - 3x_5)/2$; $x_7 = 14 -$
$(13x_2 + 14x_3 + 8x_4 + 4x_5)/3$; $x_8 = (9 + x_2 - 16x_3 - 7x_4 + 19x_5)/6$.   **A.28** Linearly dependent.
A.29 Linearly independent.   **A.30** $\lambda_1 = (3 - 2\sqrt{2})$, $\lambda_2 = (3 + 2\sqrt{2})$.   **A.31** $\lambda_1 = (3 - \sqrt{5})$,
$\lambda_2 = (3 + \sqrt{5})$.   **A.32** $\lambda_1 = (5 - \sqrt{13})/2$, $\lambda_2 = (5 + \sqrt{13})/2$, $\lambda_3 = 5$.   **A.33** $\lambda_1 = (1 - \sqrt{2})$, $\lambda_2 = 1$,
$\lambda_3 = (1 + \sqrt{2})$.   **A.34** $\lambda_1 = 0$, $\lambda_2 = (3 - \sqrt{5})$, $\lambda_3 = (3 + \sqrt{5})$.

Index

A

Acceptance criterion, 697
Acceptance–rejection (A-R)
    method, 697–698
Acceptance/rejection of trial
    design, 717
ACO. *See* Ant Colony Optimization
Adaptive numerical method for
    discrete variable
    optimization, 636–641
  continuous variable
    optimization, 636–637
  discrete variable optimization,
    637–641
Advanced first-order second
    moment method, 777–781
Agent, 727
Algebra, vector and matrix. *See*
    Vector and matrix algebra
Algorithm, for traveling salesman
    problem, 721–724
Algorithm does not converge, 217
Algorithms
  attributes of good optimization,
    588
  conceptual local-global, 699–700
  constrained problems, 417
  constraint correction, 638
  convergence of, 417
  CSD, 526–527
  Phase I, 337
  Phase II, 339–345
  robust, 587
  selection of, 587
  Simplex, 384–385
Algorithms, concepts related to
    numerical. *See* Numerical
    algorithms
Algorithms, SLP. *See* Sequential
    Linear Programming
    algorithms
Algorithms for step size
    determination, ideas,
    418–421
  alternate equal interval search,
    425

analytical method to compute
    step size, 419–421
definition of one-dimensional
    minimization subproblem,
    419
equal interval search, 423–424
example—analytical step size
    determination, 420
example—minimization of
    function by golden section
    search, 429
golden section search, 425–430
numerical methods and compute
    step size, 421–430
Alternate equal interval
    search, 425
Alternate quadratic interpolation,
    447–448
American Association of State
    Highway and Transportation
    Officials (AASHTO),
    231–232
Analyses
  engineering, 4
  operations, 702–705
Analysis, postoptimality. *See*
    Postoptimality analysis
Analysis of means (ANOM), 749
Analytical method, 419–421
Ant behavior, 718–720
  simple model/algorithm,
    719–720
Ant Colony Optimization (ACO),
    718–727
  algorithm for design
    optimization, 724–727
  algorithm for traveling salesman
    problem, 721–724
  behavior, 718–720
Application to different
    engineering fields, 52
  example problem, 724–725
  feasible solutions, finding, 725
  pheromone deposit, 726–727
  pheromone evaporation, 726
  problem definition, 724

Array operation, 276
Artificial cost function, 336, 383
Artificial variables, 334–347,
    382–383
  cost function, 336
  definition of Phase I problem,
    336–337
  degenerate basic feasible
    solution, 345–347
  example—feasible problem, 342
  example—implications of
    degenerate feasible solution,
    346
  example—unbounded problem,
    344
  example—use of artificial
    variables, 344
  example—use of artificial
    variables for equality
    constraints, 342
  example—use of artificial
    variables for $\geq$ type
    constraints, 339
  Phase I algorithm, 337
  Phase II algorithm, 339–345
  use for equality constraints, 342
Ascents, alternation of descents,
    687
Asymmetric three-bar structure,
    594–598
Augmented Lagrangian methods,
    479–481

B

Basic feasible solution, degenerate,
    346
BBM. *See* Branch and bound
    method
Beam, design of rectangular,
    174–187
Beam design problem, graphical
    solution for, 82–94
Binary variable defined, 619
Binomial crossover, 717
Bound-constrained optimization,
    549–553

Bound-constrained optimization
    (*Continued*)
  optimality conditions, 549–550
  projection methods, 550–552
  step size calculation, 552–553
Bounded objective function
    method, 675–676
Brackets
  design of two-bar, 30–36
  design of wall, 171–174
Branch and bound method (BBM),
    623–628
  basic, 623–624
  example—BBM with local
    minimizations, 626
  example—BBM with only
    discrete values allowed, 624
  for general MV-OPT, 627–628
  with local minimization,
    625–627
British versus SI units. *See* U.S.–
    British versus SI units
Broyden-Fletcher-Goldfarb-Shanno
    (BFGS) method, 470–472

                    C
Cabinet design, 37–40
Calculation of basic solution,
    314–320
  basic solutions to Ax = b,
    317–320
  pivot step, 316–317
  tableau, 314–316
Calculus concepts, 103–115
  example—calculation of gradient
    vector, 105
  example—evaluation of gradient
    and Hessian of function, 106
  example—linear Taylor's
    expansion of function, 109
  example—Taylor's expansion of
    a function of one variable,
    108
  example—Taylor's expansion of
    a function of two variables,
    108
  gradient vector, 103–105
  Hessian matrix, 105–106
  necessary and sufficient
    conditions, 115–116
  quadratic forms and definite
    matrices, 109–115
  Taylor's expansion, 106–109
Can design, 25–26

Canonical form/general solution of
    Ax = b, 308–309
Changing constraint limits, effect
    of, 153–156
Chromosome, 645, 715–716
Clustering methods, 691–694
Coefficient matrix, changes in,
    361–375
Coefficient of variation, 773
Coefficients, ranging cost, 359–361
Coil springs, design of, 43–46
Column design
  for minimum mass, 286–290
  minimum weight tubular, 40–42
Column matrix, 787–820
Columns, graphical solutions for
    minimum weight tubular,
    80–81
Column vector, 787–820
Compression members, optimum
    design of, 243–250, 244t
  discussion, 250
  example—elastic buckling
    solution, 249
  example—inelastic buckling
    solution, 247
  formulation of problem, 243–247
  formulation of problem, for
    elastic buckling, 249–250
  formulation of problem, for
    inelastic buckling, 247–248
Compromise solution, 665
Computer programs, sample, 823
  equal interval search, 823–826
  golden section search, 826–828
  modified Newton's method, 829
  steepest descent method, 829
Concepts, optimum design. *See also*
    Optimum design concepts
  duality in NLP, 201–212
  exercises, 178–180
  necessary conditions, for
    equality-constrained
    problem, 130–137
  necessary conditions, for general
    constrained problem,
    137–153
Concepts, solution. *See* Solution
    concepts
Concepts and methods, multi-
    objective optimum design.
    See Multi-objective optimum
    design concepts and
    methods

Conditions
  descent, 416
  second-ordered, 194–199
  transformation of KKT, 403–404
Conditions, alternate form of KKT
    necessary, 189–192
  example—alternate form of KKT
    conditions, 190
  example—check for KKT
    necessary conditions, 191
Conditions, concepts relating to
    optimality, 116–117
Conjugate gradient method,
    434–436, 484
  example—use of conjugate
    gradient algorithm, 435–436
Constrained design, numerical
    methods for, 491–574
  algorithms and constrained
    problems, 492–495
  basic concepts and ideas, 492–499
  constrained quasi-Newton
    methods, 573
  constraint normalization, 496–498
  constraint status at design point,
    495–496
  convergence of algorithms,
    498–499
  CSD method, 525–531
  descent function, 498
  example—constraint
    normalization and status
    at point, 497
  inexact step size determination,
    s, 0035
  linearization of constrained
    problem, 541
  miscellaneous numerical
    optimization methods,
    564–569
  potential constraint strategy,
    534–537
  QP problem, 513–514
  QP subproblem, 514–520
  SLP algorithm, 506–513
Constrained optimum design
    problems, 281–282
  example—constrained
    minimization problem using
    fmincon, 281
  example—constrained optimum
    point, 138
  example—cylindrical tank
    design, 127

example—equality constrained problem, 140
example—fmincon in Optimization Toolbox, 281
example—inequality constrained problem, 140
example—infeasible problem, 139
example—Lagrange multipliers and their geometrical meaning, 131
example—solution of KKT necessary conditions, 145, 146, 150
example—use of Lagrange multipliers, 136
example—use of necessary conditions, 140
inequality constraints, 137–139
KKT, 139–152
necessary conditions, 137–153
necessary conditions: equality constraints, 137–153
Constrained optimization, second-order conditions for, 194–199
example—check for sufficient conditions, 197
solution of KKT necessary conditions using Excel, 222
solution of KKT necessary conditions using MATLAB, 149
Constrained optimum design, numerical methods for, 533–574
bound-constrained optimization, 549–553
inexact step size calculation, 537–549
potential constraints trategy, 534–537
QP subproblem, 514–520
quasi-Newton Hessian approximation, 557–558
search direction calculation, 514–520
SQP, 513–514, 553–563
step size calculation subproblem, 520–525
Constrained problems, concepts related to algorithms for, 492–495

Constrained problems, linearization of, 499–506
example—definition of linearized subproblem, 500–506
example—linearization of rectangular beam design problem, 504
Constrained quasi-Newton methods. *See also* Sequential quadratic programming
descent functions, 563
deviation of QP subproblem, 554–557
example—use of constrained quasi-Newton method, 560
observations on, 561–563
quasi-Newton Hessian approximation, 557–558
Constrained steepest-descent (CSD) method, 513, 525–527
algorithm, 526–527
algorithm, with inexact step size, 542–549
descent function, 538–542
example—calculation of descent function, 540
example—golden section search, 429
example—use of CSD algorithm, 542
step size determination, 444–450
Constrained variable metric (CVM). *See* Sequential quadratic programming
Constraint correction (CC), algorithm for, 638
Constraint limits, effect of changing, 155
Constraint normalization, 496–498
Constraints, 300
linear, 23
notation for, 8–9
Constraints, formulation of, 22–25
equality and inequality constraints, 23
feasibility design, 23
implicit constraints, 23–25
linear and nonlinear constraints, 23
Constraint status at design point, 495–496
Constraint strategy, potential, 534–537, 587

example—determination of potential constraint set, 534
example—search direction and potential constraint strategy, 536
Constraint tangent hyperplane, 194
Continuous variable optimization, 608–609, 636–637
Contours
plotting of function, 75–77
plotting of objective function, 74
Control, optimal, 6
Control effort problem, minimum, 608–609
Controlled random search (CRS), 694–697
Control of systems by nonlinear programming. *See* Nonlinear programming, control of systems by
Control problems
minimum time, 609–610
prototype optimal, 598–602
Conventional versus optimum design, 4–5
Convergence of algorithms, 417
Convergence ratio, 482
Convex functions, 162–164
Convex programming problem, 164–170
Convex sets, 160–161
Correction algorithm, constraint, 638
Correlation coefficient, 773
Cost
algorithm for constraint correction at constant, 638
algorithm for constraint correction at specified increase in, 638
constraint correction with minimum increase in, 638
Cost coefficients, ranging, 359–361
Cost function, 300
Cost function, artificial, 336
Cost function scaling, effect on Lagrange multipliers, 156–157
Covariance, 773
Covering methods, 684–685
Criterion, acceptance, 697
Criterion method, weighted global, 673–674

Criterion space and design space, 660–662
Crossover operation to generate trial design, 716–717
CRS. *See* Controlled random search
CSD, 527–531, 572
  constrained quasi-Newton methods, 573
  CSD method, 530–531
  linearization of constrained problem, 528–529
  QP subproblem, 529
  SLP algorithm, 529
CSD method. *See* Constrained steepest-descent method
Cumulative distribution function, 770
Curve fitting, quadratic, 444–447
Cylindrical tank design, minimum cost, 42–43

D

Davidon-Fletcher-Powell (DFP) method, 467–469
DE. *See* Domain elimination (DE)
DE algorithm, 717–718
  notation and terminology for, 715t
Definite matrices, quadratic forms and, 109–115
Definitions, standard LP, 300–302
Degenerate basic feasible solution, 345–347
Derivative-based methods, 214
Derivative-free methods, 215
Derivatives of functions, 12–13
  first partial derivatives, 12
  partial derivatives, of vector functions, 13
  second partial derivatives, 13
Descent, methods of generalized, 686–688
Descent algorithm, 432–434
Descent condition, 417, 538–542
Descent direction, 415–417
  descent step, 415–417
  orthogonality of steepest, 454–455
  rate of convergence, 417
Descent function, 498, 520–522, 563
  example, 522
Descent method, steepest, 431–434, 451–455, 482–483

example—verification of properties of gradient vector, 453
properties of gradient vector, 451–454
Descents and ascents, alternation of, 687
Descent search, steepest, 829
Descent step, 415–417
Design, 714
  of cabinet, 37–40
  of can, 25–26
  of column, 286–290
  of flywheel, 290–298
  of insulated spherical tank, 26–28
  of minimum cost cylindrical tank, 42–43
  of minimum weight tubular column, 40–42
  multiple optimum, 77
  of rectangular beam, 547
  of two-bar bracket, 30–36
  of wall bracket, 171–178
Design, GA for optimum. *See* Genetic algorithms (GA) for optimum design
Design, global optimization concepts and methods for. *See* Global optimization concepts and methods
Design, introduction to, 1–16
  basic terminology and notation, 6–13
  conventional versus optimum design process, 4–5
  design process, 2–4
  engineering design versus engineering analysis, 4
  optimum design versus optimal control, 6
Design, linear programming methods for. *See* Linear programming methods for optimum design
Design, mathematical model for optimum. *See* Mathematical model for optimum design
Design, numerical methods for constrained. *See* Constrained design, numerical methods for
Design, numerical methods for constrained optimum. *See*

Constrained optimum design, numerical methods for
Design, numerical methods for unconstrained optimum. *See* Unconstrained optimum design, numerical methods for
Design concepts, optimum. *See* Optimum design concepts
Design concepts and methods, discrete variable. *See* Discrete variable optimum design concepts and methods
Design concepts and methods, multi-objective. *See* Multi-objective optimum design concepts and methods
Design examples, engineering, 171–178
Design examples with MATLAB, optimum, 284–298
Design of experiments for response surface generation, 741–748
  example—generation of a response surface using an orthogonal array, 744
  example—optimization using RSM, 746
Design optimization
  applications with implicit functions, 576–582
  practical applications with implicit functions, 575–618
Design optimization, issues in practical. *See* Practical design optimization, issues in
Design optimization applications with implicit functions
  adaptive numerical method for discrete variable optimization, 636–641
  general-purpose software, 589–590
  gradient evaluation for implicit functions, 582–587
  issues in practical design optimization, 587–588
  multiple performance requirements, 592–598
  optimal control of systems by nonlinear programming, 598–612

optimum design of three-bar structure, 592—598
optimum design of two-member frame, 590—591
out-of plane loads, 590—591
practical design optimization problems, 576—582
Design point, 714
constraint status at, 578—582
Design problem formulation, optimum, 17—64
design of cabinet, 37—40
design of can, 25—26
design of coil springs, 43—46
design of two-bar bracket, 30—36
general mathematical model for optimum design, 50—64
insulated spherical tank design, 26—28
minimum cost cylindrical tank design, 42—43
minimum weight design of symmetric three-bar truss, 46—50
minimum weight tubular column design, 40—42
problem formulation process, 18—25
saw mill operation, 28—30
Design problems
classification of mixed variable optimum, 621—622
graphical solutions for rectanglar beam, 82—94
with multiple solutions, 77—78
sufficiency check for rectangular beam, 199—201
Design problems, constrained optimum. See Constrained optimum design problems
Design problems, unconstrained optimum, 116—129
Design process, 2—4
Design representation, 645—646
Design space, 660—662
Design variables, scaling of, 456—459
example—effect of scaling of design variables, 456
Design vector, 714
Desirable direction, 415
Determination, search direction, 431—436

Deterministic methods, 684—689
covering methods, 684—685
methods of generalized descent, 686—688
tunneling method, 688—689
zooming method, 685—686
Diagonal matrix, 791—821
Differential evolution algorithm, 714—718
A-R of trial design, 717
crossover operation to generate trial design, 716—717
DE algorithm, 717—718
generation of donor design, 716
generation of initial population, 715—716
Digital human modeling, 614—617
Direct Hessian updating, 470—472
Directions
descent, 415—417
desirable, 415—417
method of feasibility, 564—565
orthogonality of steepest descent, 454—455
Direct search methods, 214—215, 412, 485—489, 713
Hooke-Jeeves method, 486—489
univariate search, 485—486
Discrete design with orthogonal arrays, 749—753
example—discrete design with an orthogonal array, 752
Discrete variable optimization, 609—610, 636—641
Discrete variable optimum design concepts and methods, 619—642
adaptive numerical method for, 607—608
basic concepts and definitions, 620—623
BBM, 623—628
dynamic rounding-off method, 632—633
IP, 628—629
methods for linked discrete variables, 633—635
neighborhood search method, 633
SA, 630—632
selection of methods, 635
sequential linearization methods, 629

Domain elimination (DE), 707—708
method, 700—702
Dominance, efficiency and, 664—665
Duality in nonlinear programming, 201—212
local duality, equality constraints case, 201—206
local duality, inequality constraints case, 206—212
Dynamic rounding-off method, 632—633

E
Efficiency and dominance, 664—665
Eigenvalues and eigenvectors, 816—818
example—calculation of eigenvalues and eigenvectors, 816—818
Eigenvectors, eigenvalues and, 816—818
Elements, off-diagonal, 791—821
Elimination, Gauss-Jordan, 800—803
Elimination domain, 700—702
Engine, optimization, 667
Engineering applications of unconstrained methods, 472—477
Engineering design examples, 171—178
design of rectangular beam, 174—187
design of wall bracket, 171—174
Engineering design optimization using Excel Solver, 231—238
data and information collection, 233—234
definition of design variables, 234
formulation of constraints, 234—235
identification of criterion to be optimized, 234
project/problem statement, 231—233
solution, 238
Solver dialog box, 237—238
spreadsheet layout, 235—237
Engineering design versus engineering analysis, 4
Equal interval search, 423—424, 823—826
alternate, 425

Equality-constrained problem,
    necessary conditions,
    130–137
  Lagrange multipliers, 131–135
  Lagrange multiplier theorem,
    135–137
Equality constraints case, local
    duality, 201–206
Equations
  general solution of $m \times n$ linear,
    792–803
  solution of m linear, 804–809
Errors, minimization of,
    602–608
Evaluation, gradient, 575–576
Excel Solver, 218–223
  for LP problems, 225–227
  for NLP, optimum design of
    springs, 227–231
  roots of a set of nonlinear
    equations, 222–223
  roots of a nonlinear equation,
    219–221
  for unconstrained optimization
    problems, 224
Excel Solver, optimum design of
    plate girders using. See also
    Plate girders, optimum
    design using Excel Solver
  data and information collection,
    233–234
  identification/definition of
    design variables, 234
  identification of constraints,
    234–235
  identification of criterion to be
    optimized, 234
  project/problem statement,
    231–233
  solution, 235–237
  Solver dialog box, 237–238
  spreadsheet layout, 235–237
Excel Solver, optimum design with,
    213–274. See also Optimum
    design, with Excel Solver
  for LP problems, 225–227
  for NLP, optimum design of
    springs, 227–231
  numerical methods for optimum
    design, 213–218
  optimum design of compression
    members, 243–250
  optimum design of members for
    flexure, 250–263

optimum design of plate girders
    using excel solver, 231–238
optimum design of
    telecommunication poles,
    263–273
optimum design of tension
    members, 238–243
  for unconstrained optimization
    problems, 224
Excel worksheet, 222–223
Expansion, Taylor's. See Taylor's
    expansion
Expected value, 772–774
Expressions, variables and,
    275–276

F

Feasible directions, method of,
    564–565
Feasible points, finding, 216
Feasible region, identification
    of, 73
Feasible solution, degenerate basic,
    345–347
Feasible solutions, finding,
    725–726
  initial link, selection, 726
  link from layer R, 726
  solution for all ants, 726
Filters, Pareto-set, 670
First-order reliability method
    (FORM), 781
Fitness functions, Pareto, 669
Fitting, quadratic curve, 444–447
Flywheel design for minimum
    mass, 290–298
  data and information collection,
    290–292
  definition of design variables,
    292
  formulation of constraints, 292
  optimization criterion, 292
  project/problem statement, 290
Formulation, design problem. See
    Design problem formulation
Formulation process, problem. See
    Problem formulation process
Formulations, comparison of three,
    611–612
Function contours
  plotting, 75–77
  plotting of objective, 74
Functions
  artificial cost, 336

descent, 498, 520–522
normalization of objective, 667
Pareto fitness, 669
plotting, 72–73
utility, 665–666
Functions, convex, 162–164
Functions, implicit, designing
    practical applications with,
    575–618
Functions, implicit, gradient
    evaluation for, 582–587
  example—gradient evaluation
    for two-member
    frame, 583
Functions of single variables,
    optimality conditions for,
    117–122

G

GA. See Genetic algorithms
Gaussian (normal) distribution,
    773–774
Gaussian elimination procedure,
    796–800
Gauss-Jordan elimination, 800–803
Gene, defined, 645
General concepts, gradient-based
    methods. See Gradient-based
    search methods
General constrained problem,
    necessary conditions,
    137–153
  KKT necessary conditions,
    139–152
  role of inequalities, 137–139
  summary of KKT solution
    approach, 152–153
General iterative algorithm, 413–415
Generalized descent, methods of,
    686–688
Generalized reduced gradient
    (GRG) method, 567–569
General-purpose software, use of,
    589–590
  integration of application into,
    589–590
Generation, 644, 714
Generation of donor design, 716
Generation of initial population,
    715–716
Genetic algorithms (GA),
    fundamentals of, 646–651
  amount of crossover and
    mutation, 649

crossover, 648
elitist strategy, 670
immigration, 651
leader of population, 650
multi-objective, 667–671
multiple runs for problem, 651
mutation, 648–649
niche techniques, 671
number of crossovers and
   mutations, 649
Pareto fitness function, 669
Pareto-set filter, 670
ranking, 669
reproduction procedure,
   647–648
stopping criteria, 650
tournament selection, 670–671
VEGA, 668–669
Genetic algorithms (GA), for
   optimum design, 643–656
applications, 653–655
basic concepts and definitions,
   644–646
fundamentals of, 646–651
Genetic algorithms (GA), for
   sequencing-type problems,
   651–653
example—bolt insertion
   sequence determination, 652
Global and local minima,
   definitions of, 96–103
Global criterion method, weighted,
   673–674
Global optimality, 159–170
convex functions, 162–164
convex programming problem,
   164–168
convex sets, 160–161
example—checking for convexity
   of function, 163, 164
example—checking for convexity
   of problem, 166, 167, 168,
   169
example—checking for convexity
   of sets, 161
sufficient conditions for convex
   programming problems,
   169–170
transformation of constraint,
   168–169
Global optimization concepts and
   methods, 681–712
basic concepts of solution
   methods, 682–684

deterministic methods, 684–689
numerical performance of
   methods, 705–712
stochastic methods, 689–698
two local-global stochastic
   methods, 699–705
Global optimization, of structural
   design problems, 708–712
Goal programming, 676–677
Golden section search, 425–430,
   523, 826–828
Golf methods, 688
Good optimization algorithm,
   attributes of, 588
Gradient-based and direct search
   methods, 411–412
nature-inspired search methods,
   412
Gradient-based search methods,
   411–412
basic concepts, 413
general algorithm, 415
general iterative algorithm,
   413–415
Gradient evaluation for implicit
   functions, 582–587
Gradient evaluation requires
   special procedures,
   575–576
Gradient method, conjugate,
   434–436
Gradient projection method,
   566–567
Gradient vectors, 103–105
properties of, 451–454
Graphical optimization, 65–94
design problem with multiple
   solutions, 77–78
graphical solution for beam
   design problem, 82–94
graphical solution for minimum-
   weight tubular column,
   80–81
graphical solution process,
   65–71
infeasible problem, 79–80
problem with unbounded
   solution, 79
use of Mathematica for graphical
   optimization, 71–74
use of MATLAB for graphical
   optimization, 75–77
Graphical optimization, use of
   Mathematica for, 71–74

identification and shading of
   infeasible region for
   inequality, 73
identification of feasible region,
   73–74
identification of optimum
   solution, 74
plotting functions, 72–73
plotting of objective function
   contours, 74
Graphical optimization, use of
   MATLAB for, 75–77
editing graphs, 77
plotting of function contours, 75–77
Graphical solution, for beam
   design problem, 82–94
Graphical solution, for minimum-
   weight tubular column, 80–81
Graphical solution procedure,
   step-by-step, 67–71
coordination of system set-up, 67
identification of feasible region
   for inequality, 67–68
identification of optimum
   solution, 69–71
inequality constraint boundary
   plot, 67
plotting objective function
   contours, 68–69
Graphical solution process, 65–71
profit maximization problem,
   65–66
Graphs, editing, 77

H

Hessian approximation, quasi-
   Newton, 557–558
Hessian matrix, 105–106
Hessian updating
direct, 470–472
inverse, 467–469
Hooke-Jeeves method, 486–489
algorithm, 486–489
exploratory search, 486
pattern search, 486
Hyperplane, constraint tangent, 194

I

Identity matrix, 791–821
Implicit functions, design
   applications with, 575–618
adaptive numerical method for
   discrete variable
   optimization, 636–641

Implicit functions, design
    applications with (*Continued*)
  formulation of practical design
    optimization problems,
    576–582
  general-purpose software,
    589–590
  gradient evaluation for implicit
    functions, 582–587
  issues in practical design
    optimization, 587–588
  multiple performance
    requirements, 592–598
  optimal control of systems by
    NLP, 598–612
  optimum design of three-bar
    structure, 592–598
  optimum design of two-member
    frame, 590–591
  out-of-plane loads, 590–591
Implicit functions, design practical
    applications with, 575–618
Implicit functions, gradient
    evaluation for, 582–587
  example—gradient evaluation
    for two-member frame, 583
Improving feasible direction,
    564–565
Inaccurate line search, 448–449
Inequality, identification and
    hatching of infeasible region
    for, 73
Inequality constraints case, local
    duality, 206–212
Inexact step-size calculation. *See*
    Step-size calculation, inexact
Infeasible problem, 79–80
Infeasible region, identification and
    shading of, 73
Insulated spherical tank design,
    26–28
Integer programming (IP), 628–629
Integer variable, 619
Integration, stochastic, 698
Interpolation, alternate quadratic,
    447–448
Interpolation, polynomial,
    444–448
  quadratic curve fitting, 444–447
Interval-reducing methods, 422–423
Interval search
  alternate equal, 425
  equal, 423–424, 823–826
Inverse Hessian updating, 467–469

IP. *See* Integer programming
Irregular points, 192–194
  example—check for KKT
    conditions at irregular
    points, 192

K

Karush-Kuhn-Tucker (KKT), 189
  conditions, transformation of,
    404–405
  conditions for LP problem,
    400–402
    optimality conditions, 400
    solution, 400–402
  necessary conditions, 139–152
  necessary conditions, alternate
    form of, 189–192
    example—alternate form of
    KKT conditions, 190
    example—check for KKT
    necessary conditions, 191
  necessary conditions for QP
    problem, 403–404
  solution approach, 152–153

L

Lagrange multipliers, 131–135
  effect of cost function scaling on,
    156–157
  physical meaning of, 153–159
    constraint variation sensitivity
    result, 159
    effect of changing constraint
    limit, 153–156
    example—effect of scaling
    constraint, 158
    example—effect of scaling cost
    function, 157
    example—Lagrange
    multipliers, 157, 158
    example—optimum cost
    function, 155
    example—variations of
    constraint limits, 155
    scaling cost function on
    Lagrange multipliers, 157
Lagrange multiplier theorem,
    135–137
Lagrangian methods, augmented,
    479–481
Length of vectors. *See* Norm/length
    of vectors
Lexicographic method, 674–675
Limit state equation, 774–776

Linear constraints, 23
Linear convergence, 482
Linear equations, general solution
    of $m \times n$, 804–809
Linear equations in n unknowns,
    solving n, 792–803
  determinants, 793–796
  example—determinant of matrix
    by Gaussian
    elimination, 799
  example—Gauss-Jordan
    reduction, 801
  example—Gauss-Jordan
    reduction process in tabular
    form, 809
  example—general solution by
    Gauss-Jordan reduction, 806
  example—inverse of matrix by
    cofactors, 801
  example—rank determination by
    elementary operation, 804
  example—solution of
    equations by Gaussian
    elimination, 798
  Gaussian elimination procedure,
    796–800
  Gauss-Jordan elimination, 806
  general solution of $m \times n$ linear
    equations, 804–809
  inverse of matrix, 800–803
  linear systems, 792–793
  rank of matrix, 803–804
Linear functions, 300
  constraints, 300
  cost function, 300
Linearization methods, sequential,
    629
Linearization of constrained
    problems, 499–506
  example—definition of linearized
    subproblem, 501
  example—linearization of
    rectangular beam design
    problem, 504
Linear limit state equation, 776
Linear programming (LP), duality
    in, 387–399
  alternate treatment of equality
    constraints, 391–392
  determination of primal solution
    from dual solution, 392–395
  dual LP program, 388–389
  dual variables as Lagrange
    multipliers, 398–399

example—dual of LP program, 389

example—dual of LP with equality and ≥ type constraints, 390

example—primal and dual solutions, 394

example—recovery of primal formulation from dual formulation, 391

example—use of final primal tableau to recover dual solutions, 398

standard primal LP, 387—388

treatment of equality constraints, 389—390

use of dual tableau to recover primal solution, 395—398

Linear programming methods, for optimum design, 299—376, 377—410

artificial variables, 334—347

basic concepts related to LP problems, 305—314

calculation of basic solution, 318—320

definition of standard LP problem, 300—305

duality in LP, 387—399

example—structure of tableau, 318

KKT conditions for LP problem, 400—402

linear functions, 300

postoptimality analysis, 348—375

QP problem, 402—409

two-phase Simplex method, 334—347

Linear programming problem, standard, 66, 300—305

example—conversion to standard LP form, 304

linear constraints, 23

unrestricted variables, 303

Linear programming problems, concepts related to, 299, 305—314

example—characterization of solution for LP problems, 311

example—determination of basic solutions, 311

example—profit maximization problem, 306

LP terminology, 310—313

optimum solutions to LP problems, 313—314

Linear programs (LPs), 299

Linear systems, 792—793

Line search, 522—525

Linked discrete variable, 619

Linked discrete variables, methods for, 633—635

Loads, out-of-plane, 590—591

Local duality, equality constraints case, 201—206

Local duality, inequality constraints case, 206—212

Local-global algorithm, conceptual, 699—705

Local minima, definition, 96—103

Lower triangle matrix, 791—821

M

Marquardt modification, 465—466

Mass

column design for minimum, 286

flywheel design for minimum, 290—298

Mathematica, use of, for graphical optimization. *See* Graphical optimization, use of Mathematica for

Mathematical model for optimum design, 50—64

active/inactive/violated constraints, 53—54

application to different engineering fields, 52

discrete integer design variables, 54

feasibility set, 53

important observations about standard model, 52—53

maximization problem treatment, 51

optimization problems, types of, 55—64

standard design optimization model, 50—51

treatment of greater than type constraints, 51—52

MATLAB, optimum design examples with, 284—298

column design for minimum mass, 286—290

flywheel design for minimum mass, 290—298

location of maximum shear stress, 284—285

two spherical bodies in contact, 284—285

MATLAB, optimum design with, 275—298

constrained optimum design problems, 281—282

Optimization Toolbox, 275—277

unconstrained optimum design problems, 278—280

MATLAB, use of for graphical optimization, 75—77

editing graphs, 77

plotting of function contours, 75—77

Matrices, 785—787

addition of, 787

column, 790

condition numbers of, 819—822

definition of, 785—787

diagonal, 791—821

equivalence of, 790

identity, 791—821

inverse of, 800—803

lower triangle, 791—821

multiplication of, 788—789

null, 787

partitioning of, 791—792

quadratic forms and definite, 109—110

rank of, 803—804

row, 790

scalar, 790—791

square, 791

transpose of, 790

upper triangle, 791—821

vector, 787

Matrices, norms and condition numbers of, 818—822

condition number of matrix, 819—822

norm of vectors and matrices, 818—819

Matrices, types of, 787—792

addition of matrices, 790

elementary row—column operations, 790

multiplication of matrices, 788—789

partitioning of matrices, 791—792

scalar product—dot product of vectors, 790—791

square matrices, 791

vectors, 787

Matrix, changes in coefficient, 361–375

Matrix, Hessian, 105–106

Matrix algebra, vector and, 785
  concepts related to set of vectors, 810–816
  definition of matrices, 785–787
  eigenvalues and eigenvectors, 816–818
  norm and condition number of matrix, 818–822
  solution of m linear equations in n unknowns, 792–803
  types of matrices and their operations, 787–792

Matrix operation, 276

Mechanical and structural design problems, 614

Members for flexure, optimum design of. *See* Optimum design of members for flexure

Meta-Model, 731–732
  normalization of variables, 737–739
  RSM, 733

Method of feasible directions, 564–565

Methods *See also individual method entries*
  alternate Simplex, 385–386
  A-R, 707
  augmented Lagrangian, 479–481
  BFGS, 469
  bounded objective function, 675–676
  clustering, 691–694
  conjugate gradient, 434–437
  constrained quasi-Newton, 573
  constrained steepest descent, 525–527
  covering, 684–685
  deterministic, 684–689
  DFP, 467–469
  domain elimination, 700–702
  dynamic rounding-off, 632–633
  of generalized descent, 686–688
  golf, 687
  gradient projection, 566–567
  GRG method, 567–569
  interval reducing, 423
  lexicographic, 674–675

linear programming, 299–410
modified Newton's, 829
multiplier, 479–481
multistart, 691
neighborhood search, 633
operations analysis of, 702–705
performance, 706–707
performance of stochastic zooming, 707–708
scalarization, 666
sequential linearization, 629
Simplex, 321–334
stochastic zooming, 702
tunneling, 688–689
two-phase Simplex, 334–347
unconstrained, 472–481
vector, 666
weighted global criterion, 673–674
weighted min-max, 672–673
weighted sum, 671–672
zooming, 685–686

Methods, for linked discrete variables, 633–635

Methods, miscellaneous numerical optimization, 564–569
  gradient projection method, 566–567
  GRG method, 567–569
  method of feasibility directions, 564–565

Methods, multi-objective optimum design concepts and. *See* Multi-objective optimum design concepts and methods

Methods, Newton's. *See* Newton's methods

Methods, numerical performance of, 705–712
  DE methods, 707–708
  global optimization of structural design problems, 708–712
  performance of methods using unconstrained problems, 706–707
  stochastic zooming method, 707–708
  summary, 705–706

Methods, for optimum design, global concepts and, 681–712

Methods, quasi-Newton. *See* Quasi-Newton methods

Methods, sequential quadratic programming (SQP). *See also* Sequential quadratic programming
  observations on constrained, 561–563

Methods, two local-global stochastic. *See* Stochastic methods, local-global

Methods, unconstrained optimization. *See* Unconstrained optimization methods

Minima, definitions of global and local, 96–103
  example—constrained minimum, 100
  example—constrained problem, 99
  example—existence of a global minimum, 102
  example—use of the definition of maximum point, 101
  example—using Weierstrass theorem, 102
  existence of minimum, 102–103

Minimization techniques, sequential unconstrained, 479

Minimum, existence of, 102–103

Minimum control effort problem, 608–609

Minimum mass
  column design for, 286–290
  flywheel design for, 290–298

Minimum-weight tubular column, graphical solution for, 80–81

Min-max method, weighted, 672–673

Mixed variable optimum design problems (MV-OPT), 620
  classification of, 621–622
  definition of, 620

Modifications, Marquardt, 465–466

Monte Carlo simulation (MCS), 781

Motion, optimal control of system, 611–612

Multi-objective optimum design concepts and methods, 657–680
  bounded objective function method, 675–676

criterion space and design space, 660–662
example—single-objective optimization problem, 658
example—two-objective optimization problem, 659
generation of Pareto optimal set, 666–667
goal programming, 676–677
lexicographic method, 674–675
multi-objective GA, 667–671
normalization of objective functions, 667
optimization engine, 667
preferences and utility functions, 665–666
problem definition, 657–659
scalarization methods, 666
selection of methods, 677–679
solution concepts, 662–665
terminology and basic concepts, 660–667
vector methods, 666
weighted global criterion method, 673–674
weighted min-max method, 672–673
weighted sum method, 671–672
Multi-objective GA, 667–671
elitist strategy, 670
niche techniques, 671
Pareto fitness function, 669
Pareto-set filter, 670
ranking, 669
tournament selection, 670–671
VEGA, 668–669
Multiple optimum designs, 77
Multiple performance requirements, 592–598
asymmetric three-bar structure, 594–598
comparison of solutions, 598
symmetric three-bar structure, 592–594
Multiple solutions, design problem with, 77–78
Multiplier methods, 479–481
Multipliers, physical meaning of Lagrange. See Lagrange multipliers, physical meaning of
Multistart method, 691

N

Nature-inspired search methods, 215, 412, 713–730
Ant Colony Optimization, 718–727
differential evolution algorithm, 714–718
Particle Swarm Optimization, 727–729
Necessary conditions, for equality-constrained problem, 130–137
Lagrange multipliers, 131–135
Lagrange multiplier theorem, 135–137
Necessary conditions, for general constrained problem, 137–153
Karush-Kuhn-Tucker necessary conditions, 139–152
role of inequalities, 137–139
summary of KKT solution approach, 152–153
Neighborhood search method, 633
Newton's methods. See also Quasi-Newton methods
classical, 460
example—conjugate gradient and modified Newton's methods, 465
example—use of modified Newton's method, 462, 463
Marquardt modification, 465–466
modified, 461–465, 829
Niche techniques, 671
Nonlinear equations, solution of, 475–477
Nonlinear limit state equation, 776–777
Nonlinear programming (NLP), 411
Nonlinear programming, control of systems by, 598–612
comparison of three formulations, 611–612
minimization of errors in state variables, 602–608
minimum control effort problem, 608–609
minimum time control problem, 609–610
optimal control of system motion, 611–612
prototype optimal control problem, 598–602

Nonlinear programming, duality in. See Duality in nonlinear programming
Nonquadratic case, 483
Normalization, constraint, 496–498
Normalization of variables, 737–739
example—response surface using normalization procedure, 740
example—response surface using the normalization procedure, 738
procedure, 737–741
Norm/length of vectors, 10–11
Notation
basic terminology and, 6–13
for constraints, 8–9
summation, 9–10
Null matrix, 787
Numerical algorithms, 415–417
convergence, 417
descent direction and descent step, 415–417
example—checking for descent condition, 417
general algorithm, 415
Numerical methods, to compute step size, 421–430
alternate equal-interval search, 425
equal-interval search, 423–424
general concepts, 421–423
golden section search, 425–430
Numerical methods, for constrained design. See Constrained design, numerical methods for
Numerical methods for constrained optimum design. See Constrained optimum design, numerical methods for
Numerical methods for optimum design, 213–218
search methods, classification of, 214–215
simple scaling of variables, 217–218
solution process, 215–217
Numerical methods for unconstrained optimum design. See Unconstrained optimum design, numerical methods for

Numerical optimization methods, 564—569
  gradient projection method, 566—567
  GRG method, 567—569
  method of feasibility directions, 564—565
Numerical performance of methods. *See* Methods, numerical performance of

O

Objective function contours, plotting of, 74
Objective functions, normalization of, 667
Off-diagonal elements, 791—821
Operations analysis of methods, 702—705
Optimal control, versus optimum design, 6
Optimal control of system motion, 611—612
Optimal control problem, prototype, 598—602
Optimality, global. *See* Global optimality
Optimality, Pareto, 663—664
Optimality conditions
  for bound constrained optimization, 549—550
  concepts relating to, 116—117
  for functions of single variables, 117—122
Optimality, weak Pareto, 664
Optimal set, generation of Pareto. *See* Pareto optimal set, generation of
Optimization
  continuous variable, 636—637
  discrete variable, 637—641
  engines, 667
Optimization, bound constrained, 549—553
Optimization, graphical. *See* Graphical optimization
Optimization, issues in practical design, 587—588
  attributes of good optimization algorithm, 588
  potential constraint strategy, 587
  robustness, 587
  selection of algorithm, 587

Optimization, practical applications of, 575—618
Optimization, practical applications of, 575—618
  discrete variable optimum design, 636—641
  formulation of practical design optimization problems, 576—582
  general-purpose software, use of, 589—590
  gradient evaluation for implicit functions, 582—587
  issues in practical design optimization, 587—588
  multiple performance requirements, 592—598
  optimal control of systems by NLP, 598—612
  optimum design of three-bar structure, 592—598
  optimum design of two-member frame, 590—591
  out-of-plane loads, 590—591
  structural optimization problems, alternative formulations for, 612—613
  time-dependent problems, alternative formulations for, 613—617
Optimization, second-order conditions for constrained. *See* Constrained optimization, second-order conditions for
Optimization, use of Mathematica for graphical. *See* Graphical optimization, use of Mathematica for
Optimization, use of MATLAB for graphical. *See* Graphical optimization, use of MATLAB for
Optimization algorithm, attributes of good. *See* Good optimization algorithm, attributes of
Optimization algorithms, by nature-inspired search methods, 713—730
Optimization methods, miscellaneous numerical, 564—569

gradient projection method, 566—567
GRG method, 567—569
method of feasibility directions, 564—565
Optimization methods, unconstrained, 477—481
  augmented Lagrangian, 479—481
  multiplier, 479—481
  sequential unconstrained minimization techniques, 478—479
Optimization problems, practical design. *See* Practical design problems, formulation of
Optimization problems, types of, 55—64
Optimization Toolbox, 275—277
  array operation, 276
  matrix operation, 276
  scalar operation, 276
  variables and expressions, 275—276
Optimum design, 731—784
  conventional versus, 4—5
  design of experiments for response surface generation, 741—748
  discrete design with orthogonal arrays, 749—753
  example— application of Taguchi method, 764, 766
  example— calculation of reliability index, 782
  example—discrete design with an orthogonal array, 752
  example—generation of a response surface using an orthogonal array, 744
  example—generation of quadratic response surface, 735
  example—optimization using RSM, 746
  example—reliability-based design optimization, 784
  example—response surface using normalization procedure, 738—739, 740—741
  example— robust optimization, 759
  general mathematical model for, 50—64

meta-models for design optimization, 731–741

RBDO, design under uncertainty, 767–784

robust design approach, 754–766

Optimum design, discrete variable. *See* Discrete variable optimum design concepts and methods

Optimum design, GA for. *See* Genetic algorithms (GA) for optimum design

Optimum design, global concepts and methods for, 681–712
basic concepts of solution methods, 682–684
deterministic methods, 684–689
numerical performance of methods, 705–712
stochastic methods, 689–698
two local-global stochastic methods, 699–705

Optimum design, LP methods for. *See* Linear programming methods, for optimum design

Optimum design, mathematical model for. *See* Mathematical model for optimum design

Optimum design, numerical methods for constrained. *See also* Constrained design, numerical methods for
approximate step-size determination, 572
bound-constrained optimization, 549–553
examples—constraint normalization and status at point, 497
inexact step size calculation, 537–549
linearization of constrained problem, 499–506
miscellaneous numerical optimization methods, 564–569
plate girders optimum design using Excel Solver, 231–238
potential constraints strategy, 534–537, 587
QP problem, 402–409
QP subproblem, 514–520

quasi-Newton Hessian approximation, 557–558
search direction calculation, 514–520
SQP, 513–514, 553–563
sequential quadratic programming methods, 553–563
SLP algorithm, 506–513
step-size calculation subproblem, 520–525

Optimum design, numerical methods for unconstrained. *See* Unconstrained optimum design, numerical methods for

Optimum design, with Excel Solver, 213–274
example—design of a shape for inelastic LTB, 259
example—design of a shape for elastic LTB, 261
example—design of noncompact shape, 262
example—elastic buckling solution, 249
example—inelastic buckling solution, 247
example—optimum design of pole, 268
example—optimum design with the local buckling constraint, 270
example—optimum design with the tip rotation constraint, 269
example—selection of W10 shape, 241
example—selection of W8 shape, 242
Excel Solver for LP problems, 225–227
Excel Solver for NLP, optimum design of springs, 227–231
Excel Solver for unconstrained optimization problems, 224
numerical methods for optimum design, 213–218
optimum design of compression members, 243–250
optimum design of members for flexure, 250–263
optimum design of plate girders using Excel Solver, 231–238

optimum design of telecommunication poles, 263–273
optimum design of tension members, 238–243

Optimum design concepts, 95–212
alternate form of KKT necessary conditions, 189–192
basic calculus concepts, 103–115
constrained optimum design problems, 281–282
engineering design examples, 171–178
exercises, 208–212
global optimality, 159–170
irregular points, 192–194
necessary conditions, for equality-constrained problem, 130–137
necessary conditions, for general unconstrained problem, 137–153
physical meaning of Lagrange multipliers, 153–159
postoptimality analysis, 153–159
second-order conditions for constrained optimization, 194–199
sufficiency check for rectangular beam design problem, 199–201
unconstrained optimum design problems, 278–280

Optimum design concepts and methods, discrete variable. *See* Discrete variable optimum design concepts and methods

Optimum design concepts and methods, multi-objective. *See* Multi-objective optimum design concepts and methods

Optimum design examples with MATLAB. *See* MATLAB, optimum design examples with

Optimum design of compression members, 243–250, 244t
discussion, 250
example—elastic buckling solution, 249
example—inelastic buckling solution, 247

Optimum design of compression
    members (*Continued*)
    formulation of problem, 243—247
    formulation of problem, for
        elastic buckling, 249—250
    formulation of problem, for
        inelastic buckling, 247—248
Optimum design of members for
    flexure, 250—263
    data and information collection,
        250—254
    definition of design variables,
        258
    deflection requirement, 258—262
    example—design of a compact
        shape for elastic LTB, 261
    example—design of a compact
        shape for inelastic LTB, 259
    example—design of noncompact
        shape, 262
    formulation of constraints,
        258—262
    moment strength requirement,
        254—255
    nominal bending strength of
        compact shapes, 255—256
    nominal bending strength of
        noncompact shapes,
        256—257
    optimization criterion, 258
    project/problem description, 250
    shear strength requirement,
        257—258
Optimum design of plate girders
    using Excel Solver. *See* Plate
    girders, optimum design
    using Excel Solver
Optimum design of
    telecommunication poles. *See*
    Telecommunication poles,
    optimum design of
Optimum design of tension
    members. *See* Tension
    members, optimum design of
Optimum design of three-bar
    structure. *See* Three-bar
    structure, optimum design of
Optimum design of two-member
    frame. *See* Two-member
    frame, optimum design of
Optimum design problem
    formulation, 17—64
    design of cabinet, 37—40
    design of can, 25—26

design of coil springs, 43—46
design of two-bar bracket, 30—36
general mathematical model for
    optimum design, 50—64
insulated spherical tank design,
    26—28
minimum cost cylindrical tank
    design, 42—43
minimum weight design of
    symmetric three-bar truss,
    46—50
minimum weight tubular column
    design, 40—42
problem formulation process,
    18—25
saw mill operation, 28—30
Optimum design problems,
    constrained. *See* Constrained
    optimum design problems
Optimum design problems,
    unconstrained. *See*
    Unconstrained optimum
    design problems
Optimum designs, multiple, 77
Optimum design versus optimal
    control, 6
Optimum design with MATLAB.
    See MATLAB, optimum
    design with
Optimum solution, identification
    of, 74
Optimum solutions to LP
    problems, 313—314
Order of convergence, 482
Out-of-plane loads, 590—591

P

Parameters, ranging right side,
    354—358
Pareto fitness function, 669
Pareto optimality, 663—664
    weak, 664
Pareto optimal set, generation of,
    666—667
Pareto-set filter, 670
Particle position, 728
Particle Swarm Optimization
    (PSO), 727—729
    algorithm, 728—729
    behavior and terminology, 727—728
Particle velocity, 728
Performance of methods using
    unconstrained problems,
    706—707

Performance requirements,
    multiple, 592—598
Phase I algorithm, 337
Phase II algorithm, 339—345
Phase I problem, definition of,
    336—337
Pheromone deposit, 726—727
Pheromone evaporation, 726
Physical programming, 665—666
Pivot step, 316—317
Plate girders, optimum design
    using Excel Solver, 231—238
    data and information collection,
        233—234
    definition of design variables,
        234
    formulation of constraints,
        234—235
    optimization criterion, 234
    project/problem description,
        231—233
    Solver Parameters dialog box,
        237—238
    spreadsheet layout, 235—237
Plotting
    of function contours, 75—77
    functions, 72—73
    of objective function contours, 74
Points
    constraint status at design,
        495—496
    sets and, 6—8
    utopia, 665
Points, irregular, 192—194
    example—check for KKT
        conditions at irregular
        points, 192
Polynomial interpolation,
    444—448
    alternate quadratic interpolation,
        447—448
    quadratic curve fitting, 444—447
Postoptimality analysis, 153—159,
    348—375
    changes in coefficient matrix,
        361—375
    changes in resource limits,
        348—349
    constraint variation sensitivity
        result, 159
    effect of scaling constraint on
        Lagrange multiplier, 158
    effect of scaling cost function on
        Lagrange multipliers, 157

example—= and $\geq$ type constraints, 352

example—$\leq$ type constraints, 350, 360

example—effect of scaling constraint, 158

example—effect of scaling cost function, 156–157

example—equality and $\geq$ type constraints, 357, 361

example—Lagrange multipliers, 156–157, 158

example—optimum cost function, 155

example—ranges for cost coefficients, 360, 361

example—ranges for resource limits, 356, 357

example—recovery of Lagrange multipliers for $\geq$ type constraint, 352

example—variations of constraint limits, 155

ranging cost coefficients, 359–361

ranging right-side parameters, 354–358

recovery of Lagrange multipliers for $\geq$ type constraints, 352

Potential constraint strategy, 587

Practical applications, design optimization, 575–618

alternative formulations for time-dependent problems, 613–617

Practical design optimization, issues in, 587–588

attributes of good optimization algorithm, 588

potential constraint strategy, 587

robustness, 587

selection of algorithm, 587

Practical design problems, formulation of, 576–582

example of practical design optimization problem, 577–582

example—design of two-member frame, 612–613

general guidelines, 576–577

Preferences and utility functions, 665–666

Probability density function (PDF), 769–770

Probability of failure, 770–771

Problem formulation, optimum design. See Optimum design problem formulation

Problem formulation process, 18–25

data and information collection, 19–20

definition of design variables, 20–21

formulation of constraints, 22–25

optimization criterion, 21–22

project/problem description, 18

Problems. See also Subproblems

classification of mixed variable optimum design problems, 621–622

concepts related to algorithms for constrained problems, 492–495

definition of Phase I, 336–337

example of practical design, 577–582

formulation of spring design, 46

graphical solutions for beam design, 82–94

infeasible, 79–80

integer programming, 40

linear programming, 66, 299, 377

minimum control effort, 608–609

minimum time control, 609–610

MV-OPT, 620

optimum solutions to LP problems, 313–314

profit maximization, 65–66

prototype optimal control, 598–602

solution to constrained problems, 477–481

sufficiency check for rectangular beam design, 199–201

with unbounded solutions, 79

Problems, concepts related to linear programming. See Linear programming problems, concepts related to

Problems, constrained optimum design. See Constrained optimum design problems

Problems, convex programming, 164–170

Problems, definition of standard linear programming. See Linear programming problem, standard

Problems, formulation of practical design optimization. See Practical design problems, formulation of

Problems, GA for sequencing-type. See Genetic algorithms (GA), for sequencing-type problems

Problems, global optimization of structural design. See Global optimization, of structural design problems

Problems, linearization of constrained. See Linearization of constrained problems

Problems, performance of methods using unconstrained. See Performance of methods using unconstrained problems

Problems, QP. See Quadratic programming (QP) problems

Problems, time-dependent. See Time-dependent problems

Problems, unconstrained design. See also Unconstrained optimum design problems

concepts relating to optimality conditions, 116–117

example—adding constant to function, 124

example—cylindrical tank design, 127

example—effects of scaling, 124

example—local minima for function of two variables, 125, 129

example—local minimum points using necessary conditions, 119, 120, 121

example—minimum cost spherical tank using necessary conditions, 122

example—multivariable unconstrained minimization, 279

example—numerical solution of necessary conditions, 128

example—single-variable unconstrained minimization, 278

Problems, unconstrained design
  (*Continued*)
  example—using necessary
    conditions, 119, 127
  example—using optimality
    conditions, 125, 129
  optimality conditions for
    functions of several
    variables, 122–129
  optimality conditions for
    functions of single variables,
    117–122
Procedures, Gaussian elimination,
  796–800
Procedures, gradient evaluation
  requires special, 575–576
Process, design, 2–4
Process, problem formulation. *See*
  Problem formulation process
Profit maximization problem,
  65–66
Programming
  duality in linear, 387–399
  goal of, 676–677
  physical, 665–666
Programming, control of systems
  by nonlinear. *See* Nonlinear
  programming, control of
  systems by
Programming problems
  convex, 164–170
  linear, 56, 299, 305–314
Programs, sample computer, 823
  equal interval search, 823–826
  golden section search, 826–828
  modified Newton's method, 829
  steepest-descent search, 829
Projection method, gradient,
  566–567
Prototype optimal control problem,
  598–602
Pure random search, 690–691

Q

QP. *See* Quadratic programming
  problems
Quadratic convergence, 482
Quadratic curve fitting, 444–447
Quadratic forms and definite
  matrices, 109–115
  example—calculations for gradient
    of quadratic form, 114
  example—calculations for Hessian
    of quadratic form, 114

example—determination of form
  of matrix, 112, 113
example—matrix of quadratic
  form, 110
Quadratic function, 482–483
Quadratic interpolation, alternate,
  447–448
Quadratic programming (QP)
  problems, 402–409, 514–520
  definition of, 402–403, 514–518
  derivation of, 554–557
  example—solution to QP
    subproblem, 519
  example—definition of QP
    subproblem, 515
  example—solution of QP
    problem, 406
  KKT necessary conditions for,
    403–404
  Simplex method for solving,
    405–409
  solution to, 518–520, 569–573
  transformation of KKT
    conditions, 404–405
Quasi-Newton Hessian
  approximation, 557–558
Quasi-Newton methods, 466–472,
  484–485
  BFGS method, 470–472
  DFP method, 467–469
  direct Hessian updating,
    470–472
  example—application of BFGS
    method, 471
  example—application of DFP
    method, 468
  inverse Hessian updating,
    467–469
  observations on constrained,
    561–563
Quasi-Newton methods,
  constrained. *See* Sequential
  quadratic programming

R

Random search, pure, 690–691
Ranging cost coefficients, 359–361
Ranging right-side parameters,
  354–358
Rate of convergence, 417
Rate of convergence of algorithms,
  481–485
  conjugate gradient method, 484
  definitions, 481–482

Newton's method, 483
quasi-Newton methods, 484–485
steepest-descent method,
  482–483
Rectangular beam, design of,
  174–187
Rectangular beam design problem,
  sufficiency check for,
  199–201
Recursive quadratic programming
  (RQP), 554. *See also*
  Sequential quadratic
  programming
Reducing methods, interval, 422–423
Regions
  identification and shading of
    infeasible, 73
Reliability-based design
  optimization (RBDO), under
  uncertainty, 767–784
  calculation of reliability index,
    774–781
  example– calculation of
    reliability index, 782
  example—reliability-based
    design optimization, 784
  review of background material
    for, 768–774
Reliability index, 773
Representation, design, 645–646
Reproduction, defined, 647–648
Requirements, multiple
  performance, 592–598
Response surface method (RSM),
  733
  example—generation of
    quadratic response surface,
    735
  quadratic response surface
    generation, 733–735
Right-side parameters, ranging,
  354–358
Robust algorithms, 587
Robust design approach, 754–766
  Taguchi method, 761–766
Robust optimization, 754–760
  example—robust optimization,
    759
  mean, 754–755
  PDF, 755–756
  problem definition,
    756–759
  standard deviation, 755
  variance, 755

Role of inequalities, 137–139
Roots of a set of nonlinear
    equations, 222–223
    Excel worksheet, 222–223
    solution to KKT cases with
        Solver, 223
    Solver Parameters dialog box,
        223
Roots of nonlinear equation,
    219–221
    Solver Parameters dialog box,
        220–221
Rounding-off method, dynamic,
    632–633
Row matrix, vector, 787–820

S

SA. *See* Simulated annealing
Saw mill operation, 28–30
Scalarization methods, 666
Scalar matrix, 791–821
Scalar operation, 276
Scaling of design variables,
    456–459
    example—effect of scaling of
        design variables, 456
Search direction calculation,
    514–520
    definition of QP subproblem,
        514–518
    example—definition of QP
        subproblem, 515
    example—solution to QP
        subproblem, 519
    solution to QP subproblem,
        518–520
Search direction determination,
    431–436, 459–466
Searches
    alternate equal interval, 425
    equal interval, 423–424, 823–826
    golden section, 425–430,
        826–828
    inexact line, 448–449
    line search, 522–525
    pure random, 690–691
    steepest descent, 829
Search method, neighborhood, 633
Search methods, classification of,
    214–215
    derivative-based, 214
    derivative-free, 215
    direct search, 214–215
    nature-inspired, 215

Second-order conditions for
    constrained optimization,
    194–199
Second-order information, 194
Sequencing-type problems, GA for,
    651–653
Sequential linearization methods,
    629
Sequential linear programming
    (SLP) algorithm, 506–513
    algorithm observations, 512–513
    example—sequential linear
        programming
        algorithm, 509
    example—use of sequential
        linear programming, 510
    move limits in, 506–508
    SLP algorithm, 508–512
Sequential quadratic programming
    (SQP), 513–514, 553–563,
    707
    algorithm, 558–561
    derivation of QP subproblem,
        554–557
    descent functions, 563
    example—solving spring design
        problem using SQP method,
        560
    example—use of SQP method,
        558
    observations on, 561–563
    option, 590–591
    quasi-Newton Hessian
        approximation, 557–558
Sequential unconstrained
    minimization techniques,
    478–479
Set, generation of Pareto optimal,
    666–667
Sets, convex, 160–161
Sets and points, 6–8
Simple scaling of variables,
    217–218
Simplex algorithms, 384–385
Simplex in two-dimensional space,
    321
Simplex method
    alternate, 385–386
    artificial cost function, 382–383
    canonical form/general solution
        of $\mathbf{Ax} = \mathbf{b}$, 308–309
    example—Big-M method for
        equality and $\geq$ type
        constraints, 386

example—identification of
    unbounded problem with
    Simplex method, 333
example—LP problem with
    multiple solutions, 331
example—pivot step, 316
example—solution by Simplex
    method, 328
example—solution of profit
    maximization problem, 329
general solution to $\mathbf{Ax} = \mathbf{b}$,
    377–379
interchange of basic and
    nonbasic variables, 316
pivot step, 316, 384
Simplex algorithms, 384–385
steps of, 322
tableau, 378–379
two-phase, 334–347
Simplex method, derivation of,
    377–385
    selection of basic variable,
        381–382
    selection of nonbasic variable,
        379–381
Simplex method, for solving QP
    problem, 405–409
Simulated annealing (SA), 630–632,
    706–707, 708
Single variables, optimality
    conditions for functions of,
    117–122
SI units versus U.S.–British, 13
SLP. *See* Sequential linear
    programming algorithm
Software, general-purpose,
    589–590
    integration of application into,
        589–590
    selection of, 589
Solution concepts, 622–623, 662–665
    compromise solution, 665
    efficiency and dominance,
        664–665
    Pareto optimality, 663–664
    utopia point, 665
    weak Pareto optimality, 664
Solution methods, basic concepts,
    682–684
Solution process, 215–217
    algorithm does not
        converge, 217
    feasible point cannot be
        obtained, 216

Solution process (*Continued*)
  finding, feasible points, 216
Solutions
  degenerate basic feasible, 346
  identification of optimum, 74
  multiple, with design problems, 77
  unbounded, 79
Solution to KKT cases with Solver, 223
Solver output, 221
Solver Parameters dialog box, 220–221, 223
Spaces
  criterion, 660–662
  design, 660–662
  Simplex in two-dimensional, 321
  vector, 814–816
Special procedures, required by gradient evaluation, 575–576
Spherical tank design, insulated, 26–28
Spring design problem formulation of, 46
Spring design problem, solving with SQP method, 560
Springs, design of coil, 43–46
SQP. *See* Sequential quadratic programming
SQP algorithm, 558–561
Square matrices, 791
Standard deviation, 773
Standard linear programming (SLP) problem, 300–305
Standard model, 52–53
State variables, minimization of errors in, 602–608
  discussion of results, 607–608
  effect of problem normalization, 605–607
  formulation for numerical solution, 602–604
  numerical results, 604–605
Steepest-descent directions, orthogonality of, 454–455
Steepest-descent method, 431–434, 451–455, 482–483
  example—use of steepest-descent algorithm, 432, 433
  example—verification of properties of gradient vector, 453
  properties of gradient vector, 451–454

Steepest-descent method, constrained. *See* Constrained steepest-descent method
Steepest-descent search, 829
Steps
  descent, 415–417
  pivot, 316
Step-size calculation
  for bound-constrained algorithm calculation, 552–553
Step-size calculation, inexact, 537–549
  basic concept, 537–538
  CSD algorithm with inexact step size, 542–549
  descent condition, 538–542
  example—effect of penalty parameter R on CSD algorithm, 546
  example—effect of γ on performance of CSD algorithm, 545
  example—minimum area design of rectangular beam, 547
  example—step size in constrained steepest-descent method, 540
  example—use of CSD algorithm, 542
Step-size calculation subproblem, 520–525
  descent function, 520–522
  line search, 522–525
Step-size determination, 418–421, 421–430
  analytical method, 419–421
  definition of, 418–419
  example—analytical step size determination, 420
  example—alternate quadratic interpolation, 446
  example—one-dimensional minimization, 446
  inexact line search, 448–449
  numerical methods, 421–430
  polynomial interpolation, 444–448
Step-size determination, approximate, 572
  basic idea, 537–538
  CSD algorithm with inexact step size, 542–549
  descent condition, 417
  example—calculations for step size, 540

example—constrained steepest-descent method, 540
example—effect of γ on performance of CSD algorithm, 545
example—minimum area design of rectangular beam, 174–187
example—use of constrained steepest-descent algorithm, 542
Step-size determination, ideas and algorithms for, 418–421
  alternate equal interval search, 425
  analytical method to compare step size, 419–421
  definition of one-dimensional minimization subproblem, 419
  equal interval search, 423–424
  example—analytical step size determination, 420
  example—minimization of function by golden section search, 429
  golden section search, 425–430
  numerical methods and compute step size, 421–430
Stochastic integration, 698
Stochastic methods, 689–698
  A-R methods, 697–698
  clustering methods, 691–694
  CRS method, 694–697
  multistart method, 691
  pure random search, 690–691
  stochastic integration, 698
Stochastic methods, local-global, 699–705
  conceptual local-global algorithm, 699–700
  domain elimination method, 700–702
  operations analysis of methods, 702–705
Stochastic zooming method, 702
  performance of, 707–708
Strategy, potential constraint. *See* Constraint strategy, potential
Structural design problems, optimization of, 708–712
Structural optimization problems, alternative formulations for, 612–613

Structures
asymmetric three-bar, 594—598
symmetric three-bar, 592—594
Structures, optimum design of three-bar. *See* Three-bar structure, optimum design of
Subproblems, QP, 514—520
definition of, 514—518
examples—definition of QP subproblem, 514—518
example—solution of QP subproblem, 519
solving, 518—520
Summation notation, superscripts/subscripts and, 9—10
Superlinear convergence, 482
Swarm leader, 728
Symmetric three-bar structure, 592—594
Symmetric three-bar truss, minimum-weight design of, 46—50
System motion, optimal control of, 611—612
Systems, linear, 792—793
Systems, optimal control, 598—612

T
Tableau, defined, 378—379
Taguchi method, 761—766
example—application of Taguchi method, 764, 766
Tangent hyperplane, constraint, 194
Tank design
cylindrical, 42—43
insulated spherical, 26—28
Taylor's expansion, 106—109
Techniques
niche, 671
sequential unconstrained minimization, 478—479
Telecommunication poles, optimum design of, 263—273
data and information collection, 263—267
definition of design variables, 267
example—optimum design of pole, 268
example—optimum design with the local buckling constraint, 270
example—optimum design with the tip rotation constraint, 269

formulation of constraints, 268—273
optimization criterion, 268
Tension members, optimum design of, 238—243
formulation of constraints, 239—242
optimization criterion, 239
Terminology, LP, 310—313
Terminology and notations, basic, 6—13
functions, 11
norm/length of vectors, 10—11
notation for constraints, 8—9
sets and points, 6—8
superscripts/subscripts and summation notation, 9—10
U.S.—British versus SI units, 13
Three-bar structure, asymmetric, 594—598
Three-bar structure, optimum design of, 592—598
asymmetric three-bar structure, 594—598
comparison of solutions, 598
symmetric three-bar structure, 592—594
Three-bar truss, symmetric minimum-weight design of, 46—50
Time control problem, minimum, 609—610
Time-dependent problems, alternative formulations for, 613—617
digital human modeling, 614—617
mechanical and structural design problems, 614
Toolbox, Optimization. *See* Optimization Toolbox
Triangle matrix
lower, 791—821
upper, 791—821
Truss, minimum-weight design of symmetric three-bar, 46—50
Tubular column, minimum-weight, graphical solution for, 80—81
Tubular column, minimum-weight, design of, 40—42
Tunneling method, 688—689
Two-bar bracket, design of, 30—36
example—optimum design of two-bar bracket, 30—36

Two-dimensional space, Simplex in, 321
Two-member frame, optimum design of, 590—591
alternate formulation for, 612—613
Two-phase Simplex method, 334—347

U
Unbounded solution, 79
Uncertainty, RBDO design under. *See* Reliability-based design optimization under uncertainty
Unconstrained methods, engineering applications of, 472—477
example—minimization of total potential energy of two-bar truss, 474
example—roots of nonlinear equations, 476
example—unconstrained minimization, 476
minimization of total potential energy, 473—475
solutions to nonlinear equations, 475—477
Unconstrained minimization techniques, sequential, 478—479
Unconstrained optimality conditions, 116—130
Unconstrained optimization methods, 477—481
augmented Lagrangian methods, 479—481
multiplier methods, 479—481
sequential unconstrained minimization techniques, 478—479
Unconstrained optimum design, numerical methods for, 411—490
concepts related to numerical algorithms, 411—415
conjugate gradient method, 434—436
descent direction and convergence of algorithms, 415—417
direct search methods, 485—489

Unconstrained optimum design, numerical methods for (*Continued*)
engineering applications of unconstrained methods, 472—477
gradient-based methods, 412—415
ideas and algorithms for step-size determination, 418—421
nature-inspired search methods, 412
Newton's method, 459—466
quasi-Newton methods, 466—472
rate of convergence of algorithms, 481—485
scaling of design variables, 456—459
search direction determination, 431—436, 459—472
solution to constrained problems, 477—481
steepest-descent method, 431—434, 451—455, 482—483
step-size determination, 418—430
unconstrained optimization methods, 477—481
Unconstrained optimum design problems, 278—280
Unconstrained problems, performance of methods using, 706—707
Unimodal functions, 421—422
Unknowns, solution to *m* linear equations in *n*, 803—809
Unrestricted variables, 303
Upper triangle matrix, 791—821
U.S.—British versus SI units, 13
Utility functions, preferences and, 665—666
Utopia point, 665

V

Variable optimization, continuous, 636—637
Variable optimization, discrete, 637—641
Variable optimum design, discrete, 619—642
Variables
binary, 619
discrete, 619
and expressions, 275—276
integer, 619
linked discrete, 619
methods for linked discrete, 633—635
unrestricted, 303
Variables, artificial. *See* Artificial variables
Variables, minimization of errors in state. *See* State variables, minimization of errors in
Variables, optimality conditions for functions of single. *See* Single variables, optimality conditions for functions of
Variables, scaling of design. *See* Design variables, scaling of
Vector, gradient, 103—105
Vector and matrix algebra, 785—822
concepts related to set of vectors, 810—816
definition of matrices, 785—787
eigenvalues and eigenvectors, 816—818
norm and condition number of matrix, 818—822
solution of *n* linear equations in *n* unknowns, 792—803

solution to *m* linear equations in *n* unknowns, 803—809
type of matrices and their operations, 787—792
Vector evaluated genetic algorithm (VEGA), 668—669
Vector methods, 666
Vectors, 787
column, 787—820
norm/length of, 10—11
properties of gradient, 451—454
row, 787—820
Vectors, set of, 810—816
example—checking for linear independence of vectors, 811
example—checking for vector spaces, 814
linear independence of set of vectors, 810—814
Vector spaces, 814—816

W

Wall bracket, design of, 171—174
Weak Pareto optimality, 664
Weighted global criterion method, 673—674
Weighted min-max method, 672—673
Weighted sum method, 671—672

Z

Zooming methods, 685—686
performances of stochastic, 707—708
stochastic, 702